Study Guide and Solutions

M000315958

Organic Chemistry

FIFTH ENHANCED EDITION

William Brown

Christopher Foote

Brent Iverson

Eric Anslyn

Prepared by

Brent Iverson
University of Texas, Austin

Sheila Iverson
University of Texas, Austin

BROOKS/COLE
CENGAGE Learning

Australia • Brazil • Japan • Korea • Mexico • Singapore • Spain • United Kingdom • United States

For product information and technology assistance, contact us at **Cengage Learning Customer & Sales Support, 1-800-354-9706**

For permission to use material from this text or product, submit all requests online at **www.cengage.com/permissions**
Further permissions questions can be emailed to **permissionrequest@cengage.com**

ISBN-13: 978-0-538-49765-7
ISBN-10: 0-538-49765-3

Brooks/Cole
20 Davis Drive
Belmont, CA 94002-3098
USA

Cengage Learning is a leading provider of customized learning solutions with office locations around the globe, including Singapore, the United Kingdom, Australia, Mexico, Brazil, and Japan. Locate your local office at: **www.cengage.com/global**

Cengage Learning products are represented in Canada by Nelson Education, Ltd.

To learn more about Brooks/Cole, visit **www.cengage.com/brookscole**

Purchase any of our products at your local college store or at our preferred online store **www.CengageBrain.com**

Printed in the United States of America
1 2 3 4 5 6 7 14 13 12 11 10

To Carina, Alexandra, Alanna, and Juliana
with love

This Student Study Guide is a companion to the fifth edition of *Organic Chemistry* by William Brown, Christopher Foote, Brent Iverson, and Eric Anslyn. All of the *problems* from the text have been reprinted in this guide, so there is no need to flip back and forth between the text and the guide. Detailed, stepwise *solutions* to all of the problems are provided.

Molecules are three-dimensional and understanding the three-dimensionality of organic chemistry is important for any student. This Student Study Guide has placed special emphasis on stereochemistry, an important aspect of three-dimensional molecular structure. Throughout the problems and answers, many of the molecules with chiral centers have the configuration explicitly denoted using wedges and dashes to indicate location in space. When the configuration is not specifically given, chiral centers are indicated by an asterisk (*). In addition, the stereochemical outcome of every reaction is stated, whether the question calls for consideration of stereochemistry or not.

In this enhanced edition, *each* mechanistic step in the solutions is identified as one of the relatively few fundamental mechanistic elements introduced mainly in Chapter 6. This is intended to help students more easily identify the similarities and thereby make connections between the mechanisms from different chapters. This revolutionary new approach will help students *understand* as opposed to just memorize Organic Chemistry

The material in this volume was reviewed for accuracy by Brent and Sheila Iverson. If you have any comments or questions, please direct them to Professor Brent Iverson, Department of Chemistry and Biochemistry, the University of Texas at Austin, Austin, Texas 78712. E-mail: biverson@mail.utexas.edu.

Brent and Sheila Iverson
Austin, Texas
December, 2009

CONTENTS:

CONTENTS

CHAPTER 1
Solutions to the Problems

Problem 1.1 Write and compare the ground-state electron configurations for each pair of elements:
(a) Carbon and silicon

C (6 electrons) $1s^2 2s^2 2p^2$
Si (14 electrons) $1s^2 2s^2 2p^6 3s^2 3p^2$
Both carbon and silicon have four electrons in their outermost (valence) shells.

(b) Oxygen and sulfur

O (8 electrons) $1s^2 2s^2 2p^4$
S (16 electrons) $1s^2 2s^2 2p^6 3s^2 3p^4$
Both oxygen and sulfur have six electrons in their outermost (valence) shells.

(c) Nitrogen and phosphorus

N (7 electrons) $1s^2 2s^2 2p^3$
P (15 electrons) $1s^2 2s^2 2p^6 3s^2 3p^3$
Both nitrogen and phosphorus have five electrons in their outermost (valence) shells.

Problem 1.2 Show how each chemical change leads to a stable octet.
(a) Sulfur forms S^{2-}.   (b) Magnesium forms Mg^{2+}.

S (16 electrons): $1s^2 2s^2 2p^6 3s^2 3p^4$ **Mg (12 electrons): $1s^2 2s^2 2p^6 3s^2$**

S^{2-} (18 electrons): $1s^2 2s^2 2p^6 3s^2 3p^6$ **Mg^{2+} (10 electrons): $1s^2 2s^2 2p^6$**

Problem 1.3 Judging from their relative positions in the Periodic Table, which element in each set is more electronegative?
(a) Lithium or potassium

In general, electronegativity increases from left to right across a row and from bottom to top of a column in the Periodic Table. This is because electronegativity increases with increasing positive charge on the nucleus and with decreasing distance of the valence electrons from the nucleus. Lithium is closer to the top of the Periodic Table and thus more electronegative than potassium.

(b) Nitrogen or phosphorus

Nitrogen is closer to the top of the Periodic Table and thus more electronegative than phosphorus.

(c) Carbon or silicon

Carbon is closer to the top of the Periodic Table and thus more electronegative than silicon.

Problem 1.4 Classify each bond as nonpolar covalent, or polar covalent, or state that ions are formed.
(a) S-H   (b) P-H   (c) C-F   (d) C-Cl

Recall that bonds formed from atoms with an electronegativity difference of less than 0.5 are considered nonpolar covalent and with an electronegativity difference of 0.5 or above are considered a polar covalent bond.

Bond	Difference in electronegativity	Type of bond
S-H	2.5 - 2.1 = 0.4	**Nonpolar covalent**
P-H	2.1 - 2.1 = 0	**Nonpolar covalent**
C-F	4.0 - 2.5 = 1.5	**Polar covalent**
C-Cl	3.0 - 2.5 = 0.5	**Polar covalent**

Problem 1.5 Using the symbols δ- and δ+, indicate the direction of polarity in each polar covalent bond.
(a) C-N   (b) N-O
δ+ δ-   **δ+ δ-**
C-N   **N-O**

Nitrogen is more electronegative than carbon   **Oxygen is more electronegative than nitrogen**

(c) C-Cl

$\delta+ \quad \delta-$

C-Cl

Chlorine is more electronegative than carbon

<u>Problem 1.6</u> Draw Lewis structures, showing all valence electrons, for these molecules.

(a) C_2H_6

(b) CS_2

(c) HCN

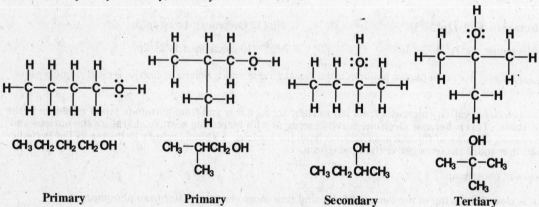

<u>Problem 1.7</u> Draw Lewis structures for these ions, and show which atom in each bears the formal charge.

(a) $CH_3NH_3^+$
    Methylammonium ion

(b) CO_3^{2-}
    Carbonate ion

(c) HO^-
    Hydroxide ion

<u>Problem 1.8</u>

Draw Lewis structures and condensed structural formulas for the four alcohols with molecular formula $C_4H_{10}O$. Classify each alcohol as primary, secondary, or tertiary.

$CH_3CH_2CH_2CH_2OH$

Primary

CH_3-CHCH_2OH
       |
       CH_3

Primary

     OH
     |
$CH_3CH_2CHCH_3$

Secondary

     OH
     |
CH_3-C-CH_3
     |
     CH_3

Tertiary

<u>Problem 1.9</u>

Draw structural formulas for the three secondary amines with molecular formula $C_4H_{11}N$.

<u>Problem 1.10</u> Draw condensed structural formulas for the three ketones with molecular formula $C_5H_{10}O$.

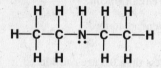
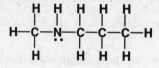
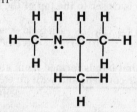

$CH_3CH_2CH_2CCH_3$

CH_3CHCCH_3
     |
     CH_3

$CH_3CH_2CCH_2CH_3$

<u>Problem 1.11</u> Draw condensed structural formulas for the two carboxylic acids with molecular formula $C_4H_8O_2$.

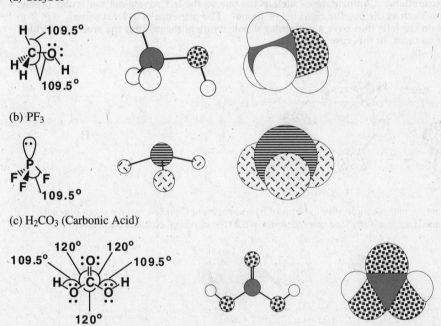

<u>Problem 1.12</u> Draw structural formulas for the four esters with molecular formula $C_4H_8O_2$.

<u>Problem 1.13</u> Predict all bond angles for these molecules.
(a) CH_3OH

(b) PF_3

(c) H_2CO_3 (Carbonic Acid)

<u>Problem 1.14</u> Which molecules are polar? For each molecule that is polar, specify the direction of its dipole moment.
(a) CH_2Cl_2

A molecular dipole moment is determined as the vector sum of the bond dipoles in three-dimensional space. Thus, by superimposing the bond dipoles on a three-dimensional drawing, the molecular dipole moment can be determined. Note that on the following diagrams, the dipole moments from the C-H bonds have been ignored because they are so small.

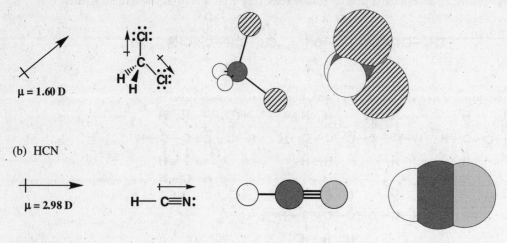

$\mu = 1.60\ D$

(b) HCN

$\mu = 2.98\ D$

(c) H_2O_2

The H_2O_2 molecule can rotate around the O-O single bond, so we must consider the molecular dipole moments in the various possible conformations. Conformations such as the one on the left have a net molecular dipole moment, while conformations such as the one the right below do not. The presence of at least some conformations (such as that on the left) that have a molecular dipole moment means that the entire molecule must have an overall dipole moment, in this case $\mu = 2.2\ D$.

$\mu = 2.2\ D$

<u>Problem 1.15</u> Describe the bonding in these molecules in terms of hybridization of C and N, and the types of bonds between carbon and nitrogen, and if there are any lone pairs, describe what type of orbital contains these electrons.

(a) $CH_3\text{-}CH{=}CH_2$

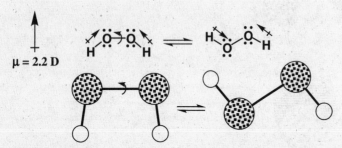

(a)

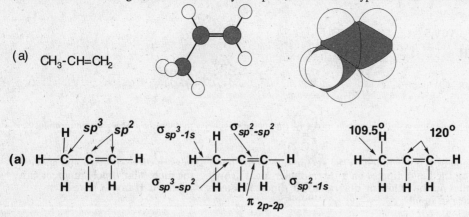

(b) CH_3-NH_2

(b) [structural diagrams showing $H-C-N-H$ with hybridization labels sp^3, σ_{sp^3-1s}, $\sigma_{sp^3-sp^3}$, σ_{sp^3-1s}, and bond angle $109.5°$]

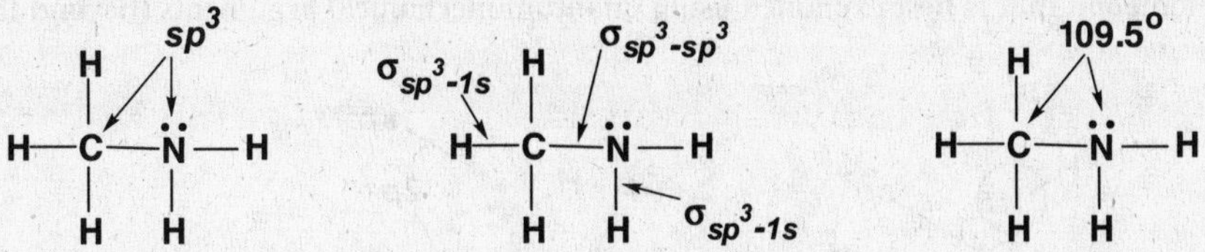

Problem 1.16 Draw the contributing structure indicated by the curved arrows. Be certain to show all valence electrons and all formal charges.

(a) [Lewis structure resonance diagrams]

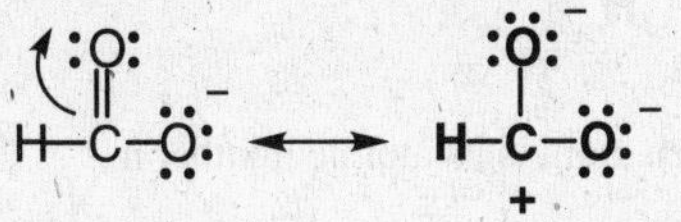

(b) [Lewis structure resonance diagrams]

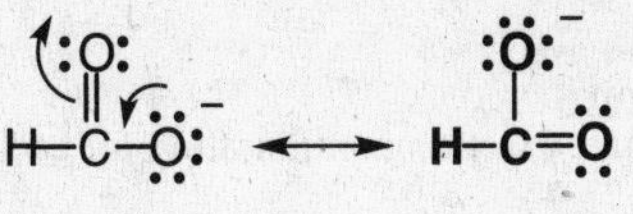

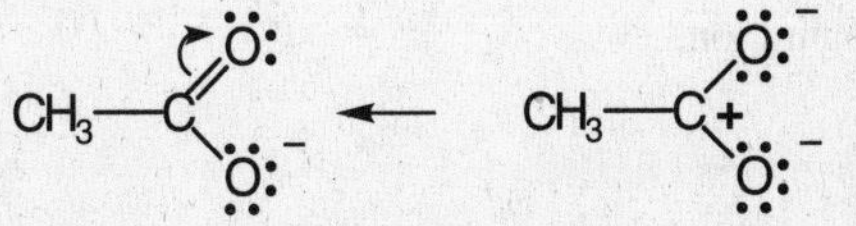

(c) $CH_3-C-O-CH_3 \longleftrightarrow CH_3-C=O-CH_3$ [Lewis structure resonance diagrams]

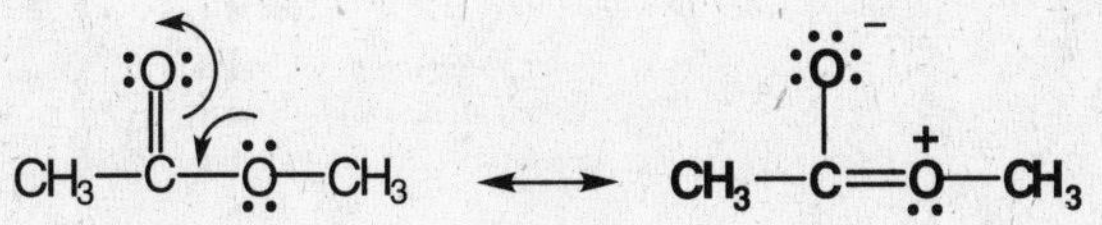

Problem 1.17 Which sets are valid pairs of contributing structures?

(a) CH_3-C ... [Lewis structure resonance diagrams]   (b) CH_3-C ... [Lewis structure resonance diagrams]

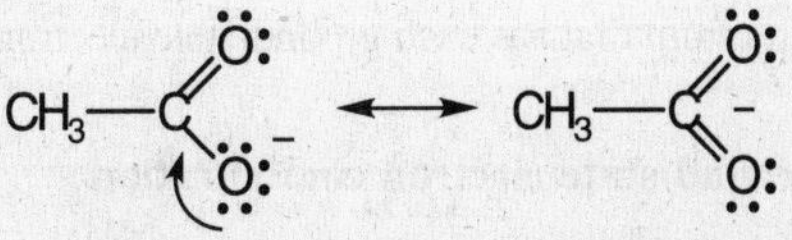

The set in (a) is a pair of contributing structures, while the set in (b) is not. The structure on the right in set (b) is not a viable contributing structure because there are five bonds to the carbon atom, implying 10 electrons in the valence shell, which can only hold a maximum of 8 electrons.

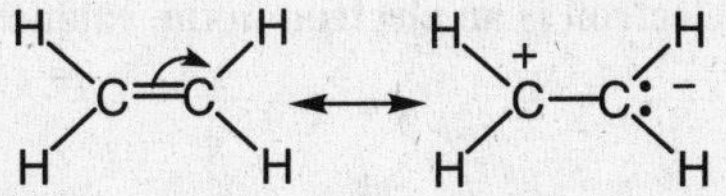

Problem 1.18 Estimate the relative contribution of the members in each set of contributing structures.

(a) [Lewis structure resonance diagrams]   (b) $H-C-O-H \longleftrightarrow H-C=O-H$ [Lewis structure resonance diagrams]

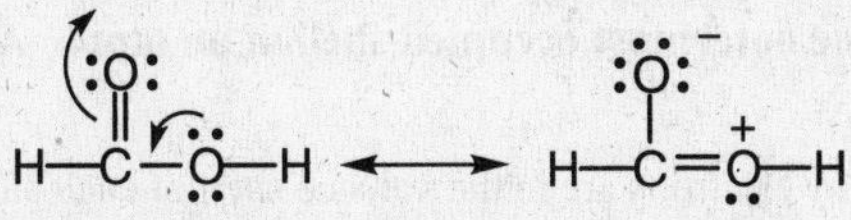

The first structure makes the greater contribution in (a) and (b). In both cases, the second contributing structure involves the disfavored creation and separation of unlike charges.

Problem 1.19 Draw three contributing structures of the following compound (called guanidine) and state the hybridization of the four highlighted atoms. In which orbitals do the three lone pairs drawn reside

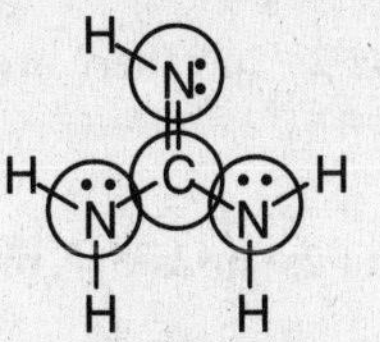

[Guanidine structure diagram]

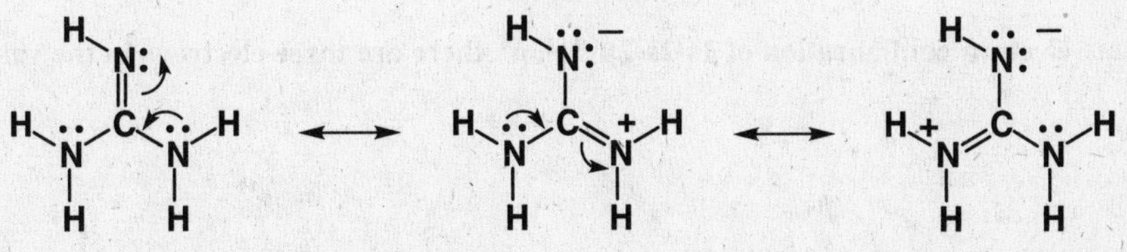

Guanidine

[Three resonance contributing structures of guanidine connected by double-headed arrows]

Remember that if any significant contributing structure contains a π bond, then the hybridization of that atom must be able to accommodate the π bond. Consideration of the three significant contributing structures indicates that all of the nitrogen atoms are sp^2 hybridized because of the π bonding. To be consistent with the contributing structures, two of the lone pairs on the original structure are in $2p$ orbitals, while the third resides in an sp^2 orbital. Guanidine is one of many examples you will encounter in which the lone pair on nitrogen is delocalized into an adjacent π bond. Such delocalization of electron density in π orbitals is stabilizing and therefore favorable, a phenomenon that is best explained using quantum mechanical arguments (beyond the scope of this text).

Electronic Structures of Atoms

Problem 1.20 Write ground-state electron configuration for each atom. After each atom is its atomic number in parentheses.

(a) Sodium (11)   **Na (11 electrons) $1s^2 2s^2 2p^6 3s^1$**

(b) Magnesium (12)   **Mg (12 electrons) $1s^2 2s^2 2p^6 3s^2$**

(c) Oxygen (8)   **O (8 electrons) $1s^2 2s^2 2p^4$**

(d) Nitrogen (7)   **N (7 electrons) $1s^2 2s^2 2p^3$**

Problem 1.21 Identify the atom that has each ground-state electron configuration.
(a) $1s^2 2s^2 2p^6 3s^2 3p^4$

Sulfur (16) has this ground-state electron configuration

(b) $1s^2 2s^2 2p^4$

Oxygen (8) has this ground-state electron configuration

Problem 1.22 Define valence shell and valence electron.

The valence shell is the outermost occupied shell of an atom. A valence electron is an electron in the valence shell.

Problem 1.23 How many electrons are in the valence shell of each atom?
(a) Carbon

With a ground-state electron configuration of $1s^2 2s^2 2p^2$, there are four electrons in the valence shell of carbon.

(b) Nitrogen

With a ground-state electron configuration of $1s^2 2s^2 2p^3$, there are five electrons in the valence shell of nitrogen.

(c) Chlorine

With a ground-state electron configuration of $1s^2 2s^2 2p^6 3s^2 3p^5$, there are seven electrons in the valence shell of chlorine.

(d) Aluminum

With a ground-state electron configuration of $1s^2 2s^2 2p^6 3s^2 3p^1$, there are three electrons in the valence shell of aluminum.

Lewis Structures and Formal Charge

Problem 1.24 Judging from their relative positions in the Periodic Table, which atom in each set is more electronegative?
(a) Carbon or nitrogen

In general, electronegativity increases from left to right across a row (period) and from bottom to top of a column in the Periodic Table. This is because electronegativity increases with increasing positive charge on the nucleus and with decreasing distance of the valence electrons from the nucleus. Nitrogen is farther to the right than carbon in Period 2 of the Periodic Table, thus nitrogen is more electronegative than carbon.

(b) Chlorine or bromine

Chlorine is higher up than bromine in column 7A of the Periodic Table, thus chlorine is more electronegative than bromine.

(c) Oxygen or sulfur

Oxygen is higher up than sulfur in column 6A of the Periodic Table, thus oxygen is more electronegative than sulfur.

Problem 1.25 Which compounds have nonpolar covalent bonds, which have polar covalent bonds, and which have ions?
(a) LiF   (b) CH_3F (c) $MgCl_2$ (d) HCl

Using the rule that ions are formed between atoms with an electronegativity difference of 1.9 or greater, the following table can be constructed:

Bond	Difference in electronegativity	Type of bond
Li-F	4.0 - 1.0 = 3.0	Ions
C-H	2.5 - 2.1 = 0.4	Nonpolar covalent
C-F	4.0 - 2.5 = 1.5	Polar covalent
Mg-Cl	3.0 - 1.2 = 1.8	Polar covalent
H-Cl	3.0 - 2.1 = 0.9	Polar covalent

Based on these values, only LiF has an ions. The other compounds have nonpolar covalent (C-H) or polar covalent (C-F, Mg-Cl, H-Cl) bonds.

Problem 1.26 Using the symbols δ- and δ+, indicate the direction of polarity, if any, in each covalent bond.

(a) C-Cl $\overset{\delta+\;\;\delta-}{C\text{-}Cl}$ **Chlorine is more electronegative than carbon.**

(b) S-H $\overset{\delta-\;\;\delta+}{S\text{-}H}$ **Sulfur is more electronegative than hydrogen.**

(c) C-S   **Carbon and sulfur have the same electronegativity so there is no polarity in a C-S bond.**

(d) P-H   **Phosphorus and hydrogen have the same electronegativity, so there is no polarity in a P-H bond.**

Problem 1.27 Write Lewis structures for these compounds. Show all valence electrons. None of them contains a ring of atoms.
(a) H_2O_2 (b) N_2H_4 (c) CH_3OH
    Hydrogen peroxide   Hydrazine   Methanol

Problem 1.28 Write Lewis structures for these ions. Show all valence electrons and all formal charges.
(a) NH_2^- (b) HCO_3^- (c) CO_3^{2-}
    Amide ion   Bicarbonate ion   Carbonate ion

(d) NO$_3^-$
Nitrate ion

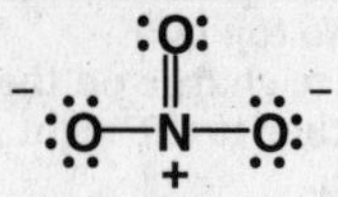

(e) HCOO$^-$
Formate ion

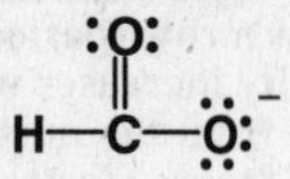

(f) CH$_3$COO$^-$
Acetate ion

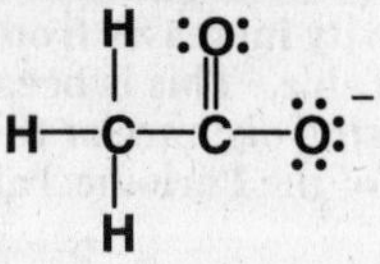

Problem 1.29 Complete these structural formulas by adding enough hydrogens to complete the tetra-valence of each carbon. Then write the molecular formula of each compound.

(a) C—C=C—C—C
with a C above the third carbon

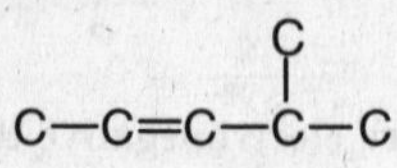

C$_6$H$_{12}$

(b) C—C—C—C—OH
with O double bonded above fourth C

C$_4$H$_8$O$_2$

(c) C—C—C—C
with O double bonded above third C

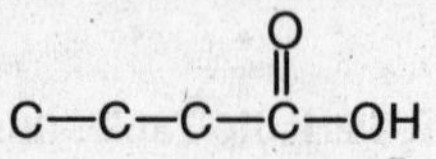

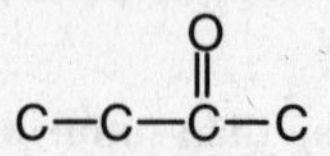

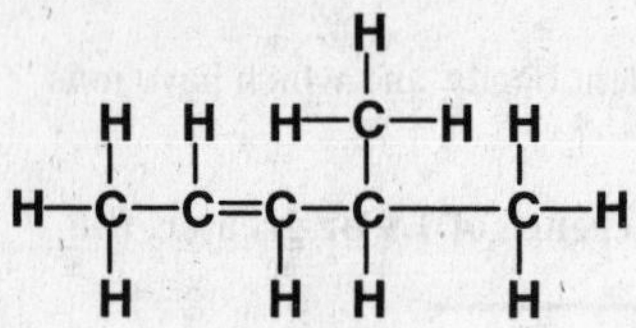

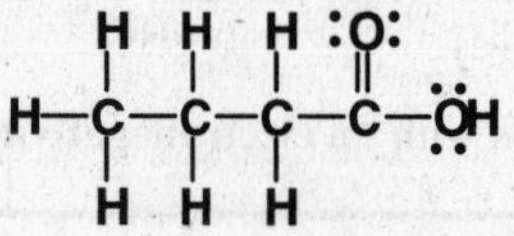

C$_4$H$_8$O

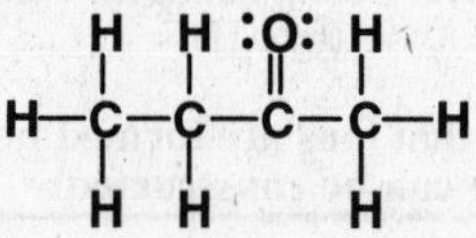

(d) C—C—C—H
with O double bonded above third C, and C below second C

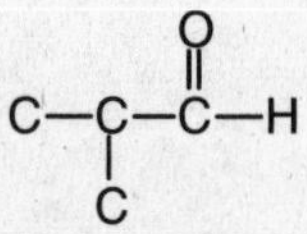

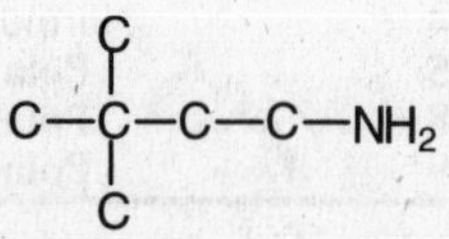

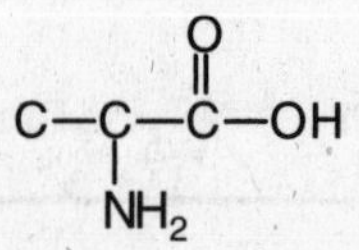

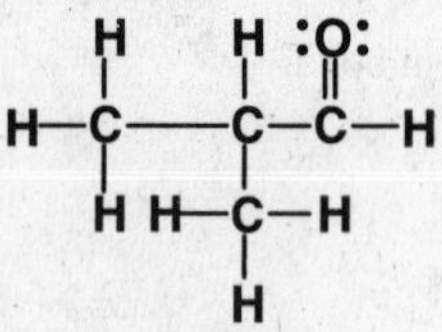

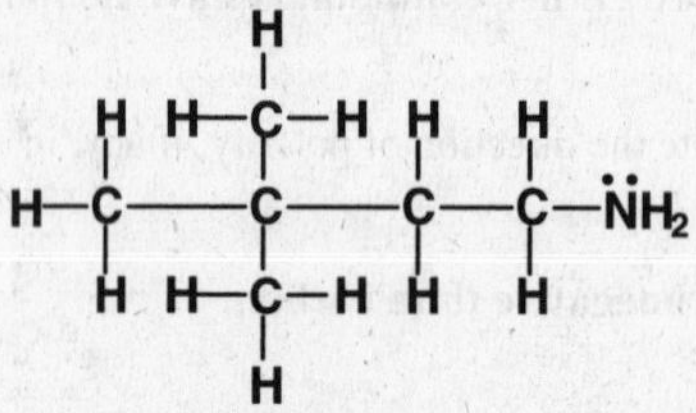

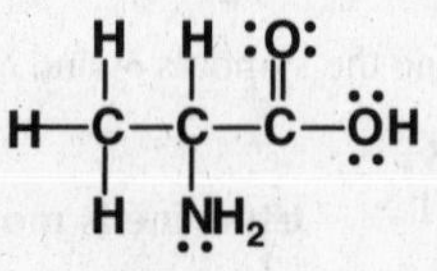

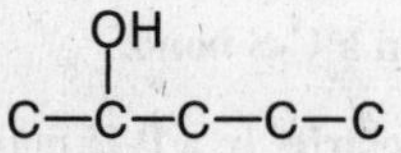

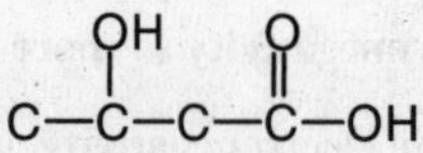

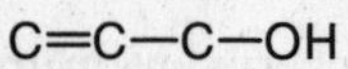

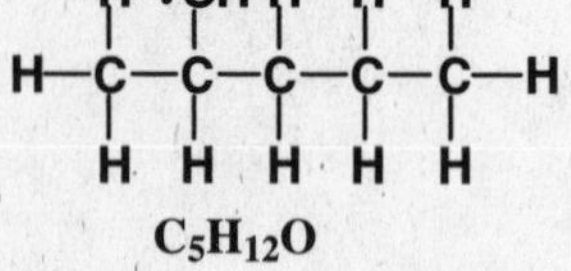

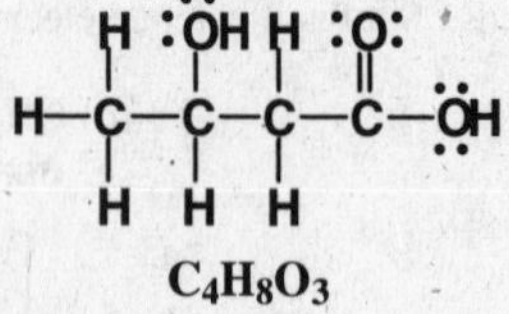

C$_4$H$_8$O

(e) C—C—C—C—NH$_2$
with C below second C

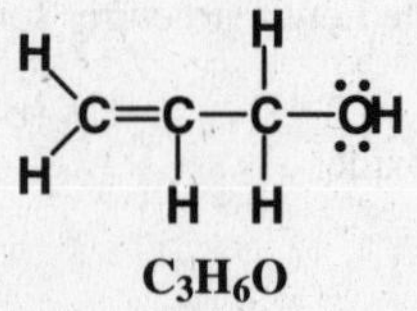

C$_6$H$_{15}$N

(f) C—C—C—OH
with O double bonded above third C, and NH$_2$ below second C

C$_3$H$_7$NO$_2$

(g) C—C—C—C—C
with OH above second C

C$_5$H$_{12}$O

(h) C—C—C—C—OH
with OH above second C, O double bonded above fourth C

C$_4$H$_8$O$_3$

(i) C=C—C—OH

C$_3$H$_6$O

<u>Problem 1.30</u> Some of these structural formulas are incorrect (i.e. they do not represent a real compound) because they have atoms with an incorrect number of bonds. Which structural formulas are incorrect, and which atoms in them have an incorrect number of bonds?

The molecules in (a), (b), (d), and (f) are incorrect, because there are five bonds to the circled carbon atom, not four.

<u>Problem 1.31</u> Following the rule that each atom of carbon, oxygen, and nitrogen reacts to achieve a complete outer shell of eight valence electrons, add unshared pairs of electrons as necessary to complete the valence shell of each atom in these ions. Then assign formal charges as appropriate.

The following structural formulas show all valence electrons and all formal charges.

<u>Problem 1.32</u> Following are several Lewis structures showing all valence electrons. Assign formal charges in each structure as appropriate.

There is a positive formal charge in parts (a), (e), and (f). There is a negative formal charge in parts (b), (c), and (d).

<u>Polarity of Covalent Bonds</u>
<u>Problem 1.33</u> Which statements are true about electronegativity?
(a) Electronegativity increases from left to right in a period of the Periodic Table.
(b) Electronegativity increases from top to bottom in a column of the Periodic Table.
(c) Hydrogen, the element with the lowest atomic number, has the smallest electronegativity.
(d) The higher the atomic number of an element, the greater its electronegativity.

Electronegativity _increases_ from left to right across a period and from bottom to top of a column in the Periodic Table. Thus, statement (a) is true, but (b), (c), and (d) are false.

<u>Problem 1.34</u> Why does fluorine, the element in the upper right corner of the Periodic Table, have the largest electronegativity of any element?

Electronegativity increases with increasing positive charge on the nucleus and with decreasing distance of the valence electrons from the nucleus. Fluorine is that element for which these two parameters lead to maximum electronegativity.

<u>Problem 1.35</u> Arrange the single covalent bonds within each set in order of increasing polarity.
(a) C-H, O-H, N-H   (b) C-H, B-H, O-H   (c) C-H, C-Cl, C-I
   C-H < N-H < O-H   **B-H < C-H < O-H**   **C-I < C-H < C-Cl**

(d) C-S, C-O, C-N   (e) C-Li, C-B, C-Mg
   C-S < C-N < C-O   **C-B < C-Mg < C-Li**

<u>Problem 1.36</u> Using the values of electronegativity given in Table 1.5, predict which indicated bond in each set is the more polar and, using the symbols δ+ and δ-, show the direction of its polarity
(a) CH_3-OH or CH_3O-H   (b) CH_3-NH_2 or CH_3-PH_2

        δ- δ+   δ+ δ-
    CH_3O-H   **H_3C-NH_2**

(c) CH_3-SH or CH_3S-H   (d) CH_3-F or H-F

      δ- δ+   δ+ δ-
   CH_3S-H   **H-F**

<u>Problem 1.37</u> Identify the most polar bond in each molecule.
(a) $HSCH_2CH_2OH$ (b) $CHCl_2F$ (c) $HOCH_2CH_2NH_2$

  The O-H bond   **The C-F bond**   **The O-H bond**
     (1.4)   **(1.5)**   **(1.4)**

The difference in electronegativities is given in parentheses underneath each answer.

<u>**Bond Angles and Shapes of Molecules**</u>
<u>Problem 1.38</u> Use VSEPR to predict bond angles about each highlighted atom.

Approximate bond angles as predicted by valence-shell electron-pair repulsion are as shown.

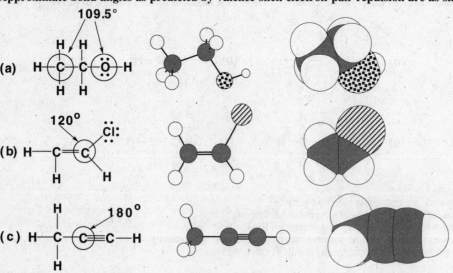

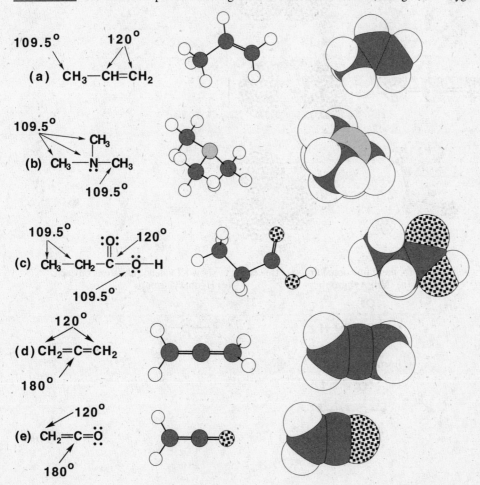

Problem 1.39 Use VSEPR to predict bond angles about each atom of carbon, nitrogen, and oxygen in these molecules.

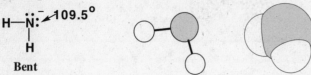

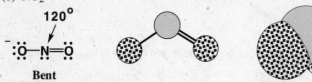

(f) $CH_3-CH=N-O-H$

109.5°   109.5°

120°

Problem 1.40 Use VSEPR to predict the geometry of these ions.
(a) NH_2^-

$H-\ddot{N}:^-$ 109.5°
  |
  H

Bent

(b) NO_2^-

120°

$^-:\ddot{O}-N=\ddot{O}$

Bent

(c) NO_2^+

180°

$\ddot{O}=N=\ddot{O}$
      +

Linear

(d) NO_3^-

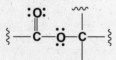

$:\ddot{O}$ 120°
  ‖
$^-:\ddot{O}-N-\ddot{O}:^-$
      +

Trigonal planar

Functional Groups
Problem 1.41 Draw Lewis structures for these functional groups. Be certain to show all valence electrons on each.
(a) Carbonyl group   (b) Carboxyl group   (c) Hydroxyl group

$\begin{array}{c}:\ddot{O}:\\ \| \\ \xi-C-\xi \end{array}$ $\begin{array}{c}:\ddot{O}:\\ \| \\ \xi-C-\ddot{O}-H \end{array}$ $\xi-\ddot{O}-H$

(d) Ester group   (e) Amide group

$\begin{array}{c}:\ddot{O}:\\ \| \\ \xi-C-\ddot{O}-C-\xi \end{array}$ $\begin{array}{c}:\ddot{O}:\\ \| \\ \xi-C-\dot{\ddot{N}}-\xi \end{array}$

Problem 1.42 Draw condensed structural formulas for all compounds with the molecular formula C_4H_8O that contain
(a) A carbonyl group (there are two aldehydes and one ketone).

Ketone

$$CH_3-\overset{\overset{O}{\|}}{C}-CH_2-CH_3 \quad \text{also written as} \quad CH_3COCH_2CH_3$$

Aldehydes

$$CH_3-CH_2-CH_2-\overset{\overset{O}{\|}}{C}-H \quad \text{also written as} \quad CH_3CH_2CH_2CHO$$

$$CH_3-\underset{\underset{CH_3}{|}}{CH}-\overset{\overset{O}{\|}}{C}-H \quad \text{also written as} \quad (CH_3)_2CHCHO$$

(b) A carbon-carbon double bond and a hydroxyl group (there are eight)

There are three separate but related things to build into this answer; the carbon skeleton (the order of attachment of carbon atoms), the location of the double bond, and the location of the -OH group. Here, as in other problems of this type, it is important to have a system and to follow it. As one way to proceed, first decide the number of different carbon skeletons that are possible. A little doodling with paper and pencil should convince you that there are only two.

$$C-C-C-C \quad \text{and} \quad C-\overset{\overset{C}{|}}{C}-C$$

Next locate the double bond on these carbon skeletons. There are three possible locations for it.

$$C=C-C-C \quad \text{and} \quad C-C=C-C \quad \text{and} \quad C=\overset{\overset{C}{|}}{C}-C$$

Finally, locate the -OH group and then add the remaining seven hydrogens to complete each structural formula. For the first carbon skeleton, there are four possible locations of the -OH group; for the second carbon skeleton there are two possible locations; and for the third, there are also two possible locations. Four of these compounds (marked by a # symbol) are not stable and are in equilibrium with a more stable aldehyde or ketone. You need not be concerned, however, with this now. Just concentrate on drawing the required eight condensed structural formulas.

$$\overset{\#}{HO}-CH=CH-CH_2-CH_3 \qquad CH_2=\overset{\overset{\#OH}{|}}{C}-CH_2-CH_3 \qquad CH_2=CH-\overset{\overset{OH}{|}}{CH}-CH_3$$

$$CH_2=CH-CH_2-CH_2-OH \qquad HO-CH_2-CH=CH-CH_3 \qquad CH_3-\overset{\overset{\#OH}{|}}{C}=CH-CH_3$$

$$\overset{\#}{HO}-CH=\overset{\overset{CH_3}{|}}{C}-CH_3 \qquad CH_2=\overset{\overset{CH_3}{|}}{C}-CH_2-OH$$

Problem 1.43 What is the meaning of the term tertiary (3°) when it is used to classify alcohols? Draw a structural formula for the one tertiary (3°) alcohol with the molecular formula $C_4H_{10}O$.

A tertiary alcohol is one in which the -OH group is on a tertiary carbon atom. A tertiary carbon atom is one that is bonded to three other carbon atoms.

$$\begin{array}{ccccc} & & H & :\!\overset{..}{O}\!H & H \\ & & | & | & | \\ H & - & C & - C - & C - H \\ & & | & | & | \\ & & H & & H \\ & & & | & \\ & & H - C - H & \\ & & | & \\ & & H & \end{array}$$

<u>Problem 1.44</u> What is the meaning of the term tertiary (3°) when it is used to classify amines? Draw a structural formula for the one tertiary (3°) amine with the molecular formula $C_4H_{11}N$.

A tertiary amine is one in which the nitrogen atom is bonded to three carbon atoms.

<u>Problem 1.45</u> Draw structural formulas for
(a) The four primary (1°) amines with the molecular formula $C_4H_{11}N$.

(b) The three secondary (2°) amines with the molecular formula $C_4H_{11}N$.

(c) The one tertiary (3°) amine with the molecular formula $C_4H_{11}N$.

<u>Problem 1.46</u> Draw structural formulas for the three tertiary (3°) amines with the molecular formula $C_5H_{13}N$.

Problem 1.47 Draw structural formulas for
(a) The eight alcohols with the molecular formula $C_5H_{12}O$.

To make it easier for you to see the patterns of carbon skeletons and functional groups, only carbon atoms and hydroxyl groups are shown in the following solutions. To complete these structural formulas, you need to supply enough hydrogen atoms to complete the tetravalence of each carbon.

There are three different carbon skeletons on which the -OH group can be placed:

$$C-C-C-C-C \qquad \overset{\displaystyle C-C-C-C}{\underset{\displaystyle C}{|}} \qquad \overset{\displaystyle C}{\underset{\displaystyle C}{C-C-C}}$$

Three alcohols are possible from the first carbon skeleton, four from the second carbon skeleton, and one from the third carbon skeleton.

$$HO-C-C-C-C-C \qquad \overset{OH}{C-C-C-C-C} \qquad \overset{OH}{C-C-C-C-C}$$
$$(1) \qquad\qquad (2) \qquad\qquad (3)$$

$$HO-\underset{\displaystyle C}{C}-C-C-C \quad \overset{OH}{\underset{\displaystyle C}{C-C-C-C}} \quad \overset{OH}{\underset{\displaystyle C}{C-C-C-C}} \quad \underset{\displaystyle C}{C-C-C-C-OH} \quad \overset{\displaystyle C}{\underset{\displaystyle C}{C-C-C-OH}}$$
$$(4) \qquad (5) \qquad (6) \qquad (7) \qquad (8)$$

(b) The eight aldehydes with the molecular formula $C_6H_{12}O$.

Following are structural formulas for the eight aldehydes of molecular formula $C_6H_{12}O$. They are drawn starting with the aldehyde group and then attaching the remaining five carbons in a chain (structure 1), then four carbons in a chain and one carbon as a branch on the chain (structures 2, 3, and 4) and finally three carbons in a chain and two carbons as branches (structures 5, 6, 7, and 8).

$$\overset{\displaystyle O}{C-C-C-C-C-\overset{\|}{C}-H} \quad \underset{\displaystyle C}{C-C-C-C-\overset{\displaystyle O}{\overset{\|}{C}}-H} \quad \underset{\displaystyle C}{C-C-C-C-\overset{\displaystyle O}{\overset{\|}{C}}-H}$$
$$(1) \qquad\qquad (2) \qquad\qquad (3)$$

$$\underset{\displaystyle C}{C-C-C-C-\overset{\displaystyle O}{\overset{\|}{C}}-H} \quad \underset{\displaystyle C}{\overset{\displaystyle C}{C-C-C-\overset{O}{\overset{\|}{C}}-H}} \quad \underset{\displaystyle C}{\overset{\displaystyle C}{C-C-C-\overset{O}{\overset{\|}{C}}-H}}$$
$$(4) \qquad\qquad (5) \qquad\qquad (6)$$

$$\underset{\displaystyle C}{\overset{\displaystyle C\ \ O}{C-C-\overset{\|}{C}-H}} \quad \underset{\displaystyle C-C}{C-C-\overset{\displaystyle O}{\overset{\|}{C}}-H}$$
$$(7) \qquad\qquad (8)$$

(c) The six ketones with the molecular formula $C_6H_{12}O$.

Following are structural formulas for the six ketones of molecular formula $C_6H_{12}O$. They are drawn first with all combinations of one carbon to the left of the carbonyl group and four carbons to the right (structures 1, 2, 3, and 4) and then with two carbons to the left and three carbons to the right (structures 5 and 6).

$$C-\overset{\overset{\displaystyle O}{\|}}{C}-C-C-C-C \qquad C-\overset{\overset{\displaystyle O}{\|}}{C}-\underset{\underset{\displaystyle C}{|}}{C}-C-C \qquad C-\overset{\overset{\displaystyle O}{\|}}{C}-C-\underset{\underset{\displaystyle C}{|}}{C}-C \qquad C-\overset{\overset{\displaystyle O}{\|}}{C}-\underset{\underset{\displaystyle C}{|}}{\overset{\overset{\displaystyle C}{|}}{C}}-C$$

      (1)   **(2)**   **(3)**   **(4)**

$$C-C-\overset{\overset{\displaystyle O}{\|}}{C}-C-C-C \qquad C-C-\overset{\overset{\displaystyle O}{\|}}{C}-\underset{\underset{\displaystyle C}{|}}{\overset{\overset{\displaystyle C}{|}}{C}}-C$$

            (5)   **(6)**

(d) The eight carboxylic acids with the molecular formula $C_6H_{12}O_2$.

There are eight carboxylic acids of molecular formula $C_6H_{12}O_2$. They have the same carbon skeletons as the eight aldehydes of molecular formula $C_6H_{12}O$ shown in part (b) of this problem. In place of the aldehyde group, substitute a carboxyl group.

$$C-C-C-C-C-\overset{\overset{\displaystyle O}{\|}}{C}-OH \quad C-\underset{\underset{\displaystyle C}{|}}{C}-C-C-\overset{\overset{\displaystyle O}{\|}}{C}-OH \quad C-C-\underset{\underset{\displaystyle C}{|}}{C}-C-\overset{\overset{\displaystyle O}{\|}}{C}-OH \quad C-C-C-\underset{\underset{\displaystyle C}{|}}{C}-\overset{\overset{\displaystyle O}{\|}}{C}-OH$$

     (1)   **(2)**   **(3)**   **(4)**

$$C-\underset{\underset{\displaystyle C}{|}}{\overset{\overset{\displaystyle C}{|}}{C}}-C-\overset{\overset{\displaystyle O}{\|}}{C}-OH \quad C-\underset{\underset{\displaystyle C}{|}}{\overset{\overset{\displaystyle C}{|}}{C}}-\underset{\underset{\displaystyle C}{|}}{C}-\overset{\overset{\displaystyle O}{\|}}{C}-OH \quad C-C-\underset{\underset{\displaystyle C}{|}}{\overset{\overset{\displaystyle C}{|}}{C}}-\overset{\overset{\displaystyle O}{\|}}{C}-OH \quad C-C-\underset{\underset{\displaystyle C-C}{|}}{C}-\overset{\overset{\displaystyle O}{\|}}{C}-OH$$

     (5)   **(6)**   **(7)**   **(8)**

(e) The nine carboxylic esters with the molecular formula $C_5H_{10}O_2$.

Start with unbranched carbon chains of all possible lengths, then add branching to complete the set.

$$C-C-C-\overset{\overset{\displaystyle O}{\|}}{C}-O-C \quad C-C-\overset{\overset{\displaystyle O}{\|}}{C}-O-C-C \quad C-\overset{\overset{\displaystyle O}{\|}}{C}-O-C-C-C \quad \overset{\overset{\displaystyle O}{\|}}{C}-O-C-C-C-C$$

       (1)   **(2)**   **(3)**   **(4)**

$$C-\underset{\underset{\displaystyle C}{|}}{C}-\overset{\overset{\displaystyle O}{\|}}{C}-O-C \quad C-\overset{\overset{\displaystyle O}{\|}}{C}-O-\underset{\underset{\displaystyle C}{|}}{C}-C \quad \overset{\overset{\displaystyle O}{\|}}{C}-O-\underset{\underset{\displaystyle C}{|}}{C}-C-C \quad \overset{\overset{\displaystyle O}{\|}}{C}-O-C-\underset{\underset{\displaystyle C}{|}}{C}-C \quad \overset{\overset{\displaystyle O}{\|}}{C}-O-\underset{\underset{\displaystyle C}{|}}{\overset{\overset{\displaystyle C}{|}}{C}}-C$$

      (5)   **(6)**   **(7)**   **(8)**   **(9)**

<u>Problem 1.48</u> Identify the functional groups in each compound.

           Hydroxyl group (2°)   **Hydroxyl group (1°)**   **Hydroxyl group (1°)**

(a) $CH_3-\underset{\underset{\displaystyle\boxed{OH}}{|}}{CH}-\boxed{\overset{\overset{\displaystyle O}{\|}}{C}-OH}$ —— Carboxyl group   (b) $\boxed{HO}-CH_2-CH_2-\boxed{OH}$

           Lactic acid   **Ethylene glycol**

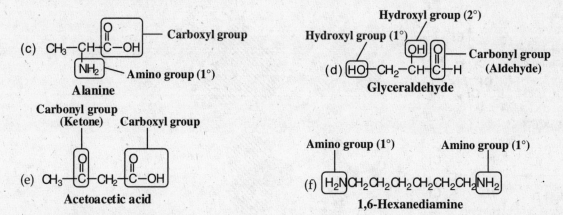

(c) CH₃–CH–C–OH — **Carboxyl group**
           | ‖
          NH₂ O — **Amino group (1°)**
      Alanine

Hydroxyl group (2°)
Hydroxyl group (1°)
(d) HO–CH₂–CH–C–H — **Carbonyl group (Aldehyde)**
              OH O
      Glyceraldehyde

Carbonyl group (Ketone)   **Carboxyl group**
(e) CH₃–C–CH₂–C–OH
          ‖   ‖
          O   O
      Acetoacetic acid

Amino group (1°)   **Amino group (1°)**
(f) H₂N–CH₂CH₂CH₂CH₂CH₂CH₂–NH₂
      1,6-Hexanediamine

Polar and Nonpolar Molecules

Problem 1.49 Draw a three-dimensional representation for each molecule. Indicate which ones have a dipole moment and in what direction it is pointing.

In the following diagrams, the C-H bond dipole moment has been left out because it is a nonpolar covalent bond. The listed dipole moments were looked up in the chemical literature and are only added for reference. You will not be expected to calculate these.

(a) CH₃F

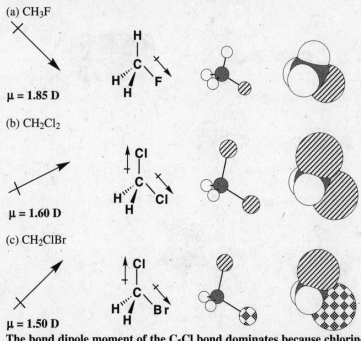

μ = 1.85 D

(b) CH₂Cl₂

μ = 1.60 D

(c) CH₂ClBr

μ = 1.50 D
The bond dipole moment of the C-Cl bond dominates because chlorine is the more electronegative element.

(d) CFCl₃

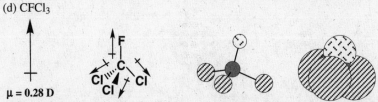

μ = 0.28 D
The bond dipole moment of the C-F bond dominates because of the higher electronegativity of fluorine.

(e) CCl$_4$

No molecular dipole moment

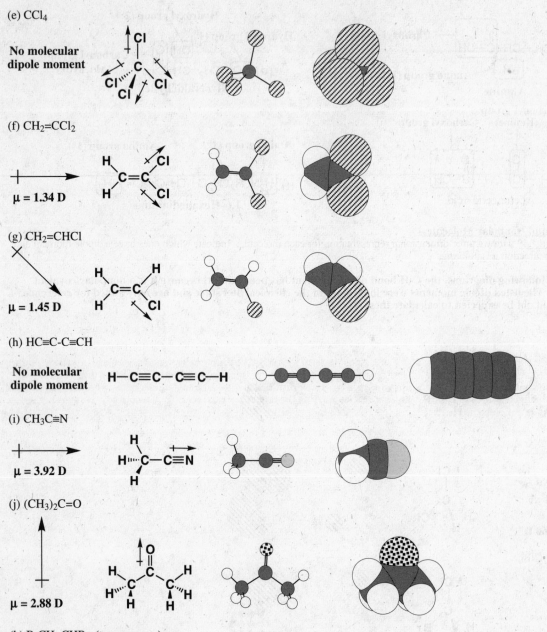

(f) CH$_2$=CCl$_2$

μ = 1.34 D

(g) CH$_2$=CHCl

μ = 1.45 D

(h) HC≡C-C≡CH

No molecular dipole moment

(i) CH$_3$C≡N

μ = 3.92 D

(j) (CH$_3$)$_2$C=O

μ = 2.88 D

(k) BrCH=CHBr (two answers)

The two bromine atoms can either be on opposite sides or on the same side of the double bond. Recall that double bonds do not rotate.

No molecular dipole moment

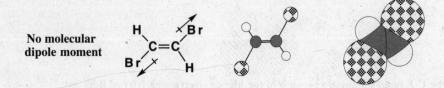

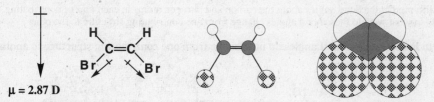

μ = 2.87 D

<u>Problem 1.50</u> Tetrafluoroethylene, C_2F_4, is the starting material for the synthesis of the polymer polytetrafluoroethylene (PTFE), one form of which is known as Teflon. Tetrafluoroethylene has a dipole moment of zero. Propose a structural formula for this molecule.

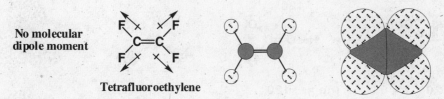

No molecular
dipole moment

Tetrafluoroethylene

Resonance and Contributing Structures
<u>Problem 1.51</u> Which statements are true about resonance contributing structures?
(a) All contributing structures must have the same number of valence electrons.
(b) All contributing structures must have the same arrangement of atoms.
(c) All atoms in a contributing structure must have complete valence shells.
(d) All bond angles in sets of contributing structures must be the same.

For sets of contributing structures, electrons (usually π electrons or lone pair electrons) move, but the atomic nuclei maintain the same arrangement in space. The atoms are arranged the same with the same bond angles among them, so statements (b) and (d) are true. In addition, the total number of electrons, valence and inner shell electrons, in each contributing structure must be the same, so statement (a) is also true. However, the movement of electrons often leaves one or more atoms without a filled valence shell in a given contributing structure, so statement (c) is false.

<u>Problem 1.52</u> Draw the contributing structure indicated by the curved arrow(s). Assign formal charges as appropriate.

(a) H—Ö—C(=O:)(Ö:⁻) ⟷ H—Ö—C(:Ö:⁻)(O:)

(b) H—Ö—C(=O:)(Ö:⁻) ⟷ H—Ö⁺=C(:Ö:⁻)(:Ö:)

(c) CH₃—Ö—C(=O:)(Ö:⁻) ⟷ CH₃—Ö—C⁺(:Ö:⁻)(:Ö:⁻)

(d) :Ö=C=Ö: ⟷ :Ö=C—Ö:⁻

(e) H—Ö—N=Ö: ⟷ H—Ö—N⁺—Ö:⁻

(f) H—Ö—N=Ö: ⟷ H—O—N⁺—Ö:⁻

Problem 1.53 Using VSEPR, predict the bond angles about the carbon and nitrogen atoms in each pair of contributing structures in problem 1.52. In what way do these bond angles change from one contributing structure to the other?

As stated in the answer to Problem 1.51, bond angles do not change from one contributing structure to another.

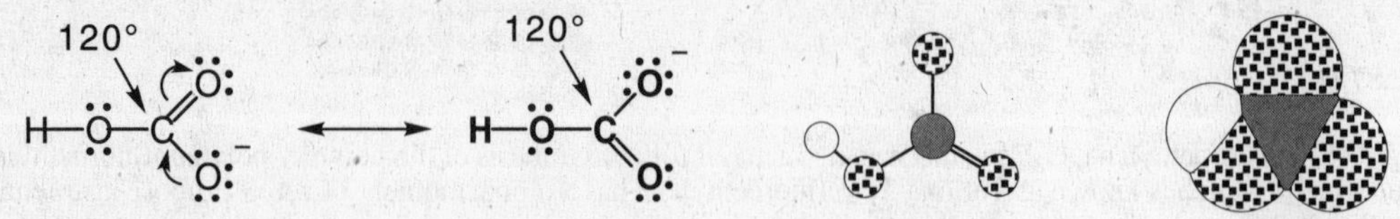

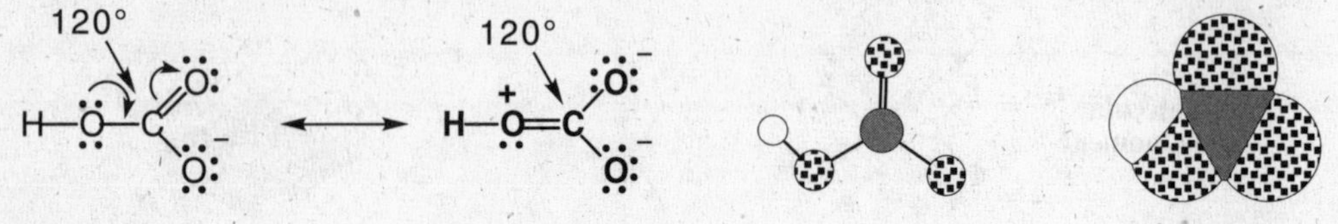

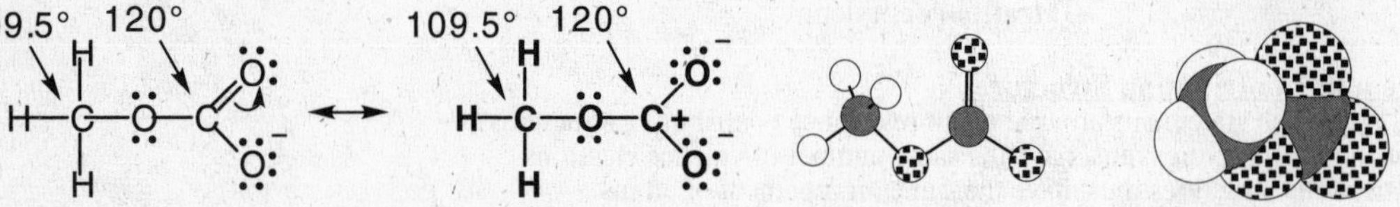

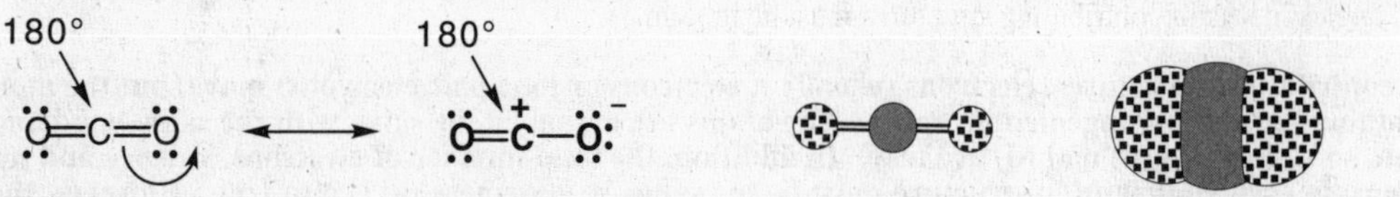

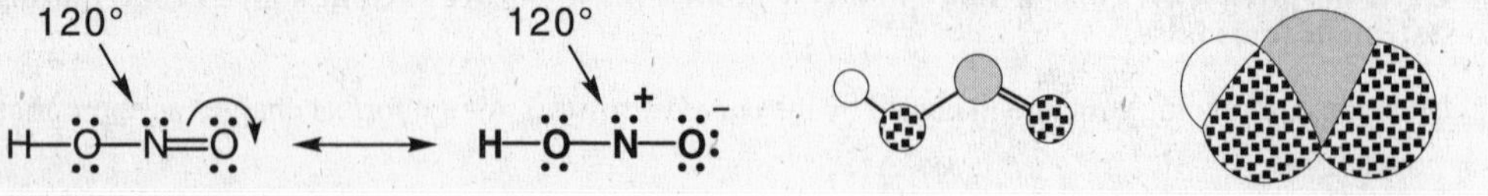

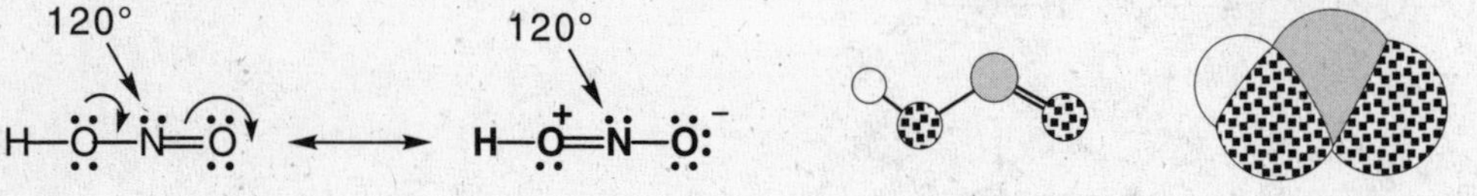

Problem 1.54 In the Problem 1.52 you were given one contributing structure and asked to draw another. Label pairs of contributing structures that are equivalent. For those sets in which the contributing structures are not equivalent, label the more important contributing structure.

(a) The two structures are equivalent because each involves a similar separation of charge.
(b, c, d, e, f) The first structure is more important, because the second involves creation and separation of unlike charges.

Problem 1.55 Are the structures in each set valid contributing structures?

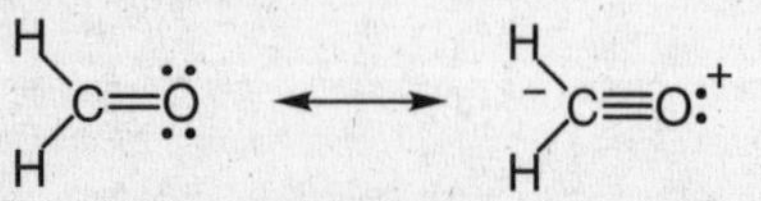

The structure on the right is not a valid contributing structure because there are 10 electrons in the valence shell of the carbon atom.

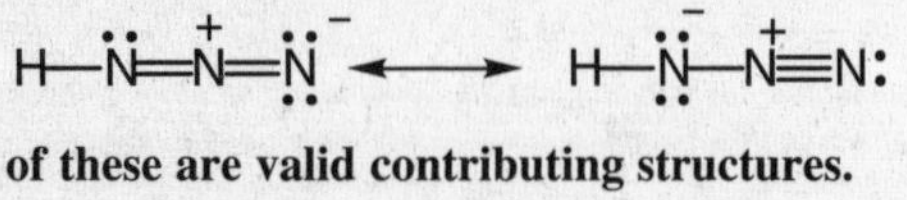

Both of these are valid contributing structures.

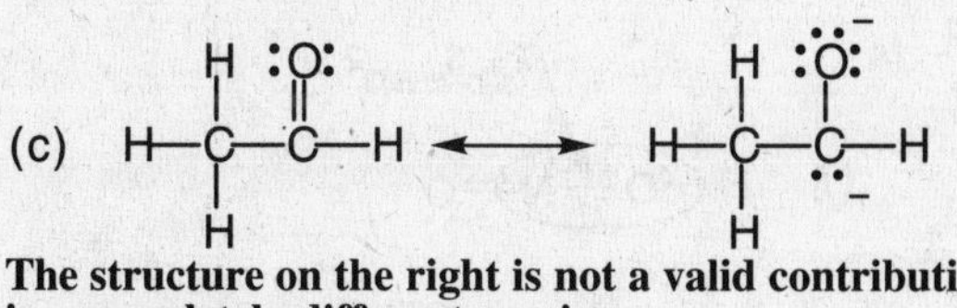

(c)

The structure on the right is not a valid contributing structure because there are two extra electrons and thus it is a completely different species.

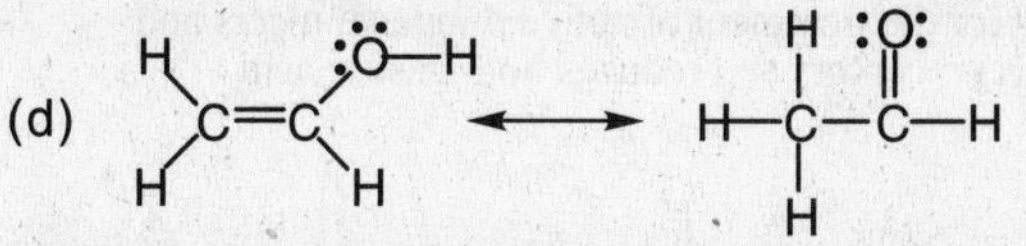

(d)

Although each is a valid Lewis structure, they are not valid contributing structures for the same resonance hybrid. An atomic nucleus, namely a hydrogen, has changed position. Later you will learn that these two molecules are related to each other and are called tautomers.

Valence Bond Theory

Problem 1.56 State the orbital hybridization of each highlighted atom.

Each circled atom is either sp, sp^2, or sp^3 hybridized.

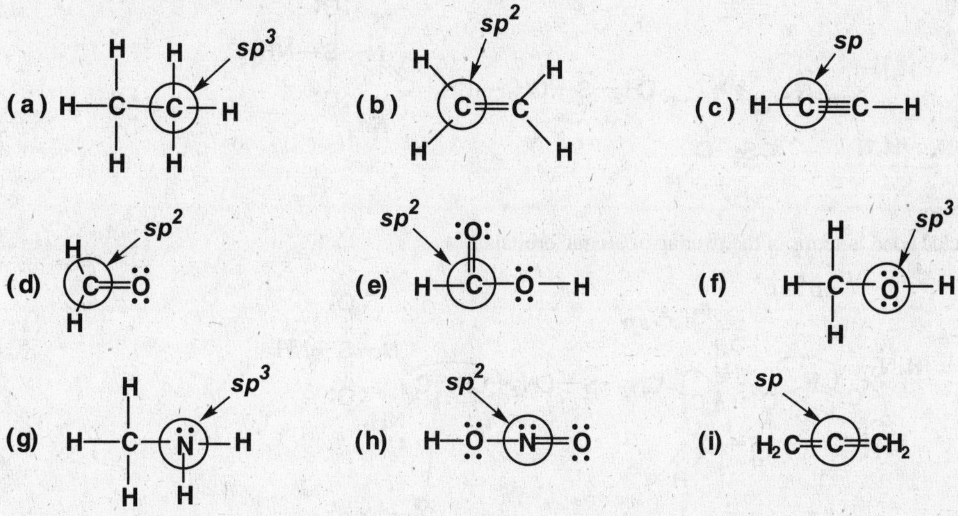

(a) (b) (c)

(d) (e) (f)

(g) (h) (i)

Problem 1.57 Describe each highlighted bond in terms of the overlap of atomic orbitals.

Shown is whether the bond is σ or π, as well as the orbitals used to form it.

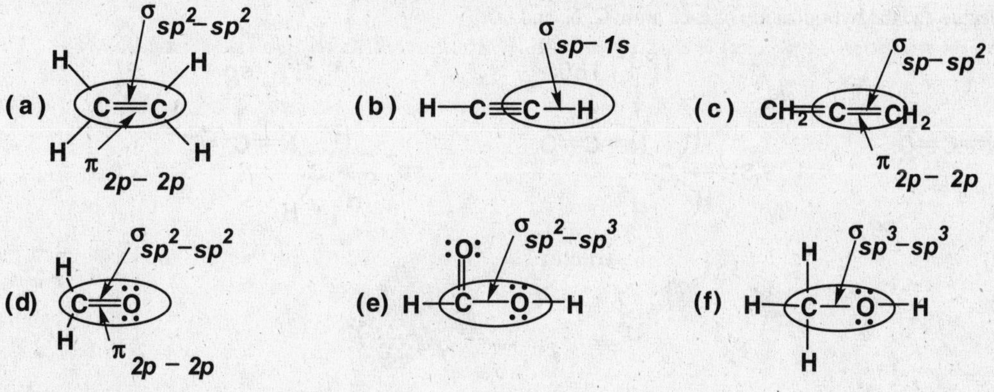

(a) (b) (c)

(d) (e) (f)

(g) $\sigma_{sp^3-sp^3}$ structure with H and C, N, H atoms

(h) σ_{sp^2-1s} structure with H, C, O, C, H atoms

(i) $\sigma_{sp^3-sp^2}$ structure with H, O, N, O atoms

Problem 1.58 Following is a structural formula of the prescription drug famotidine, manufactured by Merck Sharpe & Dohme under the name Pepcid. The primary clinical use of Pepcid is for the treatment of active duodenal ulcers and benign gastric ulcers. Pepcid is a competitive inhibitor of histamine H_2 receptors and reduces both gastric acid concentration and the volume of gastric secretions.

(a) Complete the Lewis structure of famotidine showing all valence electrons and any positive or negative charges.

(b) Describe each circled bond in terms of the overlap of atomic orbitals.

Problem 1.59 Draw a Lewis structure for methyl isocyanate, CH_3NCO, showing all valence electrons. Predict all bond angles in this molecule and the hybridization of each atom C, N, and O.

Combined MO/VB Theory

Problem 1.60 What is the hybridization of the highlighted atoms in the following structures, and what are your estimates for the bond angles around these highlighted atoms? In each case, in what kind of orbital does the lone pair of electrons on the nitrogen reside.

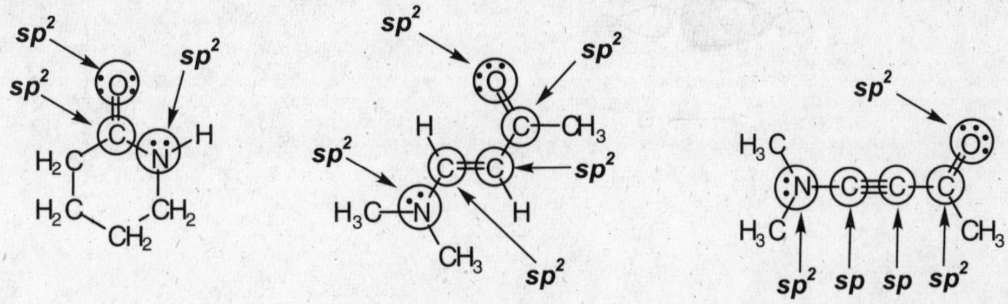

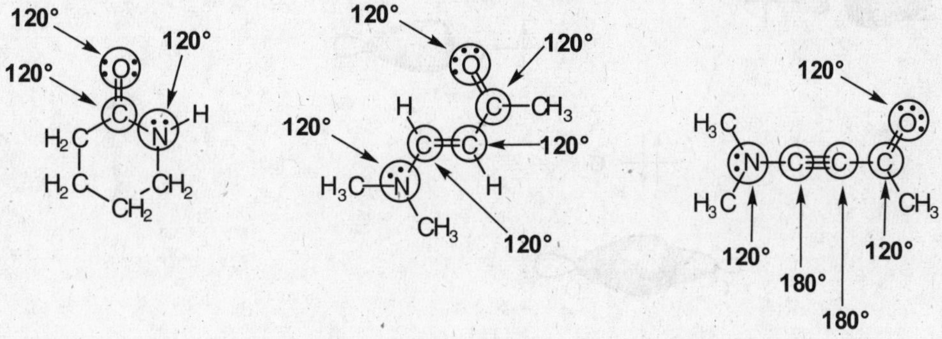

In each case there are significant contributing structures that have a π bond involving nitrogen.

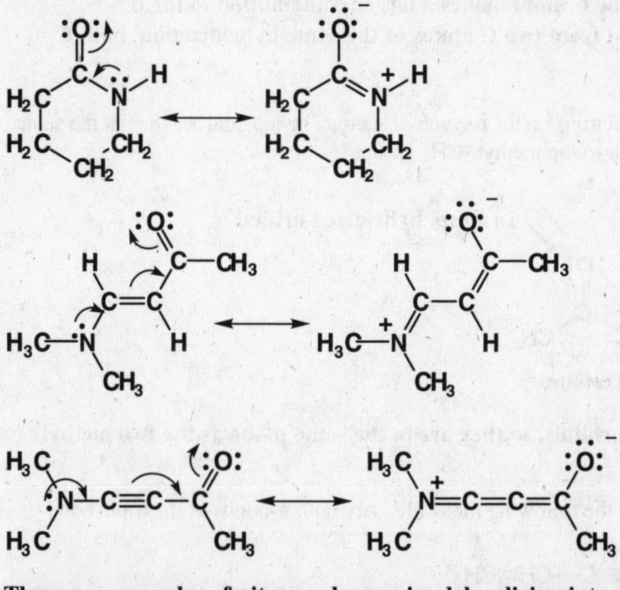

These are examples of nitrogen lone pairs delocalizing into adjacent π bonds, a common feature of many organic molecules you will come across. For this to happen, the nitrogen atoms must be *sp²* hybridized, so the lone pairs on nitrogen are best thought of as being in *2p* orbitals. Such delocalization of electron density in π orbitals is stabilizing and therefore favorable, a phenomenon that is best explained using quantum mechanical arguments (beyond the scope of this text).

Problem 1.61 Using cartoon representations, draw a molecular orbital mixing diagram for a C-O σ-bond. In your picture, consider the relative energies of C and O, and how this changes the resulting bonding and antibonding molecular orbitals relative to a C-C σ-bond.

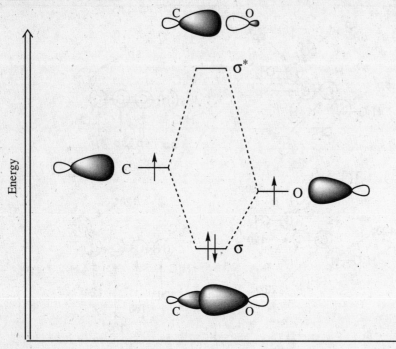

The O atom, being more electronegative, is of lower energy than the C atom. This means the O orbital makes a larger contribution to the σ-bonding orbital, while the C atom makes a larger contribution to the σ*-antibonding orbital. For a σ-bonding orbital formed from two C atoms of the same hybridization, both C orbitals make equal contributions.

Problem 1.62 In what kind of orbitals do the lone-pair electrons on the oxygen of acetone reside, and are they in the same plane as the methyl -CH$_3$ groups or are they perpendicular to the methyl -CH$_3$ groups?

In an *sp²* hybridized orbital   **In an *sp²* hybridized orbital**

$$:\overset{\cdot\cdot}{O}:$$
$$\underset{H_3C}{\overset{\|}{C}}\underset{CH_3}{}$$

Acetone

In acetone, both lone pairs reside in *sp²* hybridized orbitals, so they are in the same plane as the two methyl groups.

Problem 1.63 Draw the delocalized molecular orbitals for the following molecule. Are both π-bonds of the triple bond involved in the delocalized orbitals?

$$H_3C—C≡C—CH=CH_2$$

Shown below are the *2p* orbitals involved with delocalized π-bonding.

H₃C—C≡C—CH=CH₂

    sp   sp   sp² sp²

The delocalized molecular orbital involves only the four parallel *2p* orbitals as shown below. The perpendicular *2p* orbitals of the two *sp* hybridized carbons only overlap with each other, so they are not involved with delocalized bonding.

H₃C—C≡C—CH=CH₂

Additional Problems

Problem 1.64 Why are the following molecular formulas impossible?
(a) CH₅

Carbon atoms can only accommodate 8 electrons in their valence shell, and each hydrogen atom can only accommodate one bond. Thus, there is no way for a stable bonding arrangement to be created that utilizes one carbon atom and all five hydrogen atoms.

(b) C₂H₇

Because hydrogen atoms can only accommodate one bond each, no single hydrogen atom can make stable bonds to both carbon atoms. Thus, the two carbon atoms must be bonded to each other. This means that each of the bonded carbon atoms can accommodate only three more bonds. Therefore, only six hydrogen atoms can be bonded to the carbon atoms, not seven hydrogen atoms.

Problem 1.65 Each compound contains both ions and covalent bonds. Draw the Lewis structure for each compound, and show by dashes which are covalent bonds and show by charges which are ions.
(a) CH₃ONa   (b) NH₄Cl   (c) NaHCO₃
    Sodium methoxide   Ammonium chloride   Sodium bicarbonate

(d) NaBH₄   (e) LiAlH₄
    Sodium borohydride   Lithium aluminum hydride

In naming these compounds, the cation is named first followed by the name of the anion.

<u>Problem 1.66</u> Predict whether the carbon-metal bond in these organometallic compounds is nonpolar covalent, polar covalent, or ionic. For each polar covalent bond, show the direction of its polarity by the symbols δ+ and δ-.

$$
\begin{array}{ccc}
(0.7) & (1.3) & (0.6) \\
\delta - \quad \delta + & \delta - \quad \delta + & \delta - \quad \delta +
\end{array}
$$

(a) CH_3CH_2——Pb-CH_2CH_3 above: CH_2CH_3, below: CH_2CH_3

(b) CH_3—Mg—Cl

Methylmagnesium chloride

(c) CH_3—Hg—CH_3

Dimethylmercury

Tetraethyllead

All of these carbon-metal bonds are polar covalent because the difference in electronegativities is between 0.5 and 1.9. In each case, carbon is the more electronegative element so it has the partial negative charge. The difference in electronegativities is given above the carbon-metal bond in each answer.

<u>Problem 1.67</u> Silicon is immediately under carbon in the Periodic Table. Predict the geometry of silane, SiH_4.

Silicon is in Group 4 of the Periodic Table, and like carbon, has four valence electrons. In silane, SiH_4, silicon is surrounded by four regions of electron density. Therefore, you should predict all H-Si-H bond angles to be 109.5°, so the molecule is tetrahedral around Si.

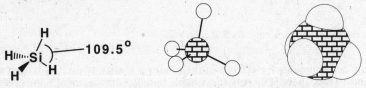

<u>Problem 1.68</u> Phosphorus is immediately under nitrogen in the Periodic Table. Predict the molecular formula for phosphine, the compound formed by phosphorus and hydrogen. Predict the H-P-H bond angle in phosphine.

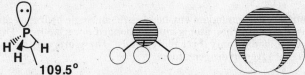

Like nitrogen, phosphorus has five valence electrons, so you should predict that phosphine has the molecular formula of PH_3 in analogy to ammonia, NH_3. In phosphine, the phosphorus atom is surrounded by four regions of electron density; one lone pair of electrons and single bonds to three hydrogen. Therefore, predict all H-P-H bond angles to be roughly 109.5°, meaning the molecule is pyramidal.

<u>Problem 1.69</u> Draw a Lewis structure for the azide ion, N_3^-. (The order of attachment is N-N-N and they do not form a ring). How does the resonance model account for the fact that the lengths of the N-N bonds in this ion are identical.

It is not possible to draw a single Lewis structure that adequately describes the azide ion. Rather, it can be drawn as the hybrid of three contributing structures.

$$:N{\equiv}N{-}\ddot{N}:^{2-} \quad \longleftrightarrow \quad \ddot{N}{=}N{=}\ddot{N}: \quad \longleftrightarrow \quad \ddot{N}{-}N{\equiv}N:$$

Taken together, the three contributing structures present a symmetric picture of azide ion bonding, thus explaining why both N-N bonds are identical.

<u>Problem 1.70</u> Cyanic acid, HOCN, and isocyanic acid, HNCO, dissolve in water to yield the same anion on loss of H^+.
(a) Write a Lewis structure for cyanic acid.   (b) Write a Lewis structure for isocyanic acid.

$$H{-}\ddot{O}{-}C{\equiv}N: \qquad\qquad \ddot{O}{=}C{=}\ddot{N}{-}H$$

(c) Account for the fact that each acid gives the same anion on loss of H^+.

Loss of an H^+ from the two different acids gives the same anion that can best be described by drawing the following two contributing structures.

$$H-\overset{..}{\underset{..}{O}}-C\equiv N:$$ $$\overset{..}{\underset{..}{O}}=C=\overset{..}{N}-H$$

loss of H⁺   **loss of H⁺**

$$:\overset{-}{\underset{..}{O}}-C\equiv N: \longleftrightarrow \overset{..}{\underset{..}{O}}=C=\overset{..}{N}:^{-}$$

Looking Ahead
Problem 1.71 In Chapter 6, we study a group of organic cations called carbocations. Following is the structure of one such carbocation, the *tert*-butyl cation.

$$\begin{array}{c} H_3C \\ \overset{+}{C}-CH_3 \\ H_3C \end{array}$$

tert-Butyl cation

(a) How many electrons are in the valence shell of the carbon bearing the positive charge?

There are six valence shell electrons on the carbon atom bearing the positive charge, two contained in each of the three single bonds.

(b) Using VSEPR predict the bond angles about this carbon.

According to VSEPR, there are three regions of electron density around the central carbon atom, so you should predict a trigonal planar geometry and C-C-C bond angles of 120°.

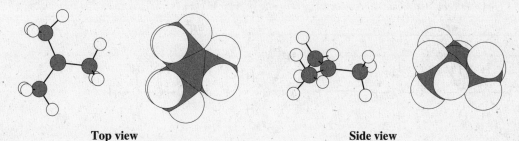

Top view   **Side view**

(c) Given the bond angle you predicted in (b), what hybridization do you predict for this carbon?

Given the trigonal planar geometry predicted in (b), so you should predict sp^2 hybridization of this carbon atom.

Problem 1.72 Many reactions involve a change in hybridization of one or more atoms in the starting material. In each reaction, identify the atoms in the organic starting material that change hybridization and indicate what the change is. We examine these reactions in more detail later in the course.

(a) $$\begin{array}{c} sp^2 \\ H H \\ C=C \\ H H \\ sp^2 \end{array} + Cl_2 \longrightarrow \begin{array}{c} sp^3 :\overset{..}{\underset{..}{Cl}}: H \\ H-C-C-H \\ H :\overset{..}{\underset{..}{Cl}}: sp^3 \end{array}$$

(b)

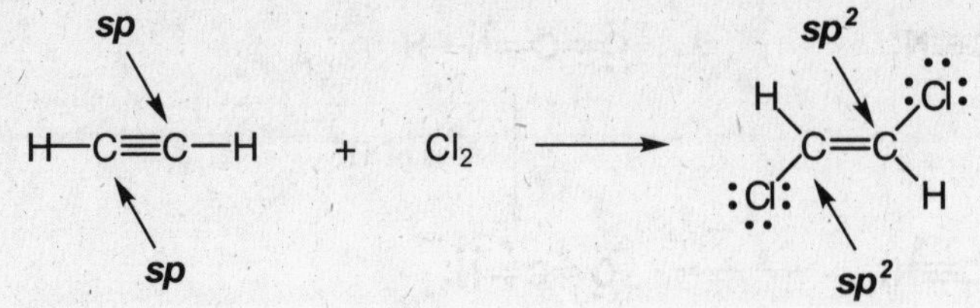

(c)

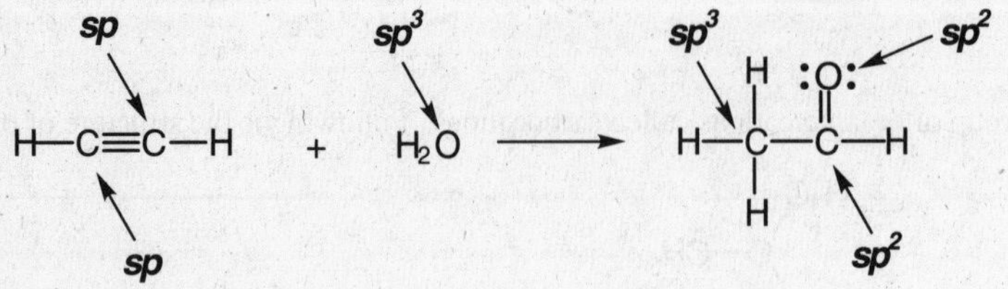

(d)

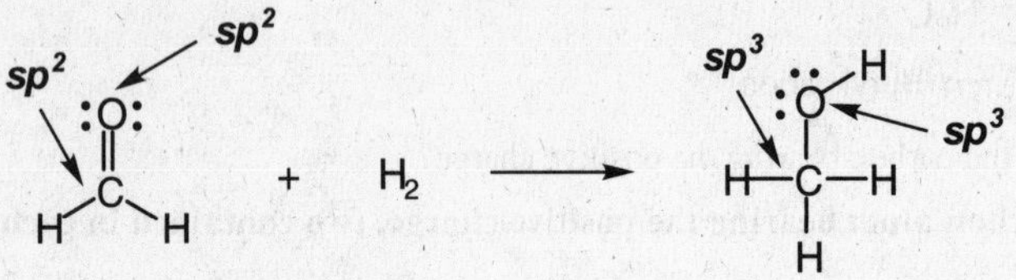

(e)

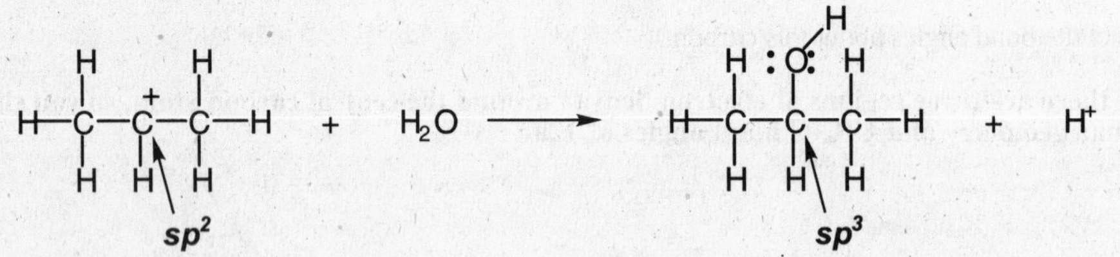

(f)

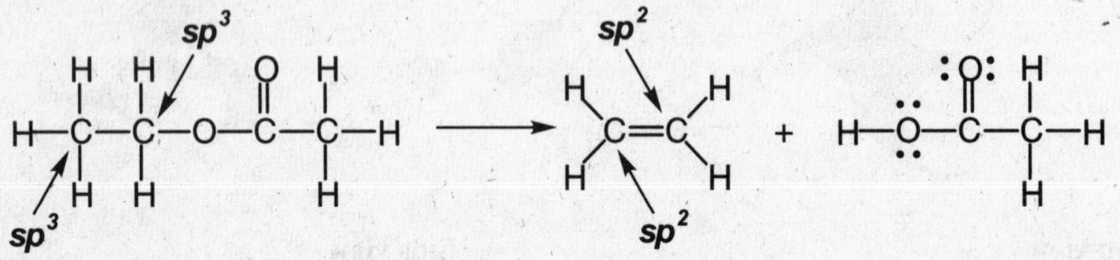

<u>Problem 1.73</u> Following is a structural formula of benzene, C_6H_6, which we study in Chapter 21

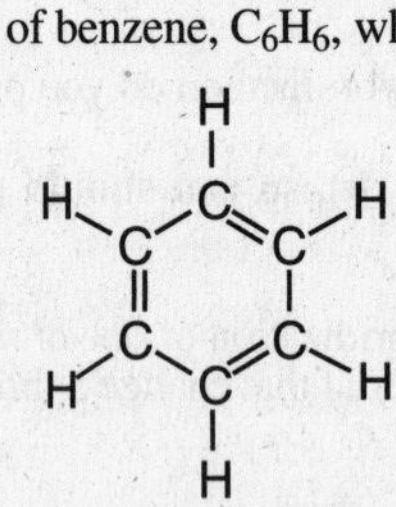

(a) Using VSEPR, predict each H-C-C and C-C-C bond angle in benzene.

Each carbon atom in benzene has three regions of electron density around it, so according to VSEPR, the carbon atoms are trigonal planar. You should predict each H-C-C bond angle to be 120° and each C-C-C bond angle to be 120°.

(b) State the hybridization of each carbon atom in benzene.

Each carbon atom is sp^2 hybridized because each one makes three σ bonds and one π bond.

(c) Predict the shape of a benzene molecule.

Because all of the carbon atoms in the ring are *sp²* hybridized and thus trigonal planar, predict carbon atoms in benzene to form a flat hexagon in shape, with the hydrogen atoms in the same plane as the carbon atoms.

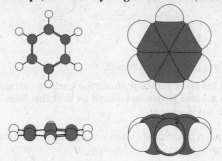

(d) Draw important resonance contributing structures.

Problem 1.74 Following are three contributing structures for diazomethane, CH_2N_2. This molecule is used to make methyl esters from carboxylic acids (Section 17.7C).

(a) Using curved arrows, show how each contributing structure is converted to the one on its right.

The arrows are indicated on the above structures.

(b) Which contributing structure makes the largest contribution to the hybrid?

The middle and left structures have filled valence shells, so these will make a larger contribution to the hybrid than the structure on the right, in which the terminal nitrogen atom has an unfilled valence shell. The structure in the middle has the negative charge on the more electronegative atom, N, compared with the structure on the left (negative charge on C), so the structure in the middle will make the largest contribution to the resonance hybrid.

Problem 1.75 Draw a Lewis structure for the ozone molecule, O_3. (The order of atom attachment is O-O-O and they do not form a ring). How does the resonance model account for the fact that the length of each O-O bond in ozone (128 pm) is shorter than the O-O single bond in hydrogen peroxide (HOOH 147 pm), but longer than the O-O double bond in the oxygen molecule (123 pm).

It is not possible to draw a single Lewis structure that adequately describes the ozone molecule. Rather, it is better to draw ozone as a hybrid of four contributing structures, each with a separation of charges.

Taken together, the two contributing structures present a symmetric picture of the bonding in which each O-O bond is intermediate between a single bond and a double bond. Recall that bonds become shorter as bond order increases. As a result, the bonds in ozone are shorter than the single O-O bond in HOOH, but longer than the O=O double bond in the oxygen molecule.

Molecular Orbitals

Problem 1.76 The following two compounds are isomers; that is, they are different compounds with the same molecular formula. We discuss this type of isomerism in Chapter 5.

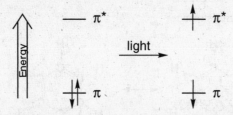

(a) Why are these different molecules that do not interconvert?

Interconversion of the two isomers involves rotation about the carbon-carbon double bond. This cannot occur without breaking the π bond. The π bond is strong enough so that this does not happen spontaneously at room temperature and the isomers do not interconvert.

(b) Absorption of light by a double bond in a molecule excites one electron from a π molecular orbital to a π* molecular orbital. Explain how this absorption can lead to interconversion of the two isomers.

Energy ↑

—— π*   ⥮ π*

light →

⥯ π   ↿ π

Putting electron density into an antibonding (*) orbital of a bond weakens that bond. Excitation of the electron from the π bond to the π* orbital upon absorption of light weakens the π bond, allowing the molecule to rotate about the carbon-carbon bond. This rotation interconverts the two isomers. A similar alkene rotation reaction is responsible for the mammalian photoreceptor molecules that allow us to see visible light.

Problem 1.77 In future chapters we will encounter carbanions-ions in which a carbon atom has three bonds and alone pair of electrons and bears a negative charge. Draw another contributing structure for the allyl anion. Now using cartoon representations, draw the three orbitals that represent the delocalized π system (look at Figure 1.26 for a hint). Which of the three orbitals are populated with electrons?

Allyl anion

Below is drawn a cartoon representation of the allyl anion π molecular orbitals.

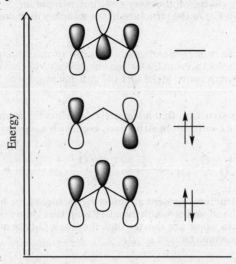

The lowest two molecular orbitals are filled with a pair of electrons each. Notice that filling of the middle orbital, with lobes on only the two terminal carbon atoms, indicates the negative charge will be found on these two atoms consistent with the contributing structures.

<u>Problem 1.78</u> Describe the bonding in PCl_5 without using d orbitals. As a hint, the geometry of PCl_5 is as shown:

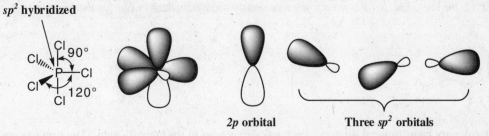

sp² **hybridized**

2p orbital   Three *sp²* orbitals

Based on the bond angles, the bonding in PCl_5 can be explained if the P atom is *sp²* hybridized. The three *sp²* hybridized orbitals would overlap with Cl orbitals to form the three "equatorial" σ bonds spaced at 120°, while the unhybridized *2p* orbital would overlap with Cl orbitals to form the two "axial" σ bonds.

Bandit angelfish *Holacanthus arcuatus,*
Lanai, Hawaii

CHAPTER 2
Solutions to the Problems

<u>Problem 2.1</u> Do the line-angle formulas in each pair represent the same compound or constitutional isomers.

(a) 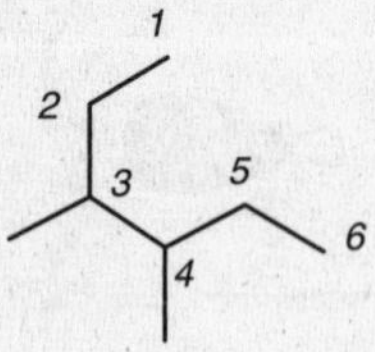 and

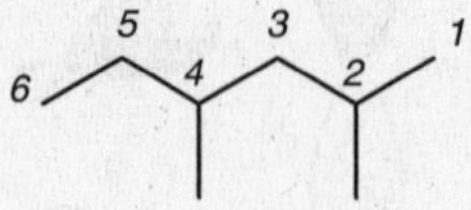

These molecules are constitutional isomers. Each has six carbons in the longest chain. The first has one-carbon branches on carbons 3 and 4 of the chain; the second has one-carbon branches on carbons 2 and 4 of the chain.

(b) 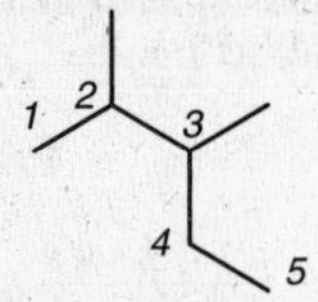 and

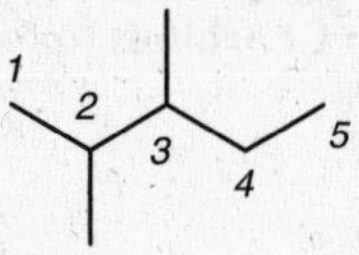

These molecules are identical. Each has five carbons in the longest chain, and one-carbon branches on carbons 2 and 3 of the chain.

<u>Problem 2.2</u> Draw line-angle formulas for the three constitutional isomers with the molecular formula C_5H_{12}.

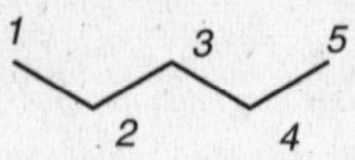

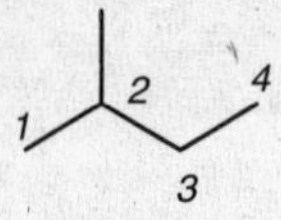

<u>Problem 2.3</u> Write IUPAC names for these alkanes.

(a)
Methyl group

1-Methylethyl group

Methyl group   **1-Methylethyl group**

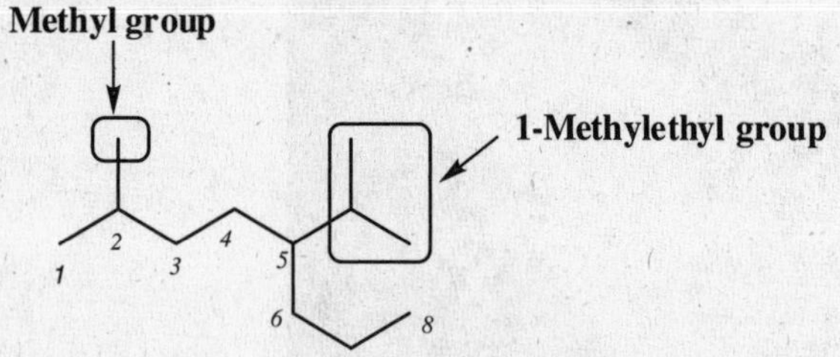

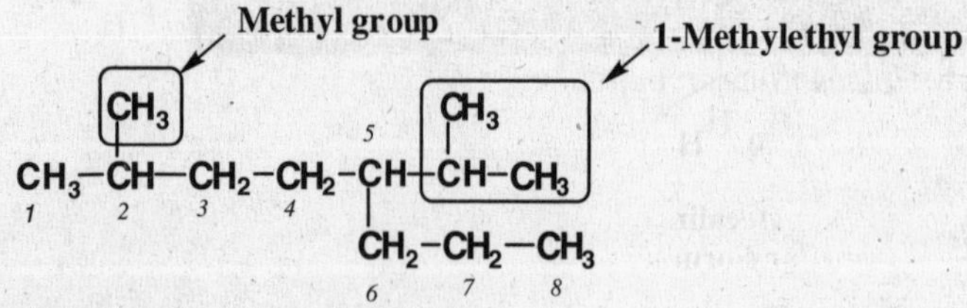

2-Methyl-5-(1-methylethyl)octane

(b)

6   8
5   7
1
2   3
4

← Propyl group

← 1-Methylethyl group

5   6   7   8
CH₂−CH₂−CH₂−CH₃

1   2   3   4
CH₃−CH₂−CH₂−C−CH₂−CH₂−CH₃   ← Propyl group

CH₃−CH−CH₃   ← 1-Methylethyl group

4-(1-Methylethyl)-4-propyloctane

Problem 2.4 Combine the proper prefix, infix, and suffix and write the IUPAC name for each compound.

$$\text{(a)} \quad CH_3\overset{\displaystyle O}{\overset{\|}{C}}CH_3 \qquad \text{(b)} \quad CH_3(CH_2)_3\overset{\displaystyle O}{\overset{\|}{C}}H$$

(c) —OH

(d)

Propanone   **Pentanal**   **Cyclopentanol**   **Cycloheptene**

Problem 2.5 Write the molecular formula, IUPAC name, and common name for each cycloalkane.

(a)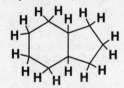

 Molecular Formula C₉H₁₈
 (2-Methylpropyl)cyclopentane (IUPAC)
 Isobutylcyclopentane (Common)

(b)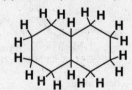

 Molecular Formula C₁₁H₂₂
 (1-Methylpropyl)cycloheptane (IUPAC)
 ***sec*-Butylcycloheptane (Common)**

(c)

 Molecular Formula C₆H₁₂
 1-Ethyl-1-methylcyclopropane (IUPAC and Common)

Problem 2.6 Write molecular formulas for each bicycloalkane, given its number of carbon atoms.
(a) Hydrindane (9 carbons)   (b) Decalin (10 carbons)   (c) Norbornane (7 carbons)

Hydrindane
Molecular Formula C₉H₁₆

Decalin
Molecular Formula C₁₀H₁₈

Norbornane
Molecular Formula C₇H₁₂

Problem 2.7 Following are the structural formulas and names of four bicycloalkanes. Write the molecular formula of each compound. Which of these compounds are constitutional isomers?

(a)   (b)   (c)   (d)

Thujane   **Carane**   **Pinane**   **Bornane**

Molecular   Molecular   Molecular   Molecular
formula   formula   formula   formula
$C_{10}H_{18}$ $C_{10}H_{18}$ $C_{10}H_{18}$ $C_{10}H_{19}$

As shown by comparing molecular formulas, the first three bicycloalkanes are constitutional isomers.

Problem 2.8 For 1,2-dichloroethane,
(a) Draw Newman projections for all eclipsed conformations formed by rotation from 0° to 360° about the carbon-carbon single bond.

Higher
Energy   Lower
                                            Energy

(b) Which eclipsed conformation(s) has (have) the lowest energy? Which has (have) the highest energy?

The chlorine atoms are the largest by far. As a result, when the chlorine atoms are eclipsed with each other (structure on the left), their steric interaction causes the higher overall energy.

(c) Which, if any, of these eclipsed conformations are related by reflection?

The two lower energy conformations (structures on the right) are related by reflection as they represent "mirror images" of each other.

Problem 2.9 Following is a chair conformation of cyclohexane with the carbon atoms numbered 1 through 6.

(a) Draw hydrogen atoms that are above the plane of the ring on carbons 1 and 2 and below the plane of the ring on carbon 4.
(b) Which of these hydrogens are equatorial? Which are axial?
(c) Draw the alternative chair conformation. Which hydrogens are equatorial? Which are axial? Which are above the plane of the ring? Which are below it?

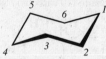

In the above figure (a) = axial and (e) = equatorial.

Problem 2.10 Draw the alternative chair conformation for the trisubstituted cyclohexane given in Example 2.10. Label all CH_3/H 1,3-diaxial interactions in this chair conformation.

Diaxial interactions between this methyl group and the circled axial hydrogens

These two methyl groups are equatorial, so they do not have any diaxial interactions.

Note how in the above equilibrium, the new chair conformation (on the right) is the more stable due to having fewer diaxial interactions (See Problem 2.13)

Problem 2.11 Draw a chair conformation of 1,4-dimethylcyclohexane in which one methyl group is equatorial and the other is axial. Draw the alternative chair conformation and calculate the ratio of the two conformations at 25° C.

Diaxial interactions between this methyl group and the circled axial hydrogens

Diaxial interactions between this methyl group and the circled axial hydrogens

Each chair conformation has diaxial interactions between the circled axial methyl group and circled axial hydrogen atoms. Because each chair has the same number of diaxial interactions, they are of the same energy. The ratio of these two conformations must therefore be 1:1.

Problem 2.12 Which cycloalkanes show *cis, trans* isomerism? For each that does, draw both isomers.

(a)

cis-1,3-Dimethyl-cyclopentane

trans-1,3-Dimethyl-cyclopentane

1,3-Dimethylcyclopentane shows *cis, trans* isomerism. Here, the ring is drawn as a planar pentagon with substituents above and below the plane of the pentagon.

(b)

Ethylcyclopentane does not show *cis, trans* isomerism.

(c)

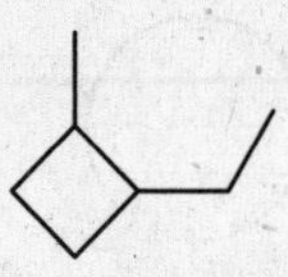

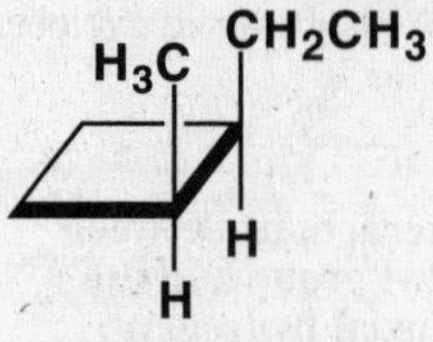

**cis-1-Ethyl-2-methyl-
cyclobutane**

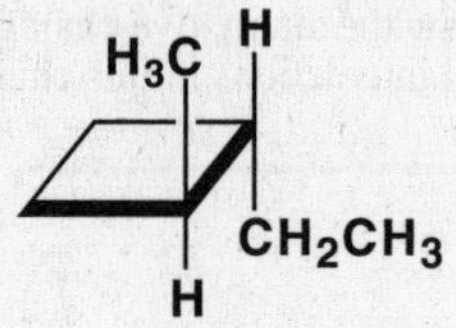

**trans-1-Ethyl-2-methyl-
cyclobutane**

1-Ethyl-2-methylcyclobutane shows *cis, trans* isomerism.

Problem 2.13 Following is a planar hexagon representation for one isomer of 1,2,4-trimethylcyclohexane. Draw the alternative chair conformations of this compound, and state which of the two is the more stable.

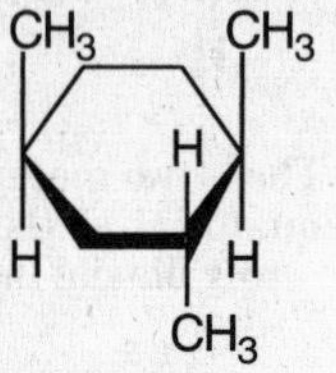

Following are alternative chair conformations for this isomer of 1,2,4-trimethylcyclohexane. The alternative chair conformation on the right is the more stable because it has only one axial methyl group.

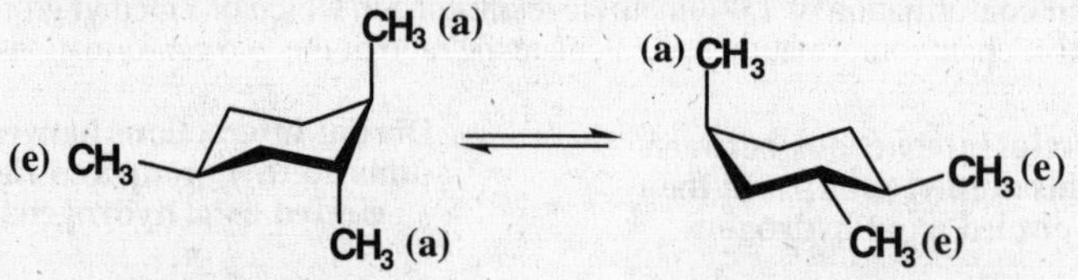

**less stable chair
(two methyl groups axial)**

**more stable chair
(one methyl group axial,
one gauche interaction)**

Problem 2.14 Here is one *cis, trans* isomer of 3,5-dimethylcyclohexanol. Complete the alternative chair conformations.

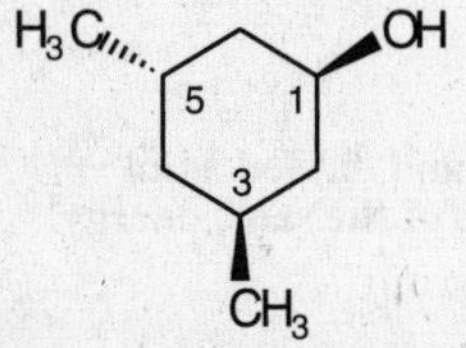

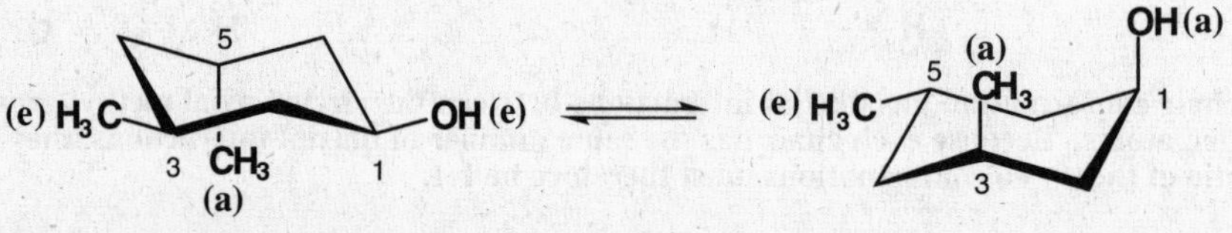

More stable

Problem 2.15 Arrange the alkanes in each set in order of increasing boiling point.
(a) 2-Methylbutane, 2,2-dimethylpropane, and pentane

**2,2-Dimethylpropane   2-Methylbutane   Pentane
(bp 9.5ºC)   (bp 29ºC)   (bp 36ºC)**

(b) 3,3-Dimethylheptane, 2,2,4-trimethylhexane, and nonane

**2,2,4-Trimethylhexane   3,3-Dimethylheptane   Nonane
(bp 130ºC)   (bp 138ºC)   (bp 151ºC)**

<u>Problem 2.16</u> Write a line-angle formula for each condensed structural formula.

(a) $CH_3CH_2CHCHCH_2CHCH_3$
with substituents CH_2CH_3, CH_3 (above) and $CH(CH_3)_2$ (below)

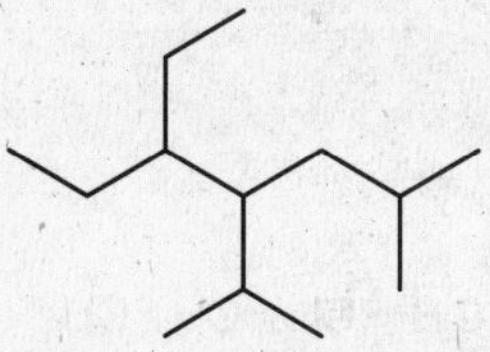

(b) CH_3CCH_3 with CH_3 above and CH_3 below

(c) $(CH_3)_2CHCH(CH_3)_2$

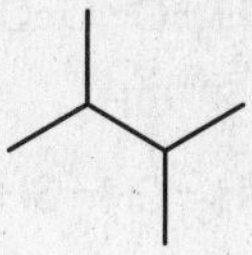

(d) $CH_3CH_2CCH_2CH_3$ with CH_2CH_3 above and CH_2CH_3 below

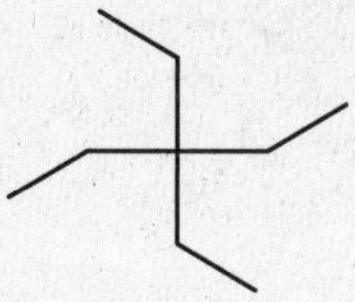

(e) $(CH_3)_3CH$

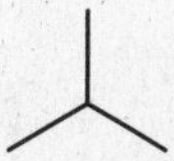

(f) $CH_3(CH_2)_3CH(CH_3)_2$

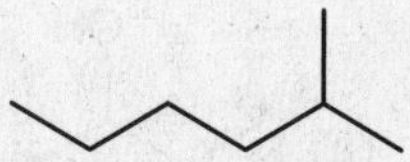

<u>Problem 2.17</u> Write the molecular formula of each alkane.

(a) $C_{10}H_{22}$

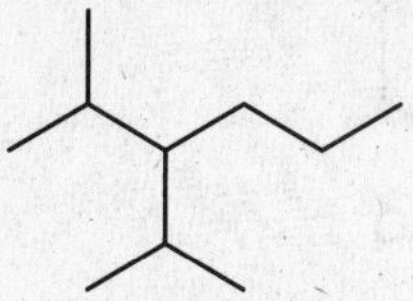

(b) C_8H_{18}

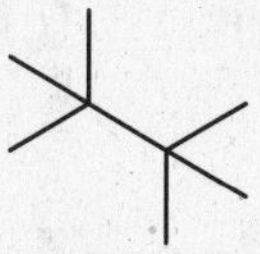

(c) $C_{11}H_{24}$

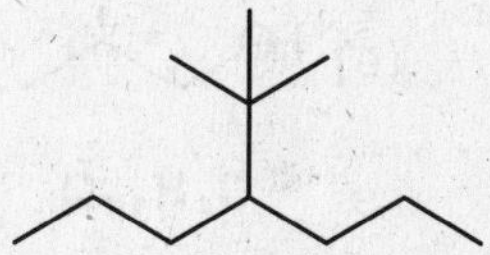

<u>Problem 2.18</u> Provide an even more abbreviated formula for each structural formula, using parentheses and subscripts..

(a) $CH_3CH_2CH_2CH_2CH_2CHCH_3$ with CH_3 above

(b) $HCCH_2CH_2CH_3$ with $CH_2CH_2CH_3$ above and $CH_2CH_2CH_3$ below

(c) $CH_3CCH_2CH_2CH_2CH_2CH_3$ with $CH_2CH_2CH_3$ above and $CH_2CH_2CH_3$ below

$CH_3(CH_2)_4CH(CH_3)_2$

$HC(CH_2CH_2CH_3)_3$

$CH_3C(CH_2CH_2CH_3)_2(CH_2)_4CH_3$

Constitutional Isomerism
<u>Problem 2.19</u> Which statements are true about constitutional isomers?
(a) They have the same molecular formula.
(b) They have the same molecular weight.
(c) They have the same order of attachment of atoms.
(d) They have the same physical properties.

Statements (a) and (b) are true, statements (c) and (d) are false.

Problem 2.20 Indicate whether the compounds in each set are constitutional isomers.

(a) CH_3-CH_2-OH and CH_3-O-CH_3

(b) $CH_3-\overset{\overset{O}{\|}}{C}-CH_3$ and $CH_3-CH_2-\overset{\overset{O}{\|}}{C}-H$

(c) $CH_3-\overset{\overset{O}{\|}}{C}-O-CH_3$ and $CH_3-CH_2-\overset{\overset{O}{\|}}{C}-OH$

(d) $CH_3-\overset{\overset{OH}{|}}{CH}-CH_2-CH_3$ and $CH_3-\overset{\overset{O}{\|}}{C}-CH_2-CH_3$

(e) ⬠ and $CH_3-CH_2-CH_2-CH_2-CH_3$

(f) ⬠ and $CH_2{=}CH-CH_2-CH_2-CH_3$

Sets (a), (b), (c), and (f) are constitutional isomers; sets (d) and (e) are not.

Problem 2.21 Each member of the following set of compounds is an alcohol; that is, each contains an -OH (hydroxyl group, Section 1.3A). Which structural formulas represent the same compound, and which represent constitutional isomers?

(a)   (b)   (c)   (d)

$C_4H_{10}O$ C_4H_8O C_4H_8O $C_4H_{10}O$

(e) HO   (f)   (g)   (h)

$C_4H_{10}O$ $C_4H_{10}O$ $C_4H_{10}O$ $C_7H_{14}O$

Structural formulas (d) and (e) are the same structural formula, and structural formulas (a) and (g) are also the same structural formula. Constitutional isomers have the same molecular formula but different connectivities between atoms. Structural formulas (a)/(g), (d)/(e) and (f) are constitutional isomers. Structural formulas (b) and (c) represent another set of constitutional isomers.

Problem 2.22 Each of the following compounds is an amine (Section 1.3B). Which structural formulas represent the same compound, and which represent constitutional isomers?

(a)   (b)   (c)   (d)

$C_4H_{11}N$ C_4H_9N $C_4H_{11}N$ $C_4H_{11}N$

(e)   (f)   (g)   (h)

$C_4H_{11}N$ $C_4H_{11}N$ $C_4H_{11}N$ $C_5H_{11}N$

Structural formulas (a) and (g) represent the same structural formula. Constitutional isomers have the same molecular formula but different connectivities between atoms.   Structural formulas (a)/(g), (c), (d), (e), and (f) are constitutional isomers.

<u>Problem 2.23</u> Each of the following compounds is either an aldehyde or a ketone (Section 1.3C). Which structural formulas represent the same compound, and which represent constitutional isomers?

(a)   (b)   (c)   (d)

C_4H_8O C_5H_8O $C_5H_{10}O$ C_4H_8O

(e)   (f)   (g)   (h)

C_4H_8O $C_5H_{10}O$ $C_6H_{10}O$ $C_6H_{10}O$

All of these molecules are different. Constitutional isomers have the same molecular formula but different connectivities between atoms. Compounds (a), (d), and (e) are constitutional isomers. Compounds (c) and (f) represent another set of constitutional isomers. A third set of constitutional isomers is composed of (g) and (h).

<u>Problem 2.24</u> Draw structural formulas, and write IUPAC names for the nine constitutional isomers with the molecular formula C_7H_{16}.

$CH_3CH_2CH_2CH_2CH_2CH_2CH_3$
Heptane
(bp 94.8°C)

$$CH_3\underset{\overset{|}{CH_3}}{CH}CH_2CH_2CH_2CH_3$$
2-Methylhexane
(bp 90.0°C)

$$CH_3CH_2\underset{\overset{|}{CH_3}}{CH}CH_2CH_2CH_3$$
3-Methylhexane
(bp 92.0°C)

$$CH_3\underset{\underset{CH_3}{\overset{|}{|}}}{\overset{\overset{CH_3}{|}}{C}}CH_2CH_2CH_3$$
2,2-Dimethylpentane
(bp 79.2°C)

$$CH_3\overset{\overset{CH_3}{|}}{CH}\underset{\overset{|}{CH_3}}{CH}CH_2CH_3$$
2,3-Dimethylpentane
(bp 89.8°C)

$$CH_3\overset{\overset{CH_3}{|}}{CH}CH_2\underset{\overset{|}{CH_3}}{CH}CH_3$$
2,4-Dimethylpentane
(bp 80.5°C)

$$CH_3CH_2\underset{\underset{CH_3}{\overset{|}{|}}}{\overset{\overset{CH_3}{|}}{C}}CH_2CH_3$$
3,3-Dimethylpentane
(bp 86.1°C)

$$CH_3CH_2\overset{\overset{CH_2CH_3}{|}}{CH}CH_2CH_3$$
3-Ethylpentane
(bp 93.5°C)

$$CH_3\overset{\overset{H_3C}{|}}{CH}\underset{\underset{CH_3}{\overset{|}{|}}}{\overset{\overset{CH_3}{|}}{C}}CH_3$$
2,2,3-Trimethylbutane
(bp 80.9°C)

<u>Problem 2.25</u> Draw structural formulas for all following.
(a) Alcohols with the molecular formula $C_4H_{10}O$.

$CH_3-CH_2-CH_2-CH_2-OH$ $CH_3-CH_2-\underset{\overset{|}{OH}}{CH}-CH_3$ $CH_3-\underset{\underset{CH_3}{|}}{\overset{\overset{CH_3}{|}}{C}}-OH$ $CH_3-\overset{\overset{CH_3}{|}}{CH}-CH_2-OH$

(b) Aldehydes with the molecular formula C_4H_8O.

$CH_3-CH_2-CH_2-\overset{\overset{O}{\|}}{C}-H$ $CH_3-\underset{\overset{|}{CH_3}}{CH}-\overset{\overset{O}{\|}}{C}-H$

(c) Ketones with the molecular formula $C_5H_{10}O$.

$$CH_3-CH_2-\overset{\displaystyle O}{\overset{\|}{C}}-CH_2-CH_3 \qquad CH_3-CH_2-CH_2-\overset{\displaystyle O}{\overset{\|}{C}}-CH_3 \qquad CH_3-\underset{\underset{\displaystyle CH_3}{|}}{CH}-\overset{\displaystyle O}{\overset{\|}{C}}-CH_3$$

(d) Carboxylic acids with the molecular formula $C_5H_{10}O_2$.

$$CH_3-\underset{\underset{\displaystyle CH_3}{|}}{CH}-CH_2-\overset{\displaystyle O}{\overset{\|}{C}}-OH \qquad CH_3-CH_2-CH_2-CH_2-\overset{\displaystyle O}{\overset{\|}{C}}-OH \qquad CH_3-\underset{\underset{\displaystyle CH_3}{|}}{\overset{\overset{\displaystyle CH_3}{|}}{C}}-\overset{\displaystyle O}{\overset{\|}{C}}-OH$$

$$CH_3-CH_2-\underset{\underset{\displaystyle CH_3}{|}}{CH}-\overset{\displaystyle O}{\overset{\|}{C}}-OH$$

Nomenclature of Alkanes and Cycloalkanes

Problem 2.26 Write IUPAC names for these alkanes and cycloalkanes.

(a)

2-Methylpentane

(b)

2,5-Dimethylhexane

(c)

3-Ethyloctane

(d)

2,2,5-Trimethylhexane

(e)

(2-Methylpropyl)cyclopentane

Problem 2.27 Write structural formulas for the following alkanes and cycloalkanes.

(a) 2,2,4-Trimethylhexane

$$CH_3\underset{\underset{\displaystyle CH_3}{|}}{\overset{\overset{\displaystyle CH_3}{|}}{C}}CH_2\underset{\underset{\displaystyle CH_3}{|}}{C}HCH_2CH_3$$

(b) 2,2-Dimethylpropane

$$CH_3\underset{\underset{\displaystyle CH_3}{|}}{\overset{\overset{\displaystyle CH_3}{|}}{C}}CH_3$$

(c) 3-Ethyl-2,4,5-trimethyloctane

$$CH_3\underset{\underset{\displaystyle CH_3}{|}}{C}H\underset{\underset{\displaystyle CH_2}{|}\;\overset{\displaystyle CH_3}{}}{C}HCHCHCH_2CH_2CH_3$$

(d) 5-Butyl-2,2-dimethylnonane

$$CH_3\underset{\underset{\displaystyle CH_3}{|}}{\overset{\overset{\displaystyle CH_3}{|}}{C}}CH_2CH_2\underset{\underset{\displaystyle CH_2CH_2CH_2CH_3}{|}}{C}HCH_2CH_2CH_2CH_3$$

(e) 4-(1-Methylethyl)octane

$$CH_3CH_2CH_2\underset{\underset{\displaystyle CH_3CHCH_3}{|}}{C}HCH_2CH_2CH_2CH_3$$

(f) 3,3-Dimethylpentane

$$CH_3CH_2\underset{\underset{\displaystyle CH_3}{|}}{\overset{\overset{\displaystyle CH_3}{|}}{C}}CH_2CH_3$$

(g) *trans*-1,3-Dimethylcyclopentane

(h) *cis*-1,2-Diethylcyclobutane

Problem 2.28 Explain why each is an incorrect IUPAC name, and write the correct IUPAC name for the intended compound.

(a) 1,3-Dimethylbutane

CH_3
$CH_3CHCH_2CH_2CH_3$

The longest chain is pentane.
Its IUPAC name is 2-methylpentane.

(b) 4-Methylpentane

CH_3
$CH_3CHCH_2CH_2CH_3$

The pentane is numbered incorrectly.
Its IUPAC name is 2-methylpentane.

(c) 2,2-Diethylbutane

CH_2CH_3
$CH_3CH_2CCH_2CH_3$
CH_3

The longest chain is pentane.
Its IUPAC name is 3-ethyl-3-methylpentane.

(d) 2-Ethyl-3-methylpentane

CH_3
$CH_3CH_2CH\ CHCH_2CH_3$
CH_3

The longest chain is hexane.
Its IUPAC name is 3,4-dimethylhexane.

(e) 2-Propylpentane

CH_3
$CH_3CH_2CH_2CHCH_2CH_2CH_3$

The longest chain is heptane.
Its IUPAC name is 4-methylheptane.

(f) 2,2-Diethylheptane

CH_2CH_3
$CH_3CH_2CCH_2CH_2CH_2CH_2CH_3$
CH_3

The longest chain is octane.
Its IUPAC name is 3-ethyl-3-methyloctane.

(g) 2,2-Dimethylcyclopropane

CH_3
CH_3

The ring is numbered incorrectly.
Its IUPAC name is 1,1-dimethylcyclopropane.

(h) 1-Ethyl-5-methylcyclohexane

H_3C CH_2CH_3

The ring is numbered incorrectly.
Its IUPAC name is 1-ethyl-3-methylcyclohexane.

Problem 2.29 For each IUPAC name, draw the corresponding structural formula.

(a) Ethanol

CH_3-CH_2-OH

(b) Butanal

$CH_3-CH_2-CH_2-\overset{\displaystyle O}{\overset{\|}{C}}-H$

(c) Butanoic acid

$CH_3-CH_2-CH_2-\overset{\displaystyle O}{\overset{\|}{C}}-OH$

(d) Ethanoic acid

$CH_3-\overset{\displaystyle O}{\overset{\|}{C}}-OH$

(e) Heptanoic acid

$CH_3(CH_2)_5-\overset{\displaystyle O}{\overset{\|}{C}}-OH$

(f) Propanoic acid

$CH_3-CH_2-\overset{\displaystyle O}{\overset{\|}{C}}-OH$

(g) Octanal

$CH_3-(CH_2)_6-\overset{\displaystyle O}{\overset{\|}{C}}-H$

(h) Cyclopentene

(i) Cyclopentanol

OH

(j) Cyclopentanone

(k) Cyclohexanol

OH

(l) Propanone

$CH_3-\overset{\displaystyle O}{\overset{\|}{C}}-CH_3$

Problem 2.30 Write the IUPAC name for each compound.

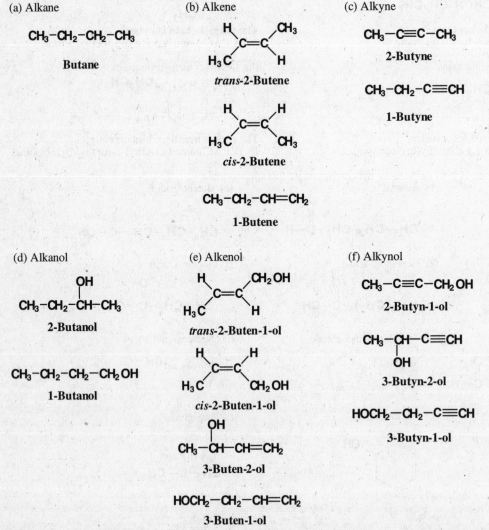

(a) $CH_3-CH_2-\overset{\overset{\displaystyle O}{\|}}{C}-CH_3$
Butanone

(b) $CH_3-CH_2-\overset{\overset{\displaystyle O}{\|}}{C}-H$
Propanal

(c) $CH_3-CH_2-CH_2-CH_2-CH_2-\overset{\overset{\displaystyle O}{\|}}{C}-OH$
Hexanoic acid

(d) $CH_3-\overset{\overset{\displaystyle OH}{|}}{C}H-CH_3$
2-Propanol

(e) Cyclohexanone

(f) Cyclopropanol

(g) $CH_3-CH=CH_2$
Propene

(h) Cyclohexene

Problem 2.31 Assume for the purposes of this problem that, to be an alcohol (-ol) or an amine (-amine), the hydroxyl or amino group must be bonded to a tetrahedral (*sp³* hybridized) carbon atom. Write the structural formula of a compound with an unbranched chain of four carbon atoms that is an:

The IUPAC names are only provided for your reference. You do not yet know how to name all of these.

(a) Alkane

$CH_3-CH_2-CH_2-CH_3$

Butane

(b) Alkene

trans-2-Butene

cis-2-Butene

$CH_3-CH_2-CH=CH_2$

1-Butene

(c) Alkyne

$CH_3-C{\equiv}C-CH_3$

2-Butyne

$CH_3-CH_2-C{\equiv}CH$

1-Butyne

(d) Alkanol

$CH_3-CH_2-\overset{\overset{\displaystyle OH}{|}}{C}H-CH_3$

2-Butanol

$CH_3-CH_2-CH_2-CH_2OH$

1-Butanol

(e) Alkenol

trans-2-Buten-1-ol

cis-2-Buten-1-ol

$CH_3-\overset{\overset{\displaystyle OH}{|}}{C}H-CH=CH_2$

3-Buten-2-ol

$HOCH_2-CH_2-CH=CH_2$

3-Buten-1-ol

(f) Alkynol

$CH_3-C{\equiv}C-CH_2OH$

2-Butyn-1-ol

$CH_3-\overset{\overset{\displaystyle OH}{|}}{C}H-C{\equiv}CH$

3-Butyn-2-ol

$HOCH_2-CH_2-C{\equiv}CH$

3-Butyn-1-ol

Note: you will learn later why the OH group cannot be bonded directly to a double bond or a triple bond.

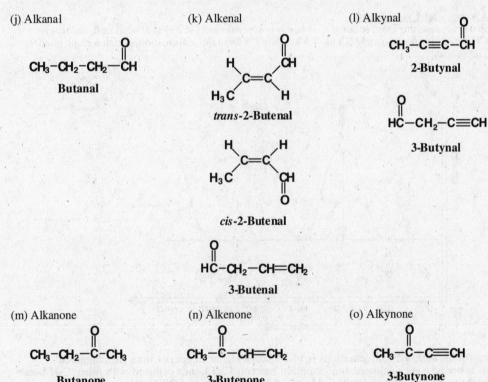

(g) Alkanamine

2-Butanamine

1-Butanamine

(h) Alkenamine

***trans*-2-Buten-1-amine**

***cis*-2-Buten-1-amine**

3-Buten-2-amine

3-Buten-1-amine

(i) Alkynamine

2-Butyn-1-amine

3-Butyn-2-amine

3-Butyn-1-amine

Note: you will learn later why the NH$_2$ group cannot be attached directly to a double bond or a triple bond.

(j) Alkanal

Butanal

(k) Alkenal

***trans*-2-Butenal**

***cis*-2-Butenal**

3-Butenal

(l) Alkynal

2-Butynal

3-Butynal

(m) Alkanone

Butanone

(n) Alkenone

3-Butenone

(o) Alkynone

3-Butynone

(p) Alkanoic acid   **(q) Alkenoic acid**   **(r) Alkynoic acid**

$$CH_3-CH_2-CH_2-\overset{\overset{\textstyle O}{\|}}{C}OH$$

Butanoic acid

trans-2-Butenoic acid

cis-2-Butenoic acid

$$HO\overset{\overset{\textstyle O}{\|}}{C}-CH_2-CH=CH_2$$

3-Butenoic acid

$$CH_3-C\equiv C-\overset{\overset{\textstyle O}{\|}}{C}OH$$

2-Butynoic acid

$$HO\overset{\overset{\textstyle O}{\|}}{C}-CH_2-C\equiv CH$$

3-Butynoic acid

(*Note*: Only one structural formula is possible for some parts of this problem. For other parts, two or more structural formulas are possible. Where two are more are possible, we will deal with how the IUPAC system distinguishes between them when we come to the chapters on those particular functional groups.)

Conformations of Alkanes and Cycloalkanes

<u>Problem 2.32</u> Torsional strain resulting from eclipsed C-H bonds is approximately 4.2 kJ (1.0 kcal)/mol, and that for eclipsed C-H and C-CH₃ bonds is approximately 6.3 kJ (1.5 kcal)/mol. Given this information, sketch a graph of energy versus dihedral angle for propane.

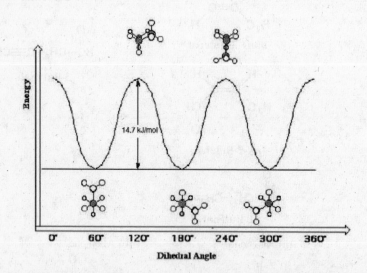

Notice that the energy of the eclipsed conformations is 14.7 kJ/mol higher in energy than the staggered conformations. This is because each eclipsed conformation has two C-H bonds eclipsed with other C-H bonds (worth 4.2 kJ/mol each) and one C-H bond eclipsed to a C-CH₃ bond (worth 6.3 kJ/mol).

<u>Problem 2.33</u> How many different staggered conformations are there for 2-methylpropane? How many different eclipsed conformations are there?

Looking down any of the carbon-carbon bonds, there is one staggered and one eclipsed conformation of 2-methylpropane.

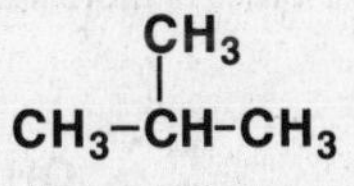

CH₃
CH₃–CH–CH₃
2-Methylpropane

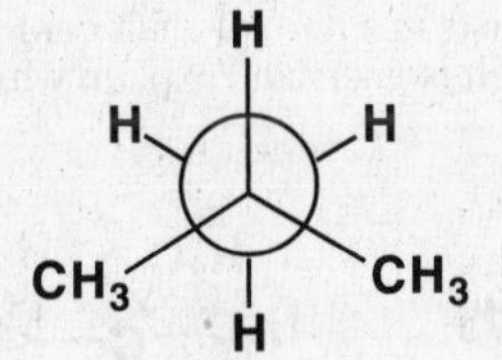

Staggered 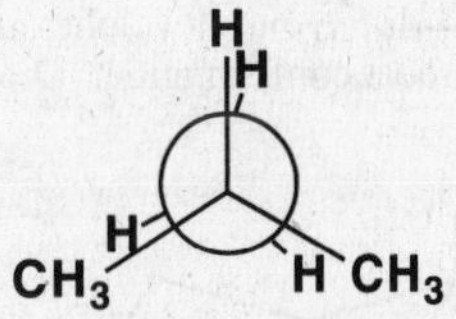   **Eclipsed**

<u>Problem 2.34</u> Consider 1-bromopropane, $CH_3CH_2CH_2Br$.
(a) Draw a Newman projection for the conformation in which -CH_3 and -Br are anti (dihedral angle 180°).

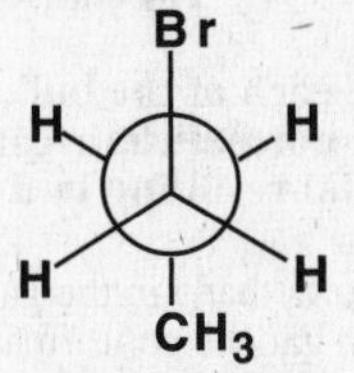

lowest in energy

(b) Draw Newman projections for the conformations in which -CH_3 and -Br are gauche (dihedral angles 60° and 300°).

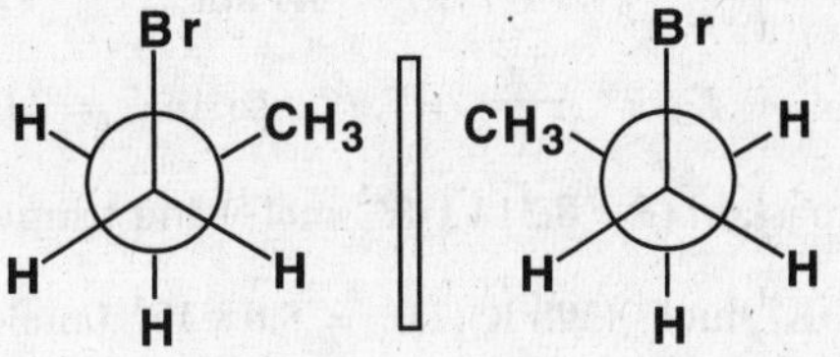

related by reflection

(c) Which of these is the lowest energy conformation.

The anti (dihedral angle 180°) is the lowest energy conformation.

(d) Which of these conformations, if any, are related by reflection?

The two gauche conformations are of equal energy and are related by reflection.

<u>Problem 2.35</u> Consider 1-bromo-2-methylpropane and draw the following.
(a) The staggered conformation(s) of lowest energy.

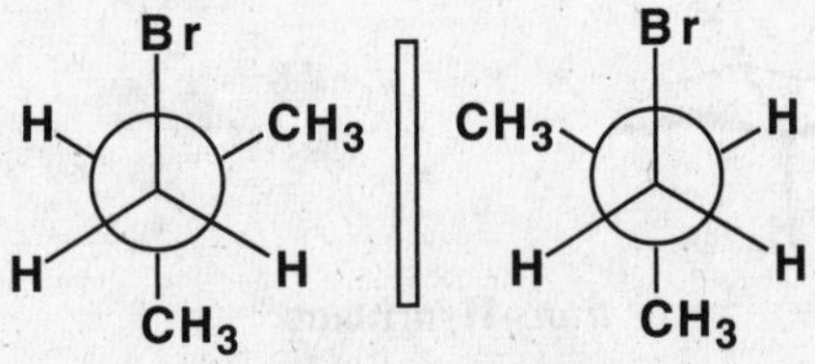

lowest in energy
(related by reflection)

(b) The staggered conformation(s) of highest energy.

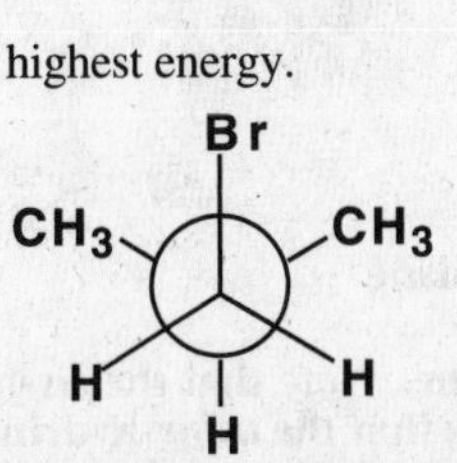

highest in energy

The lower energy staggered conformations have one methyl group anti (dihedral angle 180°) to the bromine and are related by reflection. The staggered conformation with methyl groups at dihedral angles of both 60° and 300° to the bromine have more nonbonded interaction strain and are thus higher in energy.

<u>Problem 2.36</u> *Trans*-1,4-di-*tert*-butylcyclohexane exists in a normal chair conformation. *Cis*-1,4-di-*tert*-butylcyclohexane, however, adopts a twist-boat conformation. Draw both isomers and explain why the *cis* isomer is more stable in the twist-boat conformation.

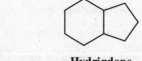

The *trans* isomer in the chair form   **The *cis* isomer in the twist-boat form**

The *cis* isomer adopts a twist-boat conformation because each of the bulky *tert*-butyl groups can be in an pseudo-equatorial position. If the *cis* isomer existed in a normal chair conformation, then one *tert*-butyl group would be equatorial, while the other would be forced axial resulting in a large nonbonded interaction strain.

<u>Problem 2.37</u> From studies of the dipole moment of 1,2-dichloroethane in the gas phase at room temperature (25°C), it is estimated that the ratio of molecules in the anti conformation to gauche conformation is 7.6 to 1. Calculate the difference in Gibbs free energy between these two conformations.

$$\Delta G° = -RT\ln K_{eq}$$

$$K_{eq} = \frac{7.6}{1} = 7.6 \quad \text{so} \quad \ln K_{eq} = 2.0$$

Plugging in the gas constant ($R = 8.314$ J·K^{-1}·mol^{-1}) and temperature ($T = 298$ K)

$$\Delta G° = -(8.314 \text{ J·K}^{-1}\text{·mol}^{-1})(298 \text{ K})(2.0) = 5.0 \times 10^3 \text{ J/mol} = \boxed{5.0 \text{ kJ/mol}}$$

<u>Problem 2.38</u> Draw structural formulas for the *cis* and *trans* isomers of hydrindane. Show each ring in its most stable conformation. Which of these isomers is the more stable?

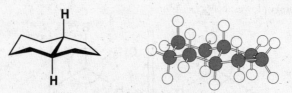

Hydrindane

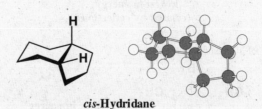

trans-**Hydridane**

cis-**Hydridane**

Shown above are the *trans* and *cis* isomers of hydrindane. Note that the *cis*-hydrindane displays some distortion from the ideal cyclohexane ring structure. The result is that the *trans*-hydrindane is the more stable.

<u>Problem 2.39</u> Following are the alternative chair conformations for *trans*-1,2-dimethylcylohexane.

trans-1,2-Dimethylcyclohexane

(a) Estimate the difference in free energy between these two conformations.

As described in Example 2.11, the difference in energy between a diaxial and diequatorial dimethyl cyclohexane conformation is 14.56 kJ (3.5 kcal)/mol. In the case of *trans*-1,2-dimethylcyclohexane, in the diequatorial conformation, there is also a gauche interaction between the two methyl groups that must be considered. Estimate the gauche interaction to be 3.8 kJ (0.91 kcal)/mol based on the gauche interaction in butane (Figure 2.9). This gauche interaction introduces a small amount of steric strain into the more stable diequatorial conformation reducing the absolute value of the total $\Delta G°$ as follows:

$$\Delta G° = \text{-14.56 kJ (3.5 kcal)/mol} + \text{3.8 kJ (0.91 kcal)/mol} = \boxed{\text{-10.76 kJ (2.59 kcal)/mol}}$$

(b) Given your value in (a), calculate the percent of each chair present in an equilibrium mixture of the two at 25°C.

$$\Delta G° = -RT\ln K_{eq}$$

Plugging in the gas constant (R = 8.314 J•K^{-1}•mol^{-1}), temperature (T = 298 K) and value of $\Delta G°$ for the equilibrium (-10.76 kJ (2.59 kcal)/mol) gives the following

$$\ln K_{eq} = \frac{-(-10{,}760 \text{ J/mol})}{8.314 \text{ J•K}^{-1}\text{•mol}^{-1} \times 298 \text{ K}} = 4.343$$

$$K_{eq} = \frac{76.9}{1} = \frac{\text{diequatorial}}{\text{diaxial}}$$

Based on this calculation, at equilibrium, there is 1.3% in the diaxial chair conformation, and 98.7% in the diequatorial chair conformation.

Cis, trans Isomerism in Cycloalkanes
<u>Problem 2.40</u> What structural feature of cycloalkanes makes *cis, trans* isomerism possible?

Because the atoms are connected in a ring, the C-C bonds cannot rotate around all 360°. As a result, groups on the ring have a fixed relationship with respect to each other, either *cis* or *trans*.

<u>Problem 2.41</u>   Is *cis, trans* isomerism possible in alkanes?

The C-C bonds in alkanes can rotate 360°, so *cis, trans* isomerism in not possible.

<u>Problem 2.42</u> Draw structural formulas for the *cis* and *trans* isomers of 1,2-dimethylcyclopropane.

cis-1,2-Dimethyl-cyclopropane   **trans-1,2-Dimethyl-cyclopropane**

<u>Problem 2.43</u> Name and draw structural formulas for all cycloalkanes with molecular formula C_5H_{10}. Be certain to include *cis* and *trans* isomers as well as constitutional isomers.

Cyclopentane   **Methylcyclo-butane**   **1,1-Dimethyl-cyclopropane**   ***cis*-1,2-Dimethyl-cyclopropane**

***trans*-1,2-Dimethyl-cyclopropane**   **Ethylcyclopropane**

<u>Problem 2.44</u> Using a planar pentagon representation for the cyclopentane ring, draw structural formulas for the *cis* and *trans* isomers of the following.
(a) 1,2-Dimethylcyclopentane   (b) 1,3-Dimethylcyclopentane.

***cis*-1,2-Dimethyl-cyclopentane**

***trans*-1,2-Dimethyl-cyclopentane**

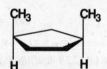
***cis*-1,3-Dimethyl-cyclopentane**

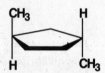
***trans*-1,3-Dimethyl-cyclopentane**

<u>Problem 2.45</u> Gibbs free energy differences between axial-substituted and equatorial-substituted chair conformations of cyclohexane were given in Table 2.4.
(a) Calculate the ratio of equatorial to axial *tert*-butylcyclohexane at 25°C.

According to the value given in Table 2.4, the equatorial *tert*-butylcyclohexane is 21 kJ/mol more stable than the axial conformation.

$$\text{Rearranging} \quad \Delta G° = -RT\ln K_{eq} \quad \text{gives} \quad \ln K_{eq} = \frac{-\Delta G°}{RT}$$

Plugging in the gas constant ($R = 8.314$ J•K^{-1}•mol^{-1}) and temperature ($T = 298$ K) as well as the value for $\Delta G°$ converted to J/mol gives

$$\ln K_{eq} = \frac{21{,}000 \text{ J/mol}}{(8.314 \text{ J•K}^{-1}\text{•mol}^{-1})(298 \text{ K})} = 8.5$$

$$K_{eq} = e^{8.5} = 4.9 \times 10^3$$

So the ratio of equatorial to axial is 4.9×10^3:1

(b) Explain why the conformational equilibria for methyl, ethyl, and isopropyl substituents are comparable but the conformational equilibrium for *tert*-butylcyclohexane lies considerably farther toward the equatorial conformation.

Rotation is possible about the single bond connecting the axial substituent to the ring. Axial methyl, ethyl, and isopropyl groups can assume a conformation where a hydrogen creates the 1,3-diaxial interactions. With a *tert*-butyl substituent, however, a bulkier -CH$_3$ group must create the 1,3-diaxial interaction. Because of the increased steric strain (nonbonded interactions) created by the axial *tert*-butyl, the energy of the axial conformation is considerably greater than that for the equatorial conformation.
As seen below, an axial isopropyl group can adopt a conformation with only a minimal 1,3 diaxial interaction:

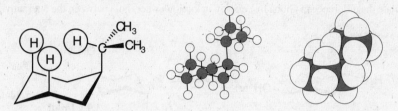

On the other hand, an axial *tert*-butyl group leads to a very severe 1,3 diaxial interaction:

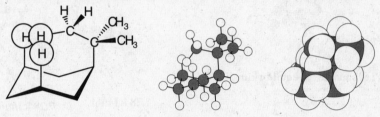

Problem 2.46 When cyclohexane is substituted by an ethynyl group, -C≡CH, the energy difference between axial and equatorial conformations is only 1.7 kJ/mol (0.41 kcal/mol). Compare the conformational equilibrium for methylcyclohexane with that for ethynylcyclohexane and account for the difference between the two.

For the ethynyl case, using the same equations as in part (a) of 2.37 gives:

$$\ln K_{eq} = \frac{1,700 \text{ J/mol}}{(8.314 \text{ J} \cdot \text{K}^{-1} \cdot \text{mol}^{-1})(298 \text{ K})} = 0.69$$

$$K_{eq} = e^{0.69} = 2.0$$

So the ratio of equatorial to axial ethynyl is 2:1

Using the value of 7.28 kJ/mol given in Table 2.4 for the methyl case gives:

$$\ln K_{eq} = \frac{7,280 \text{ J/mol}}{(8.314 \text{ J} \cdot \text{K}^{-1} \cdot \text{mol}^{-1})(298 \text{ K})} = 2.94$$

$$K_{eq} = e^{2.94} = 18.9$$

So the ratio of equatorial to axial methyl is 18.9:1

The above ratios make sense because as can be seen with the following structures and models, the bulkier methyl group is expected to have more severe 1,3 diaxial interactions than the linear -C≡CH group.

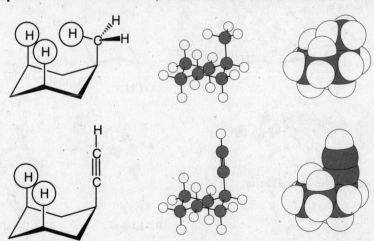

Problem 2.47 Calculate the difference in Gibbs free energy in kilojoules per mole between the alternative chair conformations of:
(a) *trans*-4-Methylcyclohexanol

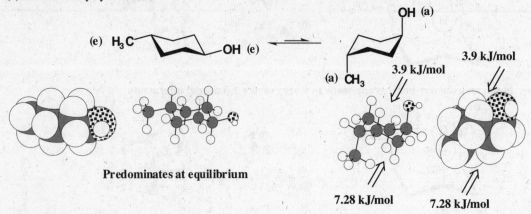

Predominates at equilibrium

The chair on the left is favored because both the methyl and hydroxyl groups are equatorial. Using the data in Table 2.4, the chair on the right is less stable by 7.28 kJ/mol (methyl group axial) + 3.9 kJ/mol (-OH axial) = 11.2 kJ/mol.

(b) *cis*-4-Methylcyclohexanol

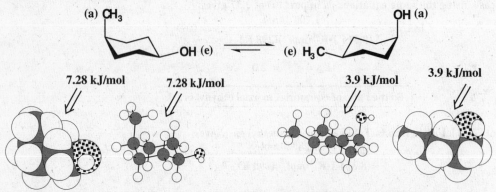

Predominates at equilibrium

The chair on the right is favored because an axial -OH group is more favorable than an axial -CH₃ group. Using the data in Table 2.4, the chair on the left has an axial -CH₃ group that costs 7.28 kJ/mol, while the chair on the right has an axial -OH that costs 3.9 kJ/mol. The chair on the right is thus more stable by 7.28 kJ/mol - 3.9 kJ/mol = 3.4 kJ/mol.

(c) *trans*-1,4-Dicyanocyclohexane

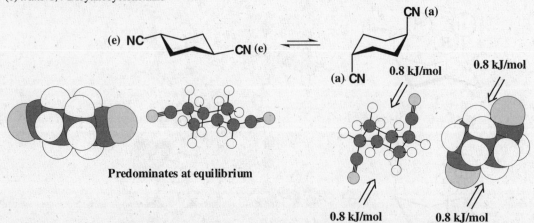

Predominates at equilibrium

The chair on the left is slightly favored because both cyano groups are equatorial. Using the data in Table 2.4, the chair on the right is less stable by 2 x 0.8 kJ/mol (cyano group axial) = 1.6 kJ/mol.

Problem 2.48 Draw the alternative chair conformations for the *cis* and *trans* isomers of 1,2-dimethylcyclohexane, 1,3-dimethylcyclohexane, and 1,4-dimethylcyclohexane.
(a) Indicate by a label whether each methyl group is axial or equatorial.
(b) For which isomer(s) are the alternative chair conformations of equal stability?
(c) For which isomer(s) is one chair conformation more stable than the other?

Cis and trans isomers are drawn as pairs. The more stable chair is labeled in cases where there is a difference.

cis-1,2-Dimethylcyclohexane
(chairs of equal stability)

trans-1,2-Dimethylcyclohexane

cis-1,3-Dimethylcyclohexane

trans-1,3-Dimethylcyclohexane
(chairs of equal stability)

cis-1,4-Dimethylcyclohexane
(chairs of equal stability)

trans-1,4-Dimethylcyclohexane

<u>Problem 2.49</u> Use your answers from problem 2.48 to complete the table showing correlations between *cis, trans* and axial, equatorial for the disubstituted derivatives of cyclohexane.

These relationships are summarized in the following table.

Position of Substitution	*cis*	*trans*
1,4	a,e or e,a	e,e or a,a
1,3	e,e or a,a	a,e or e,a
1,2	a,e or e,a	e,e or a,a

<u>Problem 2.50</u> There are four *cis, trans* isomers of 2-isopropyl-5-methylcyclohexanol.

2-Isopropyl-5-methylcyclohexanol

(a) Using a planar hexagon representation for the cyclohexane ring, draw structural formulas for the four *cis, trans* isomers.
(b) Draw the more stable chair conformation for each of your answers in part (a).
(c) Of the four *cis, trans* isomers, which is the most stable? (If you answered this part correctly, you picked the isomer found in nature and given the name menthol)

Following are planar hexagon representations for the four *cis, trans* isomers. In each, the isopropyl group is shown by the symbol R-. One way to arrive at these structural formulas is to take one group as a reference and then arrange the other two groups in relation to it. In these drawings, -OH is taken as the reference and placed above the plane of the ring. Once -OH is fixed, there are only two possible arrangements for the isopropyl group on carbon 2; either *cis* or *trans* to -OH. Similarly, there are only two possible arrangements for the methyl group on carbon-5; either *cis* or *trans* to -OH. Note that even if you take another substituent as a reference, and even if you put the reference below the plane of the ring, there are still only four *cis, trans* isomers for this compound.

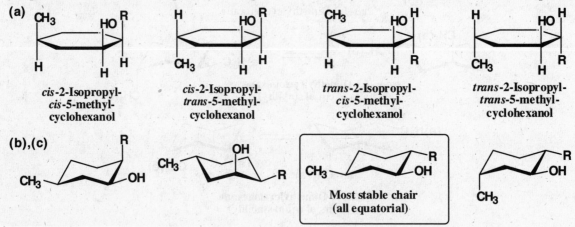

(a)

cis-2-Isopropyl-
cis-5-methyl-
cyclohexanol

cis-2-Isopropyl-
trans-5-methyl-
cyclohexanol

trans-2-Isopropyl-
cis-5-methyl-
cyclohexanol

trans-2-Isopropyl-
trans-5-methyl-
cyclohexanol

(b),(c)

Most stable chair
(all equatorial)

<u>Problem 2.51</u> Draw alternative chair conformations for each substituted cyclohexane and state which chair is more stable:

(a)

(Chairs of equal stability)

(b)

More stable chair

(c)

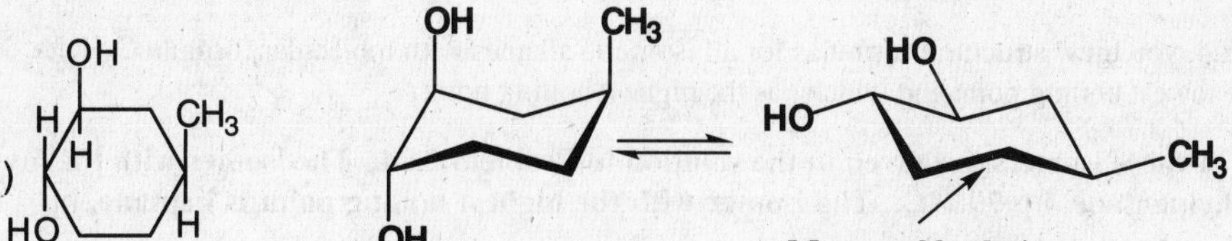

More stable chair

(d)

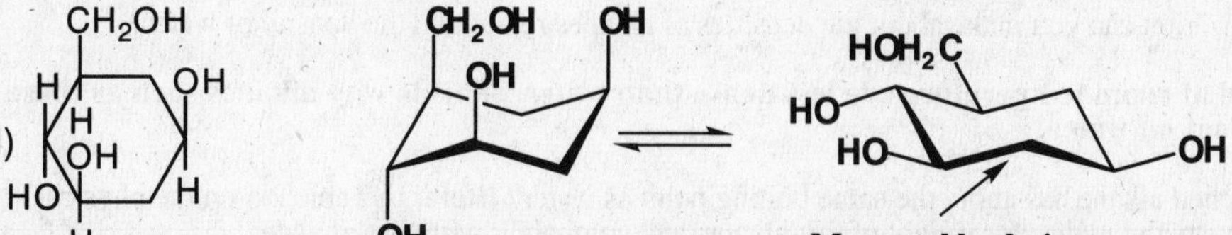

More stable chair

Problem 2.52 1,2,3,4,5,6-Hexachlorocyclohexane shows *cis, trans* isomerism. At one time a crude mixture of these isomers was sold as an insecticide. The insecticidal properties of the mixture arise from one isomer known as lindane, which is *cis*-1,2,4,5-*trans*-3,6-hexachlorocyclohexane.
(a) Draw a structural formula for 1,2,3,4,5,6-hexachlorocyclohexane disregarding for the moment the existence of *cis, trans* isomerism. What is the molecular formula of this compound?

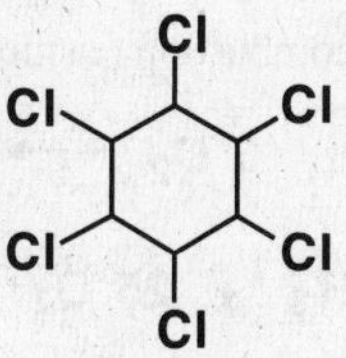

$C_6H_6Cl_6$

(b) Using a planar hexagon representation for the cyclohexane ring, draw a structural formula for lindane.

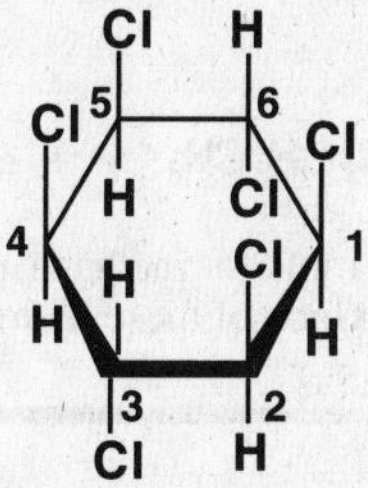

(c) Draw a chair conformation for lindane and label which chlorine atoms are axial and which are equatorial.
(d) Draw the alternative chair conformation of lindane, and again label which chlorine atoms are axial and which are equatorial.

(e) Which of the alternative chair conformations of lindane is more stable? Explain.

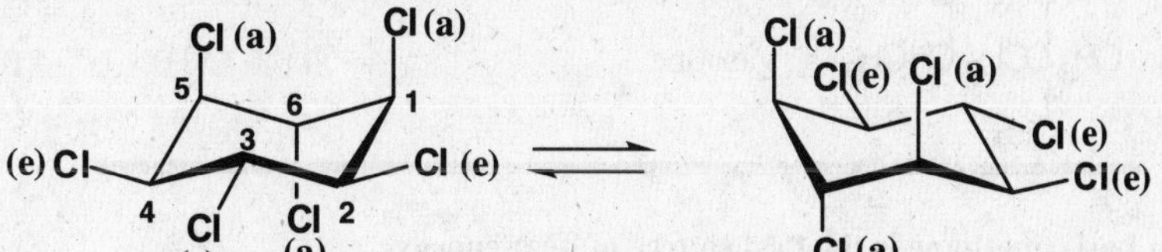

The two chairs are of equal stability; in each
three -Cl atoms are axial and three are equatorial

Physical Properties

Problem 2.53 In Problem 2.24, you drew structural formulas for all isomeric alkanes with molecular formula C_7H_{16}. Predict which isomer has the lowest boiling point and which has the highest boiling point.

Names and boiling points of these isomers are given in the solution to Problem 2.24. The isomer with the lowest boiling point is 2,2-dimethylpentane, bp 79.2°C. The isomer with the highest boiling point is heptane, bp 94.8°C.

Problem 2.54 What generalization can you make about the densities of alkanes relative to the density of water?

All alkanes that are liquid at room temperature are less dense than water. This is why alkanes such as those in gasoline and petroleum float on water.

Problem 2.55 What unbranched alkane has about the same boiling point as water? (Refer to Table 2.5 on the physical properties of alkanes.) Calculate the molecular weight of this alkane, and compare it with that of water.

Heptane, C_7H_{16}, has a boiling point of 98.4°C and a molecular weight of 100. Its molecular weight is approximately 5.5 times that of water. Although considerably smaller, the water molecules are held together by hydrogen bonding while the much larger heptane molecules are held together only by relatively weak dispersion forces.

Reactions of Alkanes

Problem 2.56 Complete and balance the following combustion reactions. Assume that each hydrocarbon is converted completely to carbon dioxide and water.

(a) Propane + O_2 $\longrightarrow$ $CH_3CH_2CH_3$ + 5 O_2 $\longrightarrow$ 3 CO_2 + 4 H_2O

(b) Octane + O_2 $\longrightarrow$ 2 $CH_3(CH_2)_6CH_3$ + 25 O_2 $\longrightarrow$ 16 CO_2 + 18 H_2O

(c) Cyclohexane + O_2 $\longrightarrow$ ⬡ + 9 O_2 $\longrightarrow$ 6 CO_2 + 6 H_2O

(d) 2-Methylpentane + O_2 $\longrightarrow$ 2 $CH_3CHCH_2CH_2CH_3$ (with CH_3 branch) + 19 O_2 $\longrightarrow$ 12 CO_2 + 14 H_2O

Problem 2.57 Following are heats of combustion per mole for methane, propane, and 2,2,4-trimethylpentane. Each is a major source of energy. On a gram-for-gram basis, which of these hydrocarbons is the best source of heat energy.

Hydrocarbon	Component of	$\Delta H°$ [kJ(kcal)/mol)]
CH_4	natural gas	-891 (-213)
$CH_3CH_2CH_3$	LPG	-2220 (-531)
$CH_3CCH_2CHCH_3$ (with two CH_3 and CH_3 groups)	gasoline	-5452(-1304)

On a gram-per-gram basis, methane is the best source of heat energy.

Hydrocarbon	Molecular Weight	Heat of Combustion (kJ/ mol)	Heat of Combustion (kJ/ gram)
Methane	16.04	-891	- 55.5
Propane	44.09	-2220	-50.4
2,2,4-Trimethylpentane	114.2	-5452	-47.7

<u>Problem 2.58</u> Following are structural formulas and heats of combustion of acetaldehyde and ethylene oxide. Which of these compounds is the more stable. Explain.

Acetaldehyde   **Ethylene oxide**

-1164 kJ (-278.8 kcal)/mol   -1264 kJ (-302.1 kcal)/mol

These molecules are constitutional isomers, so their heats of combustion are directly comparable. The molecule with the smaller (less negative) heat of combustion is the more stable, having less energy to release during the combustion process. As a result, acetaldehyde is the more stable molecule.

<u>Problem 2.59</u> Without consulting tables, arrange these compounds in order of decreasing (less negative) heat of combustion: hexane, 2-methylpentane, and 2,2-dimethylbutane.

Branching increases stability of an alkane. Therefore, the more highly branched the isomer, the smaller (less negative) the heat of combustion. The molecules listed in order of decreasing heat of combustion turn out to be exactly as they were listed in the question:
Hexane, 2-Methylpentane, 2,2-Dimethylbutane

<u>Problem 2.60</u> Which would you predict to have the larger (more negative) heat of combustion, *cis*-1,4-dimethylcyclohexane or *trans*-1,4-dimethylcyclohexane.

The less stable molecule will have the larger (more negative) heat of combustion. Because these molecules are constitutional isomers with virtually identical ring strain, any difference in energy between them must be the result of differences in conformational stability. As listed in the answer to Problem 2.39, the *cis* isomer has two chair conformations of equal energy, each one with one axial and one equatorial methyl group. The *trans* isomer has chair conformations of different stability, the more stable of which is the diequatorial conformation that has no diaxial interactions. By virtue of having diaxial interactions in both chair conformations, the *cis* isomer is higher in energy and thus will have the larger (more negative) heat of combustion.

Looking Ahead
<u>Problem 2.61</u> Following are structural formulas for 1,4-dioxane and piperidine. 1,4-Dioxane is a widely used solvent for organic compounds. Piperidine is found in small amounts in black pepper (*Piper nigrum*).

1,4-Dioxane   **Piperidine**

(a) Complete the Lewis structure of each compound by showing all unshared electron pairs.

See structures.

(b) Predict bond angles about each carbon, oxygen, and nitrogen atom.

Each carbon, oxygen, and nitrogen atom is *sp³* hybridized, so predict bond angles near 109.5° for all of them.

(c) Describe the most stable conformation of each ring, and compare these conformations with the chair conformation of cyclohexane.

Both molecules have conformations analogous to chair cyclohexane, because in both cases this conformation minimizes torsional and angle strain.

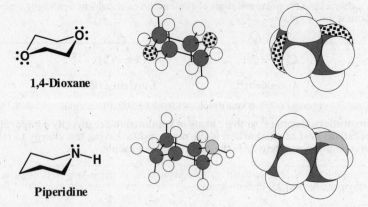

1,4-Dioxane

Piperidine

<u>Problem 2.62</u> Following is a planar hexagon representation of L-fucose, a sugar component of the determinants of the A, B, O blood group typing. For more on this system of blood typing, see Connections to Biological Chemistry "A,B,AB, and O Blood Types" in Chapter 25.

L-Fucose

(a) Draw the alternative chair conformations of L-fucose.

More stable

(b) Which of them is the more stable? Explain.

The structure on the right is more stable because it has only two axial -OH groups and three equatorial groups, the -CH₃ group and two -OH groups. The structure on the left has three axial groups so it will have the greater non-bonded interaction strain.

<u>Problem 2.63</u> On the left is a stereorepresentation of glucose (we discuss the structure and chemistry of glucose in Chapter 25):

Glucose   **(a)**   **(b)**

(a) Convert the stereorepresentation on the left to a planar hexagon representation.

See structure above.

(b) Convert the stereorepresentation on the left to a chair conformation. Which substituent groups in the chair conformation are equatorial? Which are axial?

See structure above. All of the substituents are equatorial, making this a particularly stable chair conformation.

Problem 2.64 Following is the structural formula and a ball-and-stick model of cholestanol. The only difference between this compound and cholesterol (Section 26.4) is that cholesterol has a carbon-carbon double bond in ring B.

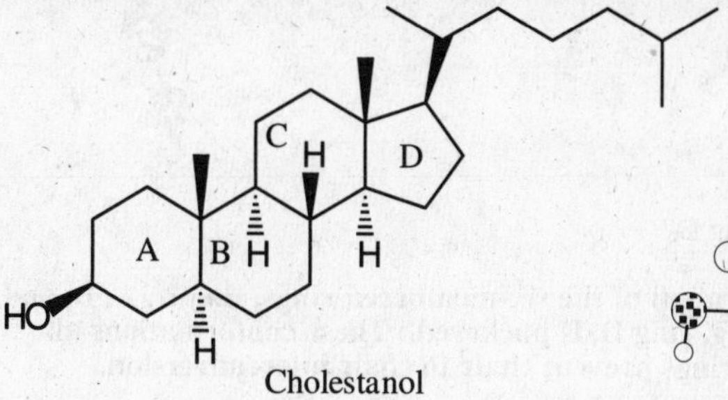

Cholestanol

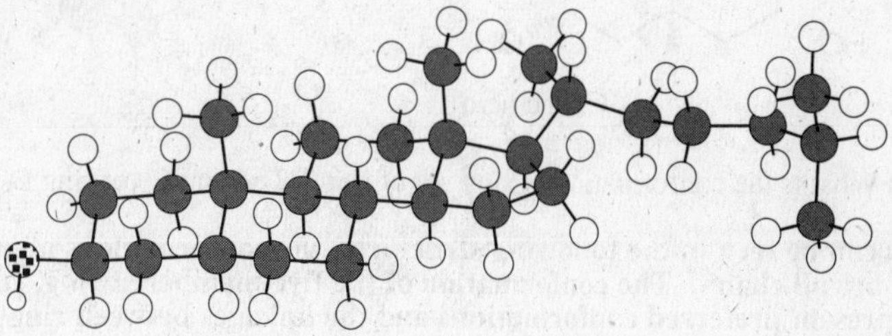

(a) Describe the conformations in rings A, B, C, and D in cholestanol.

As can be seen in the following structures, the conformations of the six-membered rings, that is, A, B, and C, are all chairs. The conformation of the five-membered ring, ring D, is puckered (envelope). These conformations all represent preferred conformations and the ring fusions between rings prevent chair-to-chair interconversion. The result is that the A, B, C, D ring system of steroids such as cholestanol are relatively rigid molecular frameworks.

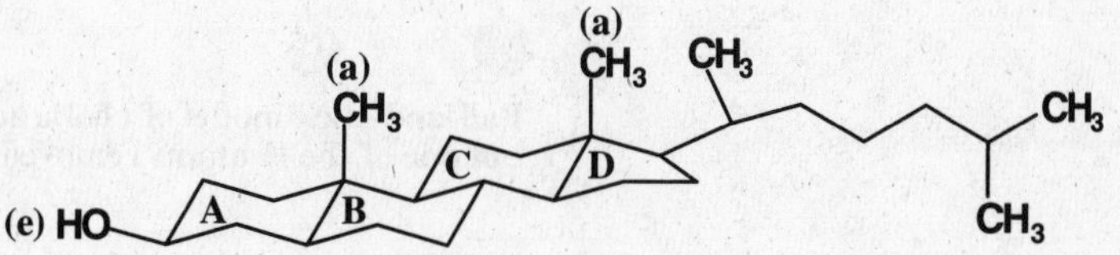

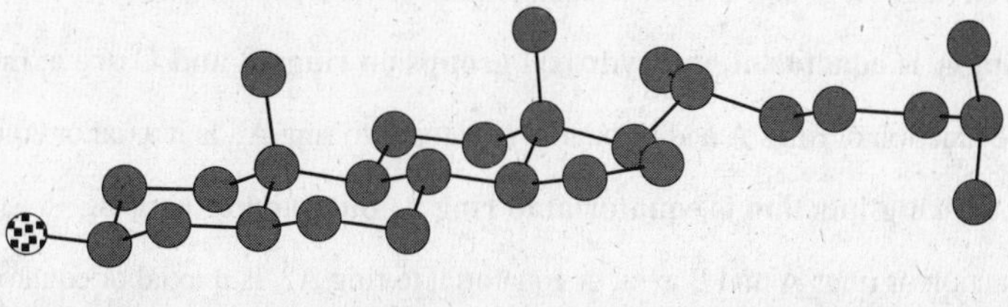

Ball-and-stick model of cholestanol with the H atoms removed for clarity.

(b) Is the hydroxyl group on ring A axial or equatorial?

The hydroxyl group is equatorial.

(c) Consider the methyl group at the junction of rings A and B. Is it axial or equatorial to ring A? Is it axial or equatorial to ring B?

The methyl group at the A,B junction is axial to both rings A and B.

(d) Is the methyl group at the junction of rings C and D axial or equatorial to ring C?

The methyl group at the C, D junction is axial to ring C.

Problem 2.65 Following is the structural formula and a ball-and-stick model of cholic acid (Chapter 26), a component of human bile whose function is to aid in the absorption and digestion of dietary fats.

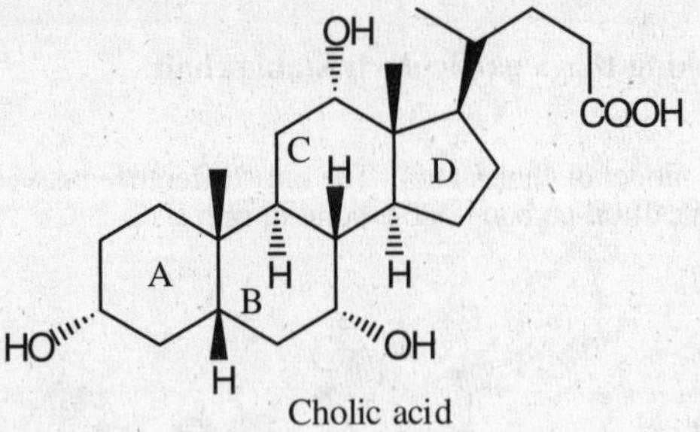

Cholic acid

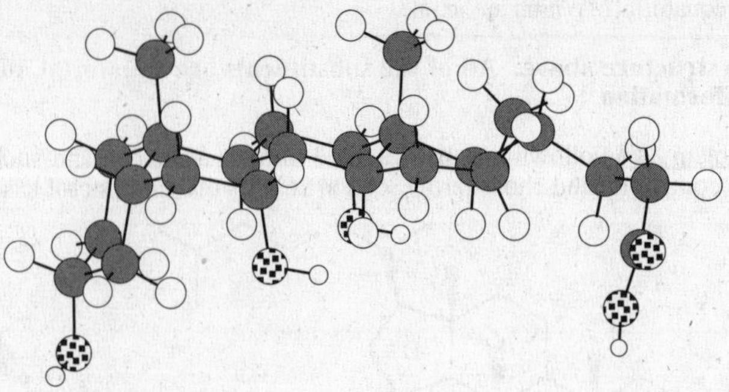

(a) What is the conformation of ring A? of ring B? of ring C? of ring D?

As can be seen in the following structures, the conformations of all of the six-membered rings, that is, A, B, and C, are all chairs. The conformation of the five-membered ring, ring D, is puckered. These conformations all represent preferred conformations and the linkages between rings prevent chair to chair interconversion.

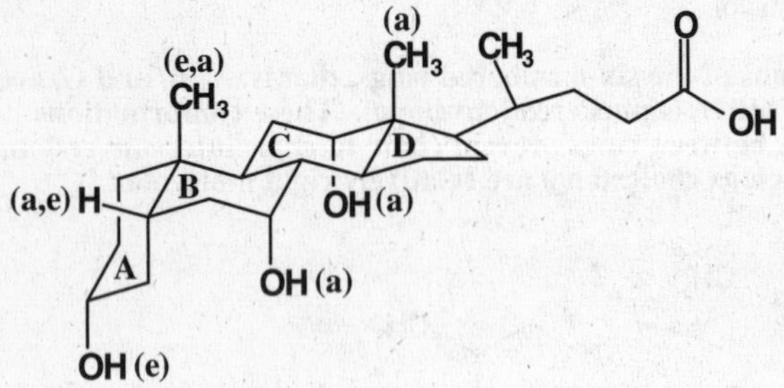

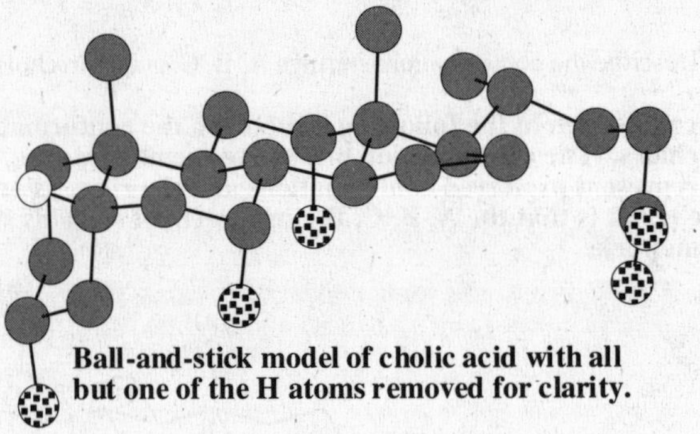

Ball-and-stick model of cholic acid with all but one of the H atoms removed for clarity.

(b) Are the hydroxyl groups on rings A, B, and C axial or equatorial to their respective rings?

The hydroxyl group on ring A is equatorial, the hydroxyl groups on rings B and C are axial.

(c) Is the methyl group at the junction of rings A and B axial or equatorial to ring A. Is it axial or equatorial to ring B?

The methyl group at the A,B ring junction is equatorial to ring A, but axial to ring B.

(d) Is the hydrogen at the junction of rings A and B axial or equatorial to ring A? Is it axial or equatorial to ring B?

The hydrogen atom at the junction of rings A and B is axial to ring A, but equatorial to ring B.

(e) Is the methyl group at the junction of rings C and D axial or equatorial to ring C?

The methyl group at the C, D junction is axial to ring C.

CHAPTER 3
Solutions to the Problems

<u>Problem 3.1</u> Each molecule has one chiral center. Draw stereorepresentations for the enantiomers of each.

Each part has a tetrahedral chiral center. The chiral centers are labeled with an asterisk.

(a)

(b)

<u>Problem 3.2</u> Assign priorities to the groups in each set.
(a) $-CH_2OH$ and $-CH_2CH_2OH$

The $-CH_2OH$ group has higher priority because the FIRST point of difference is the underlined O atom of $-CH_2\underline{O}H$ that takes priority over the underlined C atom of $-CH_2\underline{C}H_2OH$.

(b) $-CH_2OH$ and $-CH=CH_2$

$$-CH = CH_2 \xrightarrow{\text{is treated as}} -\overset{\displaystyle C}{\underset{\displaystyle H}{\overset{\displaystyle |_1}{C}}} - \overset{\displaystyle C}{\underset{\displaystyle H}{\overset{\displaystyle |_2}{C}}} - H$$

The FIRST point of difference is the underlined O atom of $-CH_2\underline{O}H$ that takes priority over any of the atoms bonded to carbon 1 of $-CH=CH_2$. Thus, the $-CH_2OH$ group takes priority over the $-CH=CH_2$ group.

(c) $-CH_2OH$ and $-C(CH_3)_3$

The FIRST point of difference is the underlined O atom of $-CH_2\underline{O}H$ that takes priority over any of the carbon atoms bonded to the central carbon atom of $-C(CH_3)_3$.

<u>Problem 3.3</u> Assign an *R* or *S* configuration to the chiral center in each molecule.

The drawings underneath each molecule show the order of priority, the perspective from which to view the molecule, and the *R,S* designation for the configuration.

(a)

view from this
perspective

(*S*)-3,3-Dimethylcyclohexanol

If you view from the perspective
shown, this is what you see

(b)

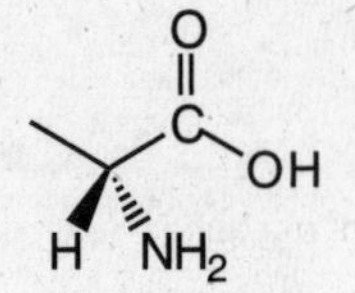

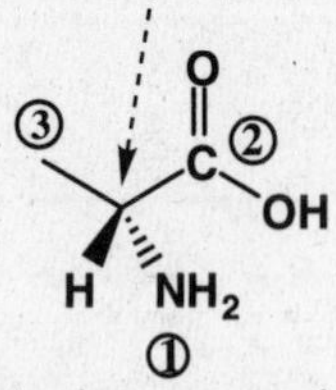

view from this
perspective

(R)-2-Aminopropanoic acid

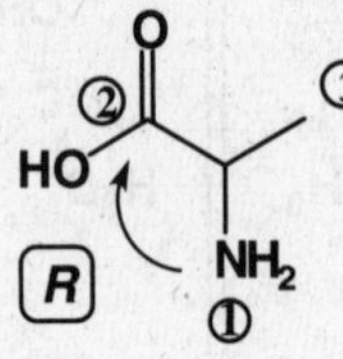

If you view from the perspective
shown, this is what you see

(c)

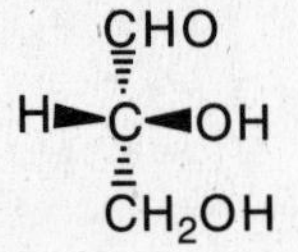

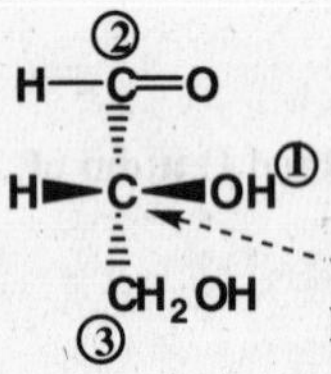

CHO
H—C—OH
CH₂OH

view from this
perspective

(R)-2,3-Dihydroxypropanal

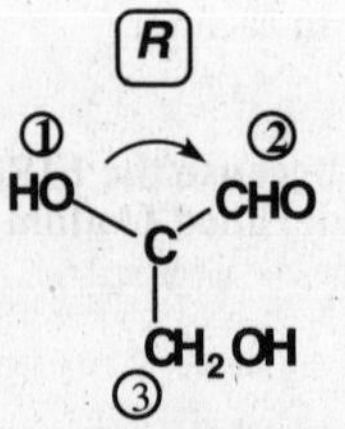

If you view from the perspective
shown, this is what you see

<u>Problem 3.4</u> Following are stereorepresentations for the four stereoisomers of 3-chloro-2-butanol.

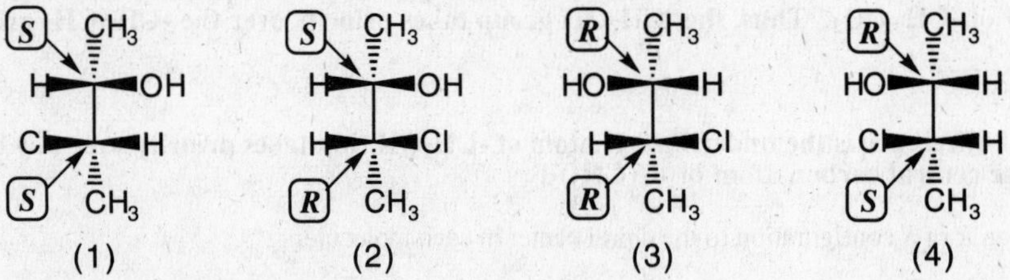

(1)   (2)   (3)   (4)

(a) Assign an R or S configuration of each chiral center.

The configuration of each chiral center is labeled on the above structures. This labeling often helps when trying to establish stereochemical relationships among molecules.

(b) Which compounds are enantiomers?

Enantiomers are stereoisomers that are mirror images of each other. The pairs of enantiomers are structures (1) and (3) (S,S and R,R) as well as structures (2) and (4) (S,R and R,S).

(c) Which compounds are diastereomers?

Diastereomers are stereoisomers that are not mirror images of each other. Therefore, there are several sets of diastereomers. The following chart describes the relationship between any pair of molecules

Problem 3.5 Following are four Newman projection formulas for tartaric acid.

(a) Which represent the same compound?

Compounds (3) and (4) are the same compound having the configuration (2S,3S). Compound (2) has the configuration of (2R,3S) while compound (1) is (2R,3R).

(b) Which represent enantiomers?

Compounds (3) and (1) or (4) and (1) represent pairs of enantiomers (recall that (3) and (4) are the same).

(c) Which represent a meso compound?

The meso compound of tartaric acid has the (2R,3S) configuration so compound (2) is the meso compound.

(d) Which are diastereomers?

The best way to analyze this is in terms of the various possible pairs of diastereomers. Compounds (3)/(4) and (2) are diastereomers. Compounds (1) and (2) are also a pair of diastereomers.

Problem 3.6 Give a complete stereochemical name for the following molecule, which is a 1,2,3-butanetriol.

(2R,3S)-1,2,3-Butanetriol

Problem 3.7 How many stereoisomers exist for 1,3-cyclopentanediol?

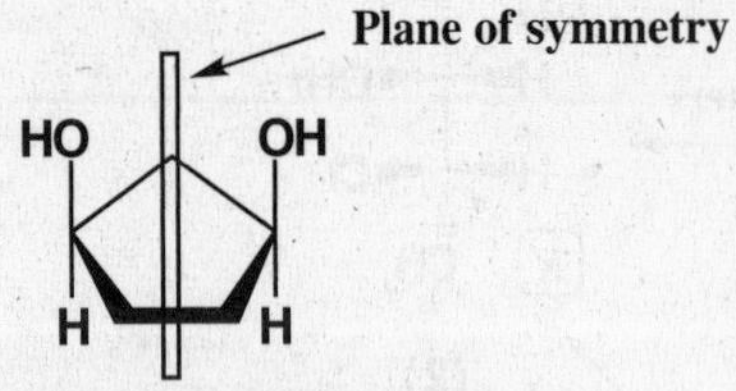

cis-1,3-Cyclopentanediol
(achiral, a meso compound)

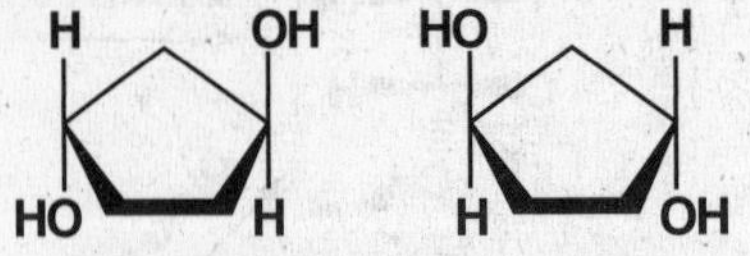

trans-1,3-Cyclopentanediol
(a pair of enantiomers)

1,3-Cyclopentanediol has three stereoisomers. The two *trans* isomers are enantiomers, the *cis* isomer is a meso compound. *Cis*-1,3-cyclopentanediol can be recognized as a meso compound because it is superimposable upon its mirror image. It also has a plane of symmetry that bisects it into two mirror halves.

Problem 3.8 How many stereoisomers exist for 1,4-cyclohexanediol?

1,4-Cyclohexanediol can exist as a pair of *cis, trans* isomers. Each is achiral because of a plane of symmetry that bisects each molecule into two mirror halves. In the figure below, the plane of symmetry in each molecule is in the plane of the paper. Because each isomer is achiral, there are only two stereoisomers of 1,4-cyclohexanediol.

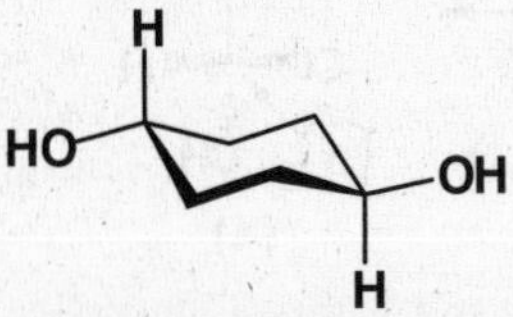

trans-1,4-Cyclohexanediol 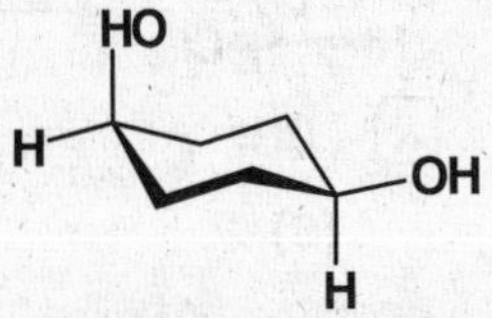   **cis-1,4-Cyclohexanediol**

Problem 3.9 The specific rotation of progesterone, a female sex hormone, is +172. Calculate the observed rotation for a solution prepared by dissolving 300 mg of progesterone in 15.0 mL of dioxane and placing it in a sample tube 10.0 cm long.

The concentration of progesterone, expressed in grams per milliliter is:

$$300 \text{ mg} / 15 \text{ mL} = 0.02 \text{ g/mL}$$

$$\text{specific rotation} = \frac{\text{observed rotation (degrees)}}{\text{length (dm) x concentration (g/mL)}}$$

Rearranging this formula to solve for observed rotation gives:

$$\text{observed rotation (degrees)} = \text{specific rotation x length (dm) x concentration (g/mL)}$$

Plugging in the experimental values gives the final answer.

$$\text{observed rotation (degrees)} = +172 \text{ x } 1.00 \text{ dm x } 0.02 \text{ g/mL} = \boxed{+3.44}$$

Problem 3.10 One commercial synthesis of naproxen (the active ingredient in Aleve and a score of other over-the-counter and prescription nonsteroidal anti-inflammatory drug preparations) gives the enantiomer shown in 97% enantiomeric excess.

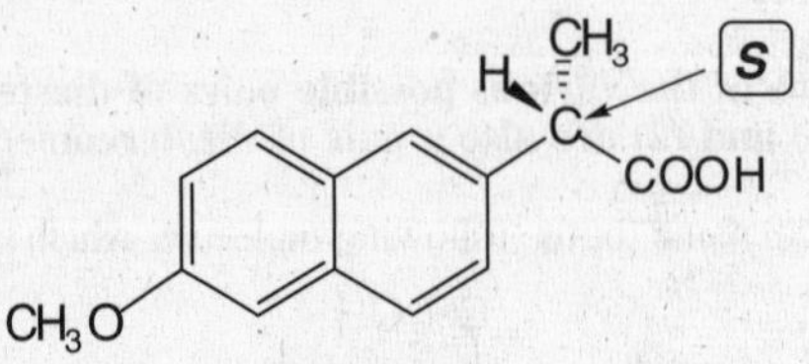

Naproxen
(A nonsteroidal antiinflammatory drug)

(a) Assign an *R* or *S* configuration to this enantiomer of naproxen.

The chiral center has the *S* configuration.

(b) What are the percentages of R and S enantiomers in the mixture?

$$\text{Enantiomeric excess (ee)} = \frac{S - R}{S + R} \times 100 = \%S - \%R = 97\%$$

solving the above equation gives

$$\boxed{98.5\% \ S \text{ and } 1.5\% \ R}$$

Chirality

Problem 3.11 Think about the helical coil of a telephone cord or a spiral binding and suppose that you view the spiral from one end and find that it is a left-handed twist. If you view the same spiral from the other end, is it a right-handed or a left-handed twist?

A helical coil has the same handedness viewed from either end.

Problem 3.12 Next time you have the opportunity to view a collection of sea shells that have a helical twist, study the chirality of their twists. Do you find an equal number of left-handed and right-handed spiral shells or mostly all of the same chirality? What about the handedness of different species of spiral shells?

This question was just meant to make you think about chirality in nature, but if you do know the answer please share it with your class.

Problem 3.13 One reason we can be sure that sp^3-hybridized carbon atoms are tetrahedral is the number of stereoisomers that can exist for different organic compounds.
(a) How many stereoisomers are possible for $CHCl_3$, CH_2Cl_2, and $CHClBrF$ if the four bonds to carbon have a tetrahedral arrangement?

Both tetrahedral $CHCl_3$ and tetrahedral CH_2Cl_2 are achiral, so no stereoisomers are possible.

On the other hand, tetrahedral CHBrClF is chiral so there are two stereoisomers possible.

(b) How many stereoisomers would be possible for each of these compounds if the four bonds to the carbon had a square planar geometry?

Even with a square planar geometry (the H and three Cl atoms are in the same plane as the C atom), there are no stereoisomers possible.

There are two possible stereoisomers of CH_2Cl_2, one with the Cl atoms adjacent to each other, and another with the Cl atoms that are opposite each other.

There are three possible stereoisomers of a square planar CHBrClF as shown.

$$
\begin{array}{c}
Br \\
| \\
H-C-Cl \\
| \\
F
\end{array}
\qquad
\begin{array}{c}
Cl \\
| \\
H-C-Br \\
| \\
F
\end{array}
\qquad
\begin{array}{c}
Br \\
| \\
H-C-F \\
| \\
Cl
\end{array}
$$

Enantiomers

Problem 3.14 Which compounds contain chiral centers?

(a) 2-Chloropentane

$$CH_3-\overset{*}{\underset{|}{C}}-CH_2CH_2CH_3$$

with Cl above the starred C and H below

(b) 3-Chloropentane

$$CH_3CH_2-\underset{|}{C}-CH_2CH_3$$

with Cl above the C and H below

(c) 3-Chloro-1-pentene

$$CH_2=CH-\overset{*}{\underset{|}{C}}-CH_2CH_3$$

with Cl above the starred C and H below

(d) 1,2-Dichloropropane

$$ClCH_2-\overset{*}{\underset{|}{C}}-CH_3$$

with Cl above the starred C and H below

Chiral centers are present in (a), (c), and (d). They are marked with an asterisk (*).

Problem 3.15 Using only C, H, and O, write structural formulas for the lowest molecular weight chiral

(a) Alkane

$$CH_3CH_2CH_2-\overset{H}{\underset{CH_3}{\overset{|}{C^*}}}-CH_2CH_3$$

3-Methylhexane

(b) Alcohol

$$HO-\overset{H}{\underset{CH_3}{\overset{|}{C^*}}}-CH_2CH_3$$

2-Butanol

(c) Aldehyde

$$CH_3CH_2-\overset{H}{\underset{CH_3}{\overset{|}{C^*}}}-\overset{O}{\overset{\|}{CH}}$$

2-Methylbutanal

(d) Ketone

$$CH_3CH_2-\overset{H}{\underset{CH_3}{\overset{|}{C^*}}}-\overset{O}{\overset{\|}{C}}-CH_3$$

3-Methyl-2-pentanone

(e) Carboxylic acid

$$CH_3CH_2-\overset{H}{\underset{CH_3}{\overset{|}{C^*}}}-\overset{O}{\overset{\|}{C}}-OH$$

2-Methylbutanoic acid

(f) Carboxylic ester

$$H-\overset{O}{\overset{\|}{C}}-O-\overset{H}{\underset{CH_3}{\overset{|}{C^*}}}-CH_2CH_3$$

2-Butyl formate

Problem 3.16 Draw mirror images for these molecules. Are they different from the original molecule

The mirror images are shown in bold.

(a)
$$CH_3\cdots\overset{OH}{\underset{H}{C}}COOH \quad \| \quad HOOC\overset{OH}{\underset{H}{C}}\cdots CH_3$$

Different

(b)
$$H-\overset{CHO}{\underset{CH_2OH}{C}}-OH \quad \| \quad HO-\overset{CHO}{\underset{CH_2OH}{C}}-H$$

Different

(c)

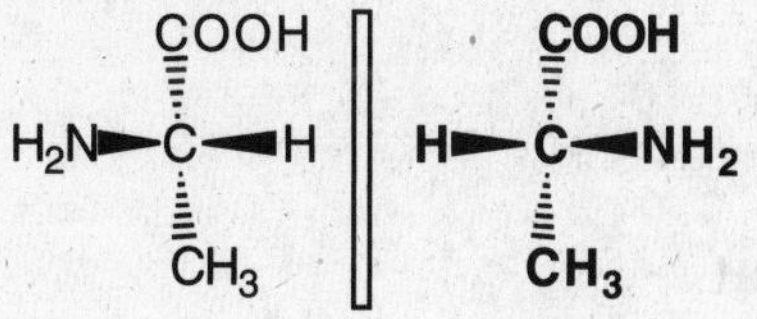

Different

(d)

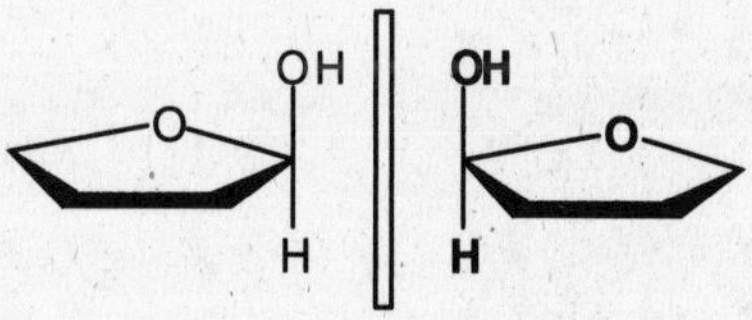

Different

(e)

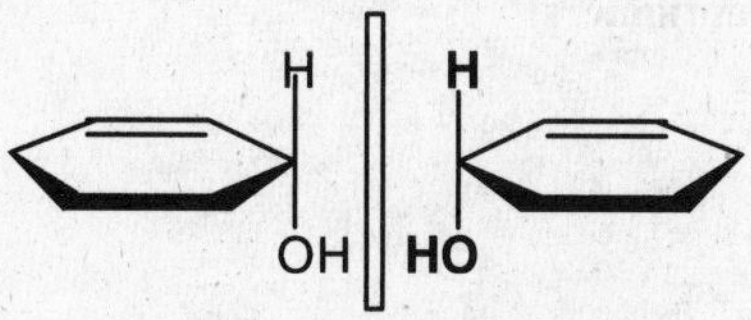

Different

(f)

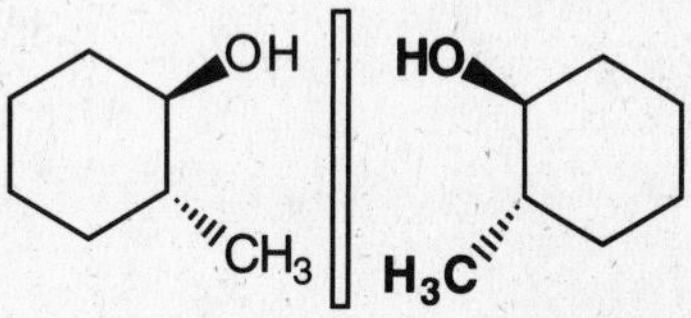

Different

(g)

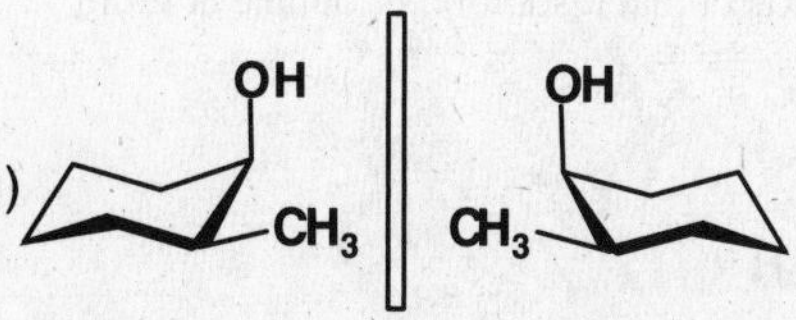

Different

(h)

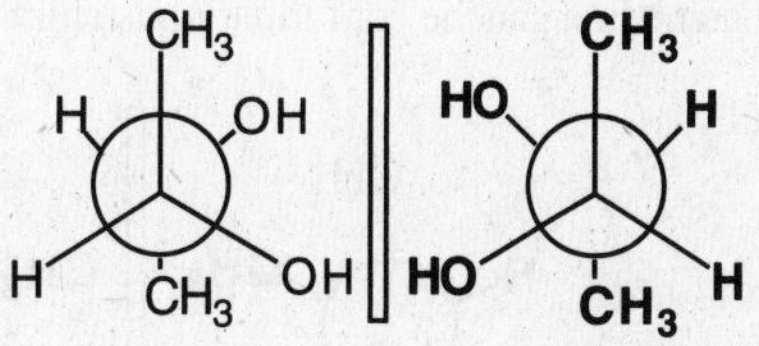

Different

<u>Problem 3.17</u> Following are several stereorepresentations for lactic acid. Take (a) as a reference structure. Which of the stereorepresentations are identical with (a) and which are mirror images of (a)?

(a) 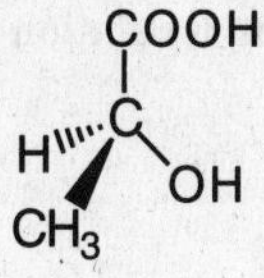 　　(b) 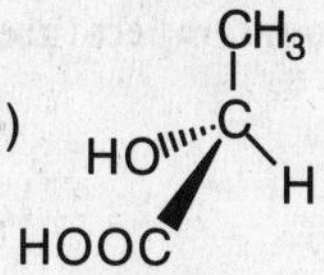 　　(c) 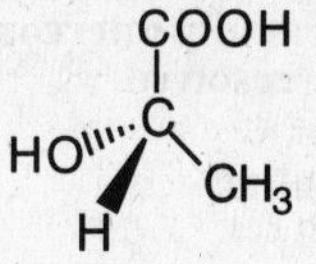 　　(d)

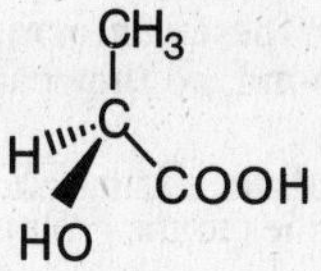

All of the above stereorepresentations have the (S)-configuration so they are identical.

<u>Problem 3.18</u> Mark each chiral center in the following molecules with an asterisk. How many stereoisomers are possible for each molecule?

(a) 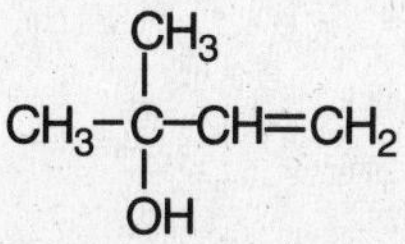 $CH_3-C-CH=CH_2$
　　　CH_3 (top), OH (bottom)

No chiral centers

(b)

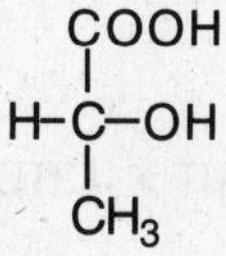

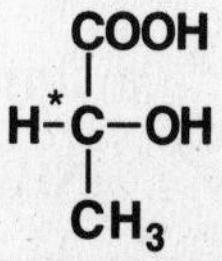

2 Stereoisomers
(a pair of enantiomers)

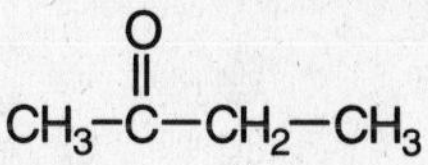

(c) 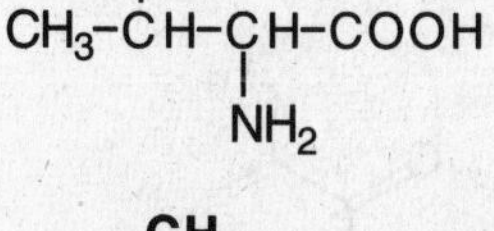 $CH_3-CH-CH-COOH$.
　　　　　　CH_3 (top), NH_2 (bottom)

(d)

$CH_3-C-CH_2-CH_3$ with $=O$

No chiral centers

$CH_3-CH-\overset{*}{CH}-COOH$
　CH_3 (top), NH_2 (bottom)

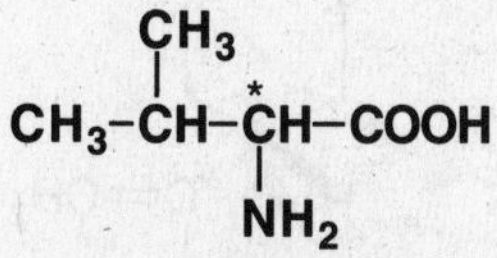

2 stereoisomers
(a pair of enantiomers)

(e)

$$CH_2OH$$
$$H-C-OH$$
$$CH_2OH$$

No chiral centers

(f)

$$CH_3-CH_2-CH-CH=CH_2$$
$$OH$$

$$CH_3-CH_2-\overset{*}{C}H-CH=CH_2$$
$$OH$$

2 stereoisomers
(a pair of enantiomers)

(g)

$$CH_2-COOH$$
$$HO-C-COOH$$
$$CH_2-COOH$$

No chiral centers

<u>Problem 3.19</u> Show that butane in a gauche conformation is chiral. Do you expect that resolution of butane at room temperature is possible?

As can be seen from the above Newman projections, the two gauche conformations are non-superimposable mirror images. However, these conformations rapidly interconvert at room temperature through rotation around the central C-C bond, so they cannot be resolved.

Designation of Configuration - The *R-S* Convention
<u>Problem 3.20</u> Assign priorities to the groups in each set.

The groups are ranked from highest to lowest under each problem. Remember that priority is assigned at the <u>first</u> point of difference.

(a) -H -CH$_3$ -OH -CH$_2$OH   (b) -CH$_2$CH=CH$_2$ -CH=CH$_2$ -CH$_3$ -CH$_2$COOH

-OH > -CH$_2$OH > -CH$_3$ > -H   **-CH=CH$_2$ > -CH$_2$COOH > -CH$_2$CH=CH$_2$ > -CH$_3$**

(c) -CH$_3$ -H -COO$^-$ -NH$_3$$^+$ (d) -CH$_3$ -CH$_2$SH -NH$_3$$^+$ -CHO

-NH$_3$$^+$ > -COO$^-$ > -CH$_3$ > -H   **-NH$_3$$^+$ > -CH$_2$SH > -CHO > -CH$_3$**

<u>Problem 3.21</u> Following are structural formulas for the enantiomers of carvone. Each has a distinctive odor characteristic of the source from which it is isolated. Assign an *R* and *S* configuration to the single chiral center in each enantiomer. Why do they smell different when they are so similar in structure?

(-)-Carvone
Spearmint oil

(+)-Carvone
Caraway and
dill seed oil

Following are *R,S* designations for each enantiomer.

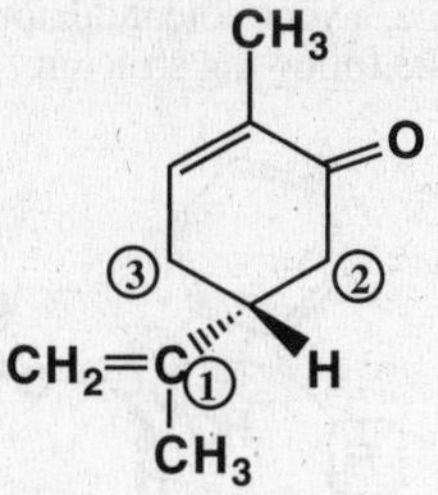

(*R*)-(-)-Carvone 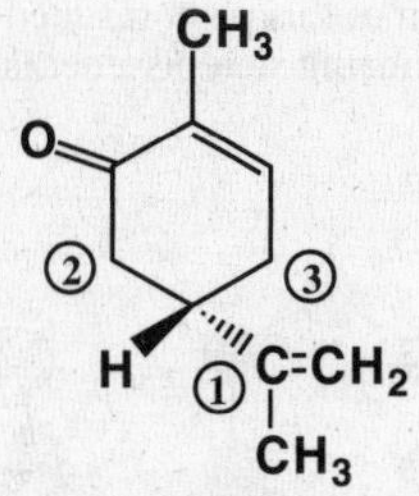   (*S*)-(+)-Carvone

Our noses operate because of receptor molecules that bind and thus detect the presence of molecules in the air we breath. The receptor then sends a signal to the brain that indicates a molecule was detected. Different receptors respond to different types of molecules. Depending on the pattern of receptors triggered, we sense different odors and learn to identify specific odors with their source, such as a delicious meal or a pretty flower. The nose receptor molecules are themselves chiral, so they can distinguish chiral molecules such as (*R*)-(-)-Carvone and (*S*)-(+)-Carvone and these enantiomers smell very different to us.

Problem 3.22 Following is a staggered conformation for one of the enantiomers of 2-butanol.

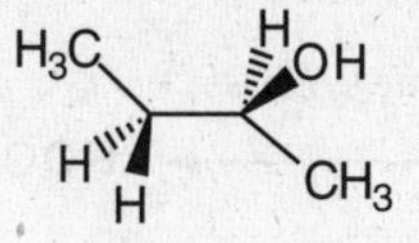

(a) Is this (*R*)-2-butanol or (*S*)-2-butanol?

The structure drawn is (*S*)-2-butanol.

(b) Draw a Newman projection for this staggered conformation, viewed along the bond between carbons 2 and 3.

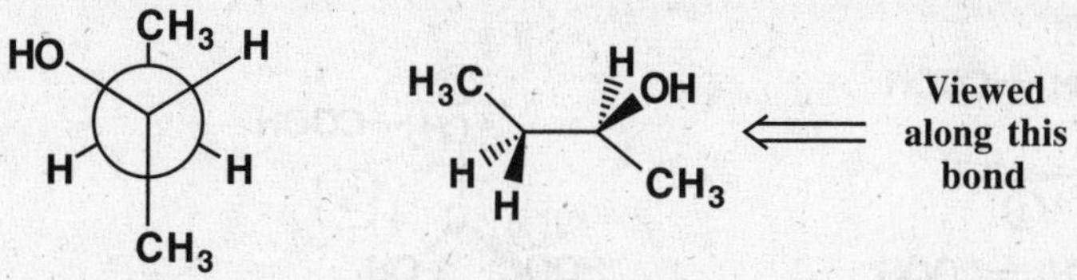

Viewed along this bond

(c) Draw a Newman projection for two more staggered conformations of this molecule. Which of your conformations is the most stable? Assume that -OH and -CH$_3$ are comparable in size.

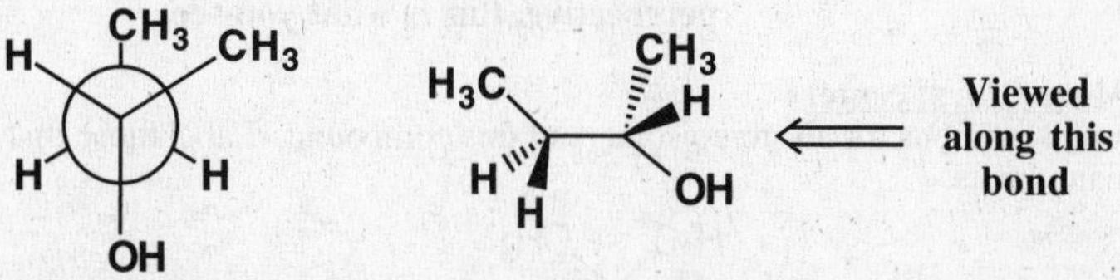

More stable

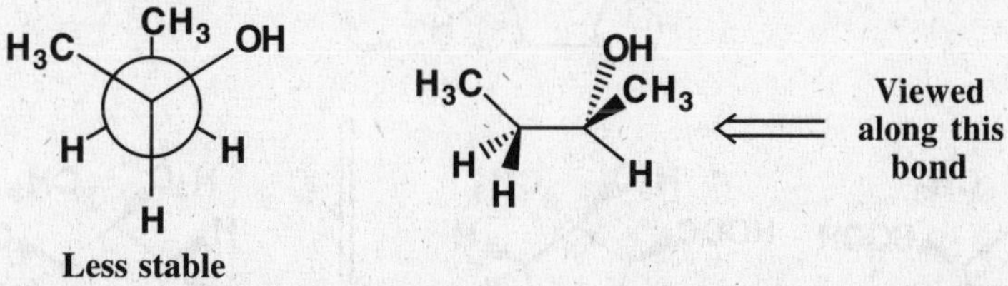

Viewed along this bond

Less stable

Viewed along this bond

Assuming that -OH and -CH$_3$ are the same size, then the structure drawn in part (b) and the upper structure shown in part (c) are of equal stability, and these are more stable than the lower structure shown in part (c). This lower structure is less stable because both the -OH and -CH$_3$ groups are adjacent to the -CH$_3$ group that is on the rear carbon atom.

<u>Problem 3.23</u> For centuries, Chinese herbal medicine has used extracts of *Ephedra sinica* to treat asthma. Ephedra as an "herbal supplement" has been implicated in the deaths of several athletes, and has recently been banned as a dietary supplement. Phytochemical investigation of this plant resulted in isolation of ephedrine, a very potent dilator of the air passages of the lungs. The naturally occurring stereoisomer is levorotatory and has the following structure. Assign *R* or *S* configuration to each chiral center.

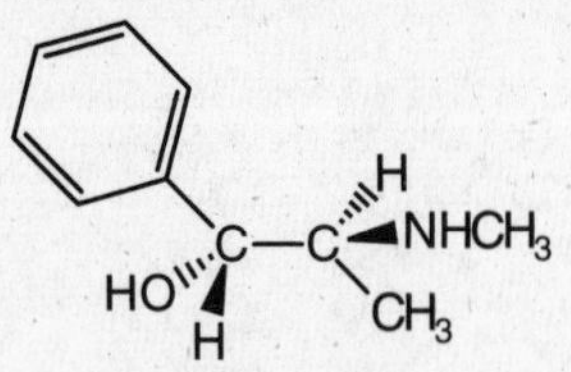

Ephedrine

$[\alpha]_D^{21} = -41$

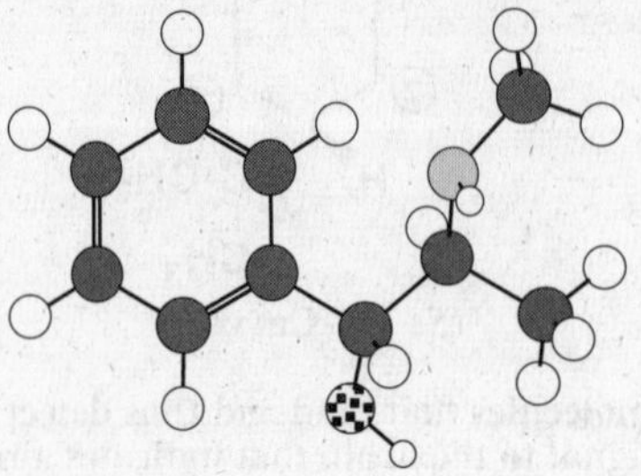

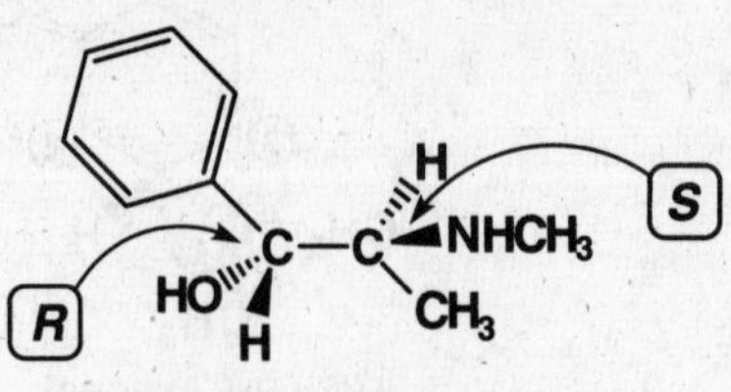

(1R,2S)-(-)-Ephedrine

<u>Problem 3.24</u> When oxaloacetic acid and acetyl-coenzyme A (acetyl-CoA) labeled with radioactive carbon-14 in position 2 are incubated with citrate synthase, an enzyme of the tricarboxylic acid cycle, only the following enantiomer of [2-^{14}C]-citric acid is formed stereoselectively. Note that citric acid containing only ^{12}C is achiral. Assign an R or S configuration to this enantiomer of [2-^{14}C] citric acid. *Note:* Carbon-14 has a higher priority than carbon-12.

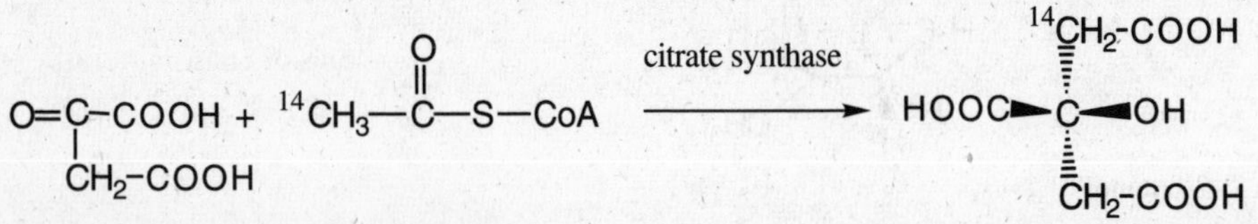

Oxaloacetic acid   **Acetyl-CoA**   **[2-^{14}C]Citric acid**

This enantiomer is (*S*)-[2-^{14}C]citric acid.

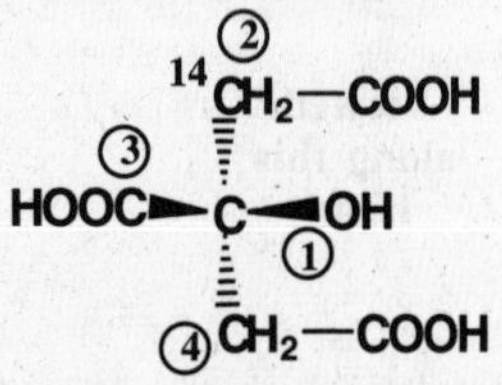

(*S*)-[2-^{14}C]Citric acid

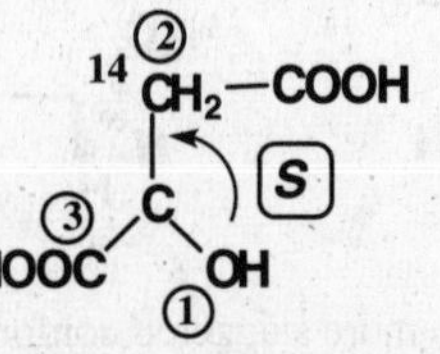

If you view from the proper perspective, this is what you see

Molecules With Two Or More Chiral centers

<u>Problem 3.25</u> Draw stereorepresentations for all stereoisomers of this compound. Label those that are meso compounds and those that are pairs of enantiomers.

Meso

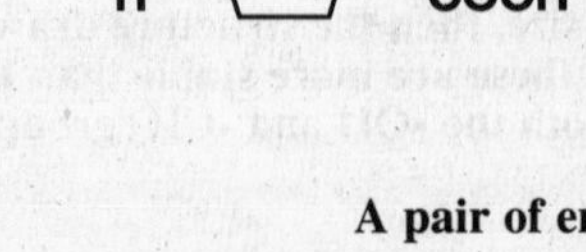

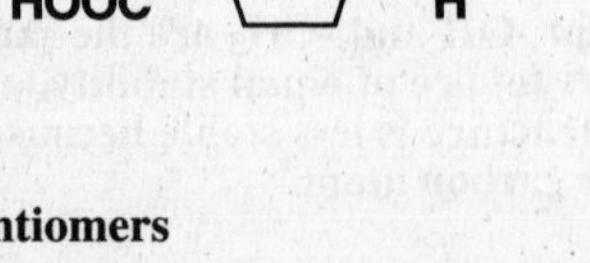

A pair of enantiomers

Problem 3.26 Mark each chiral center in the following molecules with an asterisk. How many stereoisomers are possible for each molecule?

(a) $CH_3-\overset{*}{C}H-\overset{*}{C}H-COOH$
        $\quad\quad\;\; \overset{|}{OH}\;\; \overset{|}{OH}$

2^2 = 4 Stereoisomers
(two pairs of enantiomers)

(b)
CH_2-COOH
$|$
$\overset{*}{C}H$-COOH
$|$
$HO^{*}\overset{}{C}H$-COOH

2^2 = 4 Stereoisomers
(two pairs of enantiomers)

(c)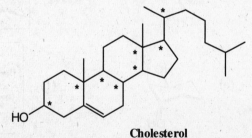

2^2 = 4 Stereoisomers
(two pairs of enantiomers)

(d)

2 Stereoisomers
(one pair of enantiomers)

(e)

2^3 = 8 Stereoisomers
(four pairs of enantiomers)

(f)

2^2 = 4 Stereoisomers
(two pairs of enantiomers)

(g)

2^2 = 4 Stereoisomers
(two pairs of enantiomers)

(h)

2^2 = 4 Stereoisomers
(two pairs of enantiomers)

Problem 3.27 Label the eight chiral centers in cholesterol. How many stereoisomers are possible for a molecule with this many chiral centers?

Cholesterol
2^8 = 256 Stereoisomers (128 pairs of enantiomers)

Problem 3.28 Label the four chiral centers in amoxicillin, which belongs to the family of semisynthetic penicillins.

Amoxicillin
2^4 = 16 Stereoisomers (8 pairs of enantiomers)

<u>Problem 3.29</u> If the optical rotation of a new compound is measured and found to have a specific rotation of +40, how can you tell if the actual rotation is not really +40 plus some multiple of +360? In other words, how can you tell if the rotation is not actually a value such as +400 or +760?

You could dilute the solution by a factor of two and then remeasure the rotation. If the new rotation is 20°, then the original rotation was 40°. If the new rotation is 200°, then the original rotation was 400°.

<u>Problem 3.30</u> Are the formulas within each set identical, enantiomers, or diastereomers?

A good approach to answering this kind of question is to label each chiral center as R or S, then compare the configurations of the different molecules.

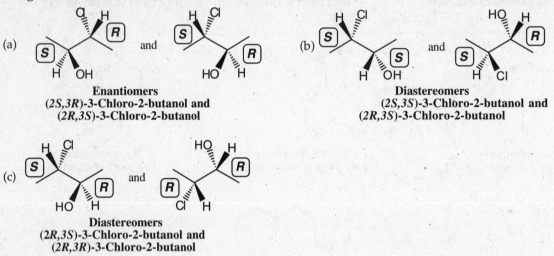

(a)   and   (b)   and

Enantiomers
(2S,3R)-3-Chloro-2-butanol and
(2R,3S)-3-Chloro-2-butanol

Diastereomers
(2S,3S)-3-Chloro-2-butanol and
(2R,3S)-3-Chloro-2-butanol

(c)   and

Diastereomers
(2R,3S)-3-Chloro-2-butanol and
(2R,3R)-3-Chloro-2-butanol

<u>Problem 3.31</u> Which of the following are meso compounds?

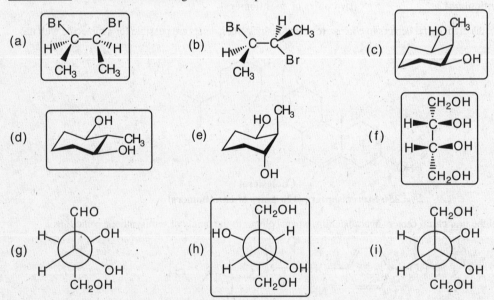

(a)   (b)   (c)

(d)   (e)   (f)

(g)   (h)   (i)

The meso compounds have a plane of symmetry: (a), (c), (d), (f), and (h).

<u>Problem 3.32</u> Vigorous oxidation of the following bicycloalkene breaks the carbon-carbon double bond and converts each carbon of the double bond to a COOH group. Assume that the conditions of oxidation have no effect on the configuration of either the starting bicycloalkene or the resulting dicarboxylic acid. Is the dicarboxylic acid produced from this oxidation one enantiomer, a racemic mixture, or a meso compound?

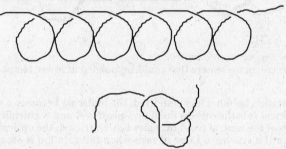

The two carboxyl groups, derived from oxidation of the double bond, must be *cis* to each other. Therefore, the compound is meso with the following configuration.

**meso-2,2-Dimethylcyclopentane-
1,3-dicarboxylic acid**

<u>Problem 3.33</u> A long polymer chain, such as polyethylene ($-CH_2CH_2-$)$_n$, can potentially exist in solution as a chiral object. Give two examples of chiral structures that a polyethylene chain could adopt.

Although there are no chiral centers in polyethylene, a long polyethylene chain could exist in chiral conformations such as a chiral helix or some kind of chiral knot.

<u>Problem 3.34</u> Which of the following compounds are chiral? Which, if any, are meso? Which, if any, does not have a possible diastereomer?

      (a)   (b)   (c)   (d)

To identify chiral molecules, look for planes or centers of symmetry. Compound (c) and (d) each have a plane of symmetry, so only molecules (a) and (b) are chiral. Because (d) has no chiral centers, it cannot have diastereomers. Molecules (a), (b), and (c) all have chiral centers (marked with asterisks (*)), so they do have diastereomers.

<u>Problem 3.35</u> Will the following compound show any optical activity if there is restricted rotation along the central C-C bond? What will happen to the optical activity at elevated temperatures as the rotation becomes less restricted.

Atropisomers are stereoisomers that occur when rotation restriction around a central bond is sufficient to allow isolation of individual conformers. The two atropisomers shown below do not have a plane of symmetry , so pure samples of each would rotate the plane of plane polarized light.

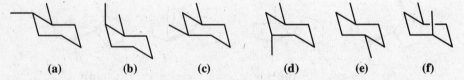

Two different atropisomers that could be isolated at lower temperature

At higher temperature, as rotation becomes less restricted, the molecule becomes a meso compound because the two chiral centers have identical substituents on them and possess *R* and *S* chirality, respectively (Section 3.4B). Therefore, as rotation of the central bond becomes less restricted, the optical rotation of either isolated conformer will be reduced until it reaches a value of zero when full rotation is observed.

<u>Problem 3.36</u> Are the following structures chiral as drawn? When placed in solution at 298 K, which structures will show an optical rotation? Explain.

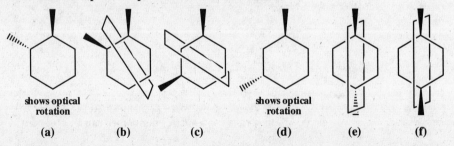

(a)   (b)   (c)   (d)   (e)   (f)

The above structures are chiral *as drawn* if the chair conformation shown has no plane or center of symmetry. Therefore, structures (a), (b), (d), and (f) are chiral. Structure (c) has a plane of symmetry, and structure (e) has a center of symmetry. To show optical rotation when placed in solution, the molecule must retain chirality even when considering both of the equilibrating chair structures. The easiest way to identify the chiral molecules is to examine the planar representations as shown below.

shows optical   shows optical
rotation   rotation

(a)   (b)   (c)   (d)   (e)   (f)

Analyzing the planar structures reveals that compounds (b), (c), (e), and (f) all have planes of symmetry and are therefore not chiral. Only compounds (a) and (d) are chiral, so only (a) and (d) will show optical rotation.

<u>Problem 3.37</u> To the following statements, answer True or False and explain your answer.
(a) All chiral centers are also stereocenters.

True. A chiral center is a tetrahedral atom with four different groups attached, and a stereocenter is an atom about which exchange of two groups produces a different stereoisomer. Because exchange of any two groups around a chiral center produces a different stereoisomer, this statement is true.

(b) All stereocenters are also chiral centers.

False. The sp^2 carbon of a double bond can be a stereocenter (*cis-trans* isomerism), but it cannot be a chiral center.

(c) All chiral molecules are optically active when pure.

True. The assumption here is that the term "pure" refers to a sample containing only a single enantiomer. It is theoretically possible that a pure chiral compound (by coincidence) might happen to not rotate the plane of plane polarized light. However, for all practical purposes, samples of a pure chiral molecule will be optically active so this statement should be labeled True.

(d) All mixtures of chiral molecules are optically active.

False. A racemic mixture is not optically active.

(e) To be optically active, a molecule must have a chiral center.

False. Atropisomers, molecules that exhibit stereoisomerism due to bond rotation restriction, can be chiral without a chiral center and they do exhibit optical activity.

(f) To be meso, a molecule must have at least two chiral centers.

True. A meso compound must contain at least two chiral centers. Meso compounds with two chiral centers occur when both chiral centers have identical groups on them and one is R while the other is S.

<u>**Looking Ahead**</u>
<u>Problem 3.38</u> The chiral catalyst (R)-BINAP-Ru is used to hydrogenate alkenes to give alkanes (Section 6.7C). The products are produced with high enantiomeric excess. An example is the formation of (S)-naproxen, a pain reliever.

BINAP

(a) What kind of isomers are the enantiomers of BINAP?

The isomers are called atropisomers.

(b) How can one enantiomer of naproxen be formed in such high yield?

Because the catalyst is chiral, the hydrogenation reaction takes place in a chiral environment. When the chiral groups on the catalyst are adjusted just right, high yields of a single enantiomer can be obtained.

<u>Problem 3.39</u> In Section 10.5D, the following reactions are discussed. Ts is the toluenesulfonate group.

Toluene sulfonate group (Ts)

(a)

(S)-2-Octanol   + Cl-Ts $\xrightarrow{\text{pyridine}}$ **(S)-2-Octyl tosylate**   + HCl

(b)

Sodium acetate   +   **(S)-2-Octyl tosylate** $\xrightarrow[\text{ethanol}]{S_N2}$ **(R)-2-Octyl acetate**   + Na$^+$OTs$^-$

In reaction (a), and *S* compound gives an *S* product. In reaction (b), an *S* compound gives an *R* product. Explain what is probably going on. (*Hint:* The oxygen atom in the starting material and produce is the same in one reaction but not in the other.) What might this say about the second reaction?

In reaction (a), the hydroxyl H atom of (S)-2-octanol is replaced by the Ts group. The reaction occurs on the oxygen atom, not the chiral center. As a result, not change in stereochemistry of the chiral center is seen. In reaction (b), the acetate replaces the -OTs group. In this case, the reaction *is* taking place on the carbon atom that is the chiral center. In particular, the reaction inverts the stereochemistry of the chiral center. You will later learn that this is characteristic for a specific type of reaction called an S$_N$2 reaction.

Gray reef shark *Carcharhinus ambylrhynchos*
Rangiroa, French Polynesia

CHAPTER 4
Solutions to the Problems

<u>Problem 4.1</u> For each conjugate acid-base pair, identify the first species as an acid or base and the second species as its conjugate acid or conjugate base. In addition, draw Lewis structures for each species, showing all valence electrons and any formal charges.

(a) H_2SO_4, HSO_4^-

Acid   Conjugate base

(b) NH_3, NH_2^-

Acid   Conjugate base

(c) CH_3OH, CH_3O^-

Acid   Conjugate base

<u>Problem 4.2</u> Write these reactions as proton-transfer reactions. Label which reactant is the acid and which is the base; which product is the conjugate base of the original acid and which is the conjugate acid of the original base. In addition, write Lewis structures for each reactant and product, and use curved arrows to show the flow of electrons in each reaction.

(a) CH_3-S-H + OH^- $\longrightarrow$ CH_3-S^- + H_2O

$CH_3-\overset{..}{\underset{..}{S}}-H$ + $^-\!:\overset{..}{O}-H$ $\rightleftharpoons$ $CH_3-\overset{..}{\underset{..}{S}}:^-$ + $H-\overset{..}{\underset{..}{O}}-H$

Acid   Base   Conjugate base   Conjugate acid

(b) $H_2C=O$ + HCl $\longrightarrow$ $H_2C=OH^+$ + Cl^-

Base   Acid   Conjugate acid   Conjugate base

<u>Problem 4.3</u> Following is a structural formula for guanidine, the compound by which migratory birds excrete excess metabolic nitrogen. The hydrochloride salt of this compound is a white crystalline powder, freely soluble in water and ethanol.

(a) Write a Lewis structure for guanidine showing all valence electrons.

Guanidine

(b) Does proton transfer occur preferentially on one of its -NH_2 groups (cation A) or to its =NH group (cation B)? Explain.

$H_2N-C-NH_2$ + HCl $\longrightarrow$ $H_2N-C-NH_3^+$ or $H_2N-C-NH_2$

Guanidine   A   B

The proton goes on the =NH group because this is the most basic nitrogen atom, and the resulting conjugate acid has 2 H atoms on each nitrogen. Distributing the H atoms evenly in this way is the most stable possible arrangement due to charge delocalization as indicated by the following resonance contributing structures.

Distributing the charge over multiple atoms is always stabilizing, so the arrangement of atoms that allows the greatest possible distribution of charge will generally be the most stable one for a given ion.

Problem 4.4 Write an equation to show the proton-transfer between each alkene or cycloalkene and HCl. Where two carbocations are possible, show each.

(a) CH_3CH_2CH═$CHCH_3$

   2-Pentene

Note how there are two different
cations produced upon protonation
of the alkene

(b)

   Cyclohexene

Problem 4.5 For each value of K_a, calculate the corresponding value of pK_a. Which compound is the stronger acid?
(a) Acetic acid, $K_a = 1.74 \times 10^{-5}$ (b) Chloroacetic acid $K_a = 1.38 \times 10^{-3}$

The pK_a is equal to $-\log_{10}K_a$. The pK_a of acetic acid is **4.76** and the pK_a of chloroacetic acid is **2.86**. Chloroacetic acid, with the smaller pK_a value, is the stronger acid.

Problem 4.6 Predict the position of equilibrium and calculate the equilibrium constant, K_{eq}, for each acid-base reaction.

(a) CH_3NH_2 + CH_3COOH ⇌ $CH_3NH_3^+$ + CH_3COO^-
      Methylamine   Acetic acid   Methylammonium   Acetate
                                              ion   ion

Acetic acid is the stronger acid; equilibrium lies to the right

$$CH_3NH_2 \quad + \quad CH_3COOH \quad \rightleftharpoons \quad CH_3NH_3^+ \quad + \quad CH_3COO^-$$

$$pK_a \; 4.76 \qquad\qquad\qquad\qquad pK_a \; 10.64$$

(stronger	(stronger	(weaker	(weaker
base)	acid)	acid)	base)

$$K_{eq} = \frac{K_a \text{ of acid on left side of equation}}{K_a \text{ of conjugate acid on right side of equation}} = \frac{10^{-4.76}}{10^{-10.64}} = \frac{1.74 \times 10^{-5}}{2.29 \times 10^{-11}}$$

$$\boxed{K_{eq} = 7.59 \times 10^5}$$

(b) $CH_3CH_2O^- + NH_3 \rightleftharpoons CH_3CH_2OH + NH_2^-$

Ethoxide ion	Ammonia	Ethanol	Amide ion

Ethanol is the stronger acid; equilibrium lies to the left.

$$CH_3CH_2O^- + NH_3 \;\rightleftharpoons\; CH_3CH_2OH + NH_2^-$$

$$pK_a \; 38 \qquad\qquad\qquad pK_a \; 15.9$$

(weaker	(weaker	(stronger	(stronger
base)	acid)	acid)	base)

$$K_{eq} = \frac{K_a \text{ of acid on left side of equation}}{K_a \text{ of conjugate acid on right side of equation}} = \frac{10^{-38}}{10^{-15.9}} = \frac{1.0 \times 10^{-38}}{1.3 \times 10^{-16}}$$

$$\boxed{K_{eq} = 7.9 \times 10^{-23}}$$

<u>Problem 4.7</u> Calculate K_{eq} for a reaction with $\Delta G° = -17.1$ kJ (-4.09 kcal)/mol at 328 K. Compare this value to the 1×10^3 seen at 298 K.

$$\Delta G° = -RT \ln K_{eq} \quad \text{can be rearranged to give} \quad \ln K_{eq} = \frac{\Delta G°}{-RT}$$

$$\ln K_{eq} = \frac{-17{,}100 \text{ J /mol}}{-(8.31 \text{ J/K mol}) \times 328 \text{ K}} = 6.27$$

$$\boxed{K_{eq} = e^{6.27} = 530}$$

<u>Problem 4.8</u> Write an equation for the reaction between each Lewis acid-base pair, showing electron flow by means of curved arrows.

(a) $(CH_3CH_2)_2B + OH^-$ $\longrightarrow$

(b) $CH_3Cl + AlCl_3 \longrightarrow$

$$CH_3\!-\!\overset{..}{\underset{..}{Cl}}: \ + \ \overset{\overset{\displaystyle Cl}{|}}{\underset{\underset{\displaystyle Cl}{|}}{Al\!-\!Cl}} \longrightarrow CH_3\!-\!\overset{+}{\underset{..}{Cl}}\!-\!\overset{-}{\underset{\underset{\displaystyle Cl}{|}}{\overset{\overset{\displaystyle Cl}{|}}{Al\!-\!Cl}}}$$

Lewis   **Lewis**
base   **acid**

Problem 4.9 For each conjugate acid-base pair, identify the first species as an acid or base and the second species as its conjugate acid or base. In addition, draw Lewis structures for each species, showing all valence electrons and any formal charge.

a) $HCOOH, HCOO^-$

$$H\!-\!\overset{\overset{\displaystyle :O:}{\|}}{C}\!-\!\overset{..}{\underset{..}{O}}\!-\!H \qquad\qquad H\!-\!\overset{\overset{\displaystyle :O:}{\|}}{C}\!-\!\overset{..}{\underset{..}{O}}\!:^-$$

           Acid   **Conjugate**
                                        base

(b) NH_4^+, NH_3

$$H\!-\!\overset{\overset{\displaystyle H}{|}}{\underset{\underset{\displaystyle H}{|}}{\overset{+}{N}}}\!-\!H \qquad\qquad H\!-\!\overset{..}{\underset{\underset{\displaystyle H}{|}}{N}}\!-\!H$$

        Acid   **Conjugate**
                                       base

(c) $CH_3CH_2O^-, CH_3CH_2OH$

$$CH_3CH_2\!-\!\overset{..}{\underset{..}{O}}:^- \qquad\qquad CH_3CH_2\!-\!\overset{..}{\underset{..}{O}}\!-\!H$$

           Base   **Conjugate**
                                         acid

(d) HCO_3^-, CO_3^{2-}

$$H\!-\!\overset{..}{\underset{..}{O}}\!-\!\overset{\overset{\displaystyle :O:}{\|}}{C}\!-\!\overset{..}{\underset{..}{O}}\!:^- \qquad\qquad {}^-\!:\!\overset{..}{\underset{..}{O}}\!-\!\overset{\overset{\displaystyle :O:}{\|}}{C}\!-\!\overset{..}{\underset{..}{O}}\!:^-$$

              Acid   **Conjugate**
                                            base

(e) $H_2PO_4^-, HPO_4^{2-}$

$$H\!-\!\overset{..}{\underset{..}{O}}\!-\!\overset{\overset{\displaystyle :O:}{\|}}{\underset{\underset{\displaystyle :O\!-\!H}{|}}{P}}\!-\!\overset{..}{\underset{..}{O}}\!:^- \qquad\qquad H\!-\!\overset{..}{\underset{..}{O}}\!-\!\overset{\overset{\displaystyle :O:}{\|}}{\underset{\underset{\displaystyle :O:^-}{|}}{P}}\!-\!\overset{..}{\underset{..}{O}}\!:^-$$

           Acid   **Conjugate**
                                              base

(f) $CH_3CH_3, CH_3CH_2^-$

$$H\!-\!\overset{\overset{\displaystyle H}{|}}{\underset{\underset{\displaystyle H}{|}}{C}}\!-\!\overset{\overset{\displaystyle H}{|}}{\underset{\underset{\displaystyle H}{|}}{C}}\!-\!H \qquad\qquad H\!-\!\overset{\overset{\displaystyle H}{|}}{\underset{\underset{\displaystyle H}{|}}{C}}\!-\!\overset{\overset{\displaystyle H}{|}}{\underset{\underset{\displaystyle H}{|}}{C}}\!:^-$$

           Acid   **Conjugate**
                                              base

(g) CH_3S^-, CH_3SH

$$CH_3\!-\!\overset{..}{\underset{..}{S}}\!:^- \qquad\qquad CH_3\!-\!\overset{..}{\underset{..}{S}}\!-\!H$$

        Base   **Conjugate**
                                       acid

Problem 4.10 Complete a net ionic equation for each proton-transfer reaction using curved arrows to show the flow of electron pairs in each reaction. In addition, write Lewis structures for all starting materials and products. Label the original acid and its conjugate base; label the original base and its conjugate acid. If you are uncertain about which substance in each equation is the proton donor, refer to Table 4.1 for the relative strengths of proton acids.

(a) $NH_3 + HCl \longrightarrow$

$$H\!-\!\overset{..}{\underset{\underset{\displaystyle H}{|}}{N}}\!: \ + \ H\!-\!\overset{..}{\underset{..}{Cl}}: \ \rightleftharpoons \ H\!-\!\overset{+}{\underset{\underset{\displaystyle H}{|}}{N}}\!-\!H \ + \ :\!\overset{..}{\underset{..}{Cl}}\!:^-$$

   Base   **Acid**   **Conjugate**   **Conjugate**
                                          acid   **base**

(b) $CH_3CH_2O^-$ + HCl $\longrightarrow$

$CH_3CH_2-\ddot{\underset{..}{O}}:^-$ + H$-\ddot{\underset{..}{Cl}}:$ $\rightleftharpoons$ $CH_3CH_2-\ddot{\underset{..}{O}}-H$ + $:\ddot{\underset{..}{Cl}}:^-$

Base   **Acid**   **Conjugate**   **Conjugate**
                                                  acid   **base**

(c) HCO_3^- + OH^- $\longrightarrow$

$:\ddot{\underset{..}{O}}-\overset{\overset{\displaystyle :O:}{\|}}{C}-\ddot{\underset{..}{O}}-H$ + $^-:\ddot{\underset{..}{O}}-H$ $\rightleftharpoons$ $^-:\ddot{\underset{..}{O}}-\overset{\overset{\displaystyle :O:}{\|}}{C}-\ddot{\underset{..}{O}}:^-$ + $H-\ddot{\underset{..}{O}}-H$

Acid   **Base**   **Conjugate**   **Conjugate**
                                                  base   **acid**

(d) CH_3COO^- + NH_4^+ $\longrightarrow$

$CH_3-\overset{\overset{\displaystyle :O:}{\|}}{C}-\ddot{\underset{..}{O}}:^-$ + $H-\overset{\overset{\displaystyle H}{|}}{\underset{\underset{\displaystyle H}{|}}{N^{\pm}}}-H$ $\rightleftharpoons$ $CH_3-\overset{\overset{\displaystyle :O:}{\|}}{C}-\ddot{\underset{..}{O}}-H$ + $H-\overset{\overset{\displaystyle H}{|}}{N}:$
                                                                                      $\underset{\displaystyle H}{|}$

Base   **Acid**   **Conjugate**   **Conjugate**
                                                  acid   **base**

Problem 4.11 Complete a net ionic equation for each proton-transfer reaction using curved arrows to show the flow of electron pairs in each reaction. Label the original acid and its conjugate base; then label the original base and its conjugate acid.

(a) NH_4^+ + OH^- $\longrightarrow$

$H-\ddot{\underset{..}{O}}:^-$ + $H-\overset{\overset{\displaystyle H}{|}}{\underset{\underset{\displaystyle H}{|}}{N^+}}-H$ $\rightleftharpoons$ $H-\ddot{\underset{..}{O}}-H$ + $:N-H$
                                                                                 $\overset{\displaystyle H}{|}$
                                                                                 $\underset{\displaystyle H}{|}$

Base   **Acid**   **Conjugate**   **Conjugate**
                                        acid   **base**

(b) CH_3COO^- + $CH_3NH_3^+$ $\longrightarrow$

$CH_3-\overset{\overset{\displaystyle :O:}{\|}}{C}-\ddot{\underset{..}{O}}:^-$ + $H-\overset{\overset{\displaystyle H}{|}}{\underset{\underset{\displaystyle H}{|}}{N^+}}-CH_3$ $\rightleftharpoons$ $CH_3-\overset{\overset{\displaystyle :O:}{\|}}{C}-\ddot{\underset{..}{O}}-H$ + $:N-CH_3$
                                                                                                  $\underset{\displaystyle H}{|}$

Base   **Acid**   **Conjugate**   **Conjugate**
                                                  acid   **base**

(c) $CH_3CH_2O^-$ + NH_4^+ $\longrightarrow$

$CH_3CH_2-\ddot{\underset{..}{O}}:^-$ + $H-\overset{\overset{\displaystyle H}{|}}{\underset{\underset{\displaystyle H}{|}}{N^+}}-H$ $\rightleftharpoons$ $CH_3CH_2-\ddot{\underset{..}{O}}-H$ + $:N-H$
                                                                                             $\overset{\displaystyle H}{|}$
                                                                                             $\underset{\displaystyle H}{|}$

Base   **Acid**   **Conjugate**   **Conjugate**
                                                  acid   **base**

(d) $CH_3NH_3^+$ + OH^- $\longrightarrow$

$$H-\overset{..}{\underset{..}{O}}:^- \quad + \quad H-\overset{\overset{H}{|}}{\underset{\underset{H}{|}}{\overset{+}{N}}}-CH_3 \quad \rightleftharpoons \quad H-\overset{..}{\underset{..}{O}}-H \quad + \quad :\overset{\overset{H}{|}}{\underset{\underset{H}{|}}{N}}-CH_3$$

Base   **Acid**   **Conjugate acid**   **Conjugate base**

Problem 4.12 Each molecule or ion can function as a base. Write a structural formula of the conjugate acid formed by reaction of each with HCl.

(a) CH_3CH_2OH

$$CH_3-CH_2-\overset{..}{\underset{..}{O}}-H$$
Base

$$CH_3-CH_2-\overset{H}{\underset{..}{\overset{|}{O}}^+}H \qquad Cl^-$$
Conjugate acid

(b) $H\overset{\overset{O}{\|}}{C}H$

$$H-\overset{\overset{:\overset{..}{O}:}{\|}}{C}-H$$
Base

$$H-\overset{\overset{:\overset{..}{O}^+{-H}}{\|}}{C}-H \qquad Cl^-$$
Conjugate acid

(c) $(CH_3)_2NH$

$$CH_3-\overset{\overset{H}{|}}{\underset{\underset{CH_3}{|}}{N}}:$$
Base

$$CH_3-\overset{\overset{H}{|}}{\underset{\underset{CH_3}{|}}{\overset{+}{N}}}-H \qquad Cl^-$$
Conjugate acid

(d) HCO_3^-

$$H-\overset{..}{\underset{..}{O}}-\overset{\overset{:\overset{..}{O}:}{\|}}{C}-\overset{..}{\underset{..}{O}}:^-$$
Base

$$H-\overset{..}{\underset{..}{O}}-\overset{\overset{:\overset{..}{O}:}{\|}}{C}-\overset{..}{\underset{..}{O}}-H \quad + \quad Cl^-$$
Conjugate acid

Problem 4.13 In acetic acid, CH_3CO_2H, the O-H hydrogen is more acidic than the CH_3 hydrogens. Explain.

Within a row, acidity increases with increasing electronegativity of the atom attached to hydrogen. Oxygen is more electronegative than carbon, so the hydrogen on the oxygen atom is more acidic. When the proton is removed from the oxygen atom of a carboxylic acid and a negatively-charged oxygen atom results, the negative charge can be shared with the adjacent oxygen atom of the carboxylate anion via the following contributing resonance structures:

$$H-\overset{\overset{H}{|}}{\underset{\underset{H}{|}}{C}}-\overset{\overset{:\overset{..}{O}:}{}}{\underset{\overset{..}{O}-H}{C}} \quad \longrightarrow \quad H-\overset{\overset{H}{|}}{\underset{\underset{H}{|}}{C}}-\overset{\overset{\overset{..}{O}:}{\|}}{\underset{\overset{..}{\underset{..}{O}}:^-}{C}} \quad \longleftrightarrow \quad H-\overset{\overset{H}{|}}{\underset{\underset{H}{|}}{C}}-\overset{\overset{\overset{..}{O}:^-}{}}{\underset{\overset{..}{O}:}{C}}$$

Contributing structures

In BOTH contributing structures, the negative charge is on a relatively electronegative oxygen atom.

When the same type of analysis is carried out on the anion produced by deprotonation of the -CH_3 group, it can be seen that this anion is less stable because one of the important contributing structures places the negative charge on the less electronegative carbon atom.

Quantitative Measure of Acid and Base Strength

Problem 4.14 Which has the larger numerical value?
(a) The pK_a of a strong acid or the pK_a of a weak acid?

The weaker acid will have the pK_a with a larger numerical value.

(b) The K_a of a strong acid or the K_a of a weak acid?

The stronger acid will have the K_a with a larger numerical value.

Problem 4.15 In each pair, select the stronger acid:
(a) Pyruvic acid (pK_a 2.49) or lactic acid (pK_a 3.85)

The stronger acid is the one with the smaller pK_a and, therefore, the larger value of K_a. Pyruvic acid is the stronger acid.

(b) Citric acid (pK_{a1} 3.08) or phosphoric acid (pK_{a1} 2.10)

Phosphoric acid is the stronger acid.

Problem 4.16 Arrange the compounds in each set in order of increasing acid strength. Consult Table 4.1 for pK_a values of each acid.

(a) CH_3CH_2OH $HOCO^-$ C_6H_5COH

           Ethanol   Bicarbonate ion   Benzoic acid
pK_a: 15.9   10.33   4.19

The compounds are already in order of increasing acid strength. Ethanol is the weakest acid, benzoic acid is the strongest acid, and bicarbonate ion is in between.

(b) $HOCOH$ CH_3COH HCl

    Carbonic acid   Acetic acid   Hydrogen chloride
pK_a: 6.36   4.76   -7

Again, the compounds are already in order of increasing acid strength. Carbonic acid is the weakest acid, hydrogen chloride is the strongest acid, and acetic acid is in between.

Problem 4.17 Arrange the compounds in each set in order of increasing base strength. Consult Table 4.1 for pK_a values of the conjugate acid of each base.

The weaker the conjugate acid (higher pK_a), the stronger the base.

(a) NH_3 $HOCO^-$ $CH_3CH_2O^-$

     9.24   6.36   15.9   pK_a of conjugate acid
Base strength increases in the order:

       $HOCO^-$ < NH_3 < $CH_3CH_2O^-$

(b) OH^- $HOCO^-$ CH_3CO^-

     15.7   6.36   4.76   pK_a of conjugate acid
Base strength increases in the order:

(b)

$pK_a = 20.2$ $pK_a = 15.9$

In the direction written, $pK_{eq} = 20.2 - 15.9 = 4.3$, so $K_{eq} = 10^{-4.3} = \boxed{5.0 \times 10^{-5}}$
As written, this reaction is endergonic.

(c)

$pK_a = 24$ $pK_a = 71$

In the direction written, $pK_{eq} = 24 - 71 = -47$, so $K_{eq} = \boxed{1.0 \times 10^{47}}$
As written, this reaction is exergonic.

<u>Problem 4.20</u> Answer True or False to the following statements about energy diagrams and reactions.
(a) A reaction coordinate diagram is used to visualize the change in the internal energy of chemical structures that occurs during chemical reactions.

True

(b) Thermodynamics is the study of the energies of structures that are represented by wells on reaction coordinate diagrams.

True

(c) Kinetics is the study of rates of the chemical reactions.

True

(d) One part of a reaction mechanism would be the understanding of which bonds break and form during a reaction.

True

(e) Thermal reactions occur via collisions between molecules, and the more energy in those collisions, the greater the rate of the reactions.

True

(f) The enthalpy of a reaction is the sole determinant of whether it will or will not occur.

False. The total change in free energy for a reaction is a function of not only enthalpy, but also temperature and entropy.

(g) An exergonic reaction will always occur during the lifespan of the standard human being.

False. Reactions that are exergonic might have a very high energy barrier, so that the occur too slowly (they take greater than 100 years) for a human being to observe over their lifetime.

<u>Problem 4.21</u> Answer True or False to the following statements about the mechanism of acid-base reactions.
(a) The acid and base must encounter each other by a collision in order for the proton to transfer.

True

(b) All collisions between acids and bases result in proton transfer

False. The acid and base must collide with the proper trajectory and geometry (linear) and energy for the proton to transfer. This does not always happen.

(c) During an acid-base reaction the lone pair on the base fills the A-H antibonding sigma orbital.

True

Problem 4.22 In each of the following three reaction coordinate diagrams state whether
(a) The reaction is exothermic or endothermic.
(b) The reaction is the slowest, the fastest or intermediate in rate.
(c) If all three reactions have the same entropy change between the reactant and product, which reaction has the largest favorable $\Delta G°$.

(a)

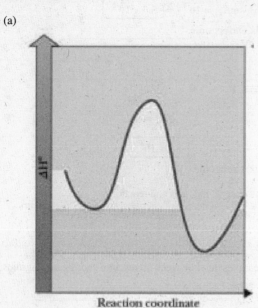

Reaction coordinate

This reaction is exothermic because the product is of lower enthalpy than the starting material. This reaction is of intermediate rate because the energy barrier (yellow shading in book) is intermediate between the other two reactions. Assuming similar entropy changes, this reaction has an intermediate favorable $\Delta G°$ because its overall change in enthalpy (pink shading in book) is favorable and intermediate among the other two reaction.

(b)

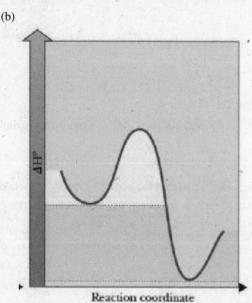

Reaction coordinate

This reaction is exothermic because the product is of lower enthalpy than the starting material. This reaction will have the fastest rate because the energy barrier (yellow shading in book) is the smallest. Assuming similar

entropy changes, this reaction will have the most favorable $\Delta G°$ because its overall change in enthalpy (pink shading in book) is favorable and the largest.

(c)

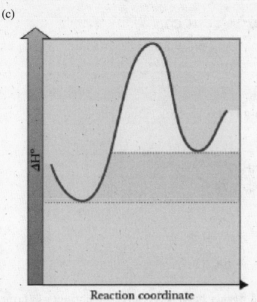

Reaction coordinate

This reaction is endothermic because the product is of higher enthalpy than the starting material. This reaction will have the slowest rate because the energy barrier is the largest. Because this reaction is endothermic, the energy barrier for this reaction is the sum of the enthalpies indicated by adding the pink and yellow areas shown in the book Assuming similar entropy changes, this reaction will have the least favorable $\Delta G°$ because its overall change in enthalpy (pink shading in book) is unfavorable and therefore the largest.

Problem 4.23 The acid-base chemistry reaction of barium hydroxide ($Ba(OH)_2$) with ammonium thiocyanate (NH_4SCN) in water creates barium thiocyanate, ammonia, and water. The reaction is highly favorable, but also so endothermic that the solution cools to such an extent that a layer of frost forms on the reaction vessel. Explain how an endothermic reaction can be favorable.

To be favorable, the total free energy of the reaction ($\Delta G°$) must be negative. If a reaction is endothermic, then the $\Delta H°$ term must be unfavorable (have a positive value). The only way for endothermic reactions to be favorable is for there to be a large positive entropy term ($\Delta S°$) associated with the reaction. Positive entropy implies that the products will have an increase in disorder compared to starting materials. Indeed, in the barium hydroxide-ammonium thiocyanate reaction, each two molecules of starting material are converted into three product molecules (barium thiocyanate, ammonia, and water) thereby increasing disorder (*i.e.* a third molecule is "released" in the process).

Position of Equilibrium in Acid-Base Reactions
Problem 4.24 Unless under pressure, carbonic acid (H_2CO_3) in aqueous solution breaks down into carbon dioxide and water, and carbon dioxide is evolved as bubbles of gas. Write an equation for the conversion of carbonic acid to carbon dioxide and water.

$$HO\overset{\displaystyle O}{\overset{\displaystyle \|}{C}}OH \longrightarrow H_2O + CO_2\uparrow$$

This relationship explains why carbonated drinks evolve CO_2 gas when they are opened and the pressure is released.

Problem 4.25 Will carbon dioxide be evolved when sodium bicarbonate is added to an aqueous solution of each compound? Explain.
(a) Sulfuric acid   (b) Ethanol   (c) Ammonium chloride

In order for carbon dioxide to be evolved, the sodium bicarbonate must be protonated to give carbonic acid (Problem 4.17). The pK_a of carbonic acid is 6.36. The pK_a's for sulfuric acid, ethanol, and ammonium chloride are -5.2, 15.9, and 9.24, respectively. Thus, sulfuric acid is the only acid strong enough to protonate sodium bicarbonate and evolve carbon dioxide.

<u>Problem 4.26</u> Acetic acid, CH_3COOH, is a weak organic acid, pK_a 4.76. Write an equation for the equilibrium reaction of acetic acid with each base. Which equilibria lie considerably toward the left? Which lie considerably toward the right?
(a) $NaHCO_3$

$$CH_3COOH \;+\; HCO_3^- \; Na^+ \;\rightleftharpoons\; CH_3COO^- Na^+ \;+\; H_2CO_3$$

pK_a 4.76   pK_a 6.36

(b) NH_3

$$CH_3COOH \;+\; NH_3 \;\rightleftharpoons\; CH_3COO^- \;+\; NH_4^+$$

pK_a 4.76   pK_a 9.24

(c) H_2O

$$CH_3COOH \;+\; H_2O \;\rightleftharpoons\; CH_3COO^- \;+\; H_3O^+$$

pK_a 4.76   pK_a -1.74

(d) NaOH

$$CH_3COOH \;+\; HO^- Na^+ \;\rightleftharpoons\; CH_3COO^- Na^+ \;+\; H_2O$$

pK_a 4.76   pK_a 15.7

Equilibrium favors the direction that gives the weaker acid and weaker base. Therefore, equilibrium will favor formation of an acid with a pK_a value higher than 4.76, or formation of acetic acid if that is the weaker acid. Based on the pK_a values shown, reactions (a), (b), and (d) have equilibria that lie considerably to the right, while reaction (c) has an equilibrium that lies considerably to the left.

<u>Problem 4.27</u> Benzoic acid, C_6H_5COOH (pK_a 4.19), is only slightly soluble in water, but its sodium salt, $C_6H_5COO^-Na^+$, is quite soluble in water. In which solutions will benzoic acid dissolve?
(a) Aqueous NaOH?   (b) Aqueous $NaHCO_3$?   (c) Aqueous Na_2CO_3?

The pK_a of benzoic acid is 4.19. The pK_a values for the conjugate acids of sodium hydroxide, sodium bicarbonate ($NaHCO_3$), and sodium carbonate (Na_2CO_3) are 15.7, 6.36, and 10.33, respectively. Thus, equilibrium will favor reaction of benzoic acid with all three of these bases to give the soluble $C_6H_5COO^-Na^+$. Therefore, benzoic acid will dissolve in aqueous solutions of all three bases.

<u>Problem 4.28</u> 4-Methylphenol, $CH_3C_6H_4OH$ (pK_a 10.26), is only slightly soluble in water, but its sodium salt, $CH_3C_6H_4O^-Na^+$, is quite soluble in water. In which solution(s) will 4-methyphenol dissolve?
(a) Aqueous NaOH?   (b) Aqueous $NaHCO_3$?   (c) Aqueous Na_2CO_3?

The pK_a of 4-methylphenol is 10.26. The pK_a values for the conjugate acids of sodium hydroxide, sodium bicarbonate($NaHCO_3$), and sodium carbonate (Na_2CO_3) are 15.7, 6.36, and 10.33, respectively. Thus, equilibrium will favor reaction of 4-methylphenol with sodium hydroxide (by a large amount) and sodium carbonate (only slightly) to give the soluble $CH_3C_6H_5O^- Na^+$. 4-Methylphenol will dissolve in aqueous solutions of these two bases. Sodium bicarbonate is not a strong enough base to deprotonate 4-methylphenol, so 4-methylphenol will not dissolve in an aqueous solution of sodium bicarbonate.

<u>Problem 4.29</u> One way to determine the predominant species at equilibrium for an acid-base reaction is to say that the reaction arrow points to the acid with the higher value of pK_a. For example:

$$NH_4^+ \;+\; H_2O \;\longleftarrow\; NH_3 \;+\; H_3O^+$$

pK_a 9.24   pK_a -1.74

$$NH_4^+ \;+\; OH^- \;\longrightarrow\; NH_3 \;+\; H_2O$$

pK_a 9.24   pK_a 15.7

Explain why this rule works.

In acid-base reactions, the position of equilibrium favors reaction of the stronger acid and stronger base to give the weaker acid and weaker base. The acid with the higher pK_a is the weaker acid, so the arrow will point toward it.

<u>Problem 4.30</u> Will acetylene react with sodium hydride according to the following equation to form a salt and hydrogen, H_2? Using pK_a values given in Table 4.1, calculate K_{eq} for this equilibrium..

$$HC\equiv CH \quad + \quad Na^+ H^- \quad \rightleftharpoons \quad HC\equiv C^- Na^+ \quad + \quad H_2$$

Acetylene	Sodium hydride	Sodium acetylide	Hydrogen
pK_a 25			pK_a 35

Since the pK_a for hydrogen is larger than the pK_a for acetylene, equilibrium will be to the right, favoring formation of the salt and hydrogen.

$$K_{eq} = \frac{K_a(\text{Acetylene})}{K_a(\text{Hydrogen})} = \frac{1 \times 10^{-25}}{1 \times 10^{-35}} = \boxed{1 \times 10^{10}}$$

<u>Problem 4.31</u> Using pK_a values given in Table 4.1, predict the position of equilibrium in this acid-base reaction and calculate its K_{eq}.

$$H_3PO_4 \quad + \quad CH_3CH_2OH \quad \rightleftharpoons \quad H_2PO_4^- \quad + \quad CH_3CH_2OH_2^+$$

pK_a 2.1 pK_a -2.4

Since the pK_a for the ethyloxonium ion ($CH_3CH_2OH_2^+$) is smaller than the pK_a for phosphoric acid (H_3PO_4), equilibrium will be to the left, favoring formation of the phosphoric acid and ethanol.

$$K_{eq} = \frac{K_a(H_3PO_4)}{K_a(CH_3CH_2OH_2^+)} = \frac{7.9 \times 10^{-3}}{2.5 \times 10^2} = \boxed{3.2 \times 10^{-5}}$$

Note the pK_a value for the ethyloxonium ion ($CH_3CH_2OH_2^+$) is not given in Table 4.1. A good estimate would be the pK_a value given for the similar species H_3O^+ (-1.74), and using this value would lead to the same conclusion that equilibrium lies to the left. However, the actual pK_a value for the ethyloxonium ion ($CH_3CH_2OH_2^+$) is -2.4 so that was used in the calculation.

<u>**Lewis Acids and Bases**</u>
<u>Problem 4.32</u> For each equation, label the Lewis acid and the Lewis base. In addition, show all unshared pairs of electrons on the reacting atoms, and use curved arrows to show the flow of electrons in each reaction.

<u>Problem 4.33</u> Complete the equation for the reaction between each Lewis acid-base pair. In each equation, label which starting material is the Lewis acid and which is the Lewis base; use curved arrows to show the flow of electrons in each reaction. In doing this problem, it is essential that you show valence electrons for all atoms participating in each reaction.

(b)

CH₃—C⁺ + H—Ö—H ⟶ CH₃—C—Ö⁺—H
with CH₃ groups, labeled **Lewis Acid** and **Lewis Base**, product with **H** and **H**

(c) CH₃-CH-CH₃ + :Br:⁻ ⟶ CH₃—CH—CH₃
Lewis Acid   **Lewis Base**   :Br:

(d) CH₃-CH-CH₃ + CH₃-Ö—H ⟶ CH₃—CH—CH₃
Lewis Acid   **Lewis Base**   H₃C—Ö⁺—H

Problem 4.34 Each of these reactions can be written as a Lewis acid/Lewis base reaction. Label the Lewis acid and the Lewis base; use curved arrows to show the flow of electrons in each reaction. In doing this problem, it is essential that you show valence electrons for all atoms participating in each reaction.

(a) CH₃—CH=CH₂ + H—Cl: ⇌ CH₃—CH—CH₂ + :Cl:⁻
Lewis Base   **Lewis Acid**

(b) CH₃—C=CH₂ + :Br—Br: ⟶ CH₃—C—CH₂—Br: + :Br:⁻
with CH₃   **Lewis Acid**   with CH₃
Lewis Base

Additional Problems

Problem 4.35 The *sec*-butyl cation can react as both a Brønsted-Lowry acid (a proton donor) and a Lewis acid (an electron pair receptor) in the presence of a water-sulfuric acid mixture. In each case, however the product is different. The two reactions are:

(1) CH₃-CH-CH₂-CH₃ + H₂O ⟶ CH₃-CH-CH₂-CH₃
sec-Butyl cation   with H—O⁺—H

(2) CH₃-CH-CH₂-CH₃ + H₂O ⟶ CH₃-CH=CH-CH₃ + H₃O⁺
sec-Butyl cation

(a) In which reaction(s) does this cation react as a Lewis acid? In which does it react as a Brønsted-Lowry acid?

As shown by the curved arrows that indicate the flow of electrons, in reaction (1), the *sec*-butyl cation is acting as a Lewis acid by reacting with the Lewis base (water). In reaction (2), the *sec*-butyl cation is acting as a Brønsted-Lowry acid by donating a proton to the water molecule.

(b) Write Lewis structures for reactants and products, and show by the use of curved arrows how each reaction occurs.

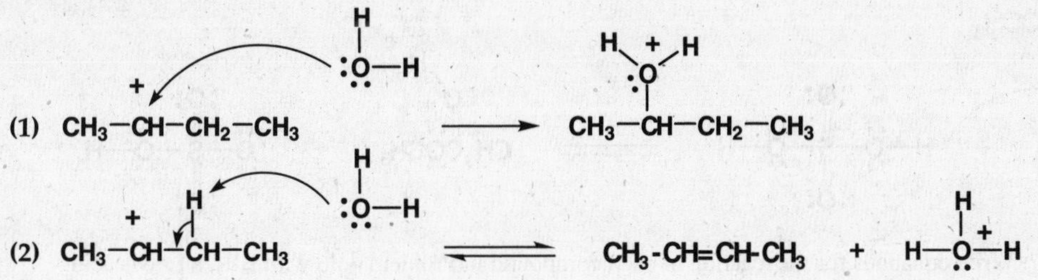

(1) $CH_3-CH-CH_2-CH_3$ $\longrightarrow$ $CH_3-CH-CH_2-CH_3$

(2) $CH_3-CH-CH-CH_3$ $\rightleftharpoons$ $CH_3-CH=CH-CH_3$ + $H-O^+-H$

Problem 4.36 Write equations for the reaction of each compound with H_2SO_4, a strong protic acid.

(a) CH_3OCH_3

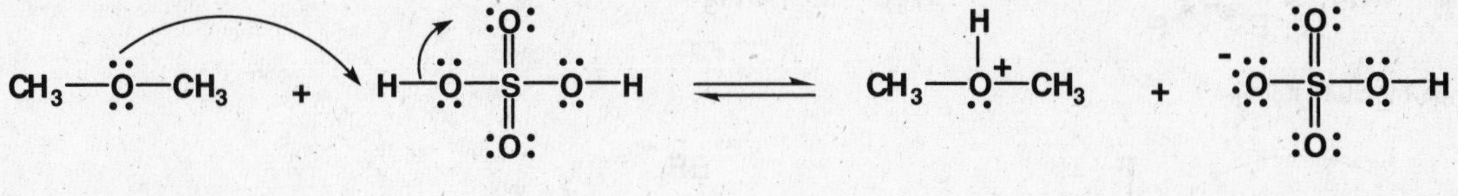

CH_3-O-CH_3 + $H-O-S-O-H$ $\rightleftharpoons$ $CH_3-O^+-CH_3$ + $^-O-S-O-H$

(b) $CH_3CH_2SCH_2CH_3$

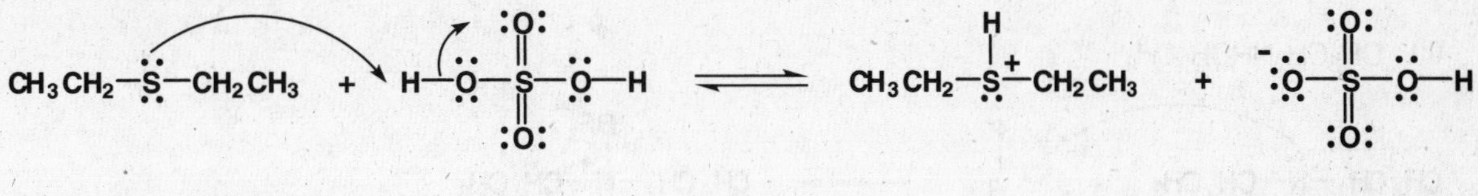

$CH_3CH_2-S-CH_2CH_3$ + $H-O-S-O-H$ $\rightleftharpoons$ $CH_3CH_2-S^+-CH_2CH_3$ + $^-O-S-O-H$

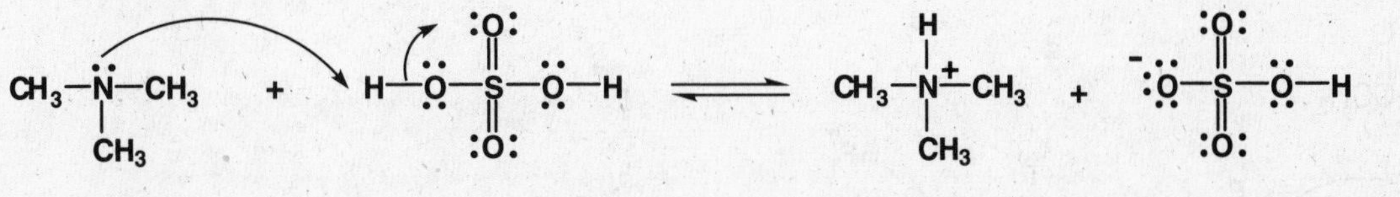

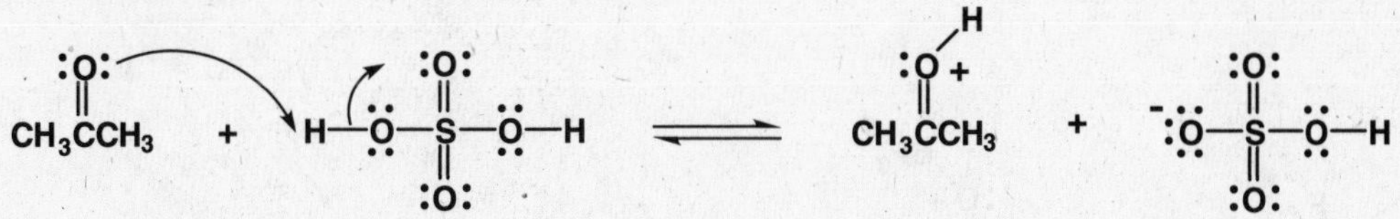

(c) $CH_3CH_2NHCH_2CH_3$

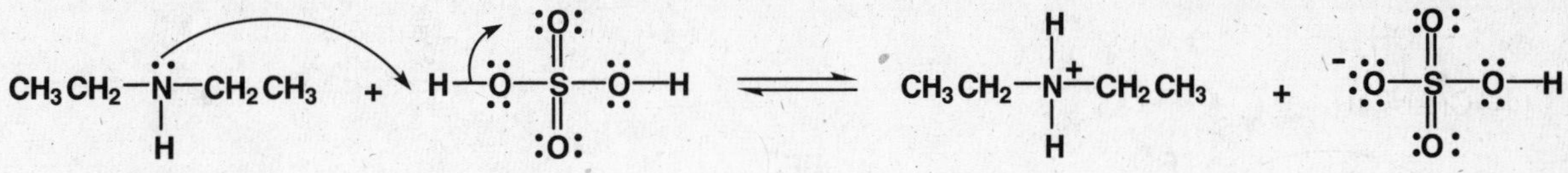

$CH_3CH_2-N-CH_2CH_3$ + $H-O-S-O-H$ $\rightleftharpoons$ $CH_3CH_2-N^+-CH_2CH_3$ + $^-O-S-O-H$

(d) CH_3NCH_3 (with CH_3 substituent)

CH_3-N-CH_3 + $H-O-S-O-H$ $\rightleftharpoons$ $CH_3-N^+-CH_3$ + $^-O-S-O-H$

(e) CH_3CCH_3 (with O double bond)

CH_3CCH_3 + $H-O-S-O-H$ $\rightleftharpoons$ CH_3CCH_3 + $^-O-S-O-H$

(f) CH_3COCH_3 (with $=O$)

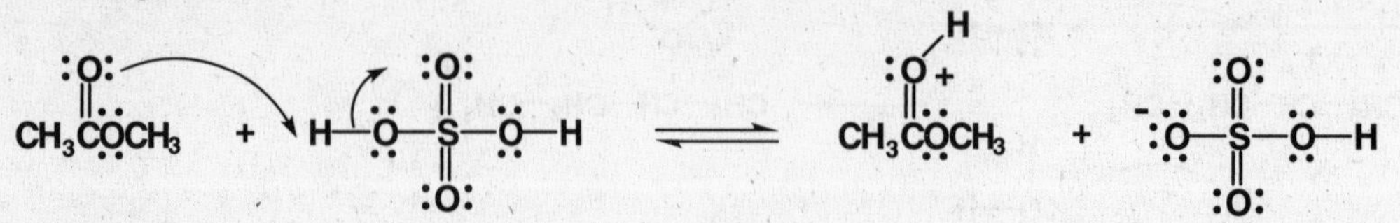

Problem 4.37 Write equations for the reaction of each compound in Problem 4.36 with BF_3, a Lewis acid.

(a) CH_3OCH_3

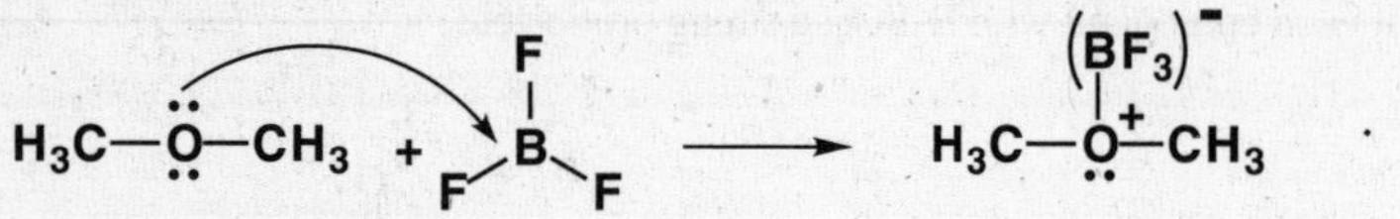

(b) $CH_3CH_2SCH_2CH_3$

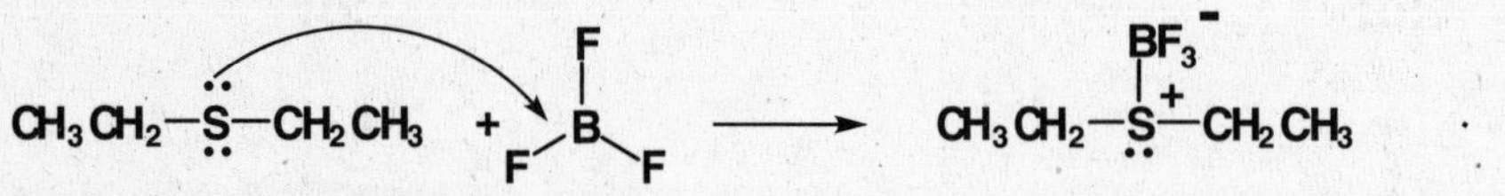

(c) $CH_3CH_2NHCH_2CH_3$

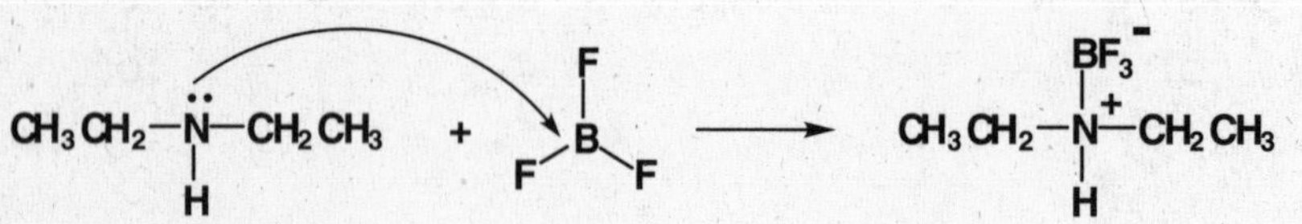

(d) CH_3NCH_3 (with CH_3)

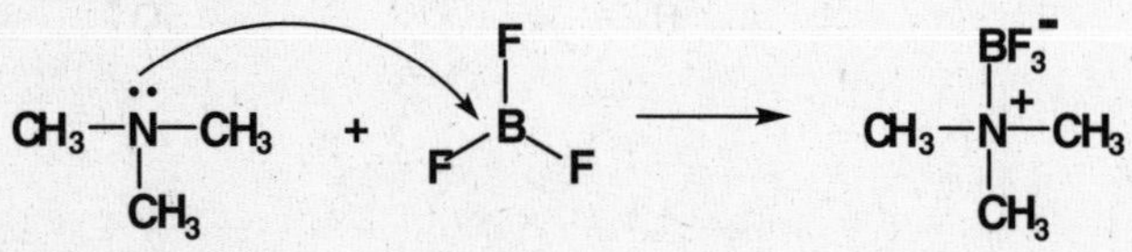

(e) CH_3CCH_3 (with $=O$)

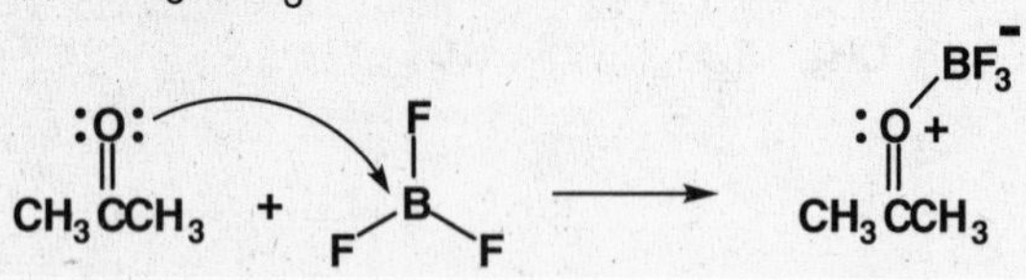

(f) CH_3COCH_3 (with $=O$)

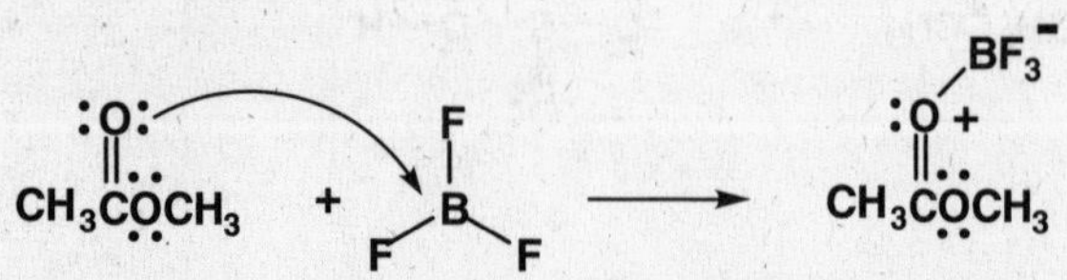

Problem 4.38 Label the most acidic hydrogen in each molecule, and justify your choice by using appropriate pK_a values.

This problem is difficult because the molecules contain more than one functional group that can be deprotonated. In each case, use the entries in Table 4.1 to determine which functional group is the more acidic. pK_a values have been added below each acidic hydrogen and the most acidic is underlined. Because exact

matches could not be found, approximate pK_a values were used corresponding to the most similar functional group entry in the Table.

(a) $\underline{H}OCH_2CH_2NH_2$
~16   ~38

(b) $\underline{H}SCH_2CH_2NH_2$
~7   ~38

(c) $\underline{H}OCH_2CH_2C\equiv CH$
~16   ~25

(d) $\underline{H}O\overset{O}{\overset{\|}{C}}CH_2CH_2SH$
~4.5   ~7

(e) $CH_3\underset{~16}{\overset{H\underset{}{O}}{CH}}\overset{O}{\overset{\|}{C}}O\underline{H}$
~4.5

(f) $H_3\overset{+}{N}CH_2CH_2\overset{O}{\overset{\|}{C}}O\underline{H}$
~9   ~4.5

(g) $\underline{H}_3\overset{+}{N}CH_2CH_2\overset{O}{\overset{\|}{C}}O^-$
~9

(h) $\underline{H}SCH_2CH_2O\underline{H}$
~7   ~16

<u>Problem 4.39</u> Explain why the hydronium ion, H_3O^+, is the strongest acid that can exist in aqueous solution. What is the strongest base that can exist in aqueous solution?

When a strong acid (HA) is placed in aqueous solution, the following equilibrium is established:

$$HA \;+\; H_2O \;\rightleftharpoons\; A^- \;+\; H_3O^+$$

Since equilibrium favors formation of the weaker acid and weaker base, if HA is a stronger acid than H_3O^+, equilibrium will favor the right side of the equation, forming H_3O^+. As a result, the strongest acid than can exist in aqueous solution is H_3O^+.
In an analogous way, when a strong base (Base) is added to water, the following equilibrium is established:

$$Base \;+\; H_2O \;\rightleftharpoons\; Base\overset{+}{-}H \;+\; HO^-$$

Equilibrium favors formation of the weaker acid and weaker base. Any base stronger than HO^- will drive the equilibrium toward formation of HO^-. As a result, the strongest base that can exist in aqueous solution is HO^-.

<u>Problem 4.40</u> What is the strongest base that can exist in liquid ammonia as a solvent?

When a strong base (Base) is added to liquid ammonia, the following equilibrium is established:

$$Base \;+\; H_3N \;\rightleftharpoons\; Base\overset{+}{-}H \;+\; H_2N^-$$

Equilibrium favors formation of the weaker acid and weaker base. Any base stronger than H_2N^- will drive the equilibrium toward formation of H_2N^-. As a result, the strongest base that can exist in liquid ammonia is H_2N^-.

<u>Problem 4.41</u> For each pair of molecules or ions, select the stronger base and write its Lewis structure.

The stronger base will have the weaker conjugate acid. For each pair, choose the base having a conjugate acid with the higher pKa value in Table 4.1.

(a) CH_3S^- or $\boxed{CH_3O^-}$

(b) $\boxed{CH_3NH^-}$ or CH_3O^-

(c) CH_3COO^- or $\boxed{OH^-}$

(d) $CH_3CH_2O^-$ or $\boxed{H^-}$

(e) NH_3 or $\boxed{OH^-}$

(f) $\boxed{NH_3}$ or H_2O

(g) CH_3COO^- or $\boxed{HOO_3^-}$

(h) HSO_4^- or $\boxed{OH^-}$

(i) $\boxed{OH^-}$ or Br^-

Problem 4.42 Account for the fact that nitroacetic acid, O_2NCH_2COOH (pK_a 1.68) is a considerably stronger acid than acetic acid, CH_3COOH (pK_a 4.76).

In general, the more stable the conjugate base, the stronger the acid. The nitro group is much more electronegative than a hydrogen atom, so the nitro group stabilizes the negative charge of the carboxylate anion through the inductive effect analogous to halogen atoms as described in the text.

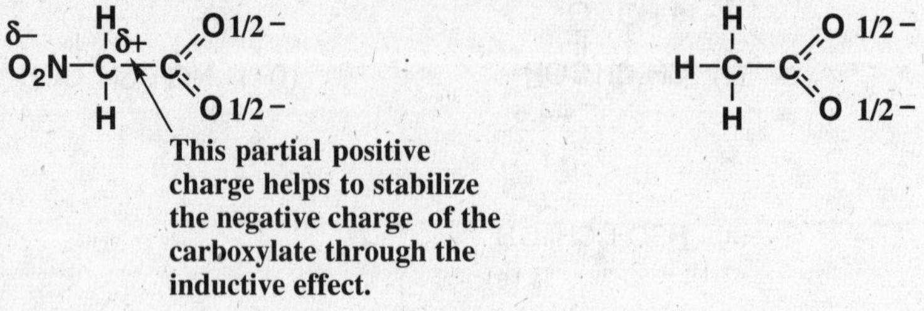

This partial positive charge helps to stabilize the negative charge of the carboxylate through the inductive effect.

Problem 4.43 Sodium hydride, NaH, is available commercially as a gray-white powder. It melts at 800°C with decomposition. It reacts explosively with water, and ignites spontaneously on standing in moist air.
(a) Write a Lewis structure for the hydride ion and for sodium hydride. Is your Lewis structure consistent with the fact that this compound is a high-melting solid? Explain.

The hydride anion   **Sodium hydride**

The Lewis structure of sodium hydride indicates that is should be thought of as being composed of the hydride anion and sodium cation. Ionic compounds such as NaCl are generally high-melting solids.

(b) When sodium hydride is added very slowly to water, it dissolves with the evolution of a gas. The resulting solution is basic to litmus. What is the gas evolved? Why has the solution become basic?

When NaH is placed in water, the following reaction takes place:

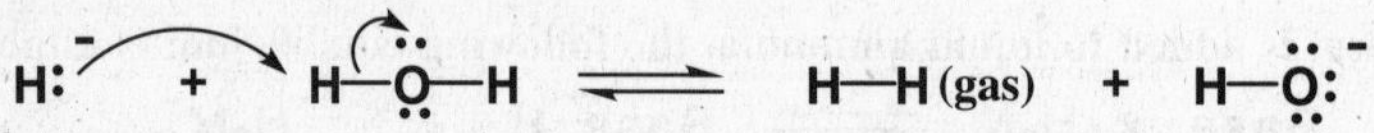

Because the hydride ion is a stronger base than HO⁻, reaction goes to the right. Formation of H_2 gas, which bubbles from the reaction, also drives the reaction to the right. Formation of HO⁻ drives up the pH of the solution.

(c) Write an equation for the reaction between sodium hydride and 1-butyne, $CH_3CH_2C{\equiv}CH$. Use curved arrows to show the flow of electrons in this reaction.

Problem 4.44 Methyl isocyanate, CH_3-N=C=O, is used in the industrial synthesis of a type of pesticide and herbicide known as a carbamate. As a historical note, an industrial accident in Bhopal, India in 1984 resulted in leakage of an unknown quantity of this chemical into the air. An estimated 200,000 persons were exposed to its vapors and over 2000 of these people died.
(a) Write a Lewis structure for methyl isocyanate and predict its bond angles. What is the hybridization of its carbonyl carbon? Of its nitrogen atom?

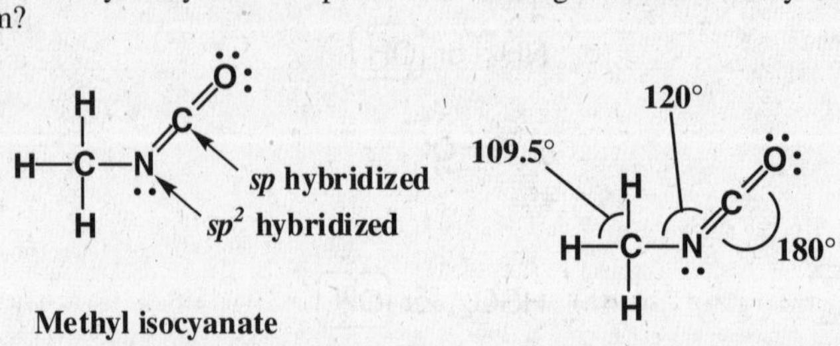

Methyl isocyanate

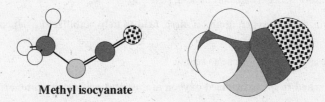

Methyl isocyanate

(b) Methyl isocyanate reacts with strong acids, such as sulfuric acid, to form a cation. Will this molecule undergo protonation more readily on its oxygen or nitrogen atom? In considering contributing structures to each hybrid, do not consider structures in which more than one atom has an incomplete octet.

Remember that protonation of a neutral species occurs most readily in the location where the positive charge can be stabilized most effectively, *i.e.* by resonance. As can be seen below, protonation at either the oxygen atom or the nitrogen atom will lead to cations that have three major contributing structures. The middle contributing structures in both cases contain a carbon atom with an unfilled octet, so these middle structures make a relatively small contribution to their respective resonance hybrids. Oxygen is more electronegative than nitrogen, so oxygen is less able to accommodate a full positive charge compared to nitrogen. As a result, the contributing structure on the right also makes a relatively small contribution to the overall resonance hybrid in the case of nitrogen protonation. This means that for the cation produced upon protonation of nitrogen, there is relatively little in the way of overall resonance stabilization.

Protonation on nitrogen - less resonance stabilization

These make relatively small contributions to the resonance hybrid

Protonation on oxygen - more resonance stabilization - will predominate

This makes relatively small contribution to the resonance hybrid

This makes more of a contribution than the analogous structure shown above for nitrogen protonation

On the other hand, the analogous contributing structure shown on the right for oxygen protonation makes more of a contribution to the resonance hybrid of that species. Greater contribution from a contributing structure leads to greater overall resonance stabilization. <u>Therefore, protonation on oxygen will predominate for methyl isocyanate.</u>

Problem 4.45 Offer an explanation for the following observations.
(a) H_3O^+ is a stronger acid than NH_4^+.

Oxygen is more electronegative than nitrogen, so the proton is more acidic on H_3O^+.

(b) Nitric acid, HNO_3, is a stronger acid than nitrous acid, HNO_2.

There is one more electronegative oxygen atom on nitric acid to help stabilize the deprotonated anion through a resonance effect.

(c) Ethanol and water have approximately the same acidity.

Both have the proton attached to sp^3 hybridized oxygen atoms, and neither deprotonated species can be resonance stabilized.

(d) Trifluoroacetic acid, CF_3COOH, is a stronger acid than trichloroacetic acid, CCl_3COOH.

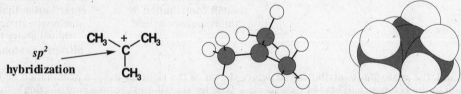

The larger this partial positive charge, the greater the stabilization of the carboxylate anion through the inductive effect.

Fluorine is more electronegative than chlorine, so the above anion is more stable when X = F compared to when X = Cl due to increased inductive effects with X = F.

Looking Ahead
Problem 4.46 Following is a structural formula for the *tert*-butyl cation. (We discuss the formation, stability, and reactions of cations such as this one in Chapter 6.)

$$CH_3 \!-\! \overset{+}{\underset{\underset{\displaystyle CH_3}{|}}{C}} \!-\! CH_3$$

tert-Butyl cation
(a carbocation)

(a) Predict all C-C-C bond angles in this cation.

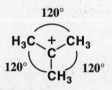

Trigonal planar geometry

According to VSEPR, the geometry of the central carbon atom will be trigonal planar because there are three regions of electron density.

(b) What is the hybridization of the carbon bearing the positive charge?

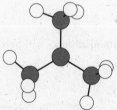
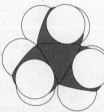

The positively-charged carbon atom is sp^2 hybridized. The three sp^2 hybrid orbitals take part in σ bonding to the three carbon atoms, and the unhybridized $2p$ orbital is empty.

(c) Write a balanced equation to show its reaction as a Lewis acid with water.

As a Lewis acid

(d) Write a balanced equation to show its reaction as a Brønsted-Lowry acid with water.

As a Brønsted-Lewis acid

Problem 4.47 Alcohols (Chapter 10) are weak organic acids, pK_a 15-18. The pK_a of ethanol, CH_3CH_2OH, is 15.9. Write equations for the equilibrium reaction of ethanol with each base. Which equilibria lie considerably toward the right? Which lie considerably toward the left?

(a) NaHCO₃

$$CH_3CH_2OH \; + \; HCO_3^- \; \rightleftharpoons \; CH_3CH_2O^- \; + \; H_2CO_3$$
pK_a 15.9 pK_a 6.36

(b) NaOH

$$CH_3CH_2OH \; + \; OH^- \; \rightleftharpoons \; CH_3CH_2O^- \; + \; H_2O$$
pK_a 15.9 pK_a 15.7

(c) NaNH₂

$$CH_3CH_2OH \; + \; NH_2^- \; \rightleftharpoons \; CH_3CH_2O^- \; + \; NH_3$$
pK_a 15.9 pK_a 38

(d) NH₃

$$CH_3CH_2OH \; + \; NH_3 \; \rightleftharpoons \; CH_3CH_2O^- \; + \; NH_4^+$$
pK_a 15.9 pK_a 9.24

Equilibrium favors the direction that gives the weaker acid and weaker base. Therefore, equilibrium will favor formation of an acid with a pK_a value higher than 15.9, or formation of ethanol if that is the weaker acid. Based on the pK_a values shown, only reaction (c) has an equilibrium that lies considerably to the right, while reactions (a) and (d) have equilibria that lie considerably to the left. Reaction (b) has an equilibrium that lies only slightly to the left because the pK_a for ethanol is only slightly higher than that for water.

Problem 4.48 As we shall see in Chapter 19, hydrogens on a carbon adjacent to a carbonyl group are far more acidic than those not adjacent to a carbonyl group. The anion derived from acetone, for example, is more stable than is the anion derived from ethane. Account for the greater stability of the anion from acetone.

Acetone   **Ethane**

$pK_a = 22$ $pK_a = 51$

The anion of acetone can be stabilized by resonance with the pi bond of the adjacent carbonyl group as shown in the following contributing structures.

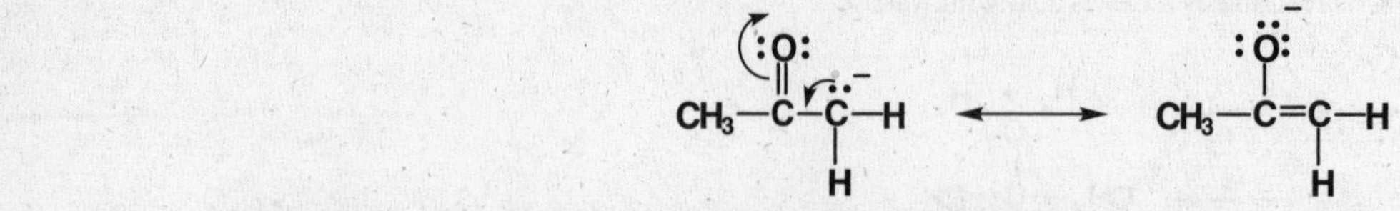

In this way, the negative charge is delocalized significantly, leading to an anion in which the charge is delocalized as indicated by the two resonance contributing structures. Note how the negative charge is placed on the more electronegative oxygen atom in the structure on the right. The anion of deprotonated ethane does not have any opportunities for this type of charge delocalization. The increased anion stability because of greater charge delocalization makes acetone more acidic.

Problem 4.49 2,4-Pentanedione is a considerably stronger acid than acetone (Chapter 19). Write a structural formula for the conjugate base of each acid, and account for the greater stability of the conjugate base from 2,4-pentanedione.

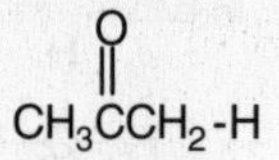

**Propanone
(Acetone)**
$pK_a = 22$

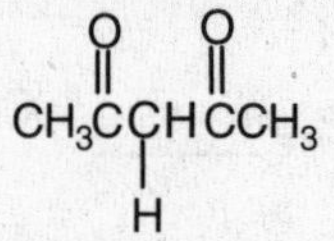

2,4-Pentanedione
$pK_a = 9$

A carbonyl group adjacent to a negatively charged carbon atom, an arrangement that is referred to as an enolate ion, can lead to resonance stabilization by virtue of a resonance contributing structure that places the negative charge on oxygen along with formation of a carbon-carbon double bond. Thus, there are two important contributing structures that describe the resonance stabilization of the conjugate base of propanone.

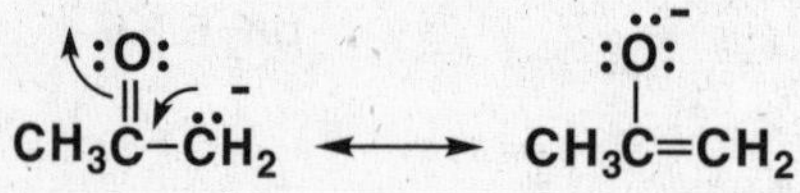

However, there are two adjacent carbonyl groups and thus three important contributing structures that describe the resonance stabilization of the conjugate base of 2,4-pentanedione.

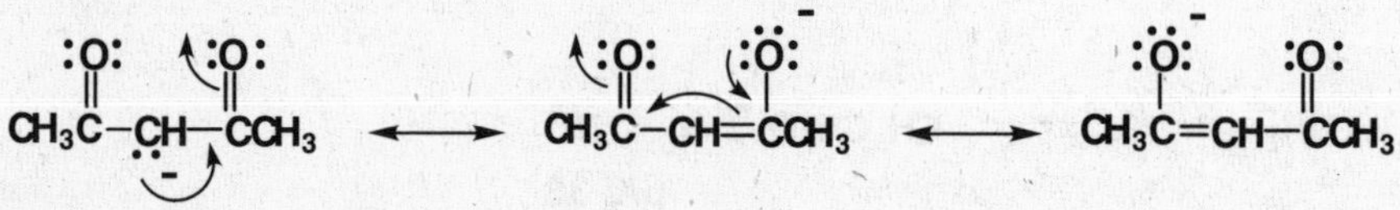

The added contributing structure provides additional resonance stabilization for the conjugate base of 2,4-pentanedione compared to propanone, making 2,4-pentanedione the stronger acid.

Problem 4.50 Write an equation for the acid-base reaction between 2,4-pentanedione and sodium ethoxide and calculate its equilibrium constant, K_{eq}. The pK_a of 2,4-pentanedione is 9; that of ethanol is 15.9.

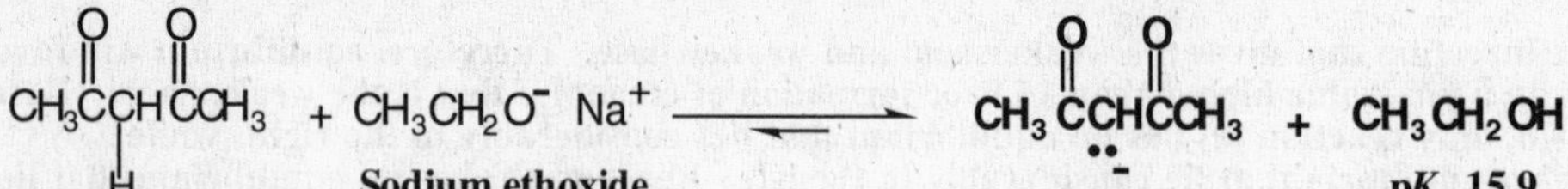

2,4-Pentanedione   **Sodium ethoxide** Na^+ **Weaker acid**
pK_a 9   **Stronger base**   **Weaker base**

pK_a 15.9

Stronger acid

Because the pK_a for 2,4-pentanedione is smaller than the pK_a for ethanol, equilibrium will favor the right side, favoring formation of the 2,4-pentanedione salt and ethanol.

$$K_{eq} = \frac{K_a(\text{2,4-Pentanedione})}{K_a(\text{Ethanol})} = \frac{1.0 \times 10^{-9}}{1.3 \times 10^{-16}} = \boxed{7.7 \times 10^6}$$

Problem 4.51 An ester is a derivative of a carboxylic acid in which the hydrogen of the carboxyl group is replaced by an alkyl group (Section 1.3E). Draw the structural formula of methyl acetate, which is derived from acetic acid by replacement of the H of its -OH group by a methyl group. Determine if proton transfer to this compound from HCl occurs preferentially on the oxygen of the C=O group or the oxygen of the OCH$_3$ group.

Remember that protonation of a neutral species occurs most readily in the location where the positive charge can be stabilized most effectively. Protonation of an ester occurs preferentially on the oxygen atom of the C=O group. Only protonation at this site leads to a product that is resonance stabilized as indicated by the two contributing structures shown. In later chapters you will learn that the C=O group is especially polarized, placing significant partial negative charge on the oxygen atom, a factor that also contributes to basicity.

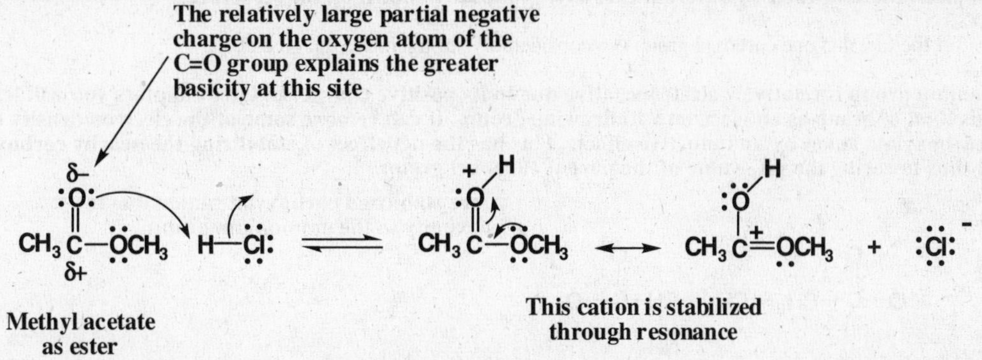

The relatively large partial negative charge on the oxygen atom of the C=O group explains the greater basicity at this site

CH$_3$ C—OCH$_3$ H—Cl: ⇌   CH$_3$ C—OCH$_3$ ⟷   CH$_3$ C=OCH$_3$ +   :Cl:⁻

**Methyl acetate
as ester**

**This cation is stabilized
through resonance**

Problem 4.52 Alanine is one of the 20 amino acids (it contains both an amino and a carboxyl group) found in proteins (Chapter 27). Is alanine better represented by the structural formula A or B? Explain.

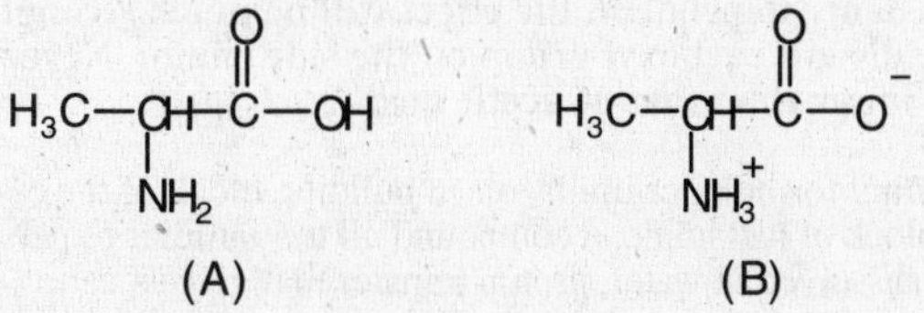

H$_3$C—CH—C—OH   H$_3$C—CH—C—O⁻
        | |
        NH$_2$ NH$_3$⁺

        (A)   (B)

Using the values in Table 4.1, you would predict the pK_a for the carboxyl group will be around 4, similar to acetic acid. The pK_a of the carboxyl groups of amino acids are actually between 2 and 3 due largely to the inductive effect of the adjacent ammonium group (see the next problem). The pK_a for the ammonium groups of amino acids around 9-10, similar to the value for the ammonium ion in Table 4.1. When conversion of (A) to (B) is written as a proton transfer reaction equilibrium, the weaker acid, namely the ammonium ion, is on the side favored at equilibrium. Thus, species (B) is the better representation of alanine.

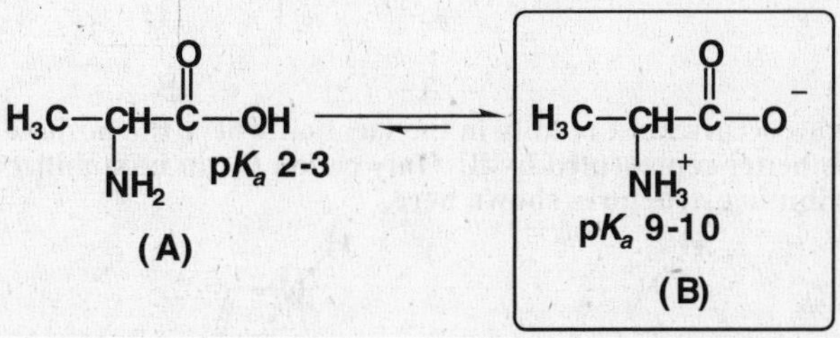

H$_3$C—CH—C—OH   ⟶   H$_3$C—CH—C—O⁻
        | pK_a 2-3 |
        NH$_2$ NH$_3$⁺
                                   pK_a 9-10
        (A)
                                      (B)

Another way to think about this problem is using the rule of thumb that if an acid has a pK_a that is at least two units lower than the pH of the solution, it will be mostly deprotonated. Similarly, if a base has a pK_a for its conjugate acid that is at least two units higher than the pH of the solution, it will be mostly protonated. At neutral pH (pH 7.0), amino acids are in their doubly ionized form (B), called a zwitterion. See the Connections to Biological Chemistry box in this chapter for a description of other biologically important functional groups that are ionized at neutral pH.

Problem 4.53 Glutamic acid is another of the amino acids found in proteins (Chapter 27). Glutamic acid has two carboxyl groups, one with pK_a 2.10 and the other with pK_a 4.07.

Glutamic acid   HO—C—CH$_2$—CH$_2$—CH—C—OH

pK_a 4.07   NH$_3^+$ **pK_a 2.10**

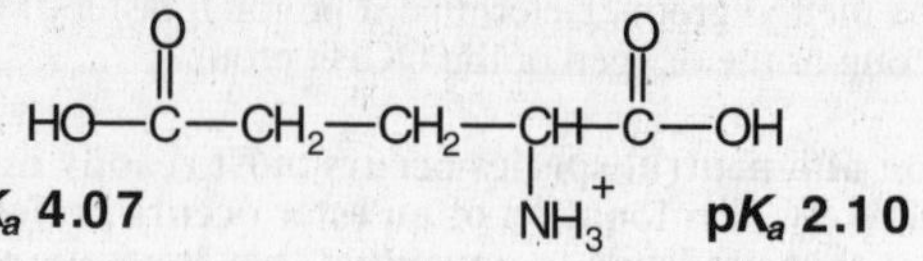

(a) Which carboxyl group has which pK_a?

The actual pK_a values for the two different carboxyl groups are shown on the structure.

(b) Account for the fact that one carboxyl group is a considerably stronger acid than the other.

The ammonium group is relatively electronegative due to its positive charge. In later chapters you will learn to refer to this kind of group as an electron withdrawing group. It can remove some of the electron density of an adjacent carboxylate anion by an inductive effect. This has the net effect of stabilizing the nearby carboxylate anion and thus lowering the pK_a value of the parent carboxyl group.

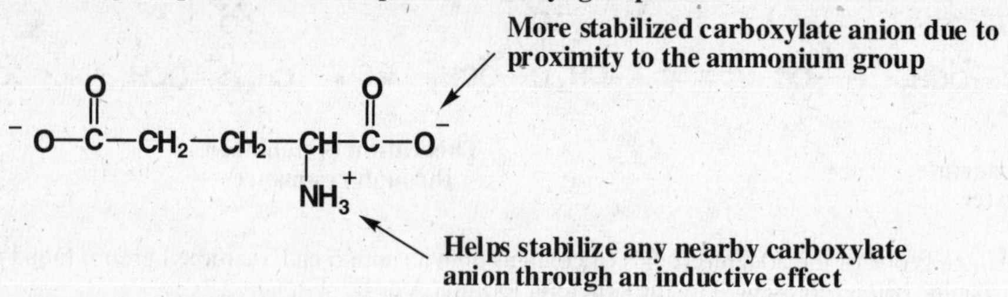

More stabilized carboxylate anion due to proximity to the ammonium group

O—C—CH$_2$—CH$_2$—CH—C—O$^-$

NH$_3^+$

Helps stabilize any nearby carboxylate anion through an inductive effect

Because an inductive effect is distance dependent, the effect will be much stronger for the nearby carboxyl group, compared with the more distant carboxyl group on the side chain. Nevertheless, even the latter carboxyl group has a pK_a that is slightly lower than that of acetic acid.

Problem 4.54 Following is a structural formula for imidazole, a building block of the essential amino acid histidine (Chapter 27). It is also a building block of histamine, a compound all too familiar to persons with allergies and takers of antihistamines. When imidazole is dissolved in water, proton transfer to it gives a cation. Is this cation better represented by structure A or B? Explain.

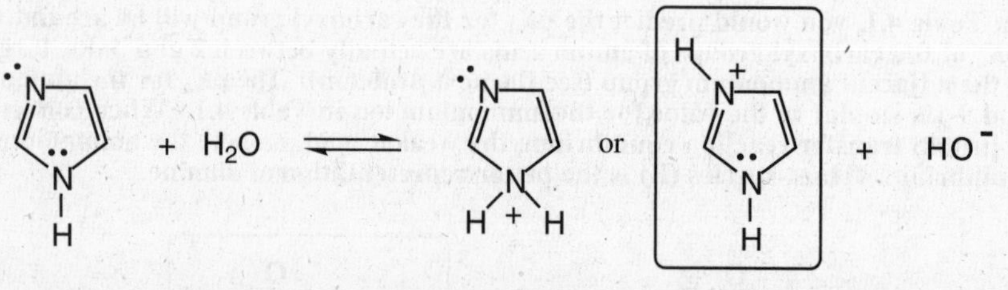

A   B

Protonation of a neutral species occurs most readily in the location where the positive charge can be stabilized most effectively. The cation is better represented by B. Only cation B can be stabilized by charge delocalization as indicated by the two contributing structures shown here.

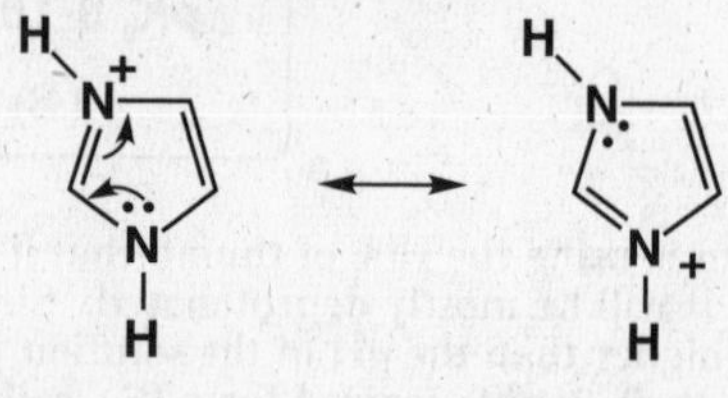

CHAPTER 5
Solutions to the Problems

<u>Problem 5.1</u> Calculate the index of hydrogen deficiency of cyclohexene, C_6H_{10}, and account for this deficiency by reference to its structural formula.

The molecular formula of the reference acyclic alkane of six carbons is C_6H_{14}. The index of hydrogen deficiency of cyclohexane (14-10)/2 = 2 and is accounted for by the combination of one ring and one π bond in cyclohexene.

Cyclohexene

<u>Problem 5.2</u> The index of hydrogen deficiency of niacin is 5. Account for this index of hydrogen deficiency by reference to the structural formula of niacin.

**Nicotinamide
(Niacin)**

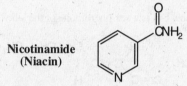

The index of hydrogen deficiency of 5 can be accounted for by the presence of the ring plus the three π bonds in the ring, as well as the π bond of the carbonyl group.

<u>Problem 5.3</u> Write the IUPAC name of each alkane.

(a)

(b)

3-Methyl-1-butene   **2,3-Dimethyl-2-butene**

<u>Problem 5.4</u> Which alkenes show *cis, trans* isomerism? For each alkene that does, draw the *trans* isomer.
(a) 2-Pentene   (b) 2-Methyl-2-pentene   (c) 3-Methyl-2-pentene

CH_3CH_2 H
          C=C
H CH_3

No *cis,trans* isomers because there are two methyl groups on one end of the double bond.

CH_3CH_2 H
          C=C
CH_3 CH_3

trans-**2-Pentene**   *trans*-**3-Methyl-2-pentene**

<u>Problem 5.5</u> Name each alkene and specify its configuration by the E-Z system.

(a)   (b)   (c)

(*E*)-1-Chloro-2,3-dimethyl-2-pentene   **(*Z*)-1-Bromo-1-chloropropene**   **(*E*)-2,3,4-Trimethyl-3-heptene**

<u>Problem 5.6</u> Write the IUPAC name of each cycloalkene

(a)   (b)   (c)

1,3-Dimethyl-   **Cyclooctene**   **4-(1,1-Dimethylethyl)-**
cyclopentene   **cyclohexene (IUPAC)**
                                                       4-*tert*-Butylcyclohexene (Common)

<u>Problem 5.7</u> Draw structural formulas for the other two stereoisomers of 2,4-heptadiene.

cis,trans-2,4-Heptadiene   *cis,cis*-2,4-Heptadiene

<u>Problem 5.8</u> (10*E*,12*Z*)-10,12-hexadecadiene-1-ol is a sex pheromone of the silkworm. Draw a structural formula for this compound.

(10*E*,12*Z*)-10,12-Hexadecadiene-1-ol

Structure of Alkenes
<u>Problem 5.9</u> Predict all approximate bond angles about each highlighted carbon atom. To make these predictions, use valence shell electron-pair repulsion (Section 1.4).

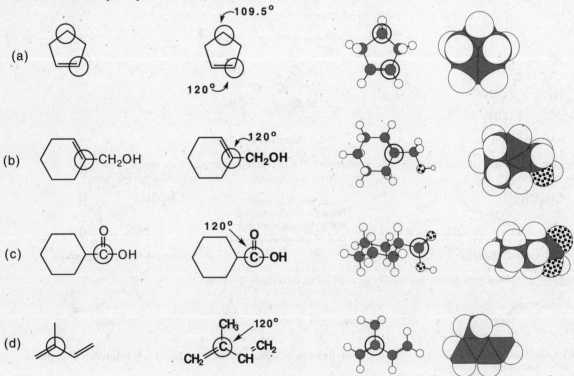

(a)

(b)   —CH₂OH   —CH₂OH

(c)

(d)

(e)   HC≡C-CH=CH₂

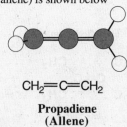

180°   **120°**

HC≡C-C(H)-C(H)(H)

<u>Problem 5.10</u> For each highlighted carbon atom in Problem 5.9, identify which atomic orbitals are used to form each σ bond and which are used to form each π bond.

Each bond is labeled σ or π and the orbitals overlapping to form each bond are shown.

(a) σ sp³-1s ; σ sp³-sp³ ; σ sp²-sp³ ; π 2p-2p ; σ 1s-sp² ; σ sp²-sp²

(b) σ sp²-sp³ ; CH₂OH ; π 2p-2p ; σ sp²-sp²

(c) π 2p-2p ; σ sp²-sp² ; C-OH ; σ sp³-sp² ; σ sp²-sp³

(d) σ sp²-sp³ ; CH₃ ; σ sp²-sp² ; π 2p-2p ; C=CH₂ ; CH₂= ; CH ; σ sp²-sp²

(e) σ sp-sp² ; π 2p-2p ; HC≡C-C ; π 2p-2p ; σ 1s-sp² ; H ; C ; H ; σ 1s-sp² ; σ sp-sp ; π 2p-2p ; σ sp²-sp²

<u>Problem 5.11</u> The structure of 1,2-propadiene (allene) is shown below

CH₂=C=CH₂

Propadiene (Allene)

(a) Predict all approximate bond angles in this molecule.
(b) State the orbital hybridization of each carbon.
(c) Explain the three-dimensional geometry of allene, and explain it in terms of the orbitals used.

180° ; **120°** ; H-C=C=C-H

sp² ; sp ; sp² ; CH₂=C=CH₂

π 2p-2p ; H-C=C=C-H ; σ sp²-sp

The central carbon atom of allene is *sp* hybridized with bond angles of 180° about it. The terminal carbons are *sp²* hybridized with bond angles of 120° about each. The planes created by H-C-H bonds at the ends of the molecule are perpendicular to each other.

<u>Problem 5.12</u> Following are lengths for a series of C-C single bonds. Propose an explanation for the differences in bond lengths.

Structure	Length of C-C single bond (pm)
CH_3-CH_3	153.7
$CH_2=CH-CH_3$	151.0
$CH_2=CH-CH=CH_2$	146.5
$HC\equiv C-CH_3$	145.9

The *s* electrons are on average held closer to atomic nuclei than *p* electrons. Thus, hybrid orbitals with higher *s* character have the electrons held closer to the nucleus and thus make bonds that are shorter. As shown in the table, a σ bond formed from overlap of an *sp³* orbital with an *sp²* orbital is shorter than a σ bond formed from overlap of an *sp³* orbital with another *sp³* orbital. Similarly, *sp³-sp* overlap produces a bond that is shorter than that produced by *sp³-sp²* overlap.

Nomenclature of Alkenes
<u>Problem 5.13</u> Draw structural formulas for these alkenes.
(a) *trans*-2-Methyl-3-hexene   (b) 2-Methyl-2-hexene   (c) 2-Methyl-1-butene

(d) 3-Ethyl-3-methyl-1-pentene   (e) 2,3-Dimethyl-2-butene   (f) *cis*-2-Pentene

(g) (Z)-1-Chloropropene   (h) 3-Methylcyclohexene   (i) 1-Isopropyl-4-methylcyclohexene

(j) (*E*)-2,6-Dimethyl-2,6-octadiene

(k) 3-Cyclopropyl-1-propene

(l) Cyclopropylethene

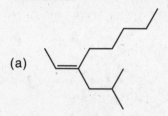

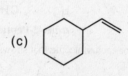 CH₂CH=CH₂

CH=CH₂

(m) 2-Chloropropene

(n) Tetrachloroethylene

(o) 1-Chlorocyclohexene

<u>Problem 5.14</u> Name these alkenes and cycloalkenes.

(a)

(*E*)-3-(2-methylpropyl)-2-Octene

(b)

4-Chloro-1,4-
dimethylcyclopentene

(c)

Ethenylcyclohexane
Vinylcyclohexane

(d)

2,4-Dimethyl-2-pentene

(e)

(*E*)-1,4-Dichloro-2-butene
trans-1,4-Dichloro-2-butene

(f)

Tetrafluoroethylene

(g)

5-Chloro-5-ethyl-
1,3-cyclopentadiene

(h)

1,4-Cyclohexadiene

<u>Problem 5.15</u> Arrange the following groups in order of increasing priority.
(a) -CH₃ -H -Br -CH₂CH₃
   -H < -CH₃ < -CH₂CH₃ < -Br

(b) -OCH₃ -CH(CH₃)₂ -B(CH₂CH₃)₂ -H
   -H < -B(CH₂CH₃)₂ < -CH(CH₃)₂ < -OCH₃

(c) -CH₃ -CH₂OH -CH₂NH₂ -CH₂Br
   -CH₃ < -CH₂NH₂ < -CH₂OH < -CH₂Br

<u>Problem 5.16</u> Assign an *E* or *Z* configuration to these dicarboxylic acids, each of which is an intermediate in the tricarboxylic acid cycle. Under each is its common name.

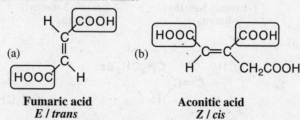

(a)

Fumaric acid
E* / *trans

(b)

Aconitic acid
Z* / *cis

The highest priority group on each *sp²* carbon atom is circled.

Problem 5.17 Name and draw structural formulas for all alkenes of molecular formula C_5H_{10}. As you draw these alkenes, remember that *cis* and *trans* isomers are different compounds and must be counted separately.

Four alkenes of molecular formula C_5H_{10} do not show *cis, trans* isomerism.

$CH_2=CHCH_2CH_2CH_3$

1-Pentene

$$CH_2=CCH_2CH_3 \quad (CH_3)$$

2-Methyl-1-butene

$$CH_2=CHCHCH_3 \quad (CH_3)$$

3-Methyl-1-butene

$$CH_3C=CHCH_3 \quad (CH_3)$$

2-Methyl-2-butene

One alkene of molecular formula C_5H_{10} shows *cis, trans* isomerism.

trans-2-Pentene

cis-2-Pentene

Problem 5.18 For each molecule that shows *cis, trans* isomerism, draw the *cis* isomer.

(a)   (b)   (c)   (d)

Problem 5.19 β-Ocimene, a triene found in the fragrance of cotton blossoms and several other essential oils, has the IUPAC name (Z)-3,7-dimethyl-1,3,6-octatriene. Draw a structural formula for β-ocimene.

β-Ocimene
(Z)-3,7-Dimethyl-1,3,6-octatriene

Problem 5.20 Draw the structural formula for at least one bromoalkene with the molecular formula C_5H_9Br that shows:
(a) Neither *E,Z* isomerism nor chirality.

$$H_2C=CH-CH_2CH_2CH_2Br$$ or $$H_3C-C=C-CH_2Br \;(CH_3, H)$$ or $$H-C=C-CH_2CH_2Br \;(H, CH_3)$$ or $$H-C=C-CH_2Br \;(H, CH_2CH_3)$$

5-Bromo-1-pentene   **1-Bromo-3-methyl-2-butene**   **4-Bromo-2-methyl-1-butene**   **2-(Bromomethyl)-1-butene**

(b) *E,Z* isomerism but not chirality.

$$H-C=C-CH_2CH_2Br \;(CH_3, H)$$ or $$CH_3-C=C-CH_2CH_2Br \;(H, H)$$ or $$H-C=C-CH_2Br \;(CH_3CH_2, H)$$

(E)-5-Bromo-2-pentene   **(Z)-5-Bromo-2-pentene**   **(E)-1-Bromo-2-pentene**

(Z)-1-Bromo-2-pentene   **(E)-1-Bromo-2-methyl-2-butene**   **(Z)-1-Bromo-2-methyl-2-butene**

(c) Chirality but not *E,Z* isomerism.

4-Bromo-1-pentene   **3-Bromo-1-pentene**   **3-Bromo-2-methyl-1-butene**

(d) Both chirality and *E,Z* isomerism.

(E)-4-Bromo-2-pentene   **(Z)-4-Bromo-2-pentene**

<u>Problem 5.21</u> Following are structural formulas and common names for four molecules that contain both a carbon-carbon double bond and another functional group. Give each an IUPAC name.

(a) $CH_2=CHCOH$
Acrylic acid
2-Propenoic acid

(b) $CH_2=CHCH$
Acrolein
2-Propenal

(c) Crotonic acid
(E)-2-Butenoic acid

(d) $CH_3CCH=CH_2$
Methyl vinyl ketone
3-Buten-2-one

<u>Problem 5.22</u> *Trans*-cyclooctene has been resolved, and its enantiomers are stable at room temperature. *Trans*-cyclononene has also been resolved, but it racemizes with a half-life of 4 min at 0°C. How can racemization of this cycloalkene take place without breaking any bonds? Why does *trans*-cyclononene racemize under these conditions but *trans*-cyclooctene does not? You will find it especially helpful to examine the molecular models of these cycloalkenes.

Following are enantiomers of *trans*-cyclooctene and *trans*-cyclononene. It may be helpful to construct molecular models and prove to yourself that the two different configurations of the ring are in fact non-superimposable mirror images of each other. For both *trans*-cyclooctene and *trans*-cyclononene, the enantiomers are configurational isomers. The enantiomers are interconverted by a change in configuration that is analogous to the chair flipping of chair cyclohexane. The *trans* double bond adds a considerable degree of rigidity to the ring, because the other atoms must exist in a slightly "stretched" configuration to accommodate the *trans* geometry. Nevertheless, the *trans*-cyclononene ring has more carbon atoms, so it is more flexible and can undergo the configurational interconversion more readily.

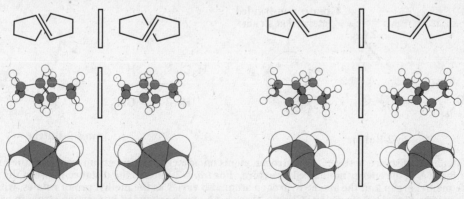

Enantiomers of *trans*-cyclooctene   **Enantiomers of *trans*-cyclononene**

<u>Problem 5.23</u> Which alkenes exist as pairs of *cis, trans* isomers? For each that does, draw the *trans* isomer.

For an alkene to exist as a pair of *cis, trans* isomers, both carbon atoms of the double bond must have two different substituents. Thus, (b), (c), and (e) exist as a pair of *cis, trans* isomers. The *trans* isomer for each alkene is drawn under its respective condensed molecular formula.

(a) $CH_2=CHBr$ (b) $CH_3CH=CHBr$ (c) $BrCH=CHBr$ (d) $(CH_3)_2C=CHCH_3$

$$\underset{H}{\overset{CH_3}{}}C=C\underset{Br}{\overset{H}{}}$$

$$\underset{H}{\overset{Br}{}}C=C\underset{Br}{\overset{H}{}}$$

(e) $(CH_3)_2CHCH=CHCH_3$

$$\underset{H}{\overset{(CH_3)_2CH}{}}C=C\underset{CH_3}{\overset{H}{}}$$

<u>Problem 5.24</u> Four stereoisomers exist for 3-penten-2-ol.

about this
double bond

$$CH_3-CH=CH-\overset{\overset{OH}{|}}{\underset{*}{CH}}-CH_3$$

3-Penten-2-ol

(a) Explain how these four stereoisomers arise.

There is one double bond that provides for *cis, trans* isomers, and one chiral center in 3-penten-2-ol.

(b) Draw the stereoisomer having the *E* configuration about the carbon-carbon double bond and the *R* configuration at the chiral center.

$$\underset{H_3C}{\overset{H}{}}C=C\underset{H}{\overset{\overset{HO}{\overset{|}{H\cdots C-CH_3}}}{}}$$

Molecular Modeling
These problems require molecular modeling programs such as Chem 3D™ or Spartan™ to solve. Pre-built models can be found at **http://now.brookscole.com/bfi4.**

<u>Problem 5.25</u> Measure the CH_3, CH_3 distance in the energy-minimized model of *cis*-2-butene, and the CH_3, H distance in the energy-minimized model of *trans*-2-butene. In which isomer is the nonbonded interaction strain greater?

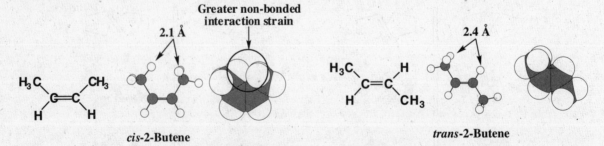

cis-2-Butene   *trans*-2-Butene

For *cis*-2-butene, the distance between the hydrogen atoms on adjacent methyl groups varies as the methyl group rotates, but is 2.1 Å in the energy minimized structure. For *trans*-2-butene, the distance between the hydrogen atoms of the methyl group and the alkene hydrogen atom also varies as the methyl group rotates, with a value of 2.4 Å in the energy minimized structure. Clearly, there is greater nonbonded interaction strain in *cis*-2-butene.

<u>Problem 5.26</u> Measure the C=C-C bond angles in the energy-minimized models of the *cis* and *trans* isomers of 2,2,5,5-tetramethyl-3-hexene. In which case is the deviation from VSEPR predictions greater?

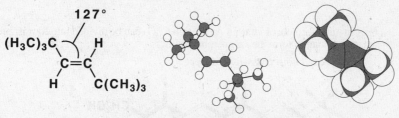

cis-2,2,5,5-Tetramethyl-3-hexene

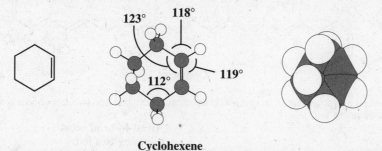

trans-2,2,5,5-Tetramethyl-3-hexene

VSEPR predicts 120° bond angles around the double bond. The minimized *cis* isomer has a 135° bond angle due to the severe non-bonded interaction strain present. The *trans* isomer has less non-bonded interaction strain and a 127° bond angle. Thus, the *cis* isomer is the one with significantly more deviation from the VSEPR predictions.

<u>Problem 5.27</u> Measure the C-C-C and C-C-H bond angles in the energy-minimized model of cyclohexene and compare them with those predicted by VSEPR. Explain any differences.

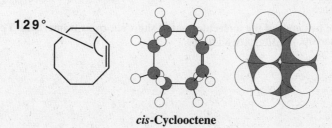

Cyclohexene

Representative measurements have been added to the above structure. Clearly, the angles predicted by Chem3D are not exactly 120° as predicted by VSEPR for an *sp²* hybridized carbon atom. This difference is the result of strain introduced by the ring of cyclohexene.

<u>Problem 5.28</u> Measure the C-C-C and C-C-H bond angles in the energy-minimized models of *cis* and *trans* isomers of cyclooctene. Compare these values with those predicted by VSEPR. In which isomer are deviations from VSEPR predictions greater?

cis-Cyclooctene

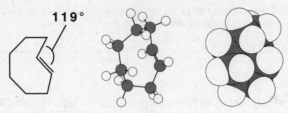

trans-Cyclooctene

Only the key C-C=C bond angles are shown on the above structures. VSEPR predicts 120° bond angles around the double bond. The angles deviate more in the *cis* isomer compared to the *trans* isomer due to increased ring strain in the *cis* isomer. Note how the *trans* isomer can more easily accommodate staggered conformations along the entire chain.

Terpenes

<u>Problem 5.29</u> Show that the structural formula of vitamin A (Section 5.3G) can be divided into four isoprene units bonded head-to-tail and cross-linked at one point to form the six-membered ring.

Isoprene chain cross-linked here

CH₂OH

<u>Problem 5.30</u> Following is the structural formula of lycopene C₄₀H₅₆, a deep-red compound that is partially responsible for the red color of ripe fruits, especially tomatoes. Approximately 20 mg of lycopene can be isolated from 1 kg of ripe tomatoes. Lycopene is an important antioxidant that may help prevent oxidative damage in atherosclerosis.

H₃C CH₃   CH₃   CH₃   H₃C

                                                        H₃C   CH₃

      CH₃   CH₃   CH₃

Lycopene

(a) Show that lycopene is a terpene, that is, its carbon skeleton can be divided into two sets of four isoprene units with the units in each set joined head-to-tail.

Head-to-head bond joining two four isoprene units

Lycopene

(b) How many of the carbon-carbon double bonds in lycopene have the possibility for *cis, trans* isomerism? Of these, which are *trans* and which are *cis*?

The double bonds on the two ends of the molecule cannot show *cis, trans* isomerism. The other 11 double bonds can show *cis, trans* isomerism, <u>but they are all *trans*</u>.

<u>Problem 5.31</u> As you might suspect, β-carotene, $C_{40}H_{56}$, precursor to vitamin A, was first isolated from carrots. Dilute solutions of β-carotene are yellow, hence its use as a food coloring. In plants, it is almost always present in combination with chlorophyll to assist in the harvesting of the energy of sunlight and to protect the plant against reactive species produced in photosynthesis. As tree leaves die in the fall, the green of their chlorophyll molecules is replaced by the yellow and reds of carotene and carotene-related molecules. Compare the carbon skeletons of β-carotene and lycopene. What are the similarities? What are the differences?

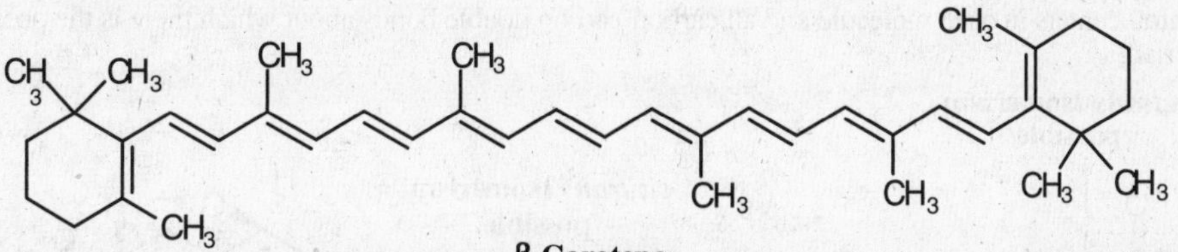

β-Carotene

The main structural difference between β-carotene and lycopene is that β-carotene has six-membered rings on the ends, not an open chain. In addtion, lycopene has two more double bonds compared to β-carotene . On the other hand, both β-carotene and lycopene can be divided into two sets of four isoprene units as shown below, and all of the double bonds are *E* in both molecules.

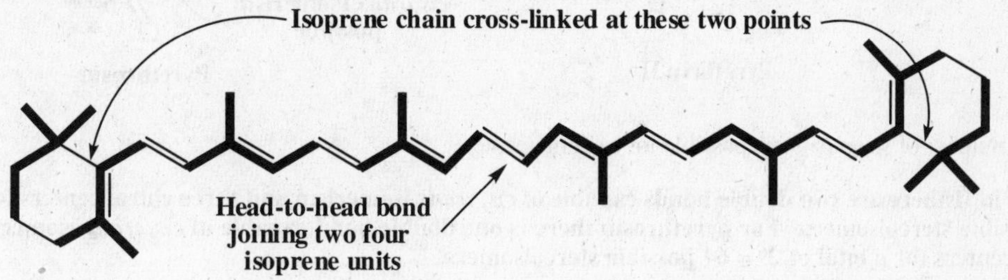

Isoprene chain cross-linked at these two points

Head-to-head bond joining two four isoprene units

<u>Problem 5.32</u> Calculate the index of hydrogen deficiency for β-carotene and lycopene.

Both molecules have an index of hydrogen deficiency of 13. β-Carotene has eleven π bonds and two rings. Lycopene has no rings, but thirteen π bonds.

<u>Problem 5.33</u> α-Santonin, isolated from the flower heads of certain species of Artemisia, is an anthelminthic (meaning against intestinal worms). This terpene is used in oral doses of 60 mg to rid the body of roundworms such as *Ascaris lumbricoides*. It has been estimated that over one third of the world's population is infested with these slender, thread-like parasites.

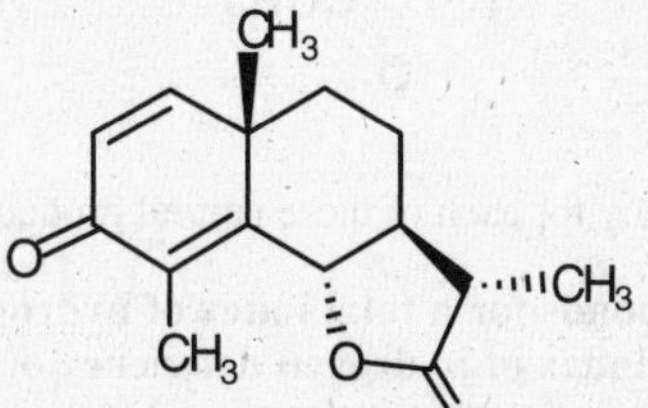

(a) Locate the three isoprene units in santonin, and show how the carbon skeleton of farnesol might be coiled and then cross-linked to give santonin. Two different coiling patterns of the carbon skeleton of farnesol can lead to santonin. Try to find them both.

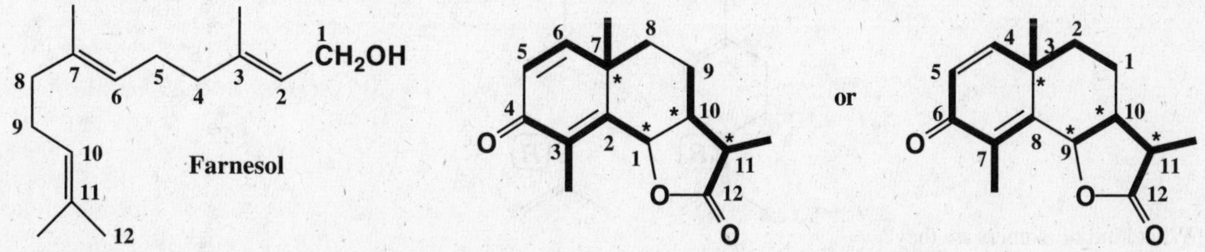

Farnesol

or

(b) Label all chiral centers in santonin. How many stereoisomers are possible for this molecule?

The four chiral centers of santonin are marked on the structures above. There are 2^4 = 16 stereoisomers possible for santonin.

(c) Calculate the index of hydrogen deficiency for santonin.

Santonin has three rings and four π bonds, so it has an index of hydrogen deficiency of seven.

<u>Problem 5.34</u> Pyrethrin II and pyrethrosin are two natural products isolated from plants of the chrysanthemum family. Pyrethrin II is a natural insecticide and is marketed as such.
(a) Label all chiral centers in each molecule and all carbon-carbon double bonds about which there is the possibility for *cis,trans* isomerism.

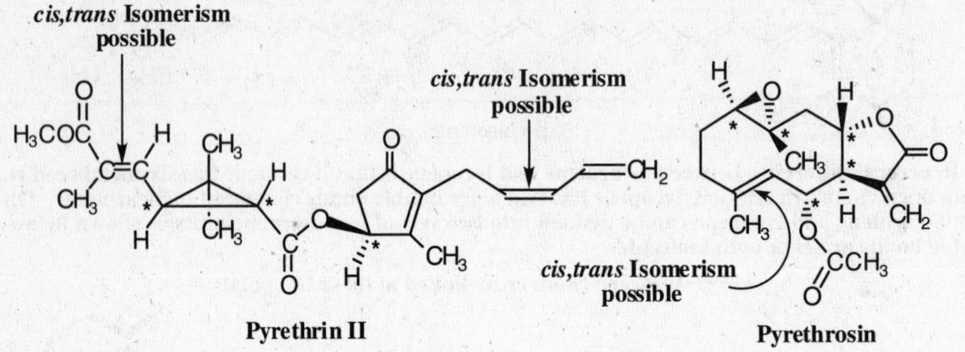

Pyrethrin II   Pyrethrosin

(b) State the number of stereoisomers possible for each molecule.

For pyrethrin II there are two double bonds capable of *cis, trans* isomerism and three chiral centers for a total of $2^5 = 32$ possible stereoisomers. For pyrethrosin there is one double bond capable of *cis, trans* isomerism and five chiral centers for a total of $2^6 = 64$ possible stereoisomers.

(c) Show that the bicyclic ring system of pyrethrosin is composed of three isoprene units.

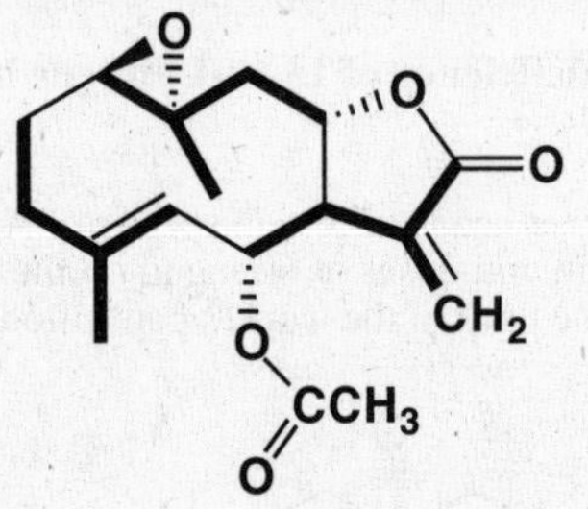

(d) Calculate the index of hydrogen deficiency for each of these natural products.

Pyrethrin II has two rings and seven π bonds for a total index of hydrogen deficiency of 9, and pyrethrosin has three rings and four π bonds for a total index of hydrogen deficiency of 7.

<u>Problem 5.35</u> Limonene is one of most common inexpensive fragrances. Two isomers of limonene can be isolated from natural sources. They are shown below. The one on the left has the odor of lemons, and the one on the right has the odor of oranges.

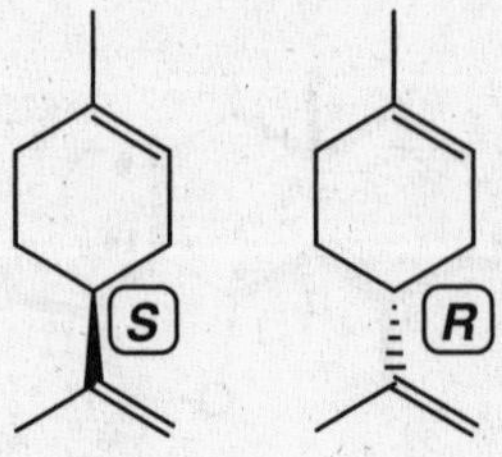

(a) What kind of isomers are they?

The two isomers differ only in the configuration of the single chiral center so these are enantiomers.

(b) Are *E,Z* isomers possible in limonene?

There are no *E,Z* isomers possible for limonene. One alkene is part of a cyclohexene ring, which is too small to accommodate a *trans* geometry. The other is a terminal alkene which also does not have *E,Z* isomer possibilities.

(c) Why do they smell different?

Many of the molecules of our body are chiral, present as single enantiomers. This includes the receptor molecules present in our noses that are responsible for our sense of smell. Thus, different enantiomers often smell different to us, because they interact differently with the chiral receptor molecules within our noses.

Problem 5.36 Nepetelactone is the active ingredient of catnip. It is isolated as an oil from the plant *Nepeta cataria*. Show that it is a terpene, that is, that its carbon skeleton can be divided into isoprene units. Is the molecule chiral? How many stereoisomers are possible?

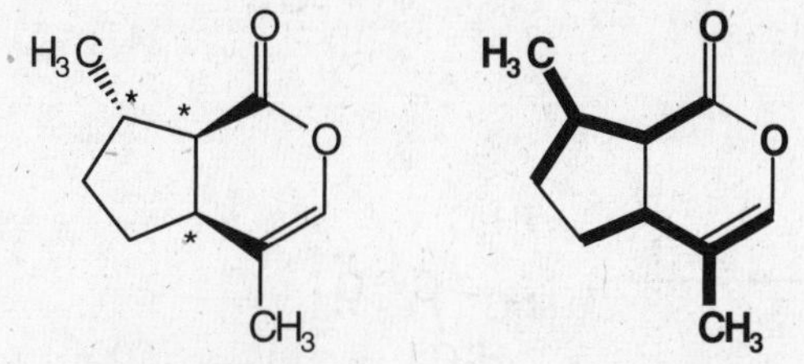

Nepetalactone
(oil of catnip)

The isoprene units are indicated on the structure to the right, and the chiral centers are indicated with asterisks (*). By virtue of the three chiral centers, the molecule is chiral and there are $2^3 = 8$ possible stereoisomers.

Looking Ahead
Problem 5.37 Bromine adds to *cis* and *trans*-2-butene to give different diastereomers of 2,3-dibromobutane. What does this say about the mode of addition of bromine to this alkene.

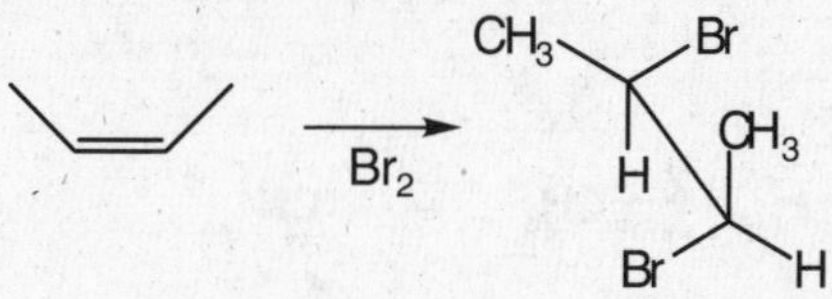

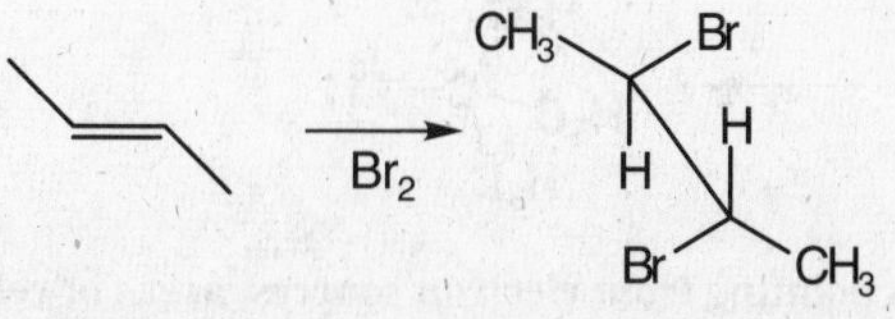

In this reaction, one Br atom of Br_2 adds to each of the carbon atoms of the alkene. The interesting feature of this process is that two new chiral centers are created in the product. Because different diastereomers are created from different isomeric starting materials, this reaction is classified as being stereospecific. In Section 6.3D you will learn that this occurs because the two bromine atoms add to opposite faces of the original alkene. This so-called anti-stereoselective addition is caused by formation of a three-membered ring intermediate. The point to make now is that when new chiral centers are created in reactions, it is important to keep track of how the mechanism of the reaction dictates which of the possible stereoisomers are produced.

CHAPTER 6
Solutions to the Problems

Problem 6.1 The following two sets of reactions ((a) and (b)), show possibilities for arrow pushing in individual reaction steps. Identify which is wrong and explain why. Next, using the correct arrow pushing, label which molecule is the nucleophile and which is the electrophile.

(a)

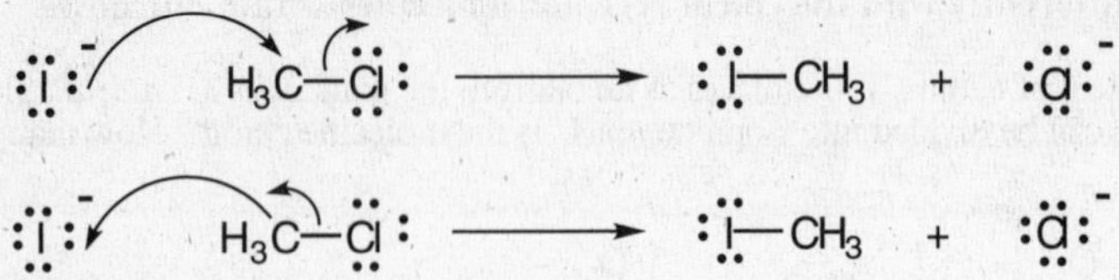

(b)

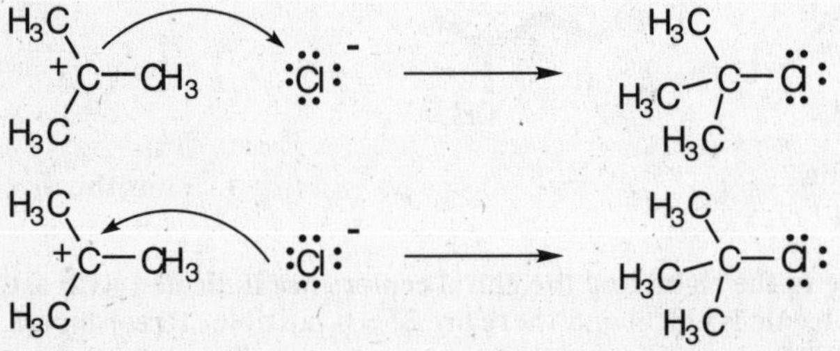

The incorrect scenarios have arrows pointing exactly the wrong way, namely from electron sinks to electron sources.

Incorrect

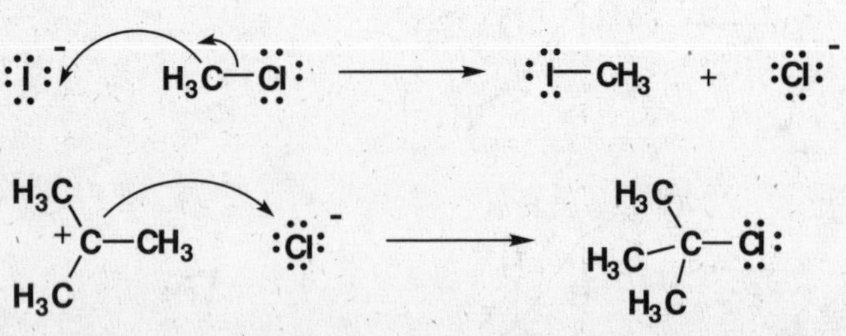

The correct scenarios have arrows pointing from electron sources, areas of relatively high electron density, to electron sinks, atoms of relatively low electron density. In particular, in each case the arrows labeled as "a" indicate a lone pair on the halide anion (nucleophile) reacting with an electrophilic carbon atom, a carbon with a full or partial postive charge. The arrow labeled "b" on methyl chloride is needed to satisfy valence.

Correct

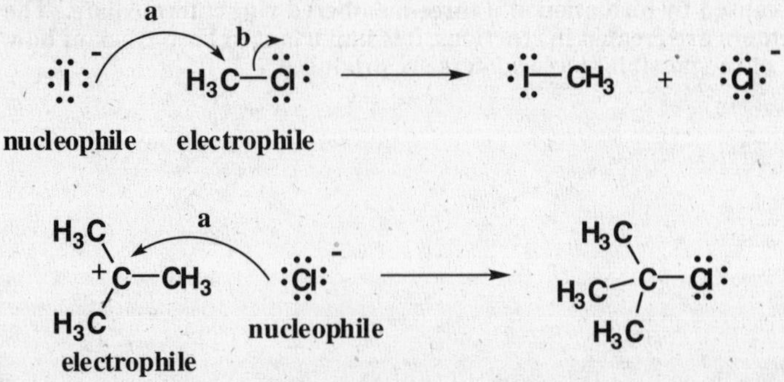

Problem 6.2 Name and draw the structural formula for the product of each alkene addition reaction.

(a) [structure]—CH₃ + HBr ⟶ [structure] CH₃ / Br

**1-Bromo-1-methyl-
cyclohexane**

(b) [structure]=CH₂ + HI ⟶ [structure] CH₃ / I

**1-Iodo-1-methyl-
cyclohexane**

Note that both alkenes give products with the halogen at the more substituted position.

Problem 6.3 Arrange these carbocations in order of increasing stability.

(a) [structure] $\overset{+}{C}$—CH₃   (b) [structure] $\overset{+}{C}$—CH₃   (c) [structure] —$\overset{+}{C}H_2$

The order of increasing stability of carbocations is methyl < primary < secondary < tertiary. Thus the three carbocations can be ranked as follows:

(c) [structure] —$\overset{+}{C}H_2$ < **(b)** [structure] $\overset{+}{C}$—CH₃ < **(a)** [structure] $\overset{+}{C}$—CH₃

Primary carbocation   Secondary carbocation   Tertiary carbocation

Problem 6.4 Propose a mechanism for addition of HI to 1-methylcyclohexene to give 1-iodo-1-methylcyclohexane. Which step in your mechanism is rate determining?

Step 1: Electrophilic addition-add a proton **A rate-determining proton transfer from HI to the carbon-carbon double bond to give a 3° carbocation intermediate.**

[reaction mechanism diagram] Slow, rate determining

Step 2: Make a new bond between an nucleophile and electrophile **Reaction of the carbocation intermediate with iodide ion completes the valence shell of carbon and gives the product.**

[reaction mechanism diagram] Fast

Problem 6.5 Draw a structural formula for the product of each alkene hydration reaction:

(a) [structure] + H₂O $\xrightarrow{H_2SO_4}$ [structure] OH

2-Methyl-2-butanol

(b) [structure] + H₂O $\xrightarrow{H_2SO_4}$ [structure] OH

2-Methyl-2-butanol

Problem 6.6 Propose a mechanism for the acid-catalyzed hydration of 1-methylcyclohexene to give 1-methylcyclohexanol. Which step in your mechanism is rate determining?

Step 1: Electrophilic addition-add a proton. **Proton transfer from the acid catalyst to the alkene gives a 3° carbocation intermediate in the rate-determining step.**

Step 2: Make a new bond between a nucleophile and an electrophile.

Step 3: Take a proton away.

Problem 6.7 The acid-catalyzed hydration of 3,3-dimethyl-1-butene gives 2,3-dimethyl-2-butanol as the major product. Propose a mechanism for formation of this alcohol.

3,3-Dimethyl-1-butene   **2,3-Dimethyl-2-butanol**

Step 1: Electrophilic addition-add a proton. **Proton transfer from the acid catalyst to the alkene gives a 2° carbocation intermediate in the rate-determining step.**

Step 2: 1,2 Shift. **The 2° carbocation rearranges to the more stable 3° carbocation.**

Step 3: Make a new bond between a nucleophile and an electrophile.

Step 4: Take a proton away

Problem 6.8 Complete these reactions.

(a) $CH_3CCH=CH_2$ + Br_2 $\xrightarrow{CH_2Cl_2}$ $CH_3CCHBrCH_2Br$

A new chiral center is created, so the product is a racemic mixture of two enantiomers.

(b) + Cl_2 $\xrightarrow{CH_2Cl_2}$

Problem 6.9 Draw the structure of the chlorohydrin formed by treating 1-methylcyclohexene with Cl_2/H_2O.

+ Cl_2 / H_2O $\longrightarrow$

Two new chiral centers are created and due to the *trans* addition geometry of the reaction, the product will be a racemic mixture of the two enantiomers shown.

Problem 6.10 Draw structural formulas for the alkene that gives each alcohol on hydroboration-oxidation.

(a) $\xrightarrow[\text{2. } H_2O_2, \text{ NaOH}]{\text{1. } BH_3}$

(b) $\xrightarrow[\text{2. } H_2O_2, \text{ NaOH}]{\text{1. } BH_3}$

Problem 6.11 Use a balanced half-reaction to show that each transformation involves a reduction.

(a)

Two hydrogens are required to produce the product alcohol from the ketone. Therefore, the balanced half-reaction needs two protons and two electrons (for charge balance) on the left-hand side. Since the electrons are on the left-hand side of the equation, the reaction is a two-electron reduction.

+ $2H^+$ + $2e^-$ $\longrightarrow$

(b) $CH_3-CH_2-\overset{\overset{\displaystyle O}{\|}}{C}OH$ ⟶ $CH_3CH_2CH_2OH$

Two hydrogens are required to produce the product alcohol from the carboxylic acid. Therefore, the balanced half-reaction needs two protons and two electrons (for charge balance) on the left-hand side. Additionally, the product alcohol has one less oxygen atom than the carboxylic acid starting material, so there must be an H_2O molecule added to the right side of the equation to balance the oxygen atoms. This H_2O molecule has two more hydrogens that must be balanced by adding two more protons and electrons to the left-hand side of the equation, giving a total of four protons and four electrons on the left-hand side. Since the electrons are on the left-hand side of the equation, the reaction is a four-electron reduction.

$$CH_3-CH_2-\overset{\overset{\displaystyle O}{\|}}{C}OH \ + \ 4H^+ \ + \ 4e^- \ \longrightarrow \ CH_3CH_2CH_2OH \ + \ H_2O$$

<u>Problem 6.12</u> Tell which of these transformations are oxidations and which are reductions based on whether there is addition or removal of O or H.

(a) $CH_3\overset{\overset{\displaystyle O}{\|}}{C}H$ ⟶ CH_3CH_2OH (b)

(c) HS〜〜〜SH ⟶

(a) Reduction; there is a gain of two hydrogens.
(b) Oxidation; there is a gain of one oxygen.
(c) Oxidation; there is a loss of two hydrogens.

<u>Problem 6.13</u> What alkene of molecular formula C_6H_{12}, when treated with ozone and then dimethyl sulfide, gives the following product(s)?

(a)   or

C_6H_{12}

$\xrightarrow[\textbf{2. (CH}_3)_2S]{\textbf{1. O}_3}$

(only product)

(b)   or

C_6H_{12}

$\xrightarrow[\textbf{2. (CH}_3)_2S]{\textbf{1. O}_3}$

+

(equal moles of each)

(c)

C_6H_{12}

$\xrightarrow[\textbf{2. (CH}_3)_2S]{\textbf{1. O}_3}$

(only product)

Energetics of Chemical Reactions

Problem 6.14   Using the table of average bond dissociation enthalpies at 25°C, determine which of the following reactions are energetically favorable at room temperature.   Assume that $\Delta S = 0$.

Bond	Bond dissociation energy [kJ(kcal)/mol]	Bond	Bond dissociation energy [kJ(kcal)/mol]
H-H	435(104)	C-I	238(57)
O-H	439(105)	C-Si	301(72)
C-H (-CH₃)	422(101)	C=C	727(174)
C-H (=CH₂)	464(111)	C=O (aldehyde)	728(174)
C-H (≡CH)	556(133)	C=O (CO₂)	803(192)
N-H	391(93)	C≡O	1075(257)
Si-H	318(76)	N≡N	950(227)
C-C	376(90)	C≡C	966(231)
C-N	355(85)	O=O	498(119)
C-O	385(92)		

The following reactions can only occur to a significant extent as written if they are exothermic, that is, if the bonds that are formed are stronger than the ones that are broken in the reaction. Recall that a catalyst increases the rate, but does not change the overall thermodynamics of a reaction. To find out if a reaction is exothermic, the dissociation enthalpy of all the bonds in the molecules on each side of the equation are added together. If the bond dissociation enthalpy total from the right side of the equation is higher than the total from the left side of the equation, then the reaction is exothermic ($\Delta H°$ for the reaction is negative).

(a) $CH_2{=}CH_2 + 2H_2 + N_2 \longrightarrow H_2N{-}CH_2{-}CH_2{-}NH_2$

The bond dissociation enthalpies from the left side of the equation:
727 + (4 x 464) + (2 x 435) + 950 = 4403 kJ/mol
    (C=C) (4 =C-H)   (H-H)   (N≡N)

The bond dissociation enthalpies from the right side of the equation:
(4 x 391) + (2 x 355) + 376 + (4 x 422) = 4338 kJ/mol
    (4 N-H)   (2 C-N)   (C-C)   (4 -C-H)

> **This reaction is endothermic because 4403 kJ/mol is larger than 4338 kJ/mol.**

(b) $CH_2{=}CH_2 + CH_4 \longrightarrow CH_3{-}CH_2{-}CH_3$

The bond dissociation enthalpies from the left side of the equation:
727 + (4 x 464) + (4 x 422) = 4271 kJ/mol
(C=C) (4 =C-H)   (4 -C-H)

The bond dissociation enthalpies from the right side of the equation:
(2 x 376) + (8 x 422) = 4128 kJ/mol
    (2 C-C)   (8 -C-H)

> **This reaction is endothermic because 4271 kJ/mol is larger than 4128 kJ/mol.**

(c) $CH_2{=}CH_2 + (CH_3)_3SiH \longrightarrow H{-}CH_2{-}CH_2{-}Si(CH_3)_3$

The bond dissociation enthalpies from the left side of the equation:
727 + (4 x 464) + (9 x 422) + (3 x 301) + 318 = 7602 kJ/mol
(C=C) (4 =C-H)   (9 -C-H)   (3 C-Si)   (Si-H)

The bond dissociation enthalpies from the right side of the equation:
(376) + (5 x 422) + (9 x 422) + (4 x 301) = 7488 kJ/mol
(C-C) (5 -C-H)   (9 -C-H)   (4 C-Si)

> **This reaction is endothermic because 7602 kJ/mol is larger than 7488 kJ/mol.**

(d) $CH_2{=}CH_2 + CHI_3 \longrightarrow H{-}CH_2{-}CH_2{-}C(I)_3$

The bond dissociation enthalpies from the left side of the equation:
727 + (4 x 464) + (422) + (3 x 238) = 3719 kJ/mol
(C=C) (4 =C-H) (1 -C-H) (3 C-I)

The bond dissociation enthalpies from the right side of the equation:
        (2 x 376) + (5 x 422) + (3 x 238) = 3576 kJ/mol
        (2 C-C)   (5 C-H)   (3 C-I)

> This reaction is endothermic because 3719 kJ/mol is larger than 3576 kJ/mol.

(e) $CH_2=CH_2$ + CO + H_2 ⟶ $H\text{-}CH_2\text{-}CH_2\text{-}\overset{\overset{O}{\|}}{C}H$

The bond dissociation enthalpies from the left side of the equation:
        727 + (4 x 464) + 1075 + 435 = 4093 kJ/mol
        (C=C) (4 =C-H)   (C≡O) (H-H)

The bond dissociation enthalpies from the right side of the equation:
        (2 x 376) + (5 x 422) + 464 + 728 = 4054 kJ/mol
        (2 C-C)   (5 C-H) (=C-H)(C=O)

> This reaction is endothermic because 4093 kJ/mol is larger than 4054 kJ/mol.

(f) + $CH_2=CH_2$ ⟶

The bond dissociation enthalpies from the left side of the equation:
        (3 x 727) + 376 + (10 x 464) = 7197 kJ/mol
        (3 C=C)   (C-C)   (10 =C-H)

The bond dissociation enthalpies from the right side of the equation:
        727 + (5 x 376) + (8 x 422) + (2 x 464) = 6911 kJ/mol
        (C=C)   (5 C-C)   (8 -C-H)   (2 =C-H)

> This reaction is endothermic because 7197 kJ/mol is larger than 6911 kJ/mol.

(g) + ⟶

The bond dissociation enthalpies from the left side of the equation:
        (2 x 727) + 376 + (2 x 803) + (6 x 464) = 6220 kJ/mol
        (2 C=C)   (C-C)   (2 C=O)   (6 =C-H)

The bond dissociation enthalpies from the right side of the equation:
        727 + (3 x 376) + (4 x 422) + (2 x 464) + (2 x 385) + 728 = 5969 kJ/mol
        (C=C)   (3 C-C)   (4 -C-H)   (2 =C-H)   (2 C-O) (C=O)

> This reaction is endothermic because 6220 kJ/mol is larger than 5969 kJ/mol.

(h) $HC≡CH$ + O_2 ⟶ $H\text{-}\overset{\overset{O}{\|}}{C}\text{-}\overset{\overset{O}{\|}}{C}\text{-}H$

The bond dissociation enthalpies from the left side of the equation:
        966 + (2 x 556) + 498 = 2576 kJ/mol
        (C≡C) (2 ≡C-H) (O=O)

The bond dissociation enthalpies from the right side of the equation:
        376 + (2 x 728) + (2 x 464) = 2760 kJ/mol
        (C-C)   (2 C=O)   (2 =C-H)

> This reaction is exothermic because 2760 kJ/mol is larger than 2576 kJ/mol.

(i) $2CH_4$ + O_2 ⟶ $2CH_3OH$

The bond dissociation enthalpies from the left side of the equation:
  2(4 x 422) + 498 = 3874 kJ/mol
  2(4 -C-H)   (O=O)

The bond dissociation enthalpies from the right side of the equation:
  2(3 x 422) + 2(385) + 2(439) = 4180 kJ/mol
  2(3 -C-H)   2(C-O)   2(O-H)

> **This reaction is exothermic because 4180 kJ/mol is larger than 3874 kJ/mol.**

Electrophilic Additions

<u>Problem 6.15</u> Draw structural formulas for the isomeric carbocation intermediates formed on treatment of each alkene with HCl. Label each carbocation 1º, 2º, or 3º, and state which of the isomeric carbocations forms more readily.

(a) $CH_3-CH_2-C=CH-CH_3$ (with CH₃ on C)

  $CH_3-CH_2-\overset{+}{C}-CH_2-CH_3$ (with CH₃)   + $CH_3-CH_2-\overset{+}{CH}-CH-CH_3$ (with CH₃)
  Tertiary   **Secondary**
  (Formed more readily)

(b) $CH_3-CH_2-CH=CH-CH_3$

  $CH_3-CH_2-\overset{+}{CH}-CH_2-CH_3$ + $CH_3-CH_2-CH_2-\overset{+}{CH}-CH_3$
  Both secondary carbocations
  (Formed at equal rates)

(c) (cyclopentene with CH₃)

  ($\overset{+}{}$—CH₃) + (—CH₃, +)
  Tertiary   **Secondary**
  (Formed more readily)

(d) (cyclohexane ring =CH₂)

  (ring—$\overset{+}{C}H_2$) + (ring $\overset{+}{}$—CH₃)
  Primary   **Tertiary**
                    (Formed more readily)

<u>Problem 6.16</u> Arrange the alkenes in each set in order of increasing rate of reaction with HI, and explain the basis for your ranking. Draw the structural formula of the major product formed in each case.

(a) $CH_3-CH=CH-CH_3$ and $CH_3-C=CH-CH_3$ (with CH₃ on C)

$CH_3-CH=CH-CH_3$ ⟶ $CH_3-CH_2-\overset{+}{CH}-CH_3$ ⟶ $CH_3-CH_2-\overset{*}{\underset{I}{CH}}-CH_3$
  2-Butene   **A secondary**   **2-Iodobutane**
                          carbocation   **(*sec*-Butyl iodide)**
                                                       New chiral center created
                                                       so racemic mixture

$CH_3-\underset{CH_3}{C}=CH-CH_3$ ⟶ $CH_3-\underset{CH_3}{\overset{+}{C}}-CH_2-CH_3$ ⟶ $CH_3-\underset{I}{\overset{CH_3}{C}}-CH_2-CH_3$
  2-Methyl-2-butene   **A tertiary**   **2-Iodo-2-methylbutane**
                               carbocation   **(Major product)**

The reaction of 2-methyl-2-butene is the only one that can form a tertiary carbocation, so 2-methyl-2-butene is the compound that reacts faster with HI.

(b)

1-Methylcyclohexene   A tertiary
carbocation

1-Iodo-1-methylcyclo-
hexane
(Major product)

Cyclohexene   A secondary
carbocation

Iodocyclohexane
(Only product)

Only 1-methylcyclohexene can form a tertiary carbocation, so 1-methylcyclohexene reacts faster with HI.

<u>Problem 6.17</u> Predict the organic product(s) of the reaction of 2-butene with each reagent.

(a) H_2O (H_2SO_4)   (b) Br_2 (c) Cl_2

$CH_3-\overset{*}{C}H-CH_2-CH_3$
$\quad\quad\; |$
$\quad\quad OH$

$CH_3-\overset{*}{C}H-\overset{*}{C}H-CH_3$
$\quad\quad |\quad\quad\; Br$
$\quad\quad Br$

$CH_3-\overset{*}{C}H-\overset{*}{C}H-CH_3$
$\quad\quad |\quad\quad\; Cl$
$\quad\quad Cl$

(d) Br_2 in H_2O (e) HI   (f) Cl_2 in H_2O

$CH_3-\overset{*}{C}H-\overset{*}{C}H-CH_3$
$\quad\quad |\quad\quad OH$
$\quad\quad Br$

$CH_3-\overset{*}{C}H-CH_2-CH_3$
$\quad\quad |$
$\quad\quad I$

$CH_3-\overset{*}{C}H-\overset{*}{C}H-CH_3$
$\quad\quad |\quad\quad OH$
$\quad\quad Cl$

(g) $Hg(OAc)_2$, H_2O (h) product (g) + $NaBH_4$

$CH_3-\overset{*}{C}H-\overset{*}{C}H-CH_3$
$\quad\quad |\quad\quad HgOAc$
$\quad\quad OH$

$CH_3-\overset{*}{C}H-CH_2-CH_3$
$\quad\quad |$
$\quad\quad OH$

In parts (a), (e), and (h) a single chiral center is created so the products are racemic mixtures. In parts (b), (c), (d), (f), and (g) two chiral centers are created. Racemic mixtures will be the result, the exact identity of which depends on whether the 2-butene starting material is *cis* or *trans* (not given in the problem).

<u>Problem 6.18</u> Draw a structural formula of an alkene that undergoes acid-catalyzed hydration to give each alcohol as the major product (more than one alkene may give each alcohol as the major product).

(a) 3-Hexanol   (b) 1-Methylcyclobutanol

$CH_3CH_2CH{=}CHCH_2CH_3$

(*cis* or *trans*)

or

(c) 2-Methyl-2-butanol   (d) 2-Propanol

$\overset{CH_3}{\underset{}{|}}$
$H_2C{=}CCH_2CH_3$ or $CH_3\overset{CH_3}{\underset{}{C}}{=}CHCH_3$

$CH_3CH{=}CH_2$

Problem 6.19 Reaction of 2-methyl-2-pentene with each reagent is regioselective. Draw a structural formula for the product of each reaction, and account for the observed regioselectivity.

In each case, the reaction mechanism involves formation of a carbocation [(a), (b), and (c)] or a positively charged, three-membered ring intermediate [(d), (e)] resulting from an electrophlic addition step. These intermediates then react with a nucleophile at the site of greatest positive charge to give the product shown (make a new bond between a nucleophile and electrophile). Tertiary carbocations are more stable then secondary carbocations, even in the context of the three-membered rings, so the reactions will give predominantly the regioisomer that has the nucleophile at the tertiary center.

(a) HI

$$CH_3-\underset{\underset{I}{|}}{\overset{\overset{CH_3}{|}}{C}}-CH_2-CH_2-CH_3$$

(b) HBr

$$CH_3-\underset{\underset{Br}{|}}{\overset{\overset{CH_3}{|}}{C}}-CH_2-CH_2-CH_3$$

(c) H_2O in the presence of H_2SO_4

$$CH_3-\underset{\underset{OH}{|}}{\overset{\overset{CH_3}{|}}{C}}-CH_2-CH_2-CH_3$$

(d) Br_2 in H_2O

$$CH_3-\underset{\underset{HO}{|}}{\overset{\overset{H_3C \quad Br}{\diagup \quad |_*}}{C}}-CH-CH_2-CH_3$$

(e) $Hg(OAc)_2$ in H_2O

$$CH_3-\underset{\underset{HO}{|}}{\overset{\overset{H_3C \quad HgOAc}{\diagup \quad |_*}}{C}}-CH-CH_2-CH_3$$

In parts (d) and (e) one new chiral center is created, so racemic mixtures are produced.

Problem 6.20 Account for the regioselectivity and stereoselectivity observed when 1-methylcyclopentene is treated with each reagent.
(a) BH_3

Attack of the borane occurs in a concerted fashion, simultaneously forming both the new C-H and C-B bonds on the same face of the double bond, a process referred to as syn addition (electrophilic addition with simultaneous bond formation to H). Largely for steric reasons, the H atom ends up on the more hindered carbon atom (the one with more/bulkier substituents) and the B atom ends up on the less hindered carbon atom.

Bond forming and bond breaking is concerted. Note that the borane could approach from either the top or the bottom of the alkene, leading to a racemic mixture of the two syn addition products as shown.

(b) Br_2 in H_2O

Attack by H_2O on the carbon of the bromonium ion intermediate bearing the methyl group (make a new bond between a nucleophile and electrophile) followed by loss of a proton (take a proton away) gives the *trans* bromohydrin. Note that the bromonium ion could form on either face of the alkene, leading to a racemic mixture of the two enantiomers shown.

**A bridged bromonium
ion intermediate**

(c) $Hg(OAc)_2$ in H_2O

Attack by water on the bridged mercurinium ion intermediate (make a new bond between a nucleophile and electrophile) followed by loss of a proton (take a proton away) results in -OH *trans* to -HgOAc. Note that the mercurinium ion could form on either face of the alkene, leading to a racemic mixture of the two enantiomers shown.

**A bridged mercurinium
ion intermediate**

<u>Problem 6.21</u> Draw a structural formula for an alkene with the indicated molecular formula that gives the compound shown as the major product (more than one alkene may give the same compound as the major product).

(a) C_5H_{10} + H_2O $\xrightarrow{H_2SO_4}$

(b) C_5H_{10} + Br_2 $\longrightarrow$

Note that a chiral center is created so the product is actually a racemic mixture.

(c) C_7H_{12} + HCl $\longrightarrow$

Problem 6.22 Account for the fact that addition of HCl to 1-bromopropene gives exclusively 1-bromo-1-chloropropane.

$$CH_3CH=CHBr \quad + \quad HCl \quad \longrightarrow \quad CH_3CH_2CHBrCl$$

1-Bromopropene   **1-Bromo-1-chloropropane**

The exclusive product must be derived from the significantly more stable carbocation. In this case, the significantly more stable carbocation is the one with the positive charge on the carbon atom attached to the bromine atom, despite the fact that this carbocation is primary versus the alternative secondary carbocation. Thus, the bromine atom must be able to stabilize an adjacent cationic carbon atom. It turns out that the stabilization is primarily a resonance effect, involving the lone pairs of the bromine atom as shown. Note how the resonance structure on the right illustrates how the positive charge is partially delocalized onto the bromine atom.

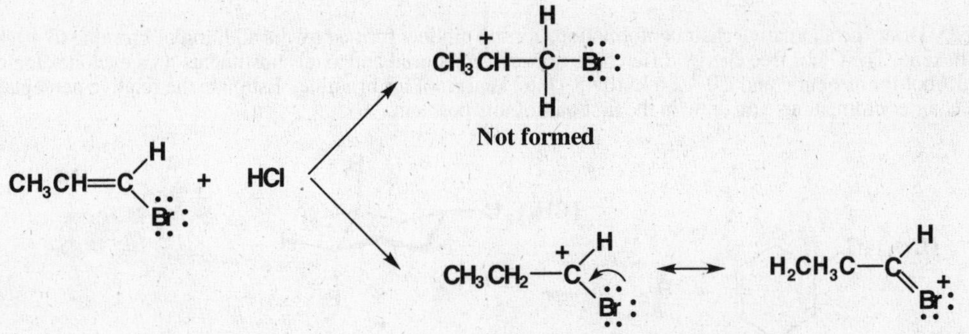

Not formed

This is the significantly more stable cation due to resonance stabilization as shown

Problem 6.23 Account for the fact that treating propenoic acid (acrylic acid) with HCl gives only 3-chloropropanoic acid.

Propenoic acid   **3-Chloropropenoic acid**   **2-Chloropropenoic acid**
(Acrylic acid)   **(this product is not formed)**

The exclusive product must be derived from the significantly more stable carbocation. In this case, the significantly more stable carbocation is the one with the positive charge on the terminal carbon atom, despite the fact that this carbocation is primary versus the alternative secondary carbocation. Thus, the carbonyl group attached to the internal carbon atom must be destabilizing to an adjacent cationic carbon atom. It turns out that the destabilization is primarily an inductive effect, based on the fact that a carbonyl group is electron withdrawing. An electron withdrawing group is destabilizing since removing charge density from a carbocation increases the charged character and thus the enthalpy of the carbocation even further.

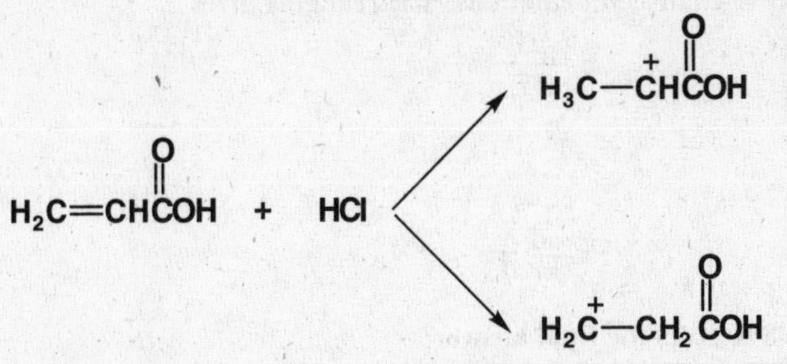

Not formed because of the destabilizing inductive effect of the carbonyl group

This is the significantly more stable carbocation

Problem 6.24 Draw a structural formula for the alkene of molecular formula C_5H_{10} that reacts with Br_2 to give each product.

(a)

(b)

(c)

A chiral center is created in each case, so the products of these bromine addition reactions are actually racemic mixtures.

Problem 6.25 Draw the alternative chair conformations for the product formed by the addition of bromine to 4-*tert*-butylcyclohexene. The Gibbs free energy differences between equatorial and axial substituents on a cyclohexane ring are 21 kJ (4.9 kcal)/mol for *tert*-butyl and 2.0 – 2.6 kJ (0.48 - 0.62 kcal)/mol for bromine. Estimate the relative percentages of the alternative chair conformations you drew in the first part of this problem.

Note that the bromine atoms are *trans* with respect to each other in the product due to anti addition geometry of the bromination reaction. Recall that large substituents are sterically disfavored in axial positions. The upper product has the large *tert*-butyl group in the strongly favored equatorial position along with both bromine atoms in the somewhat disfavored axial positions. The lower product has the large *tert*-butyl in the strongly disfavored axial position along with both bromine atoms in the somewhat more favored equatorial positions. The conformational energy difference based on the axial vs. equatorial *tert*-butyl group is 21 kJ/mol, and we will use an intermediate value of 2.3 kJ/mol for the conformational energy difference for the axial vs. equatorial bromine atoms. Thus, the relative conformation energy can be estimated as being **favorable for the upper structure** by an amount equal to the value that is favorable for the *tert*-butyl group (21 kJ/mol) minus the disfavorable contributions of the two bromine atoms (2 x +2.3 kJ/mol) for a total of 16.4 kJ/mol.

At equilibrium the relative amounts of each form are given by the equation:

$$\Delta G° = -RT \ln K_{eq}$$

Here K_{eq} refers to the ratio of the alternative chair conformations. Rearranging gives

$$\ln K_{eq} = \frac{-\Delta G°}{RT}$$

Solving gives:

$$K_{eq} = e^{\left(\frac{-\Delta G°}{RT}\right)}$$

Plugging in the values for $\Delta G°$, R, and 298 K gives the final answer:

$$K_{eq} = e^{\left(\frac{-(-16.4\,kJ/mol)}{(8.314 \times 10^{-3}\,kJ/mol\,K)(298\,K)}\right)} = e^{6.62} = 7.50 \times 10^2$$

Thus, the structure with the *tert*-butyl group equatorial will be favored by about 750 to 1 at equilibrium at room temperature.

Problem 6.26 Draw a structural formula for the cycloalkene with the molecular formula C_6H_{10} that reacts with Cl_2 to give each compound.

(a) (b) (c) (d)

Note that the chlorine addition reactions in parts (a), (b), and (c) actually a produce racemic mixtures of enantiomers, only one of which is shown.

Problem 6.27 Reaction of this bicycloalkene with bromine in carbon tetrachloride gives a *trans* dibromide. In both (a) and (b), the bromine atoms are *trans* to each other. However, only one of these products is formed.

(a)   (b)

Which *trans* dibromide is formed? How do you account for the fact that it is formed to the exclusion of the other *trans* dibromide?

Product (a) is formed. Electrophilic addition of bromine to an alkene occurs via a bridged bromonium ion intermediate and anti addition of the two bromine atoms. In a cyclohexane ring, anti addition corresponds to *trans* and diaxial addition. Only in formula (a) are the two added bromines *trans* and diaxial. In (b) they are *trans*, but diequatorial, so this isomer cannot be formed. In this case, the starting material is a single enantiomer, so only one enantiomer is produced as the product.

(a) *trans*-Diaxial   (b) *trans*-Diequatorial

<u>Problem 6.28</u> Terpin, prepared commercially by the acid-catalyzed hydration of limonene, is used medicinally as an expectorant for coughs.

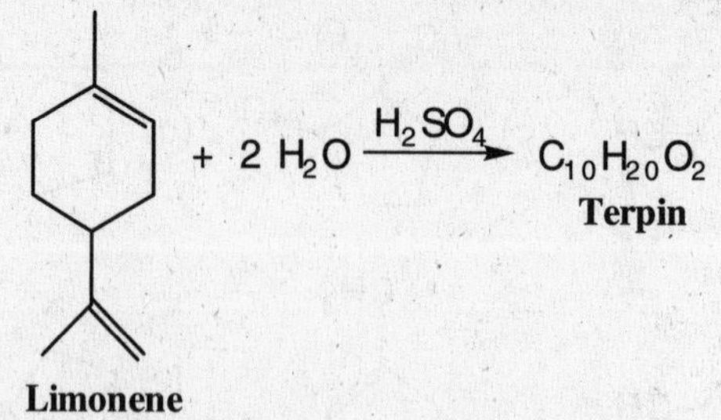

Limonene

(a) Propose a structural formula for terpin and a mechanism for its formation.

Add water to each double bond by protonation to give a 3° carbocation, reaction of each carbocation with water, and loss of the protons to give terpin hydrate. Since 3° carbocations are produced in either case, it is not clear which double bond would actually react first.

Step 1: Electrophilic addition-add a proton.

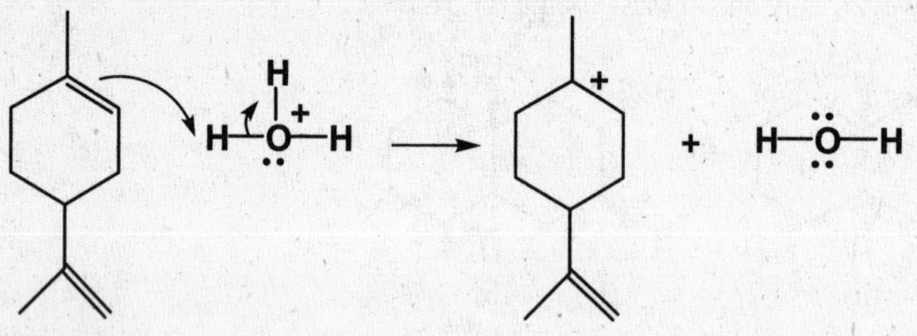

Limonene

Step 2: Make a new bond between a nucleophile and an electrophile.

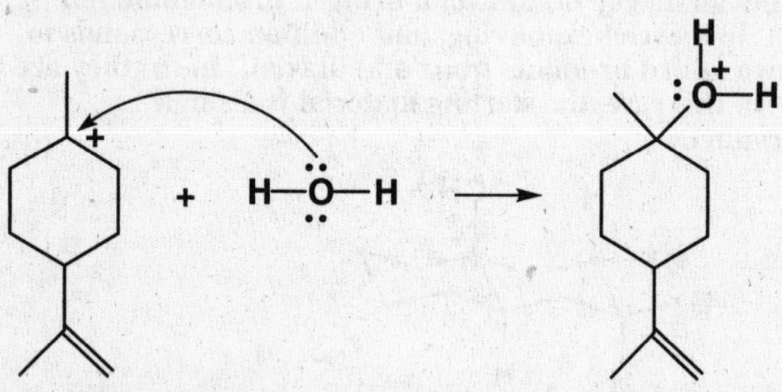

Step 3: Take a proton away.

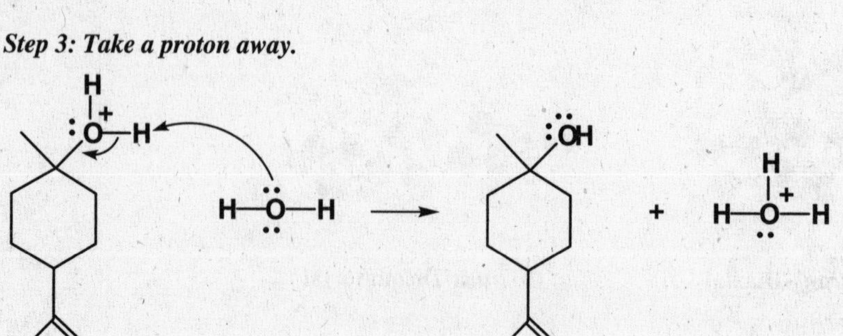

Step 4: Electrophilic addition-add a proton.

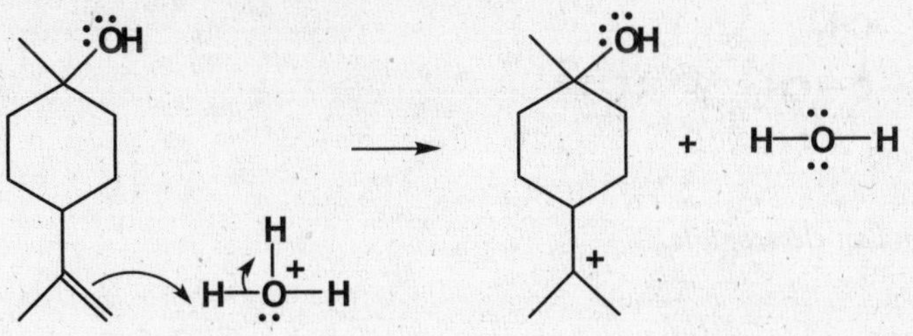

Step 5: Make a new bond between a nucleophile and an electrophile

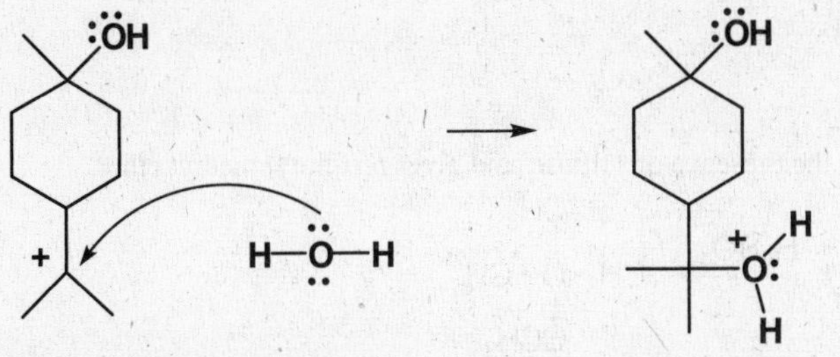

Step 6: Take a proton away

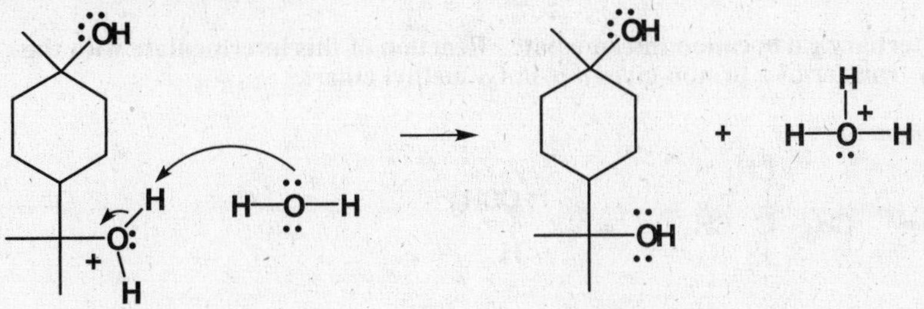

Terpin

(b) How many *cis, trans* isomers are possible for the structural formula you propose?

There are two *cis, trans* isomers, shown here as chair conformations with the (CH$_3$)$_2$COH- side chain equatorial.

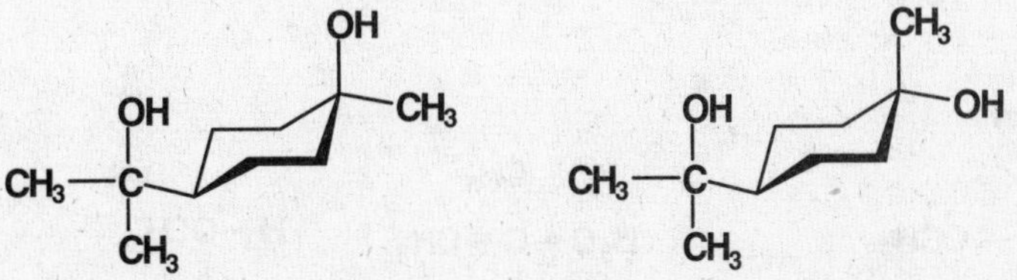

<u>Problem 6.29</u> Propose a mechanism for this reaction, and account for its regioselectivity.

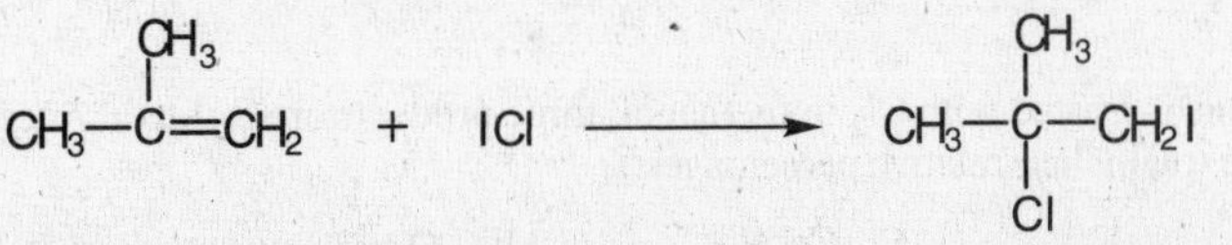

In the addition reaction of an unsymmetrical electrophilic reagent to a double bond, the "positive" portion of the reagent adds to the carbon atom of the double bond so as to yield the more stable carbocation as an intermediate. Iodine is less electronegative (by 0.5 unit on the Pauling scale) than chlorine, so iodine is added first as shown.

Step 1: Electrophilic addition.

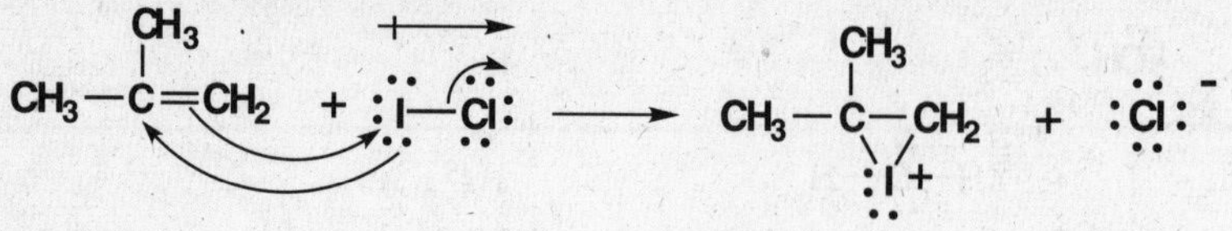

Step 2: Make a new bond between a nucleophile and an electrophile.

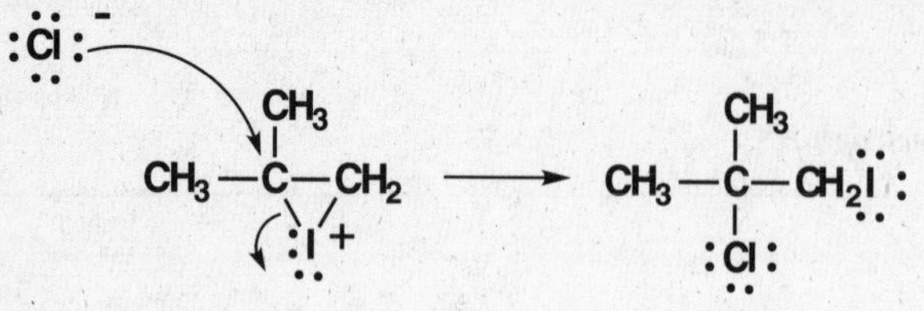

<u>Problem 6.30</u> Treating 2-methylpropene with methanol in the presence of sulfuric acid gives *tert*-butyl methyl ether.

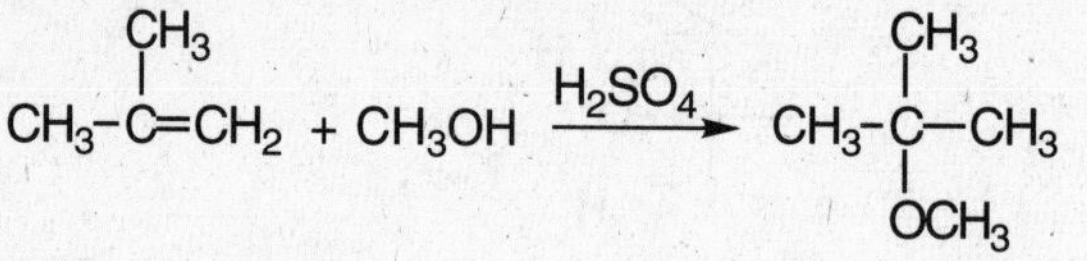

Propose a mechanism for the formation of this ether.

Proton transfer to the alkene gives a tertiary carbocation intermediate. Reaction of this intermediate with the oxygen atom of methanol followed by transfer of a proton gives *tert*-butyl methyl ether.

Step 1: Electrophilic addition-add a proton.

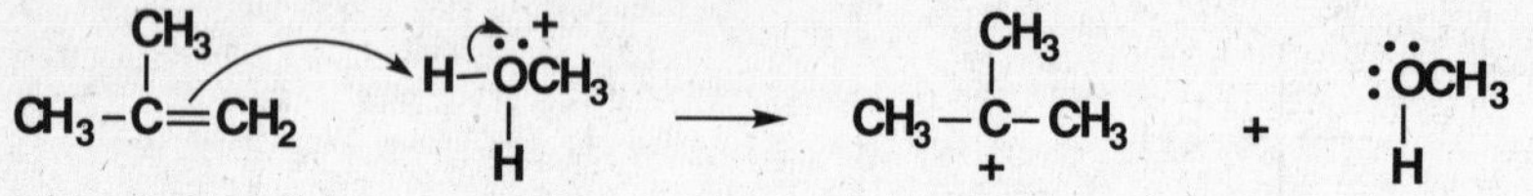

Step 2: Make a new bond between a nucleophile and an electrophile.

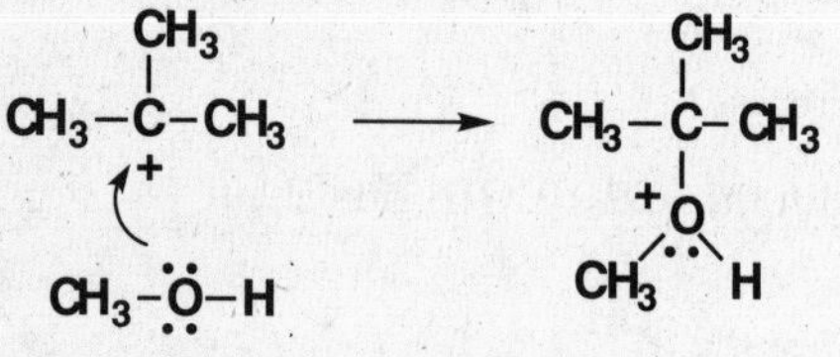

Step 3: Take a proton away.

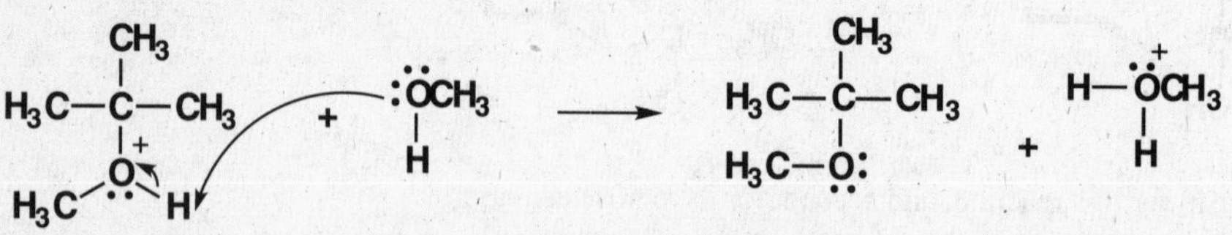

<u>Problem 6.31</u> When 2-pentene is treated with Cl_2 in methanol, three products are formed. Account for the formation of each product. (you need not explain their relative percentages.)

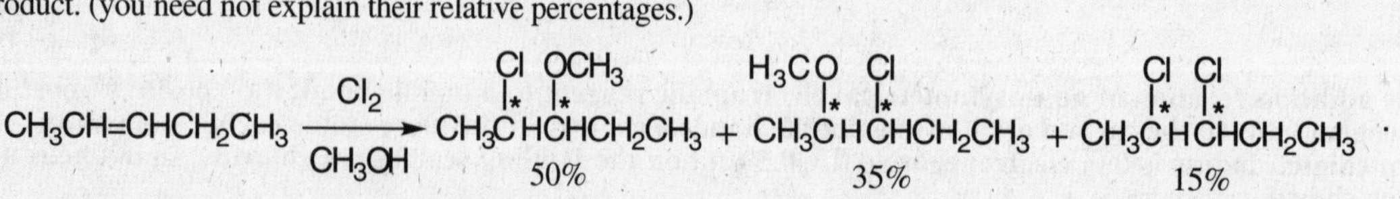

In this reaction, the chlorine reacts with the alkene to produce the chloronium ion intermediate (*electrophilic addition*) that can then react at either carbon atom (*make a new bond between a nucleophile and an electrophile*) to give the three different products shown. Note that chlorine is a poor nucleophile, therefore 2,3-dichloropentane is only a minor product. The reactions produce two new chiral centers in each case. The structures shown

actually represent a pair of enantiomers, the identity of which depends on whether the starting alkene is *cis* or *trans* (not specified).

$$CH_3CH{=}CHCH_2CH_3 \; + \; :\ddot{Cl}{-}\ddot{Cl}: \; \longrightarrow \; CH_3\overset{*}{C}H\overset{*}{C}HCH_2CH_3 \; + \; :\ddot{Cl}:^-$$

(with the cyclic chloronium intermediate shown as $\overset{+}{\underset{}{\ddot{Cl}}}$)

$$CH_3\overset{*}{C}H\overset{*}{C}HCH_2CH_3 \longrightarrow CH_3\overset{*}{C}H\overset{*}{C}HCH_2CH_3 \xrightarrow{(-H^+)} CH_3\overset{*}{C}H\overset{*}{C}HCH_2CH_3$$

with $:\ddot{O}CH_3$ / H groups as shown

$$CH_3\overset{*}{C}H\overset{*}{C}HCH_2CH_3 \longrightarrow CH_3\overset{*}{C}H\overset{*}{C}HCH_2CH_3$$

with $:\ddot{Cl}:^-$

$$CH_3\overset{*}{C}H\overset{*}{C}HCH_2CH_3 \longrightarrow CH_3\overset{*}{C}H\overset{*}{C}HCH_2CH_3 \xrightarrow{(-H^+)} CH_3\overset{*}{C}H\overset{*}{C}HCH_2CH_3$$

with $:\ddot{O}CH_3$ / H groups as shown

$$CH_3\overset{*}{C}H\overset{*}{C}HCH_2CH_3 \longrightarrow CH_3\overset{*}{C}H\overset{*}{C}HCH_2CH_3$$

with $:\ddot{Cl}:^-$

Problem 6.32 Treating cyclohexene with HBr in the presence of acetic acid gives bromocyclohexane (85%) and cyclohexyl acetate (15%).

Cyclohexene   + HBr $\xrightarrow{CH_3{-}\overset{O}{\overset{\|}{C}}{-}OH}$ Bromocyclohexane   +   Cyclohexyl acetate
                                                       (85%)   (15%)

Propose a mechanism for the formation of the latter product.

Step 1: Electrophilic addition-add a proton. **Reaction of cyclohexene with a proton gives a secondary carbocation intermediate.**

Step 2: Make a new bond between a nucleophile and an electrophile Reaction of this intermediate with an oxygen atom of acetic acid, followed by proton transfer gives cyclohexyl acetate.

Step 3: Take a proton away

Problem 6.33 Propose a mechanism for this reaction:

| 1-Pentene | | | 1-Bromo-2-pentanol |
| (a racemic mixture) |

Step 1: Electrophilic addition. Reaction of 1-pentene with bromine gives a bridged bromonium ion intermediate.

Step 2: Make a new bond between a nucleophile and an electrophile. Anti attack of water on this intermediate at the more substituted secondary carbon, followed by proton transfer, gives 1-bromo-2-pentanol.

Step 3: Take a proton away.

A new chiral center is formed, so the product is a racemic mixture.

Problem 6.34 Treating 4-penten-1-ol with bromine in water forms a cyclic bromoether.

4-Penten-1-ol

(a racemic mixture)

Account for the formation of this product rather than a bromohydrin as was formed in Problem 6.32.

Step 1: Electrophilic addition. **Reaction of the alkene with bromine gives a bridged bromonium ion intermediate.**

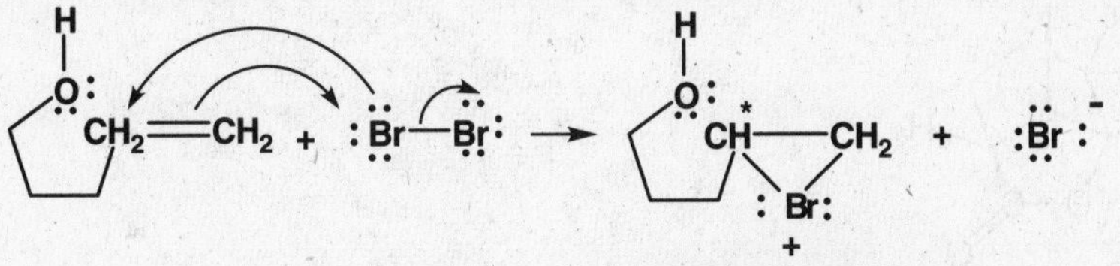

**A bridged bromonium
ion intermediate**

Step 2: Make a new bond between a nucleophile and an electrophile. **Reaction of this intermediate with the oxygen atom of the hydroxyl group followed by proton transfer gives the observed cyclic ether, a derivative of tetrahydrofuran.**

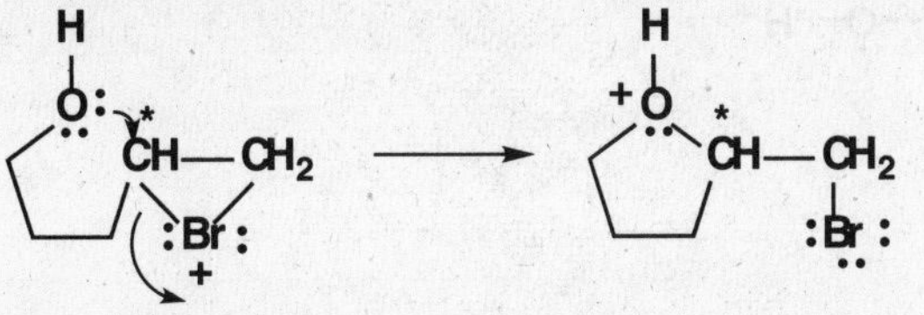

Step 3: Take a proton away.

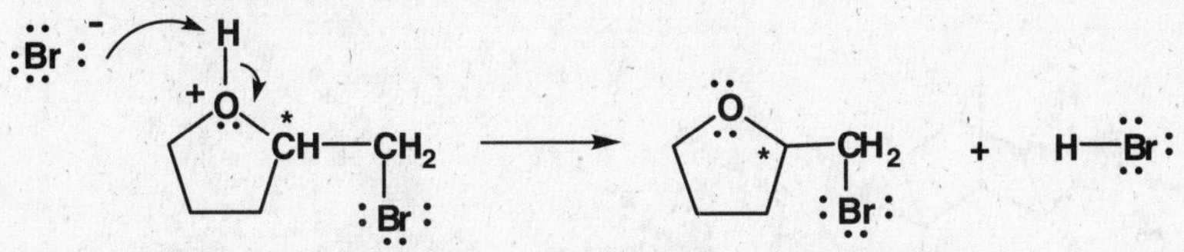

A chiral center is created, so a racemic mixture is produced.

Problem 6.35 Provide a mechanism for each reaction:

(a)

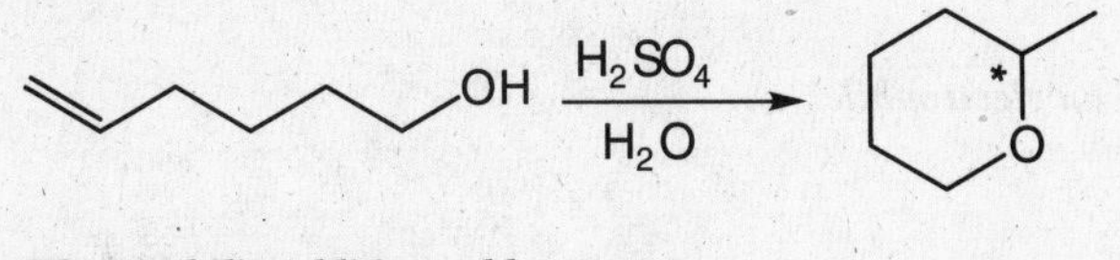

Step 1: Electrophilic addition-add a proton.

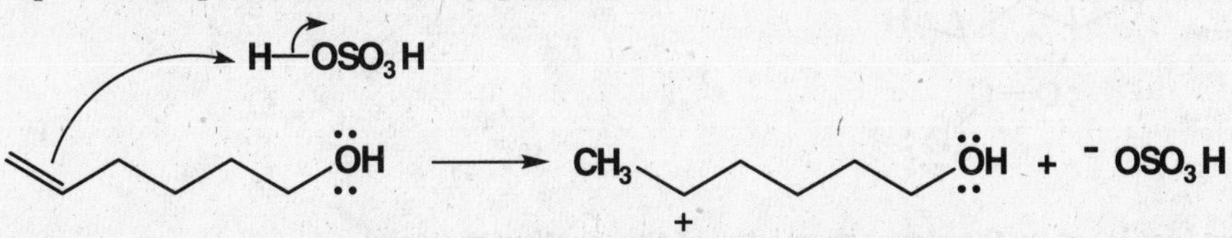

Step 2: Make a new bond between a nucleophile and an electrophile.

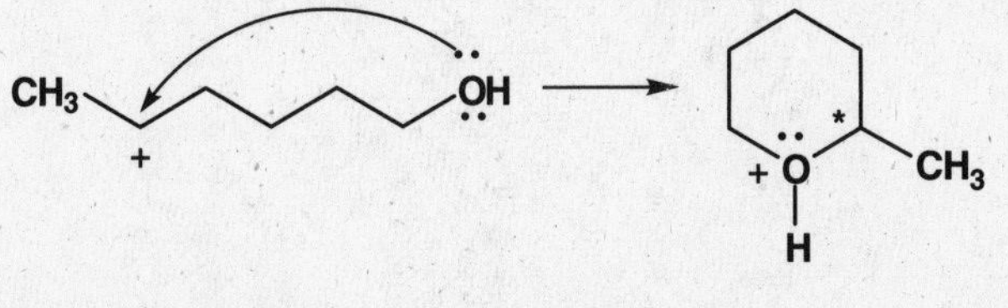

Step 3: Take a proton away.

A chiral center is created, so a racemic mixture is produced.

(b)

Step 1: Take a proton away.

Step 2: Electrophilic addition.

Step 3: Make a new bond between a nucleophile and an electrophile.

The stereochemistry of the product is set by the stereochemistry of the starting material.

<u>Problem 6.36</u> Treating 1-methyl-1-vinylcyclopentane with HCl gives mainly 1-chloro-1,2-dimethylcyclohexane.

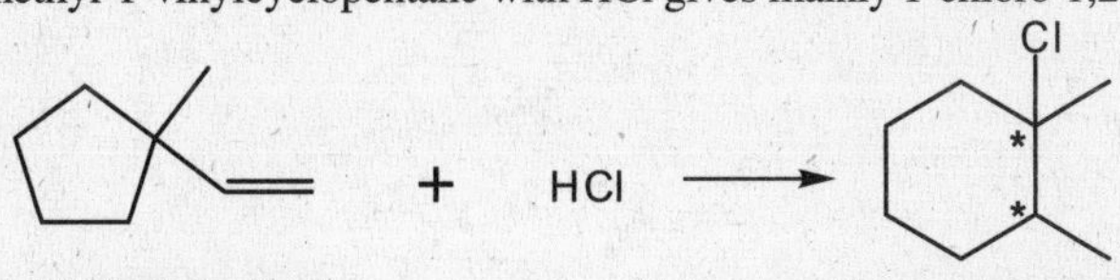

**1-Methyl-1-vinyl-
cyclopentane**

**1-Chloro-1,2-dimethyl-
cyclohexane**
(racemic)

Propose a mechanism for the formation of this product.

Step 1: Electrophilic addition-add a proton.

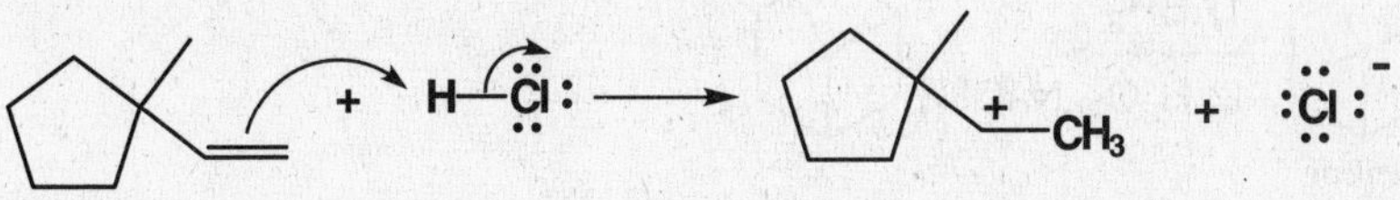

Step 2: 1,2-Shift. **The initially formed secondary carbocation can rearrange to form a more stable tertiary carbocation and a six-membered ring as shown.**

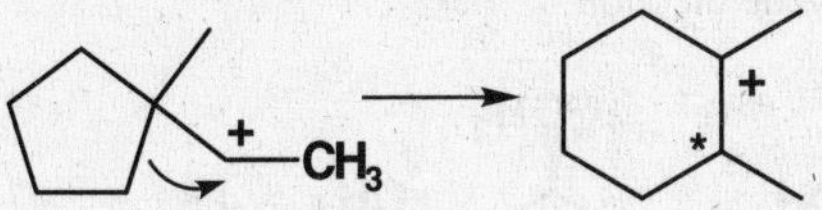

Step 3: Make a new bond between a nucleophile and an electrophile. **This new cation reacts with Cl⁻ to give the product.**

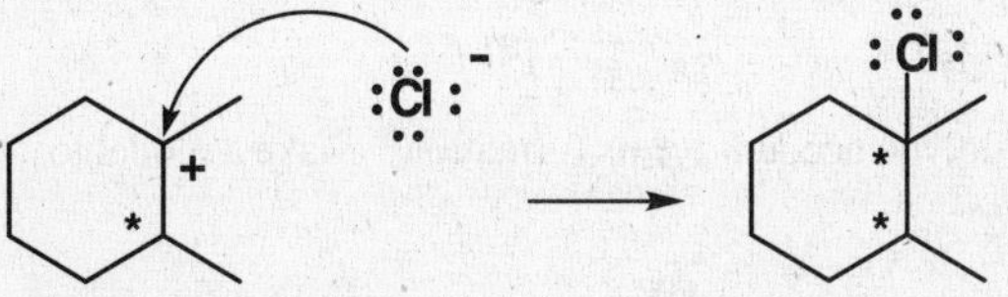

Note that two chiral centers were created, so a total of four stereoisomers as two racemic pairs are produced.

<u>**Hydroboration**</u>
<u>Problem 6.37</u> Draw a structural formula for the alcohol formed by treating each alkene with borane in tetrahydrofuran (THF) followed by hydrogen peroxide in aqueous sodium hydroxide, and specify stereochemistry where appropriate.

(a)

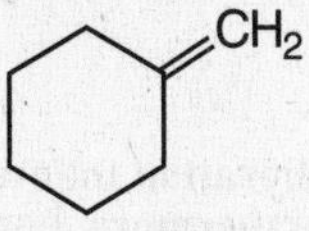

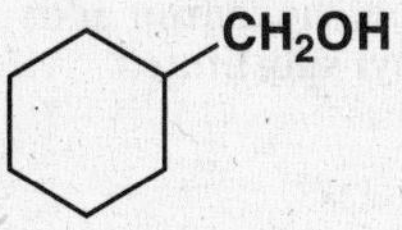

(b)

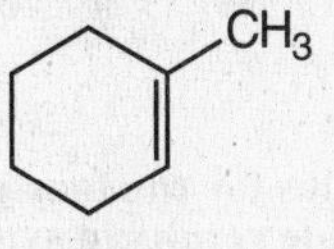

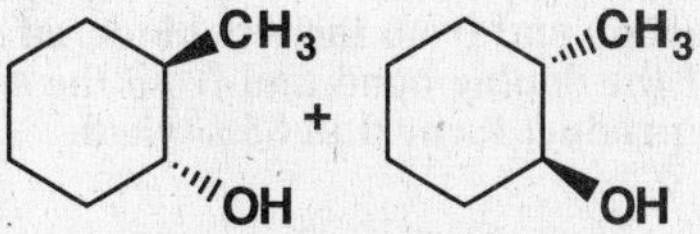

(Racemic mixture)

(c)

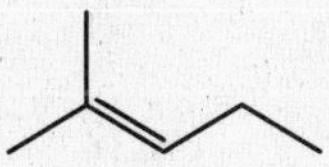

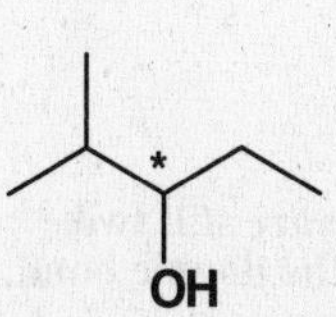

(Racemic mixture)

(d)

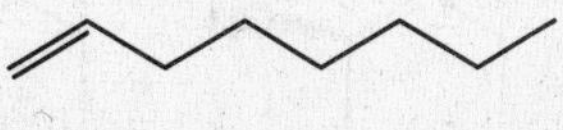

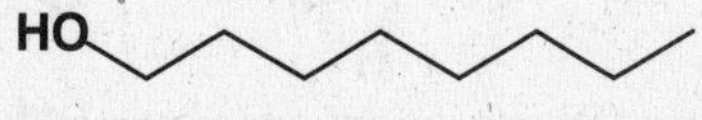

(e)

Problem 6.38 Reaction of α-pinene with borane followed by treatment of the resulting trialkylborane with alkaline hydrogen peroxide gives the following alcohol.

1) BH$_3$

2) H$_2$O$_2$, NaOH

α-Pinene

Of the four possible *cis, trans* isomers, one is formed in over 85% yield.

(a) Draw structural formulas for the four possible *cis, trans* isomers of the bicyclic alcohol.

CH$_3$ CH$_3$ CH$_3$ CH$_3$

OH   OH   OH   OH

[1]   [2]   [3]   [4]

(b) Which is the structure of the isomer formed in 85% yield? How do you account for its formation? Make a model to help you make this prediction.

CH$_3$

OH

[4]

Shown in part (a) are perspective formulas for the four possible *cis, trans* isomers. Hydroboration followed by treatment with alkaline hydrogen peroxide results in syn (*cis*) addition of -H and -OH. Furthermore, boron adds to the less substituted carbon and from the less hindered side. In hydroboration of α-pinene, boron adds to the disubstituted carbon of the double bond and from the side opposite the bulky dimethyl substituted bridge. Compound [4] is the product formed in 85% yield.

Oxidation
Problem 6.39 Write structural formulas for the major organic product(s) formed by reaction of 1-methylcyclohexene with each oxidizing agent.
(a) OsO$_4$/ H$_2$O$_2$

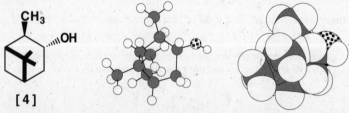

CH$_3$ OsO$_4$
                          ⟶
                           H$_2$O$_2$

CH$_3$ OH
       OH   CH$_3$
  +
       OH   H
  H   OH

Note that even though the two OH groups are added syn with respect to each other, there are still two enantiomers produced in the reaction. The OsO$_4$ reagent can attack from either face of the double bond.

(b) O₃ followed by (CH₃)₂S

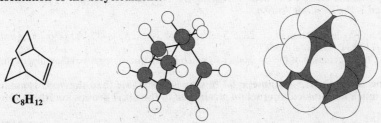

Problem 6.40 Draw the structural formula of the alkene that reacts with ozone followed by dimethyl sulfide to give each product or set of products.

(a) C₇H₁₂ $\xrightarrow[\text{2. (CH}_3)_2\text{S}]{\text{1. O}_3}$

(b) C₁₀H₁₈ $\xrightarrow[\text{2. (CH}_3)_2\text{S}]{\text{1. O}_3}$

(c) C₁₀H₁₈ $\xrightarrow[\text{2. (CH}_3)_2\text{S}]{\text{1. O}_3}$

Problem 6.41 Consider the following reaction.

C₈H₁₂ $\xrightarrow[\text{2. (CH}_3)_2\text{S}]{\text{1. O}_3}$

Cyclohexane-1,4-dicarbaldehyde

(a) Draw a structural formula for the compound with the molecular formula C₈H₁₂.

Following is a representation of the bicycloalkene.

C₈H₁₂

(b) Do you predict the product to be the *cis* isomer, the *trans* isomer, or a mixture of *cis* and *trans* isomers? Explain.

The product is the *cis* isomer, since the alkene is on one face of the starting bicycloalkane.

(c) Draw a suitable stereorepresentation for the more stable chair conformation of the dicarbaldehyde formed in this oxidation.

In either of the alternative chair conformations of the product, one carbaldehyde group is axial and the other is equatorial.

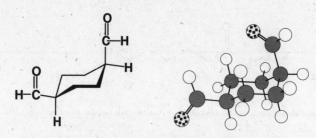

***cis*-Cyclohexane-1,4-dicarbaldehyde**

Reduction

Problem 6.42 Predict the major organic product(s) of the following reactions, and show stereochemistry where appropriate.

(a)

Geraniol   + 2H$_2$ $\xrightarrow{\text{Pt}}$ **3,7-Dimethyl-1-octanol (a racemic pair of enantiomers)**

Reduction of geraniol adds hydrogen atoms to each carbon-carbon double bond.

(b)

α-Pinene   + H$_2$ $\xrightarrow{\text{Pt}}$ **Major product**

Reduction of α-pinene adds hydrogen atoms preferentially from the less hindered side of the double bond, namely the side opposite the one-carbon bridge bearing the two methyl groups. Predict, therefore, that the major isomer formed is the first one shown.

Problem 6.43 The heat of hydrogenation of *cis*-2,2,5,5-tetramethyl-3-hexene is -154 kJ (-36.7 kcal)/mol while that of the *trans* isomer is only -113 kJ (-26.9 kcal)/mol.
(a) Why is the heat of hydrogenation of the *cis* isomer so much larger (more negative) than that of the *trans* isomer?

A larger value means the *cis*-2,2,5,5-tetramethyl-3-hexene is less stable than the *trans* isomer. This makes sense because there is so much nonbonded interaction strain due to the alkyl groups smashing into each other in the *cis* isomer.

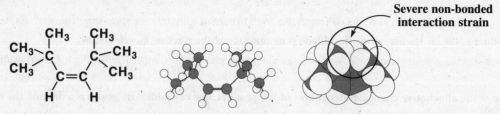

Severe non-bonded
interaction strain

***cis*-2,2,5,5-Tetramethyl-3-hexene**

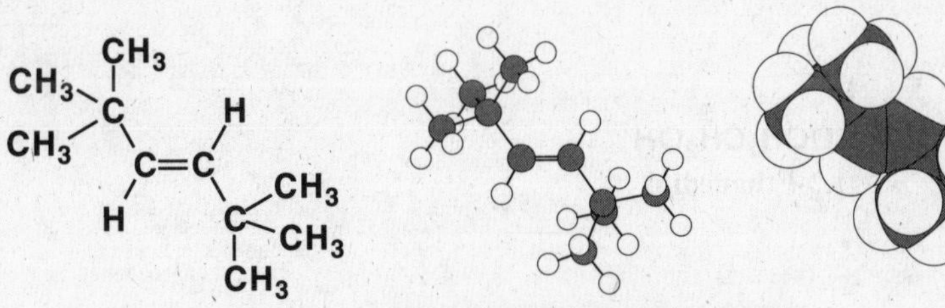

trans-2,2,5,5-Tetramethyl-3-hexene

(b) If a catalyst could be found that allowed equilibration of the *cis* and *trans* isomers at room temperature (such catalysts do exist), what would be the ratio of *trans* to *cis* isomers?

The difference in energy between the two isomers is -154 - (-113) = -41 kJ/mol. Using the equation derived in the answer to problem 6.24 gives:

$$K_{eq} = e^{\left(\frac{-\Delta G^\circ}{RT}\right)} = e^{\left(\frac{-(-41\ \text{kJ/mol})}{(8.314 \times 10^{-3}\ \text{kcal/mol K})(298\ \text{K})}\right)} = e^{16.5} = 1.5 \times 10^{7}$$

Thus, at room temperature, the ratio would be 1.5×10^7 to 1 in favor of the *trans* isomer.

Synthesis

Problem 6.44 Show how to convert ethylene to these compounds.

(a) Ethane

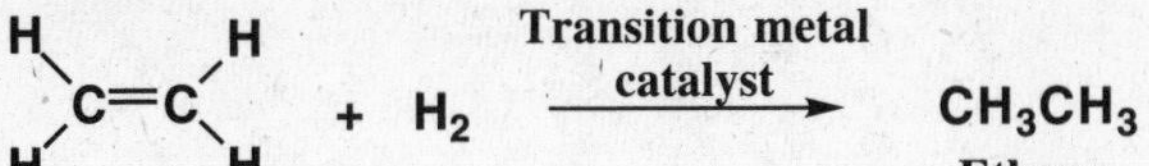

Ethane

(b) Ethanol

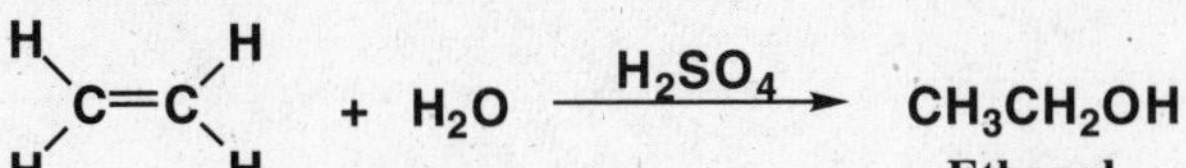

Ethanol

or

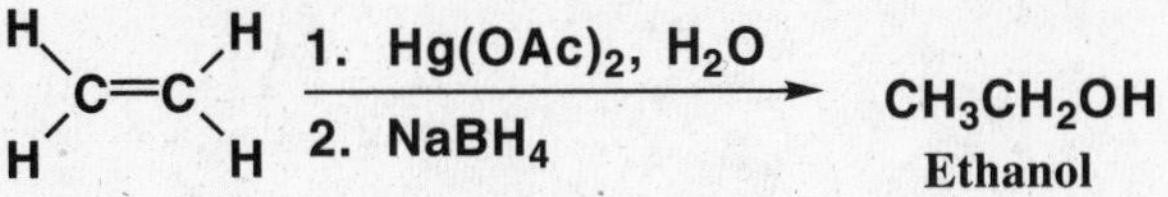

Ethanol

or

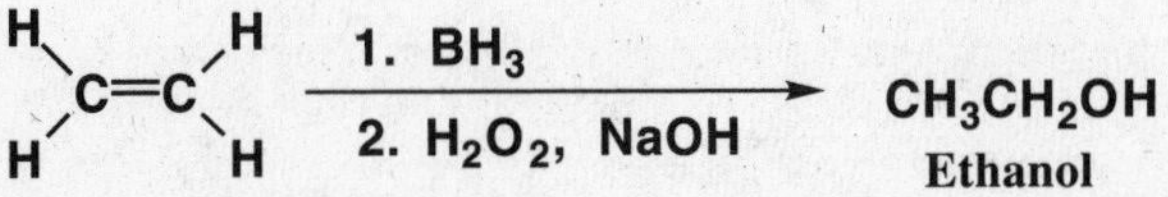

Ethanol

(c) Bromoethane

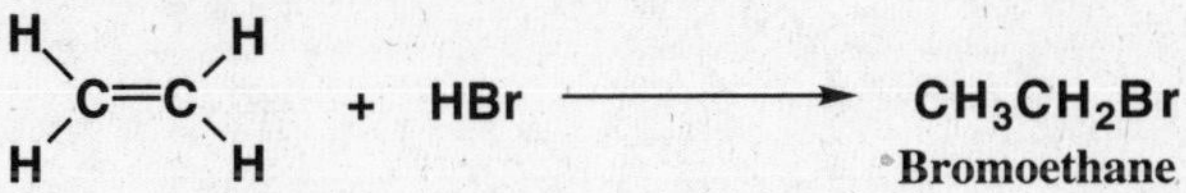

Bromoethane

(d) 2-Chloroethanol

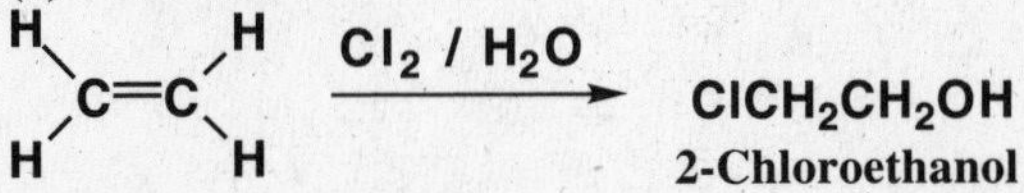

2-Chloroethanol

(e) 1,2-Dibromoethane

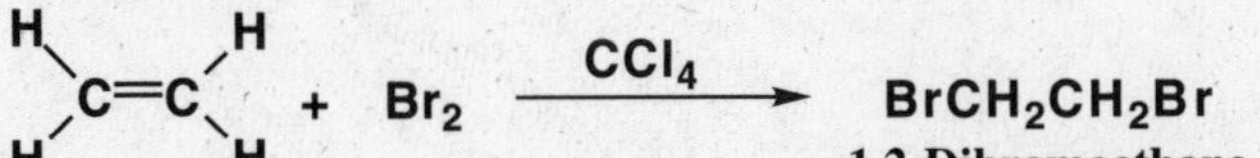

1,2-Dibromoethane

(f) 1,2-Ethanediol

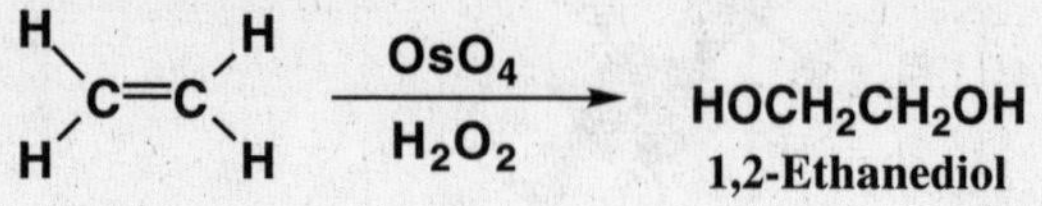

$$\underset{\text{H}}{\overset{\text{H}}{>}}C=C\underset{\text{H}}{\overset{\text{H}}{<}} \xrightarrow[\text{H}_2\text{O}_2]{\text{OsO}_4} \text{HOCH}_2\text{CH}_2\text{OH}$$
1,2-Ethanediol

(g) Chloroethane

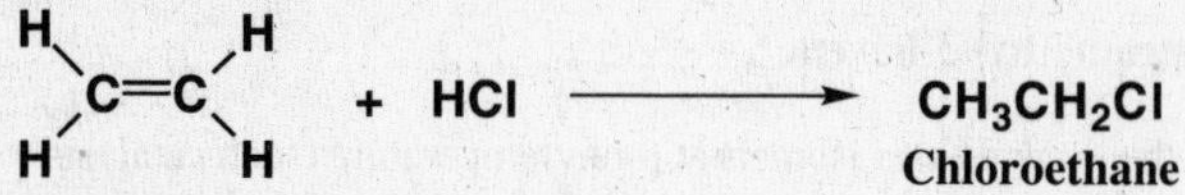

$$\underset{\text{H}}{\overset{\text{H}}{>}}C=C\underset{\text{H}}{\overset{\text{H}}{<}} + \text{HCl} \longrightarrow \text{CH}_3\text{CH}_2\text{Cl}$$
Chloroethane

Problem 6.45 Show how to convert cyclopentene into these compounds.
(a) *trans*-1,2-Dibromocyclopentane

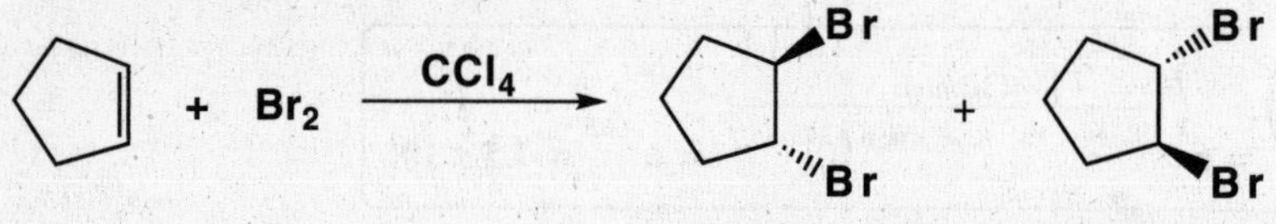

+ Br$_2$ $\xrightarrow{\text{CCl}_4}$ [Br / Br] + [Br / Br]

trans-1,2-Dibromocyclopentane

Note that the product of the reaction is actually a pair of enantiomers.

(b) *cis*-1,2-Cyclopentanediol

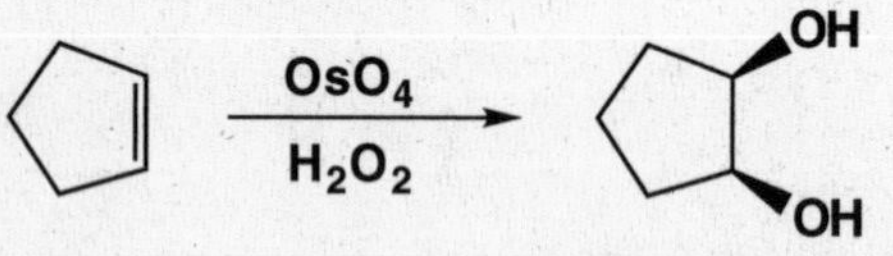

$\xrightarrow[\text{H}_2\text{O}_2]{\text{OsO}_4}$ [OH / OH]

cis-1,2-Cyclopentanediol

(c) Cyclopentanol

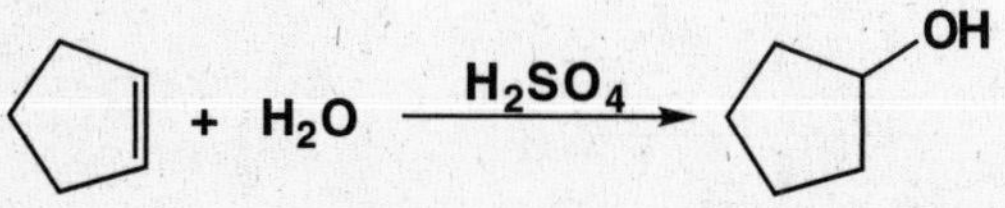

+ H$_2$O $\xrightarrow{\text{H}_2\text{SO}_4}$ [OH]

Cyclopentanol

or

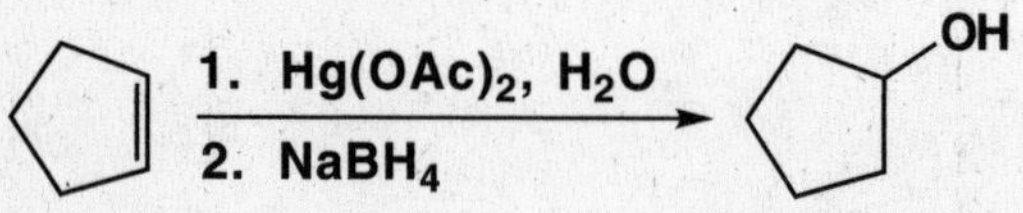

$\xrightarrow[\text{2. NaBH}_4]{\text{1. Hg(OAc)}_2,\ \text{H}_2\text{O}}$ [OH]

Cyclopentanol

or

$\xrightarrow[\text{2. H}_2\text{O}_2,\ \text{NaOH}]{\text{1. BH}_3}$ [OH]

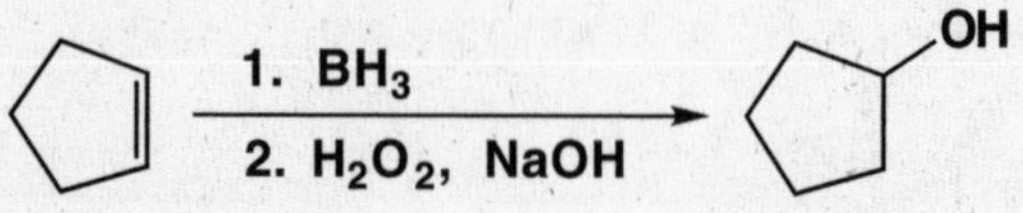

Cyclopentanol

(d) Iodocyclopentane

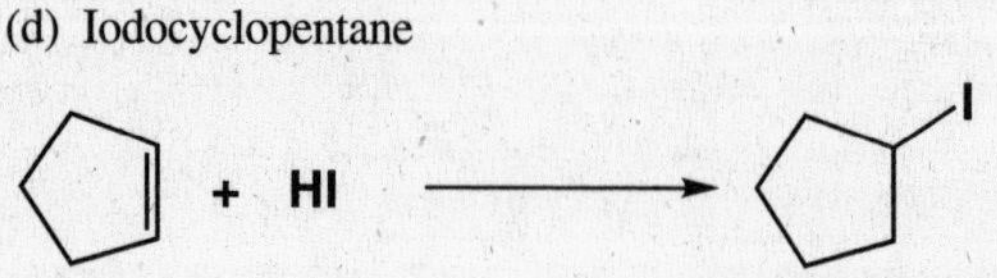

+ HI $\longrightarrow$ [I]

Iodocyclopentane

(e) Cyclopentane

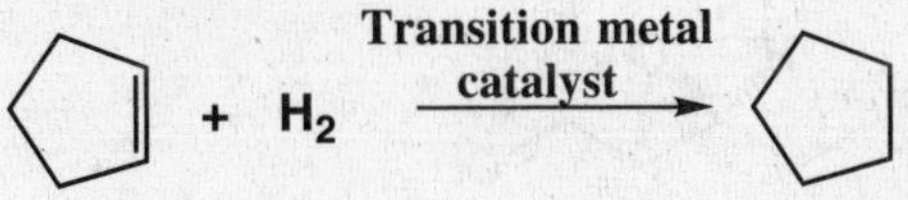

Cyclopentane

(f) Pentanedial

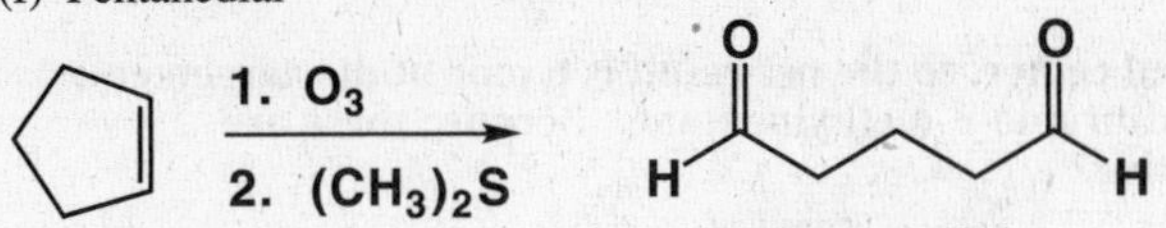

Pentanedial

Reactions that Produce Chiral Compounds

<u>Problem 6.46</u> State the number and kind of stereoisomers formed when (*R*)-3-methyl-1-pentene is treated with these reagents. Assume that the starting alkene is enantiomerically pure and optically active. Will each product be optically active or inactive?

(*R*)-3-Methyl-1-pentene

(a) Hg(OAc)$_2$, H$_2$O followed by NaBH$_4$

The alcohol produced in this reaction has a new chiral center, so the net result is a pair of diastereomers; (*2S,3R*)-3-methyl-2-pentanol and (*2R,3R*)-3-methyl-2-pentanol. Because these are diastereomers, the product solution will be optically active.

(2S,3R)-3-Methyl-2-pentanol

+

(2R,3R)-3-Methyl-2-pentanol

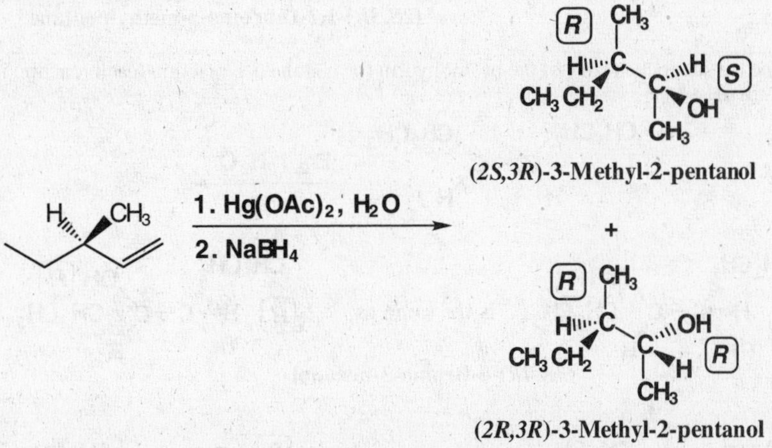

(b) H$_2$/Pt

The alkane produced in this reaction does not have any new chiral centers, and the only product, 3-methylpentane, is not chiral. Because the product is not chiral, the production solution will not be optically active.

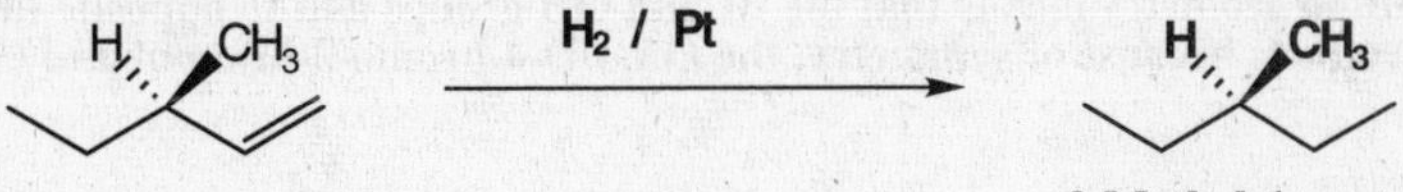

3-Methyl-1-pentane (achiral)

(c) BH$_3$ followed by H$_2$O$_2$ in NaOH

These reagents give the non-Markovnikov product, so the primary alcohol produced does not have any new chiral centers, and the only product is (*R*)-3-methyl-1-pentanol. Because there is only one enantiomer in the product solution, it will be optically active.

(R)-3-Methyl-1-pentanol

(d) Br_2 in CCl_4

The dibromide produced in this reaction has a new chiral center, so the net result is a pair of diastereomers; (2R,3R)-1,2-dibromo-3-methylpentane and (2S,3R)-1,2-dibromo-3-methylpentane. Because these are diastereomers, the product solution will be optically active.

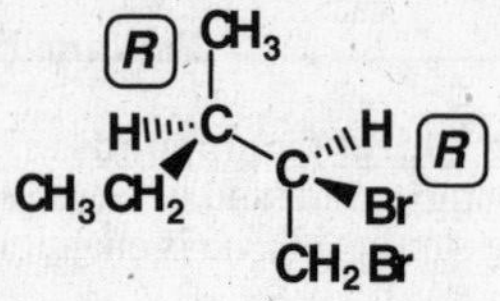

(2R,3R)-1,2-Dibromo-3-methylpentane

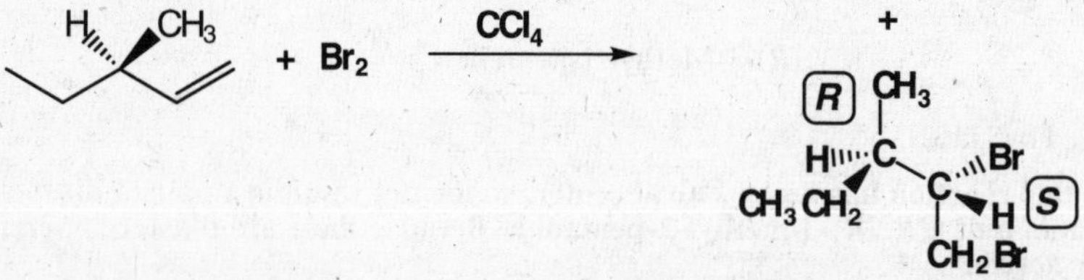

(2S,3R)-1,2-Dibromo-3-methylpentane

<u>Problem 6.47</u> Describe the stereochemistry of the bromohydrin formed in each reaction (each reaction is stereospecific).
(a) *cis*-3-Hexene + Br_2/H_2O

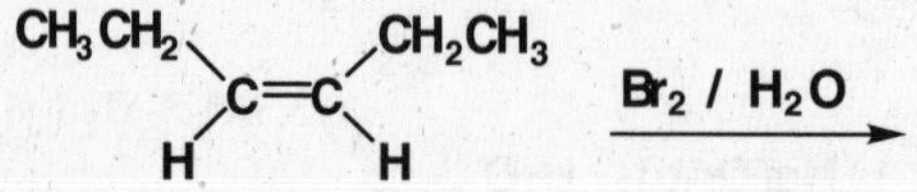

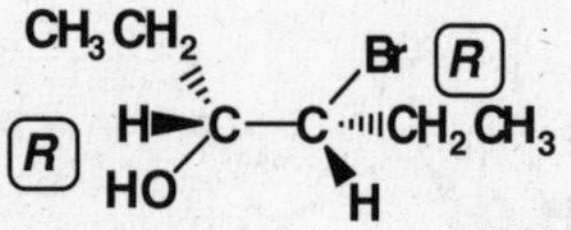

is the same as

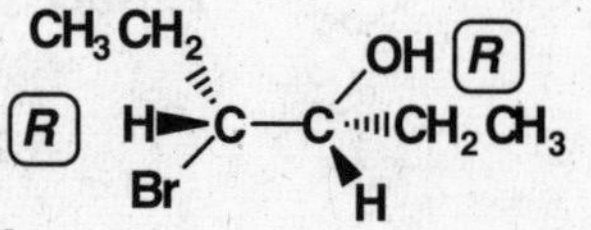

(3R,4R)-4-Bromo-3-hexanol

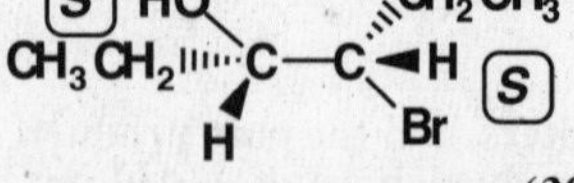

is the same as

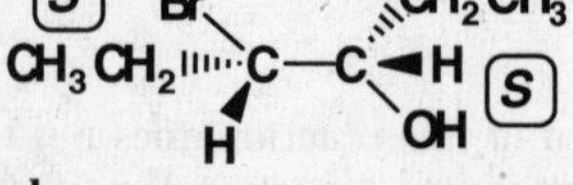

(3S,4S)-4-Bromo-3-hexanol

Remember that the anti addition geometry means that the Br and OH groups add to opposite faces of the double bond. There are only two products because of symmetry, the (3R,4R)-4-bromo-3-hexanol and (3S,4S)-4-bromo-3-hexanol

(b) *trans*-3-Hexene + Br$_2$/H$_2$O

(3R,4S)-4-Bromo-3-hexanol

(3S,4R)-4-Bromo-3-hexanol

Remember that the anti addition geometry means that the Br and OH groups add to opposite faces of the double bond. There are only two products because of symmetry, the (3R,4S)-4-bromo-3-hexanol and (3S,4R)-4-bromo-3-hexanol. Note that *cis* and *trans*-3-hexene give different stereoisomer products, explaining why the bromination reaction is referred to as being stereospecific as well as just stereoselective.

Problem 6.48 In each of these reactions, the organic starting material is achiral. The structural formula of the product is given. For each reaction determine the following.
(1) How many stereoisomers are possible for the product?
(2) Which of the possible stereoisomers is/are formed in the reaction shown?
(3) Whether the product is optically active or optically inactive?

(a)

The product has no chiral centers so no stereoisomerism is possible. The product mixture will therefore be optically inactive.

(b)

The product molecule has two new chiral centers for a total of three possible stereoisomers, a meso compound (*cis* addition) and a pair of enantiomers (*trans* addition). However, due to the anti relationship of the added bromine atoms, only the R,R and S,S isomers are formed. These are enantiomers and since they will be produced in equal amounts (racemic mixture), the product solution will be optically inactive even though each enantiomer by itself would display optical activity.

(c)

The product molecule has two new chiral centers for a total of four possible stereoisomers. However, due to the anti relationship of the added bromine atoms, only the R,R and S,S isomers can be formed. These are enantiomers and since they will be produced in equal amounts (racemic mixture), the product solution will be optically inactive even though each enantiomer by itself would display optical activity.

(d)

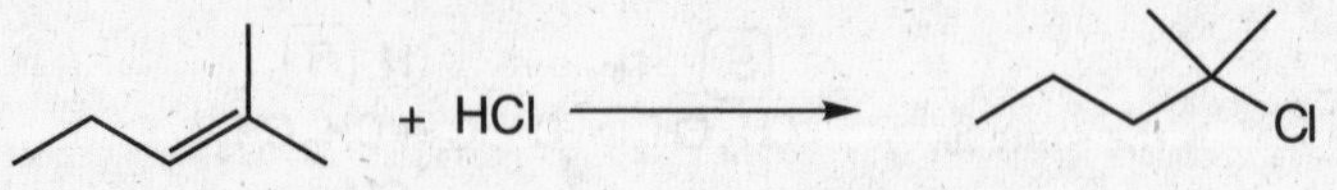

The product molecule has no chiral centers so no stereoisomerism is possible. The product solution will therefore be optically inactive.

(e)

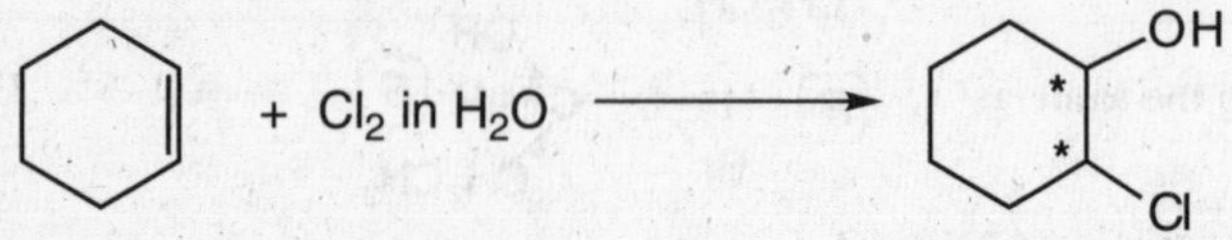

The product molecule has two new chiral centers for a total of four possible stereoisomers. However, due to the anti relationship of the added chlorine and -OH groups, only the *trans* isomers can be formed. These are enantiomers and since they will be produced in equal amounts (racemic mixture), the product solution will be optically inactive even though each enantiomer by itself would display optical activity.

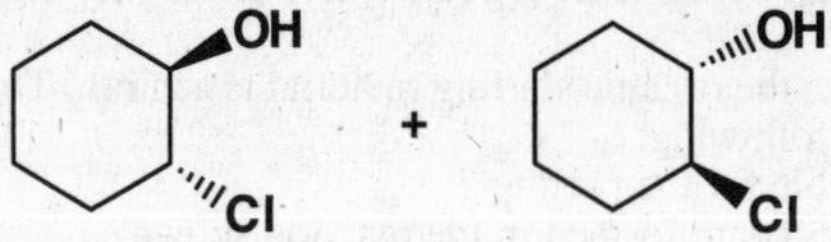

(f)

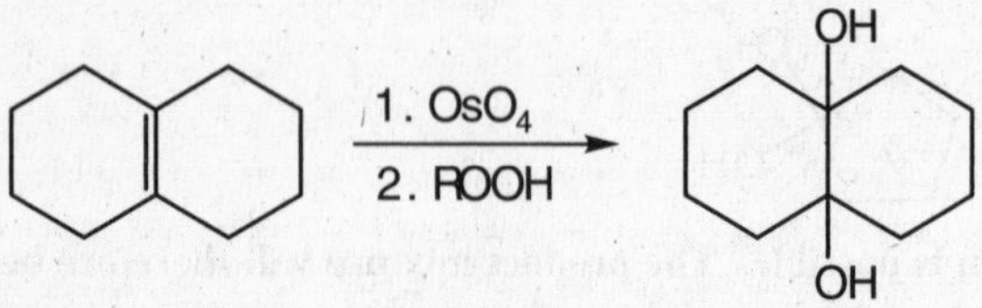

Because of symmetry in the molecule, the product has no chiral centers. However, there is still the possibility of *cis, trans* isomers, so two products are possible. Because of the mechanism of the OsO₄ reaction, only the *cis* isomer will be formed. The product is achiral so there is no optical activity.

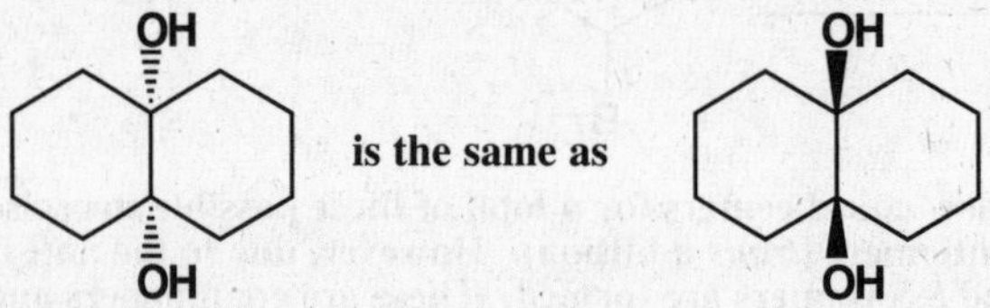

is the same as

(g)

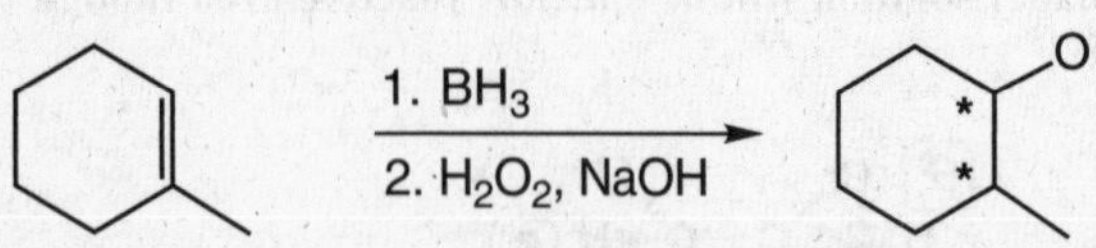

The product molecule has two new chiral centers for a total of four possible stereoisomers. However, due to the syn addition geometry observed with the reaction of borane, the -OH and methyl groups must be *trans* leading to only the two products shown. These are enantiomers and since they will be produced in equal amounts (racemic mixture), the product solution will be optically inactive even though each enantiomer by itself would display optical activity.

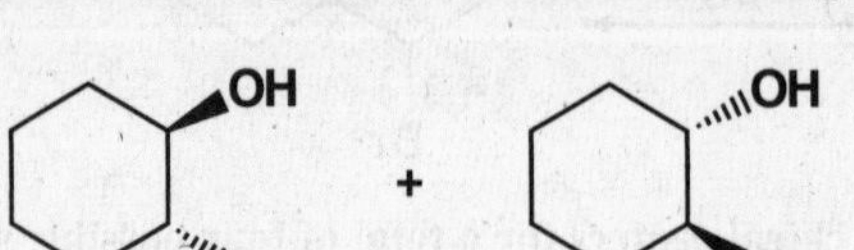

Looking Ahead

<u>Problem 6.49</u> The 2-propenyl cation appears to be a primary carbocation, and yet it is considerably more stable than a 1° carbocation such as the 1-propyl cation.

$$H_2C=CH-CH_2^+ \qquad\qquad H_3C-CH_2-CH_2^+$$

2-Propenyl cation   **1-Propyl cation**

How would you account for the differences in the stability of the two carbocations?

The 2-propenyl cation, also called the allyl cation, is stabilized by resonance with the adjacent π bond.

$$H_2C=CH-CH_2^+ \longleftrightarrow H_2C^+-CH=CH_2$$

The resonance distributes the positive charge onto both terminal carbon atoms. The greater the degree of charge distribution, the more stable the ion, so the 2-propenyl cation is considerably more stable than the 1-propyl cation in which the positive charge is localized on one primary carbon atom.

<u>Problem 6.50</u> Treating 1,3-butadiene with one mole of HBr gives a mixture of two isomeric products.

$$H_2C=CH-CH=CH_2 + H-Br \longrightarrow \overset{\overset{\displaystyle Br}{|}}{H_2C=CH-CH-CH_3} + H_3C-CH=CH-CH_2-Br$$

1,3-Butadiene   **3-Bromo-1-butene**   **1-Bromo-2-butene**

Propose a mechanism that accounts for the formation of these two products.

Step 1: Electrophilic addition-add a proton. **The key idea is that protonation of 1,3-butadiene on either π bond generates a resonance-stabilized carbocation analogous to the resonance-stabilized cation described in Problem 6.48.**

$$H_2C=CH-CH=CH_2 \longrightarrow H_3C-CH-CH=CH_2 \longleftrightarrow H_3C-CH=CH-CH_2^+ + :Br:^-$$

1,3-Butadiene   **A resonance-stabilized carbocation**

Step 2: Make a new bond between a nucleophile and an electrophile. **In the second step of the mechanism, the Br⁻ can attack either carbon with a partial positive charge to generate the two observed products.**

$$H_3C-CH-CH=CH_2 \longleftrightarrow H_3C-CH=CH-CH_2^+$$

$$:Br:^- \qquad\qquad\qquad\qquad :Br:^-$$

$$H_3C-CH-CH=CH_2 \qquad\qquad H_3C-CH=CH-CH_2-Br:$$

3-Bromo-1-butene   **1-Bromo-2-butene**

<u>Problem 6.51</u> In this chapter, we studied the mechanism of the acid-catalyzed hydration of an alkene. The reverse of this reaction is the acid-catalyzed dehydration of an alcohol.

$$\overset{\overset{\displaystyle OH}{|}}{CH_3-CH-CH_3} \xrightarrow{H_2SO_4} H_3C-CH=CH_2 + H_2O$$

2-Propanol   **Propene**
(Isopropyl alcohol)

Propose a mechanism for the acid-catalyzed dehydration of 2-propanol to propene.

The interesting feature of this mechanism is that each step is exactly the reverse of the steps of acid-catalyzed alkene hydration. Compare the following mechanism to that in Section 6.3B.

Step 1: Add a proton. The dehydration reaction begins with protonation of the hydroxyl group.

Step 2: Break a bond to give stable molecules or ions This creates a positively charged oxonium ion, that then decomposes by ejecting water to generate a carbocation.

Step 3: Take a proton away The carbocation loses a proton to water to complete the reaction.

You will see several similar conversions of hydroxyl groups into species that can depart in future chapters and it is a central theme of the chemistry of alcohols. Note that in the first step of the mechanism, H_3O^+ is shown as the acid. This is created from the equilibrium of H_2SO_4 with the water produced during the dehydration reaction. In the initial stage of the reaction, however, before significant water is produced, the acid would more appropriately be listed as H_2SO_4. The fact that this mechanism is just the reverse of alkene hydration brings up the important point that alkenes reacting with water in acid solution is a reversible process, and the position of equilibrium depends on the reaction conditions used.

<u>Problem 6.52</u> As we have seen in this chapter, carbon-carbon double bonds are electron-rich regions and are attacked by electrophiles (for example H-Br); they are not attacked by nucleophiles (for example, diethylamine).

However, when the carbon-carbon double bond has a carbonyl group adjacent to it, the double bond reacts readily with nucleophiles by nucleophilic addition (Section 19.8).

Account for the fact that nucleophiles add to a carbon-carbon double bond adjacent to a carbonyl group, and account for the regiochemistry of the reaction.

The adjacent carbonyl polarizes the carbon-carbon double bond as indicated by the following resonance contributing structure.

this position can
be attacked by
nucleophiles

The contributing structure on the right places a positive charge on the terminal carbon atom, facilitating attack at this position by nucleophiles.

Problem 6.53 Following is an example of a type of reaction known as a Diels-Alder reaction (Chapter 24).

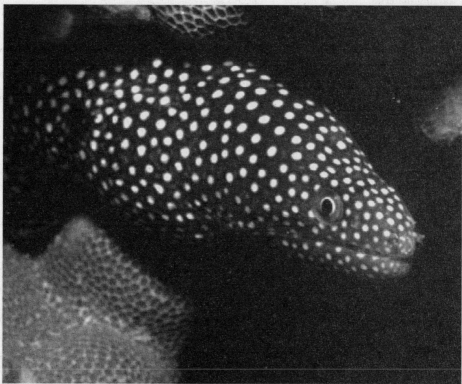

1,3-Pentadiene   Ethylene   3-Methylcyclohexene
(racemic)

The Diels-Alder reaction between a diene and an alkene is quite remarkable in that it is one of the few ways that chemists have to form two new carbon-carbon bonds in a single reaction. Given what you know about the relative strengths of carbon-carbon σ and π bonds, would you predict the Diels-Alder reaction to be exothermic or endothermic? Explain your reasoning.

The Diels-Alder reaction is exothermic as written. The reaction involves the conversion of two starting materials with a total of three π bonds into a product with only one π bond, but two new σ bonds. Since σ bonds are generally stronger than π bonds, the product 3-methylcyclohexene has overall stronger bonds than the starting materials. When products have stronger bonds compared to starting materials, the reaction is exothermic.

Whitemouth moray *Gymnothorax meleagris*
Kona coast, Hawaii

CHAPTER 7
Solutions to the Problems

Problem 7.1 Write the IUPAC name for each compound.

(a)

1-Octyne

(b)

3,3-Dimethyl-1-butyne

(c) Br

3-Bromo-3-methyl-1-butyne

Problem 7.2 Write the common name of each alkyne.

(a)

Diisopropylacetylene

(b)

Cyclohexylacetylene

(c)

Butylacetylene

Problem 7.3 Draw a structural formula for an alkene and dichloroalkane with the given molecular formula that yields the indicated alkyne by each reaction sequence.

The key to this problem is to realize that the triple bond in the product is at the same location as the double bond in the starting material.

(a) $C_6H_{12} \xrightarrow{Cl_2} C_6H_{12}Cl_2 \xrightarrow{2\ NaNH_2}$

$$CH_3CH_2CH=CHCH_2CH_3 \xrightarrow{Cl_2} CH_3CH_2\overset{*}{C}H-\overset{*}{C}HCH_2CH_3 \xrightarrow{2\ NaNH_2}$$

(cis or *trans)*
C_6H_{12}

$\underset{Cl\ \ Cl}{}$

$C_6H_{12}Cl_2$

(b) $C_7H_{14} \xrightarrow{Cl_2} C_7H_{14}Cl_2 \xrightarrow{2\ NaNH_2}$

$$CH_3-CH=CH-\overset{CH_3}{\underset{CH_3}{\overset{|}{\underset{|}{C}}}}-CH_3 \xrightarrow{Cl_2} CH_3-\overset{*}{C}H-\overset{*}{C}H-\overset{CH_3}{\underset{CH_3}{\overset{|}{\underset{|}{C}}}}-CH_3 \xrightarrow{2\ NaNH_2}$$

(cis or *trans)*
C_7H_{14}

$\underset{Cl\ \ Cl\ \ CH_3}{}$

$C_7H_{14}Cl_2$

Problem 7.4 Draw a structural formula for a hydrocarbon of the given molecular formula that undergoes hydroboration-oxidation to give the indicated product:

(a) $-C\equiv CH \xrightarrow[\text{2) } H_2O_2,\ NaOH]{\text{1) (sia)}_2BH}$

C_7H_{10}

(b) $-CH=CH_2 \xrightarrow[\text{2) } H_2O_2,\ NaOH]{\text{1) } BH_3}$

C_7H_{12}

<u>Problem 7.5</u> Hydration of 2-pentyne gives a mixture of two ketones, each with the molecular formula $C_5H_{10}O$. Propose structural formulas for these two ketones and for the enol from which each is derived.

The two ketones are 2-pentanone and 3-pentanone.

<u>Problem 7.6</u> Show how the synthetic scheme in Example 7.6 might be modified to give the following.
(a) 1-Heptanol

A reduction step can be included using the Lindlar catalyst to give an alkene, that can then be converted to the primary alcohol on the less-substituted carbon of the double bond using hydroboration.

(b) 2-Heptanol

This is the same as part (a), but the last step uses oxymercuration/reduction to place the -OH group on the more substituted carbon of the double bond. Acid catalyzed hydration (H_2SO_4/H_2O) could have also been used on the last step.

<u>Problem 7.7</u> Enanthotoxin is an extremely poisonous organic compound found in hemlock water dropwart, which is reputed to be the most poisonous plant in England. It is believed that no British plant has been responsible for more fatal accidents. The most poisonous part of the plant is the roots, which resemble small white carrots, giving the plant the name "five finger death." Also poisonous are its leaves, which look like parsley. Enanthotoxin is thought to interfere with the Na^+ current in nerve cells, which leads to convulsions and death.

Can show *cis-trans* isomerism

How many stereoisomers are possible for enanthotoxin?

There is one tetrahedral chiral center (marked with an *) and three double bonds that can show *cis-trans* isomerism, so there are 2^4 or 16 possible stereoisomers.

Preparation of Alkynes

<u>Problem 7.8</u> Show how to prepare each alkyne from the given starting material.   In part (c), D indicates deuterium. Deuterium-containing reagents such as BD_3, D_2O, and CH_3COOD are available commercially.

(a)

First treat 1-pentene with either bromine (Br_2) or chlorine (Cl_2) to form a 1,2-dihalopentane. Then carry out a double dehydrohalogenation with three moles sodium amide ($NaNH_2$) followed by an aqueous workup to form 1-pentyne.

(b)

This is an example of alkylation of a terminal alkyne. First , deprotonate the terminal alkyne with $NaNH_2$ then add propyl bromide.

(c)

First form the acetylide anion with sodium amide and then react the anion with a deuterium donor such as D_2O or CH_3COOD.

<u>Problem 7.9</u> If a catalyst could be found that would establish an equilibrium between 1,2-butadiene and 2-butyne, what would be the ratio of the more stable isomer to the less stable isomer at 25°C?

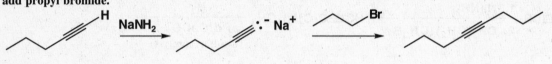

$$CH_2{=}C{=}CHCH_3 \rightleftharpoons CH_3C{\equiv}CCH_3 \qquad \Delta G° = \text{-16.7 kJ (-4.0 kcal)/mol}$$

At equilibrium the relative amounts of each form are given by the equation:

$$\Delta G° = -RT\ln K_{eq} \qquad \text{where} \quad K_{eq} = \frac{\text{[2-butyne]}}{\text{[1,2-butadiene]}}$$

Rearranging gives:

$$\ln K_{eq} = \frac{-\Delta G°}{RT}$$

Solving the equation gives:

$$K_{eq} = e^{\left(\frac{-\Delta G°}{RT}\right)}$$

Plugging in the values for $\Delta G°$, R, and 298 K gives the final answer:

$$K_{eq} = e^{\left(\frac{-(-16{,}700 \text{ J/mol})}{(8.314 \text{ J}\cdot K^{-1}\cdot mol^{-1})(298 \text{ K})}\right)} = e^{6.74} = 8.46 \times 10^2$$

Because $\Delta G°$ is negative, it means the alkyne is favored, and if a catalyst could be found for the interconversion, the ratio of alkyne to allene would be 846 to 1.

Reactions of Alkynes

Problem 7.10 Complete each acid-base reaction and predict whether the position of equilibrium lies toward the left or toward the right.

Recall that equilibrium favors formation of the weaker acid, weaker base pair.

(a) CH₃C≡CH + CH₃CH₂O⁻ Na⁺ ⇌ (CH₃CH₂OH) CH₃C≡C⁻ Na⁺ + CH₃CH₂COH

  pKₐ 25   (weaker   (stronger   pKₐ 16
  (weaker   base)   base)   (stronger
  acid)   acid)

Oxygen is more electronegative (farther to the right on the Periodic Table) than carbon so the alcohol is the stronger acid and equilibrium lies toward the left.

(b) CH₃C≡CCH₂CH₂OH + Na⁺NH₂⁻ → (NH₃ (l)) CH₃C≡CCH₂CH₂O⁻ Na⁺ + NH₃

  pKₐ ~16   (stronger   (weaker   pKₐ 38
  (stronger   base)   base)   (weaker
  acid)   acid)

Oxygen is more electronegative (farther to the right on the Periodic Table) than nitrogen, so the hydroxyl group is the stronger acid and equilibrium lies toward the right.

(c) CH₃C≡C⁻ Na⁺ + CH₃COH ⇌ CH₃C≡CH + CH₃CO⁻ Na⁺

  (stronger   pKₐ 4.76   pKₐ 25   (weaker
  base)   (stronger   (weaker   base)
                  acid)   acid)

Oxygen is more electronegative (farther to the right on the Periodic Table) than carbon, plus the carboxylate anion is resonance stabilized. The net effect is that acetic acid is by far the stronger acid and equilibrium lies toward the right.

Problem 7.11 Draw the structural formulas for the major product(s) formed by reaction of 3-hexyne with each of these reagents. (Where you predict no reaction, write NR).

(a) H₂(excess) / Pt

  CH₃CH₂CH₂CH₂CH₂CH₃

(b) H₂ / Lindlar catalyst

  CH₃CH₂ CH₂CH₃ (cis C=C)

(c) Na in NH₃ (liquid)

  CH₃CH₂ / H, H / CH₂CH₃ (trans C=C)

d) BH₃ followed by H₂O₂ / NaOH

  CH₃CH₂CCH₂CH₂CH₃ (ketone O)

(e) BH₃ followed by CH₃COOH

  CH₃CH₂ CH₂CH₃ (cis)

(f) BH₃ followed by CH₃COOD

  CH₃CH₂ CH₂CH₃ / H, D

(g) Cl₂ (one mole)

  CH₃CH₂ / Cl, Cl / CH₂CH₃

(h) NaNH₂ in NH₃ (liquid)

  No Reaction

(i) HBr (one mole)

  CH₃CH₂ / Br, H / CH₂CH₃

(j) HBr (two moles)

  CH₃CH₂CCH₂CH₂CH₃ (Br, Br)

(k) H₂O in H₂SO₄ / HgSO₄

  CH₃CH₂CCH₂CH₂CH₃ (ketone O)

Problem 7.12 Draw the structural formula of the enol formed in each alkyne hydration reaction, and then draw the structural formula of the carbonyl compound with which each enol is in equilibrium.

(a) $CH_3(CH_2)_5C{\equiv}CH + H_2O \xrightarrow[\text{H}_2\text{SO}_4]{\text{HgSO}_4}$ (an enol) $\longrightarrow$

$$CH_3(CH_2)_5\overset{\displaystyle OH}{C}{=}CH_2 \;\rightleftharpoons\; CH_3(CH_2)_5\overset{\displaystyle O}{C}{-}CH_3$$

1-Octen-2-ol   2-Octanone
(An enol)   (A ketone)

(b) $CH_3(CH_2)_5C{\equiv}CH \xrightarrow[\text{NaOH/H}_2\text{O}_2]{\text{(sia)}_2\text{BH}}$ (an enol) $\longrightarrow$.

$$CH_3(CH_2)_5\overset{\displaystyle OH}{C}H{=}CH \;\rightleftharpoons\; CH_3(CH_2)_5CH_2\overset{\displaystyle O}{C}H$$

1-Octen-1-ol   Octanal
(An enol)   (An aldehyde)

Problem 7.13 Propose a mechanism for this reaction.

$$HC{\equiv}CH \;+\; CH_3\overset{\displaystyle O}{C}OH \xrightarrow[\text{HgSO}_4]{\text{H}_2\text{SO}_4} CH_3\overset{\displaystyle O}{C}OCH{=}CH_2$$

Acetylene   Acetic acid   Vinyl acetate

Vinyl acetate is the monomer for the production of poly(vinyl acetate), the major use of which is as an adhesive in the construction and packaging industry, but it is also used in the paint and coatings industry.

Step 1: Electrophilic addition.

Bridged mercurinium
ion intermediate

Step 2: Make a new bond between a nucleophile and an electrophile.

Step 3: Take a proton away.

Step 4:

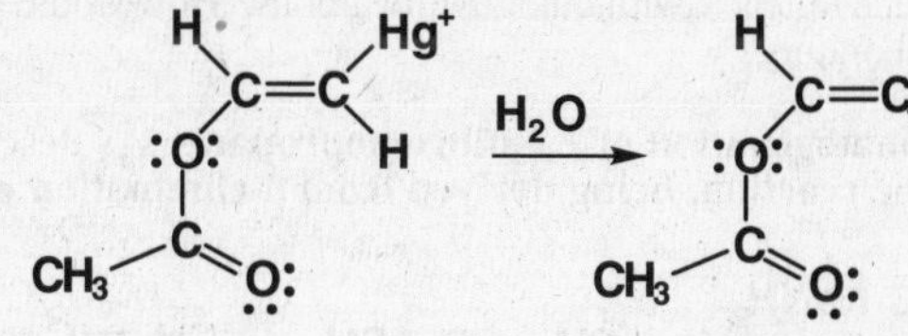

Vinyl acetate

Syntheses

Problem 7.14 Show how to convert 9-octadecynoic acid to the following:

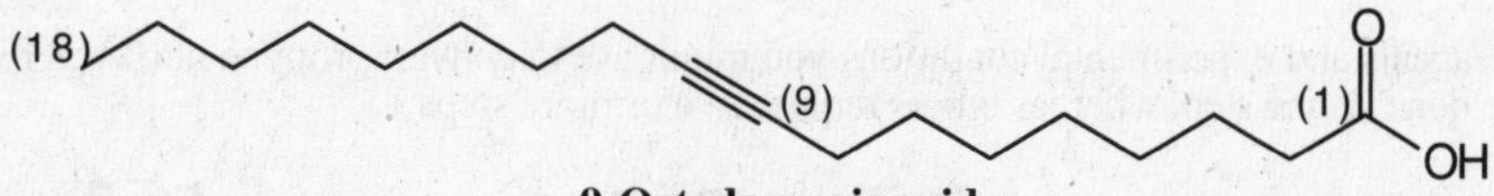

9-Octadecynoic acid

(a) (*E*)-9-Octadecenoic acid (eliadic acid)

Chemical reduction of the alkyne with two moles sodium in liquid ammonia converts the alkyne to an *E* alkene. Note that the carboxyl group is unaffected by these conditions.

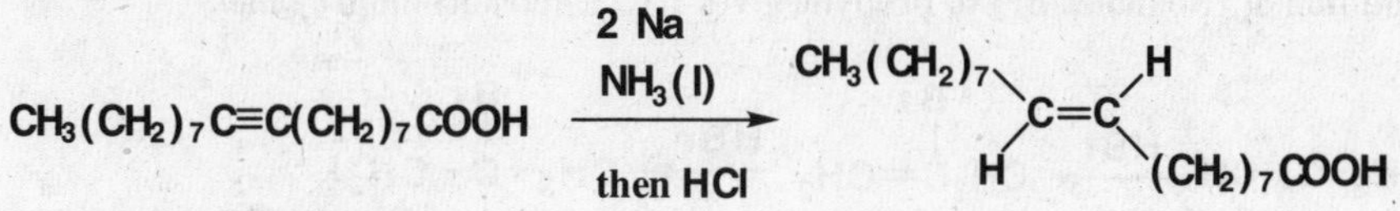

9-Octadecynoic acid

(b) (*Z*)-9-Octadecenoic acid (oleic acid)

Reduction of the alkyne with one mole of hydrogen with Lindlar's catalyst gives a *Z* alkene. The carboxyl group is unaffected by these conditions.

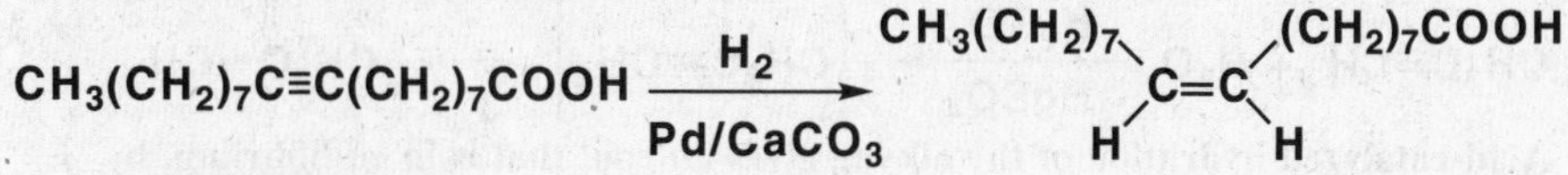

9-Octadecynoic acid

(c) 9,10-Dihydroxyoctadecanoic acid

Either the *E* alkene or the *Z* alkene can be converted to the glycol by oxidation with OsO_4/H_2O_2.

$$CH_3(CH_2)_7CH{=}CH(CH_2)_7COOH \xrightarrow[H_2O_2]{OsO_4} CH_3(CH_2)_7\overset{*}{C}H(OH){-}\overset{*}{C}H(OH)(CH_2)_7COOH$$

9-Octadecenoic acid (*cis* or *trans*)

The stereochemistry of the products depends on whether *cis* or *trans* 9-octadecanoic acid is used as starting material.

(d) Octadecanoic acid (stearic acid)

Reduction of the alkyne with two moles H_2 or reduction of either the *E* alkene or the *Z* alkene with one mole H_2 in the presence of a Ni, Pd, or Pt catalyst gives the desired product.

$$CH_3(CH_2)_7C{\equiv}C(CH_2)_7COOH \xrightarrow[Ni]{2\ H_2} CH_3(CH_2)_7CH_2{-}CH_2(CH_2)_7COOH$$

9-Octadecynoic acid

$$CH_3(CH_2)_7CH{=}CH(CH_2)_7COOH \xrightarrow[Ni]{H_2} CH_3(CH_2)_7CH_2{-}CH_2(CH_2)_7COOH$$

9-Octadecenoic acid (*cis* or *trans*)

Problem 7.15 For small-scale and consumer welding applications, many hardware stores sell cylinders of MAAP gas, which is a mixture of propyne (methylacetylene) and 1,2-propadiene (allene), with other hydrocarbons. How would you prepare the methylacetylene/allene mixture from propene in the laboratory?

This gas mixture could be prepared from a double dehydrohalogenation of 1,2-dibromopropane. As described in section 7.6, 1,2-propadiene (allene) is a side product of the reaction, being derived from β-elimination of the intermediate 2-bromopropene.

$$CH_3-CH=CH_2 \xrightarrow{Br_2} CH_3-\overset{*}{C}HBr-CH_2Br \xrightarrow[\text{2. } H_2O]{\text{1. 3 } NaNH_2} CH_3-C\equiv CH + CH_2=C=CH_2$$

Propene   **1,2-Dibromopropane**   **Propyne**   **1,2-Propadiene (Allene)**

Problem 7.16 Show reagents and experimental conditions you might use to convert propyne into each product. (Some of these syntheses can be done in one step, whereas others require two or more steps.)

(a) $CH_3-\underset{Br}{\overset{Br}{C}}-\underset{Br}{\overset{Br}{CH}}$

$$CH_3-C\equiv CH \xrightarrow{Br_2} CH_3\underset{}{\overset{Br}{C}}=\overset{Br}{CH} \xrightarrow{Br_2} CH_3-\underset{Br}{\overset{Br}{C}}-\underset{Br}{\overset{Br}{CH}}$$

Addition of two moles Br₂ to propyne gives 1,1,2,2-tetrabromopropane.

(b) $CH_3-\underset{Br}{\overset{Br}{C}}-CH_3$

$$CH_3-C\equiv CH \xrightarrow{HBr} CH_3\overset{Br}{C}=CH_2 \xrightarrow{HBr} CH_3-\underset{Br}{\overset{Br}{C}}-CH_3$$

Addition of two moles HBr occurs by electrophilic addition and gives first 2-bromopropene and then 2,2-dibromopropane.

(c) $CH_3-\overset{O}{\overset{\|}{C}}-CH_3$

$$CH_3C\equiv CH + H_2O \xrightarrow[HgSO_4]{H_2SO_4} \left[CH_3\overset{OH}{C}=CH_2 \right] \longrightarrow CH_3\overset{O}{\overset{\|}{C}}-CH_3$$

Acid-catalyzed hydration of the alkyne gives an enol that is in equilibrium, by keto-enol tautomerism, with the isomeric ketone, in this case propanone (acetone).

(d) $CH_3CH_2-\overset{O}{\overset{\|}{C}}-H$

$$CH_3C\equiv CH \xrightarrow[\text{2) } NaOH/H_2O_2]{\text{1) } (sia)_2BH} \left[CH_3CH=\overset{OH}{CH} \right] \longrightarrow CH_3CH_2\overset{O}{\overset{\|}{CH}}$$

Hydroboration with (sia)₂BH or other hindered derivative of borane followed by oxidation with alkaline hydrogen peroxide gives an enol that is in equilibrium, by keto-enol tautomerism, with the isomeric aldehyde, in this case propanal.

Problem 7.17 Show reagents and experimental conditions you might use to convert each starting material into the desired product. (Some of these syntheses can be done in one step; others require two or more steps.)

(a) $\xrightarrow[NH_3(l)]{2\ Na}$

Chemical reduction of the alkyne with sodium in liquid ammonia gives (E)-2-hexene.

(b) $\xrightarrow[\text{2) } CH_3CO_2H]{\text{1) } BH_3}$

Hydroboration of the internal alkyne followed by reaction of the organoborane with acetic acid gives (Z)-2-hexene.

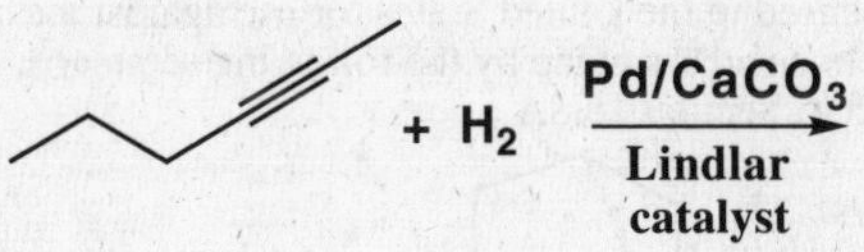

Alternatively, catalytic reduction with hydrogen in the presence of Lindlar catalyst gives the (Z)-alkene.

(c)

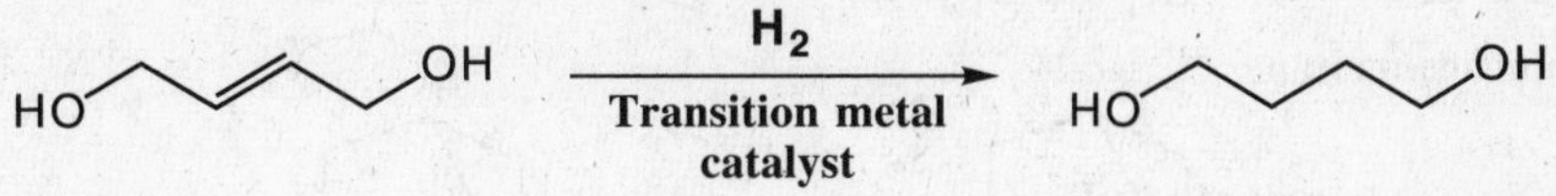

This reduction can be carried out with H_2 in the presence of a transition metal calalyst such as Pd.

(d)

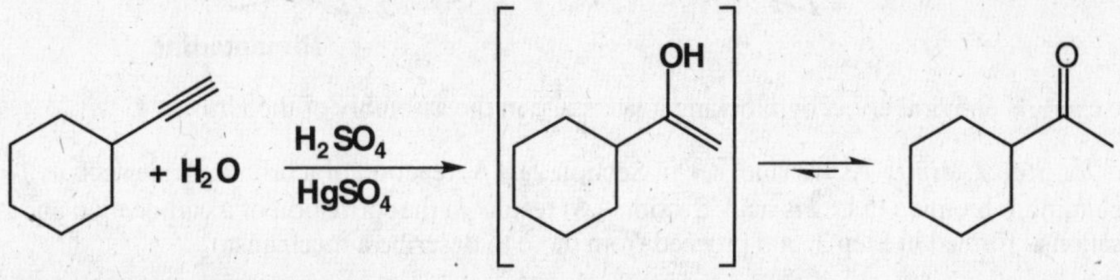

Acid-catalyzed hydration of the carbon-carbon triple bond followed by keto-enol tautomerism of the resulting enol gives the desired ketone.

Problem 7.18 Show how to convert 1-butyne to each of these compounds.

(a) $CH_3CH_2C \equiv CH$ + $NaNH_2$ ⟶ $CH_3CH_2C \equiv C^- Na^+$ + NH_3

The anion can be formed using sodium amide, $NaNH_2$.

(b) $CH_3CH_2C \equiv CH \xrightarrow{NaNH_2} CH_3CH_2C \equiv C:^- Na^+ \xrightarrow{D_2O} CH_3CH_2C \equiv C\,D$

Formation of the terminal acetylide anion followed by reaction with a deuterium donor such as D_2O gives 1-deutero-1-butyne.

(c) $CH_3CH_2C \equiv CH \xrightarrow[\text{2. } CH_3CO_2D]{\text{1. (sia)}_2BH}$

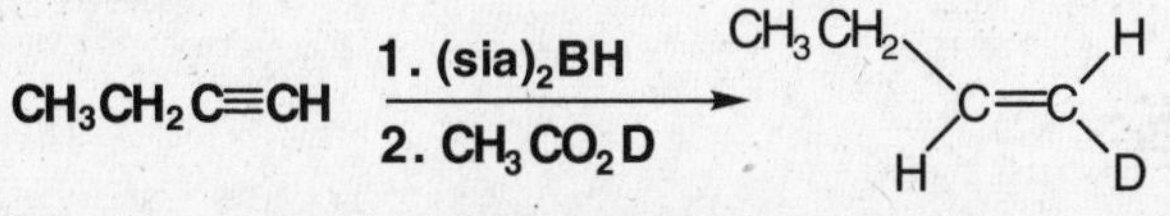

Hydroboration with this disubstituted derivative of borane followed by reaction of the organoborane with deuteroacetic acid gives the desired 1-deutero-1-butene.

(d) $CH_3CH_2C \equiv CH \xrightarrow[\text{2. } CH_3CO_2H]{\text{1. (sia)}_2BD}$

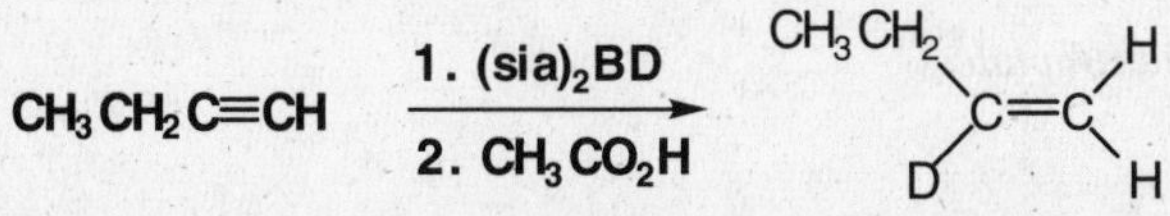

Hydroboration of the terminal alkyne with deuteroborane adds deuterium to the more substituted carbon of the alkyne. Reaction of the deuterated organoborane with acetic acid gives the 2-deutero-1-butene.

<u>Problem 7.19</u> Rimantadine was among the first antiviral drugs to be licensed in the United States for use against the influenza A virus and in treating established illnesses. It is synthesized from adamantane by the following sequence. (We discuss the chemistry of Step 1 in Chapter 8 and the chemistry of Step 5 in Section 16.8A.)

Rimantidine is thought to exert its antiviral effect by blocking a late stage in the assembly of the virus.

(a) Propose a mechanism for Step 2. *Hint:* As we shall see in Section 21.1A, reaction of a bromoalkane such as 1-bromoadamantane with aluminum bromide (a Lewis acid, Section 4.6) results in the formation of a carbocation and $AlBr_4^-$. Assume that adamantyl cation is formed in Step 2, and proceed from there to describe a mechanism.

The bromoethene pi electrons attack the admantyl cation, to create a new cation that captures a bromide from $AlBr_4^-$ to yield dibromoethyl-adamantane.

Step 1: Make a bond between a Lewis acid and Lewis base and simultaneously break a bond to give stable molecules or ions.

Step 2: Electrophilic addition.

Step 3: Make a bond between a nucleophile and an electrophile.

(b) Account for the regioselectivity of carbon-carbon bond formation in Step 2.

The carbon atom without the halogen ends up attached to the adamantane group, because this allows formation of the more stable intermediate with the cation adjacent to the halogen atom. The formation of this cation has a lower activation energy because of resonance stabilization of the cation provided by the halogen.

(c) Describe experimental conditions to bring about Step 3.

This reaction occurs via double dehydrohalogenation using a strong base such as $NaNH_2$.

(d) Describe experimental conditions to bring about Step 4.

This transformation occurs via an acid-catalyzed hydration of the alkyne using H_2O, H_2SO_4, and $HgSO_4$. The initially formed enol equilibrates to the more stable keto form.

<u>Problem 7.20</u> Show reagents and experimental conditions required to bring about the following transformations.

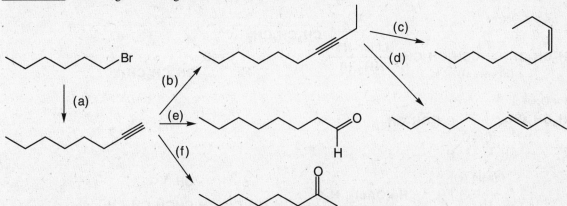

(a) $(CH_3)_3CO^-K^+$
(c) 3 $NaNH_2$ then H_2O
(e) HCl
(g) $HgSO_4$, H_2SO_4, H_2O or BH_3 then H_2O_2, NaOH
(i) H_2, $Pd/CaCO_3$ (Lindlar catalyst)

(b) Cl_2
(d) HCl
(f) $NaNH_2$ then CH_3I
(h) Li or Na in liquid NH_3
(j) Br_2

<u>Problem 7.21</u> Show reagents to bring about each conversion.

(a) $HC\equiv C^-$
(c) H_2, $Pd/CaCO_3$ (Lindlar catalyst)
(e) 1. $(sia)_2BH$; 2. H_2O_2, NaOH

(b) $NaNH_2$ then CH_3CH_2Br
(d) Li or Na in liquid NH_3
(f) $HgSO_4$, H_2SO_4, H_2O

<u>Problem 7.22</u> Propose a synthesis for (*Z*)-9-tricosene (muscalure, the sex pheromone for the common house fly (*Musca domestica*) starting with acetylene and haloalkanes as sources of carbon atoms.

Alkylation of acetylene with the two alkyl halides followed by reduction using H_2 and the Lindlar catalyst gives muscalure.

$$HC\equiv CH \xrightarrow[\text{2. } CH_3(CH_2)_6CH_2Br]{\text{1. } NaNH_2} CH_3(CH_2)_7C\equiv CH \xrightarrow[\text{2. } CH_3(CH_2)_{11}CH_2Br]{\text{1. } NaNH_2}$$

$$CH_3(CH_2)_7C\equiv(CH_2)_{12}CH_3 \xrightarrow[\substack{\text{Pd/CaCO}_3 \\ \text{(Lindlar catalyst)}}]{H_2}$$

$$\underset{\text{Muscalure}}{\underset{H \qquad\qquad H}{\overset{CH_3(CH_2)_7 \qquad (CH_2)_{12}CH_3}{C=C}}}$$

Problem 7.23 Propose a synthesis of each compound starting from acetylene and any necessary organic and inorganic reagents.

(a) 4-Octyne

$$HC\equiv CH \xrightarrow[\text{2. } CH_3CH_2CH_2Br]{\text{1. } NaNH_2} HC\equiv CCH_2CH_2CH_3 \xrightarrow[\text{2. } CH_3CH_2CH_2Br]{\text{1. } NaNH_2}$$

$$CH_3CH_2CH_2C\equiv CCH_2CH_2CH_3$$

(b) 4-Octanone

$$\underset{\text{(From (a))}}{CH_3CH_2CH_2C\equiv CCH_2CH_2CH_3} \xrightarrow[\text{HgSO}_4]{H_2O, H_2SO_4} CH_3CH_2CH_2CH_2\overset{\overset{\displaystyle O}{\|}}{C}CH_2CH_2CH_3$$

(c) cis-4-Octene

$$\underset{\text{(From (a))}}{CH_3CH_2CH_2C\equiv CCH_2CH_2CH_3} \xrightarrow[\substack{\text{Pd/CaCO}_3 \\ \text{(Lindlar catalyst)}}]{H_2} \underset{H \qquad\qquad H}{\overset{CH_3CH_2CH_2 \qquad CH_2CH_2CH_3}{C=C}}$$

(d) trans-4-Octene

$$\underset{\text{(From (a))}}{CH_3CH_2CH_2C\equiv CCH_2CH_2CH_3} \xrightarrow[NH_3 (l)]{\text{Li or Na}} \underset{H \qquad\qquad CH_2CH_2CH_3}{\overset{CH_3CH_2CH_2 \qquad\qquad H}{C=C}}$$

(e) 4-Octanol

$$\underset{\text{(From (c))}}{\underset{H \qquad\qquad H}{\overset{CH_3CH_2CH_2 \qquad CH_2CH_2CH_3}{C=C}}}$$

or

$$\xrightarrow[\text{2. } NaBH_4]{\text{1. } Hg(OAc)_2, H_2O} CH_3CH_2CH_2CH_2\overset{\overset{\displaystyle OH}{|}}{\underset{*}{C}}HCH_2CH_2CH_3$$

(racemic mixture)

$$\underset{\text{(From (d))}}{\underset{H \qquad\qquad CH_2CH_2CH_3}{\overset{CH_3CH_2CH_2 \qquad\qquad H}{C=C}}}$$

(f) meso-4,5-Octanediol

$$\underset{\text{(From (c))}}{\underset{H \qquad\qquad H}{\overset{CH_3CH_2CH_2 \qquad CH_2CH_2CH_3}{C=C}}} \xrightarrow[H_2O_2]{OsO_4} CH_3CH_2CH_2\overset{\overset{\displaystyle HO}{|}}{\underset{\underset{H}{|}}{C}}\text{—}\overset{\overset{\displaystyle OH}{|}}{\underset{\underset{H}{|}}{C}}CH_2CH_2CH_3$$

Note that only the *cis* isomer gives the meso product, the *trans* isomer gives a pair of *R,R* and *S,S* enantiomers.

Problem 7.24 Show how to prepare each compound from 1-heptene:
(a) 1,2-Dichloroheptane

$$CH_3CH_2CH_2CH_2CH_2CH=CH_2 \xrightarrow{Cl_2} CH_3CH_2CH_2CH_2CH_2\overset{\overset{\displaystyle Cl}{\overset{\displaystyle |}{*}}}{C}HCH_2Cl$$

1-Heptene   **(racemic)**

(b) 1-Heptyne

$$CH_3CH_2CH_2CH_2CH_2\overset{\overset{\displaystyle Cl}{\overset{\displaystyle |}{*}}}{C}HCH_2Cl \xrightarrow[\text{2. } H_2O]{\text{1. 3 NaNH}_2} CH_3CH_2CH_2CH_2CH_2C\equiv CH$$

(From (a))

(c) 1-Heptanol

$$CH_3CH_2CH_2CH_2CH_2CH=CH_2 \xrightarrow[\text{2. } H_2O_2, \text{ NaOH}]{\text{1. } BH_3} CH_3CH_2CH_2CH_2CH_2CH_2CH_2OH$$

(d) 2-Octyne

$$CH_3CH_2CH_2CH_2CH_2C\equiv CH \xrightarrow[\text{2. } CH_3I]{\text{1. NaNH}_2} CH_3CH_2CH_2CH_2CH_2C\equiv CCH_3$$

(From (b))

(e) *cis*-2-Octene

$$CH_3CH_2CH_2CH_2CH_2C\equiv CCH_3 \xrightarrow[\substack{\text{Pd/ CaCO}_3 \\ \text{(Lindlar catalyst)}}]{H_2}$$

(From (d))

(f) *trans*-2-Octene

$$CH_3CH_2CH_2CH_2CH_2C\equiv CCH_3 \xrightarrow[\text{NH}_3 \text{ (l)}]{\text{Li or Na}}$$

(From (d))

Problem 7.25 Show how to bring about this conversion.

The alkene is first converted to the dibromide, which is converted to the alkyne. The alkyne is reacted with base and the resulting acetylide anion treated with methyl iodide to give the alkyne with the proper number of carbon atoms. Metal reduction is used to give the desired *trans* final product.

$$\xrightarrow[\text{CCl}_4]{\text{Br}_2} \xrightarrow[\text{2. } CH_3I]{\text{1. 3 NaNH}_2} \xrightarrow[\text{NH}_3 \text{ (l)}]{\text{Li or Na}}$$

Looking Ahead

Problem 7.26 Alkyne anions react with the carbonyl groups of aldehydes and ketones to form alkynyl alcohols, as illustrated by the following sequence.

$$CH_3C\equiv C:^- \; Na^+ \;+\; H\overset{\overset{\displaystyle O}{\|}}{-C}-H \longrightarrow [\; CH_3C\equiv C-CH_2O^- \; Na^+] \xrightarrow[\text{H}_2\text{O}]{\text{HCl}} CH_3C\equiv C-CH_2OH$$

An alkynyl
alcohol

Propose a mechanism for the formation of the bracketed compound, using curved arrows to show the flow of electron pairs in the course of the reaction.

$$CH_3C\equiv C:^- \; Na^+ \;+\; H-\overset{:\overset{\displaystyle ..}{O}:}{C}-H \longrightarrow \left[\; H_3CC\equiv C-\underset{\underset{\displaystyle H}{|}}{\overset{\overset{\displaystyle :\ddot{O}:^- \; Na^+}{|}}{C}}-H \;\right]$$

The acetylide anion acts as a nucleophile and attacks the electrophilic carbonyl carbon atom (*make a new bond between a nucleophile and an electrophile*). The carbonyl pi bond is broken, generating the alkoxide anion intermediate corresponding to the species in the bracket. The carbonyl carbon atom is electrophilic because of the electronegative oxygen atom that provides for a rather polar C=O.

$$H-\underset{\delta+}{\overset{\overset{\displaystyle :\overset{\delta-}{\ddot{O}}:}{\|}}{C}}-H$$

This reaction is particularly noteworthy because a new carbon-carbon bond is formed.

Problem 7.27 Following is the structural formula of the tranquilizer meparfynol (Oblivon).

Oblivon

Propose a synthesis for this compound starting with acetylene and a ketone. (Notice the *-yn* and the *-ol* in the chemical name of this compound, indicating that it contains alkyne and hydroxyl functional groups.)

$$HC\equiv CH \xrightarrow{NaNH_2} HC\equiv C:^- Na^+ \longrightarrow \left[\; \right] \xrightarrow[\text{H}_2\text{O}]{\text{HCl}} \quad$$

Acetylene   **Oblivon**

The synthetic strategy for the construction of meparfynol (Oblivon) is based on the reaction of an acetylide anion with a carbonyl species as described in Problem 7.26. The anion of acetylene reacts with 2-butanone followed by protonation in aqueous HCl to give the desired alcohol. In later chapters you will see many other examples in which the carbonyl carbon atom acts as a nucleophile that makes a new carbon-carbon bond.

Problem 7.28 The standard procedure for synthesizing a compound is the stepwise progress toward a target molecule by forming individual bonds through single reactions. Typically, the product of each reaction is isolated and purified before the next reaction in the sequence is carried out. One of the ways Nature avoids this tedious practice of isolation and purification is by the use of a domino sequence in which each new product is built on a preexisting one in a stepwise fashion. A great example of a laboratory domino reaction is William S. Johnson's elegant synthesis of the female hormone, progesterone. Johnson first constructed the polyunsaturated monocyclic 3° alcohol (A) and then, in an acid-induced domino reaction, formed by compound B, which he then converted to progesterone.

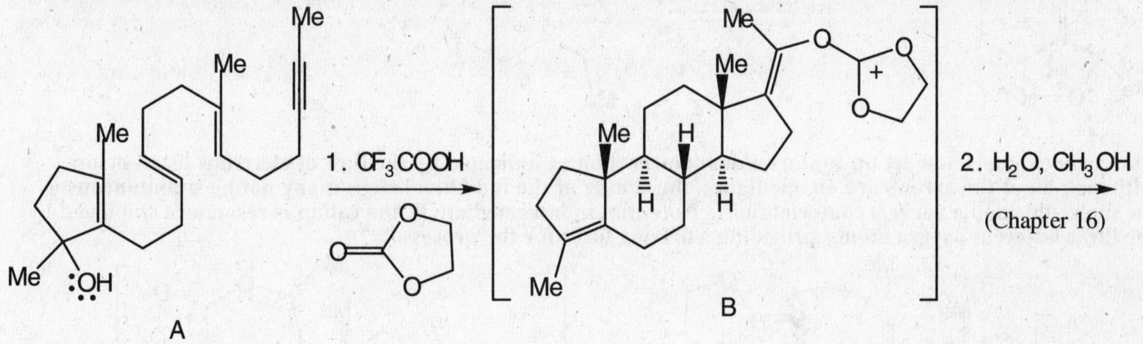

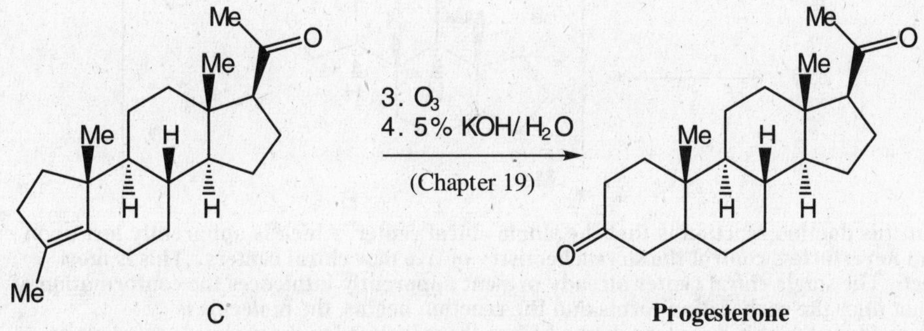

A remarkable feature of this synthesis is that compound A, which has only one chiral center, gives compound B, which has five chiral centers, each with the same configuration as those in progesterone. We will return to the chemistry of Step 2 in Section 16.7, and to the chemistry of Steps 3 and 4 in Chapter 19. In this problem, we focus on Step 1.
(a) Assume that the domino reaction in Step 1 is initiated by protonation of the 3° alcohol in compound A followed by loss of H$_2$O to give a 3° carbocation. Show how the series of reactions initiated by the formation of this cation gives compound B.

As stated in the problem, the reaction begins by protonation of the hydroxyl group (*add a proton*).

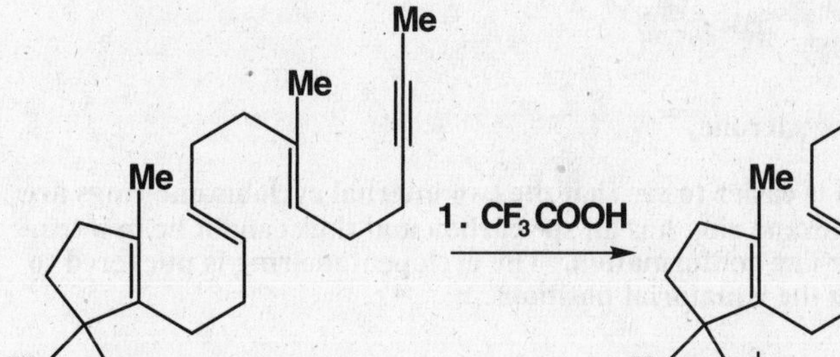

The protonated hydroxyl group makes an excellent leaving group, so H$_2$O departs, leaving behind a carbocation (*break a bond to give stable molecules or ions*). In Chapter 10, you will see many examples of reactions in which hydroxyl groups are protonated and then depart as water. This is a characteristic reaction of alcohols in acid.

The carbocation is now set up to start a domino reaction as indicated by the flow of electrons listed below. Although all of the arrows are on one figure, the timing of the individual steps many not be simultaneous as the molecule adopts various conformations. Note that in intermediate B, the cation is resonance stabilized by the three adjacent oxygen atoms, providing a driving force for the process.

The intriguing thing about this domino reaction is that the single chiral center, which is apparently lost upon carbocation formation, can nevertheless control the stereochemistry of five new chiral centers. This is most likely a conformation effect. The single chiral center already present apparently influences the conformation of the starting material, so that once the carbocation forms and the reaction begins, the molecule is conformationally set up to produce the chiral centers shown before there is time for the molecule to adopt an alternative conformation.

(b) If you have access to a large enough set of molecular models or to a computer modeling program, build a model of progesterone and describe the conformation of each ring. There are two methyl groups and three hydrogen atoms at the set of ring junctions in progesterone. Which of these five groups occupies an equatorial position? Which occupies an axial position?

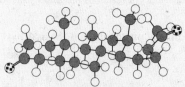

Progesterone

When only ring skeleton of progesterone is drawn, it is easier to see that the two internal cyclohexane rings are each in the chair conformation. The terminal cyclohexane ring has an sp² carbon and thus cannot be in a true chair conformation, but it does adopt a similar chair-like conformation. The cyclopentane ring is puckered to relieve ring strain. Notice that the rings are fused in the equatorial positions.

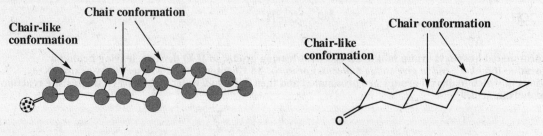

Because the ring junctions occupy all of the equatorial positions, the methyl groups and hydrogen atoms occupy only axial positions at the junctions.

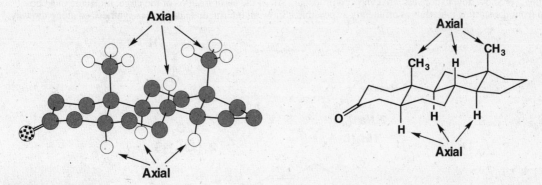

Multistep Synthesis

<u>Problem 7.29</u> Show how to convert acetylene and 1-bromoethane into 1-butene. All of the carbon atoms of the target molecule must be derived from the given starting materials. Show all intermediate molecules synthesized along the way.

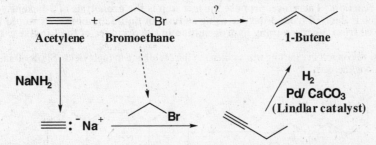

For this synthesis, recognize that a carbon-carbon bond must be formed between both two-carbon starting materials to make the 1-butene product. The best way to accomplish this is to deprotonate acetylene with sodium amide, followed by reaction with bromoethane. Reduction with hydrogen and the Lindlar catalyst completes the synthesis of 1-butene.

<u>Problem 7.30</u> Show how to convert ethylene into 1-butene. All of the carbon atoms of the target molecule must be derived from ethylene. Show all intermediate molecules synthesized along the way.

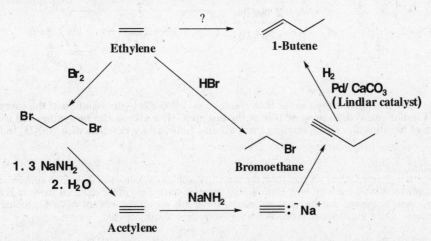

Recognize that this is just an extension of the previous synthesis (Problem 7.29). The last three steps are the same, involving deprotonation of acetylene followed by reaction with bromoethane then hydrogenation using the Lindlar catalyst. The problem then, is to use ethylene to make acetylene and bromoethane. Synthesis of acetylene is accomplished using the two reaction sequence of bromination followed by reaction with NaNH$_2$. Notice that because acetylene is a terminal alkyne, using 3 equivalents of NaNH$_2$ followed by reprotonation in H$_2$O is required. Bromoethane is generated by reaction of ethylene with HBr.

Problem 7.31 Show how to convert 3-hexyne into propanal. All of the carbon atoms of the target molecule must be derived from the starting materials as efficiently as possible. Show all intermediate molecules synthesized along the way.

3-Hexyne

Propanal

2 Na
NH₃(l)

1. O₃
2. (CH₃)₂S

Recognize that the propanal product has three carbon atoms, exactly half of the number of carbons in the 3-hexyne starting material, so a carbon-carbon bond must be broken. Recognize further that propanal is an aldehyde, exactly the type of product formed using the only reaction covered so far that breaks carbon-carbon bonds; the ozonolysis reaction. Therefore, propose the last step is the ozonolysis of 2-hexene. Shown is the use of *trans* 2-hexene, which is derived from 3-hexyne using sodium in liquid ammonia. It would have been fine to use *cis* 2-hexene, derived from 3-hexyne using hydrogenation in the presence of the Lindlar catalyst.

Problem 7.32 Show how to convert the starting *trans* alkene to the *cis* alkene in high yield. Show all intermediate molecules synthesized along the way.

Br₂

2 NaNH₂
NH₃(l)

H₂
Pd/ CaCO₃
(Lindlar catalyst)

(meso)

Recognize that the only reaction covered so far that creates a *cis* alkene is hydrogenation of the corresponding alkyne using the Lindlar catalyst, so propose this as the last step. The alkyne can be synthesized using the two reaction sequence of bromination of the starting *trans* alkene, followed by reaction with NaNH₂ in liquid ammonia.

Reactions in Context

Problem 7.33 Functional groups such as alkynes react the same in complex molecules as they do in simpler structures. The following examples of alkyne reactions were taken from syntheses carried out in the research group of E.J. Corey at Harvard University. You can assume that the reactions listed involve only the alkyne, not any of the functional groups present in the molecules. Draw the expected products for the following transformations.

(a)

(racemic)

H≡Li

DMSO (solvent)

(racemic)

(b)

$(sia)_2BH$, THF, 0°C
K_2CO_3, H_2O_2, H_2O

(c)

reactive alkyne

H_2
Lindlar Catalyst

cis alkene in product

Manta ray *Manta birostris*
Bora Bora, French Polynesia

CHAPTER 8
Solutions to the Problems

<u>Problem 8.1</u> Write the IUPAC name, and where possible, the common name of each compound. Show stereochemistry where relevant.

(a)

1-Chloro-2-methylpropane
(Isobutyl chloride)

(b)

(Z)-2-Chloro-2-butene

(c)

Chlorocyclohexane

(d)

2-Chloro-1,3-butadiene

<u>Problem 8.2</u> Name and draw structural formulas for all monochlorination products formed by treatment of 2-methylpropane with Cl_2. Predict the major product based on the regioselectivity of the reaction of Cl_2 with alkanes.

$$CH_3CHCH_3 \;+\; Cl_2 \;\xrightarrow[\text{or light}]{\text{heat}}\; \text{monochloroalkanes} \;+\; HCl$$

with CH_3 above the central carbon.

$$\underset{Cl}{\overset{CH_3}{CH_3CCH_3}} \qquad\qquad \overset{CH_3}{CH_3CHCH_2Cl}$$

2-Chloro-2-methylpropane   **1-Chloro-2-methylpropane**

There is 1 tertiary hydrogen atoms and 9 primary hydrogen atoms on the molecule. The ratio of reactivity for 3°:1° chlorination is 5:1. Therefore, the predominant product will be the 1-chloro-2-methylpropane, formed in approximately:

$$\frac{9 \times 1}{(9 \times 1) + (1 \times 5)} \times 100 \;=\; \boxed{64\%}$$

<u>Problem 8.3</u> Using tables of bond dissociation enthalpies in Appendix 3, calculate $\Delta H°$ for bromination of propane to give 1-bromopropane and hydrogen bromide.

$$CH_3CH_2CH_3 \;+\; Br_2 \;\longrightarrow\; CH_3CH_2CH_2Br \;+\; HBr$$

$\Delta H°$ **equals the difference between the bond dissociation enthalpies of bonds made vs. bonds broken in the reaction. One C-H bond [+ 422(101) kJ(kcal)/mol)] and one Br-Br bond [+ 192(46) kJ(kcal)/mol)] were broken while one C-Br [-301(-72) kJ(kcal)/mol)] bond and one H-Br bond [-368(-88) kJ(kcal)/mol)] were made in this reaction. Thus, for the complete reaction** $\Delta H°$ **= 422(101) + 192(46) - 301(-72) - 368(-88) = <u>- 55(-13) kJ(kcal)/mol</u>.**

<u>Problem 8.4</u> Write a pair of chain propagation steps for the radical bromination of propane to give 1-bromopropane, and calculate $\Delta H°$ for each propagation step and for the overall reaction.

Following is a pair of chain propagation steps for this reaction. Of these steps, the first involving hydrogen abstraction, has the higher activation energy.

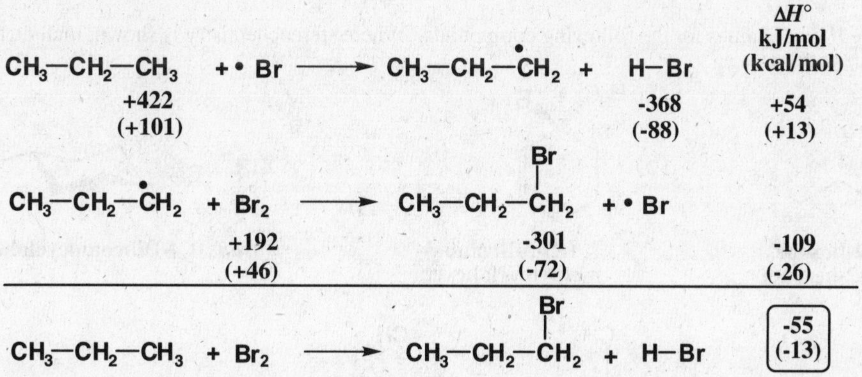

$$\Delta H° \quad \text{kJ/mol} \quad \text{(kcal/mol)}$$

$$CH_3-CH_2-CH_3 \;+\; \bullet Br \longrightarrow CH_3-CH_2-\overset{\bullet}{C}H_2 \;+\; H-Br$$

$$\begin{array}{ccc} +422 & -368 & +54 \\ (+101) & (-88) & (+13) \end{array}$$

$$CH_3-CH_2-\overset{\bullet}{C}H_2 \;+\; Br_2 \longrightarrow CH_3-CH_2-\overset{Br}{\underset{|}{C}H_2} \;+\; \bullet Br$$

$$\begin{array}{ccc} +192 & -301 & -109 \\ (+46) & (-72) & (-26) \end{array}$$

$$CH_3-CH_2-CH_3 \;+\; Br_2 \longrightarrow CH_3-CH_2-\overset{Br}{\underset{|}{C}H_2} \;+\; H-Br \qquad \boxed{\begin{array}{c} -55 \\ (-13) \end{array}}$$

<u>Problem 8.5</u> Given the solution to Example 8.5, predict the structure of the product(s) formed when 3-hexene is treated with NBS?

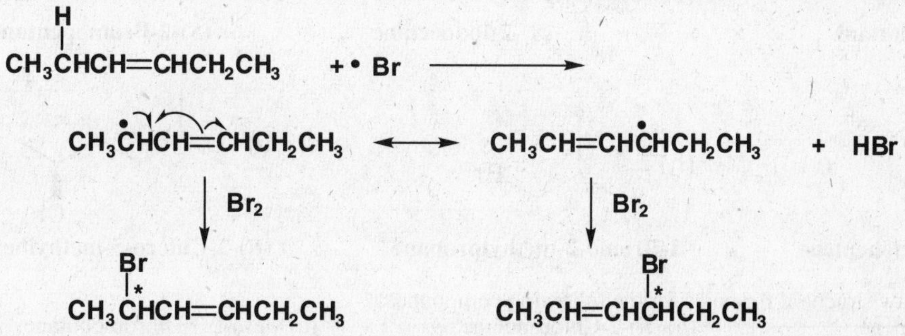

Both of the above products have a chiral center, so they will each be produced as a racemic mixture of enantiomers.

<u>Problem 8.6</u> Show the products of the following reaction and indicate the major one.

The reaction involves the initial formation of a resonance-stabilized allylic radical intermediate which reacts with O₂. The compound with the more highly substituted double bond will be the major product.

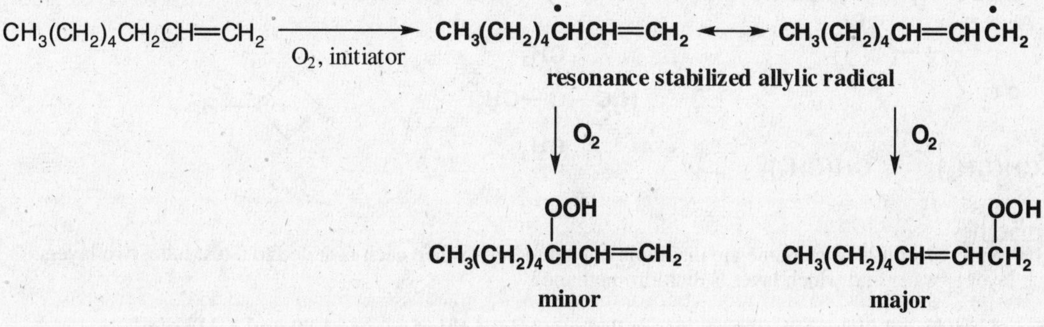

<u>Problem 8.7</u> Predict the major product of the following reaction:

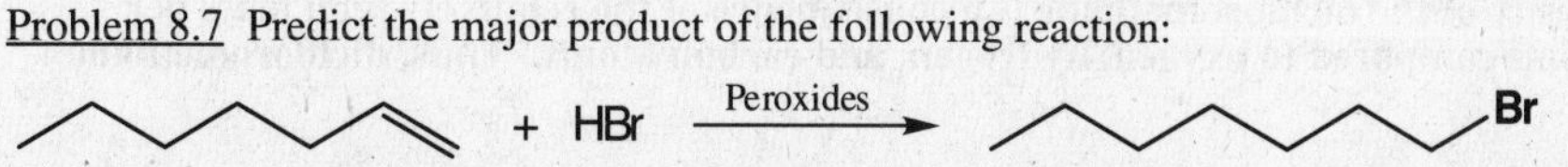

In the presence of peroxides, the reaction proceeds mostly through a radical mechanism, so the Br atom will end up predominantly on the less substituted carbon atom.

Nomenclature

Problem 8.8 Give IUPAC names for the following compounds. Where stereochemistry is shown, include a designation of configuration in your answer.

(a)

(Z)-2-Bromo-2-hexene
(*trans*-2-bromo-2-hexene)

(b)

(R)-3-Bromo-3-methylcyclohexene

(c)

***trans*-1,4-Dibromocyclohexane**

(d)

1,4-Dichlorobutane

(e)

Fischer projection

(S)-2-Iodooctane

(f)

(S)-2-Bromopentane

(g)

(R)-3-Fluorocycloheptene

(h)

1-Bromo-2-methylpropane

(i)

(R)-2-Chloro-5-methylhexane

Problem 8.9 Draw structural formulas for the following compounds.
(a) 3-Iodo-1-propene   (b) (R)-2-Chlorobutane   (c) *meso*-2,3-Dibromobutane

CH₂═CHCH₂I

(d) *trans*-1-Bromo-3-isopropylcyclohexane   (e) 1-Iodo-2,2-dimethylpropane   (f) Bromocyclobutane

or

Physical Properties

Problem 8.10 Water and dichloromethane are insoluble in each other. When each is added to a test tube, two layers form. Which layer is water and which layer is dichloromethane?

The densities of water and dichloromethane (also called methylene chloride) are 1.00 and 1.327 g/mL, respectively. The increased density of the dichloromethane is a consequence of the relatively high mass per volume ratio of the chlorine atoms compared to oxygen, hydrogen, and carbon atoms. Thus, dichloromethane will be the bottom layer.

Problem 8.11 The boiling point of methylcyclohexane (C_7H_{14}, MW 98.2) is 101°C. The boiling point of perfluoromethylcyclohexane (C_7F_{14}, MW 350) is 76°C. Account for the fact that although the molecular weight of perfluoromethylcyclohexane is over three times that of methylcyclohexane, its boiling point is lower than that of methylcyclohexane.

This difference is due to the low polarizability of fluorine that is attributed to its small size and the tightness with which its electrons are held. Low polarizability will limit the attractive interactions (i.e. dispersion forces) between molecules, lowering the boiling point.

<u>Problem 8.12</u> Account for the fact that, among the chlorinated derivatives of methane, chloromethane has the largest dipole moment and tetrachloromethane has the smallest dipole moment.

Name	Molecular Formula	Dipole Moment (debyes: D)
Chloromethane	CH_3Cl	1.87
Dichloromethane	CH_2Cl_2	1.60
Trichloromethane	$CHCl_3$	1.01
Tetrachloromethane	CCl_4	0

Each C-Cl bond is polar covalent with carbon bearing a partial positive charge and chlorine bearing a partial negative charge. Recall that molecular dipole moments are the vector sum of all the individual bond dipole moments. As shown on the following structures, adjacent C-Cl bond dipoles actually cancel each other to some extent in dichloromethane and trichloromethane, and completely in tetrachloromethane.

Vector sum   **1.87 D**   **1.60 D**   **1.01 D**

Chloromethane   Dichloromethane   Trichloromethane   Tetrachloromethane

Halogenation of Alkanes

<u>Problem 8.13</u> Name and draw structural formulas for all possible monohalogenation products that might be formed in the following reactions.

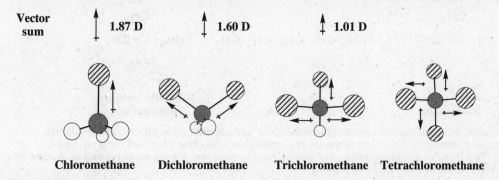

(a)   + Cl_2 $\xrightarrow{\text{light}}$

Chlorocyclopentane

(b)   + Cl_2 $\xrightarrow{\text{light}}$

CH₂Cl
|*
CH₃CHCH₂CH₂CH₃
1-Chloro-2-methyl-pentane

CH₃
|
CH₃CCH₂CH₂CH₃
|
Cl
2-Chloro-2-methyl-pentane

CH₃ *
CH₃CHCHCH₂CH₃
|
Cl
3-Chloro-2-methyl-pentane

CH₃   *
CH₃CHCH₂CHCH₃
|
Cl
2-Chloro-4-methyl-pentane

CH₃
|
CH₃CHCH₂CH₂CH₂Cl
1-Chloro-4-methyl-pentane

(c)   + Br_2 $\xrightarrow{\text{light}}$

CH₂Br
|*
CH₃CHCHCH₃
|
CH₃
1-Bromo-2,3-dimethyl-butane

CH₃
|
CH₃C—CHCH₃
| |
Br CH₃
2-Bromo-2,3-dimethyl-butane

In the above reactions, the products with chiral centers (*) are created as racemic mixtures.

Problem 8.14 Which compounds can be prepared in high yield by halogenation of an alkane?
(a) 2-Chloropentane   (b) Chlorocyclopentane
(c) 2-Bromo-2-methylheptane   (d) (R)-2-Bromo-3-methylbutane
(e) 2-Bromo-2,4,4-trimethylpentane   (f) Iodoethane

To be made in high yield, the compound must be the only monohalogentation product possible because of symmetry in the starting alkane, or, alternatively, the product must have the halogen on the single most-substituted carbon atom. Thus, (b), (c), and (e) can be prepared in high yield. (f) Cannot be prepared because it would be an endothermic reaction. The other products would be produced along with unsatisfactory amounts of other monohalogenation products. In particular, 3-chloropentane and 2-bromo-2-methylbutane would be major contaminants in preparations of (a) 2-chloropentane and (d) 2-bromo-3-methylbutane, respectively. Also, for (d), the product would contain just as much (S)-2-Bromo-3-methylbutane as the desired R isomer.

Problem 8.15 There are three constitutional isomers of molecular formula C_5H_{12}. When treated with chlorine at 300°C, isomer A gives a mixture of four monochlorination products. Under the same conditions, isomer B gives a mixture of three monochlorination products, and isomer C gives only one monochlorination product. From this information, assign structural formulas to isomers A, B, and C.

Structural formulas for the three alkanes are:

$$CH_3-CH-CH_2-CH_3 \quad CH_3-CH_2-CH_2-CH_2-CH_3 \quad CH_3-C-CH_3$$

A	**B**	**C**
2-Methylbutane	**Pentane**	**2,2-Dimethylpropane**
(Isopentane)		**(Neopentane)**

To arrive at the correct assignments of structural formulas, first write formulas for all monochloroalkanes possible from each structural formula. Then compare these numbers with those observed for A, B, and C. Because isomer B gives three monochlorination products, it must be pentane. By the same reasoning, A must be 2-methylbutane, and C must be 2,2-dimethylpropane.

Problem 8.16 Following is a balanced equation for bromination of toluene.

$$C_6H_5CH_3 \quad + \quad Br_2 \longrightarrow C_6H_5CH_2Br \quad + \quad HBr$$

Toluene   **Benzyl bromide**

(a) Using the values for bond dissociation enthalpies given in Appendix 3, calculate $\Delta H°$ for this reaction.

Formation of benzyl bromide by radical bromination of toluene is *exothermic* by 51 (12) kJ(kcal)/mol.

$$C_6H_5-CH_2 \quad + \quad Br_2 \longrightarrow C_6H_5-CH_2 \quad + \quad H-Br$$

				$\Delta H°$ kJ/mol (kcal/mol)
+376	+192		-263	-368	-63
(+90)	(+46)		(-63)	(-88)	(-15)

(b) Propose a pair of chain propagation steps and show that they add up to the observed reaction.

Following is a pair of chain propagation steps for this reaction.

$$C_6H_5-\overset{\overset{\displaystyle H}{|}}{CH_2} \;+\; \cdot Br \;\longrightarrow\; C_6H_5-\overset{\displaystyle \cdot}{CH_2} \;+\; H-Br$$

			ΔH° kJ/mol (kcal/mol)
+376 (+90)		-368 (-88)	8 (2)

$$C_6H_5-\overset{\displaystyle \cdot}{CH_2} \;+\; Br_2 \;\longrightarrow\; C_6H_5-\overset{\overset{\displaystyle Br}{|}}{CH_2} \;+\; \cdot Br$$

+192 (+46)	-263 (-63)	-71 (-17)

sum of ΔH° for chain propagation steps: | -63 (-15) |

Following is an alternative pair of chain propagation steps. Because of the considerably higher activation energy of the first of these steps, the rate of chain propagation by this mechanism is so low that it is not competitive with the chain mechanism first proposed.

$$C_6H_5-\overset{\overset{\displaystyle H}{|}}{CH_2} \;+\; \cdot Br \;\longrightarrow\; C_6H_5-\overset{\overset{\displaystyle Br}{|}}{CH_2} \;+\; \cdot H$$

		ΔH° kJ/mol (kcal/mol)
+376 (+90)	-263 (-63)	+113 (+27)

$$\cdot H \;+\; Br_2 \;\longrightarrow\; H-Br \;+\; \cdot Br$$

+192 (+46)	-368 (-88)	-176 (-42)

sum of ΔH° for chain propagation steps: | -63 (-15) |

(c) Calculate ΔH° for each chain propagation step.

See answer to part (b)

(d) Which propagation step is rate-determining?

The rate-determining step corresponds to the step with the higher potential energy barrier. Thus, the first step involving hydrogen abstraction is rate-determining because it has the higher activation energy.

<u>Problem 8.17</u> Write a balanced equation and calculate ΔH° for reaction of CH_4 and I_2 to give CH_3I and HI. Explain why this reaction cannot be used as a method of preparation of iodomethane.

The reaction is not a useful preparation method, because formation of iodomethane (methyl iodide) by radical iodination is *endothermic* by 55(13) kJ(kcal)/mol. It will not occur spontaneously.

$$CH_4 \;+\; I_2 \;\longrightarrow\; CH_3I \;+\; HI$$

				ΔH° kJ/mol (kcal/mol)
+439 (+105)	+151 (+36)	-242 (-58)	-297 (-71)	+51 (+12)

<u>Problem 8.18</u> Following are balanced equations for fluorination of propane to produce a mixture of 1-fluoropropane and 2-fluoropropane.

$$CH_3CH_2CH_3 \;+\; F_2 \;\longrightarrow\; CH_3CH_2CH_2F \;+\; HF$$
Propane   **1-Fluoropropane**

$$CH_3CH_2CH_3 \;+\; F_2 \;\longrightarrow\; CH_3\overset{\overset{\displaystyle F}{|}}{CH}CH_3 \;+\; HF$$
Propane   **2-Fluoropropane**

Assume that each product is formed by a radical chain mechanism.
(a) Calculate $\Delta H°$ for each reaction.

Formation of 1-fluoropropane (propyl fluoride) and 2-fluoropropane (isopropyl fluoride) are both exothermic.

$$CH_3-CH_2-CH_3 \;+\; F_2 \;\longrightarrow\; CH_3-CH_2-CH_2-F \;+\; HF$$

				$\Delta H°$ kJ/mol (kcal/mol)
+422 (+101)	+159 (+38)	-472 (-113)	-568 (-136)	-459 (-110)

$$CH_3-CH_2-CH_3 \;+\; F_2 \;\longrightarrow\; CH_3-\overset{\displaystyle F}{\underset{|}{CH}}-CH_3 \;+\; HF$$

+414 (+99)	+159 (+38)	-464 (-111)	-568 (-136)	-459 (-110)

(b) Propose a pair of chain propagation steps for each reaction, and calculate $\Delta H°$ for each step.

$$CH_3-CH_2-CH_3 \;+\; \bullet F \;\longrightarrow\; CH_3-CH_2-\overset{\bullet}{C}H_2 \;+\; HF$$

			$\Delta H°$ kJ/mol (kcal/mol)
+422 (+101)		-568 (-136)	-146 (-35)

$$CH_3-CH_2-\overset{\bullet}{C}H_2 \;+\; F_2 \;\longrightarrow\; CH_3-CH_2-CH_2-F \;+\; \bullet F$$

	+159 (+38)	-472 (-113)	-313 (-75)

sum of $\Delta H°$ for chain propagation steps: **-459 (-110)**

$$CH_3-CH_2-CH_3 \;+\; \bullet F \;\longrightarrow\; CH_3-\overset{\bullet}{C}H-CH_3 \;+\; HF$$

			$\Delta H°$ kJ/mol (kcal/mol)
+414 (+99)		-568 (-136)	-154 (-37)

$$CH_3-\overset{\bullet}{C}H-CH_3 \;+\; F_2 \;\longrightarrow\; CH_3-\overset{\displaystyle F}{\underset{|}{CH}}-CH_3 \;+\; \bullet F$$

	+159 (+38)	-464 (-111)	-305 (-73)

sum of $\Delta H°$ for chain propagation steps: **-459 (-110)**

(c) Reasoning from Hammond's postulate, predict the regioselectivity of radical fluorination relative to that of radical chlorination and bromination.

Because the hydrogen abstraction step in each radical fluorination sequence is highly exothermic, the transition state is reached very early in hydrogen abstraction and the intermediate in this step has very little radical character. Therefore, the relative stabilities of primary versus secondary radicals is of little importance in determination of product. Accordingly predict very low regioselectivity for fluorination of hydrocarbons.

Problem 8.19 As you demonstrated in Problem 8.18, fluorination of alkanes is highly exothermic. As per Hammond's postulate, assume that the transition state for radical fluorination is almost identical to the starting material. With this assumption, estimate the fraction of each monofluoro product formed in the fluorination of 2-methylbutane.

If it is assumed that the transition state for radical fluorination is almost identical to starting materials, then relative radical stabilities are not important. Thus, all types of carbon atoms (3°, 2°, and 1°) will react with approximately equal rate. The fraction of each monofluoro product will then be determined by the number of each kind of hydrogen present as shown.

$$50\% \ (6/12) \quad\quad\quad 16.7\% \ (2/12)$$

$$\text{H}_3\text{C}$$

$$\text{H}_3\text{C}-\text{CH}-\text{CH}_2-\text{CH}_3$$

$$8.3\% \ (1/12) \quad\quad 25\% \ (3/12)$$

Problem 8.20 Cyclobutane reacts with bromine to give bromocyclobutane, but bicyclobutane reacts with bromine to give 1,3-dibromocyclobutane. Account for the differences between the reactions of these two compounds.

Cyclobutane   **Bromocyclobutane**

Bicyclobutane   **1,3-Dibromocyclobutane**

The first reaction follows the normal radical chain mechanism. The propagation steps consist of hydrogen atom abstraction followed by reaction with Br₂ to generate the product.

The lower reaction can be explained by a mechanism in which the highly strained bridging bond reacts with the Br radical during the first propagation step. Note that the extreme ring strain of the bicyclic molecule provides the driving force for the first propagation step.

Problem 8.21 The first chain propagation step of all radical halogenation reactions we considered in Section 8.5B was abstraction of hydrogen by the halogen atom to give an alkyl radical and HX, as for example

$$\text{CH}_3\text{CH}_3 \ + \ \cdot\ddot{\text{B}}\text{r}\!: \ \longrightarrow \ \text{CH}_3\text{CH}_2\cdot \ + \ \text{H}\ddot{\text{B}}\text{r}\!:$$

Suppose, instead, that radical halogenation occurs by an alternative pair of chain propagation steps, beginning with this step:

$$\text{CH}_3\text{CH}_3 \ + \ \cdot\ddot{\text{B}}\text{r}\!: \ \longrightarrow \ \text{CH}_3\text{CH}_2\ddot{\text{B}}\text{r}\!: \ + \ \text{H}\cdot$$

(a) Propose a second chain propagation step. Remember that a characteristic of chain propagation steps is that they add to the observed reaction.

(b) Calculate the heat of reaction, $\Delta H°$, for each propagation step.

$$\begin{array}{c} \Delta H° \\ \text{kJ/mol} \\ \text{(kcal/mol)} \end{array}$$

$$CH_3CH_3 \;+\; \bullet Br \longrightarrow CH_3CH_2Br \;+\; H\bullet$$

+422	-301	+121
(+101)	(-72)	(+29)

$$Br—Br \;+\; \bullet H \longrightarrow HBr \;+\; Br$$

+192	-368	-176
(+46)	(-88)	(-42)

$$CH_3CH_3 \;+\; Br_2 \longrightarrow CH_3CH_2Br \;+\; HBr$$

	-55
	(-13)

(c) Compare the energetics and relative rates of the set of chain propagation steps in Section 8.5B with the set proposed here.

Here, the first propagation step has a very high activation energy barrier, so it would not compete with the first propagation step proposed in Section 8.5B.

Allylic Halogenation

Problem 8.22 Following is a balanced equation for the allylic bromination of propene.

$$CH_2=CHCH_3 \;+\; Br_2 \longrightarrow CH_2=CHCH_2Br \;+\; HBr$$

(a) Calculate the heat of reaction, $\Delta H°$, for this conversion.

See the answer to part (b)

(b) Propose a pair of chain propagation steps and show that they add up to the observed stoichiometry.

$$\begin{array}{c} \Delta H° \\ \text{kJ/mol} \\ \text{(kcal/mol)} \end{array}$$

$$CH_2=CH—CH_3 \;+\; \bullet Br \longrightarrow CH_2=CH—\overset{\bullet}{C}H_2 \;+\; H—Br$$

+372	-368	4
(+89)	(-88)	(1)

$$CH_2=CH—\overset{\bullet}{C}H_2 \;+\; Br_2 \longrightarrow CH_2=CH—CH_2Br \;+\; \bullet Br$$

+192	-247	-55
(+46)	(-59)	(-13)

sum of $\Delta H°$ for chain propagation steps:

$$\boxed{\begin{array}{c} -51 \\ (-12) \end{array}}$$

(c) Calculate the $\Delta H°$ for each chain propagation step and show that they add up to the observed $\Delta H°$ for the overall reaction.

See the answer to part (b).

Problem 8.23 Using the table of bond dissociation enthalpies (Appendix 3), estimate the BDE of each indicated bond in cyclohexene.

372(89)kJ(kcal)/mol (b)

414(99)kJ(kcal)/mol (a)

(c) 464(111)kJ(kcal)/mol

Estimate (a) the bond dissociation enthalpy (BDE) of this secondary site to be 414(99) kJ (kcal)/mol, (b) BDE of this allylic site to be 372(89) kJ(kcal)/mol, and (c) BDE of this vinylic site to be 464(111) kJ (kcal)/mol.

<u>Problem 8.24</u> Propose a series of chain initiation, propagation, and termination steps for this reaction and estimate its heat of reaction.

(racemic)

Initiation:

$$Br_2 \xrightarrow{\text{light}} 2\ Br\cdot$$

Propagation:

	ΔH° kJ/mol (kcal/mol)

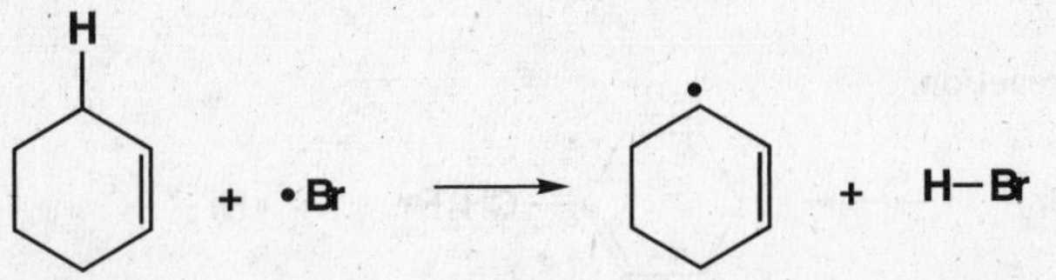

+372 (+89)   -368 (-88)   4 (1)

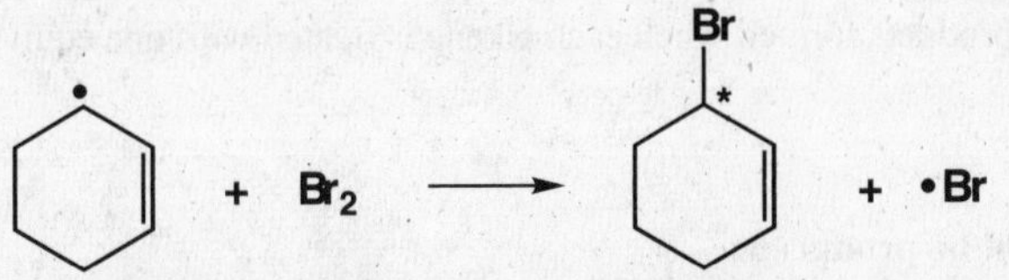

+192 (+46)   -247 (-59)   -55 (-13)

sum of ΔH° for chain propagation steps: -51 (-12)

Termination:

$$\cdot Br + \cdot Br \longrightarrow Br_2$$

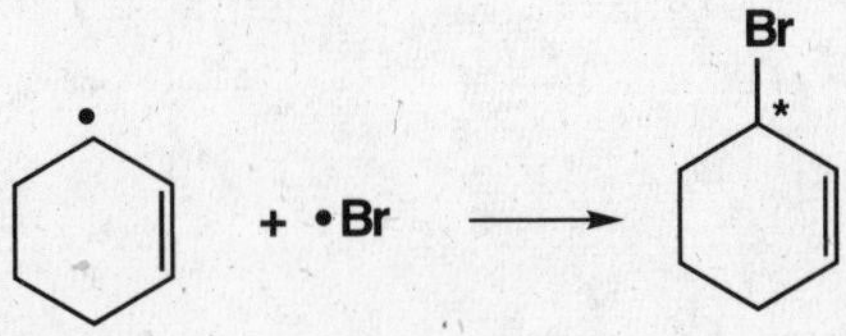

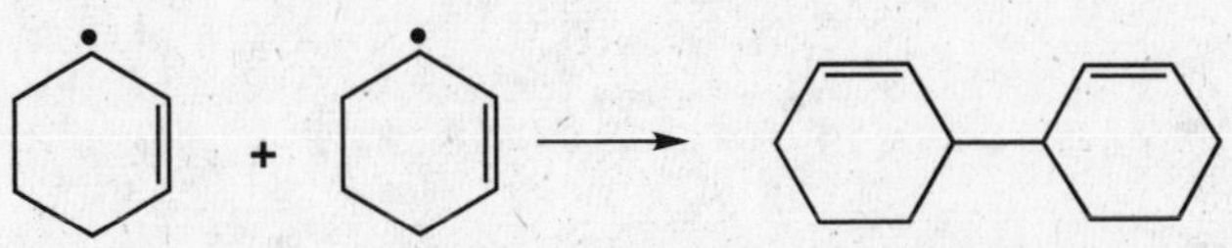

<u>Problem 8.25</u> The major product formed when methylenecyclohexane is treated with NBS in dichloromethane is 1-(bromomethyl)-cyclohexene. Account for the formation of this product.

Recall that NBS can be considered a source of Br radicals and Br₂.

Methylene-cyclohexane   **1-(Bromomethyl)-cyclohexene**

The above reaction can be explained by a first propagation step involving a hydrogen atom abstraction.

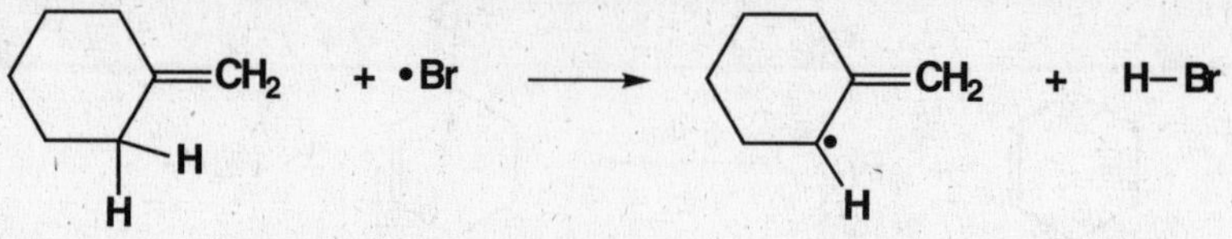

The resulting allylic radical can be represented as the hybrid of two contributing structures. The one on the right is the major contributor because it contains the more stable trisubstituted carbon-carbon double bond.

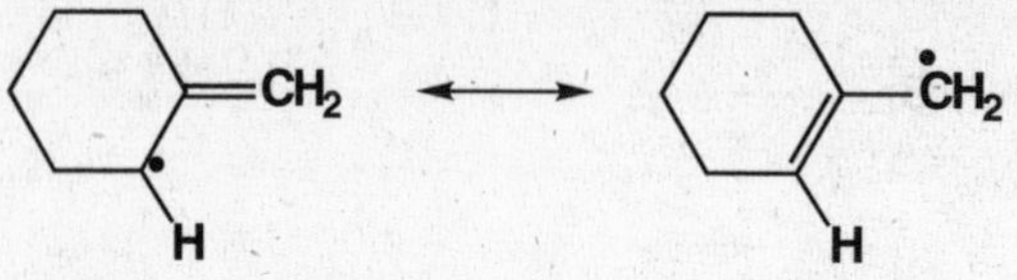

The second propagation step completes the reaction.

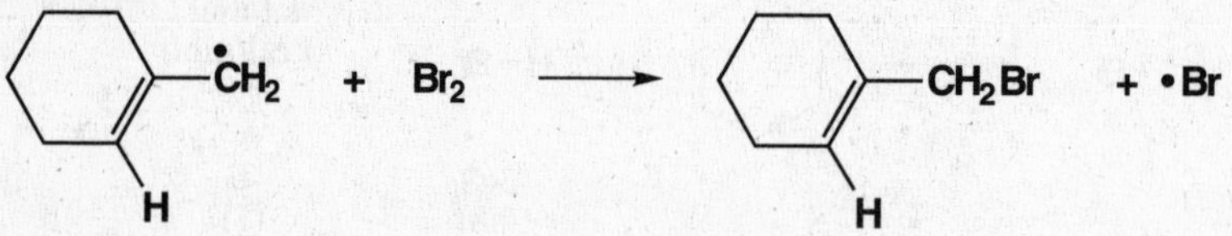

<u>Problem 8.26</u> Draw the structural formula of the products formed when each alkene is treated with one equivalent of NBS in CH_2Cl_2 in the presence of light.

(a) $CH_3CH=CHCH_2CH_3$

There are two possible allylic radicals that could be produced:

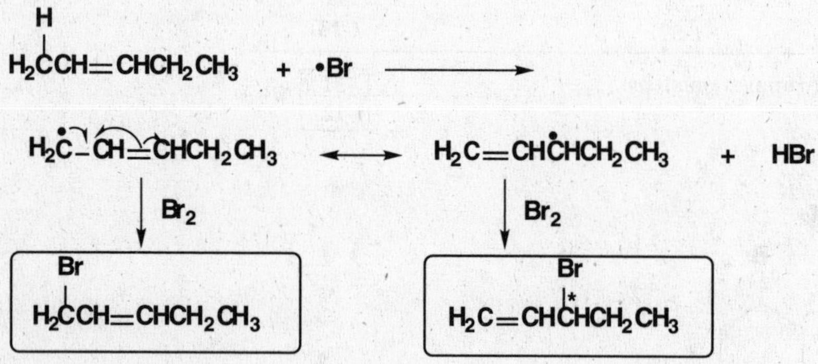

$$H_2\overset{H}{\underset{|}{C}}CH=CHCH_2CH_3 \quad + \quad \cdot Br \longrightarrow$$

$$H_2\overset{\cdot}{C}-CH \overset{\frown}{=} CHCH_2CH_3 \longleftrightarrow H_2C=CH\overset{\cdot}{C}HCH_2CH_3 \quad + \quad HBr$$

$$\downarrow Br_2 \qquad\qquad\qquad\qquad \downarrow Br_2$$

$$\boxed{\overset{Br}{\underset{|}{}} H_2CCH=CHCH_2CH_3} \qquad \boxed{\overset{Br}{\underset{|*}{}} H_2C=CHCHCH_2CH_3}$$

or alternatively:

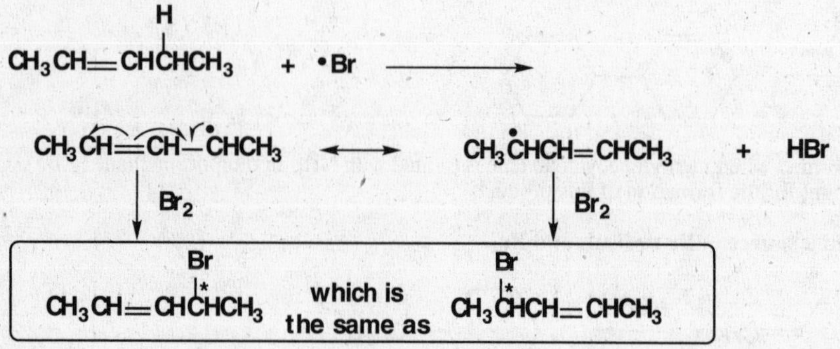

$$CH_3CH=CH\overset{H}{\underset{|}{C}}HCH_3 \quad + \quad \cdot Br \longrightarrow$$

$$CH_3\overset{\frown}{CH=CH}-\overset{\cdot}{C}HCH_3 \longleftrightarrow CH_3\overset{\cdot}{C}HCH=CHCH_3 \quad + \quad HBr$$

$$\downarrow Br_2 \qquad\qquad\qquad\qquad \downarrow Br_2$$

$$\boxed{\overset{Br}{\underset{|*}{}} CH_3CH=CHCHCH_3 \quad \text{which is} \atop \text{the same as} \quad \overset{Br}{\underset{|*}{}} CH_3CHCH=CHCH_3}$$

(b)

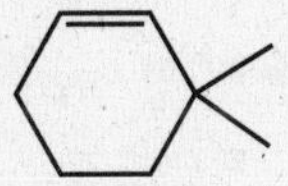

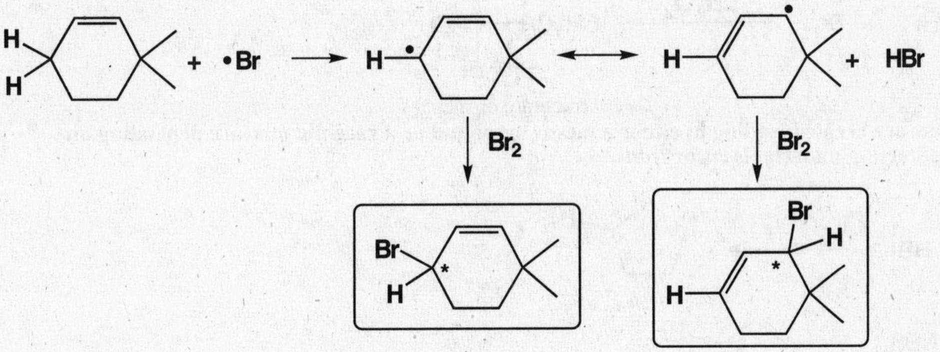

(c)

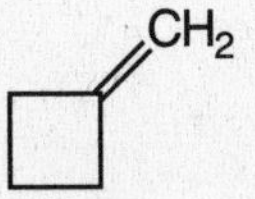

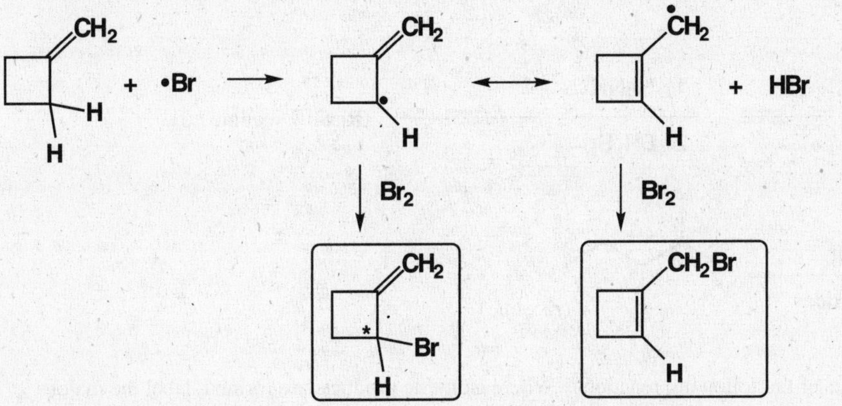

Note that the products shown above having a chiral center (*) will be created as a racemic mixture.

Problem 8.27 Calculate the $\Delta H°$ for the following reaction step. What can you say regarding the possibility of bromination at a vinylic hydrogen?

			$\Delta H°$ kJ/mol (kcal/mol)

$$CH_2{=}CH_2 \ + \ Br\bullet \longrightarrow CH_2{=}CH\bullet \ + \ H{-}Br$$

+464		-368		96
(+111)		(-88)		(23)

This reaction step has such a high $\Delta H°$, 96 kJ/mol, that it will preclude bromination at a vinylic hydrogen.

Synthesis
Problem 8.28 Show reagents and conditions to bring about these conversions, which may require more than one step.

(a)

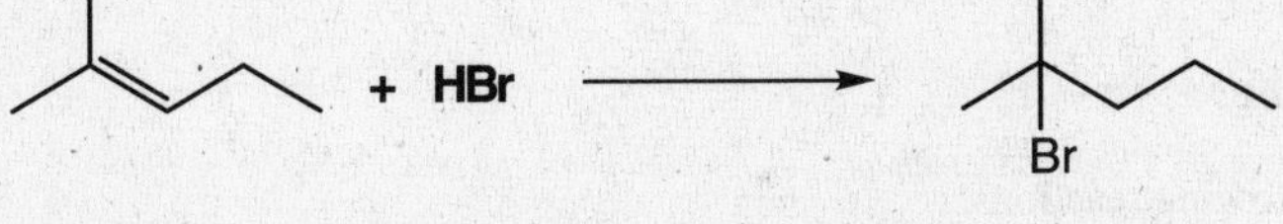

(b) $CH_3CH{=}CHCH_3$ $\xrightarrow[\text{light}]{\begin{array}{c}\text{NBS}\\\text{CH}_2\text{Cl}_2\end{array}}$ $CH_3CH{=}CHCH_2Br$

(c) $CH_3CH{=}CHCH_3$ + Br_2 $\xrightarrow{CH_2Cl_2}$ $CH_3\overset{*}{C}H{-}\overset{*}{C}HCH_3$
                                                                                          | |
                                                                                        Br   Br

(racemic or meso)

Two new chiral centers are created leading to either a meso compound or a racemic mixture depending on whether the 2-butene starting material is *cis* or *trans*.

(d)   ⬠ + **HBr** ⟶ ⬠—Br

(e)   ⬠ $\xrightarrow[\text{light}]{\begin{array}{c}\text{NBS}\\\text{CH}_2\text{Cl}_2\end{array}}$ ⬠—Br
                                                                                  *

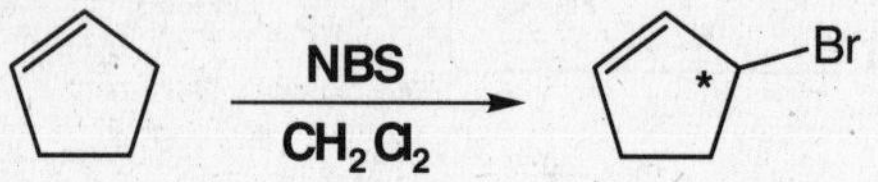

(racemic)

(f)   **2 CH$_4$** $\xrightarrow[\text{light}]{Br_2}$ **2 CH$_3$Br**

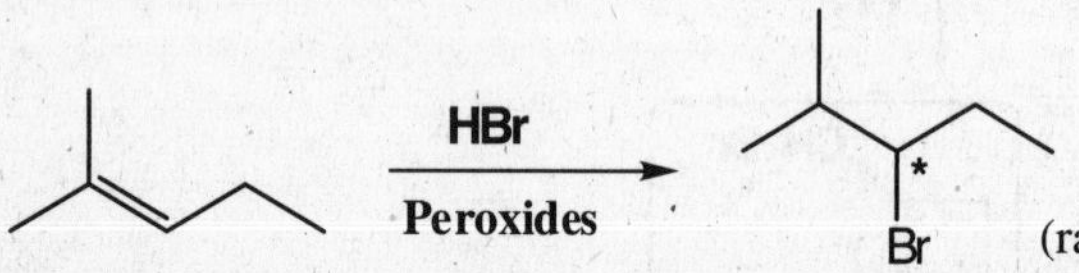

$HC{\equiv}CH$ $\xrightarrow[\text{2. CH}_3\text{Br}]{\text{1. NaNH}_2}$ ——≡—— $\xrightarrow[\text{2. CH}_3\text{Br}]{\text{1. NaNH}_2}$ ——≡—— (Review section 7.5)

(g)   ⟍⟍⟋⟍ $\xrightarrow[\text{Peroxides}]{\text{HBr}}$ ⟍⟋⟍⟋ (racemic)
                                                                Br

Autoxidation

Problem 8.29 Predict the products of the following reactions. Where isomeric products are formed, label the major product.

(a)   ⟍⟋=⟍⟋ $\xrightarrow{O_2,\text{ initiator}}$ [resonance structures] $\xrightarrow{\text{R-H}}$ Major product   Minor product

Resonance stabilized allylic radical

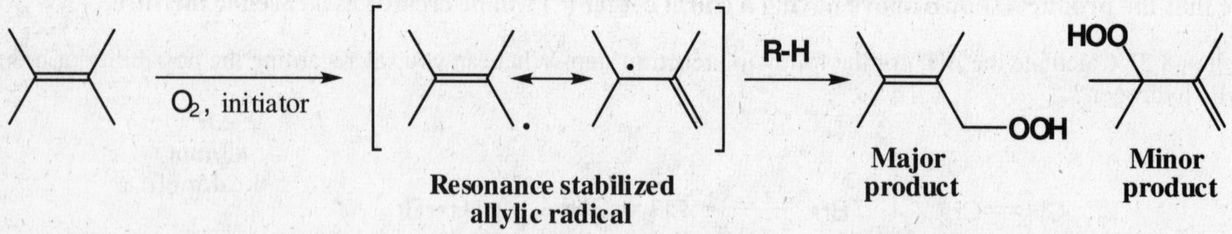

Note that the intermediate allylic radical formed by abstracting a H• from a methyl group can react at either end with O$_2$. The predicted major product is the one with the more substituted double bond.

(b)   ⬡ $\xrightarrow{O_2,\text{ initiator}}$ ⬡—OOH

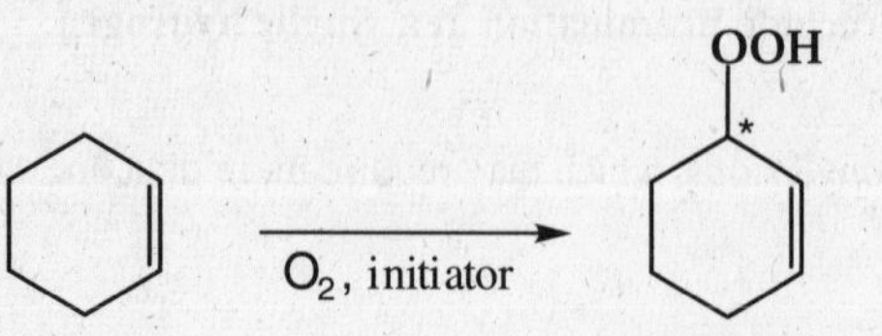

(**racemic**)

Problem 8.30 Give the major product of the following reactions.

In the presence of peroxides, the HBr adds in a non-Markovnikov fashion so that the Br atom ends up on the less substituted carbon atom.

(a) 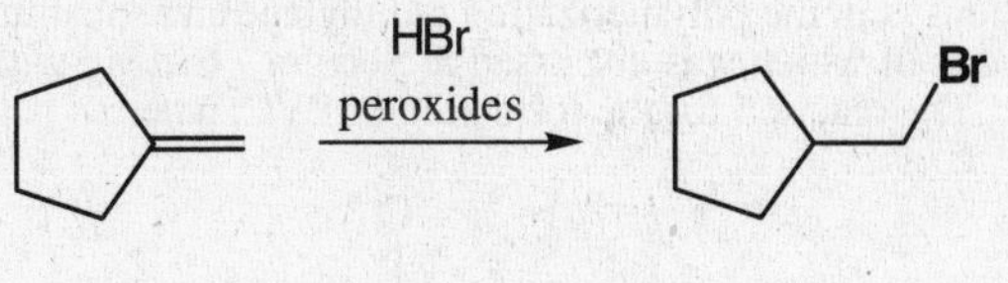 [structure: methylenecyclopentane] $\xrightarrow[\text{peroxides}]{\text{HBr}}$ [structure: (bromomethyl)cyclopentane]

(b) 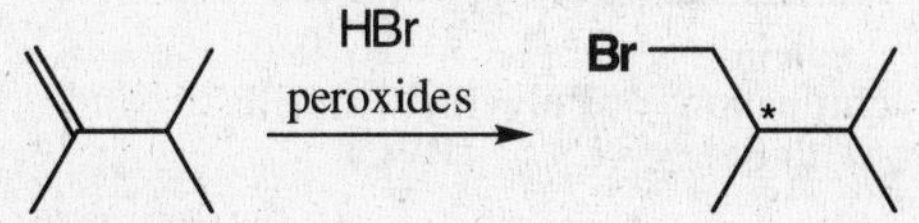 [structure: 2,3-dimethyl-1-butene] $\xrightarrow[\text{peroxides}]{\text{HBr}}$ [structure: product with Br, * marked carbon]

Looking Ahead

Problem 8.31 A major use of the compound cumene is in the industrial preparation of phenol and acetone in the two-step synthesis, shown below.

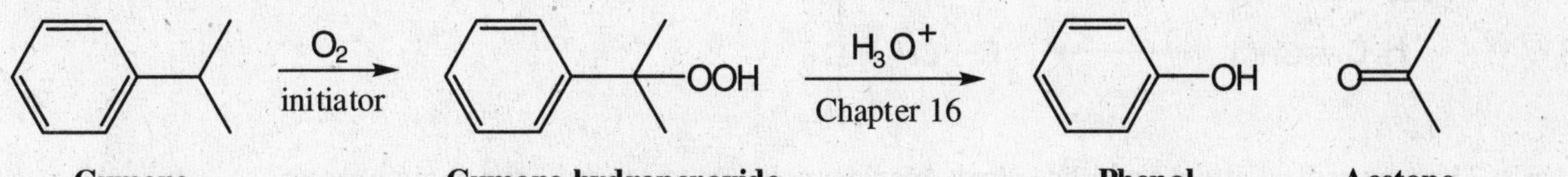

[structure: Cumene] $\xrightarrow[\text{initiator}]{O_2}$ [structure: Cumene hydroperoxide, –OOH] $\xrightarrow[\text{Chapter 16}]{H_3O^+}$ [structure: Phenol, –OH]   [structure: Acetone]

Cumene   **Cumene hydroperoxide**   **Phenol**   **Acetone**

Write a mechanism for the first step. We will see in Problem 16.63 how to complete the synthesis.

Step 1: The reaction begins when the initiator, a radical species X•, attacks cumene to make a highly resonance stabilized radical adjacent to the benzene ring, a species given the special name of benzylic radical.

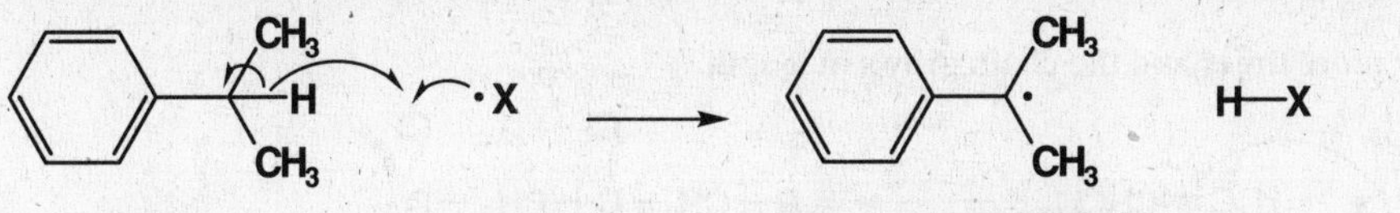

[structure: Cumene with CH_3, H, CH_3 and ·X arrow] $\longrightarrow$ [structure: Resonance stabilized benzylic radical with CH_3, CH_3]   H—X

Cumene   **Resonance stabilized
benzylic radical**

Step 2: The benzylic radical then reacts with oxygen to give a peroxy radical

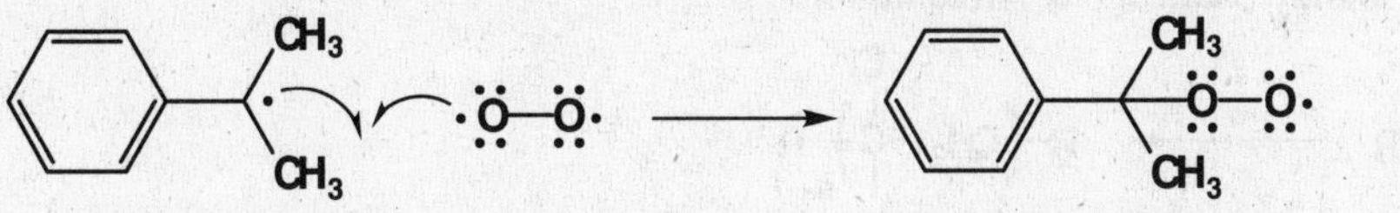

[structure: benzylic radical with CH_3, CH_3 and ·Ö—Ö·] $\longrightarrow$ [structure: peroxy radical with CH_3, CH_3, Ö—Ö·]

A peroxy radical

Step 3: The peroxy radical then reacts with another cumene molecule to give the hydroperoxide product and continue the radical chain process.

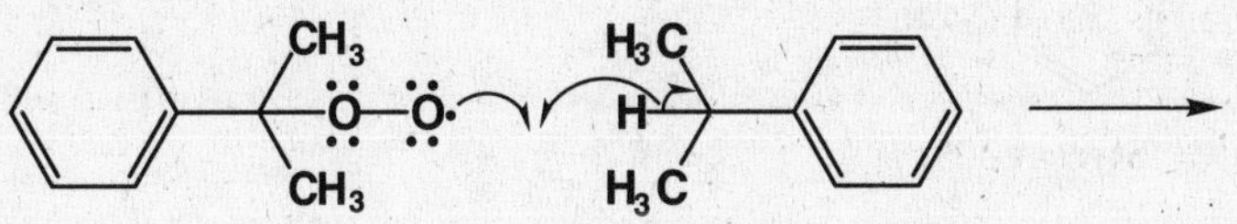

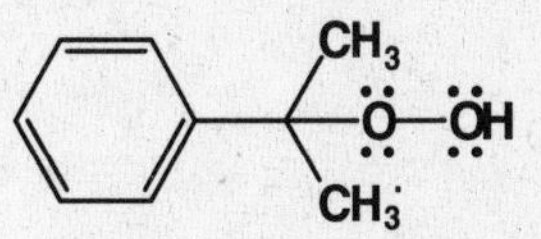

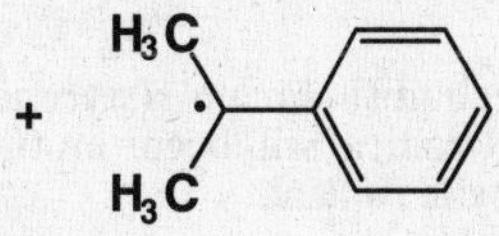

[structure: peroxy radical with CH_3, CH_3, Ö—Ö· reacting with cumene H_3C, H, H_3C] $\longrightarrow$ [structure: Cumene hydroperoxide with CH_3, CH_3, Ö—ÖH]

Cumene hydroperoxide

+ [structure: H_3C, H_3C benzylic radical]

Problem 8.32 An important use of radical chain reaction reactions is in the polymerization of ethylene and substituted ethylene monomers such as propene, vinyl chloride (the synthesis of which was discussed in Section 7.6 along with its use in the synthesis of poly(vinyl chloride), PVC, and styrene. The reaction for the formation of PVC, where n is the number of repeating units and is very large, follows.

Vinyl chloride   **Chain termination with R•'**   **Polyvinyl chloride (PVC)**

(a) Give a mechanism for this reaction (see Chapter 29).

Step 1: The R• could attack either carbon of the double bond, but it turns out that the chlorine atom is stabilizing to an adjacent radical. As a result, the predominant reaction occurs on the -CH_2 side of the double bond to give a new radical species.

Step 2: The new radical attacks another vinyl chloride molecule to continue the chain process.

Step n: This continues many more times and the chain grows in length.

Termination: The chain reaction and therefore growth of the polymer ends in a termination step when the radical at the end of the chain encounters another radical (•R') in solution.

(b) Give a similar mechanism for the formation of poly(styrene) from styrene. Which end of the styrene double bond would you expect R• to attack? Why?

Styrene

Step 1: The mechanism for poly(styrene) formation is exactly as above for vinyl chloride. Once again, the initial attack by the R radical will occur on the -CH_2 side of the double bond, since this produces the highly resonance stabilized benzylic radical.

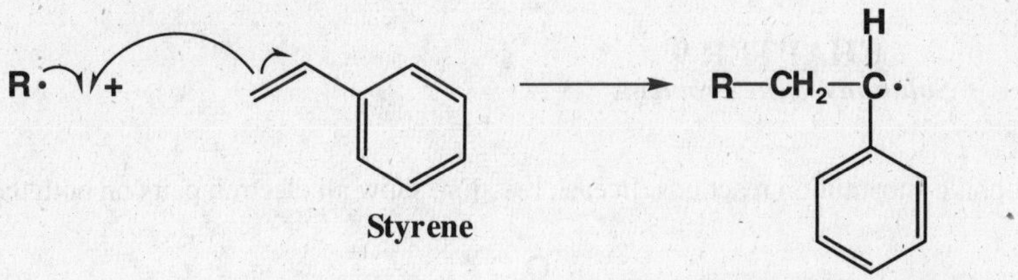

Styrene

**Resonance stabilized
benzylic radical**

Step 2: The new radical continues the chain reaction in a second step.

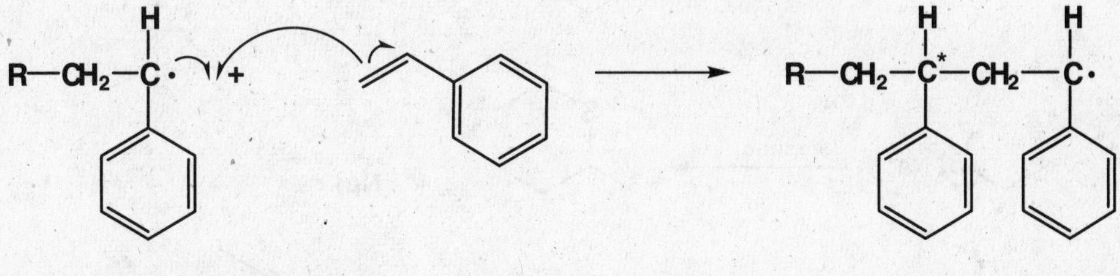

Step n: The chain continues to grow by repeated reaction of the benzylic radical at the end of the chain with a new styrene molecule.

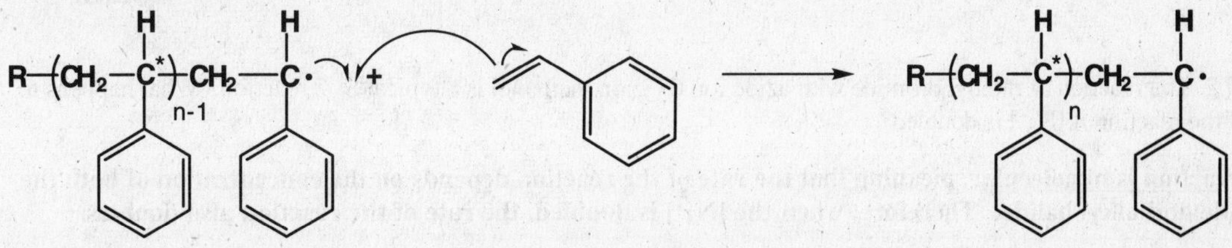

Termination: The chain reaction and therefore growth of the polymer ends in a termination step when the radical at the end of the chain encounters another radical (•R') in solution.

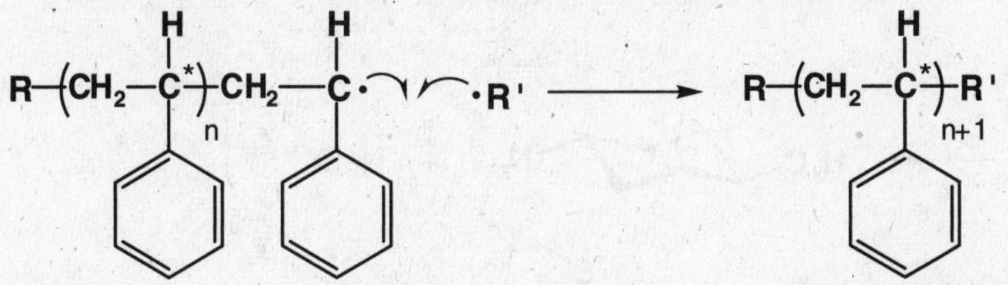

Poly(styrene)

CHAPTER 9
Solutions to the Problems

Problem 9.1 Complete the following nucleophilic substitution reactions. In each reaction, show all electron pairs on both the nucleophile and the leaving group.

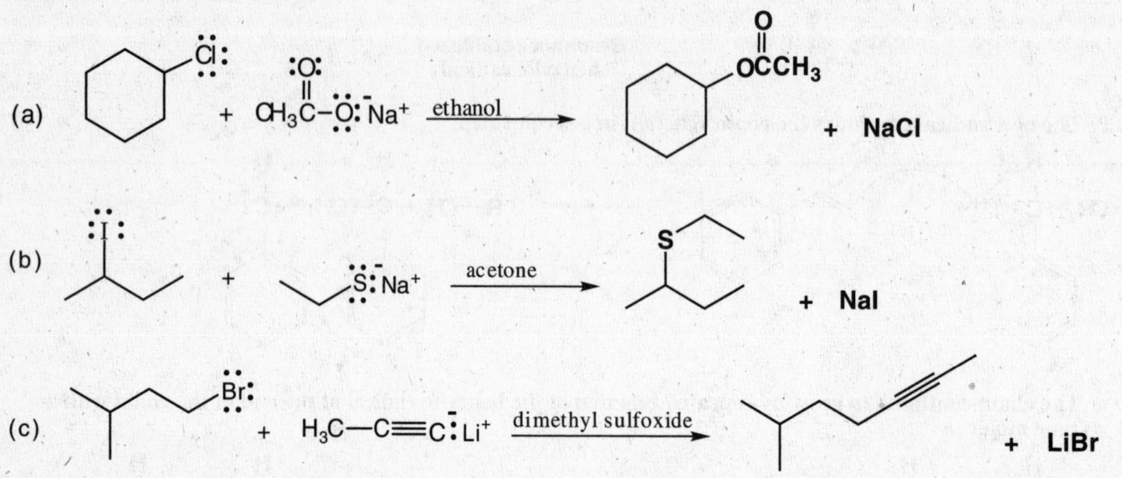

(a)

(b)

(c)

Problem 9.2 The reaction of methyl bromide with azide ion (N_3^-) in methanol is a typical S_N2 reaction. What happens to the rate of the reaction if $[N_3^-]$ is doubled?

An S_N2 reaction is bimolecular, meaning that the rate of the reaction depends on the concentration of both the nucleophile and alkyl halide. Therefore, when the $[N_3^-]$ is doubled, the rate of the reaction also doubles.

Problem 9.3 Complete these S_N2 reactions, showing the configuration of each product.

In both cases, the stereochemistry at the site of reaction is due to the nucleophile's backside attack that occurs during an S_N2 reaction.

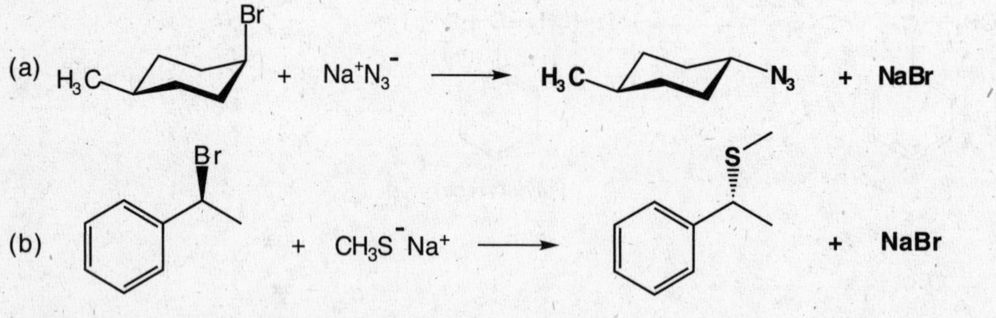

(a)

(b)

Problem 9.4 Write an additional resonance contributing structure for each carbocation, and state which of the two makes the greater contribution to the resonance hybrid.

A more highly substituted carbocation is more stable, so the contributing structure that has the more highly substituted carbocation will make the greater contribution to the resonance hybrid.

(a)

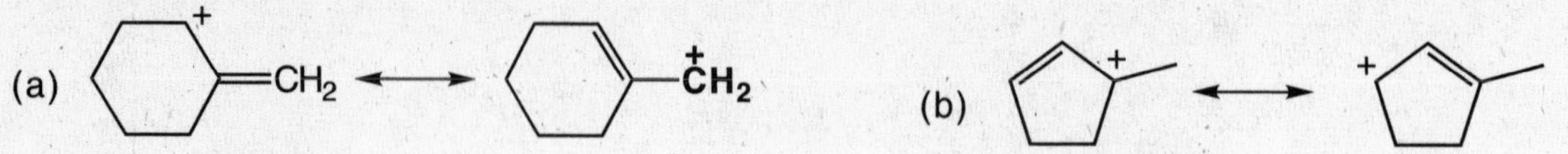

**Greater contribution
(2° carbocation)**

(b)

**Greater contribution
(3° carbocation)**

<u>Problem 9.5</u> Write the expected substitution product(s) for each reaction and predict the mechanism by which each product is formed.

(a)

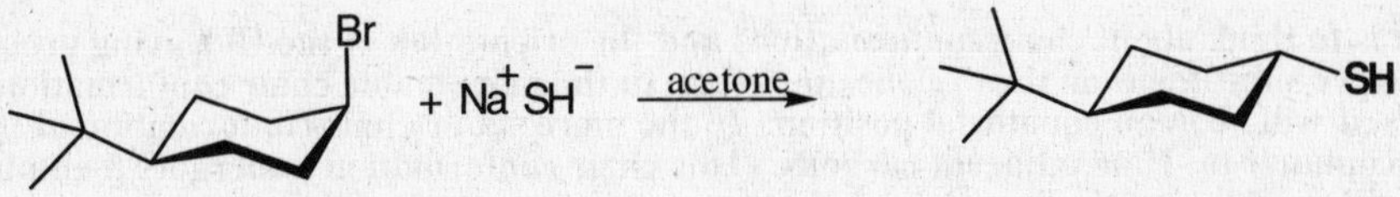

The HS⁻ is a good nucleophile and, since the reaction involves a secondary haloalkane with a good leaving group, the reaction mechanism is S_N2 so inversion of configuration is observed.

(b)

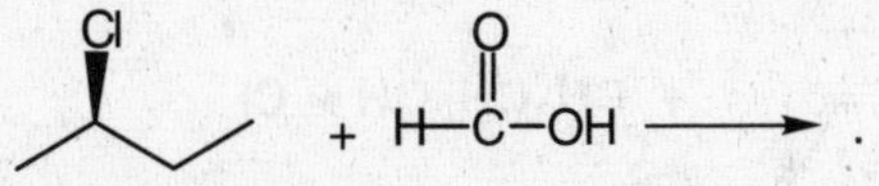

(R)-2-chlorobutane

The haloalkane is secondary and chloride is a good leaving group. Formic acid is an excellent ionizing solvent and a poor nucleophile. Therefore, substitution takes place by an S_N1 mechanism and leads to a mixture of enantiomers.

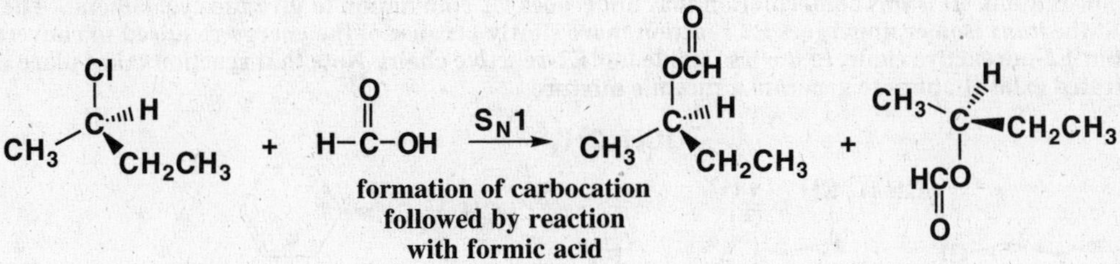

formation of carbocation
followed by reaction
with formic acid

<u>Problem 9.6</u> Predict the β-elimination product(s) formed when each chloroalkane is treated with sodium ethoxide in ethanol. If two or more products might be formed, predict which is the major product.

When there is a choice, the more highly substituted alkene will be the major product, as predicted by Zaitsev's rule.

(a)

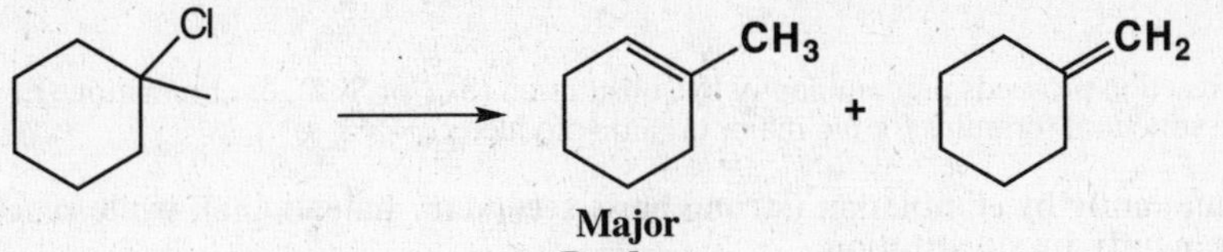

Major Product

(b)

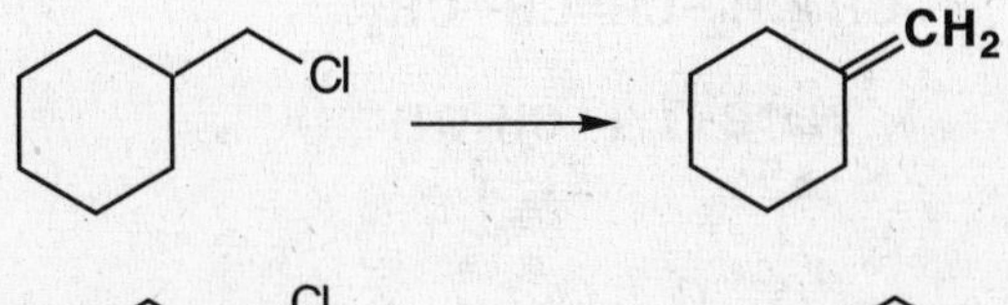

(c)

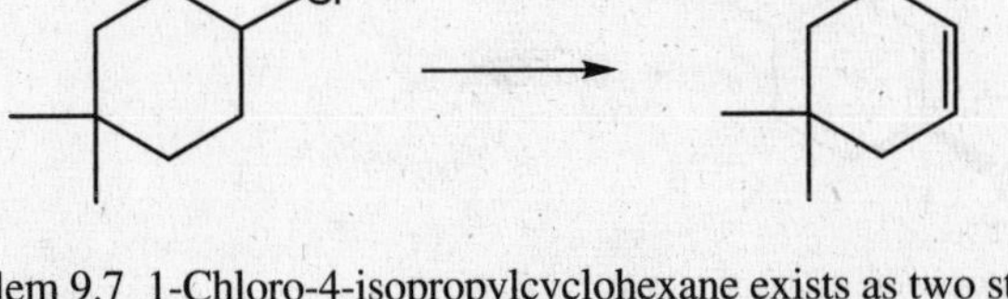

<u>Problem 9.7</u> 1-Chloro-4-isopropylcyclohexane exists as two stereoisomers: one *cis* and one *trans*. Treatment of either isomer with sodium ethoxide in ethanol gives 4-isopropylcyclohexene by an E2 reaction.

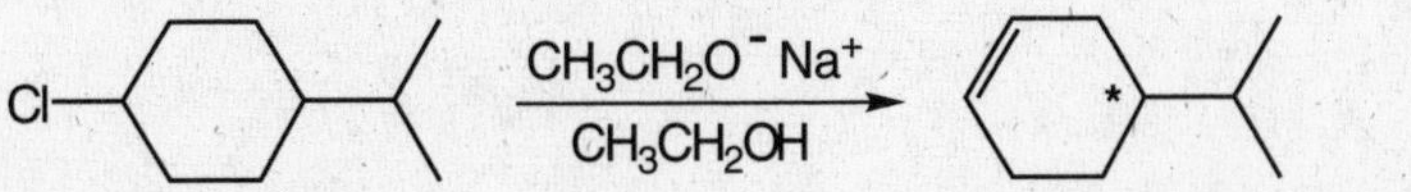

1-Chloro-4-isopropylcyclohexane   **4-Isopropylcyclohexene**

The *cis* isomer undergoes E2 reaction several orders of magnitude faster than the *trans* isomer. How do you account for this experimental observation?

The key to this problem is to think about chair conformations and the orientation of the Cl leaving group. The isopropyl group is the larger substituent on the cyclohexane ring. In the more stable chair conformation of both the *cis* and *trans* isomers, it will be in an equatorial position. In the more stable chair conformation of the *cis* isomer, -Cl is axial and coplanar to -H on adjacent carbons. This chair conformation undergoes β-elimination by an E2 mechanism at either of the two indicated axial H atoms to give a racemic mixture.

In the more stable chair conformation of the *trans* isomer, chlorine is equatorial and not coplanar to either -H on an adjacent carbon. Interconversion from this chair to the less stable chair results in the -Cl becoming axial and coplanar to an -H atom. It is this conformation that undergoes E2 elimination to give the cycloalkene. The bottom line is that the *trans* isomer undergoes E2 reaction more slowly because of the energy required to convert the more stable but E2-unreactive chair, to the less stable but E2-reactive chair. Note that reaction takes place at either of the indicated axial H atoms to generate a racemic mixture.

more stable chair
conformation

less stable chair
conformation

Problem 9.8 Predict whether each reaction proceeds predominantly by substitution (S_N1 or S_N2), or elimination (E1 or E2), or whether the two compete. Write structural formulas for the major organic product(s).

Reaction (a) will proceed predominantly by elimination (strong base, secondary haloalkane), while reactions (b) and (c) will proceed predominantly by substitution.

(a)

(b)

(c)

<u>Problem 9.9</u> Knowing what you do about the regioselectivity of S_N2 reactions, predict the product of hydrolysis of this compound.

Step 1: Make a new bond between a nucleophile and an electrophile and simultaneously break a bond to give stable molecules or ions.

Step 2: Make a new bond between a nucleophile and an electrophile and simultaneously break a bond to give stable molecules or ions **The nucleophilic attack by water on the three-membered ring intermediate will occur on the less-hindered carbon atom.**

Step 3: Take a proton away

A chiral center is created so a racemic mixture is produced.

Nucleophilic Aliphatic Substitution
<u>Problem 9.10</u> Draw a structural formula for the most stable carbocation with each molecular formula.

For (a), (b), and (c) the most stable cations are the most highly substituted alkyl cations. For (d), the most stable cation is resonance stabilized by an adjacent oxygen atom as shown. Spreading the charge over more than one atom has a stabilizing influence on charged species.

(a) $C_4H_9^+$

(b) $C_3H_7^+$

(c) $C_5H_{11}^+$

(d) $C_3H_7O^+$

<u>Problem 9.11</u> The reaction of 1-bromopropane and sodium hydroxide in ethanol occurs by an S_N2 mechanism. What happens to the rate of this reaction under the following conditions?
(a) The concentration of NaOH is doubled?

The rate of a bimolecular reaction such as the S_N2 reaction is proportional to the concentrations of both the hydroxide and haloalkane. Thus, if the concentration of hydroxide is doubled, the rate doubles.

(b) The concentrations of both NaOH and 1-bromopropane are doubled?

The rate of a bimolecular reaction, such as the S_N2 reaction, is proportional to the concentrations of both the hydroxide and haloalkane. Thus, if the concentration of both hydroxide and haloalkane are doubled, then the rate quadruples.

(c) The volume of the solution in which the reaction is carried out is doubled?

Doubling the volume lowers the concentration of each reactant by a factor of two, so the rate is slower by a factor of four.

<u>Problem 9.12</u> From each pair, select the stronger nucleophile.

(a) H_2O or OH^- (b) CH_3COO^- or OH^- (c) CH_3SH or CH_3S^-

   $OH^- > H_2O$ $OH^- > CH_3COO^-$ $CH_3S^- > CH_3SH$

(d) Cl^- or I^- (e) Cl^- or I^-
     in DMSO   in methanol   (f) CH_3OCH_3 or CH_3SCH_3

   $Cl^- > I^-$ $I^- > Cl^-$ $CH_3SCH_3 > CH_3OCH_3$

<u>Problem 9.13</u> Draw a structural formula for the product of each S_N2 reaction. Where configuration of the starting material is given, show the configuration of the product.

(a) $CH_3CH_2CH_2Cl + CH_3CH_2O^- Na^+ \xrightarrow{\text{ethanol}} CH_3CH_2CH_2OCH_2CH_3 + NaCl$

(b) $(CH_3)_3N: + CH_3I \xrightarrow{\text{acetone}} (CH_3)_4N^+ I^-$

(c) $-CH_2Br + NaCN \xrightarrow{\text{acetone}}$ $-CH_2CN + NaBr$

(d)

(e) $CH_3CH_2CH_2Cl + CH_3C{\equiv}C:^- Li^+ \longrightarrow CH_3CH_2CH_2C{\equiv}CCH_3 + LiCl$

(f)

(g)

(h) $CH_3CH_2CH_2Br$ + $Na^+ CN^-$ $\xrightarrow{\text{acetone}}$ $CH_3CH_2CH_2CN$ + Na^+Br^-

Problem 9.14 You were told that each reaction in Problem 9.13 is a substitution reaction but were not told the mechanism. Describe how you could conclude from the structure of the haloalkane, the nucleophile, and the solvent that each reaction is in fact an S_N2 reaction.

(a) **A primary halide, good nucleophile/strong base in ethanol, a moderately ionizing solvent all favor S_N2.**
(b) **Trimethylamine is a moderate nucleophile. A methyl halide in acetone, a weakly ionizing solvent, all work together to favor S_N2.**
(c) **Cyanide is a good nucleophile. A primary halide in acetone, a weakly ionizing solvent, all work together to favor S_N2.**
(d) **The alkyl chloride is secondary, so either an S_N1 or S_N2 mechanism is possible. Methanethiolate ion is a good nucleophile, but weak base. It therefore reacts by an S_N2 pathway.**
(e) **The lithium salt of the terminal alkyne is a moderate nucleophile, but also a strong base. Because the halide is primary, an S_N2 pathway is favored.**
(f) **Ammonia is a weak base and moderate nucleophile, and the halide is primary. Therefore S_N2 is favored.**
(g) **The major factor here favoring an S_N2 pathway is that the leaving group is halide and on a primary carbon.**
(h) **The cyanide anion is a good nucleophile and bromide is a good leaving group on a primary carbon. Therefore S_N2 is favored.**

Problem 9.15 Treatment of 1,3-dichloropropane with potassium cyanide results in the formation of pentanedinitrile. The rate of this reaction is about 1000 times greater in DMSO than it is in ethanol. Account for this difference in rate.

1,3-Dichloropropane   **Pentanedinitrile**

The hydroxyl H atom of ethanol is a hydrogen bond donor, so in ethanol the CN^- is strongly solvated via hydrogen bonding. This strong solvation slows down the CN^- reaction with the haloalkane. DMSO cannot act as a hydrogen bond donor, so the CN^- is not strongly solvated thereby allowing faster reaction with the haloalkane.

Problem 9.16 Treatment of 1-aminoadamantane, $C_{10}H_{17}N$, with methyl 2,4-dibromobutanoate in the presence of a non-nucleophilic base, R_3N, involves two successive S_N2 reactions and gives compound A. Propose a structural formula for compound A.

     1-Aminoadamantane   Methyl 2,4-dibromobutanoate

There are two successive S_N2 displacement reactions to give the four-membered ring.

The stereochemistry of the product will depend on the stereochemistry of the 2,4-dibromobutanoate starting material

Problem 9.17 Select the member of each pair that shows the greater rate of S_N2 reaction with KI in acetone.

The relative rates of S_N2 reactions for pairs of molecules in this problem depend on two factors: (1) bromine is a better leaving group than chlorine, and (2) a primary carbon without β-branching is less hindered and more reactive toward S_N2 substitution than a primary carbon with one, two, or three branches on the β-carbon atom. The molecule that reacts faster is circled.

(a) [structure] or [structure]   (b) [structure] or [structure]

(c) [structure] or [structure]   (d) [structure with Br*] or [structure with Br*]

<u>Problem 9.18</u> Select the member of each pair that shows the greater rate of S_N2 reaction with KN_3 in acetone.

The compound with the least steric hindrance will reacts with the greater rate. In both pairs, it is the molecule on the left.

(a) [cyclohexyl bromide] or [1-methylcyclohexyl bromide with CH₃] **This methyl group causes steric hindrance**

(b) [dimethylcyclohexyl bromide] or [structure with Br] **These methyl groups cause steric hindrance**

<u>Problem 9.19</u> What hybridization best describes the reacting carbon in the S_N2 transition state?

The hybridization state of the reacting carbon is best described as sp^2 in the S_N2 transition state.

<u>Problem 9.20</u> Each carbocation is capable of rearranging to a more stable carbocation. Limiting yourself to a single 1,2-shift, suggest a structure for the rearranged carbocation.

(a) [structure]

$H_3C-\overset{H_3C}{\underset{H_3C}{C}}-\overset{+}{\underset{H}{C}}-CH_3 \longrightarrow H_3C-\overset{+}{\underset{H_3C}{C}}-\overset{H}{\underset{H}{C}}-CH_3$

2° Carbocation   More stable 3° carbocation

(b) [structure]

$H_3C-\overset{H_3C}{\underset{H_3C}{C}}-\overset{+}{\underset{H}{C}}-CH_3 \longrightarrow H_3C-\overset{+}{\underset{H_3C}{C}}-\overset{CH_3}{\underset{H}{C}}-CH_3$

2° Carbocation   More stable 3° carbocation

(c)

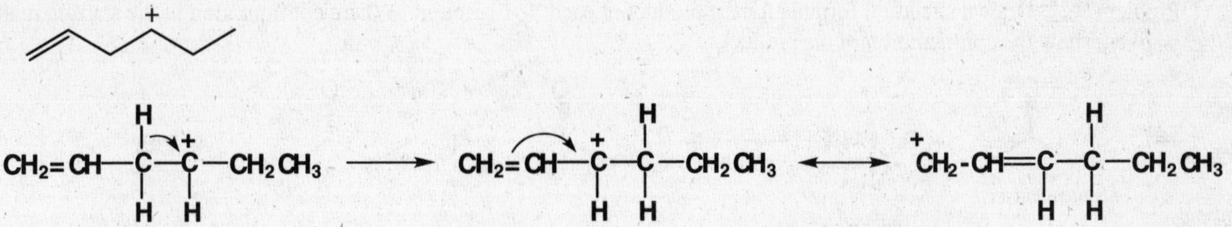

2° Carbocation

More stable
2° allylic carbocation

(d)

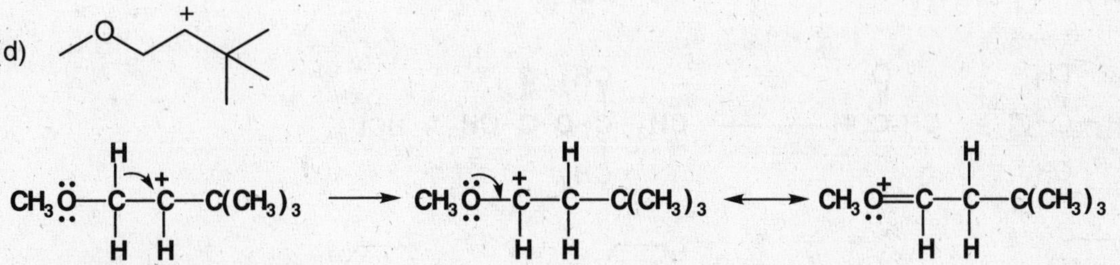

2° Carbocation

Stabilized by resonance

(e)

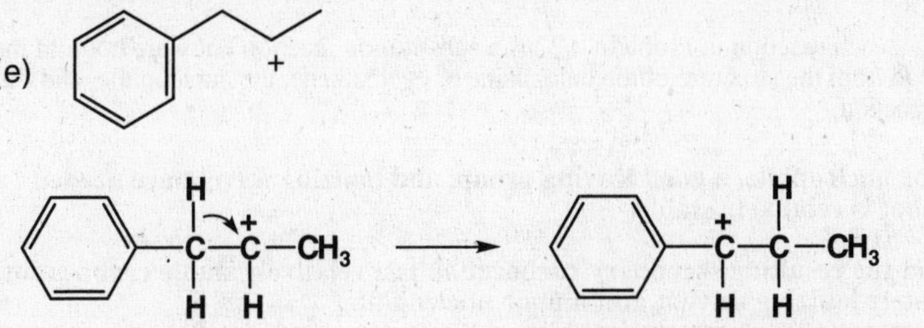

2° Carbocation

More stable 2° benzylic carbocation
(stabilized by resonance with
adjacent benzene ring)

(f)

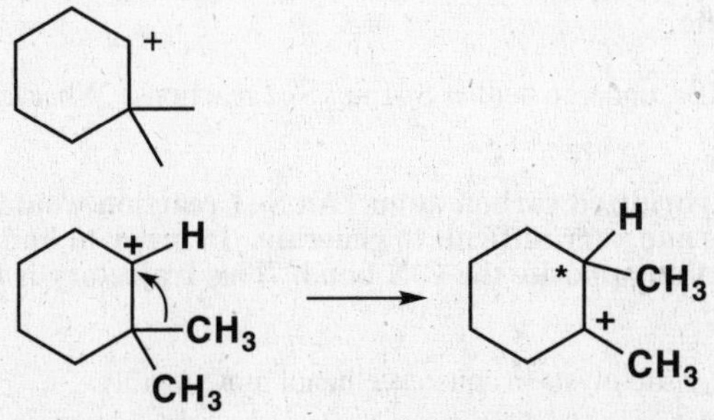

2° Carbocation   More stable 3° carbocation

Problem 9.21 Attempts to prepare optically active iodides by nucleophilic displacement on optically active bromides using I⁻ normally produce racemic iodoalkanes. Why are the product iodoalkanes racemic?

Iodide is a good nucleophile as well as a good leaving group. The iodoalkane that is formed will therefore react with other iodide ions according to an S_N2 mechanism. The resulting repeated inversion of configuration leads to full racemization of the product.

<u>Problem 9.22</u> Draw a structural formula for the product of each S_N1 reaction. Where configuration of the starting material is given, show the configuration of the product.

(a)

S enantiomer

(b)

(c)

(d)

<u>Problem 9.23</u> Suppose that you were told that each reaction in Problem 9.22 is a substitution reaction but were not told the mechanism. Describe how you could conclude from the structure of the haloalkane or cycloalkene, the nucleophile, and the solvent that each reaction is in fact an S_N1 reaction.

For an S_N1 reaction to be favored, a poor nucleophile, a good leaving group, and ionizing solvent are needed along with a carbocation intermediate that is relatively stable.

(a) **Chlorine is a good leaving group and the resulting secondary carbocation is a relatively stable carbocation intermediate. Ethanol is a moderately ionizing solvent and a poor nucleophile.**
(b) **Methanol is a moderately ionizing solvent and a poor nucleophile. Chlorine is a good leaving group and the resulting carbocation is tertiary.**
(c) **Acetic acid is a strongly ionizing solvent and a poor nucleophile. Chlorine is a good leaving group and the resulting carbocation is tertiary.**
(d) **Methanol is a moderately ionizing solvent and a poor nucleophile. Bromine is a good leaving group, and the resulting carbocation is both secondary and allylic.**

<u>Problem 9.24</u> Alkenyl halides such as vinyl bromide, $CH_2=CHBr$, undergo neither S_N1 nor S_N2 reactions. What factors account for this lack of reactivity?

In vinyl bromide, the bromine atom is bonded to an sp^2 hybridized carbon atom. An S_N1 reaction would give a vinyl carbocation, but such a carbocation is high in energy and very difficult to generate. In order to undergo an S_N2 reaction, the nucleophile must approach in a direction opposite the C-X bond. This trajectory is not possible for a vinyl halide.

<u>Problem 9.25</u> Select the member of each pair that undergoes S_N1 solvolysis in aqueous ethanol more rapidly.

Relative rates for each pair of compounds listed in this problem depend on a combination of two factors: (1) bromine is a better leaving group than chlorine and (2) the stability of the resulting carbocation. The molecule that reacts faster is circled.

(a)

Formation of a tertiary carbocation is feasible, while formation of a primary carbocation is not.

(b) 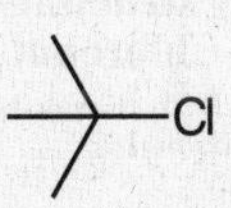 or

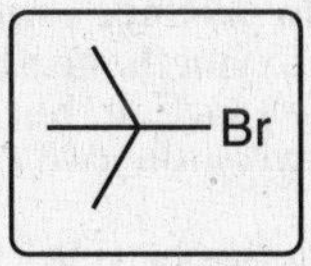

Bromine is a better leaving group than chlorine.

(c) 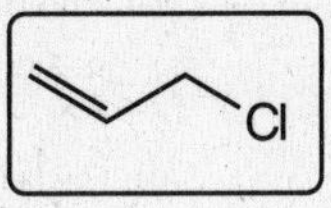 or

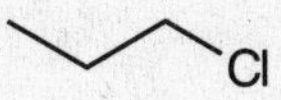

Formation of a resonance-stabilized 1° allylic carbocation is feasible, while formation of a primary carbocation is not.

(d) 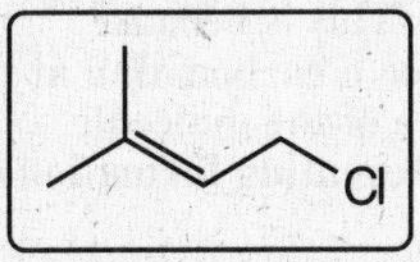 or

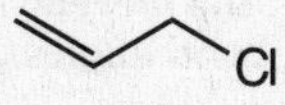

The activation energy for formation of the dialkyl allylic carbocation is lower than that for formation of an unsubstituted allylic carbocation.

(e) 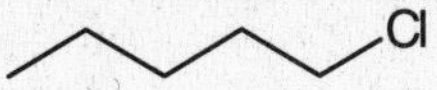 or

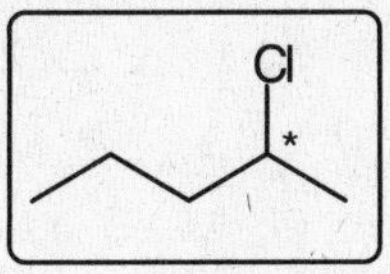

The activation energy for formation of a secondary carbocation is lower than that for formation of a primary carbocation.

(f) 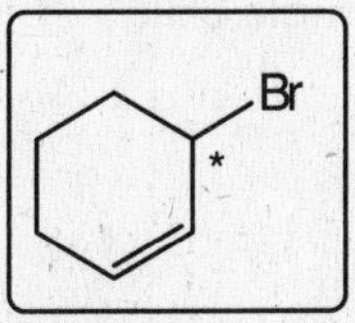 or

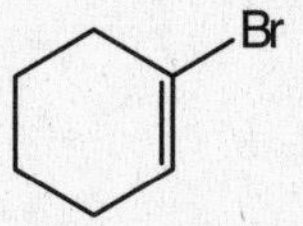

The activation energy for formation of an allylic carbocation is lower than that for formation of a vinylic carbocation.

Problem 9.26 Account for the following relative rates of solvolysis under experimental conditions favoring S_N1 reaction.

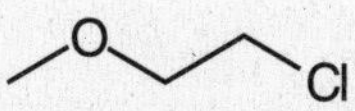

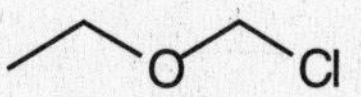

Relative rate of   0.2   1 10^9
solvolysis (S_N1)

1-Chloro-1-ethoxymethane (chloromethyl ethyl ether) reacts the fastest by far in a solvolysis reaction because the carbocation produced by loss of the chlorine atom is stabilized by the adjacent ether oxygen atom. Thus, the activation energy for this reaction is significantly lower than for the other two molecules. The most important contributing structures are shown below for the stabilized carbocation.

$$CH_3CH_2\,\ddot{O}\,CH_2\overset{\frown}{Cl} \longrightarrow CH_3CH_2\,\overset{+}{\ddot{O}}CH_2 \longleftrightarrow CH_3CH_2\,\ddot{O}\overset{+}{=}CH_2$$

1-Chloro-2-methoxyethane (2-chloroethyl methyl ether) reacts the slowest because the carbocation produced during the reaction is somewhat destabilized by the ether oxygen atom that is two carbon atoms away. This is because oxygen is more electronegative than carbon, so there is a partial positive charge on the carbon atoms bonded to the oxygen. Thus, the carbocation produced by departure of the chlorine atom is adjacent to this partially positive carbon atom; a destabilizing arrangement.

destabilizing

$$\overset{\delta+}{CH_3}-\overset{\delta-}{\underset{\cdot\cdot}{\ddot{O}}}-\overset{\delta+}{CH_2}CH_2\overset{\frown}{Cl} \longrightarrow \overset{\delta+}{CH_3}-\overset{\delta-}{\underset{\cdot\cdot}{\ddot{O}}}-\overset{\delta+}{CH_2}\overset{+}{CH_2}$$

Please note that in the case of 1-chloro-2-methoxyethane, there is no way to produce contributing structures with any positive charge on the oxygen atom like that shown for 1-chloro-1-ethoxymethane above. 1-Chlorobutane does not have similar destabilization of the carbocation formed during the reaction, but is does not have any stabilization either. Its relative reaction rate is therefore intermediate between the two ether molecules.

<u>Problem 9.27</u> Not all tertiary halides undergo S_N1 reactions readily. For example, the bicyclic compound shown here is very unreactive under S_N1 conditions. What feature of this molecule is responsible for such lack of reactivity? You will find it helpful to examine a model of this compound.

bridgehead
carbon atom

1-Iodobicyclooctane

In order to form a cation, great angle strain would have to be produced in the molecule. This is because carbocations prefer to be trigonal planar (sp^2 hybridized), and loss of iodine would place a carbocation at the bridgehead position with a preferred bond angle of 120°. However, the bicyclic structure of the molecule enforces a tetrahedral geometry at the bridgehead position (109.5° bond angles), thus preventing formation of the carbocation.

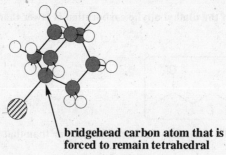

bridgehead carbon atom that is
forced to remain tetrahedral

<u>Problem 9.28</u> Show how you might synthesize the following compounds from an haloalkane and a nucleophile:

(a)

Treatment of a halocyclohexane with sodium cyanide.

+ NaCN ⟶ + NaBr

(b)

Treatment of chloromethylcyclohexane with sodium cyanide.

+ NaCN ⟶ + NaCl

(c)

Treatment of a halocyclohexane with the sodium salt of acetic acid.

$$CH_3\overset{\displaystyle O}{C}O^-Na^+ + NaBr$$

(d)

Treatment of a 1-halopentane with sodium hydrosulfide.

$$CH_3(CH_2)_3CH_2Br + HS^-Na^+ \longrightarrow CH_3(CH_2)_3CH_2SH + NaBr$$

(e)

Treatment of a 1-halohexane with the sodium salt of acetylene.

$$CH_3(CH_2)_4CH_2Br + HC\equiv C^-Na^+ \longrightarrow CH_3(CH_2)_5C\equiv CH + NaBr$$

(f)

Treatment of a haloethane with sodium or potassium ethoxide in ethanol.

$$CH_3CH_2O^-Na^+ + CH_3CH_2I \xrightarrow[CH_3CH_2OH]{} CH_3CH_2OCH_2CH_3 + NaI$$

(g)

Treatment of the appropriate *trans*-halocyclopentane enantiomer with sodium hydrosulfide.

$$+ HS^-Na^+ \longrightarrow + NaBr$$

Problem 9.29 3-Chloro-1-butene reacts with sodium ethoxide in ethanol to produce 3-ethoxy-1-butene. The reaction is second order; first order in 3-chloro-1-butene and first order in sodium ethoxide. In the absence of sodium ethoxide, 3-chloro-1-butene reacts with ethanol to produce both 3-ethoxy-1-butene and 1-ethoxy-2-butene. Explain these results.

For the second order reaction with sodium ethoxide, the mechanism is S_N2 so the -OCH$_2$CH$_3$ group ends up only where the -Cl leaving group was attached. The stereochemistry of the product depends on the stereochemistry of the starting material.

$$CH_2=CH-\overset{*}{\underset{\underset{Cl}{|}}{C}}-CH_3 + {}^-OCH_2CH_3 \xrightarrow{S_N2} CH_2=CH-\overset{OCH_2CH_3}{\underset{\underset{H}{|}}{\overset{|}{\underset{*}{C}}}}-CH_3$$

3-Chloro-1-butene   **3-Ethoxy-1-butene**

For the second reaction, the absence of a strong nucleophile allows the S_N1 mechanism to operate. The allylic carbocation intermediate can be attacked at either the 1 or 3 positions, a fact that is readily explained by considering the two predominant contributing structures of this intermediate. Note that both enantiomers of 3-ethoxy-1-butene are formed.

$$CH_2=CH-\overset{\overset{H}{|}}{\underset{\underset{Cl}{|}}{C}}{}^*-CH_3 \xrightarrow{S_N1} CH_2=CH-\overset{\overset{H}{|}}{\underset{+}{C}}-CH_3 \longleftrightarrow CH_2-CH=\overset{\overset{H}{|}}{C}-CH_3$$

3-Chloro-1-butene

EtOH ↓   EtOH ↓

$$CH_2=CH-\overset{\overset{H}{|}}{\underset{\underset{OCH_2CH_3}{|}}{C}}{}^*-CH_3 \qquad\qquad CH_2-CH=\overset{\overset{H}{|}}{\underset{\underset{OCH_2CH_3}{|}}{C}}-CH_3$$

3-Ethoxy-1-butene   **1-Ethoxy-2-butene**

<u>Problem 9.30</u> 1-Chloro-2-butene undergoes hydrolysis in warm water to give a mixture of these allylic alcohols. Propose a mechanism for their formation.

$$CH_3CH=CHCH_2Cl \xrightarrow{H_2O} CH_3CH=CHCH_2OH + \overset{\overset{OH}{|}}{CH_3CHCH=CH_2}$$

1-Chloro-2-butene   **2-Buten-1-ol**   **3-Buten-2-ol**

The two products can be explained by an S_N1 mechanism that produces an allylic cation that can react with water at either the 1 or 3 position. Note that 3-buten-2-ol is actually formed as a racemic mixture of enantiomers.

$$CH_3CH=CH-\underset{\underset{Cl}{|}}{CH_2} \xrightarrow{S_N1} CH_3CH=CH-\underset{+}{CH_2} \longleftrightarrow CH_3CH-CH=CH_2$$

H₂O ↓   H₂O ↓

$$CH_3CH=CHCH_2OH \qquad\qquad \overset{\overset{*}{CH_3CHCH=CH_2}}{\underset{\underset{OH}{|}}{}}$$

<u>Problem 9.31</u> The following nucleophilic substitution occurs with rearrangement. Suggest a mechanism for formation of the observed product. If the starting material has the *S* configuration, what is the configuration of the chiral center in the product.

Step 1: *Make a new bond between a nucleophile and an electrophile and simultaneously break a bond to give stable molecules or ions.* An internal nucleophilic reaction leads to the three-membered ring intermediate $C_8H_{18}NCl$ shown.

Step 2: *Make a new bond between a nucleophile and an electrophile and simultaneously break a bond to give stable molecules or ions.* The intermediate is then attacked by hydroxide at the less-hindered site to yield the product.

If the starting material has the *S* configuration, the product will have the *R* configuration as shown.

Problem 9.32 Propose a mechanism for the formation of these products in the solvolysis of this bromoalkane.

Make a new bond between a nucleophile and an electrophile and simultaneously break a bond to give stable molecules or ions followed by a 1,2 shift. The bromide leaves to generate a 2° carbocation, that then rearranges by ring expansion to give the more stable 3° carbocation. Note how this new 3° carbocation with a five-membered ring also has much less ring strain than the strained four-membered ring 2° carbocation.

2° carbocation   More stable
                                   3° carbocation

The new 3° carbocation either loses the proton shown to complete an E1 reaction, or reacts with ethanol to complete the S$_N$1 reaction.

More stable
3° carbocation

or alternatively,

More stable
3° carbocation

Note that the stereochemistry of the product depends on the stereochemistry of the starting material.

Problem 9.33 Solvolysis of the following bicyclic compound in acetic acid gives a mixture of products, two of which are shown. The leaving group is the anion of a sulfonic acid, ArSO$_3$H. A sulfonic acid is a strong acid, and its anion, ArSO$_3^-$, is a weak base and a good leaving group. Propose a mechanism for this reaction.

(racemic; 2 enantiomers)   (racemic; 4 diastereomers)

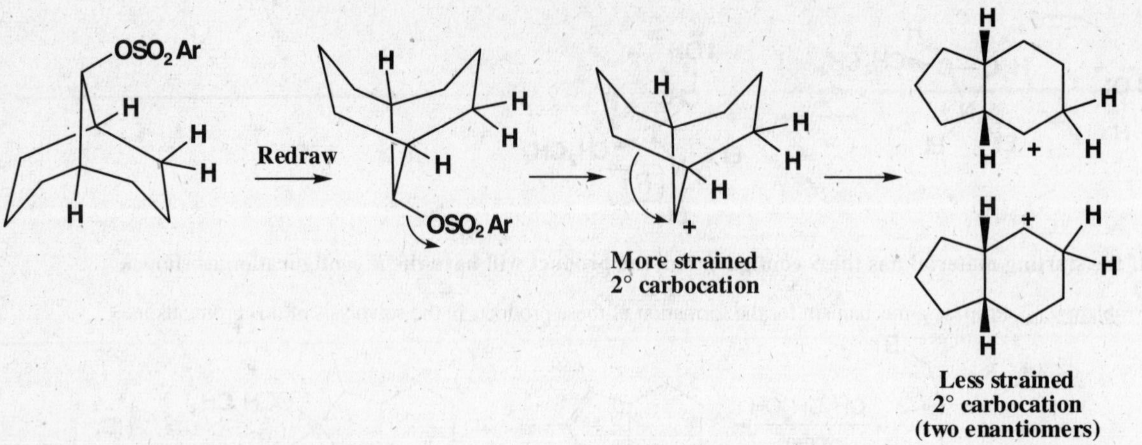

More strained
2° carbocation

Less strained
2° carbocation
(two enantiomers)

The sulfonate anion leaves to generate a 2° carbocation, that then rearranges to give a new 2° carbocation with less ring strain. The new 2° carbocation either loses the proton shown to complete an E1 reaction, or reacts with acetic acid to complete the S_N1 reaction.

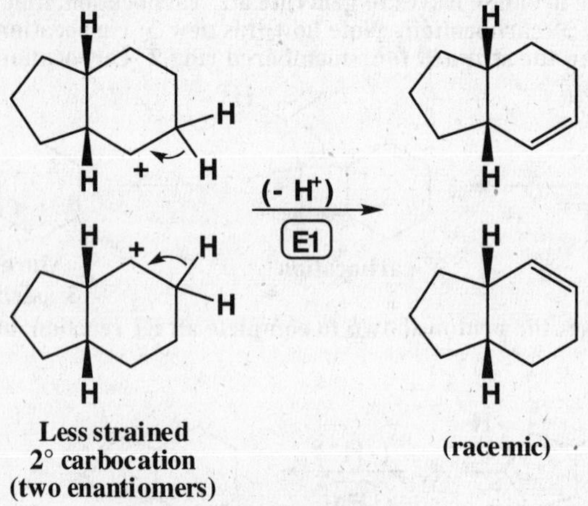

(− H⁺)

E1

Less strained
2° carbocation
(two enantiomers)

(racemic)

or alternatively,

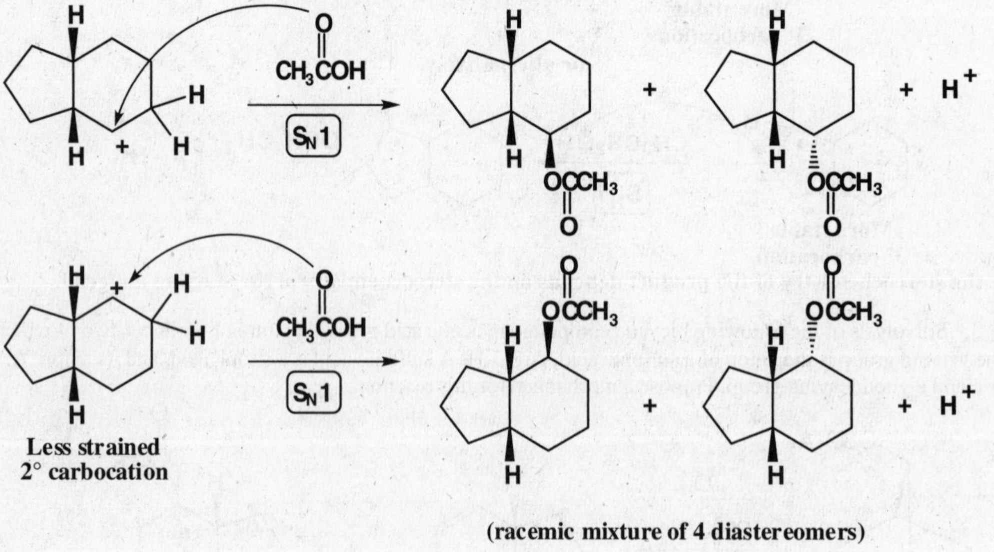

CH₃COH

S_N1

CH₃COH

S_N1

Less strained
2° carbocation

(racemic mixture of 4 diastereomers)

Problem 9.34 Which compound in each set undergoes more rapid solvolysis when refluxed in ethanol? Show the major product formed from the more reactive compound.

More rapid solvolysis will occur for the molecule that can produce the more stable carbocation or all things being equal, the one with the better leaving group.

(a) ⌐Br⌐ or Br   ⟶ **Allylic carbocation**

The molecule on the left can make a more stable allylic carbocation.

(b) Cl or Br   ⟶

Bromide is a better leaving group than chloride.

(c) Br or Br   CH₃ / Br ⟶ CH₃ / +   **3° Carbocation**

The molecule on the left can make a more stable 3° carbocation.

(d) Cl or Cl   Cl ⟶ +   **3° Carbocation**

The molecule on the right can make a stable 3° carbocation. In fact, the molecule on the left cannot adopt the favored trigonal planar geometry so it reacts much slower than expected.

Problem 9.35 Account for the relative rates of solvolysis of these compounds in aqueous acetic acid.

$(CH_3)_3CBr$ Br   Br   Br

1   10^{-2} 10^{-7} 10^{-12}

More rapid solvolysis will occur for the molecule that can produce the more stable carbocation.

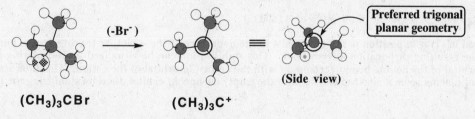

$(-Br^-)$ ≡   **Preferred trigonal planar geometry**

(Side view)

$(CH_3)_3CBr$ $(CH_3)_3C^+$

The *tert*-butyl carbocation produced upon solvolysis of *tert*-butyl bromide can easily adopt the preferred trigonal planar geometry as shown above. A trigonal planar geometry is preferred because this allows for maximum stabilization (maximum orbital overlap) due to hyperconjugation with the three adjacent methyl groups. As shown below, it is much more difficult for the three bicyclic bromides to produce the preferred trigonal planar geometries due to ring strain. Thus, the activation energies for these reactions are much higher, and the reactions are slower.

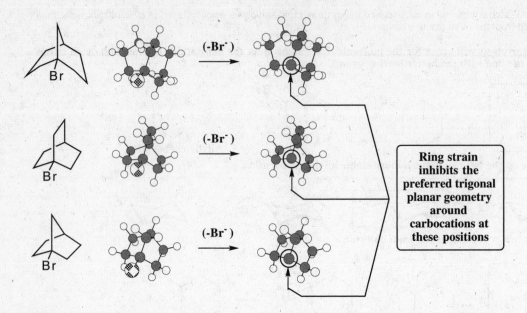

<u>Problem 9.36</u> A comparison of the rates of S$_N$1 solvolysis of the bicyclic compounds (1) and (2) in acetic acid shows that compound (1) reacts 10^{11} times faster than compound (2). Furthermore, solvolysis of (1) occurs with complete retention of configuration: The nucleophile occupies the same position on the one-carbon bridge as did the leaving -OSO$_2$Ar group.

(a) Draw structural formulas for the products of solvolysis of each compound.

(b) Account for difference in rate of solvolysis of (1) and (2).

The pi bond of (1) is in position to not only assist in the departure of the -OSO$_2$Ar leaving group, but also stabilize the resulting carbocation intermediate. This stabilization can be visualized by considering the filled π bonding orbital of the double bond overlapping with the empty *2p* orbital of the cation. This interaction has the effect of placing some π electron density into the empty carbon *2p* orbital, thereby stabilizing the positive charge.

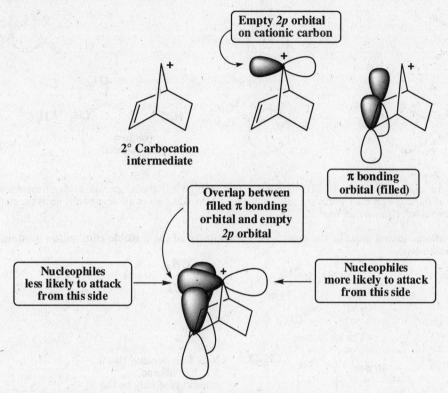

Empty *2p* orbital on cationic carbon

2° Carbocation intermediate

π bonding orbital (filled)

Overlap between filled π bonding orbital and empty *2p* orbital

Nucleophiles less likely to attack from this side

Nucleophiles more likely to attack from this side

(c) Account for complete retention of configuration in the solvolysis of (1).

As shown on the figure above, the pi bond will block a nucleophile from attacking the cation on one face. Thus, the nucleophile will approach from the other side, the side that the -OSO$_2$Ar departed from, leading to predominantly retention of configuration.

β-Eliminations
<u>Problem 9.37</u> Draw structural formulas for the alkene(s) formed by treatment of each haloalkane or halocycloalkane with sodium ethoxide in ethanol. Assume that elimination occurs by an E2 mechanism.

The major and minor products for each E2 reaction are shown. Where there is more than one combination of leaving groups *anti* and coplanar, the major product is the more substituted alkene (the so-called Zaitzev product).

(a)

(b)

CH_3 + CH_2

(major)

(c)

CH_3

(the other elimination cannot occur)

(d) (e) (f)

(major)

Problem 9.38 Draw structural formulas of all chloroalkanes that undergo dehydrohalogenation when treated with KOH to give each alkene as the major product. For some parts, only one chloroalkane gives the desired alkene as the major product. For other parts, two chloroalkanes may work.

Recall that the alkene shown must be the most highly substituted of the possible elimination products for the starting chloroalkane(s).

(a)

The *cis* isomer

(b)   Note that because this is a 1° haloalkane, substitution may be the predominant reaction.

(c)   Note that because this is a 1° haloalkane, substitution may be the predominant reaction.

(d)

(e)

Note that for (c), (d), and (e) any of the possible stereoisomers would work.

Problem 9.39 Following are diastereomers (A) and (B) of 3-bromo-3,4-dimethylhexane. On treatment with sodium ethoxide in ethanol, each gives 3,4-dimethyl-3-hexene as the major product. One diastereomer gives the *E*-alkene, and the other gives the *Z*-alkene. Which diastereomer gives which alkene? Account for the stereoselectivity and stereospecificity of each β-elimination.

(A)   (B)

Rotate the given conformation of each stereoisomer into a conformation in which Br and H are anti and coplanar and then made to undergo E2 elimination. You will find that diastereomer (A) gives the (*E*)-isomer and diastereomer (B) gives the (*Z*)-isomer.

(A)   (E) Product

(B)   (Z) Product

Problem 9.40 Treatment of the following stereoisomer of 1-bromo-1,2-diphenylpropane with sodium ethoxide in ethanol gives a single stereoisomer of 1,2-diphenylpropene. Predict whether the product has the E configuration or the Z configuration.

1-Bromo-1,2-diphenylpropane   1,2-Diphenylpropene

(E) Product

Problem 9.41 Elimination of HBr from 2-bromonorbornane gives only 2-norbornene and no 1-norbornene. How do you account for the regioselectivity of this dehydrohalogenation? In answering this question, you will find it helpful to look at molecular models of both 1-norbornene and 2-norbornene and analyze the strain in each.

2-Bromonorbornane   2-Norbornene   1-Norbornane

1-Norbornene is not formed for at least two reasons. First, it is not possible to achieve the preferred *anti* and coplanar geometry of the hydrogen and Br atom that would lead to 1-norbornene. Second, the alkene in 1-norbornene has considerably more angle strain compared with 2-norbornene.

Problem 9.42 Which isomer of 1-bromo-3-isopropylcyclohexane reacts faster when refluxed with potassium *tert*-butoxide, the *cis* isomer or the *trans* isomer? Draw the structure of the expected product from the faster reacting compound.

The isopropyl group is the larger substituent on the cyclohexane ring. In the more stable chair conformation of both the *cis* and *trans* isomers, it will be in an equatorial position. In the more stable chair conformation of the *trans* isomer, -Br is axial and coplanar to -H on adjacent carbons. This chair conformation readily undergoes β-elimination by an E2 mechanism so this is the faster reaction.

More stable chair

In the more stable chair conformation of the *cis* isomer, bromine is equatorial and therefore not coplanar to either -H on adjacent carbon atoms. Interconversion from this chair to the less stable chair is required for the -Br to become axial and coplanar to an adjacent -H atom. It is this conformation that undergoes E2 elimination to give the cycloalkene. The bottom line is that the *cis* isomer undergoes E2 reaction more slowly because of the energy required to convert the more stable chair, but E2-unreactive chair, to the less stable, but E2-reactive, chair. Note this reaction will lead to the same products as above.

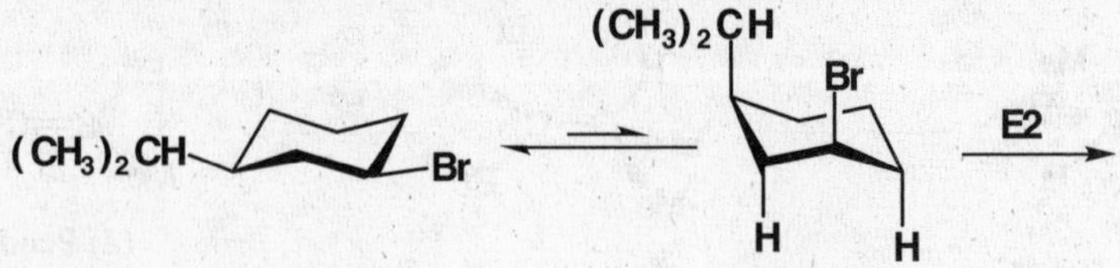

Less stable chair

Note that in the above, only one of the possible enantiomers was shown as starting material for simplicity.

Substitution Versus Elimination

Problem 9.43 Consider the following statements in reference to S_N1, S_N2, E1, and E2 reactions of haloalkanes. To which mechanism(s), if any, does each statement apply?

(a) Involves a carbocation intermediate **S_N1, E1**
(b) Is first order in haloalkane and first order in nucleophile **S_N2**
(c) Involves inversion of configuration at the site of substitution **S_N2**
(d) Involves retention of configuration at the site of substitution **None. S_N1, however, may proceed with predominantly racemization, but some retention.**
(e) Substitution at a chiral center gives predominantly a racemic product **S_N1**
(f) Is first order in haloalkane and zero order in base **E1**
(g) Is first order in haloalkane and first order in base **E2**
(h) Is greatly accelerated in protic solvents of increasing polarity **S_N1, E1**
(i) Rearrangements are common **S_N1, E1**
(j) Order of reactivity of haloalkanes is 3° > 2° > 1° **S_N1, E2, E1**
(k) Order of reactivity of haloalkanes is methyl > 1° > 2° > 3° **S_N2**

Problem 9.44 Arrange these haloalkanes in order of increasing ratio of E2 to S_N2 products observed on reaction of each with sodium ethoxide in ethanol.

(a) CH_3CH_2Br (b) $CH_3CH(CH_3)CH_2Br$ (c) $CH_3C(CH_3)(Cl)CH_2CH_3$ (d) $CH_3CH(CH_3)CH_2CH_2Br$

More E2 will occur with a more hindered site of reaction and/or a more substituted (stable) product alkene. Thus, listed in order from least to most E2 product:

$$CH_3CH_2Br < CH_3CH(CH_3)CH_2CH_2Br < CH_3CH(CH_3)CH_2Br < CH_3C(CH_3)(Cl)CH_2CH_3$$

Problem 9.45 Draw a structural formula for the major organic product of each reaction and specify the most likely mechanism by which each is formed.

The substitution and elimination products for the reactions are given in bold. In each case, the different parameters discussed in the chapter are considered including the type of haloalkane (primary, secondary, tertiary, etc.) and the relative strength of the nucleophile/base.

(a)

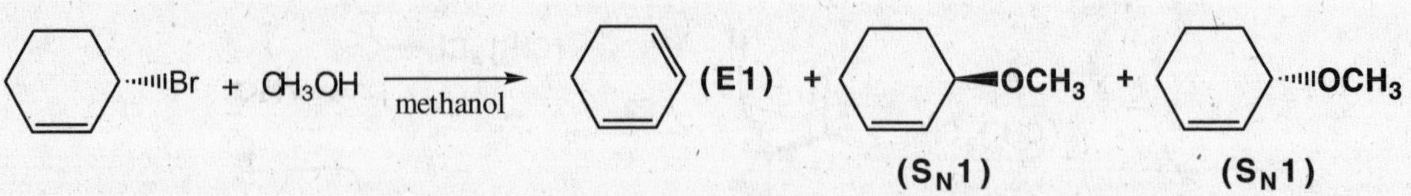

(b) 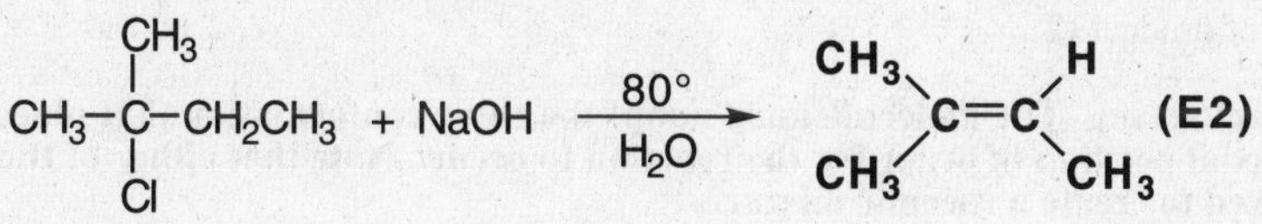 (E2)

(c)

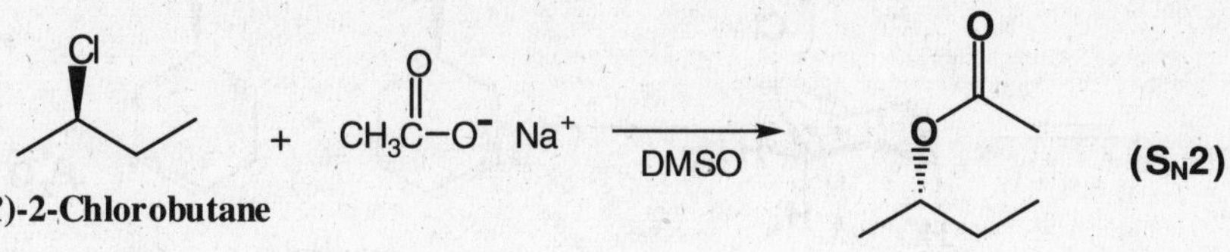

(R)-2-Chlorobutane   (S$_N$2)

(d) 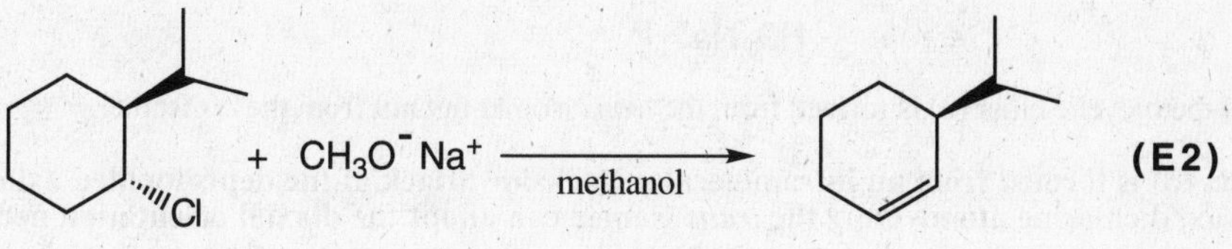 (E2)

(e) 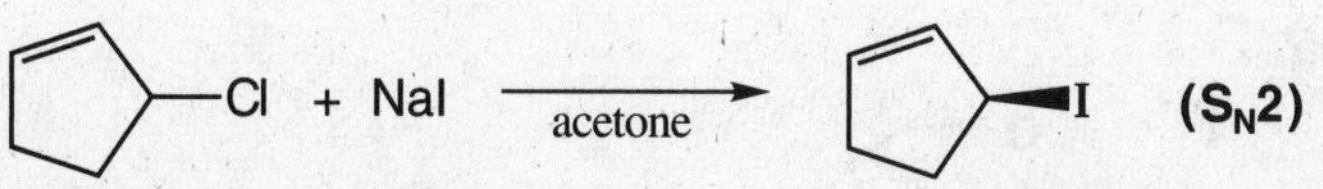 (S$_N$2)

R Enantiomer   *S* Enantiomer

(f)

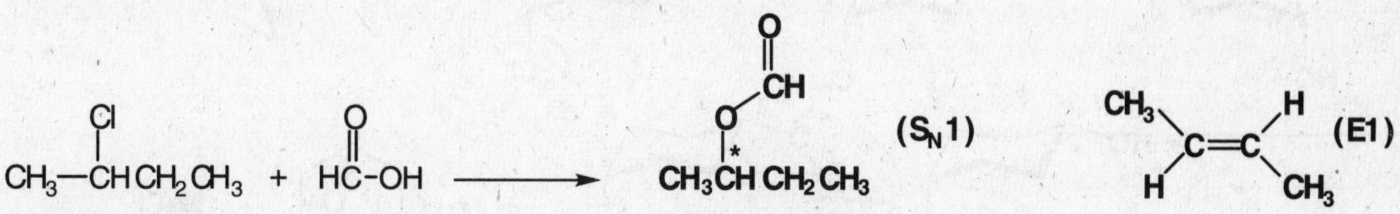

R enantiomer   **R,S enantiomers** (S$_N$1)   (E1)

(g) CH$_3$CH$_2$O$^-$ Na$^+$ + CH$_2$=CHCH$_2$Cl $\xrightarrow{\text{ethanol}}$ CH$_3$CH$_2$OCH$_2$CH=CH$_2$ (S$_N$2)

Problem 9.46 When *cis*-4-chlorocyclohexanol is treated with sodium hydroxide in ethanol, it gives mainly the substitution product *trans*-1,4-cyclohexanediol (1). Under the same reaction conditions, *trans*-4-chlorocyclohexanol gives 3-cyclohexenol (2) and the bicyclic ether (3).

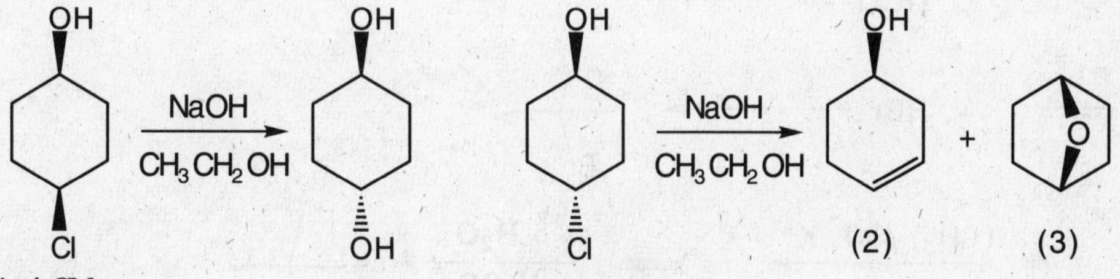

cis-**4-Chloro-**
cyclohexanol   (1)   *trans*-**4-Chloro-**
cyclohexanol   (2)   (3)

(a) Propose a mechanism for formation of product (1), and account for its configuration.

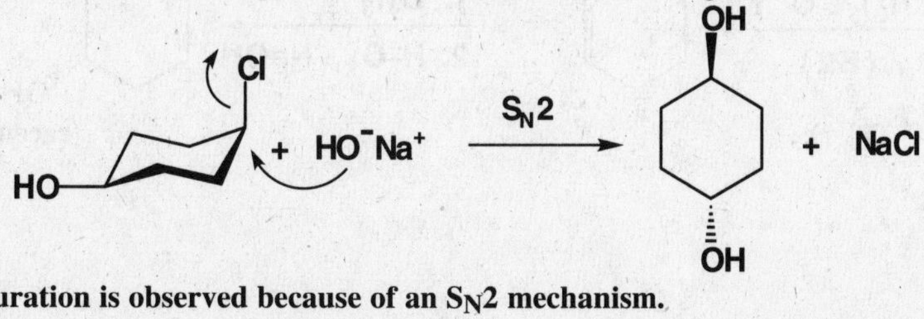

Inversion of configuration is observed because of an S$_N$2 mechanism.

(b) Propose a mechanism for formation of product (2).

The reaction takes place by an E2 mechanism. The molecule must adopt the chair conformation that places both the HO- and Cl- groups in the axial position in order for the reaction to occur. Note that either of the two axial H atoms indicated can be removed to create a racemic mixture.

(c) Account for the fact that the bicyclic ether (3) is formed from the *trans* isomer but not from the *cis* isomer.

The bicyclic ether product (3) is formed from an intramolecular backside attack of the deprotonated axial hydroxyl group upon an axial chlorine atom. Only the *trans* isomer can adopt the diaxial orientation necessary for this process.

Synthesis
Problem 9.47 Show how to convert the given starting material into the desired product. Note that some syntheses require only one step, whereas others require two or more.

(e)

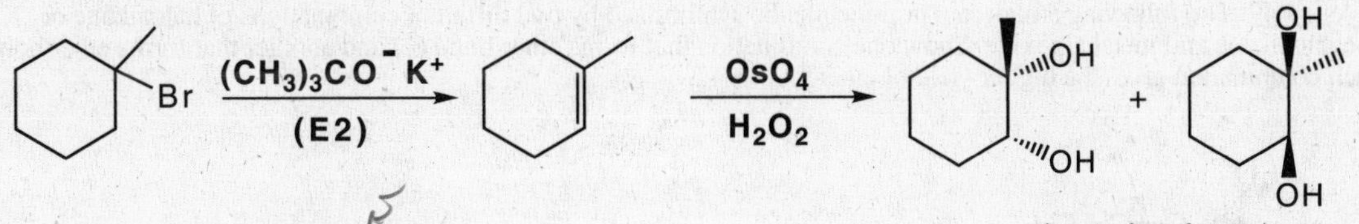

(racemic mixture)

(f)

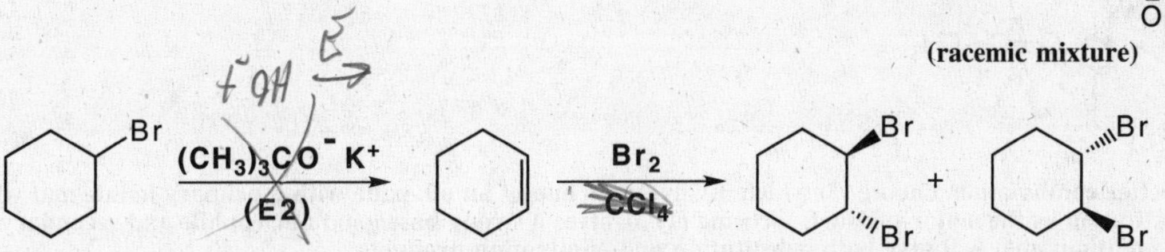

(racemic mixture)

(g)

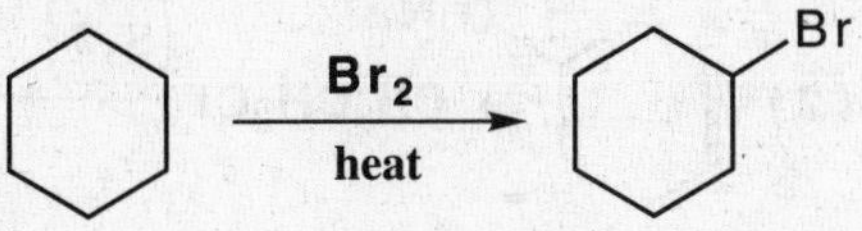

(h)

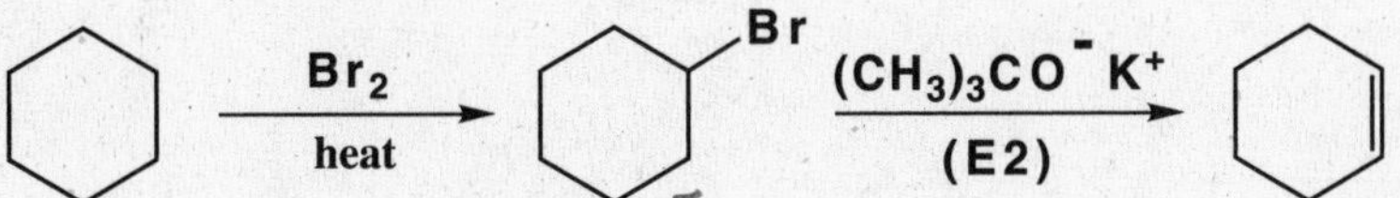

(i)

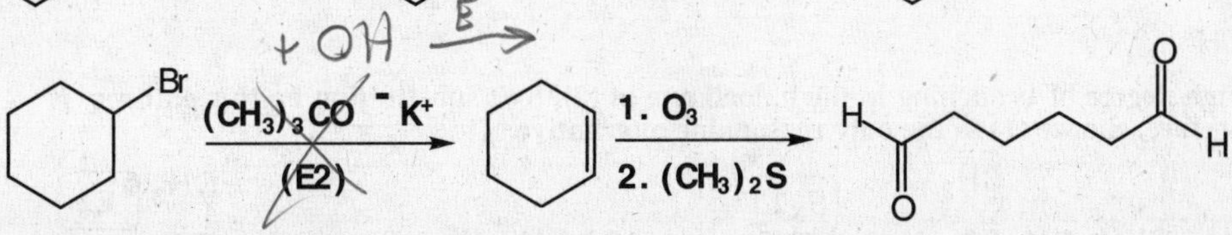

Problem 9.48 The Williamson ether synthesis involves treatment of a haloalkane with a metal alkoxide. Following are two reactions intended to give benzyl *tert*-butyl ether. One reaction gives the ether in good yield, the other reaction does not. Which reaction gives the ether? What is the product of the other reaction? How do you account for its formation?

(a)

(b)

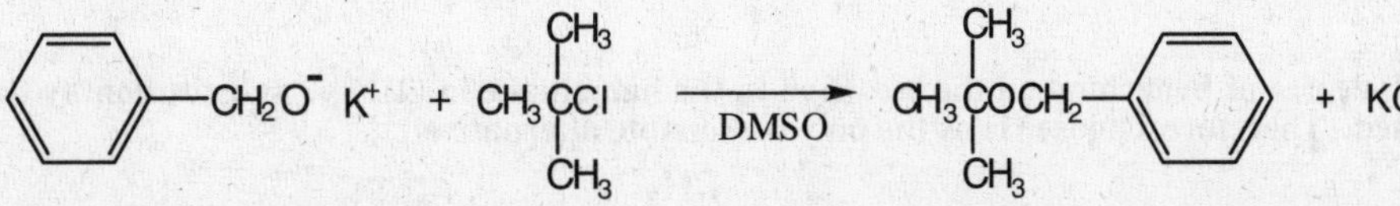

The only reaction that will give the desired ether product in good yield is the one shown in (a). In (b), the major product will be the E2 elimination product 2-methylpropene (CH₂=C(CH₃)₂), because the halogen atom is on a tertiary carbon atom and ethoxide is a strong base, both of which favor E2 reaction.

Problem 9.49 The following ethers can, in principle, be synthesized by two different combinations of haloalkane or halocycloalkane and metal alkoxide. Show one combination that forms ether bond (1) and another that forms ether bond (2). Which combination gives the higher yield of ether?

(a)

As the better combination, choose (2) which involves reaction of an alkoxide with a primary halide and will give substitution as the major product. Scheme (1) involves a strong base/good nucleophile and secondary halide, conditions that will give both substitution and elimination products.

(1) [cyclohexenyl-Cl] + CH$_3$CH$_2$O$^-$ Na$^+$ $\xrightarrow{\text{S}_N\text{2 and E2}}$ (2) [cyclohexenyl-O$^-$Na$^+$] + CH$_3$CH$_2$Cl $\xrightarrow{\text{S}_N\text{2}}$

(b) CH$_3$–O–C(CH$_3$)$_2$–CH$_3$ with CH$_3$

Because of the high degree of branching in the haloalkane in (2), S$_N$2 substitution by this pathway is impossible. Therefore, choose (1) as the only reasonable alternative.

(1) CH$_3$–Cl + CH$_3$C(CH$_3$)$_2$O$^-$ Na$^+$ $\xrightarrow{\text{S}_N\text{2}}$ (2) CH$_3$–O$^-$ Na$^+$ + CH$_3$C(CH$_3$)$_2$Cl $\xrightarrow{\text{E2}}$

(c) H$_2$C=CHCH$_2$–O–CH$_2$C(CH$_3$)$_2$CH$_3$

Because of the high degree of branching on the β-carbon in the haloalkane in (2), S$_N$2 substitution by this pathway is prevented. Therefore, choose (1) as the only reasonable alternative.

(1) H$_2$C=CHCH$_2$Cl + CH$_3$C(CH$_3$)$_2$CH$_2$O$^-$ Na$^+$ $\xrightarrow{\text{S}_N\text{2}}$

(2) H$_2$C=CHCH$_2$O$^-$ Na$^+$ + CH$_3$C(CH$_3$)$_2$CH$_2$Cl $\xrightarrow{\text{No reaction}}$

Problem 9.50 Propose a mechanism for this reaction.

$$ClCH_2CH_2OH \xrightarrow{\text{Na}_2\text{CO}_3,\ \text{H}_2\text{O}} H_2C \overset{O}{\underset{\diagdown}{\diagup}} CH_2$$

The mechanism of this reaction involves an initial deprotonation of the hydroxyl group, followed by an intramolecular S_N2 reaction to give the epoxide.

Step 1: Take a proton away

$$ClCH_2CH_2OH \ + \ Na_2CO_3 \ \rightleftharpoons \ ClCH_2CH_2O^-Na^+ \ + \ NaHCO_3$$

Step 2: Make a new bond between a nucleophile and an electrophile and simultaneously break a bond to give stable molecules or ions

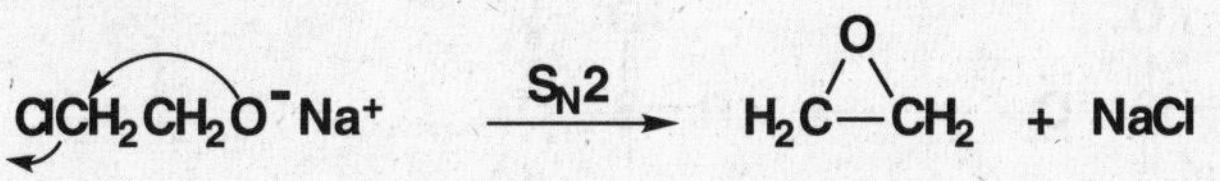

$$ClCH_2CH_2O^-Na^+ \xrightarrow{S_N2} H_2C \overset{O}{\underset{\diagdown}{\diagup}} CH_2 \ + \ NaCl$$

Problem 9.51 Each of these compounds can be synthesized by an S_N2 reaction. Suggest a combination of haloalkane and nucleophile that will give each product.

In the following reactions, the Br atom could be replaced with Cl or I, except for (f) and (k) in which Cl is required.

(a) $CH_3OCH_3 \longleftarrow CH_3O^-Na^+ \ + \ CH_3Br$

(b) $CH_3SH \longleftarrow HS^-Na^+ \ + \ CH_3Br$

(c) $CH_3CH_2CH_2PH_2 \longleftarrow PH_3 \ + \ CH_3CH_2CH_2Br$

(d) $CH_3CH_2CN \longleftarrow Na^+CN^- \ + \ CH_3CH_2Br$

(e) $CH_3SCH_2C(CH_3)_3 \longleftarrow (CH_3)_3CCH_2S^-Na^+ \ + \ CH_3Br$

(f) $(CH_3)_3NH^+Cl^- \longleftarrow (CH_3)_2NH \ + \ CH_3Cl$

(g) $C_6H_5\overset{O}{\overset{\|}{C}}OCH_2C_6H_5 \longleftarrow C_6H_5\overset{O}{\overset{\|}{C}}O^-Na^+ \ + \ C_6H_5CH_2Br$

(h) $(R)\text{-}CH_3\overset{N_3}{\underset{|}{C}}HCH_2CH_2CH_3 \longleftarrow Na^+N_3^- \ + \ (S)\text{-}CH_3\overset{Br}{\underset{|}{C}}HCH_2CH_2CH_3$

(i) $CH_2=CHCH_2OCH(CH_3)_2 \longleftarrow (CH_3)_2CHO^-Na^+ \ + \ CH_2=CHCH_2Br$

(j) $CH_2=CHCH_2OCH_2CH=CH_2 \longleftarrow CH_2=CHCH_2O^-Na^+ \ + \ CH_2=CHCH_2Br$

(k) 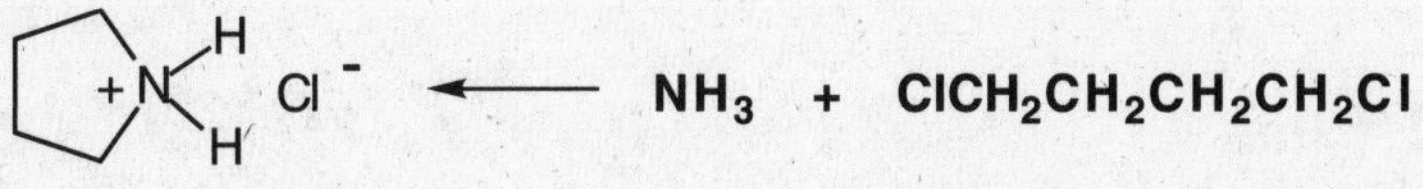 $\underset{H}{\overset{H}{\underset{\displaystyle \diagup}{\overset{\displaystyle \diagdown}{+N}}}}\ Cl^- \longleftarrow NH_3 \ + \ ClCH_2CH_2CH_2CH_2Cl$

(l) 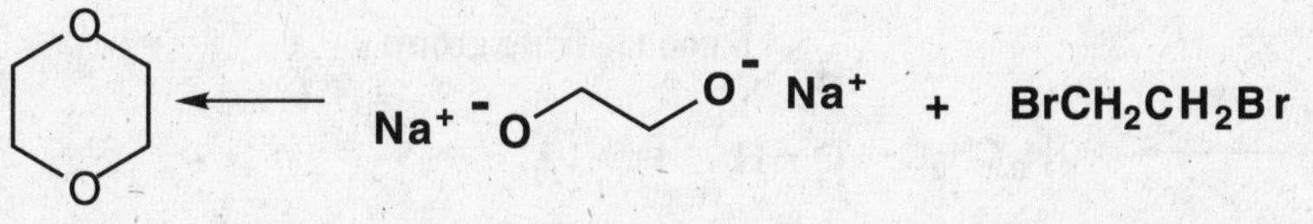 (1,4-dioxane ring) $\longleftarrow Na^+\ ^-O \diagdown\diagup O^-Na^+ \ + \ BrCH_2CH_2Br$

Looking Ahead

<u>Problem 9.52</u> OH⁻ is a very poor leaving group. However, many alcohols react with alkyl or aryl sulfonyl chlorides to give sulfonate esters.

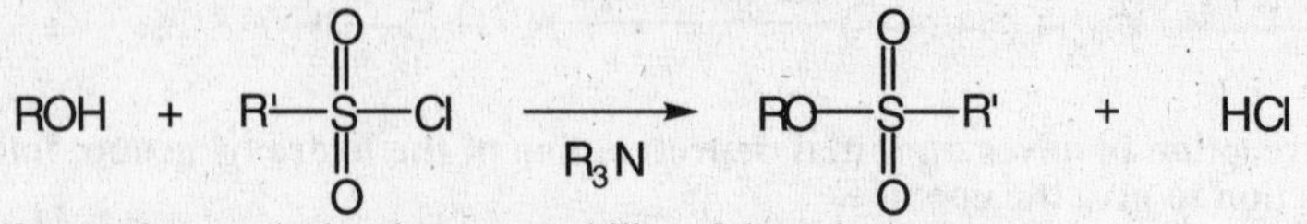

$$\text{ROH} + \text{R'}\text{SO}_2\text{Cl} \xrightarrow{R_3N} \text{RO}\text{SO}_2\text{R'} + \text{HCl}$$

(a) Explain what this change does to the leaving group ability of the substituent.

Better leaving group ability is correlated with anion stability. Anion stability, in turn, is related to pK_a in that stronger acids have generally more stable anions as conjugate bases. Water has a pK_a of 15.7 and sulfonic acids have pK_a values in 0-1 range. Thus, the sulfonate anions are significantly more stable anions, and therefore significantly better leaving groups compared to OH⁻. The increased stability of sulfonate anions derives from resonance stabilization that serves to distribute the negative charge over all three oxygen atoms as shown below.

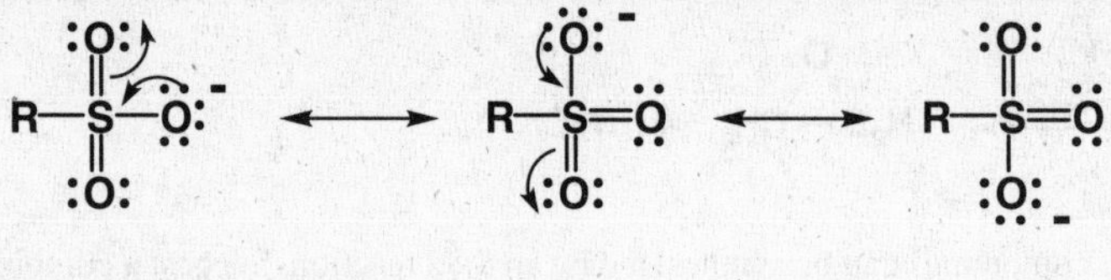

$$R-\overset{O}{\underset{O}{S}}-O^{-} \longleftrightarrow R-\overset{O^{-}}{\underset{O}{S}}=O \longleftrightarrow R-\overset{O}{\underset{O^{-}}{S}}=O$$

(b) Suggest the product of the following reaction.

$$\text{CH}_3\text{CH}_2\text{O}-\overset{O}{\underset{O}{S}}-\text{C}_6\text{H}_5 + \text{CH}_3\text{S}^-\text{Na}^+ \xrightarrow{\text{DMSO}} \text{CH}_3\text{CH}_2\text{SCH}_3 + \text{C}_6\text{H}_5-\overset{O}{\underset{O}{S}}-\text{O}^-$$

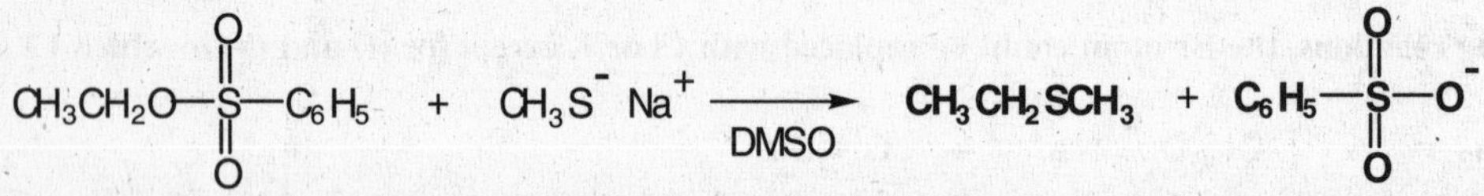

In this S_N2 reaction, the good nucleophile CH₃S⁻ displaces the sulfonate leaving group. The net process of reacting an -OH group with a sulfonyl chloride can thus be thought of as a good way turn the -OH group, a poor leaving group, into a good leaving group that can take part in many S_N2 reactions (Section 10.5D). This transformation is useful in a number of synthesis applications.

<u>Problem 9.53</u> Suggest a product of the following reaction. HI is a very strong acid.

$$\text{CH}_3\text{CH}_2\text{OCH}_2\text{CH}_3 + 2\ \text{HI} \longrightarrow 2\ \text{CH}_3\text{CH}_2\text{I} + \text{H}_2\text{O}$$

The key to this problem is to realize that the strong acid can protonate the ether oxygen and an -OH group to create better leaving groups (Section 10.5A) that can be displaced by the strong nucleophile I⁻ in two sequential S_N2 reactions.

Step 1: Add a proton

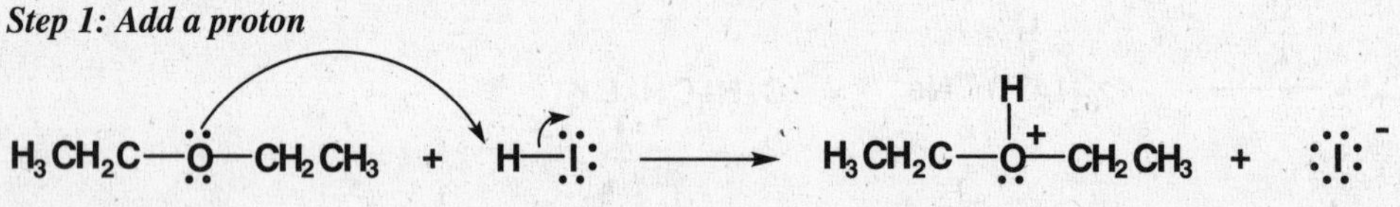

$$\text{H}_3\text{CH}_2\text{C}-\overset{..}{\underset{..}{O}}-\text{CH}_2\text{CH}_3 + \text{H}-\overset{..}{\underset{..}{I}}: \longrightarrow \text{H}_3\text{CH}_2\text{C}-\overset{H}{\underset{}{O^+}}-\text{CH}_2\text{CH}_3 + :\overset{..}{\underset{..}{I}}:^-$$

Step 2: Make a new bond between a nucleophile and an electrophile and simultaneously break a bond to give stable molecules or ions

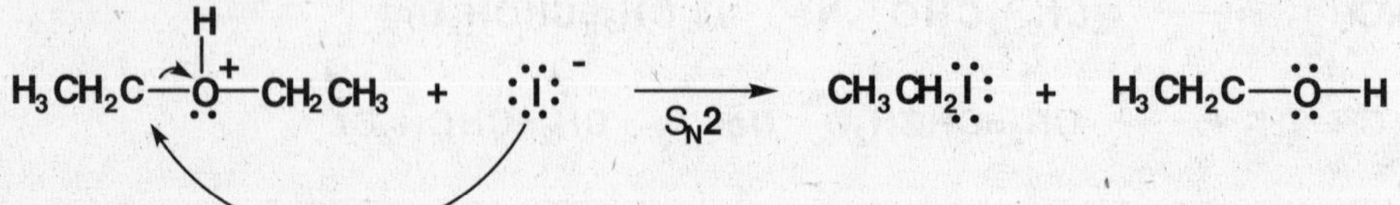

$$\text{H}_3\text{CH}_2\text{C}-\overset{H}{\underset{}{O^+}}-\text{CH}_2\text{CH}_3 + :\overset{..}{\underset{..}{I}}:^- \xrightarrow{S_N2} \text{CH}_3\text{CH}_2\overset{..}{\underset{..}{I}}: + \text{H}_3\text{CH}_2\text{C}-\overset{..}{\underset{..}{O}}-\text{H}$$

Step 3: Add a proton

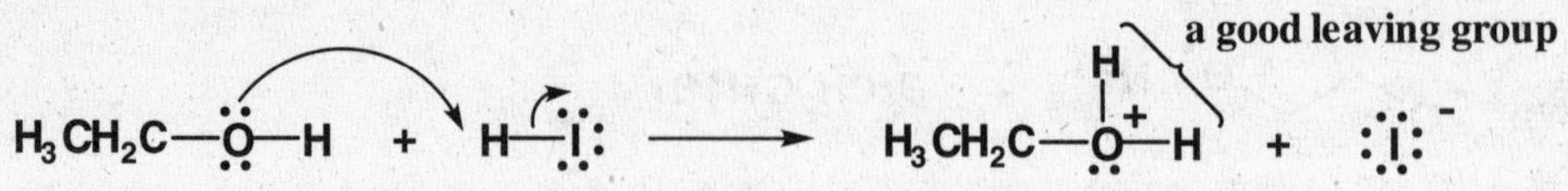

a good leaving group

$$\text{H}_3\text{CH}_2\text{C}-\overset{..}{\underset{..}{O}}-\text{H} + \text{H}-\overset{..}{\underset{..}{I}}: \longrightarrow \text{H}_3\text{CH}_2\text{C}-\overset{H}{\underset{}{O^+}}-\text{H} + :\overset{..}{\underset{..}{I}}:^-$$

Step 4: Make a new bond between a nucleophile and an electrophile and simultaneously break a bond to give stable molecules or ions

$$H_3CH_2C-\overset{+}{\underset{..}{O}}{-}H \;+\; :\ddot{I}:^- \;\xrightarrow{S_N2}\; CH_3CH_2\ddot{I}: \;+\; H-\ddot{O}-H$$

Multistep Synthesis Problems

Some reaction sequences are more useful than others in organic synthesis. Among the reactions you have a learned thus far, a particularly useful sequence involves the combination of free radical halogenation of an alkane to give a haloalkane, which is then subjected to an E2 elimination to give an alkene. The alkene is then converted to a variety of possible functional groups. Note that free radical halogenation is the only reaction you have seen that uses an alkane as a starting material.

Problem 9.54 Show how to convert butane into 2-butyne. Show all reagents and all molecules synthesized along the way.

Recognize that the product alkyne has the same number of carbons as the butane starting material, so no carbon-carbon bonds need to be made or broken. A good way to make an alkyne is from the vicinal dihalide. The vicinal dihalide is conveniently made from the corresponding alkene, which can be derived from the E2 elimination of a secondary haloakane as shown. The haloalkene, in turn, is easily made from the butane starting material using free radical halogenation. Notice that conversion of an alkane into an alkene using free radical halogenation followed by an E2 reaction is a very useful reaction sequence.

Problem 9.55 Show how to convert 2-methylbutane into racemic 3-bromo-2-methyl-2-butanol. Show all reagents and all molecules synthesized along the way.

2-Methylbutane   **3-Bromo-2-methyl-2-butanol**
                                           (racemic)

Recognize that the product halohydrin has the same number of carbons as the 2-methylbutane starting material, so no carbon-carbon bonds need to be made or broken. Halohydrins are made from the corresponding alkene treated with X_2 in H_2O, to the same sequence of free radical halogenation followed by an E2 elimination completes the synthesis as shown.

<u>Problem 9.56</u> Show how to convert cyclohexane into hexanedial. Show all reagents and all molecules synthesized along the way.

Recognize that the product hexanedial has the same number of carbons as the cyclohexane starting material, so no carbon-carbon bonds need to be made. However, the cyclohexane ring must be opened, so one carbon-carbon bond needs to be broken. The only reaction you have learned that can break a carbon-carbon bond is ozonolysis, and the dial product is consistent with the last step being ozonolysis starting with cyclohexene. The cyclohexene is conveniently prepared from the familiar free radical halogenation followed by E2 reaction sequence.

<u>Problem 9.57</u> Show how to convert cyclohexane into racemic 3-bromocyclohexene. Show all reagents and all molecules synthesized along the way.

Recognize that the racemic 3-bromocyclohexene product has the same number of carbons as the cyclohexane starting material, so no carbon-carbon bonds need to be made. Further, notice that the product is an allylic bromide, so propose the last step as an allylic halogenation using NBS. The appropriate starting material for an allylic halogenation reaction is cyclohexene, which is conveniently prepared from the familiar free radical halogenation followed by E2 reaction sequence.

Problem 9.58 Another important pattern in organic synthesis is the construction of C-C bonds. Show how to convert propane into hex-1-en-4-yne. You must use propane as the source of all of the carbon atoms in the hex-1-en-4-yne product. Show all reagents needed and all molecules synthesized along the way.

Recognize that the hex-1-en-4-yne product has six carbon atoms, so it must be derived by joining two three-carbon units. The only way you have learned to accomplish this is to react a primary haloalkane with an acetylide anion. The required alkyne can be made from propane through the familiar free radical halogenation followed by E2, then halogenation, and finally alkyne formation using NaNH$_2$. The required primary haloalkane can be made through the allylic halogenation of propene using NBS.

Problem 9.59 Show how to convert propane into butyronitrile. You must use propane and sodium cyanide as the source of all of the carbon atoms in butyronitrile. Show all reagents and all molecules synthesized along the way.

Recognize that the butyronlitrile product has four carbon atoms, the same number as the two starting materials combined. The new carbon-carbon bond in the product nitrile is most conveniently prepared through an S$_N$2 reaction using sodium cyanide and 1-bromopropane. The 1-bromopropane could be obtained as the non-Markovnikov product by treating propene with HBr using free radical conditions. The propene could be derived as in Problem 9.58 using the familiar free radical halogenation followed by an E2 reaction sequence. An alternative approach is also possible in which the propene is reacted with NBS to give 3-bromopropene followed by hydrogenation (H$_2$/Pd) to generate the 1-bromopropane.

Reactions in Context

Problem 9.60 Fluticasone is a glucocorticoid drug that has been used to treat asthma. In the synthesis of fluticasone, the following transformation is used that involves a limiting amount of sodium iodide. Analyze the structure using the chemistry you have learned in this chapter, and draw the product of the reaction.

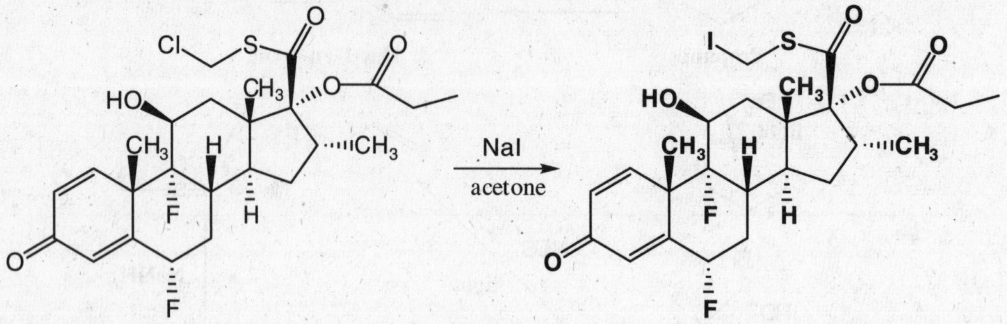

Iodide is a good nucleophile and the best leaving group on the starting material is the chlorine atom, so substitution will take place there.

Problem 9.61 The following reaction sequence was used in the synthesis of several derivatives prostaglandin C2. Analyze the structure using the chemistry you have learned in this chapter, and draw the structures of the synthetic intermediates A and B.

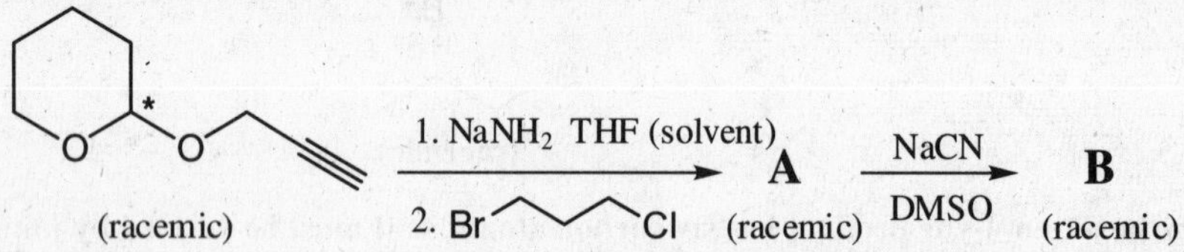

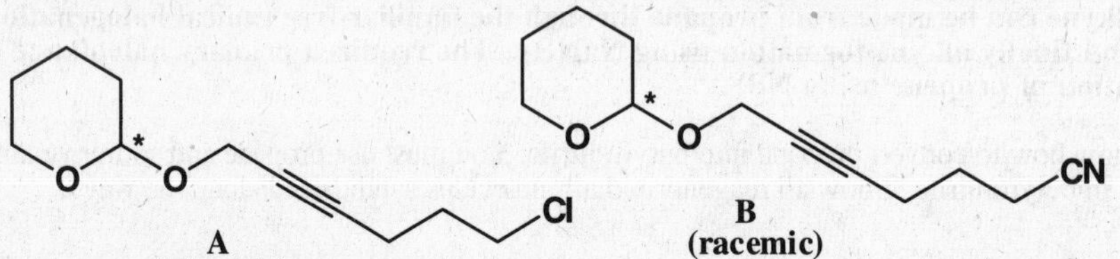

You should be able to recognize the useful reaction sequence of deprotonating a terminal alkyne followed by reaction with a primary haloalkane to give a new carbon-carbon bond. The reaction occurs at the Br atom, because bromide is a better leaving group than chloride. The second reaction is an S$_N$2 reaction using the cyanide anion as the nucleophile to displace the Cl leaving group.

Problem 9.62 The following reaction was used in the synthesis of various prostaglandin derivatives. Analyze the structure using the chemistry you have learned in this chapter, and draw the product of the reaction.

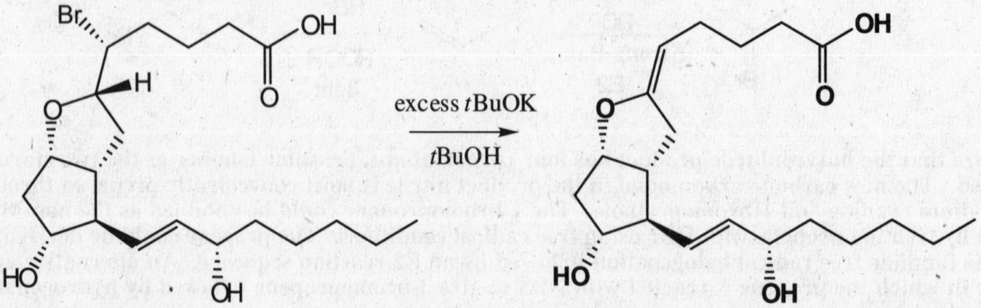

This reaction uses standard E2 conditions, and elimination occurs to give the more highly substituted alkene in agreement with Zaitsev's rule.

CHAPTER 10
Solutions to the Problems

Problem 10.1 Write IUPAC names for these alcohols, including configuration for (a):

(a)

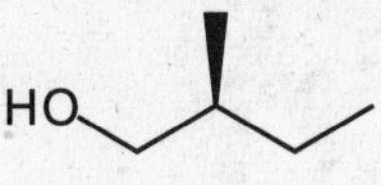

(b)

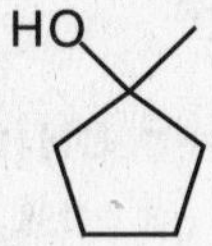

(S)-2-Methyl-1-butanol   **1-Methylcyclopentanol**

Problem 10.2 Classify each alcohol as primary, secondary, or tertiary.

(a) 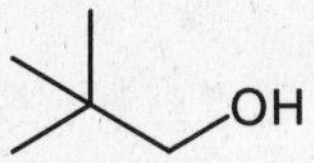   (b) 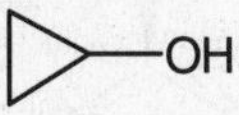   (c) 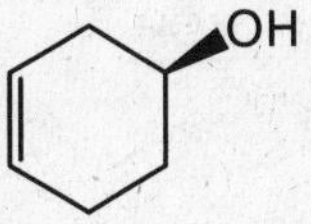   (d)

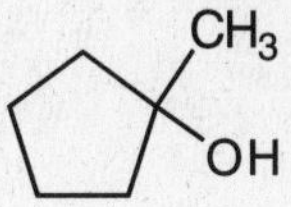

Primary   **Secondary**   **Secondary**   **Tertiary**

Problem 10.3 Write IUPAC names for these unsaturated alcohols.

(a)    (b)

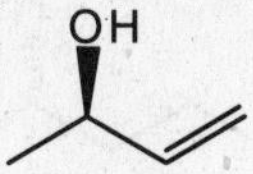

3-Buten-1-ol   **(R)-3-Buten-2-ol**

Problem 10.4 Arrange these compounds in order of increasing boiling point.

Heptane   **1,5-Pentanediol**   **1-Hexanol**

In order of increasing boiling point they are:

Heptane   **1-Hexanol**   **1,5-Pentanediol**
bp 98.4°C   **bp 157°C**   **bp 242°C**

Hydrogen bonding, or lack of it, is the key. Both 1-hexanol and 1,4-pentanediol can associate by hydrogen bonding, so their boiling points are substantially higher than heptane. Because 1,5-pentanediol has more sites for hydrogen bonding, it has a higher boiling point than 1-hexanol.

Problem 10.5 Arrange these compounds in order of increasing solubility in water.

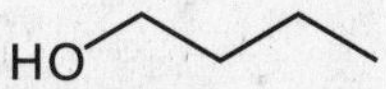

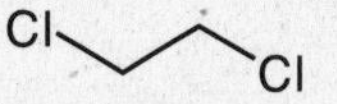

1-Butanol   **1-Propanol**   **1,2-Dichloroethane**

In order of increasing solubility in water, they are:

1,2-Dichloroethane   **1-Butanol**   **1-Propanol**
Slightly soluble   **8 g/100 g H$_2$O**   **Soluble in all**
proportions

Problem 10.6 Predict the position of equilibrium for this acid-base reaction.

$$CH_3CH_2O^- \ Na^+ \ + \ CH_3-\overset{\overset{\displaystyle O}{\|}}{C}-OH \ \rightleftharpoons \ CH_3CH_2OH \ + \ CH_3-\overset{\overset{\displaystyle O}{\|}}{C}-O^- \ Na^+$$

Acetic acid is the stronger acid; equilibrium lies to the right.

$$CH_3CH_2O^- \ Na^+ \ + \ CH_3-\overset{\overset{\displaystyle O}{\|}}{C}-OH \ \rightleftharpoons \ CH_3CH_2OH \ + \ CH_3-\overset{\overset{\displaystyle O}{\|}}{C}-O^- \ Na^+$$

	pK_a 4.76	pK_a 15.9	
(stronger base)	**(stronger acid)**	**(weaker acid)**	**(weaker base)**

Problem 10.7 Show how to convert (R)-2-pentanol to (S)-2-pentanethiol via a tosylate:

(R)-2-Pentanol   (S)-2-Pentanethiol

Step 1: **In the first step, which is really several steps, the -OH group is converted to the good leaving group -OTs by treatment with tosyl chloride (Ts-Cl) in pyridine.**

Step 2: Make a bond between a nucleophile and an electrophile and simultaneously break a bond to give stable molecules or ions. **Next the good nucleophile HS⁻ is used to carry out an S_N2 reaction and displace the -OTs group. Note that the inversion of stereochemistry seen with S_N2 reactions insures that the desired (S) isomer of 2-pentanethiol is produced.**

Problem 10.8 Draw structural formulas for the alkenes formed by acid-catalyzed dehydration of each alcohol. Where isomeric alkenes are possible, predict which is the major product.

(a) 2-Methyl-2-butanol

(b) 1-Methylcyclopentanol

<u>Problem 10.9</u> Propose a mechanism to account for this acid-catalyzed dehydration.

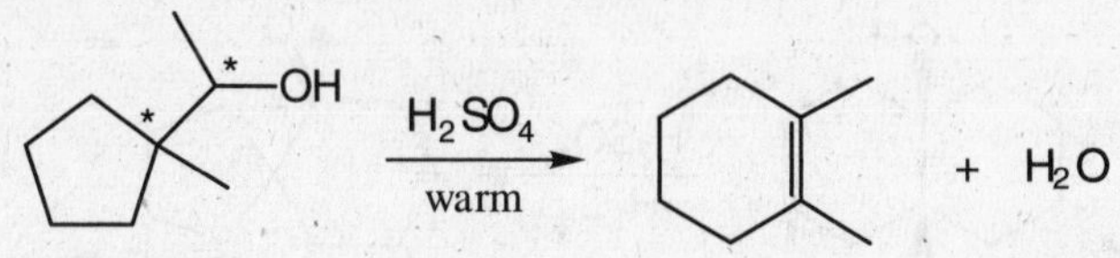

(racemic)

This reaction can best be explained as an acid catalyzed E1 reaction with a carbocation rearrangement. The first steps involve protonation of the -OH group, followed by loss of H_2O to give a 2° carbocation.

Step 1: Add a proton.

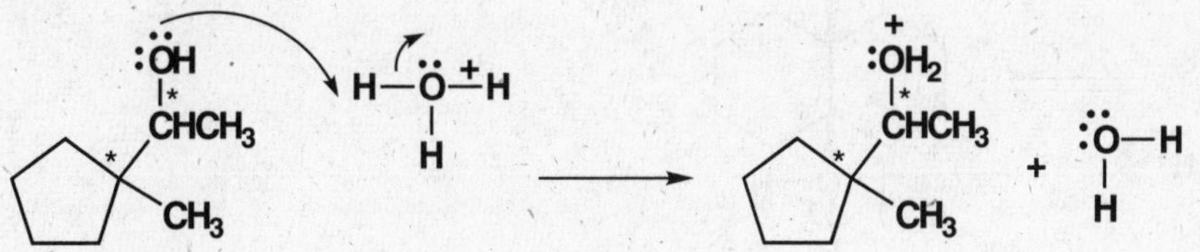

Step 2: Break a bond to give stable molecules or ions.

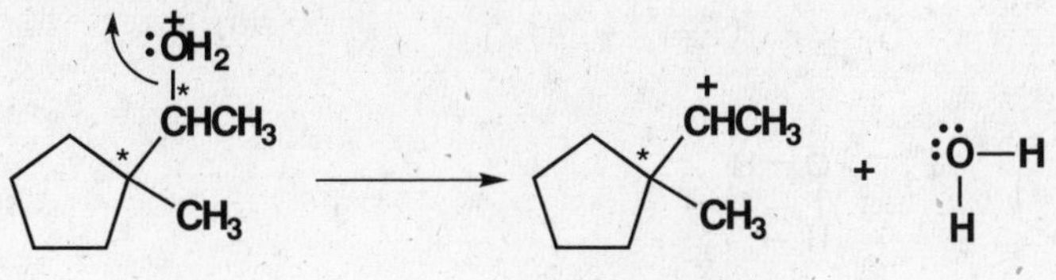

2° Carbocation

Step 3: 1,2 Shift. **This 2° carbocation then rearranges to a more stable 3° carbocation by expanding to a six-membered ring.**

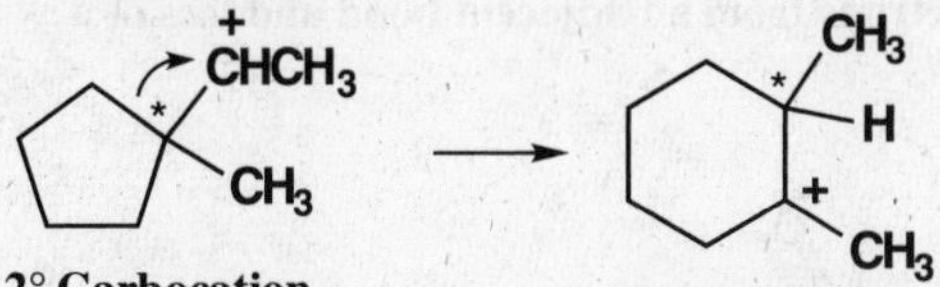

2° Carbocation

More stable
3° carbocation

Step 4: Take a proton away. **The 3° carbocation completes the E1 reaction by losing a proton to generate the product alkene.**

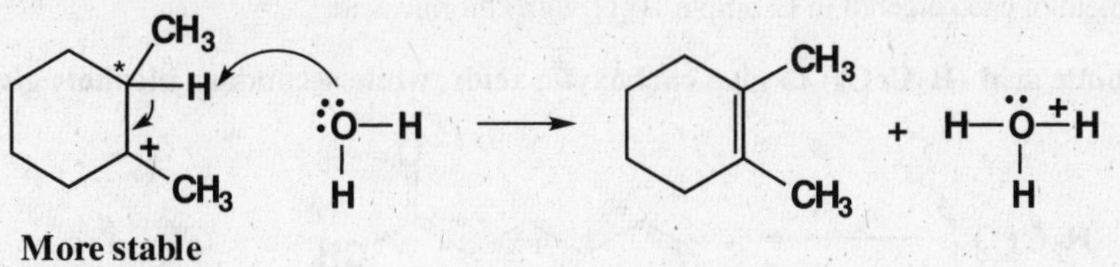

More stable
3° carbocation

Problem 10.10 Propose a mechanism to account for the following transformation:

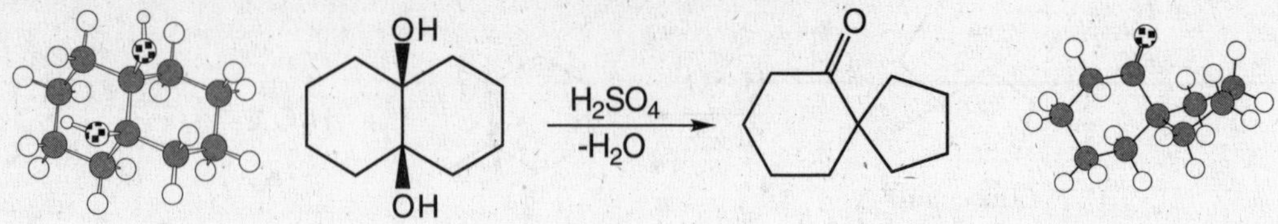

Step 1: Add a proton. **Protonation of either hydroxyl group followed by departure of water gives a tertiary carbocation in the first two steps.**

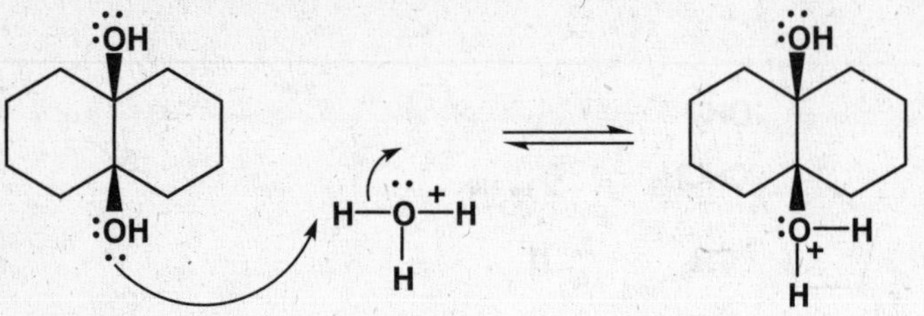

Step 2: Break a bond to give stable molecules or ions.

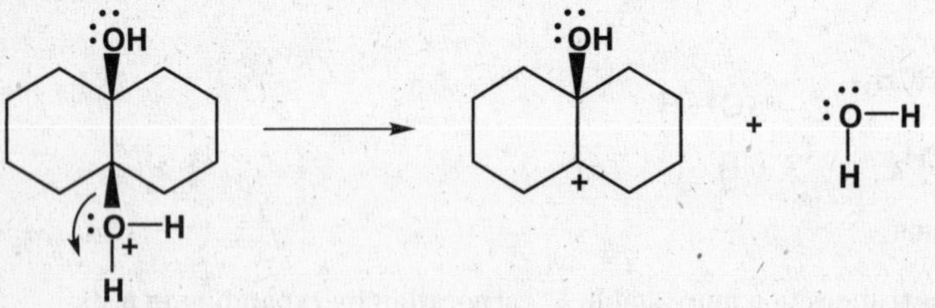

Step 3: Take a proton away and 1,2 shift. **Migration of a pair of electrons from an adjacent bond and loss of a proton to give a ketone gives the observed product.**

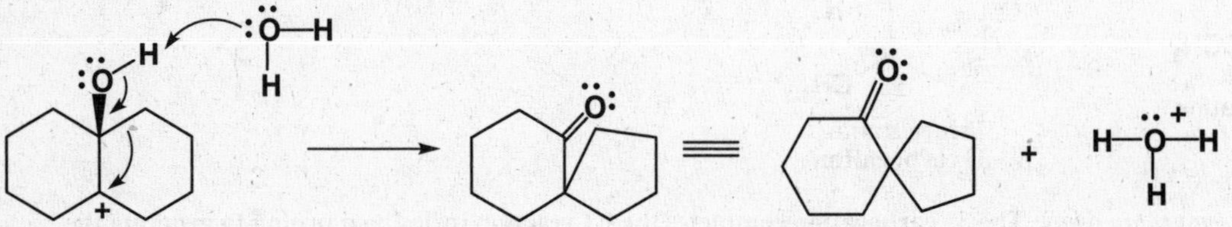

Problem 10.11 Draw the product of treatment of each alcohol in Example 10.11 with chromic acid.

Primary alcohols are oxidized by chromic acid (H_2CrO_4) to give carboxylic acids, while secondary alcohols give ketones.

(a)

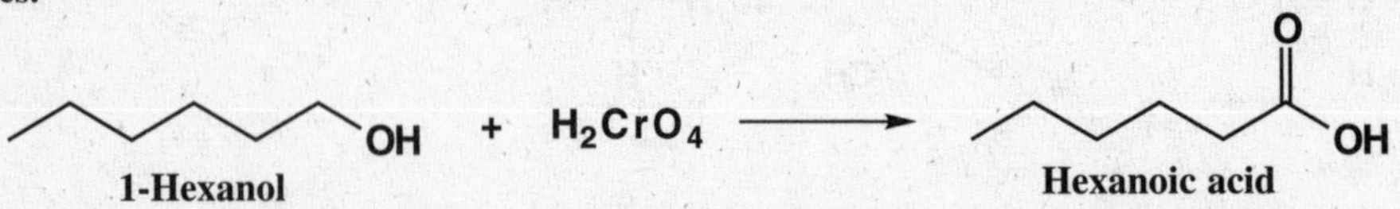

1-Hexanol **Hexanoic acid**

(b)

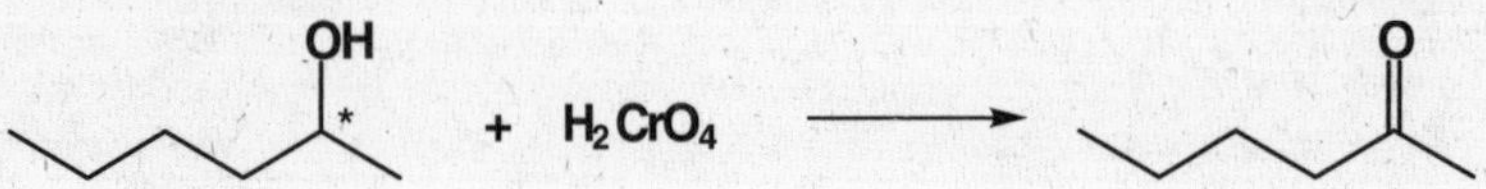

2-Hexanol **2-Hexanone**

(c)

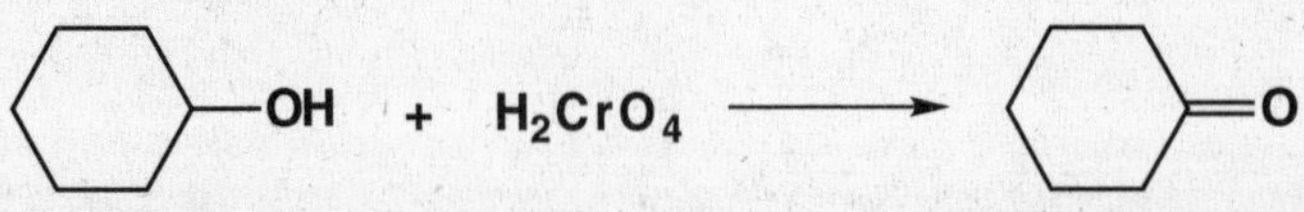

Cyclohexanol **Cyclohexanone**

Problem 10.12 α-Hydroxyketones and α-hydroxyaldehydes are also oxidized by treatment with periodic acid.

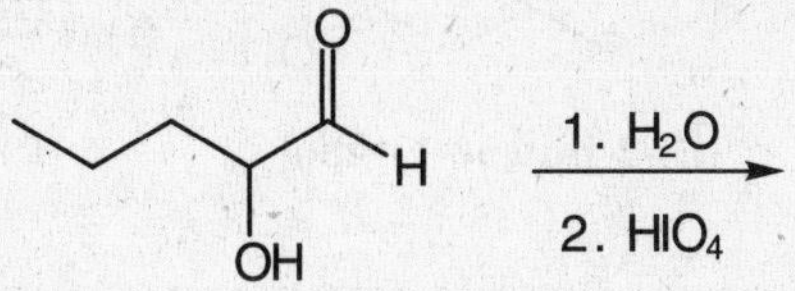

It is not the α-hydroxyketone or aldehyde, however, that undergoes reaction with periodic acid, but the hydrate formed by addition of water to the carbonyl group of the α-hydroxyketone or aldehyde. Write a mechanism for the oxidation of this α-hydroxyaldehyde with HIO_4.

Step 1: **Which is actually several steps, the aldehyde group reacts with water to give the following molecule, a geminal diol.**

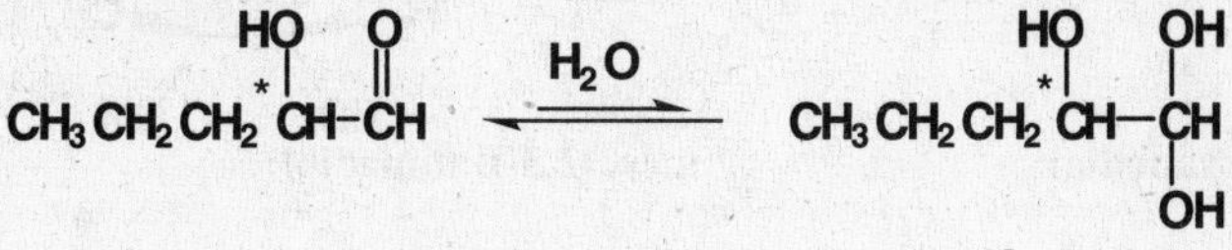

Step 2: **Which is also actually several steps, this geminal diol then reacts with HIO_4 according to the usual mechanism to give butanal plus formic acid.**

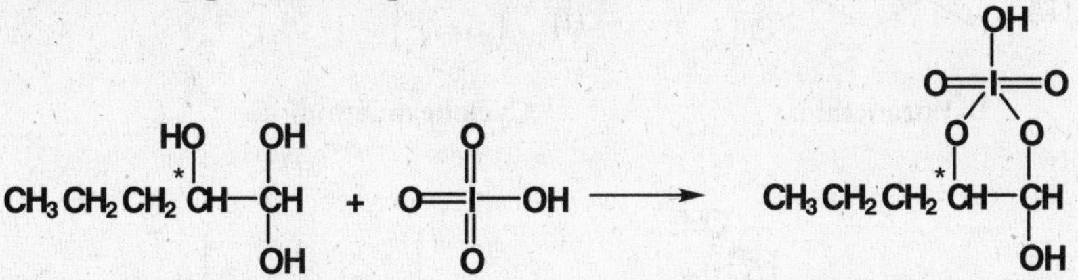

Step 3:

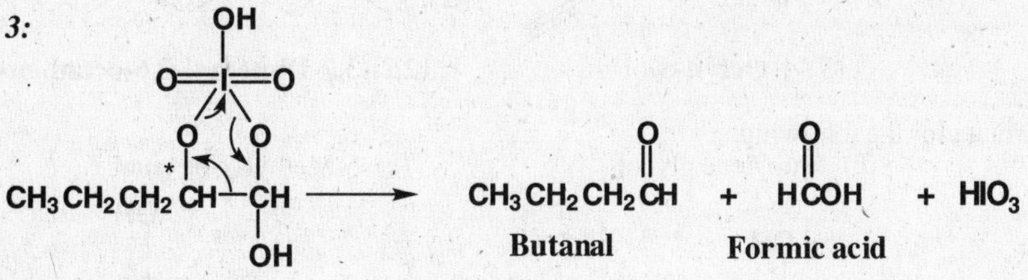

Problem 10.13 Write IUPAC names for these thiols.

(a)

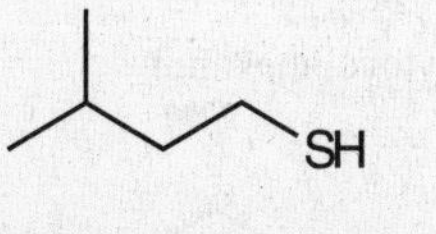

3-Methyl-1-butanethiol

(b)

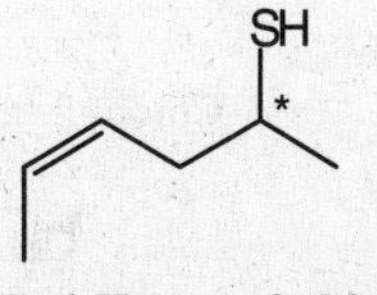

(Z)-4-Hexene-2-thiol

Structure and Nomenclature
Problem 10.14 Which are secondary alcohols?

(a) 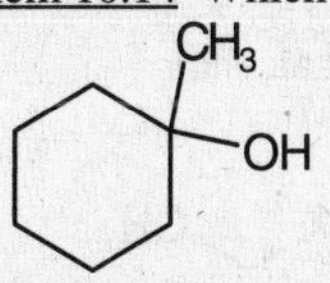   (b) $(CH_3)_3COH$ (c) 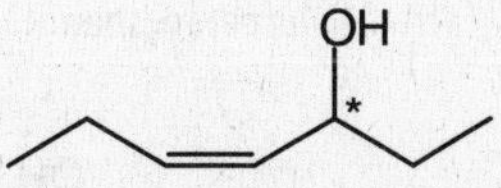   (d)

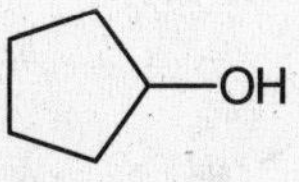

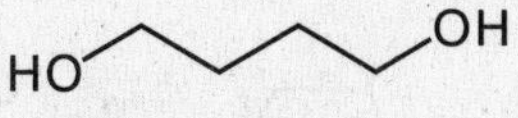

The secondary alcohols are (c) and (d). Molecules (a) and (b) are tertiary alcohols.

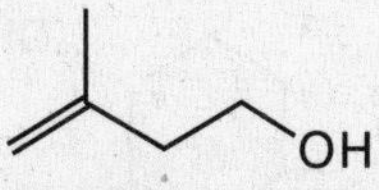

Problem 10.15 Name each compound.

(a) 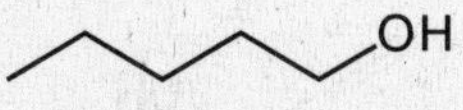   (b)   (c)

1-Pentanol   **1,4-Butanediol**   **3-Methyl-3-buten-1-ol**

(d) **(R)-1-Chloro-2-propanol**

(e) **(E)-2-Butene-1,4-diol**

(f) **(R)-1,3-butanediol**

(g) **(R)-3-Cyclohexenol**

(h) **cis-1,2-Cyclohexanediol**

(i) **meso-2,3-Butanediol**

(j) **(1R,2R)-trans-2-Bromocyclohexanol**

(k) **1-Butanethiol**

(l) **Cyclohexanethiol**

(m) **1,2-Ethanedithiol**

(n) **(S)-1-Octyn-3-ol**

(o) **(2E)-3,7-Dimethyl-2,6-octadien-1-ol**

<u>Problem 10.16</u> Write structural formula for the following:

(a) Isopropyl alcohol

(b) Propylene glycol

(c) 5-Methyl-2-hexanol

(d) 2-Methyl-2-propyl-1,3-propanediol

(e) 1-Chloro-2-hexanol

(f) cis-3-Isobutylcyclohexanol

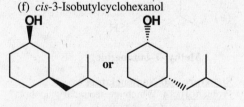

(g) 2,2-Dimethyl-1-propanol

(h) 2-Mercaptoethanol

(i) Allyl alcohol

(j) trans-2-Vinylcyclohexanol

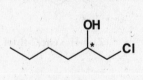

(k) (Z)-5-Methyl-2-hexen-1-ol

(l) 2-Propyn-1-ol

(m) 3-Chloro-1,2-propanediol   (n) *cis*-3-Pentene-1-ol

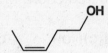

Problem 10.17 Name and draw structural formulas for the eight constitutional isomeric alcohols with molecular formula $C_5H_{12}O$. Classify each as alcohol as primary, secondary, or tertiary. Which are chiral?

The eight isomeric alcohols of molecular formulas $C_5H_{12}O$ are grouped by carbon skeleton. First, the three alcohols derived from pentane, then the four derived from 2-methylbutane (isopentane), and then the single alcohol derived from 2,2-dimethylpropane (neopentane). Each is given an IUPAC name. Common names, where appropriate, are given in parentheses.

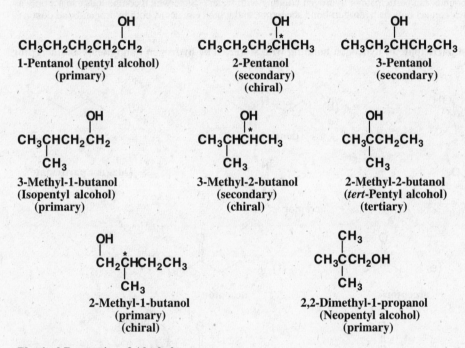

1-Pentanol (pentyl alcohol)
(primary)

2-Pentanol
(secondary)
(chiral)

3-Pentanol
(secondary)

3-Methyl-1-butanol
(Isopentyl alcohol)
(primary)

3-Methyl-2-butanol
(secondary)
(chiral)

2-Methyl-2-butanol
(*tert*-Pentyl alcohol)
(tertiary)

2-Methyl-1-butanol
(primary)
(chiral)

2,2-Dimethyl-1-propanol
(Neopentyl alcohol)
(primary)

Physical Properties of Alcohols

Problem 10.18 Arrange these compounds in order of increasing boiling point. (Values in °C are -42, 78, 138, and 198.)

(a) ⌇⌇⌇OH   (b) ⌇OH   (c) HO⌇⌇OH   (d) ⌇⌇

In order of increasing boiling point they are:

Propane
bp -42°C

Ethanol
bp 78°C

1-Pentanol
bp 138°C

Ethylene glycol
bp 198°C

The keys for this problem are hydrogen bonding and size. Propane cannot make any hydrogen bonds, so it has the lowest boiling point by far. Ethanol and 1-pentanol can each make hydrogen bonds through their single -OH group. However, 1-pentanol is larger (greater dispersion forces, higher molecular weight), so it will have a higher boiling point than ethanol. Ethylene glycol has two -OH groups per molecule with which to make hydrogen bonds, so it will have the highest boiling point of this set.

Problem 10.19 Arrange these compounds in order of increasing boiling point. (Values in °C are -42, -24, 78, and 118.)
(a) CH_3CH_2OH (b) CH_3OCH_3 (c) $CH_3CH_2CH_3$ (d) CH_3COOH

In order of increasing boiling point they are:

$CH_3CH_2CH_3$	CH_3OCH_3	CH_3CH_2OH	CH_3CO_2H
Propane	Dimethyl ether	Ethanol	Acetic acid
bp -42°C	bp -24°C	bp 78°C	bp 118°C

The keys for this problem are hydrogen bonding, polarity, and size. We know from the last problem that propane and ethanol have boiling points of -42°C and 78°C, respectively. Dimethyl ether is polar, but cannot make hydrogen bonds. Therefore, it makes sense that dimethyl ether has a boiling point (-24°C) that is higher than propane, but lower than ethanol. Acetic acid can make strong hydrogen bonds and has a higher molecular weight than ethanol, so it makes sense that acetic acid has the highest boiling point of this set.

Problem 10.20 Which compounds can participate in hydrogen bonding with water? State which compounds can act only as hydrogen-bond donors, which can act only as hydrogen-bond acceptors, and which can act as both hydrogen-bond donor and hydrogen bond acceptors.

The molecules in bold can participate in hydrogen bonding with water. The hydrogen bond donor and acceptor sites are labeled.

Problem 10.21 From each pair of compounds, select the one that is more soluble in water.

(a) CH_2Cl_2 or CH_3OH

Methanol, CH_3OH, is soluble in all proportions in water. Dichloromethane, CH_2Cl_2, is insoluble. The highly polar -OH group of methanol is capable of participating both as a hydrogen bond donor and hydrogen bond acceptor with water and, therefore, interacts strongly with water by intermolecular association. No such interaction is possible with dichloromethane.

(b) $CH_3\overset{O}{\overset{\|}{C}}CH_3$ or $CH_3\overset{CH_2}{\overset{\|}{C}}CH_3$

Propanone (acetone), CH_3COCH_3, is soluble in water in all proportions. 2-Methylpropene (isobutylene) is insoluble in water. Acetone has a large dipole moment and can function as a hydrogen bond acceptor from water.

(c) CH_3CH_2Cl or $NaCl$

NaCl is the more soluble. Chloroethane is insoluble in water. Following is a review of some of the general water solubility rules developed in General Chemistry. For these rules, <u>soluble</u> is defined as dissolving greater than 0.10 mol/L. <u>Slightly soluble</u> is dissolving between 0.01 mol/L and 0.10 mol/L.
1. Sodium, potassium, and ammonium salts of halogens or nitrates are soluble.
2. Silver, lead, and mercury(I) salts of halogens are insoluble.
Thus, applying Rule 1, NaCl is soluble in water. Chloroethane (ethyl chloride) is a nonpolar organic compound and insoluble in water.

(d)   or

The alcohol is more soluble in water. Sulfur is less electronegative than oxygen, so an S-H bond is less polarized than an O-H bond. Hydrogen bonding is therefore weaker with thiols than alcohols, so the alcohol is more able to interact with water molecules through hydrogen bonding.

(e)   or

The alcohol is more able to interact with water through hydrogen bonding, and is more soluble in water. The hydroxyl group has both a hydrogen bond donor and acceptor (the oxygen and hydrogen atoms of the -OH group, respectively), while the carbonyl group has only a hydrogen bond acceptor (the oxygen atom).

<u>Problem 10.22</u> Arrange the compounds in each set in order of decreasing solubility in water.
(a) Ethanol, butane, diethyl ether

CH_3CH_2OH $CH_3CH_2OCH_2CH_3$ $CH_3C H_2CH_2CH_3$

soluble in all   8 g/100 mL water   insoluble in water
proportions

In general, the more strongly a molecule can take part in hydrogen bonding with water, the greater the molecule is able to interact with the water molecules and dissolve. Only ethanol can act both as a donor and acceptor of hydrogen bonds with water. Diethyl ether can act as an acceptor of hydrogen bonds. Butane can act as neither a donor nor an acceptor of hydrogen bonds.

(b) 1-Hexanol, 1,2-hexanediol, hexane

$CH_2CHCH_2CH_2CH_2CH_3$ $CH_3CH_2CH_2CH_2CH_2CH_2OH$ $CH_3CH_2CH_2CH_2CH_2CH_3$
$OH \ OH$

1,2-Hexanediol molecules can take part in more hydrogen bonds with water than the 1-hexanol, because the diol has two -OH groups. Hexane has no polar bonds and thus cannot take part in any dipole-dipole interactions with water molecules.

<u>Problem 10.23</u> Each compound given in this problem is a common organic solvent. From each pair of compounds, select the solvent with the greater solubility in water.

Solubility in water increases with increasing hydrogen bonding ability and decreases with increasing surface area of hydrophobic groups such as alkyl groups.

(a) CH_2Cl_2 or CH_3CH_2OH (b) Et_2O or $EtOH$

   CH_3CH_2OH EtOH

(c)   or   (d)   or

Problem 10.24 The decalinols A and B can be equilibrated using aluminum isopropoxide in 2-propanol (isopropyl alcohol) containing a small amount of acetone.

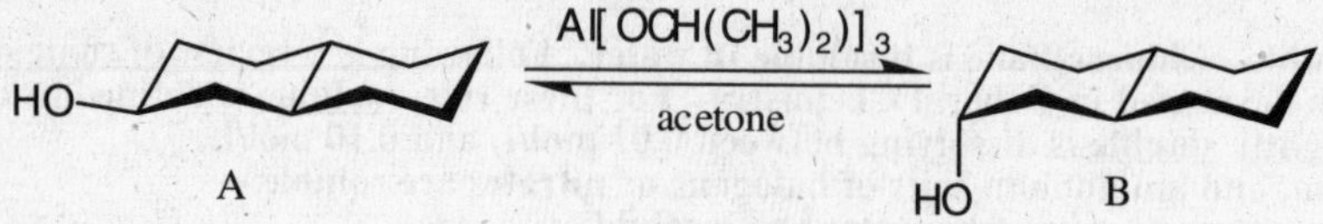

Assume a value of $\Delta G°$ (equatorial - axial) for cyclohexanol is 4.0 kJ(0.95 kcal)/mol, calculate the percent of each decalinol in the equilibrium mixture at 25°C.

At equilibrium, the relative amounts of A and B are given by the equation:

$$\Delta G° = -RT \ln K_{eq} \qquad \text{where} \quad K_{eq} = \frac{[B]}{[A]}$$

$$\ln K_{eq} = \frac{-\Delta G°}{RT}$$

Rearranging gives:

$$K_{eq} = e^{\left(\frac{-\Delta G°}{RT}\right)}$$

Plugging in the values for $\Delta G°$, R, and 298 K gives the final answer:

$$K_{eq} = e^{\left(\frac{-4,000 \text{ J/mol}}{(8.314 \text{ J} \cdot \text{K}^{-1} \cdot \text{mol}^{-1})(298 \text{ K})}\right)} = e^{-1.61} = 0.20$$

Thus, the molecule with an equatorial hydroxyl group (A) will predominate in a 5 to 1 ratio.

Acid-Base Reactions of Alcohols

Problem 10.25 Complete the following acid-base reactions. In addition, show all valence electrons on the interacting atoms and show by the use of curved arrows the flow of electrons in each reaction.

(a) $CH_3CH_2{-}O{-}H$ + H_3O^+ ⟶

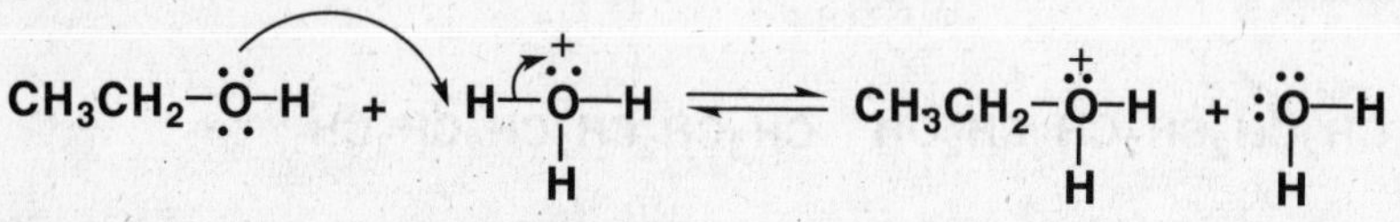

Brønsted-Lowry Acid

(b) 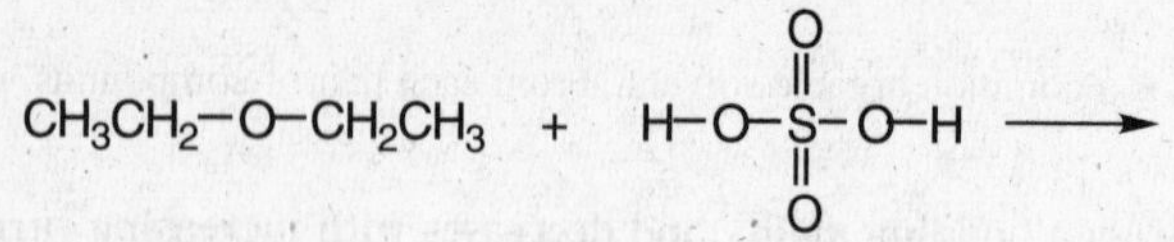 $CH_3CH_2{-}O{-}CH_2CH_3$ +   ⟶

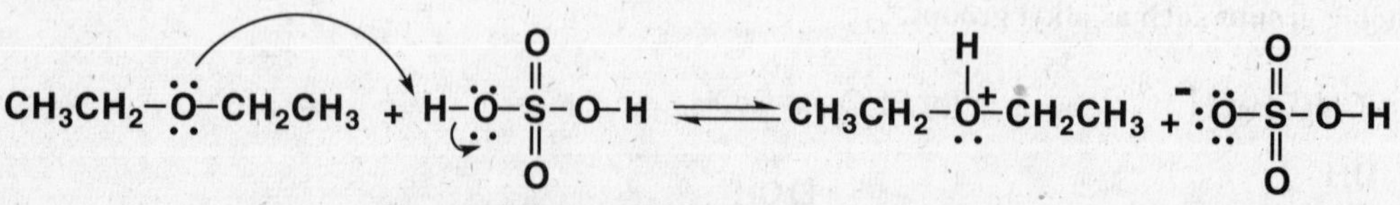

Brønsted-Lowry Acid

(c) 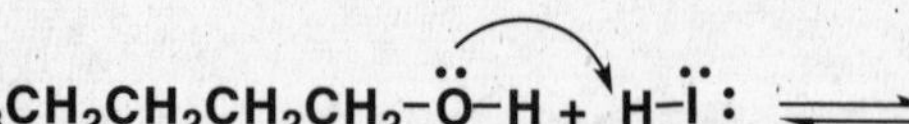   OH + HI ⟶

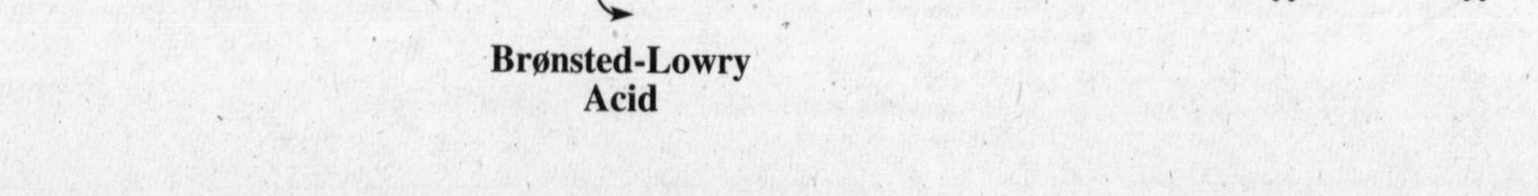

Brønsted-Lowry Acid

(d)

CH₃CH₂CH₂C–O–H + H–Ö–S–O–H → CH₃CH₂CH₂C–O–H + :Ö–S–O–H

Brønsted-Lowry Acid

(e)

Lewis acid

(f)

CH₃CH=CHCHCH₃ + H–Ö–H ⇌ CH₃CH=CHCHCH₃

Lewis acid

<u>Problem 10.26</u> Select the stronger acid from each pair and explain your reasoning. For each stronger acid, write a structural formula for its conjugate base.

(a) H_2O or $\boxed{H_2CO_3}$ (b) CH_3OH or $\boxed{CH_3COOH}$

(c) $\boxed{CH_3CH_2OH}$ or $CH_3C{\equiv}CH$ (d) CH_3CH_2OH or $\boxed{CH_3CH_2SH}$

Relative acidity rankings can be predicted by examining the relative stability of the anions produced upon deprotonation of the different acidic groups. *The more stable anion will be derived from deprotonation of the stronger acid.*

For (a), the carbonyl group of carbonic acid allows for resonance stabilization of the bicarbonate anion relative to the hydroxide anion.

For (b), the carbonyl group of acetic acid allows for resonance stabilization of the carboxylate anion relative to the methoxide anion.

For (c), oxygen is a more electronegative element (farther to the right on the Periodic Table) than carbon, adding greater stability to the alkoxide anion relative to the deprotonated alkyne.

For (d), sulfur is below oxygen in the Periodic Table and is a larger atom. Larger atoms are more able to carry negative charge, so the thiolate anion is more stable than the alkoxide anion.

Under each acid is given its pK_a. Recall that acid strength increases with decreasing values of pK_a.

	weaker acid	stronger acid	conjugate base of stronger acid
(a)	H_2O pK_a 15.7	H_2CO_3 pK_a 6.36	HCO_3^-
(b)	CH_3OH pK_a 15.5	CH_3COOH pK_a 4.76	CH_3COO^-
(c)	$CH_3C{\equiv}CH$ pK_a 25	CH_3CH_2OH pK_a 15.9	$CH_3CH_2O^-$
(d)	CH_3CH_2OH pK_a 15.9	CH_3CH_2SH pK_a 8.5	$CH_3CH_2S^-$

<u>Problem 10.27</u> From each pair, select the stronger base. For each stronger base, write a structural formula of its conjugate acid.

Recall that the stronger base will have the weaker conjugate acid.

(a) OH^- or CH_3O^- (each in H_2O)

$$HO^- \longrightarrow HOH$$
stronger   weaker
base   acid

This is a close one, but HOH is a weaker acid than CH_3OH because HOH has a less-electronegative H atom bonded to the OH group.

(b) $CH_3CH_2O^-$ or $CH_3C{\equiv}C^-$

$$CH_3C{\equiv}C^- \longrightarrow CH_3C{\equiv}CH$$
stronger   weaker
base   acid

Carbon is less electronegative (farther to the left on the Periodic Table) than oxygen, so the alkyne is the weaker acid.

(c) $CH_3CH_2S^-$ or $CH_3CH_2O^-$

$$CH_3CH_2O^- \longrightarrow CH_3CH_2OH$$
stronger   weaker
base   acid

Oxygen is above sulfur in the Periodic Table, so the alcohol is the weaker acid.

(d) $CH_3CH_2O^-$ or NH_2^-

$$NH_2^- \longrightarrow NH_3$$
stronger   weaker
base   acid

Nitrogen is less electronegative (farther to the left on the Periodic Table) than oxygen, so ammonia is the weaker acid.

<u>Problem 10.28</u> In each equilibrium, label the stronger acid, the stronger base, the weaker acid, and the weaker base. Also estimate the position of each equilibrium.

(a) $CH_3CH_2O^-$ + $CH_3C{\equiv}CH$ $\rightleftharpoons$ CH_3CH_2OH + $CH_3C{\equiv}C^-$

$$CH_3CH_2O^- + CH_3C{\equiv}CH \underset{}{\overset{}{\rightleftharpoons}} CH_3CH_2OH + CH_3C{\equiv}C^- \quad K_{eq} = 10^{-9.1}$$
                    pK_a 25 pK_a 15.9
weaker base   weaker acid   stronger acid   stronger base

Oxygen is a more electronegative element (farther to the right on the Periodic Table) than carbon, so the alcohol is the stronger acid.

(b) $CH_3CH_2O^-$ + HCl $\rightleftharpoons$ CH_3CH_2OH + Cl^-

$$CH_3CH_2O^- + HCl \underset{}{\overset{}{\rightleftharpoons}} CH_3CH_2OH + Cl^- \quad K_{eq} = 10^{22.9}$$
                 pK_a -7 pK_a 15.9
stronger   stronger   weaker   weaker
base   acid   acid   base

Chlorine is more electronegative (farther to the right on the Periodic Table) than oxygen, so HCl is the stronger acid.

(c) CH_3CO_2H + $CH_3CH_2O^-$ $\rightleftharpoons$ $CH_3CO_2^-$ + CH_3CH_2OH

$$CH_3CO_2H + CH_3CH_2O^- \underset{}{\overset{}{\rightleftharpoons}} CH_3CO_2^- + CH_3CH_2OH \quad K_{eq} = 10^{11.1}$$
pK_a 4.76 pK_a 15.9
stronger   stronger   weaker   weaker
acid   base   base   acid

The carbonyl group of acetic acid inductively weakens the O-H bond and stabilizes the anion through resonance.

Reactions of Alcohols

<u>Problem 10.29</u> Write equations for the reaction of 1-butanol with each reagent. Where you predict no reaction, write NR.
(a) Na metal

$$2 \ CH_3CH_2CH_2CH_2OH \ + \ 2 \ Na \longrightarrow 2 \ CH_3CH_2CH_2CH_2\bar{O} \ Na^+ \ + \ H_2$$

(b) HBr, heat

$$CH_3CH_2CH_2CH_2OH \ + \ HBr \xrightarrow{heat} CH_3CH_2CH_2CH_2Br \ + \ H_2O$$

(c) HI heat

$$CH_3CH_2CH_2CH_2OH \ + \ HI \xrightarrow{heat} CH_3CH_2CH_2CH_2I \ + \ H_2O$$

(d) PBr_3

$$3 \ CH_3CH_2CH_2CH_2OH \ + \ PBr_3 \longrightarrow 3 \ CH_3CH_2CH_2CH_2Br \ + \ P(OH)_3$$

(e) $SOCl_2$, pyridine

$$CH_3CH_2CH_2CH_2OH \ + \ SOCl_2 \xrightarrow{pyridine} CH_3CH_2CH_2CH_2Cl \ + \ SO_2 \ + \ HCl$$

(f) $K_2Cr_2O_7$, H_2SO_4, H_2O, heat

$$CH_3CH_2CH_2CH_2OH \ + \ K_2Cr_2O_7 \xrightarrow[\substack{H_2O \\ heat}]{H_2SO_4} CH_3CH_2CH_2\overset{\overset{O}{\|}}{C}OH \ + \ Cr^{3+}$$

(g) HIO_4
    NR

(h) PCC

$$CH_3CH_2CH_2CH_2OH \ + \ PCC \longrightarrow CH_3CH_2CH_2\overset{\overset{O}{\|}}{C}H \ + \ Cr^{3+}$$

(i) CH_3SO_2Cl, pyridine

$$CH_3CH_2CH_2CH_2OH \ + \ Cl\overset{\overset{O}{\|}}{\underset{\underset{O}{\|}}{S}}CH_3 \xrightarrow{pyridine} CH_3CH_2CH_2CH_2O\overset{\overset{O}{\|}}{\underset{\underset{O}{\|}}{S}}CH_3 \ + \ HCl$$

<u>Problem 10.30</u> Write equations for the reaction of 2-butanol with each reagent listed in Problem 10.29. Where you predict no reaction, write NR.
(a) Na metal

$$2 \ CH_3CH_2\overset{OH}{\underset{|*}{C}}HCH_3 \ + \ 2 \ Na \longrightarrow 2 \ CH_3CH_2\overset{\bar{O}\,Na^+}{\underset{|*}{C}}HCH_3 \ + \ H_2$$

(b) HBr, heat

$$CH_3CH_2\overset{OH}{\underset{|*}{C}}HCH_3 \ + \ HBr \xrightarrow{heat} CH_3CH_2\overset{Br}{\underset{|*}{C}}HCH_3 \ + \ H_2O$$

(c) HI, heat

$$CH_3CH_2\overset{OH}{\underset{|*}{C}}HCH_3 \ + \ HI \xrightarrow{heat} CH_3CH_2\overset{I}{\underset{|*}{C}}HCH_3 \ + \ H_2O$$

(d) PBr_3

$$3 \ CH_3CH_2\overset{OH}{\underset{|*}{C}}HCH_3 \ + \ PBr_3 \longrightarrow 3 \ CH_3CH_2\overset{Br}{\underset{|*}{C}}HCH_3 \ + \ P(OH)_3$$

(e) SOCl$_2$, pyridine

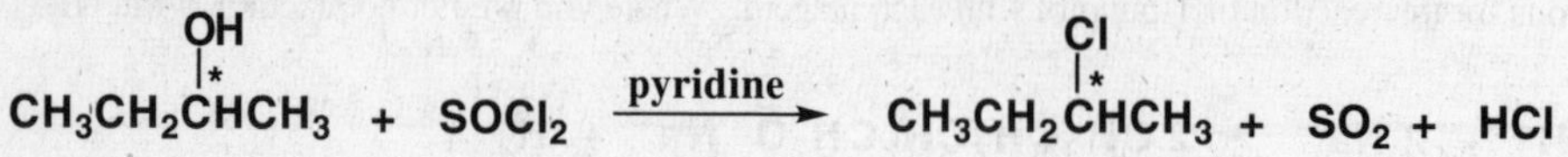

$$\underset{\overset{|}{*}}{\overset{\overset{OH}{|}}{}}$$
CH$_3$CH$_2$CHCH$_3$ + SOCl$_2$ $\xrightarrow{\text{pyridine}}$ CH$_3$CH$_2$CHCH$_3$ + SO$_2$ + HCl

(f) K$_2$Cr$_2$O$_7$, H$_2$SO$_4$, H$_2$O, heat

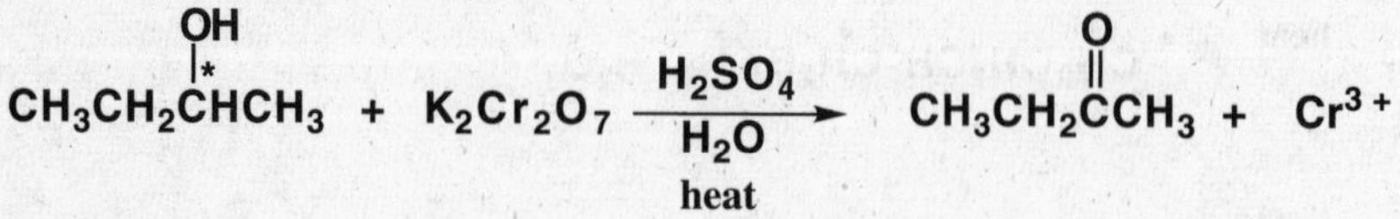

$$\overset{\overset{OH}{|}}{}$$
CH$_3$CH$_2$CHCH$_3$ + K$_2$Cr$_2$O$_7$ $\xrightarrow[\overset{H_2O}{\text{heat}}]{H_2SO_4}$ CH$_3$CH$_2$CCH$_3$ + Cr^{3+}

(g) HIO$_4$
    NR

(h) PCC

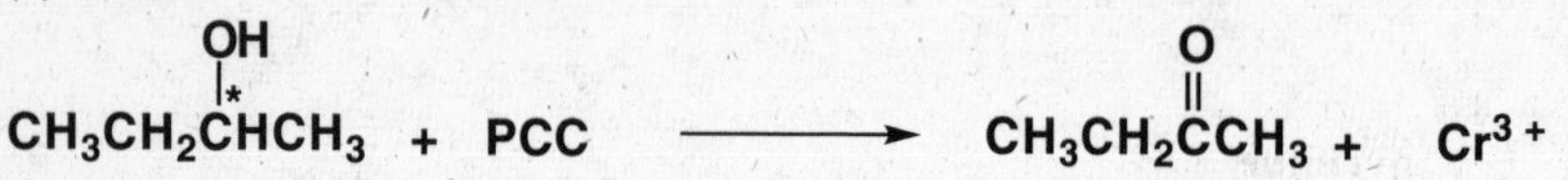

$$\overset{\overset{OH}{|}}{}$$
CH$_3$CH$_2$CHCH$_3$ + PCC $\longrightarrow$ CH$_3$CH$_2$CCH$_3$ + Cr^{3+}

(i) CH$_3$SO$_2$Cl, pyridine

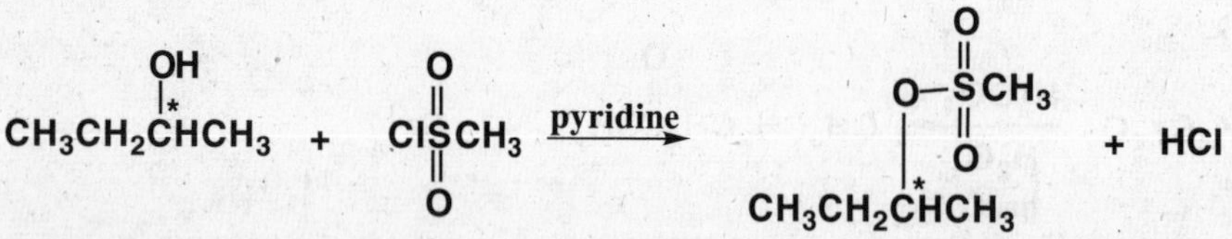

CH$_3$CH$_2$CHCH$_3$ + ClSCH$_3$ $\xrightarrow{\text{pyridine}}$ CH$_3$CH$_2$CHCH$_3$ + HCl

For the above reactions with a chiral center in the product, the stereochemistry observed depends on the stereochemistry of the starting material. In parts (a) and (i), the configuration of the starting material will be retained. In parts (d) and (e), the configuration of the starting material will be inverted. In parts (b) and (c), depending on conditions, a mixture of enantiomers might be observed even if the starting material is a single enantiomer.

<u>Problem 10.31</u> Complete these equations. Show structural formulas for the major products, but do not balance.

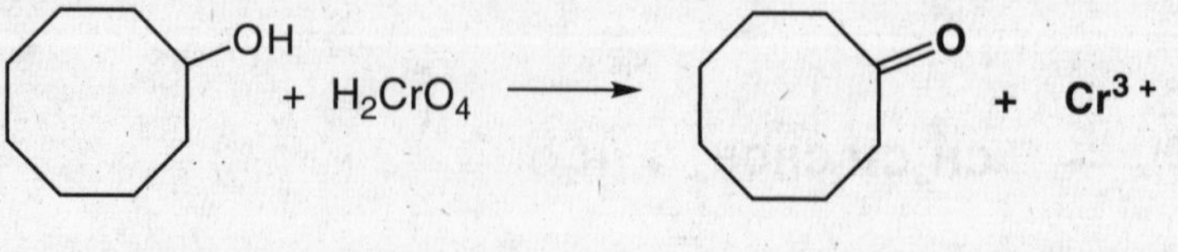

(a) + H$_2$CrO$_4$ $\longrightarrow$ + Cr^{3+}

(b) + SOCl$_2$ $\longrightarrow$ + SO$_2$ + HCl

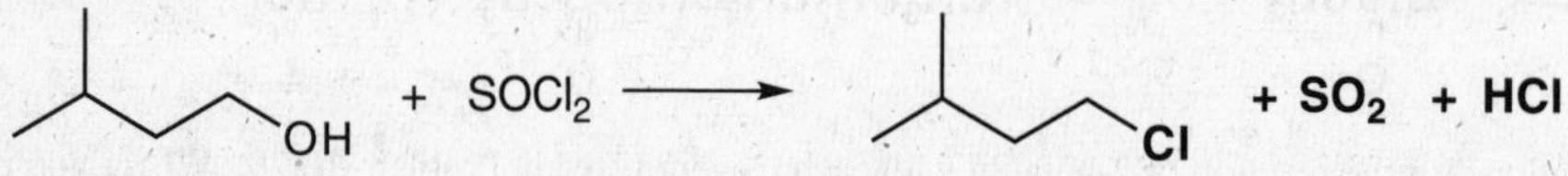

(c) + HCl $\longrightarrow$ + H$_2$O

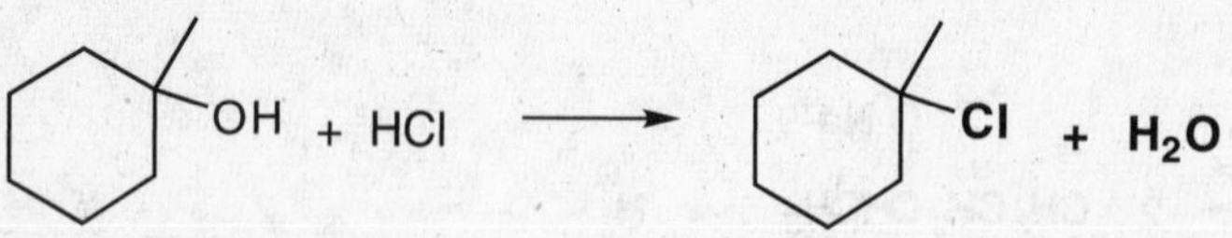

(d) HO $\diagup\diagdown\diagup$ OH + HBr $\longrightarrow$ Br $\diagup\diagdown\diagup$ Br + 2 H$_2$O
     (excess)

(e) + H$_2$CrO$_4$ $\longrightarrow$ + Cr^{3+}

(f)

(g)

(h)

Problem 10.32 When (R)-2-butanol is left standing in aqueous acid, it slowly loses its optical activity. Account for this observation.

In aqueous acid, (R)-2-butanol is in equilibrium with a small amount of the protonated form, that subsequently loses H$_2$O to create an achiral cation. The achiral cation reacts with another water molecule to give back 2-butanol. The key to this question is that the achiral cation can react with H$_2$O on either face of the ion, to give either (R)-2-butanol or (S)-2-butanol. Thus, a pure sample of (R)-2-butanol will gradually turn into a racemic mixture of (R)-2-butanol and (S)-2-butanol, thereby losing optical activity.

Problem 10.33 Two diastereomeric sets of enantiomers, A/B and C/D, exist for 3-bromo-2-butanol.

When enantiomer A or B is treated with HBr, only racemic 2,3-dibromobutane is formed; no meso isomer is formed. When enantiomer C or D is treated with HBr, only meso 2,3-dibromobutane is formed; no racemic 2,3-dibromobutane formed. Account for these observations.

The -OH group of the starting materials will be protonated as usual, and H$_2$O will depart to produce a carbocation. The key to this problem is realizing that a bromine atom adjacent to a carbocation will form a three-membered ring bromonium ion intermediate that then dictates an *anti* geometry for the incoming bromide nucleophile as shown.

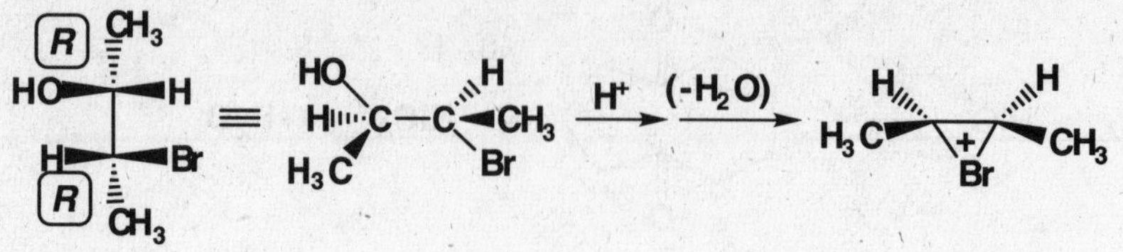

A

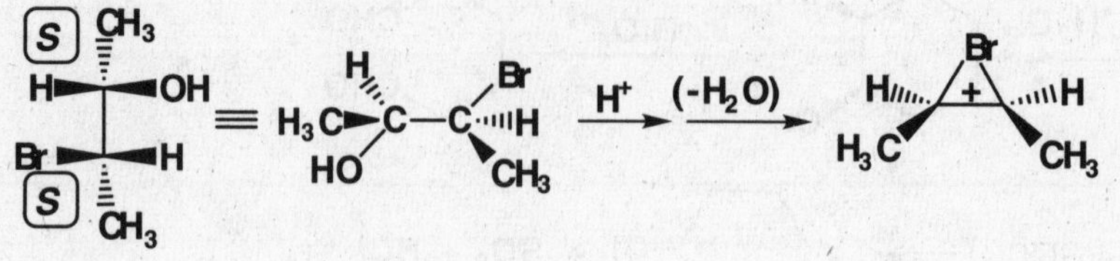

B

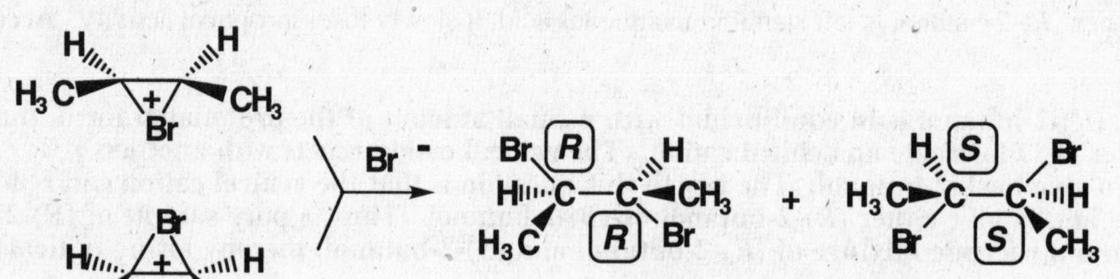

Racemic mixture of *R,R* and *S,S* 2,3-dibromobutane

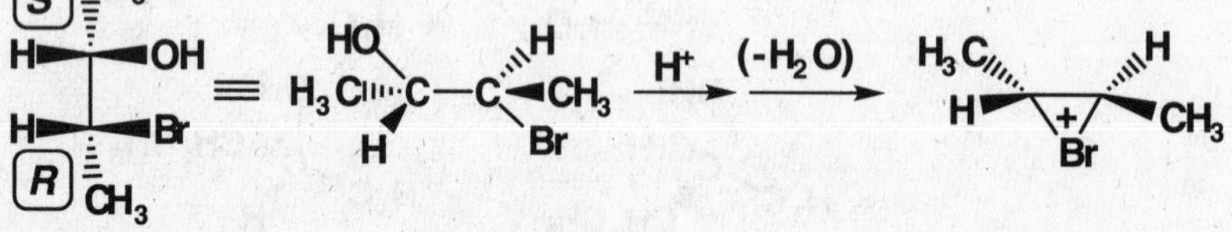

C

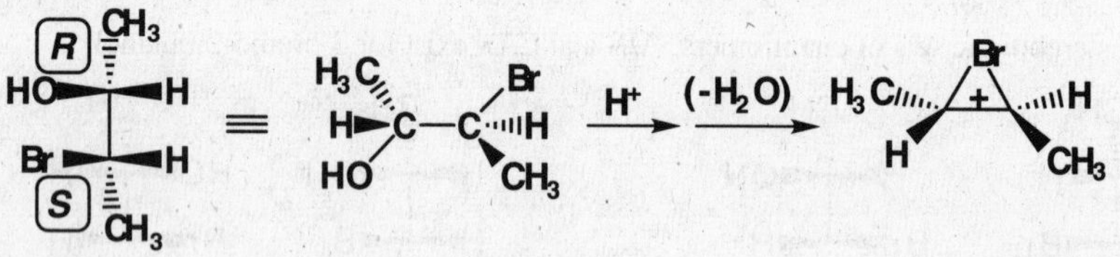

D

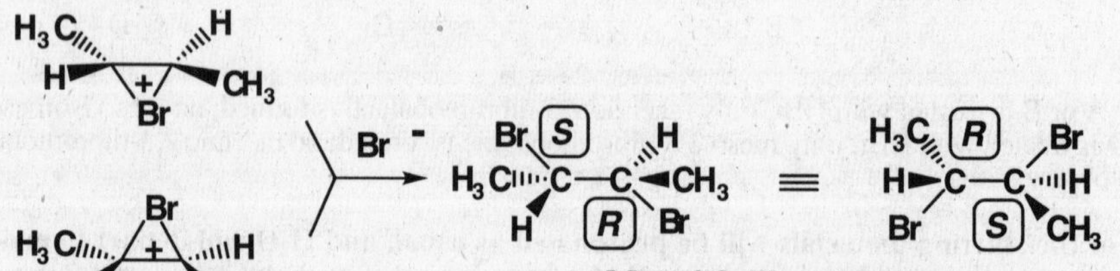

Meso 2,3-dibromobutane

Problem 10.34 Acid-catalyzed dehydration of 3-methyl-2-butanol gives three alkenes: 2-methyl-2-butene, 3-methyl-1-butene, and 2-methyl-1-butene. Propose a mechanism to account for the formation of each product.

Step 1: Add a proton. **The 3-methyl-2-butanol is protonated on the -OH group.**

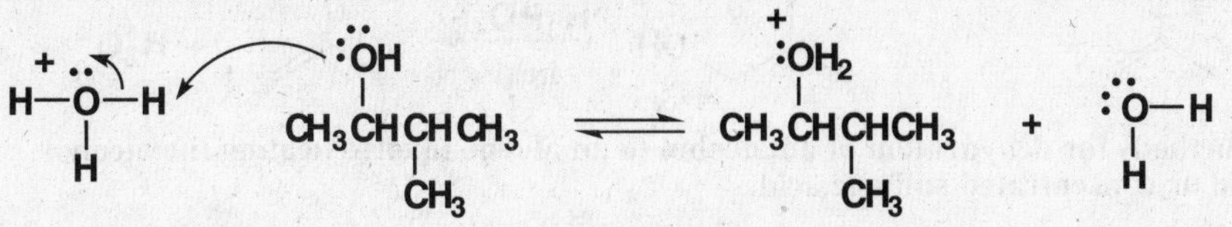

Step 2: Break a bond to give stable molecules or ions. **Loss of water to produce a 2° carbocation.**

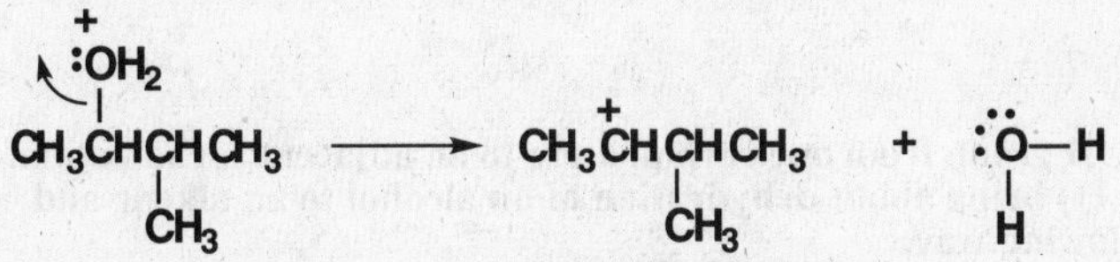

2° Carbocation

Step 3: Take a proton away. **This 2° carbocation can lose either of two protons to give 2-methyl-2-butene or 3-methyl-1-butene.**

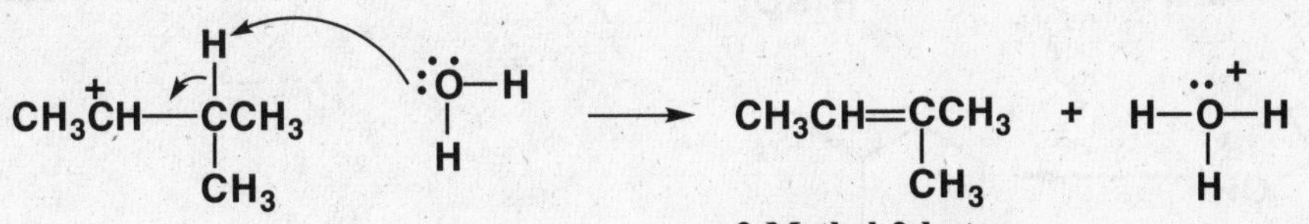

2-Methyl-2-butene

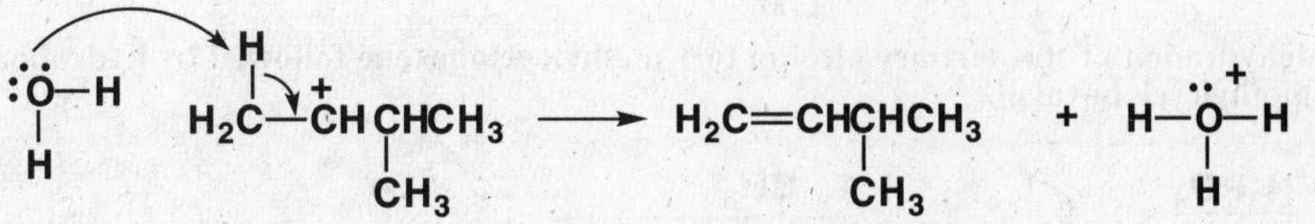

3-Methyl-1-butene

Step 4: 1,2 Shift **Alternatively, the 2° carbocation can rearrange to give a more stable 3° carbocation.**

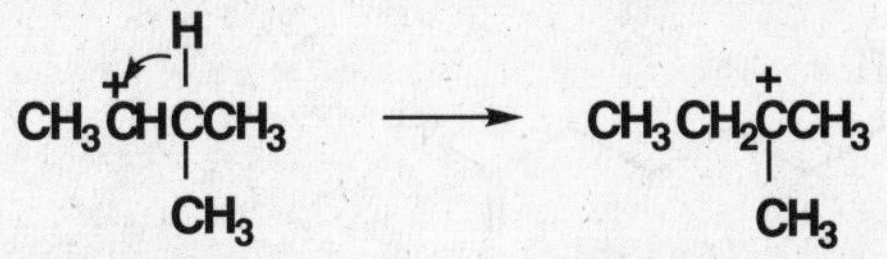

2° Carbocation   3° Carbocation

Step 5: Take a proton away. **This 3° carbocation can lose either of 2 protons to give 2-methyl-2-butene or 2-methyl-1-butene.**

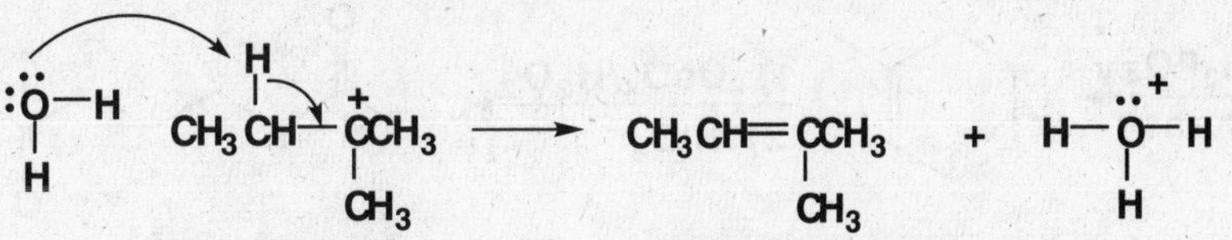

3° Carbocation   2-Methyl-2-butene

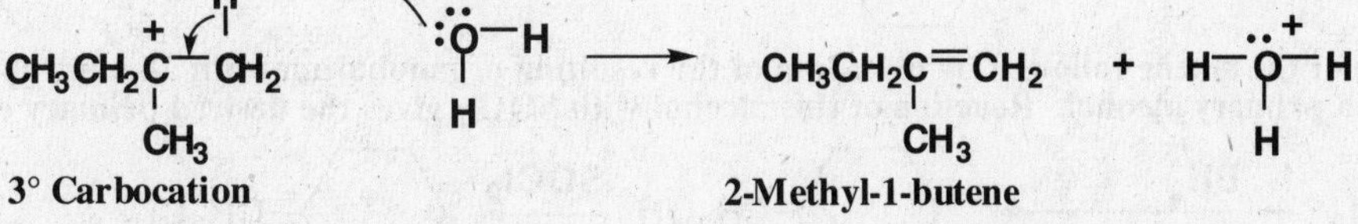

3° Carbocation   2-Methyl-1-butene

Problem 10.35 Show how you might bring about the following conversions. For any conversion involving more than one step, show each intermediate compound formed.

(a)

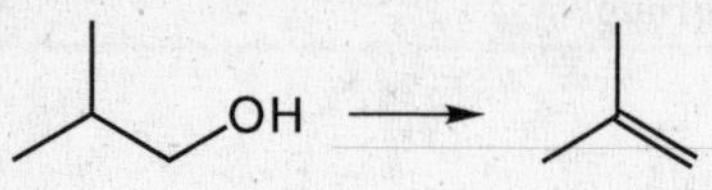

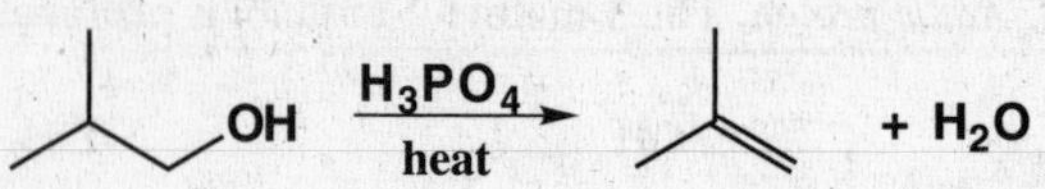

The most common laboratory methods for dehydration of an alcohol to an alkene involve heating the alcohol with either 85% phosphoric acid or concentrated sulfuric acid.

(b)

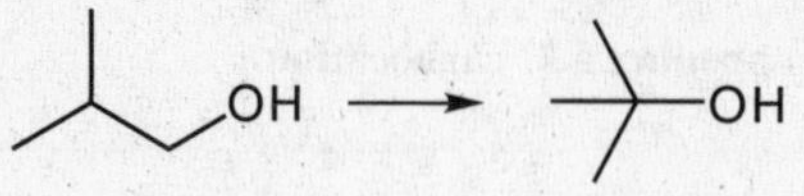

We have not seen any reaction that can switch an -OH group from one carbon atom to an adjacent carbon atom. We have seen, however, reactions by which we can (1) bring about dehydration of an alcohol to an alkene and then (2) hydrate the alkene to an alcohol in the following way.

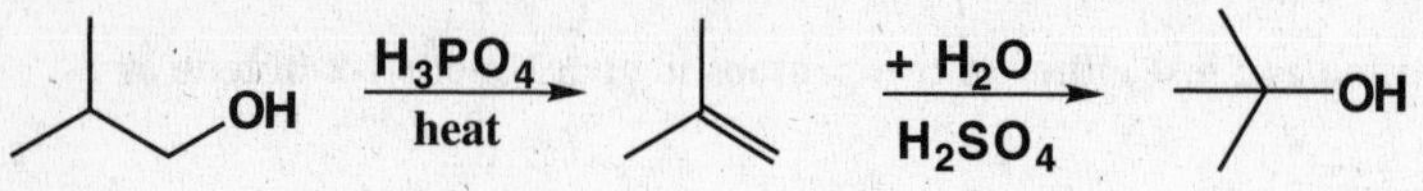

(c)

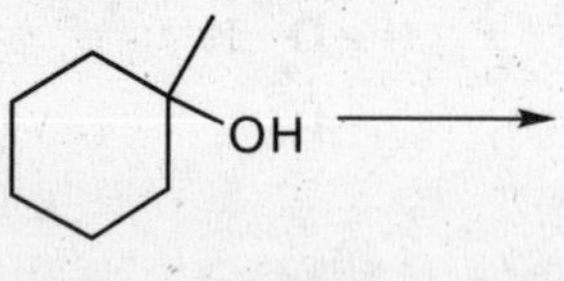

Acid-catalyzed dehydration of this tertiary alcohol to 1-methylcyclohexene followed by hydroboration/oxidation to form *trans*-2-methylcyclohexanol.

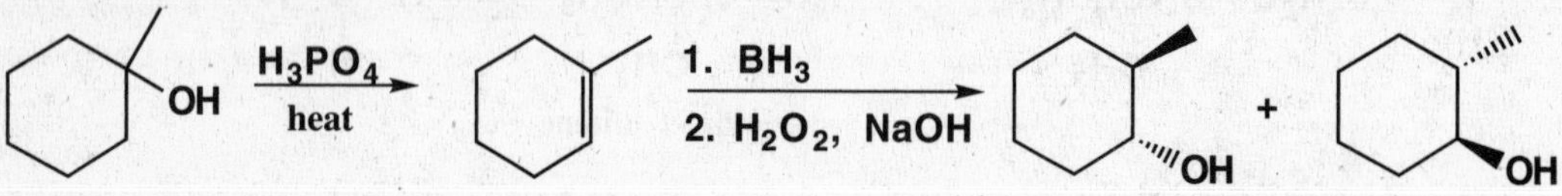

(racemic mixture)

(d)

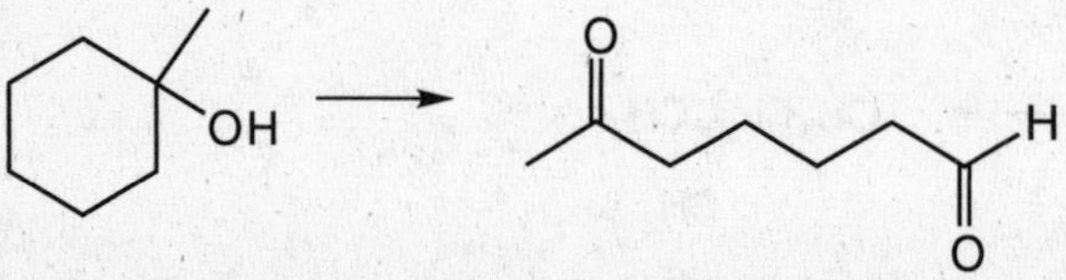

Acid-catalyzed dehydration of the tertiary alcohol as in part (c) followed by oxidative cleavage of the carbon-carbon double bond using osmium tetroxide in the presence of periodic acid, or ozonolysis followed by work-up in the presence of dimethyl sulfide.

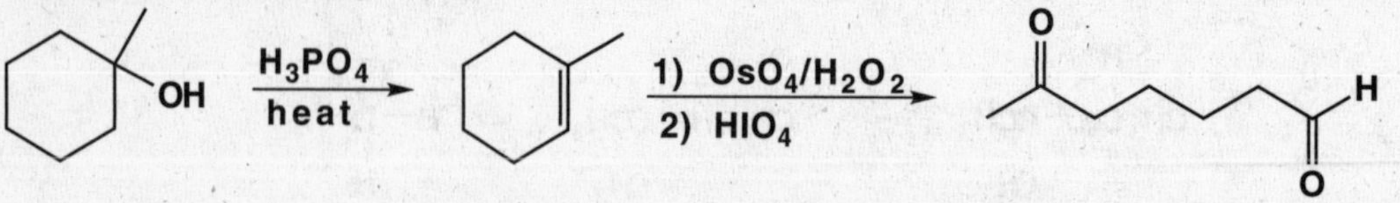

(e)

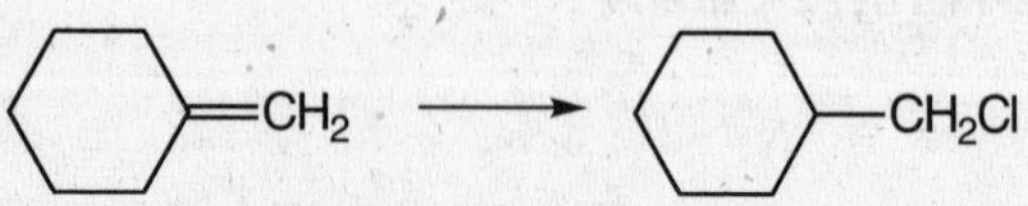

Hydroboration of the alkene followed by oxidation of the resulting organoborane with alkaline hydrogen peroxide gives a primary alcohol. Reaction of this alcohol with $SOCl_2$ gives the desired primary chloride.

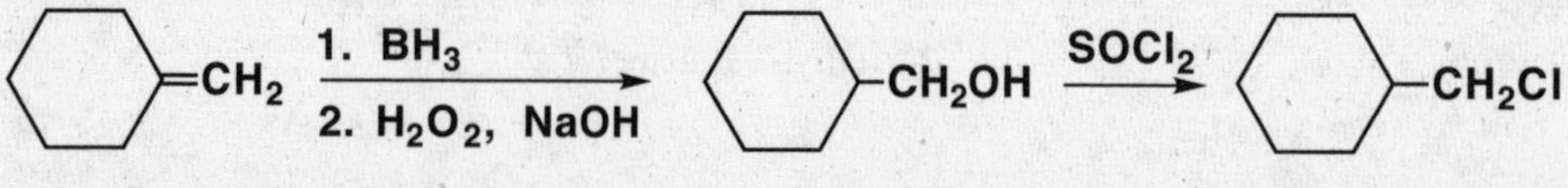

(f)

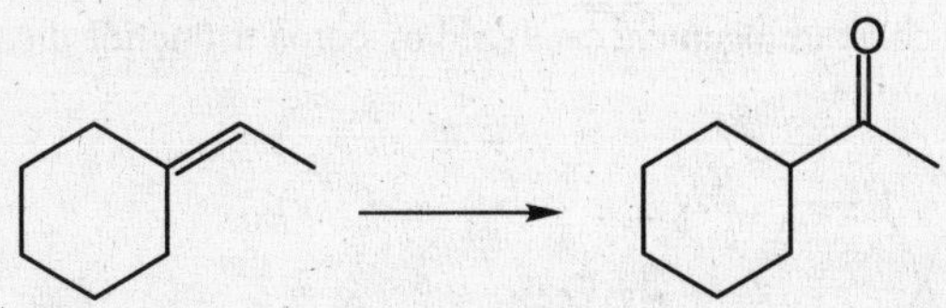

Hydroboration of the alkene followed by oxidation in alkaline hydrogen peroxide gives a secondary alcohol. Oxidation of this alcohol with chromic acid in sulfuric acid or pyridinium chlorochromate gives the desired ketone.

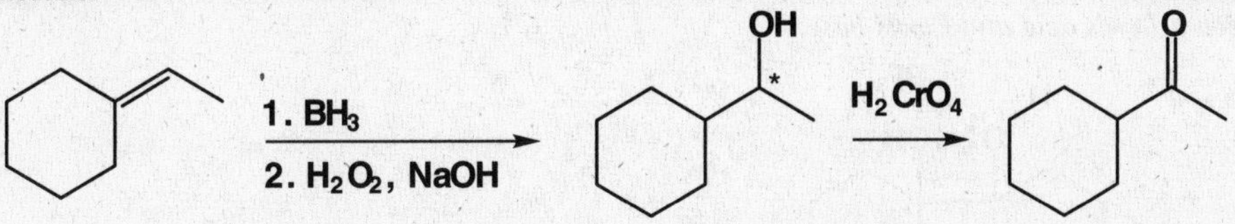

1. BH₃
2. H₂O₂, NaOH

H₂CrO₄

(g)

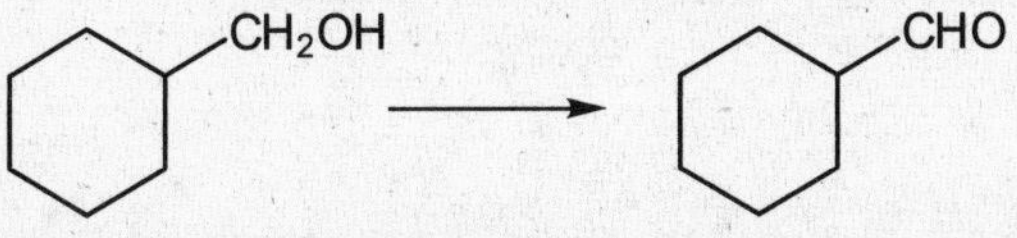

This conversion can be accomplished through oxidation with PCC.

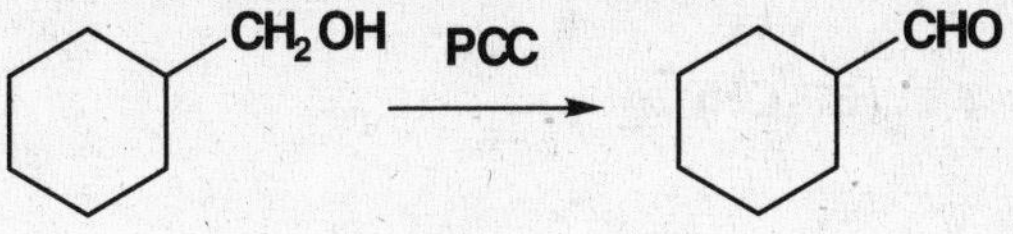

PCC

(h)

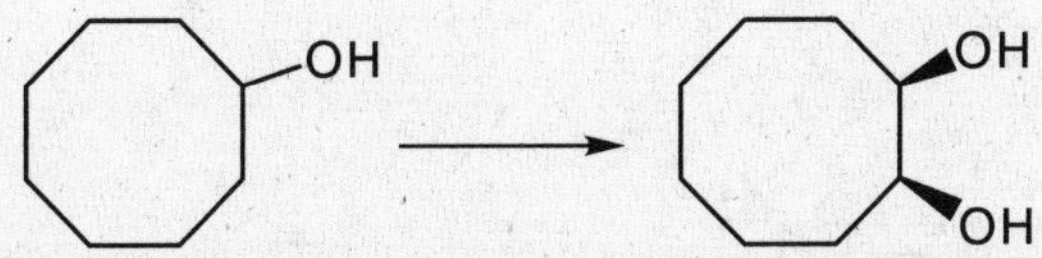

Acid-catalyzed dehydration of the secondary alcohol to an alkene followed by oxidation to the *cis*-glycol using osmium tetroxide in the presence of hydrogen peroxide.

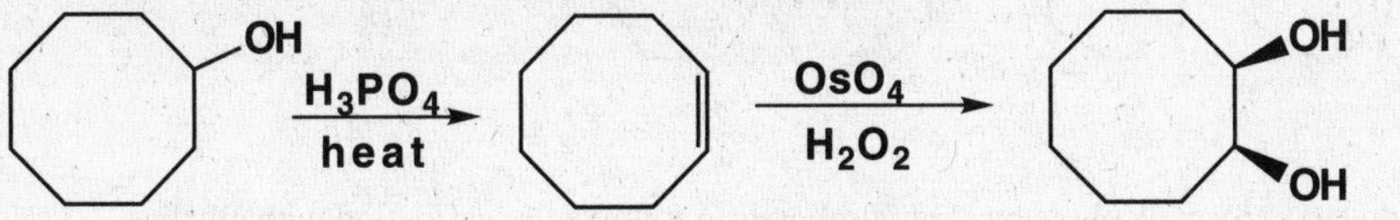

H₃PO₄
heat

OsO₄
H₂O₂

(i)

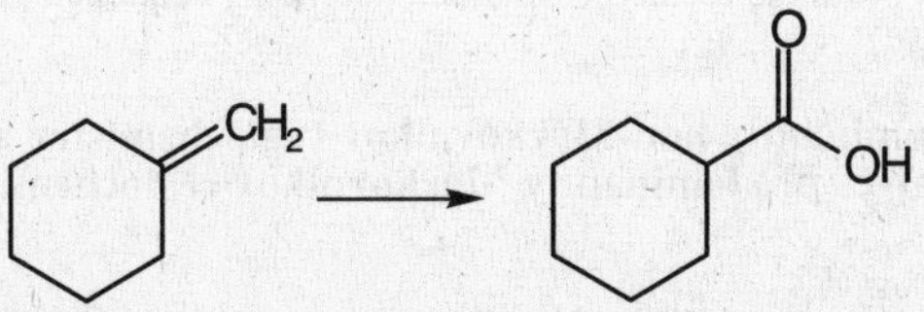

Hydroboration of the alkene followed by oxidation of the resulting organoborane with alkaline hydrogen peroxide gives a primary alcohol. Oxidation of this alcohol with H₂CrO₄ gives the desired carboxylic acid.

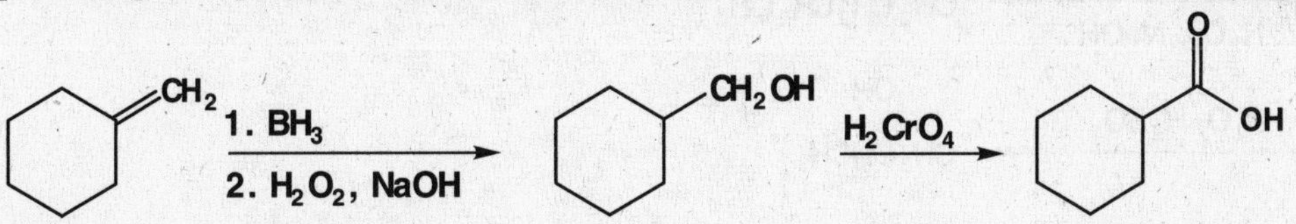

1. BH₃
2. H₂O₂, NaOH

H₂CrO₄

Pinacol Rearrangement
<u>Problem 10.36</u> Propose a mechanism for the following pinacol rearrangement catalyzed by boron trifluoride diethyl etherate:

Step 1: Make a bond between a Lewis acid and Lewis base.

Step 2: Break a bond to give stable molecules or ions.

Step 3: Take a proton away and 1,2-shift.

Synthesis
<u>Problem 10.37</u> Alkenes can be hydrated to form alcohols by (1) hydroboration followed by oxidation with alkaline hydrogen peroxide and (2) acid-catalyzed hydration. Compare the products formed from each alkene by sequence (1) with those formed from (2).

Recall that hydroboration followed by oxidation gives predominantly non-Markovnikov regiochemistry and *syn* stereochemistry of addition. Acid-catalyzed hydration gives predominantly Markovnikov regiochemistry and *anti* stereochemistry of addition.

(a) Propene

$$CH_3CH=CH_2 \xrightarrow[\text{2. } H_2O_2, \text{ NaOH}]{\text{1. } BH_3} CH_3CH_2CH_2OH$$

$$CH_3CH=CH_2 \xrightarrow{H_2O, H_2SO_4} CH_3\overset{\overset{\displaystyle OH}{|}}{C}HCH_3$$

(b) *cis*-2-Butene

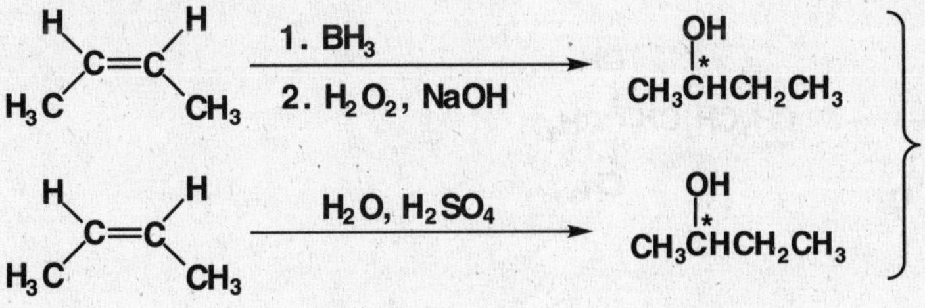
1. BH3
2. H2O2, NaOH
CH3CHCH2CH3
H2O, H2SO4
CH3CHCH2CH3

A racemic mixture is formed in both reactions

(c) *trans*-2-Butene

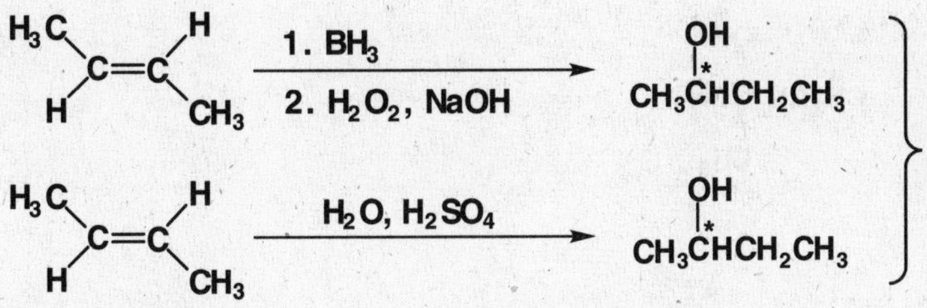
1. BH3
2. H2O2, NaOH
CH3CHCH2CH3
H2O, H2SO4
CH3CHCH2CH3

A racemic mixture is formed in both reactions

(d) Cyclopentene

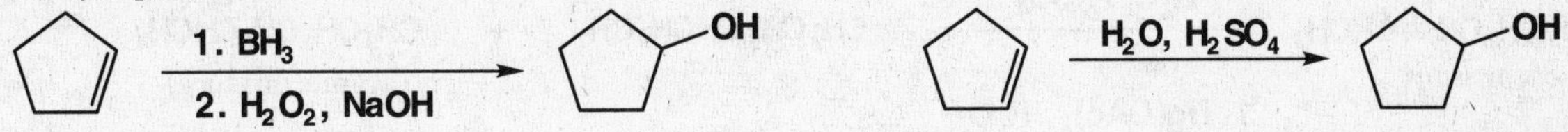
1. BH3
2. H2O2, NaOH
H2O, H2SO4

(e) 1-Methylcyclohexene

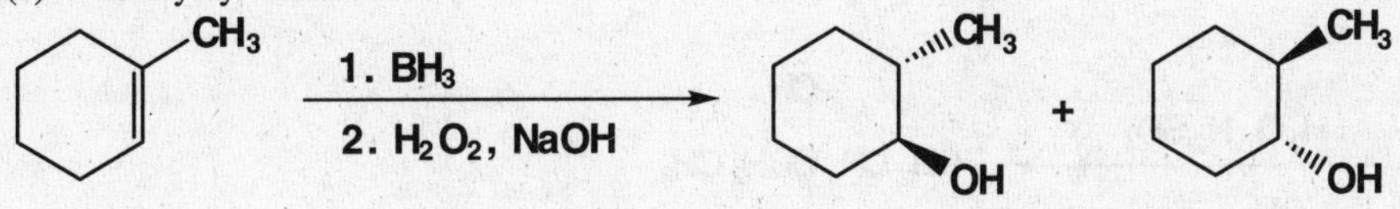
1. BH3
2. H2O2, NaOH

(racemic mixture)

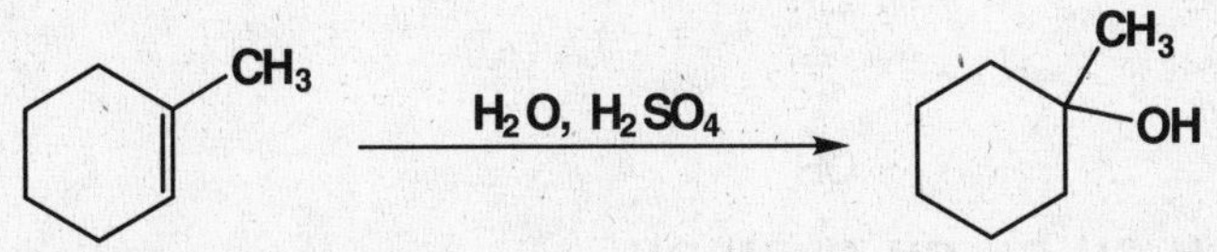
H2O, H2SO4

Problem 10.38 Show how each alcohol or diol can be prepared from an alkene.
(a) 2-Pentanol

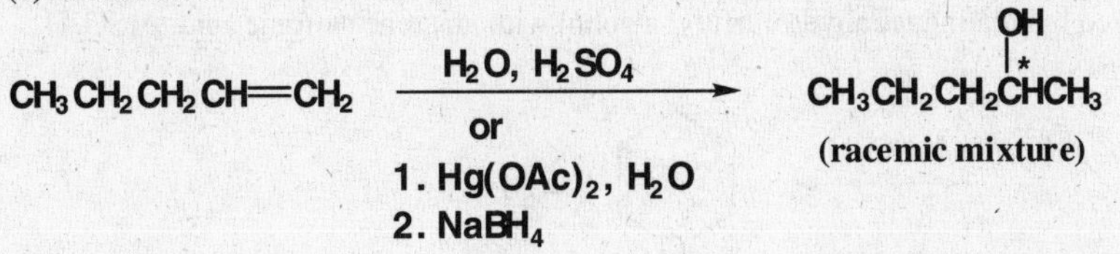
CH3CH2CH2CH=CH2
H2O, H2SO4
or
1. Hg(OAc)2, H2O
2. NaBH4
CH3CH2CH2CHCH3
(racemic mixture)

(b) 1-Pentanol

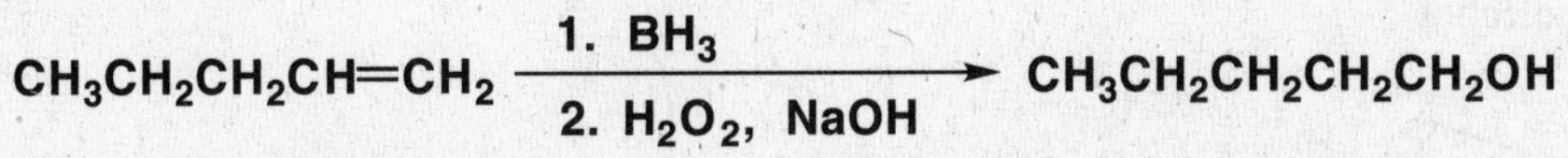
CH3CH2CH2CH=CH2
1. BH3
2. H2O2, NaOH
CH3CH2CH2CH2CH2OH

(c) 2-Methyl-2-pentanol

$CH_3CH_2CH_2\underset{\underset{\displaystyle CH_3}{|}}{C}{=}CH_2$

or

$CH_3CH_2CH{=}\underset{\underset{\displaystyle CH_3}{|}}{C}CH_3$

$\xrightarrow[\substack{\text{or} \\ \textbf{1. Hg(OAc)}_2\text{, H}_2\text{O} \\ \textbf{2. NaBH}_4}]{\text{H}_2\text{O, H}_2\text{SO}_4}$

$CH_3CH_2CH_2\underset{\underset{\displaystyle CH_3}{|}}{\overset{\overset{\displaystyle OH}{|}}{C}}CH_3$

(d) 2-Methyl-2-butanol

$CH_3CH_2\underset{\underset{\displaystyle CH_3}{|}}{C}{=}CH_2$

or

$CH_3CH{=}\underset{\underset{\displaystyle CH_3}{|}}{C}CH_3$

$\xrightarrow[\substack{\text{or} \\ \textbf{1. Hg(OAc)}_2\text{, H}_2\text{O} \\ \textbf{2. NaBH}_4}]{\text{H}_2\text{O, H}_2\text{SO}_4}$

$CH_3CH_2\underset{\underset{\displaystyle CH_3}{|}}{\overset{\overset{\displaystyle OH}{|}}{C}}CH_3$

(e) 3-Pentanol

$CH_3CH_2CH{=}CHCH_3$
cis or *trans*

$\xrightarrow[\substack{\text{or} \\ \textbf{1. Hg(OAc)}_2\text{, H}_2\text{O} \\ \textbf{2. NaBH}_4}]{\text{H}_2\text{O, H}_2\text{SO}_4}$

$CH_3CH_2\overset{\overset{\displaystyle OH}{|}}{C}HCH_2CH_3$ + $CH_3CH_2CH_2\overset{\overset{\displaystyle OH}{|}}{\underset{*}{C}}HCH_3$

(racemic mixture)

(f) 3-Ethyl-3-pentanol

$(CH_3CH_2)_2C{=}CHCH_3$

$\xrightarrow[\substack{\text{or} \\ \textbf{1. Hg(OAc)}_2\text{, H}_2\text{O} \\ \textbf{2. NaBH}_4}]{\text{H}_2\text{O, H}_2\text{SO}_4}$

$CH_3CH_2\underset{\underset{\displaystyle CH_2CH_3}{|}}{\overset{\overset{\displaystyle OH}{|}}{C}}CH_2CH_3$

(g) 1,2-Hexanediol

$CH_3CH_2CH_2CH_2CH{=}CH_2 \xrightarrow{\text{OsO}_4/\text{H}_2\text{O}_2} CH_3CH_2CH_2CH_2\overset{\overset{\displaystyle OH}{|}}{\underset{*}{C}}HCH_2OH$

(racemic mixture)

<u>Problem 10.39</u> Dihydropyran is synthesized by treating tetrahydrofurfuryl alcohol with an arene sulfonic acid, ArSO$_3$H. Propose a mechanism for this conversion.

Tetrahydrofurfuryl   Dihydropyran
alcohol

Step 1: Add a proton

Step 2: Break a bond to give stable molecules or ions.

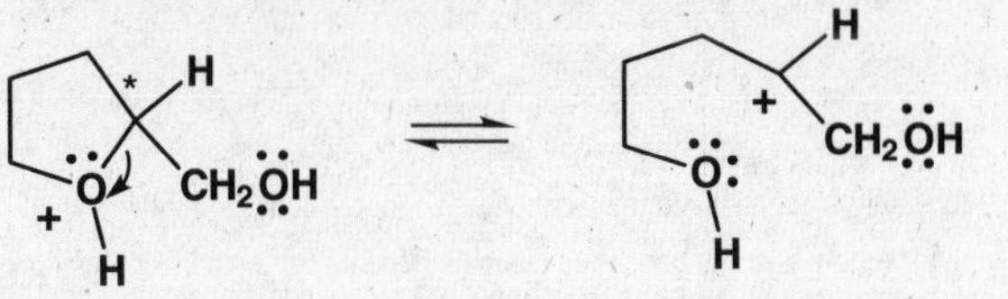

Step 3: 1,2 Shift.

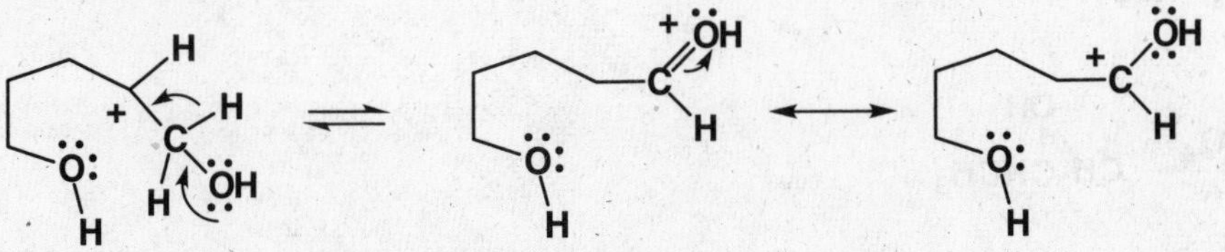

Step 4: Make a new bond between a nucleophile and an electrophile.

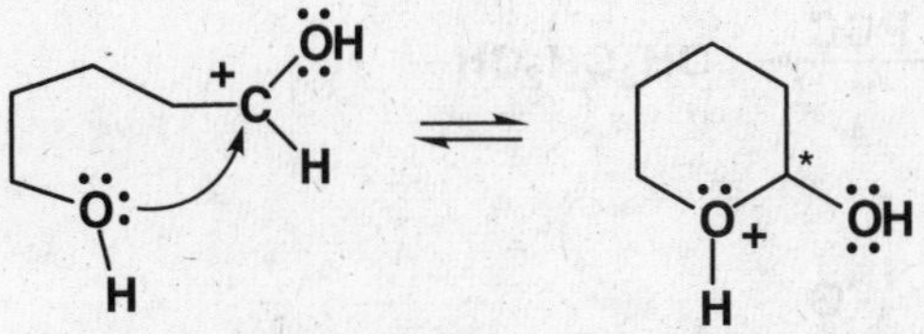

Step 5: Add a proton-take a proton away. **Transfer a proton from one part of the molecule to another.**

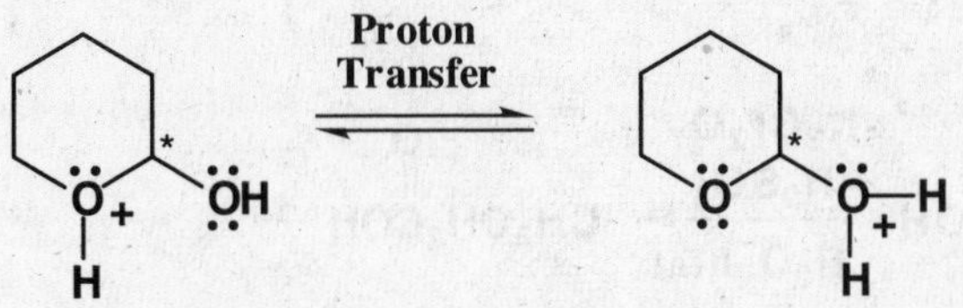

Step 6: Break a bond to give stable molecules or ions.

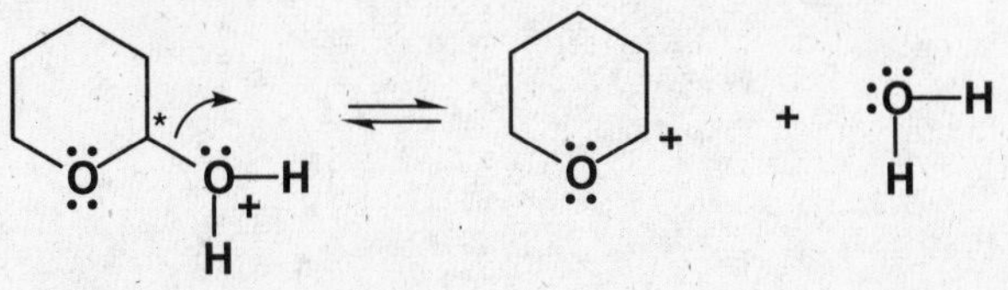

Step 7: Take a proton away.

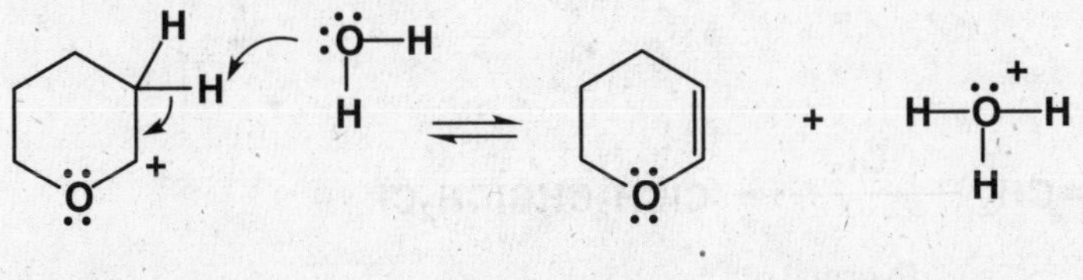

<u>Problem 10.40</u> Show how to convert propene to each of these compounds, using any inorganic reagents as necessary:
(a) Propane

$$CH_3CH\!=\!CH_2 \;+\; H_2 \;\xrightarrow[\text{catalyst}]{\text{Transition metal}}\; CH_3CH_2CH_3$$

(b) 1,2-Propanediol

$$CH_3CH=CH_2 \xrightarrow[\text{H}_2\text{O}_2]{\text{OsO}_4} \overset{\overset{\displaystyle OH}{|}}{CH_3\overset{*}{C}HCH_2OH}$$

(racemic mixture)

(c) 1-Propanol

$$CH_3CH=CH_2 \xrightarrow[\text{2. H}_2\text{O}_2,\ \text{NaOH}]{\text{1. BH}_3} CH_3CH_2CH_2OH$$

(d) 2-Propanol

$$CH_3CH=CH_2 \xrightarrow[]{\text{H}_2\text{O, H}_2\text{SO}_4} \overset{\overset{\displaystyle OH}{|}}{CH_3CHCH_3}$$

(e) Propanal

$$CH_3CH=CH_2 \xrightarrow[\text{2. H}_2\text{O}_2,\ \text{NaOH}]{\text{1. BH}_3} CH_3CH_2CH_2OH \xrightarrow{\text{PCC}} \overset{\overset{\displaystyle O}{\parallel}}{CH_3CH_2CH}$$

(f) Propanone

$$CH_3CH=CH_2 \xrightarrow{\text{H}_2\text{O, H}_2\text{SO}_4} \overset{\overset{\displaystyle OH}{|}}{CH_3CHCH_3} \xrightarrow[\text{H}_2\text{O, heat}]{\overset{\text{K}_2\text{Cr}_2\text{O}_7}{\text{H}_2\text{SO}_4}} \overset{\overset{\displaystyle O}{\parallel}}{CH_3CCH_3}$$

(g) Propanoic acid

$$CH_3CH=CH_2 \xrightarrow[\text{2. H}_2\text{O}_2,\ \text{NaOH}]{\text{1. BH}_3} CH_3CH_2CH_2OH \xrightarrow[\text{H}_2\text{O, heat}]{\overset{\text{K}_2\text{Cr}_2\text{O}_7}{\text{H}_2\text{SO}_4}} \overset{\overset{\displaystyle O}{\parallel}}{CH_3CH_2COH}$$

(h) 1-Bromo-2-propanol

$$CH_3CH=CH_2 \xrightarrow{\text{Br}_2\ /\ \text{H}_2\text{O}} \overset{\overset{\displaystyle OH}{|}}{CH_3\overset{*}{C}HCH_2Br}$$

(racemic mixture)

(i) 3-Chloropropene

$$CH_3CH=CH_2 \xrightarrow[\substack{\text{heat}\\ \text{(Allylic}\\ \text{halogenation)}}]{\text{Cl}_2} ClCH_2CH=CH_2$$

(j) 1,2,3-trichloropropane

$$CH_3CH=CH_2 \xrightarrow[\substack{\text{heat}\\ \text{(Allylic}\\ \text{halogenation)}}]{\text{Cl}_2} ClCH_2CH=CH_2 \xrightarrow[\substack{\text{(Normal}\\ \text{halogenation)}}]{\text{Cl}_2} ClCH_2CHClCH_2Cl$$

(k) 1-Chloropropane

$$CH_3CH=CH_2 \xrightarrow[\text{2. H}_2\text{O}_2,\ \text{NaOH}]{\text{1. BH}_3} CH_3CH_2CH_2OH \xrightarrow[\text{Pyridine}]{\text{SOCl}_2} CH_3CH_2CH_2Cl$$

(l) 2-Chloropropane

$$CH_3CH=CH_2 \xrightarrow{H_2O,\ H_2SO_4} CH_3\overset{\overset{\displaystyle OH}{|}}{C}HCH_3 \xrightarrow[\text{Pyridine}]{SOCl_2} CH_3\overset{\overset{\displaystyle Cl}{|}}{C}HCH_3$$

or

$$CH_3CH=CH_2 \xrightarrow{HCl} CH_3\overset{\overset{\displaystyle Cl}{|}}{C}HCH_3$$

(m) 2-Propen-1-ol

$$CH_3CH=CH_2 \xrightarrow[\substack{\text{Peroxides}\\(\text{Allylic}\\\text{halogenation})}]{NBS} BrCH_2CH=CH_2 \xrightarrow[S_N2]{NaOH} HOCH_2CH=CH_2$$

(n) Propenal

$$CH_3CH=CH_2 \xrightarrow[\substack{\text{Peroxides}\\(\text{Allylic}\\\text{halogenation})}]{NBS} BrCH_2CH=CH_2 \xrightarrow[S_N2]{NaOH} HOCH_2CH=CH_2 \xrightarrow{PCC} \overset{\overset{\displaystyle O}{\|}}{H}CCH=CH_2$$

<u>Problem 10.41</u> (a) How many stereoisomers are possible for 4-methyl-1,2-cyclohexanediol?

4-Methyl-1,2-cyclohexanediol

There are 2^3 or eight stereoisomers, because there are three chiral centers in this molecule and no planes of symmetry.

(b) Which of the possible stereoisomers are formed by oxidation of (*S*)-4-methylcyclohexene with osmium tetroxide?

There are two *cis-trans* isomers possible for the diol formed by oxidation of 4-methylcyclohexene. Because of the stereospecificity of oxidation by osmium tetroxide under these conditions, the two -OH groups are *cis* to each other, and both are either *cis* or *trans* to the methyl group.

(*S*)-4-Methyl-1,2-
cyclohexanediol

both -OH groups
cis to -CH₃

both -OH groups
trans to -CH₃

(c) Is the product formed in part (b) optically active or optically inactive?

Because a single enantiomer is used as the starting material, it is reasonable to expect that the product will also be optically active. Although hard to predict which one predominates, it is safe to assume the two products will not be formed in exactly equal amounts starting from (*S*)-4-methylcyclohexene.

Problem 10.42 Show how to bring about this conversion in good yield.

The alcohol is first converted to the alkene by treatment with acid, then the resulting alkene is treated with O₃.
Alternatively, (not shown) the alkene could be treated with OsO₄ / H₂O₂ followed by HIO₄.

Problem 10.43 The tosylate of a primary alcohol normally undergoes an S_N2 reaction with hydroxide ion to give a primary alcohol. Reaction of this tosylate, however, gives a compound of molecular formula $C_7H_{12}O$.

Propose a structural formula for this compound and a mechanism for its formation.

In this reaction, the HO⁻ is acting as a base, not a nucleophile. This makes sense because proton transfer reactions are generally so fast. The alkoxide produced by deprotonation of the alcohol carries out an intramolecular S_N2 attack to give the bicyclic product, $C_7H_{12}O$.

Step 1: Take a proton away.

Step 2: Make a new bond between a nucleophile and an electrophile and simultaneously break a bond to give stable molecules or ions.

The stereochemistry of the product will be the same as that of the starting material because neither chiral center is involved in bond making or bond breaking during the reaction.

Problem 10.44 Chrysanthemic acid occurs as a mixture of esters in flowers of the chrysanthemum (pyrethrum) family. Reduction of chrysanthemic acid to its alcohol (Section 17.6A) followed by conversion of the alcohol to its tosylate gives chrysanthemyl tosylate. Solvolysis of the tosylate (Section 9.3) gives a mixture of artemesia and yomogi alcohols.

Chrysanthemic acid

Chrysanthemyl alcohol

Chrysanthemyl tosylate

Artemisia alcohol

Yomogi alcohol

Propose a mechanism for the formation of these alcohols from chrysanthemyl tosylate.

Step 1: Break a bond to give stable molecules or ions. In the first step, the tosylate leaves along with bond migration to create a resonance stabilized allylic cation.

Chrysanthemyl tosylate

Allylic cation

Step 2: Make a new bond between a nucleophile and an electrophile. The allyl cation can react with H_2O at either end to add an OH group to create the Artemisia or Yomogi alcohols.

Allylic cation

Artemisia alcohol (racemic mixture)

Yomogi alcohol

Problem 10.45 Show how to convert cyclohexene to each compound in good yield.

(a) **NBS and light or peroxides (allylic halogenation)**
(b) **Solvolysis in H_2O (S_N1); note that there will likely be some E1 elimination observed in this step.**
(c) **Acid catalyzed dehydration in H_3PO_4**
(d) **Solvolysis in CH_3OH (S_N1); note that there will likely be some E1 elimination observed in this step.**
(e) **Oxidation with H_2CrO_4 or PCC**

(f) Acid catalyzed hydration (H₂O, H₂SO₄), oxymercuration-reduction (1. Hg(OAc)₂, H₂O; 2. NaBH₄), or hydroboration-oxidation (1. BH₃; 2. H₂O₂, NaOH).
(g) Oxidation with H₂CrO₄ or PCC
(h) OsO₄, H₂O₂
(i) HIO₄
(j) Oxidation with H₂CrO₄

<u>Problem 10.46</u> Hydroboration of the following bicycloalkane followed by oxidation in alkaline hydroperoxide is both stereoselective and regioselective. The product is a single alcohol in better than 95% yield.

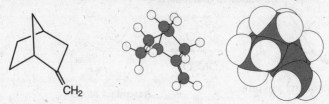

Propose a structural formula for this alcohol and account for the stereo- and regioselectivity of its formation. *Hint*: Examine a molecular model of this alkene and see if you can determine which face of the double bond is more accessible to hydroboration.

Following is the structural formula of the bicyclic alcohol formed by hydroboration/oxidation

First, given the known regioselectivity of hydroboration/oxidation, predict that -OH adds to the less substituted carbon atom of the double bond. To account for the stereoselectivity, predict that the concerted syn addition of H and BH₂ will occur from the less hindered side. That is, the same side as the one-carbon bridge (from the top of the molecule as it is drawn in the problem) rather than on the same side as the two-carbon bridge (from the bottom side as the molecule is drawn in the problem). This puts the H atom on the same side of the ring as the one carbon bridge.

Side view:

2. H₂O₂, NaOH

Looking Ahead

Problem 10.47 Compounds that contain an N-H group associate by hydrogen bonding.
(a) Do you expect this association to be stronger or weaker than that of compounds containing an O-H group?

Weaker. The O-H bond is more polar, because the difference in electronegativity between N and H is less than the difference between O and H. Thus, the degree of intermolecular interaction between compounds containing an N-H group is less than that between compounds containing an -OH group.

Bond	Difference in electronegativity
N-H	3.0 - 2.1 = 0.9
O-H	3.5 - 2.1 = 1.4

(b) Based on your answer to part (a), which would you predict to have the higher boiling point, 1-butanol or 1-butanamine?

$$CH_3CH_2CH_2CH_2OH \qquad CH_3CH_2CH_2CH_2NH_2$$
1-Butanol   **1-Butanamine**
bp 117°C   **bp 78°C**

The stronger the hydrogen bonds, the higher the boiling point because hydrogen bonds in the liquid state must be broken upon boiling. Therefore, 1-butanol, with the stronger hydrogen bonds, will have the higher boiling point.

Problem 10.48 Ethanol (CH_3CH_2OH) and dimethyl ether (CH_3OCH_3) are constitutional isomers. (a) Predict which of the two has the higher boiling point. (b) Predict which of the two is the more soluble in water.

The -OH group of ethanol can act as both a hydrogen bond donor and acceptor, so it has a relatively high boiling point (78° C) and an infinite solubility in water. The oxygen atom of dimethyl ether, on the other hand, is only a hydrogen bond acceptor, so it cannot hydrogen bond with itself. As a result, it has a relatively low boiling point (-24°C). Dimethyl ether does have moderate solubility in water (7.8 g/100g).

Problem 10.49 Following are structural formulas for phenol and cyclohexanol along with the acid dissociation constants for each.

Phenol   **Cyclohexanol**
pK_a 9.96 pK_a 18

Propose an explanation for the fact that phenol is a considerably stronger acid than cyclohexanol.

The key to predicting relative acid strength is to compare relative stabilities of the conjugate base anions. More directly, the more stable anion is the one with greater delocalization (*i.e.* spread out over more atoms) of the negative charge. Cyclohexanol has a pK_a that is normal for alcohols. It's conjugate base anion has the negative charge localized on the deprotonated O atom. The pK_a for phenol is a significantly stronger acid because its conjugate base, the phenoxide anion, has the negative charge delocalized into the adjacent aromatic ring.

Conjugate base

Conjugate base stabilized through resonance

Mixed Synthesis

Alcohols are important for organic synthesis, especially in situations involving alkenes. The alcohol might be the desired product, or the OH group might be transformed into another functional group via halogenation, oxidation, or perhaps conversion to a sulfonic ester derivative. Formation of an alcohol from an alkene is particularly powerful because conditions can be chosen to produce either the Markovnikov or non-Markovnikov product from an unsymmetrical alkene.

<u>Problem 10.50</u> Show how to convert 4-methyl-1-pentene into 5-methylhexanenitrile. You must use 4-methyl-1-pentene and sodium cyanide as the source of all carbon atoms in the target molecule. Show all reagents needed and all molecules synthesized along the way.

4-Methyl-1-pentene   +   NaCN →?   **5-Methylhexanenitrile**

1. BH$_3$
2. H$_2$O$_2$ / NaOH

HBr, ROOR, light

NaCN, S$_N$2

OH → PBr$_3$ →   Br

Recognize that the 5-methylhexanenitrile product has seven carbon atoms, so it must be derived by joining the two starting materials with a new carbon-carbon bond. The obvious way to accomplish this is to react a primary haloalkane with sodium cyanide in an S$_N$2 reaction. The required haloalkane can be made a couple of different ways. One approach involves formation of an alcohol with non-Markovnikov regiochemistry using hydroboration/oxidation, followed by conversion to the haloalkane using PBr$_3$. Alternatively, HBr could be used with peroxides in a free radical process to give the haloalkane directly from the alkene.

<u>Problem 10.51</u> Show how to convert butane into 2-butanone. Show all reagents and all molecules synthesized along the way.

Butane →?   **2-Butanone**

Br$_2$, light

Br / Br (racemic)

RO⁻ Strong Base E2

H$_2$O, H$_2$SO$_4$

OH (racemic)

PCC or H$_2$CrO$_4$

Recognize that the 2-butanone product has the same number of carbons as the butane starting material so carbon-carbon bond formation is not required. Recognize also that a good way to make a ketone product is to oxidize a secondary alcohol using either PCC or H$_2$CrO$_4$. The required 2-butanol can be synthesized from 2-butene using the acid-catalyzed hydration reaction or oxymercuration (not shown). 2-Butene can be prepared from butane using the familiar sequence of free radical halogenation followed by an E2 reaction in strong base.

Problem 10.52 Show how to convert butane into butanal. Show all reagents needed and all molecules synthesized along the way.

Recognize that the 2-butanal product has the same number of carbons as the butane starting material so carbon-carbon bond formation is not required. Recognize also that the product aldehyde can be derived from the PCC oxidation of 1-butanol. The difficult aspect of this problem is to find a way to functionalize the terminal position of butane. A good way to accomplish terminal functionalization is to carry out an allylic halogenation of the 2-butene using NBS, followed by hydrogenation to give 1-bromobutane. An S_N2 reaction on 1-bromobutane (a primary haloalkane) using hydroxide or even water as the nucleophile gives 1-butanol. 2-Butene can be prepared from butane using the familiar sequence of free radical halogenation followed by an E2 reaction in strong base.

Problem 10.53 Show how to convert 1-propanol into 2-hexyne. You must use 1-propanol as the source of all carbon atoms in the target molecule. Show all reagents needed and all molecules synthesized along the way.

Recognize that the 2-hexyne product has six carbons, twice the number of the 1-propanol starting material. Assume that a new carbon-carbon bond must be made between two three-carbon units. The best way to accomplish this

involves reaction of the appropriate acetylide anion with 1-bromopropane. The required propyne is derived from dehydration of propanol in anhydrous sulfuric acid to give propene, followed by halogenation then reaction with NaNH$_2$. Note that dehydration of primary alcohols in acid to give alkenes is generally not a useful process due to competing carbocation rearrangements, but in the case of 1-propanol, no rearrangements are possible. The 1-bromopropane required for the final step is conveniently prepared through the treatment of 1-propanol with PBr$_3$.

Problem 10.54 Show how to convert 2-methylpentane into 2-methyl-3-pentanone. Show all reagents needed and all molecules synthesized along the way.

Recognize that the 2-methyl-3-pentanone product has the same number of carbons as the 2-methylpentane starting material so carbon-carbon bond formation is not required. Recognize also that a good way to make a ketone product is to oxidize a secondary alcohol using either PCC or H$_2$CrO$_4$. The required secondary alcohol is conveniently prepared using hydroboration/oxidation of 2-methyl-2-pentene . Note that the non-Markovnikov regiochemistry is needed for this step. 2-Methyl-2-pentene can be prepared from 2-methylpentane using the familiar sequence of free radical halogenation (reaction predominantly at the tertiary carbon) followed by an E2 reaction in strong base to give the more highly substituted alkene in agreement with Zaitsev's rule.

Reactions in Context
Problem 10.55 Atorvistatin (Lipitor) is used to decrease patient serum cholesterol levels. It works by inhibiting an enzyme called HMG-CoA reductase. See the Connections to Biological Chemistry box in Section 10.2 for more information about the action of atorvistatin. In one synthesis of atorvistatin that produces the desired single enantiomer of the final product, the following reagents are used. Draw the structures of synthetic intermediates A and B.

The first reaction occurs on the primary alcohol group of the starting material to give the corresponding sulfonic ester (A), which is displaced by the nucleophilic cyanide anion in an S$_N$2 reaction in the second step to give the nitrile product (B).

Problem 10.56 Paroxetine (Paxil) is an antidepressant that is a member of a family of drugs known as *Selective Serotonin Reuptake Inhibitors* (SSRIs). This family of drugs also includes fluoxetine (Prozac) and sertraline (Zoloft). SSRIs work by inhibiting the reuptake of the neurotransmitter serotonin in the synapses of the central nervous system following release of serotonin during excitation of individual nerve cells. Between firings, the serotonin is taken back up by a nerve cell in preparation for firing again. Inhibition of reuptake has the effect of increasing the time serotonin molecules remain in the synapses following excitation, leading to a therapeutic effect. In one synthesis of paroxetine, the following reagents are used. Draw the structures of synthetic intermediates A and B.

A   **B**

The first reaction generates a primarily alkyl chloride (A) from the primary alcohol, which is displaced by the nucleophilic OH group in an S$_N$2 reaction to give ether product B.

Problem 10.57 Tartaric acid is an inexpensive and readily available chiral starting material for the synthesis of chiral molecules. In a well-known prostaglandin synthesis, the (*S,S*)-tartaric acid enantiomer was used to prepare the chiral diol in several steps. The chiral diol was isolated as a synthetic intermediate and the following reagents are used. Draw the structures of synthetic intermediates A and B.

A   **B**

The first reaction occurs on the two primary alcohol groups to give the corresponding double sulfonic ester (A), which are both displaced by the nucleophilic iodide anion in an S$_N$2 reaction in the second step to give the diiodo product (B).

CHAPTER 11
Solutions to the Problems

Problem 11.1 Write IUPAC and common names for these ethers.

(a)

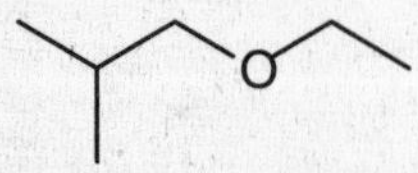

(b)

(c)

1-Ethoxy-2-methylpropane
(Ethyl isobutyl ether)

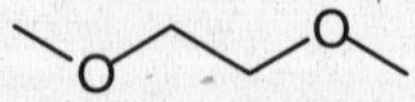

1,2-Dimethoxyethane
(Ethylene glycol dimethyl ether)

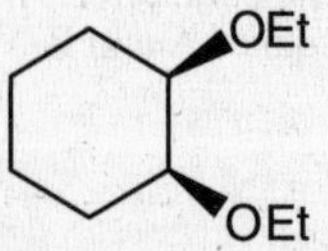

cis-**1,2-Diethoxycyclohexane**

Problem 11.2 Arrange these compounds in order of increasing boiling point.

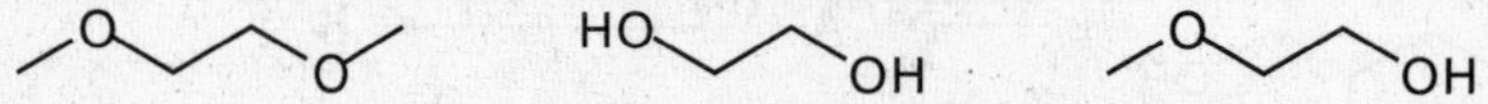

Boiling point increases with an increasing number of hydrogen bonding groups. In order of increasing boiling point, they are:

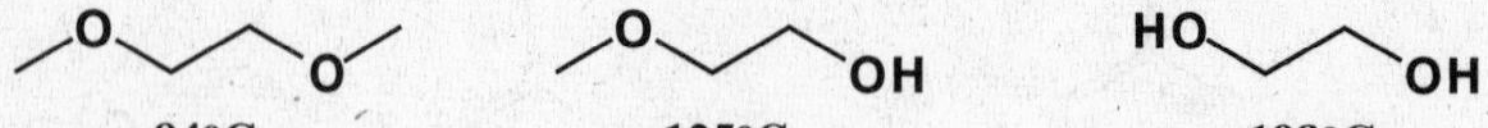

84ºC   **125ºC**   **198ºC**

Problem 11.3 Show how you might use the Williamson ether synthesis to prepare each ether:

(a)

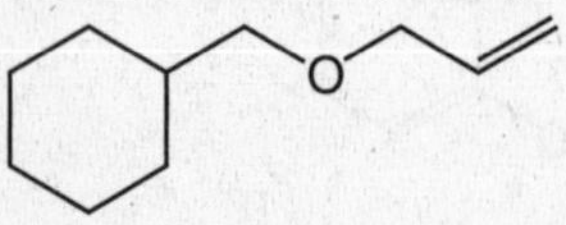

There are two combinations of reagents that could produce the desired ether in high yield because both involve primary haloalkanes and are thus unlikely to eliminate via an E2 pathway

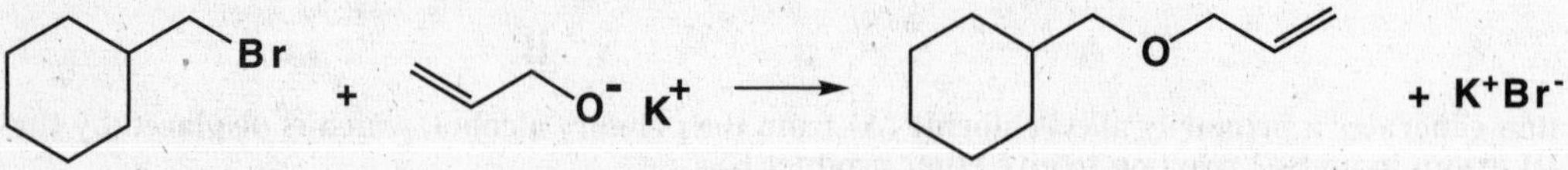

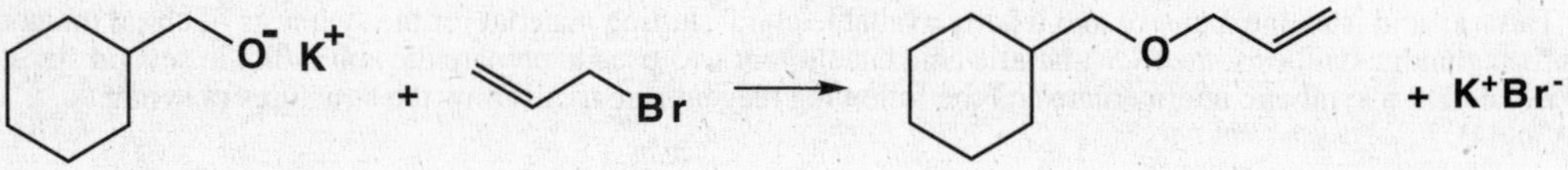

(b)

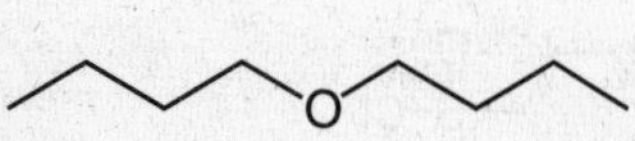

Treatment of 1-bromobutane with sodium butoxide gives dibutyl ether.

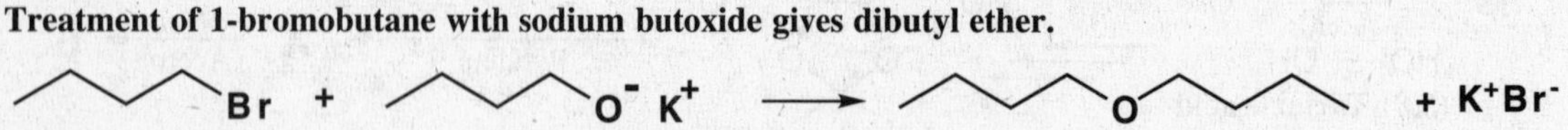

Problem 11.4 Show how ethyl hexyl ether might be prepared by a Williamson ether synthesis.

Using the Williamson ether synthesis, ethyl hexyl ether could be synthesized by either of the two following routes:

$$CH_3CH_2Br \; + \; CH_3(CH_2)_4CH_2O^-Na^+$$

or $\longrightarrow CH_3CH_2OCH_2(CH_2)_4CH_3$

$$CH_3(CH_2)_4CH_2Br \; + \; CH_3CH_2O^-Na^+$$

Problem 11.5 Account for the fact that treatment of *tert*-butyl methyl ether with a limited amount of concentrated HI gives methanol and *tert*-butyl iodide rather than methyl iodide and *tert*-butyl alcohol.

The first step of the reaction involves protonation of the ether oxygen to give an oxonium ion intermediate. Cleavage of the oxonium ion intermediate on one side gives methanol and the *tert*-butyl cation. Cleavage on the other side gives a methyl cation and *tert*-butyl alcohol. Because of the greater stability of the tertiary carbocation, cleavage to give methanol and *tert*-butyl cation is favored. Reaction of the tertiary cation with the iodide completes the reaction.

There is an alternative pathway for formation of product, namely reaction of the oxonium ion with iodide ion by an S_N2 pathway on the less-hindered methyl carbon to give iodomethane and the 2-methyl-2-propanol (*tert*-butyl alcohol). The fact is that the S_N1 pathway by way of a tertiary carbocation has a lower activation energy (a faster rate) than reaction of the oxonium ion with iodide ion by an S_N2 pathway so the S_N1 reaction predominates.

Problem 11.6 Draw structural formulas for the major products of each reaction.

Problem 11.7 The trimethylsilyl protecting group is easily removed in aqueous solution containing a trace of acid. Propose a mechanism for this reaction. (Note that a TBDMS protecting group is stable under these conditions because of the greater steric crowding around silicon created by the *t*-butyl group.)

Step 1: Add a proton. **Protonation in acid will occur on oxygen, making the alcohol a better leaving group.**

Step 2: Make a bond between a nucleophile and an electrophile and simultaneously break a bond to give stable molecules or ions. **Water attacks the silyl group in an S_N2 process to generate the free alcohol and silanol products (after a proton transfer back to water). Note that the S_N2 reaction on a tertiary Si is possible because Si atoms are larger than carbon, reducing steric hindrance enough to allow a backside attack.**

Step 3: Take a proton away.

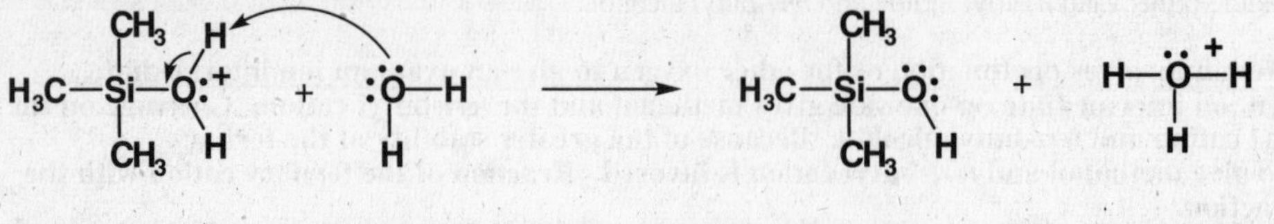

<u>Problem 11.8</u> Consider the possibilities for stereoisomerism in the bromohydrin and epoxide formed from *trans*-2-butene. (a) How many stereoisomers are possible for the bromohydrin? Which of the possible bromohydrin stereoisomers are formed by treating *trans*-2-butene with bromine in water?

This chlorohydrin has two chiral centers and, by the 2^n rule, there are four possible stereoisomers - two pairs of enantiomers.

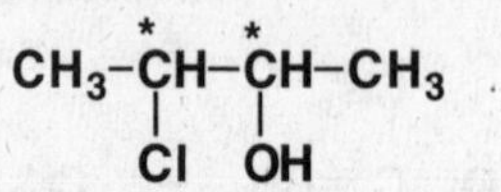

However, given the anti stereoselectivity of halohydrin formation, only one pair of these enantiomers is formed from *trans*-2-butene as a racemic mixture.

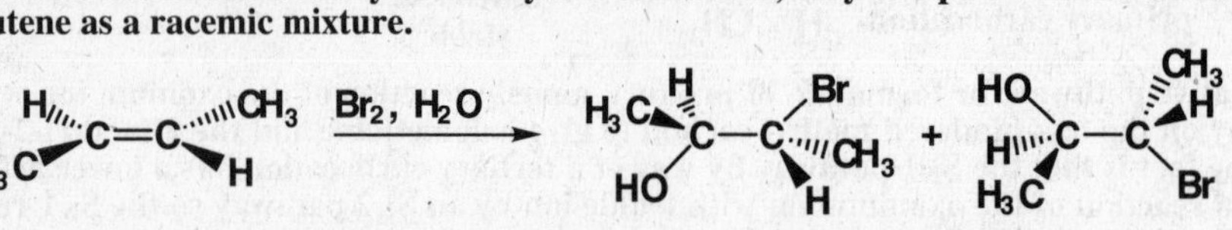

(b) How many stereoisomers are possible for the epoxide? Which of the possible stereoisomers is/are formed in this two step sequence?

There are three stereoisomers possible for this epoxide; a pair of enantiomers and a meso compound. However, because of the stereoselectivity (requirement for backside attack) of the S_N2 reaction, it is the pair of enantiomers that will be produced in the reaction sequence that began with *trans*-2-butene.

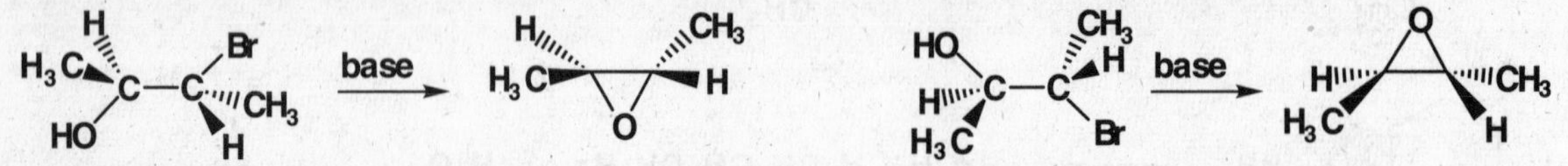

<u>Problem 11.9</u> Draw the expected products of Sharpless epoxidation of each allylic alcohol using (+)-diethyl tartrate as the chiral catalyst.

(a)

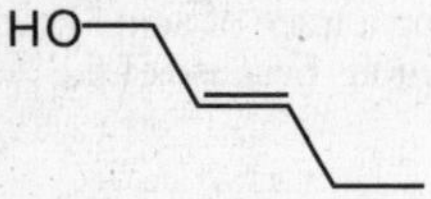

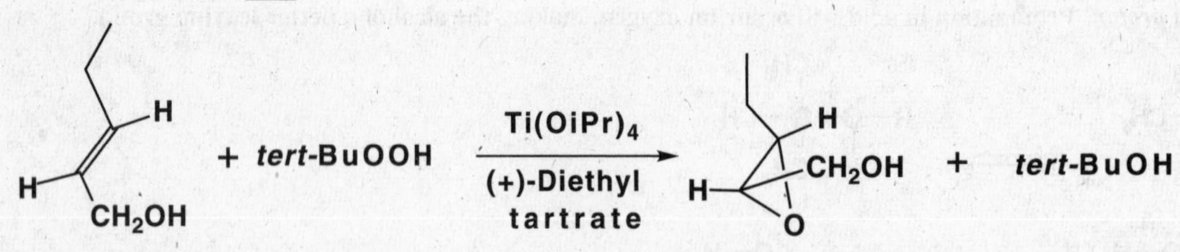

(b)

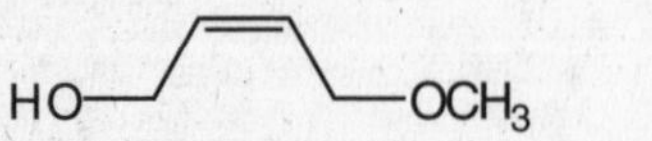

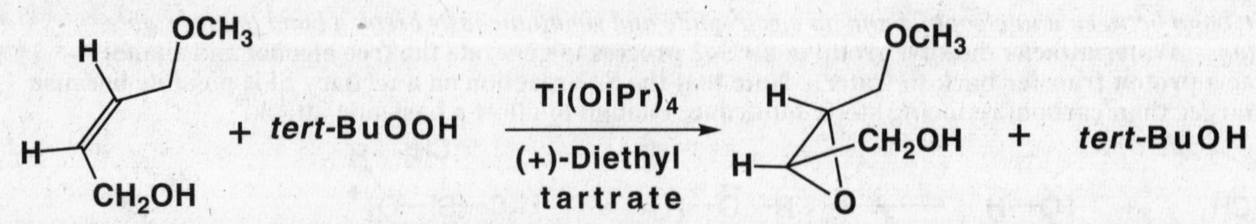

Structure and Nomenclature
Problem 11.10 Write names for these compounds. Where possible, write both IUPAC names and common names.

Common names are listed in parentheses.

(a)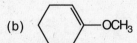

Cyclopentoxycyclopentane
(Dicyclopentyl ether)

(b)

1-Methoxycyclohexene
(1-Cyclohexenyl methyl ether)

(c)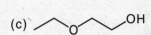

2-Ethoxyethanol

(d)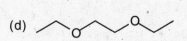

1,2-Diethoxyethane
(Ethylene glycol diethyl ether)

(e)

Oxalane
(Tetrahydrofuran)

(f)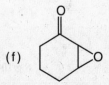

2,3-Epoxycyclohexanone

(g)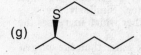

(R)-2-Ethylsulfanylhexane
((R) Ethyl 2-hexyl sulfide)

(h)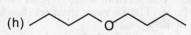

Butoxybutane
(Dibutyl ether)

Problem 11.11 Draw structural formulas for these compounds.

(a) 2-(1-Methylethoxy)propane

(b) *trans*-2,3-Diethyloxirane

(c) *trans*-2-Ethoxy-cyclopentanol

$$CH_3CH-O-CHCH_3$$
with CH_3 above each CH

$$CH_3CH_2-\overset{H}{\underset{O}{C}}-\overset{CH_2CH_3}{\underset{}{C}}H$$

OH / OCH₂CH₃ cyclopentane

or

(d) Ethenyloxyethene

(e) Cyclohexene oxide

OH / OCH₂CH₃ cyclopentane

$$CH_2=CH-O-CH=CH_2$$

cyclohexane with H, O, H epoxide

(f) 3-Cyclopropyloxy-1-propene

(g) (R)-2-Methyloxirane

(h) 1,1-Dimethoxycyclohexane

$$H_2C \diagdown CH-O-CH_2-CH=CH_2$$
$$H_2C \diagup$$

$$H-\overset{CH_3}{\underset{O}{C}}-CH_2$$

cyclohexane with OCH₃ / OCH₃

Physical Properties
Problem 11.12 Each compound given in this problem is a common organic solvent. From each pair of compounds, select the solvent with the greater solubility in water.

In each case select the molecule that can make better hydrogen bonds with water [(a) and (b)] or is more polar [(c) and (d)].

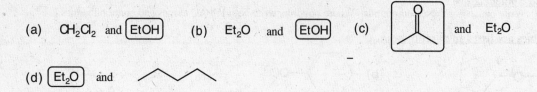

(a) CH_2Cl_2 and $\boxed{EtOH}$ (b) Et_2O and $\boxed{EtOH}$ (c) 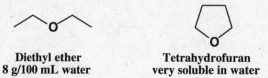 and Et_2O

(d) $\boxed{Et_2O}$ and ⌁⌁⌁

<u>Problem 11.13</u> Account for the fact that tetrahydrofuran (THF) is very soluble in water, whereas the solubility of diethyl ether in water is only 8 g/100 mL water.

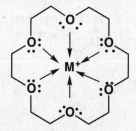

Diethyl ether   **Tetrahydrofuran**
8 g/100 mL water   **very soluble in water**

There are two factors to be considered:
(1) The shape of the hydrocarbon is important in determining water solubility. For example, 2-methyl-2-propanol (*tert*-butyl alcohol) is considerably more soluble in water than 1-butanol. A *tert*-butyl group is much more compact than a butyl group and, consequently, there is less disruption of water hydrogen bonding when *tert*-butyl alcohol is dissolved in water compared to when 1-butanol is dissolved in water. Similarly, the hydrocarbon portion of tetrahydrofuran (THF) is more compact than that of diethyl ether which increases the solubility of THF in water relative to diethyl ether.
(2) The oxygen atom of THF is more accessible for hydrogen bonding and solvation by water than the oxygen atom of diethyl ether. This greater accessibility arises because the hydrocarbon chains bonded to oxygen in THF are "tied back" whereas those on the oxygen atom of diethyl ether have more degrees of freedom and consequently present more steric hindrance to solvation.

<u>Problem 11.14</u> Because of the Lewis base properties of ether oxygen atoms, crown ethers are excellent complexing agents for Na^+, K^+, and NH_4^+. What kind of molecule might serve as a complexing agent for Cl^- or Br^-?

18-crown-6 with a generic metal M^+

To build an analogous system for complexation of Cl^- it is necessary to first realize that the chloride ion is a Lewis base. Instead of a cavity with electron pair donors (as in crown ethers) we need a system with electron pair acceptors (or Lewis acid sites), such as the hydrogen atoms of protonated amines.

The size of the cavity will determine which anions will be complexed preferentially.

Preparation of Ethers

Problem 11.15 Write equations to show a combination of reactants to prepare each ether. Which ethers can be prepared in good yield by a Williamson ether synthesis? If there are any that cannot be prepared by the Williamson method, explain why not.

In each case, choose reagents that utilize a methyl halide (best), primary alkyl halide (2nd best), or a secondary alkyl halide (3rd best).

(a)

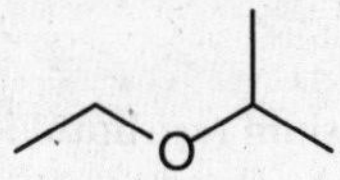

$$Na^+ \ ^-OCHCH_3 + CH_3CH_2Br \longrightarrow CH_3CH_2OCHCH_3 + Na^+Br^-$$

(b)

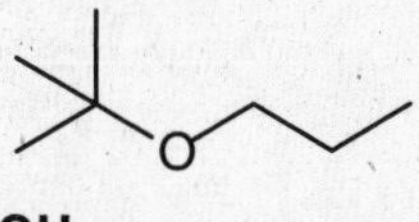

$$CH_3CO^-K^+ + CH_3CH_2CH_2Br \longrightarrow CH_3COCH_2CH_2CH_3 + K^+Br^-$$

(c)

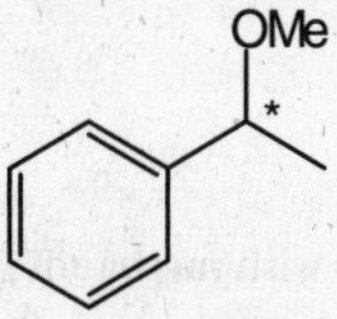

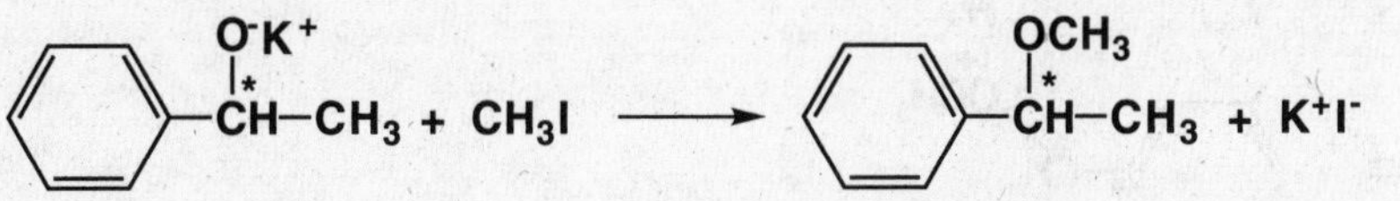

$$\underset{}{\overset{O^-K^+}{\underset{*}{C}H-CH_3}} + CH_3I \longrightarrow \underset{}{\overset{OCH_3}{\underset{*}{C}H-CH_3}} + K^+I^-$$

(d)

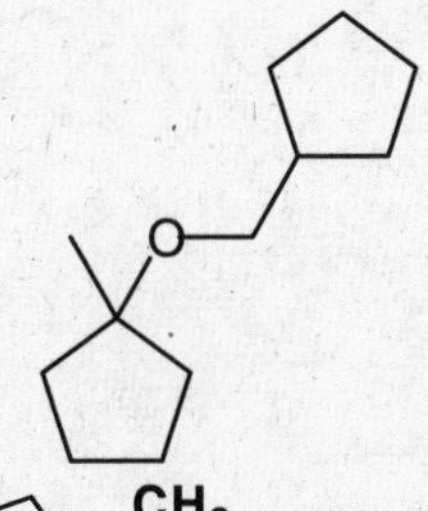

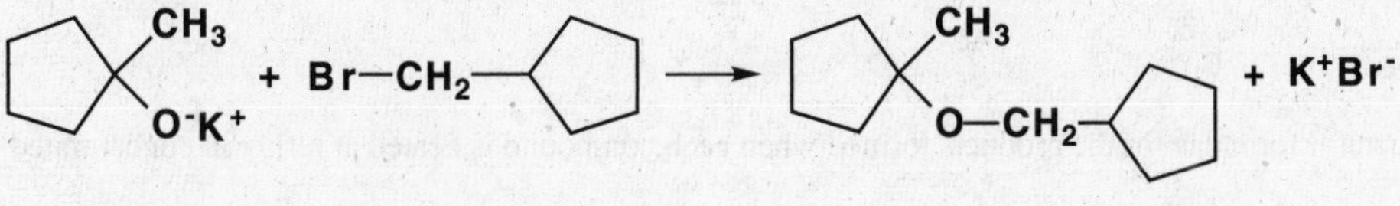

$$\overset{CH_3}{\underset{O^-K^+}{C}} + Br-CH_2 \longrightarrow \overset{CH_3}{\underset{O-CH_2}{C}} + K^+Br^-$$

(e)

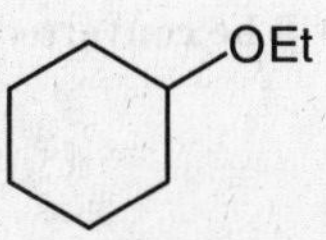

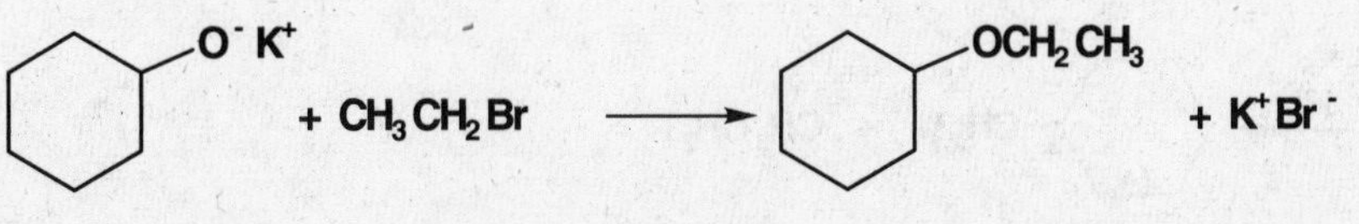

$$+ CH_3CH_2Br \longrightarrow OCH_2CH_3 + K^+Br^-$$

(f)

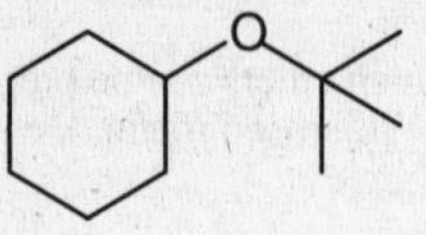

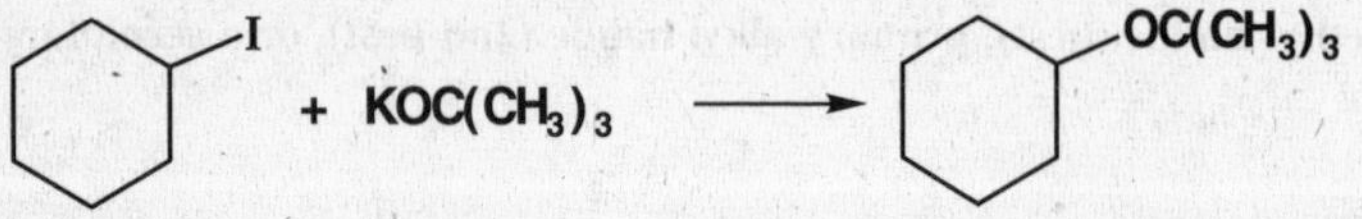

Significant elimination will be observed in (f) because a secondary alkyl iodide and potassium *tert*-butoxide are being used.

Problem 11.16 Propose a mechanism for this reaction.

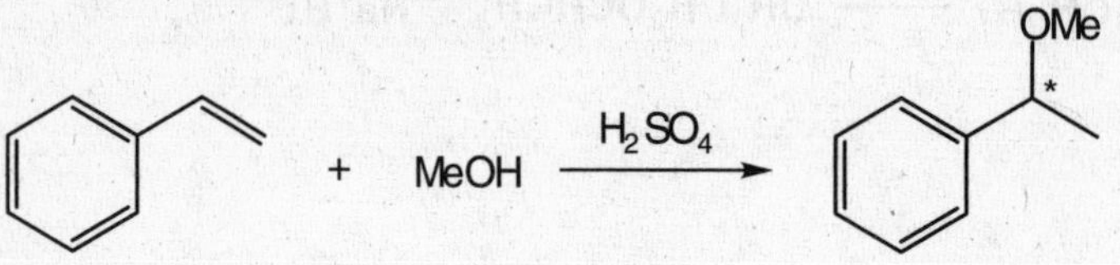

(racemic)

Step 1: Add a proton. **Protonation of the alkene gives a resonance-stabilized carbocation.**

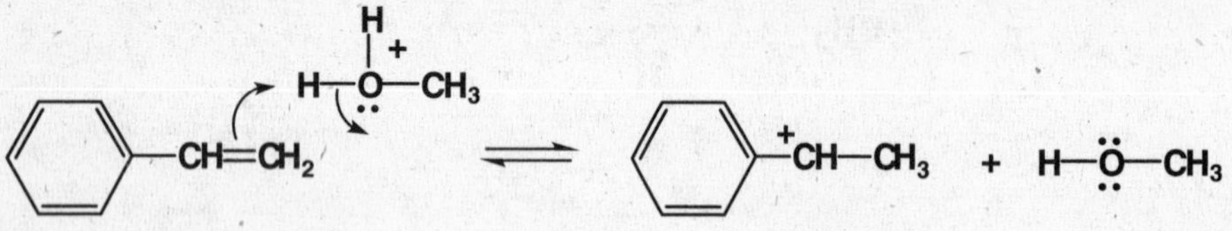

Step 2: Make a bond between a nucleophile and an electrophile. **Reaction of the carbocation with methanol gives an oxonium ion.**

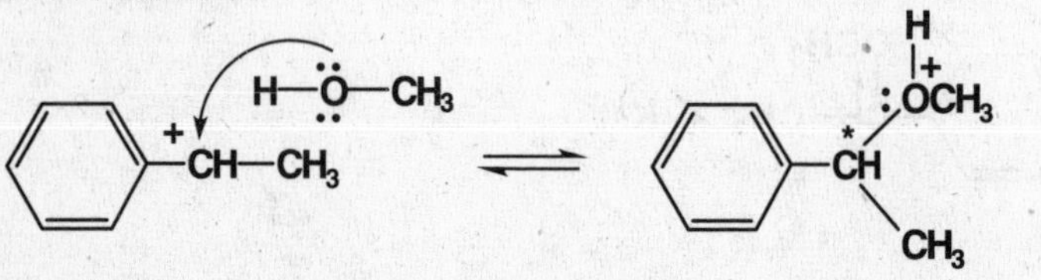

Step 3: Take a proton away. **Loss of a proton to gives the ether.**

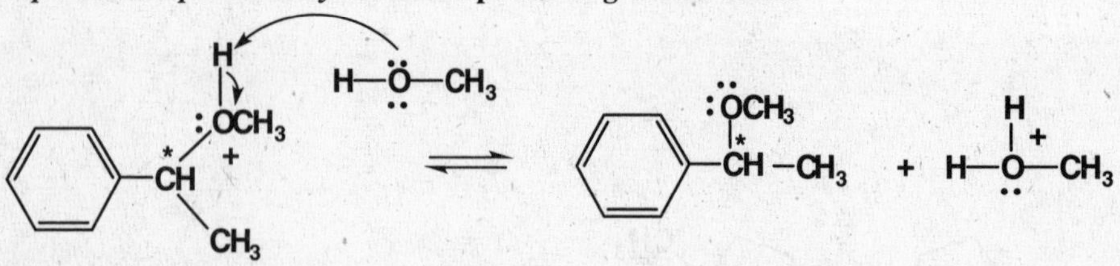

Reactions of Ethers
Problem 11.17 Draw structural formulas for the products formed when each compound is heated at reflux in concentrated HI.

Since an excess of HI is used, you can assume that the alcohol products initially produced will be converted to alkyl iodides under these conditions.

(a)

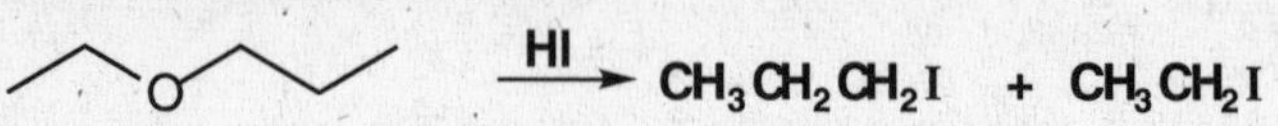

$$\xrightarrow{HI} \quad CH_3CH_2CH_2I \quad + \quad CH_3CH_2I$$

(b)

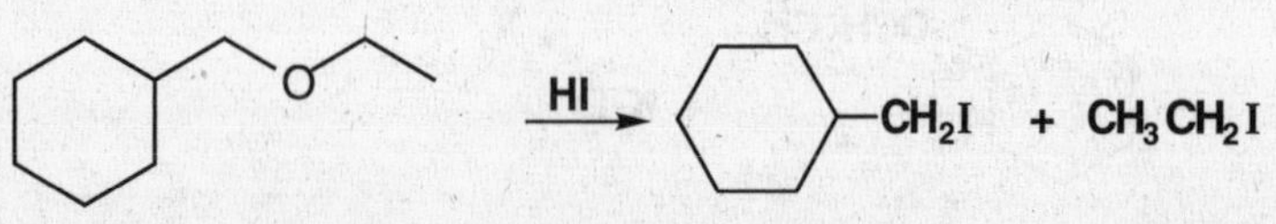

$$\xrightarrow{HI} \quad \text{(cyclohexyl)}-CH_2I \quad + \quad CH_3CH_2I$$

(c)

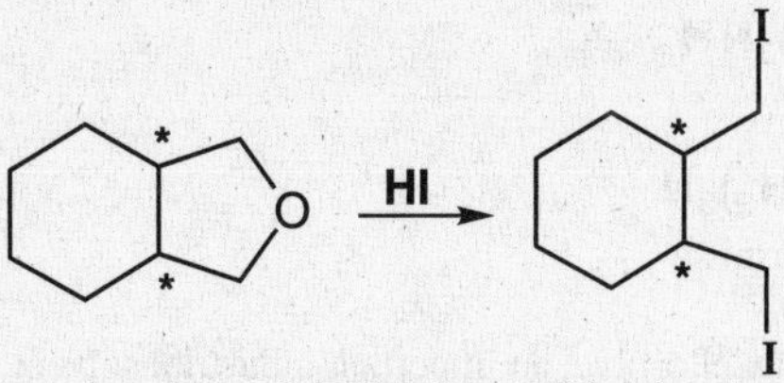

$$\text{HI} \longrightarrow$$

The stereochemistry of the product will be the same as that in the starting material (*cis* or *trans*, etc.).

(d)

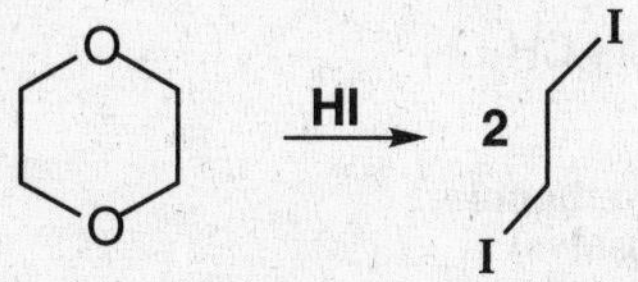

$$\xrightarrow{\text{HI}} 2$$

Problem 11.18 Following is an equation for the reaction of diisopropyl ether and oxygen to form a hydroperoxide.

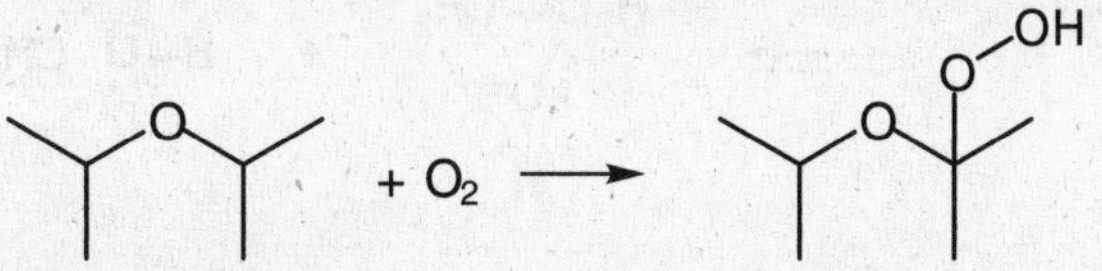

Diisopropyl ether   (A hydroperoxide)

Formation of an ether hydroperoxide is a radical chain reaction.
(a) Write a pair of chain propagation steps that accounts for the formation of this ether hydroperoxide. Assume that initiation is by a radical, R•.

Initiation:

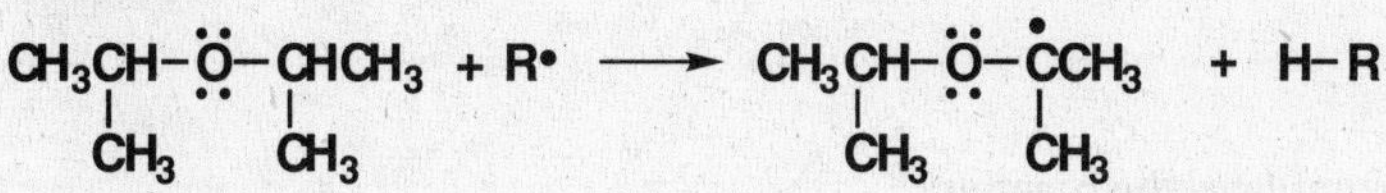

Propagation:

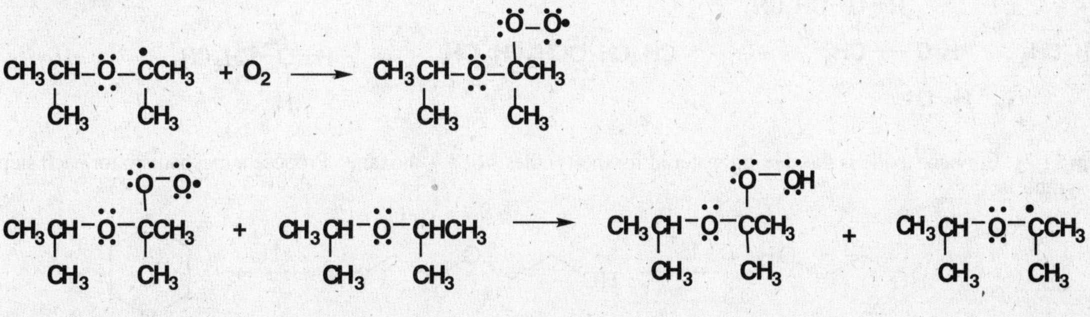

(b) Account for the fact that hydroperoxidation of ethers is regioselective; that is, reaction occurs preferentially at a carbon adjacent to the ether oxygen.

The radical formed is secondary rather than primary and it is stabilized by a resonance interaction with the adjacent oxygen atom.

Synthesis and Reactions of Epoxides
Problem 11.19 Triethanolamine, $(HOCH_2CH_2)_3N$, is a widely used biological buffer, with maximum buffering capacity at pH 7.8. Propose a synthesis of this compound from ethylene oxide and ammonia.

Triethanolamine can be prepared from ammonia by three successive reactions with ethylene oxide.

$HOCH_2CH_2NH_2$ + H_2C-CH_2 (epoxide, O) $\longrightarrow$ $(HOCH_2CH_2)_2NH$

$(HOCH_2CH_2)_2NH$ + H_2C-CH_2 (epoxide, O) $\longrightarrow$ $(HOCH_2CH_2)_3N$
Triethanolamine

<u>Problem 11.20</u> Ethylene oxide is the starting material for the synthesis of Cellosolve, an important industrial solvent. Propose a mechanism for this reaction.

H_2C-CH_2 (epoxide, O) + CH_3CH_2OH $\xrightarrow{H_2SO_4}$ $CH_3CH_2OCH_2CH_2OH$

    Oxirane   **2-Ethoxyethanol**
(Ethylene oxide)   **(Cellosolve)**

Step 1: Add a proton. **protonation of the epoxide gives a cyclic oxonium ion.**

Step 2: Make a bond between a nucleophile and an electrophile. **Reaction of this oxonium ion with the alcohol opens the epoxide ring.**

Step 3: Take a proton away. **Loss of a proton completes the reaction.**

<u>Problem 11.21</u> Ethylene oxide is the starting material for the synthesis of 1,4-dioxane. Propose a mechanism for each step in this synthesis.

                                                           1,4-dioxane

Step 1: Add a proton. **The mechanism for this reaction involves an initial protonation on the epoxide O atom.**

Step 2: Make a bond between a nucleophile and an electrophile. **There is a nucleophilic attack by the diol on the protonated epoxide.**

$$H_2C-CH_2 \ (\overset{+}{O}H) \ + \ HOCH_2CH_2OH \longrightarrow CH_2CH_2\overset{+}{O}CH_2CH_2OH$$

Step 3: Take a proton away.

$$\overset{OH}{CH_2}CH_2\overset{+}{O}CH_2CH_2OH \longrightarrow HOCH_2CH_2OCH_2CH_2OH \ + \ H-\overset{+}{O}-R$$

Step 4: Add a proton. **Next, one of the terminal hydroxyl groups is protonated.**

$$HOCH_2CH_2OCH_2CH_2OH \ + \ H-\overset{+}{O}-R \ \rightleftharpoons \ [\text{ring structure}] \ + \ H-O-R$$

Step 5: Make a bond between a nucleophile and an electrophile and simultaneously break a bond to give stable molecules or ions. **The terminal -OH group displaces water to form a six-member ring.**

$$[\text{structure}] \longrightarrow [\text{six-membered ring}] \ + \ H-O-H$$

Step 6: Take a proton away.

$$[\text{protonated ring}] \ + \ H-O-R \longrightarrow [\text{ring}] \ + \ H-\overset{+}{O}-R$$

Problem 11.22 Propose a synthesis for 18-crown-6. If a base is used in your synthesis, does it make a difference whether it is lithium hydroxide or potassium hydroxide ?

$$[\text{diCl ether}] \ + \ [\text{diOH ether}] \ \overset{KOH}{\longrightarrow} \ [\text{18-crown-6 with } K^+]$$

18-Crown-6

Because 18-crown-6 binds K$^+$ the best, KOH should be used so that the crown ether can form around the K$^+$ ion. This type of strategy is called the template approach and is used in a variety of similar situations. The above two pieces can be synthesized as follows:

$$HOCH_2CH_2OCH_2CH_2OH \quad + \quad H_2C\overset{O}{-}CH_2 \quad \xrightarrow{H^+}$$

(From Problem 11.21)

$$\xrightarrow{SOCl_2}$$

Problem 11.23 Predict the structural formula of the major product of the reaction of 2,2,3-trimethyloxirane with each set of reagents.

Nucleophiles in the absence of acid catalysis react via an S$_N$2 mechanism, so the nucleophile ends up on the less hindered, that is less substituted, carbon atom of the epoxide. The acid-catalyzed reaction of epoxides occurs in an S$_N$1-like fashion, so substitution occurs at the site of the more stable cation, namely the more substituted carbon atom of the epoxide. In both cases, the stereochemistry of addition is *trans* to the epoxide oxygen atom. Note that because 2,2,3-trimethyloxirane is actually a pair of enantiomers, the products of each reaction are a pair of enantiomers as well.

2,2,3-Trimethyloxirane (a pair of enantiomers)

(a) MeOH/ MeO$^-$ Na$^+$

Methoxide (MeO$^-$) is a nucleophilic reagent so the reaction takes place on the less hindered carbon atom.

(b) MeOH/ H$^+$

This reaction involves protonation of the epoxide, so the MeOH attacks the tertiary side that has more positive charge character

(c) Me$_2$NH

Dimethylamine (Me$_2$NH) is a nucleophilic reagent so the reaction takes place on the less hindered carbon atom.

<u>Problem 11.24</u> The following equation shows the reaction of *trans*-2,3-diphenyloxirane with hydrogen chloride in benzene to form 2-chloro-1,2-diphenylethanol.

trans-2,3-Diphenyloxirane   **2-Chloro-1,2-diphenylethanol**

(a) How many stereoisomers are possible for 2-chloro-1,2-diphenylethanol?

There are two chiral centers in the molecule so there are 2^2 or 4 possible stereoisomers.

A pair of enantiomers
(from addition of HCl to *trans*-2,3-diphenyloxirane)

A pair of enantiomers
(from addition of HCl to *cis*-2,3-diphenyloxirane)

(b) Given that opening of the epoxide ring in this reaction is stereoselective, predict which of the possible stereoisomers of 2-chloro-1,2-diphenylethanol is/are formed in the reaction.

Only the first (upper) pair of enantiomers is formed.

<u>Problem 11.25</u> Propose a mechanism to account for this rearrangement.

Tetramethyloxirane   3,3-Dimethyl-2-butanone

Reaction between boron trifluoride, a Lewis acid, and the epoxide oxygen, a Lewis base, forms an oxonium ion.

Step 1: Make a bond between a Lewis acid and Lewis base.

Step 2: Break a bond to give stable molecules or ions. **The ring then opens to give a carbocation intermediate.**

Step 3: 1,2 Shift. **Rearrangement of the resulting secondary carbocation then occurs.**

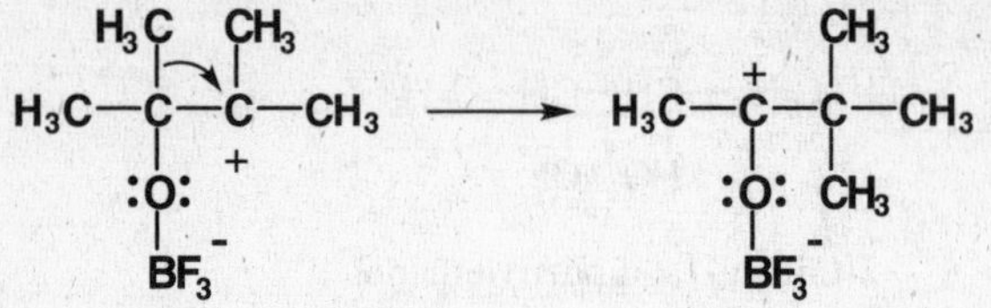

Step 4: Break a bond to give stable molecules or ions. **Loss of boron trifluoride gives the observed ketone product.**

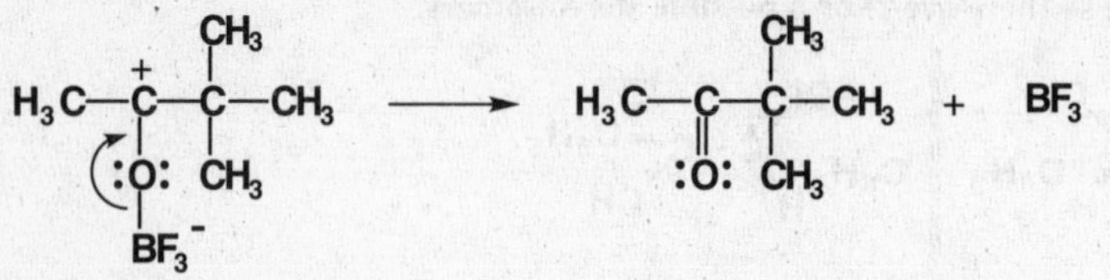

<u>Problem 11.26</u> Acid-catalyzed hydrolysis of the following epoxide gives a *trans* diol.

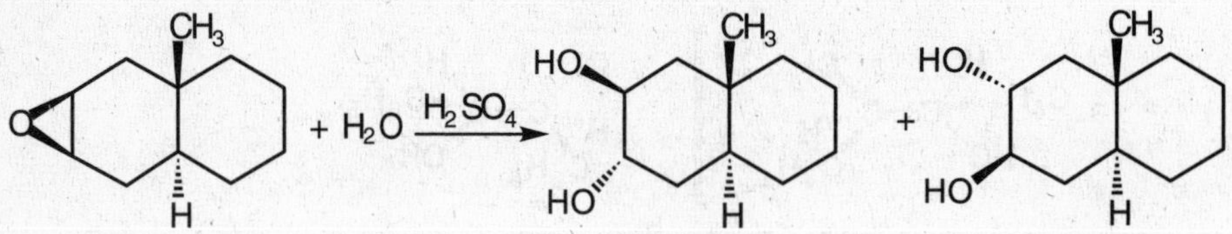

Only this glycol
is formed

This glycol is
not formed

Of the two possible *trans* diols, only one is formed. How do you account for this stereoselectivity?

The key to this problem is that the incoming nucleophile and the leaving protonated epoxide oxygen must be anti coplanar. In a cyclohexane ring, anti coplanar corresponds to *trans* and diaxial. An accurate model of the glycol formed will show that in it, the two -OH groups are diaxial and, therefore, *trans* and coplanar. In the alternative glycol, the -OH groups are also *trans*, but because they are diequatorial, they are not coplanar and this is not formed.

-OH groups
are *trans*
and coplanar

-OH groups
are *trans* but
not coplanar

<u>Problem 11.27</u> Following are two reaction sequences for converting 1,2-diphenylethylene into 2,3-diphenyloxirane.

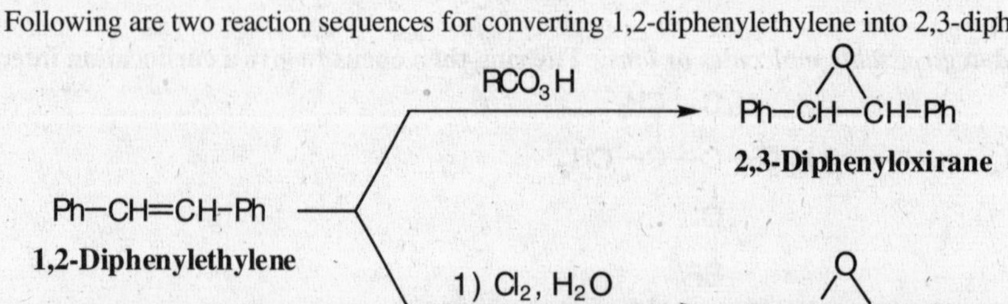

Ph–CH=CH–Ph

1,2-Diphenylethylene

RCO$_3$H

Ph–CH–CH–Ph

2,3-Diphenyloxirane

1) Cl$_2$, H$_2$O
2) CH$_3$O⁻Na⁺

Ph–CH–CH–Ph

2,3-Diphenyloxirane

Suppose that the starting alkene is *trans*-1,2-diphenylethylene.
(a) What is the configuration of the oxirane formed in each sequence?

In each case, the oxirane has the *trans* configuration.

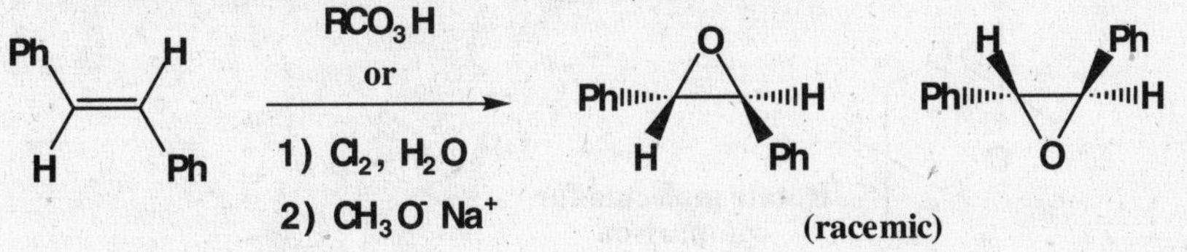

(b) Will the oxirane formed in either sequence rotate the plane of polarized light? Explain.

Neither product will rotate the plane of polarized light. The *trans* isomer can exist as a pair of enantiomers. In each reaction, the product formed is a racemic mixture and will not be optically active.

Problem 11.28 The following enantiomer of a chiral epoxide is an intermediate in the synthesis of the insect pheromone frontalin.

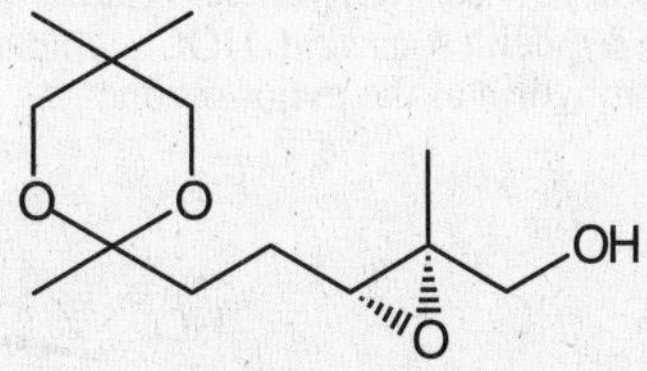

Show how can this enantiomer be prepared from an allylic alcohol, using the Sharpless epoxidation.

The key to this problem is to use the following diagram of how the Sharpless epoxidation operates, then identify which face of the alkene must react to give the indicated epoxide stereoisomer.

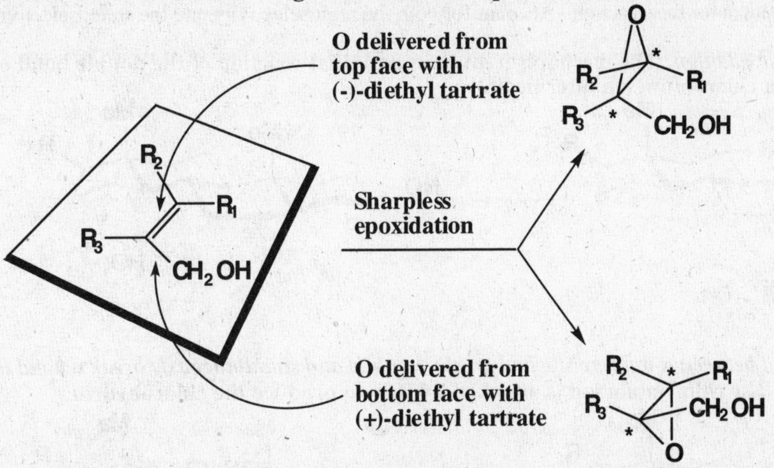

The starting alkene can be drawn, and it is helpful to rotate it so that it can be more easily compared with the diagram. You should use models to make sure you can visualize this rotation.

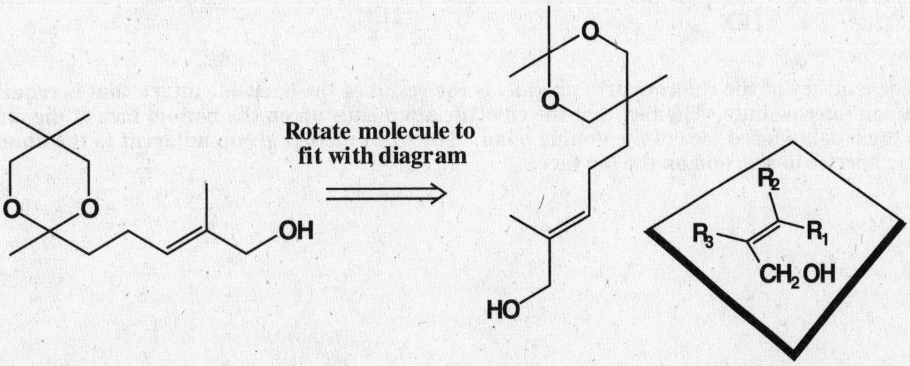

The Sharpless epoxidation reaction run with (-)-diethyl tartrate delivers the O atom to the top face of the alkene as drawn. When rotated back for comparison with the stereoisomer given in the problem, it can be seen that (-)-diethyl tartrate is the correct choice for this epoxidation reaction. You should make a model of the epoxide product to confirm it is the desired enantiomer.

Problem 11.29 Human white cells produce an enzyme called myeloperoxidase. This enzyme catalyzes the reaction between hydrogen peroxide and chloride ion to produce hypochlorous acid, HOCl, which reacts as if it were Cl^+OH^-. When attacked by white cells, cholesterol gives a chlorohydrin as the major product.

Cholesterol

Cholesterol chlorohydrin

(a) Propose a mechanism for this reaction. Account for both the regioselectivity and the stereoselectivity.

Step 1: Electrophilic addition. **The mechanism involves an initial reaction of the double bond of cholesterol with HOCl to produce a chloronium ion intermediate.**

Step 2: Make a bond between a nucleophile and an electrophile and simultaneously break a bond to give stable molecules or ions. **The chloronium ion is attacked by HO⁻ to produce the chlorohydrin.**

The *trans* stereochemistry of the chlorohydrin product is the result of the backside attack that is required by the chloronium ion intermediate. The fact that the chlorine atom ends up on the bottom face of the molecule is because this is the less-hindered face of the double bond. The axial methyl group adjacent to the double bond provides a steric barrier to reaction on the top face.

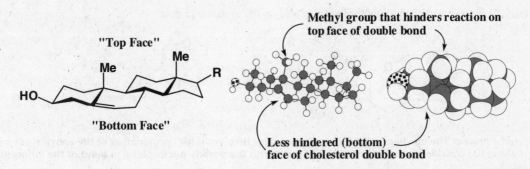

"Top Face"

Me
Me
R
HO

"Bottom Face"

Methyl group that hinders reaction on
top face of double bond

Less hindered (bottom)
face of cholesterol double bond

The HO⁻ nucleophile attacks the secondary carbon atom of the chloronium ion, as opposed to the tertiary
carbon atom, again because of steric hindrance caused by the axial methyl group preventing access to the
tertiary carbon atom. Note that attack by HO⁻ must be from the top face of the molecule, anti to the chlorine
atom.

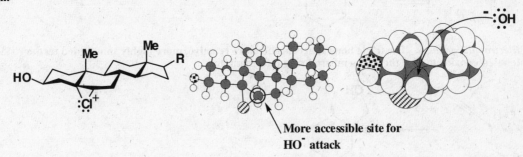

More accessible site for
HO⁻ attack

(b) On standing or (much more rapidly) on treatment with base, the chlorohydrin is converted to an epoxide. Show the
structure of the epoxide and a mechanism for its formation. This epoxide is believed to be involved in induction of certain
cancers.

Step 1: Take a proton away. The OH group of the chlorohydrin is deprotonated by the base.

*Step 2: Make a bond between a nucleophile and an electrophile and simultaneously break a bond to give stable
molecules or ions.* The alkoxide intermediate leads epoxide formation as chloride anion departs. The backside
attack of the oxygen atom dictates that the epoxide product will have the oxygen atom on what we are calling
the top face of the molecule.

Problem 11.30 Propose a mechanism for the following acid-catalyzed rearrangement.

Step 1: Add a proton. The mechanism of this rearrangement involves initial protonation of the epoxide oxygen atom, making the epoxide electrophilic enough to react with the weakly nucleophilic pi bond of the adjacent triple bond.

Step 2: Electrophlic addition. The triple bond reacts at the more reactive, more highly substituted tertiary side of the protonated epoxide to give the six-membered ring intermediate.

Step 3: Make a bond between a nucleophile and an electrophile. Attack of the alcohol upon the carbocation of this intermediate creates the five-member ring.

Step 4: Take a proton away. Loss of a proton completes the reaction.

Synthesis
Problem 11.31 Show reagents and experimental conditions to synthesize the following compounds from 1-propanol (any derivative of 1-propanol prepared in an earlier part of this problem may then be used for another part of the problem).
(a) Propanal

(b) Propanoic acid

(c) Propene

$$CH_3CH_2CH_2OH \xrightarrow[\text{heat}]{H_3PO_4} CH_3CH=CH_2$$

(d) 2-Propanol

$$CH_3CH=CH_2 \xrightarrow[\text{H}_2\text{SO}_4]{\text{H}_2\text{O}} \overset{\overset{\displaystyle OH}{|}}{CH_3CHCH_3}$$

From (c)

(e) 2-Bromopropane

$$\overset{\overset{\displaystyle OH}{|}}{CH_3CHCH_3} \quad or \quad H_2C=CHCH_3 \; + \; HBr \xrightarrow{\text{heat}} \overset{\overset{\displaystyle Br}{|}}{CH_3CHCH_3}$$

From (d)   From (c)

(f) 1-Chloropropane

$$CH_3CH_2CH_2OH \xrightarrow[\text{Pyridine}]{SOCl_2} CH_3CH_2CH_2Cl \; + \; SO_2 \; + \; HCl$$

(g) 1,2-Dibromopropane

$$CH_3CH=CH_2 \xrightarrow[\text{CCl}_4]{\text{Br}_2} \overset{\overset{\displaystyle Br}{|*}}{CH_3CHCH_2Br}$$

From (c)   **(racemic mixture)**

(h) Propyne

$$\overset{\overset{\displaystyle Br}{|}}{CH_3CHCH_2Br} \xrightarrow[\text{2. H}_2\text{O}]{\text{1. 3NaNH}_2} H_3CC\equiv CH$$

From (g)

(i) 2-Propanone

$$\overset{\overset{\displaystyle OH}{|}}{CH_3CHCH_3} \xrightarrow[\text{H}_2\text{O, heat}]{\overset{\displaystyle K_2Cr_2O_7}{H_2SO_4}} CH_3\overset{\overset{\displaystyle O}{||}}{C}CH_3$$

From (d)

(j) 1-Chloro-2-propanol

$$CH_3CH=CH_2 \xrightarrow{Cl_2 / H_2O} \overset{\overset{\displaystyle OH}{|*}}{CH_3CHCH_2Cl}$$

(racemic mixture)

(k) Methyloxirane

$$\overset{\overset{\displaystyle OH}{|*}}{CH_3CHCH_2Cl} \xrightarrow{NaOH} CH_3\overset{*}{CH}\overset{O}{\triangle}CH_2$$

From (j)

or

(racemic mixture assuming racemic starting material)

$$CH_3CH=CH_2 \xrightarrow{RCO_3H} CH_3\overset{*}{CH}\overset{O}{\triangle}CH_2$$

(racemic mixture)

From (c)

(l) Dipropyl ether

$$CH_3CH_2CH_2Cl \;+\; CH_3CH_2CH_2O^-Na^+ \longrightarrow CH_3CH_2CH_2OCH_2CH_2CH_3$$

  From (f)   From 1-propanol
                       and Na

(m) Isopropyl propyl ether

$$\underset{\text{From (f)}}{CH_3CH_2CH_2Cl} \;+\; \underset{\substack{\text{From (d)}\\ \text{and Na}}}{CH_3\overset{\overset{O^-Na^+}{|}}{C}HCH_3} \longrightarrow CH_3\overset{\overset{OCH_2CH_2CH_3}{|}}{C}HCH_3$$

(n) 1-Mercapto-2-propanol

$$\underset{\text{From (k)}}{CH_3\overset{\overset{*}{\triangle}}{CH}-CH_2} \xrightarrow[\;H_2O\;]{HS^-Na^+} \underset{\text{(racemic mixture)}}{CH_3\overset{\overset{OH}{|}}{\underset{*}{C}}HCH_2SH}$$

(o) 1-Amino-2-propanol

$$\underset{\text{From (k)}}{CH_3\overset{\overset{*}{\triangle}}{CH}-CH_2} \xrightarrow{NH_3} \underset{\text{(racemic mixture)}}{CH_3\overset{\overset{OH}{|}}{\underset{*}{C}}HCH_2NH_2}$$

(p) 1,2-Propanediol

$$\underset{\text{From (c)}}{CH_3CH{=}CH_2} \xrightarrow[\;H_2O_2\;]{OsO_4} \underset{\text{(racemic mixture)}}{CH_3\overset{\overset{HO}{|}}{\underset{*}{C}}H-\overset{\overset{OH}{|}}{C}H_2}$$

or

$$\underset{\text{From (k)}}{CH_3\overset{\overset{*}{\triangle}}{CH}-CH_2} \xrightarrow[\;H_3O^+\;]{HO^-\ \text{or}} \underset{\text{(racemic mixture)}}{CH_3\overset{\overset{HO}{|}}{\underset{*}{C}}H-\overset{\overset{OH}{|}}{C}H_2}$$

Note that the products in g, j, k, n, o, and p will be racemic mixtures.

<u>Problem 11.32</u> Starting with *cis*-3-hexene, show how to prepare the following diols.
(a) Meso 3,4-hexanediol

The *syn* geometry of addition observed with OsO$_4$ leads to meso 3,4-hexanediol.

(b) Racemic 3,4-hexanediol

The *anti* geometry produced via epoxidation followed by hydrolysis gives a racemic mixture of enantiomers.

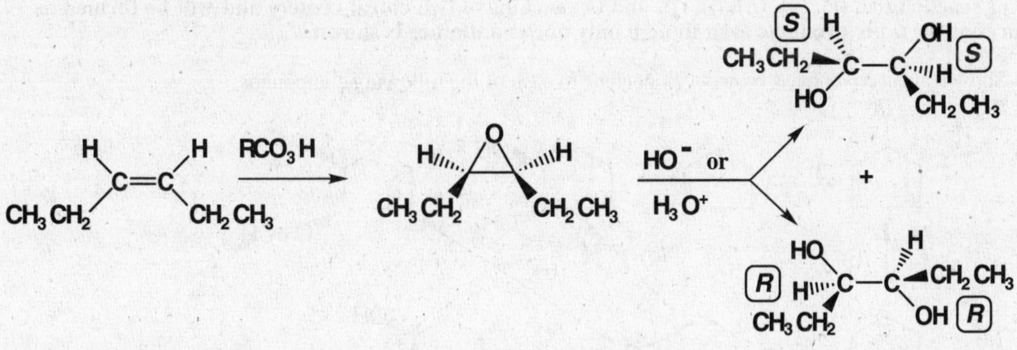

Problem 11.33 Show reagents to convert cycloheptene to the following.

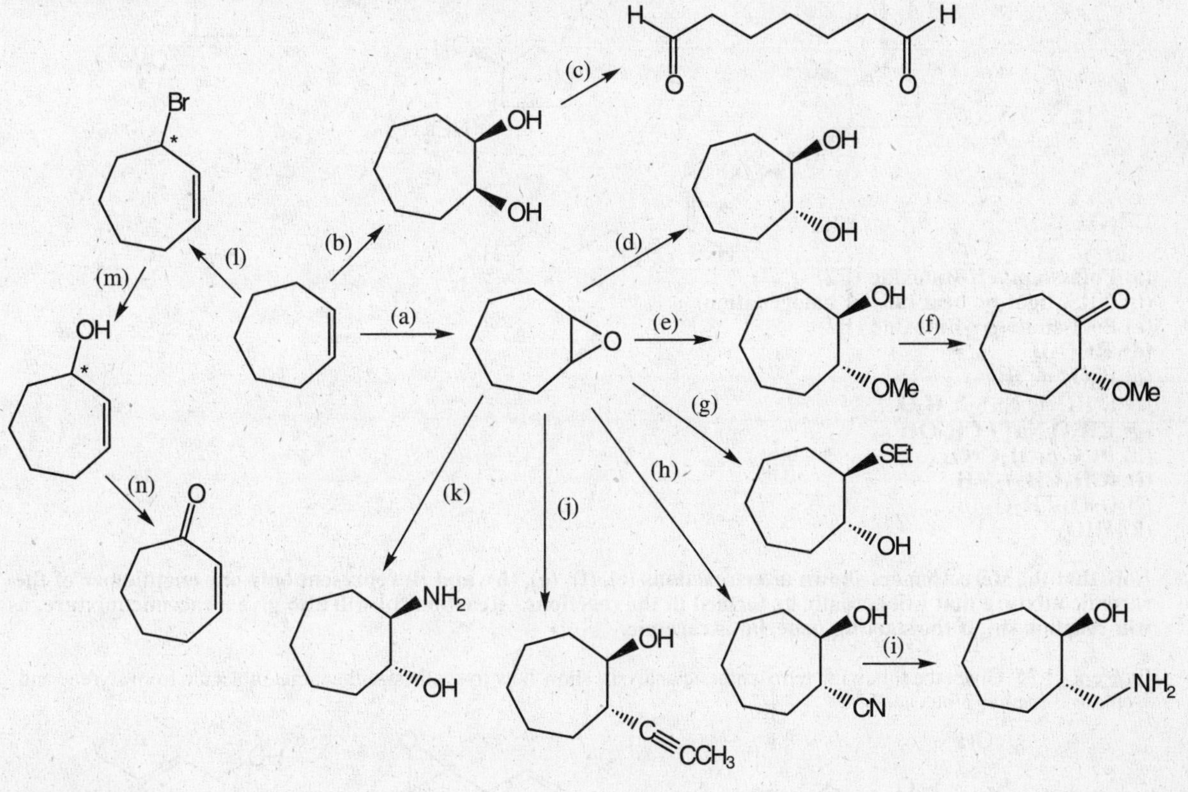

(a) RCO$_3$H
(b) OsO$_4$ / H$_2$O$_2$
(c) HIO$_4$
(d) H$_3$O$^+$ or HO$^-$
(e) CH$_3$O$^-$Na$^+$ / CH$_3$OH or CH$_3$OH/H$^+$
(f) PCC or H$_2$CrO$_4$
(g) 1. EtS$^-$Na$^+$ 2. H$_2$O or EtSH
(h) 1. NaCN 2. H$_2$O
(i) H$_2$/Pd will reduce the C≡N bond just as it does the C≡C bond.
(j) 1. H$_3$CC≡C$^-$Na$^+$ 2. H$_2$O
(k) NH$_3$
(l) NBS, light or heat (allylic halogenation)
(m) NaOH (Some E2 will occur as well here)
(n) PCC or H$_2$CrO$_4$

In the above scheme, the epoxide product of reaction (a), the *cis* diol product of reaction (b), the dialdehyde product of (c), and the ketone product of reaction (n) are not chiral. All of the other products are chiral and will be formed as racemic mixtures. The products of reaction (f),(l), and (m) each have a single chiral center.

The products of reaction (d), (e), (g), (h), (i), (j), and (k) each have two chiral centers and will be formed as racemic mixtures of the *trans* products even though only one enantiomer is shown.

Problem 11.34 Show reagents to convert bromocyclopentane to each of the following compounds,

(a) **Potassium *tert*-Butoxide (E2)**
(b) **NBS, light or heat (allylic halogenation)**
(c) **Potassium *tert*-Butoxide (E2)**
(d) **RCO₃H**
(e) **H₃O⁺ or HO⁻**
(f) **1. HC≡C⁻Na⁺ 2. H₂O**
(g) **CH₃O⁻Na⁺ / CH₃OH**
(h) **PCC or H₂CrO₄**
(i) **(CH₃CH₂)₂NH**
(j) **OsO₄ / H₂O₂**
(k) **HIO₄**

Note that the stereoisomers shown after reactions (e), (f), (g), (h), and (i) represent only one enantiomer of the racemic mixture that will actually be formed in the reactions. Reaction (b) will also give a racemic mixture, as will reaction (h), if the starting material is racemic.

Problem 11.35 Given the following retrosynthesic analysis, show how to synthesize the target molecule from styrene and 1-chloro-3-methyl-2-butene.

Styrene   1-Chloro-3-methyl-
                                              2-butene

The 1-chloro-3-methyl-2-butene is converted to the alcohol in base then the alkoxide by reaction with Na metal. The styrene is converted to the epoxide, then reacted with the alkoxide followed by an aqueous workup to give the final product. Notice that the last step requires reaction of the nucleophile at the less-hindered carbon of the epoxide, so an acid-catalyzed step could not be used. The final product is actually a racemic mixture.

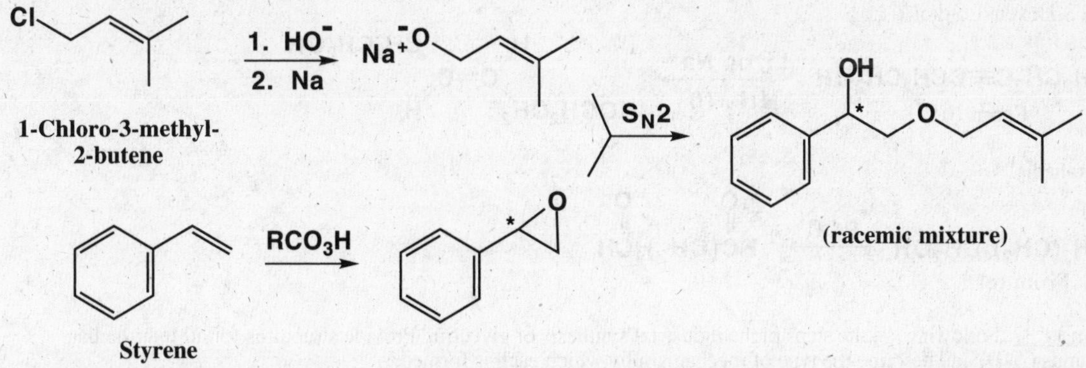

1-Chloro-3-methyl-2-butene

Styrene

(racemic mixture)

Problem 11.36 Starting with acetylene and ethylene oxide as the only sources of carbon atoms, show how to prepare these compounds.
(a) 3-Butyn-1-ol

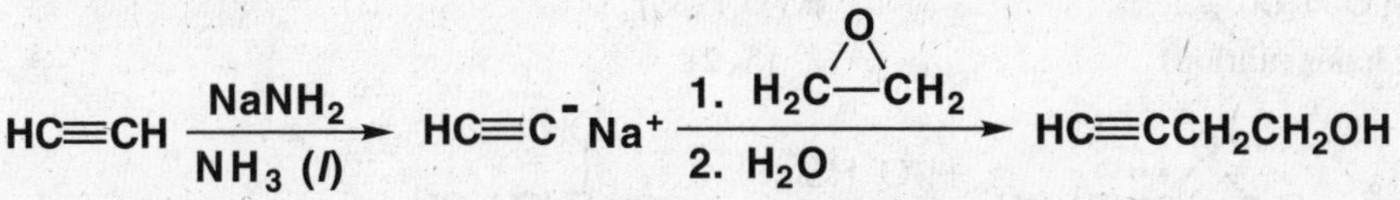

(b) 3-Hexyn-1,6-diol

Because the acidity of the -OH group will interfere with formation of the acetylide anion in base, it must be protected by a group such as the trimethylsilyl or *tert*-butyldimethylsilyl (TBDMS). Note how the silyl ether is created from reaction of the alcohol with the corresponding silyl chloride.

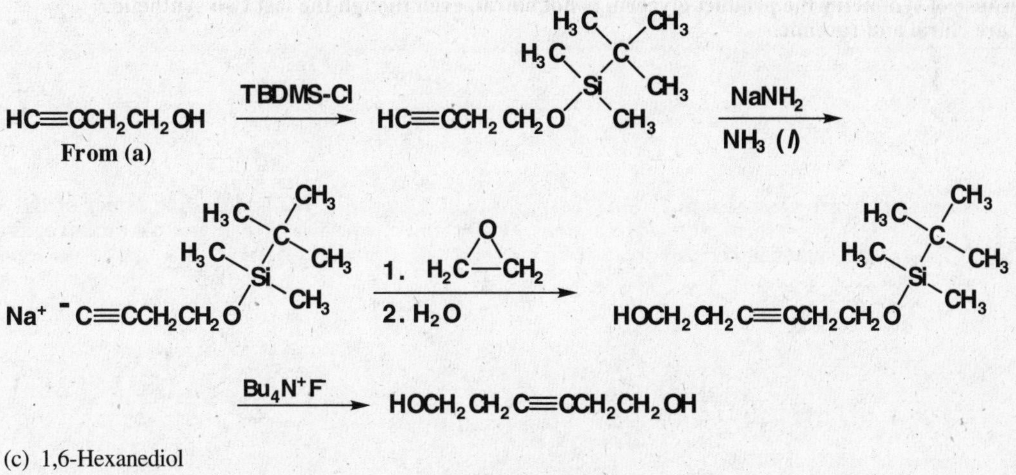

(c) 1,6-Hexanediol

$$HOCH_2CH_2C \equiv CCH_2CH_2OH \ + \ 2 \ H_2 \ \xrightarrow[\text{catalyst}]{\text{Transition metal}} \ HOCH_2(CH_2)_4CH_2OH$$

From (b)

(d) (Z)-3-Hexen-1,6-diol

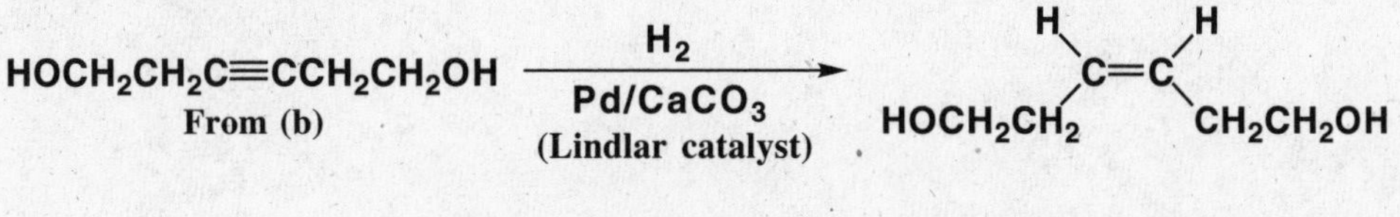

(e) (*E*)-3-Hexen-1,6-diol

$$\text{HOCH}_2\text{CH}_2\text{C}\equiv\text{CCH}_2\text{CH}_2\text{OH} \xrightarrow[\text{NH}_3 \ (l)]{\text{Li or Na}}$$

From (b)

Product: a cis/trans alkene drawn with

$$\begin{array}{c} \text{H} \qquad\qquad \text{CH}_2\text{CH}_2\text{OH} \\ \text{C}=\text{C} \\ \text{HOCH}_2\text{CH}_2 \qquad\qquad \text{H} \end{array}$$

(f) Hexanedial

$$\text{HOCH}_2(\text{CH}_2)_4\text{CH}_2\text{OH} \xrightarrow{\text{PCC}} \overset{\text{O}}{\overset{\|}{\text{HC}}}(\text{CH}_2)_4\overset{\text{O}}{\overset{\|}{\text{CH}}}$$

From (c)

<u>Problem 11.37</u> Following are the steps in the industrial synthesis of glycerin. Provide structures for all intermediate compounds (**A-D**) and describe the type of mechanism by which each is formed.

$$\text{CH}_2\text{=CHCH}_3 \xrightarrow[\text{heat}]{\text{Cl}_2} \text{CH}_2\text{=CHCH}_2\text{Cl} \xrightarrow{\text{NaOH, H}_2\text{O}} \text{CH}_2\text{=CHCH}_2\text{OH} \xrightarrow{\text{Cl}_2, \text{H}_2\text{O}}$$

Propene   **A** ($\text{C}_3\text{H}_5\text{Cl}$)   **B** ($\text{C}_3\text{H}_6\text{O}$)

                              (**Allylic halogenation**)   ($\text{S}_\text{N}2$)

$$\begin{array}{c}\overset{\text{OH}}{\underset{*}{|}}\\ \text{ClCH}_2\text{CHCH}_2\text{OH}\end{array} \xrightarrow[\text{heat}]{\text{Ca(OH)}_2} \overset{\text{O}}{\underset{*}{\text{H}_2\text{C}-\text{CHCH}_2\text{OH}}} \xrightarrow{\text{H}_2\text{O, HCl}} \begin{array}{c}\overset{\text{OH}}{|}\\ \text{HOCH}_2\text{CHCH}_2\text{OH}\end{array}$$

C ($\text{C}_3\text{H}_7\text{ClO}_2$)   **D** ($\text{C}_3\text{H}_6\text{O}_2$)   **1,2,3-Propanetriol**
(**Halohydrination**)   (**Internal displacement**   (glycerol, glycerin)
                                  by an alkoxide)
                                                          (**Acid-catalyzed**
                                                          epoxide hydrolysis)

Notice that because of symmetry the product glycerin is not chiral, even though the last two synthetic intermediates are chiral and racemic.

Problem 11.38 Gossyplure, the sex pheromone of the pink bollworm, is the acetic ester of 7,11-hexadecadien-1-ol. The active pheromone has the Z configuration at the C7-C8 double bond and is a mixture of E,Z isomers at the C11-C12 double bond. Shown here is the Z,E isomer.

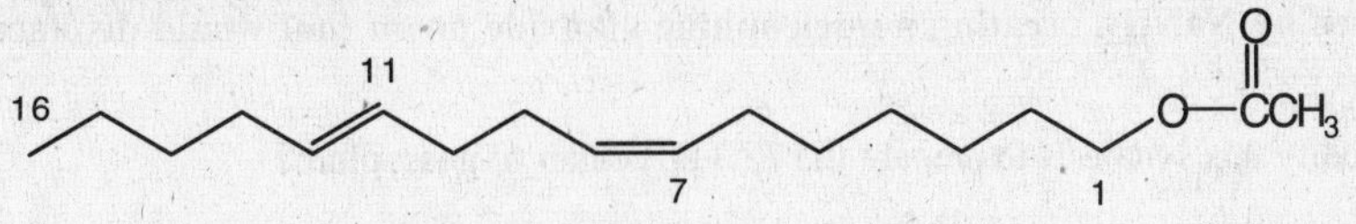

(7Z,11E)-7,11-hexadecadienyl acetate

Following is a retrosynthetic analysis for (7Z, 11E)-7,11-hexadecadien-1-ol, which then led to a successful synthesis of gossyplure.

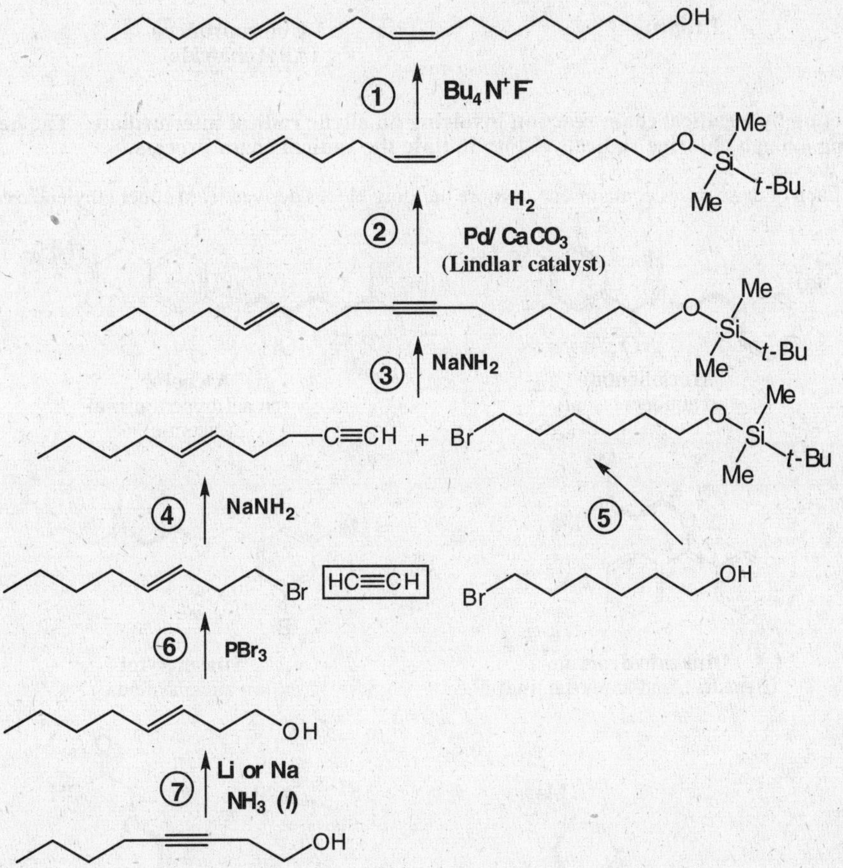

(a) Suggest reagents and experimental conditions for each step in this synthesis.

Reasonable reagents are shown on the above scheme.

(b) Why is it necessary to protect the -OH group of 6-bromo-1-hexanol?

A hydroxyl group is considerably more acidic than an alkyne. Thus, left unprotected, the hydroxyl group would be deprotonated by $NaNH_2$, creating a nucleophilic alkoxide anion that would displace bromine to give an unwanted ether.

(c) How might you modify this synthesis to prepare the 7Z,11Z isomer of gossyplure?

Lindlar's catalyst would be used in place of the Li or Na in NH_3 for the reduction of the 3-octyn-2-ol.

<u>Problem 11.39</u> Epichlorhydrin (Section 11.10) is a valuable synthetic intermediate because each of its three carbons contains a reactive group. Following is the first step in its synthesis from propene. Propose a mechanism for this step.

Propene is reacting by a radical chain reaction involving an allylic radical intermediate. The heat breaks the chlorine bond, producing enough chlorine radicals (Cl·)to initiate the radical chain process.

<u>Problem 11.40</u> Each of these drugs contains one or more building blocks derived from either ethylene oxide or epichlorohydrin.

Identify the part of each molecule that can be derived from one or the other of these building blocks and propose structural formulas for the nucleophile(s) that can be used along with either ethylene oxide or epichlorohydrin to synthesize each molecule. We will learn about the actual syntheses of each molecule in later chapters.

The pieces of the molecules that could be synthesized from either ethylene oxide or epichlorohydrin are shown as bold bonds in the structures. For a potential ethylene oxide building block, look for two methylene units separating two heteroatoms, especially O and N. These units can be seen in (a), (c), (d), (e), and (f). For epichlorohydrin, look for a three carbon unit, with a heteroatom attached to each. This can be found in (b).

One of the exciting and challenging aspects of organic chemistry is that there are many ways to synthesize a complex molecule. There are thus a variety of ways in which ethylene oxide or epichlorohydrin may be used to synthesize the molecules in this problem. Shown below are just representative examples for each target structure. You may come up with many more, as ethylene oxide and epichlorohydrin are very versatile synthetic building blocks.

(a)

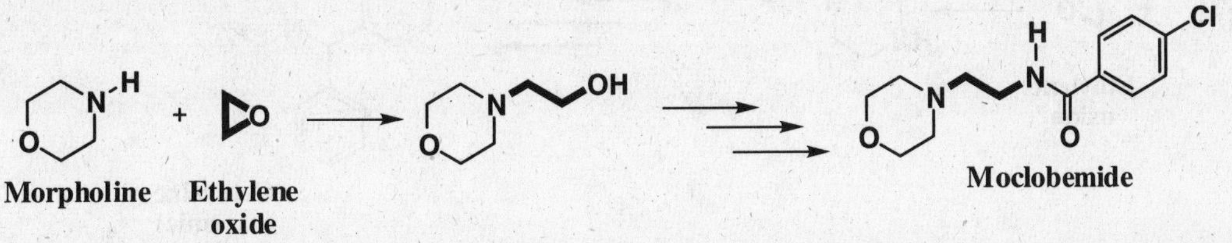

Morpholine   Ethylene oxide

Moclobemide

(b) $(CH_3)_2CHNH_2$ +

Isopropyl-amine   Epichloro-hydrin

Atenolol (racemic)

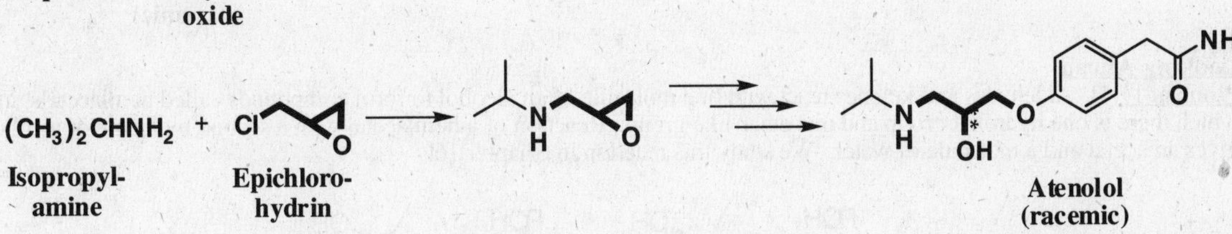

(c) $(CH_3)_2NH$ +

Dimethyl-amine   Ethylene oxide

Diphenhydramine

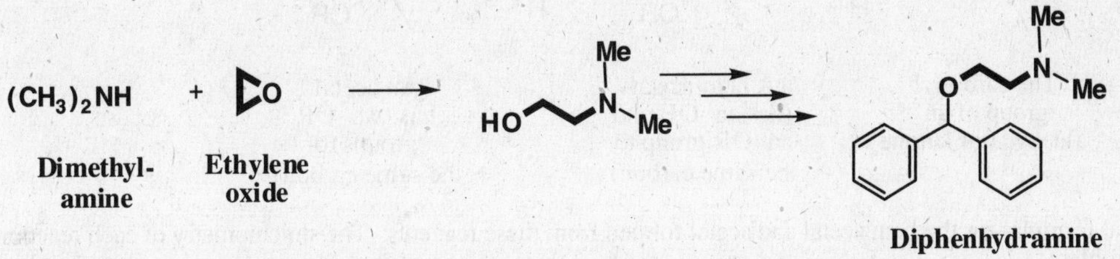

(d) $(Et)_2NH$ +

Diethyl-amine   Ethylene oxide

Spasmolytol

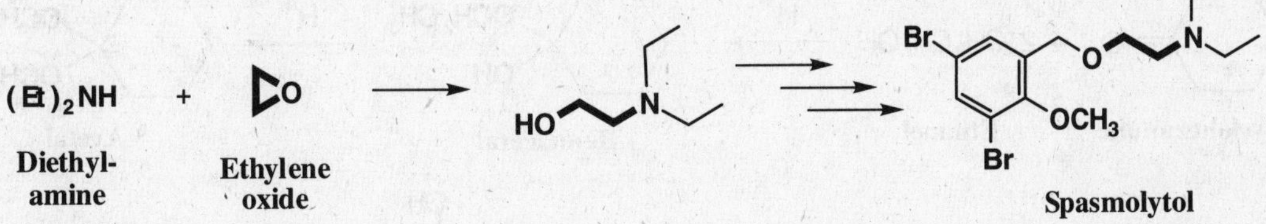

(e) H_3C NH_2 +

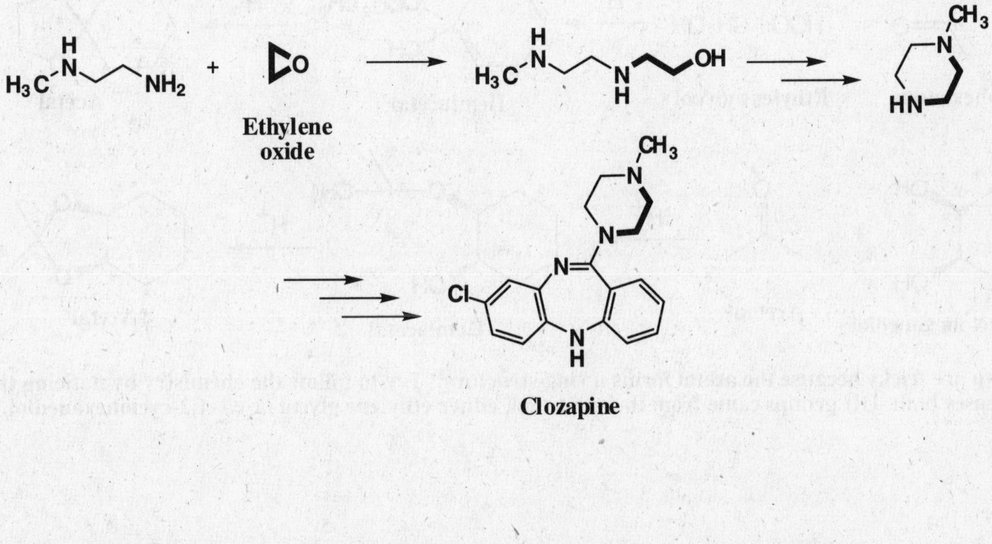

Ethylene oxide

Clozapine

(f)

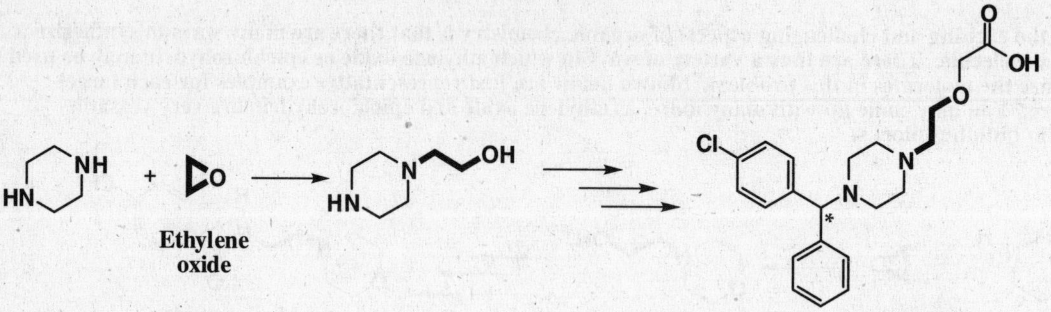

Ethylene
oxide

Cetirizine
(racemic)

Looking Ahead

Problem 11.41 Aldehydes and ketones react with one molecule of an alcohol to form compounds called hemiacetals, in which there is one hydroxyl group and one ether-like group. Reaction of a hemiacetal with a second molecule of alcohol gives an acetal and a molecule of water. We study this reaction in Chapter 16.

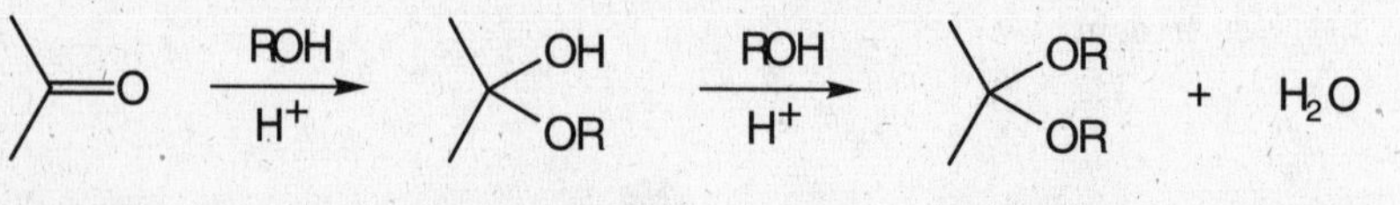

The carbonyl
group of an
aldehyde or ketone

A hemiacetal
(has an -OH and
an -OR group to
the same carbon)

An acetal
(has two -OR
groups to
the same carbon)

Draw structural formulas for the hemiacetal and acetal formed from these reagents. The stoichiometry of each reaction is given in the problem.

(a)

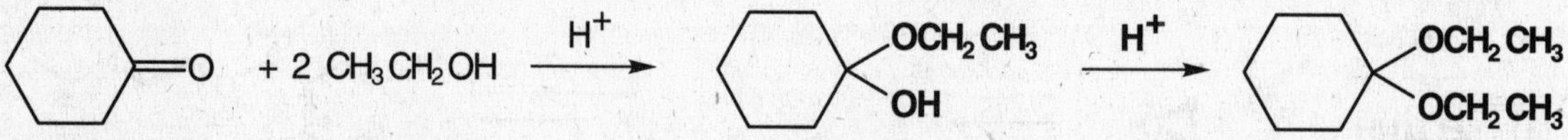

Cyclohexanone   Ethanol   Hemiacetal   Acetal

(b)

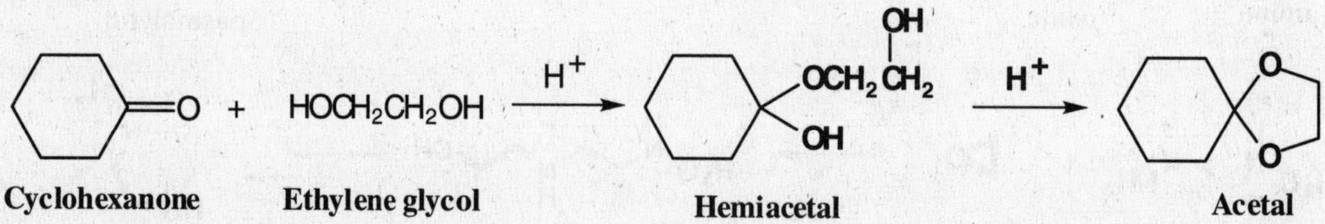

Cyclohexanone   Ethylene glycol   Hemiacetal   Acetal

(c)

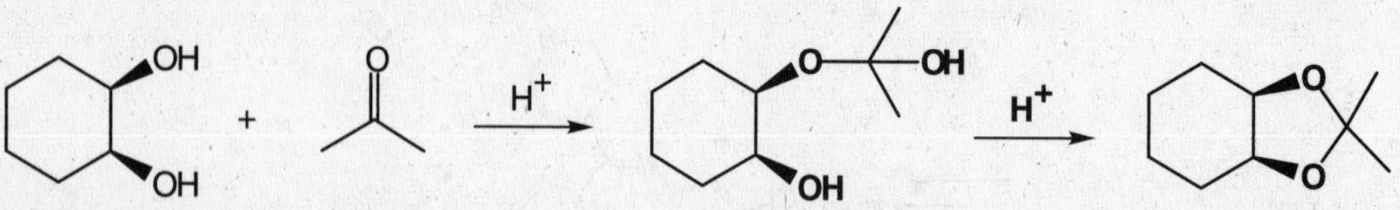

cis-1,2-Cyclohexanediol   Acetone   Hemiacetal   Acetal

The last two are tricky because the acetal forms a ring structure. Try to follow the chemistry by noticing that in these two cases both -OH groups came from the same diol, either ethylene glycol or *cis*-1,2-cyclohexanediol.

Mixed Synthesis

Problem 11.42 Show how to convert 1-butene into dibutyl ether. You must use 1-butene as the source of all carbon atoms in the dibutyl ether. Show all required reagents and all molecules synthesized along the way.

1-Butene   **Dibutyl ether**

1. BH$_3$

2. H$_2$O$_2$ / NaOH

PBr$_3$ S$_N$2

Na°

OH   O$^-$ Na$^+$

Recognize that the dibutyl ether product must be constructed from two four-carbon pieces. The Williamson ether synthesis is a logical choice for ether construction in this case because there is no branching. The required alkoxide can be created using hydroboration/oxidation of 1-butene (non-Markovnikov regiochemistry is needed) followed by reaction with metallic sodium. The required 1-bromobutane can be synthesized conveniently by treating 1-butanol with PBr$_3$.

Problem 11.43 Show how to convert cyclohexanol into racemic *trans*-1,2-cyclohexanediol. Show all required reagents and all molecules synthesized along the way.

OH

Cyclohexanol   ***Trans*-1,2-cyclohexanediol**
                                      (racemic)

H$_2$SO$_4$ H$_3$O or NaOH

RCO$_3$H   O   O

(racemic)

Recognize that the vicinal diol product has *trans* stereochemistry, so that epoxide ring-opening is a logical reaction to use as the last step. The epoxide can be conveniently prepared by reacting cyclohexene with a peracid such as peracetic acid or MCPBA. The required cyclohexene can be made though the dehydration of the starting cyclohexanol using treatment with anhydrous H$_2$SO$_4$.

Problem 11.44 Show how to convert 1-butanol and ethanol into racemic 2-ethoxy-1-butanol. You must use 1-butanol and ethanol as the source of all carbon atoms in the ether product. Show all required reagents and all molecules synthesized along the way.

OH

1-Butanol   **2-Ethoxy-1-butanol**
                                  (racemic)

+

OH

Ethanol

PBr$_3$ OH   H$_2$SO$_4$ (catalytic amount)

Br   O   O

t-BuO$^-$ RCO$_3$H   (racemic)
Strong Base
E2

Recognize that the hydroxy ether product can be constructed by attacking ethyloxirane with ethanol using an acid catalyst. Using acid assures the proper regiochemistry, since in the presence of acid, the ethanol will attack the more hindered side of the epoxide thereby giving the desired product. The ethyloxirane can be conveniently prepared by reacting 1-butene with a peracid such as peracetic acid or MCPBA. 1-Butene can be conveniently prepared by first converting 1-butanol to 1-bromobutane, followed by an E2 reaction using a hindered strong base such as *tert*-butoxide. Note that using acid-catalyzed dehydration of 1-butanol directly to give 1-butene would not proceed in high yield because of competing carbocation rearrangement.

Reactions in Context

Problem 11.45 During the synthesis of the antiasthmatic drug montelukast (Singulair), a silyl ether protecting group is used to mask the reactivity of an OH group. The silyl group chosen is the *tert*-butyldimethylsilyl (TBDMS) group. Draw the product of the following transformation, assuming the TBDMS-Cl reagent reacts only once with the starting material. Briefly explain your answer.

Reaction will occur exclusively at the less-hindered secondary alcohol, rather than the tertiary alcohol. Note that the stereochemistry of the secondary alcohol is not changed during the reaction.

Problem 11.46 The Sharpless epoxidation is used when a single enantiomer product is required. Predict the structure of the predominant product of the following transformation.

Using (-)-diethyl tartrate will lead to the stereoisomer shown.

CHAPTER 12
Solutions to the Problems

Problem 12.1 Which is higher in energy?
(a) Infrared radiation of 1715 cm^{-1} or 2800 cm^{-1}?

The higher the wavenumber, the higher the energy. As a result, 2800 cm^{-1} is higher energy than 1705 cm^{-1}.

(b) Radiofrequency radiation of 300 MHz or 60 Hz?

Energy is directly proportional to frequency, so 300 MHz is higher energy than 60 Hz.

Problem 12.2 Without doing the calculation, which member of each pair do you expect to occur at the higher frequency?
(a) C=O or C=C stretching?

The atomic weight of O is slightly larger than that of C. However, the C=O bond is much stronger than C=C and, hence, has a substantially larger force constant. Thus, C=O stretching occurs at higher frequency.

(b) C=O or C-O stretching?

Double bonds have higher force constants than single bonds, so the C=O bond will have a stretching frequency that occurs at a higher frequency than C-O.

(c) C≡C or C=O stretching?

Triple bonds have higher force constants than double bonds, so the C≡C bond will have a stretching frequency that occurs at a higher frequency than C=O.

(d) C-H or C-Cl stretching?

Assuming that C-H and C-Cl have similar force constants, then the C-H will have an absorbance at a higher frequency because the atomic weight of H is much smaller than Cl.

Problem 12.3 A compound shows strong, very broad IR absorption in the region 3300-3600 cm^{-1} and strong, sharp absorption at 1715 cm^{-1}. What functional group accounts for both of these absorptions?

The very broad IR absorption centered at 3300-3600 cm^{-1} corresponds to O-H stretching and the strong absorption at 1715 cm^{-1} corresponds to a carbonyl C=O stretch. The functional group responsible for these absorptions is the -COOH group.

Problem 12.4 Propanoic acid and methyl ethanoate are constitutional isomers. Show how to distinguish between them by IR spectroscopy.

$$CH_3CH_2\overset{\displaystyle O}{\overset{\displaystyle \|}{C}}OH \qquad\qquad CH_3\overset{\displaystyle O}{\overset{\displaystyle \|}{C}}OCH_3$$

      Propanoic acid   **Methyl ethanoate**
                                     (Methyl acetate)

The key difference between these constitutional isomers is the OH group of propanoic acid that is not present in methyl ethanoate. The -OH group will show a strong, broad O-H stretch in the IR spectrum of propanoic acid between 2500-3300 cm^{-1} that will be absent in the spectrum of methyl ethanoate. Also, the strong C=O stretching absorption will be between 1700-1725 cm^{-1} for propanoic acid, but 1735-1800 cm^{-1} for methyl ethanoate. Both spectra will have C-H stretching absorptions near 3000 cm^{-1} and C-O stretching absorptions near 1100 cm^{-1}.

Infrared Spectra

Problem 12.5 Following are infrared spectra of methylenecyclopentane and 2,3-dimethyl-2-butene. Assign each compound its correct spectrum.

Methylenecyclopentane   **2,3-Dimethyl-2-butene**

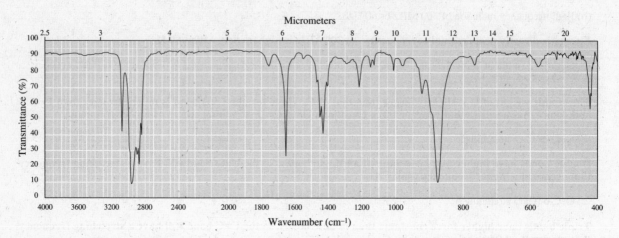

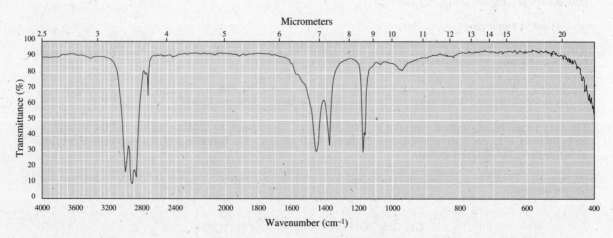

Both molecules have several C-H bonds and thus both spectra have C-H stretches and C-H bending vibrations at 2900 cm⁻¹ and 1450 cm⁻¹, respectively. The alkene in methylenecyclopentane is unsymmetrical and therefore has a permanent dipole, so the C=C stretching will have a prominent band at 1654 cm⁻¹ as seen in the upper spectrum. On the other hand, 2,3-dimethyl-2-butene has a symmetrically substituted carbon-carbon double bond with no permanent dipole, so no C=C stretching should be prominent. In addition, the four methyl groups of 2,3-dimethyl-2-butene should give a prominent CH₃ bending band at 1375 cm⁻¹ and 1450 cm⁻¹ as is seen in the lower spectrum.

Thus, the first (upper) spectrum corresponds to methylenecyclopentane and the second (lower) spectrum corresponds to 2,3-dimethyl-2-butene.

<u>Problem 12.6</u> Following are infrared spectra of nonane and 1-hexanol. Assign each compound its correct spectrum.

Nonane   **1-Hexanol**

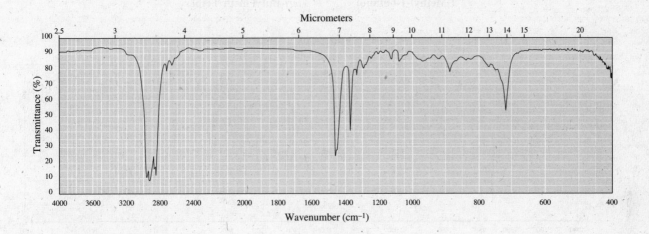

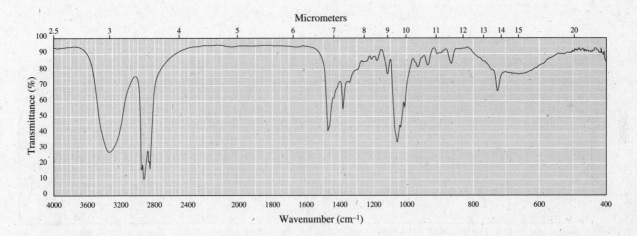

Both compounds have C-H bonds, so both spectra have C-H stretches and bends at 2900 cm^{-1} and 1450 cm^{-1}, respectively. Furthermore, both compounds have methyl groups so both spectra have methyl bending vibrations at 1375 cm^{-1} and 1450 cm^{-1}. On the other hand, the 1-hexanol has an OH group, that will give rise to an O-H and C-O stretching vibrations at 3340 cm^{-1} and 1050 cm^{-1}, respectively.

The 3340 cm^{-1} and 1050 cm^{-1} features are in the second (lower) spectrum, so this second spectrum must correspond to the 1-hexanol and the first (upper) spectrum must correspond to nonane.

<u>Problem 12.7</u> Following are infrared spectra of 2-methyl-1-butanol and *tert*-butyl methyl ether. Assign each compound its correct spectrum.

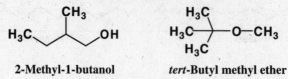

2-Methyl-1-butanol   ***tert*-Butyl methyl ether**

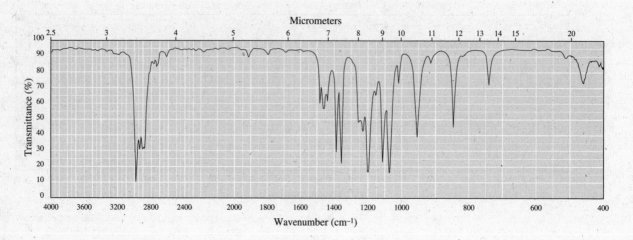

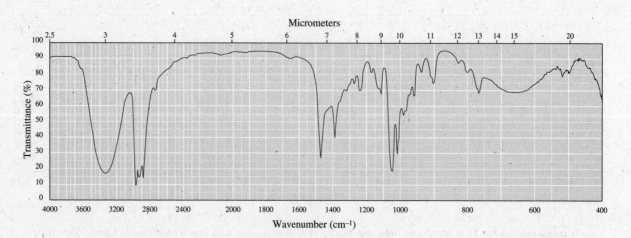

The molecules are extremely similar except for the -OH group present in the 2-methyl-1-butanol. The characteristic O-H stretch is present at 3300 - 3600 cm⁻¹ in the lower spectrum.

The broad peak at 3300 - 3600 cm⁻¹ verifies that the second (lower) spectrum corresponds to 2-methyl-1-butanol. Therefore, the first (upper) spectrum corresponds to *tert*-butyl methyl ether.

Problem 12.8 The IR C≡C stretching absorption in symmetrical alkynes is usually absent. Why is this so?

For a bond vibration to have a corresponding absorption of high intensity in the IR, that bond must have a dipole moment, and absorption of energy must result in a large change in that dipole moment. For symmetrical alkynes, there is no dipole moment associated with the C≡C bond, and thus no large change in dipole moment associated with bond stretching. Thus, the IR band is usually absent.

Problem 12.9 Explain the fact that the C-O stretch in ethers and esters occurs at 1000-1100 cm^{-1} when the C is sp^3 hybridized, but at 1250 cm^{-1} when it is sp^2 hybridized.

When the C atom is sp^2 hybridized, it has more s character compared to an sp^3 hybridized atom. As can be seen in the table of bond dissociation enthalpies(Appendix 3), all things being equal, sp^2 hybridized atoms will make stronger bonds than sp^3 hybridized atoms. Stronger bonds absorb infrared radiation at higher energy, thus higher frequency.

Problem 12.10 A compound has strong infrared absorption bands at the following frequencies. Suggest likely functional groups that may be present.

(a) 1735, 1250, and 1100 cm^{-1}.   (b) 1745 cm^{-1} but not 1000-1250 cm^{-1} (c) 1710 and 2500-3400 (broad) cm^{-1}.
   An ester   **A ketone**   **A carboxyl group**

(d) A single band at about 3300 cm^{-1} (e) 3600 and 1050 cm^{-1}.   (f) 1100 but not 3300-3650 cm^{-1}
   A secondary amine   **An alcohol**   **An ether**

Problem 12.11 Show how IR spectroscopy can be used to distinguish between the compounds in each set.

The key to using IR spectroscopy for distinguish between molecules is to look for characteristic peaks that arise from functional groups unique to one of the structures. Characteristic peaks are listed under each molecule. Note that all of the following will have C-H stretching peaks at 2850-3000 cm^{-1} and bending peaks near 1450 cm^{-1}, so these have been ignored. Also, the methyl group bending peaks at 1375 cm^{-1} and 1450 cm^{-1} are generally ignored, unless they will be useful as in (d) and (f).

(a) OH   and   O
   1000-1250 cm^{-1} 1°(C-O)   1000-1100 cm^{-1} (C-O)
   3200-3500 cm^{-1} (O-H)

The alcohol will be identifiable by the broad O-H peak at 3200-3500 cm^{-1}.

(b) H   and
   1630-1820 cm^{-1} (C=O)   1600-1680 cm^{-1} (C=C)

The aldehyde will be identifiable by the C=O peak at 1630-1820 cm^{-1}, while the alkene will have a characteristic peak at 1600-1680 cm^{-1}. The aldehyde will also have a characteristic C-H stretch at 2720 cm^{-1}.

(c) OH   and
   1000-1250 cm^{-1} 2°(C-O)   1600-1680 cm^{-1} (C=C)
   3200-3500 cm^{-1} (O-H)

The alcohol will be identifiable by the broad O-H peak at 3200-3500 cm^{-1} and C-O peak at 1100 cm^{-1}, while the alkene will have a characteristic C=C stretch at 1600-1680 cm^{-1}.

(d) and

1745 cm⁻¹ (C=O)

$1745 \text{ cm}^{-1} \text{ (C=O)}$

1000-1100 cm⁻¹ (C-O)
1375, 1450 cm⁻¹ (-CH₃)

The ketone will be identifiable by the C=O peak at 1630-1820 cm⁻¹, while the ether will have a characteristic C-O peak at 1000-1100 cm⁻¹ and methyl bending peaks at 1375 cm⁻¹ and 1450 cm⁻¹.

(e) and

1000-1100 cm⁻¹ (*sp³* C-O)
1450-1600 cm⁻¹ (C=C) aromatic
1735-1800 cm⁻¹ (C=O)

1000-1100 cm⁻¹ (*sp³* C-O)
1200-1250 cm⁻¹ (*sp²* C-O) 2x
1450-1600 cm⁻¹ (C=C) aromatic
1735-1800 cm⁻¹ (C=O)
2500-3300 cm⁻¹ (-OH)

Both spectra will have the characteristic C=O, *sp³* C-O, and *sp²* C-O peaks of an alkyl ester, but only the molecule on the right will have an additional *sp²* C-O peak at 1200-1250 cm⁻¹.

(f) and

1375, 1450 cm⁻¹ (-CH₃)
1600-1680 cm⁻¹ (C=C)

1000-1100 cm⁻¹ (C-O)

The alkene will be identifiable by the C=C peak at 1600-1680 cm⁻¹ and methyl bending peaks at 1375 cm⁻¹ and 1450 cm⁻¹, while the cyclic ether will have a characteristic C-O peak at 1000-1100 cm⁻¹.

(g) and

1450-1600 cm⁻¹ (C=C) aromatic
1630-1820 cm⁻¹ (C=O)
1600-1680 cm⁻¹ (C=C)

1200-1250 cm⁻¹ (*sp²* C-O)
1450-1600 cm⁻¹ (C=C) aromatic
1600-1680 cm⁻¹ (C=C)
1700-1725 cm⁻¹ (C=O)
2500-3300 cm⁻¹ (O-H)

The main difference will be the broad O-H peak at 2500-3300 cm⁻¹ of the carboxylic acid that is not present in the aldehyde.

(h) and

3000-3100 cm⁻¹ (=C-H)
1600-1680 cm⁻¹ (C=C)

3300 cm⁻¹ (≡C-H)
2100-2250 cm⁻¹ (C≡C)

The alkene will have a C=C peak at 1600-1680 cm⁻¹ and the alkyne will have a C≡C peak at 3300 cm⁻¹. The C-H stretches will also help distinguish the two as the C-H stretch for the alkene will be between 3000-3100 cm⁻¹ and the C-H stretch for the alkyne will be near 3300 cm⁻¹.

(i)

1000-1100 cm⁻¹ (C-O)   1000-1250 cm⁻¹ (C-O)
                            3200-3500 cm⁻¹ (O-H)

The main difference will be the broad OH peak of the alcohol at 3200-3500 cm⁻¹.

(j)   and

1000-1250 cm⁻¹ (C-O)   1700-1725 cm⁻¹ (C=O)
3200-3500 cm⁻¹ (O-H)   2500-3300 cm⁻¹ (O-H)

The main difference will be the strong C=O peak of the carboxylic acid at 1700-1725 cm⁻¹.

Whitetip reef shark *Triaenodon obesus*
Maui, Hawaii

CHAPTER 13
Solutions to the Problems

<u>Problem 13.1</u> Calculate the ratio of nuclei in the higher spin state to those in the lower spin state, N_h/N_l, for ^{13}C at 25°C in an applied field strength of 7.05 T. The difference in energy between the higher and lower nuclear spin states in this applied field is approximately 0.030 J (0.00715 cal)/mol.

The important equation relates the change in energy of two spin states to their equilibrium concentrations:

$$\Delta G° = -RT\ln \frac{N_h}{N_l}$$

Rearranging this expression in terms of N_h/N_l gives:

$$\ln \frac{N_h}{N_l} = \frac{-\Delta G°}{RT}$$

Substituting in the appropriate values for R (8.314 J•deg^{-1}•mol^{-1}), T (298 K) and $\Delta G°$ (0.030 J•mol^{-1}) gives:

$$\ln \frac{N_h}{N_l} = \frac{-0.030 \text{ J•mol}^{-1}}{(8.314 \text{ J•deg}^{-1}\text{•mol}^{-1})(298 \text{ deg})} = -1.21 \times 10^{-5}$$

$$\frac{N_h}{N_l} = \boxed{0.9999879 = \frac{1.0000000}{1.0000121}}$$

<u>Problem 13.2</u> State the number of sets of equivalent hydrogens in each compound and the number of hydrogens in each set.

(a) 3-Methylpentane
Numbers have been added to the carbon atoms of the structures to aid in referring to specific hydrogens. Use the "test atom" approach if you have trouble understanding the answers.

There are four sets of equivalent hydrogens. <u>Set a:</u> 6 hydrogens from the methyl groups of carbon atoms 1 and 5. <u>Set b:</u> 4 hydrogens from the -CH$_2$- groups of carbon atoms 2 and 4. <u>Set c:</u> 3 hydrogens from the methyl group of carbon atom 5. <u>Set d:</u> 1 hydrogen from the -CH- group of carbon atom 3.

(b) 2,2,4-Trimethylpentane

There are four sets of equivalent hydrogens. <u>Set a:</u> 9 hydrogens from the methyl groups of carbon atoms 1, 7, and 8. <u>Set b:</u> 6 hydrogens from the methyl groups of carbon atoms 5 and 6. <u>Set c:</u> 2 hydrogens from the -CH$_2$- group of carbon atom 3. <u>Set d:</u> 1 hydrogen from the -CH- group of carbon atom 4.

Problem 13.3 Each compound gives only one signal in its ^{1}H-NMR spectrum. Propose a structural formula for each compound.

In order for these molecules to give a single absorption peak, each of the hydrogen nuclei must be in an identical environment. This will only occur in symmetrical molecules.

(a) C_3H_6O

$$\underset{CH_3}{\overset{\overset{\displaystyle O}{\|}}{C}}CH_3$$

(b) C_5H_{10}

(c) C_5H_{12}

$$CH_3-\underset{CH_3}{\overset{CH_3}{C}}-CH_3$$

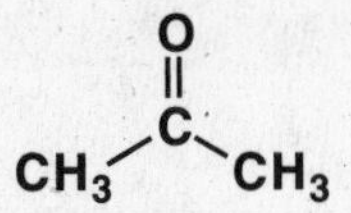

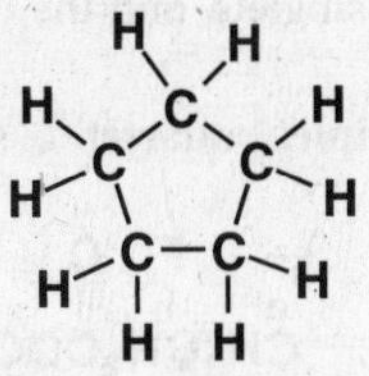

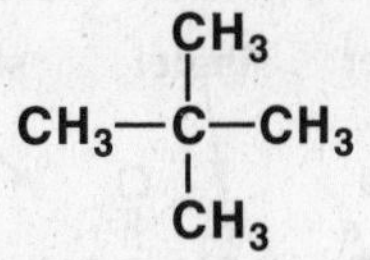

(d) $C_4H_6Cl_4$

CH_3-CCl_2-CCl_2-CH_3

Problem 13.4 The line of integration of the two signals in the ^{1}H-NMR spectrum of a ketone with the molecular formula $C_7H_{14}O$ rises 62 and 10 chart divisions, respectively. Calculate the number of hydrogens giving rise to each signal, and propose a structural formula for this ketone.

The ratio of signals is approximately 6:1, which corresponds to a 12:2 ratio of the 14 hydrogens. Thus, the larger signal represents 12 hydrogens and the smaller signal represents 2 hydrogens. A structure consistent with this assignment is 2,4-dimethyl-3-pentanone as shown below:

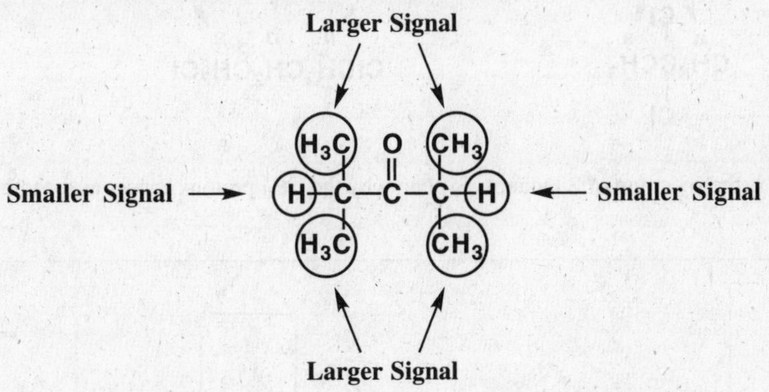

Larger Signal

Smaller Signal ⟶   **⟵ Smaller Signal**

Larger Signal

Problem 13.5 Following are two constitutional isomers of molecular formula $C_4H_8O_2$.

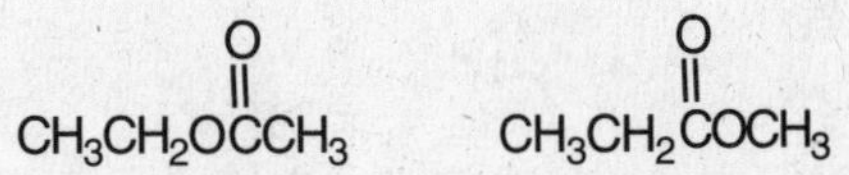

$$\underset{(1)}{CH_3CH_2O\overset{\overset{\displaystyle O}{\|}}{C}CH_3}\qquad\qquad \underset{(2)}{CH_3CH_2\overset{\overset{\displaystyle O}{\|}}{C}OCH_3}$$

(a) Predict the number of signals in the ^{1}H-NMR spectrum of each isomer.

Each isomer will have three signals.

(b) Predict the ratio of areas of the signals in each spectrum.

In each spectrum, the ratio of areas of the three signals will be 3:2:3.

(c) Show how to distinguish between these isomers on the basis of chemical shift.

The -CH_3 singlet signals will be diagnostic. Isomer (1) is the only one with a -CH_3 attached to the carbonyl carbon atom, while isomer (2) is the only one with a -CH_3 attached to an ester sp^3 oxygen atom. Therefore, the spectrum of isomer (1) will be the one with a -CH_3 singlet at δ 2.1-2.3, and the spectrum of isomer (2) will be the one with a -CH_3 singlet at δ 3.7-3.9.

<u>Problem 13.6</u> Following are pairs of constitutional isomers. Predict the number of signals in the ^{1}H-NMR spectrum of each isomer and the splitting pattern of each signal.

(a) $CH_3OCH_2\overset{\overset{\displaystyle O}{\|}}{C}CH_3$ and $CH_3CH_2\overset{\overset{\displaystyle O}{\|}}{C}OCH_3$

The molecule on the left will have three signals that are all singlets, and the molecule on the right will have three signals with splitting patterns as indicated.

singlet   singlet   singlet   triplet quartet   singlet

$\underset{\text{a}}{CH_3}O\underset{\text{b}}{CH_2}\overset{\overset{\displaystyle O}{\|}}{\underset{\text{c}}{C}}CH_3$ $\underset{\text{a}}{CH_3}\underset{\text{b}}{CH_2}\overset{\overset{\displaystyle O}{\|}}{C}O\underset{\text{c}}{CH_3}$

(b) $CH_3\overset{\overset{\displaystyle Cl}{|}}{\underset{\underset{\displaystyle Cl}{|}}{C}}CH_3$ and $ClCH_2CH_2CH_2Cl$

The molecule on the left will have one signal and the molecule on the right will have two signals with splitting patterns as indicated.

singlet   triplet   quintet   triplet

$\underset{\text{a}}{CH_3}\overset{\overset{\displaystyle Cl}{|}}{\underset{\underset{\displaystyle Cl}{|}}{C}}\underset{\text{a}}{CH_3}$ $\underset{\text{a}}{Cl CH_2}\underset{\text{b}}{CH_2}\underset{\text{a}}{CH_2 Cl}$

<u>Problem 13.7</u> Following is the spectrum of 2-butanol. Explain why the CH$_2$ protons appear as a complex multiplet rather than as a simple quintet.

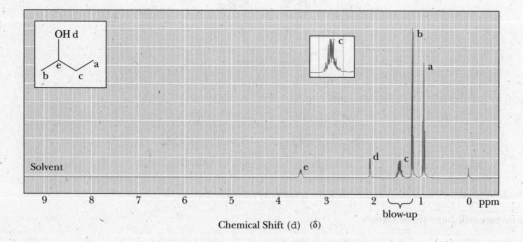

Chemical Shift (d) (δ)

$\underset{\underset{\underset{\text{(d)}}{\displaystyle OH}}{\displaystyle |}}{\overset{4}{CH_3}-\overset{3}{CH_2}-\overset{2}{\underset{}{CH}}-\overset{1}{CH_3}}$ with labels a, c, e, b

Carbon atom 2 is a chiral center and the two hydrogens labeled as group c on carbon atom 3 are diastereotopic. Thus, these two protons are in different environments and give rise to different chemical shifts, greatly complicating the corresponding signals in the observed spectra.

Problem 13.8 Explain how to distinguish between the members of each pair of constitutional isomers based on the number of signals in the proton-decoupled ^{13}C-NMR spectrum of each member.

(a) 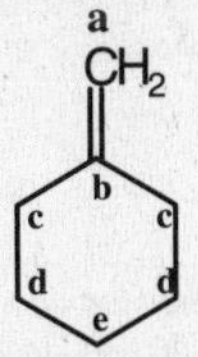 and

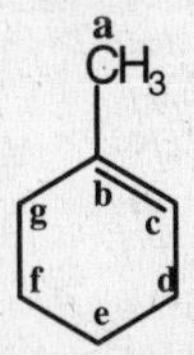

These molecules can be distinguished because they have different numbers of nonequivalent carbon nuclei and thus will have different numbers of ^{13}C-NMR signals. Different signals are indicated by different letters on the above structures. The molecule on the left has higher symmetry and will have 5 signals corresponding to the carbon atoms labeled as a - e, while the molecule on the right has less symmetry and will have 7 signals corresponding to the carbon atoms labeled as a - g.

(b) 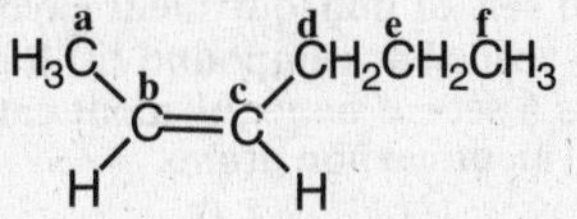 and

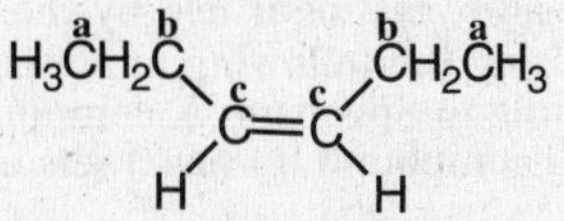

These molecules can also be distinguished because they have different numbers of nonequivalent carbon nuclei and thus will have different numbers of ^{13}C-NMR signals. Different signals are indicated by different letters on the above structures. The molecule on the right has higher symmetry and will only have 3 signals corresponding to the carbon atoms labeled as a - c, while the molecule on the left has less symmetry and will have 6 signals corresponding to the carbon atoms labeled as a - f.

Problem 13.9 Calculate the index of hydrogen deficiency of these compounds:

(a) Aspirin, $C_9H_8O_4$

$(20-8)/2 = 6$

(b) Ascorbic acid (vitamin C), $C_6H_8O_6$

$(14-8)/2 = 3$

(c) Pyridine, C_5H_5N

$(13-5)/2 = 4$
(nitrogen correction)

(d) Urea, CH_4N_2O

$(6-4)/2 = 1$
(nitrogen correction)

(e) Cholesterol, $C_{27}H_{46}O$

$(56-46)/2 = 5$

(f) Dopamine, $C_8H_{11}NO_2$

$(19-11)/2 = 4$
(nitrogen correction)

Interpretation of ^1H-NMR and ^{13}C-NMR Spectra

Problem 13.10 Complete the following table. Which nucleus requires the least energy to flip its spin at this applied field? Which nucleus requires the most energy?

Nucleus	Applied field (T)	Radio frequency (MHz)	Energy (J/mol)
^1H	7.05	300	0.110
^{13}C	7.05	75.5	0.0301
^{19}F	7.05	282	0.113

Based on the entries in the table, the ^{13}C requires the least energy to flip its spin and the ^1H requires the most.

Problem 13.11 The natural abundance of ^{13}C is only 1.1%. Furthermore, its sensitivity in NMR spectroscopy (a measure of the energy difference between a spin aligned with or against an applied magnetic field) is only 1.6% that of ^1H. What are the relative signal intensities expected for the ^1H-NMR and ^{13}C-NMR spectra of the same sample of $Si(CH_3)_4$?

A given ^{13}C signal is $(0.011)(0.016) = 0.000176$ as strong as a given ^1H signal. There are three times as many H atoms as C atoms in $Si(CH_3)_4$, so overall the ratio of H to C signals is 1 : (0.000176/3) = 1 : 0.000059. (Note that this is the same as 17,000 to 1)

Problem 13.12 Following are structural formulas for three constitutional isomers with the molecular formula $C_7H_{16}O$ and three sets of ^{13}C-NMR spectral data. Assign each constitutional isomer its correct spectral data.

(a) $CH_3CH_2CH_2CH_2CH_2CH_2CH_2OH$

	Spectrum 1	Spectrum 2	Spectrum 3
	74.66	70.97	62.93
	30.54	43.74	32.79
	7.73	29.21	31.86
		26.60	29.14
		23.27	25.75
		14.09	22.63
			14.08

(b) $CH_3CCH_2CH_2CH_2CH_3$ with OH and CH_3

(c) $CH_3CH_2CCH_2CH_3$ with OH and CH_2CH_3

These constitutional isomers are most readily distinguished by the number of sets of nonequivalent carbon atoms and thus different ^{13}C signals. Using the following analysis, it can be seen that compound (a) has 7 nonequivalent carbon atoms corresponding to Spectrum 3, compound (b) has 6 sets of nonequivalent carbon atoms corresponding to Spectrum 2, and compound (c) has 3 sets of nonequivalent carbon atoms corresponding to Spectrum 1.

g f e d c b a
$CH_3CH_2CH_2CH_2CH_2CH_2CH_2OH$

e a b c d f
$CH_3CCH_2CH_2CH_2CH_3$ / CH_3 / e

c b a b c
$CH_3CH_2CCH_2CH_3$ / CH_2CH_3 / b c

Problem 13.13 Following are structural formulas for the *cis* isomers of 1,2-, 1,3-, and 1,4-dimethylcyclohexane and three sets of ^{13}C-NMR spectral data. Assign each constitutional isomer its correct spectral data.

	Spectrum 1	Spectrum 2	Spectrum 3
	31.35	34.20	44.60
	30.67	31.30	35.14
	20.85	23.56	32.88
		15.97	26.54
			23.01

These constitutional isomers are most readily distinguished by the number of sets of nonequivalent carbon atoms and thus different ^{13}C signals. Using the following analysis, it can be seen that compound (a) has 4 sets of nonequivalent carbon atoms corresponding to Spectrum 2, compound (b) has 5 sets of nonequivalent carbon atoms corresponding to Spectrum 3, and compound (c) has 3 sets of nonequivalent carbon atoms corresponding to Spectrum 1. The different sets of equivalent carbon atoms are indicated by the letters.

Problem 13.14 Following are structural formulas, dipole moments, and ^{1}H-NMR chemical shifts for acetonitrile, fluoromethane, and chloromethane.

$$CH_3—C≡N \qquad\qquad CH_3—F \qquad\qquad CH_3—Cl$$

Acetonitrile	**Fluoromethane**	**Chloromethane**
3.92 D	1.85 D	1.87 D
δ 1.97	δ 4.26	δ 3.05

(a) How do you account for the fact that the dipole moments of fluoromethane and chloromethane are almost identical even though fluorine is considerably more electronegative than chlorine?

Recall that dipole moment is proportional to the partial charge times the distance of charge separation. Fluorine is a much smaller atom than chlorine, so it makes shorter bonds leading to relatively short charge separation distances. The differences in bond lengths happens to almost exactly offset the differences in electronegativities between fluorine and chlorine and the dipole moments come out almost the same.

(b) How do you account for the fact that the dipole moment of acetonitrile is considerably greater than that of either fluoromethane or chloromethane?

Again, the key is distance. The acetonitrile has partial charge distributed over more atoms and thus a larger distance than fluoromethane or chloromethane.

(c) How do you account for the fact that the chemical shift of the methyl hydrogens in acetonitrile is considerably less than that for either fluoromethane or chloromethane?

A magnetic field is induced in the π system of the nitrile that is against the applied field, thus decreasing the chemical shift.

Problem 13.15 Following are three compounds with the molecular formula $C_4H_8O_2$, and three ^{1}H-NMR spectra. Assign each compound its correct spectrum and assign all signals to their corresponding hydrogens.

$$\begin{array}{ccc} \overset{\textstyle O}{\overset{\textstyle \|}{CH_3COCH_2CH_3}} & \overset{\textstyle O}{\overset{\textstyle \|}{HCOCH_2CH_2CH_3}} & \overset{\textstyle O}{\overset{\textstyle \|}{CH_3OCCH_2CH_3}} \\ (1) & (2) & (3) \end{array}$$

For the spectral interpretations in the rest of this chapter the chemical shift (δ) is shown on the structure adjacent to the appropriate hydrogen atom.

Compound A:

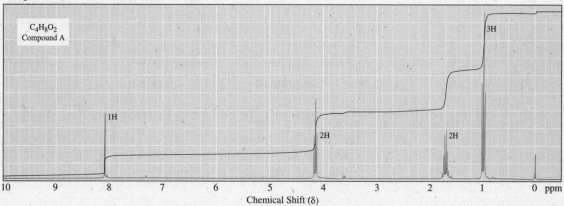

The spectrum for Compound A corresponds to compound 2: ^{1}H-NMR δ 8.1 (1H, singlet, H-C(O)-), 4.2 (2H, triplet, -O-CH$_2$-), 1.7 (2H, multiplet, -CH$_2$-), 1.0 (3H, triplet, -CH$_3$).

$$\overset{\textstyle O}{\overset{\textstyle \|}{\underset{\textstyle 8.1}{HC}-O-\underset{\textstyle 4.2}{CH_2}\underset{\textstyle 1.7}{CH_2}\underset{\textstyle 1.0}{CH_3}}}$$

(2)
Compound A

Compound B:

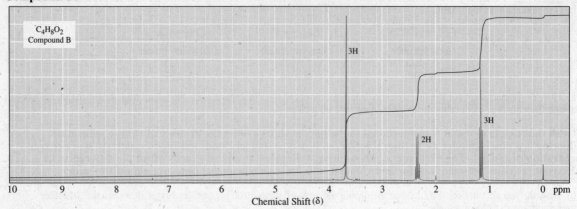

The spectrum for Compound B corresponds to compound 3: ¹H-NMR δ 3.7 (3H, singlet, CH₃-C(O)-), 2.3 (2H, quartet, -C(O)-CH₂-), 1.2 (3H, triplet, -CH₃).

$$
\overset{3.7}{CH_3}-O-\overset{\overset{\displaystyle O}{\displaystyle \|}}{C}\overset{2.3}{CH_2}\overset{1.2}{CH_3}
$$

(3)
Compound B

Compound C:

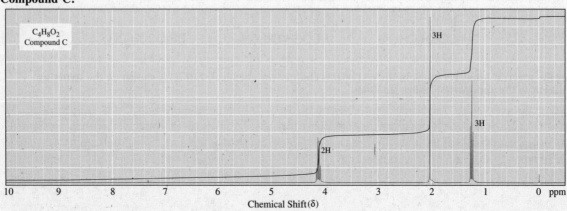

The spectrum for Compound C corresponds to compound 1: ¹H-NMR δ 4.1 (2H, quartet, -O-CH₂-), 2.0 (3H, singlet, CH₃-C-), 1.2 (3H, triplet, -CH₃).

$$
\overset{2.0}{CH_3}\overset{\overset{\displaystyle O}{\displaystyle \|}}{C}-O-\overset{4.1}{CH_2}\overset{1.2}{CH_3}
$$

(1)
Compound C

<u>Problem 13.16</u> Following are ¹H-NMR spectra for compounds D, E, and F, each with molecular formula C₆H₁₂. Each readily decolorizes a solution of Br₂ in CCl₄. Propose structural formulas for compounds D, E, and F, and account for the observed patterns of signal splitting.

Each of the compounds has an index of hydrogen deficiency of 1, in the form of a double bond as evidenced by the reaction with Br₂. The rest of the detailed structures can be deduced from the spectra.

Compound D:

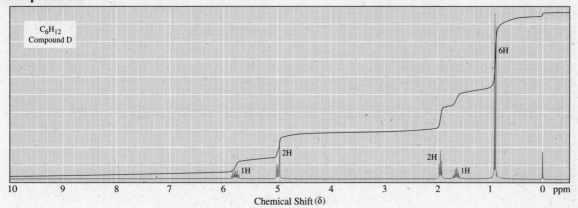

¹H-NMR δ 5.8 (1H, multiplet; this is more complex than expected because the adjacent vinylic hydrogens are not equivalent, -CH=), 4.95-5.0 (2H, multiplet, =CH₂; this is asymmetric because these two vinylic hydrogens are not equivalent and the hydrogen *trans* to the hydrogen on the other vinylogous carbon has the larger signal splitting so it is the signal at 5.0), 1.9 (2H, multiplet; doublet of doublets, -CH₂-), 1.6 (1H, multiplet, -CH-), 0.9 (6H, one doublet, -CH₃).

Compound E:

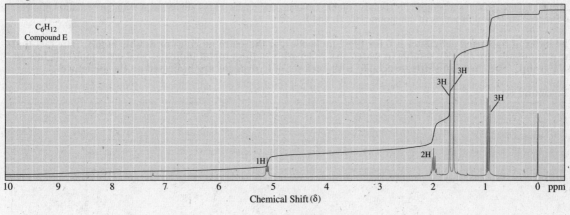

¹H-NMR δ 5.1 (1H, triplet, -CH=), 2.0 (2H, multiplet, -CH₂-), 1.6 and 1.7 (6H, two singlets, =C(CH₃)₂), 0.9 (3H, triplet, -CH₃)

Compound F:

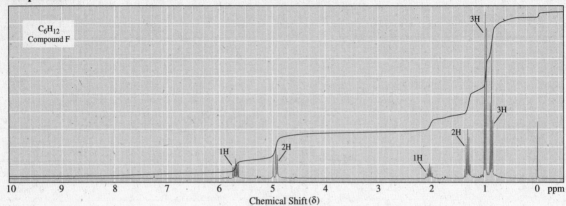

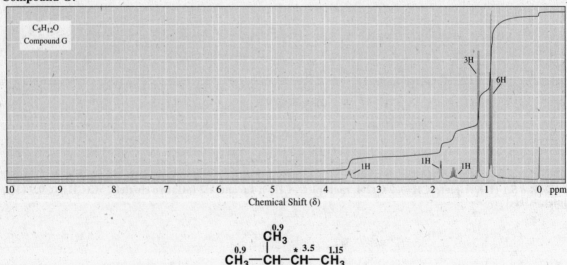

¹H-NMR δ 5.7 (1H, multiplet; this is more complex than expected because the adjacent vinylic hydrogens are not equivalent, -CH=), 4.9-5.0 (2H, multiplet, =CH₂; this is asymmetric because these two vinylic hydrogens are not equivalent and the hydrogen *trans* to the hydrogen on the other vinylogous carbon has the larger signal splitting so it is the signal at 5.0), 2.0 (1H, multiplet, -CH-), 1.3 (2H, multiplet, -CH₂-), 1.0 (3H, doublet,-CH-CH₃), 0.8 (3H, triplet, -CH₂-CH₃)

<u>Problem 13.17</u> Following are ¹H-NMR spectra for compounds G, H, and I, each with the molecular formula C₅H₁₂O. Each is a liquid at room temperature, is slightly soluble in water, and reacts with sodium metal with the evolution of a gas. (a) Propose structural formulas of compounds G, H, and I.

The index of hydrogen deficiency is 0 for these molecules, so there are no rings or double bonds. The fact that the compounds are slightly soluble in water and react with sodium metal indicates that each molecule has an -OH group. The chemical shifts associated with each set of hydrogens are indicated on the structures.

Compound G:

^{1}H-NMR δ 3.5 (1H, multiplet, -CH-OH-), 1.85 (1H, doublet, -OH), 1.6 (1H, multiplet, -CH-(CH$_3$)$_2$), 1.15 (3H, doublet, -C(OH)-CH$_3$), 0.9 (6H, overlapping doublets; this is more complex than expected because it is adjacent to a chiral center -CH-(CH$_3$)$_2$).

Compound H:

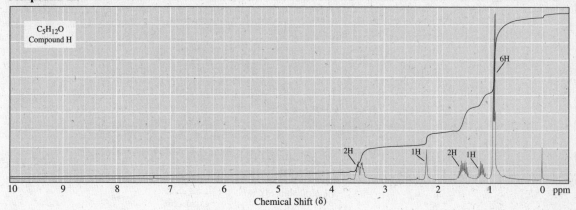

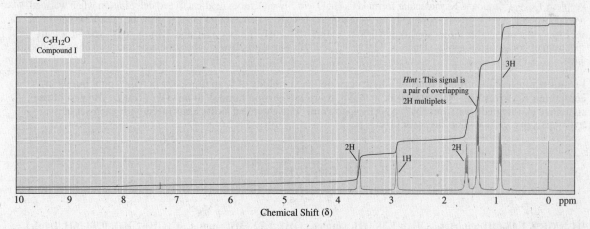

0.8-0.9
CH$_3$
0.8-0.9 1.4-1.6 | *   3.4-3.5
CH$_3$CH$_2$CHCH$_2$OH
1.1   2.2

^{1}H-NMR δ 3.4-3.5 (2H, multiplet; this is more complex than expected because it is adjacent to a chiral center -CH$_2$-OH), 2.2 (1H, broad triplet, -OH), 1.4-1.6 (2H, multiplet; this is more complex than expected because it is adjacent to a chiral center CH$_3$-CH$_2$-), 1.1 (1H, multiplet, -CH-), 0.8-0.9 (6H, broad multiplet, both -CH$_3$ groups).

Compound I:

0.9   1.4   1.4   1.55   3.6   2.9
CH$_3$CH$_2$CH$_2$CH$_2$CH$_2$OH

^{1}H-NMR δ 3.6 (2H, broad multiplet, -CH$_2$-OH), 2.9 (1H, broad peak, -OH), 1.55 (2H, multiplet, -CH$_2$-CH$_2$-OH), 1.4 (4H, multiplet, CH$_3$-CH$_2$-CH$_2$- and CH$_3$-CH$_2$-CH$_2$-), 0.9 (3H, triplet, -CH$_3$)

(b) Explain why there are four lines between δ 0.86 and 0.90 for Compound G.

Carbon atom 2 (with the -OH group) is a chiral center. That makes the two methyl groups diastereotopic, so they have different chemical shifts. The four lines are actually two doublets.

(c) Explain why the 2-H multiplets at δ 1.5 and 3.5 ppm for compound H are so complex.

Carbon atom 2 is a chiral center. The chiral center makes the adjacent -CH$_2$ protons diastereotopic, so the signals are complex.

<u>Problem 13.18</u> Propose a structural formula for compound J, molecular formula C_3H_6O, consistent with the following ^1H-NMR spectrum:

Compound J:

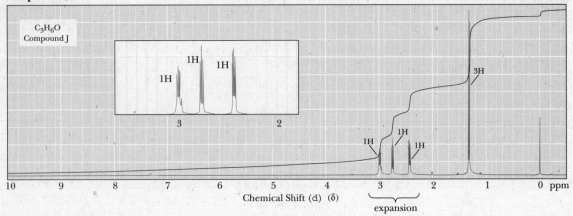

From the molecular formula, there is an index of hydrogen deficiency of 1 indicating that there is one ring or π bond.

^1H-NMR δ 3.0 (1H, multiplet, H_c), 2.4 and 2.75 (2H, multiplets; these hydrogens are not equivalent because the carbon-carbon bonds cannot rotate freely), 1.3 (3H, doublet, $-CH_3$)

<u>Problem 13.19</u> Compound K, molecular formula $C_6H_{14}O$, readily undergoes acid-catalyzed dehydration when warmed with phosphoric acid to give compound L, molecular formula C_6H_{12}, as the major organic product. The ^1H-NMR spectrum of compound K shows signals at δ 0.90 (t, 6H), 1.12 (s, 3H), 1.38 (s, 1H), and 1.48 (q, 4H). The ^{13}C-NMR spectrum of compound K shows signals at 72.98, 33.72, 25.85, and 8.16. Deduce the structural formulas of compounds K and L.

From the molecular formula, there is a hydrogen deficiency index of 0, so there are no rings or π bonds in compound K. From the ^{13}C-NMR peak at 72.98, there is a carbon bonded to an -OH group. The rest of the structure can be deduced from the ^1H-NMR spectrum. The chemical shifts associated with each set of hydrogens are indicated on the structure.

$$\begin{array}{c} 1.38 \\ \overset{\displaystyle OH}{\underset{\displaystyle |}{}} \\ \overset{0.90 \quad 1.48 \,|\, 1.48 \quad 0.90}{CH_3CH_2CCH_2CH_3} \\ \underset{\displaystyle CH_3}{\overset{\displaystyle |}{}} \\ {\scriptstyle 1.12} \end{array}$$

^1H-NMR δ 1.48 (4H, quartet, $-CH_2$), 1.38 (1H, singlet, -OH), 1.12 (3H, singlet, $-C(OH)-CH_3$), 0.90 (6H, triplet, both $-CH_2-CH_3$ groups)

Dehydration of compound K gives the following alkene as compound L:

$$\underset{H}{\overset{CH_3}{}}C=C\overset{CH_3}{\underset{CH_2CH_3}{}}$$

<u>Problem 13.20</u> Compound M, molecular formula $C_5H_{10}O$, readily decolorizes Br_2 in CCl_4, and is converted by H_2/Ni into compound N, molecular formula $C_5H_{12}O$. Following is the ^1H-NMR spectrum of compound M. The ^{13}C-NMR spectrum of compound M shows signals at δ 146.12, 110.75, 71.05, and 29.38. Deduce the structural formulas of compounds M and N.

Compound M:

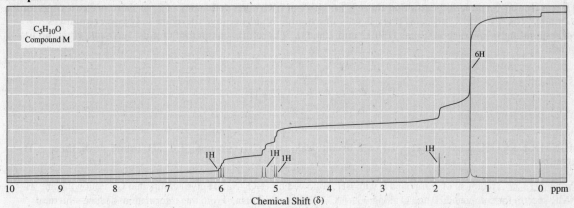

From the reaction with Br_2 and H_2/Ni it is clear the compound M has a carbon-carbon double bond. These conclusions are supported by the ^{13}C-NMR signals corresponding to the sp^2 carbons at δ 146.12 and 110.75. There is also a carbon bonded to an -OH group judging from the signal at δ 71.05. The rest of the structure is deduced from the ^1H-NMR spectrum.

^1H-NMR δ 6.0 (1H, doublet of doublets), 5.0 and 5.2, (2H, doublet of doublets; the vinylic hydrogens are not equivalent and the hydrogen *trans* to the hydrogen on the other vinylogous carbon has the larger signal splitting so it is the signal at 5.2), 1.9 (1H, singlet, -OH), 1.3 (6H, singlet, C-(CH$_3$)$_2$)

Upon hydrogenation, compound M is reduced to the alcohol shown below as compound N:

Problem 13.21   Following is the ^1H-NMR spectrum of compound O, molecular formula C_7H_{12}. Compound O reacts with bromine in carbon tetrachloride to give a compound with the molecular formula $C_7H_{12}Br_2$. The ^{13}C-NMR spectrum of compound O shows signals at δ 150.12, 106.43, 35.44, 28.36, and 26.36. Deduce the structural formula of compound O.

Compound O:

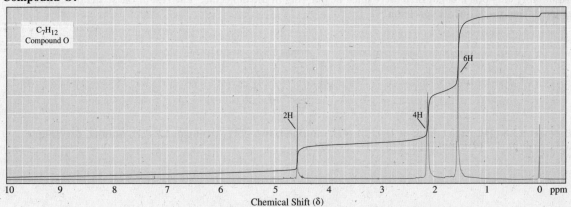

The molecular formula indicates that there is an index of hydrogen deficiency of 2, so there are two rings and/or π bonds. The ^{13}C-NMR indicates there is only one double bond because there are only two resonances corresponding to sp^2 carbon atoms (150.12 and 106.43). Therefore, compound O must have one ring and one π bond.

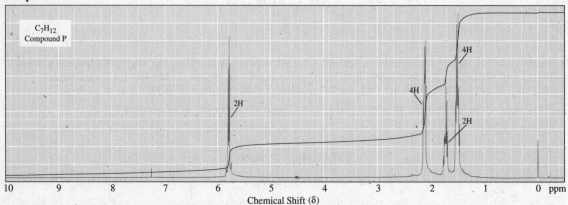

^{1}H-NMR δ 4.6 (2H, singlet, =CH$_2$), 2.1 (4H, broad peak, two -CH$_2$- groups), 1.6 (6H, broad peak, three -CH$_2$-groups)

Problem 13.22 Treatment of compound P with BH$_3$ followed by H$_2$O$_2$/NaOH gives compound Q. Following are ^{1}H-NMR spectra for compounds P and Q along with ^{13}C-NMR spectral data. From this information, deduce structural formulas for compounds P and Q

		^{13}C-NMR	
		(P)	(Q)
C$_7$H$_{12}$	1) BH$_3$ → C$_7$H$_{14}$O	132.38	72.71
(P)	2) H$_2$O$_2$, NaOH (Q)	32.12	37.59
		29.14	28.13
		27.45	22.68

Compound P:

The molecular formula for P indicates that it has an index of hydrogen deficiency of 2 so that is has two rings and/or π bonds. The ^{13}C-NMR spectral data shows that there is an sp^2 carbon atom (132.38). Since there must be two sp^2 carbon atoms to make a π bond, then the molecule must be symmetric so that both sp^2 carbon atoms are equivalent. This also explains why there are so few other ^{13}C-NMR signals. Since there is presumably only one π bond, then there must be one ring in the molecule. The rest of the structure can be deduced from the ^{1}H-NMR spectrum.

^{1}H-NMR δ 5.8 (2H, triplet, both =CH-), 2.1 (4H, multiplet, two -CH$_2$- groups), 1.7 (2H, quintet, the unique -CH$_2$- group), 1.5 (4H, multiplet, two -CH$_2$- groups).

Compound Q:

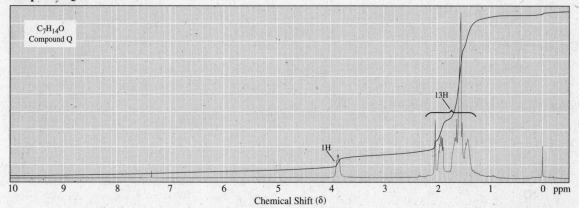

Given the structural formula for P, it is clear that compound Q would be the hydroboration/oxidation product, namely the alcohol shown below. This structure is consistent with the ^{13}C-NMR spectral data provided as well as the ^1H-NMR spectrum.

^1H-NMR δ 3.8 (1H, broad peak, -C(OH)H-), 2.0 (1H, sharp singlet, -OH), 1.4-1.9 (12H, broad multiplets, all the remaining hydrogens on the ring. The peaks are so broad and the patterns so complex because this ring does not have a double bond to hold it rigid, so it has a great deal of flexibility)

<u>Problem 13.23</u> The ^1H-NMR of Compound R, $C_6H_{14}O$, consists of two signals: δ 1.1 (doublet) and δ 3.6 (septet) in the ratio 6:1. Propose a structural formula for compound R consistent with this information.

There is a hydrogen deficiency index of 0, so there are no rings or pi bonds in compound R. The simplicity of the ^1H-NMR spectrum indicates a highly level of symmetry in the molecule, with each methyl group being bonded to a carbon with a single hydrogen atom. The only structure consistent with all of this information is the following ether. The chemical shifts associated with each set of hydrogens are indicated on the structure.

Compound R

<u>Problem 13.24</u> Write structural formulas for the following compounds:
(a) $C_2H_4Br_2$: δ 2.5 (d 3H) and 5.9 (q 1H)

$CH_3\text{-}CHBr_2$

(b) $C_4H_8Cl_2$: δ 1.60 (d 3H), 2.15 (m, 2H), 3.72 (t, 2H), and 4.27 (m 1H)

$CH_3\text{-}CHCl\text{-}CH_2\text{-}CH_2Cl$

(c) $C_5H_8Br_4$: δ 3.6 (s, 8H)

$$\underset{3.6}{CH_2Br}-\underset{\underset{\underset{3.6}{CH_2Br}}{|}}{\overset{\overset{3.6}{CH_2Br}}{|}}{C}-\underset{3.6}{CH_2Br}$$

(d) C_4H_8O: δ 1.0 (t, 3H), 2.1 (s, 3H), and 2.4 (q, 2H)

$$\underset{1.0}{CH_3}\underset{2.4}{CH_2}-\overset{\overset{O}{||}}{C}-\underset{2.1}{CH_3}$$

(e) $C_4H_8O_2$: δ 1.2 (t 3H), 2.1 (s, 3H) and 4.1 (q, 2H); contains an ester group

$$\underset{2.1}{CH_3}-\overset{\overset{O}{||}}{C}-O-\underset{4.1}{CH_2}-\underset{1.2}{CH_3}$$

(f) $C_4H_8O_2$: δ 1.2 (t, 3H), 2.3 (q, 2H) and 3.6 (s, 3H); contains an ester group

$$\underset{1.2}{CH_3}-\underset{2.3}{CH_2}-\overset{\overset{O}{||}}{C}-O-\underset{3.6}{CH_3}$$

(g) C_4H_9Br: δ 1.1 (d, 6H), 1.9 (m, 1H), and 3.4 (d, 2H)

$$\underset{1.1}{CH_3}-\underset{\underset{1.9}{CH}}{\overset{\overset{1.1}{CH_3}}{|}}-\underset{3.4}{CH_2Br}$$

(h) $C_6H_{12}O_2$: δ 1.5 (s, 9H) and 2.0 (s, 3H)

$$\underset{2.0}{CH_3}-\overset{\overset{O}{||}}{C}-O-\underset{\underset{\underset{1.5}{CH_3}}{|}}{\overset{\overset{1.5}{CH_3}}{|}}{C}-\underset{1.5}{CH_3}$$

(i) $C_7H_{14}O$: δ 0.9 (t, 6H), 1.6 (s, 4H), and 2.4 (t, 4H)

$$\underset{0.9}{CH_3}-\underset{1.6}{CH_2}-\underset{2.4}{CH_2}-\overset{\overset{O}{||}}{C}-\underset{2.4}{CH_2}-\underset{1.6}{CH_2}-\underset{0.9}{CH_3}$$

(j) $C_5H_{10}O_2$: δ 1.2 (d, 6H), 2.0 (s, 3H) and 5.0 (septet, 1H)

$$\underset{2.0}{CH_3}-\overset{\overset{O}{||}}{C}-O-\overset{\overset{5.0}{H}}{\underset{\underset{1.2}{CH_3}}{|}}{C}-\underset{1.2}{CH_3}$$

(k) $C_5H_{11}Br$: δ 1.1 (s, 9H) and 3.2 (s, 2H)

$$\underset{1.1}{CH_3}-\overset{\overset{1.1}{CH_3}}{\underset{\underset{1.1}{CH_3}}{|}}{C}-\underset{3.2}{CH_2Br}$$

(l) $C_7H_{15}Cl$: δ 1.1 (s, 9H) and 1.6 (s, 6H)

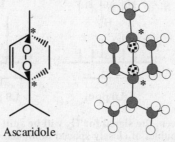

Problem 13.25 The percent *s*-character of carbon participating in a C-H bond can be established by measuring the ^{13}C-^{1}H coupling constant and using the relationship

$$\text{Percent } s\text{-character} = 0.2 \, J(^{13}C\text{-}^{1}H)$$

The ^{13}C-^{1}H coupling constant observed for methane, for example, is 125 Hz, which gives 25% *s*-character, the value expected for an sp^3 hybridized carbon atom.
(a) Calculate the expected ^{13}C-^{1}H coupling constant in ethylene and acetylene.

For ethylene and acetylene the carbon atoms are sp^2 and sp hybridized and thus 33% and 50% *s*-character, respectively. Using the above equation gives coupling constants of 165 Hz and 250 Hz, respectively.

(b) In cyclopropane, the ^{13}C-^{1}H coupling constant is 160 Hz. What is the hybridization of carbon in cyclopropane?

The carbon atoms in cyclopropane are (0.2)(160) = 32% *s*-character. This corresponds roughly to an sp^2 hybridized carbon atom.

Problem 13.26 Ascaridole is a natural product that has been used to treat intestinal worms. Explain why the two methyls on the isopropyl group in ascaridole appear in its ^{1}H-NMR spectrum as four lines of equal intensity, with two sets of two each separated by 7 Hz.

Ascaridole

Ascaridole is a chiral molecule, having two chiral centers as shown. As a result, the NMR spectrum is complex and the two methyl groups of the isopropyl groups are split. The best way to think about it is that the two methyl groups are themselves diastereotopic, thus giving rise to split signals.

Problem 13.27 The ^{13}C-NMR spectrum of 3-methyl-2-butanol shows signals at δ 17.88 (CH_3), 18.16 (CH_3), 20.01 (CH_3), 35.04 (carbon-3), and 72.75 (carbon-2). Account for the fact that each methyl group in this molecule gives a different signal.

Because of the chiral center at carbon-2, the two methyl groups of carbon-4 and carbon-5 are diastereotopic and thus have two different chemical shifts. In addition, carbon-1 is not equivalent to 4 or 5, so it also has a unique chemical shift.

Problem 13.28 Sketch the NMR spectrum you would expect from a partial molecule with the following parameters.

$$H_a = 1.0 \text{ ppm}$$
$$H_b = 3.0 \text{ ppm}$$
$$H_c = 6.0 \text{ ppm}$$
$$J_{ab} = 5.0 \text{ Hz}$$
$$J_{bc} = 8.0 \text{ Hz}$$
$$J_{ac} = 1.0 \text{ Hz}$$

The J_{ac} coupling of 1 Hz is so small that it can be safely ignored since it will not show up on normal resolution spectra. Assuming the R groups are not involved with signal splitting, expect the signal for H_a at 1.0 ppm, integrating to a single hydrogen atom, to be split into a doublet with coupling constant $J_{ab} = 5.0$ Hz. Expect the signal for H_c at 6.0 ppm, integrating to two hydrogen atoms, to be split into a doublet with coupling constant $J_{bc} = 8.0$ Hz. Expect the signal for H_b at 3.0 ppm, integrating to a single hydrogen atom, to be split into a doublet of triplets with coupling constants $J_{ab} = 5.0$ Hz and $J_{bc} = 8.0$ Hz. The following is drawn as expanded views around the signals.

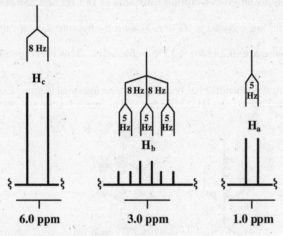

Note that if the $J_{ac} = 1.0$ Hz coupling is seen, the signal for H_a will be split into a doublet of closely spaced triplets, and the signal for H_c will be a doublet of closely spaced doublets.

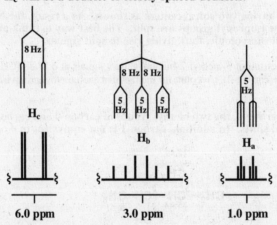

CHAPTER 14
Solutions to the Problems

<u>Problem 14.1</u> Calculate the nominal mass of each ion. Unless otherwise indicated, use the mass of the most abundant isotope of each element.

(a) $[CH_3Br]^{\ddot{+}}$ (b) $[CH_3{}^{81}Br]^{\ddot{+}}$ (c) $[{}^{13}CH_3Br]^{\ddot{+}}$

  94   96   95

<u>Problem 14.2</u> Propose a structural formula for the cation of m/z 41 observed in the mass spectrum of methylcyclopentane.

The cation of m/z 41 must have a molecular formula of C_3H_5. This corresponds to the stable allyl cation:

$$CH_2=CH-CH_2^+$$

<u>Problem 14.3</u> The low-resolution mass spectrum of 2-pentanol shows 15 peaks. Account for the formation of the peaks at m/z 73, 70, 55, 45, 43, and 41.

2-Pentanol has a molecular formula of $C_5H_{12}O$ and thus a nominal mass of 88.

$$CH_3\text{-}CH_2\text{-}CH_2\text{-}\overset{\overset{\displaystyle OH}{|}}{CH}\text{-}CH_3$$
2-Pentanol

Loss of a methyl radical would leave a fragment of mass 73:

$$CH_3\text{-}CH_2\text{-}CH_2\text{-}\overset{+OH}{\overset{||}{CH}}$$

Loss of water results in a fragment of mass 70:

$$[CH_3\text{-}CH_2\text{-}CH_2\text{-}CH=CH_2]^{\ddot{+}} \quad [CH_3\text{-}CH_2\text{-}CH=CH\text{-}CH_3]^{\ddot{+}}$$

The above alkene radical cations could then lose a methyl radical to give the following fragments of mass 55:

$$[CH_2\text{-}CH_2\text{-}CH=CH_2]^+ \quad [CH_2\text{-}CH=CH\text{-}CH_3]^+$$

An alkyl radical could break off from the alcohol to give a fragment of mass 45:

$$\overset{+OH}{\overset{||}{CH}}\text{-}CH_3$$

The original alcohol cation radical could have fragmented in such as way as to generate the following cation with mass 43:

$$CH_3\text{-}CH_2\text{-}CH_2^+$$

The allyl cation has a mass of 41:

$$CH_2=CH-CH_2^+$$

<u>Problem 14.4</u> Draw acceptable Lewis structures for the molecular ion (radical cation) formed from the following molecules when each is bombarded by high-energy electrons in a mass spectrometer.

(a)

(b) [structure: 1,3-butadiene] → [structure: cyclic cation with H-C, C-H atoms, C+ and ·C]

(c) [structure: O, C-N-H, with H, H] → [structure: O, C+-N-H, with H, H]

(d) $H-C\equiv C-H$ → $H-\dot{C}=\overset{+}{C}-H$.

Problem 14.5 The molecular ion for compounds containing only C, H, and O always has an even mass-to-charge value. Why is this so? What can you say about mass-to-charge ratio of ions that arise from fragmentation of one bond in the molecular ion? From fragmentation of two bonds in the molecular ion?

Stable organic molecules composed of C, H, and O with filled valences will always have an even number of hydrogen atoms because carbon and oxygen make an even number of bonds (4 and 2, respectively). Try making some molecular formulas for yourself to verify this. This fact, combined with the even-numbered atomic weights of carbon and oxygen, guarantee an even mass-to-charge ratio of molecular ions. If one bond is broken in the molecular ion, then a cation and a radical are formed. The cation is the only species observed in the mass spectrum, and this will have an odd mass-to-charge ratio. If the molecular ion fragments so that two bonds are broken, then a new radical cation is formed, and the new radical cation will have an even mass-to-charge ratio.

Problem 14.6 For which compounds containing a heteroatom (an atom other than carbon or hydrogen) does the molecular ion have an even-numbered mass and for which does it have an odd-numbered mass?
(a) A chloroalkane with the molecular formula $C_nH_{2n+1}Cl$.

The molecular ion has an even-numbered mass. The C and H atoms will add up to an odd-numbered mass, because there will be an odd number of hydrogen atoms. The chlorine provides an additional odd-numbered mass as either of the two most abundant isotopes (35 and 37, respectively).

(b) A bromoalkane with the molecular formula $C_nH_{2n+1}Br$.

The molecular ion has an even-numbered mass. Again, the C and H atoms add up to an odd-numbered mass, and bromine provides an additional odd-numbered mass as either of the two most abundant isotopes (79 and 81, respectively).

(c) An alcohol with the molecular formula $C_nH_{2n+1}OH$.

The molecular ion has an even-numbered mass. There is an even number of hydrogen atoms, so the C and H atoms add up to an even-numbered mass. The most abundant isotope of oxygen (16) is also even, so the entire molecular ion has an even-numbered mass.

(d) A primary amine with the molecular formula $C_nH_{2n-1}NH_2$.

The molecular ion has an odd-numbered mass. These compounds have an odd number of hydrogens, so the C and H atoms add up to an odd-numbered mass. The nitrogen contributes an even-numbered mass (14) so the molecular ion will have an odd-numbered mass.

(e) A thiol with the molecular formula $C_nH_{2n+1}SH$.

The molecular ion has an even-numbered mass. These compounds have an even number of hydrogens, so the C and H atoms add up to an even-numbered mass. The sulfur contributes an even-numbered mass (32 or 34) so the entire molecular ion has an even-numbered mass.

Problem 14.7 The so-called nitrogen rule states that if a compound has an odd number of nitrogen atoms, the value of m/z for its molecular ion will be an odd number. Why is this so?

Nitrogen atoms have an even-numbered mass, but one lone pair of electrons in the neutral state. Thus, they make an odd number of bonds, namely three, to other atoms. As a result, compounds with an odd number of nitrogens in the neutral state will have an odd-numbered m/z ratio for the molecular ion.

Problem 14.8 Both $C_6H_{10}O$ and C_7H_{14} have the same nominal mass, namely 98. Show how these compounds can be distinguished by the m/z ratio of their molecular ions in high resolution mass spectrometry.

The values in Table 14.1 can be used to calculate precise masses for each molecular ion.
    $C_6H_{10}O$: 6(12.0000) + 10(1.00783) + 15.9949 = 98.0732
    C_7H_{14}: 7(12.0000) + 14(1.00783) = 98.1096
A high resolution mass spectrum will distinguish these compounds based on this small difference in the expected m/z ratio of the molecular ions.

Problem 14.9 Show how the compounds with the molecular formula C_6H_9N and C_5H_5NO can be distinguished by the m/z ratio of their molecular ions in high resolution mass spectrometry.

The values in Table 14.1 can be used to calculate precise masses for each molecular ion.
    C_6H_9N: 6(12.0000) + 9(1.00783) + 14.0031 = 95.0736
    C_5H_5NO: 5(12.0000) + 5(1.00783) + 14.0031 + 15.9949= 95.0372
A high resolution mass spectrum will distinguish these compounds based on this small difference in the expected m/z ratio of the molecular ions.

Problem 14.10 What rule would you expect for the m/z values of fragment ions resulting from the cleavage of one bond in a compound with an odd number of nitrogen atoms?

As stated in Problem 14.7, molecules having an odd number of nitrogen atoms have an odd-numbered m/z ratio for the molecular ion. Cleavage of one bond of a molecular ion would generate a cation and a radical, only the cation of which would be observed in the mass spectrum. If the cation fragment retained an odd number of nitrogen atoms, then the fragment would have an even m/z ratio. If the cation fragment contained an even number of nitrogen atoms (or zero), then the fragment would have an odd m/z ratio.

Problem 14.11 Determine the probability of the following in a natural sample of ethane.
(a) One carbon in an ethane molecule is ^{13}C?

The relative abundance of ^{13}C listed in Table 14.1 is 1.11 atoms of ^{13}C for every 100 atoms of ^{12}C. This corresponds to a probability that any one atom is a ^{13}C atom of 1.11/(100 + 1.11) = 0.0110. The corresponding probability that a given carbon atom is ^{12}C is 100/(100 + 1.11) = 0.0989. Therefore the probability of having one carbon ^{12}C and the other ^{13}C is (0.989)(0.011) = 0.0109. There are two ways to have this situation since either of the carbon atoms could be the ^{13}C so the final answer is 2(0.0109) = 0.0218. Converting to percent by multiplying by 100 means that there is a 2.18% chance that one of the carbon atoms in ethane is ^{13}C.

(b) Both carbons in an ethane molecule are ^{13}C?

The probability that both of the two carbon atoms in ethane is ^{13}C is $(0.0110)^2 = 1.21 \times 10^{-4}$. Converting to percent by multiplying by 100 means that there is a 0.012% chance that both of the carbon atoms in ethane is ^{13}C.

(c) Two hydrogens in an ethane molecule are replaced by deuterium atoms?

The relative abundance of ^2H listed in Table 14.1 is 0.016 atoms of ^2H for every 100 atoms of ^1H. This corresponds to a probability of $0.016/(100 + 0.016) = 1.6 \times 10^{-4}$ that any given H atom is a ^2H atom. The corresponding probability that any given H atom is ^1H is 0.99984. The probability that a given two H atoms in ethane are ^2H is given by $(0.99984)^4(0.00016)^2 = 2.55 \times 10^{-8}$. There are 15 possible combinations of having two ^2H atoms in ethane, so the overall probability is $(15)(2.55 \times 10^{-8}) = 3.8375 \times 10^{-7}$. Converting to percent by multiplying by 100 means that there is a $3.84 \times 10^{-5}\%$ chance that two of the hydrogen atoms in ethane are ^2H.

Problem 14.12 The molecular ions of both $C_5H_{10}S$ and $C_6H_{14}O$ appear at m/z 102 in low-resolution mass spectrometry. Show how determination of the correct molecular formula can be made from the appearance and relative intensity of the M + 2 peak of each compound.

As shown in Table 14.1, ^{16}O occurs in greater than 99.7% abundance, so no (M + 2) peak will be observable for $C_6H_{14}O$. On the other hand, sulfur has one isotope, ^{32}S, that is 95% abundant and another isotope 2 amu higher, namely ^{34}S, that has an abundance of 4.2%.

Thus, if the low-resolution mass spectrum has an (M + 2) peak that is 4.2% the height of the molecular ion peak, the compound must be $C_5H_{10}S$.

Problem 14.13 In Section 14.3, we saw several examples of fragmentation of molecular ions to give resonance-stabilized cations. Make a list of these resonance-stabilized cations, and write important contributing structures of each. Estimate the relative importance of the contributing structures in each set.

$$H_2C=CH-\overset{+}{C}H_2 \longleftrightarrow H_2\overset{+}{C}-CH=CH_2 \qquad HC\equiv C-\overset{+}{C}H_2 \longleftrightarrow HC=C=CH_2$$

Equal Contributor   **Equal Contributor**   **Major Contributor**   **Minor Contributor**

Allyl cation   **Propargyl cation**

The major contributor of the propargyl cation has the positive charge on the less-electronegative sp² hybridized carbon atom.

$$\overset{\diagdown}{C}=\overset{\cdot\cdot}{\overset{+}{O}} \longleftrightarrow \overset{\diagdown}{\overset{+}{C}}-\overset{\cdot\cdot}{\overset{\cdot\cdot}{O}}$$

Major Contributor   Minor Contributor

Oxonium ion

The more stable contributor of the oxonium ion is predicted to be the one that has the positive charge on oxygen due to having filled valence shells on both carbon and oxygen.

Problem 14.14 Carboxylic acids often give a strong fragment ion at m/z (M - 17). What is the likely structure of this cation? Show by drawing contributing structures that this cation is stabilized by resonance.

Loss of 17 results from losing the -OH group of the carboxylic acid. This produces an acylium ion:

$$R-\overset{+}{C}=\overset{\cdot\cdot}{O} \longleftrightarrow R-C\equiv\overset{\cdot\cdot}{O}\colon$$

These ions are probably derived from fragmentation of the molecular ion as follows:

$$R-\overset{\overset{\displaystyle O:}{\|}}{C}-\overset{\cdot\cdot}{O}\text{-H} \xrightarrow{-e^-} R-\overset{\overset{\displaystyle +\;\;O:}{\|}}{C}-\overset{\cdot\cdot}{O}\text{-H} \longrightarrow R-\overset{+}{C}=\overset{\cdot\cdot}{\overset{\cdot\cdot}{O}} + \cdot\overset{\cdot\cdot}{O}H$$

molecular ion

Problem 14.15 For primary amines with no branching on the carbon bearing the nitrogen, the base peak occurs at m/z 30. What cation does this peak represent ? How is it formed? Show by drawing contributing structures that this cation is stabilized by resonance.

$$\left[\begin{array}{c}\overset{\displaystyle |}{\underset{\displaystyle |}{C}}-\overset{\displaystyle H}{\underset{\displaystyle H}{C}}-NH_2\end{array}\right]^{+} \longrightarrow \overset{\displaystyle |}{\underset{\displaystyle |}{C}}\cdot \;+\; \overset{H}{\underset{H}{C}}=\overset{+}{\underset{H}{N}}\overset{H}{} \longleftrightarrow \overset{H}{\underset{H}{\overset{+}{C}}}-\overset{\cdot\cdot}{\underset{H}{N}}\overset{H}{}$$

A radical   **This resonance stabilized cation has m/z = 30**

Problem 14.16 The base peak in the mass spectrum of propanone (acetone) occurs at m/z 43. What cation does this peak represent?

$$\left[\begin{array}{c}\overset{\displaystyle O}{\underset{\displaystyle \|}{}}\\ H_3C\overset{\displaystyle C}{}CH_3\end{array}\right]^{+} \longrightarrow H-\overset{\displaystyle H}{\underset{\displaystyle H}{C}}\cdot \;+\; \overset{\displaystyle :\overset{+}{O}:}{\underset{\displaystyle |}{\underset{\displaystyle \|}{C}}}_{H_3C} \longleftrightarrow \overset{\displaystyle :\overset{\cdot\cdot}{O}:}{\underset{\displaystyle C+}{}}_{H_3C}$$

A methyl radical   **This resonance stabilized cation has m/z = 43**

Problem 14.17 A characteristic peak in the mass spectrum of most aldehydes occurs at *m/z* 29. What cation does this peak represent? (No, it is not an ethyl cation, $CH_3CH_2^+$.)

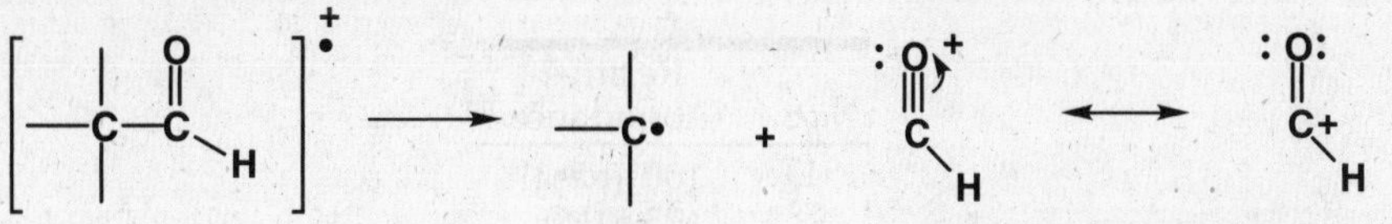

An alkyl
radical

This resonance stabilized
cation has *m/z* = 29

Problem 14.18 Predict the relative intensities of the M and M + 2 peaks for the following.

The following are based on the values in Table 14.1.

(a) CH_3CH_2Cl

100 : 31.98

(b) CH_3CH_2Br

100 : 98

(c) $BrCH_2CH_2Br$

We will assume that the M peak has two ^{79}Br atoms, and the M + 2 peak is due to molecules with one ^{79}Br atom and one ^{81}Br atom. Using the abundance given in the Table, the probability of having a given Br atom ^{79}Br is 100/(100 +98) = 0.5051 and the probability that it is ^{81}Br is 98/(100 + 98) = 0.4949. The probability that a given Br atom is ^{79}Br and the other is ^{81}Br is given by (0.5051)(0.4949) = 0.2500. There are two combinations that give this arrangement so the probability of having one ^{81}Br and one ^{79}Br is 2 x 0.2500 = 0.5000. The probability of having two ^{79}Br atoms is (0.5051)(0.5051) = .2551. Thus the ratio of M to M+2 is 0.2551 to 0.5000, which normalizes to 51:100

(d) CH_3CH_2SH

100 : 4.40

Problem 14.19 The mass spectrum of Compound A shows the molecular ion at *m/z* 85, an M + 1 peak at *m/z* 86 of approximately 6% abundance relative to M, and an M + 2 peak at *m/z* 87 of less than 0.1% abundance relative to M.
(a) Propose a molecular formula for Compound A.

The odd mass indicates an odd number of nitrogen atoms. Assuming only C, H, and N atoms, the 6% abundance of an M + 1 peak is primarily due to ^{13}C. Therefore guess 5 carbon atoms and 1 nitrogen atom giving a molecular formula of $C_5H_{11}N$.

(b) Draw at least 10 possible structural formulas for this molecular formula.

This molecular formula contains either one pi bond or one ring. The following structures are representative examples out of the large number of possible stable molecules that are consistent with this formula.

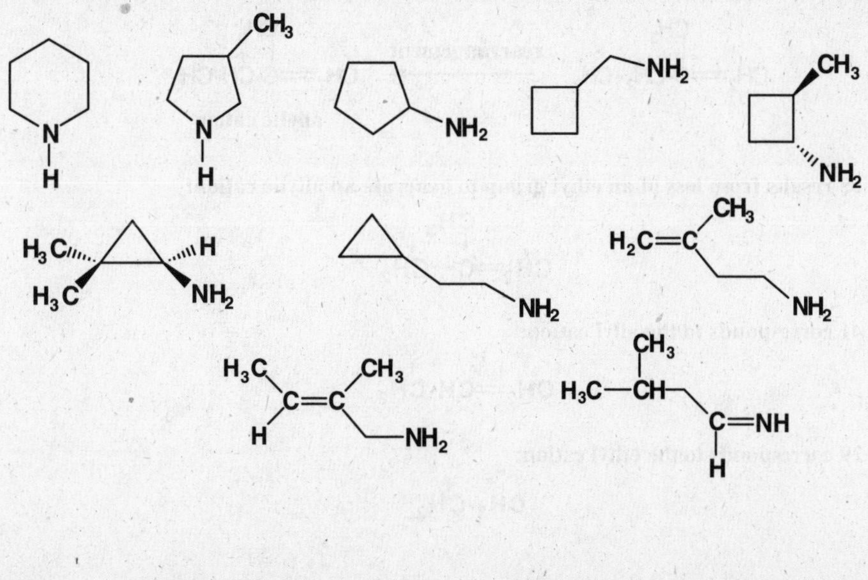

__Problem 14.20__ The mass spectrum of Compound B, a colorless liquid, shows these peaks in its mass spectrum. Determine the molecular formula of Compound B and propose a structural formula for it.

m/z	Relative Abundance
43	100 (base)
78	23.6 (M)
79	1.00
80	7.55
81	0.25

The ratio of the M to M + 2 peaks is 23.6/7.55 = 3.13 to 1. This is approximately the same ratio as seen with ^{35}Cl to ^{37}Cl, so there must be a chlorine atom in the molecule. The base peak at 43 corresponds to either a propyl cation or isopropyl cation. Therefore, the molecular formula for compound E must be C_3H_7Cl. The two structures consistent with this formula are:

$$CH_3CH_2CH_2Cl$$

$$\overset{\displaystyle Cl}{\underset{\displaystyle}{CH_3CHCH_3}}$$

__Problem 14.21__ Write molecular formulas for the five possible molecular ions of m/z 88 containing the elements C, H, N, and O.

$C_3H_4O_3 = 36 + 4 + 48 = 88$ $\quad$ $C_3H_8N_2O = 36 + 8 + 28 + 16 = 88$ $\quad$ $C_4H_8O_2 = 48 + 8 + 32 = 88$
$C_4H_{12}N_2 = 48 + 12 + 28 = 88$ $\quad$ $C_5H_{12}O = 60 + 12 + 16 = 88$

__Problem 14.22__ Write molecular formulas for the five possible molecular ions of m/z 100 containing only the elements C, H, N, and O.

Because of the nitrogen rule, there must be an even number of nitrogen atoms or no nitrogen atoms in the molecular formulas.

$C_4H_8N_2O = 48 + 8 + 28 + 16 = 100$ $\quad$ $C_5H_8O_2 = 60 + 8 + 32 = 100$ $\quad$ $C_5H_{12}N_2 = 60 + 12 + 28 = 100$
$C_6H_{12}O = 72 + 12 + 16 = 88$ $\quad$ $C_7H_{16} = 84 + 16 = 100$

__Problem 14.23__ The molecular ion in the mass spectrum of 2-methyl-1-pentene appears at m/z 84. Propose structural formulas for the prominent peaks at m/z 69, 55, 41, and 29.

$$\overset{\displaystyle CH_3}{\underset{\displaystyle}{CH_2{=}C\text{-}CH_2\text{-}CH_2\text{-}CH_3}}$$
2-Methyl-1-pentene

The peak at m/z 69 results from loss of a methyl radical followed by rearrangement to give an allylic cation:

$$\overset{CH_3}{CH_2{=}C\text{-}CH_2\text{-}CH_2^+} \xrightarrow{\text{rearrangement}} \overset{CH_3}{CH_2{=}C\text{-}\overset{+}{CH}\text{-}CH_3}$$
allylic cation

The peak at m/z 55 results from loss of an ethyl group to generate an allylic cation:

$$\overset{CH_3}{CH_2{=}C{-}\overset{+}{CH_2}}$$

The peak at m/z 41 corresponds to the allyl cation:

$$CH_2{=}CH\text{-}\overset{+}{CH_2}$$

The peak at m/z 29 corresponds to the ethyl cation:

$$CH_3\text{-}\overset{+}{CH_2}$$

Problem 14.24 Following is the mass spectrum of 1,2-dichloroethane.
(a) Account for the appearance of an (M + 2) peak with approximately two-thirds the intensity of the molecular ion peak.

The relative abundance of ^{37}Cl is 31.98 (Table 14.1) corresponding to a probability of 31.98/(100 + 31.98) = 0.2423. The corresponding probability that a Cl atom is ^{35}Cl is 100/(100 + 31.98) = 0.7577. The probability that a given Cl atom is ^{35}Cl and the other is ^{37}Cl is (0.2423)(0.7577) = 0.1836. There are two combinations with one of each so the total probability of having one ^{35}Cl and one ^{37}Cl is 2(0.1836) = 0.3672. The probability that both Cl atoms are ^{35}Cl is (0.7577)2 = 0.5741. Therefore, expect an M to (M + 2) relative ratio of 0.57 to 0.37 = 1.54. The (M + 2) peak is therefore predicted to be a little over two-thirds as high as the M peak, just what was seen on the spectrum.

(b) Predict the intensity of the M + 4 peak.

The probability of having a ^{37}Cl at both Cl atoms is $(0.2423)^2$ = 0.0587 so the relative intensity of the (M + 4) to the M peak is predicted to be 0.059 to 0.57 or about 1 to 9.7.

(c) Propose structural formulas for the cations of m/z 64, 63, 62, 51, 49, 27, and 26.

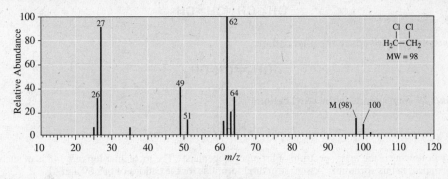

$$CH_2Cl\text{-}CH_2Cl$$
1,2-Dichloroethane

The peaks at m/z 64, 63, and 62 correspond to the following structures:

$$\left[^{37}Cl\text{—}CH\!=\!CH_2 \right]^{+\cdot} \qquad \left[^{37}Cl\text{—}CH_2\text{—}CH_2 \right]^{+\cdot} \qquad \left[^{35}Cl\text{—}CH\!=\!CH_2 \right]^{+\cdot}$$
$$m/z = 64 \qquad\qquad\qquad m/z = 63 \qquad\qquad\qquad m/z = 62$$

The peaks at m/z 51 and 49 correspond to the following structures:

$$^{37}Cl\text{—}CH_2^{+} \qquad\qquad ^{35}Cl\text{—}CH_2^{+}$$
$$m/z = 51 \qquad\qquad\qquad m/z = 49$$

The peaks at m/z 27 and 26 correspond to the following structures:

$$^{+}CH\!=\!CH_2 \qquad\qquad \left[HC\!\equiv\!CH \right]^{+\cdot}$$
$$m/z = 27 \qquad\qquad\qquad m/z = 26$$

Problem 14.25 Following is the mass spectrum of 1-bromobutane.
(a) Account for the appearance of the M + 2 peak of approximately 95% of the intensity of the molecular ion peak.

As detailed in Table 14.1, the most abundant isotope of bromine has a mass of 79 amu. Bromine has another isotope with a mass of 81 amu that has a natural abundance of 49.49%. The (M+ 2) peak results from the presence of a ^{81}Br atom in the 1-bromobutane molecule.

(b) Propose structural formulas for the cations of *m/z* 57, 41, and 29.

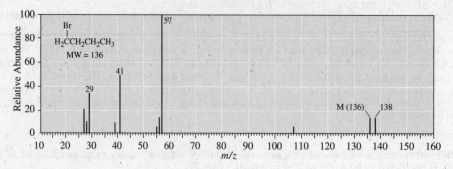

The peak at *m/z* 57 results from loss of a bromine atom:

$$\overset{+}{C}H_2\text{-}CH_2\text{-}CH_2\text{-}CH_3$$

The peak at *m/z* 41 corresponds to the allyl cation:

$$CH_2{=}CH\text{-}\overset{+}{C}H_2$$

The peak at *m/z* 29 corresponds to the ethyl cation:

$$CH_3\text{-}\overset{+}{C}H_2$$

<u>Problem 14.26</u> Following is the mass spectrum of bromocyclopentane. The molecular ion *m/z* 148 is of such low intensity that it does not appear in this spectrum. Assign structural formulas for the cations of *m/z* 69 and 41.

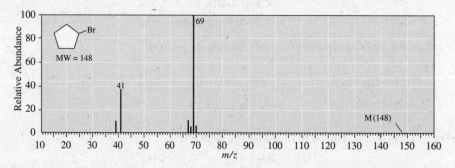

The peak at *m/z* 69 corresponds to the loss of the bromine atom to give the cyclopentyl cation:

The peak at *m/z* 41 corresponds to the allyl cation:

$$CH_2{=}CH\text{-}\overset{+}{C}H_2$$

<u>Problem 14.27</u> Following is the mass spectrum of an unknown compound. The two highest peaks are at m/z 120 and 122. Suggest a structure for this compound (data from http://webbook.nist.gov/chemistry/).

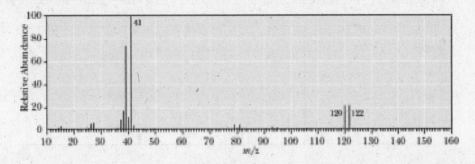

Notice that the (M + 2) peak is of approximately equal height compared to the M peak. Predict that one Br atom is present. The base peak at m/z 41 is the allyl cation, but this does not give any useful structural information. The most reasonable formula for a mass of 120/122 is C_3H_5Br. Since the allyl cation is present, propose the unknown compound is allyl bromide $CH_2=CHCH_2Br$.

<u>Problem 14.28</u> Following is the mass spectrum of 3-methyl-2-butanol. The molecular ion m/z 88 does not appear in this spectrum. Propose structural formulas for the cations of m/z 45, 43, and 41.

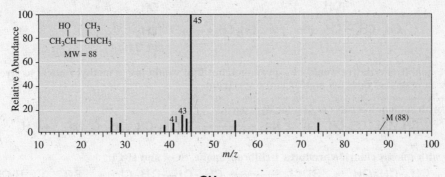

$$\underset{\substack{| \\ CH_3}}{CH_3-\overset{\overset{OH}{|}}{\underset{*}{CH}}-CH-CH_3}$$

3-Methyl-2-butanol

The peak at m/z 45 corresponds to the following resonance stabilized cation:

$$CH_3-\overset{\overset{+}{\overset{OH}{\|}}}{CH}$$

The peak at m/z 43 corresponds to the isopropyl cation:

$$\underset{CH_3}{\overset{+}{CH}-CH_3}$$

The peak at m/z 41 corresponds to the allyl cation:

$$CH_2=CH-\overset{+}{C}H_2$$

<u>Problem 14.29</u> The following is the mass spectrum of a compound C, C_3H_8O. Compound C is infinitely soluble in water, undergoes reaction with sodium metal with the evolution of a gas, and undergoes reaction with thionyl chloride to give a water-insoluble chloroalkane. Propose a structural formula for compound C, and write equations for each of its reactions.

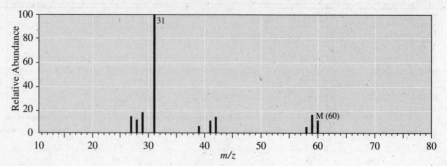

Compound C is 1-propanol.

$$CH_3\text{-}CH_2\text{-}CH_2OH$$
Compound C

In the mass spectrum, the peak at *m/z* **31 could only have come from 1-propanol due to the following fragmentation.**

$$CH_3\text{-}CH_2\overset{\displaystyle OH}{\underset{}{-CH_2}} \longrightarrow CH_3CH_2\bullet \;+\; \overset{\displaystyle OH}{\underset{m/z\ \mathbf{31}}{{}^+CH_2}}$$

Note that a resonable alternative would be 2-propanol, but that would lose a methyl radical to give a prominant peak at *m/z* **45, a mass that does not appear in the spectrum.**

The reaction with sodium liberates molecular hydrogen:

$$2\ CH_3\text{-}CH_2\text{-}CH_2OH \;+\; 2\ Na \longrightarrow 2\ CH_3\text{-}CH_2\text{-}CH_2\text{-}O^-Na^+ \;+\; H_2$$

The reaction with thionyl chloride produces 1-chloropropane, SO_2, and H-Cl.

$$CH_3\text{-}CH_2\text{-}CH_2OH \;+\; SOCl_2 \longrightarrow CH_3\text{-}CH_2\text{-}CH_2Cl \;+\; SO_2 \;+\; H\text{-}Cl$$

<u>Problem 14.30</u> Following are mass spectra for the constitutional isomers 2-pentanol and 2-methyl-2-butanol. Assign each isomer its correct spectrum.

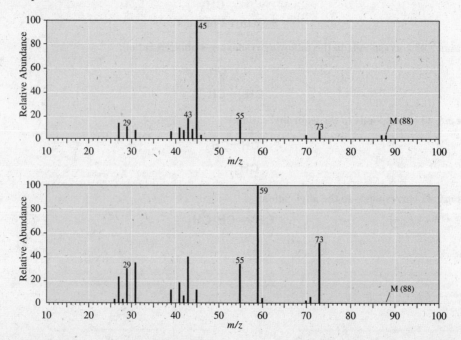

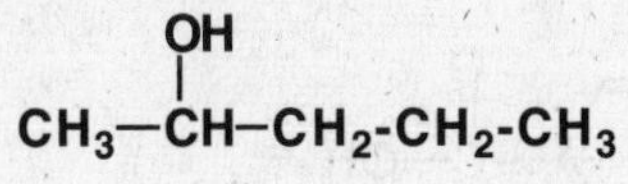

2-Pentanol 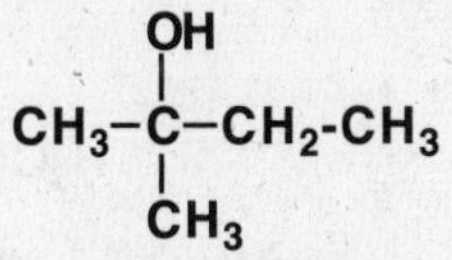   2-Methyl-2-butanol

The main difference in the above spectra is the base peak at 45 in the first spectrum that is only about 10% of the base peak in the second spectrum. This peak corresponds to the following resonance stabilized cation:

This species can be readily produced by loss of the propyl group from 2-pentanol. It is unlikely that 2-methyl-2-butanol could produce such a cation, thus the upper mass spectrum is that of 2-pentanol and the lower mass spectrum is that of 2-methyl-2-butanol. Also, the base peak in the second spectrum has a m/z of 59 that corresponds to the following fragmentation.

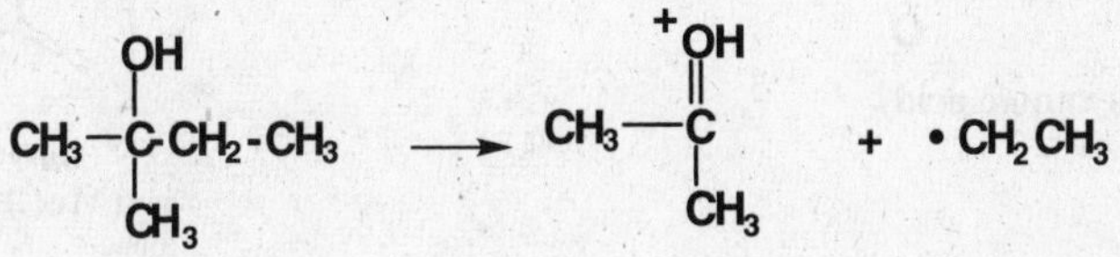

Problem 14.31 2-Methylpentanal and 4-methyl-2-pentanone are constitutional isomers with the molecular formula $C_6H_{12}O$. Each shows a molecular ion peak in its mass spectrum at m/z 100. Spectrum A shows significant peaks at m/z 85, 58, 57, 43, and 42. Spectrum B shows significant peaks at m/z 71, 58, 57, 43, and 29. Assign each compound its correct spectrum.

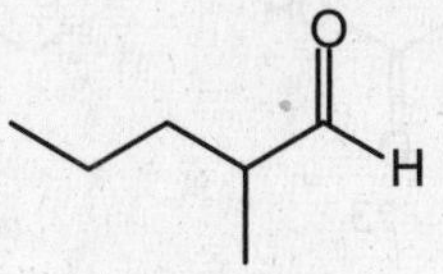

2-Methylpentanal 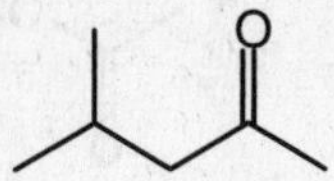   4-Methyl-2-pentanone

Some expected α–cleavage and McClafferty rearrangement fragments can be predicted as follows.

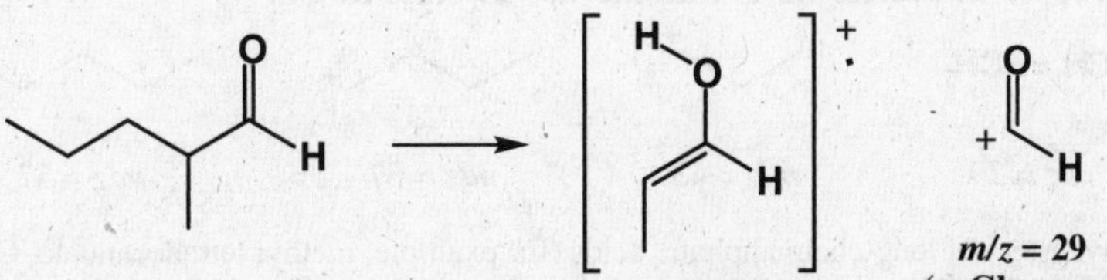

m/z = 58
(McClafferty)

m/z = 29
(α-Cleavage)

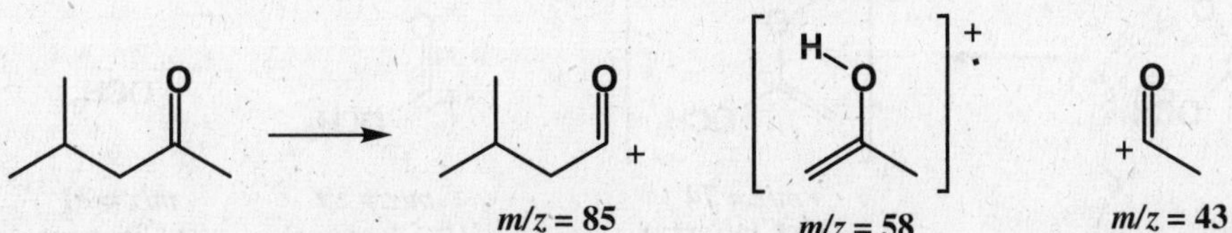

m/z = 85
(α-Cleavage)

m/z = 58
(McClafferty)

m/z = 43
(α-Cleavage)

The α-cleavage fragments at m/z = 85 (Spectrum A) and 29 (Spectrum B) are particularly diagnostic and serve to verify that Spectrum A corresponds to 4-methyl-2-pentanone, and Spectrum B corresponds to 2-methylpentanal.

<u>Problem 14.32</u> Account for the presence of peaks at *m/z* 135 and 107 in the mass spectrum of 4-methoxybenzoic acid (*p*-anisic acid).

4-Methoxybenzoic acid

$m/z = 135$ $m/z = 107$

<u>Problem 14.33</u> Account for the presence of the following peaks in the mass spectrum of hexanoic acid, $CH_3(CH_2)_4COOH$.
(a) *m/z* 60

Hexanoic acid

$m/z = 60$
(McClafferty)

(b) A series of peaks differing by 14 amu at *m/z* 45, 59, 73, and 87

The key to identifying this series of fragments is that the difference between them, namely 14 amu, corresponds to CH$_2$ groups. The series starts with the *m/z* 45 fragment that results from α-cleavage.

$m/z = 45$ $m/z = 59$ $m/z = 73$ $m/z = 87$
(α-Cleavage)

(c) a series of peaks differing by 14 amu at *m/z* 29, 43, 57, and 71

The key to identifying this series of fragments is once again that the difference between them, namely 14 amu corresponds to CH$_2$ groups. The series starts with the *m/z* 29 ethyl cation.

CH_3—CH_2^+

$m/z = 29$ $m/z = 43$ $m/z = 57$ $m/z = 57$

<u>Problem 14.34</u> All methyl esters of long-chain aliphatic acids (for example, methyl tetradecanoate, $C_{13}H_{27}COOCH_3$) show significant fragment ions at *m/z* 74, 59, and 31. What are the structures of these ions? How are they formed?

$m/z = 74$ $m/z = 59$ $m/z = 31$
(McClafferty)   **(α-Cleavage)**   **(α-Cleavage)**

<u>Problem 14.35</u> Propylbenzene, $C_6H_5CH_2CH_2CH_3$, and isopropyl benzene, $C_6H_5CH(CH_3)_2$, are constitutional isomers of molecular formula C_9H_{12}. One of these compounds shows prominent peaks in its mass spectrum at *m/z* 120 and 105. The other shows prominent peaks at *m/z* 120 and 91. Which compound has which spectrum?

The peak at *m/z* 120 is the parent ion in both cases. The peak at *m/z* 105 corresponds to the methyltropylium ion that can be most easily formed from isopropyl benzene following the loss of a single methyl group. The *m/z*

91 peak is the tropylium ion that can most easily be formed from propylbenzene following loss of an ethyl group.

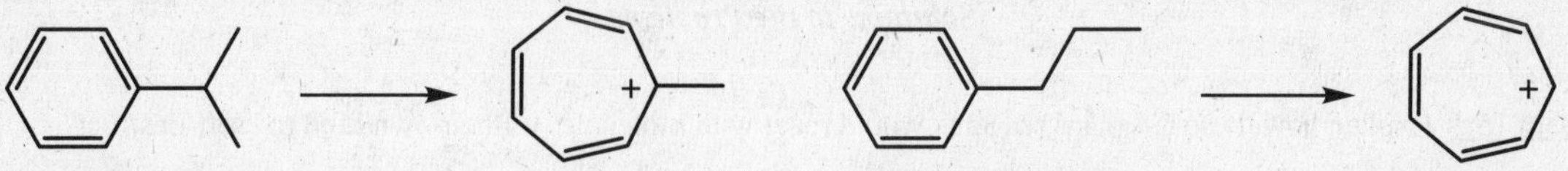

Isopropyl benzene $\longrightarrow$ $m/z = 105$ Propyl benzene $\longrightarrow$ $m/z = 91$

Problem 14.36 Account for the formation of the base peaks in these mass spectra.
(a) Isobutylamine, m/z 30   (b) Diethylamine, m/z 58

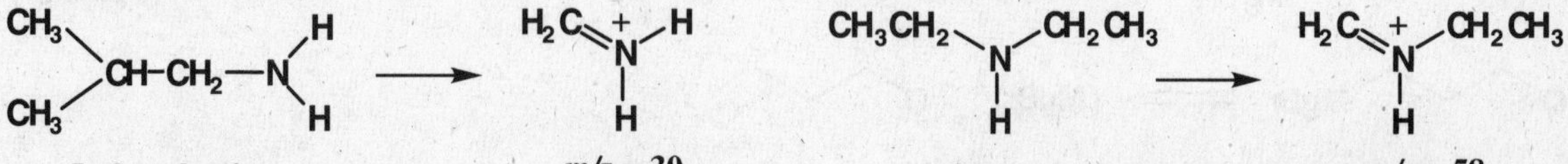

Isobutylamine $\longrightarrow$ $m/z = 30$ (β-Cleavage)   Diethylamine $\longrightarrow$ $m/z = 58$ (β-Cleavage)

Problem 14.37   Because of the sensitivity of mass spectrometry, it is often used to detect the presence of drugs in blood, urine, or other biological fluids. Tetrahydrocannabinol (nominal mass 314), a component of marijuana, exhibits two strong fragment ions at m/z 246 and 231 (the base peak). What is the likely structure of each ion?

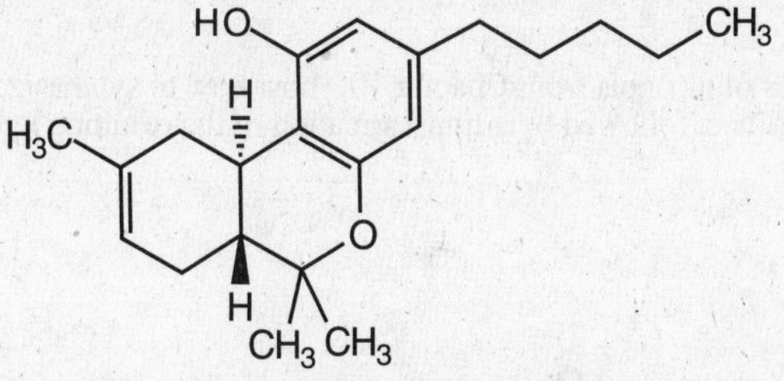

Tetrahydrocannibinol
($C_{21}H_{30}O$)

The peak at m/z 246 is probably the result of a reverse Diels-Alder reaction as described in section 14.4B. In this case, the cyclohexene ring in the top left of the structure is split from the rest of the molecule to leave the radical cation shown below. The peak at m/z 231 results from loss of a methyl group from the m/z 246 radical cation to give a tertiary cation as shown.

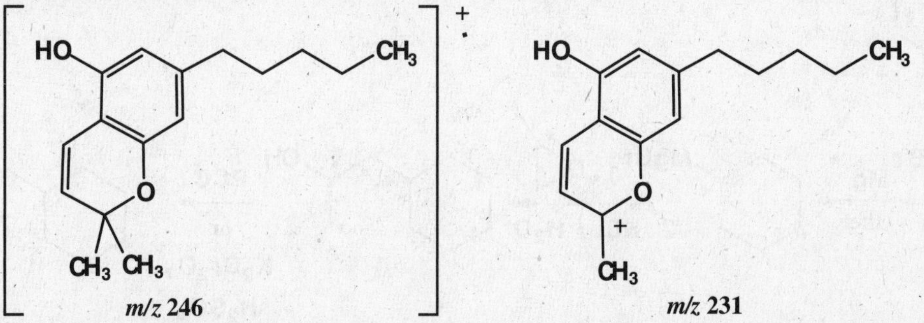

m/z 246 m/z 231

Problem 14.38 Electrospray mass spectrometry is a recently developed technique for looking at large molecules with a mass spectrometer. In this technique, molecular ions, each associated with one or more H^+ ions, are prepared under mild conditions in the mass spectrometer. As an example, a protein (P) with a molecular mass of 11,812 gives clusters of the type $(P + 8H)^{8+}$, $(P + 7H)^{7+}$, and $(P + 6H)^{6+}$.   At what mass-to-charge values do these three clusters appear in the mass spectrum?

The key to this question is to notice that these ions have multiple charges. Because ions are recorded in a mass spectrum according to their mass divided by their total charge, the mass-to-charge values for these ions are calculated by dividing their total mass (11,812 + the number of protons) by their total charge (8, 7, and 6, respectively). The final answers are thus (11,812 + 8)/8 = <u>1477.5 *m/z*.</u>, (11,812 + 7)/7 = <u>1688.4 *m/z*</u>., and (11,812 + 6)/6 = <u>1969.7 *m/z*</u>.

CHAPTER 15
Solutions to the Problems

Problem 15.1 Explain how these Grignard reagents would react with molecules of their own kind to "self-destruct".

Grignard reagents are strong bases. In both cases the Grignard reagent will "self-destruct" due to an acid-base reaction.

(a)

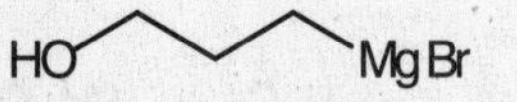

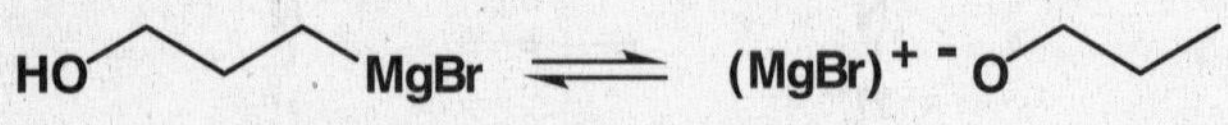

(b)

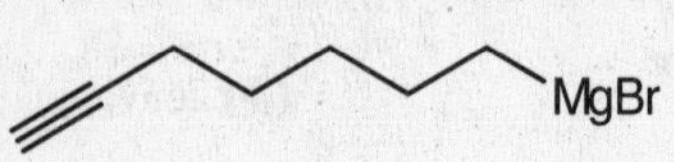

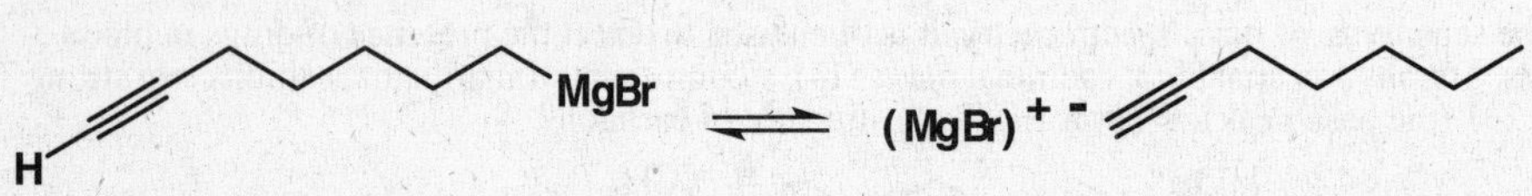

Problem 15.2 Recalling the reactions of alcohols from Chapter 10, show how to synthesize each compound from an organohalogen compound and an oxirane, followed by a transformation of the resulting hydroxyl group to the desired oxygen-containing functional group.

(a)

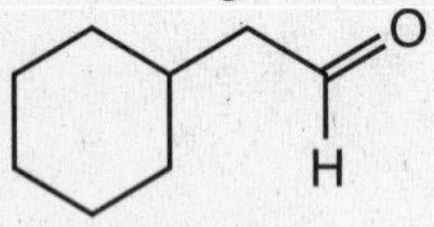

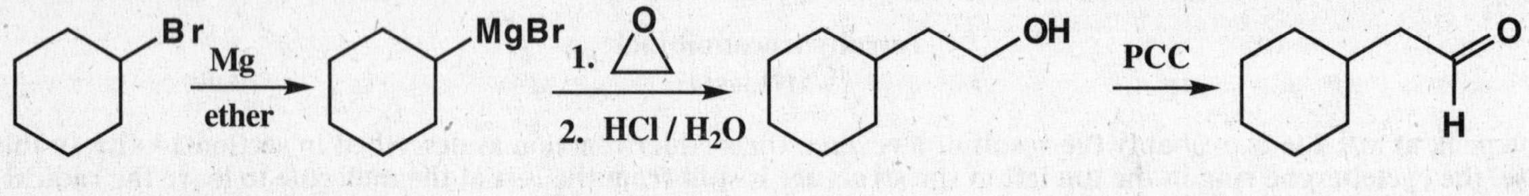

(b)

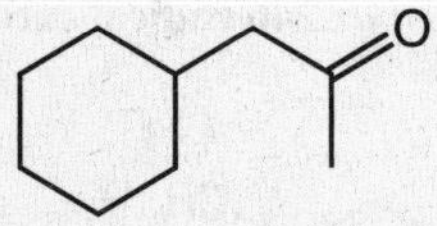

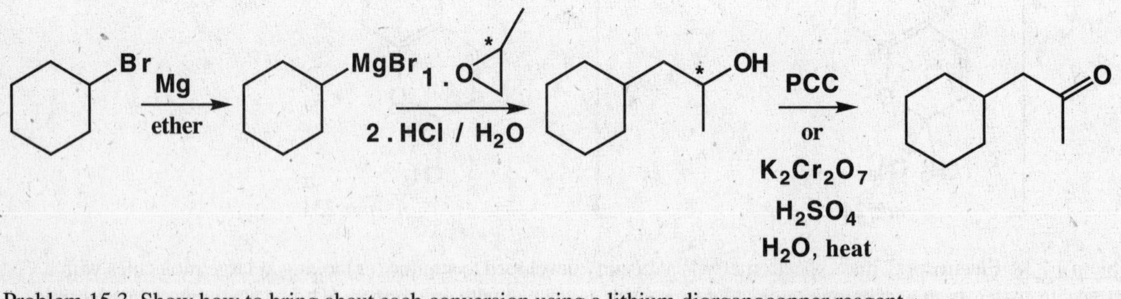

Problem 15.3 Show how to bring about each conversion using a lithium diorganocopper reagent.

(a)

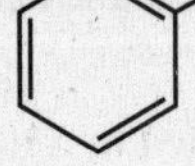

$+ \ (CH_3CH_2CH_2CH_2)_2CuLi \longrightarrow$

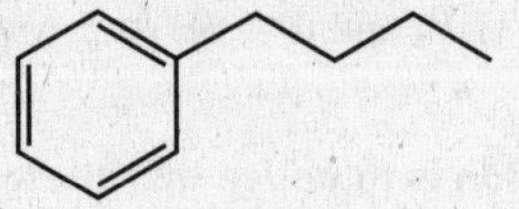

(b)

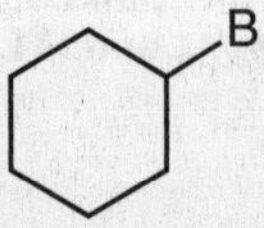

$+ \ ((CH_3)_2CHCH_2)_2CuLi \longrightarrow$

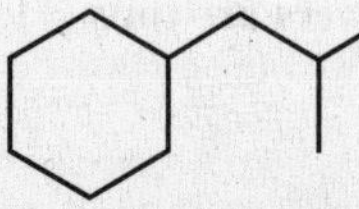

Problem 15.4 Show how to prepare each Gilman reagent in Example 15.4 from an appropriate alkyl halide and each epoxide from an appropriate alkene.

CH_3CH_2Br $\xrightarrow{2\ Li}$ CH_3CH_2Li + $LiBr$

$2\ CH_3CH_2Li$ $\xrightarrow{CuI}$ $[(CH_3CH_2)_2Cu]Li$ + $Li\bar{I}$

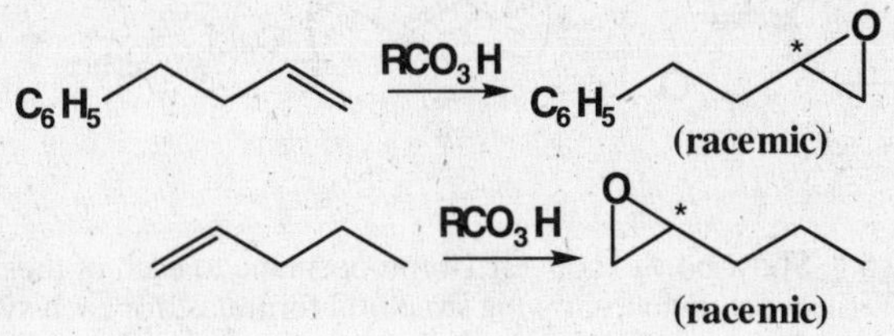

or alternatively,

$C_6H_5CH_2Br$ $\xrightarrow{2\ Li}$ $C_6H_5CH_2Li$ + $LiBr$

$2\ C_6H_5CH_2Li$ $\xrightarrow{CuI}$ $[(C_6H_5CH_2)_2Cu]Li$ + LiI

Problem 15.5 Predict the product of the following reaction.

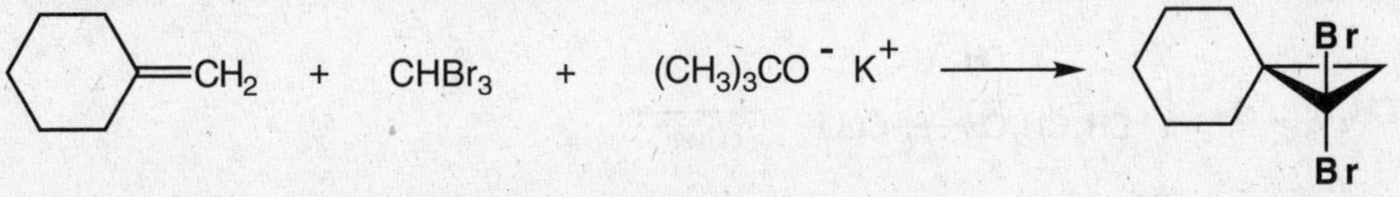

Problem 15.6 Show how the following compound could be prepared from any compound containing ten carbons or fewer.

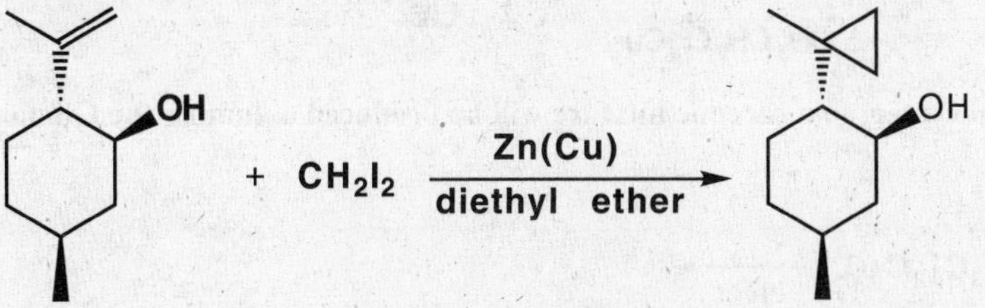

Problem 15.7 Complete these reactions involving lithium diorganocopper (Gilman) reagents.

(a)

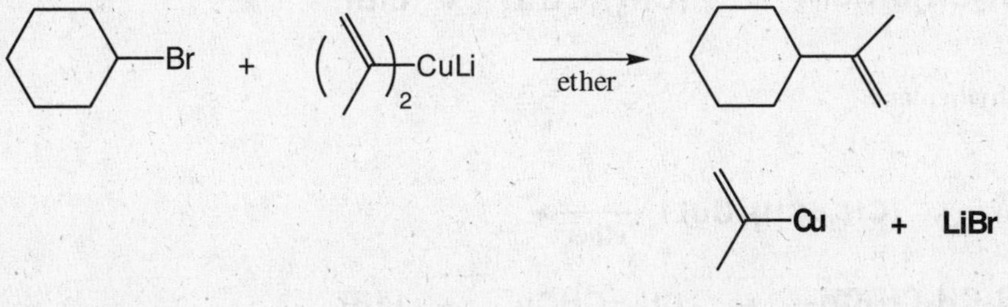

(b)

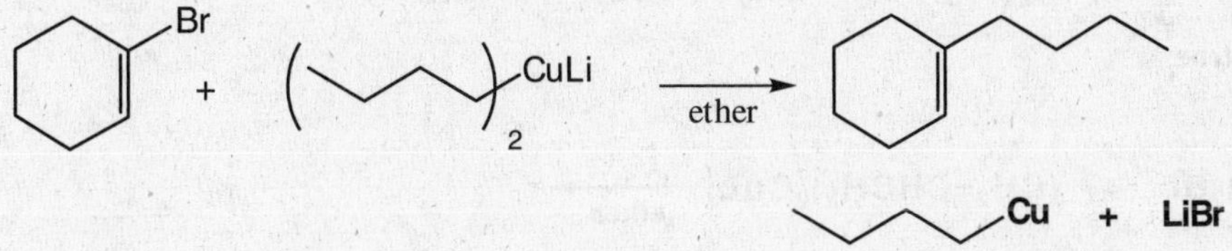

(c)

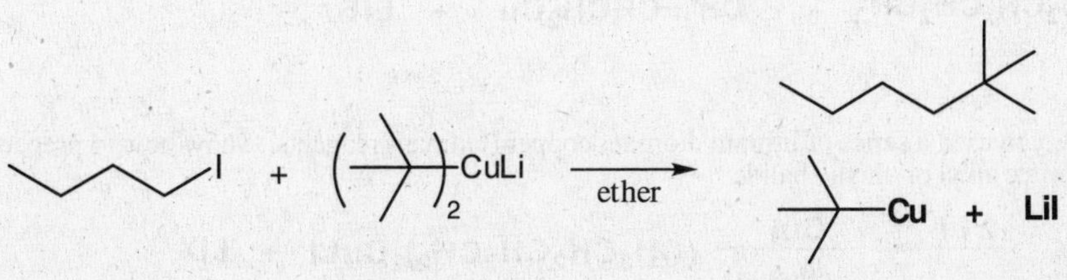

(d)

Problem 15.8 Show how to convert 1-bromopentane to each of these compounds using a lithium diorganocopper (Gilman) reagent. Write an equation, showing structural formulas, for each synthesis.

(a) Nonane

$$CH_3CH_2CH_2CH_2CH_2Br \quad + \quad (CH_3CH_2CH_2CH_2)_2CuLi \xrightarrow{\text{ether}}$$

$$CH_3(CH_2)_7CH_3 \quad + \quad CH_3CH_2CH_2CH_2Cu \quad + \quad LiBr$$
$$\text{Nonane}$$

(b) 3-Methyloctane

$$CH_3CH_2CH_2CH_2CH_2Br \quad + \quad \left(\underset{}{\overset{CH_3}{CH_3CH_2\overset{|}{\underset{*}{CH}}}} \right)_2 CuLi \xrightarrow{\text{ether}}$$

$$\underset{\text{3-Methyloctane}}{\overset{CH_3}{CH_3CH_2\overset{|}{\underset{*}{CH}}CH_2CH_2CH_2CH_2CH_3}} \quad + \quad \overset{CH_3}{CH_3CH_2\overset{|}{\underset{*}{CH}}Cu} \quad + \quad LiBr$$

Note how the 3-methyloctane has one chiral center, so a racemic mixture will be produced assuming the Gilman reagent starting material is racemic.

(c) 2,2-Dimethylheptane

$$CH_3CH_2CH_2CH_2CH_2Br \quad + \quad [(CH_3)_3C]_2CuLi \xrightarrow{\text{ether}}$$

$$\underset{\underset{CH_3}{|}}{\overset{\overset{CH_3}{|}}{CH_3\overset{}{C}CH_2CH_2CH_2CH_2CH_3}} \quad + \quad (CH_3)_3CCu \quad + \quad LiBr$$
$$\text{2,2-Dimethylheptane}$$

(d) 1-Heptene

$$CH_3CH_2CH_2CH_2CH_2Br \quad + \quad (CH_2{=}CH)_2CuLi \xrightarrow{\text{ether}}$$

$$CH_2{=}CHCH_2CH_2CH_2CH_2CH_3 \quad + \quad CH_2{=}CHCu \quad + \quad LiBr$$
$$\text{1-Heptene}$$

(e) 1-Octene

$$CH_3CH_2CH_2CH_2CH_2Br \quad + \quad (CH_2{=}CHCH_2)_2CuLi \xrightarrow{\text{ether}}$$

$$CH_2{=}CHCH_2CH_2CH_2CH_2CH_2CH_3 \quad + \quad CH_2{=}CHCH_2Cu \quad + \quad LiBr$$
$$\text{1-Octene}$$

Problem 15.9 In Problem 15.8, you used a series of lithium diorganocopper (Gilman) reagents. Show how to prepare each Gilman reagent from an appropriate alkyl or vinylic halide.

(a) $$2\ CH_3CH_2CH_2CH_2X \xrightarrow[\text{ether}]{2\ Li} \xrightarrow{CuI} (CH_3CH_2CH_2CH_2)_2CuLi \quad + \quad LiX$$

(b) $2\ CH_3CH_2\overset{CH_3}{\underset{X}{\overset{*}{C}H}}$ $\xrightarrow{2\ Li}$ $\xrightarrow[ether]{CuI}$ $\left(CH_3CH_2\overset{CH_3}{\underset{}{\overset{*}{C}H}}\right)_2 CuLi\ +\ LiX$

(c) $2\ (CH_3)_3CX$ $\xrightarrow{2\ Li}$ $\xrightarrow[ether]{CuI}$ $[(CH_3)_3C]_2CuLi\ +\ LiX$

(d) $2\ \overset{H}{\underset{H}{>}}C=C\overset{H}{\underset{X}{<}}$ $\xrightarrow{2\ Li}$ $\xrightarrow[ether]{CuI}$ $(CH_2=CH)_2CuLi\ +\ LiX$

(e) $2\ \overset{H}{\underset{H}{>}}C=C\overset{H}{\underset{CH_2X}{<}}$ $\xrightarrow{2\ Li}$ $\xrightarrow[ether]{CuI}$ $(CH_2=CHCH_2)_2CuLi\ +\ LiX$

<u>Problem 15.10</u> Show how to prepare each compound from the given starting compound through the use of a lithium diorganocopper (Gilman) reagent.
(a) 4-Methylcyclopentene from 4-bromocyclopentene

**4-Bromocyclo-
pentene**   **4-Methylcyclo-
pentene**

(b) (Z)-2-Undecene from (Z)-1-bromopropene

**(Z)-1-Bromo-
propene**   **(Z)-2-Undecene**

(c) 1-Butylcyclohexene from 1-iodocyclohexene

1-Iodocyclohexene   **1-Butylcyclohexene**

(d) 1-Decene from 1-iodooctane

$CH_3(CH_2)_6CH_2I$ + $(CH_2=CH)_2CuLi$ $\xrightarrow{ether}$ $CH_2=CH(CH_2)_7CH_3$ + $CH_2=CHCu$ + LiI
1-Iodooctane   **1-Decene**

(e) 1,8-Nonadiene from 1,5-dibromopentane

$BrCH_2(CH_2)_3CH_2Br$ + $2\ (CH_2=CH)_2CuLi$ $\xrightarrow{ether}$ $CH_2=CH(CH_2)_5CH=CH_2$ + $2\ CH_2=CHCu$
1,5-Dibromopentane   **1,8-Nonadiene**
                                                                          + $2\ LiBr$

Problem 15.11 The following is a retrosynthetic scheme for the preparation of *trans*-2-allylcyclohexanol. Show reagents to bring about the synthesis of this compound from cyclohexane.

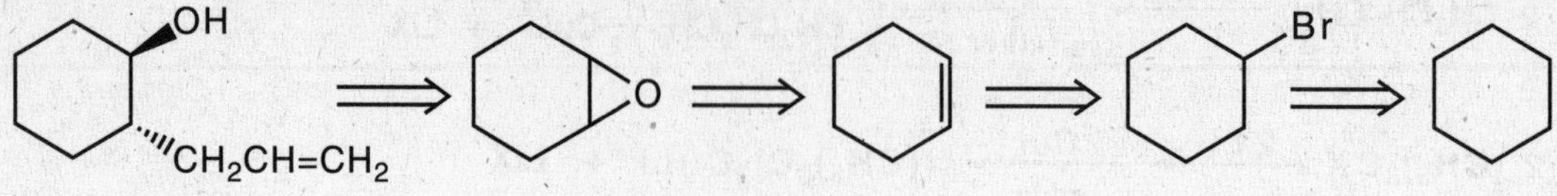

(racemic)

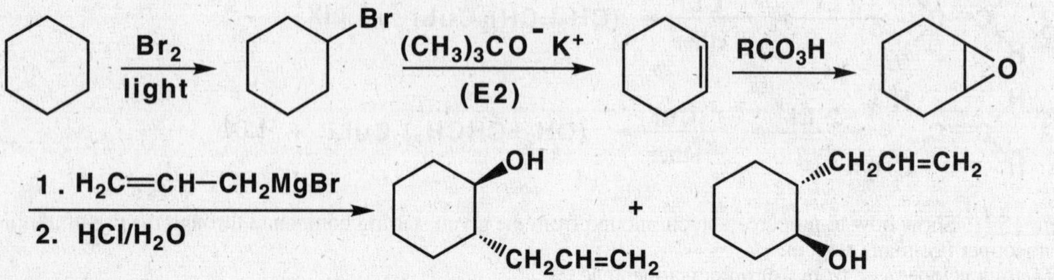

Note that in this reaction scheme two enantiomers are formed as a racemic mixture. Also, an alternative approach is possible using a Gilman reagent to react with the epoxide in the last step.

Problem 15.12 Complete these equations.

(a) $CH_3CH_2CH_2C \equiv CH$ + CH_3CH_2MgBr $\xrightarrow{\text{diethyl ether}}$ $CH_3CH_2CH_2C \equiv C^- \ MgBr^+$ + CH_3CH_3

b)

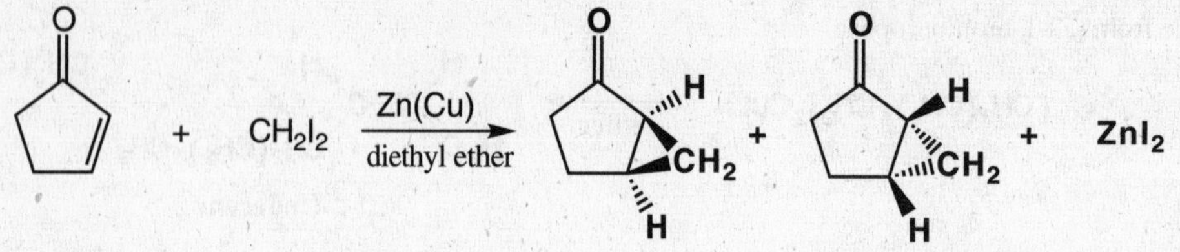

(c)

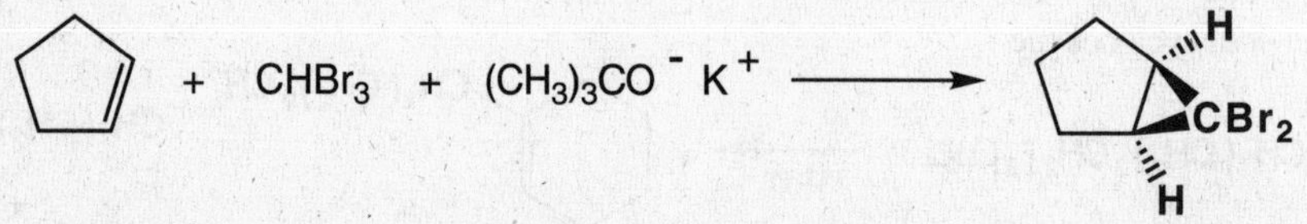

(d)

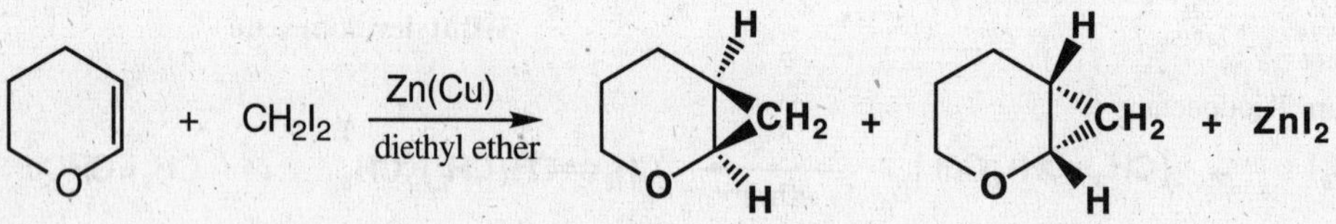

(e)

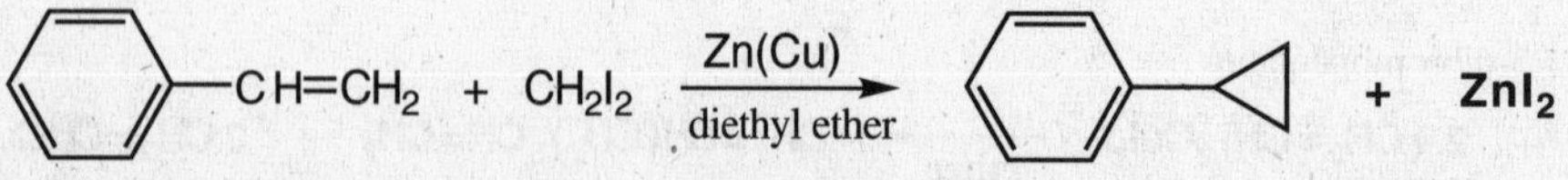

Note that in (b) and (d) racemic mixtures are produced.

Problem 15.13 Reaction of the following cycloalkene with the Simmons-Smith reagent is stereospecific and gives only the isomer shown. Suggest a reason for this stereospecificity.

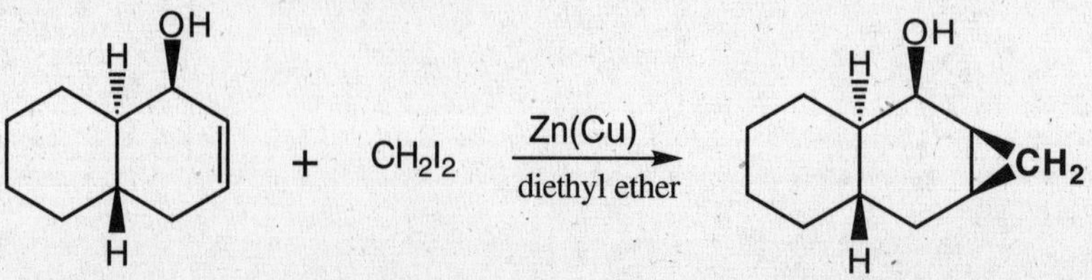

Recall that the reaction mechanism of the Simmons-Smith reaction is formation of the following Zn species.

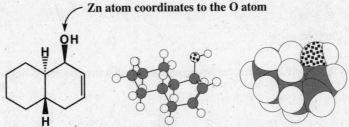

$$CH_2I_2 \xrightarrow[\text{diethyl ether}]{\textbf{Zn(Cu)}} I{-}Zn\diagdown_{CH_2}{-}I$$

As shown in the following three-dimensional models, the -OH group is suspended above one face of the double bond.

Zn atom coordinates to the O atom

The Zn atom of the reagent acts as a Lewis acid and forms a weak complex with the Lewis base oxygen atom of the -OH group. This positions the reagent above the double bond and directs attack of the carbenoid from the top face, leading to the observed product.

<u>Problem 15.14</u> Show how the following compound can be prepared in good yield.

The product will be obtained in high yield from reacting the corresponding bicycloalkene with CHCl₃ and hindered base.

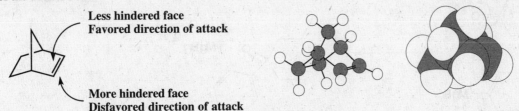

$$+ \quad CHCl_3 \quad + \quad (CH_3)_3CO^-K^+ \quad \longrightarrow$$

Note how the predominant product is the one shown, in which the dichlorocarbene attacks the less hindered face of the double bond.

Less hindered face
Favored direction of attack

More hindered face
Disfavored direction of attack

<u>Problem 15.15</u> Show the product of the following reaction (do not concern yourself with which side of the ring is attacked).

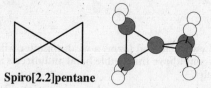

Caryophyllene

<u>Problem 15.16</u> Show how the spiro[2.2]pentane can be prepared in one step from organic compounds containing three carbons or less and any necessary inorganic reagents or solvents.

Spiro[2.2]pentane

Spiro[2.2]pentane can be prepared in one step through the reaction of allene with the Simmons-Smith reagent.

$$CH_2{=}C{=}CH_2 \quad + \quad CH_2I_2 \quad \xrightarrow{Zn(Cu)}$$
Allene   **(excess)**

Looking Ahead
<u>Problem 15.17</u> One of the most important uses for Grignard reagents is their addition to carbon compounds to give new carbon-carbon bonds (Section 16.5) In this reaction, the carbon of the organometallic compound acts as a nucleophile to add to the positive carbon of the carbonyl.

$$\xrightarrow[\text{2. } H_3O^+]{\text{1. } C_6H_5MgBr}$$

*OH

C₆H₅ → C_6H_5

(racemic)

(a) Give a mechanism for the first step of the reaction.

Step 1: Make a bond between a nucleophile and an electrophile. **The first step of the reaction involves the attack of the nucleophilic carbon atom of the Grignard reagent onto the relatively electrophilic carbon atom of the carbonyl group. The carbonyl pi bond breaks in the process, generating an alkoxide intermediate.**

$$\delta^+ \quad \overset{\cdot\cdot}{\underset{\cdot\cdot}{O}} \quad \delta^-$$

MgBr

$\delta^- \quad \delta^+$

$$\ddot{O}:^- \; (MgBr)^+$$

(racemic)

(b) Explain the function of the acid in the second step.

Step 2: Add a proton. The acid protonates the alkoxide to give the alcohol product.

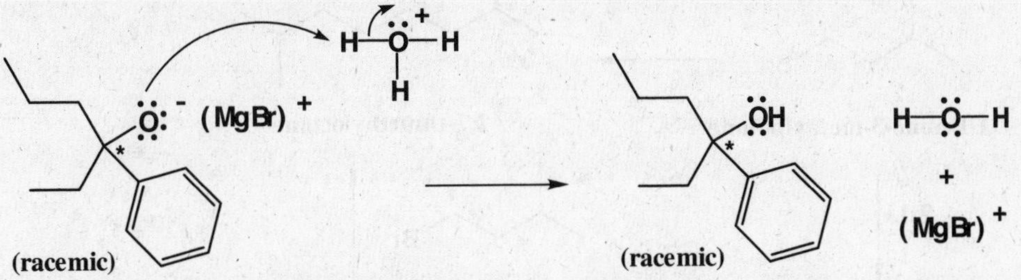

(racemic)   (racemic)

An important aspect of this process is that a new carbon-carbon is formed.

Problem 15.18 Organolithium compounds react with carbonyl compounds in a way that is similar to that of Grignard reactions. Suggest a product of the following reaction.

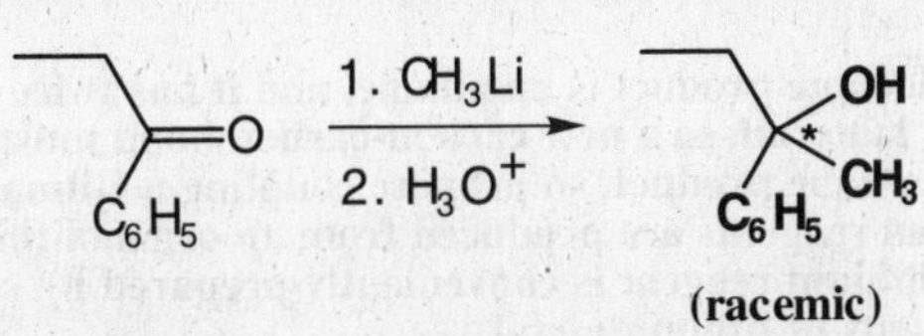

(racemic)

In analogy to the Grignard reaction of Problem 15.17, the nucleophilic methyl group of methyllithium reacts with the relatively electrophilic carbonyl carbon atom. Protonation in acid completes the reaction to give the racemic alcohol product. An important aspect of this reaction is that a new carbon-carbon bond is formed.

Problem 15.19 1-Bromobutane can be converted into either of the two products shown by a suitable choice of reagents. Give reagents and conditions for each reaction.

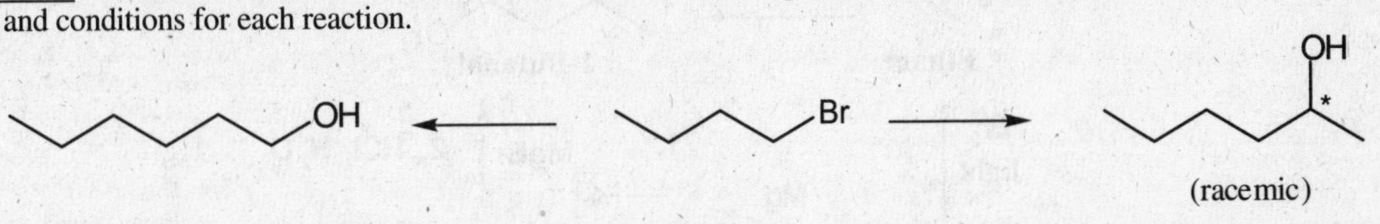

(racemic)

A Grignard reagent can be created that reacts with ethylene oxide to generate 1-hexanol. Alternatively, the Grignard reagent can react with acetaldehyde in analogy to Problem 15.17 to give racemic 2-hexanol. Note that the 1-bromobutane could be converted into butyllithium and reacted in an analogous fashion to give the same products.

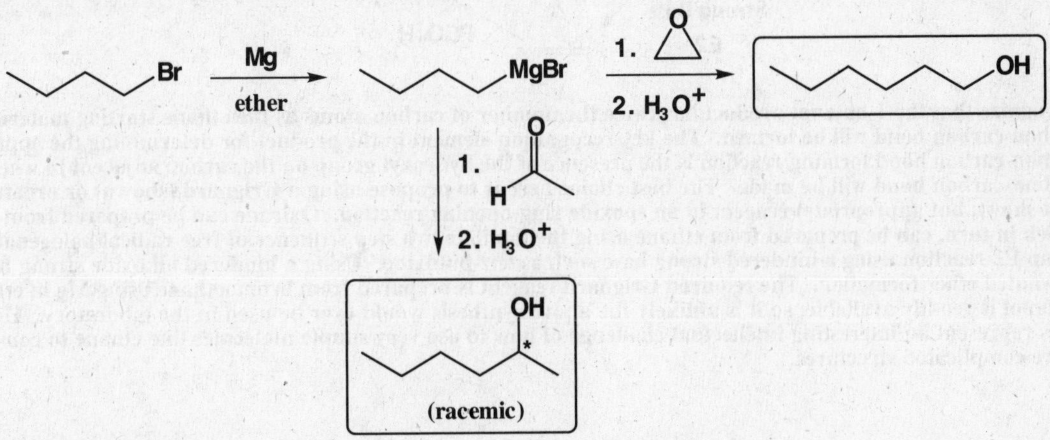

(racemic)

<u>Problem 15.20</u> Show how to convert 1-bromo-3-methylbutane into 2,7-dimethyloctane. You must use 1-bromo-3-methylbutane as the source of all carbon atoms in the target molecule. Show all reagents and all molecules synthesized along the way.

Recognize that the 2,7-dimethyloctane product is an alkane, and it has twice the number of carbon atoms as the 1-bromo-3-methylbutane starting material, so a new carbon-carbon bond must be formed. Gilman reagents create a new carbon-carbon bond to give an alkane product, so propose coupling a Gilman reagent to the starting haloalkane in the last step of the synthesis. Gilman reagents are produced from an organolithium reagent reacting with half of an equivalent of CuI. The organolithium reagent is conveniently prepared by reacting haloalkanes such as 1-bromo-3-methylbutane with two equivalents of lithium metal.

<u>Problem 15.21</u> Show how to convert ethane into 1-butanol. You must use ethane as the source of all carbon atoms in the target molecule. Show all reagents and all molecules synthesized along the way.

Recognize that the 1-butanol product has twice the number of carbon atoms as the ethane starting material, so a new carbon-carbon bond will be formed. The key recognition element in the product for determining the appropriate carbon-carbon bond forming reaction is the presence of the hydroxyl group on the carbon adjacent to where the new carbon-carbon bond will be made. The best choice here is to propose using a Grignard (shown) or organolithium (not shown but appropriate) reagent in an epoxide ring-opening reaction. Oxirane can be prepared from ethylene, which in turn, can be prepared from ethane using the familiar two step sequence of free radical halogenation followed by an E2 reaction using a hindered strong base such as *tert*-butoxide. Using a hindered alkoxide strong base avoids unwanted ether formation. The required Grignard reagent is prepared from bromoethane using Mg in ether. 1-Butanol is readily available, so it is unlikely the above synthesis would ever be used in the laboratory. However, it does represent an interesting intellectual challenge of how to use very simple molecules like ethane to construct larger, more complicated structures.

<u>Problem 15.22</u> Show how to convert 1-propanol and diiodomethane into racemic *trans*-1-methyl-2-propylcyclopropane. You must use 1-propanol and diiodomethane as the source of all carbon atoms in the target molecule. Show all molecules synthesized along the way.

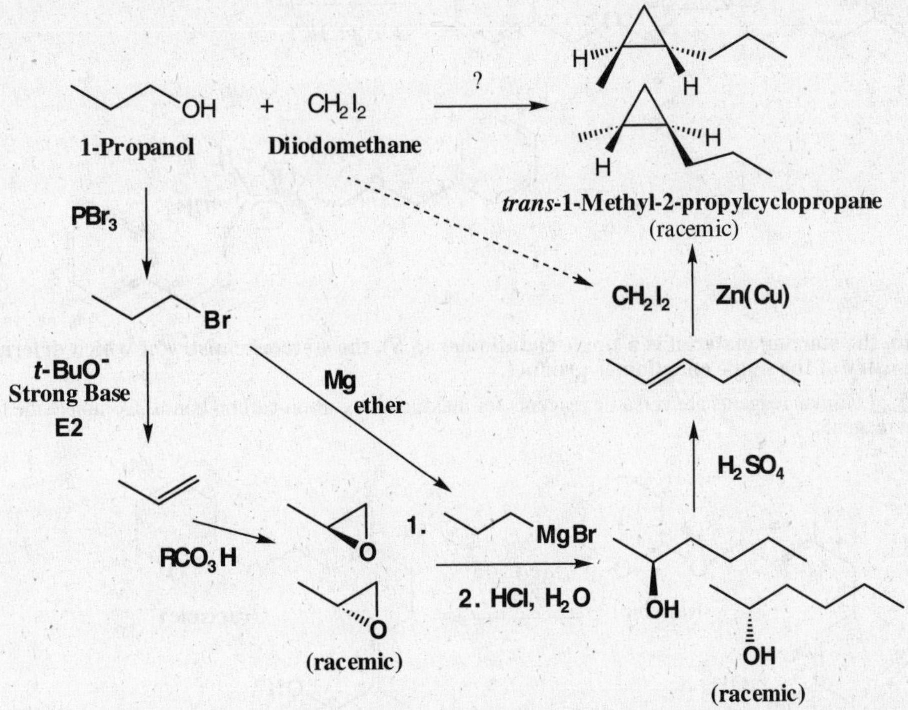

Recognize that the product has seven carbon atoms, so it must be made from a combination of two three-carbon and one single-carbon starting material units. The key recognition element of the product is the cyclopropane group, that is best made using the Simmons-Smith reaction. As confirmation of this proposal, the required diiodomethane is provided as a starting material. The alkene component required for the Simmons-Smith reaction is *trans*-2-hexene. Synthesis of *trans*-2-hexene in high yield from 1-propanol represents the real challenge of this synthesis problem. A good strategy involves formation of the carbon-carbon bond using a Grignard reagent with an epoxide to create racemic 2-hexanol. Dehydration of 2-hexanol in acid according to Zaitsev's rule will lead to the desired *trans*-2-hexene. The two components needed, the Grignard reagent and the epoxide, can both be derived from 1-bromopropane, which, in turn, can be made from 1-propanol using PBr$_3$. Following an E2 reaction using the hindered *tert*-butoxide base (to prevent ether formation), propene can be converted into the desired epoxide using a peracid such as peracetic acid or MCPBA. The Grignard reagent can be made in a single step from 1-bromopropane using Mg in ether.

Reactions in Context

<u>Problem 15.23</u> The synthesis of carbohydrates can be particularly difficult because of the large number of chiral centers and OH functional groups present. Epoxides can be useful synthetic intermediates in carbohydrate syntheses. Draw the product of the following reactions of a Gilman reagent with each epoxide.

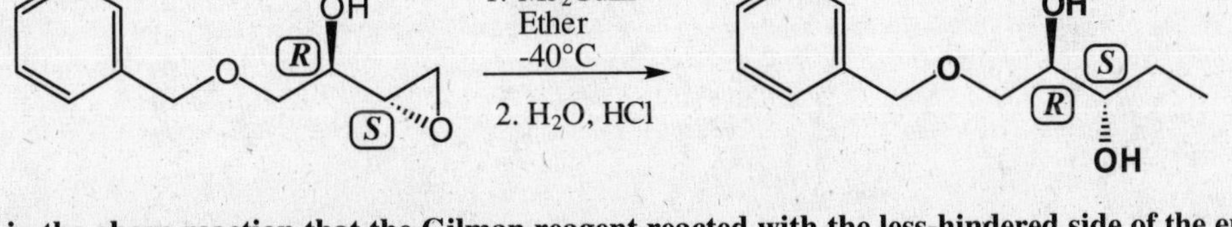

Note in the above reaction that the Gilman reagent reacted with the less-hindered side of the epoxide. Note also that the starting epoxide was a single enantiomer (*R*,*S*), the stereochemistry of which determines the stereochemistry of the product, which was therefore formed as a single enantiomer.

(b)

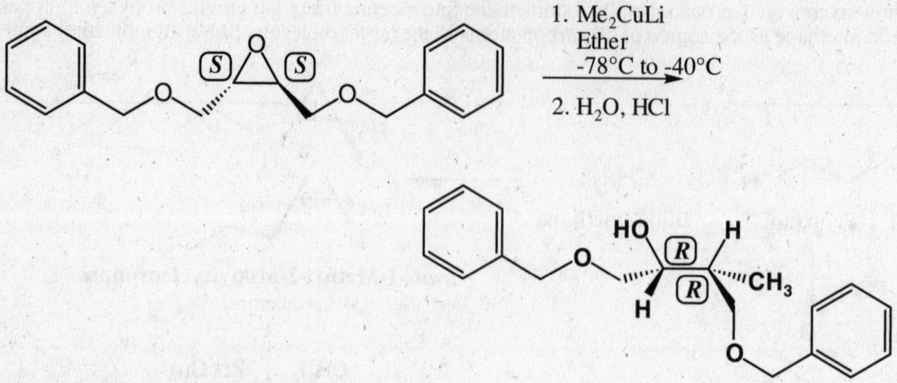

1. Me$_2$CuLi
   Ether
   $\xrightarrow{\text{-78°C to -40°C}}$
2. H$_2$O, HCl

Once again, the starting material is a single enantiomer (S,S), the stereochemistry of which determines the stereochemistry of the single enantiomer product.

Problem 15.24 Gilman reagents are versatile reagents for making new carbon-carbon bonds. Complete the following reactions that use Gilman reagents.

(a)

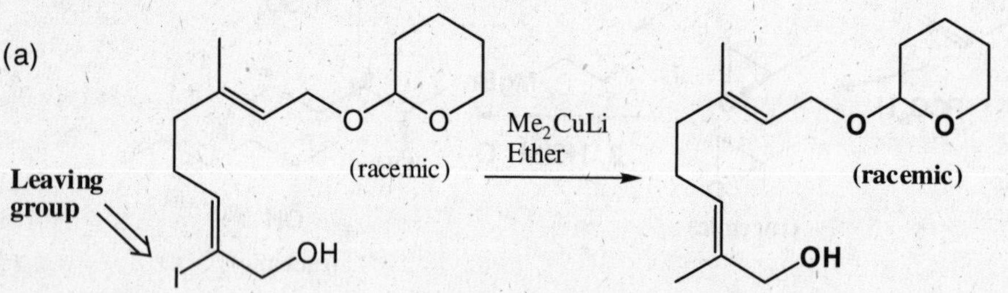

Leaving group

(racemic) $\xrightarrow[\text{Ether}]{\text{Me}_2\text{CuLi}}$ (racemic)

(b)

$\xrightarrow[\text{-78°C to -40°C}]{\begin{array}{c}\text{Me}_2\text{CuLi}\\\text{Ether}\end{array}}$

Leaving group

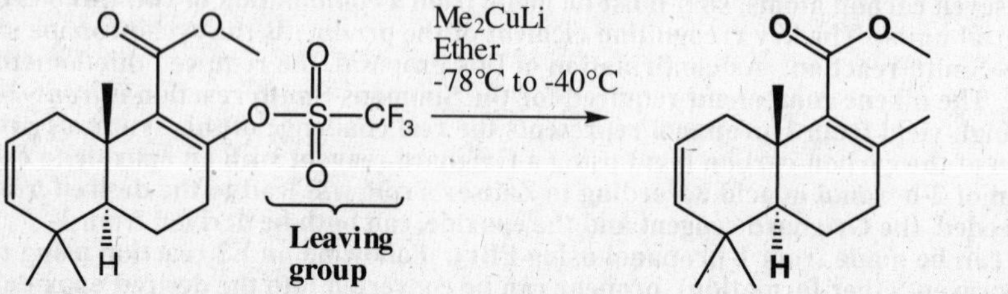

In either case, the Gilman reagent is reacting with the leaving group, either the iodo atom, or trifluoromethyl sulfonyl (triflate) group, to give a new carbon-carbon bond.

CHAPTER 16
Solutions to the Problems

Problem 16.1 Write the IUPAC names for each compound. Specify the configuration in (c).

(a)

3,3-Dimethylbutanal

(b)

1,3-Cyclohexanedione

(c)

(R)-2-Phenylpropanal

Problem 16.2 Write structural formulas for all aldehydes with the molecular formula $C_6H_{12}O$ and give each its IUPAC name. Which of these aldehydes are chiral?

$$CH_3—CH_2—CH_2—CH_2—CH_2—\overset{\overset{\displaystyle O}{\|}}{CH}$$

Hexanal

$$CH_3—\underset{\underset{\displaystyle CH_3}{|}}{CH}—CH_2—CH_2—\overset{\overset{\displaystyle O}{\|}}{CH}$$

4-Methylpentanal

$$CH_3—CH_2—\overset{*}{\underset{\underset{\displaystyle CH_3}{|}}{CH}}—CH_2—\overset{\overset{\displaystyle O}{\|}}{CH}$$

3-Methylpentanal
(chiral)

$$CH_3—CH_2—CH_2—\overset{*}{\underset{\underset{\displaystyle CH_3}{|}}{CH}}—\overset{\overset{\displaystyle O}{\|}}{CH}$$

2-Methylpentanal
(chiral)

$$CH_3—CH_2—\underset{\underset{\underset{\displaystyle CH_3}{|}}{\underset{\displaystyle CH_2}{|}}}{CH}—\overset{\overset{\displaystyle O}{\|}}{CH}$$

2-Ethylbutanal

$$CH_3—\underset{\underset{\displaystyle CH_3}{|}}{CH}—\overset{*}{\underset{\underset{\displaystyle CH_3}{|}}{CH}}—\overset{\overset{\displaystyle O}{\|}}{CH}$$

2,3-Dimethylbutanal
(chiral)

$$CH_3—\underset{\underset{\displaystyle CH_3}{|}}{\overset{\overset{\displaystyle CH_3}{|}}{C}}—CH_2—\overset{\overset{\displaystyle O}{\|}}{CH}$$

3,3-Dimethylbutanal

$$CH_3—CH_2—\underset{\underset{\displaystyle CH_3}{|}}{\overset{\overset{\displaystyle CH_3}{|}}{C}}—\overset{\overset{\displaystyle O}{\|}}{CH}$$

2,2-Dimethylbutanal

Problem 16.3 Write the IUPAC name for each compound.

(a)

1,3-Dihydroxy-2-propanone

(b)

1,2-Cyclohexanedione

(c)

4-Aminobutanal

Problem 16.4 Show how these four products can be synthesized from the same Grignard reagent.

(a) (racemic)   (b) (racemic)   (c) (racemic)   (d) (racemic)

All of the products can be obtained using 3-cyclohexenylmagnesium bromide and the other reactants shown. Note that in each case at least one chiral center is produced so that racemic mixtures will be generated in the reactions.

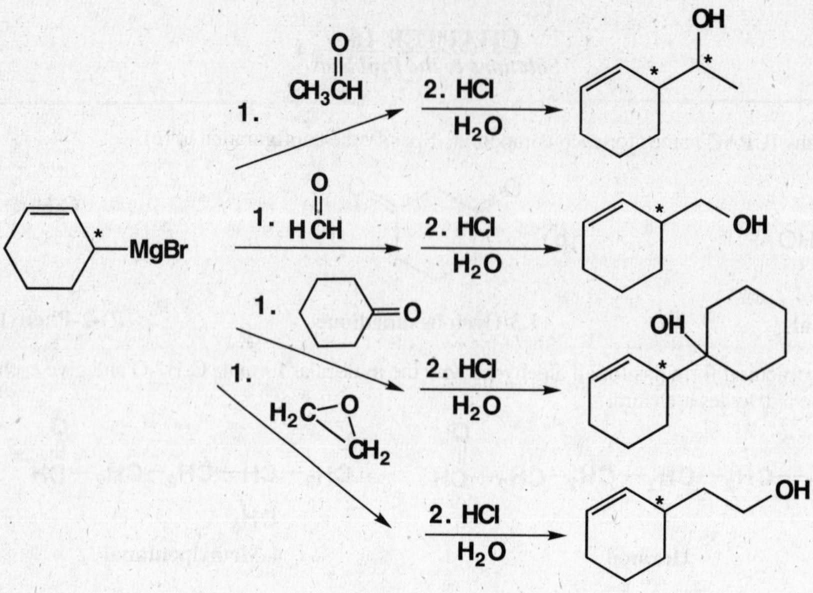

Problem 16.5 Show how each alkene can be synthesized by a Wittig reaction (there are two routes to each).

(a)

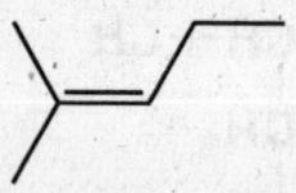

There are two combinations of Wittig reagent and carbonyl compound that could be used to make this alkene. The first uses 1-bromopropane to produce the Wittig reagent.

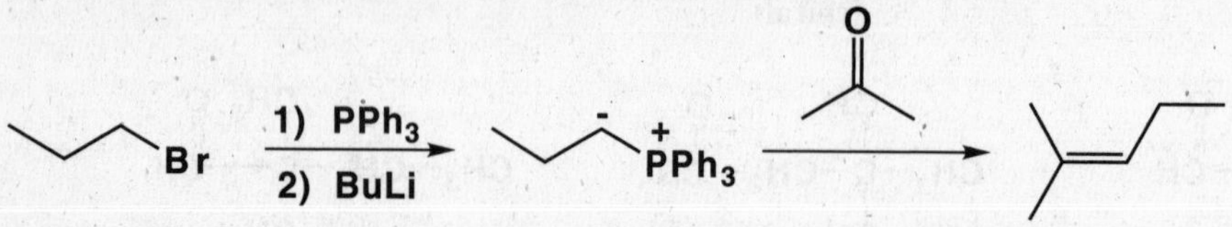

The other involves the Wittig reagent prepared from 2-bromopropane

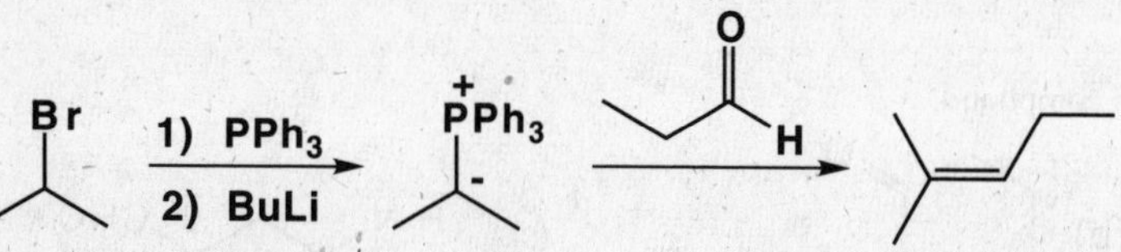

The first synthesis is better because a more reactive 1° alkyl halide is used in the S_N2 reaction with PPh₃.

(b)

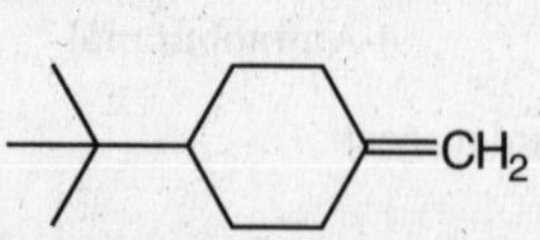

There are also two combinations of Wittig reagent and carbonyl compound that could be used to make this alkene. The first uses 1-bromo-4-*tert*-butylcyclohexane to produce the Wittig reagent.

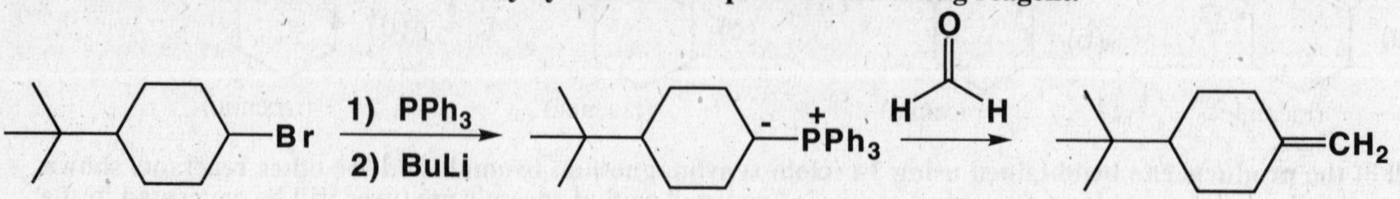

The other involves the Wittig reagent prepared from bromomethane

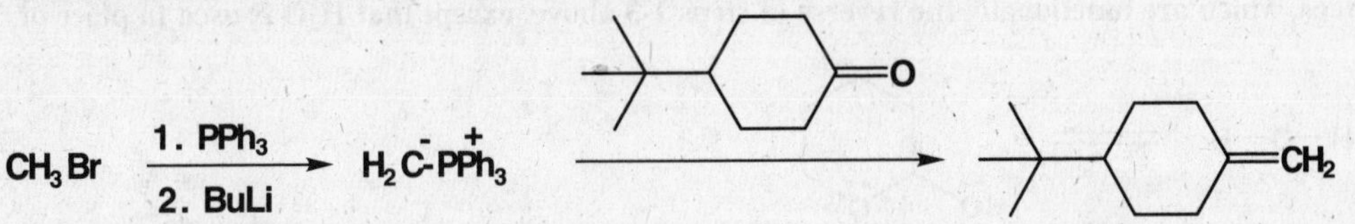

$$CH_3Br \xrightarrow[\text{2. BuLi}]{\text{1. PPh}_3} H_2\overset{-}{C}\text{-}\overset{+}{PPh_3}$$

The second synthesis is better because the more reactive CH₃Br is used in the S_N2 reaction with PPh₃.

In the above examples, alkyl bromides were used. In fact, alkyl chlorides and iodides could have also been utilized.

<u>Problem 16.6</u> Hydrolysis of an acetal in aqueous acid gives an aldehyde or ketone and two molecules of alcohol or one molecule of a diol. Draw the structural formulas for the products of hydrolysis of each in aqueous acid.

(a)

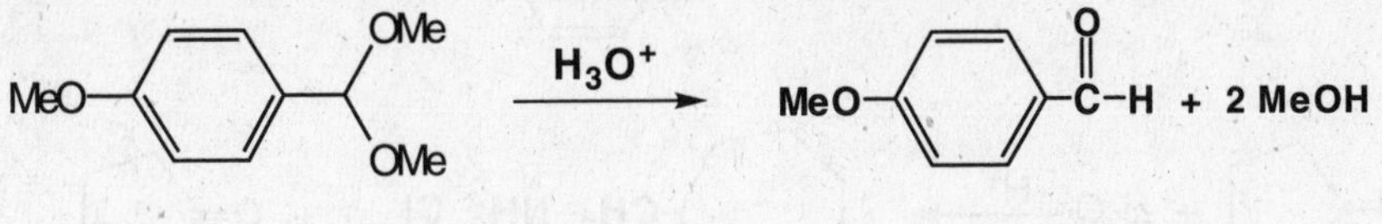

(b)

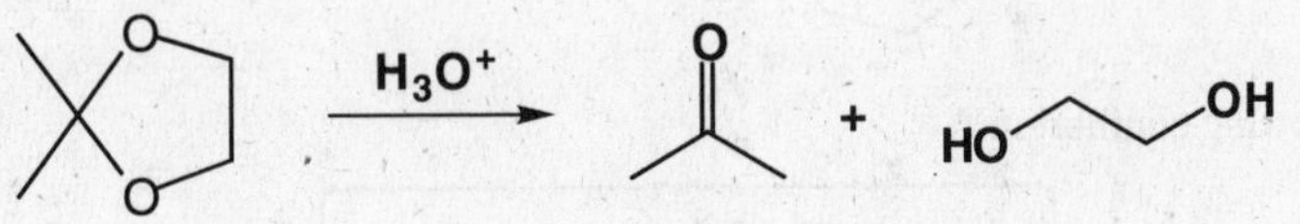

(c)

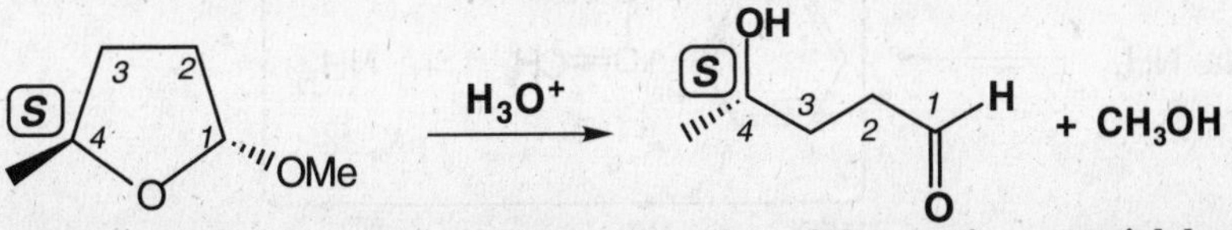

Note that stereochemistry of this product will be that of the starting material, because the configuration of the chiral center is not altered during the reaction.

<u>Problem 16.7</u> Write a mechanism for the acid-catalyzed hydrolysis of a THP ether to regenerate the original alcohol. Into what compound is the THP group converted?

The mechanism for acid-catalyzed THP ether hydrolysis is exactly the reverse of the acid-catalyzed THP ether formation.

Step 1: Add a proton.

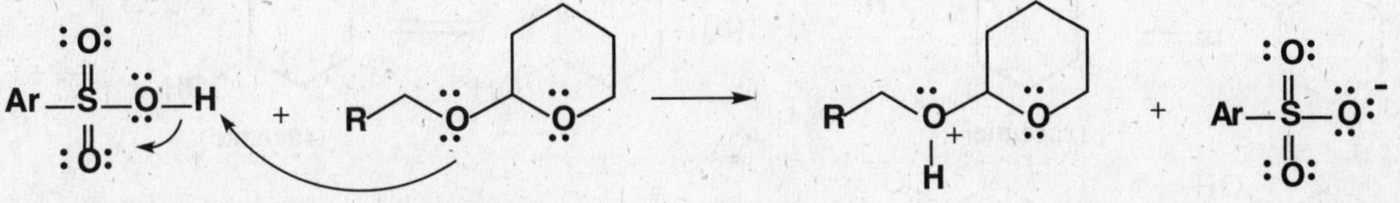

Step 2: Break a bond to give stable molecules or ions.

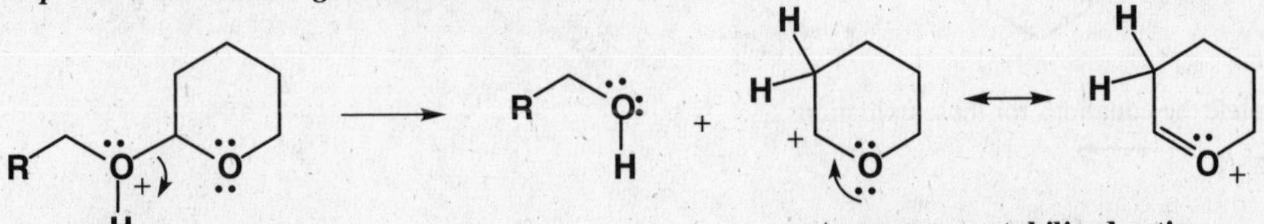

A resonance-stabilized cation

Step 3: Take a proton away.

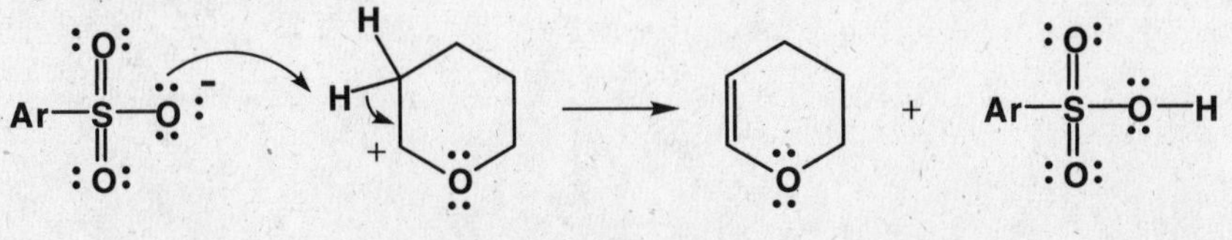

Step 4: The final step involves equilibration with water to create a cyclic hemiacetal structure. This actually involves several steps, which are functionally the reverse of steps 1-3 above, except that H_2O is used in place of the alcohol.

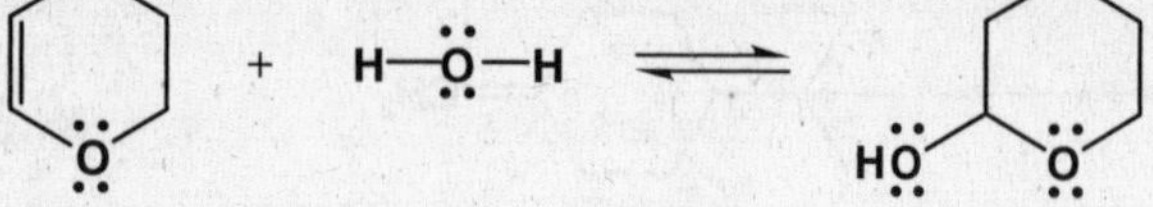

Problem 16.8 Acid-catalyzed hydrolysis of an imine gives an amine and an aldehyde or ketone. When one equivalent of acid is used, the amine is converted to an ammonium salt. Write structural formulas for the products of hydrolysis of the following imines using one equivalent of HCl:

(a)

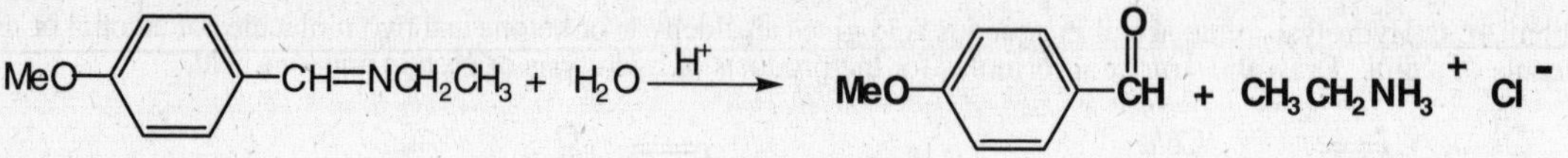

(b)

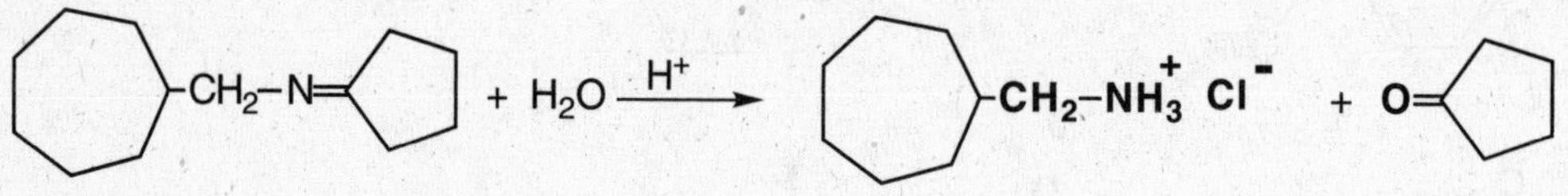

Problem 16.9 Predict the position of the following equilibrium.

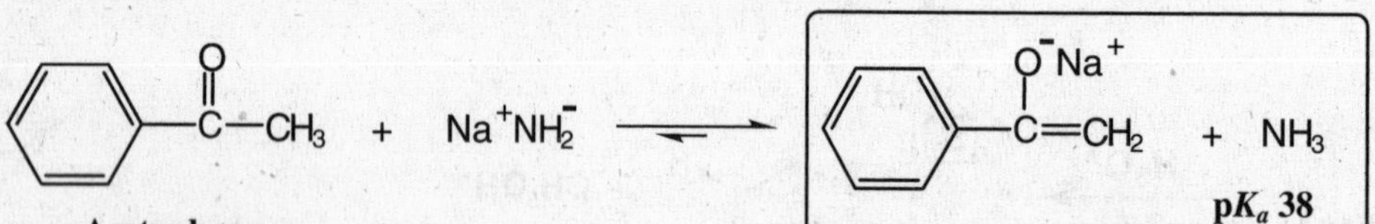

Acetophenone

$pK_a \sim 20$

Like all acid-base reactions, equilibrium favors formation of the weaker acid/weaker base. Because ammonia, with pK_a of 38, is a much weaker acid than acetophenone ($pK_a \sim 20$) the position of equilibrium will be strongly to the right.

Problem 16.10 Draw a structural formula for the keto form of each enol.

(a)

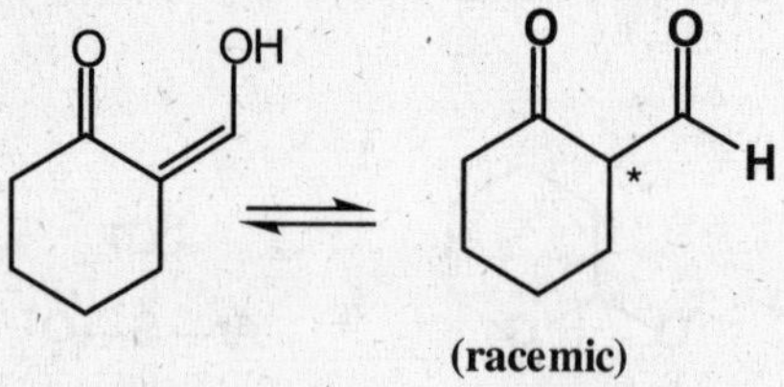

(racemic)

(b)

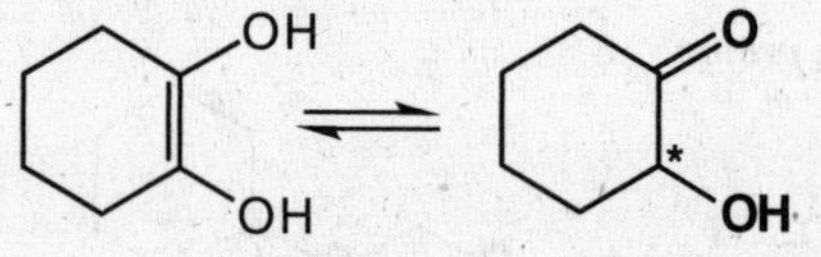

(racemic)

(c)

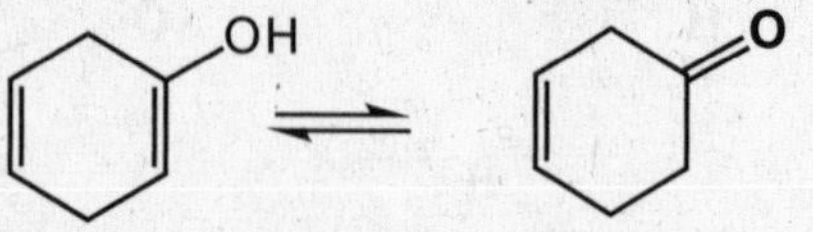

Problem 16.11 Complete the equations for these oxidations.

(a) Hexanal + H_2O_2 ⟶

$$CH_3(CH_2)_4\overset{O}{\overset{\|}{C}}H \ + \ H_2O_2 \longrightarrow CH_3(CH_2)_4\overset{O}{\overset{\|}{C}}OH \ + \ H_2O$$

(b) 3-Phenylpropanal + Tollens' reagent ⟶

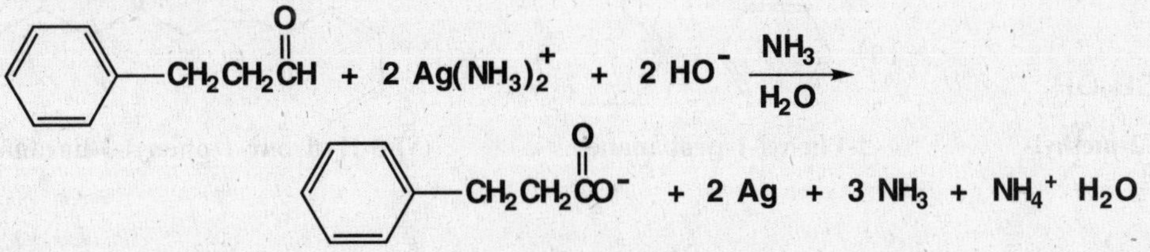

$$\text{Ph-CH}_2\text{CH}_2\overset{\displaystyle O}{\overset{\|}{\text{CH}}} + 2\ \text{Ag}(\text{NH}_3)_2^+ + 2\ \text{HO}^- \xrightarrow[\text{H}_2\text{O}]{\text{NH}_3}$$

$$\text{Ph-CH}_2\text{CH}_2\overset{\displaystyle O}{\overset{\|}{\text{C}}}\text{O}^- + 2\ \text{Ag} + 3\ \text{NH}_3 + \text{NH}_4^+\ \text{H}_2\text{O}$$

Problem 16.12 What aldehyde or ketone gives these alcohols on reduction with $NaBH_4$?

(a)

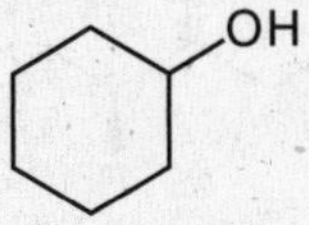

(b)

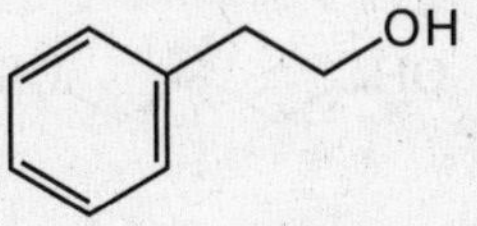

(c)

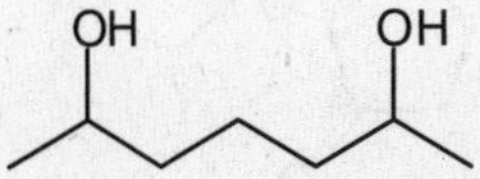

(Mixture of 3 stereoisomers)

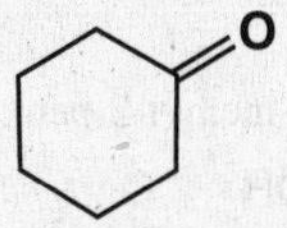

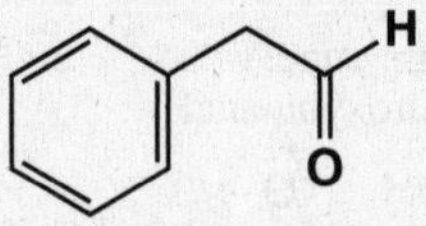

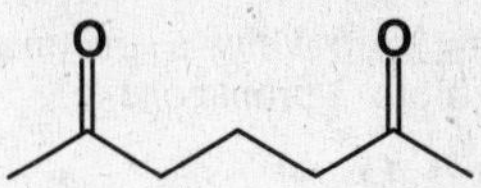

Problem 16.13 Complete the following reactions:

(a)

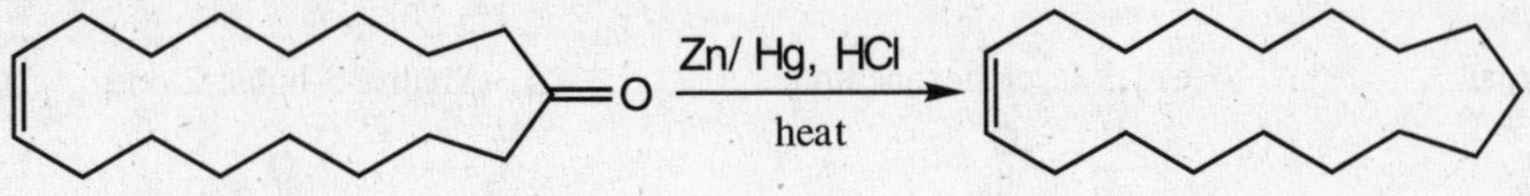

Zn/ Hg, HCl
heat

Civetone
(from the civet cat; used in perfumery)

The above reaction is an example of a Clemmensen reduction.

(b)

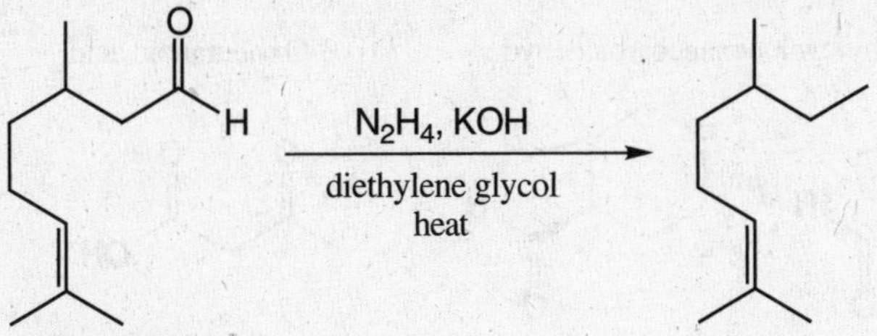

N_2H_4, KOH
diethylene glycol
heat

Citronellal
(from citronella and
lemon grass oils)

The above reaction is an example of a Wolff-Kishner reduction.

Structure and Nomenclature
Problem 16.14 Name each compound, showing stereochemistry where relevant.

(a)

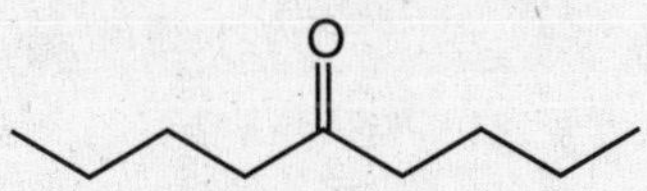

5-Nonanone
(dibutyl ketone)

(b)

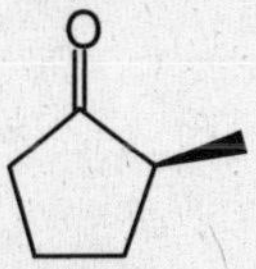

(S)-2-Methylcyclo-
pentanone

(c)

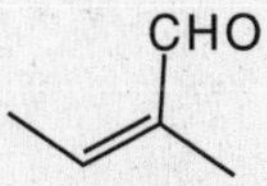

(Z)-2-Methyl-2-butenal

(d)

(R)-3-Hydroxy-2-methyl-
propanal

(e)

1-Phenyl-1-propanone

(f)

(S)-5-Hydroxy-1-phenyl-3-hexanone

(g)

2-Propyl-1,3-cyclopentanedione

(h) OHC   CHO

Pentanedial

(i)

(R)-2-Bromo-3-pentanone

<u>Problem 16.15</u> Draw a structural formulas for each compound.

(a) 1-Chloro-2-propanone

(b) 3-Hydroxybutanal

(c) 4-Hydroxy-4-methyl-2-pentanone

(d) 3-Methyl-3-phenylbutanal

(e) 1,3-Cyclohexanedione

(f) 3-Methyl-3-buten-2-one

(g) 5-Oxohexanal

(h) 2,2-Dimethylcyclohexanecarbaldehyde

(i) 3-Oxobutanoic acid

or

<u>Problem 16.16</u> The infrared spectrum of Compound A, $C_6H_{12}O$, shows a strong, sharp peak at 1724 cm^{-1}. From this information and its ^{1}H-NMR spectrum, deduce the structure of compound A.

Compound A:

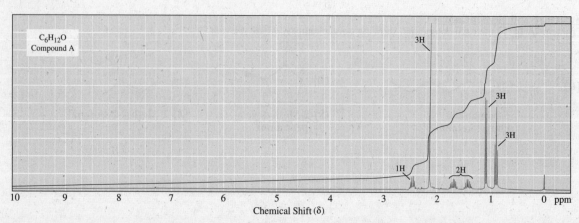

$C_6H_{12}O$
Compound A

3H

3H

3H

1H   2H

Chemical Shift (δ)

Compound A has an aldehyde or ketone function judging from the peak at 1724 cm⁻¹ in the IR spectrum. The rest of the structure can be deduced from the ¹H-NMR, with assignments as listed below.

$$\underset{0.89}{CH_3}-\underset{1.42,1.68}{CH_2}-\underset{\underset{\underset{CH_3}{|}}{1.08}}{\underset{*}{CH}}-\underset{2.45}{\overset{\overset{O}{\|}}{C}}-\underset{2.1}{CH_3}$$

The -CH₂- hydrogens with signals at δ 1.42 and δ 1.68 above are not equivalent because they are adjacent to a tetrahedral chiral center in the molecule.

<u>Problem 16.17</u> Following are ¹H-NMR spectra for compounds B ($C_6H_{12}O_2$), and C ($C_6H_{10}O$). On warming in dilute acid, compound B is converted to compound C. Deduce the structural formulas for compounds B and C.

Compound B:

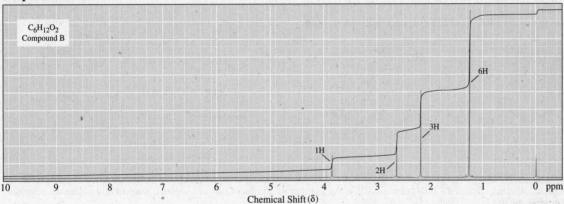

Compound C:

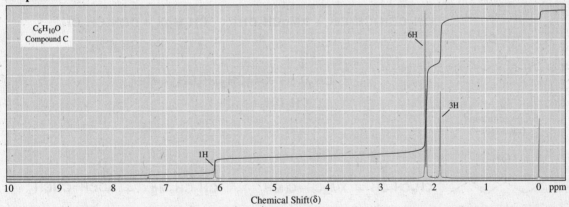

From the molecular formula it is clear that compound B undergoes acid-catalyzed dehydration to create compound C. Thus, compound B must have an -OH group. Furthermore, because its index of hydrogen deficiency is one, compound B must have one ring or π bond. However, compound C has an index of hydrogen deficiency of two, so it must have two π bonds or rings. The ¹H-NMR spectrum of compound B shows all singlets. Especially helpful are the methyl group resonances; the singlet integrating to 6H at δ 1.22 and the singlet integrating to 3H at δ 2.18. This latter signal is assigned as a methyl ketone by comparison to the previous problem. The other two methyl groups are equivalent. There is the -OH hydrogen at δ 3.85 and a -CH₂- resonance at δ 2.62. The only structure consistent with these signals is 4-hydroxy-4-methyl-2-pentanone.

$$\underset{CH_3}{\overset{\overset{\displaystyle 1.22}{\displaystyle CH_3}}{\underset{1.22}{\underset{\underset{3.85}{|}}{\overset{\displaystyle |}{C}}}}}\ \underset{2.62}{CH_2}-\underset{}{\overset{\overset{O}{\|}}{C}}-\underset{2.18}{CH_3}$$
$$\underset{OH}{}$$

Compound B

Upon dehydration, Compound B would be converted to 4-methyl-3-pentene-2-one.

$$\underset{\text{2.15}}{CH_3}-\underset{\text{6.10}}{C}=\underset{}{CH}-\underset{\text{O}}{\overset{\text{O}}{C}}-\underset{\text{1.86}}{CH_3}$$

with CH_3 at 2.15

Compound C

The structure of 4-methyl-3-pentene-2-one is entirely consistent with the ^{1}H-NMR spectrum, especially the presence of the signal at δ 6.10 (-CH=C) and the methyl singlets at δ 1.86 (integrating to 3H) and δ 2.15 (integrating to 6H).

Addition of Carbon Nucleophiles

Problem 16.18 Draw structural formulas for the product formed by treating each compound with propylmagnesium bromide followed by aqueous HCl.

The products after acid hydrolysis are given in bold.

(a) CH_2O (b) [epoxide structure]   (c) [3-pentanone structure]   (d) [2-cyclopentenone structure]

[n-butanol with OH]   [pentanol with OH]   [tertiary alcohol with OH]   [cyclopentene with OH and propyl]

Problem 16.19 Suggest a synthesis for the following alcohols starting from an aldehyde or ketone and an appropriate Grignard reagent. Below each target molecule is the number of combinations of Grignard reagent and aldehyde or ketone that might be used.

(a) [alcohol structure with OH and *]

     $CH_3\overset{O}{C}CH_2CH_2CH_3 + XMgCH_2CH_3$ $CH_3MgX + \underset{CH_2CH_3}{\overset{O}{C}}CH_2CH_2CH_3$

           $CH_3\underset{CH_2CH_3}{\overset{O}{C}} + XMgCH_2CH_2CH_3$

3 Combinations
(racemic)

(b) [alcohol structure with OH and *]

     $CH_3CH_2\overset{O}{C}H + XMgCH=CHCH_3$ $CH_3CH_2MgX + H\overset{O}{C}CH=CHCH_3$

2 Combinations
(racemic)

(c) [diaryl alcohol structure with OH, *, and OCH₃]

     $CH_3O-\text{[C}_6\text{H}_4\text{]}-\overset{O}{C}H + XMg-\text{[C}_6\text{H}_5\text{]}$

2 Combinations
(racemic)

     $CH_3O-\text{[C}_6\text{H}_4\text{]}-MgX + H\overset{O}{C}-\text{[C}_6\text{H}_5\text{]}$

Note that each product is actually a racemic mixture.

Problem 16.20 Show how to synthesize the following alcohol using 1-bromopropane, propanal, and ethylene oxide as the only sources of carbon atoms.

(racemic)

This synthesis is divided into two stages. In the first stage, 1-bromopropane is treated with magnesium to form a Grignard reagent and then with propanal followed by hydrolysis in aqueous acid to give 3-hexanol.

3-Hexanol

In the second stage, 3-hexanol is treated with thionyl chloride followed by magnesium in ether to form a Grignard reagent. Treatment of this Grignard reagent with ethylene oxide followed by hydrolysis in aqueous acid gives 3-ethyl-1-hexanol. A chiral center is created in this synthesis. The product is a racemic mixture. Note that the alcohol could have also been treated with PBr$_3$ followed by Mg to make a bromo Grignard.

3-Ethyl-1-hexanol
(racemic)

Problem 16.21 1-Phenyl-2-butanol is used in perfumery. Show how to synthesize this alcohol from bromobenzene, 1-butene, and any necessary inorganic reagents.

Bromobenzene   1-Butene   1-Phenyl-2-butanol
(racemic)

(a) Bromobenzene is treated with magnesium in diethyl ether to form phenylmagnesium bromide, in preparation for part (c).

(b) Treatment of 1-butene with a peroxycarboxylic acid gives 1,2-epoxybutane.

(c) Treatment of phenylmagnesium bromide with 1,2-epoxybutane followed by hydrolysis in aqueous acid gives 1-phenyl-2-butanol.

A chiral center is created in this synthesis. The product is a racemic mixture. A Gilman reagent could also have been used in this step.

<u>Problem 16.22</u> With organolithium and organomagnesium compounds, approach to the carbonyl carbon from the less hindered direction is generally preferred. Assuming this is the case, predict the structure of the major product formed by reaction of methylmagnesium bromide with 4-*tert*-butylcyclohexanone.

The bulky *tert*-butyl group lies in an equatorial position. Approach to the carbonyl carbon may be by way of a pseudo-axial direction or a pseudo-equatorial direction. The less hindered approach is from the pseudo-equatorial direction which then places the incoming group equatorial and the -OH axial.

Less hindered approach   **Less hindered approach**   **Less hindered approach**

Wittig Reaction
<u>Problem 16.23</u> Draw structural formulas for (1) the triphenylphosphonium salt formed by treatment of each haloalkane with triphenylphosphine, (2) the phosphonium ylide formed by treatment of each phosphonium salt with butyllithium, and (3) the alkene formed by treatment of each phosphonium ylide with acetone.

In the following, the triphenylphosphonium salts are listed as (1), the phosphonium ylides as (2), and the alkene formed upon reaction with acetone as (3).

(a)

(1) $CH_3-\overset{+}{CH}-PPh_3Br^-$
          $|$
          CH_3

(2) $CH_3-\overset{..-}{\underset{|}{C}}-\overset{+}{PPh_3}$
          CH_3

(3) $\underset{CH_3}{\overset{CH_3}{>}}C=C\underset{CH_3}{\overset{CH_3}{<}}$

(b)

(1) $CH_2=CHCH_2\overset{+}{PPh_3}Br^-$

(2) $CH_2=CH\overset{..-}{C}H\overset{+}{PPh_3}$

(3) $\underset{CH_3}{\overset{CH_3}{>}}C=C\underset{CH=CH_2}{\overset{H}{<}}$

(c)

(1) $-CH_2\overset{+}{PPh_3}Cl^-$

(2) $-\overset{..-}{C}H\overset{+}{PPh_3}$

(3) $\underset{CH_3}{\overset{CH_3}{>}}C=C\overset{H}{<}$

(d) (structure: ethyl chloroacetate, ClCH₂C(=O)OCH₂CH₃)

(1) Cl⁻ Ph₃P⁺CH₂COCH₂CH₃   (with O above)

(2) Ph₃P⁺CH⁻COCH₂CH₃   (with O above)

(3) (structure)
$$CH_3\text{—}C=C\text{—}COCH_2CH_3$$
with CH₃ groups on left carbon and H on right carbon, and C=O on the ester.

(e) (structure: 1-bromobutane)

(1) CH₃CH₂CH₂CH₂P⁺Ph₃ Br⁻

(2) CH₃CH₂CH₂CH⁻P⁺Ph₃

(3) (structure)
$$CH_3\text{—}C=C\text{—}CH_2CH_2CH_3$$
with CH₃ groups on left carbon and H on right carbon.

(f) Ph—CH=CH—CH₂Cl

(1) (phenyl)—CH=CHCH₂P⁺Ph₃ Cl⁻

(2) (phenyl)—CH=CHCH⁻P⁺Ph₃

(3) (structure)
$$CH_3\text{—}C=C\text{—}CH=CH\text{—(phenyl)}$$
with CH₃ groups on left carbon and H on right carbon.

Problem 16.24 Show how to bring about the following conversions using a Wittig reaction.

(a) (structure: acetone → 2-methyl-2-heptene)

Start with 1-halopentane and, by treatment with triphenylphosphine followed by butyllithium, convert it to a Wittig ylide. Treatment of this ylide with acetone gives the desired alkene.

$$CH_3(CH_2)_3CH_2I \xrightarrow[\text{2) BuLi}]{\text{1) PPh}_3} CH_3(CH_2)_3CH^-P^+Ph_3 \xrightarrow{CH_3CCH_3}$$

(with O above CH₃CCH₃)

$$CH_3C=CH(CH_2)_3CH_3$$
(with CH₃ above the left carbon)

(b) (structure: acetophenone → α-methylstyrene)

Treatment of acetophenone with the Wittig ylide derived from methyl iodide gives the desired alkene.

$$CH_3I \xrightarrow[\text{2) BuLi}]{\text{1) PPh}_3} CH_2^-\text{—}P^+Ph_3 \xrightarrow{\text{(acetophenone)}} \text{(product)}$$

(structure: phenyl—C(CH₃)=CH₂)

(c)

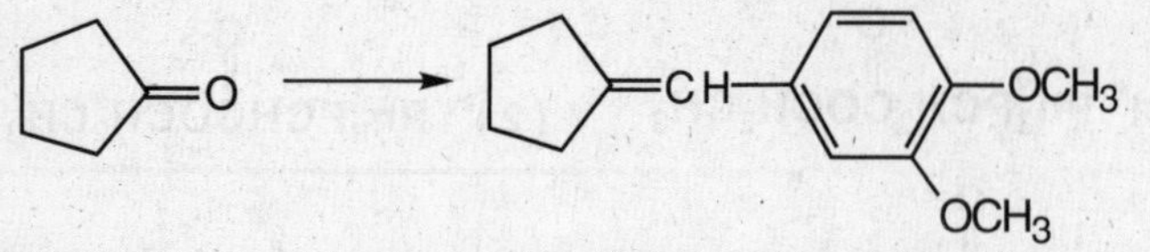

Treatment of cyclopentanone with the Wittig ylide derived from 3,4-dimethoxybenzyl bromide gives the desired alkene.

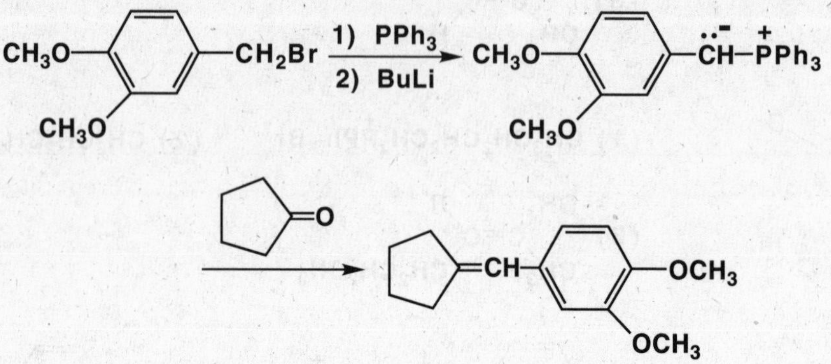

CH_3O—⟨ ⟩—CH_2Br $\xrightarrow[\text{2) BuLi}]{\text{1) PPh}_3}$ CH_3O—⟨ ⟩—$\overset{..-}{C}H$—$\overset{+}{P}Ph_3$

Problem 16.25 The Wittig reaction can be used for the synthesis of conjugated dienes, as for example, 1-phenyl-1,3-pentadiene.

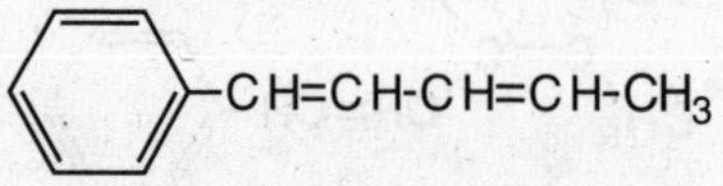

⟨ ⟩—CH=CH-CH=CH-CH₃

1-Phenyl-1,3-pentadiene

Propose two sets of reagents that might be combined in a Wittig reaction to give this conjugated diene.

Allylic halides can be used as starting materials for preparation of Wittig reagents. Below are combinations of allylic halides and aldehydes that can be used to prepare the desired diene.

⟨ ⟩—$\overset{\displaystyle O}{\overset{\|}{C}}$H + Br-CH₂-CH=CH-CH₃   ⟨ ⟩—CH=CH-CH₂Br + $\overset{\displaystyle O}{\overset{\|}{H C}}$-CH₃

Benzaldehyde   **1-Bromo-2-butene**   **3-Bromo-1-phenylpropene**   **Ethanal**

These other sets of reagents may also be used as shown below.

⟨ ⟩—CH₂Br + $\overset{\displaystyle O}{\overset{\|}{H C}}$-CH=CH-CH₃   ⟨ ⟩—CH₂=CH-$\overset{\displaystyle O}{\overset{\|}{C}}$H + BrCH₂CH₃

Benzyl bromide   **2-Butenal**   **3-Phenyl-2-propenal**   **Bromoethane**
                                                  (Cinnamaldehyde)

Problem 16.26 Wittig reactions with the following α-chloroethers can be used for the synthesis of aldehydes and ketones:

$ClCH_2OCH_3$ $Cl\overset{\overset{\displaystyle CH_3}{|}}{C}HOCH_3$

(A)   (B)

(a) Draw the structure of the triphenylphosphonium salt and Wittig reagent formed from each chloroether.

(A) $Ph_3\overset{+}{P}$-$\overset{..-}{C}HOCH_3$ **(B)** $Ph_3\overset{+}{P}$-$\overset{..-}{C}OCH_3$
                                                                 $\overset{\displaystyle |}{CH_3}$

(b) Draw the structural formula of the product formed by treatment of each Wittig reagent with cyclopentanone. Note that the functional group is an enol ether, or, alternatively, a vinyl ether.

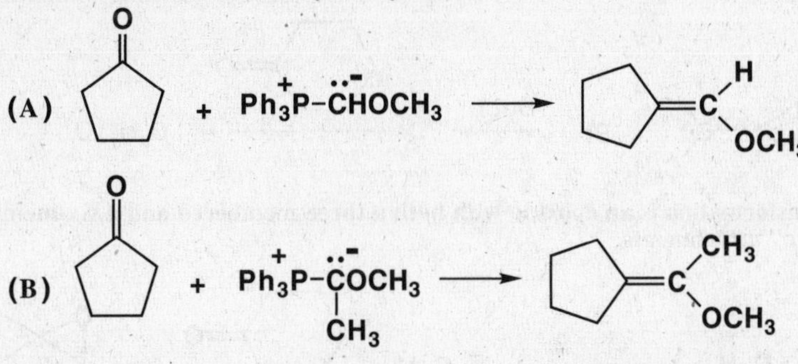

(A)

(B)

(c) Draw the structural formula of the product formed on acid-catalyzed hydrolysis of each enol ether from part (b).

The acid catalyzed hydrolysis leads to an aldehyde and methyl ketone for (A) and (B), respectively.

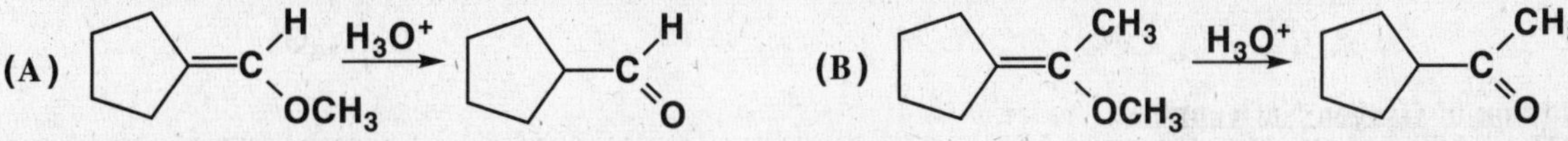

(A)   (B)

<u>Problem 16.27</u> It is possible to generate sulfur ylides in a manner similar to that used to produce phosphonium ylides. For example, treating a sulfonium salt with a strong base gives the sulfur ylide.

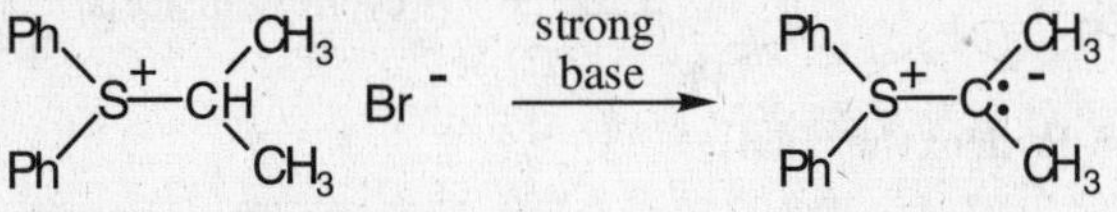

A sulfonium bromide salt   A sulfur ylide

Sulfur ylides react with ketones to give epoxides. Suggest a mechanism for this reaction.

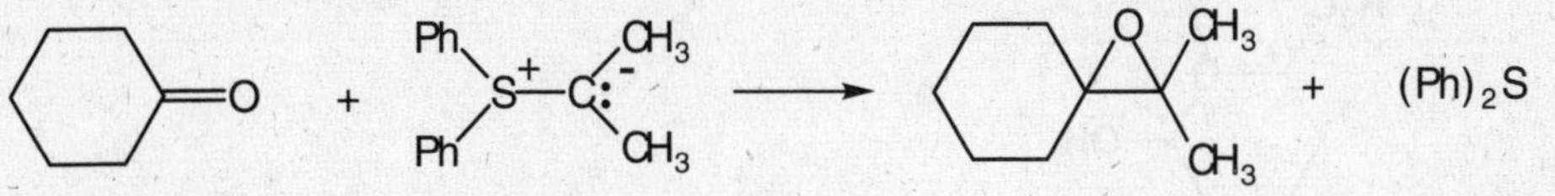

Step 1: Make a bond between a nucleophile and an electrophile.

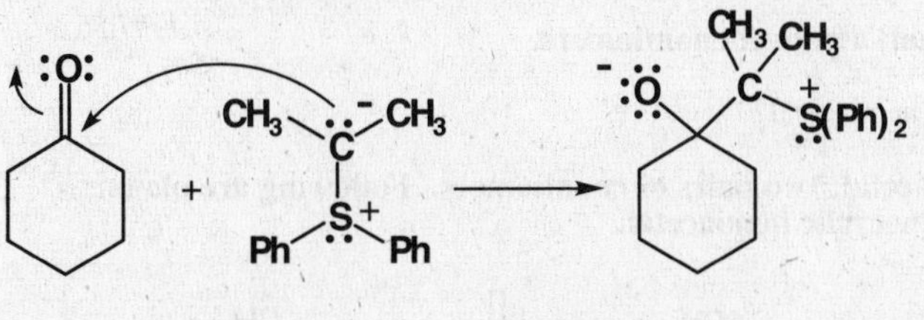

Step 2: Make a bond between a nucleophile and an electrophile and simultaneously break a bond to give stable molecules or ions.

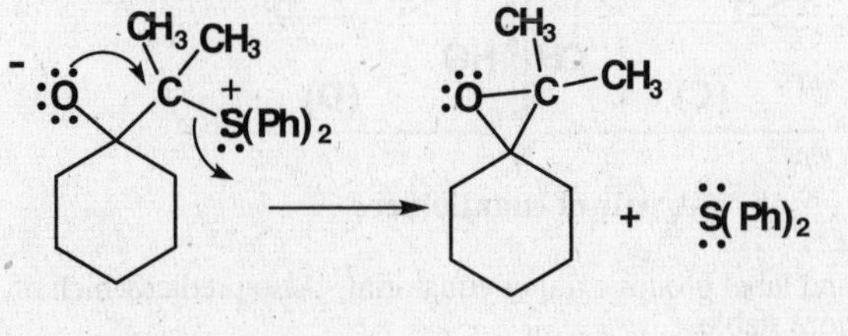

Problem 16.28 Propose a structural formula for Compound D and for the product, $C_9H_{14}O$, formed in this reaction sequence.

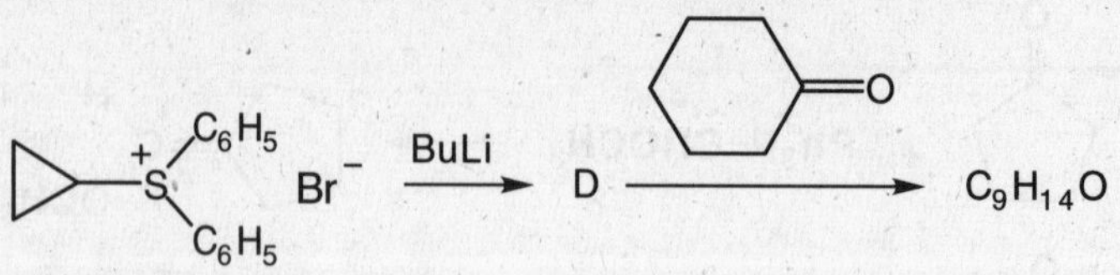

The product of this transformation is an epoxide with both a three-membered and a six-membered ring bonded through so-called "spiro" attachments.

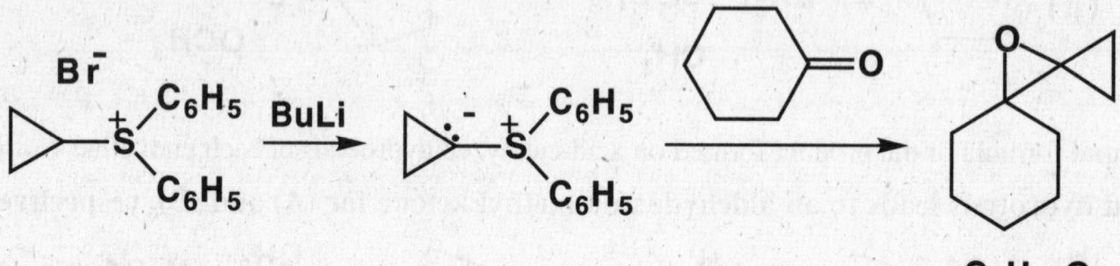

$$C_9H_{14}O$$

Addition of Oxygen Nucleophiles

Problem 16.29 5-Hydroxyhexanal forms a six-membered cyclic hemiacetal, which predominates at equilibrium in aqueous solution.

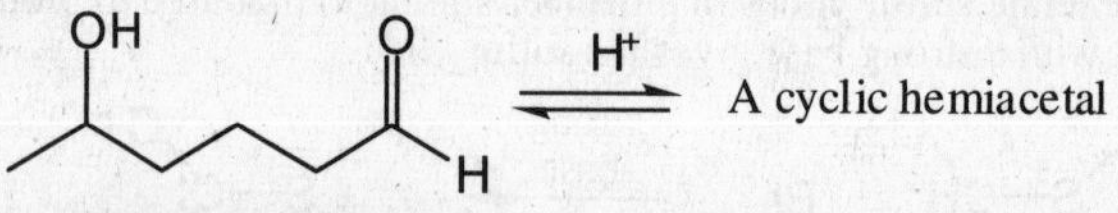

$$\xrightarrow{H^+}$$ A cyclic hemiacetal

5-Hydroxyhexanal

(a) Draw a structural formula for this cyclic hemiacetal.

5-Hydroxyhexanal forms a six-membered cyclic hemiacetal.

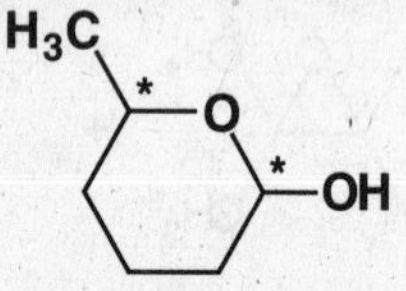

(b) How many stereoisomers are possible for 5-hydroxyhexanal?

Two stereoisomers are possible for 5-hydroxyhexanal; a pair of enantiomers.

(c) How many stereoisomers are possible for this cyclic hemiacetal?

Four stereoisomers are possible for the cyclic hemiacetal; two pairs of enantiomers. Following are planar hexagon formulas for each pair of enantiomers of the cyclic hemiacetal.

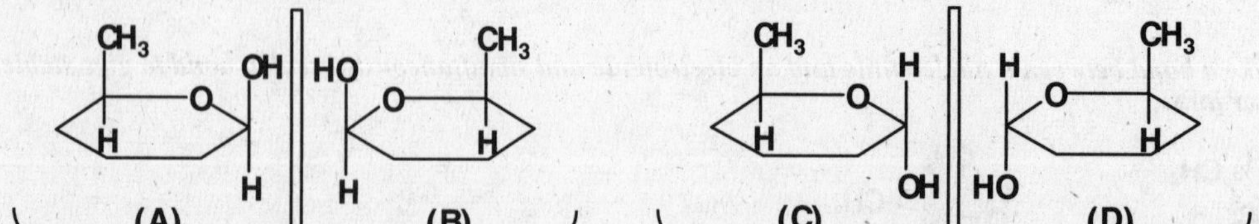

(d) Draw alternative chair conformations for each stereoisomer and label groups axial or equatorial. Also predict which of the alternative chair conformations for each stereoisomer is the more stable.

Alternative chair conformations are drawn for (A), one of the *cis* enantiomers, and for (C), one of the *trans* enantiomers. Methyl and hydroxyl groups are 7.28 kJ/mol and 3.9 kJ/mol more stable in the equatorial position, respectively. For (A), the diequatorial chair is the more stable by 7.28 + 3.9 = 11.2 kJ/mol. For (C), the chair with the methyl group is equatorial (structure on the left) is 7.28 - 3.9 = 3.4 kJ/mol more stable.

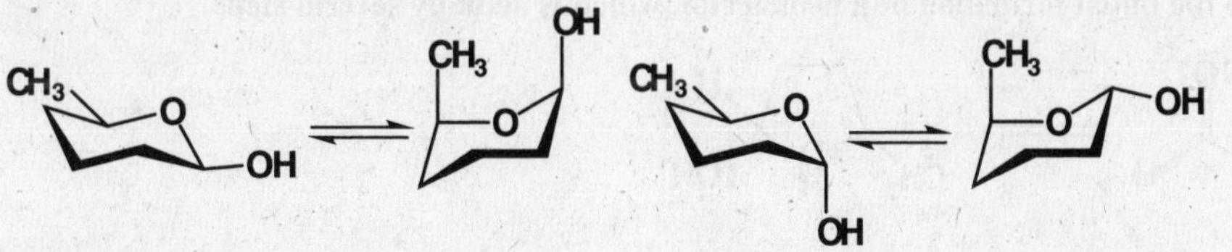

Alternative chair conformations of A   **Alternative chair conformations of C**

<u>Problem 16.30</u> Draw structural formulas for the hemiacetal and then the acetal formed from each pair of reactants in the presence of an acid catalyst.

(a)

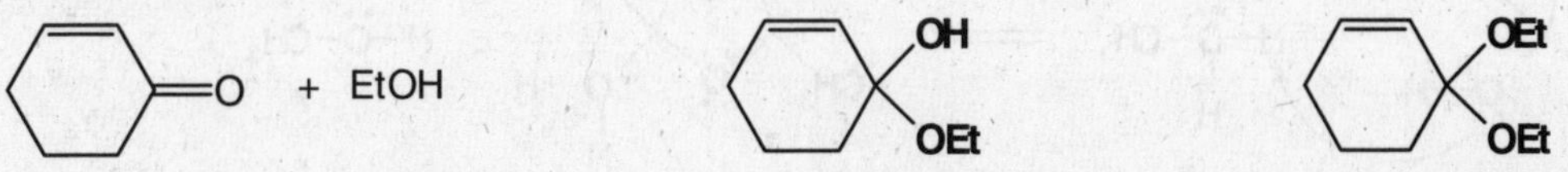

(b)

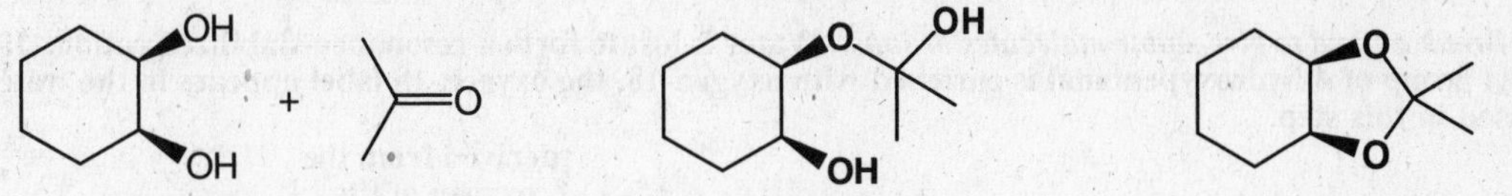

(c)

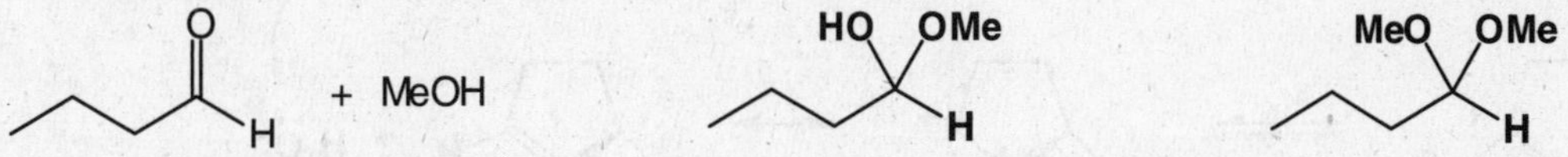

<u>Problem 16.31</u> Draw structural formulas for the products of hydrolysis of the following acetals in aqueous HCl.

(a)

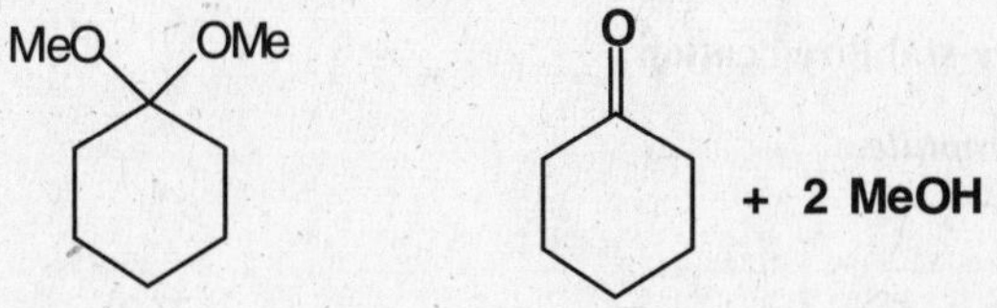

(b)

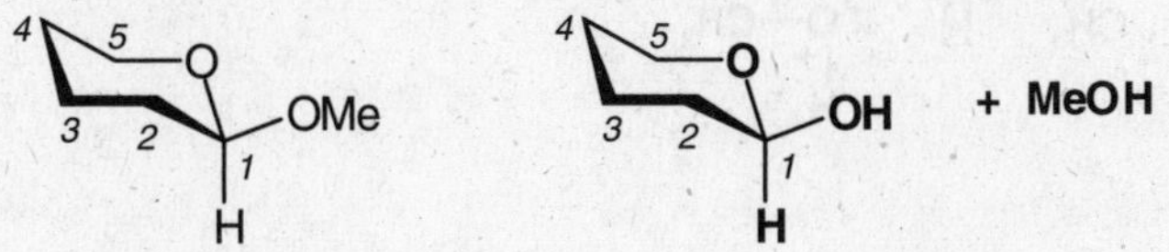

(c)

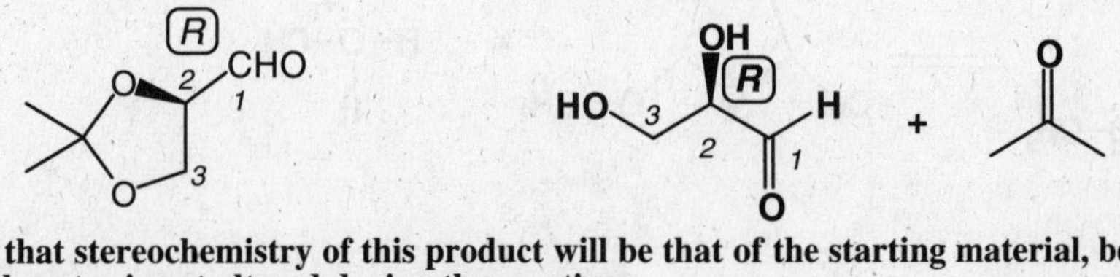

Note that stereochemistry of this product will be that of the starting material, because the configuration of the chiral center is not altered during the reaction.

<u>Problem 16.32</u> Propose a mechanism to account for the formation of a cyclic acetal from 4-hydroxypentanal and one equivalent of methanol. If the carbonyl oxygen of 4-hydroxypentanal is enriched with oxygen-18, do you predict that the oxygen label appears in the cyclic acetal or in the water?

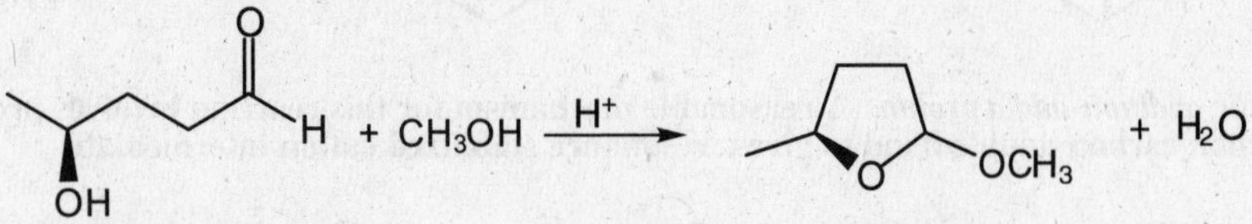

4-Hydroxypentanal   A cyclic acetal

Step 1: Propose the initial formation of a hemiacetal, which is actually several steps.

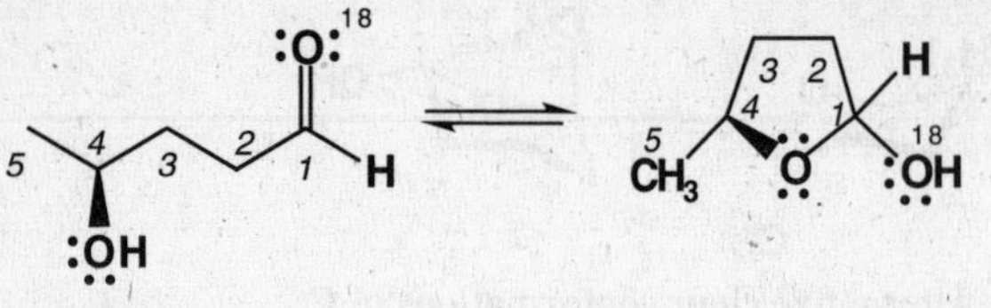

Step 2: Add a proton. The hemiacetal is protonated on the -OH group.

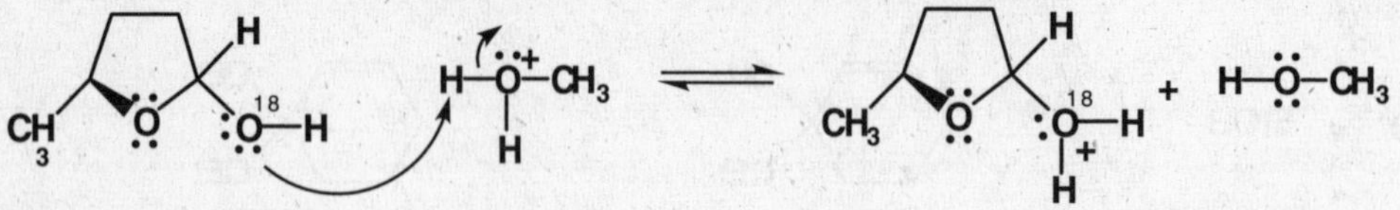

Step 3: Break a bond to give stable molecules or ions. Water is lost to form a resonance-stabilized cation. If the carbonyl group of 4-hydroxypentanal is enriched with oxygen-18, the oxygen-18 label appears in the water that is released in this step.

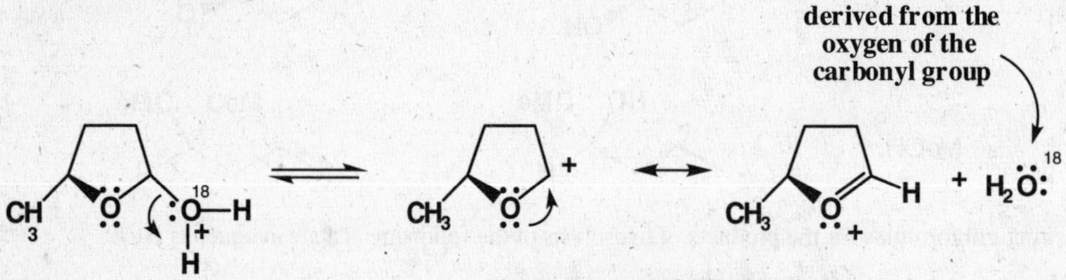

derived from the
oxygen of the
carbonyl group

A resonance-stabilized cation

Step 4: Make a bond between a nucleophile and an electrophile.

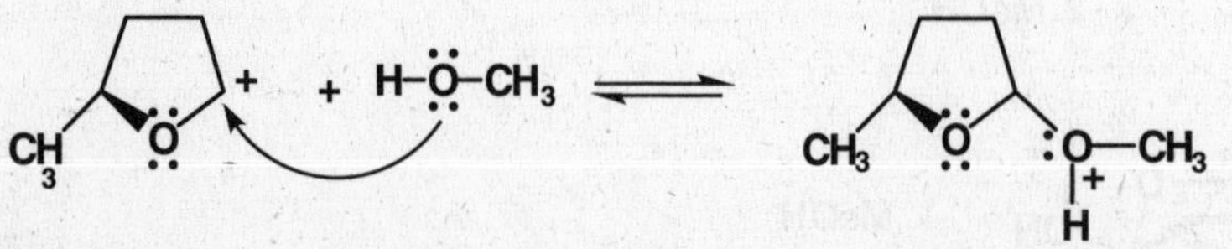

Step 5: Take a proton away.

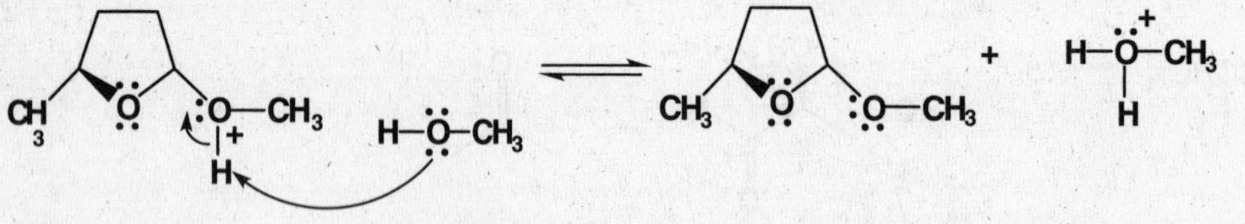

<u>Problem 16.33</u> Propose a mechanism for this acid-catalyzed hydrolysis.

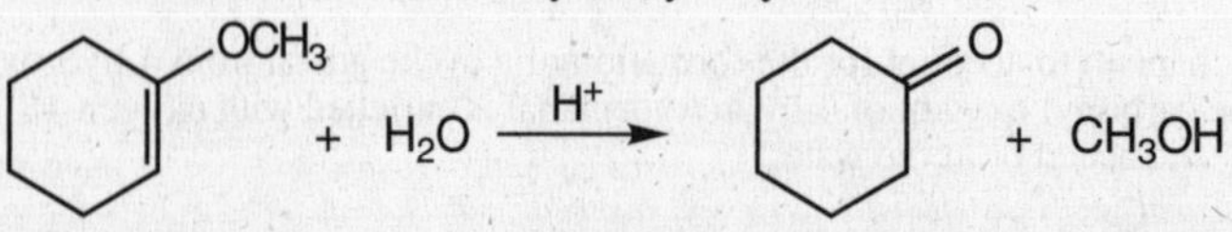

Step 1: Electrophilic addition-add a proton. A reasonable mechanism for this reaction involves protonation of the π bond of the carbon-carbon double bond to give a resonance stabilized cation intermediate.

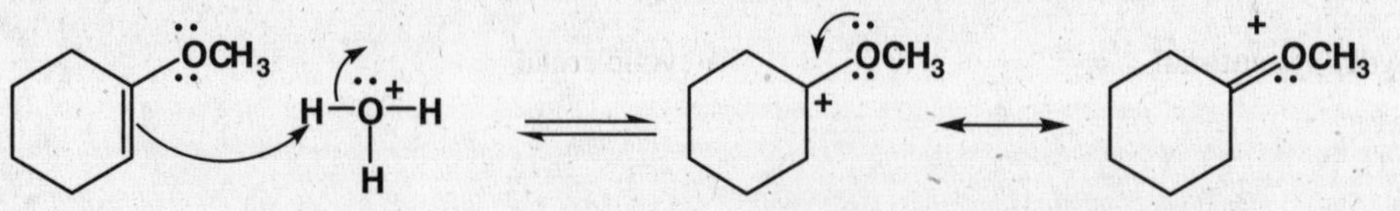

A resonance stabilized cation

Step 2: *Make a bond between a nucleophile and an electrophile.* Water reacts with the cation to form a hemiacetal after loss of a proton.

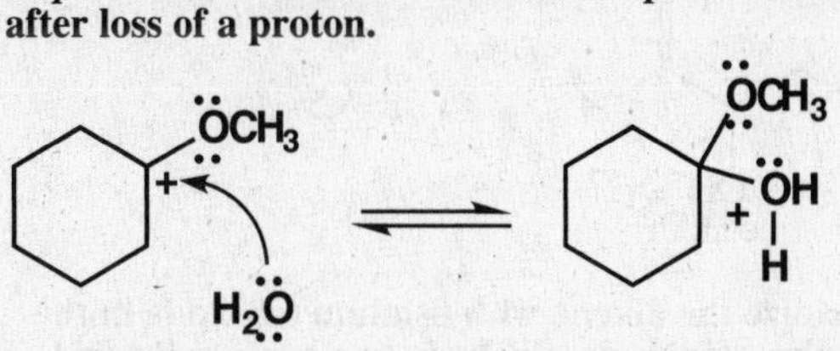

Step 3: *Take a proton away.*

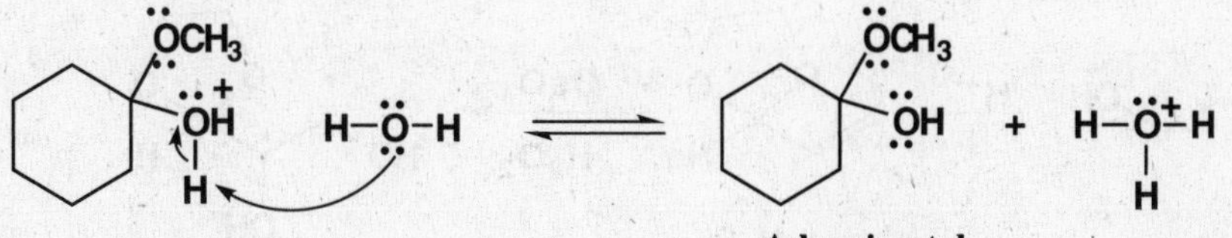

A hemiacetal

Step 4: *Add a proton.*

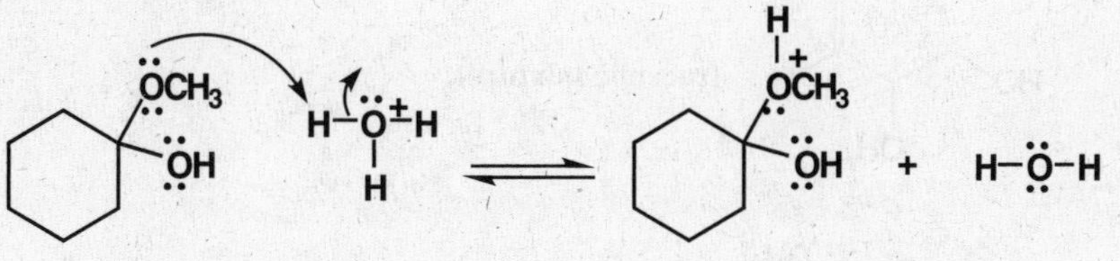

Step 5: *Break a bond to give stable molecules or ions.* Methanol departs.

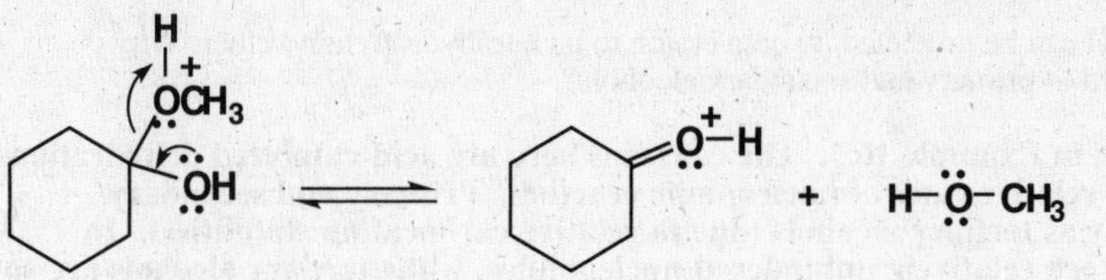

Step 6: *Take a proton away.*

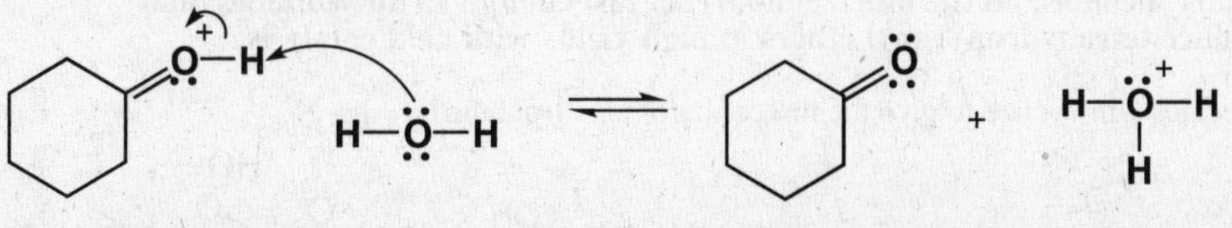

<u>Problem 16.34</u> In Section 11.5 we saw that ethers, such as diethyl ether and tetrahydrofuran, are quite resistant to the action of dilute acids and require hot concentrated HI or HBr for cleavage. However, acetals in which two ether groups are linked to the same carbon undergo hydrolysis readily, even in dilute aqueous acid. How do you account for this marked difference in chemical reactivity toward dilute aqueous acid between ethers and acetals?

The first step of the cleavage reactions in acid for both ethers and acetals is protonation of an oxygen to form an oxonium ion. For acetals, the following step is cleavage of a carbon-oxygen bond to form a resonance-stabilized cation. For ethers, similar cleavage occurs, but the cation formed has no comparable resonance stabilization. Therefore, it is the resonance-stabilization of the cation intermediate formed during the cleavage of acetals that lowers the activation energy for their hydrolysis much below that for the hydrolysis of ethers.

Problem 16.35 Show how to bring about the following conversion:

(racemic)

The most convenient way to convert an alkene to a glycol is to oxidize the alkene with osmium tetroxide in the presence of hydrogen peroxide. These conditions, however, will also oxidize an aldehyde to a carboxylic acid. Therefore, it is necessary to first protect the aldehyde by transformation to an acetal. In the following answer, ethylene glycol is the protecting agent.

+ H$_2$O

(racemic mixture)

Problem 16.36 A primary or secondary alcohol can be protected by conversion to its tetrahydropyranyl ether. Why is formation of THP ethers by this reaction limited to primary and secondary alcohols?

It will help to refer to the mechanism given in Example 16.7. The concerns here are acid-catalyzed dehydration of the alcohol as a competing reaction and relative rates of nucleophilic reaction. Primary and secondary alcohols do not dehydrate nearly as readily as tertiary alcohols (due to relative carbocation stabilities). In addition, primary and secondary alcohols are relatively unhindered nucleophiles, while tertiary alcohols are so sterically hindered that they are sluggish nucleophiles. The net result is that tertiary alcohols dehydrate too readily, and react too slowly as nucleophiles, to be converted efficiently into tetrahydropyranyl ethers in acidic conditions. Primary and secondary alcohols, on the other hand, react fast enough as nucleophiles and dehydrate slowly enough to produce tetrahydropyranyl ethers in high yields with acid catalysis.

Problem 16.37 Which of these molecules will cyclize to give the insect pheromone frontalin?

Frontalin   (A)   (B)   (C)

The answer is B. Using models may help with this answer.

Addition of Nitrogen Nucleophiles

Problem 16.38 Draw a structural formula for the product of each acid-catalyzed reaction:

(a) Phenylacetaldehyde + hydrazine ----->

$$PhCH_2\overset{\overset{\displaystyle O}{\|}}{C}H + H_2N-NH_2 \longrightarrow PhCH_2CH=N-NH_2 + H_2O$$

(b) Cyclopentanone + semicarbazide ----->

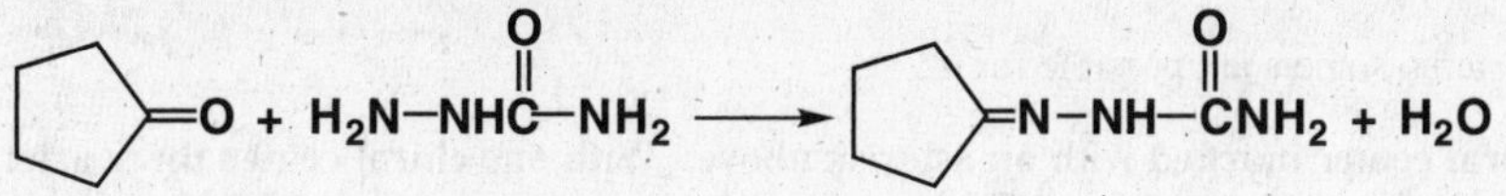

(c) Acetophenone + 2,4-dinitrophenylhydrazine ----->

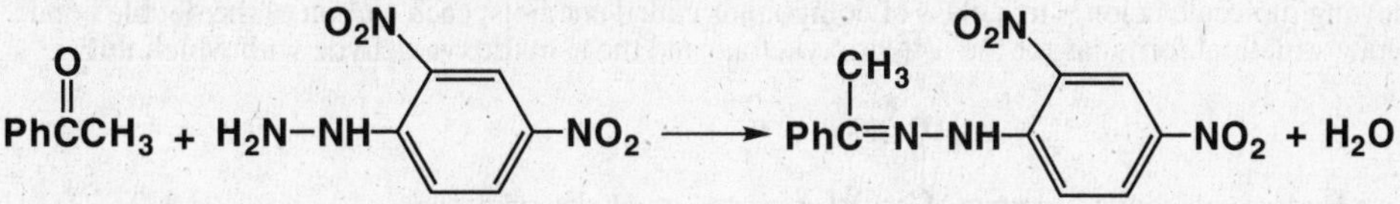

(d) Benzaldehyde + hydroxylamine ----->

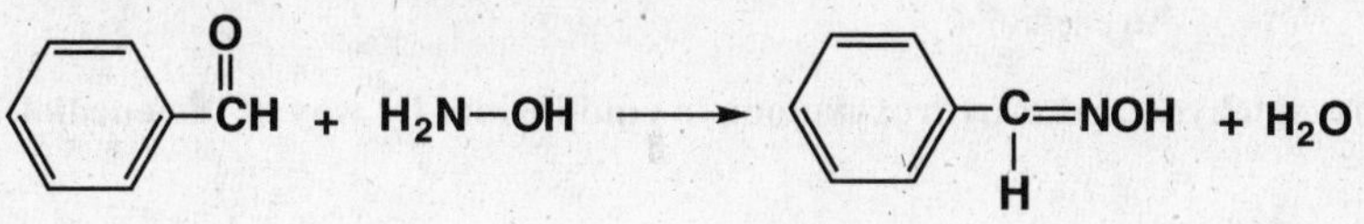

Problem 16.39 Following are structural formulas for amphetamine and methamphetamine.

(a)

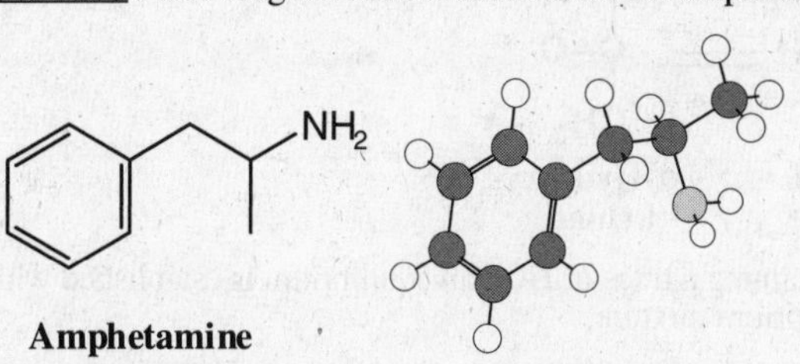

Amphetamine
(racemic)

(b)

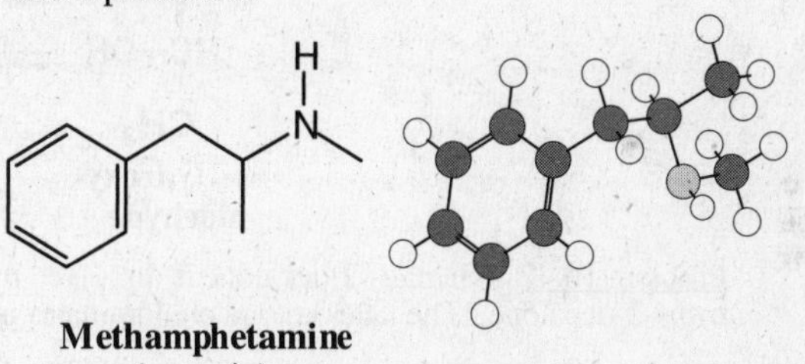

Methamphetamine
(racemic)

The major central nervous system effects of amphetamine and amphetamine-like drugs are locomotor stimulation, euphoria and excitement, stereotyped behavior, and anorexia. Show how each drug can be synthesized by reductive amination of an appropriate aldehyde or ketone and amine. For the structural formulas of several more anorexics, see 23.62 and 23.63.

Both molecules can be synthesized from the same ketone via reductive amination with either ammonia (amphetamine) or methylamine (methamphetamine). A new chiral center is created in each case. Racemic mixtures will be produced.

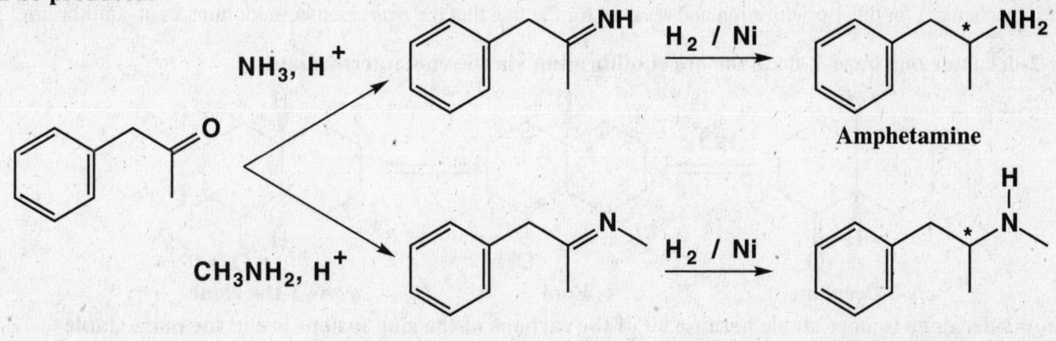

Amphetamine

Methamphetamine

Problem 16.40 Following is the final step in the synthesis of the antiviral drug Rimantadine (Problem 7.19).
(a) Describe experimental conditions to bring this conversion.

(racemic)

(b) Is Rimantidine chiral? How many stereoisomers are possible for it?.

Rimantidine is chiral due to the chiral center marked with an asterisk above. With one chiral center there are two stereoisomers possible, a racemic pair of enantiomers.

Keto-Enol Tautomerism

Problem 16.41 The following molecule belongs to a class of compounds called enediols; each carbon of the double bond carries an -OH group. Draw structural formulas for the α-hydroxyketone and the α-hydroxyaldehyde with which this enediol is in equilibrium.

α-Hydroxyaldehyde ⇌ [structure] ⇌ α-Hydroxyketone

An enediol

Following are formulas for the α-hydroxyaldehyde and α-hydroxyketone in equilibrium by way of the enediol intermediate.

α-Hydroxy-aldehyde   Enediol   α-Hydroxy-ketone

Problem 16.42 When *cis*-2-decalone is dissolved in ether containing a trace of HCl, an equilibrium is established with *trans*-2-decalone. The latter ketone predominates in the equilibrium mixture.

Propose a mechanism for this isomerization and account for the fact that the *trans* isomer predominates at equilibrium.

The *cis*-2-decalone and *trans*-2-decalone are equilibrating via the enol intermediate.

cis-2-Decalone   Enol   *trans*-2-Decalone

The *trans*-2-decalone is more stable because all of the carbons of the ring fusions are in the more stable equatorial arrangement. In *cis*-2-decalone, there is one equatorial and one less-favored axial arrangement at the ring fusion. Note that in the models of *cis*-2-decalone, the carbonyl oxygen is hidden from view because it is behind the structure.

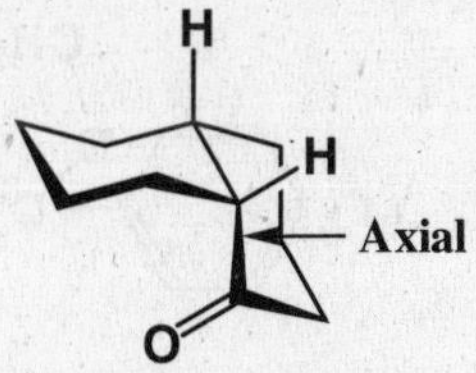

cis-2-Decalone

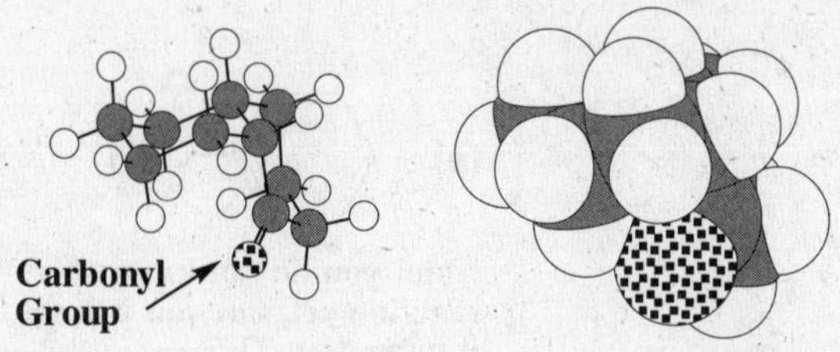

Carbonyl Group

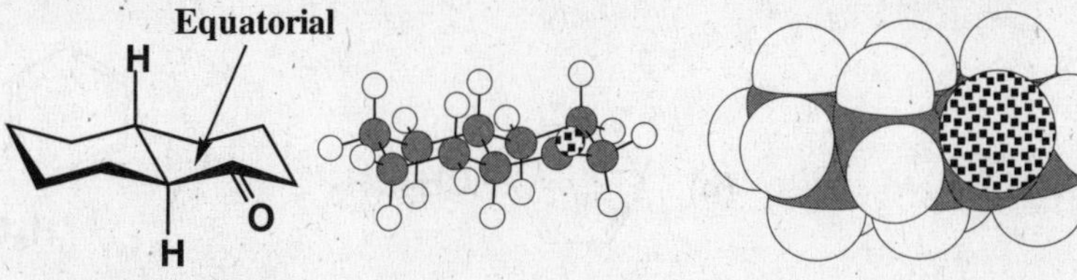

Equatorial

trans-2-Decalone

Oxidation/Reduction of Aldehydes and Ketones

<u>Problem 16.43</u> Draw a structural formula for the product formed by treating butanal with each reagent.

(a) $LiAlH_4$ followed by H_2O

$$CH_3CH_2CH_2CH_2OH$$

(b) $NaBH_4$ in CH_3OH/H_2O

$$CH_3CH_2CH_2CH_2OH$$

(c) H_2/Pt

$$CH_3CH_2CH_2CH_2OH$$

(d) $Ag(NH_3)_2^+$ in NH_3/H_2O

$$CH_3CH_2CH_2COO^-NH_4^+$$

(e) H_2CrO_4, heat

$$CH_3CH_2CH_2COOH$$

(f) $HOCH_2CH_2OH$, HCl

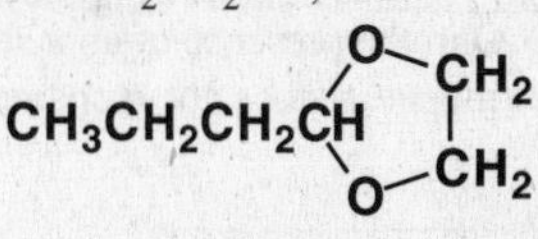

$$+ \ H_2O$$

(g) Zn(Hg)/HCl

$$CH_3CH_2CH_2CH_3$$

(h) N_2H_4, KOH at $250°C$

$$CH_3CH_2CH_2CH_3$$

(i) $C_6H_5NH_2$

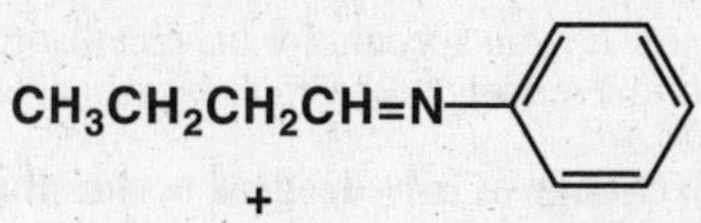

$$+$$
$$H_2O$$

(j) $C_6H_5NHNH_2$

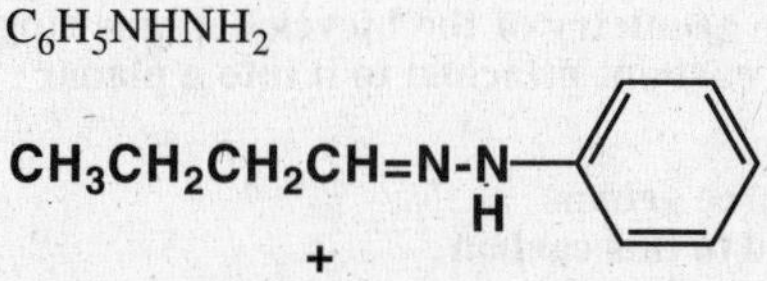

$$+$$
$$H_2O$$

<u>Problem 16.44</u> Draw a structural formula for the product of the reaction of acetophenone with each reagent given in Problem 16.43.

Following are structural formulas for each product. Note that in parts (a), (b), and (c) a new chiral center is created in the reaction so the products will actually be racemic mixtures.

(a)

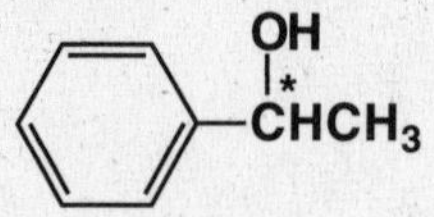

(b)

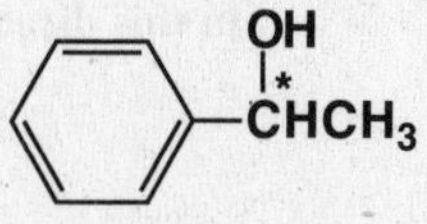

(c)

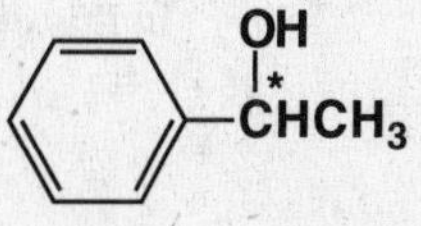

(d) No reaction

(e)

Note: you do not know this
reaction yet, but you will see
it in Section 21.5

(f)

(g)

(h)

(i)

(j)

Reactions at the α-Carbon

Problem 16.45 The following bicyclic ketone has two α-carbons and three α-hydrogens. When this molecule is treated with D_2O in the presence of an acid catalyst, only two of the three α-hydrogens exchange with deuterium. The α-hydrogen at the bridgehead does not exchange.

These two α-hydrogens
exchange

This α-hydrogen
does not exchange

 How do you account for the fact that two α-hydrogens do exchange but the third does not? You will find it helpful to build a models of the enols by which exchange of α-hydrogens occurs.

Exchange of α-hydrogens occurs through keto-enol tautomerism and an enol intermediate. The key to this problem centers on the possibility of placing a carbon-carbon double bond between carbons 2-3 of the bicyclic ring and between carbons 1-2 of the ring. Enolization in the first direction is possible with the result that the two α-hydrogens on carbon-3 exchange.
Enolization toward the bridgehead carbon is not possible because of the geometry of the bicycloheptane ring. The energy required to force the bridgehead carbon and the three other carbons attached to it into a planar conformation with bond angles of 120° is prohibitively high.

The three groups
bonded to this carbon
cannot become planar

enolization is possible
in this direction

enolization is not possible
in this direction

<u>Problem 16.46</u> Propose a mechanism for this reaction.

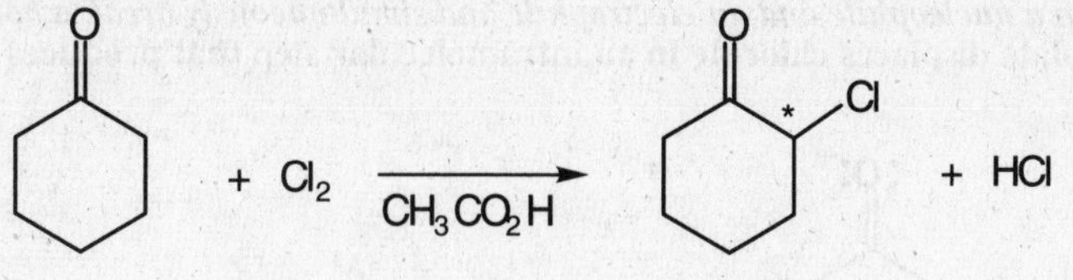

(racemic)

Step 1: Keto-enol tautomerism. **A reasonable mechanism for this reaction involves tautomerization to the enol form.**

keto-enol
tautomerism

Step 2: Electrophilic addition. **The pi bond of the enol then reacts with the Cl_2.**

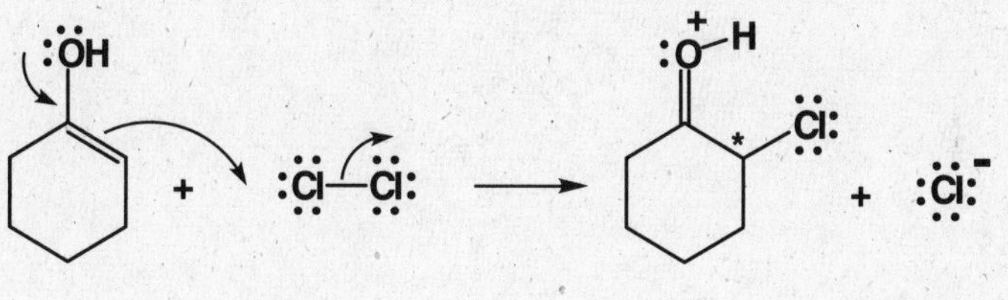

Step 3: Take a proton away. **A final proton transfer completes the reaction.**

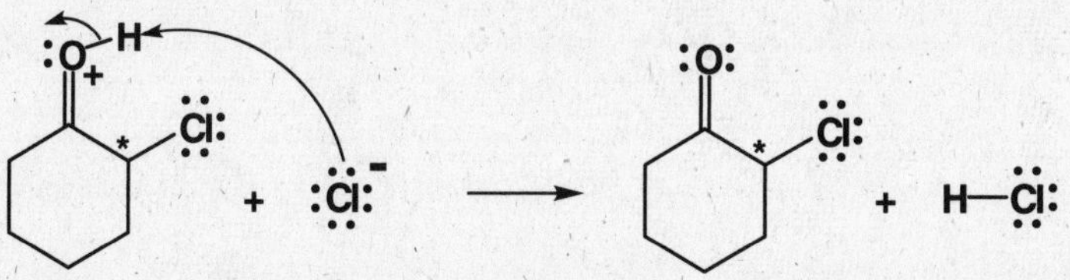

A new chiral center is created in the product. A racemic mixture is produced.

<u>Problem 16.47</u> The base-promoted rearrangement of an α-haloketone to a carboxylic acid, known as the Favorskii rearrangement, is illustrated by the conversion of 2-chlorocyclohexanone to cyclopentanecarboxylic acid.

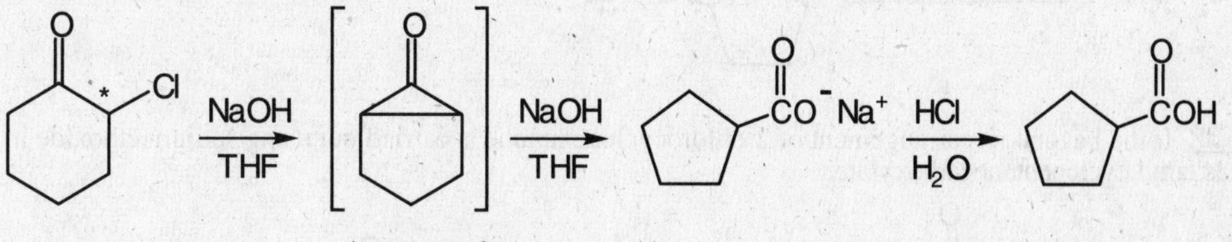

A proposed
intermediate

It is proposed that NaOH first converts the α-haloketone to the substituted cyclopropanone shown in brackets, and then to the sodium salt of cyclopentanecarboxylic acid.
(a) Propose a mechanism for base-promoted conversion of 2-chlorocyclohexanone to the proposed intermediate.

Step 1: Take a proton away. **A reasonable mechanism involves an initial formation of an enolate anion.**

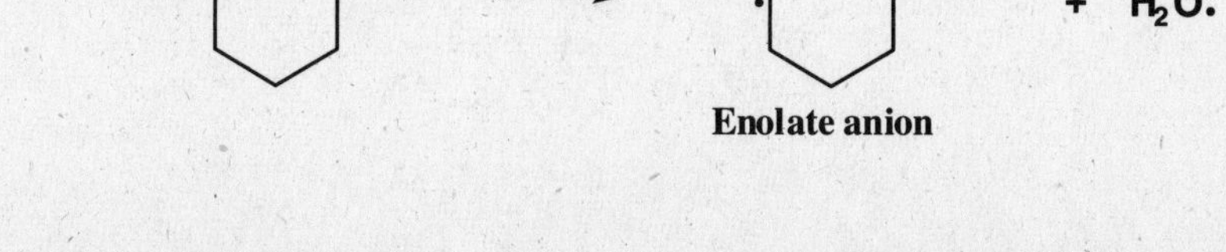

Enolate anion

Step 2: Make a bond between a nucleophile and an electrophile and simultaneously break a bond to give stable molecules or ions. The enolate displaces chloride in an intramolecular step that produces the three-membered ring intermediate.

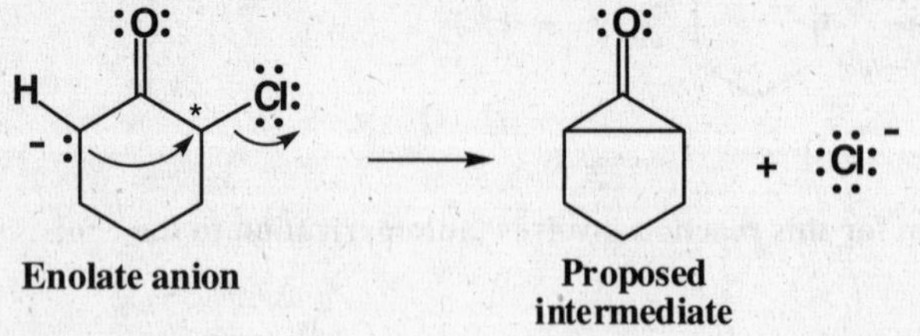

Enolate anion   Proposed
                                intermediate

(b) Propose a mechanism for base-promoted conversion of the bracketed intermediate to sodium cyclopentanecarboxylate.

Step 1: Make a bond between a nucleophile and an electrophile. As stated in the hint, a reasonable mechanism can be written that starts with formation of a tetrahedral carbonyl addition intermediate,

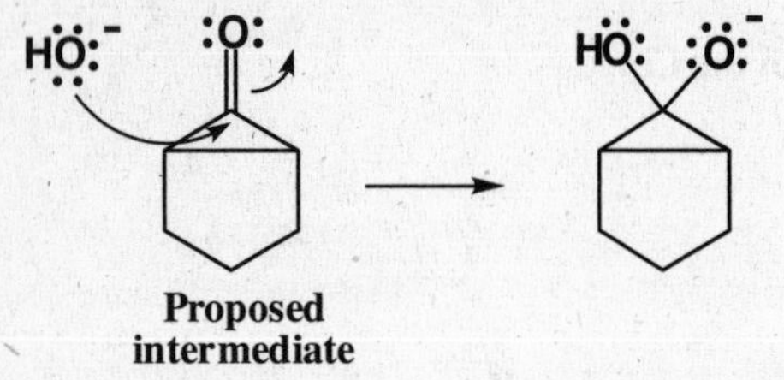

Proposed
intermediate

Step 2: Break a bond to give stable molecules or ions. The intermediate collapses by breaking of one of the cyclopropane ring bonds.

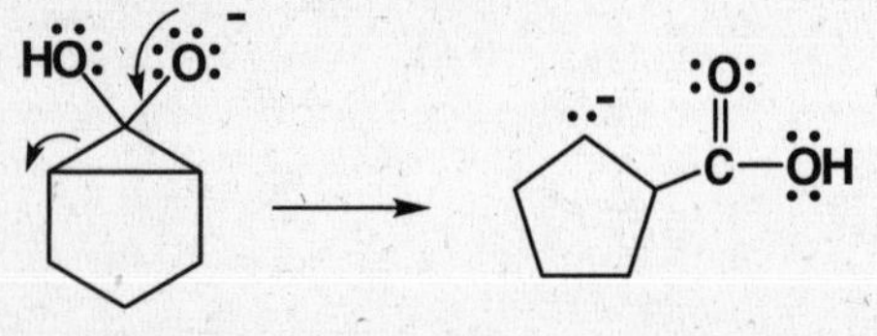

Step 3: Add a proton-take a proton away. A proton transfer completes the reaction.

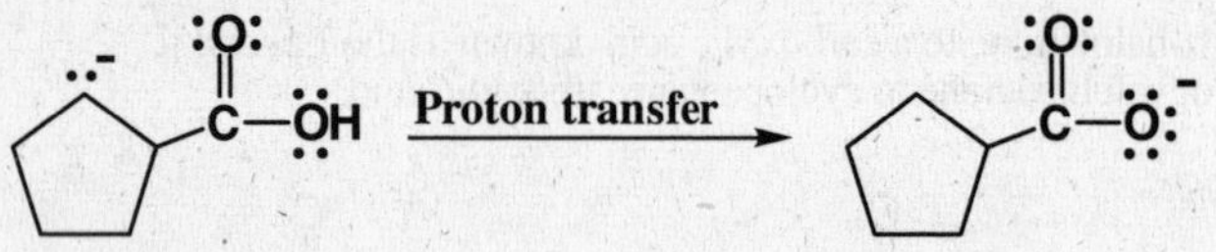

Proton transfer

Problem 16.48 If the Favorskii rearrangement of 2-chlorocyclohexanone is carried out using sodium ethoxide in ethanol, the product is ethyl cyclopentanecarboxylate.

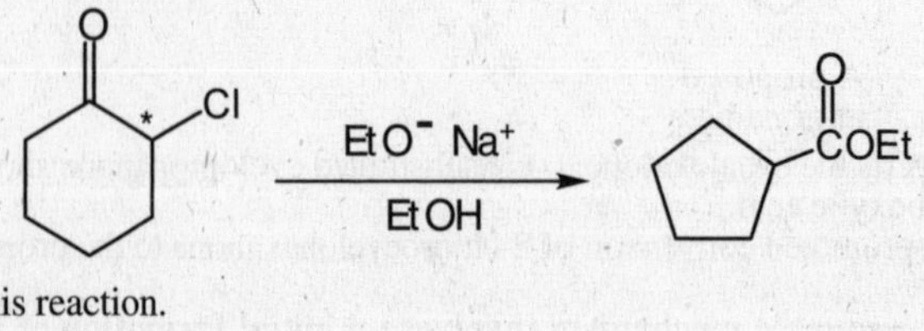

Propose a mechanism for this reaction.

This mechanism is entirely analogous to that of the previous problem, except this time ethoxide is the base and nucleophile, not hydroxide.

Step 1: Take a proton away.

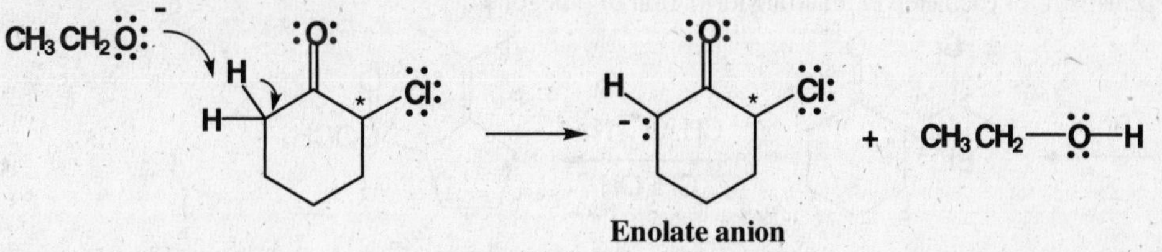

Enolate anion

Step 2: Make a bond between a nucleophile and an electrophile and simultaneously break a bond to give stable molecules or ions.

Enolate anion   Proposed intermediate

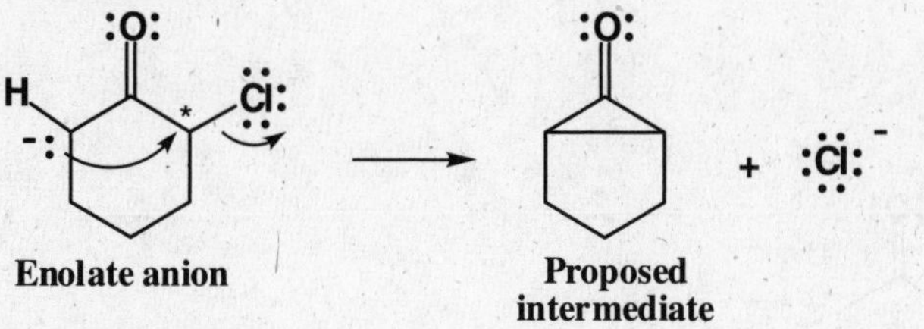

Step 3: Make a bond between a nucleophile and an electrophile.

Proposed intermediate

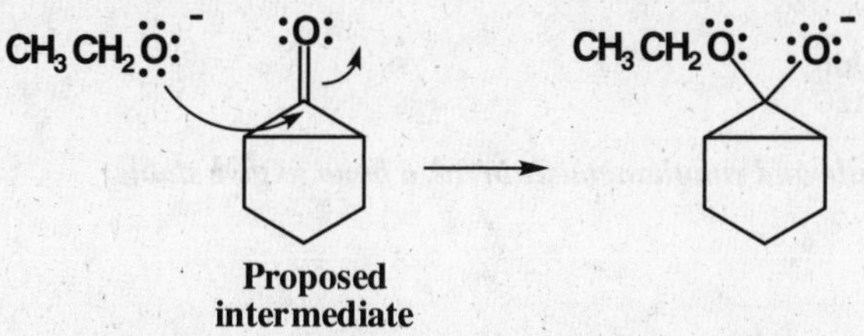

Step 4: Break a bond to give stable molecules or ions.

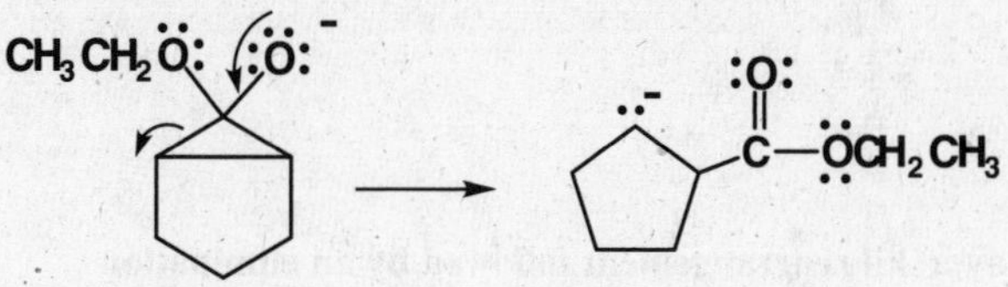

Step 5: Add a proton.

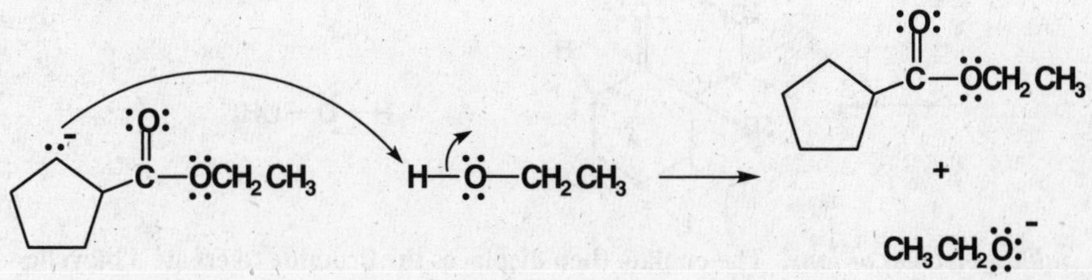

Problem 16.49 (*R*)-Pulegone, readily available from pennyroyal oil, is an important enantiopure building block for organic syntheses. Propose a mechanism for each step in this transformation of pulegone.

(*R*)-(+)-Pulegone

The first step involves bromination of the double bond, a process that proceeds through the familiar bromonium ion intermediate.

Step 1: Electrophilic addition.

Bromonium ion intermediate

Step 2: Make a bond between a nucleophile and an electrophile and simultaneously break a bond to give stable molecules or ions.

Step 3: Take a proton away. **The next steps involve the Favorskii rearrangement followed by an elimination reaction, the first step of which is enolate formation.**

Step 4: Break a bond to give stable molecules or ions. **The enolate then displaces the bromide to create a bicyclic intermediate.**

Step 5: Make a bond between a nucleophile and an electrophile. **Methoxide attacks the carbonyl to generate a tetrahedral addition intermediate.**

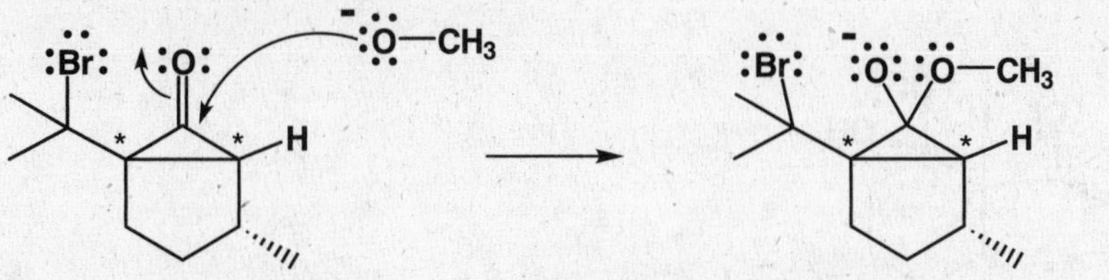

Step 6: Break a bond to give stable molecules or ions. **The tetrahedral addition intermediate collapses, the bromide anion is displaced, and the final product is generated.**

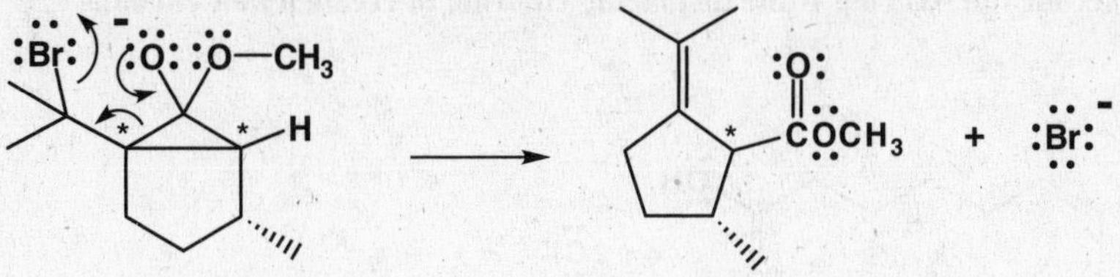

Problem 16.50 (*R*)-(+)-Pulegone is converted to (*R*)-citronellic acid by addition of HCl followed by treatment with NaOH. Propose a mechanism for each step in this transformation, including the regioselectivity of HCl addition.

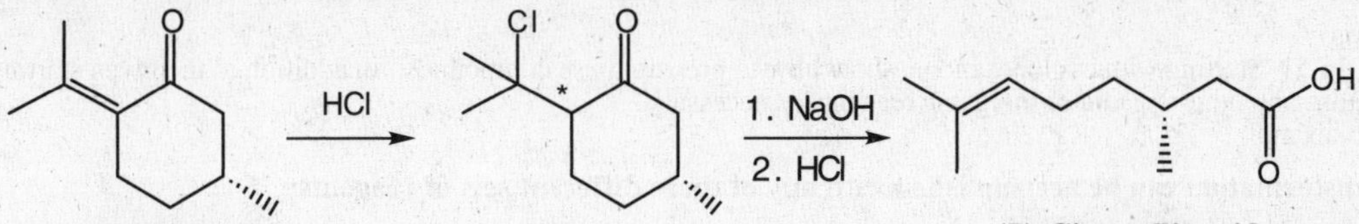

(*R*)-Pulegone **(*R*)-Citronellic acid**

Propose a mechanism for each step in this transformation, including the regioselectivity of HCl addition.

Step 1: Electrophilic addition-add a proton. **The first step of the process involves protonation of the carbon-carbon double bond to create the tertiary carbocation shown below.**

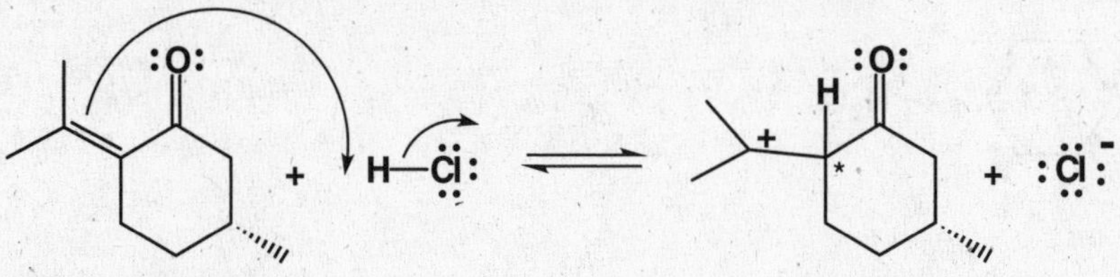

(R)-(+)-Pulegone

Step 2: Make a bond between a nucleophile and an electrophile. **Chloride reacts with this carbocation to give the chlorinated intermediate. This regiochemistry of hydrochlorination reflects the most stable carbocation that is possible. The other possible carbocation produced upon protonation would also have been a tertiary carbocation, but one that is α to the carbonyl group. Carbonyl groups are electron withdrawing and thus destabilizing to an adjacent carbocation.**

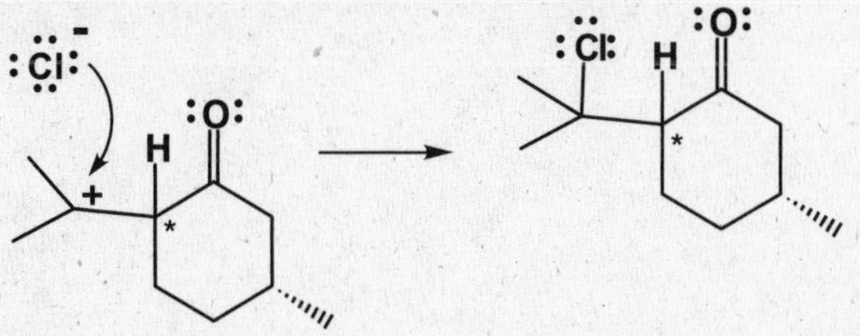

Step 3: Make a bond between a nucleophile and an electrophile. After NaOH is added to the reaction, hydroxide attacks the carbonyl group to give a tetrahedral carbonyl addition intermediate.

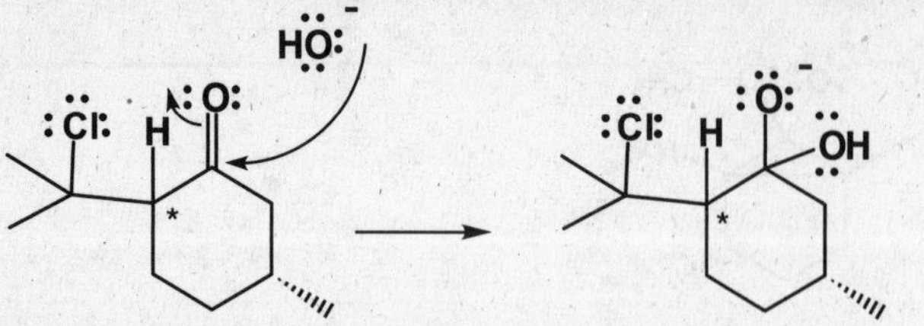

Step 4: Break a bond to give stable molecules or ions. The tetrahedral carbonyl addition intermediate then decomposes to break one bond of the six-membered ring while displacing chloride to create a new carbon-carbon double bond.

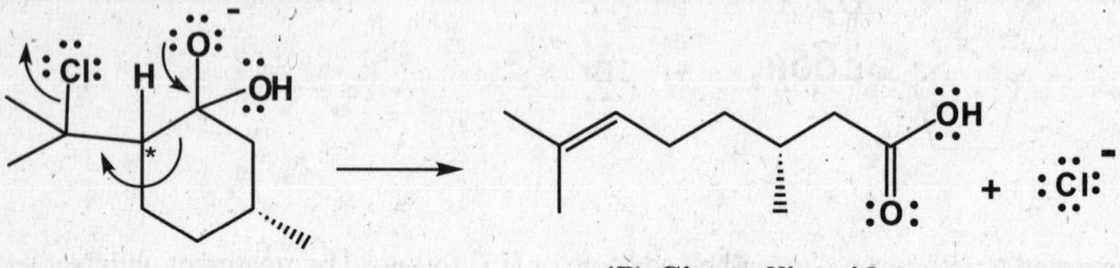

(R)-Citronellic acid

Synthesis

Problem 16.51 Starting with cyclohexanone, show how to prepare these compounds. In addition to the given starting material, use any other organic or inorganic reagents as necessary.

(a) Cyclohexanol

This transformation can be accomplished with any of three different sets of reagents:

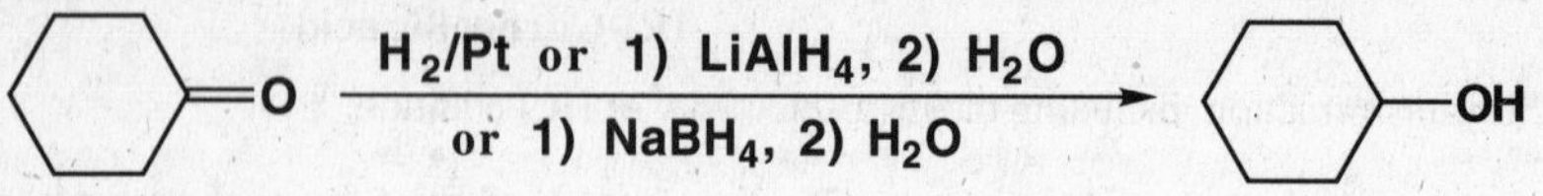

$$\text{H}_2/\text{Pt or 1) LiAlH}_4, \text{ 2) H}_2\text{O}$$
$$\text{or 1) NaBH}_4, \text{ 2) H}_2\text{O}$$

(b) Cyclohexene

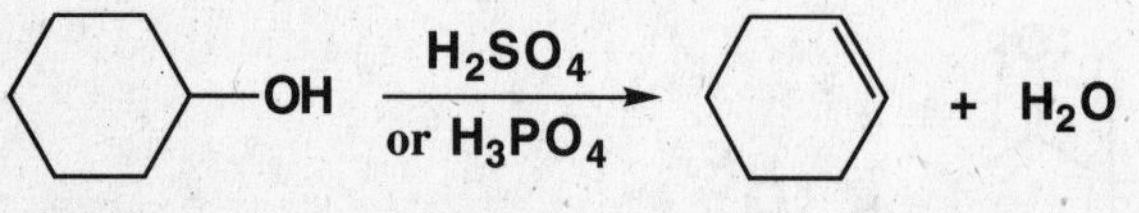

$$\xrightarrow[\text{or H}_3\text{PO}_4]{\text{H}_2\text{SO}_4} \quad + \text{ H}_2\text{O}$$

From (a)

(c) *cis*-1,2-Cyclohexanediol

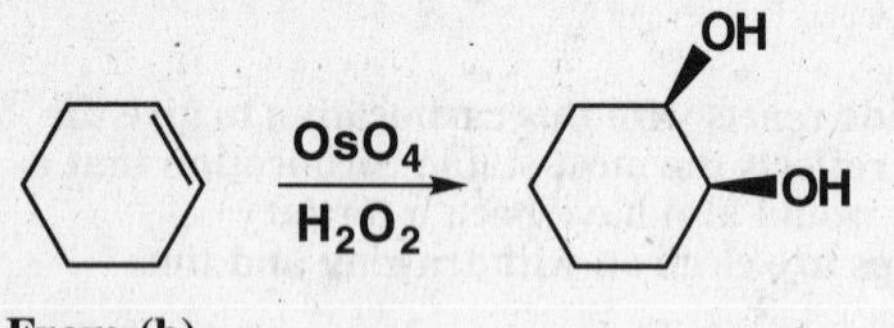

$$\xrightarrow[\text{H}_2\text{O}_2]{\text{OsO}_4}$$

From (b)

(d) 1-Methylcyclohexanol

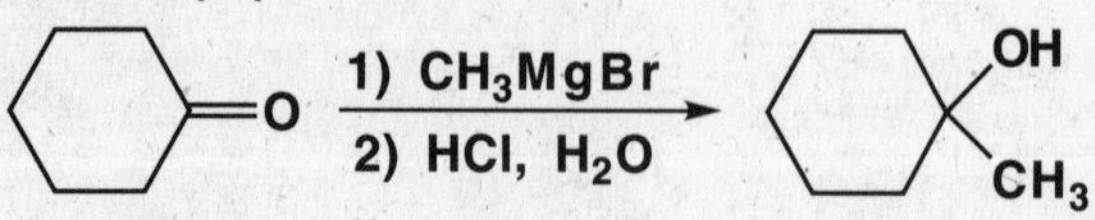

$$\xrightarrow[\text{2) HCl, H}_2\text{O}]{\text{1) CH}_3\text{MgBr}}$$

(e) 1-Methylcyclohexene

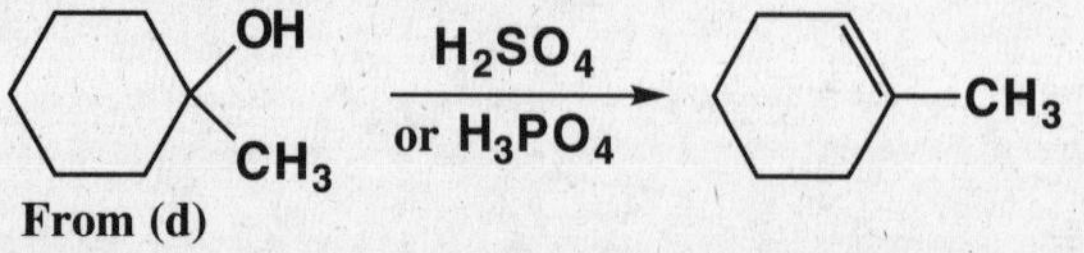

$$\xrightarrow[\text{or H}_3\text{PO}_4]{\text{H}_2\text{SO}_4}$$

From (d)

(f) 1-Phenylcyclohexanol

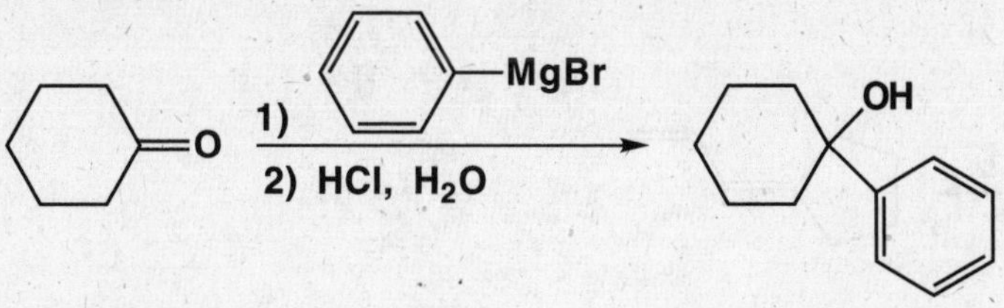

(g) 1-Phenylcyclohexene

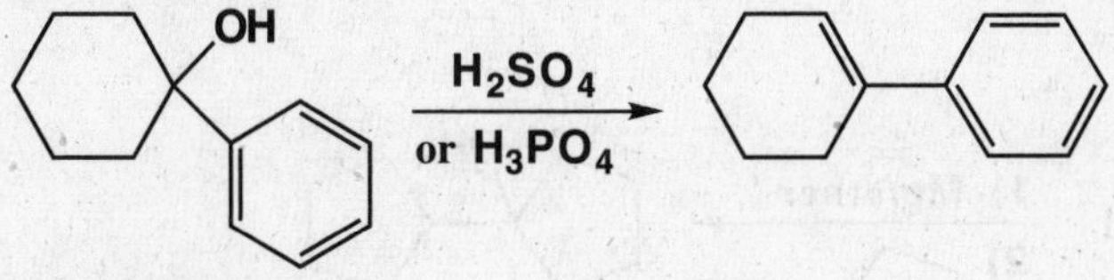

From (f)

(h) Cyclohexene oxide

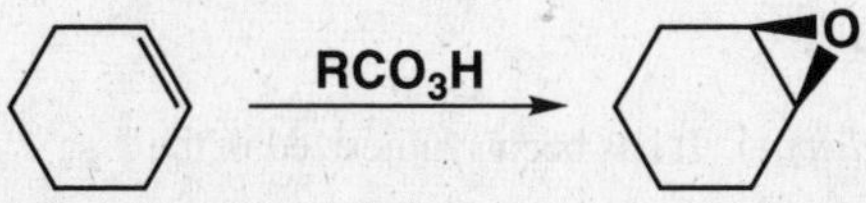

From (b)

(i) *trans*-1,2-Cyclohexanediol

Recall that ring-opening of an epoxide in acid gives the desired *trans* product. Compare this to part (c) of this problem in which the *cis* product is desired.

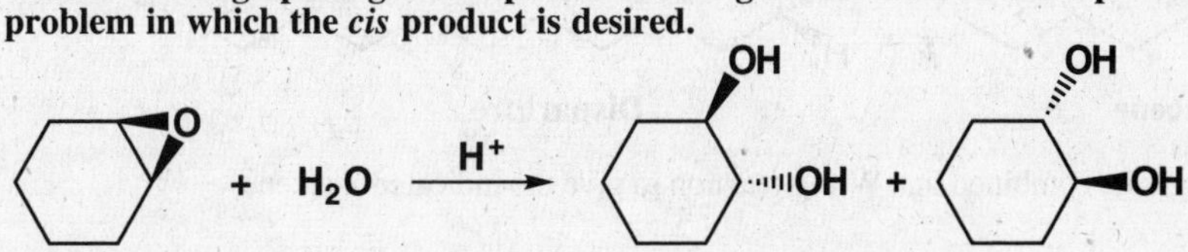

(racemic mixture)

From (h)

<u>Problem 16.52</u> Show how to convert cyclopentanone to these compounds. In addition to cyclopentanone, use any other organic or inorganic reagents as necessary.

(a)

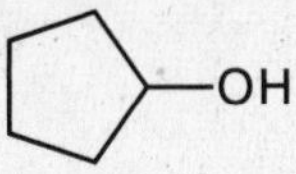

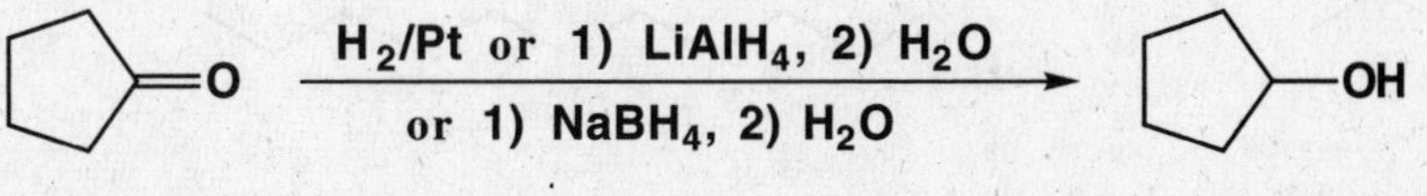

(b)

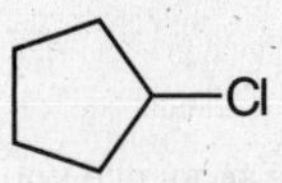

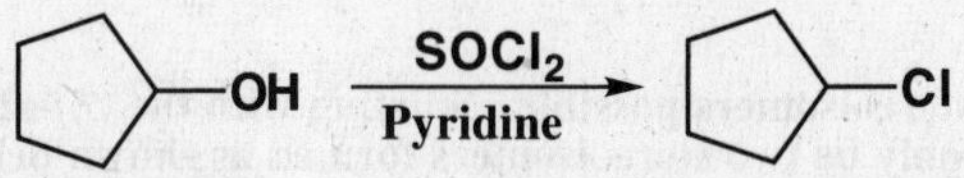

from (a)

(c)

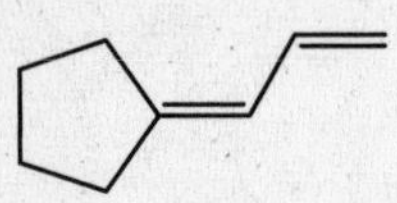

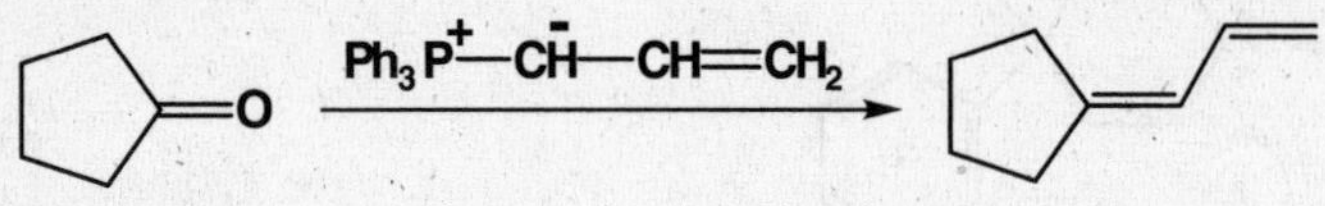

(d)

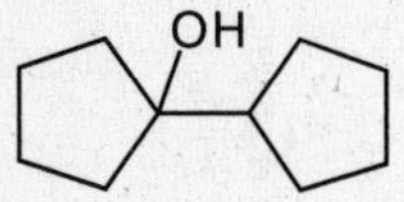

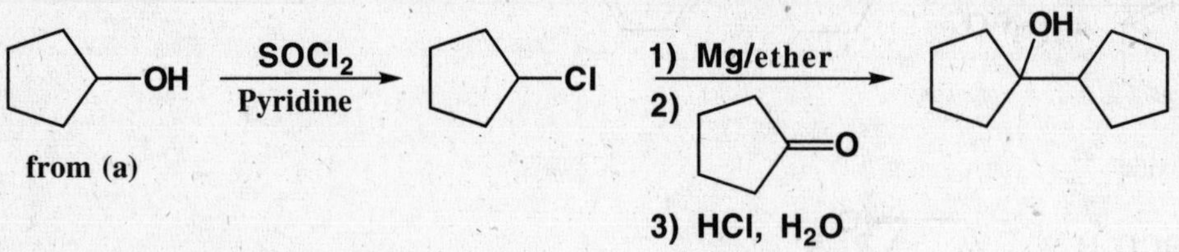

from (a)

1) Mg/ether
2)
3) HCl, H₂O

Problem 16.53 Disparlure is a sex attractant of the gypsy moth (*Porthetria dispar*). It has been synthesized in the laboratory from the following Z alkene.

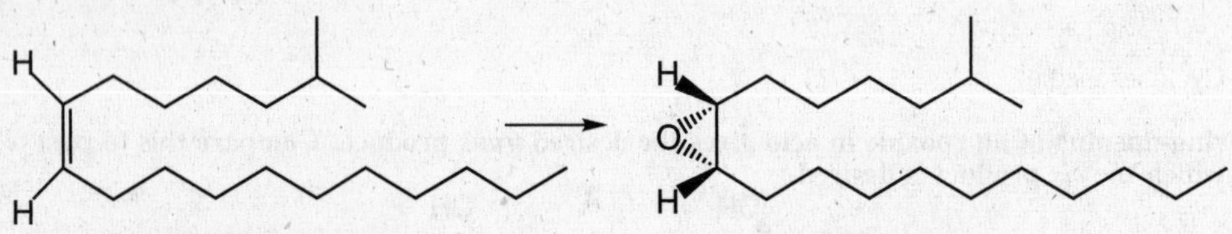

(Z)-2-Methyl-7-octadecene   **Disparlure**

(a) Propose two sets of reagents that might be combined in a Wittig reaction to give the indicated Z alkene.

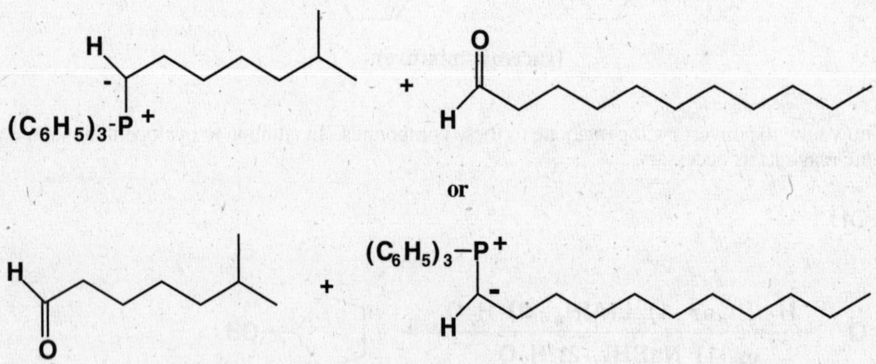

or

(b) How might the Z alkene be converted to disparlure?

The Z alkene is converted to disparlure by reaction with a peracid.

(c) How many stereoisomers are possible for disparlure? How many stereoisomers are formed in the sequence you chose?

Disparlure has two chiral centers, so there are a total of 4 stereoisomers possible. Starting with the (Z)-alkene means that only (Z) epoxides are formed. Thus, there will only be two stereoisomers formed as shown below.

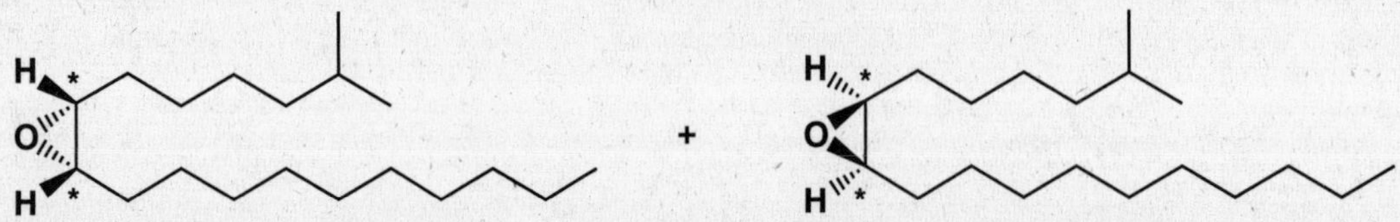

Problem 16.54 Propose structural formulas for compounds A, B, and C in the following conversion. Show also how to prepare compound C by a Wittig reaction.

Compound C could also be prepared by the following Wittig reaction:

Problem 16.55 Following is a retrosynthetic scheme for the synthesis of *cis*-3-penten-2-ol.

Write a synthesis for this compound from acetylene, acetaldehyde, and iodomethane.

The product has a chiral center, so it will be formed as a racemic mixture.

Problem 16.56 Following is the structural formula of Surfynol, a defoaming surfactant. Describe the synthesis of this compound from acetylene and a ketone. How many stereoisomers are possible for Surfynol?

Surfynol

Surfynol (surfactant containing alkyne and alcohol functional groups) is synthesized by reaction of the sodium salt of acetylene in the presence of excess sodium amide with two moles 4-methyl-2-pentanone. Surfynol has two chiral centers, so a racemic mixture of stereoisomers is produced in this synthesis.

Surfynol

There are three stereoisomers of surfynol; 2 enantiomers and a meso compound.

Problem 16.57 Propose a mechanism for this isomerization.

Step 1: Add a proton. **A reasonable mechanism for this transformation involves initial protonation of the -OH group.**

Step 2: Break a bond to give stable molecules or ions. **Loss of water generates a resonance stabilized carbocation.**

Step 3: Make a bond between a nucleophile and an electrophile. **This carbocation is attacked by water to give an intermediate.**

Step 4: Take a proton away. **The intermediate loses a proton to give an enol**

Step 5: Keto-enol tautomerism. The enol tautomerizes to produce the final product.

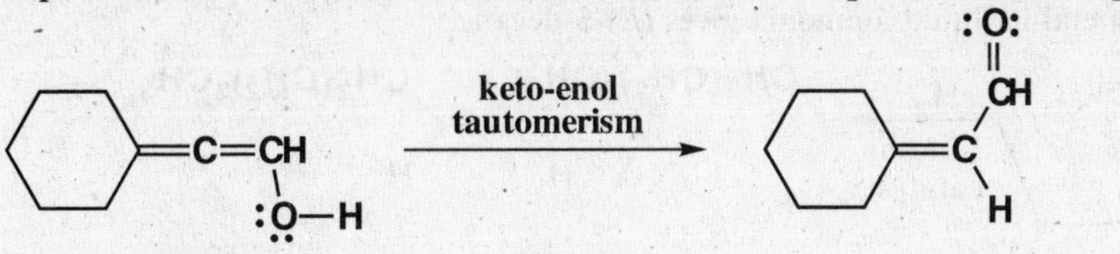

Enol

Problem 16.58 Propose a mechanism for this isomerization.

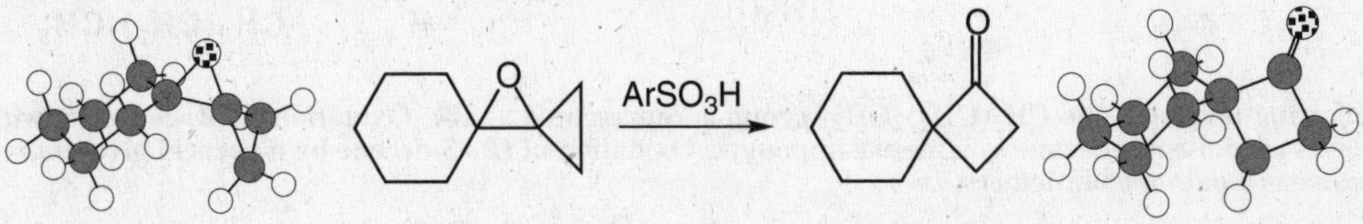

Step 1: Add a proton. A reasonable mechanism for this reaction involves an initial protonation of the epoxide.

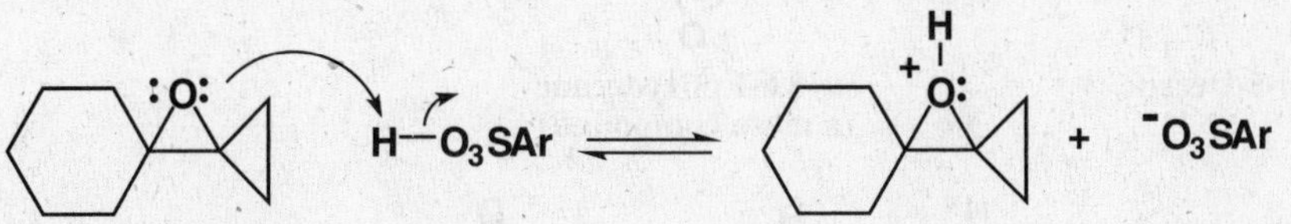

Step 2: Break a bond to give stable molecules or ions. Ring opening forms a tertiary carbocation.

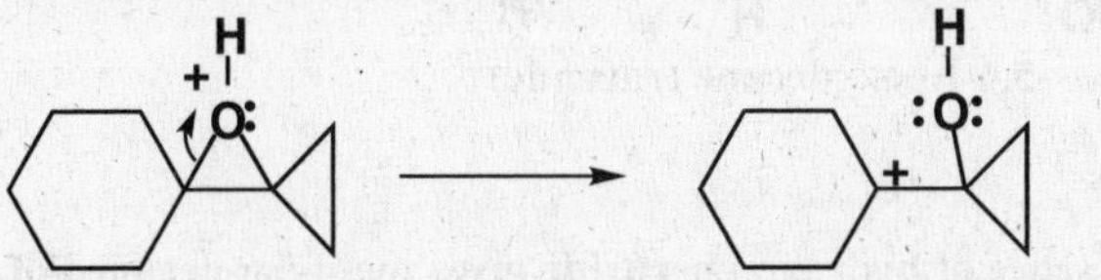

Step 3: 1,2 Shift. This undergoes a rearrangement to create a resonance-stabilized cation that contains the less strained four-membered ring instead of the original, more highly strained three-membered ring.

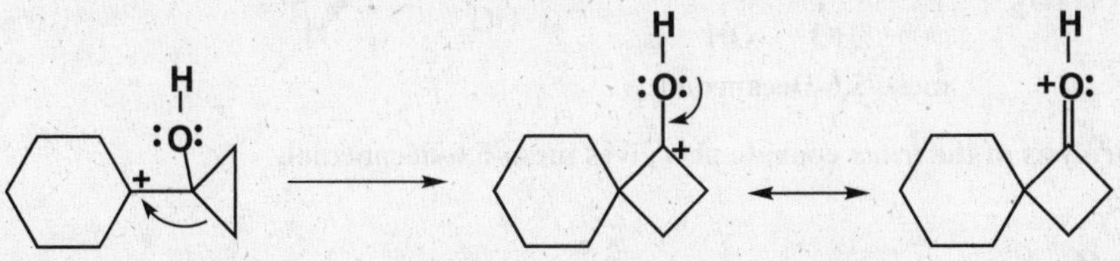

Step 4: Take a proton away. Deprotonation gives the final product.

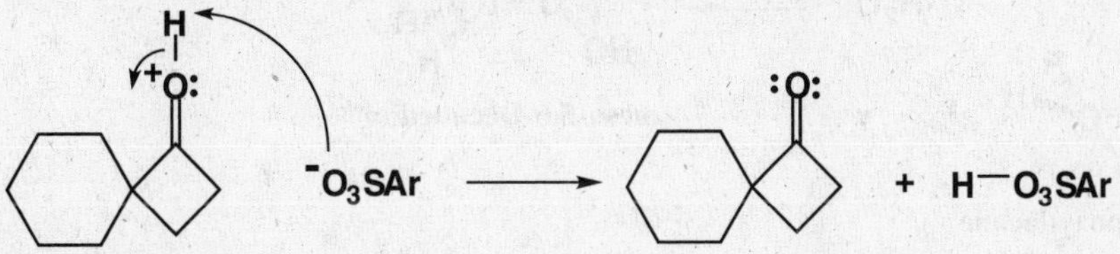

Problem 16.59 Starting with acetylene and 1-bromobutane as the only sources of carbon atoms, show how to synthesize the following:

All molecules in this problem are decanes. The way to build this carbon skeleton using the compounds given is by dialkylation of acetylene to give 5-decyne.

$$HC\equiv CH \xrightarrow[\text{2) 2 } CH_3CH_2CH_2CH_2Br]{\text{1) 2 } NaNH_2} CH_3(CH_2)_3C\equiv C(CH_2)_3CH_3$$

5-Decyne

Catalytic reduction of 5-decyne using hydrogen over a Lindlar catalyst gives (Z)-5-decene. Chemical reduction of 5-decyne using sodium or lithium metal in liquid ammonia gives (E)-5-decene.

$CH_3(CH_2)_3C≡C(CH_2)_3CH_3$

$\xrightarrow[\text{Lindlar catalyst}]{H_2}$

$CH_3(CH_2)_2CH_2 \quad CH_2(CH_2)_2CH_3$
H   H

$\xrightarrow[NH_3(l)]{Na}$

$CH_3(CH_2)_2CH_2 \quad H$
H $CH_2(CH_2)_2CH_3$

In the following formulas, the $CH_3(CH_2)_2CH_2-$ group is represented as R-. Oxidation of (Z)-5-decene with a peracid gives *cis*-5,6-epoxydecane as a meso compound. Oxidation of (E)-5-decene by a peracid gives *trans*-5,6-epoxydecane as a pair of enantiomers.

(Z)-5-Decene $\xrightarrow{RCO_3H}$ *cis*-5,6-Epoxydecane
(a meso compound)

(E)-5-Decene $\xrightarrow{RCO_3H}$ *trans*-5,6-Epoxydecane (racemic)

(a) meso-5,6-Decanediol

Oxidation of (Z)-5-decene by osmium tetroxide in the presence of hydrogen peroxide gives meso-5,6-decanediol.

(Z)-5-Decene $\xrightarrow[H_2O_2]{OsO_4}$ meso-5,6-Decanediol   ≡

Alternatively, acid-catalyzed hydrolysis of the *trans* epoxide also gives meso-5,6-decanediol.

$+ H_2O \xrightarrow{H^+}$ meso-5,6-Decanediol

trans-5,6-Epoxydecane

(b) Racemic 5,6-Decanediol

Oxidation of (E)-5-decene by osmium tetroxide in the presence of hydrogen peroxide gives racemic 5,6-decanediol.

(E)-5-Decene $\xrightarrow[H_2O_2]{OsO_4}$ 5,6-Decanediol (racemic)

Alternatively, acid-catalyzed hydrolysis of the *cis* epoxide also gives racemic 5,6-decanediol.

cis-**5,6-Epoxydecane**   **5,6-Decanediol (racemic)**

c) 5-Decanone

Acid-catalyzed hydration of 5-decyne gives 5-decanone.

$$CH_3(CH_2)_3C\equiv C(CH_2)_3CH_3 \;+\; H_2O \xrightarrow[\text{HgSO}_4]{\text{H}_2\text{SO}_4} CH_3(CH_2)_3\overset{\overset{\displaystyle O}{\|}}{C}CH_2(CH_2)_3CH_3$$

5-Decyne   **5-Decanone**

(d) 5,6-Epoxydecane

See the beginning of this solution for methods to prepare both the *cis*-epoxide and the *trans*-epoxide.

(e) 5-Decanol

Catalytic reduction of 5-decanone using hydrogen over a transition metal catalyst, or chemical reduction using NaBH₄ or LiAlH₄ gives 5-decanol as a racemic mixture.

$$CH_3(CH_2)_3\overset{\overset{\displaystyle O}{\|}}{C}(CH_2)_4CH_3 \;+\; NaBH_4 \longrightarrow CH_3(CH_2)_3\overset{\overset{\displaystyle OH}{|}}{\underset{*}{C}}H(CH_2)_4CH_3$$

5-Decanone   **5-Decanol (racemic)**

(f) Decane

Catalytic reduction of 5-decyne using hydrogen over a transition metal catalyst gives decane.

$$CH_3(CH_2)_3C\equiv C(CH_2)_3CH_3 \;+\; 2\,H_2 \xrightarrow{\text{Pt}} CH_3(CH_2)_8CH_3$$

5-Decyne   **Decane**

(g) 6-Methyl-5-decanol

Treatment of either the *cis*-epoxide or the *trans*-epoxide from part (d) with methylmagnesium iodide gives 6-methyl-5-decanol as a racemic mixture, the exact stereochemistry of which depends on whether the *cis* or *trans* epoxide is used as starting material.

cis or *trans* **5,6-Epoxydecane**   **6-Methyl-5-decanol**

(h) 6-Methyl-5-decanone

Chromic acid oxidation of 6-methyl-5-decanol from part (g) gives 6-methyl-5-decanone, the stereochemistry of which will depend on the stereochemistry of the starting material.

6-Methyl-5-decanol   **6-Methyl-5-decanone**

Problem 16.60 Following are the final steps in one industrial synthesis of vitamin A acetate:

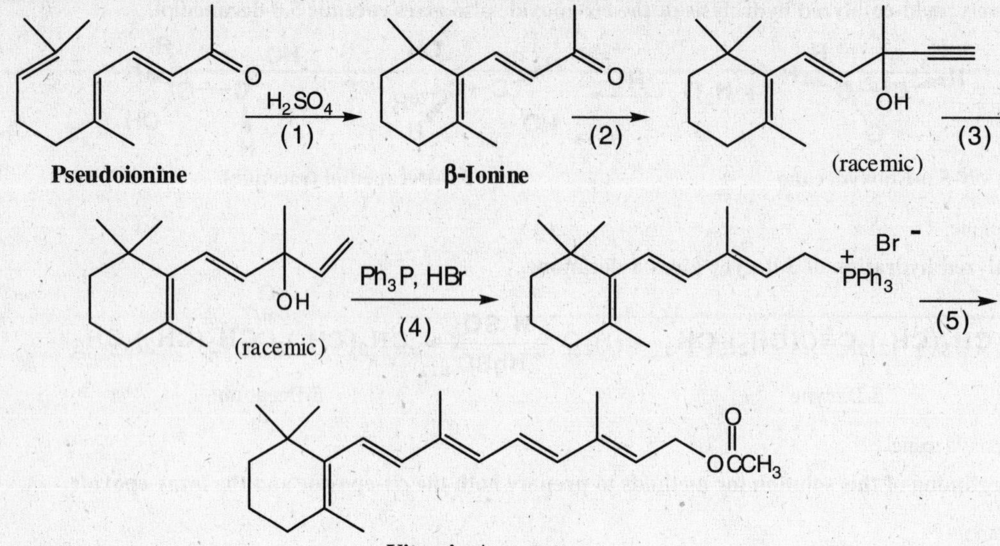

Vitamin A acetate

(a) Propose a mechanism for the acid-catalyzed cyclization in Step 1.

Step 1: Electrophilic addition-add a proton.

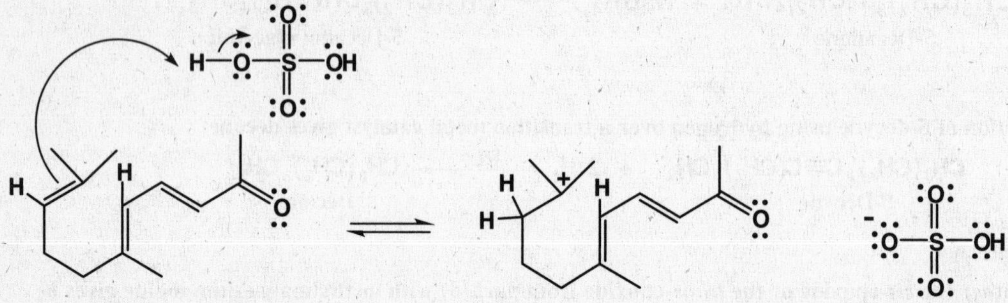

Step 2: Electrophilic addition.

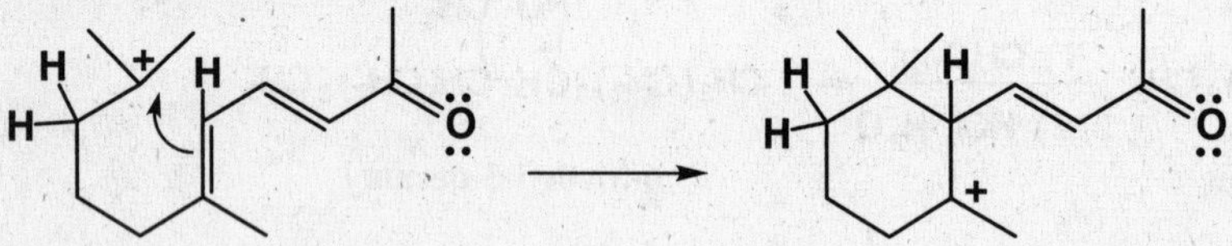

Step 3: Take a proton away.

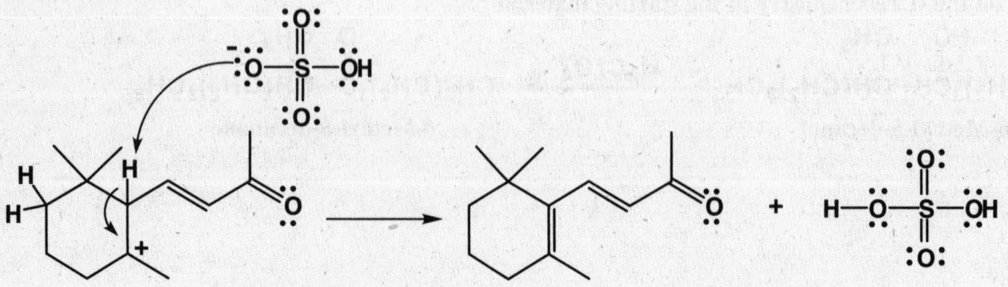

(b) Propose reagents to bring about Step 2.

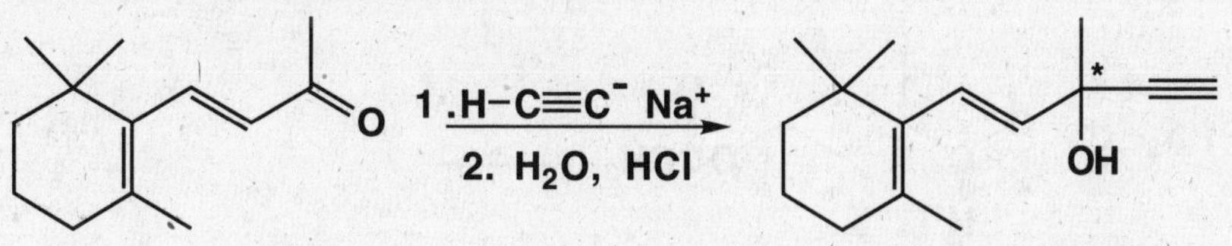

1 .H–C≡C⁻ Na⁺

2. H₂O, HCl

(racemic mixture)

(c) Propose reagents to bring about Step 3

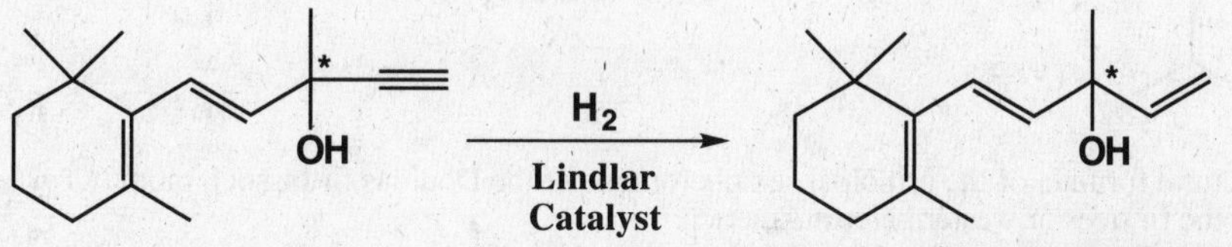

H₂

Lindlar Catalyst

(racemic mixture assuming racemic starting material)

(d) Propose a mechanism for formation of the phosphonium salt in Step 4.

Step 1: Add a proton.

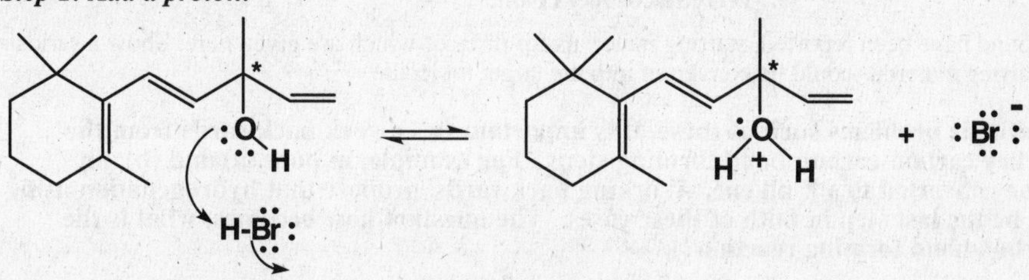

Step 2: Break a bond to give stable molecules or ions.

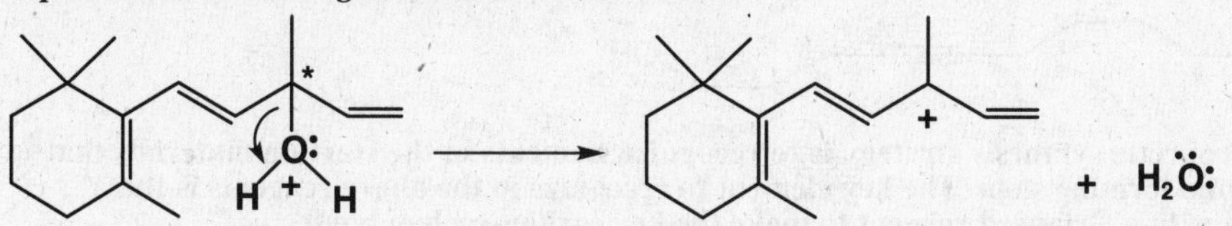

Step 3: Make a bond between a nucleophile and an electrophile.

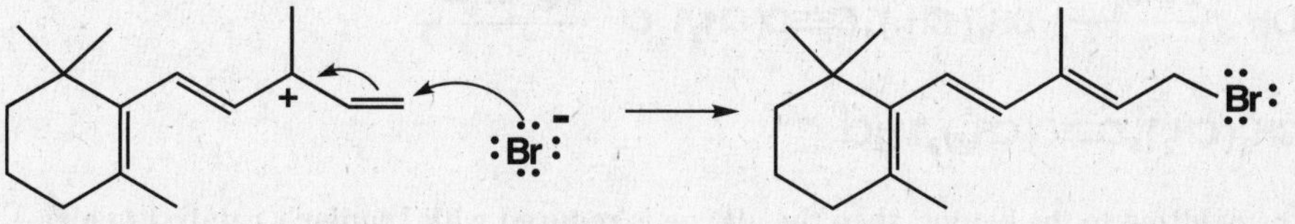

Step 4: Make a bond between a nucleophile and an electrophile and simultaneously break a bond to give stable molecules or ions.

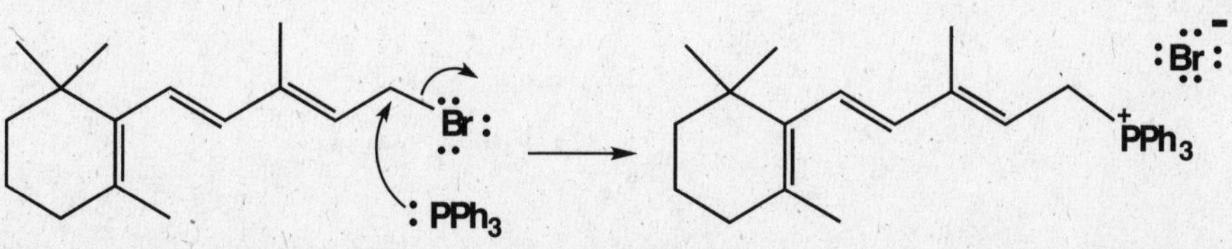

Note it is also possible the PPh₃ might attack the allylic cation directly in Step 3, without the intermediate alkyl bromide being formed.

(e) Show how Step 5 can be completed by a Wittig reaction.

Problem 16.61 Following is the structural formula of the principal sex pheromone of the Douglas fir tussock moth (*Orgyia pseudotsugata*), a severe defoliant of the fir trees of western North America.

(Z)-6-Heneicosene-11-one

Several syntheses of this compound have been reported, starting materials for three of which are given here. Show a series of steps by which each set of starting materials could be converted into the target molecule.

When analyzing difficult synthesis problems such as these, it is important to 1) work backwards from the product and 2) focus on the key carbon-carbon bond forming steps. For example, in both (a) and (b), an alkyne is present that must be converted to a Z alkene. Working backwards, propose that hydrogenation using Lindlar's catalyst will likely be the last step in both of these cases. The question now becomes, what is the nature of the key carbon-carbon bond forming reaction?

(a)

The key to deducing the appropriate synthesis strategy is to recognize elements in the starting materials that suggest the carbon-carbon bond forming step. The key element to recognize in the above reagents is the aldehyde, that can be reacted with a Grignard reagent to make the key carbon-carbon bond.

$$CH_3(CH_2)_4C{\equiv}C(CH_2)_3OH \xrightarrow{SOCl_2} CH_3(CH_2)_4C{\equiv}C(CH_2)_3Cl \xrightarrow{Mg/ether}$$

$$CH_3(CH_2)_4C{\equiv}C(CH_2)_3MgCl$$

The resulting alcohol can be oxidized to the ketone, then the alkyne is reduced with Lindlar's catalyst to give the final Z product. For the original reference, see R. Smith and D. Daves, Journal of Organic Chemistry, 40, 1593 (1975).

(Z)-6-Heneicosene-11-one

(b)

The key carbon-carbon bond forming step can be deduced from the presence of the terminal alkyne and alkyl halide. Here, alkylation of an acetylide anion with the alkyl halide will be used to construct the carbon backbone of the molecule. However, the ketone group must first be protected as an acetal, since ketones also react with acetylide anions. Reduction using Lindlar's catalyst completes the synthesis. For the original reference, see K. Mori, M. Uchida, and M. Matsui, Tetrahedron, 33, 385 (1977).

$$CH_3(CH_2)_9\overset{O}{\overset{||}{C}}(CH_2)_3C\equiv CH \xrightarrow{\text{H}^+ / \text{HO-CH}_2\text{CH}_2\text{-OH}} CH_3(CH_2)_9\overset{\text{O}\quad\text{O}}{\overset{\diagup\;\diagdown}{C}}(CH_2)_3C\equiv CH \xrightarrow{\text{NaNH}_2}$$

$$CH_3(CH_2)_9\overset{\text{O}\quad\text{O}}{\overset{\diagup\;\diagdown}{C}}(CH_2)_3C\equiv C^- \ Na^+ \xrightarrow{CH_3(CH_2)_4Br} CH_3(CH_2)_9\overset{\text{O}\quad\text{O}}{\overset{\diagup\;\diagdown}{C}}(CH_2)_3C\equiv C(CH_2)_4CH_3$$

$$\xrightarrow{H_3O^+} CH_3(CH_2)_9\overset{O}{\overset{||}{C}}(CH_2)_3C\equiv C(CH_2)_4CH_3 \xrightarrow[\text{(Lindlar catalyst)}]{\underset{Pd/CaCO_3}{H_2}} \begin{array}{c} CH_3(CH_2)_9\overset{O}{\overset{||}{C}}(CH_2)_3 \quad (CH_2)_4CH_3 \\ \diagdown\;\;\diagup \\ C=C \\ \diagup\quad\;\diagdown \\ H \qquad\quad H \end{array}$$

(Z)-6-Heneicosene-11-one

(c)

This part is by far the most challenging since it is difficult to deduce the nature of the carbon-carbon bond forming steps. The key element to recognize is that the ring structure is actually a cyclic hemiacetal, in other words, the cyclized form of 5-hydroxypentanal.

As a result, the cyclic hemiacetal can take part in a Wittig reaction like other aldehydes. The first key carbon-carbon bond forming step uses a Wittig reagent prepared from 1-bromohexane. The authors report that the desired Z isomer was the major product.

The next carbon-carbon bond forming step must add the remaining ten carbon atoms onto the end of the molecule containing the primary hydroxyl group. The synthesis continues with a Grignard reaction after first transforming the primary alcohol into an aldehyde group using PCC. Oxidation of the resulting alcohol to the ketone completes the synthesis. See M. Fetizon and C. Lazare, Journal of the Chemical Society, Perkin Transactions 1, 842, (1978).

(Z)-6-Heneicosene-11-one

Note how three very different approaches all produce the same molecule in high yield. One of the fascinating aspects of organic synthesis is that there are usually several good ways to make a molecule, proving that organic chemistry involves not only knowledge and experience, but considerable scientific creativity as well.

Problem 16.62 Both (S)-citronellal and isopulegol are naturally occurring terpenes (Section 5.4). When (S)-citronellal is treated with tin(IV) chloride (a Lewis acid) followed by neutralization with aqueous ammonium chloride, isopulegol is obtained in 85% yield.

(S)-Citronellal
($C_{10}H_{18}O$)

Isopulegol
($C_{10}H_{18}O$)

(a) Show that both compounds are terpenes.

The five carbon units characteristic of terpenes are shown in bold in the following representations.

(b) Propose a mechanism for the conversion of (*S*)-citronellal to isopulegol.

Step 1: Make a bond between a Lewis acid and Lewis base. In the first step of the reaction, the very strong Lewis acid SnCl₄ associates with the most Lewis basic site in the molecule, namely the carbonyl oxygen atom.

Step 2: Electrophilic addition. Association of the carbonyl oxygen atom with the strong Lewis acid SnCl₄ increases the electrophilic character of the carbonyl carbon atom, facilitating nucleophilic attack by the adjacent π bond. Note that a six-membered ring is formed in the process, the relative stability of six-membered rings assists in the reaction process. Note also that the Lewis acid helps stabilize the oxygen anion produced.

Step 3: Add a proton-take a proton away.

The NH₄Cl removes the SnCl₄ from the reaction.

(c) How many chiral centers are present in isopulegol? How many stereoisomers are possible for a molecule with this number of chiral centers?

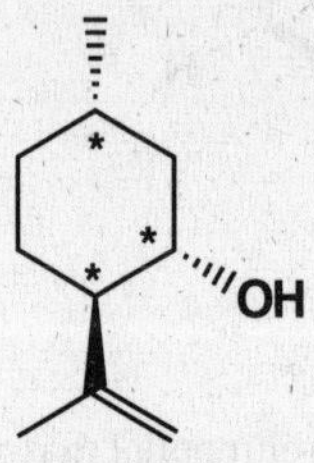

Isopulegol has three chiral centers so there are $2^3 = 8$ possible stereoisomers.

(d) Isopulegol is formed as a single stereoisomer. Account for the fact that only a single stereoisomer is formed.

The key step that sets the stereochemistry of the final product is Step 2. In this step, the π bond of the alkene acts as a nucleophile to attack the carbonyl carbon atom. A six-membered ring is formed during the step. As a result, we can approximate the transition state leading to product as a six-membered ring. Of the two possible chair conformations that could form as transition states, the more stable one places the methyl group equatorial. During the reaction, the two new chiral centers will form such that both substituents will also be equatorial. The net result is a single enantiomer, derived from the more stable chair transition state.

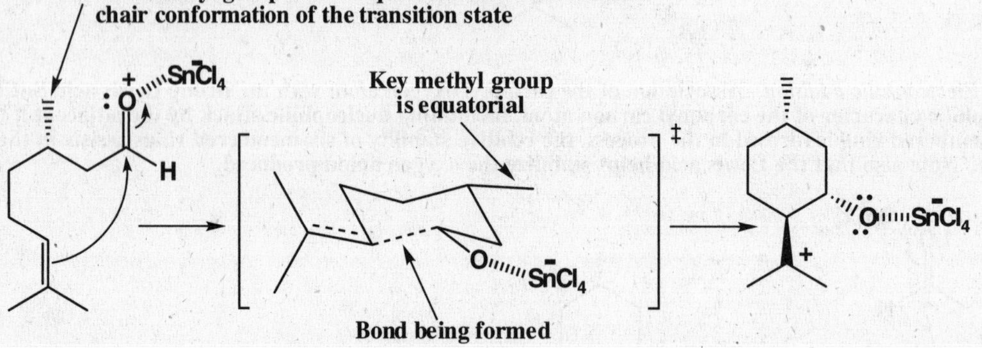

This methyl group will be equatorial in the most stable chair conformation of the transition state

Key methyl group is equatorial

Bond being formed

More stable six-membered ring transition state.

A careful evaluation of the chiral centers being created in the equatorial positions of the transition state shown above will reveal that they do accurately predict the observed stereochemistry in the isopulegol product.

Problem 16.63 At some point during the synthesis of a target molecule, it may be necessary to protect an -OH group (that is, to prevent its reacting). In addition to the trimethylsilyl, *tert*-butyldimethylsilyl, and other trialkylsilyl groups described in Section 11.6, and the tetrahydropyranyl group described in Section 16.7D, the ethoxyethyl group may also be used as a protecting group.

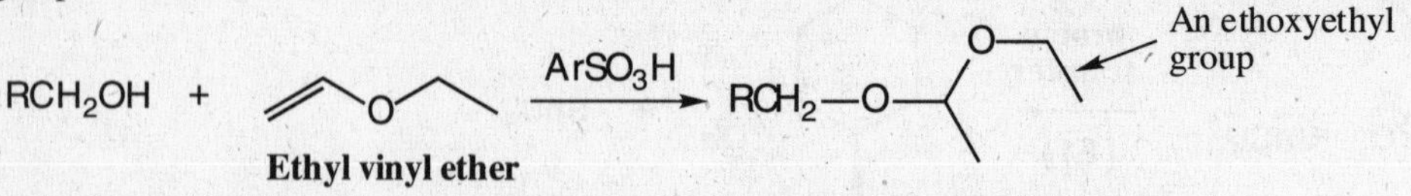

$$RCH_2OH \ + \ \text{(ethyl vinyl ether)} \ \xrightarrow{ArSO_3H} \ RCH_2{-}O{-}\text{(ethoxyethyl group)}$$

An ethoxyethyl group

Ethyl vinyl ether

(a) Propose a mechanism for the acid-catalyzed formation of the ethoxyethyl protecting group.

Step 1: Electrophilic addition-add a proton.

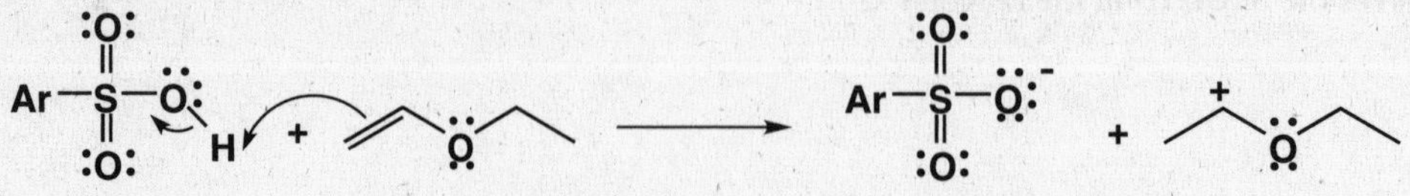

Step 2: Make a bond between a nucleophile and an electrophile.

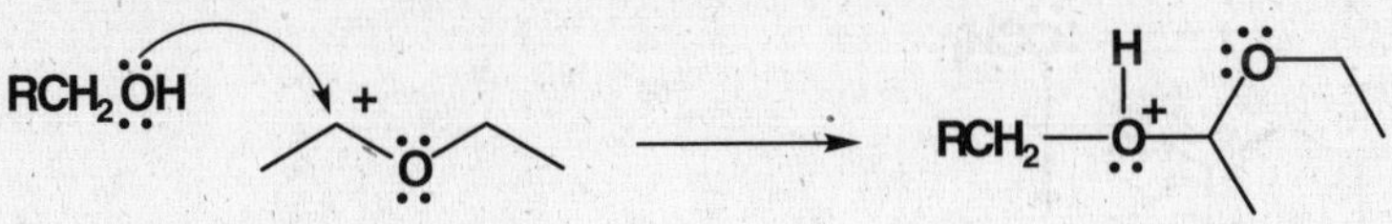

Step 3: Take a proton away.

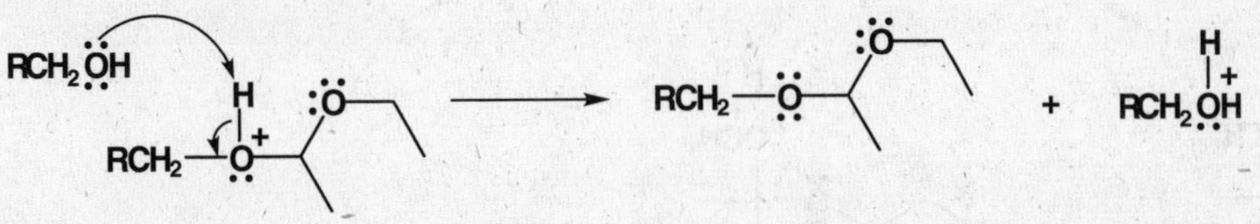

(b) Suggest an experimental procedure whereby this protecting group can be removed to regenerate the unprotected alcohol.

The key to this question is to realize that the ethoxyethyl protecting group contains an acetal functional group. An acidic solution with water present will hydrolyze the acetal and regenerate the alcohol, acetaldehyde, and ethanol.

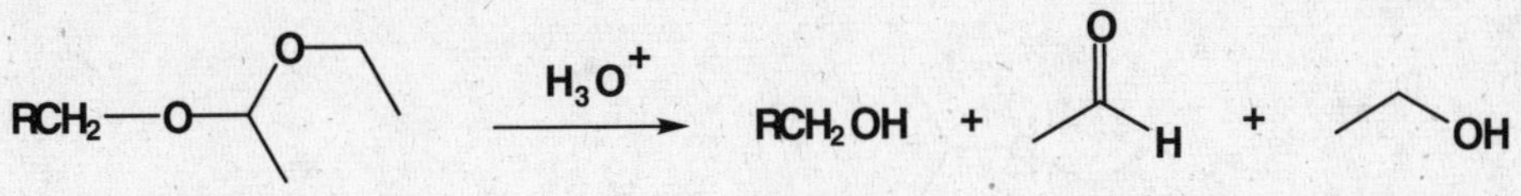

Problem 16.64 Both 1,2-diols and 1,3-diols can be protected by treatment with 2-methoxypropene according to the following reaction.

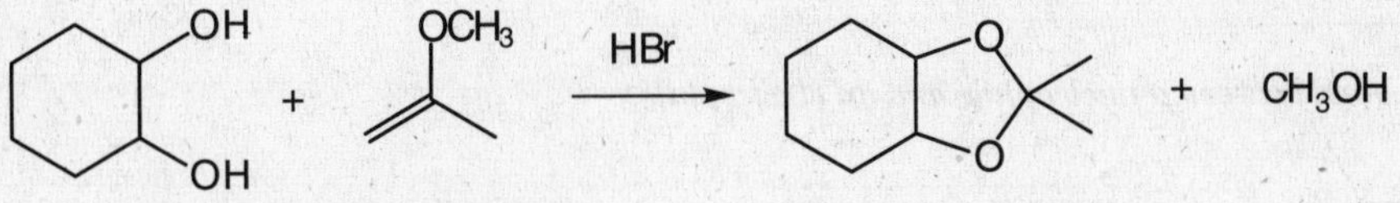

2-Methoxypropene   A protected 1,2-diol

(a) Propose a mechanism for the formation of this protected diol.

Step 1: Electrophilic addition-add a proton.

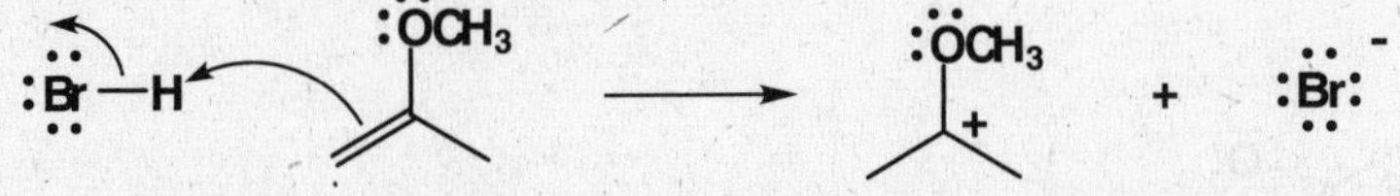

Step 2: Make a bond between a nucleophile and an electrophile.

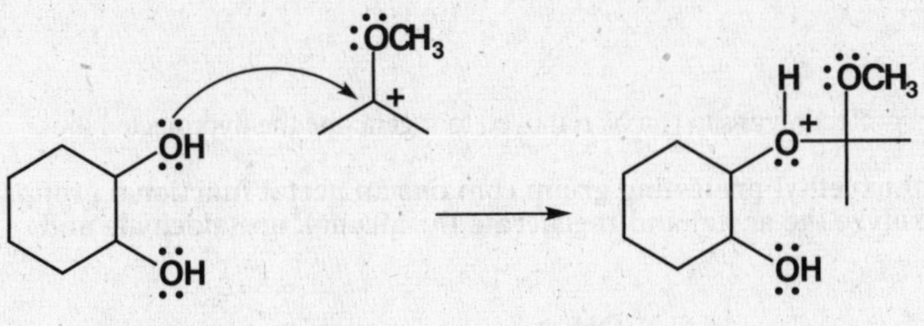

Step 3: Take a proton away.

Step 4: Add a proton.

Step 5: Break a bond to give stable molecules or ions.

Step 6: Make a bond between a nucleophile and an electrophile.

Step 7: Take a proton away.

(b) Suggest an experimental procedure whereby this protecting group can be removed to regenerate the unprotected diol.

The key to this question is to realize that the ethoxyethyl protecting group contains an acetal functional group. An acidic solution with water present will hydrolyze the acetal and regenerate the alcohol, acetaldehyde and ethanol.

Looking Ahead

<u>Problem 16.65</u> All rearrangements we have discussed so far have involved generation of an electron deficient carbon followed by a 1,2-shift of an atom or group of atoms from an adjacent atom to the electron deficient carbon. Rearrangements by a 1,2-shift can also occur following the generation of an electron-deficient oxygen. Propose a mechanism for the acid-catalyzed rearrangement of cumene hydroperoxide to phenol and acetone.

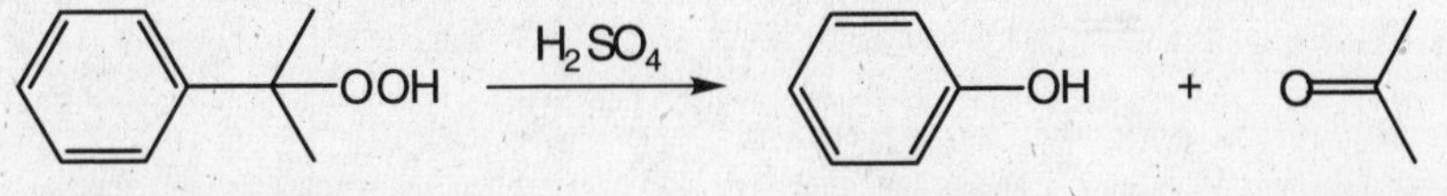

Cumene hydroperoxide   **Phenol**   **Acetone**

Step 1: Add a proton. **A reasonable mechanism for this reaction involves protonation of the terminal oxygen atom of the hydroperoxide group.**

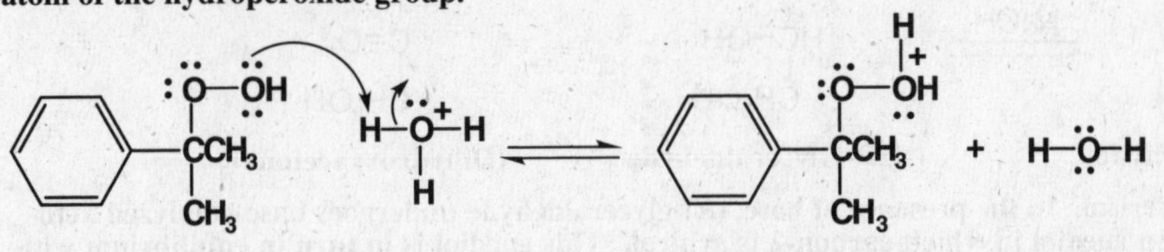

Step 2: 1,2 Shift and simultaneously break a bond to give stable molecules or ions. **A 1,2 shift of the benzene ring and simultaneous loss of water gives an intermediate stabilized by resonance delocalization.**

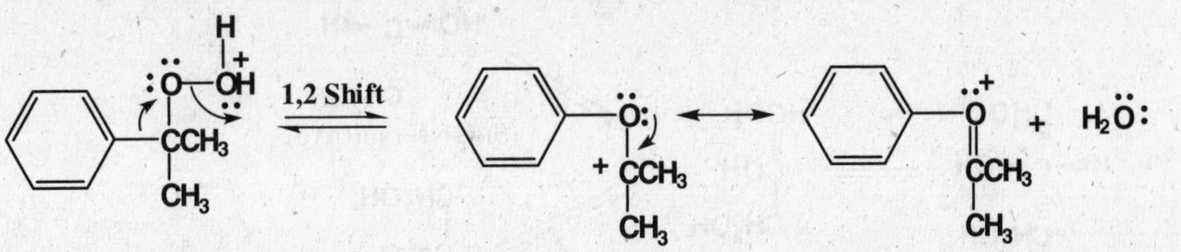

Step 3: Make a bond between a nucleophile and an electrophile. **Attack of the resulting carbocation by water and loss of a proton yields a hemiacetal.**

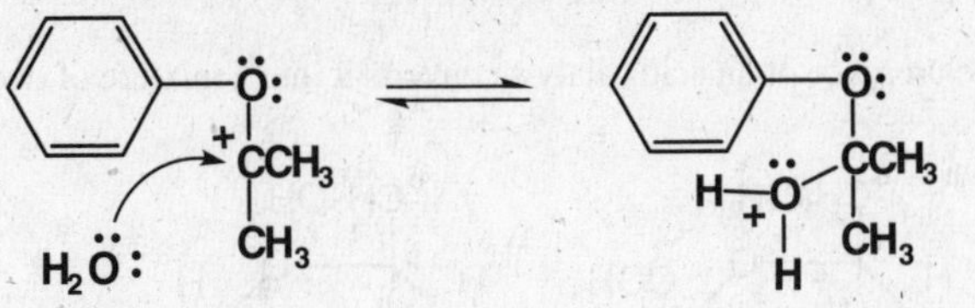

Step 4: Take a proton away.

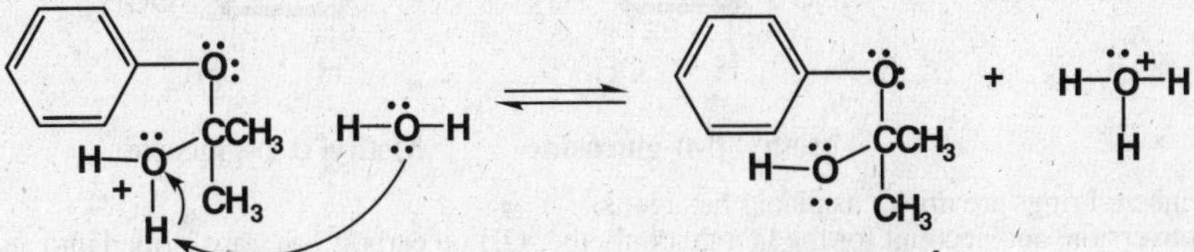

A hemiacetal

Step 5: Add a proton. **The hemiacetal decomposes to give phenol and acetone.**

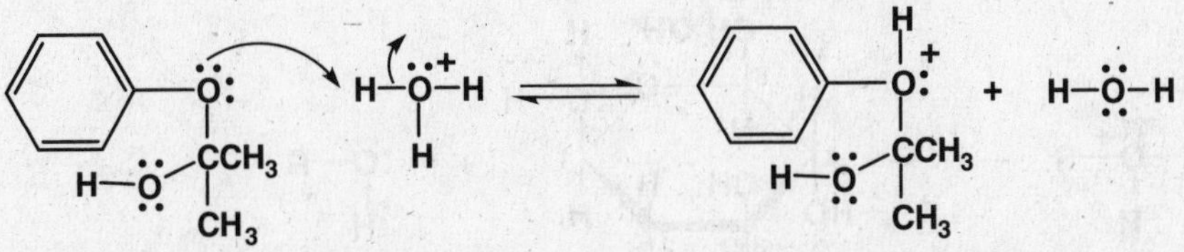

Step 6: Break a bond to give stable molecules or ions.

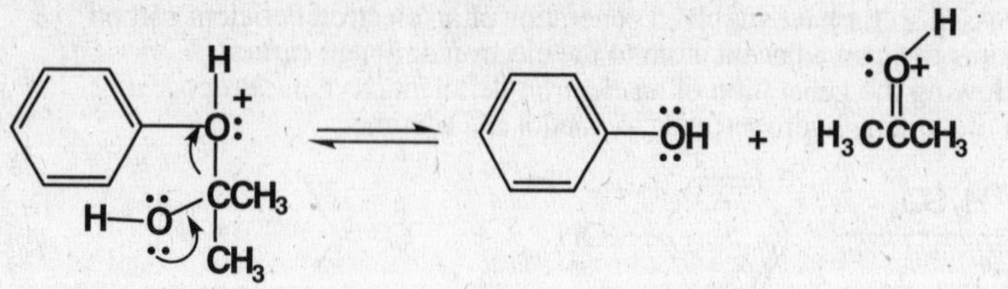

<u>Problem 16.66</u> In dilute aqueous base, (R)-glyceraldehyde is converted into an equilibrium mixture of (R,S)-glyceraldehyde and dihydroxyacetone. Propose a mechanism for this isomerization.

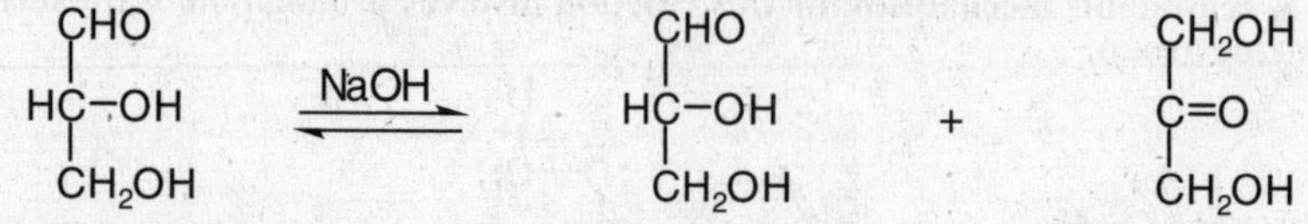

(R)-Glyceraldehyde   **(R,S)-Glyceraldehyde**   **(Dihydroxyacetone)**

The key is keto-enol tautomerism. In the presence of base, (R)-glyceraldehyde undergoes base-catalyzed keto-enol tautomerism to form an enediol in which carbon-2 is achiral. This enediol is in turn in equilibrium with (S)-glyceraldehyde and dihydroxyacetone.

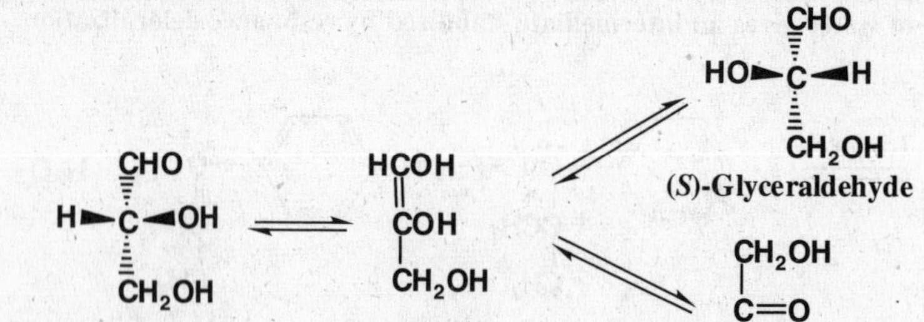

(R)-Glyceraldehyde   **An enediol**

(S)-Glyceraldehyde

Dihydroxyacetone

<u>Problem 16.67</u> Treatment of β-D-glucose with methanol in the presence of an acid catalyst converts it into a mixture of two compounds called methyl glucosides (Section 25.3A).

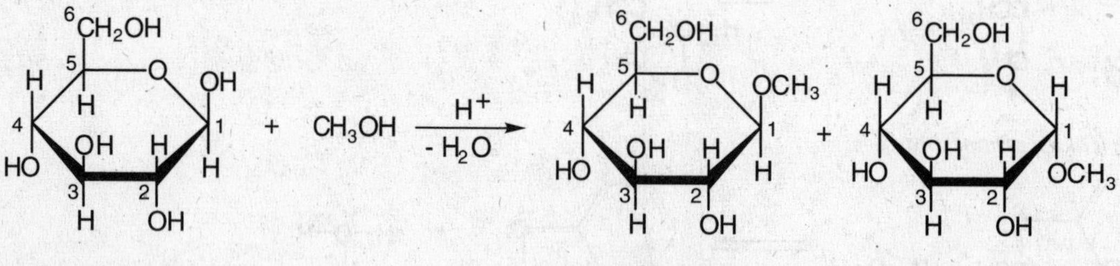

β-D-Glucose   **Methyl β-D-glucoside**   **Methyl α-D-glucoside**

In these representations, the six-membered rings are drawn as planar hexagons.
(a) Propose a mechanism for this conversion, and account for the fact that only the -OH on carbon 1 is transformed into an -OCH₃ group.

Step 1: Add a proton. The reaction begins with protonation of the -OH group on carbon 1.

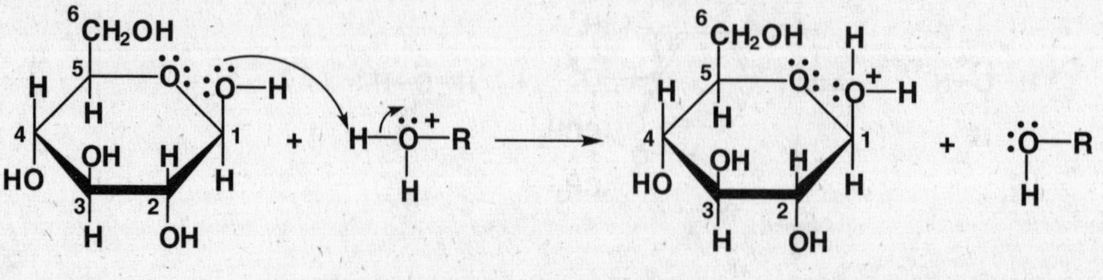

Step 2: *Break a bond to give stable molecules or ions.* Water departs to create a resonance stabilized cation intermediate. Resonance stabilization with the adjacent O atom is only possible with the departure of water from C1. That makes this the most stable intermediate by far, compared to reaction at the other -OH groups. As a result, reaction proceeds exclusively at the -OH of C1.

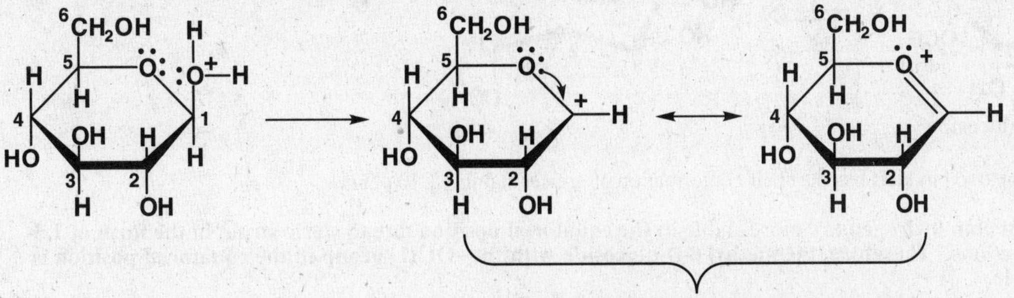

Resonance stabilized cation intermediate

Step 3: *Make a bond between a nucleophile and an electrophile then take a proton away.* The resonance stabilized cation intermediate can react with CH_3OH on either face, followed by loss of a proton to generate the two glucoside products. Note that these are not enantiomers, so they are not formed in equal amounts. The methyl β-D-glucoside, with the -OH group in the more favored equatorial position, dominates at equilibrium.

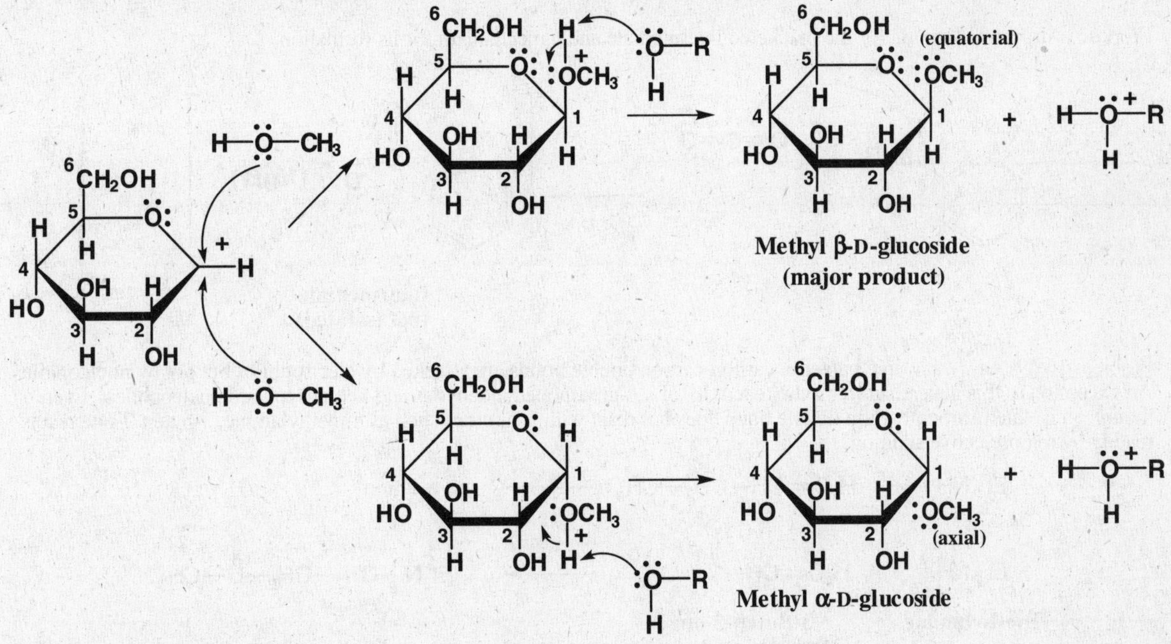

Methyl β-D-glucoside
(major product)

Methyl α-D-glucoside

(b) Draw the more stable chair conformation for each product.

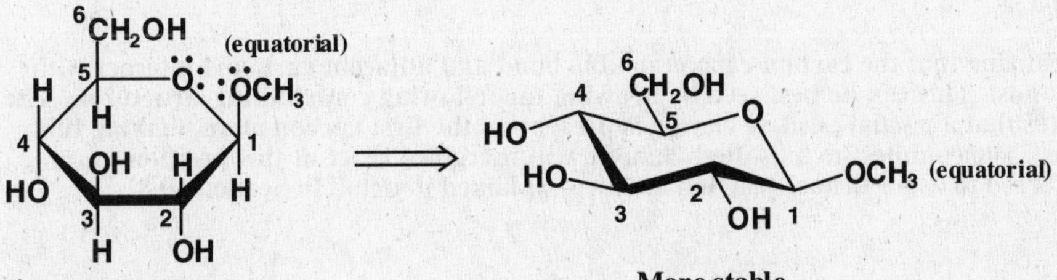

Methyl β-D-glucoside

More stable

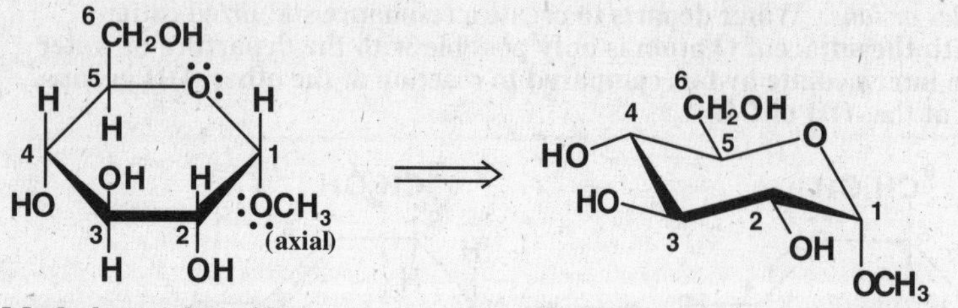

Methyl α-D-glucoside

(c) Which of the two products has the chair conformation of greater stability? Explain.

Groups larger than hydrogen are more stable in the equatorial position due to steric strain in the form of 1,3-diaxial interactions. Therefore, the methyl β-D-glucoside with the -OCH_3 group in the equatorial position is the more stable.

Problem 16.68 Treating a Grignard reagent with carbon dioxide followed by aqueous HCl gives a carboxylic acid.

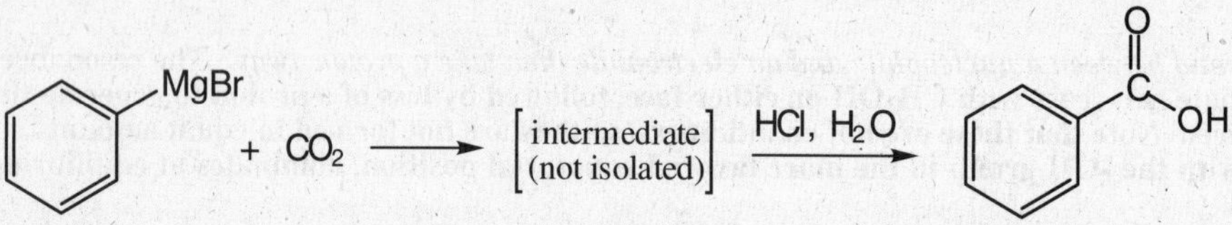

Propose a structural formula for the bracketed intermediate and a mechanism for its formation.

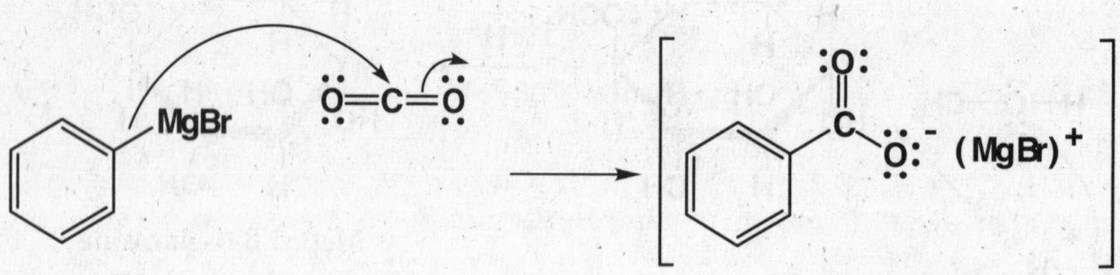

intermediate (not isolated)

Problem 16.69 As we saw in Chapter 6, carbon-carbon double bonds are attacked by electrophiles but not by nucleophiles. An exception to this generalization is the reactivity of α,β-unsaturated aldehydes and ketones toward nucleophiles. Even though an isolated carbon-carbon double bond does not react with 2° amines such as dimethylamine, 3-buten-2-one reacts readily by regioselective addition.

$$Et_2NH + H_2C{=}CH{-}CH_2{-}CH_3 \longrightarrow \text{no reaction}$$

$$Et_2NH + H_2C{=}CH{-}\overset{\overset{\displaystyle O}{\|}}{C}{-}CH_3 \longrightarrow Et_2N{-}CH_2{-}CH_2{-}\overset{\overset{\displaystyle O}{\|}}{C}{-}CH_3$$

Diethylamine **3-Buten-2-one (Methyl vinyl ketone)**

Account for the addition of nucleophiles to the carbon-carbon double bond of an α,β-unsaturated aldehyde or ketone and the regioselectivity of the addition.

The key to this problem is realizing that the carbon-carbon double bond and adjacent carbonyl interact with each other through their π bonds. This can be best seen by drawing the following contributing structures. The structure on the right indicates that a partial positive charge is present on the first carbon atom, making this position slightly electrophilic. Nucleophiles such as diethylamine will therefore react at this position. Reactions of this type are referred to as a Michael reaction and are explained in detail in Section 19.8.

Some nucleophiles such as
diethylamine attack here

Problem 16.70 Ribose, a carbohydrate with the formula shown, forms a cyclic hemiacetal, which, in principle, could contain either a four-membered, a five-membered, or a six-membered ring. When D-ribose is treated with methanol in the presence of an acid catalyst, two cyclic acetals, A and B, are formed, both with molecular formula $C_6H_{10}O_5$. These are separated, and each is treated with sodium periodate (Section 10.8C) followed by dilute aqueous acid. Both A and B yield the same three products in the same ratios.

From this information, deduce whether the cyclic hemiacetal formed by D-ribose is four membered, five membered, or six membered.

D-Ribose forms five membered ring acetals with alcohols such as methanol. There are two isomeric cyclic acetals because there are two orientations possible for the -OCH$_3$ group relative to the other substituents on the ring. The periodate cleaves between the carbons labeled as 2 and 3. The acidic aqueous solution hydrolyzes the acetal at carbon 1.

Problem 16.71 The favorite nuclide used in positron emission tomography (PET scan) to follow glucose metabolism is fluorine-18, which decays by positron emission to oxygen-18 and has a half-life of 110 minutes. Fluorine-18 is administered in the form of fludeoxyglucose F-18; the product of this molecule's decay is glucose.

Fludeoxyglucose F-18   **D-Glucose**   A positron

Draw the alternative chair conformations for fludeoxyglucose F-18, and select the more stable of the two.

The major chair conformation is the one with all but the acetal -OCH$_3$ group equatorial.

Fludeoxyglucose F-18   **Major chair**
                                      conformation

Synthesis
Problem 16.72 Show how to convert 1-butanol into racemic 4-octanol. You must use 1-butanol as the source of all carbon atoms in the target molecule. Show all reagents and all molecules synthesized along the way.

1-Butanol   **4-Octanol**
                                                 (racemic)

PCC

PBr$_3$

1.   2. HCl, H$_2$O

Mg
ether

Recognize that the product has eight carbon atoms, so it must be made from a combination of two four-carbon units. The key recognition element of the product is the hydroxyl group at the site of new carbon-carbon bond formation, which indicates a Grignard reaction. The required Grignard reagent can be made by first converting the 1-butanol into 1-bromobutane followed by treatment with Mg in ether. The required butanal can be made by reacting 1-butanol with PCC.

Problem 16.73 Show how to convert ethylene into 2-methyl-1,3-dioxolane. You must use ethylene as the source of all carbon atoms in the target molecule. Show all reagents and all molecules synthesized along the way.

Recognize that the product is actually the cyclic acetal derived from acetaldehyde and ethylene glycol. The acetaldehyde can be synthesized from ethylene by the two reaction sequence of acid-catalyzed hydration followed by oxidation with PCC. Ethylene glycol can be synthesized in one step using OsO_4 (shown) or in two steps using RCO_3H to form ethylene oxide followed by acid or base-catalyzed epoxide opening in water (not shown).

Problem 16.74 Show how to convert (2-bromoethyl)benzene into racemic 2-chloro-1-phenylethanone. Show all reagents and all molecules synthesized along the way.

Recognize that the product has the same number of carbon atoms as the starting material, so no new carbon-carbon bonds will need to be made. Further, recognize that the product can be derived though the α-halogenation of acetophenone. Acetophenone can be synthesized using the reaction sequence of E2 elimination using a hindered strong base, followed by acid-catalyzed hydration then oxidation using PCC (shown) or chromic acid (not shown).

Problem 16.75 Show how to convert acetaldehyde into racemic 3-hydroxybutanal. You must use acetaldehyde as the source of all carbon atoms in the target molecule. Show all reagents and all molecules synthesized along the way.

Recognize that the product has four carbon atoms, so a new carbon-carbon bond must be created between a couple of two-carbon units. Further, notice the key recognition element of a hydroxyl group next to the location of carbon-carbon bond formation indicating the use of a Grignard reaction. Ordinarily, the aldehyde function of the product would be incompatible with a Grignard reaction, but not if a protecting group such as a cyclic acetal is used. Therefore, propose formation of the acetal-protected Grignard reagent using the reaction sequence of α-halogenation, followed by cyclic acetal formation to mask the aldehyde function, then treatment with Mg in ether. Reaction of this Grignard reagent with acetaldehyde followed by deprotection of the cyclic acetal gives the final product. Note that the cyclic acetal might hydrolyze in the acidic workup following the Grignard reaction, or it might require more forcing acidic conditions as shown.

Reactions in Context

Problem 16.76 Cetrizine is a nonsedating antihistamine. The first step in a synthesis of cetrizine involves the following Grignard reaction.

(a) Draw the product of this Grignard reaction.

(b) Cetrizine is chiral. Like many chiral drugs, one enantiomer of cetrizine is more active than the other enantiomer or the racemic mixture. The levorotatory (S) enantiomer of cetrizine is more active, and syntheses have been developed that produce only the desired enantiomer in high yield. Label the chiral center of cetrizine with an asterisk.

Cetrizine   **(S)-Cetrizine**

<u>Problem 16.76</u> Wittig reactions are widely used in drug synthesis. Write the predominant product of the following Wittig reaction used in the first step of a synthesis of elitriptan, which is used to treat migraine headaches.

Elitriptan

Recall that a carbonyl adjacent to the anionic carbon of a Wittig reagent leads to primarily *E* products as shown.

<u>Problem 16.77</u> Complete the following Wittig reactions.

Problem 16.78 Complete the following Grignard reaction. The starting material is chiral and present as a single enantiomer. Using models, predict which product enantiomer predominates and include that stereochemical prediction in your answer.

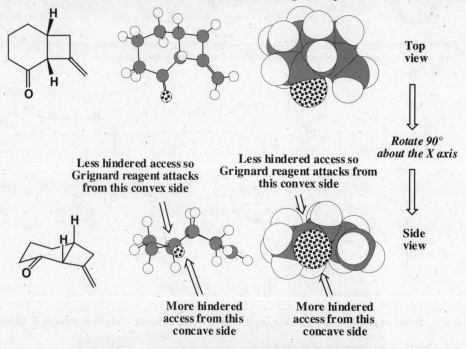

As can be seen using the models below, the starting ketone has a bent structure, with one face of the molecule being more convex, and the other side being more concave. There is much greater access to the carbonyl on the convex face of the molecule, so the Grignard reagent will react from this side, leading to the product stereoisomer shown.

Top view

Rotate 90° about the X axis

Side view

Less hindered access so Grignard reagent attacks from this convex side

Less hindered access so Grignard reagent attacks from this convex side

More hindered access from this concave side

More hindered access from this concave side

Green sea turtle *Chelonia mydas*
Kona coast, Hawaii

CHAPTER 17
Solutions to the Problems

Problem 17.1 Each of these carboxylic acids has a well-recognized common name. A derivative of glyceric acid is an intermediate in glycolysis. Maleic acid is an intermediate in the tricarboxylic acid (TCA) cycle. Mevalonic acid is an intermediate in the biosynthesis of steroids. Write the IUPAC name for each compound. Be certain to specify configuration.

(a)

Glyceric acid

(R)-2,3-Dihydroxy-propanoic acid

(b)

Maleic acid

(Z)-2-Butenedioic acid

(c)

Mevalonic acid

(R)-3,5-Dihydroxy-3-methylpentanoic acid

Problem 17.2 Which is the stronger acid in each pair?

(a) CH_3COOH or CH_3SO_3H

Acetic acid   **Methanesulfonic acid**

$pK_a = 4.8$ $pK_a = -1.8$

(b)

2-Oxopropanoic acid (Pyruvic acid)   or   **Propanoic acid**

$pK_a = 2.5$ $pK_a = 4.8$

In (a), the third oxygen atom on sulfur increases acidity by both inductive and resonance effects compared to the carboxylic acid. In (b), the electron withdrawing carbonyl group in the 2 position of pyruvic acid increases acidity due to inductive effects.

Problem 17.3 Write equations for the reaction of each acid in Example 17.3 with ammonia, and name the carboxylic salt formed.

(a) $COOH$ + NH_3 $\longrightarrow$ $COO^-\ NH_4^+$

Butanoic acid   **Ammonium butanoate**

(b) $COOH$ + NH_3 $\longrightarrow$ $COO^-\ NH_4^+$

(S)-Lactic acid   **Ammonium (S)-lactate**

Problem 17.4 Complete the equation for each Fischer esterification.

(a)   + HO $\rightleftharpoons$ + H_2O

(b) HO $\rightleftharpoons$ + H_2O

(a cyclic ester)

<u>Problem 17.5</u> Complete the equation for each reaction.

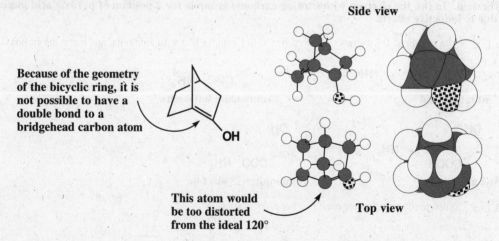

(a) [structure: benzene ring with COOH and OCH₃ substituents] + SOCl₂ ⟶ [structure: benzene ring with CCl(=O) and OCH₃] + SO₂ + HCl

(b) [structure: cyclohexane with OH] + SOCl₂ ⟶ . [structure: cyclohexane with Cl] + SO₂ + HCl

<u>Problem 17.6</u> Account for the observation that the following β-ketoacid can be heated for extended periods at temperatures above its melting point without noticeable decarboxylation.

The mechanism we have proposed for decarboxylation of a β-ketoacid involves formation of an enol intermediate and equilibrium of the enol, via keto-enol tautomerism, with the keto form. Given the geometry of a bicycloalkane, it is not possible to have a carbon-carbon double bond to a bridgehead carbon because of the prohibitively high angle strain. Therefore, because the enol intermediate cannot be formed, this β-ketoacid does not undergo thermal decarboxylation.

Side view

Because of the geometry of the bicyclic ring, it is not possible to have a double bond to a bridgehead carbon atom

This atom would be too distorted from the ideal 120°

Top view

<u>Problem 17.7</u> Write the IUPAC name of each compound, showing stereochemistry where relevant:

(a) [structure: cyclohexene with COOH]

1-Cyclohexenecarboxylic acid

(b) [structure with OH and COOH]

(R)-4-Hydroxypentanoic acid

(c) [structure with COOH]

(E)-3,7-Dimethyl-2,6-octadienoic acid

(d) [structure: cyclopentane with methyl and COOH]

**1-Methylcyclopentane-
carboxylic acid**

(e) [structure: COO⁻NH₄⁺]

Ammonium hexanoate

(f) [structure with HO, COOH, COOH]

(R)-2-Hydroxybutanedioic acid

<u>Problem 17.8</u> Draw a structural formula for each compound.

(a) Phenylacetic acid

(b) 4-Aminobutanoic acid

H_2N —————— COOH

(c) 3-Chloro-4-phenylbutanoic acid

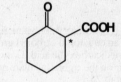

(d) Propenoic acid (acrylic acid)

(e) (Z)-3-Hexenedioic acid

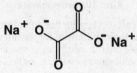
HOOC ————— COOH

(f) 2-Pentynoic acid

$CH_3CH_2C \equiv CCOOH$

(g) Potassium phenylacetate

(h) Sodium oxalate

Na^+ O^- ——— O^- Na^+

(i) 2-Oxocyclohexanecarboxylic acid

—COOH

(j) 2,2-Dimethylpropanoic acid

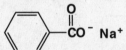

$$CH_3 \\ CH_3CCOOH \\ CH_3$$

<u>Problem 17.9</u> Megatomoic acid, the sex attractant of the female black carpet beetle, has the following structure:

————————————COOH

Megatomoic acid

(a) What is its IUPAC name?

Its IUPAC name is (3E, 5E)-3,5-tetradecadienoic acid.

(b) State the number of stereoisomers possible for this compound.

Four stereoisomers are possible; each double bond can have either an E or Z (trans or cis) configuration.

<u>Problem 17.10</u> Draw structural formulas for each salt.

(a) Sodium benzoate

—CO⁻ Na⁺

(b) Lithium acetate

$CH_3CO^- \ Li^+$

(c) Ammonium acetate

$CH_3CO^- \ NH_4^+$

(d) Disodium adipate

Na^+ O^- ———————— O^- Na^+

(e) Sodium salicylate

OH
—CO⁻ Na⁺

(f) Calcium butanoate

$\left(\quad O^- \right)_2 Ca^{2+}$

<u>Problem 17.11</u> The monopotassium salt of oxalic acid is present in certain leafy vegetables, including rhubarb. Both oxalic acid and its salts are poisonous in high concentrations. Draw the structural formula of monopotassium oxalate.

O O
HOC—CO⁻ K⁺

Monopotassium oxalate

Problem 17.12 Potassium sorbate is added as a preservative to certain foods to prevent bacteria and molds from causing food spoilage and to extend the foods' shelf life. The IUPAC name of potassium sorbate is potassium (2E,4E)-2,4-hexadienoate. Draw a structural formula for potassium sorbate.

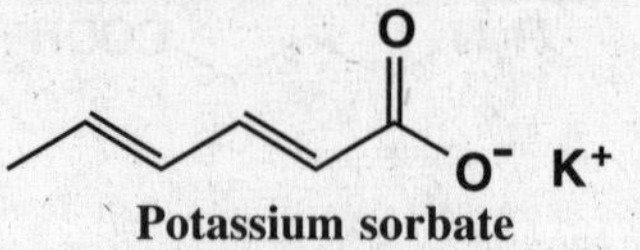

Potassium sorbate

Problem 17.13 Zinc 10-undecenoate, the zinc salt of 10-undecenoic acid, is used to treat certain fungal infections, particularly *Tinea pedis* (athlete's foot). Draw the structural formula for this zinc salt.

$$\left(CH_2{=}CH(CH_2)_8\overset{\overset{O}{\|}}{C}O^- \right)_2 Zn^{2+}$$

Zinc 10-undecenoate

Problem 17.14 On a cyclohexane ring, an axial carboxyl group has a conformational energy of 5.9 kJ (1.4 kcal)/mol relative to an equatorial carboxyl group. Consider the equilibrium for the alternative chair conformations of *trans*-1,4-cyclohexanedicarboxylic acid. Draw the less stable chair conformation on the left of the equilibrium arrows and the more stable chair on the right. Calculate ΔG° for the equilibrium as written, and calculate the ratio of more stable chair to the less stable chair at 25°C.

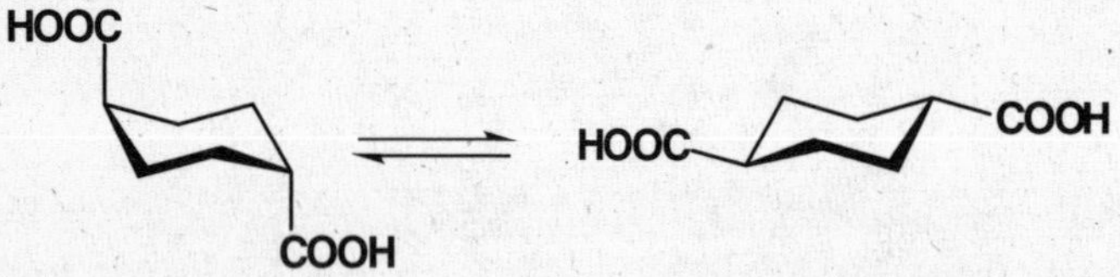

$\Delta G° = -11.8$ kJ/mol

As written, the conformation on the right is (2 x 5.9) = 11.8 kJ/mol more stable than the conformation on the left.

At equilibrium the relative amounts of each form are given by the equation:

$$\Delta G° = -RT\ln K_{eq}$$

Here K_{eq} refers to the ratio of the alternative chair conformations. Rearranging gives:

$$\ln K_{eq} = \frac{-\Delta G°}{RT}$$

Solving gives:

$$K_{eq} = e^{\left(\frac{-\Delta G°}{RT}\right)}$$

Plugging in the values for ΔG°, R, and 298 K gives the answer:

$$K_{eq} = e^{\left(\frac{-(-11.8 \text{ kJ/mol})}{(8.314 \times 10^{-3} \text{ kJ/mol K})(298 \text{ K})}\right)} = e^{4.76} = 1.17 \times 10^2$$

Physical Properties

Problem 17.15 Arrange the compounds in each set in order of increasing boiling point:

The better the hydrogen bond capability, the higher the boiling point.

(a) $CH_3(CH_2)_5COOH$ $CH_3(CH_2)_6CHO$ $CH_3(CH_2)_6CH_2OH$

The following are listed in order of increasing boiling point:

$CH_3(CH_2)_6CHO$ $CH_3(CH_2)_6CH_2OH$ $CH_3(CH_2)_5COOH$

  bp 171°C   bp 195°C   bp 223°C

(b)

The following are listed in order of increasing boiling point:

bp 35°C   bp 117°C   bp 141°C

Problem 17.16 Acetic acid has a boiling point of 118°C, whereas its methyl ester has a boiling point of 57°C. Account for the fact that the boiling point of acetic acid is higher than that of its methyl ester, even though acetic acid has a lower molecular weight.

Acetic acid can make strong hydrogen bonds, but the methyl ester lacks a hydrogen bonding hydrogen atom. Thus, acetic acid will have the much higher boiling point compared to the methyl ester.

Can form hydrogen bonds with this H atom

Acetic acid
bp 118°C

Methyl acetate
bp 57°C

Spectroscopy

Problem 17.17 Given here are ^{1}H-NMR and ^{13}C-NMR spectral data for nine compounds. Each compound shows strong absorption between 1720 and 1700 cm^{-1}, and strong, broad absorption over the region 2500 - 3500 cm^{-1}. Propose a structural formula for each compound. Refer to Appendices 4, 5, and 6 for spectral correlation tables.

(a) $C_5H_{10}O_2$

^{1}H-NMR	^{13}C-NMR
0.94 (t, 3H)	180.71
1.39 (m, 2H)	33.89
1.62 (m, 2H)	26.76
2.35 (t, 2H)	22.21
12.0 (s, 1H)	13.69

0.94 1.39 1.62 2.35   12.0
$CH_3CH_2CH_2CH_2COOH$

(b) $C_6H_{12}O_2$

^{1}H-NMR	^{13}C-NMR
1.08 (s, 9H)	179.29
2.23 (s, 2H)	47.82
12.1 (s, 1H)	30.62
	29.57

1.08
CH_3
1.08 | 2.23   12.1
CH_3CCH_2COOH
| 1.08
CH_3

(c) $C_5H_8O_4$

^{1}H-NMR	^{13}C-NMR
0.93 (t, 3H)	170.94
1.80 (m, 2H)	53.28
3.10 (t, 1H)	21.90
12.7 (s, 2H)	11.81

12.7   3.10   12.7
$HOOCCHCOOH$
| 1.80 0.93
CH_2CH_3

(d) $C_5H_8O_4$

^{1}H-NMR	^{13}C-NMR
1.29 (s, 6H)	174.01
12.8 (s, 2H)	48.77
	22.56

1.29
CH_3
12.8 | 12.8
$HOOCCCOOH$
| 1.29
CH_3

(e) $C_4H_6O_2$

^{1}H-NMR	^{13}C-NMR
1.91 (d, 3H)	172.26
5.86 (d, 1H)	147.53
7.10 (m, 1H)	122.24
12.4 (s, 1H)	18.11

1.91 7.10 5.86   12.4
$CH_3CH=CHCOOH$

(f) $C_3H_4Cl_2O_2$

^{1}H-NMR	^{13}C-NMR
2.34 (s, 3H)	171.82
11.3 (s, 1H)	79.36
	34.02

Cl
2.34 | 11.3
CH_3CCOOH
|
Cl

(g) $C_5H_8Cl_2O_2$

^{1}H-NMR	^{13}C-NMR
1.42 (s, 6H)	180.15
6.10 (s, 1H)	77.78
12.4 (s, 1H)	51.88
	20.71

(h) $C_5H_9BrO_2$

^{1}H-NMR	^{13}C-NMR
0.97 (t, 3H)	176.36
1.50 (m, 2H)	45.08
2.05 (m, 2H)	36.49
4.25 (t, 1H)	20.48
12.1 (s, 1H)	13.24

(i) $C_4H_8O_3$

^{1}H-NMR	^{13}C-NMR
2.62 (t, 2H)	177.33
3.38 (s, 3H)	67.55
3.68 (t, 2H)	58.72
11.5 (s, 1H)	34.75

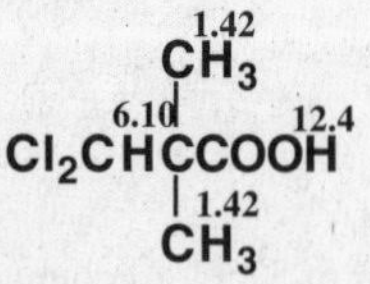

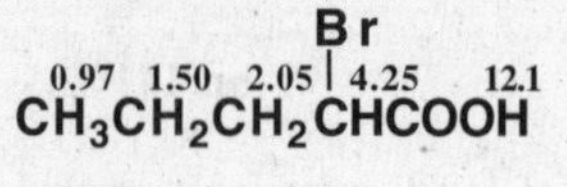

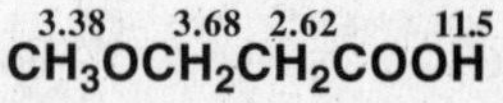

Preparation of Carboxylic Acids

Problem 17.18 Complete each reaction:

(a) 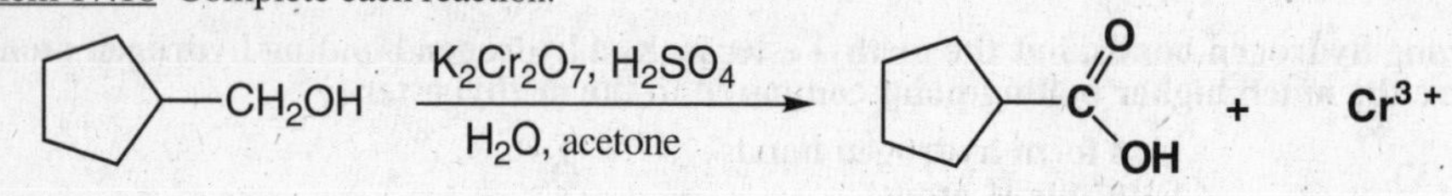

(b) 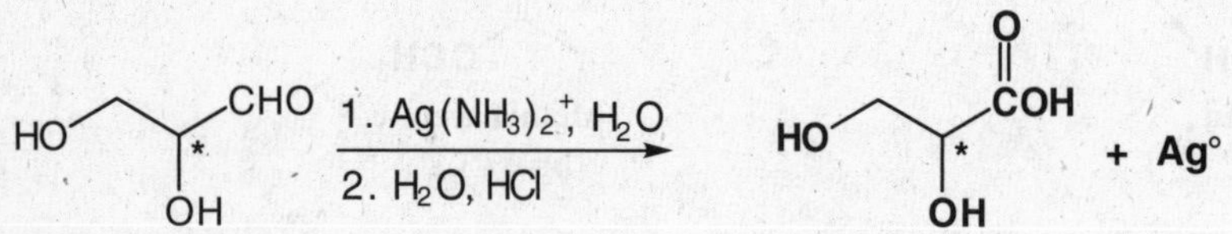

The stereochemistry of the product will be the same as the stereochemistry of the starting material.

(c) 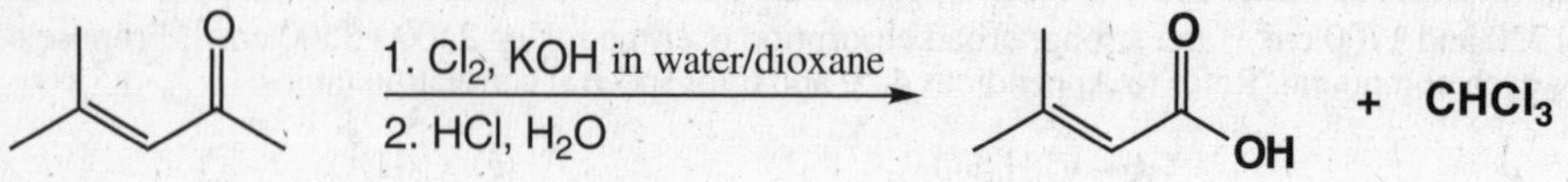

(d) 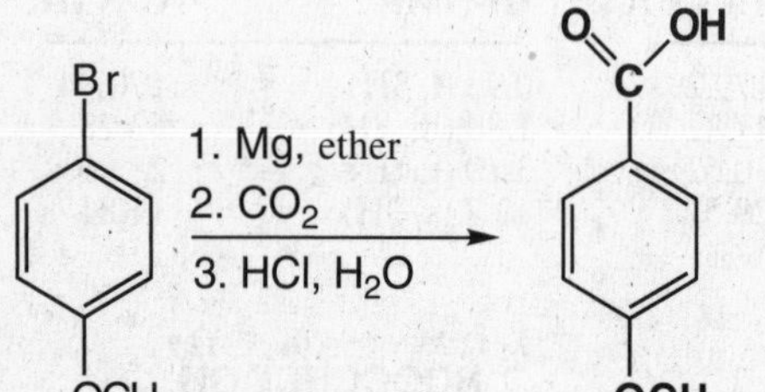

Problem 17.19 Show how to bring about each conversion in good yield.

(a) 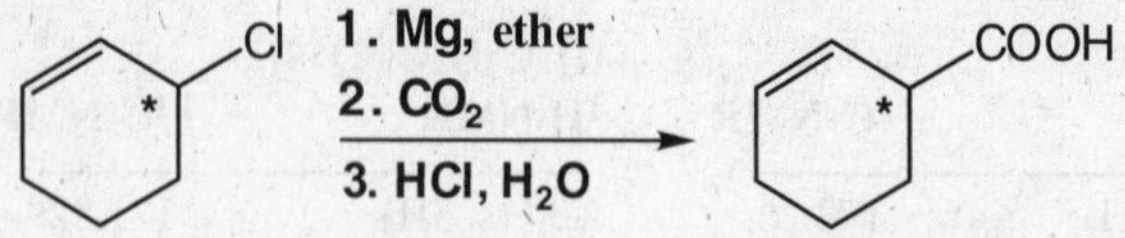

(racemic mixture)

(b) 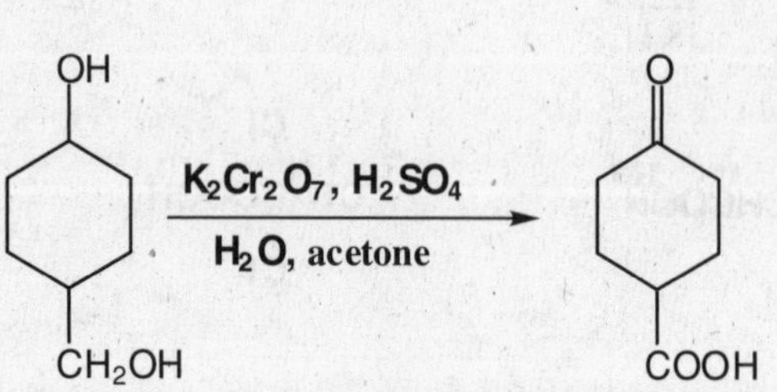

(c) 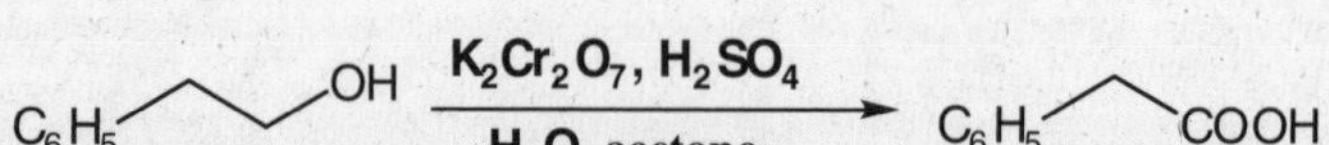

Problem 17.20 Show how to prepare pentanoic acid from each compound:
(a) 1-Pentanol

Oxidation of 1-pentanol by chromic acid gives pentanoic acid.

$$CH_3(CH_2)_3CH_2OH \xrightarrow[\text{H}_2\text{O, acetone}]{\text{K}_2\text{Cr}_2\text{O}_7,\ \text{H}_2\text{SO}_4} CH_3(CH_2)_3COOH + Cr^{3+}$$

(b) Pentanal

Oxidation of pentanal by chromic acid, Tollens' solution, or molecular oxygen gives pentanoic acid.

$$CH_3(CH_2)_3\overset{O}{\overset{\|}{C}}H + O_2 \longrightarrow CH_3(CH_2)_3\overset{O}{\overset{\|}{C}}OH$$

(c) 1-Pentene

Hydroboration/oxidation of 1-pentene gives 1-pentanol. Oxidation of 1-pentanol as in part (a) gives pentanoic acid.

$$CH_3CH_2CH_2CH=CH_2 \xrightarrow[\text{2) H}_2\text{O}_2,\ \text{NaOH}]{\text{1) BH}_3\cdot\text{THF}} CH_3(CH_2)_3CH_2OH \text{ then as in part (a)}$$

(d) 1-Butanol

Conversion of 1-butanol to 1-bromobutane, then to butyllithium or butylmagnesium bromide followed by carbonation and then acidification gives pentanoic acid.

$$CH_3(CH_2)_3OH \xrightarrow[\text{2) Mg,ether}]{\text{1) PBr}_3} CH_3(CH_2)_3MgBr \xrightarrow[\text{2) HCl,H}_2\text{O}]{\text{1) CO}_2} CH_3(CH_2)_3COOH$$

(e) 1-Bromopropane

Treatment of 1-bromopropane with magnesium in ether followed by treatment of the Grignard reagent with ethylene oxide gives 1-pentanol. Oxidation of 1-pentanol as in part (a) gives pentanoic acid.

$$CH_3CH_2CH_2Br \xrightarrow{\text{Mg, ether}} CH_3CH_2CH_2MgBr \xrightarrow[\text{2) HCl, H}_2\text{O}]{\text{1) H}_2\text{C}-\text{CH}_2 (O)} CH_3(CH_2)_3CH_2OH$$

$$\xrightarrow{\text{Oxidize as in part (a)}} CH_3(CH_2)_3COOH$$

(f) 1-Hexene

Ozonolysis of the alkene followed by oxidation of the resulting aldehyde gives the carboxylic acid product.

$$CH_3(CH_2)_3CH=CH_2 \xrightarrow[\text{2. (CH}_3)_2\text{S}]{\text{1. O}_3} CH_3(CH_2)_3\overset{O}{\overset{\|}{C}}H + H\overset{O}{\overset{\|}{C}}H$$

$$CH_3(CH_2)_3\overset{O}{\overset{\|}{C}}H \xrightarrow[\text{H}_2\text{O, acetone}]{\text{K}_2\text{Cr}_2\text{O}_7,\ \text{H}_2\text{SO}_4} CH_3(CH_2)_3\overset{O}{\overset{\|}{C}}OH$$

Problem 17.21 Draw the structural formula of a compound with the given molecular formula that, on oxidation by potassium dichromate in aqueous sulfuric acid, gives the carboxylic acid or dicarboxylic acid shown.

(a) $C_6H_{14}O$ ──oxidation──▶ [carboxylic acid structure]   [alcohol structure]

(b) $C_6H_{12}O$ ──oxidation──▶ [carboxylic acid structure]   [aldehyde structure]

(c) $C_6H_{14}O_2$ ──oxidation──▶ [dicarboxylic acid structure]   [diol structure]

Problem 17.22 Show the reagents and experimental conditions necessary to bring about each conversion in good yield.

(a)   [cyclopentanol] ───▶ [cyclopentanecarboxylic acid, COOH]

The most convenient way to add a carbon atom in the form of a carboxyl group is carbonation of an organolithium or organomagnesium compound.

[cyclopentanol, OH] 1) $SOCl_2$ [cyclopentyl MgCl] 1) CO_2 [cyclopentane COOH]
                      2) Mg, ether   2) HCl, H_2O

(b) $CH_3\overset{CH_3}{\underset{CH_3}{C}}OH$ ───▶ $CH_3\overset{CH_3}{\underset{CH_3}{C}}COOH$

Use the same set of reactions in this part as you used in part (a), except because the starting material is a tertiary alcohol, HCl is used in place of $SOCl_2$.

$CH_3\overset{CH_3}{\underset{CH_3}{C}}OH$ ──HCl──▶ $CH_3\overset{CH_3}{\underset{CH_3}{C}}Cl$ 1) Mg, ether   2) CO_2 3) HCl, H_2O ──▶ $CH_3\overset{CH_3}{\underset{CH_3}{C}}COOH$

(c) $CH_3\overset{CH_3}{\underset{CH_3}{C}}OH$ ───▶ $CH_3\overset{CH_3}{C}HCOOH$

This conversion is best accomplished by acid-catalyzed dehydration of the tertiary alcohol to an alkene followed by hydroboration/oxidation. These two reactions in sequence shift the -OH group from the tertiary carbon to a primary carbon. Oxidation of this primary alcohol with chromic acid gives 2-methylpropanoic acid.

$CH_3\overset{CH_3}{\underset{CH_3}{C}}OH$ ──H_2SO_4 (-H_2O)──▶ $CH_3\overset{CH_3}{C}=CH_2$ 1) $BH_3 \cdot THF$ 2) H_2O_2, NaOH ──▶ $CH_3\overset{CH_3}{C}HCH_2OH$

──$K_2Cr_2O_7$, H_2SO_4 / H_2O, acetone──▶ $CH_3\overset{CH_3}{C}HCOOH$

(d)   CH₃
        |
   CH₃COH   ⟶   CH₃CHCH₂COOH
        | |
        CH₃   CH₃

Repeat the first set of steps as in part (c) to convert *tert*-butyl alcohol to isobutyl alcohol. Then use the sequence of steps as in part (a) and (b) to convert this alcohol to a carboxylic acid containing one more carbon atom.

      CH₃   CH₃   CH₃
      | 1) SOCl₂ | 1) CO₂ |
CH₃CHCH₂OH   ──────────────⟶   CH₃CHCH₂MgCl ──────────⟶   CH₃CHCH₂COOH
                  2) Mg, ether   2) HCl, H₂O

From part (c)

(e) CH₃CH=CHCH₃   ⟶   CH₃CH=CHCH₂COOH

This problem is analogous to part (a) except an allylic halogenation of the alkene starting material is the first step.

                                                              1) Mg, ether
                            NBS   2) CO₂
CH₃CH=CHCH₃   ───────────────────⟶   CH₃CH=CHCH₂Br ──────────────⟶   CH₃CH=CHCH₂COOH
               Light or peroxides   3) HCl, H₂O

Problem 17.23 Succinic acid can be synthesized by the following series of reactions from acetylene. Show the reagents and experimental conditions necessary to carry out this synthesis.

| Acetylene | 2-Butyne-1,4-diol | 1,4-Butanediol | Butanedioic acid (Succinic acid) |

Two one-carbon fragments in the form of formaldehyde are added to the carbon skeleton of acetylene. To bring about this double addition, acetylene is treated with two moles of sodium amide followed by two moles of formaldehyde. Acidification of the resulting disodium salt gives 2-butyne-1,4-diol.

                  1. 2 NaNH₂
HC≡CH   ─────────────────────⟶   Na⁺O⁻   O⁻Na⁺   +   2 NH₃
                       O
                       ‖
                  2. 2 H–C–H

Na⁺O⁻   O⁻Na⁺   + 2HCl ⟶   HO   OH

                                        2-Butyne-1,4-diol

Catalytic reduction of the carbon-carbon triple bond in 2-butyne-1,4-diol over a transition metal catalyst gives 1,4-butanediol.

                                      Pt
HO   OH + 2H₂   ────────────────⟶   HO   OH
                                   or other
  2-Butyne-1,4-diol   transition metal   1,4-Butanediol
                                  catalyst

Oxidation of 1,4-butanediol by chromic acid gives succinic acid.

1,4-Butanediol

**Butanedioic acid
(Succinic acid)**

<u>Problem 17.24</u> The reaction of an α–diketone with concentrated sodium or potassium hydroxide to give the salt of an α–hydroxyacid is given the general name benzil-benzilic acid rearrangement. It is illustrated by the conversion of benzil to sodium benzilate and then to benzilic acid.

Benzil
(an α-diketone)

Sodium benzilate

Benzilic acid

Propose a mechanism for this rearrangement.

Step 1: Make a bond between a nucleophile and an electrophile. **Addition of hydroxide ion to one of the carbonyl groups forms a tetrahedral carbonyl addition intermediate.**

**Tetrahedral carbonyl
addition intermediate**

Step 2: 1,2 Shift. **Collapse of this intermediate with simultaneous regeneration of the carbonyl group and migration of a phenyl group gives sodium benzilate**

Step 3: Add a proton-take a proton away.

**Sodium
benzilate**

Acidity of Carboxylic Acids
<u>Problem 17.25</u> Select the stronger acid in each set.
(a) Phenol (pK_a 9.95) and benzoic acid (pK_a 4.19)

Recall that pK_a is the negative $\log_{10}$ of K_a. The smaller the pK_a, the stronger the acid, so benzoic acid is the stronger acid.

(b) Lactic acid (K_a 8.4 x 10^{-4}) and ascorbic acid (K_a 7.9 x 10^{-5})

The larger the value of K_a, the stronger the acid, so lactic acid is the stronger acid.

Problem 17.26 In each set, assign the acid its appropriate pK_a.

(a) and (pK_a 4.19 and 0.70)

4.19   0.70

The third oxygen on sulfur increases acidity compared to an analogous carboxylic acid.

(b) and (pK_a 3.58 and 2.49)

3.58   2.49

The electron withdrawing carbonyl group increases acidity through inductive effects, so the closer the carbonyl group to the carboxylic acid, the greater the effect and the stronger the acid.

(c) CH_3CH_2COOH and $N\equiv CCH_2COOH$ (pK_a 4.78 and 2.45)

4.78   2.45

The nitrile is electron withdrawing, so it increases acidity through inductive effects.

Problem 17.27 Low-molecular-weight dicarboxylic acids normally exhibit two different pK_a values. Ionization of the first carboxyl group is easier than the second. This effect diminishes with molecular size, and, for adipic acid and longer chain dicarboxylic acids, the two acid ionization constants differ by about one pK unit.

Dicarboxylic Acid	Structural Formula	pK_{a1}	pK_{a2}
Oxalic	HOOCCOOH	1.23	4.19
Malonic	$HOOCCH_2COOH$	2.83	5.69
Succinic	$HOOC(CH_2)_2COOH$	4.16	5.61
Glutaric	$HOOC(CH_2)_3COOH$	4.31	5.41
Adipic	$HOOC(CH_2)_4COOH$	4.43	5.41

Why do the two pK_a values differ more for the shorter chain dicarboxylic acids than for the longer chain dicarboxylic acids?

For these dicarboxylic acids, in going from the first dissociation to the second, the molecule is changing from a monoanion to a dianion. Electrostatic repulsion hinders formation of two negative charges in nearby regions of space. Thus, the second pK_a values are higher than the first, and the effect is more pronounced the shorter the chain between the two carboxyl groups.

Problem 17.28 Complete the following acid-base reactions:

(a) $-CH_2COOH$ + NaOH $\longrightarrow$ $-CH_2COO^-$ Na$^+$ + H_2O

(b) $CH_3CH=CHCH_2COOH$ + NaHCO$_3$ $\longrightarrow$ $CH_3CH=CHCH_2COO^-$ Na$^+$ + H_2O + CO_2

(c) + NaHCO$_3$ $\longrightarrow$ + H_2O + CO_2

(d)   CH₃CHCOOH + H₂NCH₂CH₂OH ⟶ CH₃CHCOO⁻ (H₃N⁺CH₂CH₂OH)

$$\text{(d)} \quad \underset{\underset{\text{OH}}{|}}{CH_3CHCOOH} + H_2NCH_2CH_2OH \longrightarrow \underset{\underset{\text{OH}}{|}}{CH_3CHCOO^-} (H_3\overset{+}{N}CH_2CH_2OH)$$

$$\text{(e)} \quad CH_3CH=CHCH_2COO^-Na^+ + HCl \longrightarrow CH_3CH=CHCH_2COOH + NaCl$$

$$\text{(f)} \quad CH_3CH_2CH_2CH_2Li + CH_3COOH \longrightarrow CH_3CH_2CH_2CH_3 + CH_3COO^-Li^+$$

$$\text{(g)} \quad CH_3CH_2CH_2CH_2MgBr + CH_3CH_2OH \longrightarrow CH_3CH_2CH_2CH_3 + CH_3CH_2O^- MgBr^+$$

<u>Problem 17.29</u> The normal pH range for blood plasma is 7.35 - 7.45. Under these conditions, would you expect the carboxyl group of lactic acid (pK_a 3.08) to exist primarily as a carboxyl group or as a carboxylate anion? Explain.

Recall from the definition of K_a that:

$$K_a = \frac{[A^-][H^+]}{[H\text{-}A]} \quad \text{so dividing both sides by } [H^+] \text{ gives} \quad \frac{K_a}{[H^+]} = \frac{[A^-]}{[H\text{-}A]}$$

Here, $[H^+]$ is concentration of H^+, $[H\text{-}A]$ is concentration of protonated acid (lactic acid in this case) and $[A^-]$ is the concentration of deprotonated acid (lactic acid carboxylate anion in this case). Therefore, if the ratio of K_a / $[H^+]$ is greater than 1, $[A^-]$ will be the predominant form, and if the ratio of K_a / $[H^+]$ is less than 1, then $[H\text{-}A]$ will be the predominant form. Recall that pH = $-\log_{10}[H^+]$, so a pH of 7.4 corresponds to a $[H^+]$ of $10^{-(pH)}$ = $10^{-(7.4)}$ = 4.0×10^{-8}. Similarly, $pK_a = -\log_{10} K_a$, so for lactic acid $K_a = 10^{-(pKa)} = 10^{-(3.07)}$ = 8.5×10^{-4}. Using these numbers:

$$\frac{[A^-]}{[H\text{-}A]} = \frac{K_a}{[H^+]} = \frac{8.5 \times 10^{-4}}{4.0 \times 10^{-8}} = 2.1 \times 10^4$$

Therefore, lactic acid will exist primarily as the carboxylate anion in blood plasma.

<u>Problem 17.30</u> The K_{a1} of ascorbic acid is 7.94×10^{-5}. Would you expect ascorbic acid dissolved in blood plasma (pH 7.35-7.45) to exist primarily as ascorbic acid or as ascorbate anion? Explain.

Using the same reasoning described in the answer to Problem 17.29:

$$\frac{[A^-]}{[H\text{-}A]} = \frac{K_a}{[H^+]} = \frac{7.9 \times 10^{-5}}{4.0 \times 10^{-8}} = 2.0 \times 10^3$$

Therefore, ascorbic acid will exist primarily as the ascorbate anion in blood plasma.

<u>Problem 17.31</u> Excess ascorbic acid is excreted in the urine, the pH of which is normally in the range 4.8 - 8.4. What form of ascorbic acid would you expect to be present in urine of pH 8.4, free ascorbic acid or ascorbate anion? Explain.

At pH 8.4, $[H^+] = 4.0 \times 10^{-9}$, therefore using the same reasoning as described in the answer to Problem 17.29 and 17.30:

$$\frac{[A^-]}{[H\text{-}A]} = \frac{K_a}{[H^+]} = \frac{7.9 \times 10^{-5}}{4.0 \times 10^{-9}} = 2.0 \times 10^4$$

Ascorbic acid will exist primarily as the ascorbate anion in urine of pH 8.4.

Reactions of Carboxylic Acids
<u>Problem 17.32</u> Give the expected organic product when phenylacetic acid, $PhCH_2COOH$, is treated with each reagent.
(a) $SOCl_2$

$$PhCH_2\overset{\overset{O}{||}}{C}OH + SOCl_2 \longrightarrow PhCH_2\overset{\overset{O}{||}}{C}Cl + SO_2 + HCl$$

(b) $NaHCO_3$, H_2O

$$PhCH_2\overset{\overset{\displaystyle O}{\|}}{C}OH + NaHCO_3 \longrightarrow PhCH_2\overset{\overset{\displaystyle O}{\|}}{C}O^- Na^+ + H_2O + CO_2$$

(c) NaOH, H_2O

$$PhCH_2\overset{\overset{\displaystyle O}{\|}}{C}OH + NaOH \longrightarrow PhCH_2\overset{\overset{\displaystyle O}{\|}}{C}O^- Na^+ + H_2O$$

(d) CH_3MgBr (1 equivalent)

$$PhCH_2\overset{\overset{\displaystyle O}{\|}}{C}OH + CH_3MgBr \longrightarrow PhCH_2\overset{\overset{\displaystyle O}{\|}}{C}O^- MgBr^+ + CH_4$$

(e) $LiAlH_4$ followed by H_2O

$$PhCH_2\overset{\overset{\displaystyle O}{\|}}{C}OH \xrightarrow[\text{2) } H_2O]{\text{1) } LiAlH_4} PhCH_2CH_2OH$$

(f) CH_2N_2

$$PhCH_2\overset{\overset{\displaystyle O}{\|}}{C}OH + CH_2N_2 \longrightarrow PhCH_2\overset{\overset{\displaystyle O}{\|}}{C}OCH_3 + N_2$$

(g) CH_3OH + H_2SO_4 (catalyst)

$$PhCH_2\overset{\overset{\displaystyle O}{\|}}{C}OH + CH_3OH \xrightarrow{H_2SO_4} PhCH_2\overset{\overset{\displaystyle O}{\|}}{C}OCH_3 + H_2O$$

Problem 17.33 Show how to convert *trans*-3-phenyl-2-propenoic acid (cinnamic acid) to each compound.

(a) C_6H_5 —CH=CH—COOH $\xrightarrow[\text{2) } H_2O]{\text{1) } LiAlH_4}$ C_6H_5—CH=CH—CH$_2$OH

(b) C_6H_5 —CH=CH—COOH $\xrightarrow[\substack{\text{Pt 25°C} \\ \text{2 atm}}]{H_2}$ C_6H_5—CH$_2$CH$_2$—COOH

(c) C_6H_5 —CH=CH—COOH $\xrightarrow[\text{2) } H_2O]{\text{1) } LiAlH_4}$ $\xrightarrow[\substack{\text{Pt 25°C} \\ \text{2 atm}}]{H_2}$ C_6H_5—CH$_2$CH$_2$CH$_2$—OH

Problem 17.34 Show how to convert 3-oxobutanoic acid (acetoacetic acid) to each compound.

(a) 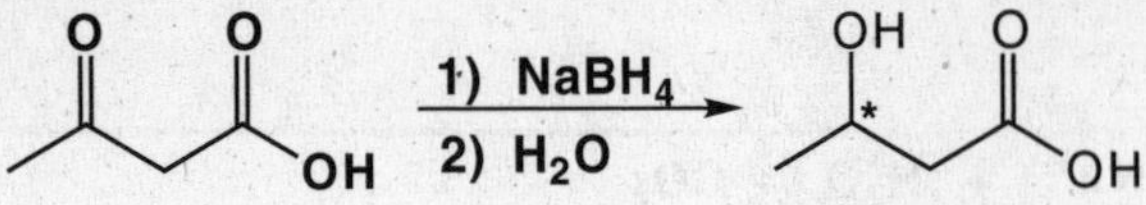
(racemic mixture)

(b) 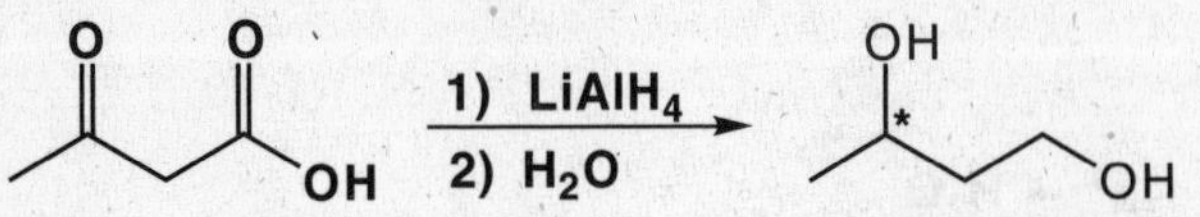
(racemic mixture)

(c)

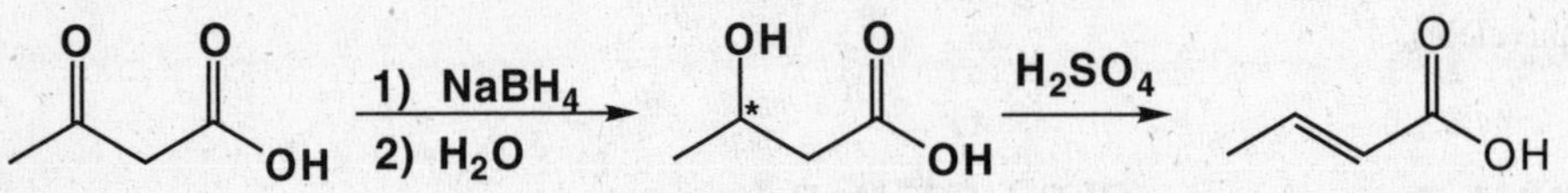

Problem 17.35 Complete these examples of Fischer esterification. Assume that alcohol is present in excess.

(a)

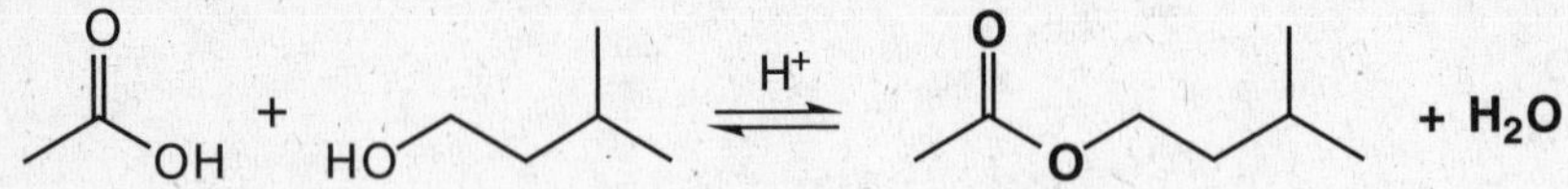

(b)

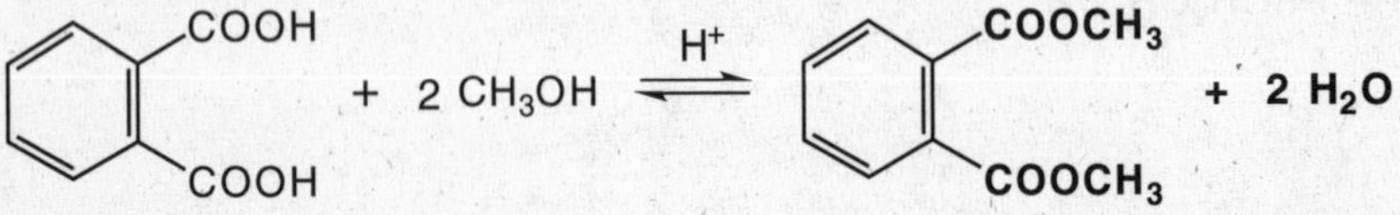

(c)

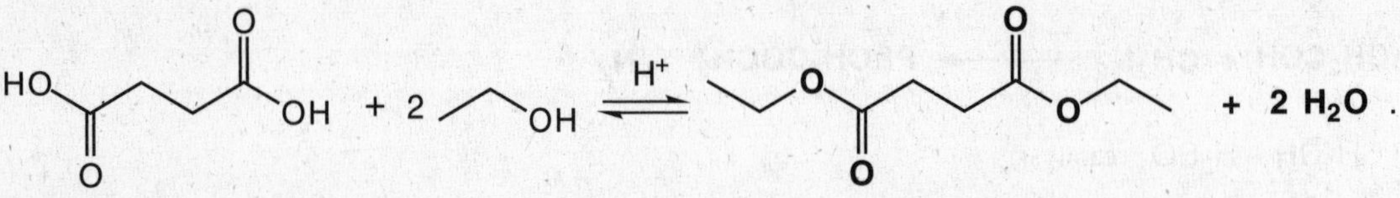

Problem 17.36 Benzocaine, a topical anesthetic, is prepared by treatment of 4-aminobenzoic acid with ethanol in the presence of an acid catalyst followed by neutralization. Draw a structural formula for benzocaine.

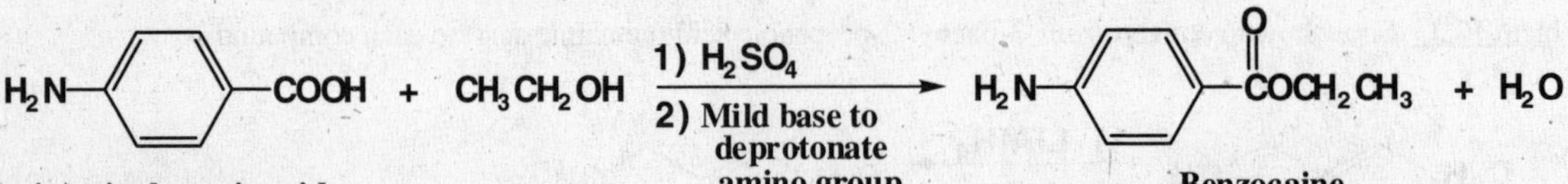

4-Aminobenzoic acid   1) H_2SO_4 2) Mild base to deprotonate amino group   Benzocaine (a topical anesthetic)

Problem 17.37 Name the carboxylic acid and alcohol from which each ester is derived.

(a) 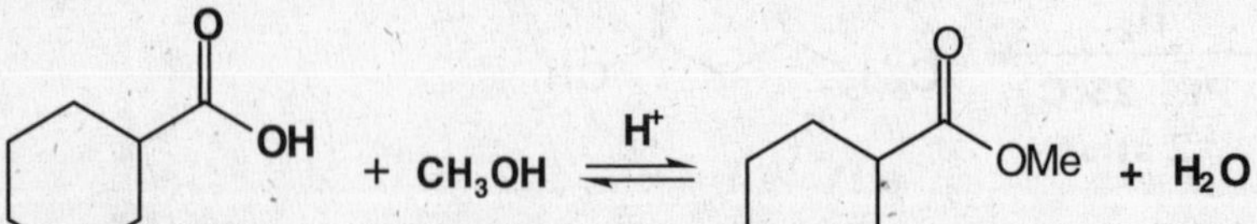
+ CH_3OH
Methanol
Cyclohexane-carboxylic acid

(b) 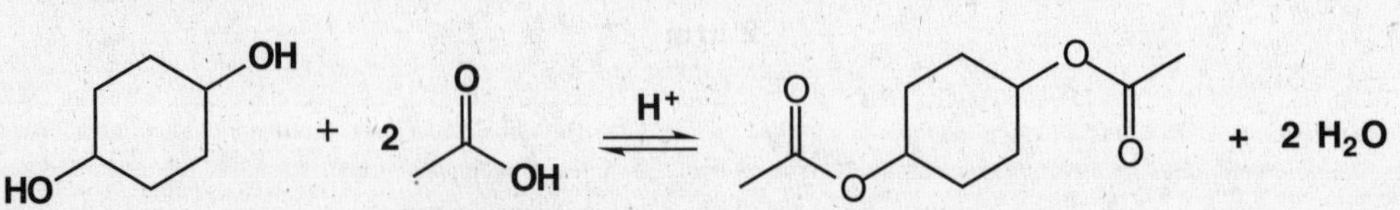
+ 2
Acetic acid
1,4-Cyclohexanediol

(c)

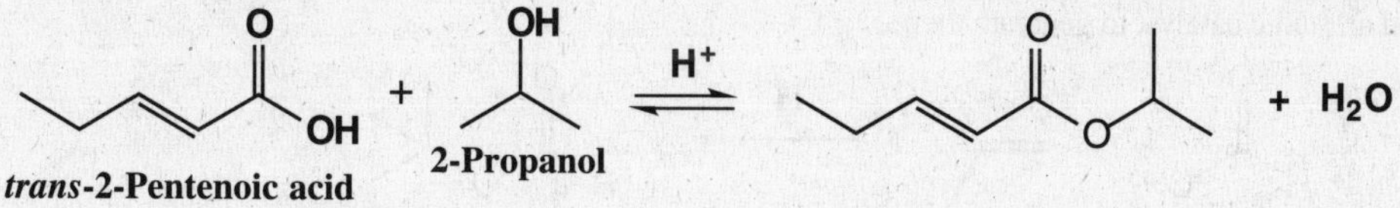

***trans*-2-Pentenoic acid**   **2-Propanol**

(d)

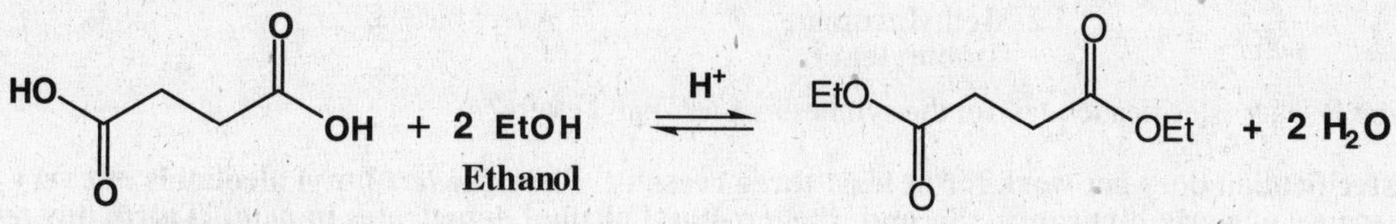

Butanedioic acid
(Succinic acid)

Problem 17.38 When 4-hydroxybutanoic acid is treated with an acid catalyst, it forms a lactone (a cyclic ester). Draw the structural formula of this lactone, and propose a mechanism for its formation.

Step 1: Add a proton.

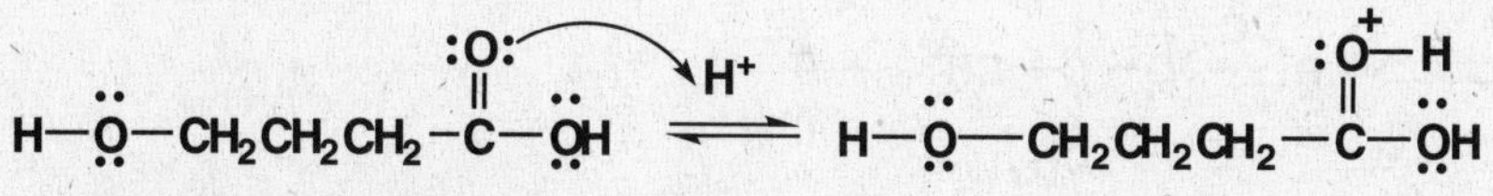

4-Hydroxybutanoic acid

Step 2: Make a bond between a nucleophile and an electrophile.

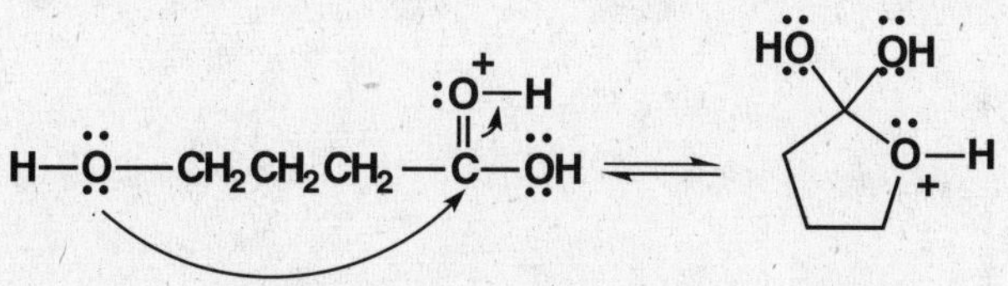

Step 3: Add a proton-take a proton away.

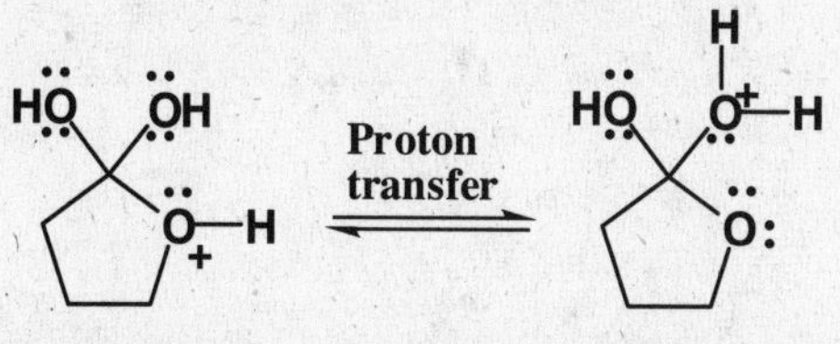

Proton
transfer

Step 4: Break a bond to give stable molecules or ions.

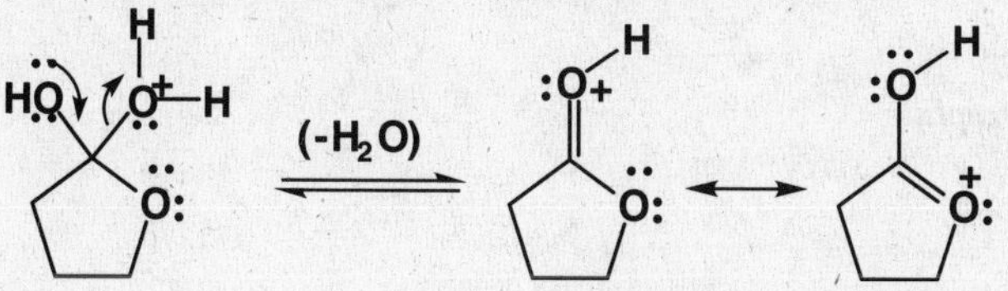

(−H₂O)

(resonance stabilized)

Step 5: Take a proton away.

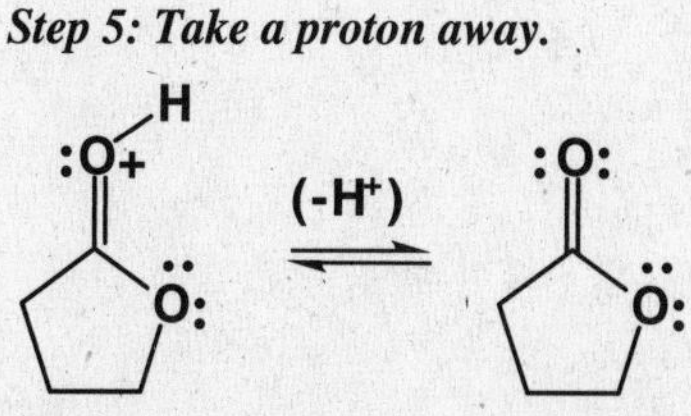

(−H⁺)

Lactone

<u>Problem 17.39</u> Fischer esterification cannot be used to prepare *tert*-butyl esters. Instead, carboxylic acids are treated with 2-methylpropene and an acidic catalyst to generate them.

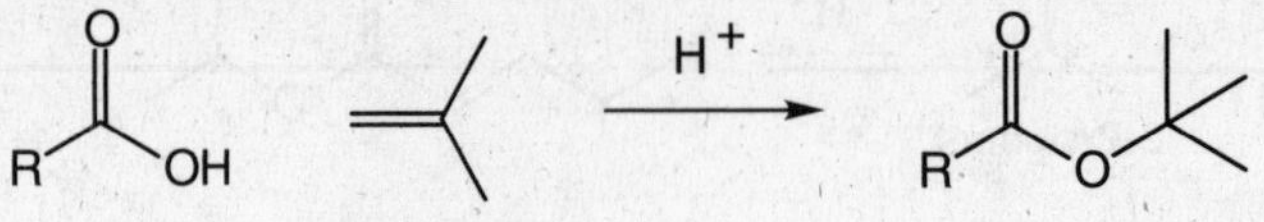

**2-Methylpropene
(Isobutylene)**   A *tert*-butyl ester

(a) Why does the Fischer esterification fail for the synthesis of *tert*-butyl esters?

The Fischer esterification does not work for at least three reasons. First, the *tert*-butyl alcohol is not very nucleophilic because of steric hindrance. Second, the *tert*-butyl alcohol dehydrates in acid. Third, any *tert*-butyl ester that forms falls apart to give the carboxylic acid and *tert*-butyl cation in acid as shown in the two part mechanism below.

Step 1: Add a proton.

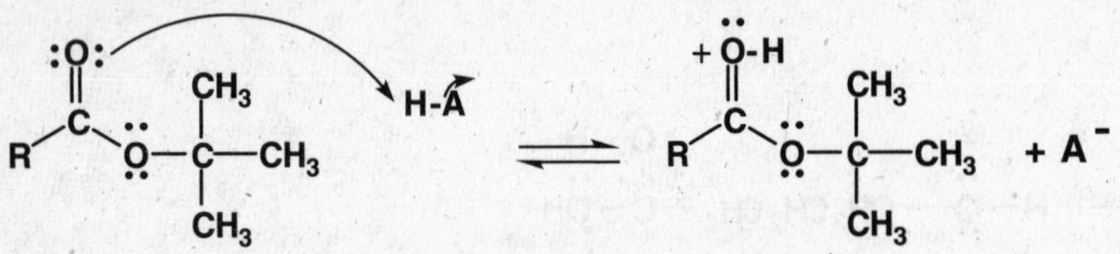

Step 2: Break a bond to give stable molecules or ions.

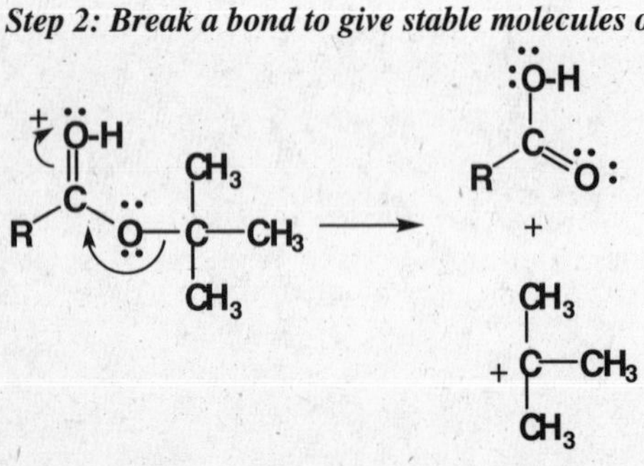

(b) Propose a mechanism for the 2-methylpropene method.

Step 1: Add a proton.

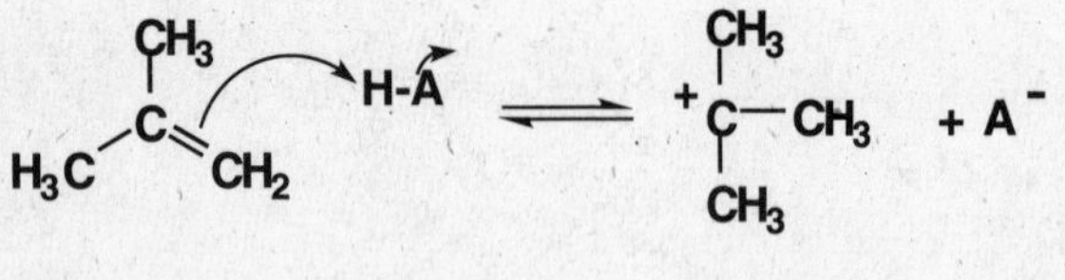

Step 2: Make a bond between a nucleophile and an electrophile.

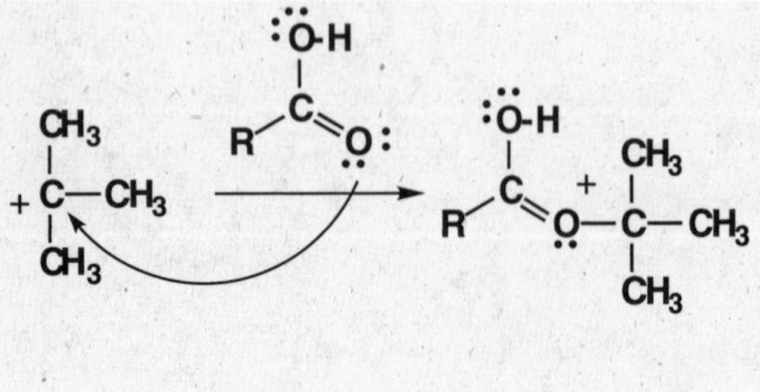

Step 3: Take a proton away.

Problem 17.40 Draw the product formed on thermal decarboxylation of each compound.

(a) $C_6H_5\overset{O}{\overset{\|}{C}}CH_2COOH$ $\xrightarrow{\text{Heat}}$ $C_6H_5\overset{O}{\overset{\|}{C}}CH_3$ + CO_2

(b) $C_6H_5CH_2\overset{COOH}{\overset{|}{C}H}COOH$ $\xrightarrow{\text{Heat}}$ $C_6H_5CH_2CH_2COOH$ + CO_2

(c) $\xrightarrow{\text{Heat}}$ + CO_2

Problem 17.41 When heated, carboxylic salts in which there is a good leaving group on the carbon beta to the carboxylate group undergo decarboxylation/elimination to give an alkene. Propose a mechanism for this type of decarboxylation/elimination. Compare the mechanism of these decarboxylations with the mechanism for decarboxylation of β-ketoacids; in what way(s) are the mechanisms similar?

The mechanisms are similar in that, in both cases, CO_2 is generated and a carbon-carbon bond is broken. As can be seen in the mechanisms above, that is where the similarity ends. In the case of the decarboxylation/elimination reaction, the starting material is the carboxylate anion and a bromide anion is lost in the decarboxylation step to give the product alkene. In the case of decarboxylation of β-ketoacids, the protonated carboxylic acid group is usually used. A proton is transferred from the carboxylic acid to the ketone to give an enol in the decarboxylation step, that tautomerizes to produce the product ketone.

<u>Problem 17.42</u> Show how cyclohexanecarboxylic acid could be synthesized from cyclohexane in good yield.

Propose a synthesis that begins with radical halogenation. Note that this is the only synthetic reaction you have learned that uses an alkane as a starting material. Next create a Grignard reagent, followed by reaction with CO$_2$ then an acid workup to give cyclohexanecarboxylic acid in high yield.

<u>Looking Ahead</u>
<u>Problem 17.43</u> In Section 17.7B, we suggested that the mechanism of Fischer esterification of carboxylic acids is a model for the reactions of functional derivatives of carboxylic acids. One of these reactions is that of an acid chloride with water (Section 18.4A).

Suggest a mechanism for this reaction.

Step 1: Make a bond between a nucleophile and an electrophile. **The reaction is initiated when the weakly nucleophilic water molecule attacks the highly electrophilic carbonyl carbon atom to create a tetrahedral addition intermediate.**

Step 2: Add a proton-take a proton away. **The tetrahedral addition intermediate undergoes proton transfer.**

Step 3: Break a bond to give stable molecules or ions. **Chloride is a good leaving group, so it departs.**

Step 4: Take a proton away. **A proton is lost to complete the reaction. Note that the chloride is such a good leaving group that the exact timing of the proton transfer and chloride departure is difficult to predict.**

Problem 17.44 We have studied Fischer esterification, in which a carboxylic acid is reacted with an alcohol in the presence of an acid catalyst to form an ester. Suppose that you start instead with a dicarboxylic acid such as terephthalic acid and a diol such as ethylene glycol. Show how Fischer esterification in this case can lead to a macromolecule with a molecular weight several thousands of times that of the starting materials.

1,4-Benzenedicarboxylic acid   **1,2-Ethandiol**   **poly(ethylene terephthalate)**
(Terephthalic acid)   **(Ethylene glycol)**   **(PET)**

poly(ethylene terephthalate)
(PET)

As we shall see in Section 29.5B, the material produced in this reaction is a high-molecular-weight polymer, which can be fabricated into Mylar films, and into the textile fiber known as Dacron polyester.

Synthesis

Problem 17.45 Show how to convert propane into propyl propanoate. You must use propane as the source of all carbon atoms in the target molecule. Show all reagents needed and all molecules synthesized along the way.

Recognize the product as an ester formed from 1-propanol and propanoic acid using a catalytic amount of acid (Fischer esterification). 1-propanol can be synthesized from propane using the three reaction sequence of free radical halogenation, E2 elimination, then hydroboration/oxidation (non-Markovnikov regiochemistry required here). Propanoic acid is made from 1-propanol using chromic acid oxidation. Alternatively, the propanoic acid could have been converted to propanoyl chloride with $SOCl_2$ then reacted with 1-propanol (not shown).

Problem 17.46 Show how to convert 4-methyl-1-pentene and carbon dioxide into 5-methylhexanoic acid. You must use 4-methyl-1-pentene and carbon dioxide as the source of all carbon atoms in the target molecule. Show all reagents and all molecules synthesized along the way.

Recognize the product as a carboxylic acid with one more carbon atom than the 4-methyl-1-pentene starting material, so propose a Grignard reaction with CO_2 as the last step. The Grignard reagent can be made from 4-methyl-1-pentene using the sequence of hydroboration/oxidation (non-Markovnikov regiochemistry required), conversion to the primary bromoalkane with PBr_3, and treatment with Mg in ether. Alternatively, the primary bromoalkane could have been synthesized from 4-methyl-1-pentene in one step using HBr with peroxides and light (free radical conditions).

Problem 17.47 Show how to convert cyclohexane into adipoyl dichloride. Show all reagents and all molecules synthesized along the way.

Cyclohexane

Adipoyl dichloride

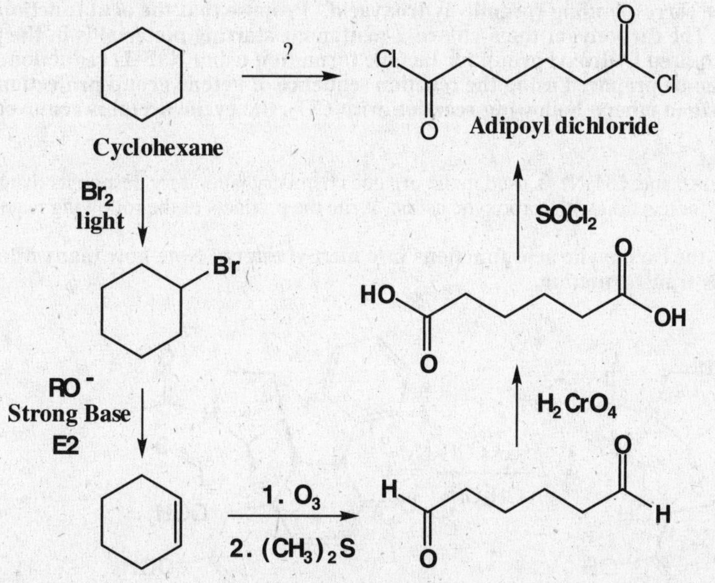

Recognize the same number of carbon atoms in both starting material and product, although the ring must be opened to give the open chain adipoyl structure. The only reaction covered so far that can break a carbon-carbon bond is ozonolysis, so assume it will be used in this synthesis. Also recognize the product as a diacid chloride that must have come from a diacid. The diacid, in turn, can be easily derived from the corresponding dialdehyde, which is the product of the ozonolysis reaction. The required cyclohexene can be synthesized from cyclohexane using the familiar sequence of free radical halogenation followed by E2 elimination in strong base.

Problem 17.48 Show how to convert 5-chloro-2-pentanone and carbon dioxide into racemic tetrahydro-6-methyl-2-pyranone. You must use 5-chloro-2-pentanone and carbon dioxide as the source of all carbon atoms in the racemic target molecule. Show all reagents and all molecules synthesized along the way.

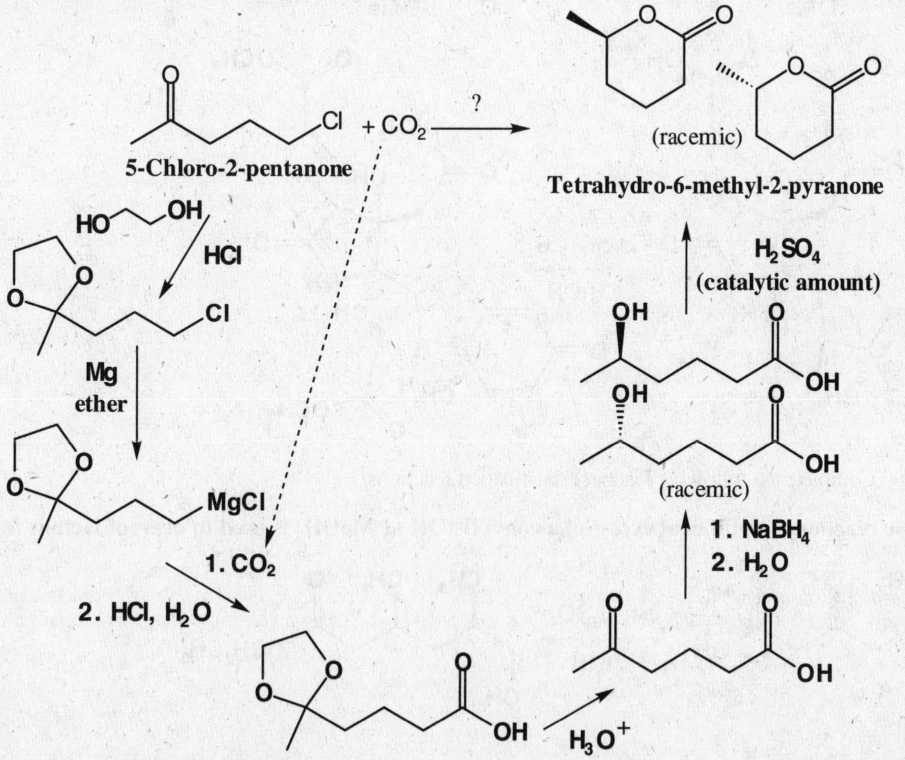

5-Chloro-2-pentanone

Tetrahydro-6-methyl-2-pyranone

Recognize that the product has six carbon atoms, so a new carbon-carbon bond must be created between the 5-chloro-2-pentanone and CO_2. Also recognize the product as a racemic lactone, that is, a cyclic ester molecule that is synthesized from the corresponding racemic hydroxyacid. Propose that the acid function is derived from Grignard reaction with CO_2. The carbonyl of the 5-chloro-2-pentanone starting material is in the proper position for a conversion to the required hydroxyl group for lactone formation using $NaBH_4$ reduction. Therefore, the required Grignard reagent can be prepared using the reaction sequence of ketone group protection as the cyclic acetal, followed by treatment with Mg in ether. Following reaction with CO_2, the cyclic acetal is removed using aqueous acid.

Reactions in Context

Problem 17.49 Diazomethane, CH_2N_2, is used in the organic chemistry laboratory despite its danger because it produces very high yields and is selective for reaction with carboxylic acids. Write the products of the following reactions.

In all cases, convert the carboxylic acid functions into methyl esters. Note how many other functional groups are compatible with this transformation.

(a)

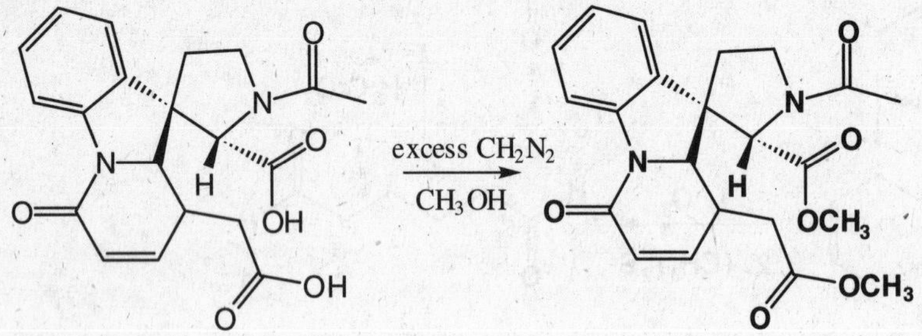

(b)

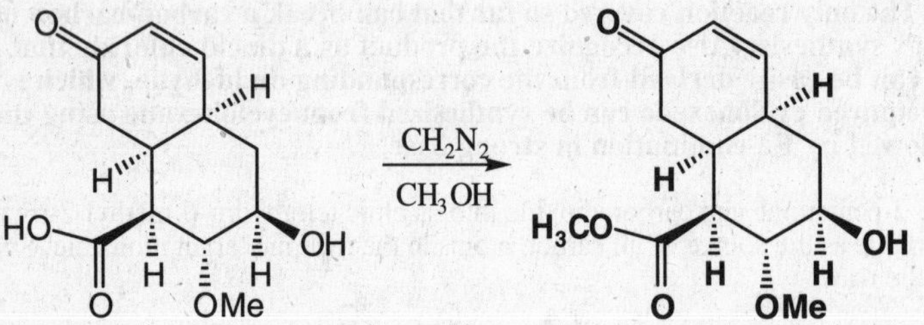

(c)

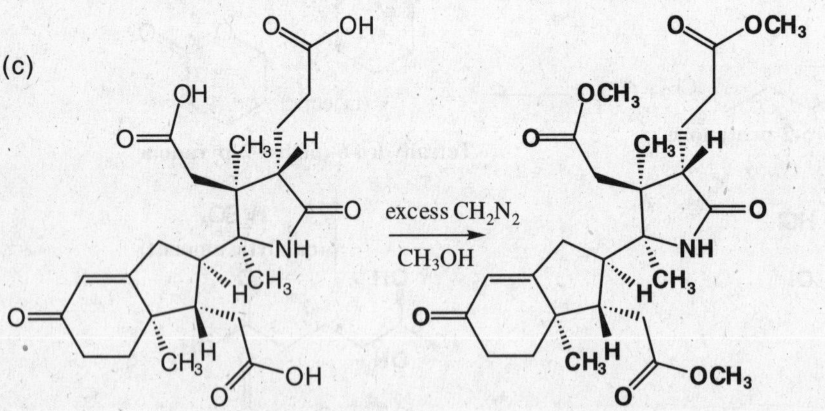

Problem 17.50 Complete the following Fischer esterification reactions.

In both these reaction, a significant excess of alcohol (EtOH or MeOH) is used to prevent lactone formation.

(a)

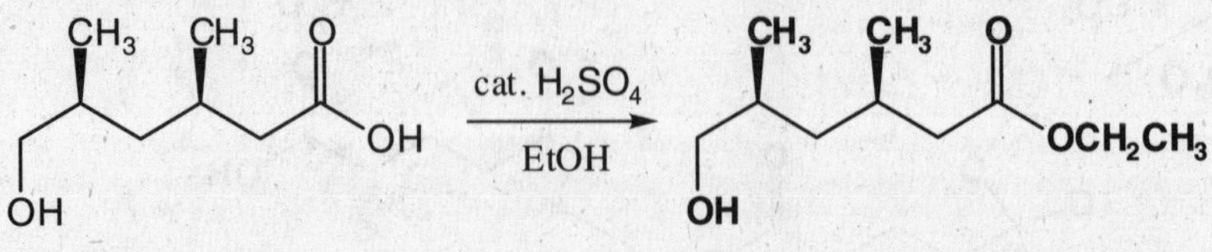

(b)

Problem 17.51 So far, you have seen a number of reducing agents used in reactions. Functional groups react differently with each of these reagents. With this in mind, complete the following reaction, which is the last step in a synthesis of fexofenadine, a nonsedating antihistamine sold under the trade name of Allegra.

Note that fexofenedine is actually sold as the hydrochloride salt to aid solubility.

1. NaBH$_4$, NaOH, H$_2$O/EtOH   2. HCl

pH 7-8, 35 hours

Fexofenedine hydrochloride (racemic)

Ornate butterflyfish *Chaetodon ornatissimus*
Maui, Hawaii

CHAPTER 18
Solutions to the Problems

Problem 18.1 Draw a structural formula for each compound.

(a) *N*-Cyclohexylacetamide.   (b) 1-Methylpropyl methanoate   (c) Cyclobutyl butanoate

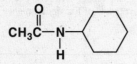

(d) *N*-(1-Methylheptyl)succinamide   (e) Diethyl adipate   (f) 2-Aminopropanamide

Problem 18.2 Will phthalimide dissolve in aqueous sodium bicarbonate?

In order to dissolve in water, the phthalimide must be deprotonated to give the negatively-charged phthalimide anion. The question becomes whether sodium bicarbonate is a strong enough base to deprotonate phthalimide according to the following equilibrium:

$$\text{phthalimide (NH)} + NaHCO_3 \rightleftharpoons \text{phthalimide (N}^-\text{ Na}^+) + H_2O + CO_2$$

$pK_a = 8.3$

The pK_a of phthalimide is 8.3, while the pK_a of carbonic acid H_2CO_3, which dissociates into H_2O and CO_2, is 6.4. Because equilibria in acid-base reactions favor formation of the weaker acid (phthalimide), the equilibrium will be to the left and the phthalimide will not dissolve.

Problem 18.3 Complete and balance equations for the hydrolysis of each ester in aqueous solution; show each product as it is ionized under the indicated experimental conditions.

(a) (benzene-1,2-dicarboxylic acid dimethyl ester, COOCH₃ / COOCH₃) + 2 NaOH $\xrightarrow{H_2O}$ (benzene-1,2-dicarboxylate, COO⁻ Na⁺ / COO⁻ Na⁺) + 2 CH₃OH

(b) (keto ester, OEt) + H₂O $\xrightarrow{HCl}$ (keto acid, OH) + EtOH

Problem 18.4 Complete equations for the hydrolysis of the amides in Example 18.4 in concentrated aqueous NaOH. Show all products as they exist in aqueous NaOH, and the number of moles of NaOH required for hydrolysis of each amide.

Each product is shown as it would exist in aqueous NaOH.

(a) $CH_3\overset{O}{\overset{\|}{C}}-\underset{\underset{CH_3}{|}}{N}-CH_3 + NaOH \xrightarrow{H_2O} CH_3CO^-Na^+ + (CH_3)_2NH$

(b) 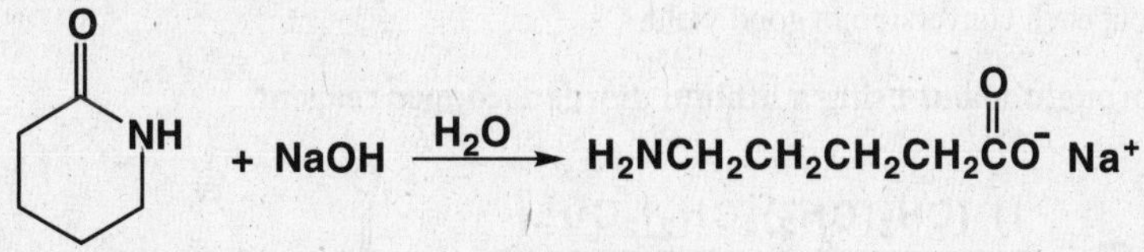   + NaOH $\xrightarrow{H_2O}$ H$_2$NCH$_2$CH$_2$CH$_2$CH$_2$CO$^-$ Na$^+$

<u>Problem 18.5</u> Synthesis of nitriles by nucleophilic displacement of halide from an alkyl halide is practical only with primary and secondary alkyl halides. It fails with tertiary alkyl halides. Why? What is the major product of the following reaction?

$\xrightarrow[\substack{\text{ethanol,}\\\text{water}}]{\text{KCN}}$ –CH$_3$ + HCN + KCl

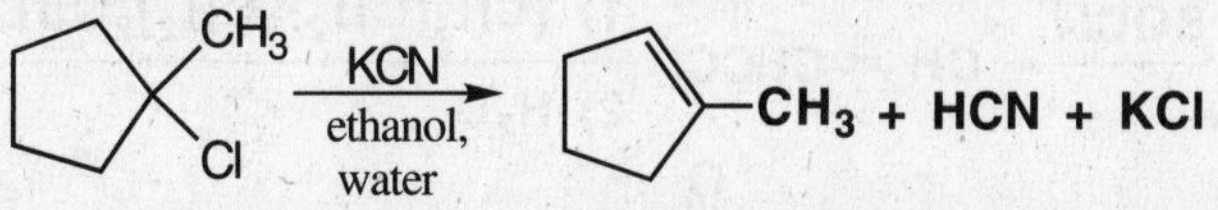

Cyanide ion is both a base and a nucleophile. With a tertiary halide (a lot of steric hindrance to backside attack) and a moderate base/moderate nucleophile such as CN⁻, E2 is the principal reaction. Thus, the major product in this instance is 1-methylcyclopentene.

<u>Problem 18.6</u> Complete the following transesterification reaction (the stoichiometry is given in the equation).

2 + HO⌣OH $\xrightarrow{H^+}$

+ 2 CH$_3$OH

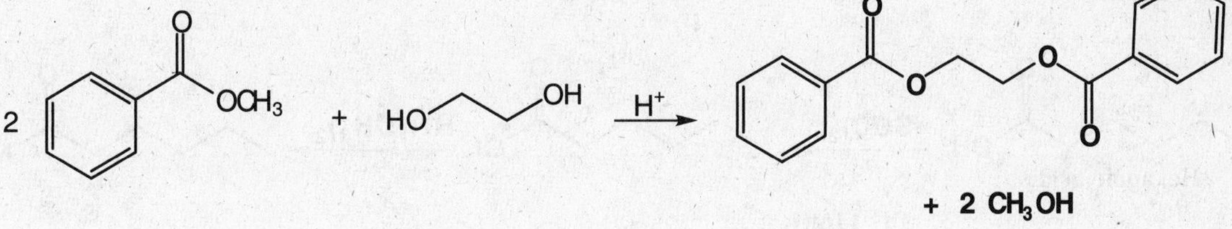

<u>Problem 18.7</u> Complete and balance equations for the following reactions (the stoichiometry of each reaction is given in the equation).

Each is an example of aminolysis of an ester.

(a) 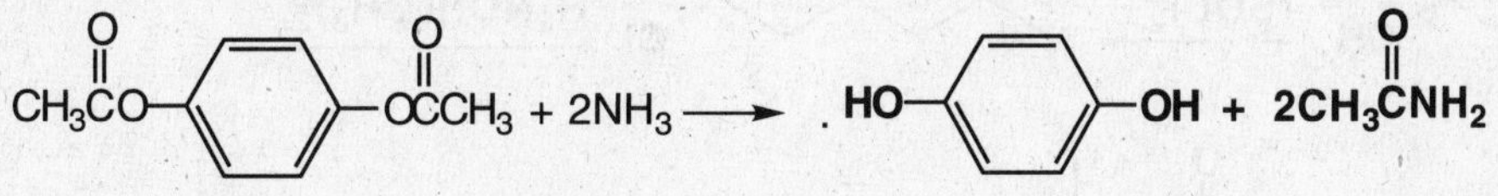 CH$_3$CO——OCCH$_3$ + 2NH$_3$ $\longrightarrow$ HO——OH + 2CH$_3$CNH$_2$

(b) 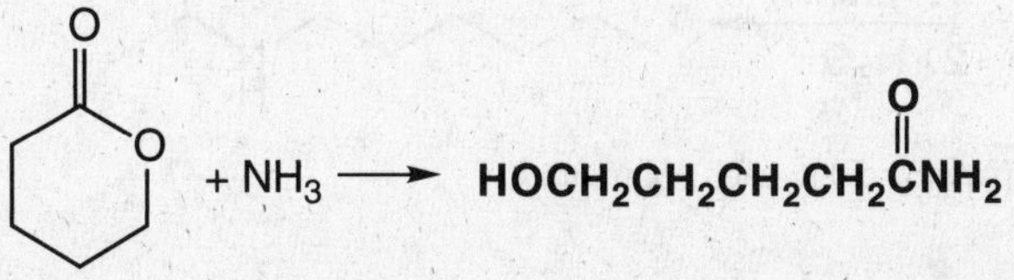 + NH$_3$ $\longrightarrow$ HOCH$_2$CH$_2$CH$_2$CH$_2$CNH$_2$

<u>Problem 18.8</u> Show how to prepare each alcohol by treating an ester with a Grignard reagent.

Each can be prepared by treatment of an ester with two moles of an organomagnesium reagent. In these solutions, the ester chosen is the ethyl ester.

(a) 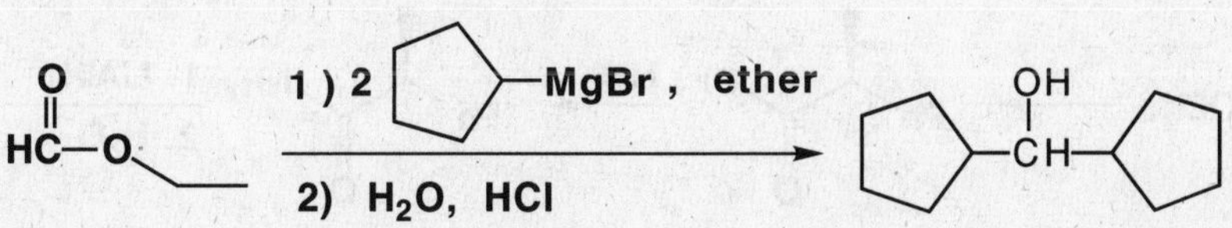 HC–O— $\xrightarrow[\text{2) H}_2\text{O, HCl}]{\text{1) 2}\,\square\text{–MgBr, ether}}$

(b) 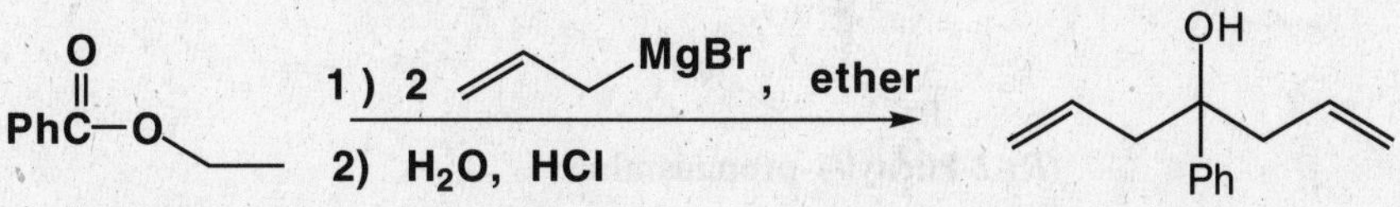 PhC–O— $\xrightarrow[\text{2) H}_2\text{O, HCl}]{\text{1) 2}\,\diagup\diagup\text{MgBr, ether}}$

Problem 18.9 Show how to bring about each conversion in good yield.

Both of these conversions can be brought about using a lithium diorganocopper reagent.

a) PhCOOH $\xrightarrow{\text{SOCl}_2}$ PhCCl $\xrightarrow[\text{2) H}_2\text{O}]{\text{1) [CH}_3\text{(CH}_2\text{)}_4\text{CH}_2\text{]}_2\text{CuLi}}$

(b) CH$_2$=CHCl $\xrightarrow[\substack{\text{1. Mg, ether} \\ \text{2. CO}_2 \\ \text{3. HCl, H}_2\text{O}}]{}$ CH$_2$=CHCOH $\xrightarrow{\text{SOCl}_2}$ CH$_2$=CHCCl $\xrightarrow[\text{2) H}_2\text{O}]{\text{1) [CH}_3\text{(CH}_2\text{)}_3\text{CH}_2\text{]}_2\text{CuLi}}$

Problem 18.10 Show how to convert hexanoic acid to each amine.

In both cases, the conversions can be carried out through formation of an amide, followed by reduction to give the desired amine.

(a) Hexanoic acid $\xrightarrow{\text{SOCl}_2}$ $\xrightarrow{\text{HN(CH}_3\text{)}_2}$ [NMe$_2$]

$\xrightarrow[\text{2. H}_2\text{O}]{\text{1. LiAlH}_4}$ [NMe$_2$]

(b) Hexanoic acid $\xrightarrow{\text{SOCl}_2}$ $\xrightarrow{\text{H}_2\text{NCH(CH}_3\text{)}_2}$

$\xrightarrow[\text{2. H}_2\text{O}]{\text{1. LiAlH}_4}$

Problem 18.11 Show how to convert (*R*)-2-phenylpropanoic acid to each compound.

(a) Ph [OH] $\xrightarrow[\text{2. H}_2\text{O}]{\text{1. LiAlH}_4}$ Ph [OH]
(*R*)-2-Phenyl-1-propanol

(b) Ph [OH] + SOCl$_2$ $\longrightarrow$ Ph [Cl] $\xrightarrow{\text{NH}_3}$ Ph [NH$_2$] $\xrightarrow[\text{2. H}_2\text{O}]{\text{1. LiAlH}_4}$
(*R*)-2-Phenylprop-
anoic acid

Ph [NH$_2$]
(*R*)-2-Phenyl-1-propanamine

Structure and Nomenclature

Problem 18.12 Draw a structural formula for each compound.

(a) Dimethyl carbonate

(b) Benzonitrile

(c) Isopropyl 3-methylhexanoate

(d) Diethyl oxalate

(e) Ethyl (Z)-2-pentenoate

(f) Butanoic anhydride

(g) Dodecanamide

(h) Ethyl 3-hydroxybutanoate

(i) Octanoyl chloride

(j) Diethyl *cis* 1,2-cyclohexane-dicarboxylate

(k) Methanesulfonyl chloride

(l) *p*-Toluenesulfonyl chloride

Problem 18.13 Write the IUPAC name for each compound.

(a)
Benzoic anhydride

(b)
Benzenesulfonamide

(c)
N-Methylhexanamide

(d)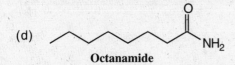
Octanamide

(e)
Diethyl propanedioate (Diethyl malonate)

(f) $CH_3O-S-OCH_3$
Dimethyl sulfate

(g)
(R)-Methyl 2-methyl-3-oxo-4-phenylbutanoate

(h)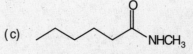
Pentanedioyl chloride (Glutaryl chloride)

(i) $CH_3(CH_2)_5CN$
Heptanenitrile

Physical Properties
Problem 18.14 Both the melting point and boiling point of acetamide are higher than those of its *N,N*-dimethyl derivative.
How do you account for these differences?

$$CH_3\overset{\overset{\displaystyle O}{\|}}{C}NH_2 \qquad\qquad CH_3\overset{\overset{\displaystyle O}{\|}}{C}N(CH_3)_2$$

Acetamide   ***N,N*-Dimethylacetamide**
mp 82.3°C, bp 221.2°C   mp -20°C, bp 165°C

Acetamide has two amide N-H atoms that can take part in hydrogen bonding with carbonyl oxygen atoms, while *N,N*-dimethylacetamide has none. Thus, due to this ability to hydrogen bond, acetamide molecules can associate with each other more strongly, leading to higher melting and boiling points compared to *N,N*-dimethylacetamide.

Spectroscopy
Problem 18.15 Each hydrogen of a primary amide typically has a separate ^{1}H-NMR resonance, as illustrated by the separate signals for the two amide hydrogens of propanamide, which fall at δ 6.22 and δ 6.58. Furthermore, each methyl group of *N,N*-dimethylformamide has a separate resonance (δ 3.88 and δ 3.98). How do you account for these observations?

δ 6.22 and 6.58   δ 3.88 and 3.98

Propanamide   ***N,N*-Dimethylformamide**

Amide resonance causes a rotation barrier around the C(O)-N bond that prevents interconversion of the two nitrogen substituents on the NMR time scale. For this reason, the two hydrogens and two methyl groups, respectively, are in different chemical environments and thus give different signals.

Problem 18.16 Propose a structural formula for compound A, $C_7H_{14}O_2$, consistent with its ^{1}H-NMR and IR spectra.

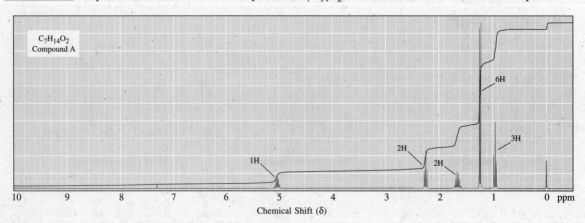

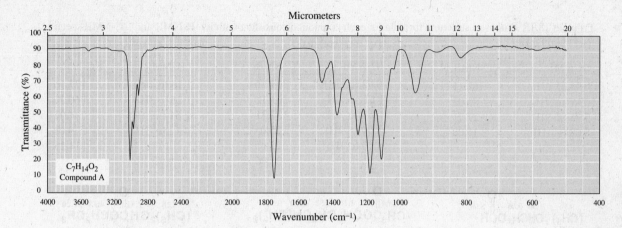

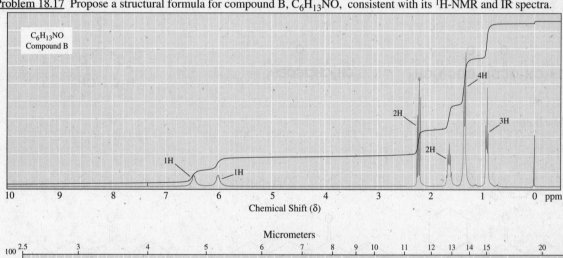

Isopropyl butanoate

Problem 18.17 Propose a structural formula for compound B, $C_6H_{13}NO$, consistent with its ^1H-NMR and IR spectra.

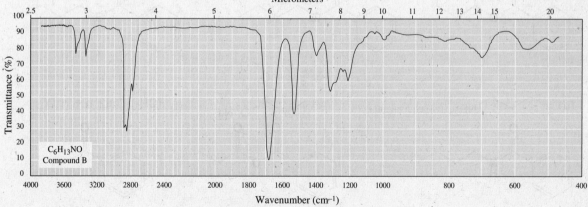

Hexanamide

Problem 18.18 Propose a structural formula for each compound consistent with its ^{1}H-NMR and ^{13}C-NMR spectra.

(a) $C_5H_{10}O_2$

^{1}H-NMR	^{13}C-NMR
0.96 (d, 6H)	161.11
1.96 (m, 1H)	70.01
3.95 (d, 2H)	27.71
8.08 (s, 1H)	19.00

$$\underset{0.96}{(CH_3)_2}\underset{1.96}{CH}\underset{3.95}{CH_2}\underset{}{O}\overset{O}{\underset{8.08}{CH}}$$

(b) $C_7H_{14}O_2$

^{1}H-NMR	^{13}C-NMR
0.92 (d, 6H)	171.15
1.52 (m, 2H)	63.12
1.70 (m, 1H)	37.31
2.09 (s, 3H)	25.05
4.10 (t, 2H)	22.45
	21.06

$$\underset{2.09}{CH_3}\overset{O}{C}O\underset{4.10}{CH_2}\underset{1.52}{CH_2}\underset{1.70}{CH}\underset{0.92}{(CH_3)_2}$$

(c) $C_6H_{12}O_2$

^{1}H-NMR	^{13}C-NMR
1.18 (d, 6H)	177.16
1.26 (t, 3H)	60.17
2.51 (m, 1H)	34.04
4.13 (q, 2H)	19.01
	14.25

$$\underset{1.18}{(CH_3)_2}\underset{2.51}{CH}\overset{O}{C}O\underset{4.13}{CH_2}\underset{1.26}{CH_3}$$

d) $C_7H_{12}O_4$

^{1}H-NMR	^{13}C-NMR
1.28 (t, 6H)	166.52
3.36 (s, 2H)	61.43
4.21 (q, 4H)	41.69
	14.07

$$\underset{1.28}{CH_3}\underset{4.21}{CH_2}O\overset{O}{C}\underset{3.36}{CH_2}\overset{O}{C}O\underset{4.21}{CH_2}\underset{1.28}{CH_3}$$

(e) $C_4H_7ClO_2$

^{1}H-NMR	^{13}C-NMR
1.68 (d, 3H)	170.51
3.80 (s, 3H)	52.92
4.42 (q, 1H)	52.32
	21.52

$$\underset{1.68}{CH_3}\underset{4.42}{CHCl}\overset{O}{C}O\underset{3.80}{CH_3}$$

(f) $C_4H_6O_2$

^{1}H-NMR	^{13}C-NMR
2.29 (m, 2H)	177.81
2.50 (t, 2H)	68.58
4.36 (t, 2H)	27.79
	22.17

Reactions

Problem 18.19 Draw a structural formula for the principal product formed when benzoyl chloride is treated with each reagent.

(g) CH₃O—⟨benzene⟩—NH₂, pyridine

(h) C₆H₅MgBr (two equivalents), then H₃O⁺

Problem 18.20 Draw a structural formula of the principal product formed when ethyl benzoate is treated with each reagent.

(a) H₂O, NaOH, heat

(b) H₂O, H₂SO₄, heat

(c) CH₃CH₂CH₂CH₂NH₂

(d) DIBALH (-78°C), then H₂O

(e) LiAlH₄, then H₂O

(f) C₆H₅MgBr (two equivalents), then HCl/ H₂O

Problem 18.21 The mechanism for hydrolysis of an ester in aqueous acid involves formation of a tetrahedral carbonyl addition intermediate. Evidence in support of this mechanism comes from an experiment designed by Myron Bender. He first prepared ethyl benzoate enriched with oxygen-18 in the carbonyl oxygen and then carried out acid-catalyzed hydrolysis of the ester in water containing no enrichment in oxygen-18. If he stopped the experiment after only partial hydrolysis and isolated the remaining ester, the recovered ethyl benzoate had lost a portion of its enrichment in oxygen-18. In other words, some exchange had occurred between oxygen-18 of the ester and oxygen-16 of water. Show how this observation bears on the formation of a tetrahedral carbonyl addition intermediate during acid-catalyzed ester hydrolysis.

This observation can only be explained by proposing reversible formation of a tetrahedral carbonyl addition intermediate, and the assumption that exchange of oxygen-18 in this manner is more rapid than collapse of the intermediate to give benzoic acid and ethanol.

Problem 18.22 Predict the distribution of oxygen-18 in the products obtained from hydrolysis of ethyl benzoate labeled in the ethoxy oxygen under the following conditions:

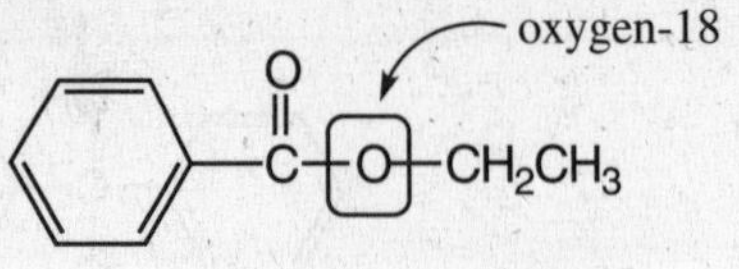

(a) In aqueous NaOH.   (b) In aqueous HCl.

Under both basic (a) and acidic (b) conditions, oxygen-18 will appear in ethanol. This is because both conditions involve a tetrahedral carbon addition intermediate that decomposes with loss of ethoxide (a) or ethanol (b). Either way, the oxygen-18 ends up in the ethanol product

(c) What distribution would you predict if the reaction were done with the *tert*-butyl ester in HCl?

The oxygen-18 will end up on the carboxylic acid because following protonation, the *tert*-butyl ester cleaves to give a carboxylic acid (containing the oxygen-18) and the *tert*-butyl cation.

Step 1: Add a proton.

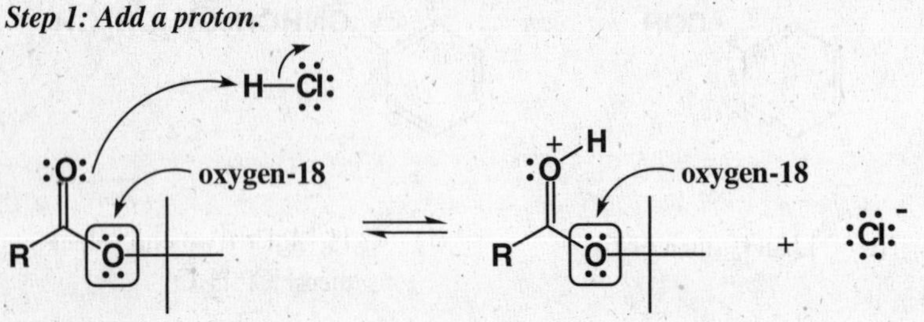

Step 2: Break a bond to give stable molecules or ions.

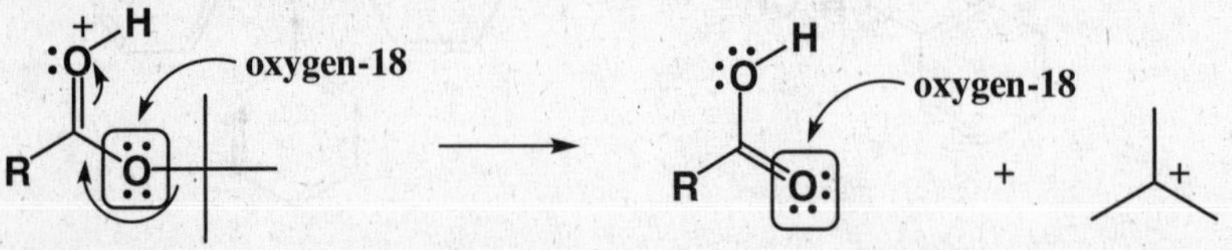

Problem 18.23 Draw a structural formula of the principal product formed when benzamide is treated with each reagent.

(a) H_2O, HCl, heat   (b) NaOH, H_2O, heat

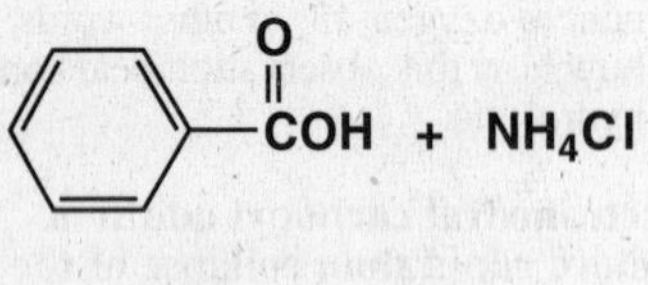

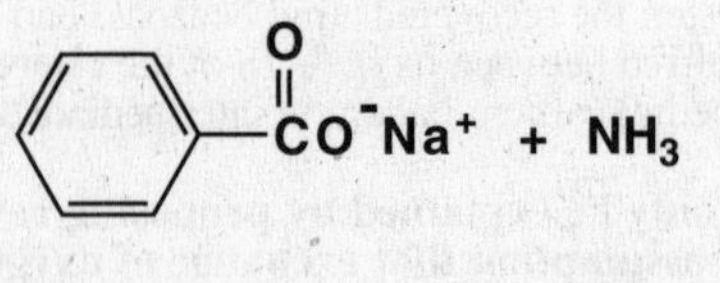

(c) $LiAlH_4$, then H_2O

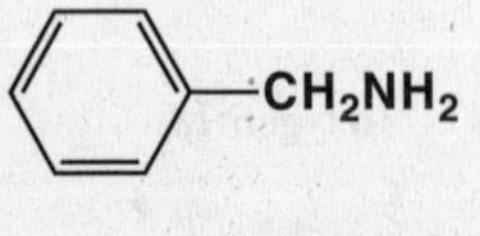

Problem 18.24 Draw a structural formula of the principal product formed when benzonitrile is treated with each reagent.

(a) H_2O (one equivalent), H_2SO_4, heat   (b) H_2O (excess), H_2SO_4, heat

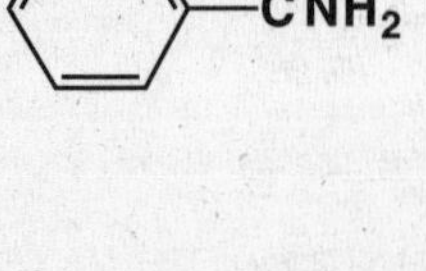

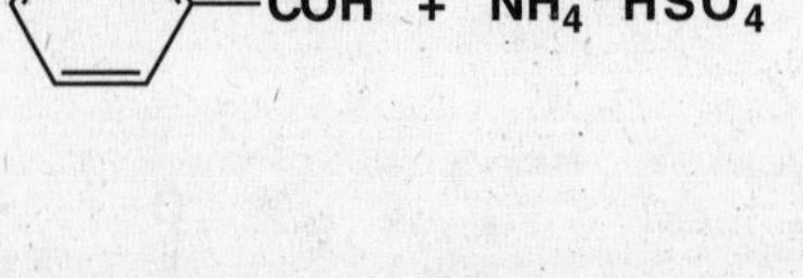

(c)   NaOH, H₂O, heat   (d)   LiAlH₄, then H₂O

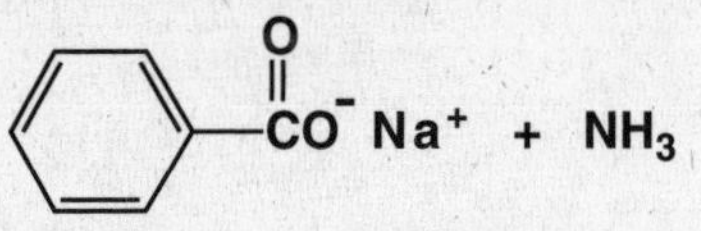

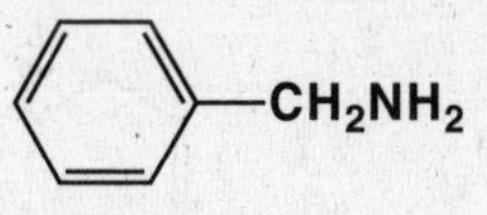

Problem 18.25 Show the product expected when the following unsaturated δ-ketoester is treated with each reagent.

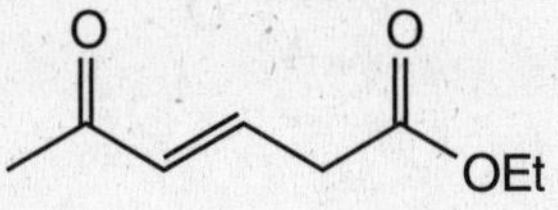

(a) $\dfrac{H_2 \ (1 \ mol)}{Pd, \ EtOH}$

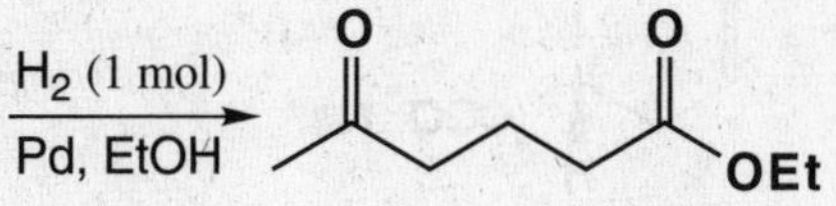

(b) $\dfrac{NaBH_4}{CH_3OH}$

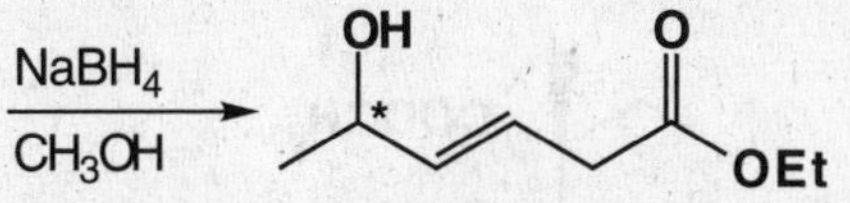

(racemic mixture)

(c) $\dfrac{1. \ LiAlH_4, \ THF}{2. \ H_2O}$

+ EtOH

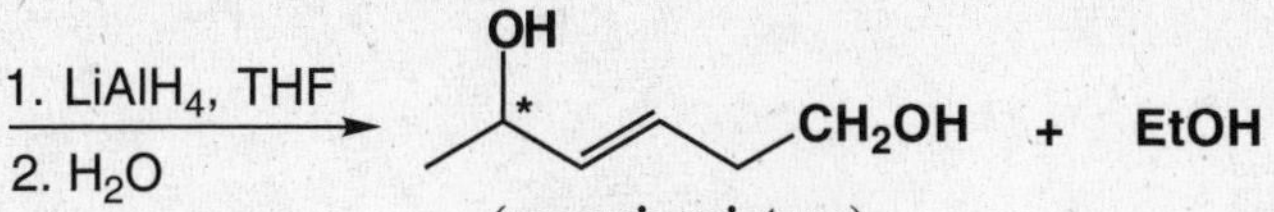

(racemic mixture)

(d) $\dfrac{1. \ DIBALH, \ -78°}{2. \ H_2O}$

+ EtOH

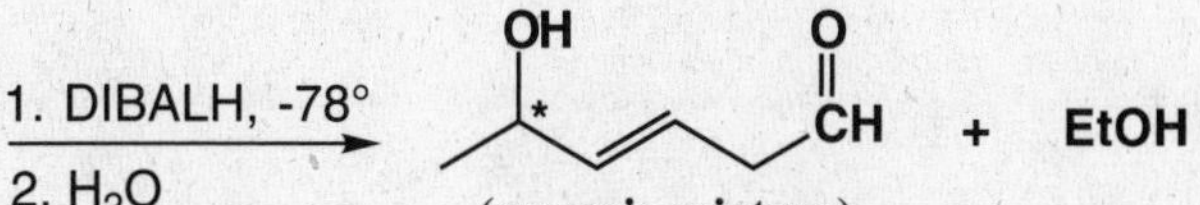

(racemic mixture)

Problem 18.26 The reagent diisobutylaluminum hydride (DIBALH) reduces esters to aldehydes. When nitriles are treated with DIBALH, followed by mild acid hydrolysis, the product is also an aldehyde. Propose a mechanism for this reaction.

A reasonable mechanism for this reaction involves an initial reaction of the hydride with the nitrile carbon atom, followed by hydrolysis of the resulting imine anion.

Step 1: Make a bond between a nucleophile and an electrophile.

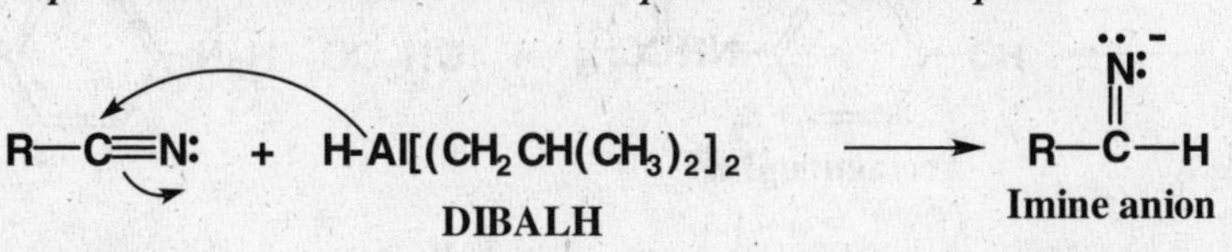

Step 2:

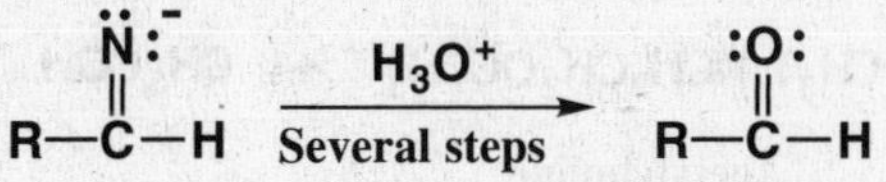

Problem 18.27 Show the product of treating this anhydride with each reagent.

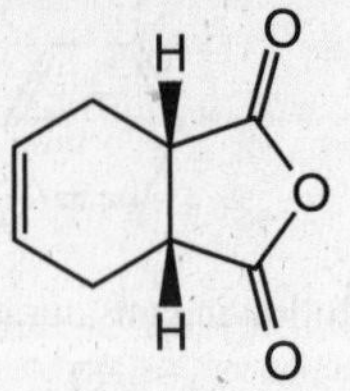

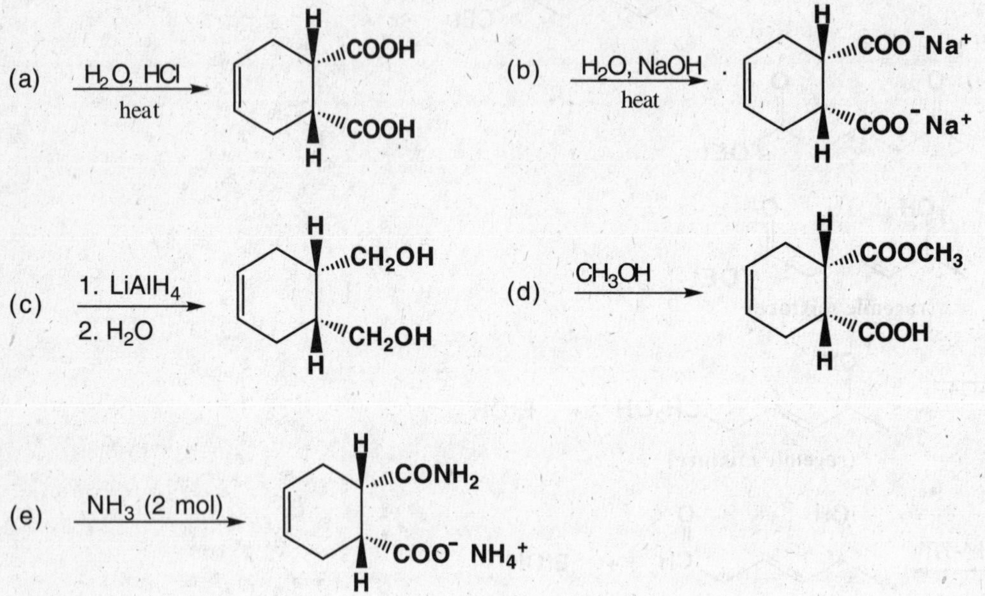

(a) $\xrightarrow[\text{heat}]{H_2O, HCl}$ [structure with ,,,,COOH and ,,,,COOH]

(b) $\xrightarrow[\text{heat}]{H_2O, NaOH}$ [structure with ,,,,COO$^-$Na$^+$ and ,,,,COO$^-$Na$^+$]

(c) $\xrightarrow{\text{1. LiAlH}_4 \quad \text{2. H}_2O}$ [structure with ,,,,CH$_2$OH and ,,,,CH$_2$OH]

(d) $\xrightarrow{CH_3OH}$ [structure with ,,,,COOCH$_3$ and ,,,,COOH]

(e) $\xrightarrow{NH_3 \text{ (2 mol)}}$ [structure with ,,,,CONH$_2$ and ,,,,COO$^-$ NH$_4^+$]

Problem 18.28 The analgesic acetaminophen is synthesized by treating 4-aminophenol with one equivalent of acetic anhydride. Draw a structural formula for acetaminophen.

Note how, in the following reaction scheme, the acylation occurs at the more nucleophilic amino group rather than the less nucleophilic hydroxyl group of 4-aminophenol.

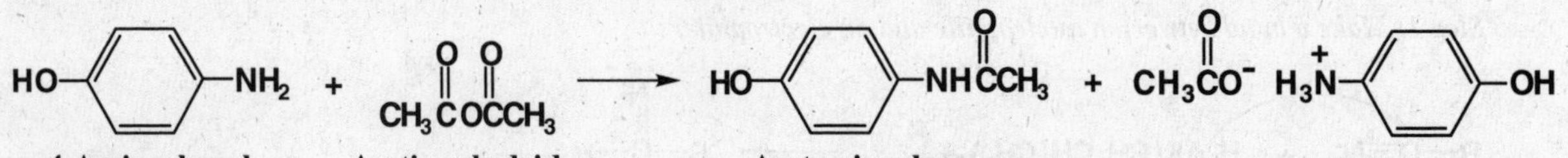

HO—⟨benzene ring⟩—NH$_2$ + CH$_3$COCCH$_3$ (acetic anhydride) $\longrightarrow$ HO—⟨benzene ring⟩—NHCCH$_3$ + CH$_3$CO$^-$ H$_3$N$^+$—⟨benzene ring⟩—OH

4-Aminophenol   **Acetic anhydride**   **Acetaminophen**

Problem 18.29 Treating choline with acetic anhydride gives acetylcholine, a neurotransmitter. Write an equation for the formation of acetylcholine.

$(CH_3)_3 \overset{+}{N}CH_2CH_2OH$ + CH$_3$COCCH$_3$ $\longrightarrow$ $(CH_3)_3 \overset{+}{N}CH_2CH_2OCCH_3$ + CH$_3$COH

Choline   **Acetylcholine**

Problem 18.30 Nicotinic acid, more commonly named niacin, is one of the B vitamins. Show how nicotinic acid can be converted to (a) ethyl nicotinate and then to (b) nicotinamide.

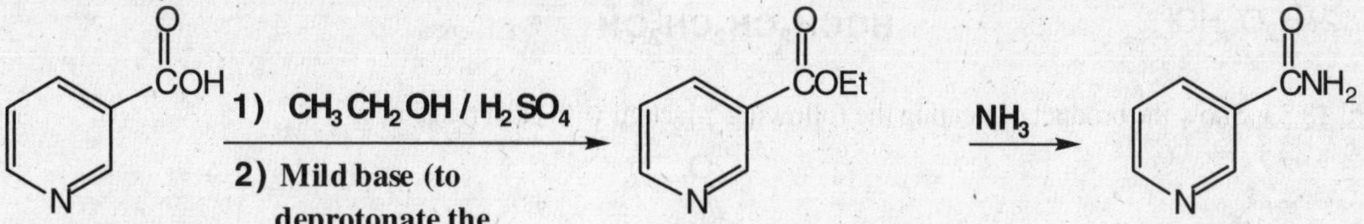
COH
1) CH_3CH_2OH / H_2SO_4
2) Mild base (to deprotonate the pyridine N atom)
COEt
NH_3
CNH_2

Nicotinic acid (Niacin) Ethyl nicotinoate Nicotinamide

Problem 18.31 Complete each reaction.

(a) 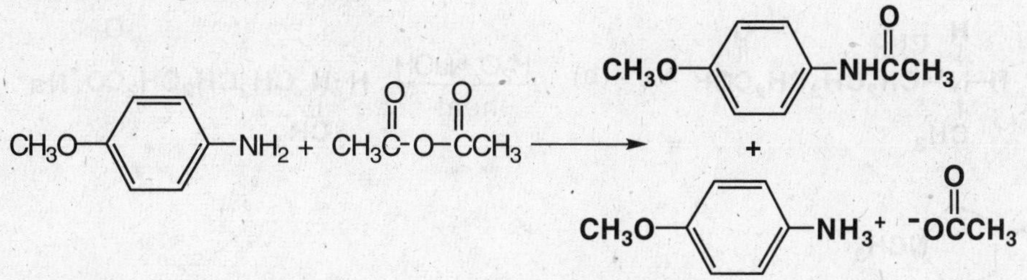

(b) 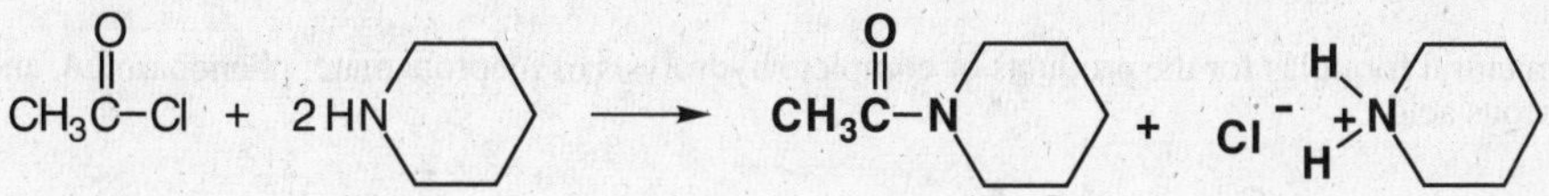

(c) 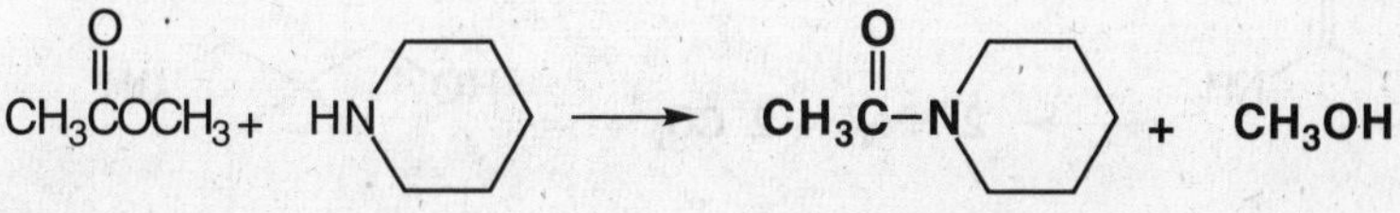

(d) 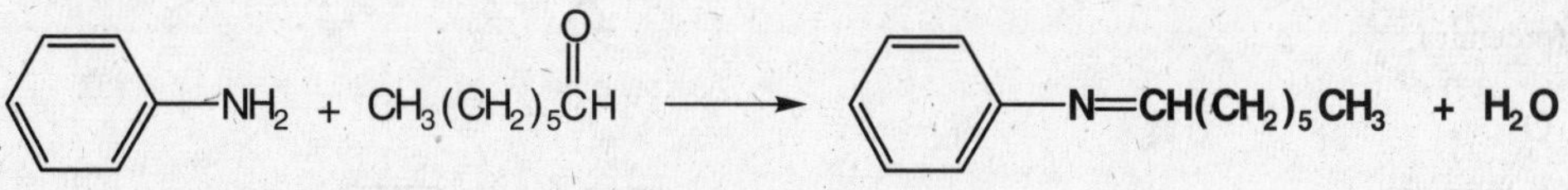

Problem 18.32 Show the product of treating γ-butyrolactone with each reagent.

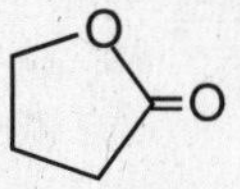

(a) $NH_3 \longrightarrow HOCH_2CH_2CH_2C(=O)NH_2$

(b) $\xrightarrow[2.\ H_2O]{1.\ LiAlH_4} HOCH_2CH_2CH_2CH_2OH$

(c) $\xrightarrow[2.\ H_2O,\ HCl]{1.\ 2\ PhMgBr,\ ether} HOCH_2CH_2CH_2C(OH)(Ph)-Ph$

(d) NaOH $\xrightarrow[heat]{H_2O} HOCH_2CH_2CH_2C(=O)O^- Na^+$

(e) $\xrightarrow[2.\ H_2O,\ HCl]{1.\ 2\ CH_3Li,\ ether} HOCH_2CH_2CH_2C(OH)(CH_3)-CH_3$

(f) $\dfrac{\text{1. DIBALH, ether, -78°C}}{\text{2. H}_2\text{O, HCl}}$ → $\text{HOCH}_2\text{CH}_2\text{CH}_2\overset{\overset{\displaystyle O}{\|}}{\text{C}}\text{H}$

<u>Problem 18.33</u> Show the product of treating the following γ-lactam with each reagent.

(a) $\dfrac{\text{H}_2\text{O, HCl}}{\text{heat}}$ → $\underset{\displaystyle \text{CH}_3}{\overset{\displaystyle H}{H-\overset{+}{N}}}\text{Cl}^- \; \text{-CH}_2\text{CH}_2\text{CH}_2\text{COH}$ (with C=O)

(b) $\dfrac{\text{H}_2\text{O, NaOH}}{\text{heat}}$ → $\underset{\displaystyle \text{CH}_3}{H-N}\text{-CH}_2\text{CH}_2\text{CH}_2\text{CO}^-\text{Na}^+$ (with C=O)

(c) $\dfrac{\text{1. LiAlH}_4}{\text{2. H}_2\text{O}}$ →

<u>Problem 18.34</u> Draw structural formulas for the products of complete hydrolysis of meprobamate, phenobarbital, and pentobarbital in hot aqueous acid.

(a)

Meprobamate
(racemic)

→ $2\ \text{NH}_4^+ + 2\ \text{CO}_2 +$

(b)

Phenobarbital
(racemic)

→ $2\ \text{NH}_4^+ + 2\ \text{CO}_2 +$

(racemic mixture)

(c)

Pentobarbital
(racemic)

→ $2\ \text{NH}_4^+ + 2\ \text{CO}_2 +$

(mixture of stereoisomers)

Meprobamate is a tranquilizer prescribed under 58 different trade names, including Equanil and Miltown. Phenobarbital is a long-acting sedative, hypnotic, and anticonvulsant. Luminal is one of over a dozen names under which it is prescribed. Pentobarbital is a short-acting sedative, hypnotic, and anticonvulsant. Nembutal is one of several trade names under which it is prescribed.

In parts (b) and (c) above there is a decarboxylation that occurs from a β-diacid that accompanies hydrolysis.

Synthesis

Problem 18.35 *N,N*-Diethyl *m*-toluamide (DEET) is the active ingredient in several common insect repellents. Propose a synthesis for DEET from 3-methylbenzoic acid.

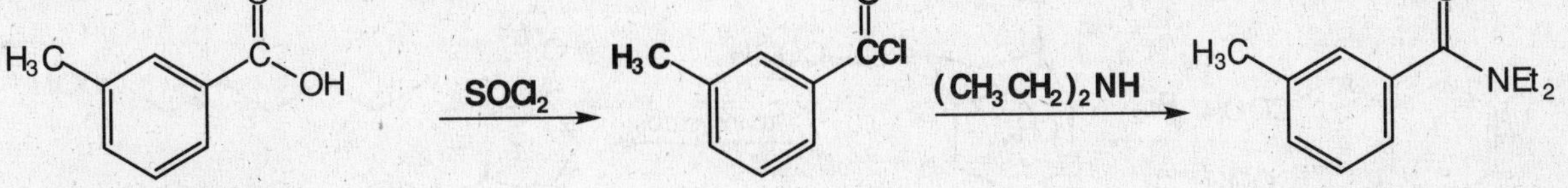

3-Methylbenzoic acid
(*m*-Toluic acid)

SOCl₂

(CH₃CH₂)₂NH

N,N-Diethyl *m*-toluamide
(DEET)

Problem 18.36 Isoniazid, a drug used to treat tuberculosis, is prepared from pyridine-4-carboxylic acid. How might this synthesis be carried out?

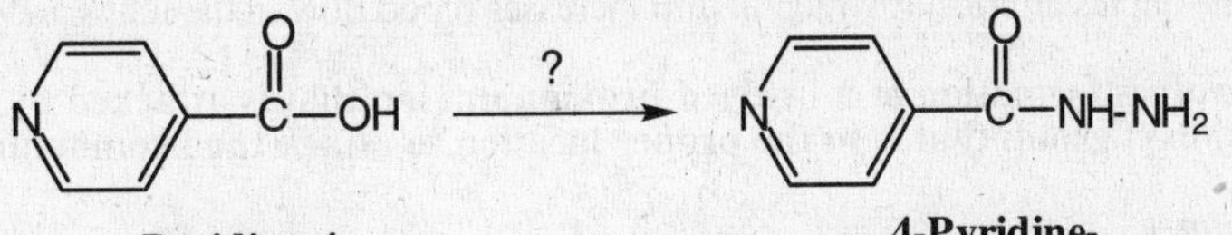

Pyridine-4-
carboxylic acid

?

4-Pyridine-
carboxylic acid hydrazide
(Isoniazid)

Isoniazid can be prepared by reaction of hydrazine (NH₂-NH₂) with a suitable carboxylic acid derivative, such as an ester.

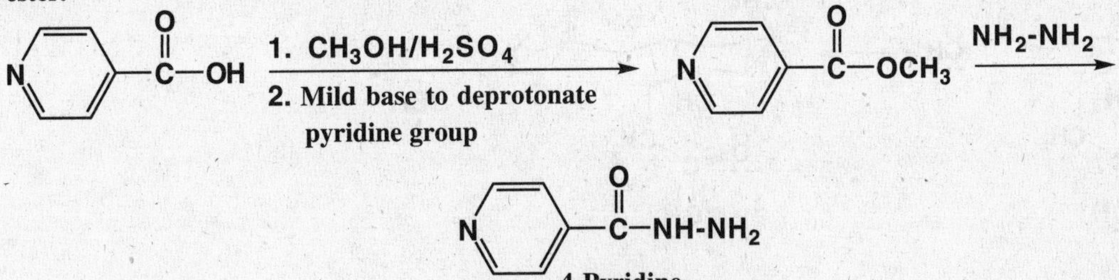

1. CH₃OH/H₂SO₄
2. Mild base to deprotonate pyridine group

NH₂-NH₂

4-Pyridine-
carboxylic acid hydrazide
(Isoniazid)

Problem 18.37 Show how to convert phenylacetylene to allyl phenylacetate.

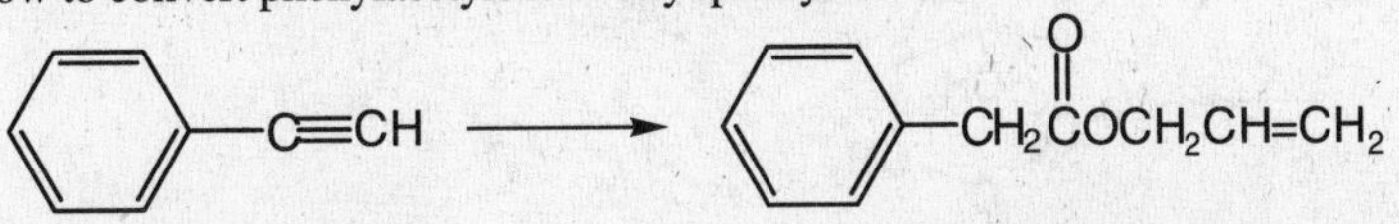

Phenylacetylene

Allyl phenylacetate

The key to this synthesis is the first step in which hydroboration / oxidation is used to make the aldehyde. Oxidation to the carboxylic acid, conversion to the acid chloride, and reaction with allyl alcohol gives the desired product.

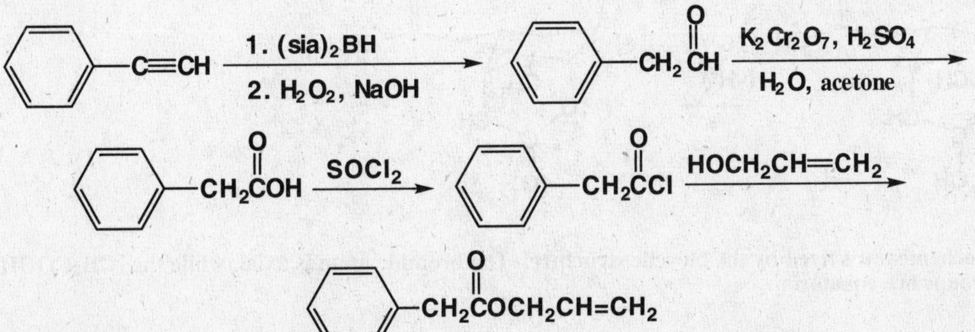

1. (sia)₂BH
2. H₂O₂, NaOH

K₂Cr₂O₇, H₂SO₄
H₂O, acetone

SOCl₂

HOCH₂CH=CH₂

<u>Problem 18.38</u> A step in a synthesis of PGE$_1$ is the reaction of a trisubstituted cyclohexene with bromine to form a bromolactone. Propose a mechanism for formation of this bromolactone and account for the observed stereochemistry of each substituent on the cyclohexane ring.

A bromolactone   PGE$_1$ (Alprostadil)

Alprostadil is used as a temporary therapy for infants born with congenital heart defects that restrict pulmonary blood flow. It brings about dilation of the ductus arteriosus, which in turn increases blood flow in the lungs and blood oxygenation.

The mechanism likely involves formation of a bridged bromonium ion that is attacked by the axial carboxyl group. It is the axial carboxyl group that is in the proper location to attack the bromonium ion as shown:

Step 1: Electrophilic addition.

Step 2: Make a bond between a nucleophile and an electrophile.

Step 3: Take a proton away.

The stereochemistry is fixed by the bicyclic structure. The bromine atom is axial, while the -CH$_2$COOH and methyl groups are equatorial.

Problem 18.39 Barbiturates are prepared by treatment of a derivative of diethyl malonate with urea in the presence of sodium ethoxide as a catalyst. Following is an equation for the preparation of barbital, a long-duration hypnotic and sedative, from diethyl diethylmalonate and urea.

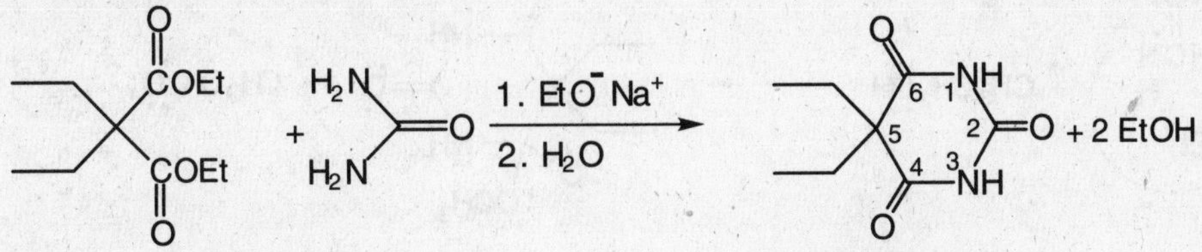

Diethyl diethylmalonate   Urea   **5,5-Diethylbarbituric acid**
                                                                          (Barbital)

Barbital is prescribed under one of a dozen or more trade names.
(a) Propose a mechanism for this reaction.

Treatment of urea with ethoxide ion gives an anion that then attacks a carbonyl group of the malonic ester to displace ethoxide ion. This second reaction is an example of nucleophilic displacement at a carbonyl carbon.

Step 1: Take a proton away.

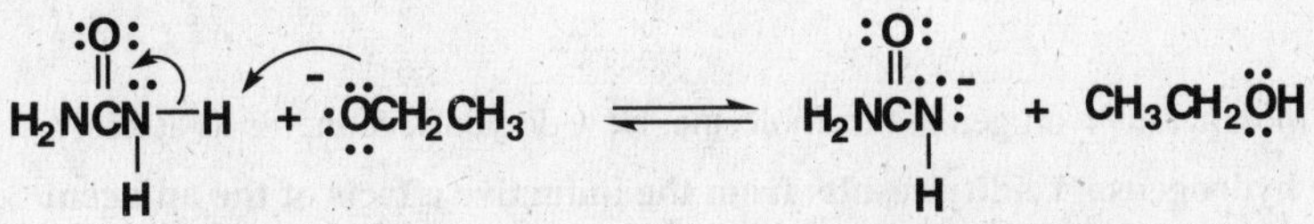

Anion from urea

Step 2: Make a bond between a nucleophile and an electrophile.

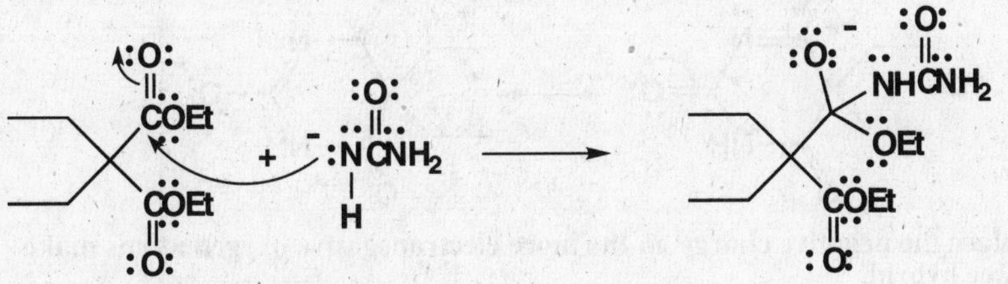

Step 3: Break a bond to give stable molecules or ions.

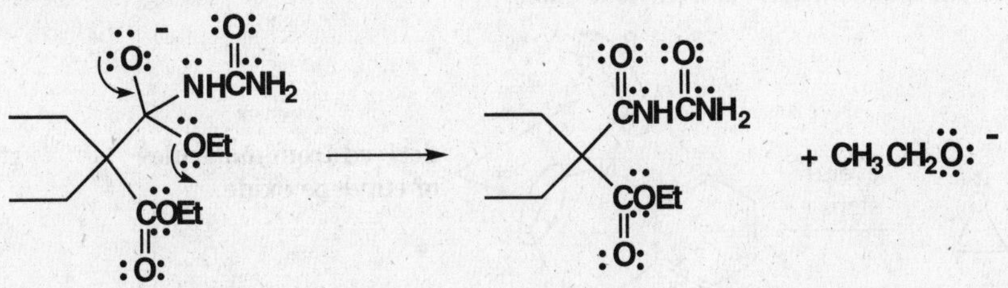

Step 4: Take a proton away.

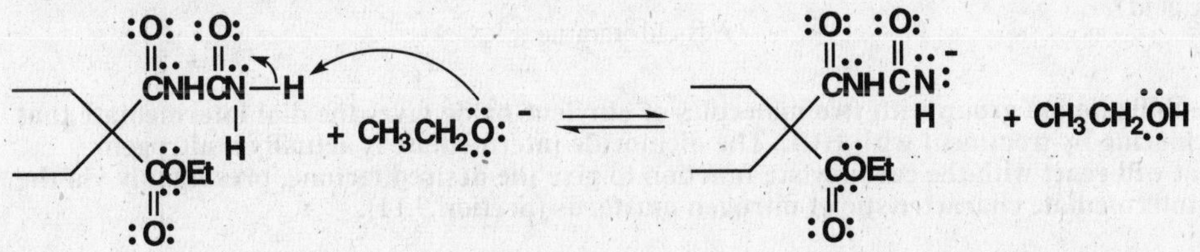

Step 5: *Make a bond between a nucleophile and an electrophile.* Nucleophilic displacement at the carbonyl group of the remaining ester to complete formation of the six-membered ring.

Step 6: *Break a bond to give stable molecules or ions.*

(b) The pK_a of barbital is 7.4. Which is the most acidic hydrogen in this molecule. How do you account for its acidity?

The most acidic hydrogens are the imide hydrogens. Acidity results from the inductive effects of the adjacent carbonyl groups and stabilization of the deprotonated anion by resonance interactions with the carbonyl groups. Following are three contributing structures for the deprotonated barbiturate anion.

The two contributing structures that place the negative charge on the more electronegative oxygen atoms make the greater contribution to the resonance hybrid.

Problem 18.40 The following compound is one of a group of β-chloroamines, many of which have antitumor activity. Describe a synthesis of this compound from anthranilic acid and ethylene oxide.

2-Aminobenzoic acid
(Anthranilic acid)

A β-chloramine

derived from molecules
of ethylene oxide

Reaction of the nucleophilic amino group with two molecules of ethylene oxide gives the diol intermediate that is converted to the dichloride by treatment with HCl. The dichloride intermediate is actually a nitrogen mustard derivative that will react with the carboxylate function to give the desired lactone, presumably via the three-membered ring intermediate characteristic of nitrogen mustards (Section 9.11).

<u>Problem 18.41</u> Show how to synthesize 5-nonanone from 1-bromobutane as the only organic starting material.

There are nine carbon atoms in the product and four in 1-bromobutane. Two 1-bromobutane molecules will be used, but that leaves one carbon atom that must be added. This is most easily accomplished through formation of a nitrile or by a Grignard reaction with CO₂.

1-Bromobutane is also converted to a Gilman reagent via the organolithium species.

The key step in the synthesis involves reaction of pentanoyl chloride, produced from pentanoic acid by reaction with SOCl₂, with the Gilman reagent to give the product 5-nonone. This is an example of a so-called "convergent synthesis" in which two pieces, themselves the product of several synthetic steps, are combined in a late step in the reaction sequence.

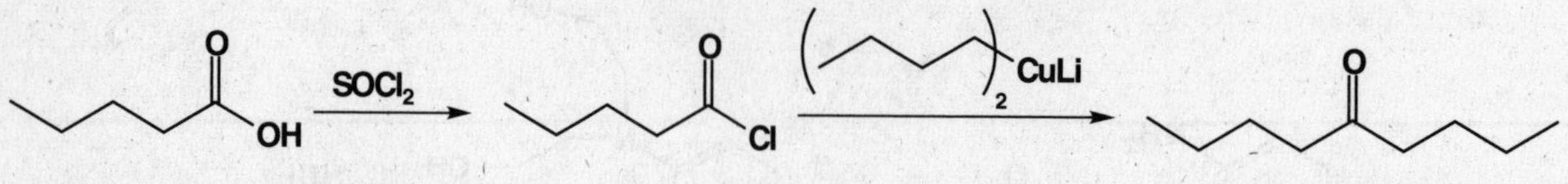

<u>Problem 18.42</u> Procaine (its hydrochloride is marketed as Novocain) was one of the first local anesthetics for infiltration and regional anesthesia. See Chemical Connections: "From Cocaine to Procaine and Beyond." According to the following retrosynthetic scheme, procaine can be synthesized from 4-aminobenzoic acid, ethylene oxide, and diethylamine as sources of carbon atoms.

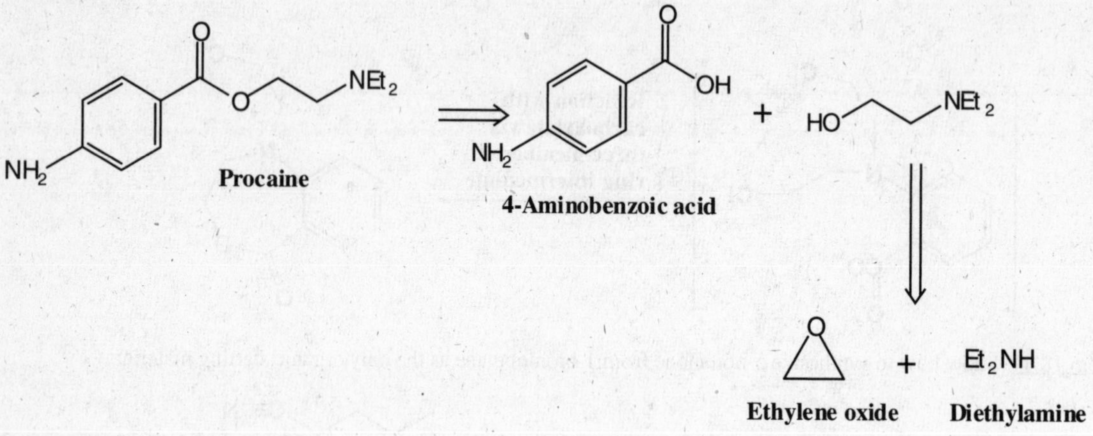

Provide reagents and experimental conditions to carry out the synthesis of procaine from these three compounds.

A reasonable synthetic scheme is shown below. First, the amine is reacted with ethylene oxide to create the corresponding β-aminoalcohol:

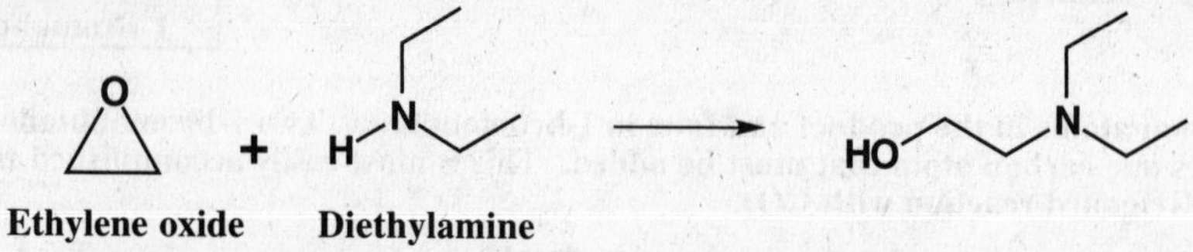

Next, *p*-aminobenzoic acid is converted to procaine by Fischer esterification.

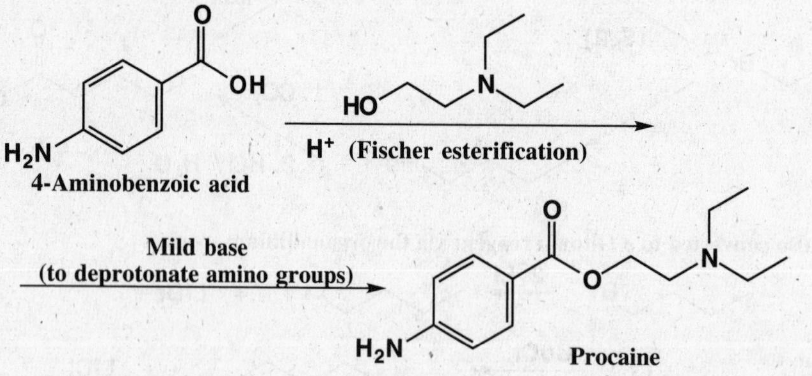

<u>Problem 18.43</u> The following sequence converts (*R*)-2-octanol to (*S*)-2-octanol.

(*R*)-2-Octanol   →[*p*-TsCl / pyridine] A →[CH₃COO⁻Na⁺ / DMSO] B →[1. LiAlH₄ / 2. H₂O] (*S*)-2-Octanol

Propose structural formulas for intermediates A and B, specify the configuration of each, and account for the inversion of configuration in this sequence.

Formation of the tosylate ester (A) occurs with retention of configuration; reaction takes place on oxygen of the alcohol and does not involve the tetrahedral chiral center. Nucleophilic substitution to form (B) occurs by an S$_N$2 pathway with inversion of configuration at the chiral center undergoing reaction. Reduction of the ester with LAH occurs at the carbonyl carbon, not at the chiral center. These relationships are shown in the following structural formulas.

(*R*)-2-Octanol   →[*p*-TsCl / pyridine] (A) →[CH₃COO⁻Na⁺ / DMSO / (S$_N$2)] (B)

→[1. LiAlH₄ / 2. H₂O] (*S*)-2-Octanol

<u>Problem 18.44</u> Reaction of a primary or secondary amine with diethyl carbonate under controlled conditions gives a carbamic ester. Propose a mechanism for this reaction.

Diethyl carbonate + Butylamine → Ethyl *N*-butylcarbamate + EtOH

A reasonable mechanism for this reaction involves initial attack by the nucleophilic nitrogen atom of butylamine on the carbonyl carbon atom of diethyl carbonate to create a tetrahedral carbonyl addition intermediate that leads to the product carbamate after a proton transfer then elimination of ethoxide.

Step 1: Make a bond between a nucleophile and an electrophile.

Step 2: Add a proton-take a proton away.

Step 3: Break a bond to give stable molecules or ions.

Step 4: Take a proton away.

<u>Problem 18.45</u> Several sulfonylureas, a class of compounds containing $RSO_2NHCONHR$, are useful drugs as orally active replacements for injected insulin in patients with adult-onset diabetes. These drugs decrease blood glucose concentrations by stimulating β cells of the pancreas to release insulin and by increasing the sensitivity of insulin receptors in peripheral tissues to insulin stimulation. Tolbutamide is synthesized by the reaction of the sodium salt of *p*-toluenesulfonamide and ethyl *N*-butylcarbamate (see Problem 18.44 for the synthesis of this carbamic ester). Propose a mechanism for this step.

**Sodium salt of
p-toluenesulfonamide**

A carbamic ester

**Tolbutamide
(Oramide, Orinase)**

A reasonable mechanism for this reaction involves initial attack by the nucleophilic nitrogen atom of the sulfonamide sodium salt on the carbonyl carbon atom of the carbamic ester to create a tetrahedral carbonyl addition intermediate that eliminates ethoxide to give the final product.

Step 1: Make a bond between a nucleophile and an electrophile.

Step 2: Break a bond to give stable molecules or ions.

**Tolbutamide
(Oramide, Orinase)**

<u>Problem 18.46</u> Following are structural formulas for two more widely used sulfonylurea hypoglycemic agents. Show how each might be synthesized by converting an appropriate amine to a carbamic ester and then treating the carbamate with the sodium salt of a substituted benzenesulfonamide.

(a)

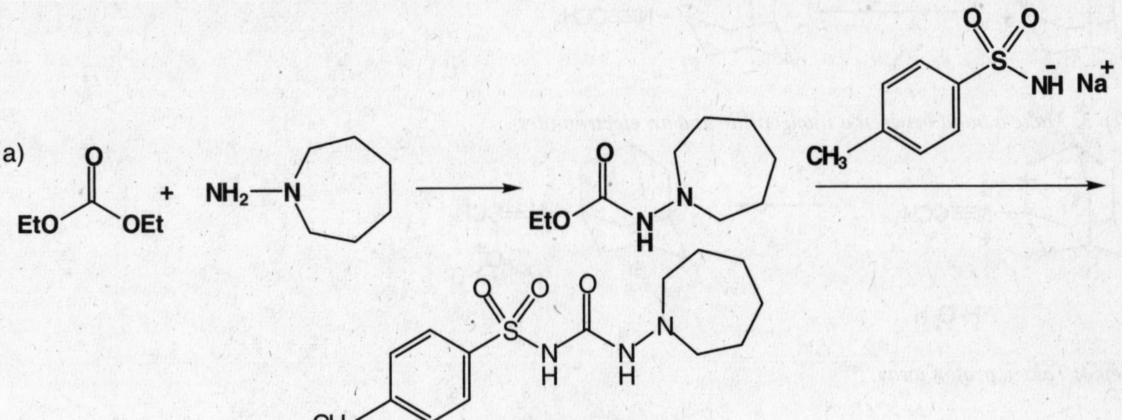

Tolazamide
(Tolamide, Tolinase)

(b)

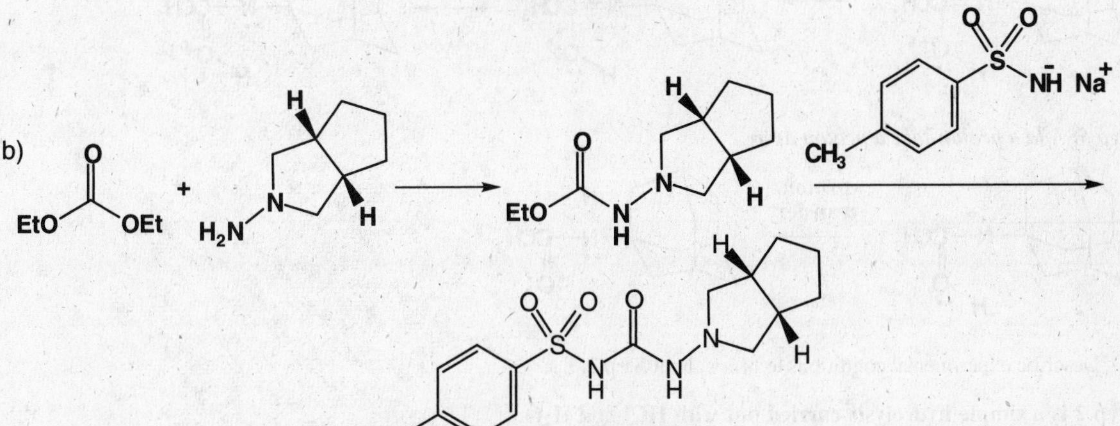

Gliclazide
(Diamicron)

<u>Problem 18.47</u> Amantadine is effective in preventing infections caused by the influenza A virus and in treating established illnesses. It is thought to block a late stage in the assembly of the virus. Amantadine is synthesized by treating 1-bromoadamantane with acetonitrile in sulfuric acid to give *N*-adamantylacetamide, which is then converted to amantadine.

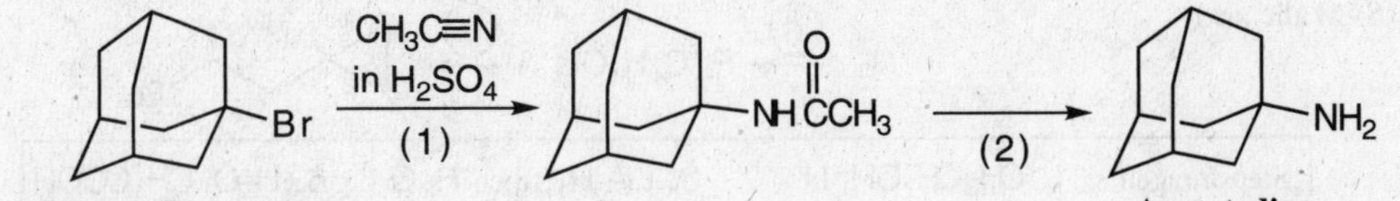

1-Bromoadamantane   **Amantadine**

(a) Propose a mechanism for the transformation in Step 1.

A reasonable mechanism is shown below. The last proton transfer may occur at the same time as formation of the carbonyl.

Step 1: Break a bond to give stable molecules or ions.

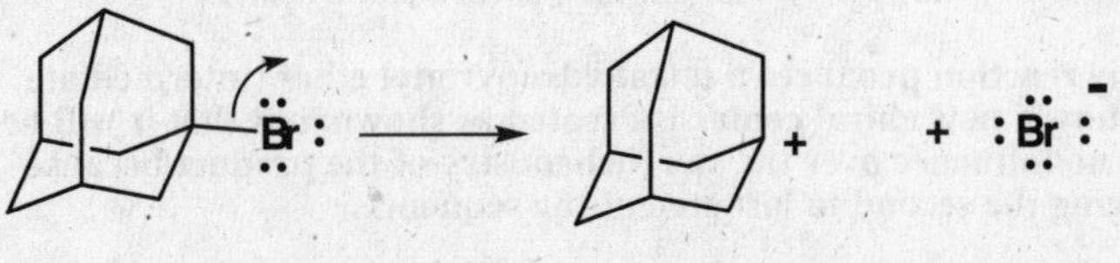

Step 2: Make a bond between a nucleophile and an electrophile.

Step 3: Make a bond between a nucleophile and an electrophile.

Step 4: Take a proton away.

Step 5: Add a proton-take a proton away.

(b) Describe experimental conditions to bring about Step 2.

Step 2 is a simple hydrolysis carried out with HCl and H₂O.

Problem 18.48 In a series of seven steps, (S)-malic acid is converted to the bromoepoxide shown on the right in 50% overall yield. This synthesis is enantioselective- of the stereoisomers possible for the bromoepoxide, only one is formed.

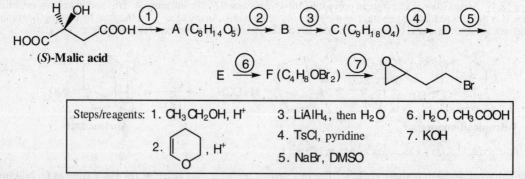

(S)-Malic acid

Steps/reagents: 1. CH₃CH₂OH, H⁺   3. LiAlH₄, then H₂O   6. H₂O, CH₃COOH

2. ⟨ ⟩, H⁺   4. TsCl, pyridine   7. KOH

5. NaBr, DMSO

In thinking about the chemistry of these steps, you will want to review the use of dihydropyran as an -OH protecting group (Section 16.7D) and the use the *p*-toluenesulfonyl chloride to convert the -OH, a poor leaving group, into a tosylate, a good leaving group (Section 10.5D).
(a) Propose structural formulas for intermediates A through F, and specify configuration at each chiral center.

Transformation 2. deserves special comment. This reaction produces a tetrahydropyranyl ether (intermediate B) that serves as an -OH protecting group. Note how a new chiral center is created as shown, but that it will be a mixture of stereoisomers (R,S). This center has no influence over the stereochemistry of the product because the tetrahydropyranyl ether is cleaved in acid during the second to last step of the sequence.

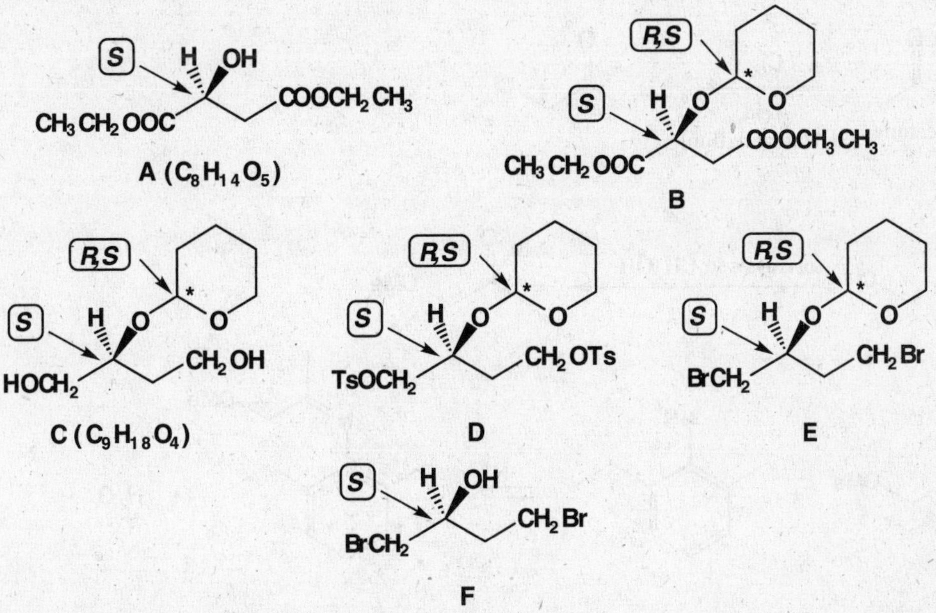

A $(C_8H_{14}O_5)$

B

C $(C_9H_{18}O_4)$

D

E

F

(b) What is the configuration of the chiral center in the bromoepoxide and how do you account for the stereoselectivity of this seven-step conversion?

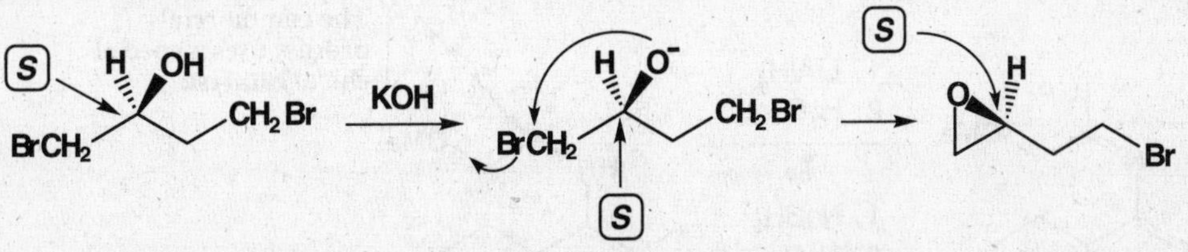

As shown on the above structures, the configuration at the 2 position during the synthesis is S**. This stereochemistry is set in the starting material and maintained throughout the synthetic sequence, because none of the reactions involves a change or loss of stereochemistry at this position.**

<u>Problem 18.49</u> Following is a retrosynthetic analysis for the synthesis of the herbicide (S)-Metolachlor from 2-ethyl-6-methylaniline, chloroacetic acid, acetone, and methanol.

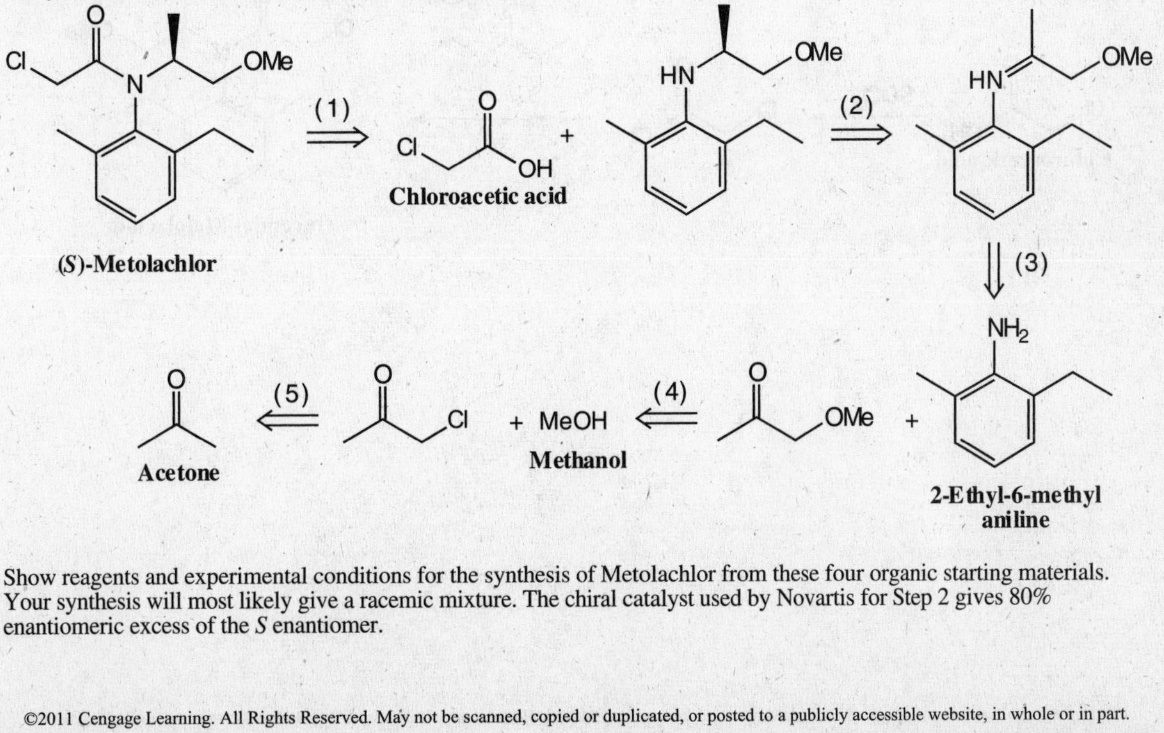

(S)-Metolachlor

Chloroacetic acid

Acetone

Methanol

2-Ethyl-6-methyl aniline

Show reagents and experimental conditions for the synthesis of Metolachlor from these four organic starting materials. Your synthesis will most likely give a racemic mixture. The chiral catalyst used by Novartis for Step 2 gives 80% enantiomeric excess of the S enantiomer.

olutions   Chapter 18: Functional Derivatives of Carboxylic Acids

(5)

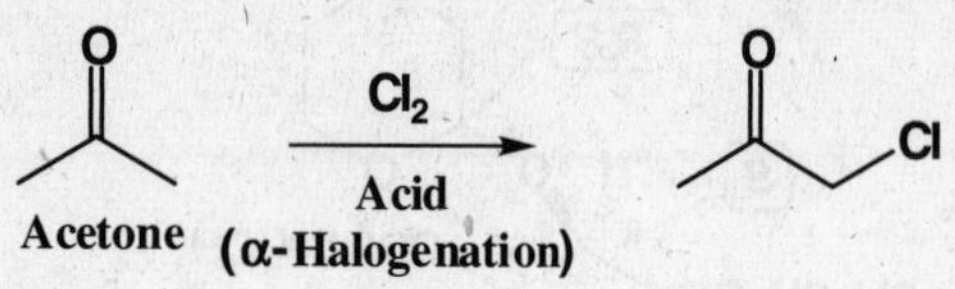

Acetone

Cl_2

Acid
(α-Halogenation)

(4)

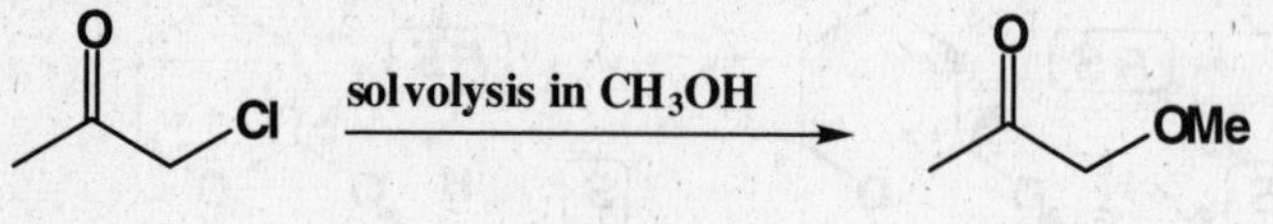

solvolysis in CH_3OH

OMe

(3)

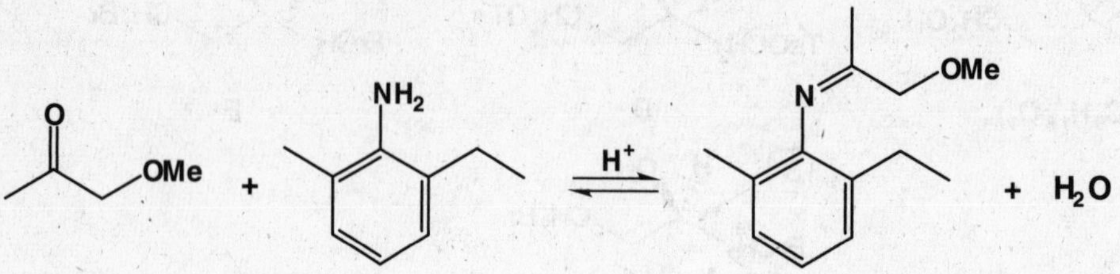

OMe   + NH_2

H^+

OMe

+ H_2O

Note how a racemic
mixture is produced
at this chiral center.
The commercial
process uses a special
chiral catalyst.

(2)

OMe

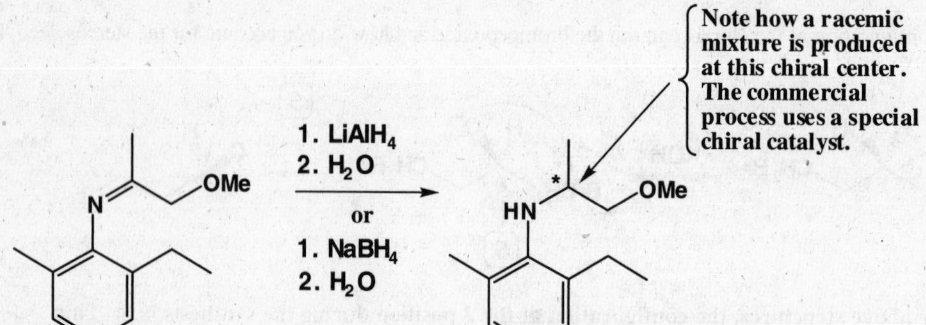

1. $LiAlH_4$
2. H_2O

or

1. $NaBH_4$
2. H_2O

OMe

(1)

Cl   OH
Chloroacetic acid

$SOCl_2$

Cl   Cl

OMe

Cl   OMe

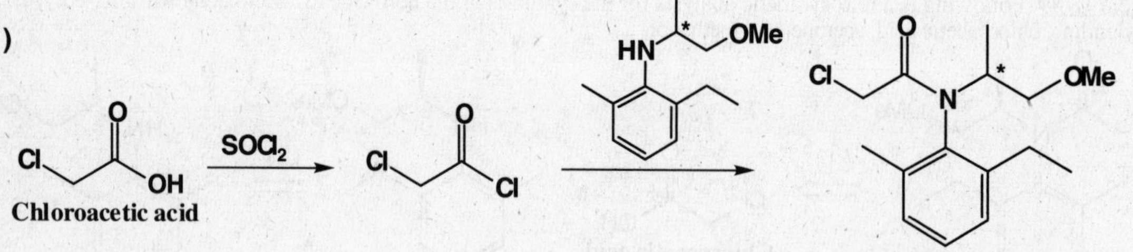

(racemic)-Metolachlor

2011 Cengage Learning. All Rights Reserved. May not be scanned, copied or duplicated, or posted to a publicly accessible website, in whole or in part.

<u>Problem 18.50</u> Following is a retrosynthetic analysis for the anthelmintic (against worms) diethylcarbamazine.

Diethylcarbamazine   (1)   (A)

(2)

**Methylamine   Ethylene   (C)   (3)   (B)   Ethyl
              oxide   chloroformate**

Diethylcarbamazine is used chiefly against nematodes, small cylindrical or slender threadlike worms such as the common roundworm, which are parasitic in animals and plants. Given this retrosynthetic analysis, propose a synthesis of diethylcarbamazine from the four named starting materials.

Treatment of methylamine with two moles of ethylene oxide results in nucleophilic opening of the highly strained epoxide ring (Section 11.9).

(4)

**Methylamine   Ethylene   (C)
              oxide**

Treatment of the diol from step 1 with two moles of thionyl chloride (Section 10.5) gives a dichloride. Treatment of the dichloride with one mole of ammonia results in nucleophilic displacement of first one chloride and then the second to form a six-membered nitrogen-containing ring. Note that the dichloride has the structural characteristics of a nitrogen mustard (Section 9.11) and displacement of chlorine most probably involves neighboring group participation by the tertiary amine of the mustard.

(3)

(C)   (B)

Nucleophilic displacement of chlorine from ethyl chloroformate by nitrogen (Section 18.6) gives the carbamic ester; that is, an derivative of carbonic acid that is both an amide and an ester. Note that in the reaction of the amine with ethyl chloroformate, it is -Cl that is displaced and not the -OEt group; chloride is a more stable anion than ethoxide ion and, therefore, is more readily displaced from ethyl chloroformate than ethoxide ion.

(2)

**(B)   Ethyl   (A)
            chloroformate**

Treatment of the carbamide by diethylamine (Section 18.6) gives diethylcarbamazine.

(1)

Diethylcarbamazine

In an alternative synthesis, *N*-methylpiperazine is then treated with phosgene, the diacid chloride of carbonic acid, to give a carbamoyl chloride. This acid chloride is then with diethylamine to give diethylcarbamazine.

Diethylcarbamazine

<u>Problem 18.51</u> Given this retrosynthetic analysis, propose a synthesis for the antidepressant moclobemide.

Moclobemide   Morpholine   Ethylene   *p*-Chlorobenzoic
                                      oxide   acid

Treatment of ethylene oxide with morpholine results in nucleophilic opening of the highly strained three-membered ring (Section 11.9) and formation of a primary alcohol. Treatment of the primary alcohol with thionyl chloride (Section 10.5) converts the alcohol to the alkyl chloride.

Morpholine   Ethylene
                oxide

The most direct way to covert the primary halide to a primary amine is displacement of chloride by ammonia (S_N2, Section 9.3). However, as will be pointed out in Section 23.7, unless reaction conditions are very carefully controlled, a complex mixture of primary, secondary, tertiary, and amines is formed as well as a quaternary ammonium salt. An alternative strategy is to use azide ion, N_3^-, as a nitrogen-containing nucleophile. The resulting alkyl azide is no longer a nucleophile. Reduction of the alkyl azide to the desired primary amine is accomplished using H_2/Pt. The carboxylic acid is converted to its acid chloride by reaction with thionyl chloride (Section 17.8). The synthesis is completed by reaction of the acid chloride and amine (Section 18.6) to form the amide group of moclobemide.

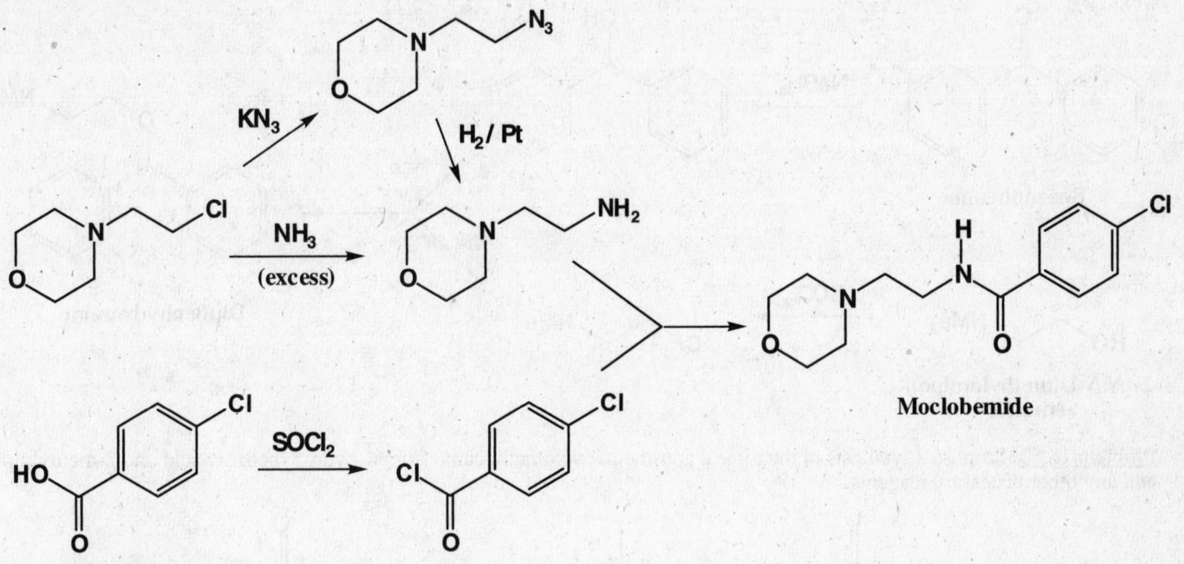

Moclobemide

p-Chlorobenzoic acid

An alternative synthesis of moclobemide begins with aziridine, the nitrogen analog of ethylene oxide. Acylation of the amine nitrogen with 4-chlorobenzoyl chloride gives an aziridine amide. Treatment of the aziridine ring with morpholine results in nucleophilic opening of the ring to give moclobemide.

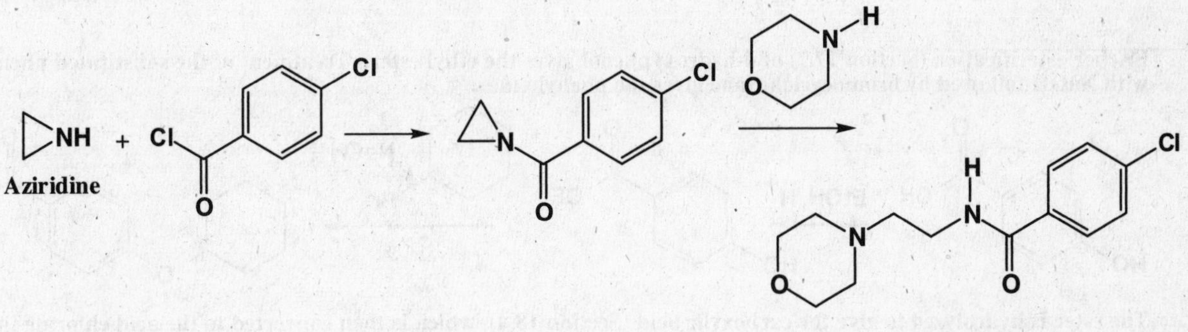

Aziridine

Moclobemide

<u>Problem 18.52</u> Propose a synthesis for diphenhydramine starting from benzene, benzoic acid, and 2-(*N,N*-dimethylamino)ethanol.

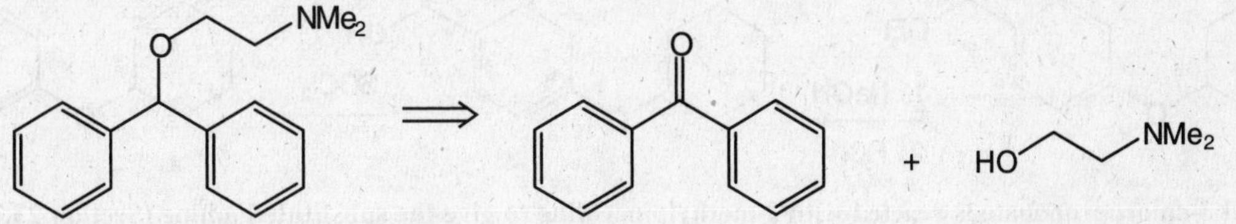

Diphenhydramine   **Benzophenone**   **2-(*N,N*-Dimethylamino)-ethanol**

The hydrochloride salt of diphenhydramine, best known by its trade name of Benadryl, is an antihistamine.

NaBH₄ reduction of the carbonyl group of the ketone gives a secondary alcohol (Section 16.11). Reaction of the primary alcohol of 2-(*N,N*-dimethylamino)ethanol with thionyl chloride (Section 10.5) gives an alkyl chloride. Because of participation by the neighboring nitrogen atom (Section 9.11), this primary chloride undergoes reaction with the secondary hydroxyl group of diphenylmethanol to give diphenhydramine.

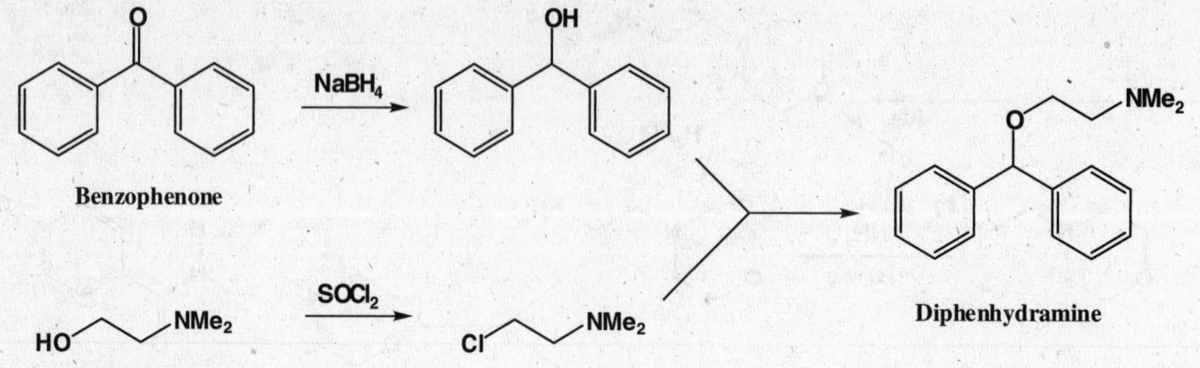

Benzophenone

2-(*N*,*N*-Dimethylamino)- ethanol

Diphenhydramine

Problem 18.53 Propose a synthesis of the topical anesthetic cyclomethycaine from 4-hydroxybenzoic acid and 2-methylpiperidine and any other necessary reagents.

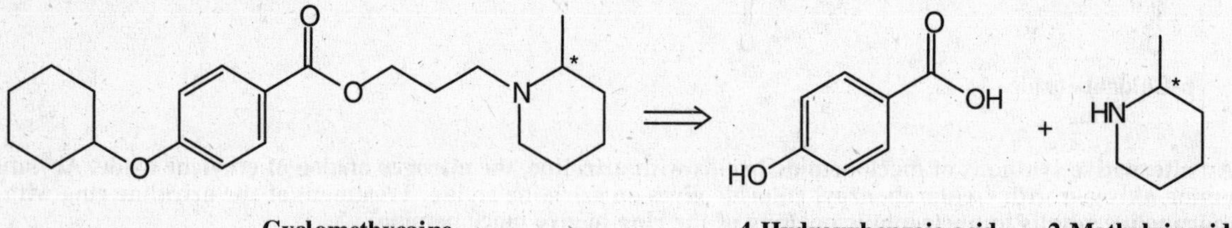

Cyclomethycaine
(racemic)

4-Hydroxybenzoic acid

2-Methylpiperidine
(racemic)

Fischer esterification (Section 17.7) of 4-hydroxyphenol gives the ethyl ester. Treatment of the substituted phenol with NaOH followed by bromocyclohexane gives the phenyl ether.

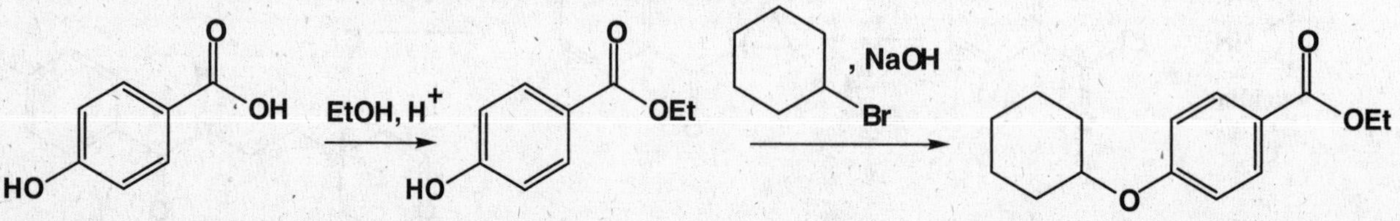

The ester is hydrolyzed to give the carboxylic acid (Section 18.4), which is then converted to the acid chloride through treatment with thionyl chloride (Section 17.9).

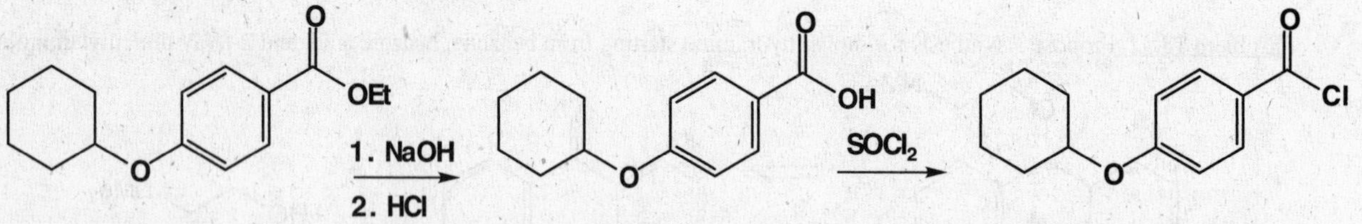

Methyl 3-chloroproponate is reacted with 2-methylpiperidine to give the substituted amine (Section 23.7). Note that in this synthesis, racemic 2-methylpiperidine is used, which will lead to racemic product. The ester is reduced to a primary amine with LiAlH₄ (Section 18.10).

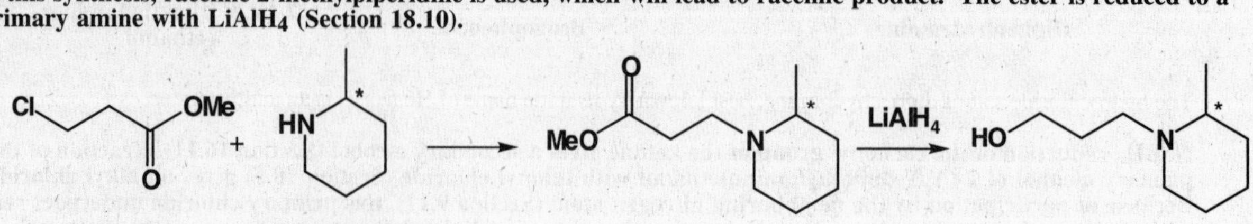

Methyl 3-chloro- proponate

2-Methylpiperidine
(racemic)

In the last step of the synthesis, the acid chloride reacts with the primary alcohol (Section 18.5) to give the racemic product cyclomethycaine.

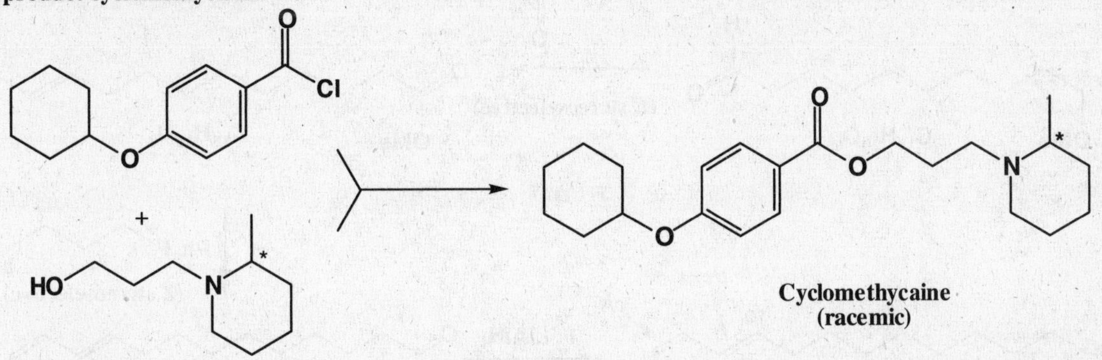

Cyclomethycaine
(racemic)

Problem 18.54 Following is an outline of a synthesis of bombykol, the sex attractant of the male silkworm moth. Of the four stereoisomers possible for this conjugated diene, the 10-*trans*-12-*cis* isomer shown here is over 10^6 times more potent as a sex attractant than any of the other three possible stereoisomers.

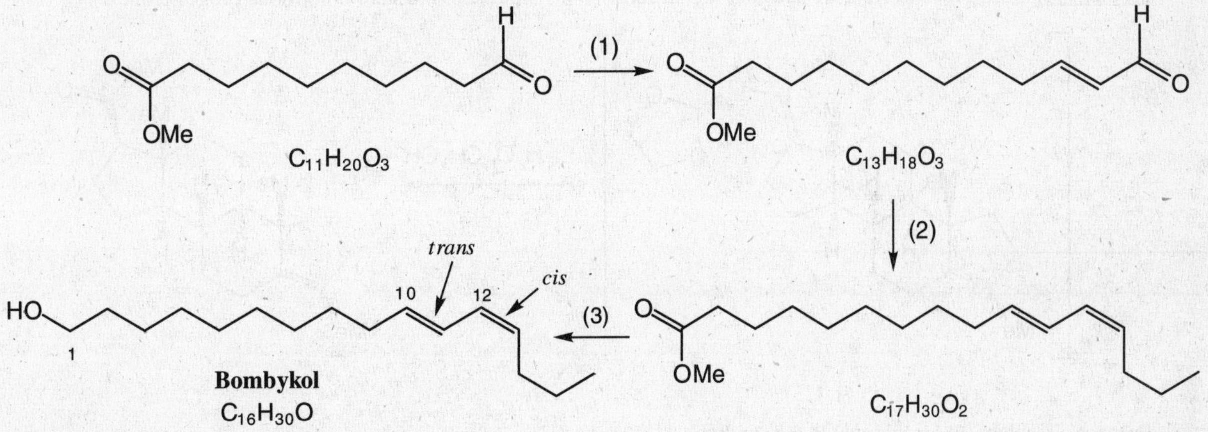
(1)
(2)
(3)
OMe
$C_{11}H_{20}O_3$
$C_{13}H_{18}O_3$
$C_{17}H_{30}O_2$
trans
cis
10
12
1
Bombykol
$C_{16}H_{30}O$

Show how this synthesis might be accomplished, and explain how your proposed synthesis is stereoselective for the 10-*trans*-12-*cis* isomer.

The key to this reaction sequence is the control of stereoselectivity of the Wittig reaction (Section 16.6). Recall that as a general rule, those Wittig reagents with anion-stabilizing substituents, such as a carbonyl group, are *E* selective. Those Wittig reagents with no anion stabilization are generally *Z* selective.
(1) The first Wittig reaction needs to be *E* selective and can be accomplished with a Wittig reagent with an anion stabilizing group, in this case an aldehyde. This reaction is possible because the product aldehyde is less reactive than the starting aldehyde, favoring reaction of the starting material and thus completion of the reaction. The double bond is conjugated (Section 20.1)to the aldehyde group in the product, which reduces its electrophilic character.
(2) The second reaction uses an alkyl Wittig reaction that gives predominantly the *Z* product. Note, that with no alternative aldehydes to react with, the conjugated aldehyde reacts sufficiently to give a high yield.
(3) The final reaction is a reduction of the ester group with 1. $LiAlH_4$, 2. H_2O to give the desired alcohol.

Problem 18.55 In Problem 7.28, we saw this step in Johnson's synthesis of the steroid hormone progesterone.

Propose a mechanism for this step in the synthesis.

Step 1: Make a bond between a nucleophile and an electrophile. **The reaction begins with attack by water on the highly electrophilic carbocation.**

Step 2: Add a proton-take a proton away. A proton is transferred, setting up the departure of the steroid enol.

Step 3: Break a bond to give stable molecules or ions.

Step 4: Keto-enol tautomerism. The product is formed following keto-enol tautomerization.

The stereoselectivity of the last step deserves special comment. During the keto-enol equilibrium, the proton is added to the C=C double bond from the bottom face of the steroid, because this is the significantly less-hindered face. The axial methyl group on the top face is the major contributor to the steric hindrance on the top face of the double bond.

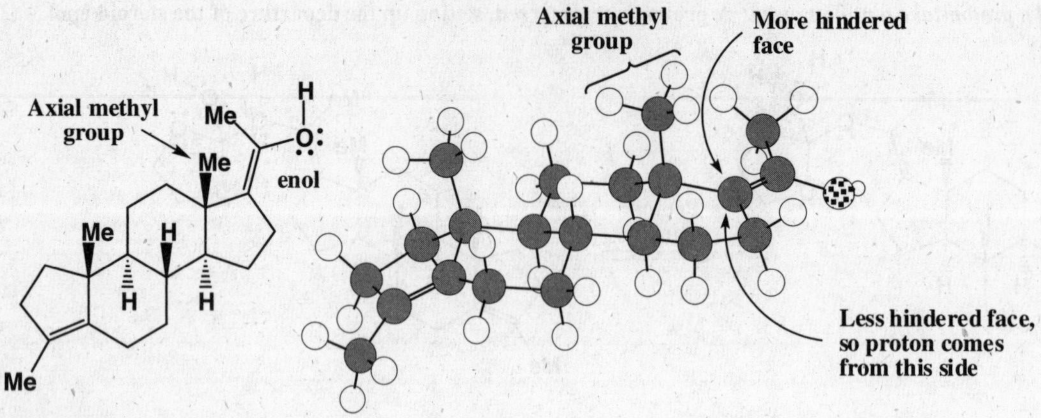

Axial methyl group

H
Me
Me
:O:
H
enol

Axial methyl group

More hindered face

Me
Me
H
H
H

Less hindered face, so proton comes from this side

Me

Mechanisms

<u>Problem 18.56</u> Using the principles for writing mechanisms and the four steps in the "How to Write Mechanisms for Interconversions of Carboxylic Acid Derivatives" box of this chapter, write mechanisms showing all electron flow arrows for the following reactions:

(a) Hydrolysis of *N,N*-dimethylacetamide in acidic water.

The following reaction occurs in acid media, so no basic intermediates are formed. The presence of strong acid predicts that the first step is addition of a proton.

Step 1: Add a proton.

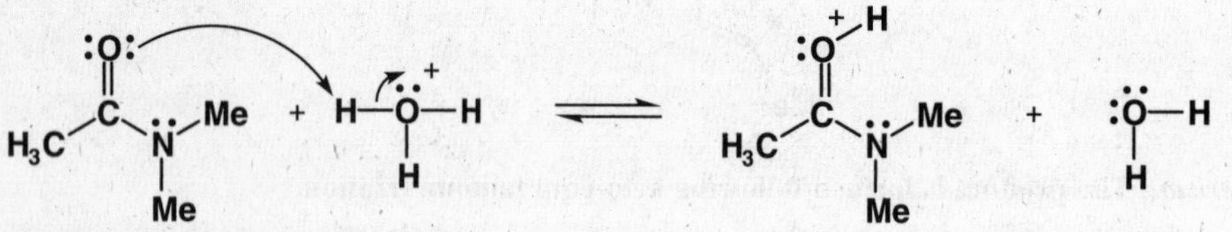

:O:
‖
H₃C—C—N—Me
      |
      Me   +   H—O—H⁺   ⇌   :Ö⁺—H   +   :Ö—H
                   | ‖ |
                   H   H₃C—C—N—Me   H
                              |
                              Me

N,N-Dimethylacetamide

Step 2: Make a bond between a nucleophile and an electrophile.

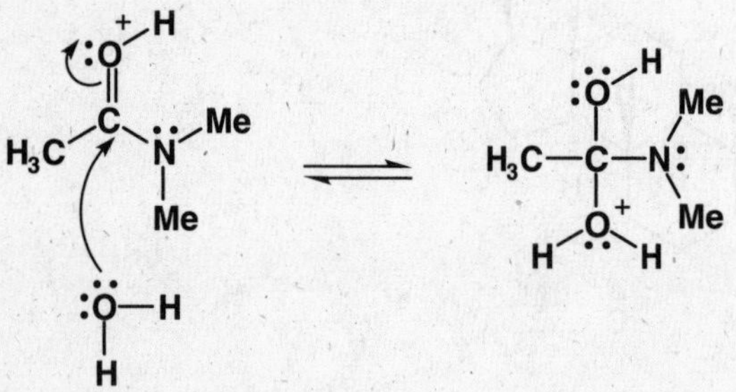

:O⁺—H
‖
H₃C—C—N—Me
      |
      Me   :Ö—H
                   |
:Ö—H   ⇌   H₃C—C—N—Me
| |
H   O⁺
                  / \
                 H   H

Step 3: Take a proton away.

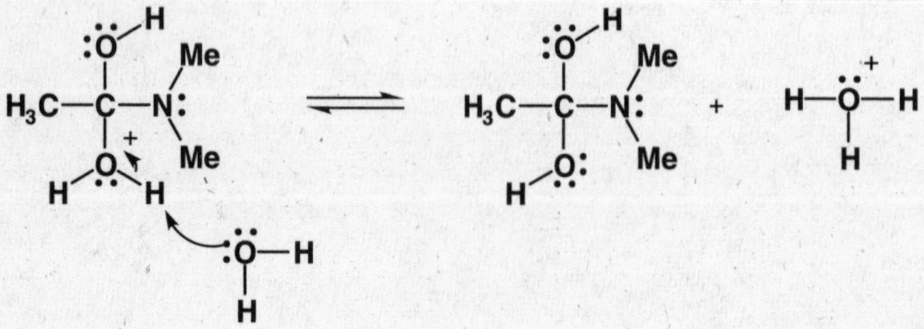

:O—H
|
H₃C—C—N:—Me
| |
O⁺   Me
/ \
H   H   :O—H   :Ö—H
            | |
:Ö—H ⇌ H₃C—C—N:—Me   +   H—O⁺—H
| | |
H   Me   H

Step 4: Add a proton.

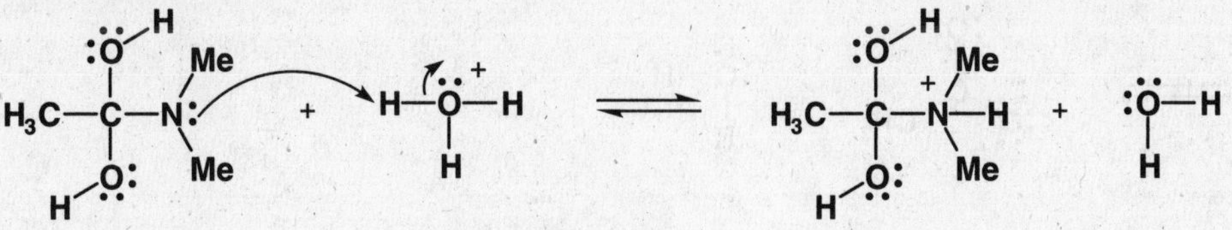

Step 5: Break a bond to give stable molecules or ions.

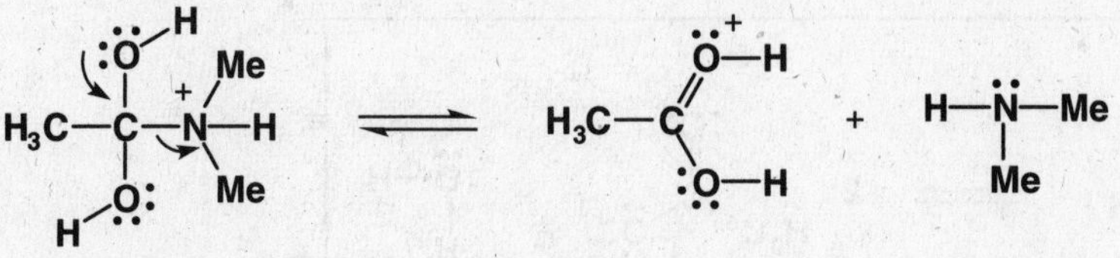

Step 6: Take a proton away.

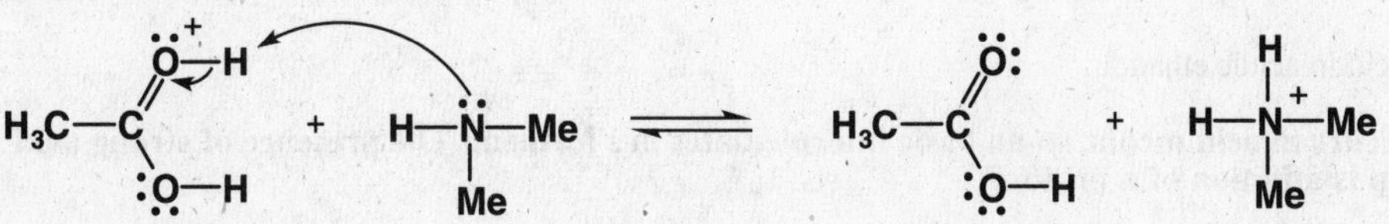

Overall

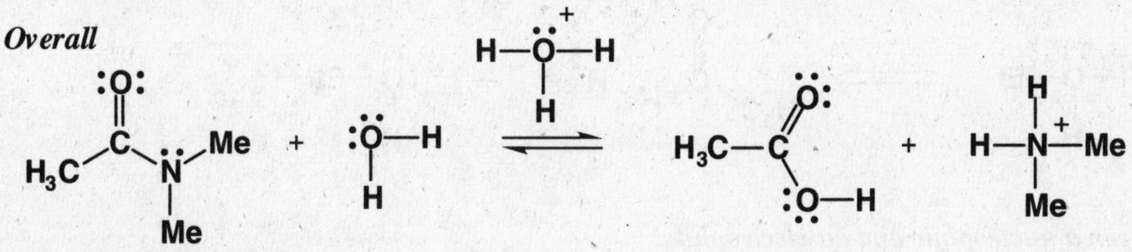

N,N-Dimethylacetamide

(b) Hydrolysis of acetic anhydride in basic water.

The following reaction occurs in basic media, so no acidic intermediates are formed.

Step 1: Make a bond between a nucleophile and an electrophile.

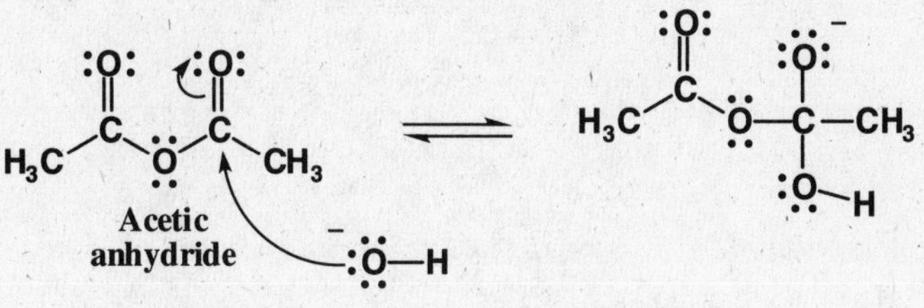

Acetic anhydride

Step 2: Break a bond to give stable molecules or ions.

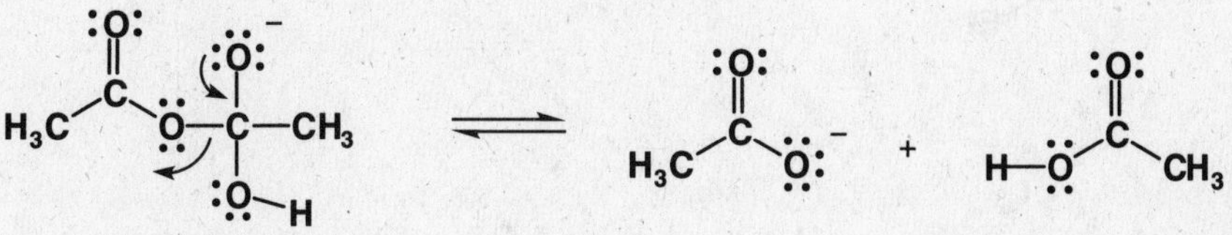

Step 3: Take a proton away.

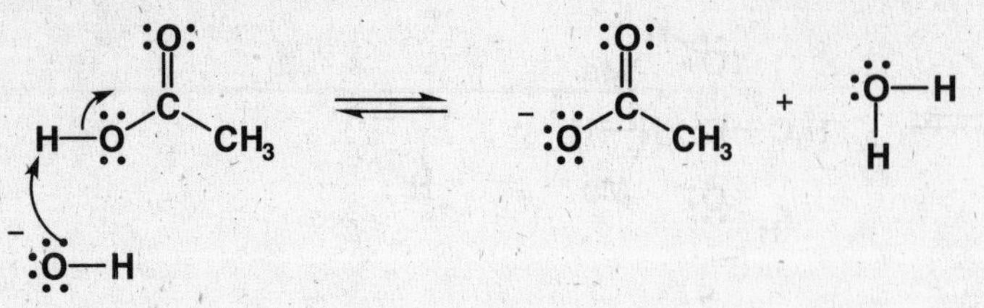

Overall

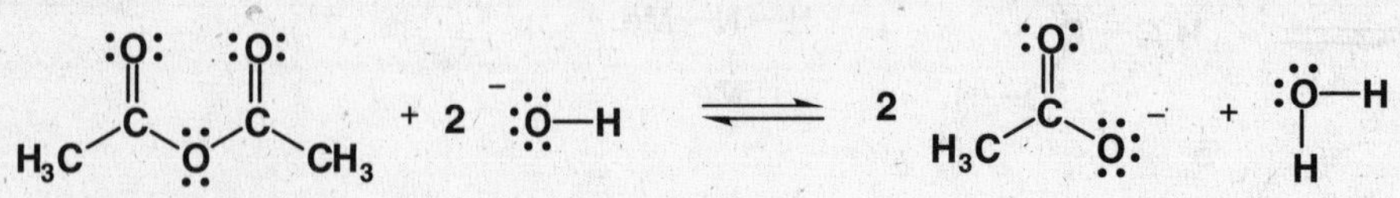

Acetic anhydride

(c) Esterification of acetic acid in acidic ethanol.

The following reaction occurs in acid media, so no basic intermediates are formed. The presence of strong acid predicts that the first step is addition of a proton.

Step 1: Add a proton.

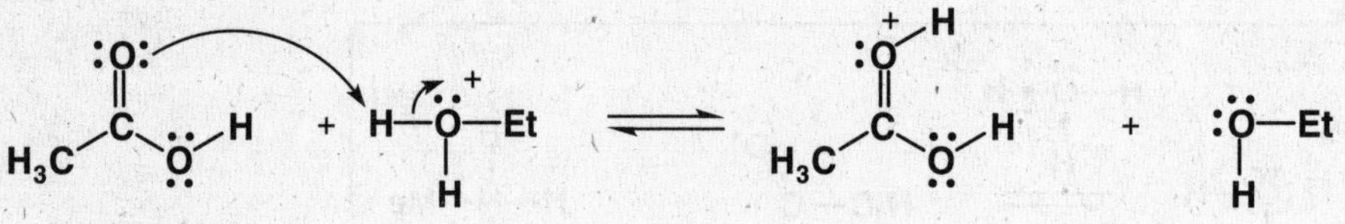

Acetic acid

Step 2: Make a bond between a nucleophile and an electrophile.

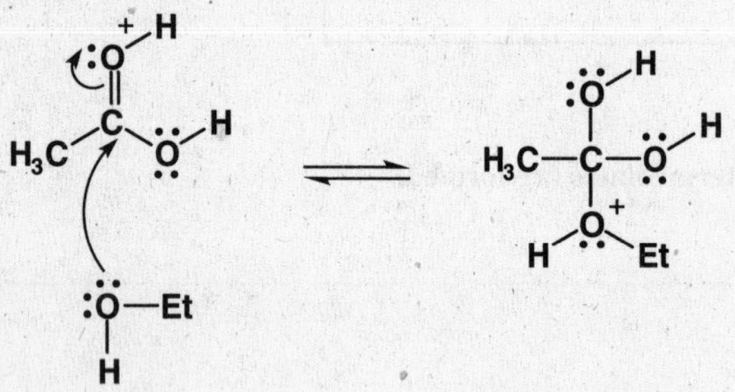

Step 3: Take a proton away.

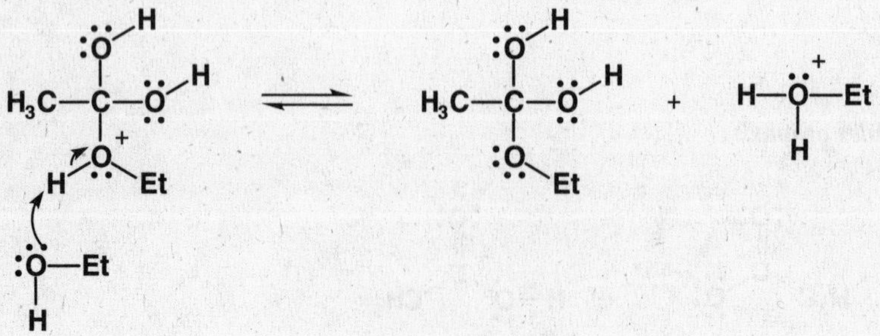

Step 4: Add a proton.

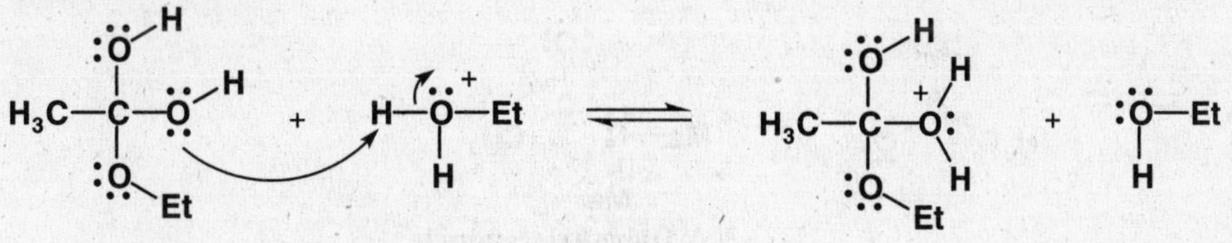

$$H_3C-C(OH)(OH)(OEt) \; + \; H-OEt_2^+ \; \rightleftharpoons \; H_3C-C(OH)(OH_2^+)(OEt) \; + \; :O-Et$$

Step 5: Break a bond to give stable molecules or ions.

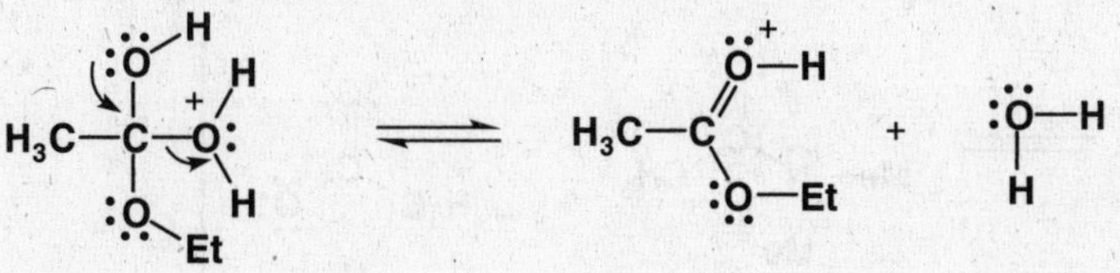

$$H_3C-C(OH)(OH_2^+)(OEt) \; \rightleftharpoons \; H_3C-C(=O^+H)(OEt) \; + \; :O-H$$

Step 6: Take a proton away.

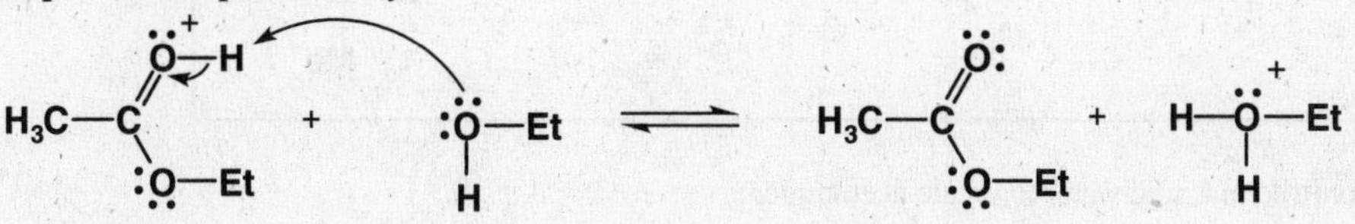

$$H_3C-C(=O^+H)(OEt) \; + \; :O-Et \; \rightleftharpoons \; H_3C-C(=O)(OEt) \; + \; H-O^+Et$$

Overall

$$H_3C-C(=O)(OH) \; + \; :O(H)-Et \; \xrightleftharpoons[H-O^+Et]{} \; H_3C-C(=O)(OEt) \; + \; :O(H)-H$$

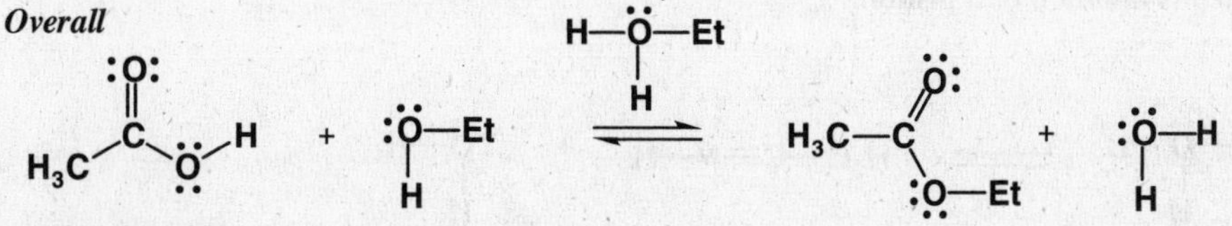

Acetic acid

(d) The reaction of dimethylamine in water with acetic anhydride to create *N,N*-dimethylacetamide.

Amines are bases, so the following reaction occurs in basic media and no acidic intermediates are formed.

Step 1: Make a bond between a nucleophile and an electrophile.

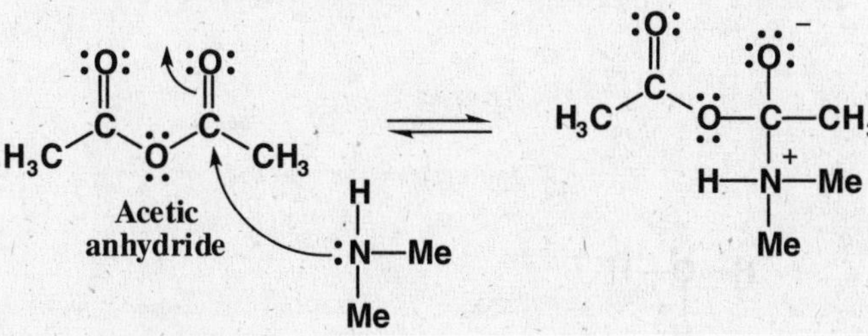

$$H_3C-C(=O)-O-C(=O)-CH_3 \;(\text{Acetic anhydride}) \; + \; :N(H)(Me)(Me) \; \rightleftharpoons \; H_3C-C(=O)-O-C(-O^-)(CH_3)(N^+H(Me)(Me))$$

Step 2: Take a proton away.

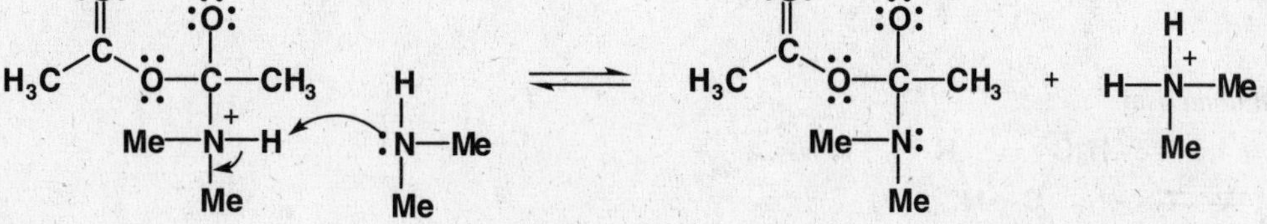

$$H_3C-C(=O)-O-C(-O^-)(CH_3)(Me-N^+(H)(Me)) \; + \; :N(H)(Me)(Me) \; \rightleftharpoons \; H_3C-C(=O)-O-C(-O^-)(CH_3)(Me-N:(Me)) \; + \; H-N^+(H)(Me)(Me)$$

Step 3: Break a bond to give stable molecules or ions.

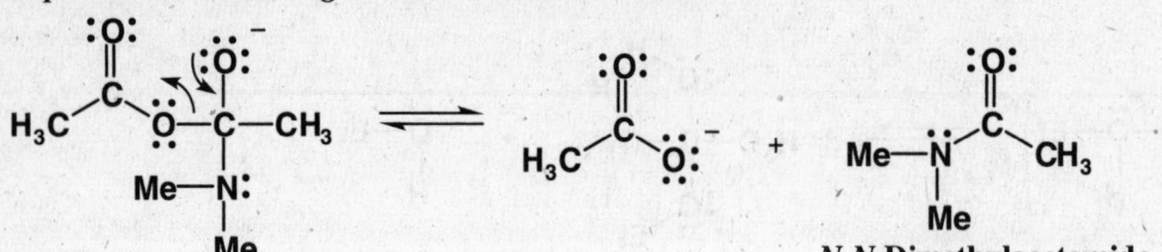

N,N-Dimethylacetamide

Overall

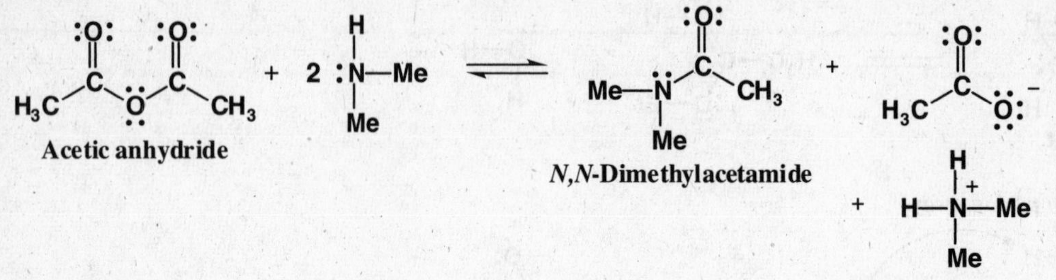

Acetic anhydride

N,N-Dimethylacetamide

(e) Partial hydrolysis of acetonitrile in acidic water to create acetamide.

The following reaction occurs in acid media, so no basic intermediates are formed. The presence of strong acid predicts that the first step is addition of a proton.

Step 1: Add a proton.

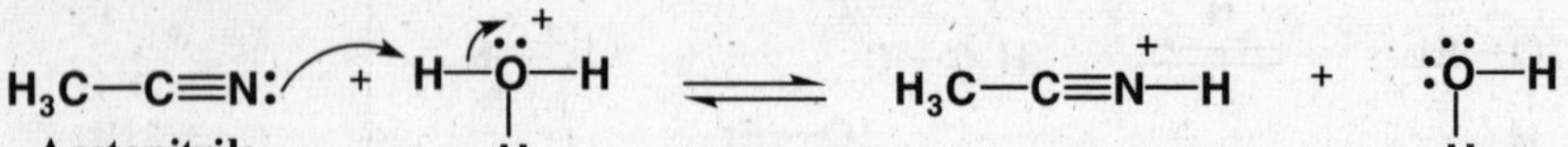

Acetonitrile

Step 2: Make a bond between a nucleophile and an electrophile.

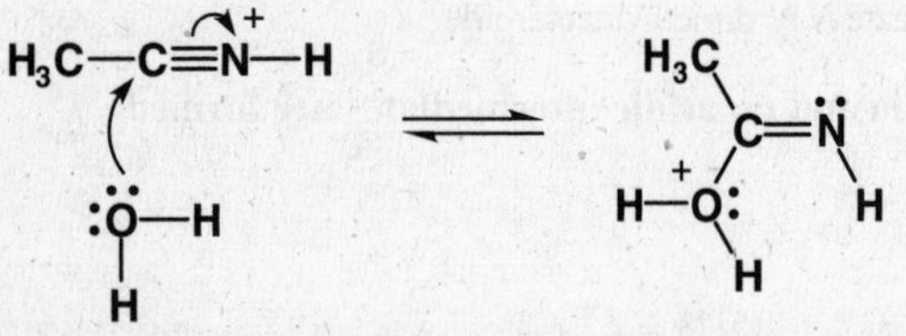

Step 3: Take a proton away.

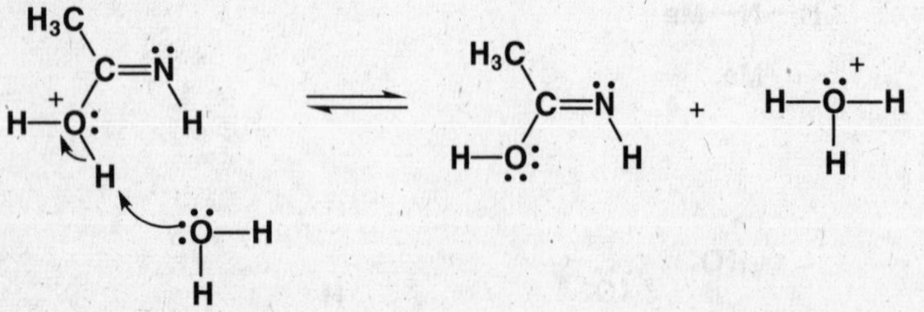

Step 4: Keto-enol tautomerism.

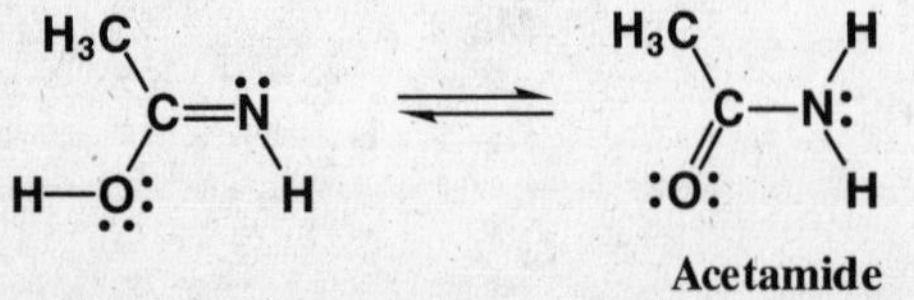

Acetamide

Problem 18.57 The following statements are true experimental observations. Explain the reason behind each observation.
(a) The reaction of acetic acid with ammonia in water does not give any amide products.

Acetic acid is acidic, and ammonia is a base. They react through proton transfer to produce the acetate anion and ammonium ion.

(b) The reaction of acetyl chloride with water causes the pH to decrease.

As acetyl chloride is hydrolyzed, acetic acid is produced causing a decrease in solution pH.

(c) The hydrolysis of an amide at neutral pH takes seven years at room temperature, while the hydrolysis of an acid chloride takes a few minutes.

The chloride anion is a much better leaving group compared with the amide anion. In addition, the extra stability of the amide group, as described in the Connections to Biological Chemistry Box "The Unique Structure of Amide Bonds" (Section 18.1) also helps amides resist hydrolysis.

Looking Ahead
Problem 18.58 We have seen two methods for converting a carboxylic acid and an amine into an amide. Suppose that you start instead with a dicarboxylic acid such as hexanedioic acid and a diamine such as 1,6-hexanediamine. Show how amide formation in this case can lead to a polymer (a macromolecule of molecular weight several thousands of times that of the starting materials).

As we shall see in Section 29.5A, the material produced in this reaction is the high molecular-weight polymer nylon-66, so named because it is synthesized from two 6-carbon starting materials.

Reaction of 1,6-hexanediamine with hexanedioic acid creates an amide bond. The free amine end can react with another molecule of diacid, and the free acid end can react with another diamine and so on. Hydrogen bonding among amide chains gives nylon its strength, and the freely rotating alkyl chains contribute to it flexibility.

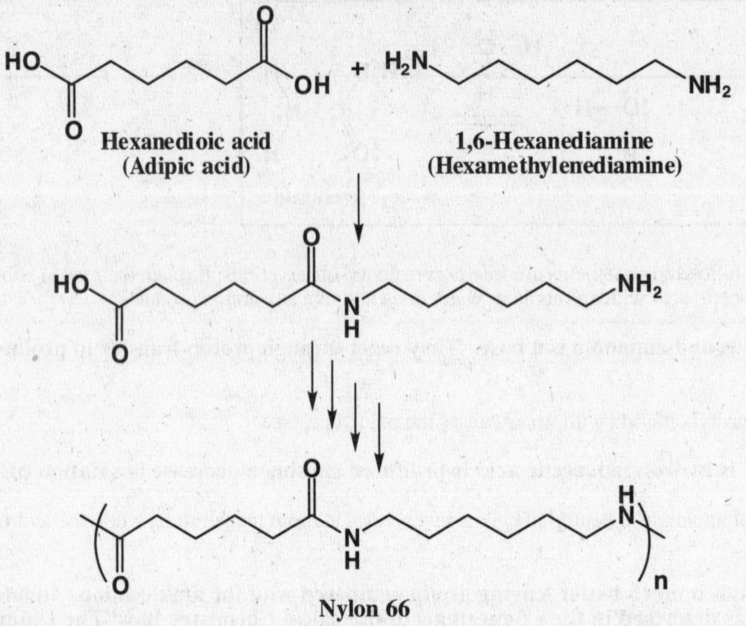

Hexanedioic acid
(Adipic acid)

1,6-Hexanediamine
(Hexamethylenediamine)

Nylon 66

Problem 18.59 Using the same reasoning as in Problem 18.56, show how amide formation between this combination of dicarboxylic acid and diamine will also lead to a polymer, in this case Kevlar.

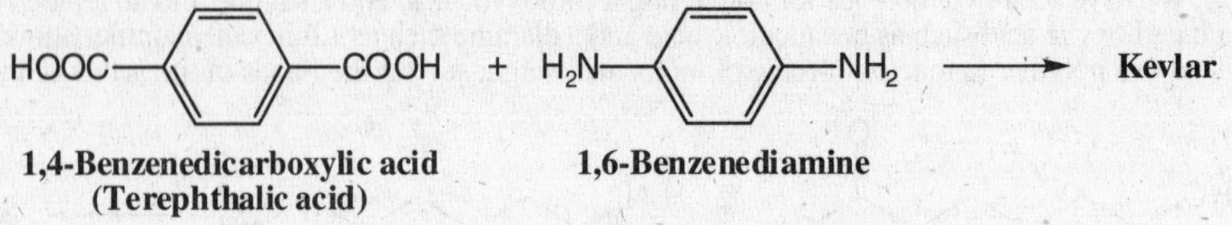

HOOC—⬡—COOH + H₂N—⬡—NH₂ ⟶ **Kevlar**

1,4-Benzenedicarboxylic acid
(Terephthalic acid)

1,6-Benzenediamine

Reaction of 1,4-benzenedicarboxylic acid with 1,6-benzenediamine creates an amide bond. The free amine end can react with another molecule of diacid, and the free acid end can react with another diamine and so on. Hydrogen bonding among amide chains gives nylon its strength, and the relatively rigid backbone make for very strong fibers.

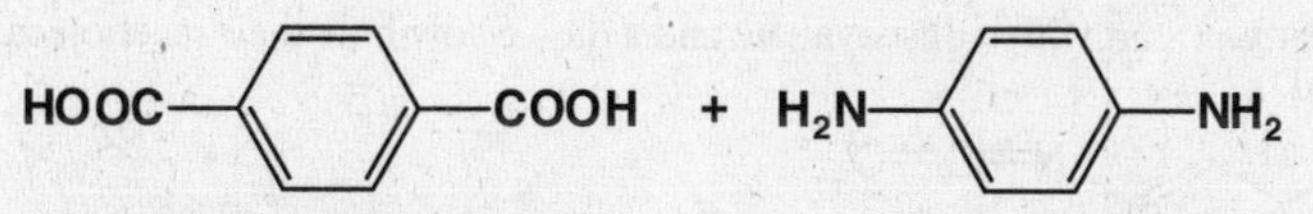

HOOC—⟨benzene⟩—COOH + H₂N—⟨benzene⟩—NH₂

1,4-Benzenedicarboxylic acid
(Terephthalic acid)   **1,6-Benzenediamine**

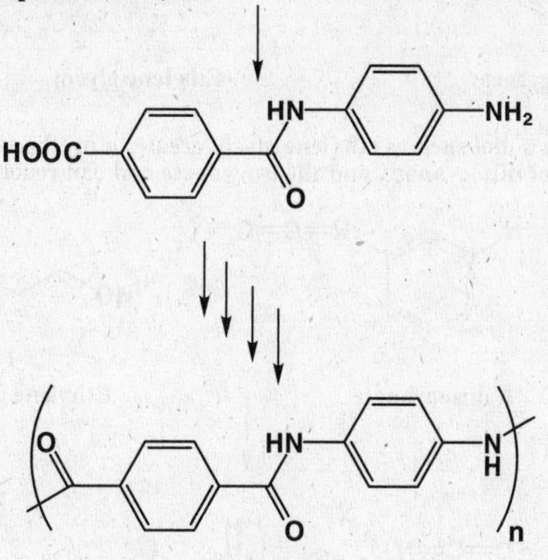

HOOC—⟨benzene⟩—C(=O)—HN—⟨benzene⟩—NH₂

⟨O⟩—⟨benzene⟩—C(=O)—HN—⟨benzene⟩—N(H)—₎ₙ

Kevlar

Problem 18.60 A urethane is a molecule in which a carbonyl group is part of an ester and an amide (it is an amide in one direction, an ester in the other direction). Propose a mechanism for the reaction of an isocyanate with an alcohol to form a urethane.

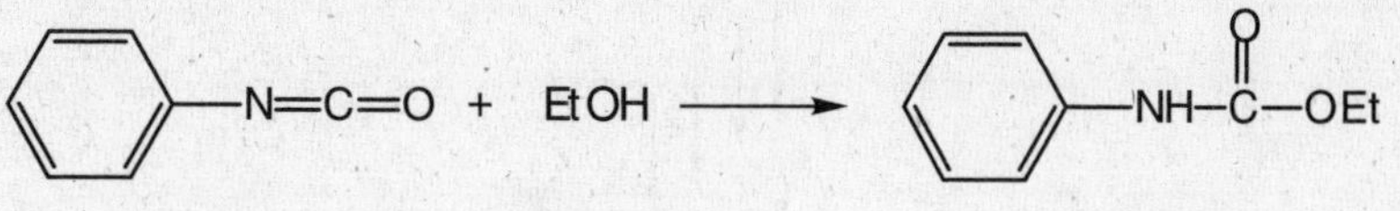

⟨benzene⟩—N=C=O + EtOH ⟶ ⟨benzene⟩—NH—C(=O)—OEt

Phenylisocyanate   Ethanol   A urethane

Step 1: Make a bond between a nucleophile and an electrophile. **Isocyanates react like other carbonyl derivatives in that a nucleophile will attack the relatively electrophilic carbon atom. The product of this step is resonance stabilized.**

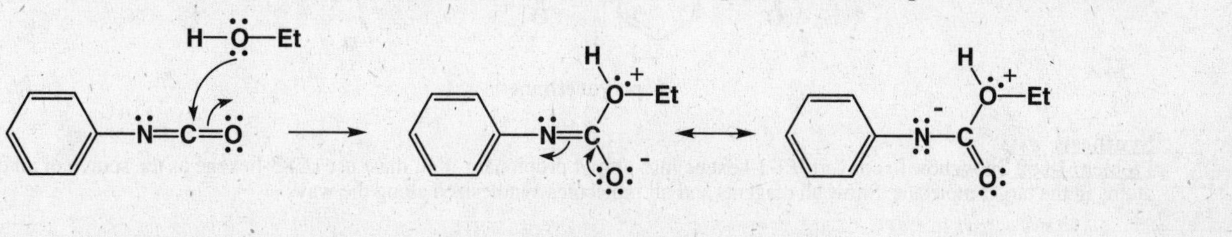

Step 2: Add a proton-take a proton away. **A proton transfer completes the reaction.**

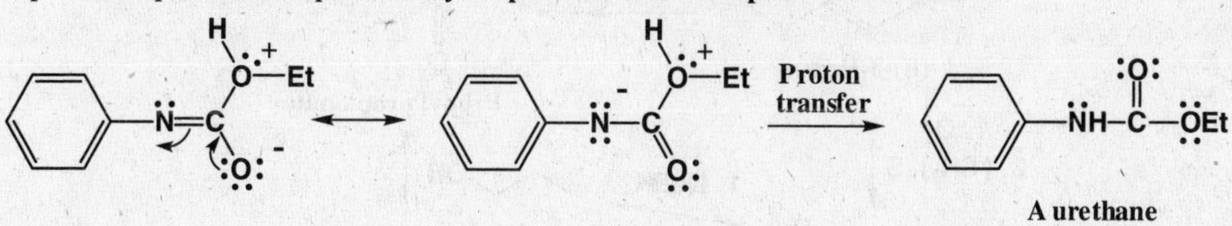

A urethane

Problem 18.61 Suppose that you start with a diisocyanate and a diol. Show how their reaction can lead to a polymer called a polyurethane (Section 29.5D).

A diisocyanate   Ethylene glycol

Reaction of diisocyanate with a diol such as ethylene glycol creates a urethane. The free hydroxyl group end can react with another molecule of diisocyanate, and the isocyanate end can react with another diol and so on.

A diisocyanate   Ethylene glycol

A polyurethane

Synthesis

Problem 18.62 Show how to convert (*E*)-3-hexene into propyl propionate. You must use (*E*)-3-hexene as the source of all carbon atoms in the target molecule. Show all reagents and all molecules synthesized along the way.

(*E*)-3-Hexene   **Propyl propionate**

Recognize that the starting alkene has six carbons in a chain, while the product is an ester with two three-carbon pieces, so a carbon-carbon bond must be broken. The only reaction covered so far that cleaves a carbon-carbon bond is ozonolysis, so propose that the 3-hexene starting material is converted to propanal using ozonolysis as the first step. Oxidation of propanal with chromic acid to give propanoic acid followed by reaction with $SOCl_2$ gives propionyl chloride. Reduction of propanal with $NaBH_4$ gives 1-propanol, which can be treated with propionyl chloride to give the desired propyl propionate product. Note that a Fisher esterification reaction using propanoic acid and 1-propanol with a catalytic amount of acid would also work to create the ester product (not shown).

Problem 18.63 Show how to convert 1-bromopentane and sodium cyanide into *N*-hexylhexanamide. You must use 1-bromopentane and sodium cyanide as the source of all carbon atoms in the target molecule. Show all reagents and all molecules synthesized along the way.

Recognize that the product amide has two six-carbon chains, while the starting 1-bromopentane contains only five carbons. Propose an initial S_N2 reaction between the starting materials to generate a six-carbon nitrile. The nitrile can be hydrolyzed in acid to give hexanoic acid followed by treatment with $SOCl_2$ to give hexanoyl chloride. The nitrile can also be reduced with $LiAlH_4$ to give 1-aminohexane. Reaction of 1-aminohexane with hexanoyl chloride gives the desired *N*-hexylhexanamide product.

Problem 18.64 Show how to convert 1-bromopropane and carbon dioxide into 4-propyl-4-heptanol. You must use 1-bromopropane and carbon dioxide as the source of all carbon atoms in the target molecule. Show all reagents and all molecules synthesized along the way.

Recognize that the product has ten carbons, and the two starting materials contain only three and one carbons, respectively. Therefore, propose that at least two carbon-carbon bonds need to be formed. Recognize further that the

product contains the key recognition element of a tertiary alcohol with at least two identical carbon chains attached, so propose that the last step involves a Grignard reagent of three carbons reacting twice with an ester containing four carbons in the parent chain (a butanoic acid derivative). The required propyl Grignard reagent can be produced by treating the 1-bromopropane starting material with Mg in ether. The four-carbon ester can be made by reacting the propyl Grignard reagent with CO_2 to give butanoic acid, which is converted to the required ester by treatment with $SOCl_2$ then an unhindered alcohol (ROH) such as ethanol. Note that a Fisher esterification reaction using butanoic acid and the alcohol with a catalytic amount of acid would also work to create the ester (not shown).

Problem 18.65 Show how to convert 1-bromopropane and carbon dioxide into 4-heptanone. You must use 1-bromopropane and carbon dioxide as the source of all carbon atoms in the target molecule. Show all reagents and all molecules synthesized along the way.

Recognize that the product 4-heptanone has seven carbons, while the starting materials contain three and one carbons, respectively. Therefore, propose that at least two different carbon-carbon bonds need to be formed. Recognize also that the product is a ketone. The only carbon-carbon bond forming reaction covered so far that produces a ketone product is a diorganocuprate (Gilman reagent) reacting with an acid chloride. Propose that the last step involves a propyl Gilman reagent reacting with butanoyl chloride. The Gilman reagent can be prepared from 1-bromopropane by reaction with Li followed by CuI. Butanoyl chloride can be derived as in the last problem, namely conversion of 1-bromopropane to the corresponding Grignard reagent by treatment with Mg in ether, reaction with CO_2 to give butanoic acid, then reaction with $SOCl_2$. An entirely different strategy that would also work for this synthesis (not shown) that involves a Grignard reagent reacting with an aldehyde followed by oxidation to give the ketone product.

Reactions in Context

Problem 18.66 Minoxidil is a molecule that causes hair growth in some people. It was originally synthesized as a vasodilator for the treatment of hypertension (high blood pressure). Most of the patients taking the drug for hypertension were seen to grow body hair. Due to other side effects, its oral use was stopped, but it became popular as a topical cream to promote hair growth.

Minoxidil

The first key reaction in one synthesis of minoxidil follows. Draw the product of this reaction.

Unhindered esters will often react with amines to give amides directly, although heating or long reaction times are usually required.

Problem 18.67 Chloroformates have the functional group ROC(O)Cl, in which R is often a *tert*-butyl or benzyl group. A chloroformate is used in the following synthesis of the antibacterial drug, linezolid (Zyvox).

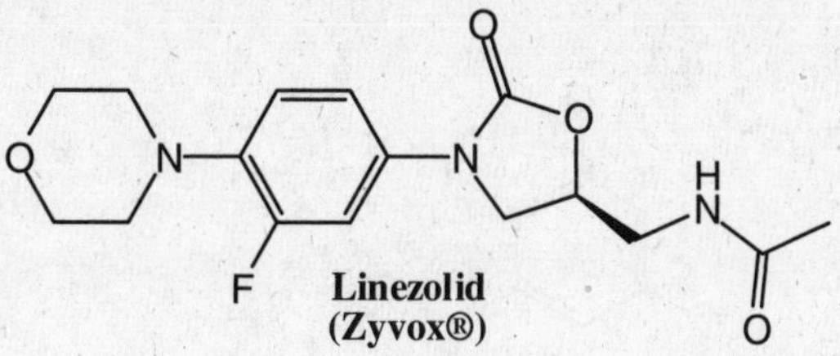

Linezolid
(Zyvox®)

Based on your knowledge of carboxylic acid derivatives, predict the product of the following transformation used in a synthesis of linezolid (Zyvox). The new functional group created is called a carbamate. Carbamates are often used as protecting groups for amine groups during complex syntheses.

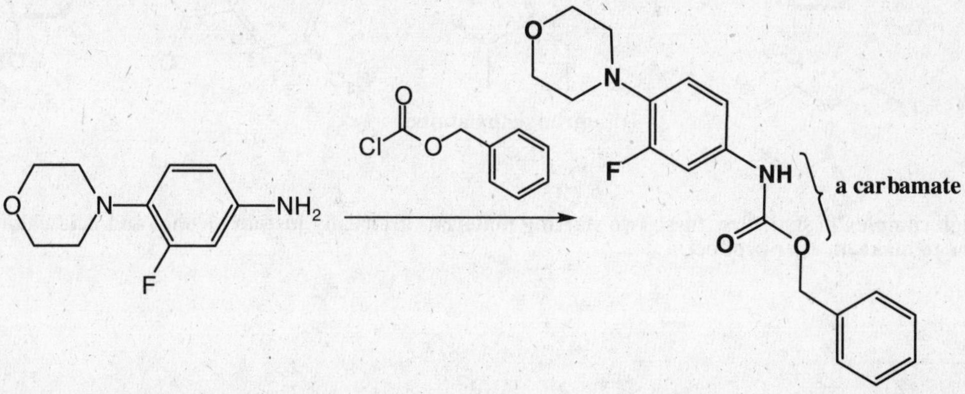

a carbamate

Problem 18.68. Acid anhydrides are often used in place of acid chlorides because a less acidic carboxylic acid, not the much stronger acid HCl, is the byproduct of the reaction. In the following reaction of a carbohydrate derivative, acetic anhydride is used to obtain the product in 99% yield as a single stereoisomer. Note that the stereochemistry of the starting anomeric carbon is not indicated. Draw the product of the following transformation in a chair form and show the single stereoisomer product of this transformation, which is also the most stable possible chair species.

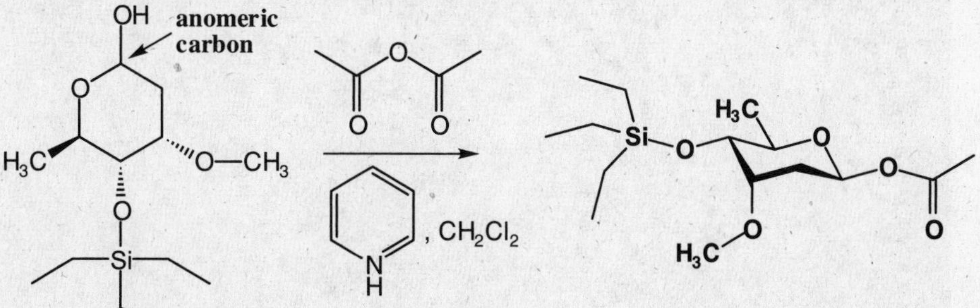

anomeric carbon

The large triethylsilyl ether will dominate the chair conformation of the molecule, remaining equatorial. This will place the methyl and methyl ether groups equatorial and axial, respectively. The anhydride will react at the only OH group on the molecule, which is on the anomeric carbon atom. As a result, the ester group could in theory be either axial or equatorial. The equatorial ester product is the more stable, so propose it as the single stereoisomer seen in the actual reaction as shown.

<u>Problem 18.69</u> The benzyl ether group (-OBn) is often used as a protecting group for OH during the synthesis of complex molecules. The following structure has a number benzyl ether groups used for this purpose. Draw the product of the following transformation. What role does the diisopropylethylamine play in this reaction?

Diisopropylethylamine

Although complex in structure, these two starting materials are really just an alcohol and acid chloride, which combine to make an ester product.

Whitetip reef shark *Triaenodon obesus*
Maui, Hawaii

CHAPTER 19
Solutions to the Problems

<u>Problem 19.1</u> Draw the product of the base-catalyzed aldol reaction of each compound.
(a) Phenylacetaldehyde   (b) Cyclopentanone

Two new chiral centers are created in each reaction, so racemic mixtures of stereoisomers will be produced.

<u>Problem 19.2</u> Draw the product of dehydration of each aldol product in Problem 19.1.

<u>Problem 19.3</u> Draw the product of the base-catalyzed crossed aldol reaction between benzaldehyde and 3-pentanone and the product formed by its dehydration.

<u>Problem 19.4</u> Show the product of Claisen condensation of ethyl 3-methylbutanoate in the presence of sodium ethoxide followed by acidification with aqueous HCl.

One new chiral center is created in the reaction, so a racemic mixture will be produced.

<u>Problem 19.5</u> Complete the equation for this crossed Claisen condensation.

(excess)

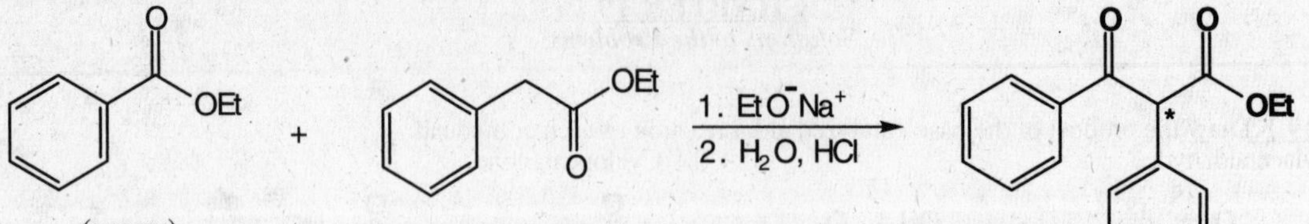

One new chiral center is created in the reaction, so a racemic mixture will be produced.

<u>Problem 19.6</u> Show how to convert benzoic acid to 3-methyl-1-phenyl-1-butanone (isobutyl phenyl ketone) by the following synthetic strategies, each of which uses a different type of reaction to form the new carbon-carbon bond to the carbonyl group of benzoic acid.

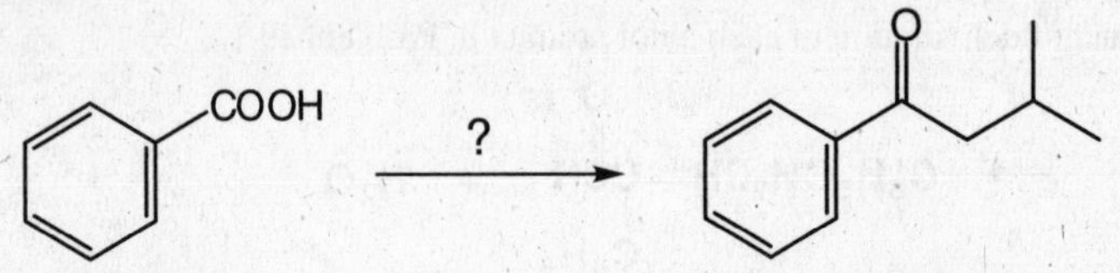

Benzoic acid   **3-Methyl-1-phenyl-1-butanone**

(a) A lithium diorganocopper (Gilman) reagent

Treatment of benzoic acid with thionyl chloride gives benzoyl chloride. Treatment of benzoyl chloride with lithium diisobutylcopper followed by hydrolysis in aqueous acid gives the desired product.

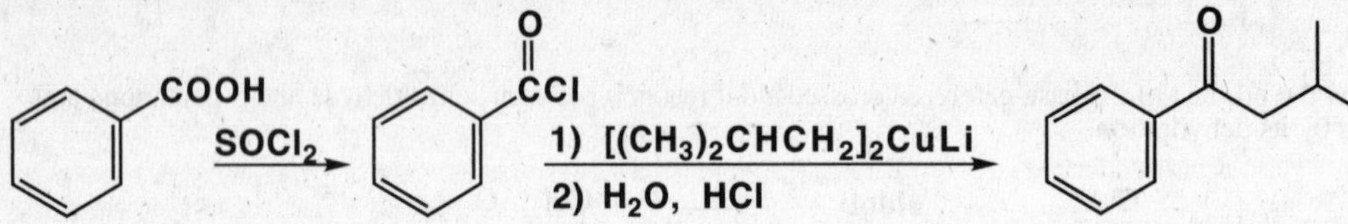

(b) A Claisen condensation

A crossed Claisen condensation of ethyl benzoate and ethyl 3-methylbutanoate gives a β-ketoester. Saponification of the ester followed by acidification gives the β-ketoacid. Heating causes decarboxylation and gives the desired product.

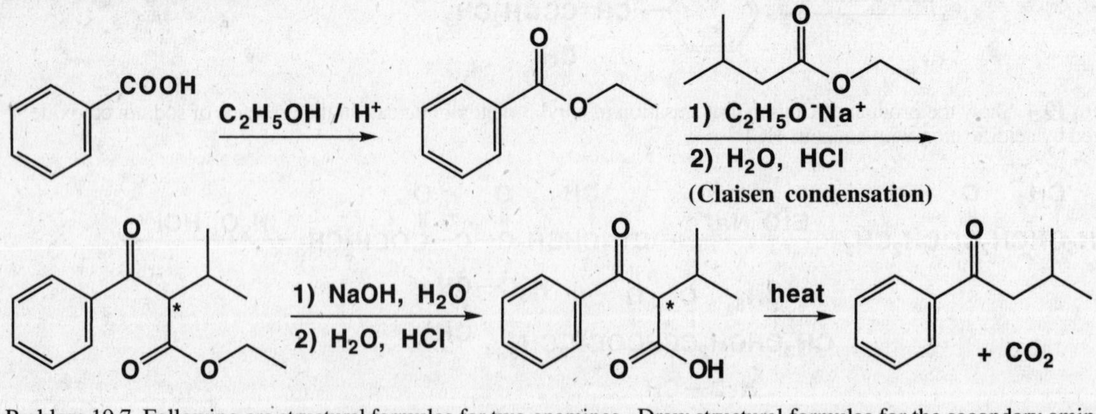

<u>Problem 19.7</u> Following are structural formulas for two enamines. Draw structural formulas for the secondary amine and carbonyl compound from which each is derived.

(a)

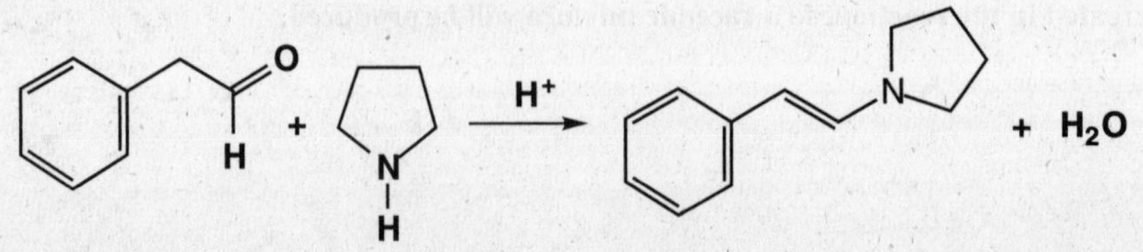

(b)

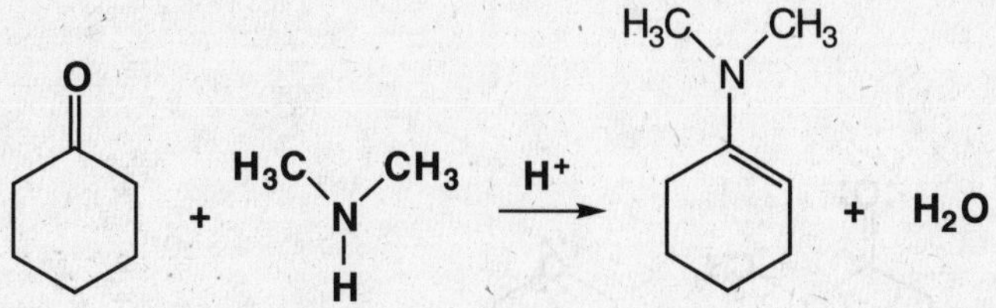

Problem 19.8 Write a mechanism for the hydrolysis of the following iminium chloride in aqueous HCl:

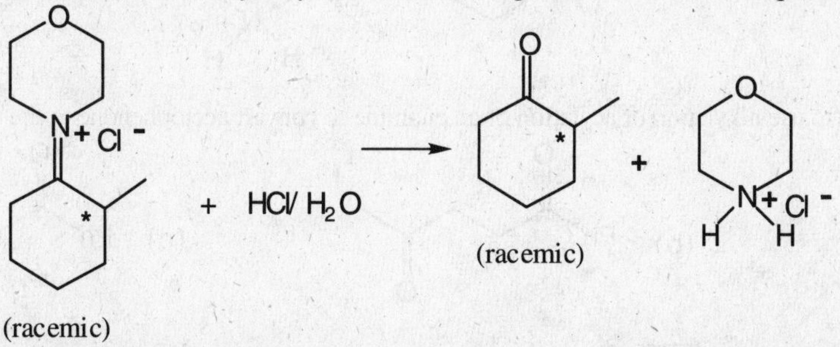

(racemic)   (racemic)

The positively charged iminium ion is attacked by water acting as a nucleophile. A proton is transferred to the nitrogen, followed by departure of the amine and formation of the carbon-oxygen double bond. The reaction is completed with a proton transfer step to produce the product ketone and protonated morpholine. Under the acidic conditions in this reaction, an enamine is not formed, so hydrolysis predominates.

Step 1: Make a bond between a nucleophile and an electrophile.

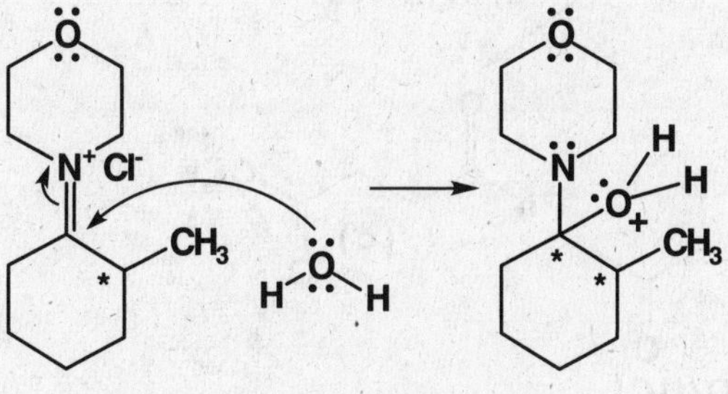

Step 2: Add a proton-take a proton away.

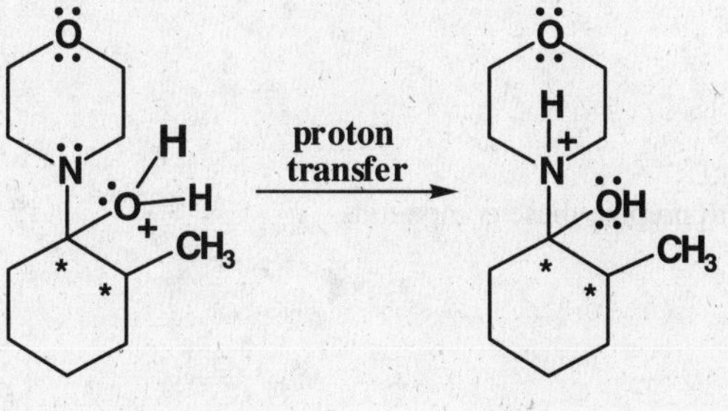

proton
transfer

Step 3: Break a bond to give stable molecules or ions.

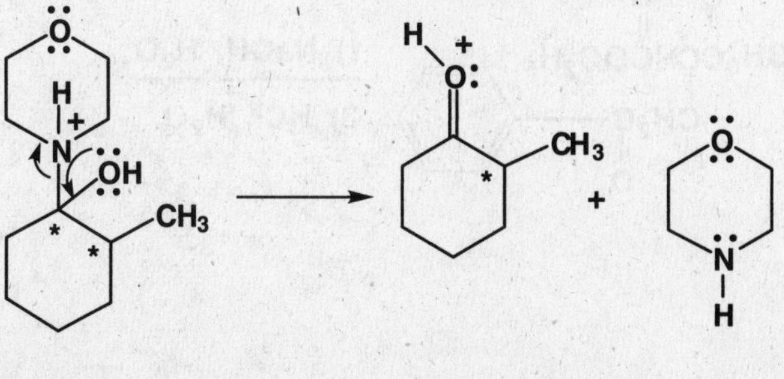

Step 4: Take a proton away.

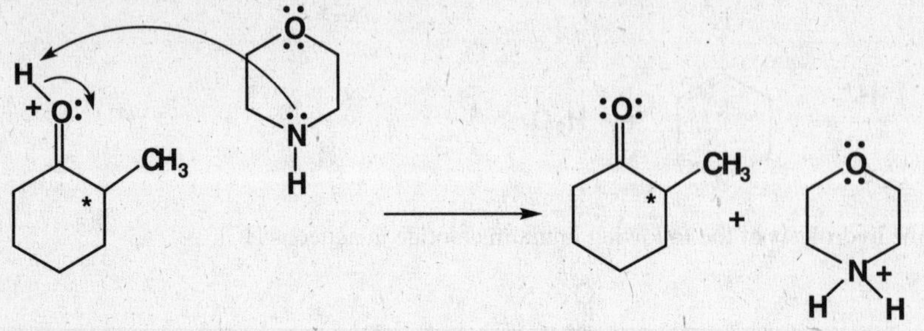

Problem 19.9 Show how to use alkylation or acylation of an enamine to convert acetophenone to the following compounds.

a) 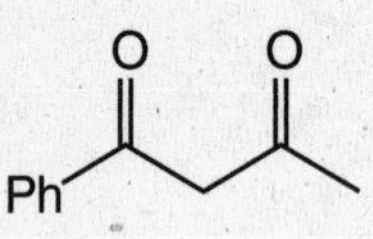 Ph

(b) 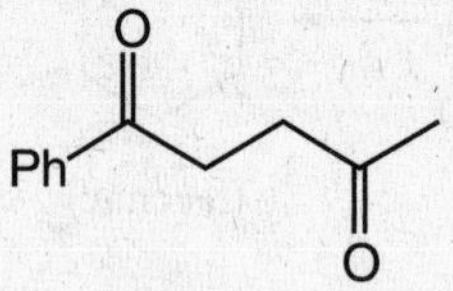 Ph

(c) 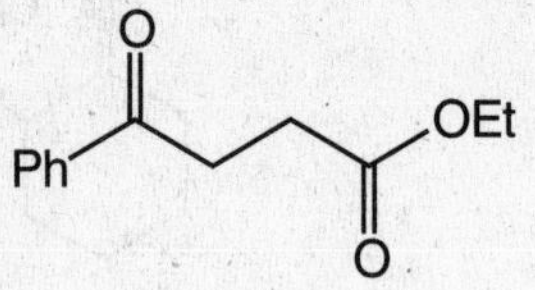 Ph   OEt

1) Cl   O   OEt

2) H₂O/HCl

1)   O   Cl

2) H₂O/HCl

1) Cl   O

2) H₂O/HCl

Ph   OEt

(c)

Ph

(a)

Ph

(b)

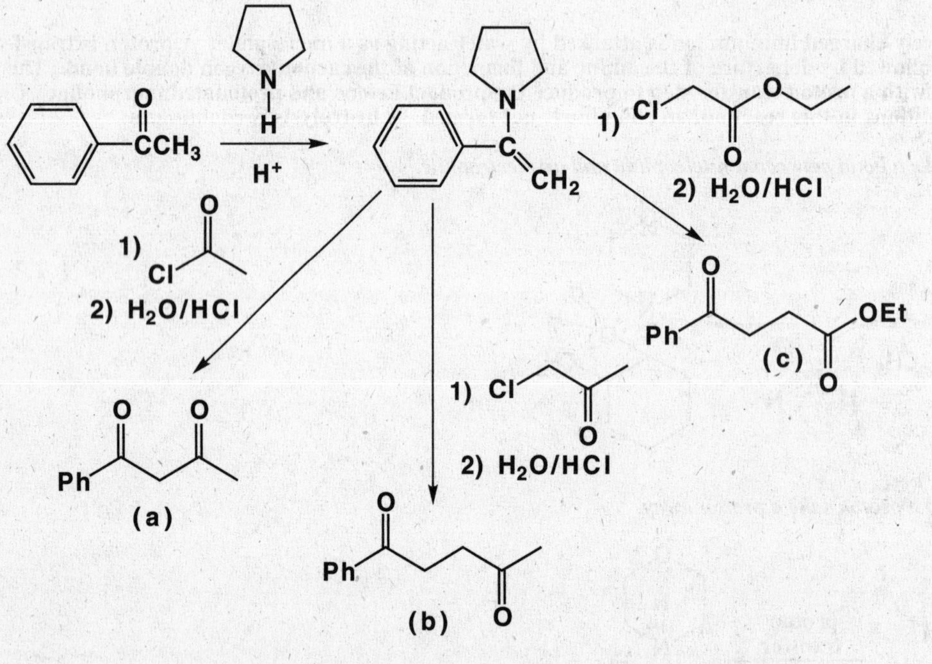

Problem 19.10 Show how the acetoacetic ester synthesis can be used to prepare these compounds.

(a)

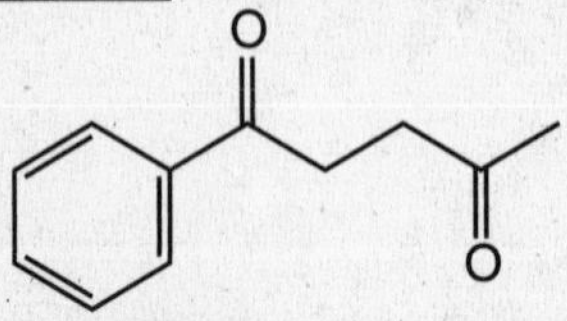

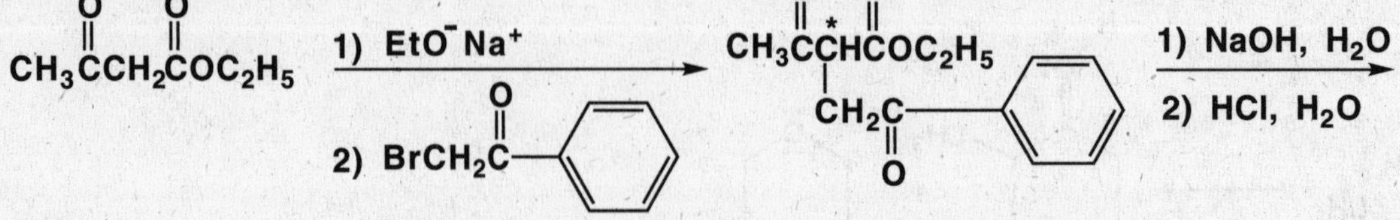

$$CH_3CCH_2COC_2H_5 \xrightarrow[\text{2) BrCH}_2C\text{Ph}]{\text{1) EtO}^- \text{Na}^+} CH_3CCHCOC_2H_5 \xrightarrow[\text{2) HCl, H}_2O]{\text{1) NaOH, H}_2O}$$

(b)

(c)

Problem 19.11 Show how to convert ethyl 2-oxocyclopentanecarboxylate to this compound.

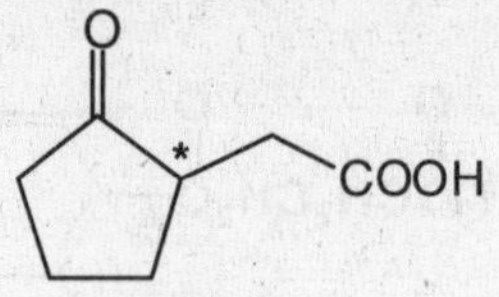

(racemic)

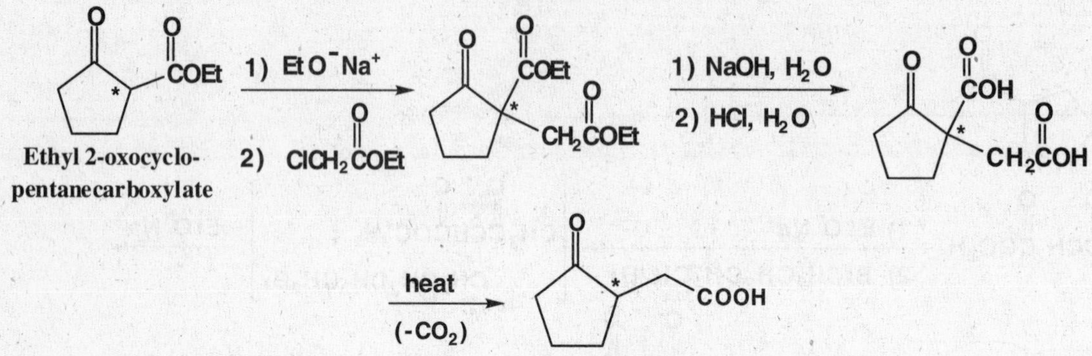

Ethyl 2-oxocyclo-
pentanecarboxylate

heat
(-CO₂)

(racemic)

Note that, in the decarboxylation step, only the carboxyl group bonded to the five-membered ring is lost. This is because this is the only carboxyl group that is β to another carbonyl, namely the ketone function on the five-membered ring. The β arrangement is required to allow for the six-membered ring transition state leading to decarboxylation. A new chiral center is created in the reaction, so a racemic mixture will be produced.

Problem 19.12 Show how the malonic ester synthesis can be used to prepare the following substituted acetic acids.

(a)

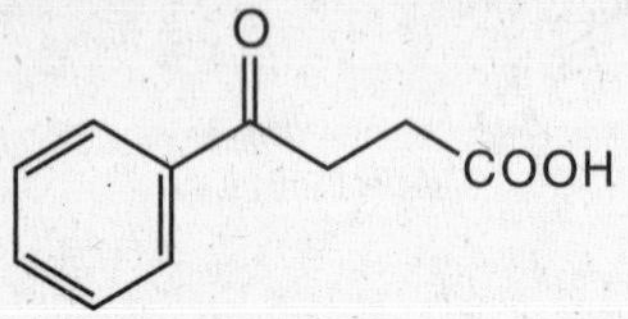

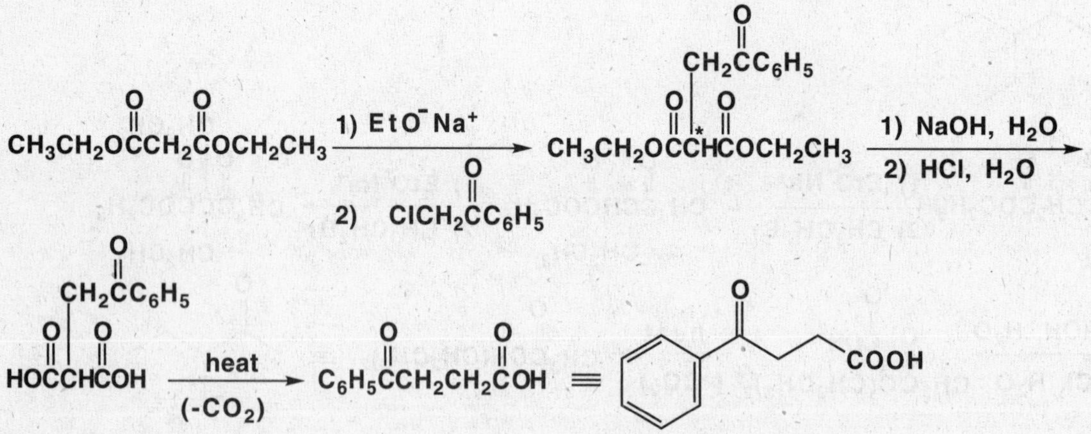

(b) —COOH

$$CH_3CH_2O\overset{O}{\underset{}{C}}CH_2\overset{O}{\underset{}{C}}OCH_2CH_3 \quad \xrightarrow[\text{2) 1 eq. } CH_3CH_2Br]{\text{1) 1 eq. EtO}^-\text{Na}^+} \quad \xrightarrow[\text{2) 1 eq. } CH_3CH_2Br]{\text{1) 1 eq. EtO}^-\text{Na}^+} \quad CH_3CH_2O\overset{O}{\underset{}{C}}\overset{CH_2CH_3}{\underset{CH_2CH_3}{C}}\overset{O}{\underset{}{C}}OCH_2CH_3$$

$$\xrightarrow[\text{2) HCl, }H_2O]{\text{1) NaOH, }H_2O} \quad HO\overset{O}{\underset{}{C}}\overset{CH_2CH_3}{\underset{CH_2CH_3}{C}}\overset{O}{\underset{}{C}}OH \quad \xrightarrow[\text{(-CO}_2)]{\text{heat}} \quad (CH_3CH_2)_2CH\overset{O}{\underset{}{C}}OH \quad \equiv \quad \text{—COOH}$$

(c) —COOH

$$CH_3CH_2O\overset{O}{\underset{}{C}}CH_2\overset{O}{\underset{}{C}}OCH_2CH_3 \quad \xrightarrow[\text{2) 1 eq. } BrCH_2CH_2CH_2CH_2Br]{\text{1) 1 eq. EtO}^-\text{Na}^+} \quad \left[\begin{array}{c} C_2H_5O\overset{O}{\underset{}{C}}\overset{-}{\underset{CH_2CH_2CH_2CH_2Br}{CH}}\overset{O}{\underset{}{C}}OC_2H_5 \end{array} \right]$$

$$\xrightarrow{\text{EtO}^-\text{Na}^+} \left[\begin{array}{c} C_2H_5O\overset{O}{\underset{}{C}}\overset{\cdot\cdot}{\underset{CH_2CH_2CH_2CH_2Br}{C}}\overset{O}{\underset{}{C}}OC_2H_5 \end{array} \right] \longrightarrow C_2H_5O\overset{O}{\underset{}{C}}\underset{}{C}\overset{O}{\underset{}{C}}OC_2H_5 \quad \xrightarrow[\text{2) HCl, }H_2O]{\text{1) NaOH, }H_2O}$$

$$HO\overset{O}{\underset{}{C}}\underset{}{C}\overset{O}{\underset{}{C}}OH \quad \xrightarrow[\text{(-CO}_2)]{\text{heat}} \quad HO\overset{O}{\underset{}{C}}\underset{}{CH} \quad \equiv \quad \text{—COOH}$$

<u>Problem 19.13</u> Show the product formed from each Michael product in the solution to Example 19.13 after (1) hydrolysis in aqueous NaOH, (2) acidification, and (3) thermal decarboxylation of each β-ketoacid or β-dicarboxylic acid. These reactions illustrate the usefulness of the Michael reaction for the synthesis of 1,5-dicarbonyl compounds.

$$CH_3\overset{O}{\underset{EtO\overset{}{\underset{O}{C}}}{C}}\overset{*}{\underset{}{CH}}CH_2CH_2\overset{O}{\underset{}{C}}OEt \quad \xrightarrow[\text{2) HCl, }H_2O]{\text{1) NaOH, }H_2O} \quad CH_3\overset{O}{\underset{\underbrace{HO\overset{}{\underset{O}{C}}}}{C}}\overset{*}{\underset{}{CH}}CH_2CH_2\overset{O}{\underset{}{C}}OH \quad \xrightarrow[\text{(-CO}_2)]{\text{3) heat}} \quad CH_3\overset{O}{\underset{}{C}}CH_2CH_2CH_2\overset{O}{\underset{}{C}}OH$$

**this group is lost upon
heating because it is β
to the ketone**

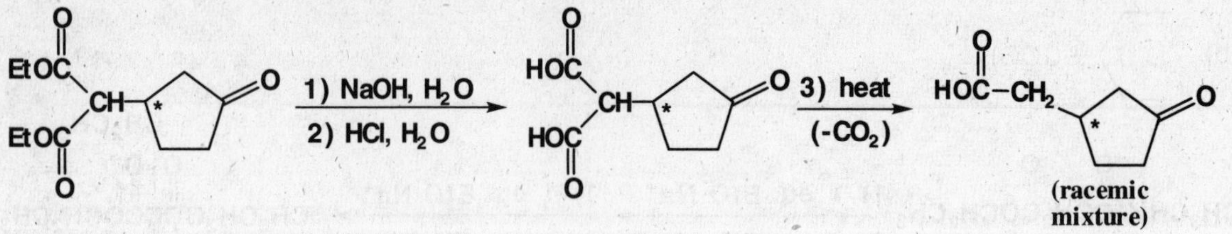

Problem 19.14 Show how the sequence of Michael reaction, hydrolysis, acidification, and thermal decarboxylation can be used to prepare pentanedioic acid (glutaric acid).

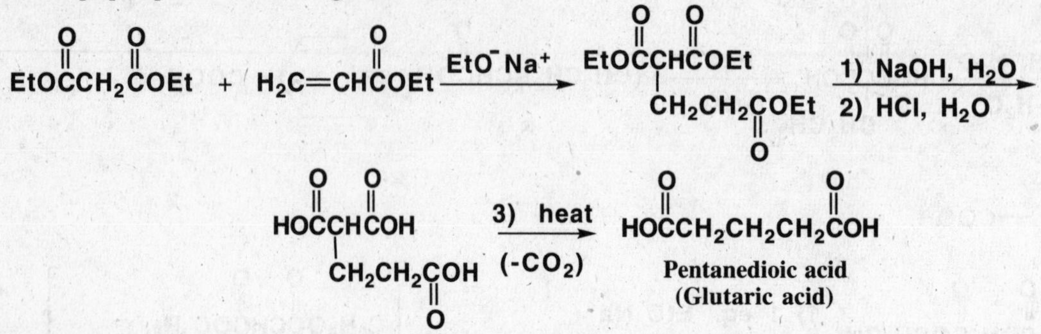

Problem 19.15 Show how to bring about the following conversion.

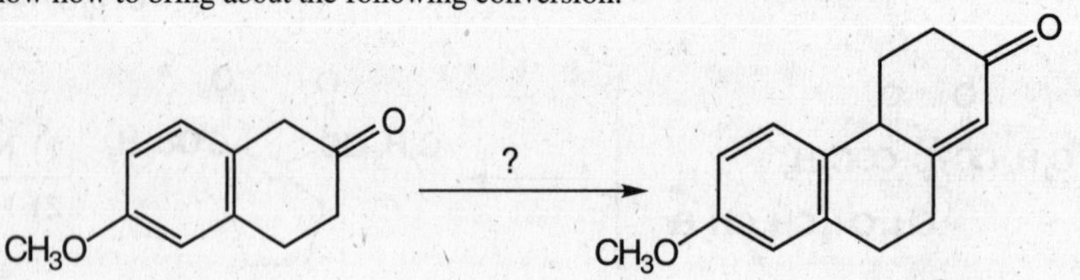

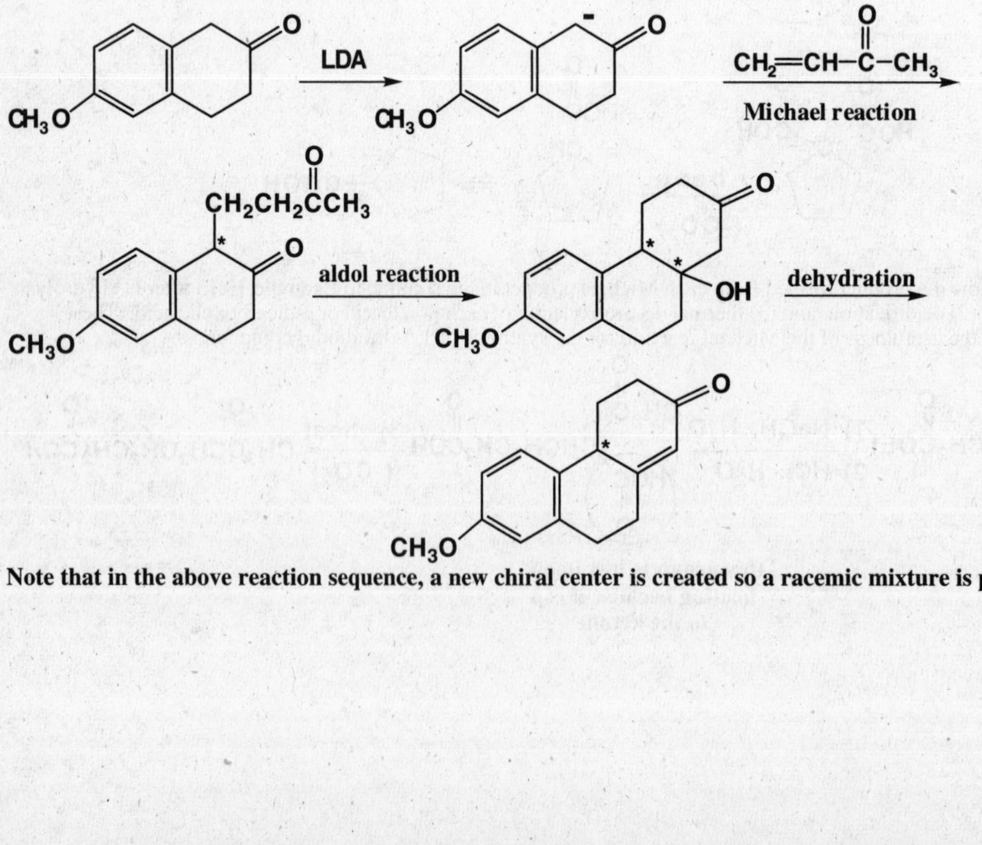

Note that in the above reaction sequence, a new chiral center is created so a racemic mixture is produced.

Problem 19.16 Propose two syntheses of 4-phenyl-2-pentanone, each involving conjugate addition of a lithium diorganocopper reagent.

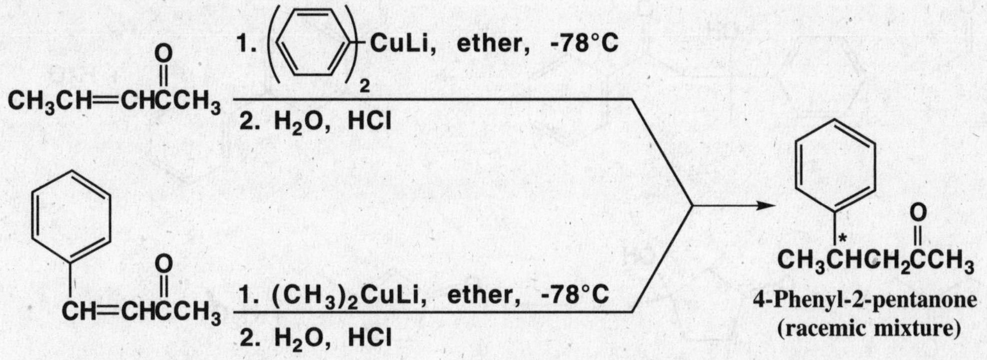

$CH_3CH=CHCCH_3$
1. (phenyl)$_2$CuLi, ether, -78°C
2. H$_2$O, HCl

$CH=CHCCH_3$ (phenyl)
1. (CH$_3$)$_2$CuLi, ether, -78°C
2. H$_2$O, HCl

$CH_3CHCH_2CCH_3$
4-Phenyl-2-pentanone
(racemic mixture)

Problem 19.17 Show how you might prepare the following compounds using directed aldol reactions.

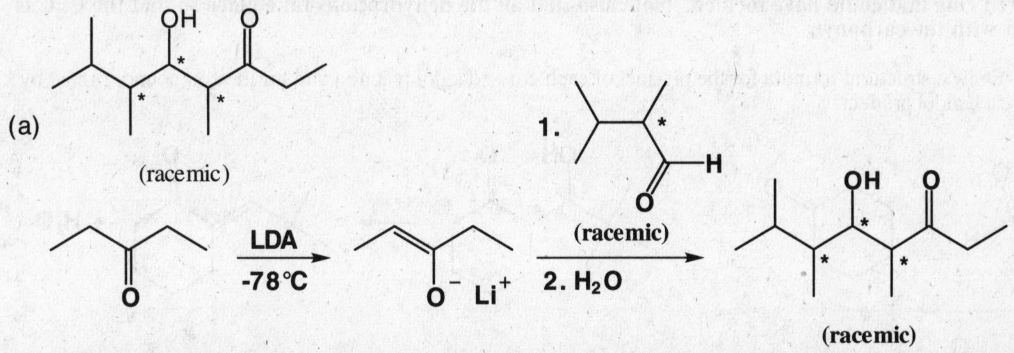

(a)

(racemic)

$\xrightarrow[\text{-78°C}]{\text{LDA}}$ $\xrightarrow[\text{2. H}_2\text{O}]{\text{1.}}$

(racemic)

(racemic)

An enolate is made from 3-pentanone, and used to carry out an aldol reaction with 2,3-dimethylbutanal.

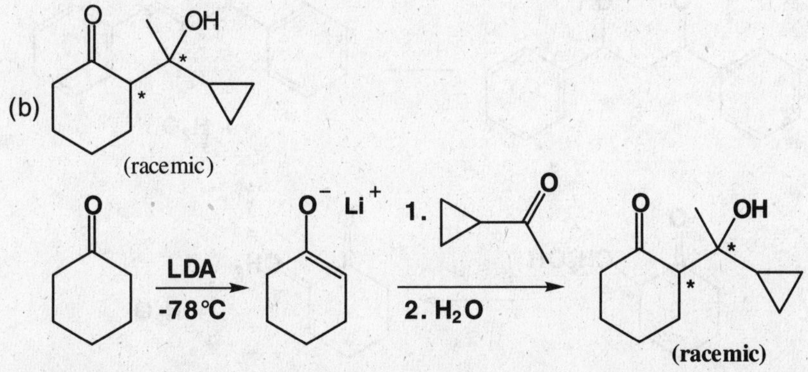

(b)

(racemic)

$\xrightarrow[\text{-78°C}]{\text{LDA}}$ $\xrightarrow[\text{2. H}_2\text{O}]{\text{1.}}$

(racemic)

An enolate is made from cyclohexanone that is used to carry out an aldol reaction with cyclopropyl methyl ketone.

The Aldol Reaction

Problem 19.18 Draw a structural formula for the product of the aldol reaction of each compound and for the α,β-unsaturated aldehyde or ketone formed from dehydration of each aldol product.

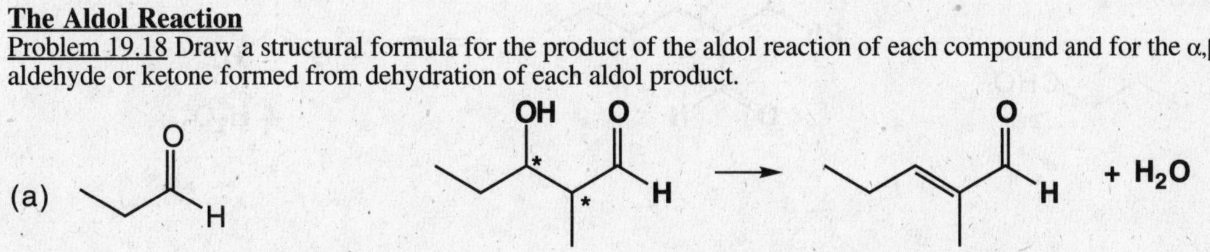

(a)

$\longrightarrow$ + H$_2$O

(b)

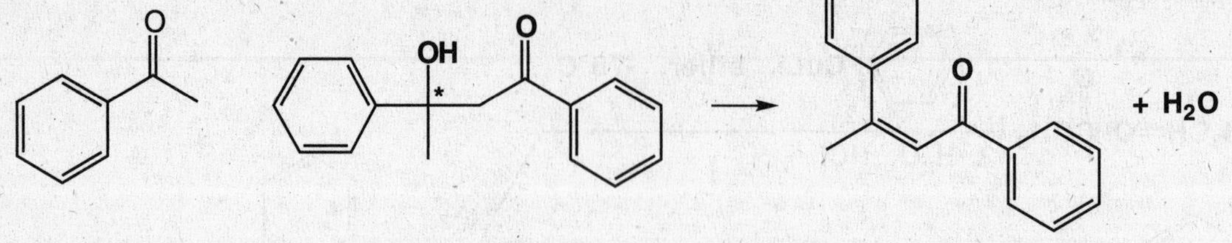

+ H₂O

(c)

+ H₂O

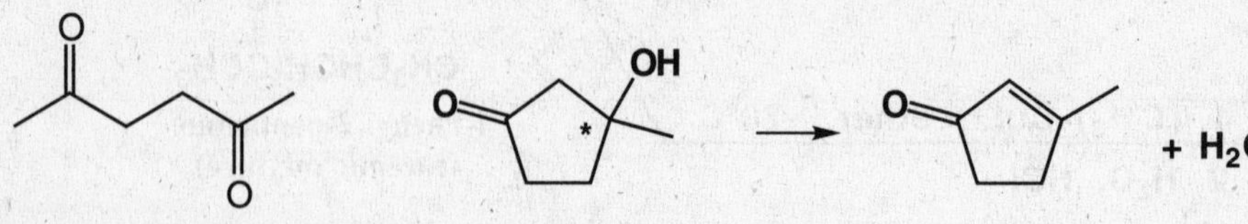

Note that in this problem the relatively stable five-membered ring was formed instead of the highly strained three-membered ring that could have formed. Note also that all the dehydrations take place so that the C=C is in conjugation with the carbonyl.

<u>Problem 19.19</u> Draw a structural formula for the product of each crossed aldol reaction and for the compound formed by dehydration of each aldol product.

(a)

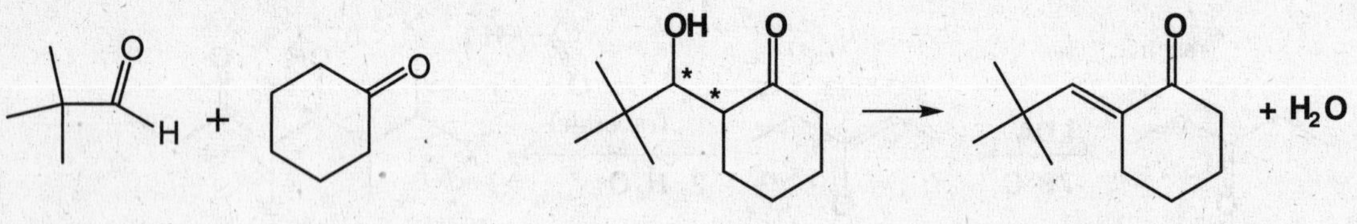

+ H₂O

(b)

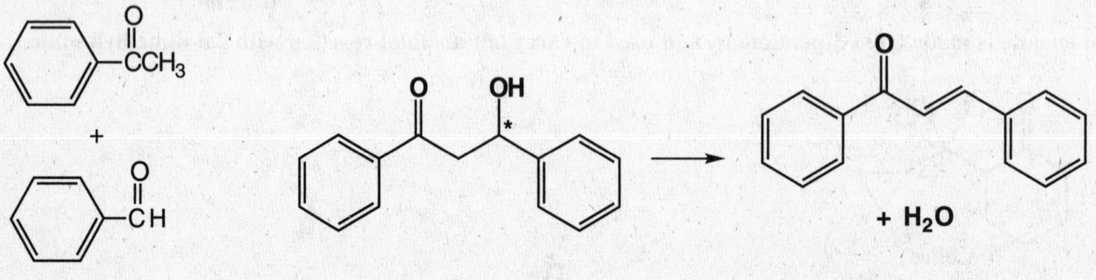

+ H₂O

(c)

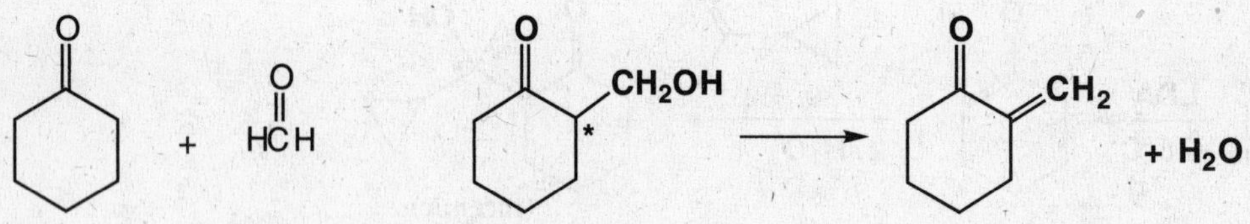

+ H₂O

(d) 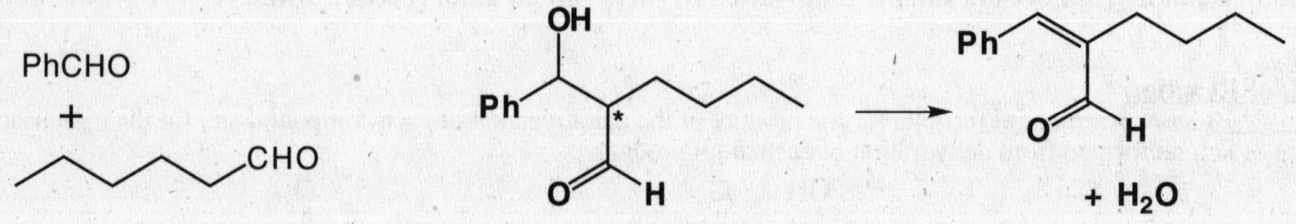 PhCHO

+

+ H₂O

Problem 19.20 When a 1:1 mixture of acetone and 2-butanone is treated with base, six aldol products are possible. Draw a structural formula for each product.

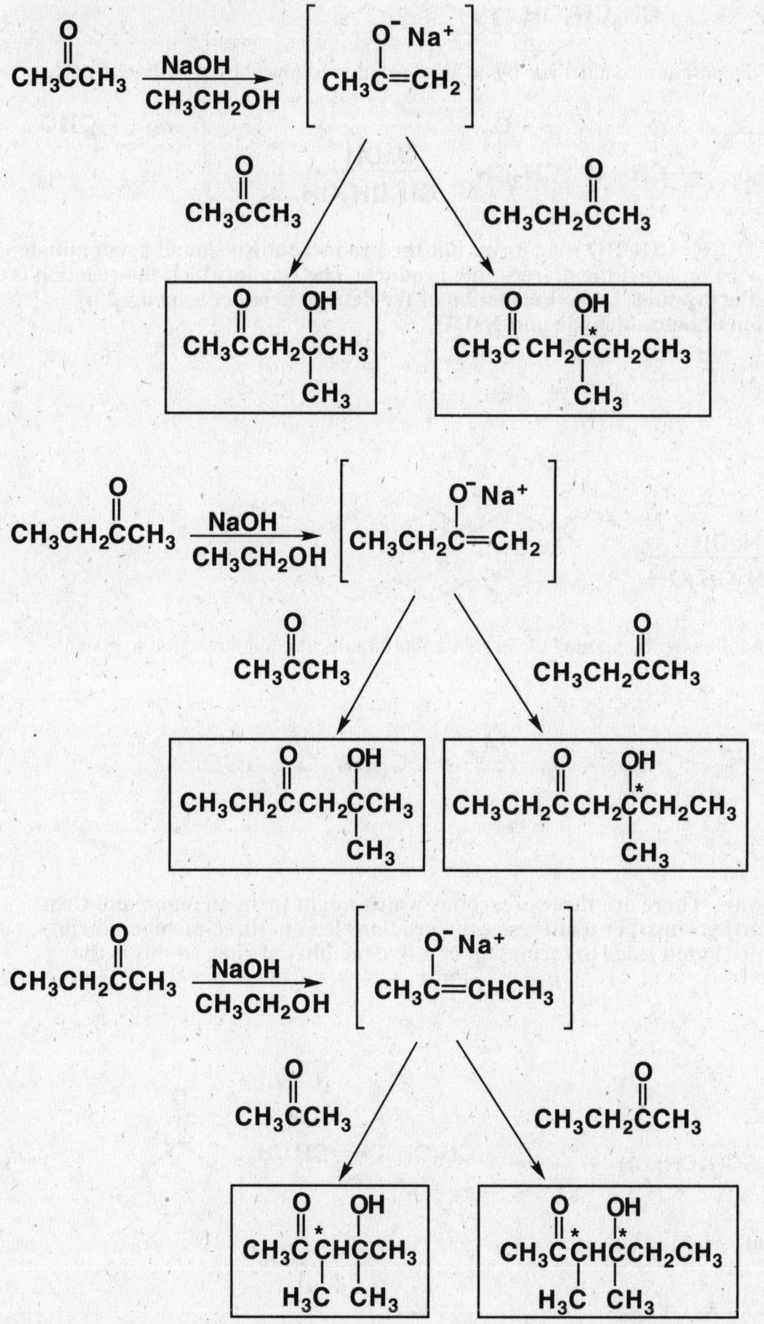

Notice four of the six products in the above scheme have chiral centers, so racemic mixtures will be formed.

Problem 19.21 Show how to prepare each α,β-unsaturated ketone by an aldol reaction followed by dehydration of the aldol product.

(a)

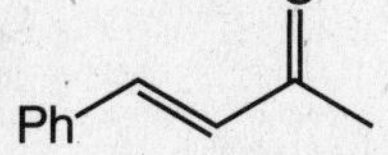

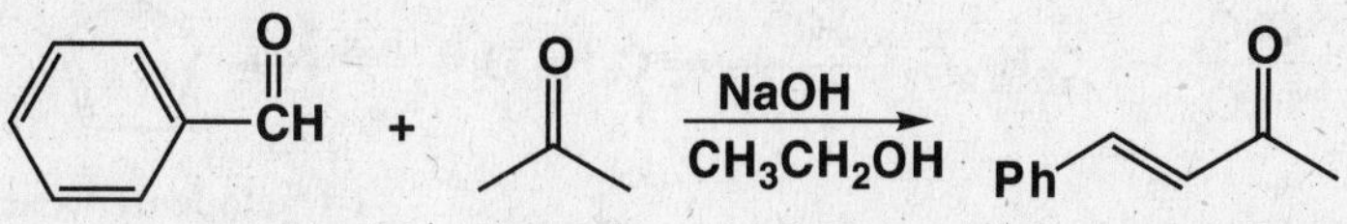

(b)

Problem 19.22 Show how to prepare each α,β-unsaturated aldehyde by an aldol reaction followed by dehydration of the aldol product.

(a)

Note that in the reaction above, some $CH_3CH=CHCHO$ may form, but the product shown should predominate because conjugation with the aromatic ring makes it the more stable product. The way in which the reaction is run can also have a major influence on the products here. Formation of the desired product is favored by adding the ethanal dropwise to a solution of benzaldehyde and NaOH.

(b)

Problem 19.23 When treated with base, the following compound undergoes an intramolecular aldol reaction to give a product containing a ring (yield 78%).

Propose a structural formula for this product.

Analyze this problem in the following way. There are three α-carbons which might form an anion and then condense with either of the other carbonyl groups. Two of these condensations lead to three-membered rings and, therefore, are not feasible. The third anion leads to formation of a five-membered ring, so this is the pathway followed.

An intramolecular aldol condensation by this α-carbon gives a five-membered ring

An intramolecular aldol condensation by these α-carbons would give highly strained three-membered rings

$$CH_3CH_2CH=CHCH_2CH_2CCH_2CH_2CH \longrightarrow CH_3CH_2CH=CHCH_2-$$

$$C_{10}H_{14}O$$

Problem 19.24 Cyclohexene can be converted to 1-cyclopentenecarbaldehyde by the following series of reactions.

$$\underset{H_2O_2}{\overset{OsO_4}{\longrightarrow}} C_6H_{12}O_2 \overset{HIO_4}{\longrightarrow} C_6H_{10}O_2 \overset{base}{\longrightarrow}$$

1-Cyclopentenecarbaldehyde

Propose a structural formula for each intermediate compound.

Oxidation of cyclohexene by osmium tetroxide gives a *cis*-glycol, which is then oxidized by periodic acid to hexanedial. Base-catalyzed aldol reaction of this dialdehyde followed by dehydration of the aldol product gives 1-cyclopentenecarbaldehyde.

| Cyclo-hexene | *cis*-1,2-Cyclo-hexanediol | Hexanedial | 1-Cyclopentene-carbaldehyde |

<u>Problem 19.25</u> Propose a structural formula for each lettered compound.

$$\text{CrO}_3, \text{pyridine} \longrightarrow \text{A } (C_{11}H_{18}O_2) \xrightarrow[\text{EtOH}]{\text{EtO}^- \text{Na}^+} \text{B } (C_{11}H_{16}O)$$

Oxidation of the starting material by CrO_3 gives the diketone, compound A. Base-catalyzed aldol reaction of this diketone could potentially give cyclic products with either a four-membered or six-membered ring. The six-membered ring is much more stable, so this product predominates. Dehydration of the aldol product gives compound B. One new chiral center is created, so assuming racemic starting material, a racemic mixture will be produced.

(assume a racemic mixture of
stereoisomers here because no
detailed stereochemistry is given)

A $(C_{11}H_{18}O_2)$

B $(C_{11}H_{16}O)$

(racemic)

<u>Problem 19.26</u> How might you bring about the following conversion?

The alkene is oxidized to a diketone by OsO_4/H_2O_2 followed by treatment with HIO_4. Note that ozonolysis could have been used for this step as well. Sodium ethoxide or other base catalyzes an intramolecular aldol reaction of the diketone to give a β-hydroxyketone. Dehydration of this aldol product under the conditions of the aldol reaction gives the desired α,β-unsaturated ketone.

$$\xrightarrow[\text{H}_2\text{O}_2]{\text{OsO}_4} \qquad \xrightarrow{\text{HIO}_4} \qquad \xrightarrow[\text{EtOH}]{\text{EtO}^- \text{Na}^+}$$

(Intramolecular
aldol reaction)

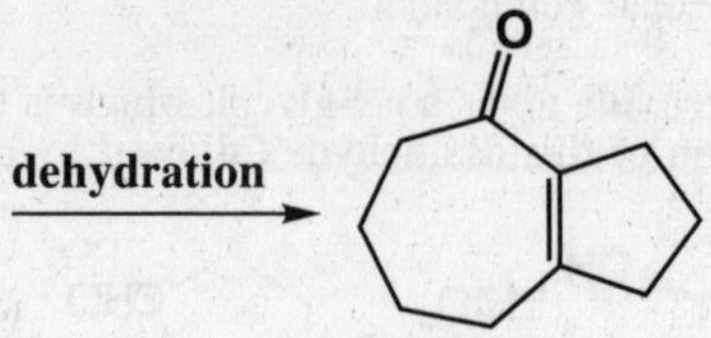

$$\xrightarrow{\text{dehydration}}$$

<u>Problem 19.27</u> Pulegone, $C_{10}H_{16}O$, a compound from oil of pennyroyal, has a pleasant odor midway between peppermint and camphor. Treatment of pulegone with steam produces acetone and 3-methylcyclohexanone.

Pulegone

3-Methyl-cyclohexanone

Acetone

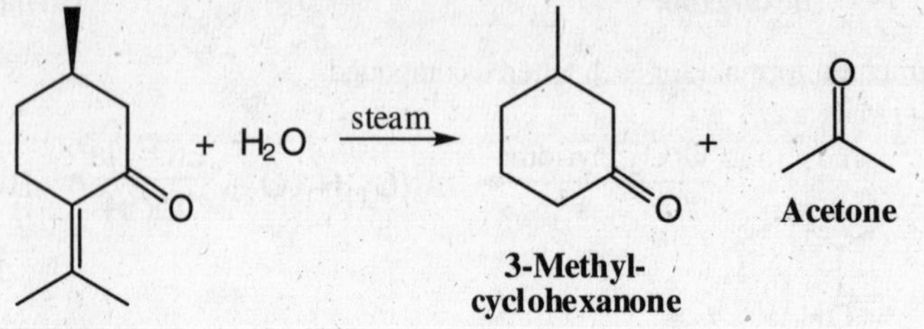

(a) Natural pulegone has the configuration shown. Assign an *R* or *S* configuration to its chiral center.

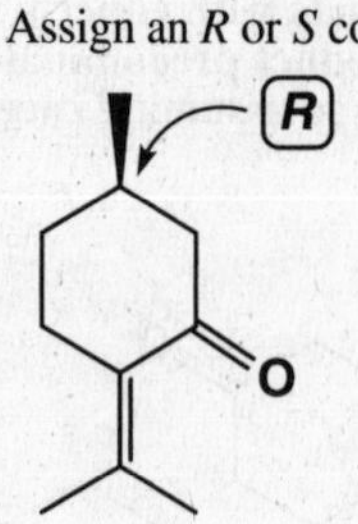

(b) Propose a mechanism for the steam hydrolysis of pulegone to the compounds shown.

An aldol reaction is reversible, and this steam hydrolysis of pulegone is formally the reverse of an aldol reaction. Thus, a reasonable mechanism is exactly the reverse of what would be written for an aldol reaction between acetone and 3-methylcyclohexanone.

Step 1: Hydration. **This actually occurs over several steps.**

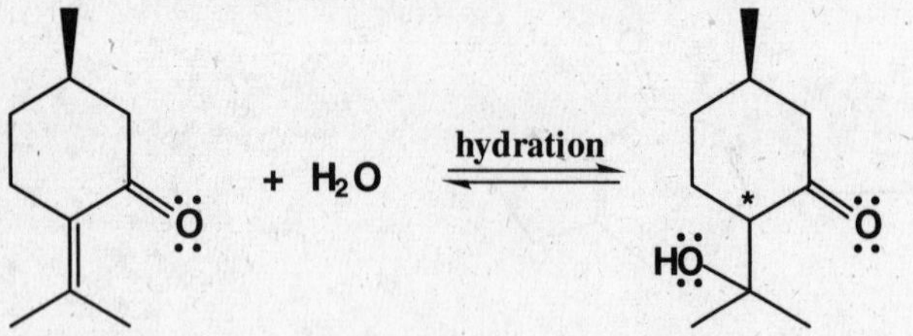

Step 2: Break a bond to give stable molecules or ions.

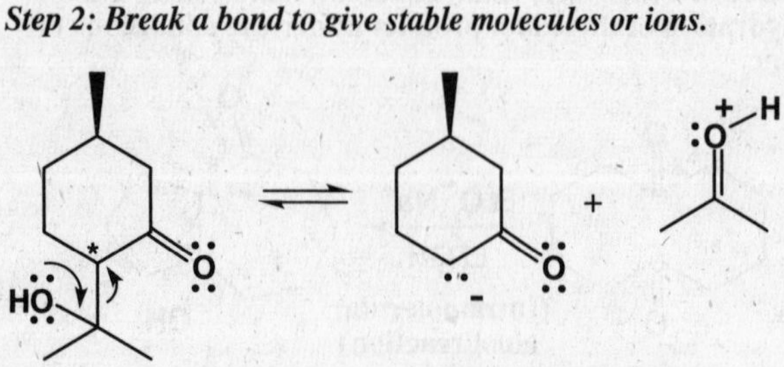

Step 3: Add a proton.

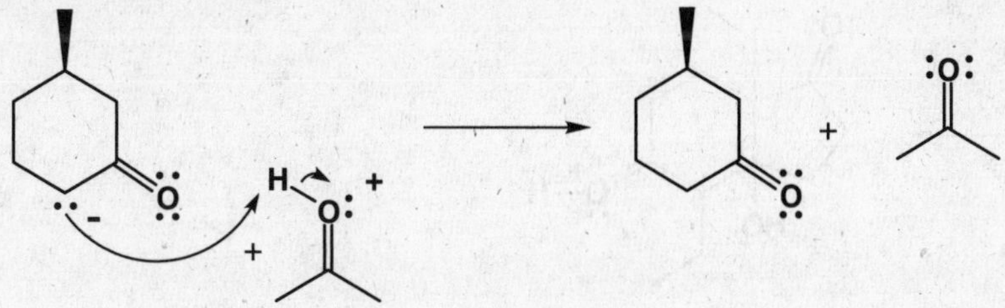

(c) In what way does this steam hydrolysis affect the configuration of the chiral center in pulegone? Assign an *R* or *S* configuration to the 3-methylcyclohexanone formed in this reaction.

This reaction mechanism does not involve the chiral center; it remains in the *R* configuration in the 3-methylcyclohexanone product.

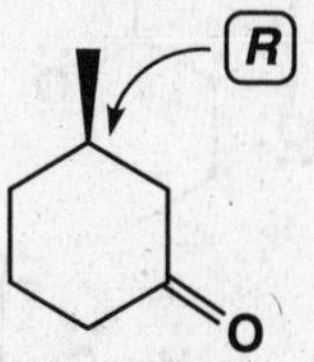

<u>Problem 19.28</u> Propose a mechanism for this acid-catalyzed aldol reaction and the dehydration of the resulting aldol product.

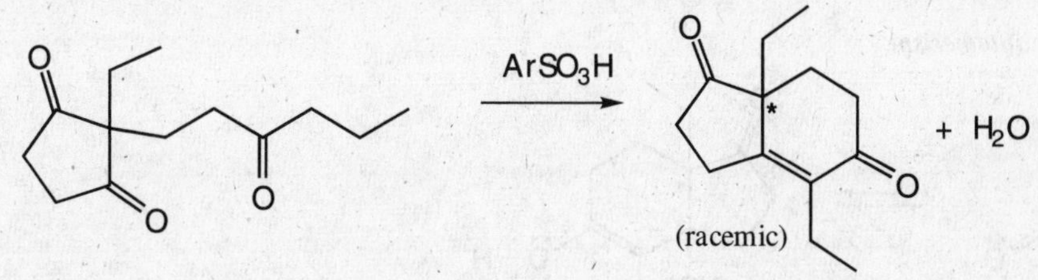

A reasonable mechanism for this acid-catalyzed aldol reaction involves attack of the enol on the protonated carbonyl group, followed by acid-catalyzed dehydration. Note how the first two steps could occur in either order.

Step 1: Keto-enol tautomerism.

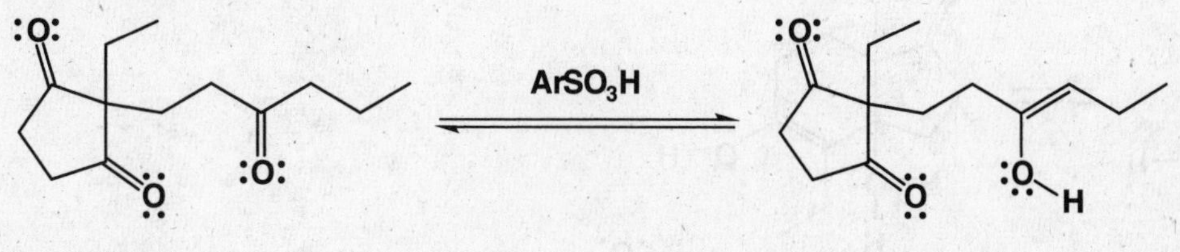

Step 2: Add a proton.

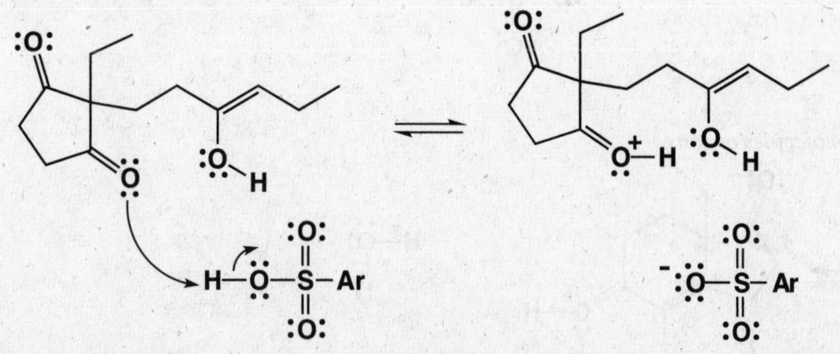

Step 3: Electrophilic addition-make a bond between a nucleophile and an electrophile.

Step 4: Take a proton away.

Step 5: Keto-enol tautomerism.

Step 6: Add a proton.

Step 7: Break a bond to give stable molecules or ions.

Step 8: Take a proton away.

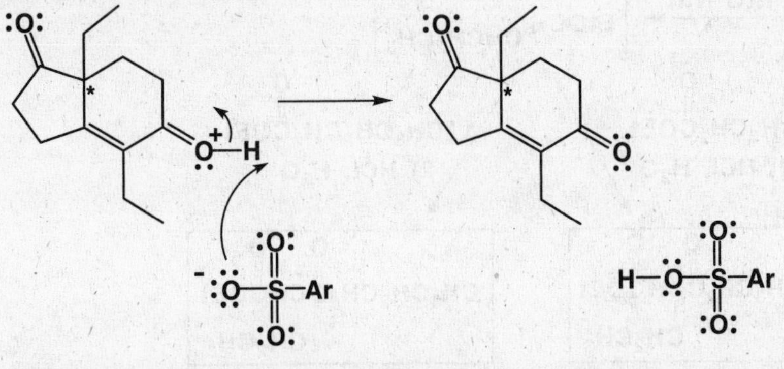

Note that steps 7 and 8 may take place more or less simultaneously. One chiral center is created, so a racemic mixture is produced in the overall reaction.

The Claisen Condensation
Problem 19.29 Show the product of Claisen condensation of each ester.
(a) Ethyl phenylacetate in the presence of sodium ethoxide.

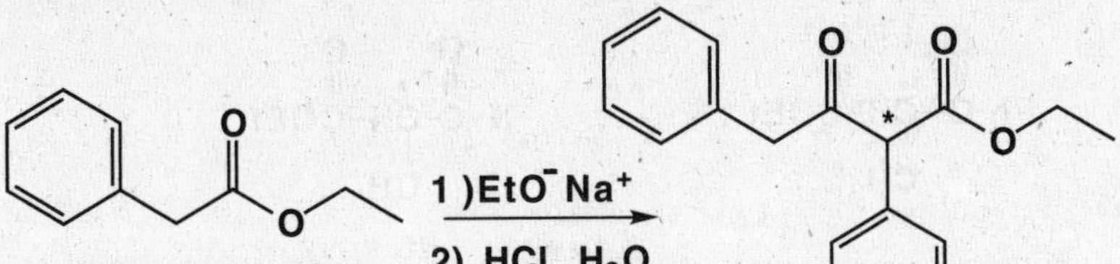

Ethyl phenylacetate

1) EtO⁻ Na⁺

2) HCl, H₂O

One chiral center is created, so a racemic mixture is produced.

(b) Methyl hexanoate in the presence of sodium methoxide.

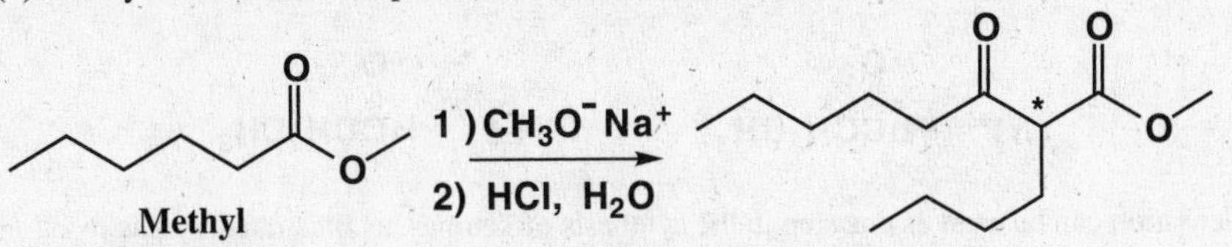

Methyl hexanoate

1) CH₃O⁻ Na⁺

2) HCl, H₂O

One chiral center is created, so a racemic mixture is produced.

Problem 19.30 When a 1:1 mixture of ethyl propanoate and ethyl butanoate is treated with sodium ethoxide, four Claisen condensation products are possible. Draw a structural formula for each product.

For the following scheme, in the cases where one chiral center is created, a racemic mixture is produced.

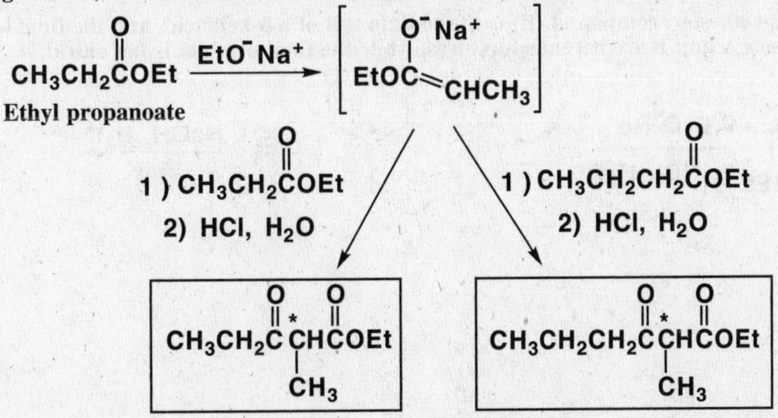

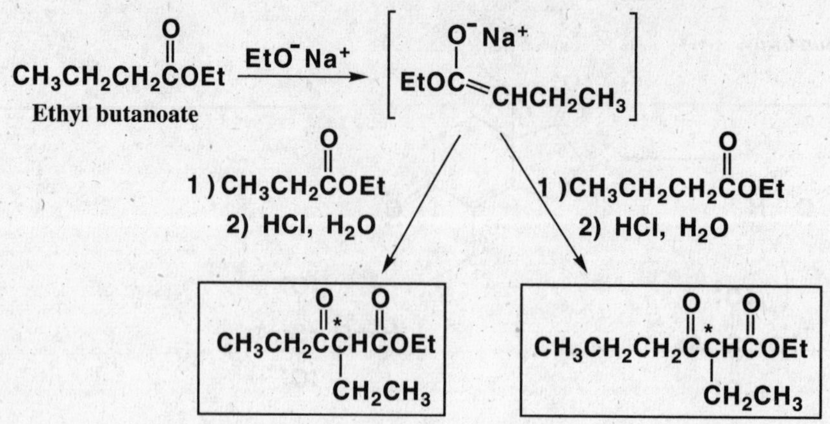

CH₃CH₂CH₂COEt Ethyl butanoate

Problem 19.31 Draw structural formulas for the β-ketoesters formed by Claisen condensation of ethyl propanoate with each ester:

(a) EtOC—COEt

(b) Ph COEt

(c) HCOEt

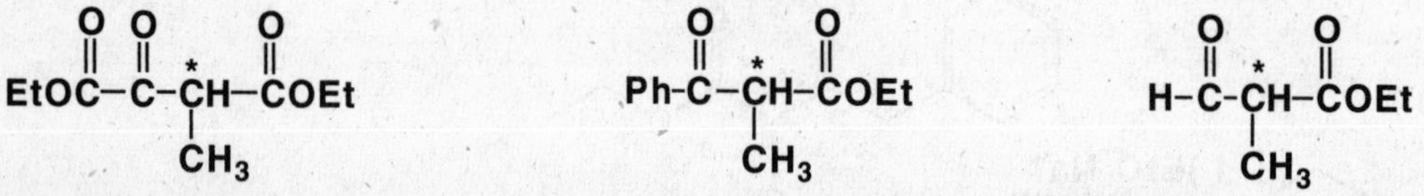

In all three of the above reactions, one chiral center is created so racemic mixtures are produced.

Problem 19.32 Draw a structural formula for the product of saponification, acidification, and decarboxylation of each β-ketoester formed in Problem 19.31.

Following are the structures for the products of saponification and decarboxylation of each β-ketoester from the previous problem.

(a) HOC—CCH₂CH₃

(b) PhCCH₂CH₃

(c) HCCH₂CH₃

Problem 19.33 The Claisen condensation can be used as one step in the synthesis of ketones, as illustrated by this reaction sequence.

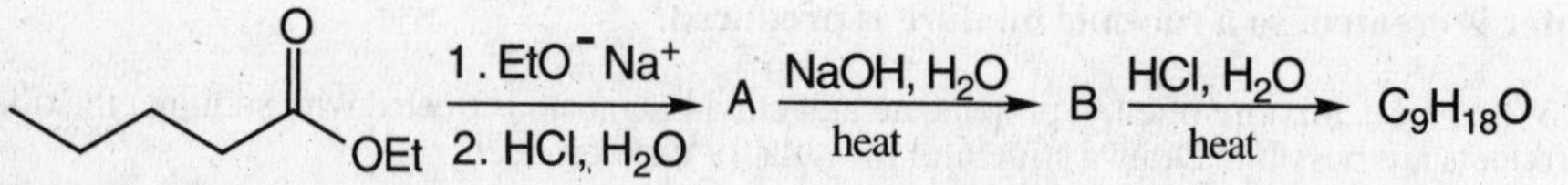

Propose structural formulas for compounds A and B and the ketone formed in this sequence.

Compound (A) is a β-ketoester, compound (B) is the sodium salt of a β-ketoacid, and the final ketone is 5-nonanone. Compounds A and B are racemic mixtures, while the final product is not chiral.

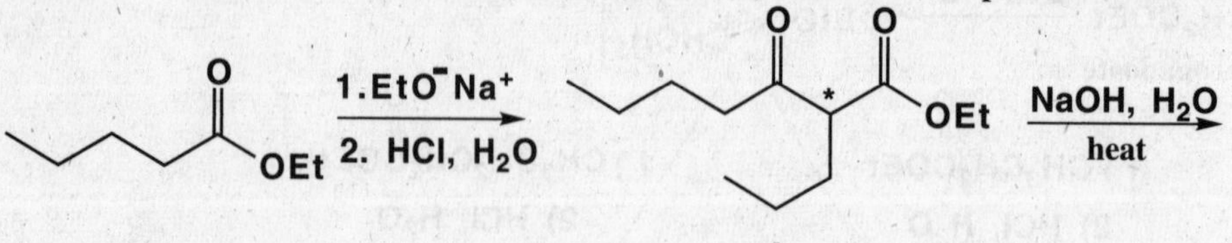

A

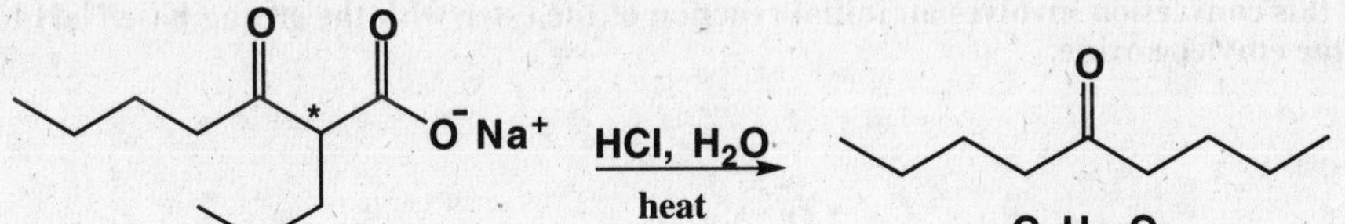

B   **C₉H₁₈O**
                                                                    5-Nonanone

Reaction: B + HCl, H₂O, heat → 5-Nonanone

Problem 19.34 Propose a synthesis for each ketone, using as one step in the sequence a Claisen condensation and the reaction sequence illustrated in Problem 19.33.

(a)

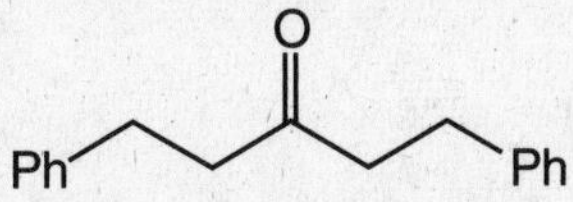

Each target molecule is synthesized using the sequence of Claisen condensation, saponification, and decarboxylation as outlined in the previous problem. The starting ester is ethyl 3-phenylpropanoate.

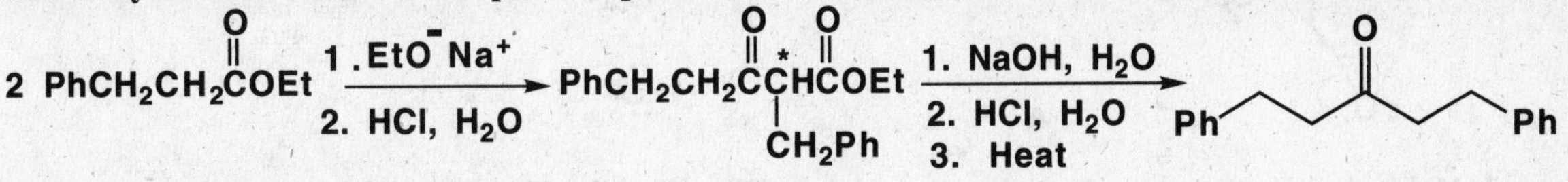

(b)

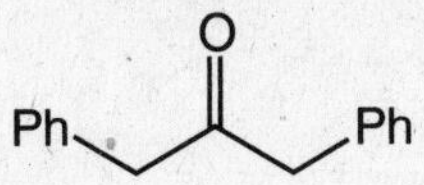

The starting ester is ethyl phenylacetate.

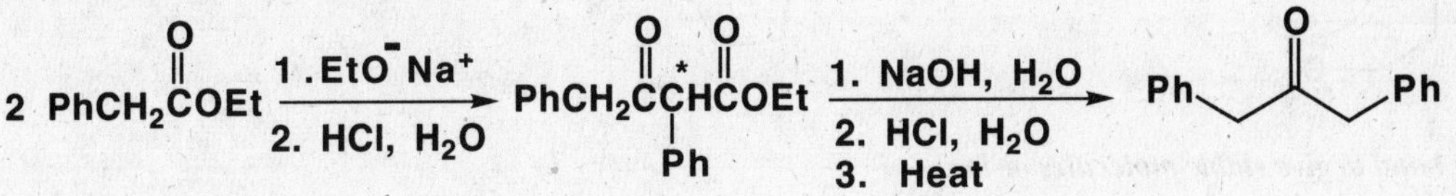

(c)

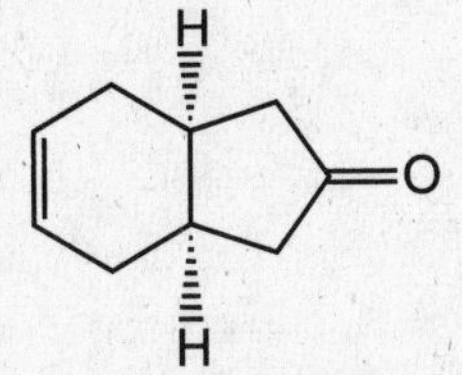

The starting diester is derived from a *cis*-4,5-disubstituted cyclohexene.

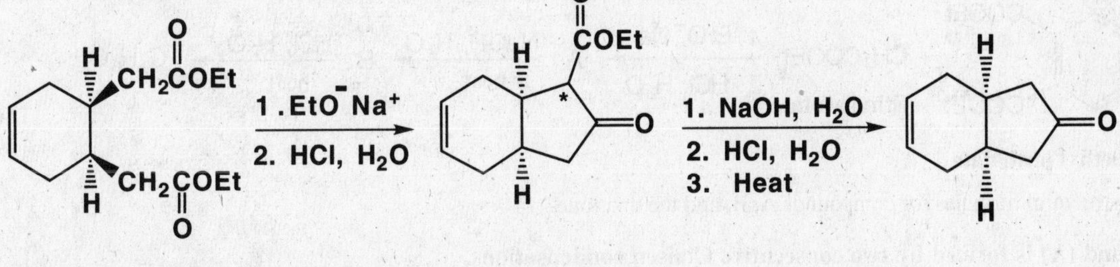

Problem 19.35 Propose a mechanism for the following conversion.

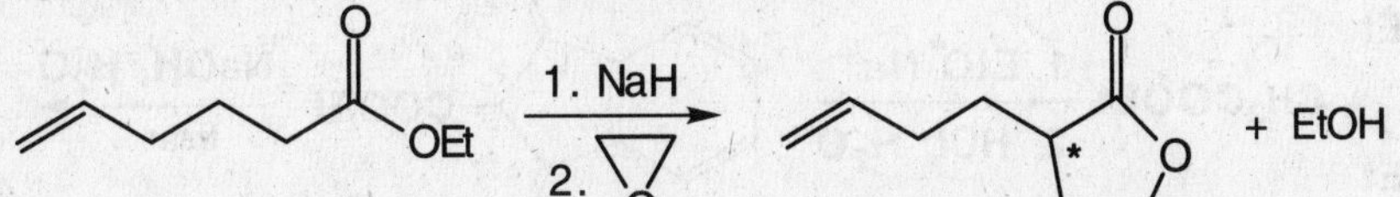

(racemic)

A reasonable mechanism for this conversion involves an initial reaction of the ester with the strong base NaH to give an enolate, that attacks the ethylene oxide.

Step 1: Take a proton away.

Step 2: Electrophilic addition-make a bond between a nucleophile and an electrophile.

Step 3: Make a bond between a nucleophile and an electrophile. The alkoxide intermediate then undergoes an intramolecular reaction with the ester function to give the product.

Step 4: Break a bond to give stable molecules or ions.

One chiral center is created, so a racemic mixture will be produced.

<u>Problem 19.36</u> Claisen condensation between diethyl phthalate and ethyl acetate followed by saponification, acidification, and decarboxylation forms a diketone, $C_9H_6O_2$.

Diethyl phthalate + CH_3COOEt (Ethyl acetate) $\xrightarrow[\text{2. HCl, H}_2\text{O}]{\text{1. EtO}^-\text{Na}^+}$ A $\xrightarrow[\text{heat}]{\text{NaOH, H}_2\text{O}}$ B $\xrightarrow[\text{heat}]{\text{HCl, H}_2\text{O}}$ $C_9H_6O_2$

Propose structural formulas for compounds A, B, and the diketone.

Compound (A) is formed by two consecutive Claisen condensations.

Diethyl phthalate + CH_3COOEt $\xrightarrow[\text{2. HCl, H}_2\text{O}]{\text{1. EtO}^-\text{Na}^+}$ A $\xrightarrow[\text{heat}]{\text{NaOH, H}_2\text{O}}$

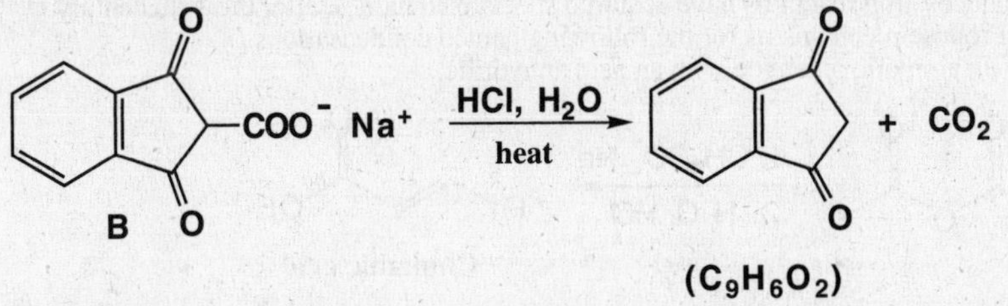

$$(C_9H_6O_2)$$

Problem 19.37 In 1887, the Russian chemist Sergei Reformatsky at the University of Kiev discovered that treatment of an α-haloester with zinc metal in the presence of an aldehyde or ketone followed by hydrolysis in aqueous acid results in formation of a β-hydroxyester. This reaction is similar to a Grignard reaction in that a key intermediate is an organometallic compound, in this case a zinc salt of an ester enolate anion. Grignard reagents, however, are so reactive that they undergo self-condensation with the ester.

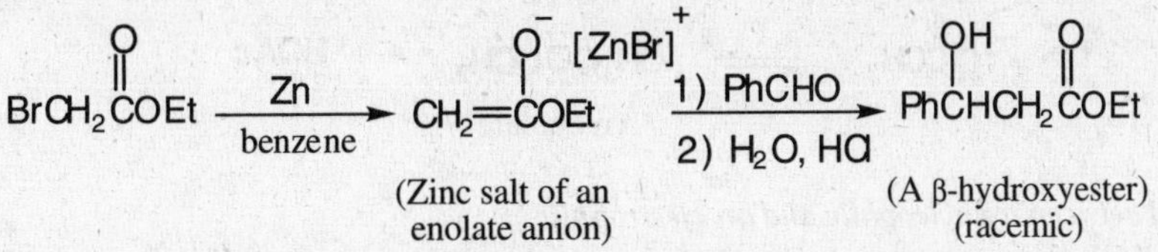

(Zinc salt of an enolate anion)   (A β-hydroxyester) (racemic)

Show how a Reformatsky reaction can be used to synthesize these compounds from an aldehyde or ketone and an α-haloester.

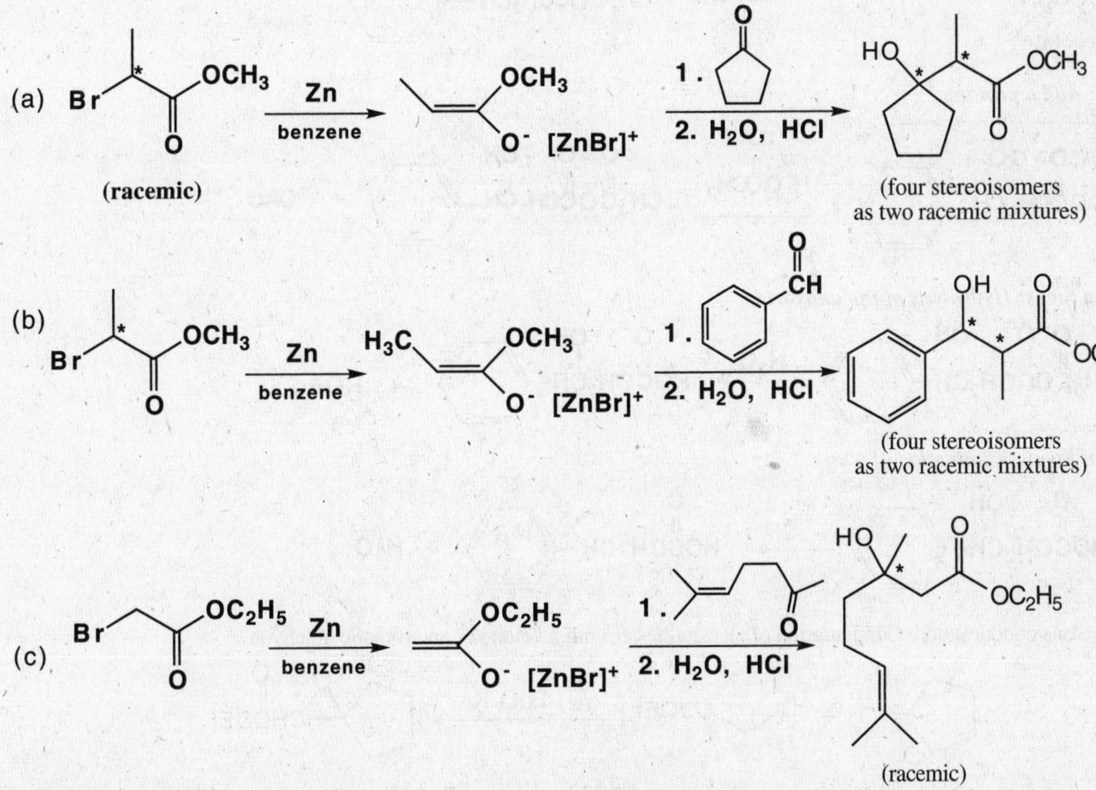

(a)

(racemic)   (four stereoisomers as two racemic mixtures)

(b)

(four stereoisomers as two racemic mixtures)

(c)

(racemic)

In parts (a) and (b), one new chiral center is created and one chiral center is present in the ester starting material, so a mixture of four stereoisomers will be produced. In part (c), one new chiral center is created, so a racemic mixture will be produced.

<u>Problem 19.38</u> Many types of carbonyl condensation reactions have acquired specialized names, after the 19th century organic chemists who first studied them. Propose mechanisms for the following named condensations.
(a) Perkin condensation: Condensation of an aromatic aldehyde with an acid anhydride.

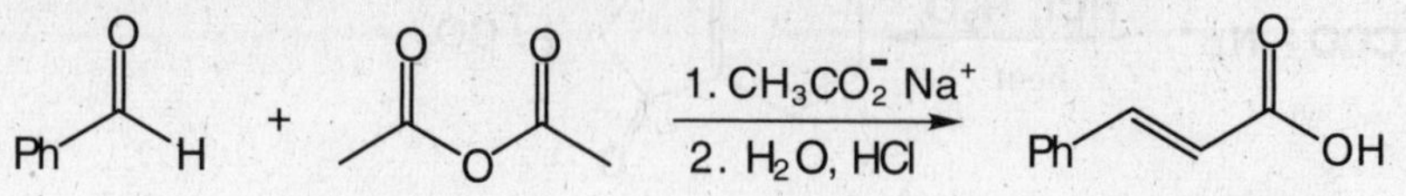

Cinnamic acid

In the Perkin reaction, an enolate of the anhydride is formed with acetate acting as the base. This enolate then attacks benzaldehyde reminiscent of an aldol reaction. Hydrolysis of the anhydride and dehydration complete the mechanism.

Step 1: Take a proton away.

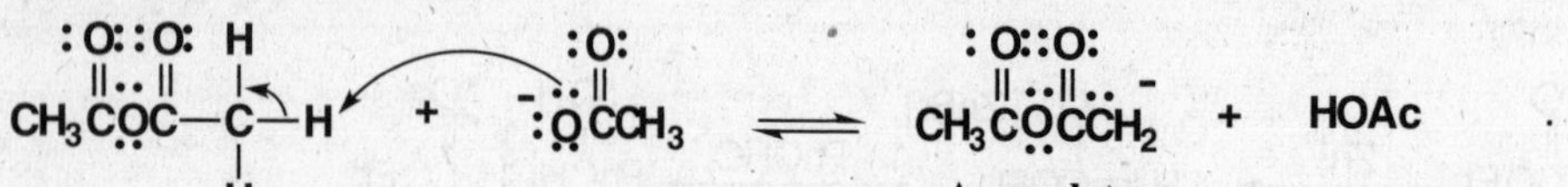

An enolate

Step 2: Make a bond between a nucleophile and an electrophile.

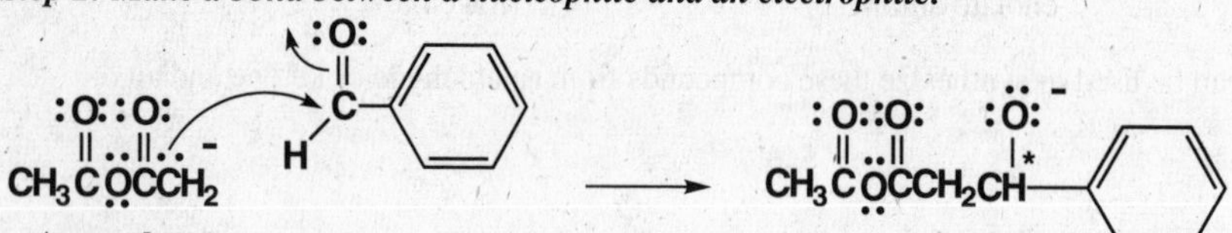

An enolate

Step 3: Add a proton.

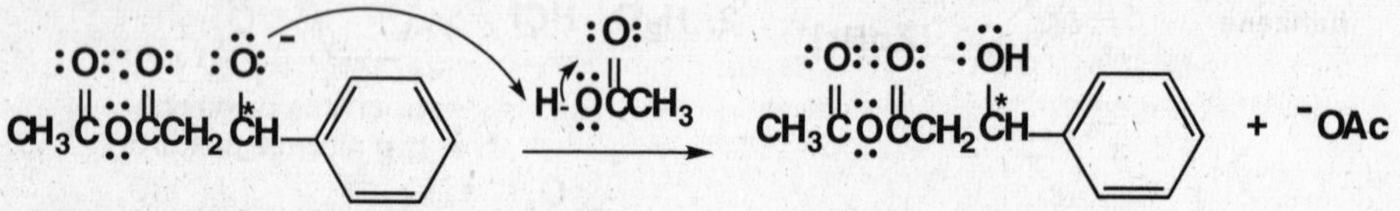

Several Steps: Hydrolysis of the anhydride

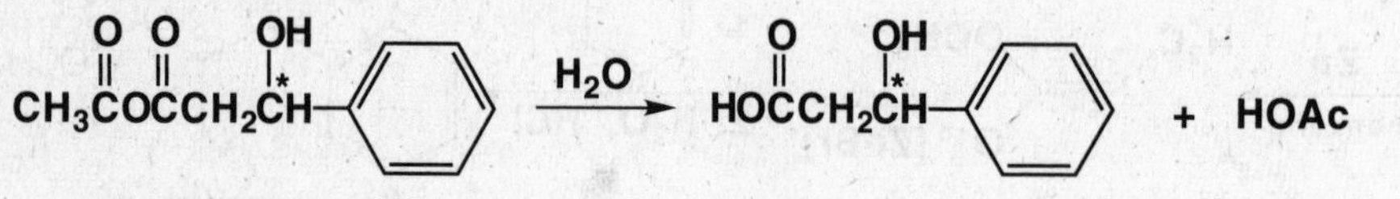

Several Steps: Elimination of water

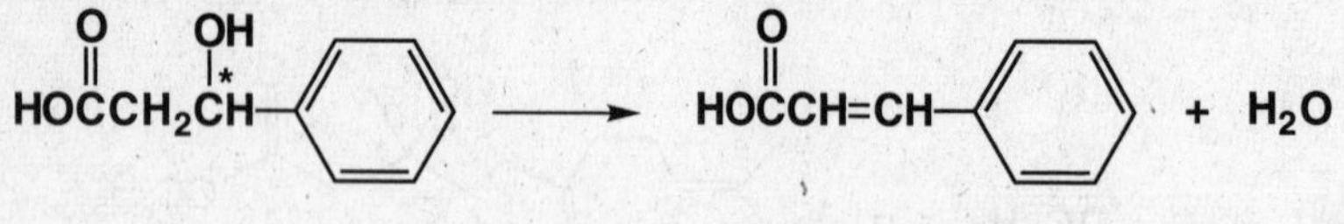

(b) Darzens condensation: Condensation of an α-haloester with a ketone or an aromatic aldehyde.

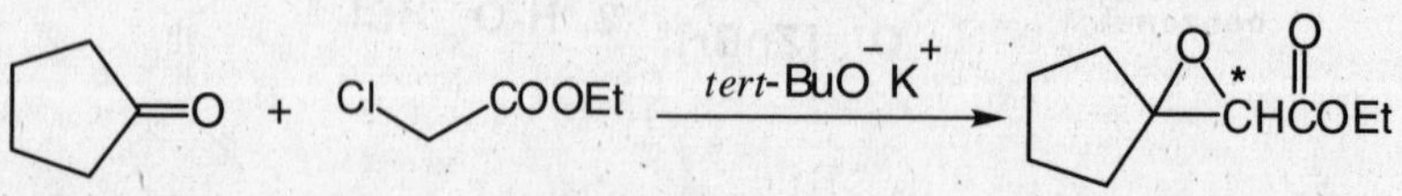

In the Darzens condensation, the enolate of the α-haloester is formed, which then attacks the ketone carbonyl group reminiscent of an aldol reaction.

Step 1: Take a proton away.

Step 2: Make a bond between a nucleophile and an electrophile.

Step 3: Make a bond between a nucleophile and an electrophile and simultaneously break a bond to give stable molecules or ions. The oxyanion displaces chloride via backside attack to create an epoxide.

One new chiral center is created, so a racemic mixture will be produced.

Enamines
Problem 19.39 When 2-methylcyclohexanone is treated with pyrrolidine, two isomeric enamines are formed.

A (85%)   B(15%)

Why is enamine A with the less substituted double bond the thermodynamically favored product? (You will find it helpful to examine the models of these two enamines.)

Remember that in the six atom system of an enamine, the nitrogen is sp^2 hybridized, so there is overlap of the $2p$ orbital of nitrogen atom with the $2p$ orbitals of the carbon-carbon double bond. This overlap allows electron delocalization and thus stabilization of the enamine.

Enamine A   this six-atom
                                  sytem is planar

steric interactions

sp^2 carbon atom   Enamine B

In enamine B, non-bonded interactions force rotation about the C-N bond and reduction in planarity of the enamine system, resulting in loss of stability. In enamine A, there is no such non-bonded interaction, so the enamine system is planar providing for maximum stabilization. Thus, enamine A predominates.

<u>Problem 19.40</u> Enamines normally react with methyl iodide to give two products: one arising from alkylation at nitrogen and the second arising from alkylation at carbon. For example,

Heating the mixture of C-alkylation and N-alkylation products gives only the product from C-alkylation. Propose a mechanism for this isomerization.

The I⁻ that is in the solution can act as a nucleophile to regenerate CH₃I from the N-alkylated product. This then reacts with the enamine to make the C-alkylated product.

Step 1: Make a bond between a nucleophile and an electrophile and simultaneously break a bond to give stable molecules or ions.

Step 2: Electrophilic addition-make a bond between a nucleophile and an electrophile and simultaneously break a bond to give stable molecules or ions.

One new chiral center is created, so a racemic mixture will be produced.

Problem 19.41 Propose a mechanism for the following conversion.

Step 1: Make a bond between a nucleophile and an electrophile and simultaneously break a bond to give stable molecules or ions.

Step 2: Take a proton away.

Step 3: Electrophilic addition-make a bond between a nucleophile and an electrophile and simultaneously break a bond to give stable molecules or ions.

Step 4: Hydrolysis This hydroysis actually involves several steps.

Due to symmetry, the product is not chiral.

Problem 19.42 The following intermediate was needed for the synthesis of tamoxifen, a widely used antiestrogen drug for treating estrogen-dependent cancers such as breast and ovarian cancer.

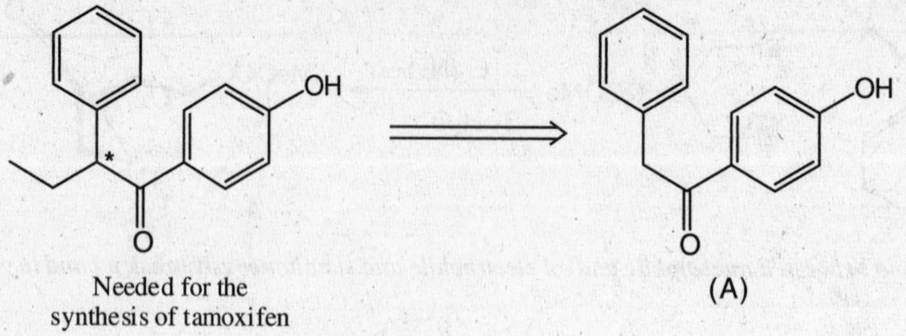

Needed for the
synthesis of tamoxifen

(A)

Propose a synthesis for this intermediate from compound A.

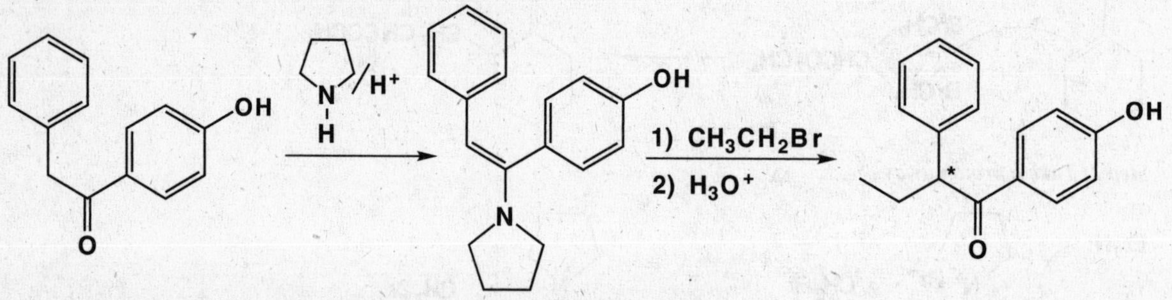

1) CH₃CH₂Br

2) H₃O⁺

One new chiral center is created, so a racemic mixture will be produced.

Problem 19.43 Propose a mechanism for the following reaction.

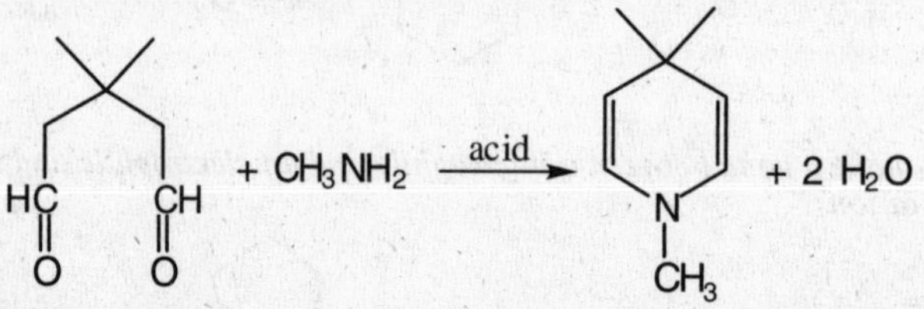

+ CH₃NH₂ $\xrightarrow{\text{acid}}$ + 2 H₂O

Note that in the following mechanism, familiar steps (amino alcohol formation, acid catalyzed dehydration) are summarized rather than shown explicitly.

Several Steps: Amino alcohol formation.

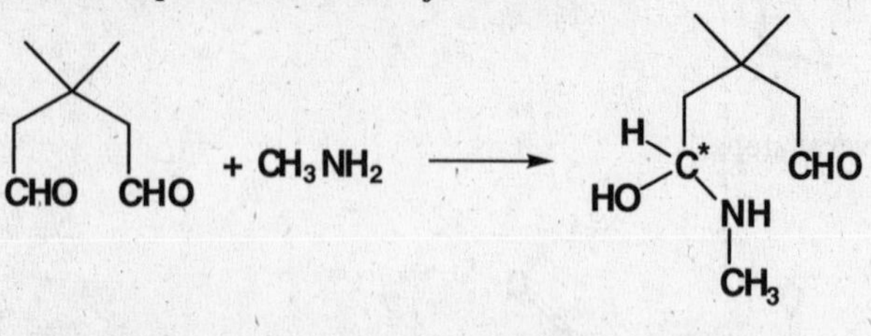

+ CH₃NH₂   ⟶

Several Steps: Amino alcohol formation.

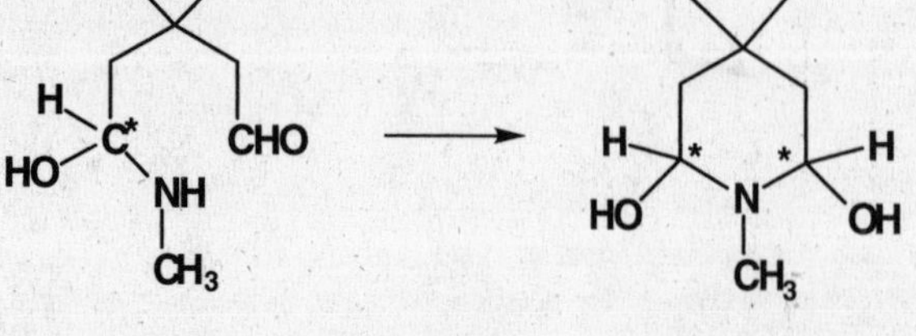

⟶

Several Steps: Dehydration

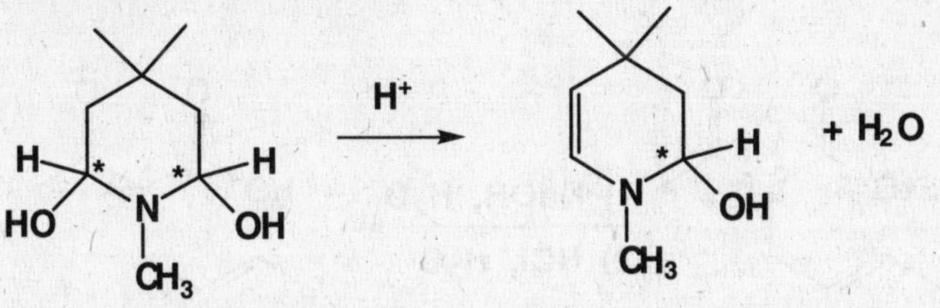

Several Steps: Dehydration

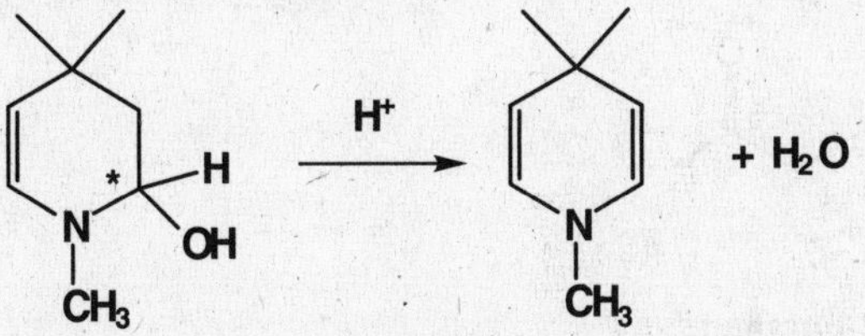

Acetoacetic Ester and Malonic Ester Syntheses

<u>Problem 19.44</u> Propose syntheses of the following derivatives of diethyl malonate, each of which is a starting material for synthesis of a barbiturate.

(a)

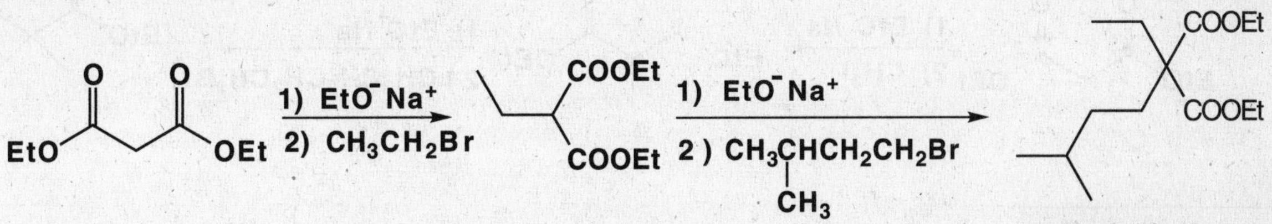

Needed for the synthesis of amobarbital

(b)

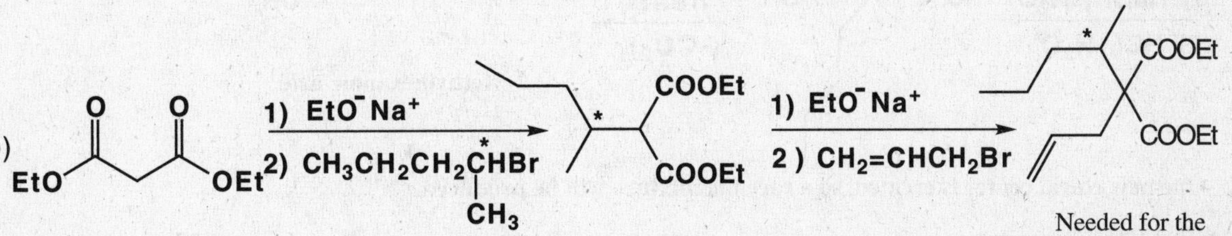

Needed for the synthesis of secobarbital

<u>Problem 19.45</u> 2-Propylpentanoic acid (valproic acid) is an effective drug for treatment of several types of epilepsy, particularly absence seizures, which are generalized epileptic seizures characterized by brief and abrupt loss of consciousness. Propose a synthesis of valproic acid starting with diethyl malonate.

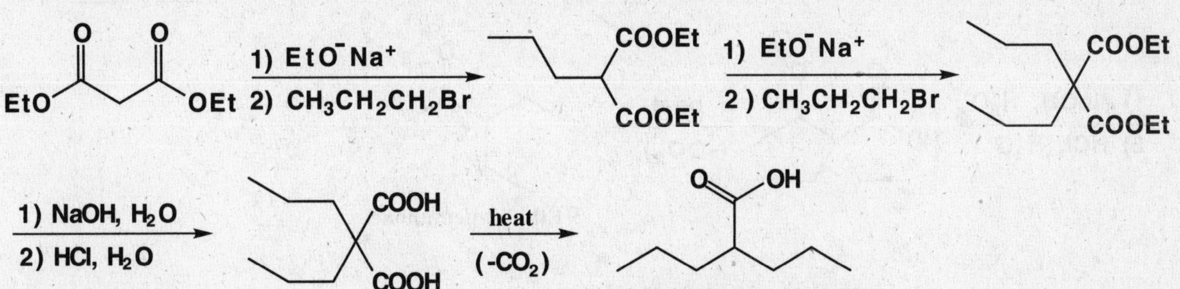

2-Propylpentanoic acid
(Valproic acid)

<u>Problem 19.46</u> Show how to synthesize the following compounds using either the malonic ester synthesis or the acetoacetic ester synthesis.

a) 4-Phenyl-2-butanone

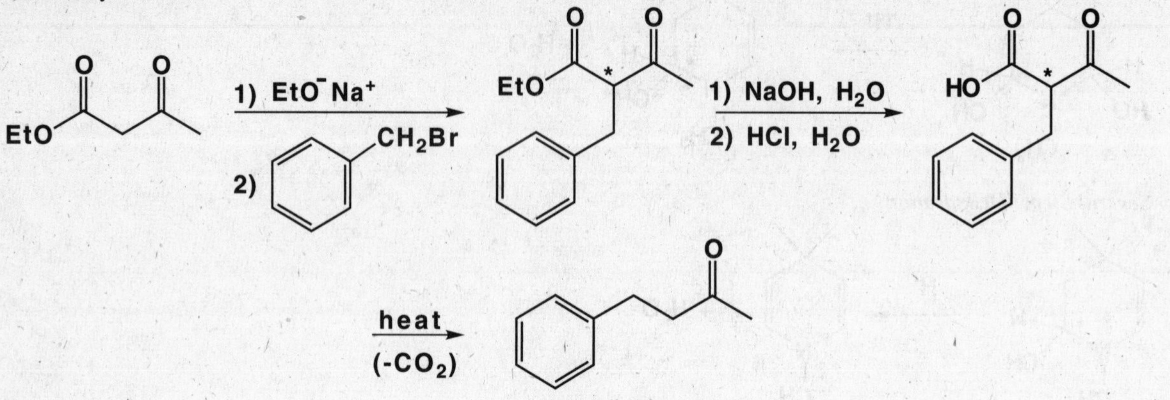

4-Phenyl-2-butanone

(b) 2-Methylhexanoic acid

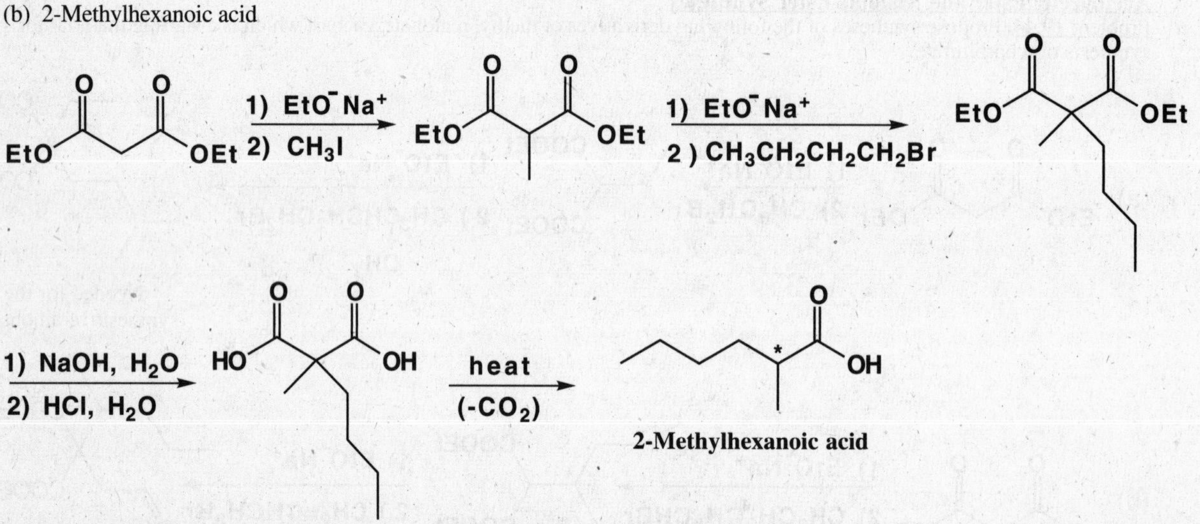

2-Methylhexanoic acid

One new chiral center is created, so a racemic mixture will be produced.

(c) 3-Ethyl-2-pentanone

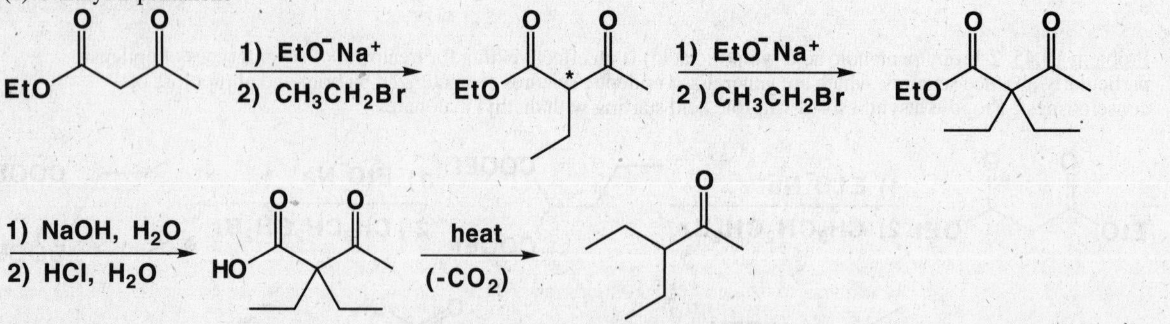

3-Ethyl-2-pentanone

(d) 2-Propyl-1,3-propanediol

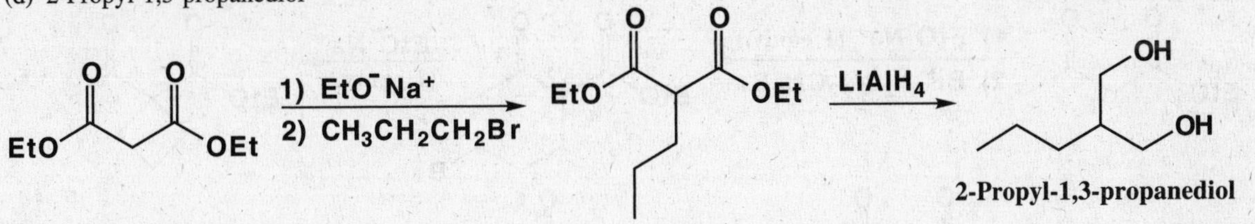

2-Propyl-1,3-propanediol

(e) 4-Oxopentanoic acid

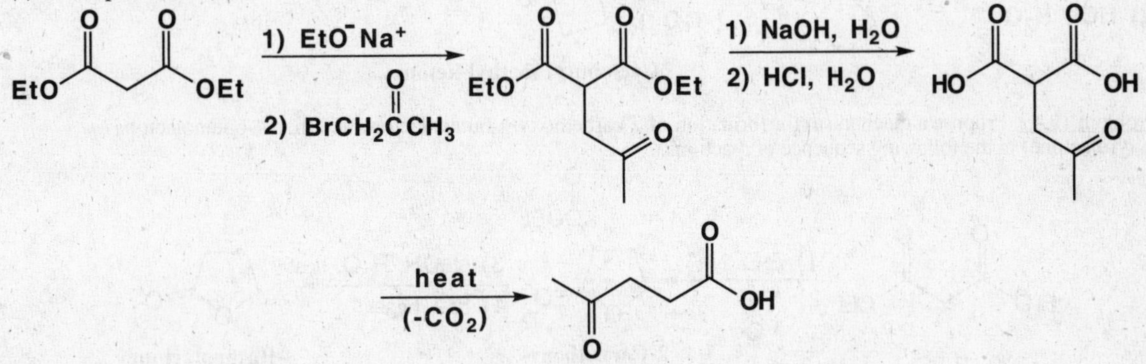

4-Oxopentanoic acid

(f) 3-Benzyl-5-hexene-2-one

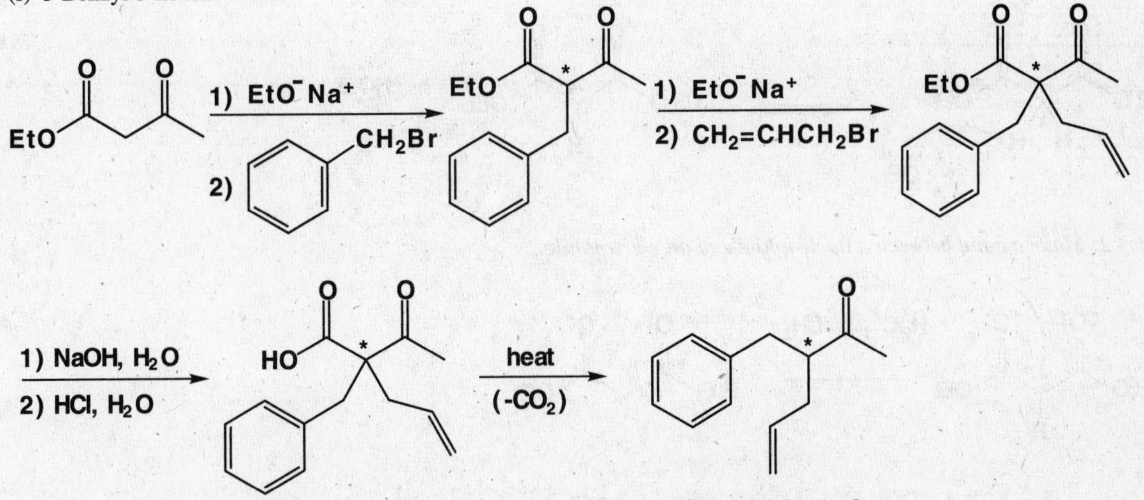

3-Benzyl-5-hexen-2-one

One new chiral center is created, so a racemic mixture will be produced.

(g) Cyclopropanecarboxylic acid

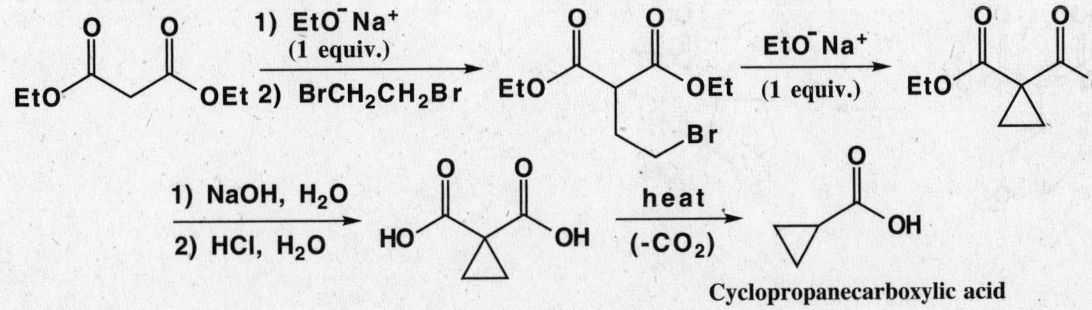

Cyclopropanecarboxylic acid

(h) Cyclobutyl methyl ketone

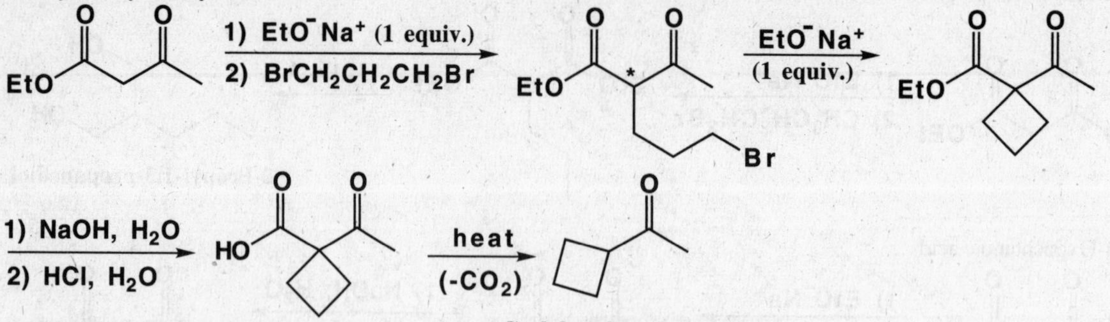

Cyclobutyl methyl ketone

<u>Problem 19.47</u> Propose a mechanism for formation of 2-carbethoxy-4-butanolactone and then 4-butanolactone (γ-butyrolactone) in the following sequence of reactions.

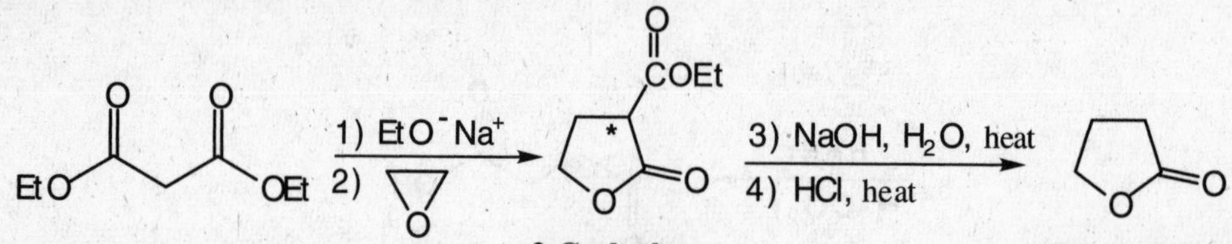

2-Carbethoxy-4-butanolactone
(racemic)

4-Butanolactone
(γ-Butyrolactone)

Step 1: Take a proton away.

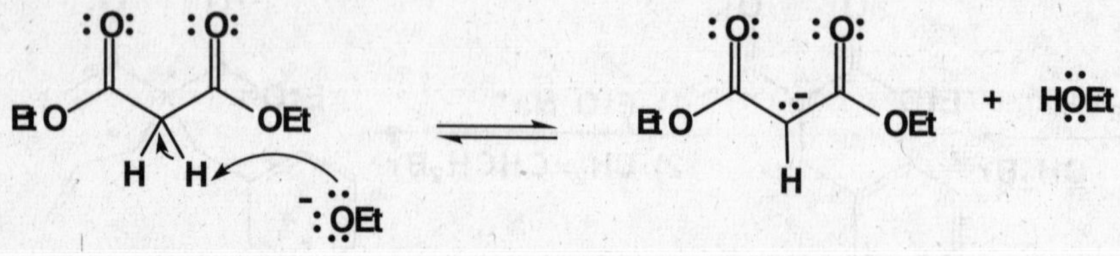

Step 2: Make a bond between a nucleophile and an electrophile.

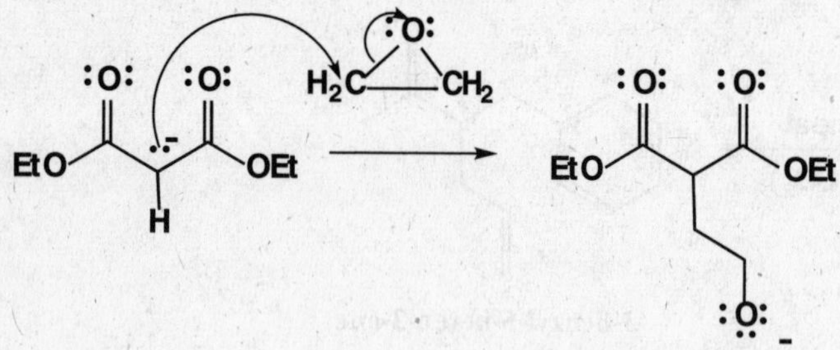

Step 3: Make a bond between a nucleophile and an electrophile.

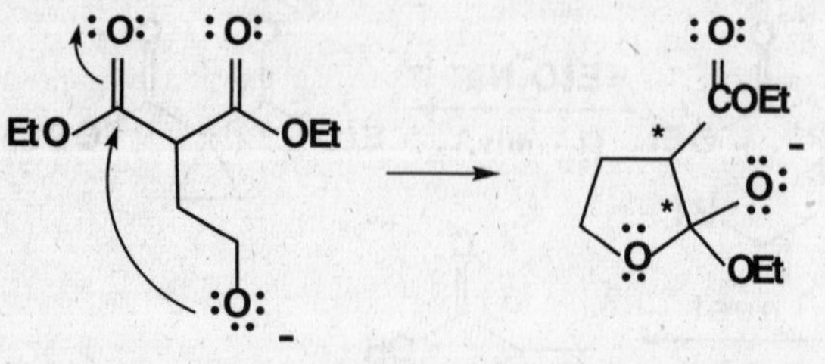

Step 4: Break a bond to give stable molecules or ions.

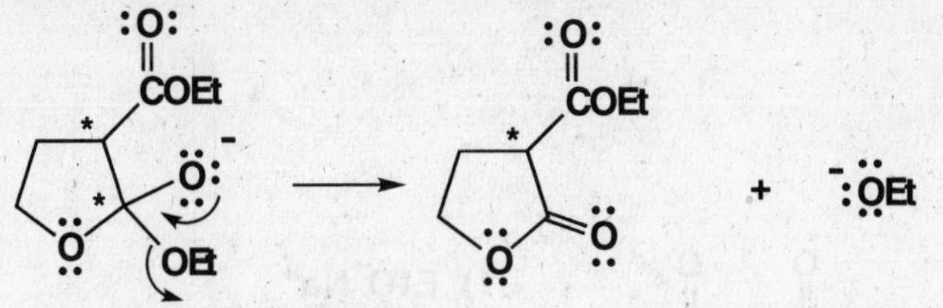

Several steps: Ester hydrolysis

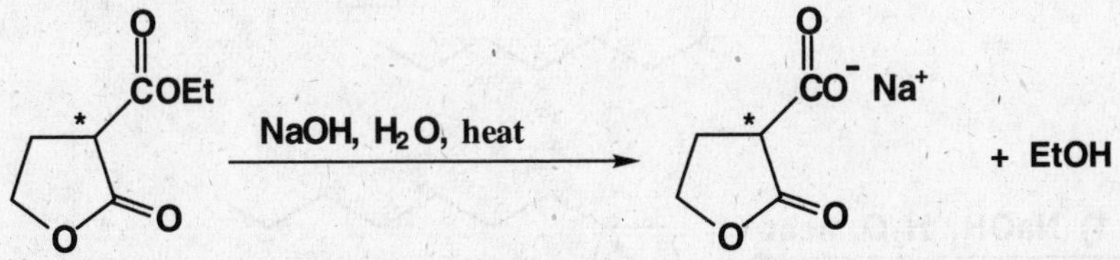

NaOH, H$_2$O, heat

\+ EtOH

Two steps: Add a proton and decarboxylation.

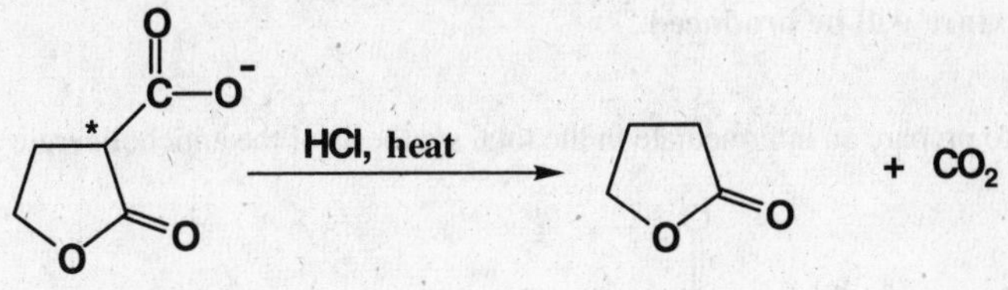

HCl, heat

\+ CO$_2$

Note that in an effort to save space, in steps 5 and 6 above there are some simple individual steps not shown.

<u>Problem 19.48</u> Show how the scheme for formation of 4-butanolactone in Problem 19.47 can be used to synthesize lactones (a) and (b), each of which has a peach odor and is used in perfumery. As sources of carbon atoms for these syntheses, use diethyl malonate, ethylene oxide, 1-bromoheptane, and 1-nonene.

(a)

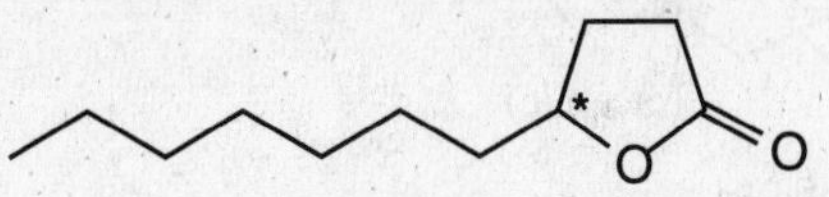

(racemic)

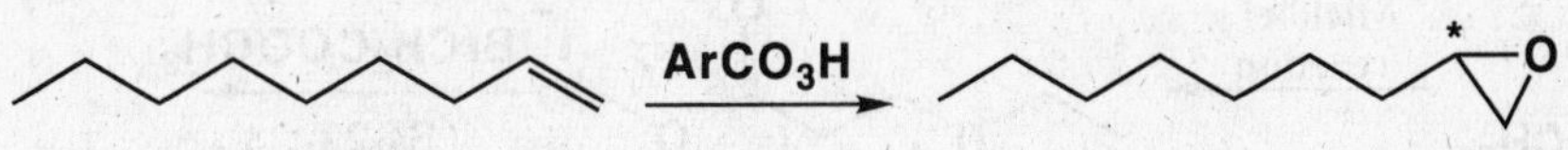

1-Nonene   ArCO$_3$H

1) EtO$^-$ Na$^+$

2)

1) NaOH, H$_2$O, heat

2) HCl, heat (-CO$_2$)

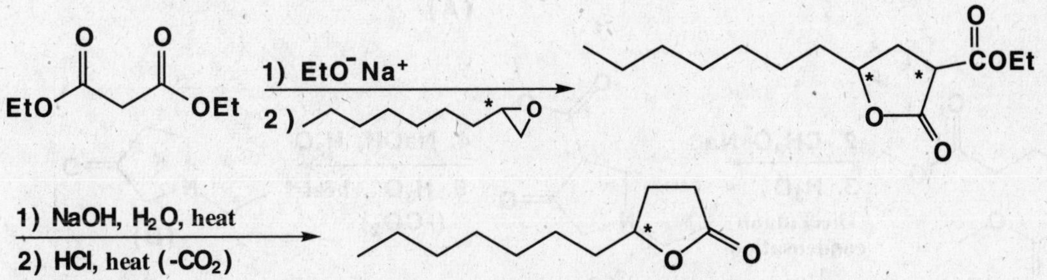

One new chiral center is created, so a racemic mixture will be produced.

(b)

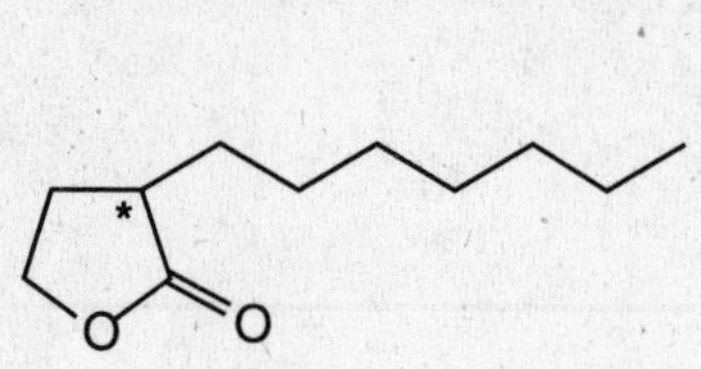

(racemic)

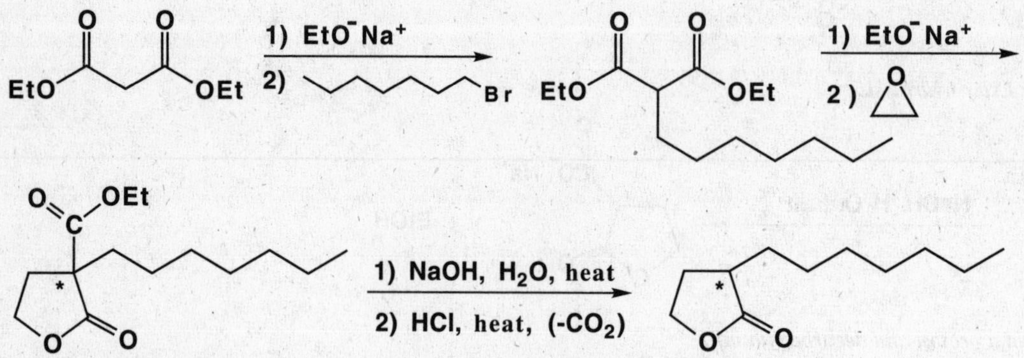

1) EtO⁻ Na⁺
2) [hexyl bromide]

1) EtO⁻ Na⁺
2) [epoxide]

1) NaOH, H_2O, heat
2) HCl, heat, ($-CO_2$)

One new chiral center is created, so a racemic mixture will be produced.

Michael Reactions

Problem 19.49 The following synthetic route is used to prepare an intermediate in the total synthesis of the anticholinergic drug benzilonium bromide.

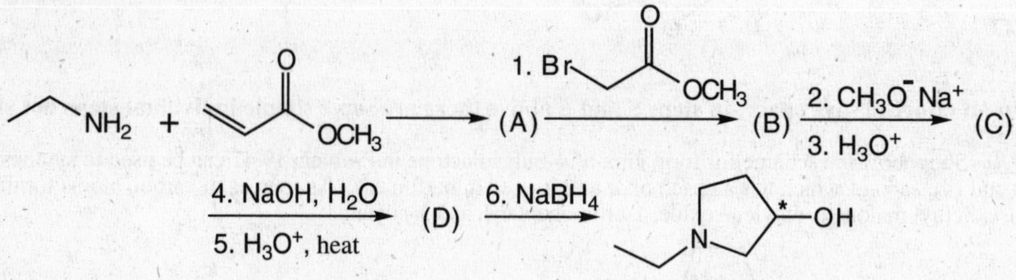

[reaction scheme]

$CH_3CH_2NH_2$ + [methyl acrylate OCH_3] ⟶ (A) ⟶ [1. BrCH₂C(O)OCH₃] ⟶ (B) ⟶ [2. CH_3O^- Na⁺ ; 3. H_3O^+] ⟶ (C)

4. NaOH, H_2O
5. H_3O^+, heat ⟶ (D) ⟶ 6. NaBH₄ ⟶ [1-ethyl-3-hydroxypyrrolidine] (racemic)

Propose structural formulas for intermediates A, B, C, and D in this synthesis.

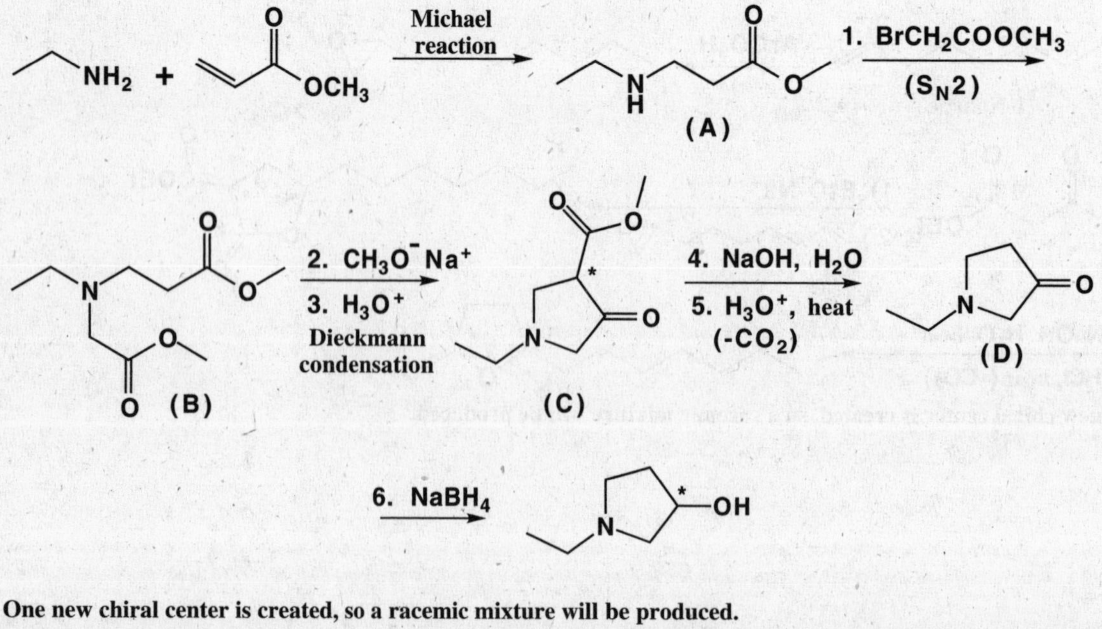

NH_2 + [methyl acrylate OCH₃] —Michael reaction→ [ethylamino ester] (A) —1. BrCH₂COOCH₃ (S_N2)→

(B) —2. CH_3O^- Na⁺ ; 3. H_3O^+ Dieckmann condensation→ (C) —4. NaOH, H_2O ; 5. H_3O^+, heat ($-CO_2$)→ (D) [1-ethyl-3-oxopyrrolidine]

6. NaBH₄ ⟶ [1-ethyl-3-hydroxypyrrolidine]

One new chiral center is created, so a racemic mixture will be produced.

Problem 19.50 Propose a mechanism for formation of the bracketed intermediate, and for the bicyclic ketone formed in the following reaction sequence.

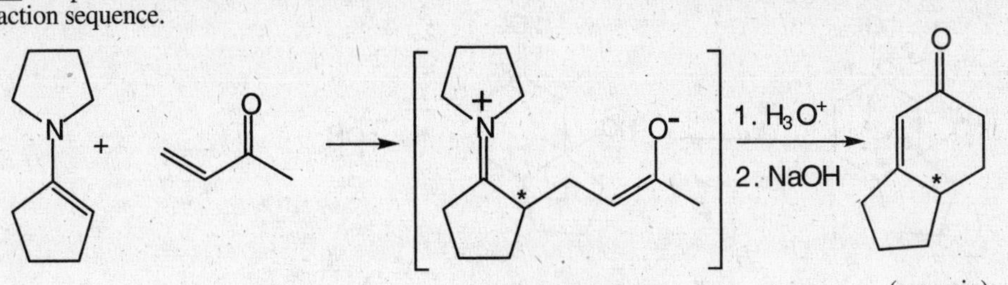

(an intermediate not isolated) (racemic)

Step 1: Electrophilic addition-make a bond between a nucleophile and an electrophile. **The reaction begins with a Michael reaction.**

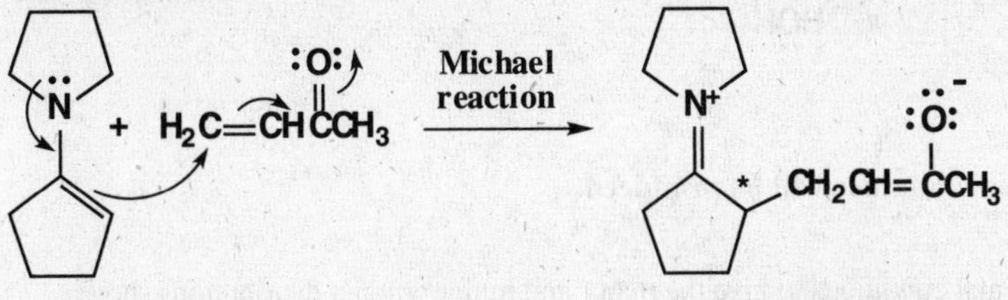

(an intermediate not isolated)

The chemist then adds aqueous acid. There will be several steps of hydrolysis of the imine, protonation of the alkoxide and tautomerization to give the ketone.

Step 2: Hydrolysis-add a proton-keto-enol tautomerism.

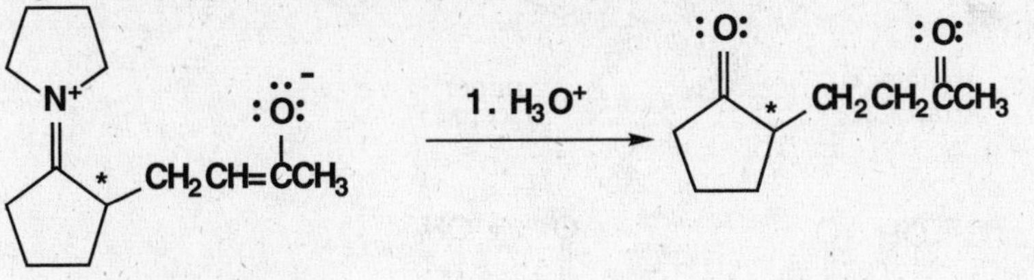

(an intermediate not isolated)

Step 3: Take a proton away.

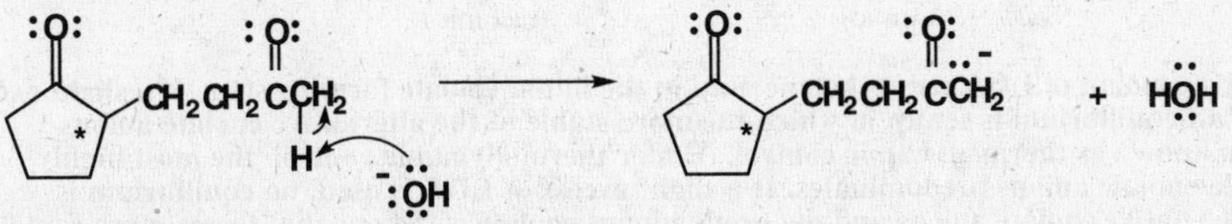

Step 4: Make a bond between a nucleophile and an electrophile.

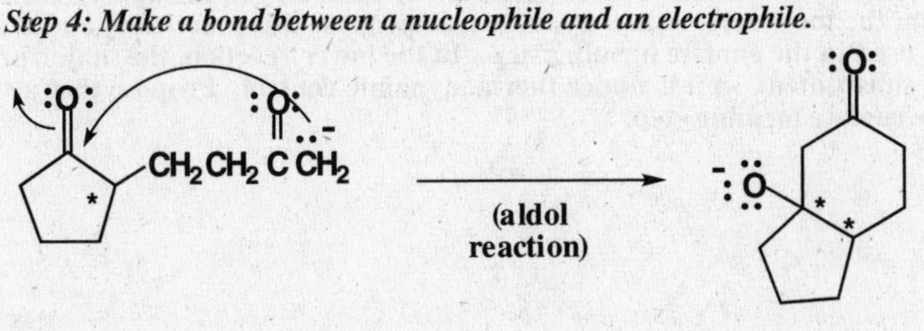

Step 5: Add a proton.

Step 6: Dehydration As always, the dehydration occurs through several steps.

One new chiral center is created, so a racemic mixture will be produced.

Directed Aldol and Alkylation

Problem 19.51 Discuss the different experimental conditions to give the major and minor product distributions shown.

The key here is the relative amount of LDA versus ketone used in the initial enolate forming step. If a slight excess of carbonyl species is used, an equilibrium is set up in which the more stable of the alternative enolate anions predominates, a situation known as thermodynamic control. Under thermodynamic control, the most highly substituted of the possible enolate anions predominates. If a slight excess of LDA is used, no equilibrium is established among the alternative enolate anions and the predominant enolate is the one that forms more rapidly, *i.e.* the one in which the α-hydrogen removed is the one that is more accessible (kinetic control). In the upper reaction, the predominant product is the one derived from the most accessible α-hydrogen, so propose this reaction is under kinetic control and that a slight excess of LDA was used in the enolate forming step. In the lower reaction, the major product derives from the more stable enolate (more substituted), so it is under thermodynamic control. Propose that a slight excess of the starting ketone was used in the enolate forming step.

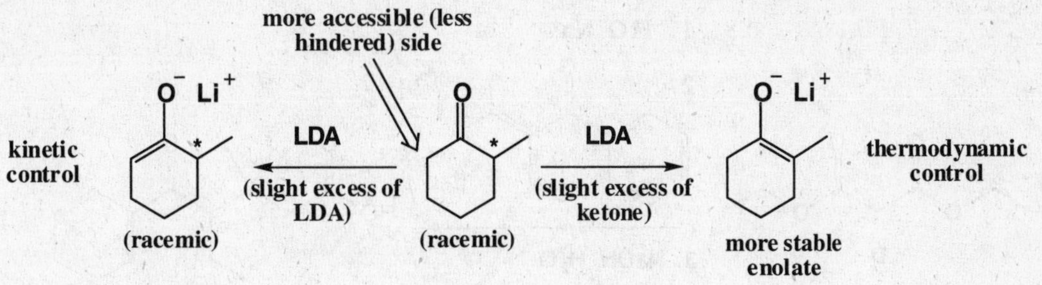

more accessible (less
hindered) side

kinetic
control

(racemic)

LDA

(slight excess of LDA)

(racemic)

LDA

(slight excess of ketone)

thermodynamic
control

more stable
enolate

Problem 19.52 Why does the following reaction give the product shown as the major product when 0.95 equivalents of LDA relative to ketone are used in the first step?

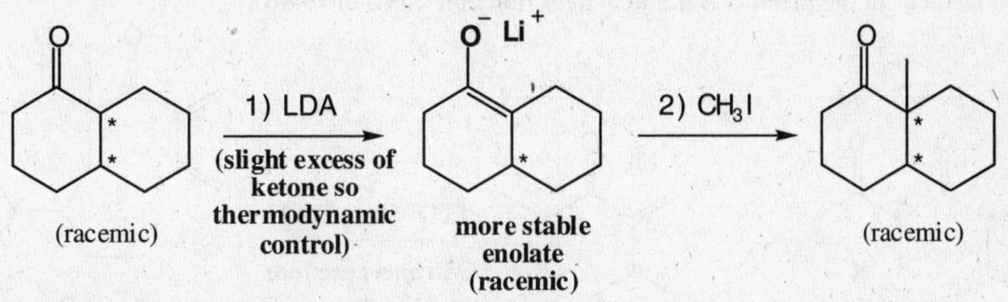

(racemic)

1) LDA

(slight excess of ketone so thermodynamic control)

more stable
enolate
(racemic)

2) CH₃I

(racemic)

Because an excess of ketone is being used, the more stable enolate will predominate (thermodynamic control). The more stable enolate is the one with greater substitution, so alkylation takes place at the ring junction.

Retrosynthetic Analysis
Problem 19.53 Using one of the reactions in this chapter, give the correct starting material (A–L) needed to produce each structure (a–f). Name the type of reaction used.

a)

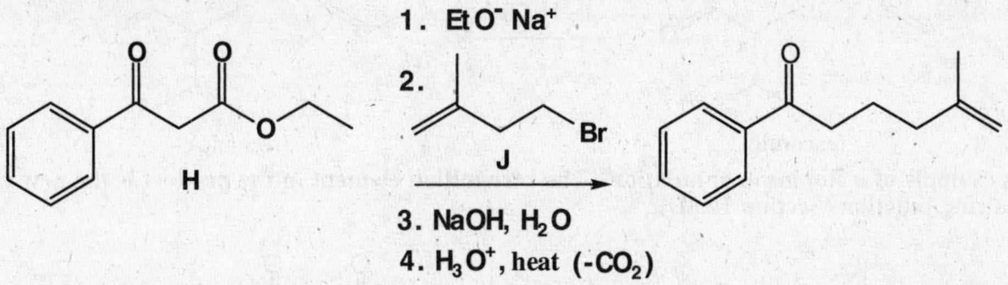

H

1. Et O⁻ Na⁺

2.

J

Br

3. NaOH, H₂O

4. H₃O⁺, heat (-CO₂)

This reaction sequence is analogous to the acetoacetic ester synthesis. The recognition element in the product is an alkylated ketone group (Section 19.6B).

b)

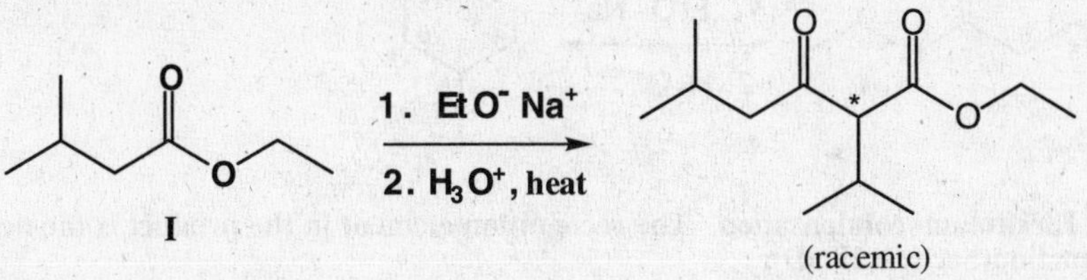

I

1. Et O⁻ Na⁺

2. H₃O⁺, heat

(racemic)

This reaction is an example of a Claisen condensation. The recognition element in the product is an β-keto ester function (Section 19.3D).

c)

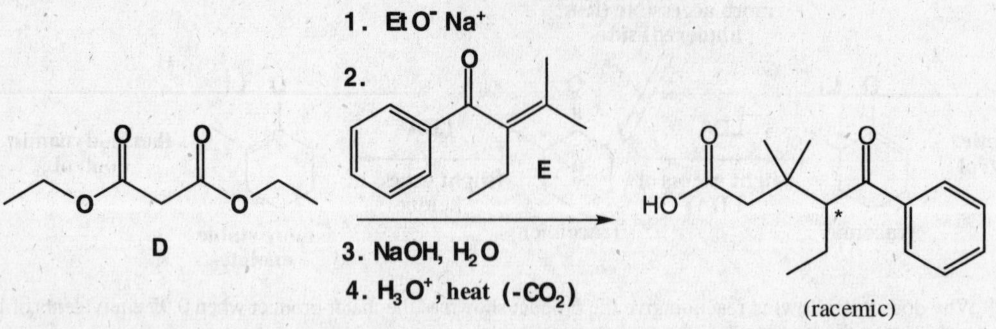

This reaction is an example of a Michael addition of an enolate anion, followed by decarboxylation. The recognition element in the product is a δ-keto acid function (Section 19.8B).

d)

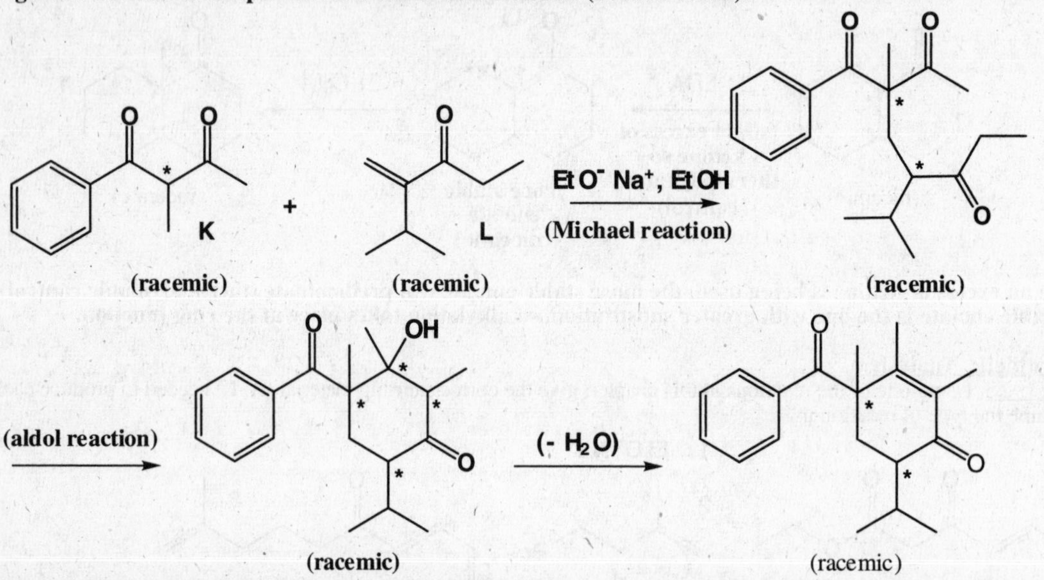

This reaction is an example of a Robinson annulation. The recognition element in the product is the new ring, with an alkene at a ring junction (Section 19.8D).

e)

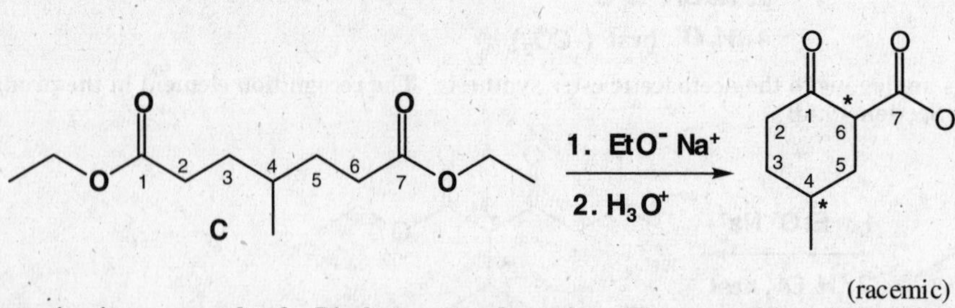

This reaction is an example of a Dieckmann condensation. The recognition element in the product is the new ring along with a β-keto ester function(Section 19.3D).

f)

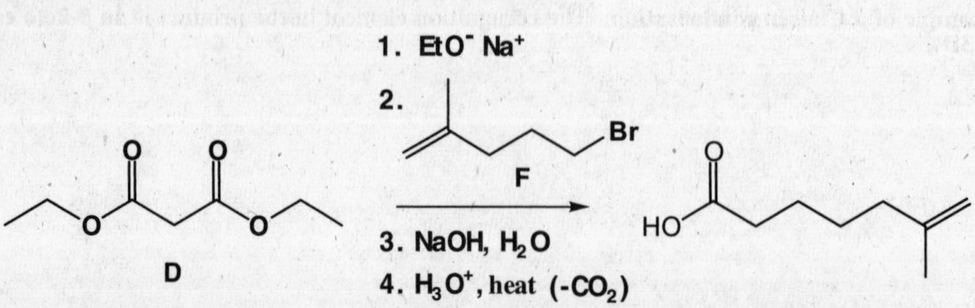

This reaction is an example of a malonic ester synthesis. The recognition element in the product is the carboxylic acid with a primary alkyl group attached to the α-carbon (Section 19.7B).

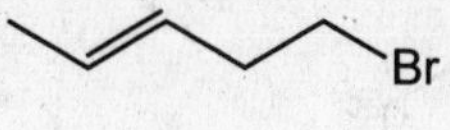
A

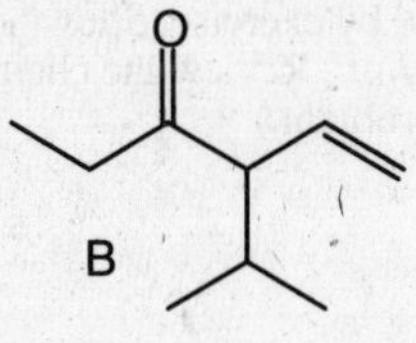
B

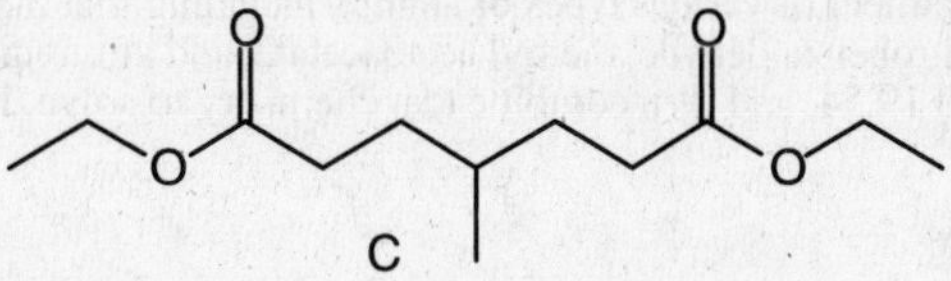
C

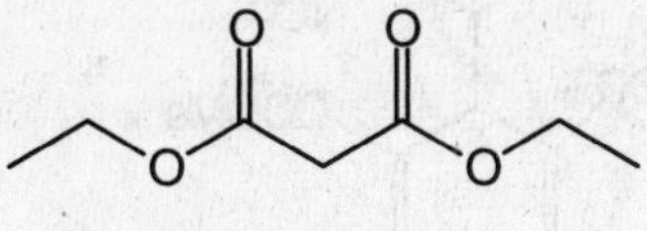
D

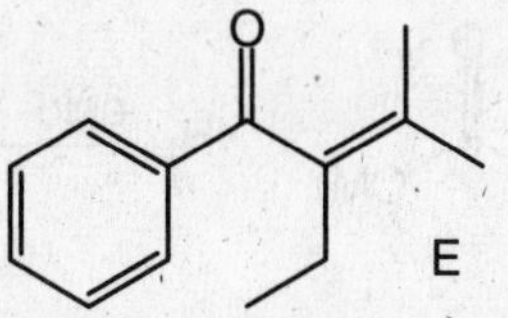
E

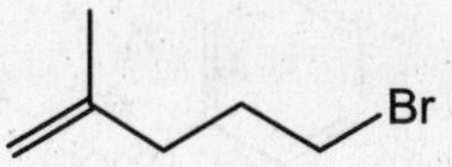
F

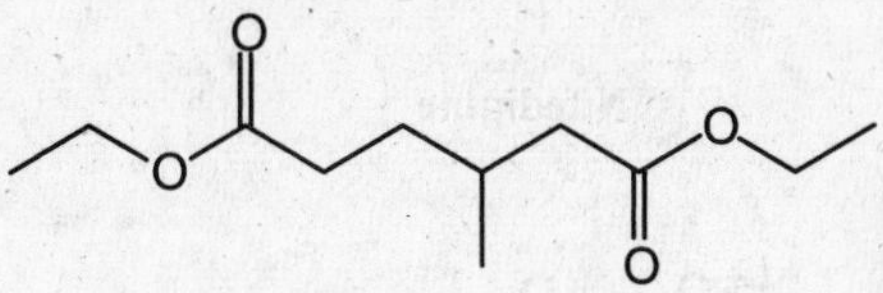
G

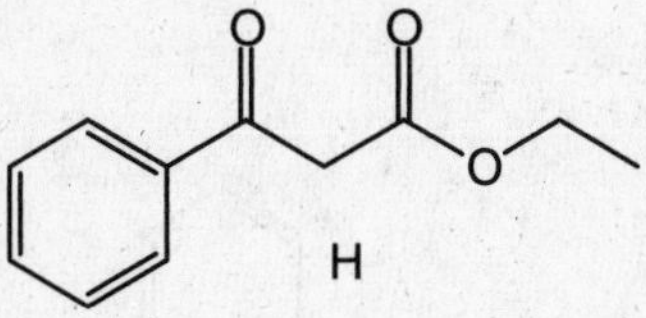
H

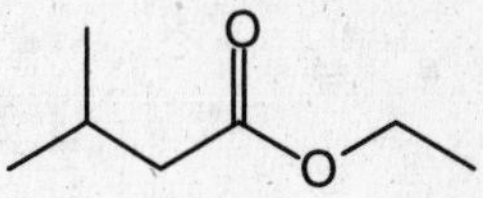
I

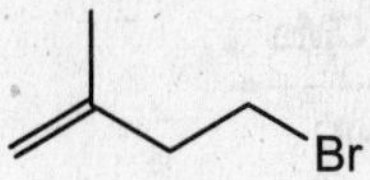
J

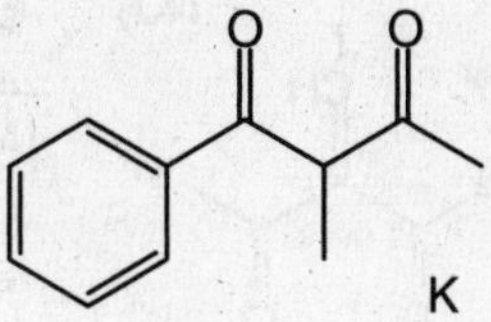
K

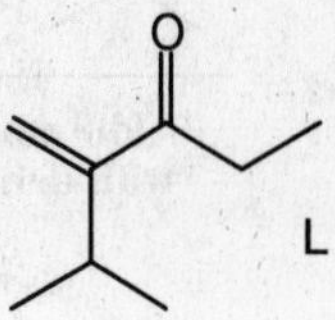
L

Synthesis

Problem 19.54 Show experimental conditions by which to carry out the following synthesis.

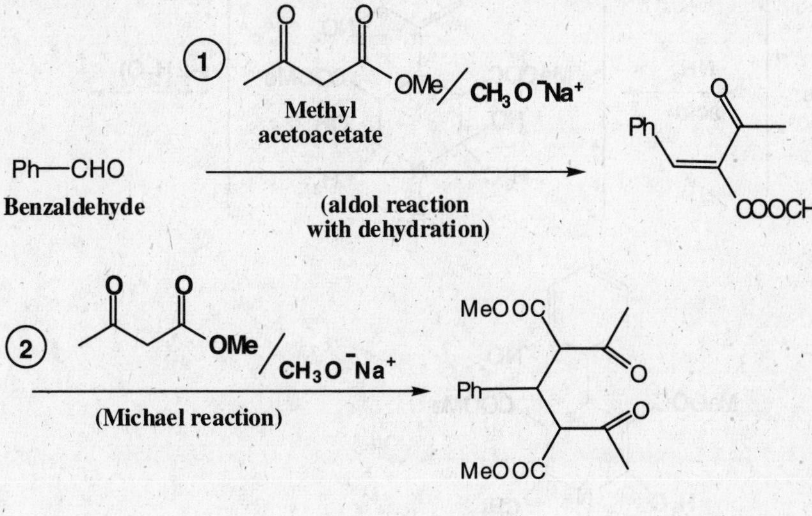
1
OMe
$CH_3O^-Na^+$
Methyl acetoacetate
Ph—CHO
Benzaldehyde
(aldol reaction with dehydration)
Ph
$COOCH_3$
2
OMe
$CH_3O^-Na^+$
(Michael reaction)
MeOOC
Ph
MeOOC

<u>Problem 19.55</u> Nifedipine (Procardia, Adalat,) belongs to a class of drugs called calcium channel blockers and is effective in the treatment of various types of angina, including that induced by exercise. Show how nifedipine can be synthesized from 2-nitrobenzaldehyde, methyl acetoacetate, and ammonia. (*Hint:* Review the chemistry of your answers to Problems 19.43 and 19.54, and then combine that chemistry to solve this problem.)

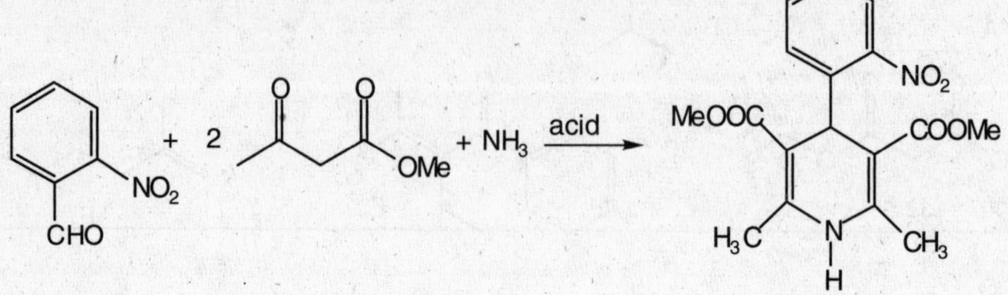

Nifedipine

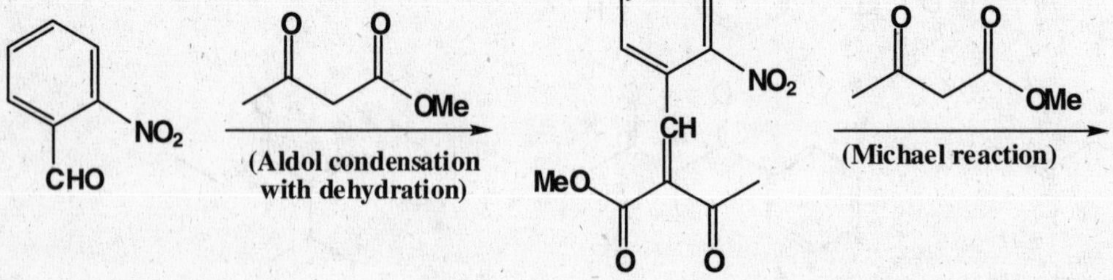

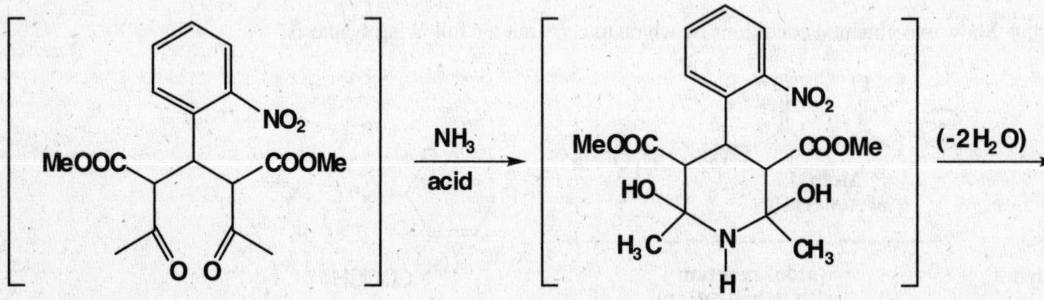

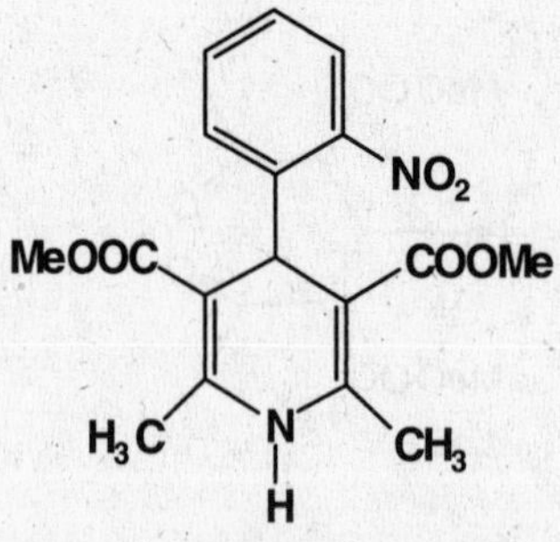

Nifedipine

Problem 19.56 The compound 3,5,5-trimethyl-2-cyclohexenone can be synthesized using acetone and ethyl acetoacetate as sources of carbon atoms. New carbon-carbon bonds in this synthesis are formed by a combination of aldol reactions and Michael reactions. Show reagents and conditions by which this synthesis might be accomplished.

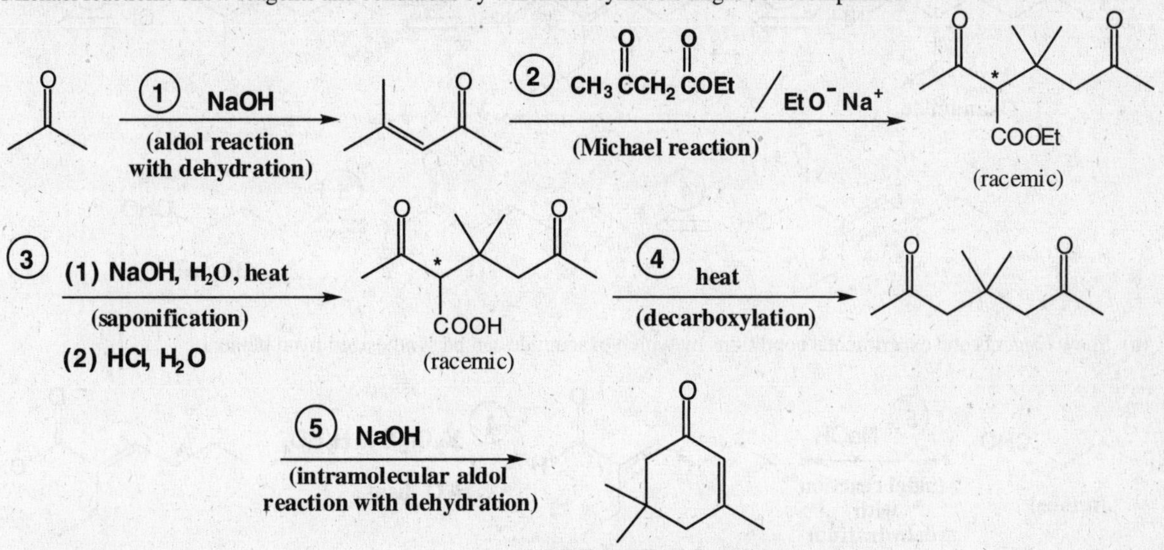

① **NaOH**
(aldol reaction with dehydration)

② CH₃CCH₂COEt / EtO⁻Na⁺
(Michael reaction)

③ (1) NaOH, H₂O, heat
(saponification)
(2) HCl, H₂O

④ heat
(decarboxylation)

⑤ **NaOH**
(intramolecular aldol reaction with dehydration)

3,5,5-Trimethyl-2-cyclohexanone

Problem 19.57 The following β-diketone can be synthesized from cyclopentanone and an acid chloride using an enamine reaction.

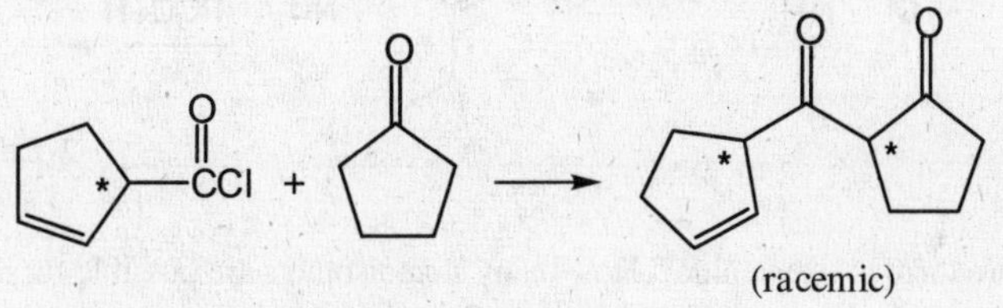

(racemic)

(a) Propose a synthesis of the starting acid chloride from cyclopentene.

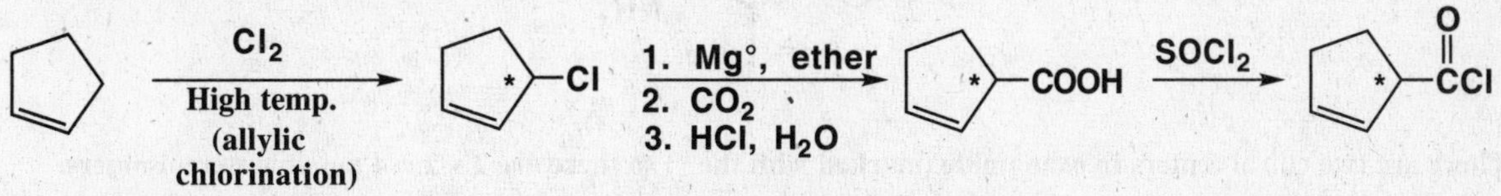

Cl₂
High temp.
(allylic chlorination)

1. **Mg°, ether**
2. **CO₂**
3. **HCl, H₂O**

SOCl₂

(b) Show the steps in the synthesis of the β-diketone using a morpholine enamine.

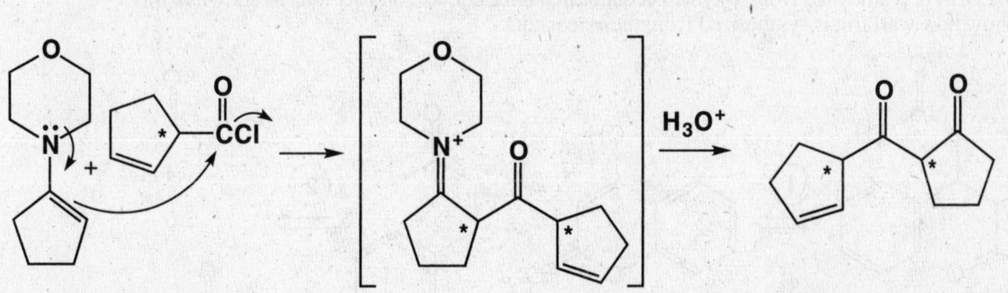

H₃O⁺

A racemic mixture of stereoisomers will be produced.

Problem 19.58 Oxanamide is a mild sedative belonging to a class of molecules called oxanamides. As seen in this retrosynthetic scheme, the source of carbon atoms for the synthesis of oxanamide is butanal.

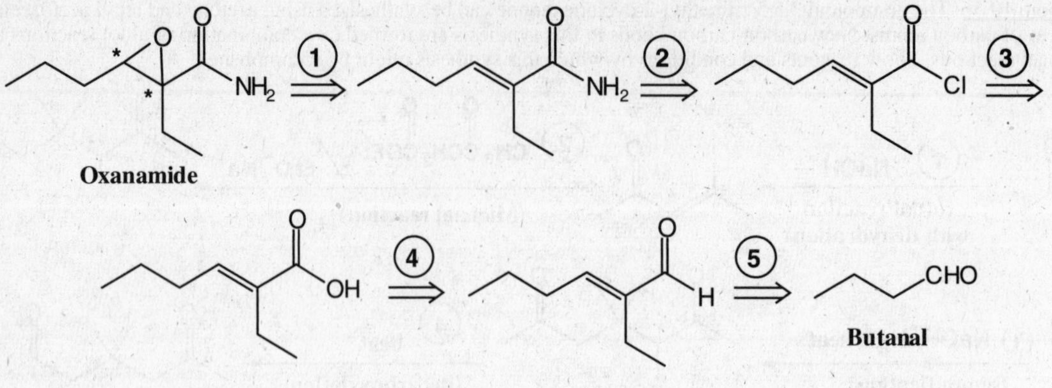

Oxanamide

(a) Show reagents and experimental conditions by which oxanamide can be synthesized from butanal.

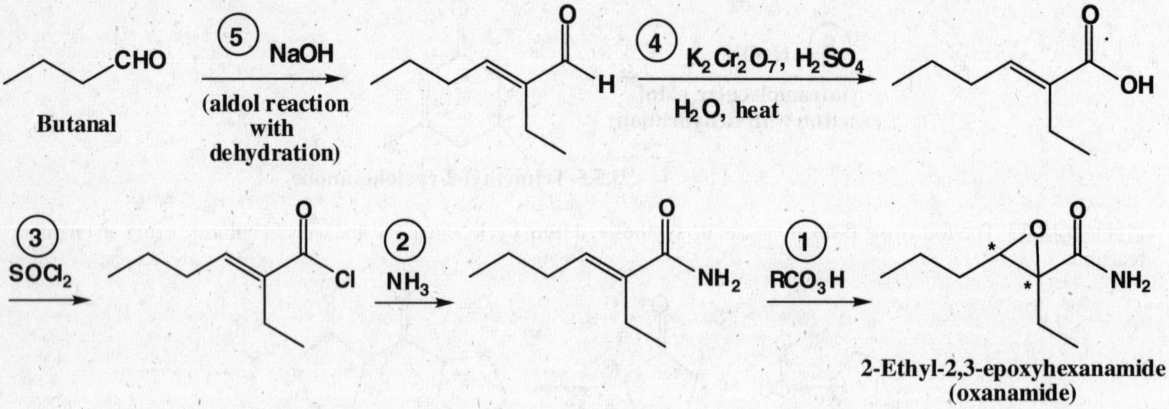

2-Ethyl-2,3-epoxyhexanamide
(oxanamide)

(b) How many chiral centers are there in oxanamide? How many stereoisomers are possible for this compound?

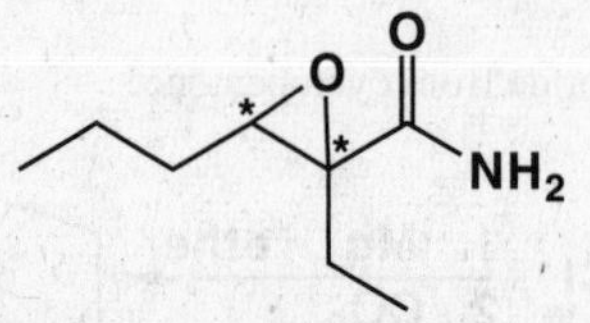

There are two chiral centers in oxanamide (marked with the *) so there are 2 x 2 = 4 possible stereoisomers.

Problem 19.59 The widely used anticoagulant warfarin (see Chemical Connections: "From Moldy Clover to a Blood Thinner" in Chapter 18) is synthesized from 4-hydroxycoumarin, benzaldehyde, and acetone as shown in this retrosynthesis. Show how warfarin is synthesized from these reagents.

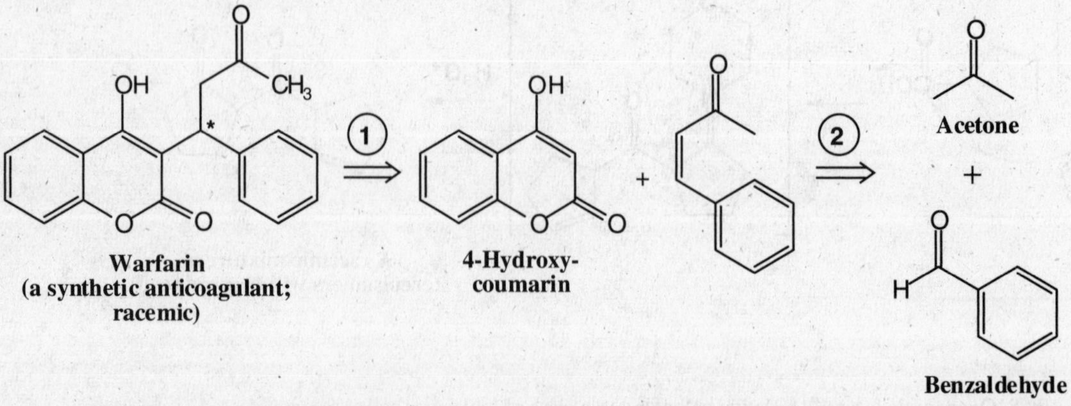

Warfarin
(a synthetic anticoagulant;
racemic)

4-Hydroxy-coumarin

Acetone

Benzaldehyde

The synthesis can be accomplished through the combination of a crossed aldol reaction with dehydration between benzaldehyde and acetone, followed by a Michael reaction with 4-hydroxycoumarin.

Benzaldehyde + **Acetone** → (2) NaOH / **crossed aldol reaction with dehydration**

4-Hydroxy-coumarin + ... (1) **Michael reaction** →

proton transfer and **keto-enol tautomerization** →

Warfarin (a synthetic anticoagulant)

One new chiral center is created, so a racemic mixture will be produced.

Problem 19.60 Following is a retrosynthetic analysis for an intermediate in the industrial synthesis of vitamin A.

(needed for the synthesis of vitamin A) (1) ⟹ ... + Cl + **Ethyl acetoacetate** COOEt

(2) ⟹ **2-Methyl-1,3-butadiene (Isoprene)** + HCl

(a) Addition of one mole of HCl to isoprene gives 4-chloro-2-methyl-2-butene as the major product. Propose a mechanism for this addition and account for its regioselectivity.

Step 1: Electrophilic addition-add a proton.

A resonance stabilized allylic cation

Step 2: Make a bond between a nucleophile and an electrophile.

The reaction mechanism involves a 1,4 conjugate addition to the alkene (Section 23.1). The key to understanding the observed regiochemistry is to analyze the relative stabilities of the two possible allylic cation intermediates produced by protonation at either end of the molecule. Protonation at the 1 position gives the following resonance stabilized allylic cation.

More stable allylic cation

Protonation at the 4 position gives the following resonance stabilized allylic cation.

Less stable allylic cation

The upper allylic cation is the more stable, because it has a considerable contribution from a 3° allylic cation contributing structure. The more stable cation leads to the major product 4-chloro-2-methyl-2-butene. Note that the chlorine atom ends up on the terminal atom, because this allows formation of the more stable internal alkene.

(b) Propose a synthesis of the vitamin A precursor from this allylic chloride and ethyl acetoacetate.

Problem 19.61 Following are the steps in one of the several published synthesis of frontalin, a pheromone of the western pine beetle.
(a) Propose reagents for Steps 1-8.

1. H_3O^+ (hydrolysis to carboxylic acid)

2. $SOCl_2$ (conversion to acid chloride)

3. $(CH_3)_2CuLi$, ether, -78°C

4. H_2O ⑥

⑦ RCO_3H

⑧ H_3O^+

cyclizes under these conditions

Two new chiral centers are created, but because of the bicyclic structure, there are only two possible stereoisomers of the product. These stereoisomers are enantiomers, so a racemic mixture of these enantiomers will be produced. Both enantiomers are drawn above.

(b) Propose a mechanism for the cyclization of the ketodiol from Step 8 to frontalin.

This reaction is an intramolecular acetal formation. In the following steps, the atoms are labeled to help keep track of them during the cyclization steps. It may be helpful to create molecular models to follow this reaction sequence.

Several steps:

intramolecular hemiacetal formation

Several steps:

acetal formation

The other enantiomer is formed by attack of the carbonyl from the other face in Step 1.

<u>Problem 19.62</u> 2-Ethyl-1-hexanol was needed for the synthesis of the sunscreen octyl *p*-methylcinnamate (See Chapter 23, Chemical Connections: "Sunscreens and Sunblocks"). Show how this alcohol could be synthesized
(a) by an aldol condensation of butanal

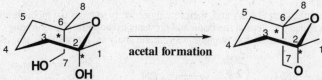

2 **Butanal**

NaOH

dehydration

H_2 / Pt

2-Ethyl-1-hexanol

(b) by a malonic ester synthesis starting with diethyl malonate.

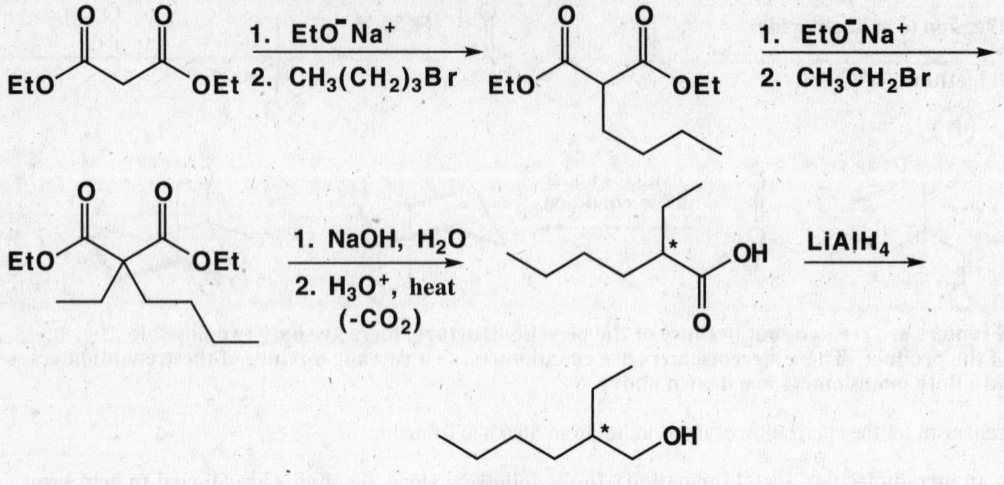

2-Ethyl-1-hexanol

Because a new chiral center is created, reactions in parts (a) and (b) both create a racemic mixture.

<u>Problem 19.63</u> Gabapentin, an anticonvulsant used in the treatment of epilepsy, is structurally related to the neurotransmitter 4-aminobutanoic acid (GABA).

Gabapentin

4-Aminobutanoic acid
(γ-Aminobutyric acid, GABA)

Gabapentin was designed specifically to be more lipophilic than GABA and, therefore more likely to cross the blood-brain barrier, the lipid-like protective membrane that surround the capillary system in the brain and prevents hydrophilic (water-loving) compounds from entering the brain by passive diffusion. Given the following retrosynthetic analysis, propose a synthesis for gabapentin.

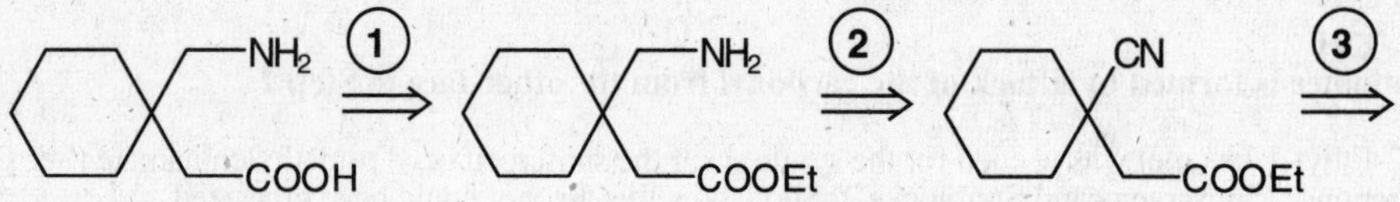

Gabapentin

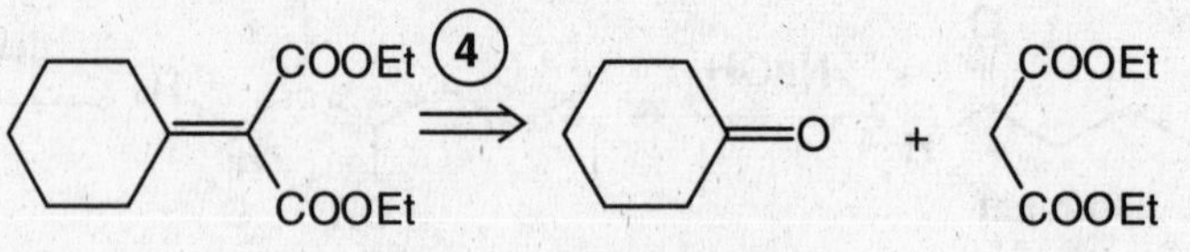

Cyclohexanone

Diethyl malonate

(4) When treated with sodium ethoxide, diethyl malonate is converted to its enolate anion (Section 19.7) and then, in a carbonyl condensation related to the aldol reaction (Section 19.2) and the Claisen condensation (Section 19.3), adds to the carbonyl carbon of cyclohexanone to form a tetrahedral carbonyl addition compound. Dehydration of this addition compound gives an α,β–unsaturated diester.

Cyclohexanone   **Diethyl malonate**

(3) This step is really four different reactions, all of which can be seen once it is realized that a decarboxylation step is required to set up a Michael reaction with NaCN. The sequence begins with hydrolysis of the diester in aqueous base followed by acidification with HCl (Section 18.4). Heating the β-dicarboxylic acid results in decarboxylation (Section 17.9) and gives an α,β-unsaturated carboxylic acid. Treatment of the carboxylic acid with ethanol in the presence of a *p*-toluenesulfonic acid catalyst (Fischer esterification, Section 17.7) coverts the carboxyl group to an ethyl ester. Michael addition (Section 19.8) of cyanide ion to the α,β-unsaturated ester gives a β-cyanoester.

(2) Treatment of the cyano group with hydrogen over a platinum-on-charcoal catalyst reduces the cyano group to a primary amine (Section 18.10).

(1) Hydrolysis of the ester group (Section 18.4) using either aqueous NaOH or HCl gives gabapentin.

Gabapentin

Note that although gabapentin is shown in the problem as containing a primary amino groups and a carboxyl group, it is better represented as an internal salt resulting from proton transfer from the carboxyl group to the amino group.

<u>Problem 19.64</u> The following three derivatives of succinimide are anticonvulsants and have found use in the treatment of epilepsy, particularly petit mal seizures.

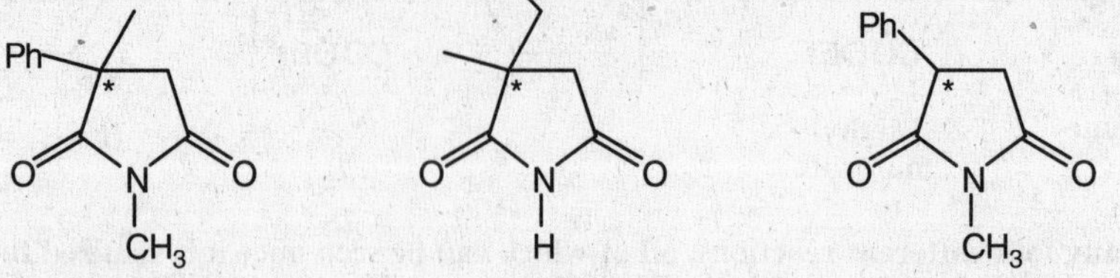

Methsuximide　　　　　**Ethosuximide**　　　　　**Phensuximide**
(racemic)　　　　　　　　(racemic)　　　　　　　　(racemic)

Following is a synthesis of phensuximide.

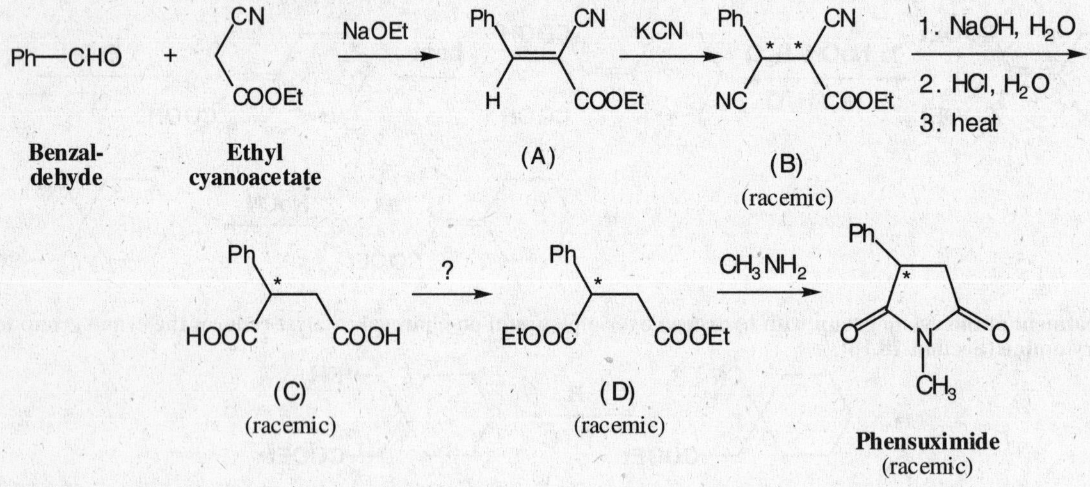

(a) Propose a mechanism for the formation of (A).

The mechanism involves formation of an enolate. Note that a cyano group has about the same enolate stabilizing capability as an ester group, so ethyl cyanoacetate has about the same pK_a as diethyl malonate. The enolate adds to the carbonyl carbon of cyclohexanone to form a tetrahedral carbonyl addition compound in analogy to a carbonyl condensation related to the aldol reaction (Section 19.2) and the Claisen condensation (Section 19.3). Dehydration of this addition compound gives the α,β-unsaturated cyanoester.

Step 1: Take a proton away.

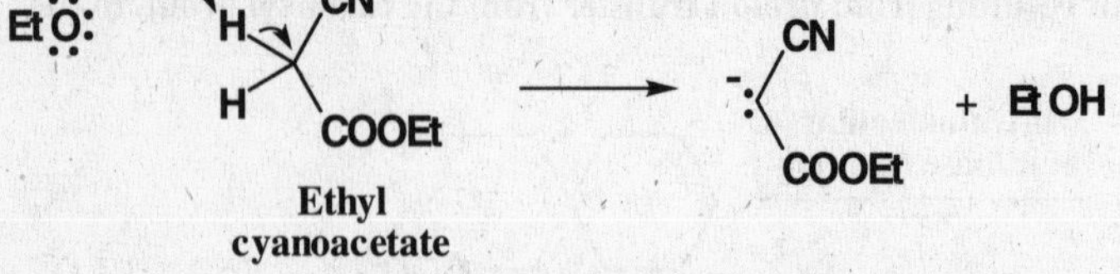

Ethyl cyanoacetate

Step 2: Make a bond between a nucleophile and an electrophile.

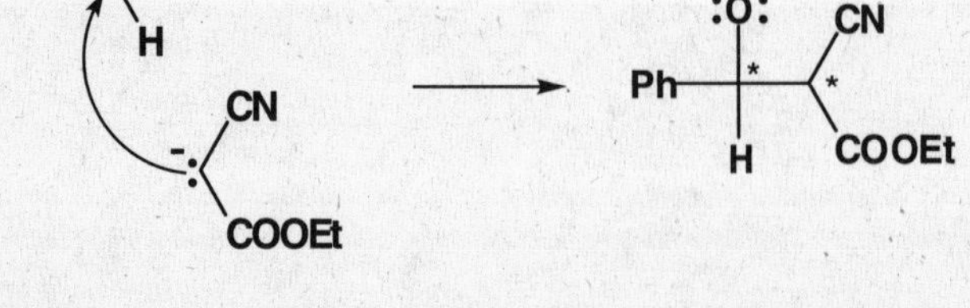

Step 3: Take a proton away.

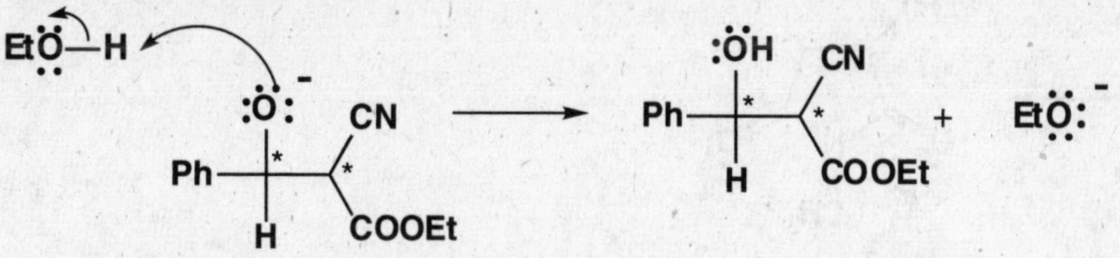

Step 4: Dehydration.

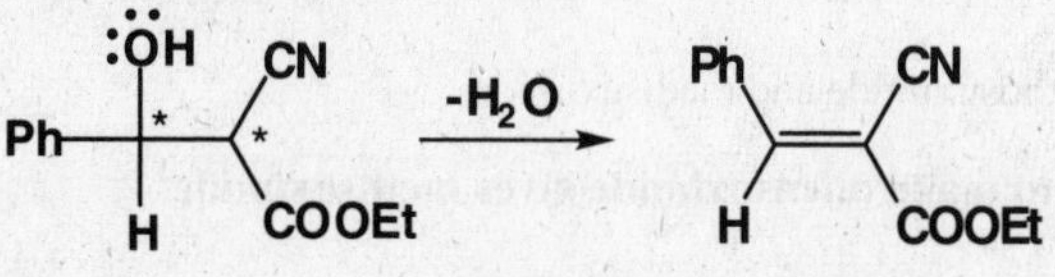

(A)

(b) What (person's) name is given to this type of reaction involved in the conversion of (A) to (B)?

Conversion of (A) to (B) is a Michael reaction in which the cyanide anion is the nucleophile.

(c) Describe the chemistry involved in the conversion of (B) to (C). You need not present detailed mechanisms. Rather, state what is accomplished by treating (B) first with NaOH and then with HCl followed by heating.

The aqueous base cleaves the ester, the acid hydrolyzes the nitriles to carboxyl groups, and heating leads to decarboxylation of the β-diacid to give intermediate (C).

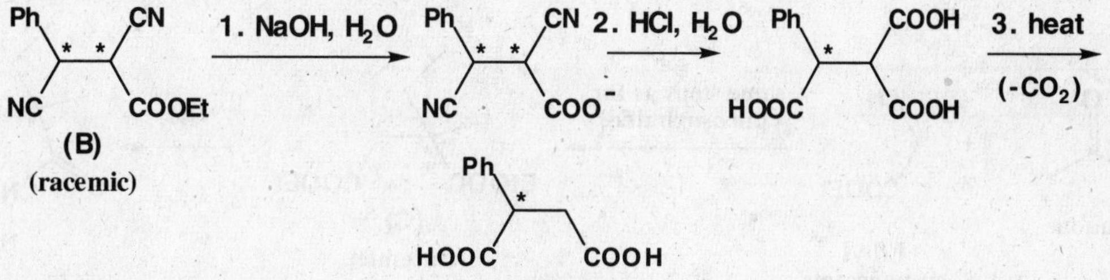

(B)
(racemic)

(C)
(racemic)

(d) Propose experimental conditions for the conversion of (C) to (D).

The (C) to (D) transformation can be accomplished with a Fischer esterification using EtOH and acid.

(e) Propose a mechanism for the conversion of (D) to phensuximide.

The mechanism involves two consecutive amide forming reactions.

Several Steps:

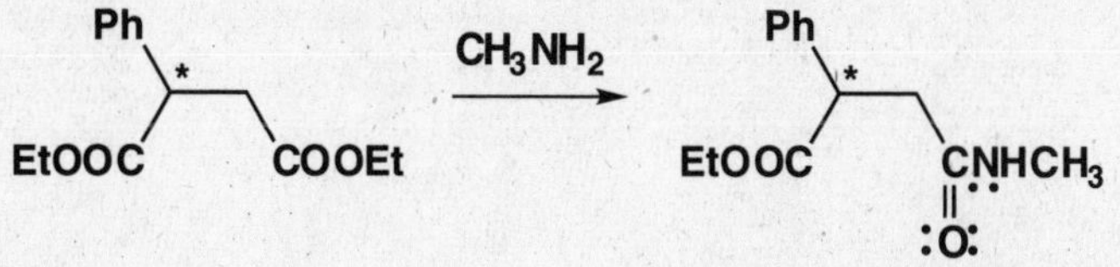

Several steps:

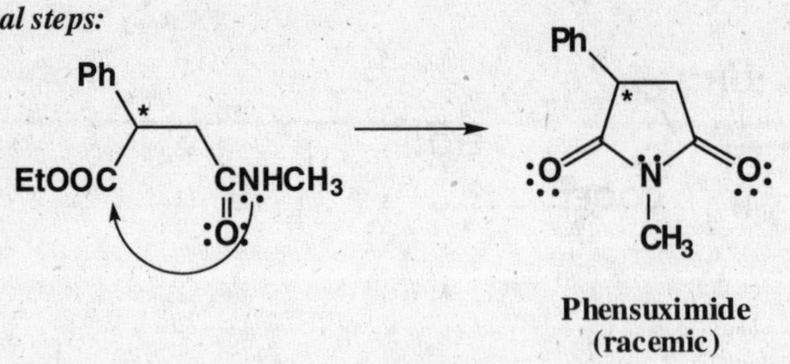

**Phensuximide
(racemic)**

(f) Show how this same synthetic strategy can be used to prepare ethosuximide and methsuximide.

The use of acetophenone, followed by the same scheme used to make phensuximide gives methsuximide.

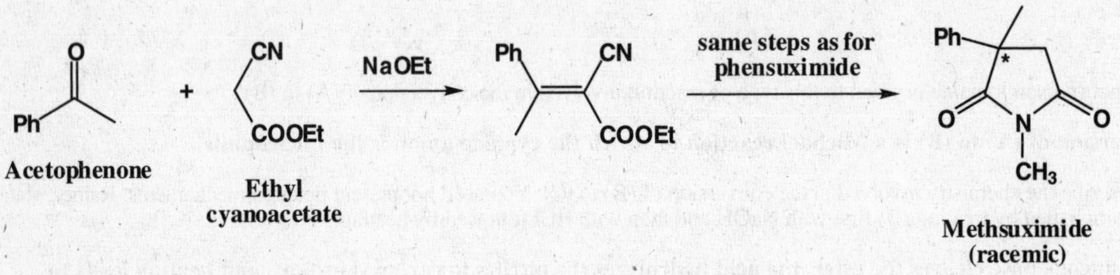

Acetophenone   **Ethyl
cyanoacetate**

**same steps as for
phensuximide**

**Methsuximide
(racemic)**

Starting with 2-butanone in the first step and ammonia in the last step gives ethosuximide.

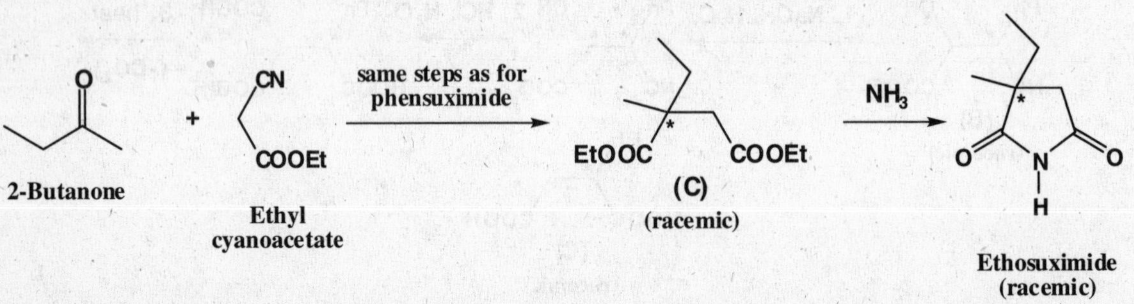

2-Butanone   **Ethyl
cyanoacetate**

**same steps as for
phensuximide**

**(C)
(racemic)**

**Ethosuximide
(racemic)**

(g) Of these three anticonvulsants, one is considerably more acidic than the other two. Which is the most acidic compound? Estimate its pKa and account for its acidity. How does its acidity compare with that of phenol? With that of acetic acid?

Only ethosuximide has a hydrogen on the imide nitrogen. It is the most acidic and has an acidity comparable to that of succinimide (pK_a 11, Section 18.2). The imide anion is stabilized by resonance interaction with the carbonyl groups on either side of it. This is about as acidic as phenol (pK_a 10-11), but not as acidic as acetic acid (pK_a 4.75).

Problem 19.65 The analgesic meperidine (Demerol) was developed in the search for analgesics without the addictive effects of morphine. As shown in these structural formulas, it represents a simplification of morphine's structure.

Morphine   **Meperidine**

Meperidine is prepared by treating phenylacetonitrile with one mole of *bis*(N-2-chloroethyl)methylamine (a nitrogen mustard) in the presence of two moles of sodium hydride to give (A). Refluxing (A) with concentrated sodium hydroxide followed by neutralization of the reaction mixture with dilute HCl gives (B). Treating (B) with ethanol in the presence of one equivalent of HCl gives meperidine as its hydrochloride salt.

Phenylacetonitrile

(a) Propose structural formulas for (A) and (B).

An α-hydrogen of a nitrile is weakly acidic, with a pK_a of approximately 25. See the solution to 19.64(a) for a discussion of the acidity of nitrile α-hydrogens. Treatment of the nitrile with sodium amide, a very strong base, gives an anion, which then displaces chlorine from the nitrogen mustard (Section 9.9). The process is then repeated to displace the second chlorine and close the six-membered ring.

Hydrolysis of the nitrile of (A) gives (B), the carboxylic acid.

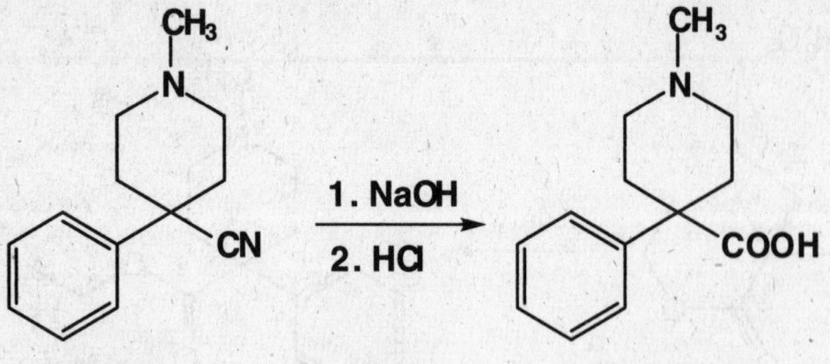

$C_{13}H_{16}N_2$

(A)

$C_{13}H_{17}NO_2$

(B)

(b) Propose a mechanism for the formation of (A).

Step 1: Take a proton away.

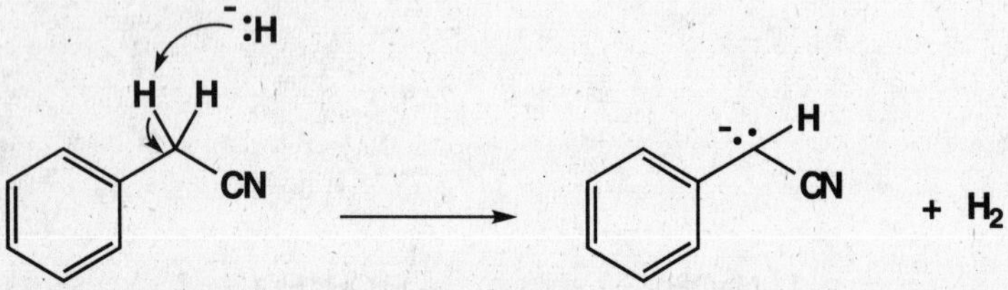

Step 2: Make a bond between a nucleophile and an electrophile.

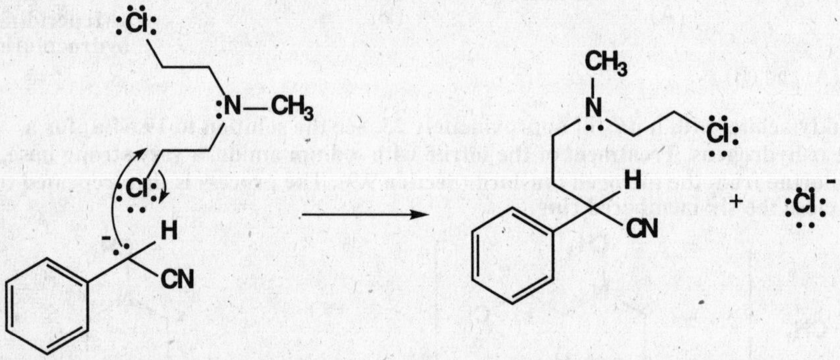

Step 3: Take a proton away.

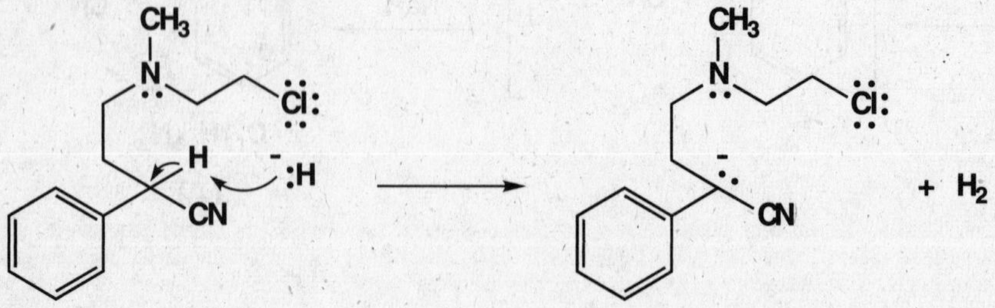

Step 4: *Make a bond between a nucleophile and an electrophile and simultaneously break a bond to give stable molecules or ions.*

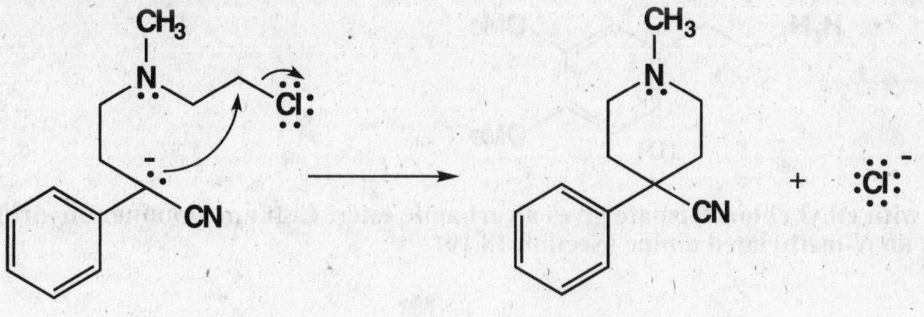

$C_{13}H_{16}N_2$

(A)

Problem 19.66 Verapamil (Effexor), a coronary artery vasodilator, is used in the treatment of angina caused by insufficient blood flow to cardiac muscle. Even though its effect on coronary vasculature tone was recognized over 30 years ago, it has only been more recently that its role as a calcium channel blocker has become understood. Below is a retrosynthetic analysis leading to a convergent synthesis: it is convergent because (A) and (B) are made separately and then combined (i.e. the route converges) to give the final product. Convergent syntheses are generally much more efficient than those in which the skeleton is built up stepwise.

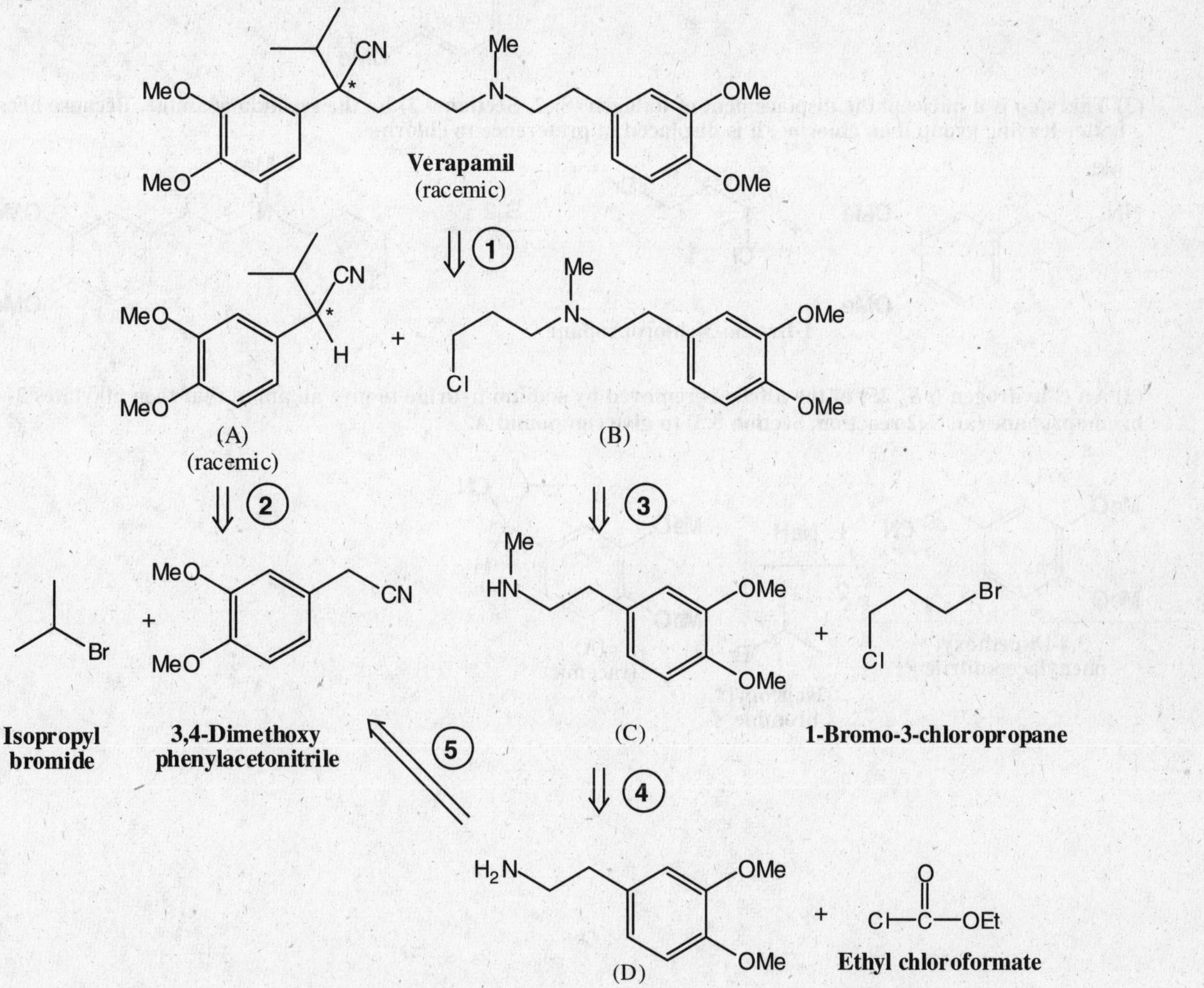

Verapamil
(racemic)

(A)
(racemic)

(B)

Isopropyl bromide

3,4-Dimethoxy phenylacetonitrile

(C)

1-Bromo-3-chloropropane

(D)

Ethyl chloroformate

(a) Given this retrosynthetic analysis, propose a synthesis for verapamil from the four named starting materials.

(5) Lithium aluminum hydride reduction of the cyano group gives a primary amine (Section 18.10)

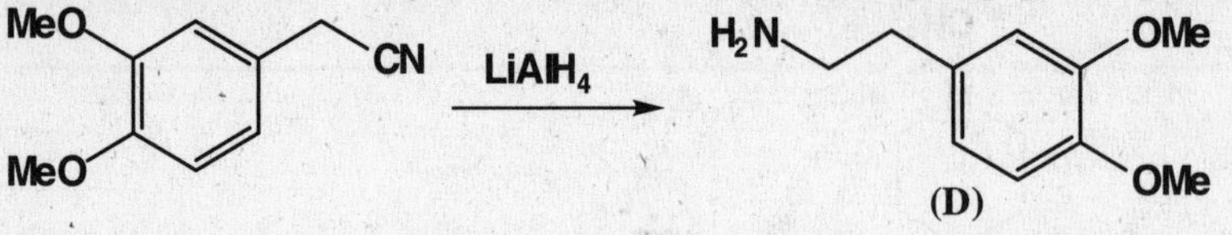

(4) Treatment of the primary amine with ethyl chloroformate gives a carbamic ester. Lithium aluminum hydride reduction of the carbamic ester gives an *N*-methylated amine (Section 18.10).

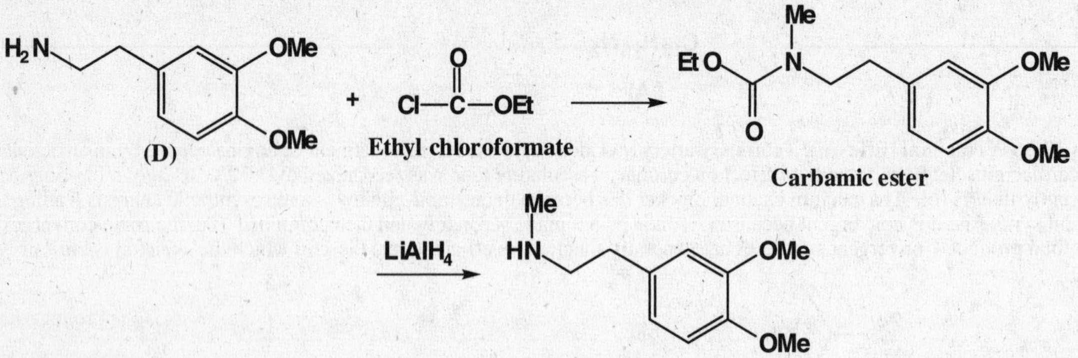

(3) This step is a nucleophilic displacement of halogen (S$_N$2, Section 9.3) by the secondary amine. Because bromine is a better leaving group than chlorine, it is displaced in preference to chlorine.

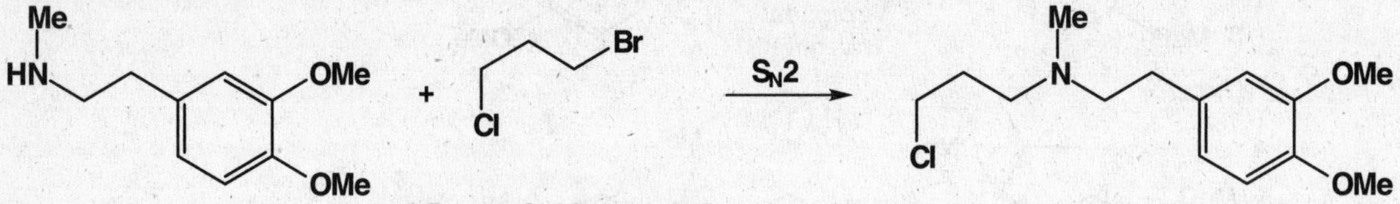

1-Bromo-3-chloropropane

(2) An α-hydrogen (*pK$_a$* 25) of the nitrile is removed by sodium hydride to give an anion that then alkylates 2-bromopropane (an S$_N$2 reaction, Section 8.3) to give compound A.

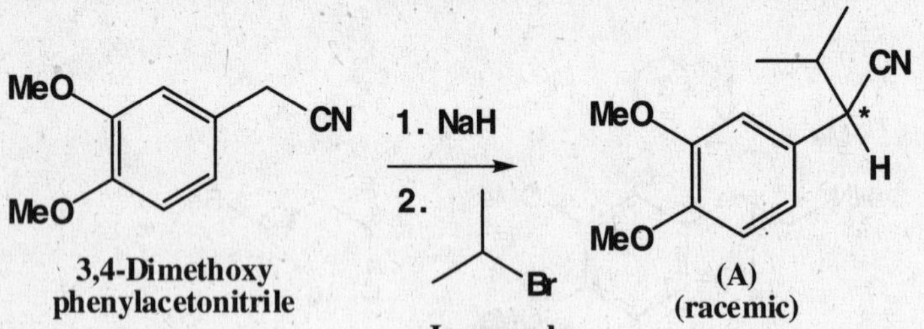

3,4-Dimethoxy
phenylacetonitrile

Isopropyl
bromide

(A)
(racemic)

(1) Treatment of (A) with sodium hydride followed by addition of (B) gives verapamil.

(b) It requires two steps to convert (D) to (C). The first is treatment of (D) with ethyl chloroformate. What is the product of this first step? What reagent can be used to convert this product to (C)?

See (4) above. Treatment of the primary amine with ethyl chloroformate gives a carbamic ester. Lithium aluminum hydride reduction of the carbamic ester gives an *N*-methylated amine (Section 18.10).

(c) How do you account for the regioselectivity of the nucleophilic displacement involved in converting (C) to (B)?

See (3) above. This step is a nucleophilic displacement of halogen (S$_N$2, Section 9.3) by the secondary amine. Because bromine is a better leaving group than chlorine, it is displaced in preference to chlorine.

<u>Problem 19.67</u> Based on this retrosynthetic analysis, propose a synthesis of the anticoagulant (inhibits blood clotting) diphenadione.

Diphenadione   **Diethyl phthalate**

Because of its anticoagulant activity for blood, this compound is used as a rodenticide. For the story of the discovery of the anticoagulant dicoumarin, see Chemical Connections: "From Moldy Clover to a Blood Thinner" in Chapter 18.

The key to this synthesis is that the methyl ketone reacts as an enolate on the methyl group, as opposed to the benzylic α-carbon, because methyl group is far more accessible to the electrophilic ethyl ester. Note also that there are no α-hydrogens on diethyl phthalate that could make alternative enolates that would interfere with the desired reaction.

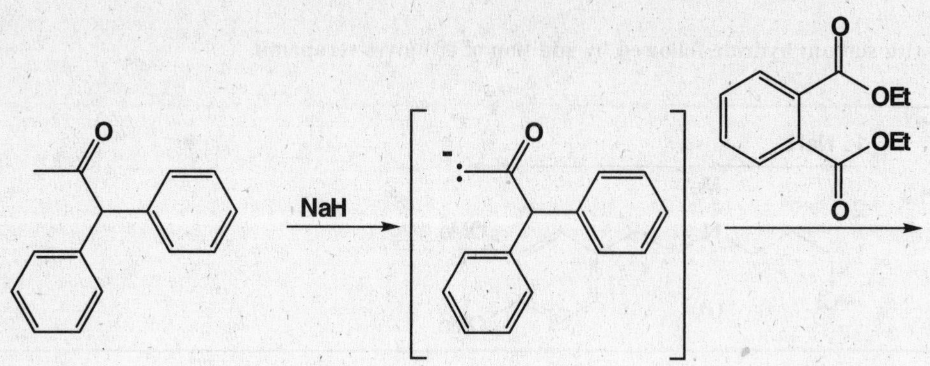

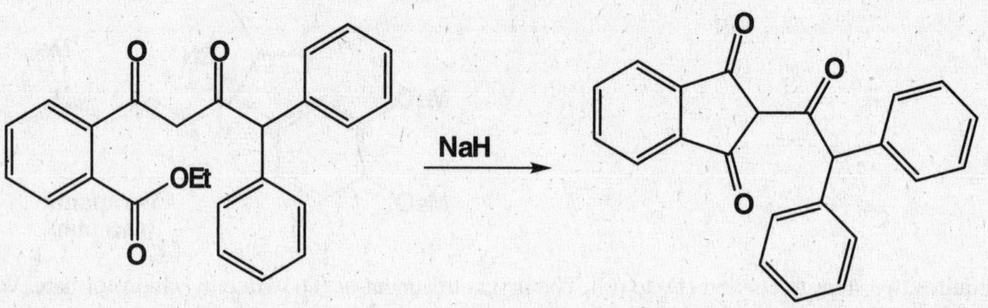

<u>Problem 19.68</u> Following are two possible retrosynthetic analyses for the anticholinergic drug cycrimine. Fill in the details of each potential synthesis.

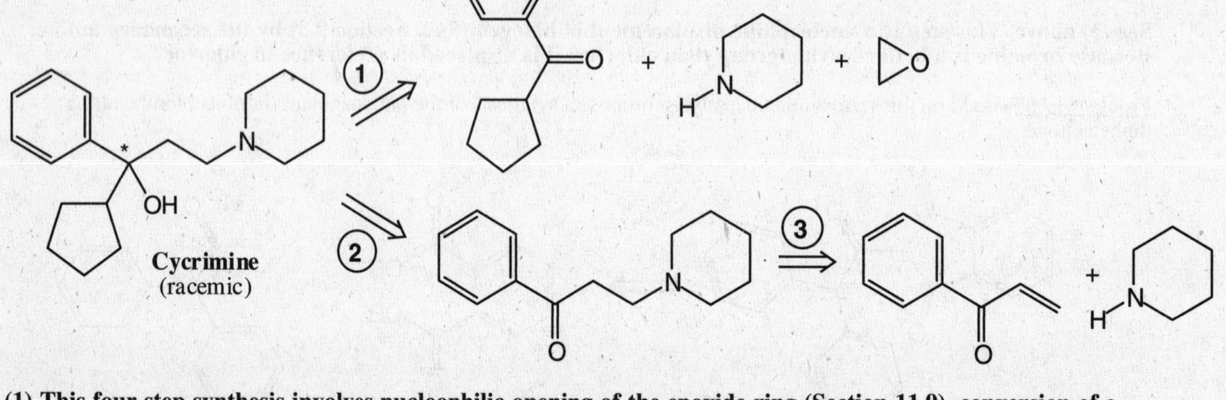

(1) This four step synthesis involves nucleophilic opening of the epoxide ring (Section 11.9), conversion of a primary alcohol to a alkyl chloride (Section 10.5), preparation of a Grignard reagent (Section 15.1), and its reaction with a ketone (Section 16.5).

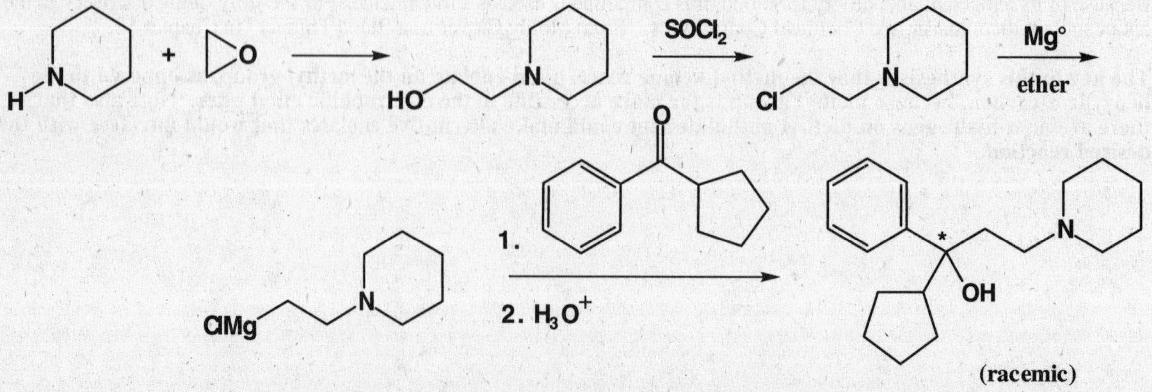

(2),(3) This two step synthesis used a Michael addition (Section 19.8) followed by a Grignard reaction (Section 16.5).

(racemic)

Problem 19.69 Show how the tranquilizer valnoctamide can be synthesized using diethyl malonate as the source of the carboxamide group.

2-Ethyl-3-methylpentanamide
(**Valnoctamide**, racemic)

The first steps result in dialkylation of diethyl malonate (Section 19.7). Base-promoted hydrolysis followed by acidification results in hydrolysis of the β-diester to a β-dicarboxylic acid. Heating the β-dicarboxylic acid brings about decarboxylation. Treatment of the carboxylic acid with thionyl chloride (Section 17.8) and then ammonia (Section 18.6) converts the carboxyl group to an amide and completes the synthesis of racemic valnoctamide.

2-Ethyl-3-methylpentanamide
(Valnoctamide, racemic)

<u>Problem 19.70</u> In Problem 7.28 we saw this two-step sequence in Johnson's synthesis of the steroid hormone progesterone. Propose a structural formula for the intermediate formed in Step 3, and a mechanism for its conversion in Step 4 to progesterone.

Ozonolysis of the one carbon-carbon double bond gives the diketone intermediate shown.

The mechanism of step 4 is an aldol reaction followed by dehydration.

Problem 19.71 Monensin, a polyether antibiotic, was isolated from a strain of *Streptomyces cinamonensis* in 1967, and its structure was determined shortly thereafter.

Monensin

This molecule exhibits a broad-spectrum anticoccidial activity and, since its introduction in 1971, has been used to treat coccidial infections in poultry and as an additive in cattle feed. In the synthesis of monensin, Y. Kishi chose to create the molecule in sections and then join them to create the target molecule. Following is an outline of the steps by which he created the seven-carbon-chain building block on the left side of the molecule. Propose a reagent or reagents for Steps 1–14. Note that this fragment contains five chiral centers. You do not have to to predict or rationalize the stereochemistry of each step, but rather only to propose a reagent or type of reagent to bring about each step.

The synthetic reagents are listed over the arrows. The key element of the synthesis are two cycles of the steps comprised of first oxidation of an alcohol to an aldehyde with PCC (4) and (11) (Section 10.8), a Wittig reaction (5) and (12) (Section 16.6), reduction of the terminal ester to an alcohol with LiAlH₄ (6) and (13) (Section 18.10), and hydroboration of a carbon-carbon double bond on the less substituted carbon with 1. BH₃, 2. H₂O₂, NaOH (8) and (14) (Section 6.4). The other steps include methylation of an enolate with CH₃I (1) (Section 9.3), hydrolysis of a nitrile in aqueous acid (2) (Section 18.4), reduction of a carboxyl group to an alcohol with LiAlH₄ (3) (Section 17.6), protection of a primary alcohol as a benzyl ether with NaH and benzyl chloride (BnCl) (7) (Section 11.4), protection of a hydroxyl group with NaH and CH₃I (9) (Section 9.3), and removal of the benzyl ether protecting group with H₂/Pd (10) (Section 21.5).

Multi-step Synthesis

Problem 19.72 Show how to convert 2-methylpropylbenzene into 4-phenyl-3-buten-2-one. You must use 2-methylpropylbenzene as the source of all carbon atoms in the target molecule. Show all reagents and all molecules synthesized along the way.

Recognize that the product has the same number of carbons as the starting material, but the connectivity pattern has changed. Therefore, propose that a carbon-carbon bond must first be broken then a new one formed. Notice also that the product contains an α,β-unsaturated ketone, the recognition element of an aldol reaction followed by dehydration. Acetone and benzaldehyde are the required carbonyl species for the appropriate crossed aldol reaction, and they can

be derived from ozonolysis of the corresponding phenyl alkene. The phenyl alkene can be derived from 2-methylpropylbenzene using the familiar free radical halogenation followed by E2 elimination sequence.

Problem 19.73 Show how to convert ethanol into 2-pentanone. You must use ethanol as the source of all carbon atoms in the target molecule. Show all reagents and all molecules synthesized along the way.

Recognize that the product has five carbons, but the starting material has only two. Therefore, propose that at least two different carbon-carbon bonds must be made, and that one must also be broken. Notice also that the product contains a methyl ketone group substituted with a primary alkyl group on one α-carbon. This is the recognition element of the acetoacetic ester synthesis sequence in which a two carbon alkyl halide is used. The required ethyl acetoacetate can be prepared from ethanol using the sequence of oxidation to acetic acid, conversion to acetyl chloride using SOCl$_2$, conversion of acetyl chloride to ethyl acetate using ethanol, then finally preparation of ethyl acetoacetate from ethyl acetate using a Claisen condensation reaction. The required ethyl halide can be derived in one step using PBr$_3$. As always, it would have been perfectly acceptable to use a Fischer esterification to make ethyl acetate from acetic acid and ethanol directly.

Problem 19.74 Show how to convert ethanol, formaldehyde, and acetone into racemic ethyl 2-acetyl-5-oxohexanoate. You must use ethanol, formaldehyde, and acetone as the source of all carbon atoms in the target molecule. Show all reagents and all molecules synthesized along the way.

Recognize that the product has ten carbons, but the required starting materials have only one, two, or three. Therefore, propose that several different carbon-carbon bonds must be made. Notice also that the product contains a δ-dicarbonyl group, the recognition element of a Michael reaction. The required components of the Michael reaction are ethyl acetoacetate (nucleophile) and 3-buten-2-one (Michael acceptor). Ethyl acetoacetate can be made from ethanol using the sequence of oxidation to acetic acid, conversion to acetyl chloride using $SOCl_2$, conversion of acetyl chloride to ethyl acetate using ethanol, then finally preparation of ethyl acetoacetate from ethyl acetate using a Claisen condensation reaction. Notice that the required 3-buten-2-one Michael acceptor is an α,β-unsaturated ketone, the recognition element of an aldol reaction followed by dehydration, which can be made from the crossed aldol of acetone and formaldehyde. As always, it would have been perfectly acceptable to use a Fischer esterification to make ethyl acetate from acetic acid and ethanol directly.

<u>Problem 19.75</u> Show how to convert cyclohexane and ethanol into racemic ethyl 2-oxocyclopentanecarboxylate. You must use ethanol and cyclohexane as the source of all carbon atoms in the target molecule. Show all reagents and all molecules synthesized along the way.

Recognize that the product has eight carbons, which is the sum of the carbons in the starting materials. Importantly, there is an ester linkage in the product linking the six and a two carbon pieces, not a new carbon-carbon bond. On the other hand, the cyclohexane ring of the starting material has been contracted to a five-membered ring in the product, indicating initial cleavage of a carbon-carbon bond of cyclohexane, that is reformed as the smaller five-membered ring. Notice also that the product contains a ring and a β-keto ester group, the recognition element of a Dieckmann condensation. The required starting material for the appropriate Dieckmann condensation is the diethyl ester of a six-carbon diacid (diethyl adipate). The six-carbon diester can be derived from the corresponding diacid chloride, which in turn, can be made by treating the six-carbon diacid (adipic acid) with $SOCl_2$. Adipic acid can be synthesized through an oxidation of the corresponding six-carbon dialdehyde. The six-carbon dialdehyde can be derived from cyclohexane using the sequence of free radical halogenation followed by E2 elimination in strong base to give cyclohexene, which undergoes ozonolysis to give the required six-carbon dialdehyde.

Problem 19.76 Show how to convert cyclohexane and ethanol into racemic 2-acetylcyclohexanone. You must use ethanol and cyclohexane as the source of all carbon atoms in the target molecule. Show all reagents and all molecules synthesized along the way.

Recognize that the product has eight carbons, which is the sum of the carbons in the starting materials, so they must be connected through a new carbon-carbon bond. Notice also that the product contains a β-diketo group, the recognition element of an enamine acylation reaction. The required starting materials for the appropriate enamine acylation reaction are cyclohexanone and acetyl chloride. Acetyl chloride can be created from ethanol through the sequence of oxidation with chromic acid to give acetic acid followed by treatment with $SOCl_2$. The required cyclohexanone can by synthesized from cyclohexane using the sequence of free radical halogenation, then substitution of the halogen with an OH group to give cyclohexanol, which is then oxidized to cyclohexanone using H_2CrO_4. Note that the desired product could also be derived by treating cyclohexanone with LDA to give the enolate, that is acylated with acetyl chloride.

Problem 19.77 Show how to convert 2-oxepane and ethanol into 1-cyclopentenecarbaldehyde. You must use 2-oxepane as the source of all carbon atoms in the target molecule. Show all reagents and all molecules synthesized along the way.

Recognize that the product has six carbons like the starting material, but there is an ester linkage in the ring of the starting material, and a five-member carbon only ring of the product. Notice also that the product contains an α,β-unsaturated aldehyde, the recognition element of an aldol condensation followed by dehydration. Propose that the product is derived from an intramolecular aldol reaction followed by dehydration starting from a six-carbon dialdehyde. The six-carbon dialdehyde can be made from the starting lactone using the sequence of $LiAlH_4$ reduction to the diol followed by oxidation with PCC.

Reactions in Context

Problem 19.78 Atorvastatin (Lipitor) is a popular treatment for high cholesterol. See "The Importance of Hydrogen Bonding in Drug-Receptor Interactions" in Chapter 10 for more information about atorvastatin. One synthesis of atorvastatin involves the following enolate reaction. Draw the predominant product of this reaction, which gives an overall yield of 90%.

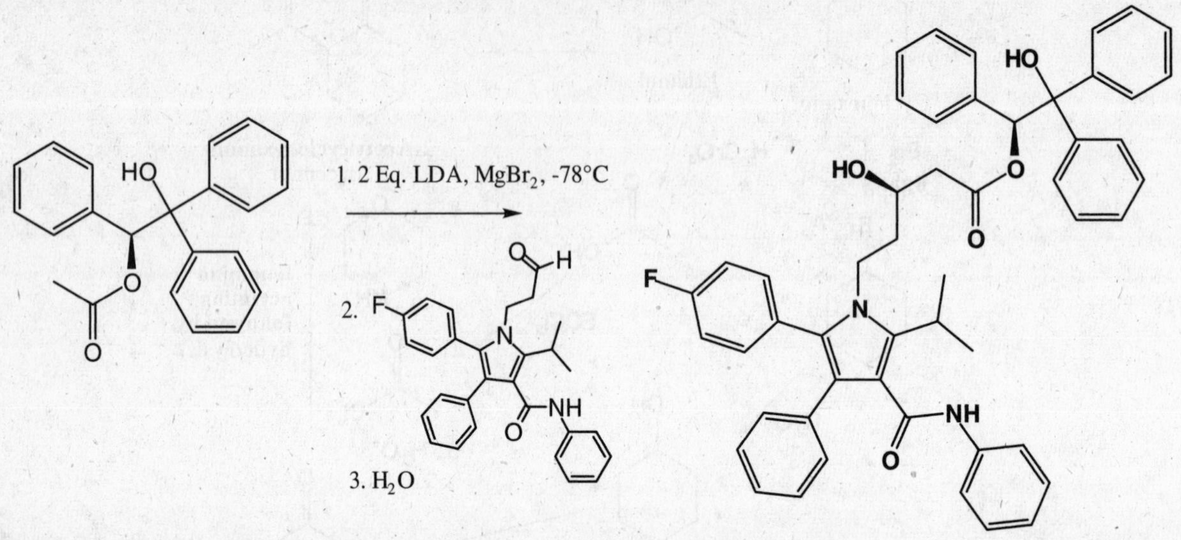

There are a couple of noteworthy aspects to this reaction. First, $MgBr_2$ is added to exchange with Li and make Mg enolates. This is helpful for controlling stereochemistry. Notice that the starting material is chiral and a single enantiomer is used. The product of this reaction is a 97:3 (94% ee) mixture of two enantiomers, not a racemic mixture. You don't have to be able to deduce which enantiomer is the predominant product, but be aware that being able to control the stereochemical outcome of a reaction by using a single enantiomer of a chiral starting material can save time and resources in the large-scale synthesis of chiral drugs.

Problem 19.79 E. J. Corey used the following reaction in a synthesis of thromboxane B_2. Predict the major product of the reaction. There are two possible products here. State why you think the pathway that creates the predominant product is favored under the conditions of the reaction.

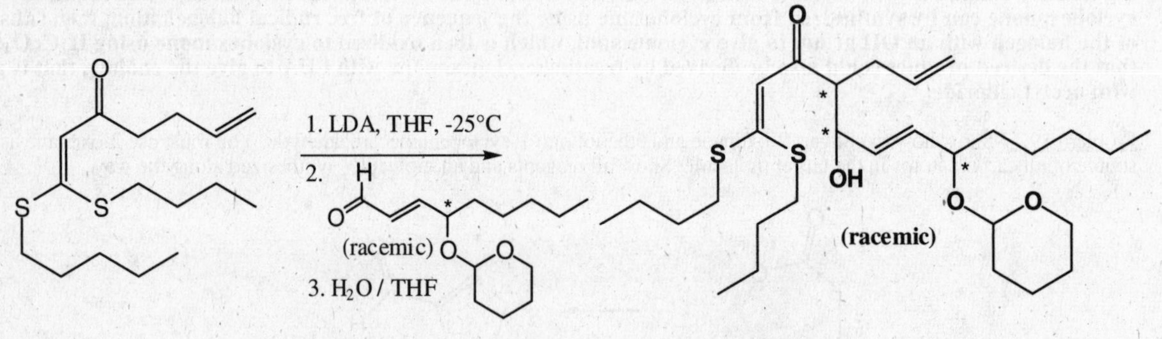

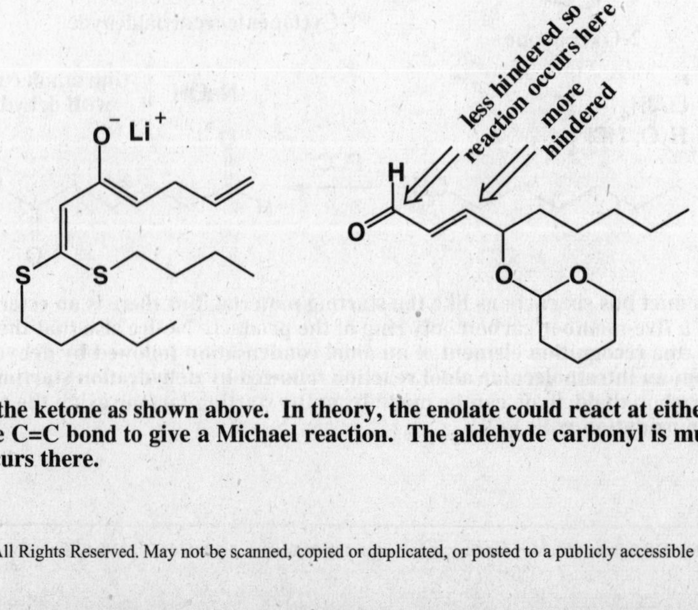

The enolate forms on the ketone as shown above. In theory, the enolate could react at either the aldehyde carbonyl to give an aldol, or at the C=C bond to give a Michael reaction. The aldehyde carbonyl is much less hindered, so reaction primarily occurs there.

Problem 19.80 The following molecule undergoes an intramolecular reaction in the presence of pyrrolidium acetate, the protonated form of pyrrolidine. Draw the product of this reaction, assuming that a dehydration reaction takes place.

Intramolecular reactions are favored that produce five or six-membered rings, as in this example.

Problem 19.81 Organocuprates predominantly react to give 1,4-addition products with α,β-unsaturated carbonyl species, while Grignard reagents often add to the carbonyl, in a process referred to as 1,2-addition. To increase the yield of 1,4-addition products CuI is added to convert an easily prepared Grignard reagent into a organocuprate reagent *in situ* (during the reaction). Predict the major product and stereochemistry of the following reaction, assuming the more stable chair product predominates.

1. ⟍⟍ MgBr
   CuI, THF, -30°C
2. Equilibration to more stable chair product in base

The predominant product chair conformation has both the methyl group and the alkyl group equatorial.

Raccoon butterflyfish *Chaetodon lunula*
Kauai, Hawaii

CHAPTER 20
Solutions to the Problems

<u>Problem 20.1</u> Which of these terpenes (Section 5.4) contains conjugated double bonds?

(a)

Geraniol

(b)

Limonene

(c)

HO

An aggregating
pheromone of
bark beetles

Only the aggregating pheromone of bark beetles (c) has conjugated double bonds.

<u>Problem 20.2</u> Estimate the stabilization gained due to conjugation when 1,4-pentadiene is converted to *trans*-1,3-pentadiene. Note that the answer is not as simple as comparing the heats of hydrogenation of 1,4-pentadiene and *trans*-1,3-pentadiene. Although the double bonds are moved from unconjugated to conjugated, the degree of substitution of one of the double bonds is also changed, in this case from a monosubstituted double bond to a *trans* disubstituted double bond. To answer this question, you must separate the effect that is the result of conjugation from that caused by a change in the degree of substitution.

1,4-Pentadiene   ***trans*-1,3-Pentadiene**

As stated in the question, there are actually two important issues to be addressed. It is best to isolate the two different interactions by considering analogous systems in Table 20.1. First, *trans*-1,3-pentadiene has an internal double bond while 1,4-pentadiene does not. Internal double bonds are lower in energy because they are more highly substituted. A good estimate of the stabilization energy provided by having an internal versus terminal double bond is 12 kJ/mol, calculated as the difference in energy between 1-butene (ΔH° = -127 kJ/mol) and *trans*-2-butene (ΔH° = -115 kJ/mol).

The overall difference in energy between 1,4-pentadiene (ΔH° = -254 kJ/mol) and *trans*-1,3-pentadiene (ΔH° = -226 kJ/mol) from Table 20.1 is 28 kJ/mol. Corrected by 12 kJ/mol for the presence of the internal double bond in *trans*-1,3-pentadiene leaves:

16 kJ/mol as the stabilization due to conjugation when 1,4-pentadiene is converted to *trans*-1,3-pentadiene.

<u>Problem 20.3</u> Predict the product(s) formed by addition of one mole of Br_2 to 2,4-hexadiene.

Predict both 1,2-addition and 1,4-addition. Two new chiral centers are created in each case leading to a racemic mixture of stereoisomers.

Br_2

Br   Br

Br   Br

1,2-addition   **1,4-addition**

<u>Problem 20.4</u> Wavelengths in ultraviolet-visible spectroscopy are commonly expressed in nanometers; wavelengths in infrared spectroscopy are sometimes expressed in micrometers. Carry out the following conversions.
(a) 2.5 μm to nanometers. **2.5 μm is equal to 2500 nm.**
(b) 200 nm to micrometers. **200 nm is equal to 0.2 μm.**

<u>Problem 20.5</u> The visible spectrum of β-carotene ($C_{40}H_{56}$, MW 536.89, the orange pigment in carrots) dissolved in hexane shows intense absorption maxima at 463 nm and 494 nm, both in the blue-green region. Because light of these wavelengths is absorbed by β-carotene, we perceive the color of this compound as that of the complement to blue-green, namely, red-orange.

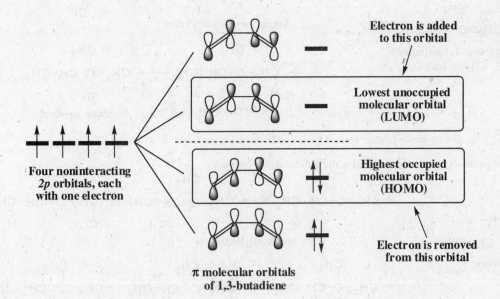

β-Carotene

λ_{max} 463 (log ε 5.10); 494 (log ε 4.77)

Calculate the concentration in milligrams per milliliter of β-carotene that gives an absorbance of 1.8 at 463 nm.

$$c = 1.8/(1.00 \text{ cm})(10^{5.10}) = 1.43 \times 10^{-5} \text{ mol/liter}$$

The molecular weight of β-carotene, $C_{40}H_{56}$, is (12 x 40) + (1 x 56) = 536 g/mol. Concentration in units of milligrams per milliliter is equal to the value of concentration of grams per liter, so:

$$c = (1.43 \times 10^{-5} \text{ mol/liter})(536 \text{ g/mol}) = 7.7 \times 10^{-3} \text{ g/liter}$$

$$\boxed{= 7.7 \times 10^{-3} \text{ milligram / milliliter.}}$$

Structure and Stability

<u>Problem 20.6</u> If an electron is added to 1,3-butadiene, into which molecular orbital does it go? If an electron is removed from 1,3-butadiene, from which molecular orbital is it taken?

When an electron is removed from a conjugated pi system, such as that in 1,3-butadiene, it is taken from the <u>h</u>ighest <u>o</u>ccupied <u>m</u>olecular <u>o</u>rbital (HOMO). When an electron is added to a conjugated π system, is added to the <u>l</u>owest <u>u</u>noccupied <u>m</u>olecular <u>o</u>rbital (LUMO).

<u>Problem 20.7</u> Draw all important contributing structures for the following allylic carbocations; then rank the structures in order of relative contributions to each resonance hybrid.

(a)

**Primary allylic
(lesser contribution)**   **Secondary allylic
(greater contribution)**

The secondary allylic cation makes the greater contribution.

(b) $CH_2=CH-CH=CH-CH_2^+$

$CH_2=CH-CH=CH-CH_2^+ \longleftrightarrow CH_2=CH-\overset{+}{CH}-CH=CH_2 \longleftrightarrow \overset{+}{CH_2}-CH=CH-CH=CH_2$

     Primary allylic   **Secondary allylic**   **Primary allylic**
   (lesser contribution)   **(greater contribution)**   **(lesser contribution)**

The secondary allylic cation makes the greater contribution.

(c)
$$CH_3-\underset{+}{\overset{CH_3}{\underset{|}{\overset{|}{C}}}}-CH=CH_2 \quad\quad CH_3-\underset{\underset{+}{}}{\overset{CH_3}{\underset{|}{\overset{|}{C}}}}-CH=CH_2 \longleftrightarrow CH_3-\overset{CH_3}{\underset{|}{\overset{|}{C}}}=CH-\overset{+}{CH_2}$$

                            Tertiary allylic   **Primary allylic**
                          (greater contribution) **(lesser contribution)**

The tertiary allylic cation makes the greater contribution.

Electrophilic Addition to Conjugated Dienes

<u>Problem 20.8</u> Predict the structure of the major product formed by 1,2-addition of HCl to 2-methyl-1,3-butadiene (isoprene).

A tertiary allylic cation is more stable than a secondary allylic cation. Assume that because the tertiary allylic cation is the more stable of the two, the activation energy is lower for its formation, and accordingly, it is formed at a greater rate than the secondary allylic cation. Therefore, the major product of 1,2-addition is 3-chloro-3-methyl-1-butene.

<u>Problem 20.9</u> Predict the major product formed by 1,4-addition of HCl to isoprene.

The major product of 1,4-addition is 1-chloro-3-methyl-2-butene.

<u>Problem 20.10</u> Predict the structure of the major 1,2-addition product formed by reaction of one mole of Br_2 with isoprene. Also predict the structure of the major 1,4-addition product formed under these conditions.

Following the reasoning developed in the previous problems, predict that the major product of 1,2-addition to isoprene is 3,4-dibromo-3-methyl-1-butene. One new chiral center is created, so a racemic mixture will be produced.

There is only one 1,4-addition product possible from the addition of bromine.

Problem 20.11 Which of the two molecules shown do you expect to be the major product formed by 1,2-addition of HCl to cyclopentadiene? Explain.

1,3-Cyclo-   **3-Chloro-**   **4-Chloro-**
pentadiene   **cyclopentene**   **cyclopentene**
                        (racemic)

The major product of 1,2-addition will be 3-chlorocyclopentene. This is because 3-chlorocyclopentene is the only product that will be derived from the more stable allylic cation intermediate. The 4-chlorocyclopentene derives from a less stable secondary cation intermediate, so it is formed in lesser amounts. A new chiral center is formed in 3-chlorocyclopentene, so it is actually formed as a racemic mixture.

Less stable
secondary cation

More stable   **Major product**
allylic cation

Problem 20.12 Predict the major product formed by 1,4-addition of HCl to cyclopentadiene.

The conjugate addition product would also be 3-chlorocyclopentene. Because of symmetry in the ring, the 1,4-addition product is the same as the predominant 1,2-addition product, both being derived from the allylic cation.

Problem 20.13 Draw structural formulas for the two constitutional isomers with the molecular formula $C_5H_6Br_2$ formed by adding one mole of Br_2 to cyclopentadiene.

The two constitutional isomers are the 1,2 and 1,4-addition products. Two new chiral centers are formed, leading to a racemic mixture of stereoisomers.

1,2-Addition   **1,4-Addition**
product   **product**

<u>Problem 20.14</u> What are the expected kinetic and thermodynamic products from addition of one mole of Br_2 to the following dienes?

Kinetic product   **Thermodynamic product**

The kinetic product will be the 1,2-addition product that is derived from the more highly substituted and thus more stable bridged bromonium ion intermediate. The thermodynamic product is the most stable one, namely the 1,4-addition product, because this has the most highly substituted alkene as shown. Two new chiral centers are created in the molecules as shown, leading to a racemic mixture of stereoisomers in either case.

Kinetic product   **Thermodynamic product**

Because of symmetry in the molecule, both possible bromonium ion intermediates are the same. Thus, there is only one possible 1,2-addition product and this is the kinetic product. The thermodynamic product is the more stable one, namely the 1,4-addition product, because this has the most highly substituted alkene as shown. Once again, two new chiral centers are created in the molecules as shown, leading to a racemic mixture of stereoisomers in either case.

Ultraviolet-Visible Spectra
<u>Problem 20.15</u> Show how to distinguish between 1,3-cyclohexadiene and 1,4-cyclohexadiene by ultraviolet spectroscopy.

1,3-Cyclohexadiene   **1,4-Cyclohexadiene**

In 1,3-cyclohexadiene, the pi bonds are conjugated so the absorption will occur near 217 nm compared with the 165 nm or so absorption that will be observed for the 1,4-cyclohexadiene that has only isolated (unconjugated) pi bonds.

<u>Problem 20.16</u> Pyridine exhibits a UV transition of the type n --> π^* at 270 nm. In this transition, one of the unshared electrons on nitrogen is promoted from a nonbonding MO to a π^*-antibonding MO. What is the effect on this UV peak if pyridine is protonated?

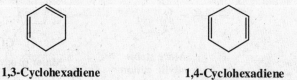

Pyridine   **Pyridinium ion**

When the pyridinium ion is protonated, the lone pair of electrons on nitrogen is tied up in a bond to hydrogen. As a result, it is lower in energy compared with the unprotonated form, so it takes more energy to promote one of these electrons into the π^* orbital. Therefore, protonation shifts the absorbance peak to a lower wavelength (higher energy).

Problem 20.17 The weight of proteins or nucleic acids in solution is commonly determined by UV spectroscopy using the Beer-Lambert law. For example, the ε of double-stranded DNA at 260 nm is 6670 M^{-1} cm^{-1}. The formula weight of the repeating unit in DNA (650 Daltons) can be used as the molecular weight. What is the weight of DNA in 2.0 mL of aqueous buffer if the absorbance, measured in a 1-cm cuvette, is 0.75?

According to the Beer-Lambert law:
$$0.75 = (6670)(1)(x)$$
where x equals the unknown concentration of DNA in mol per liter. Rearranging gives:
$$x = 0.75/(6670)(1) = 1.12 \times 10^{-4} \text{ mol/liter}$$

The molecular weight used is that of a single base pair of DNA, namely 650 grams/mol. Furthermore, there is a total of 2.0 mL of solution, so the total weight of double stranded DNA in the sample is:
$$(1.12 \times 10^{-4} \text{ mol/liter})(650 \text{ grams/mol})(2.0 \times 10^{-3} \text{ liter}) =$$
$$\boxed{1.46 \times 10^{-4} \text{ gram}}$$

Problem 20.18 A sample of adenosine triphosphate (ATP) (MW 507, ε = 14,700 M^{-1} cm^{-1} at 257 nm) is dissolved in 5.0 mL of buffer. A 250-μL aliquot is removed and placed in a 1-cm cuvette with sufficient buffer to give a total volume of 2.0 mL. The absorbance of the sample at 257 nm is 1.15. Calculate the weight of ATP in the original 5.0-mL sample.

The concentration of sample in the cuvette can be determined by using the Beer-Lambert law as follows:
$$1.15 = (14,700)(1.0 \text{ cm})(x \text{ mol/liter})$$
$$x = 1.15/14,700 = 7.82 \times 10^{-5} \text{ mol/liter}$$

The sample measured in the cuvette is actually diluted 0.250 mL in 2.0 mL or 1 in 8 compared to the original unknown sample, so the concentration of ATP in the original unknown sample is:
$$(8)(7.82 \times 10^{-5}) = 6.26 \times 10^{-4} \text{ mol/liter}$$

Since there is a total of 5.0 mL or 5 x 10^{-3} liter and ATP has a MW = 507 grams/mol then the weight of ATP in the original sample can be calculated as:
$$6.26 \times 10^{-4} \text{ mol/liter} = (x \text{ grams})/(506 \text{ grams/mol})(5 \times 10^{-3} \text{ liters})$$
$$x = (6.26 \times 10^{-4})(5 \times 10^{-3} \text{ liters})(506) =$$
$$\boxed{1.59 \times 10^{-3} \text{ grams} = 1.59 \text{ mg}}$$

Problem 20.19 The following equilibrium was discussed in Section 20.1.

2-Cyclohexenone   **3-Cyclohexenone**
(more stable)   (less stable)

(a) Give a mechanism for this reaction under either acidic or basic conditions.

In acid, protonation on the ketone followed by loss of a proton on the 3 position gives a conjugated enol intermediate.

Step 1: Add a proton.

2-Cyclohexenone

Step 2: Take a proton away.

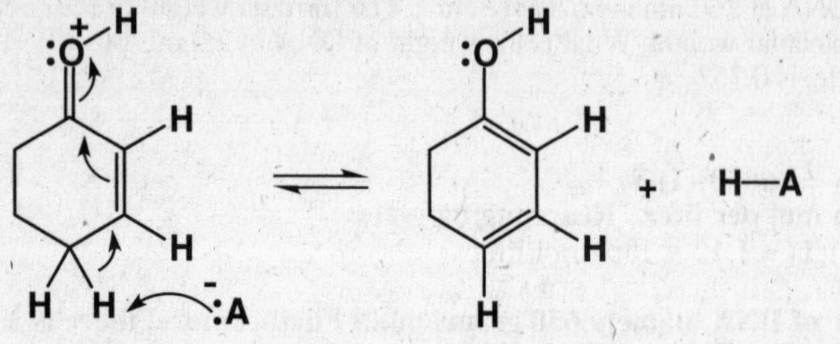

Step 3: Keto-enol tautomerism. Keto-enol tautomerism of the enol gives 3-cyclohexenone.

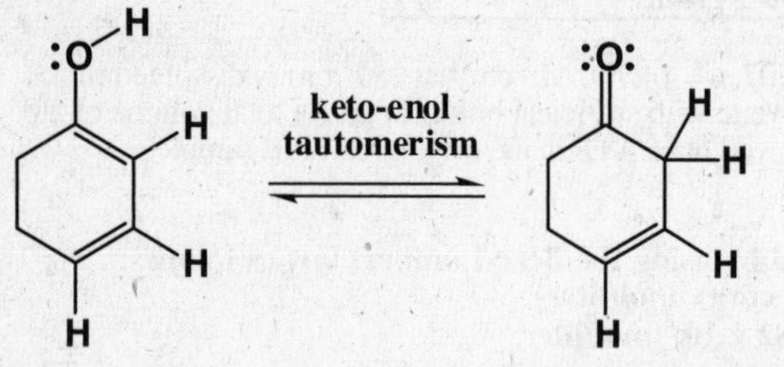

3-Cyclohexenone

In base, deprotonation at the 3-position gives an enolate.

Step 1: Take a proton away.

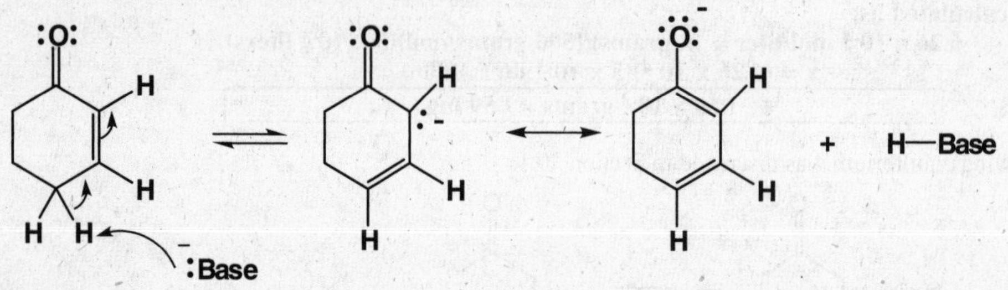

Step 2: Add a proton. Protonation of the enolate followed by tautomerization gives 3-cyclohexenone.

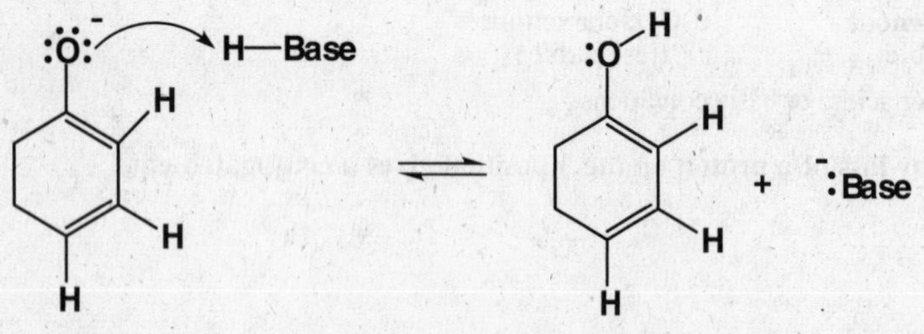

Step 3: Keto-enol tautomerism. Tautomerism of the enol gives 3-cyclohexenone.

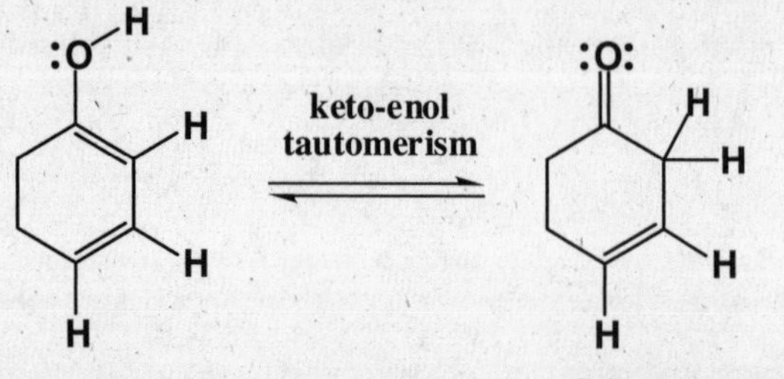

3-Cyclohexenone

(b) Explain the position of equilibrium.

Conjugation is stabilizing, so the conjugated 2-cyclohexenone is favored at equilibrium compared to 3-cyclohexenone, which is not conjugated.

Orangefin anemonefish *Amphiprion chrysopterus*
in a *Stichodactyla kenti* anenome
Bora Bora, French Polynesia

Reticulated butterflyfish *Chaetodon reticulatus*
Bora Bora, French Polynesia

CHAPTER 21
Solutions to the Problems

<u>Problem 21.1</u> Construct a Frost circle for a planar eight-membered ring with one $2p$ orbital on each atom of the ring, and show the relative energies of its eight π molecular orbitals. Which are bonding MOs, which are antibonding, and which are nonbonding?

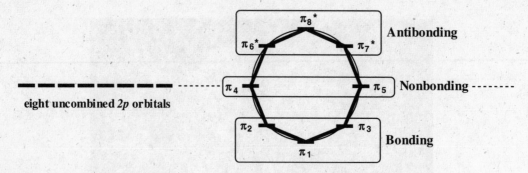

<u>Problem 21.2</u> Which compound gives a signal in the ^{1}H-NMR spectrum with a larger chemical shift, furan or cyclopentadiene? Explain.

Furan   Cyclopentadiene

Furan is aromatic by virtue of a lone pair on the oxygen atom, when combined with the four π electrons of the carbon-carbon double bonds give six π electrons. Being aromatic, the hydrogens on this aromatic ring are deshielded by the induced ring current and appear farther downfield compared to those of the nonaromatic cyclopentadiene.

<u>Problem 21.3</u> Describe the ground-state electron configuration of the cyclopentadienyl cation and radical. Assuming each species is planar, would you expect it to be aromatic or antiaromatic?

Cyclopentadienyl cation:

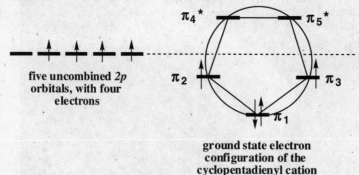

ground state electron
configuration of the
cyclopentadienyl cation

Cyclopentadienyl radical:

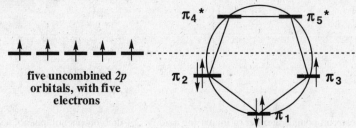

five uncombined *2p*
orbitals, with five
electrons

ground state electron
configuration of the
cyclopentadienyl anion

The cyclopentadienyl cation has 4 π electrons so it is antiaromatic. The cyclopentadienyl radical has five π electrons, so it is neither aromatic (4n + 2) nor antiaromatic (4n).

<u>Problem 21.4</u> Describe the ground-state electron configuration of the cycloheptatrienyl radical and anion. Assuming that each species is planar, would you expect it to be aromatic or antiaromatic?

Cycloheptatrienyl radical:

seven uncombined *2p*
orbitals, with seven
electrons

ground state electron
configuration of the
cycloheptatrienyl radical

Cycloheptatrienyl anion:

seven uncombined *2p*
orbitals, with eight
electrons

ground state electron
configuration of the
cycloheptatrienyl anion

The cycloheptatrienyl anion has 8 π electrons so it is antiaromatic. The cycloheptatrienyl radical has seven π electrons, so it is neither aromatic (4n + 2) or antiaromatic (4n).

<u>Problem 21.5</u> Write names for these molecules.

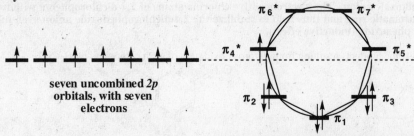

(a)

2-Phenyl-2-propanol

(b)

N-Phenylbutanamide

(c)

**3-Nitrobenzoyl
chloride**

(d)

COOH

COOH

**1,2-Benzenedicarboxylic
acid
(Phthalic acid)**

<u>Problem 21.6</u> Arrange these compounds in order of increasing acidity: 2,4-dichlorophenol, phenol, cyclohexanol.

The following compounds are ranked from least to most acidic:

$$\text{Cyclohexanol} \quad < \quad \text{Phenol} \quad < \quad \text{2,4-Dichlorophenol}$$
$$pK_a \sim 18 \qquad\qquad pK_a \ 9.95 \qquad\qquad pK_a \sim 7.5$$

**A good way to predict relative acidities between related compounds is to keep track of the anionic conjugate
bases produced upon deprotonation. In general, the more stable the conjugate base anion, the stronger the acid.
Anions become increasingly stabilized as the negative charge is more delocalized around the molecule. Thus,
phenol is more acidic than an aliphatic alcohol like cyclohexanol, because resonance involving the aromatic
ring of phenol leads to increased charge delocalization and thus stabilization of the phenoxide anion compared
to the cyclohexylalkoxide anion. The electronegative chlorine atoms of 2,4-dichlorophenol withdraw electron
density from the aromatic ring and thus help to stabilize the 2,4-dichlorophenoxide anion even further than
what is seen with phenoxide (inductive effect).**

<u>Problem 21.7</u> Predict the products resulting from vigorous oxidation of each compound by H_2CrO_4.

(a)

(b)

O_2N

NO_2

COOH

COOH

O_2N COOH

NO_2

Nomenclature and Structural Formulas
<u>Problem 21.8</u> Name the following compounds and ions.

(a)

NO_2

Cl

1-Chloro-4-nitrobenzene

(b)

CH_3

Br

**2-Bromotoluene
(*o*-Bromotoluene)**

(c)

OH

3-Phenyl-1-propanol

(d)

NO_2

NO_2

1,5-Dinitronaphthalene

(e)

OH

(*S*)-2-Phenyl-3-buten-2-ol

(f)

O_2N

**3-Nitrophenylethyne
(*m*-Nitrophenylacetylene)**

(g)

2-Phenylphenol
(*o*-Phenylphenol)

(h) CH₃O—⟨ ⟩—CH₂⁺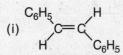

4-Methoxybenzyl cation

(i)

C₆H₅   H
  \   /
   C = C
  /   \
 H   C₆H₅

(*E*)-1,2-Diphenylethene
(*trans*-1,2-Diphenylethylene)

(j)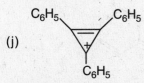

C₆H₅   C₆H₅

C₆H₅

Triphenylcyclopropenyl cation

<u>Problem 21.9</u> Draw structural formulas for these compounds.
(a) 1-Bromo-2-chloro-4-ethylbenzene   (b) *m*-Nitrocumene   (c) 4-Chloro-1,2-dimethylbenzene

Br

Cl

NO₂

CH₃

CH₃

Cl

(d) 3,5-Dinitrotoluene   (e) 2,4,6-Trinitrotoluene   (f) (2*S*,4*R*)-4-Phenyl-2-pentanol

CH₃

O₂N   NO₂

CH₃

O₂N   NO₂

NO₂

OH

(g) *p*-Cresol   (h) Pentachlorophenol   (i) 1-Phenylcyclopropanol

OH

CH₃

OH

Cl   Cl

Cl   Cl

Cl

OH

(j) Triphenylmethane   (k) Phenylethylene (styrene)   (l) Benzyl bromide

CH

Br

(m) 1-Phenyl-1-butyne

(n) (*E*)-3-Phenyl-2-propen-1-ol

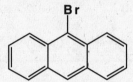

<u>Problem 21.10</u> Draw a structural formulas for each compound.

(a) 1-Nitronaphthalene

NO₂

(b) 1,6-Dichloronaphthalene

Cl

Cl

(c) 9-Bromoanthracene

Br

(d) 2-Methylphenanthrene

CH₃

<u>Problem 21.11</u> Molecules of 6,6'-dinitrobiphenyl-2,2'-dicarboxylic acid have no tetrahedral chiral center, and yet they can be resolved to a pair of enantiomers. Account for this chirality. You will find it helpful to examine the model on the CD of this compound.

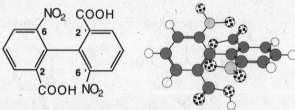

6,6'-Dinitrobiphenyl-2,2'-dicarboxylic acid

The key here is that the central bond between the benzene rings cannot rotate freely at room temperature due to the steric hindrance provided by the nitro and carboxyl groups. In other words, these groups run into each other as the molecule attempts to rotate around the central bond, so it is prevented from rotating. As a result, the molecule is chiral because, like a propeller, there are two different orientations possible. The two orientations represent non-superimposable mirror images so they are a pair of enantiomers.

Resonance in Aromatic Compounds

Problem 21.12 Following each name is the number of Kekulé structures that can be drawn for it. Draw these Kekulé structures, and show, using curved arrows, how the first contributing structure for each molecule is converted to the second and so forth.

(a) Naphthalene (3)

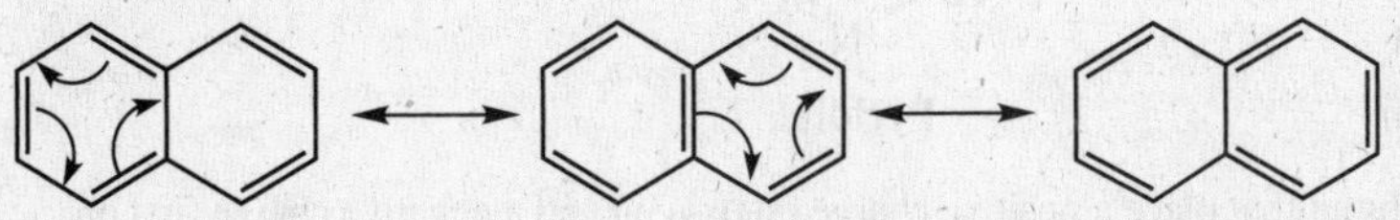

(b) Phenanthrene (5)

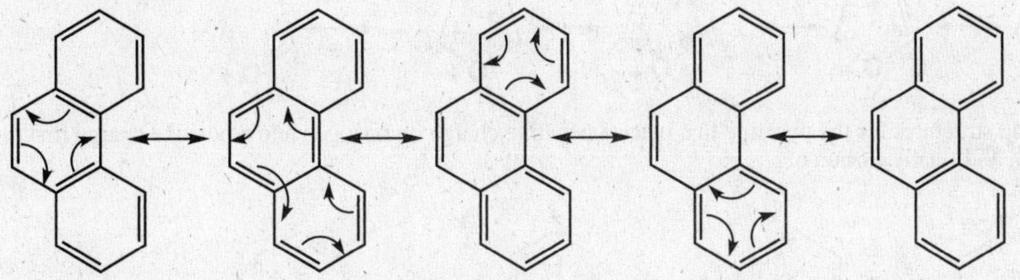

Problem 21.13 Each molecule in this problem can be drawn as a hybrid of five contributing structures: two Kekulé structures and three that involve creation and separation of unlike charges. Draw these five contributing structures for each molecule.

(a) Chlorobenzene

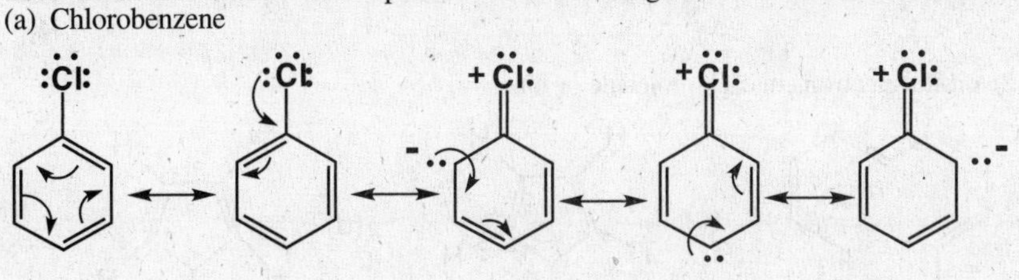

(b) Phenol

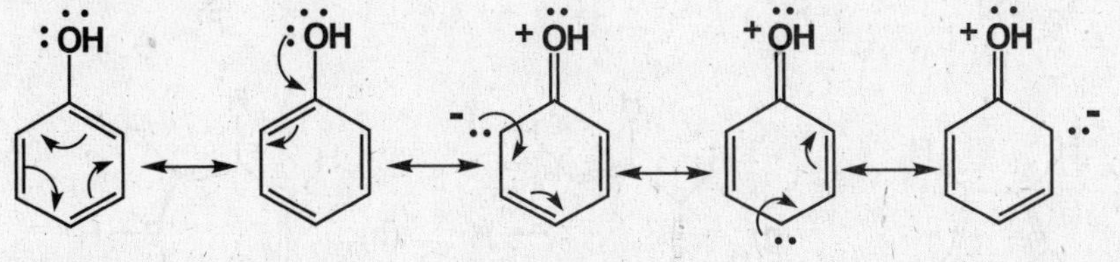

(c) Nitrobenzene

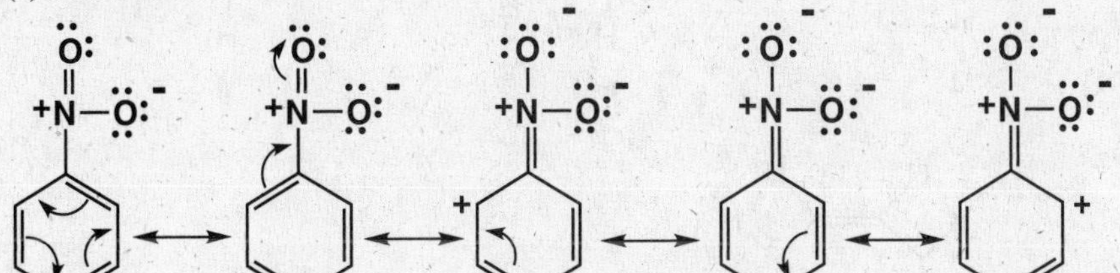

Problem 21.14 Following are structural formulas for furan and pyridine.

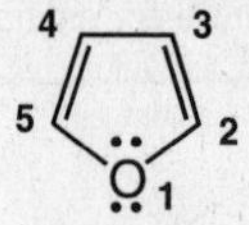

Furan 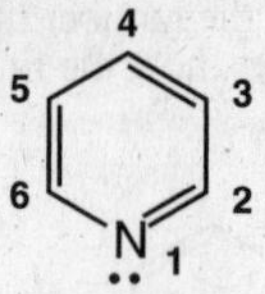   **Pyridine**

(a) Draw four contributing structures for the furan that place a positive charge on oxygen and a negative charge first on carbon 3 of the ring and then on each other carbon of the ring.

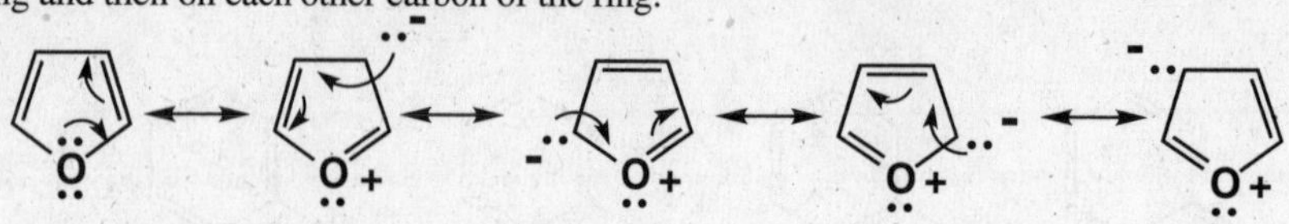

(b) Draw three contributing structures for the pyridine that place a negative charge on nitrogen and a positive charge first on carbon 2, then on carbon 4, and finally carbon 6.

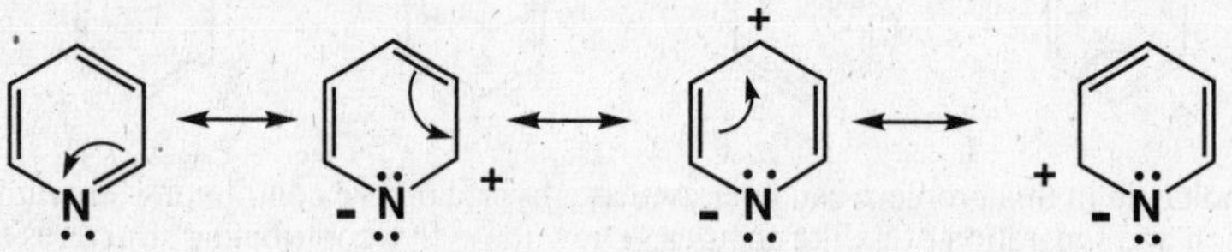

The Concept of Aromaticity
Problem 21.15 State the number of $2p$ orbital electrons in each molecule or ion.

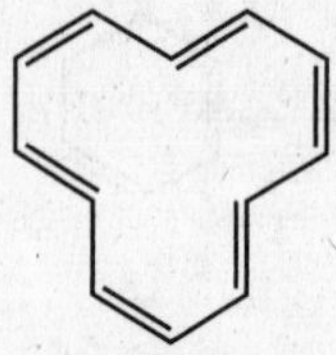

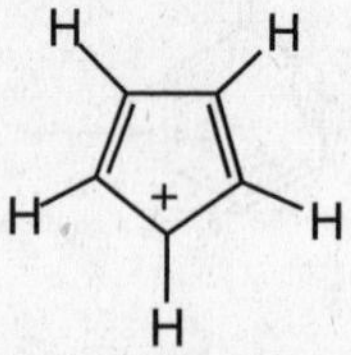

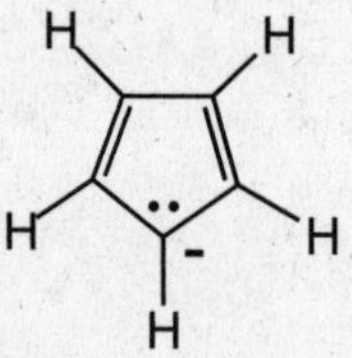

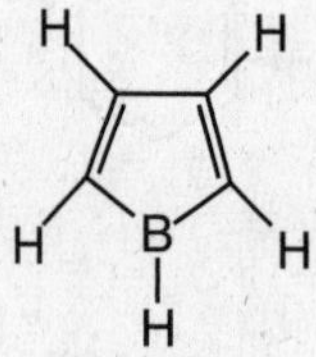

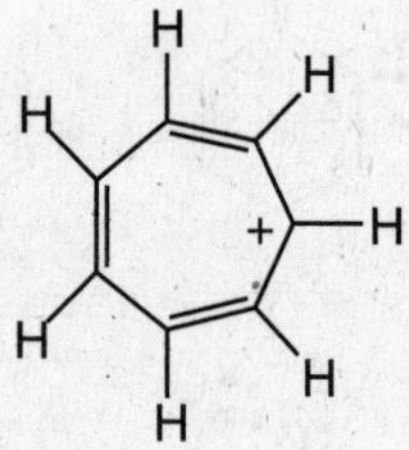

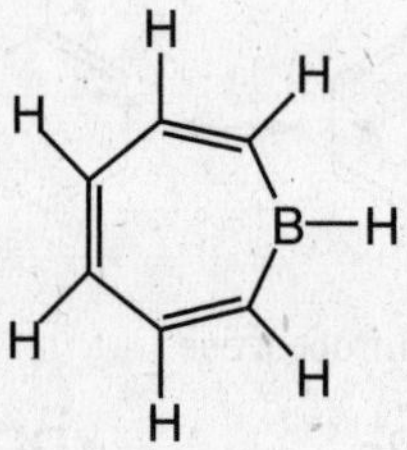

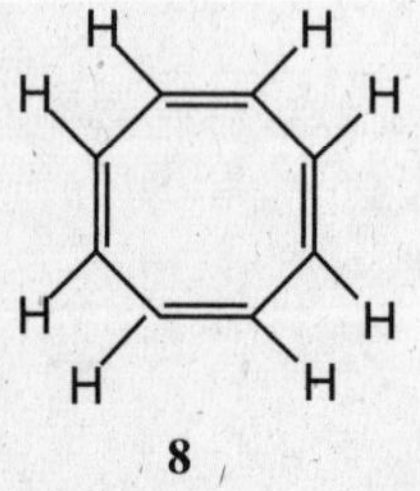

(a)   (b)   (c)   (d)

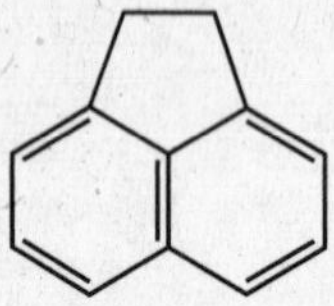

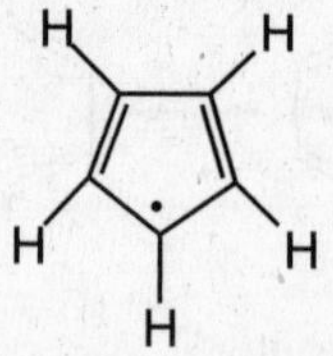

10   12   4   5

(e)   (f)   (g)   (h)

6   4   6   6

(i)   (j)

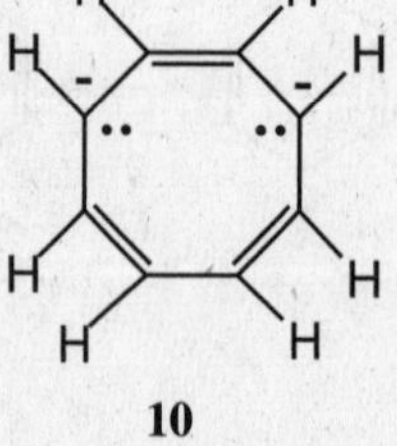

8   10

Problem 21.16 Which of the molecules and ions given in Problem 21.15 are aromatic according to the Hückel criteria? Which, if planar, would be antiaromatic?

The following molecules are aromatic because they have 4n + 2 π electrons: a, e, g, h, and j.
The following molecules would be antiaromatic if planar because they have 4n π electrons: b, c, f, and i.

<u>Problem 21.17</u> Construct MO energy diagrams for the cyclopropenyl cation, radical, and anion. Which of these species is aromatic according to the Hückel criteria?

Cyclopropenyl cation:

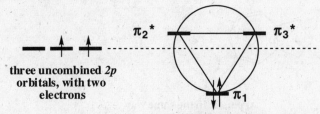

three uncombined *2p*
orbitals, with two
electrons

ground state electron
configuration of the
cyclopropenyl cation

Cyclopropenyl radical:

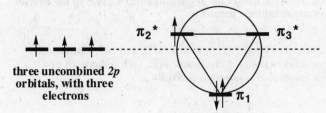

three uncombined *2p*
orbitals, with three
electrons

ground state electron
configuration of the
cyclopropenyl radical

Cyclopropenyl anion:

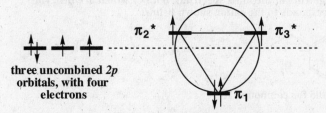

three uncombined *2p*
orbitals, with four
electrons

ground state electron
configuration of the
cyclopropenyl anion

Of these species, the only one that satisfies the Hückel criteria is the first one, namely the cyclopropenyl cation with 2 π electrons, a Hückel number (4n + 2, here n = 0).

<u>Problem 21.18</u> Naphthalene and azulene are constitutional isomers of molecular formula $C_{10}H_8$.

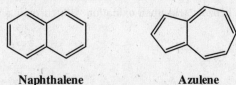

Naphthalene   **Azulene**

Naphthalene is a colorless solid with a dipole moment of zero. Azulene is a solid with an intense blue color and a dipole moment of 1.0 D. Account for the difference in dipole moments of these constitutional isomers.

Naphthalene has no permanent dipole moment because it possesses a high degree of symmetry. Azulene has a remarkably large permanent dipole moment (1.0 D) for a hydrocarbon. The dipole moment of azulene can be explained using the contributing structures such as the one shown below on the right:

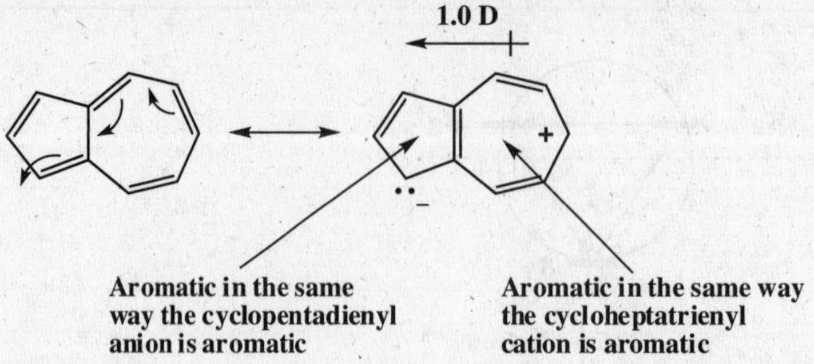

1.0 D

Aromatic in the same
way the cyclopentadienyl
anion is aromatic

Aromatic in the same way
the cycloheptatrienyl
cation is aromatic

Neither a cyclopentadiene or cycloheptatriene ring is aromatic in the neutral state. However, by transferring electron density from the seven-membered ring π system to the five-membered ring π system of azulene, the aromatic cyclopentadienyl anion and cycloheptatrienyl cation π systems are formed. This aromatic stabilization explains why the resonance structure on the right makes an important contribution to the overall resonance hybrid, resulting in the large observed permanent dipole moment.

Spectroscopy
Problem 21.19 Compound A (C_9H_{12}) shows prominent peaks in its mass spectrum at m/z 120 and 105. Compound B (C_9H_{12}) shows prominent peaks at m/z 120 and 91. On vigorous oxidation with chromic acid, both compounds give benzoic acid. From this information, deduce the structural formulas of compounds A and B.

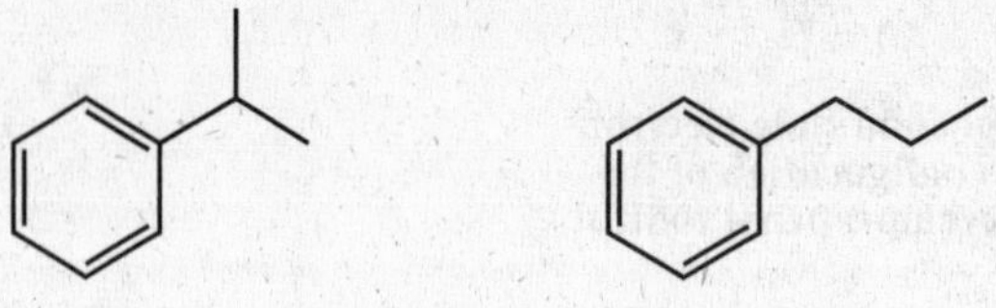

A   B

The compounds can be identified by the major fragments in the mass spectrum, both of which are benzylic cations. For compound A, the peak at m/z 105 corresponds to the cation shown below:

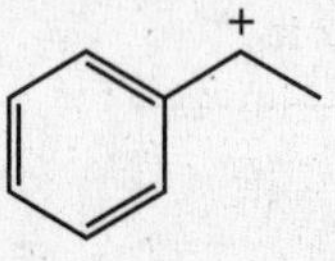

Peak at m/z 105 for compound A

For compound B, the peak at m/z 91 corresponds to the tropylium ion that was produced by rearrangement of the benzyl cation (Section 15.4)

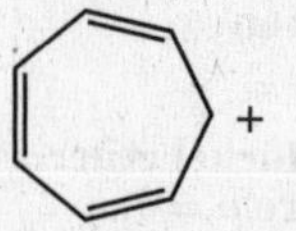

Peak at m/z 91 for compound B

Both of the compounds will produce benzoic acid upon oxidation with chromic acid.

Problem 21.20 Compound C shows a molecular ion at *m/z* 148, and other prominent peaks at *m/z* 105, and 77. Following are its infrared and ^{1}H-NMR spectra.

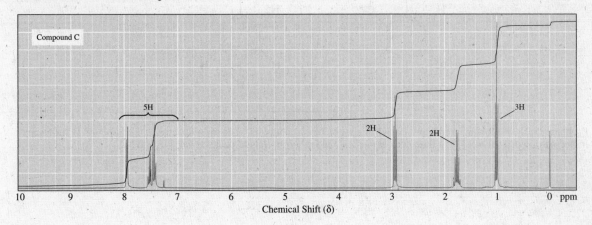

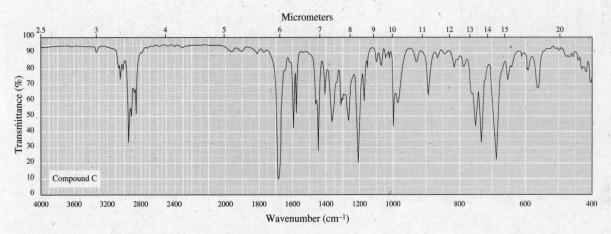

(a) Deduce the structural formula of Compound C.

Compound C

This compound, $C_{10}H_{12}O$, has the correct molecular formula of 148. This compound also has a carbonyl corresponding to the peak at 1680 cm^{-1} in the IR spectrum and the correct pattern of hydrogens to explain the ^{1}H-NMR spectrum.

(b) Account for the appearance of peaks in its mass spectrum at *m/z* 105 and 77.

The peaks at *m/z* 105 and 77 correspond to the following fragments produced by α-cleavage on either side of the carbonyl group.

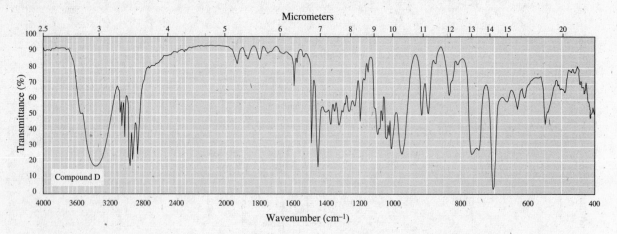

m/z 105 m/z 77

<u>Problem 21.21</u> Following are IR and ^{1}H-NMR spectra of compound D. The mass spectrum of compound D shows a molecular ion peak at m/z 136, a base peak at m/z 107, and other prominent peaks at m/z 118 and 59.

(a) Propose a structural formula for compound D.

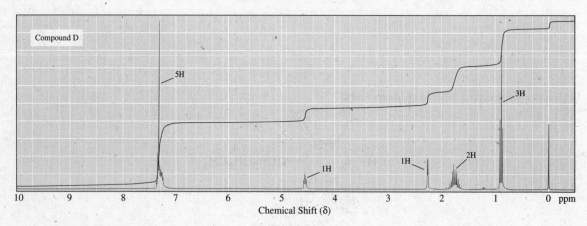

Compound D

This compound, $C_9H_{12}O$, has the correct molecular weight of 136. The hydroxyl group gives the broad peak at 3300 cm^{-1} in the IR spectrum and the pattern of hydrogens present explains the ^{1}H-NMR spectrum.

(b) Propose structural formulas for ions in the mass spectrum at *m/z* 118, 107 and 59.

The peaks at *m/z* 118, 107 and 77 correspond to the following fragments:

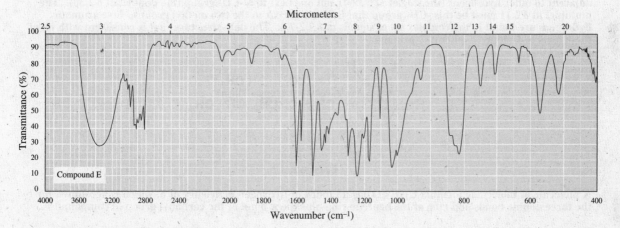

| *m/z* 118 | *m/z* 107 | *m/z* 59 |

<u>Problem 21.22</u> Compound E ($C_8H_{10}O_2$) is a neutral solid. Its mass spectrum shows a molecular ion at *m/z* 138 and prominent peaks at M-1 and M-17. Following are IR and ¹H-NMR spectra of compound E. Deduce the structure of compound E.

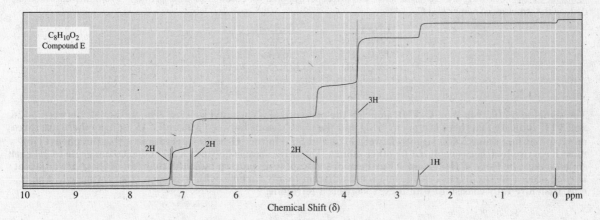

A molecule of molecular formula $C_8H_{10}O_2$ has an index of hydrogen deficiency of 4, which is accounted for by the three double bonds and ring of the benzene ring of compound E. Compound E also has an alcohol function corresponding to the broad peak at 3350 cm^{-1} in the IR spectrum, and the correct pattern of hydrogens to explain the ^{1}H-NMR spectrum. Notice especially the distinctive *para* pattern near δ 7.0 and the methyl group signal at δ 3.85.

Problem 21.23 Following are ^{1}H-NMR and ^{13}C-NMR spectral data for compound F ($C_{12}H_{16}O$). From this information, deduce the structure of compound F.

^{1}H-NMR	^{13}C-NMR	
0.83 (d, 6H)	207.82	50.88
2.11 (m, 1H)	134.24	50.57
2.30 (d, 2H)	129.36	24.43
3.64 (s, 2H)	128.60	22.48
7.2 - 7.4 (m, 5H)	126.86	

The signal in the ^{13}C-NMR spectrum at δ 217.82 indicates the presence of a carbonyl group. The signals around δ 130 indicate there is an aromatic ring. The doublet in the ^{1}H-NMR at δ 0.83 that integrates to 6H indicates two methyl groups adjacent to a -CH- group. There are also two -CH$_2$- groups; one that is not adjacent to other hydrogens (the singlet at δ 3.64) and one next to a -CH- group (the doublet at δ 2.30). The multiplet at δ 2.11 must be this -CH- group that is also adjacent to the two methyl groups. Five aromatic hydrogens are found in the complex set of signals at δ 7.2-7.4. The only structure that is consistent with all of these facts is 4-methyl-1-phenyl-2-pentanone.

Compound F

A molecule of molecular formula $C_{12}H_{16}O$ has an index of hydrogen deficiency of 5, which is accounted for by the three double bonds and ring of the benzene ring plus the π bond of the carbonyl group in compound F.

Problem 21.24 Following are ^{1}H-NMR and ^{13}C-NMR spectral data for compound G ($C_{10}H_{10}O$). From this information, deduce the structure of compound G.

^{1}H-NMR	^{13}C-NMR	
2.50 (t, 2H)	210.19	126.82
3.05 (t, 2H)	136.64	126.75
3.58 (s, 2H)	133.25	45.02
7.1 to 7.3 (m, 4H)	128.14	38.11
	127.75	28.34

The signal at δ 210.19 in the ^{13}C-NMR indicate the presence of a carbonyl species. The six signals between δ 126 and δ 137 indicate that there is a phenyl ring and the three signals between δ 28 and δ 45 indicate there are three more sp^3 carbon atoms. The signals between δ 7.1 and δ 7.3 in the ^{1}H-NMR integrate to 4H so the phenyl ring must have two hydrogens replaced by other atoms. The three signals between δ 2.6 and δ 3.6 integrate to 2H each so these must be three -CH$_2$- groups. Furthermore, the splitting pattern indicates that two of these are adjacent to each other (the two triplets) while one is not adjacent to any carbons with hydrogen atoms attached. The only structure fully consistent with these spectra is β-tetralone.

Compound G

A molecule of molecular formula $C_{10}H_{10}O$ has an index of hydrogen deficiency of 6, which is accounted for by the three double bonds and ring of the benzene ring, the π bond of the carbonyl group, and the cyclohexane ring in compound G.

<u>Problem 21.25</u> Compound H ($C_8H_6O_3$), gives a precipitate when treated with hydroxylamine in aqueous ethanol, and a silver mirror when treated with Tollens' solution. Following is its ^{1}H-NMR spectrum. Deduce the structure of compound H.

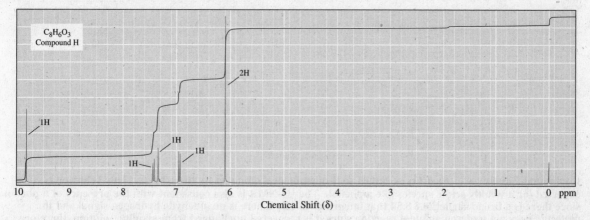

The hydroxylamine reaction indicates the presence of a carbonyl species, and the positive Tollens' test confirms the presence of an aldehyde. The aldehyde is also indicated by the singlet in the ^{1}H-NMR at δ 9.8. The splitting pattern of the aromatic hydrogen signals between δ 6.95 and 7.5 indicates that the phenyl ring only has three hydrogens, two are adjacent to each other (the two doublets), while the third is not adjacent to any hydrogens. The singlet at δ 6.1, integrating to 2H, indicates the presence of a -CH_2- group bound to two very electronegative atoms (oxygen). Given the molecular formula, the only structure that agrees with this information is piperonal.

Compound H

A molecule of molecular formula $C_8H_6O_3$ has an index of hydrogen deficiency of 6, which is accounted for by the three double bonds and ring of the benzene ring, the π bond of the carbonyl group, and the five-membered ring in compound H.

<u>Problem 21.26</u> Compound I, $C_{11}H_{14}O_2$, is insoluble in water, aqueous acid, and aqueous $NaHCO_3$ but dissolves readily in 10% Na_2CO_3 and 10% NaOH. When these alkaline solutions are acidified with 10% HCl, compound I is recovered unchanged. Given this information and its ^{1}H-NMR spectrum, deduce the structure of compound I.

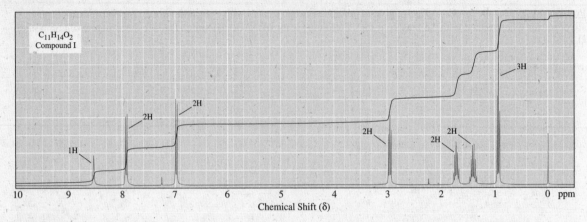

Compound I dissolves in Na_2CO_3 or NaOH so it must be deprotonated by these bases. This behavior is expected for a weakly acidic species like a phenol. The ^{1}H-NMR is also consistent with the presence of a phenol since there is a broad singlet at δ 8.54 that integrates to 1H. There is no aldehyde hydrogen signal and the molecule does not undergo an aldol reaction since it is recovered unchanged from alkaline solution, therefore there is no aldehyde in the molecule. The two doublets at δ 6.9 and δ 7.9, each integrating to 2H, indicate a 1,4 disubstituted phenyl ring. The signals between δ 0.9 and δ 3.0 indicate a -CH_2-CH_2-CH_2-CH_3 group. Finally, the molecular formula indicates an index of hydrogen deficiency of 5. The phenyl ring accounts for 4 so there must be one other π bond (or ring) in the molecule. The only structure consistent with the molecular formula and the spectrum is 1-(4-hydroxyphenyl)-1-pentanone (4-hydroxy-valerophenone).

Compound I

<u>Problem 21.27</u> Propose a structural formula for compound J ($C_{11}H_{14}O_3$) consistent with its ^{1}H-NMR and infrared spectra.

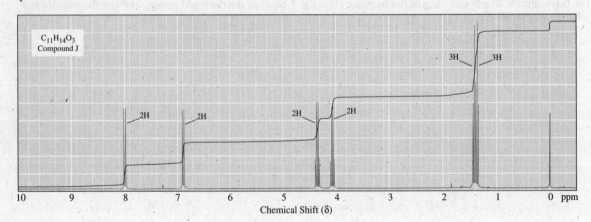

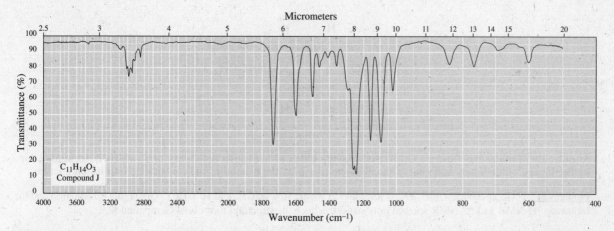

The strong peak at 1721 cm^{-1} in the IR indicates the presence of a carbonyl group. The two sets of doublets integrating to 2H each between δ 6.9 and 8.0 indicate the presence of a 1,4 disubstituted benzene ring. The two quartets integrating to 2H each at δ 4.35 and δ 4.05 indicate there are two -CH$_2$- groups attached to oxygen atoms as well methyl groups. The molecular formula has three oxygen atoms, so the -OCH$_2$- groups are likely part of one ester and one ether. The two triplets integrating to a total of 6H at δ 1.4 are the signals from the two methyl groups, confirming the presence of two ethyl groups. The only structure that is consistent with the molecular formula C$_{11}$H$_{14}$O$_3$ and the spectra is ethyl 4-ethoxybenzoate.

6.9   8.0
H   H

1.4 4.07   O   4.35 1.4
CH$_3$CH$_2$O   COCH$_2$CH$_3$

H   H
6.9   8.0

Compound J

A molecule of molecular formula C$_{11}$H$_{14}$O$_3$ has an index of hydrogen deficiency of 5, which is accounted for by the three double bonds and ring of the benzene ring plus the π bond of the carbonyl group in compound J.

<u>Problem 21.28</u> Propose a structural formula for the analgesic phenacetin, molecular formula C$_{10}$H$_{13}$NO$_2$, based on its ^{1}H-NMR spectrum.

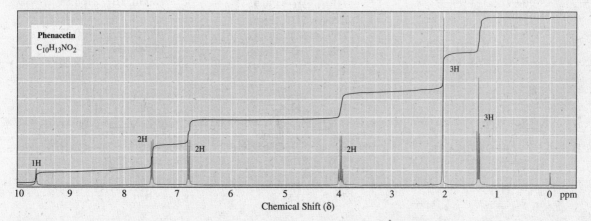

This structure is not only consistent with the molecular formula, but also with the ^{1}H-NMR spectrum. The characteristic two doublets centered at δ 6.8 and 7.5 indicate the presence of a 1,4-disubstituted benzene ring. The singlet at δ 9.65 integrating to 1H indicates a primary amide, and the singlet integrating to 3H at δ 2.05 indicates this is an acetamide. Finally, the typical ethyl splitting pattern for the signals at δ 1.32 and δ 3.95 indicates the presence of an ethyl group. These signals are shifted so far downfield that they must be part of an ethoxy group.

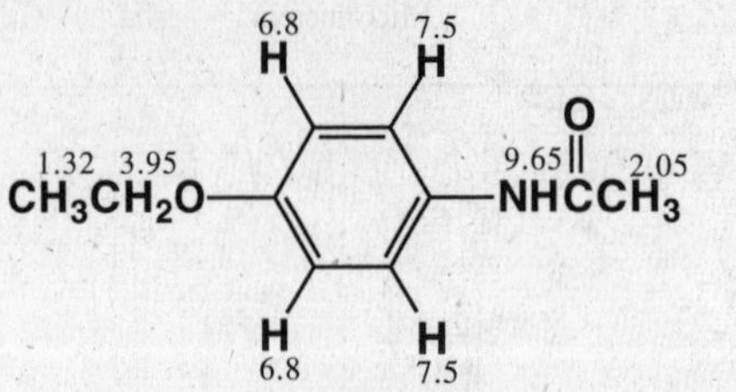

Phenacetin

A molecule of molecular formula $C_{10}H_{13}NO_2$ has an index of hydrogen deficiency of 5, which is accounted for by the three double bonds and ring of the benzene ring plus the π bond of the carbonyl group in phenacetin.

<u>Problem 21.29</u> Compound K, $C_{10}H_{12}O_2$, is insoluble in water, 10% NaOH, and 10% HCl. Given this information and the following ^{1}H-NMR and^{13}C-NMR spectral information, deduce the structural formula of Compound K.

^{1}H-NMR	^{13}C-NMR	
2.10 (s, 3H)	206.51	114.17
3.61 (s, 2H)	158.67	55.21
3.77 (s, 3H)	130.33	50.07
6.86 (d, 2H)	126.31	29.03
7.12 (d, 2H)		

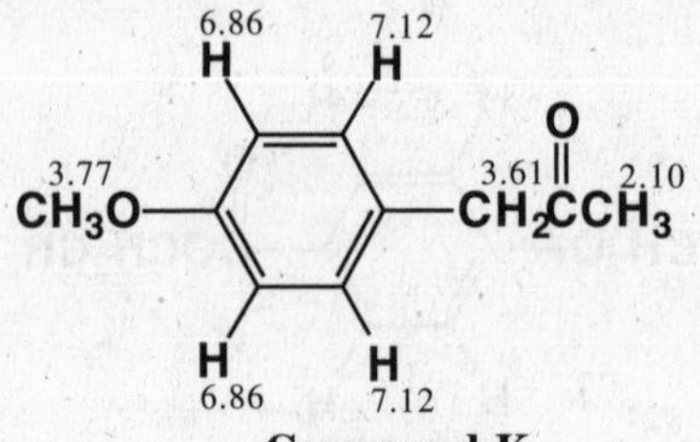

Compound K

<u>Problem 21.30</u> Propose a structural formula for each compound given these NMR data.

(a) $C_9H_9BrO_2$

^{1}H-NMR	^{13}C-NMR
1.39 (t, 3H)	165.73
4.38 (q, 2H)	131.56
7.57 (d, 2H)	131.01
7.90 (d, 2H)	129.34
	127.81
	61.18
	14.18

(b) C_8H_9NO

^{1}H-NMR	^{13}C-NMR
2.06 (s, 3H)	168.14
7.01 (t, 1H)	139.24
7.30 (m, 2H)	128.51
7.59 (d, 2H)	122.83
9.90 (s, 1H)	118.90
	23.93

(c) $C_9H_9NO_3$

^{1}H-NMR	^{13}C-NMR
2.10(s, 3H)	168.74
7.72 (d, 2H)	166.85
7.91 (d, 2H)	143.23
10.3 (s, 1H)	130.28
12.7 (s, 1H)	124.80
	118.09
	24.09

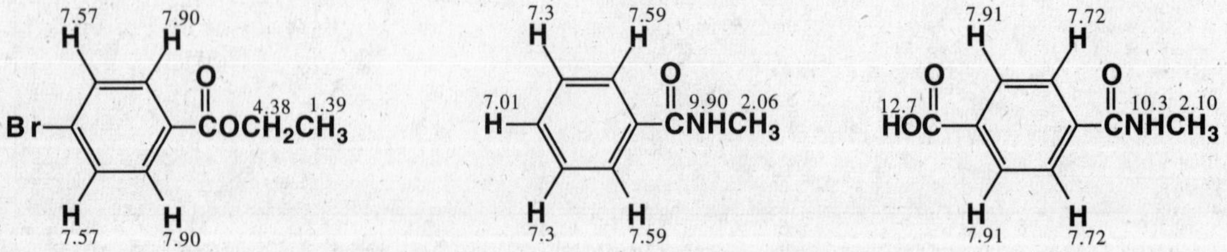

Problem 21.31 Given here are ^{1}H-NMR and ^{13}C-NMR spectral data for two compounds. Each shows strong, sharp absorption between 1700 and 1721 cm^{-1}, and strong, broad absorption over the region 2500 - 3000 cm^{-1}. Propose a structural formula for each compound.

(a) $C_{10}H_{12}O_3$

^{1}H-NMR	^{13}C-NMR
2.49 (t, 2H)	173.89
2.80 (t, 2H)	157.57
3.72 (s, 3H)	132.62
6.78 (d, 2H)	128.99
7.11 (d, 2H)	113.55
12.4 (s 1H)	54.84
	35.75
	29.20

(b) $C_{10}H_{10}O_2$

^{1}H-NMR	^{13}C-NMR
2.34 (s, 3H)	167.82
6.38 (d, 1H)	143.82
7.18 (d, 1H)	139.96
7.44 (d, 2H)	131.45
7.56 (d, 2H)	129.37
12.0 (s 1H)	127.83
	111.89
	21.13

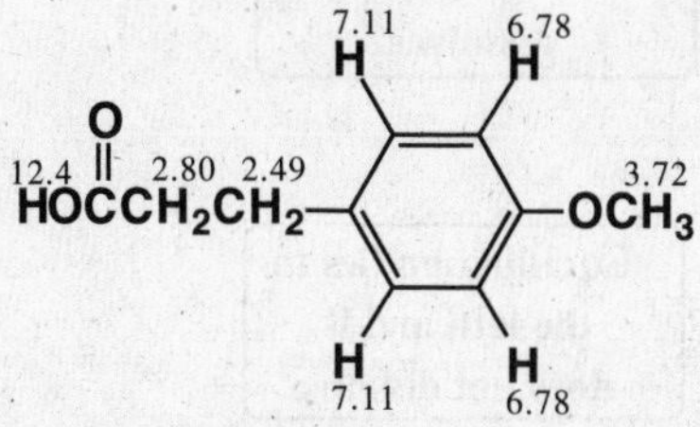

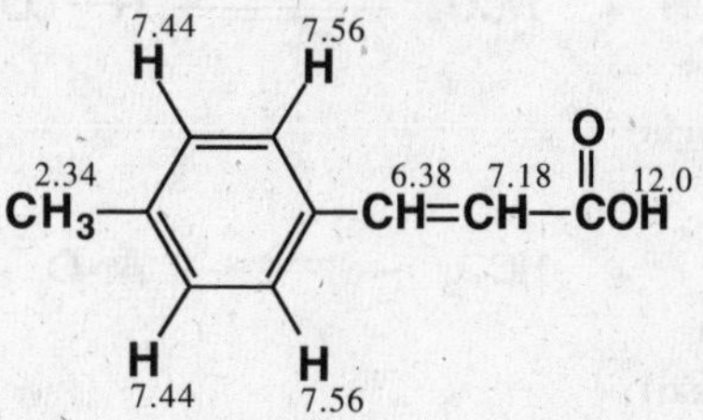

Acidity of Phenols

Problem 21.32 Account for the fact that *p*-nitrophenol ($K_a = 7.0 \times 10^{-8}$) is a stronger acid than phenol ($K_a = 1.1 \times 10^{-10}$).

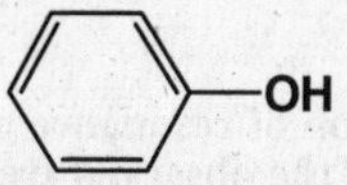

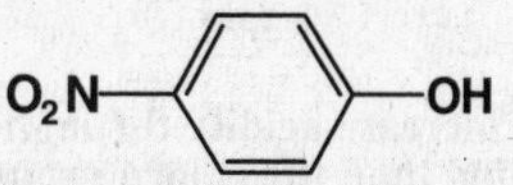

$$K_a = 1.1 \times 10^{-10} \qquad K_a = 7.0 \times 10^{-8}$$

As seen from the acid ionization constants, *p*-nitrophenol is the stronger acid. To account for this fact, draw contributing structures for each anion and compare the degree of delocalization of negative charge (*i.e.*, the resonance stabilization of each anion). Phenoxide ion is a resonance hybrid of five important contributing structures, two of which place the negative charge on the phenoxide oxygen, and three of which place the negative charge on the atoms of the ring.

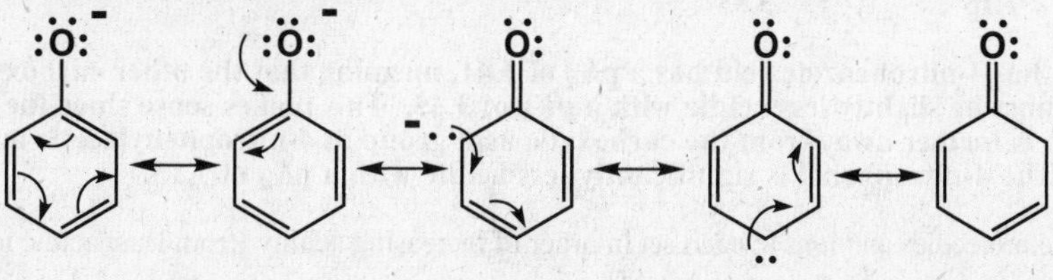

The *p*-nitrophenoxide ion is a hybrid of six important contributing structures. In addition to the five similar to those drawn above for the phenoxide ion, there is a sixth that places the negative charge on the oxygen atoms of the *p*-nitro group.

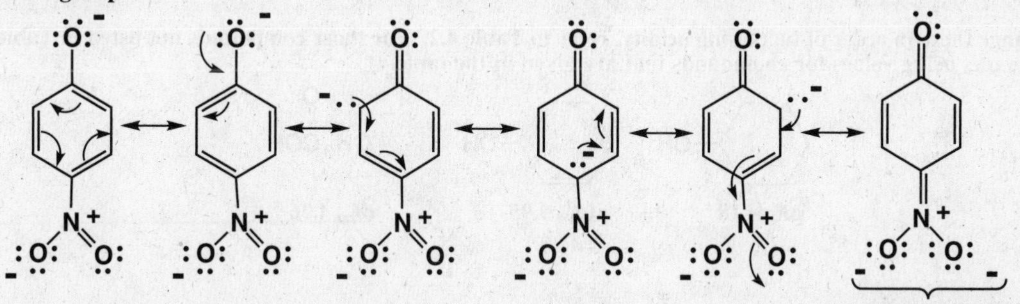

negative charge
delocalized to oxygens
of nitro group

Thus, because of the greater delocalization of the negative charge onto the more electronegative oxygen atoms of the nitro group, *p*-nitrophenol is a stronger acid than phenol.

In addition to the resonance effect discussed above, the nitro group is highly electron-withdrawing because nitrogen and oxygen are both more electronegative than carbon. The inductive effect also helps stabilize the phenoxide anion by pulling electron density away from the anionic oxygen atom.

Problem 21.33 Account for the fact that water-insoluble carboxylic acids (pK_a 4-5) dissolve in 10% aqueous sodium bicarbonate (pH 8.5) with the evolution of a gas but water-insoluble phenols (pK_a 9.5-10.5) do not dissolve in 10% sodium bicarbonate.

In order to dissolve, the carboxylic acid or phenol must be deprotonated.

pK_a = 4-5
(stronger acid)

pK_a = 6.36
(weaker acid)

Equilibrium lies to
the right, and it
dissolves

$$Ar\text{-}OH + HCO_3^- \rightleftharpoons Ar\text{-}O^- + H_2CO_3$$

pK_a = 10
(weaker acid)

pK_a = 6.36
(stronger acid)

Equilibrium lies to
the left, and it
does not dissolve

Problem 21.34 Match each compound with its appropriate pK_a value.
(a) 4-Nitrobenzoic acid, benzoic acid, 4-chlorobenzoic acid pK_a = 4.19, 3.98, and 3.41
pK_a 3.41   4.19   3.98

Electron-withdrawing groups increase acidity through a combination of resonance and inductive effects. The nitro group is more withdrawing than the chloro group, explaining the observed trend.

(b) Benzoic acid, cyclohexanol, phenol pK_a = 18.0, 9.95, and 4.19
pK_a 4.19   18.0   9.95

Acidity increases in the order of aliphatic alcohols, phenols, and carboxylic acids.

(c) 4-Nitrobenzoic acid, 4-nitrophenol, 4-nitrophenylacetic acid pK_a = 7.15, 3.85, and 3.41
pK_a 3.41   7.15   3.85

From part (a) it is clear that 4-nitrobenzoic acid has a pK_a of 3.41, meaning that the other carboxylic acid, 4-nitrophenylacetic acid, must be slightly less acidic with a pK_a of 3.85. This makes sense since the electron-withdrawing nitro group is farther away from the carboxylic acid group in 4-nitrophenylacetic acid compared to 4-nitrobenzoic acid. The 4-nitrophenol is significantly less acidic with a pK_a of 7.15.

Problem 21.35 Arrange the molecules and ions in each set in order of increasing acidity (from least acidic to most acidic).

(a)

To arrange these in order of increasing acidity, refer to Table 4.2. For those compounds not listed in Table 4.2, estimate pK_a using values for compounds that are given in the table.

pK_a ~ 18 pK_a 9.95 pK_a 4.76

(b) [benzene ring]—OH HCO_3^- H_2O

H_2O HCO_3^- [benzene ring]—OH

pK_a 15.7 pK_a 10.33 pK_a 9.95

(c) [benzene ring]—C≡CH   [benzene ring]—OH   [benzene ring]—CH_2OH

[benzene ring]—C≡CH   [benzene ring]—CH_2OH [benzene ring]—OH

pK_a ~ 25 pK_a ~ 18 pK_a 9.95

<u>Problem 21.36</u> Explain the trends in the acidity of phenol and the monofluoro derivatives of phenol.

[phenol, OH]   [2-fluorophenol, OH with F ortho]   [3-fluorophenol, OH with F meta]   [4-fluorophenol, OH with F para]

pK_a = 10.0 pK_a = 8.81 pK_a = 9.28 pK_a = 9.81

The electronegative fluoro substituent increases the acidity of the phenol through an inductive effect, so all of the monofluoro isomers of phenol are more acidic than phenol itself. Because this is an inductive effect, the closer the fluorine atom to the phenolic OH group, the stronger the effect and the greater the acidity. Thus, the *ortho* fluoro derivative is most acidic, followed by the *meta* fluoro derivative, then finally the *para* fluoro derivative.

<u>Problem 21.37</u> Suppose you wish to determine the inductive effects of a series of functional groups, for example Cl, Br, CN, COOH, and C_6H_5. Is it best to use a series of ortho-, meta-, or para-substituted phenols? Explain.

The question to be addressed involves inductive effects only. It would be best to use the derivatives with the substituents in the *meta* position, because this would minimize any contributions from resonance effects. Resonance effects are maximal when substituents are in the *ortho* and *para* positions.

<u>Problem 21.38</u> From each pair, select the stronger base.

To estimate which is the stronger base, first determine which conjugate acid is the weaker acid. The weaker the acid, the stronger its conjugate base.

(a) [benzene ring]—O⁻   or   OH⁻   OH⁻

**Stronger base
(anion of weaker acid)**

(b) [benzene ring]—O⁻   or   [cyclohexane ring]—O⁻   [cyclohexane ring]—O⁻

**Stronger base
(anion of weaker acid)**

(c) 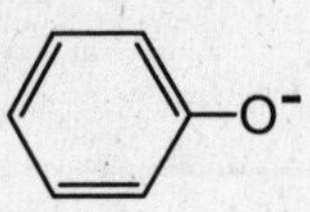 or HCO_3^- 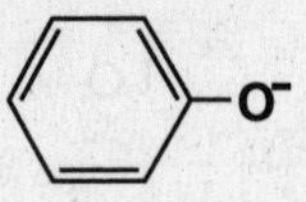

Stronger base (anion of weaker acid)

(d) 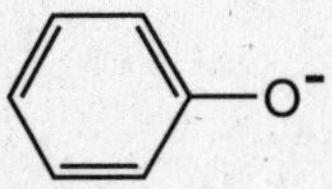 or CH_3COO^- 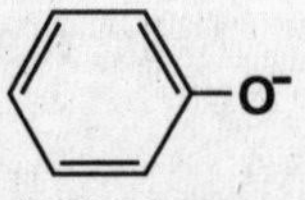

Stronger base (anion of weaker acid)

Problem 21.39 Describe a chemical procedure to separate a mixture of benzyl alcohol and *o*-cresol and recover each in pure form.

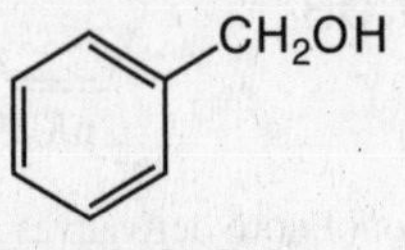
CH2OH

Benzyl alcohol

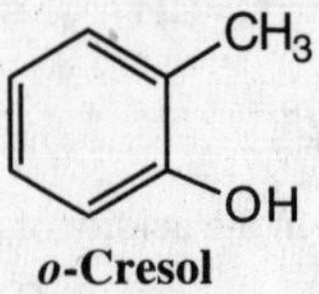
CH3
OH

***o*-Cresol**

Following is a flow chart for an experimental method for separating these two compounds. Separation is based on the facts that each is insoluble in water, soluble in diethyl ether, and that *o*-cresol reacts with 10% NaOH to form a water-soluble phenoxide salt while benzyl alcohol does not.

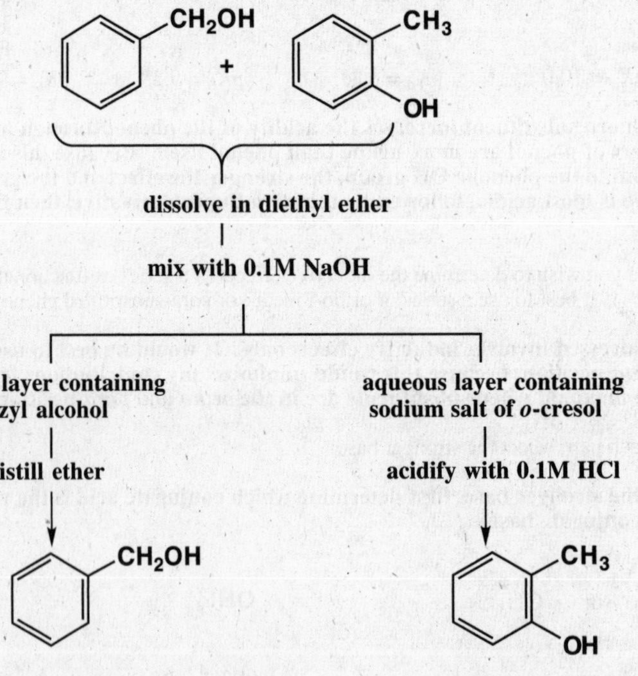
CH2OH
+
CH3
OH
dissolve in diethyl ether
mix with 0.1M NaOH
ether layer containing benzyl alcohol
aqueous layer containing sodium salt of o-cresol
distill ether
acidify with 0.1M HCl
CH2OH
CH3
OH
Benzyl alcohol
o-Cresol

Problem 21.40 The compound 2-hydroxypyridine, a derivative of pyridine, is in equilibrium with 2-pyridone. 2-Hydroxypyridine is aromatic. Does 2-pyridone have comparable aromatic character? Explain.

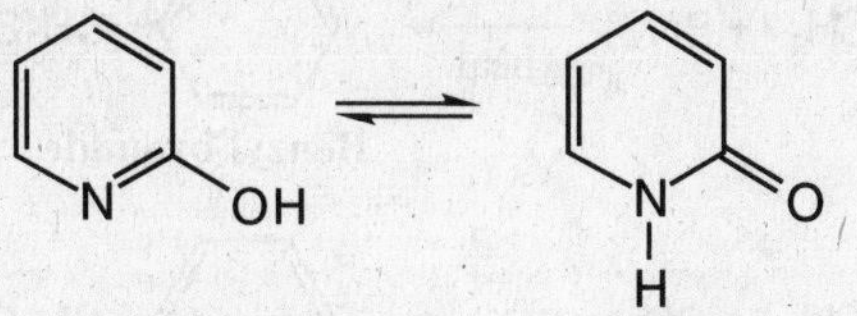

2-Hydroxypyridine   **2-Pyridone**

2-Pyridones have aromatic character because of the contributing structure shown on the right.

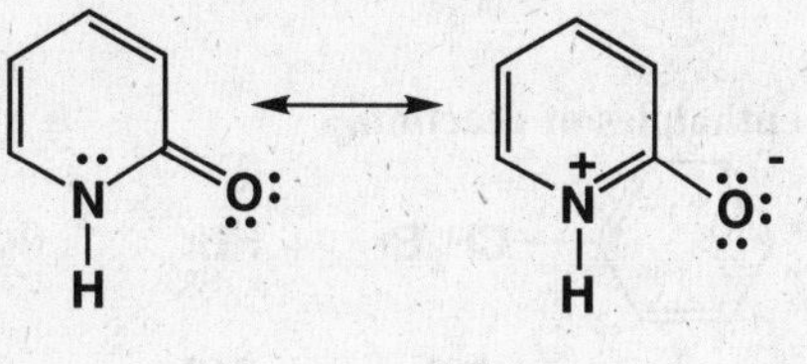

Reactions at the Benzylic Position

Problem 21.41 Write a balanced equation for the oxidation of *p*-xylene to 1,4-benzenedicarboxylic acid (terephthalic acid) using potassium dichromate in aqueous sulfuric acid. How many milligrams of $H_2Cr_2O_4$ are required to oxidize 250 mg of *p*-xylene to terephthalic acid?

Following are balanced equations for the oxidation half-reaction and the reduction half-reaction. Because oxidation takes place in aqueous acid, the reactions are balanced with H_2O and H^+.

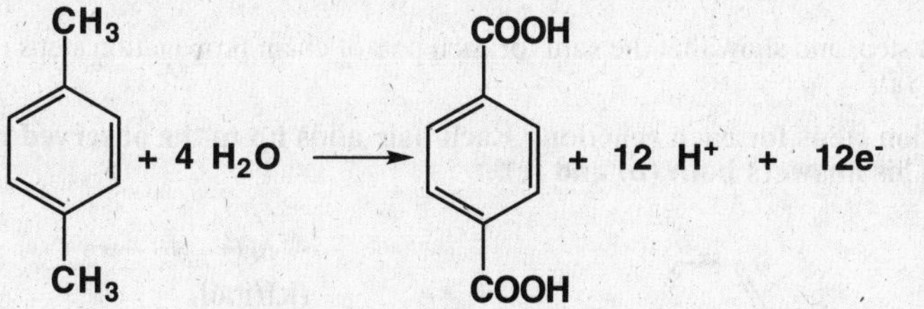

+ 4 H_2O ⟶ + 12 H^+ + 12e⁻   **oxidation half-reaction**

4 H_2CrO_4 + 24 H^+ + 12e⁻ ⟶ 4 Cr^{3+} + 16 H_2O **reduction half-reaction**

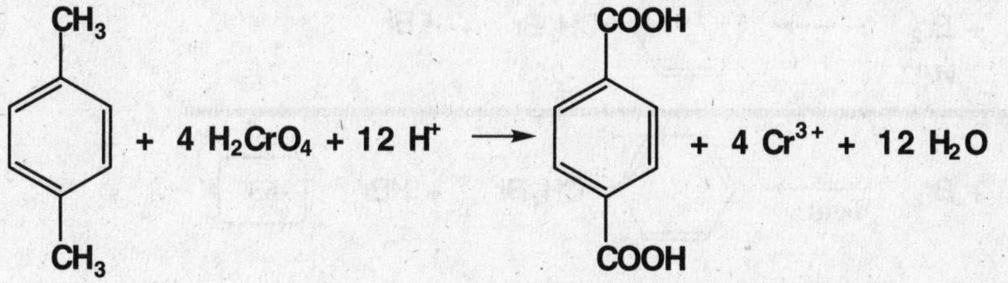

+ 4 H_2CrO_4 + 12 H^+ ⟶ + 4 Cr^{3+} + 12 H_2O **balanced redox equation**

$$0.250 \text{ g Xyl} \times \left(\frac{\text{mol Xyl}}{106 \text{ g Xyl}} \right) \times \left(\frac{4 \text{ mol } H_2CrO_4}{1 \text{ mol Xyl}} \right) \times \left(\frac{118 \text{ g } H_2CrO_4}{\text{mol } H_2CrO_4} \right)$$

$$= 1.11 \text{ g } H_2CrO_4 = 1110 \text{ mg } H_2CrO_4$$

Problem 21.42 Each of the following reactions occurs by a radical chain mechanism.

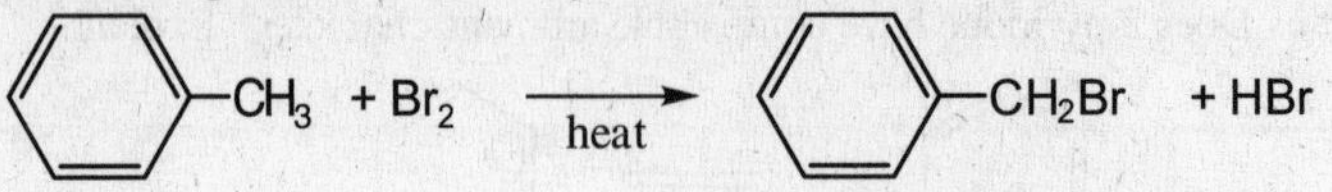

Toluene   **Benzyl bromide**

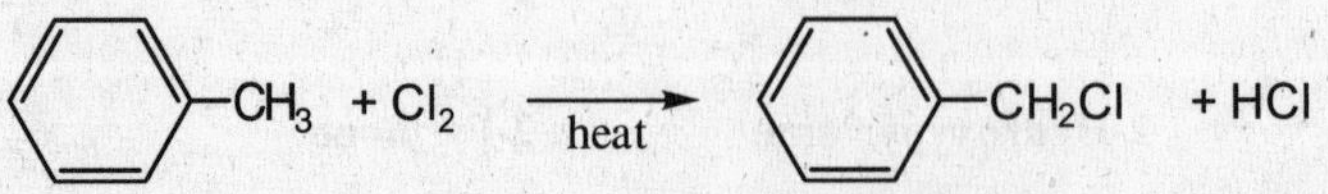

Toluene   **Benzyl chloride**

(a) Calculate the heat of reaction, $\Delta H°$, in kilocalories per mol for each reaction. (Consult Appendix 3 for bond dissociation enthalpies.)

Following are the calculations of the enthalpies of reaction.

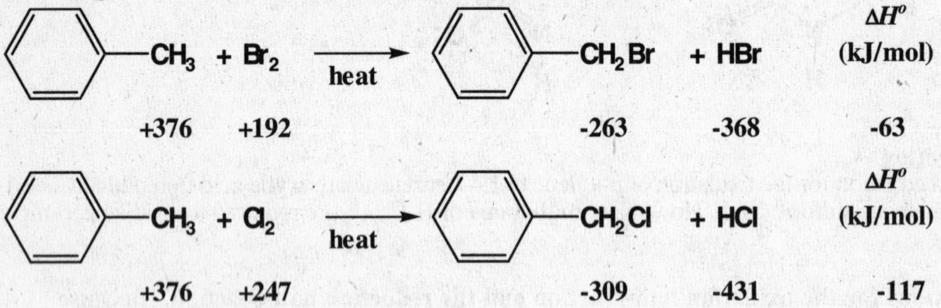

(b) Write a pair of chain propagation steps for each mechanism, and show that the net result of the chain propagation steps is the observed reaction.

(c) Calculate $\Delta H°$ for each chain propagation step, and show that the sum for each pair of chain propagation steps is identical with the $\Delta H°$ value calculated in part (a).

Shown below are pairs of chain propagation steps for each reaction. Each pair adds up to the observed reaction and the observed enthalpy of reaction. This answers both (b) and (c).

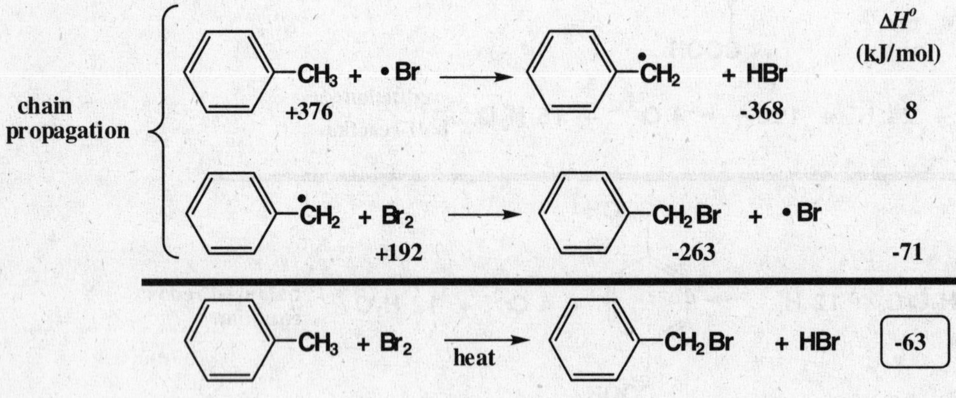

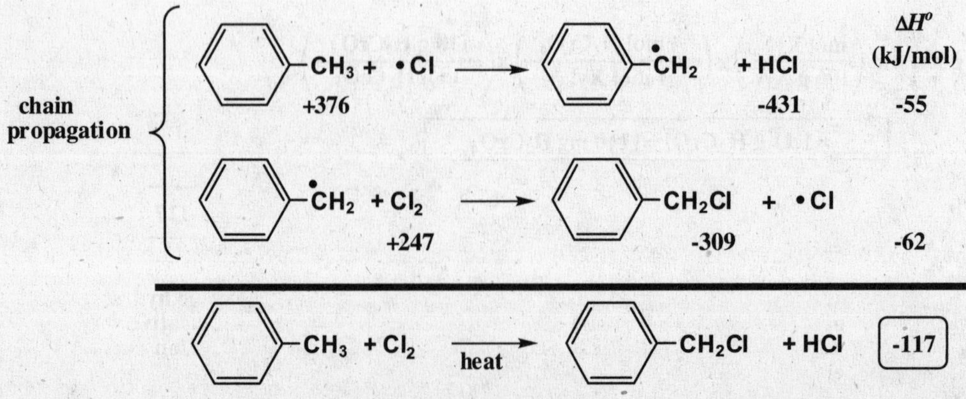

Problem 21.43 Following is an equation for iodination of toluene.

Toluene   **Benzyl iodide**

This reaction does not take place. All that happens under experimental conditions for the formation of radicals is initiation to form iodine radicals, I•, followed by termination to reform I_2. How do you account for these observations?

Reaction of toluene and iodine to form benzyl iodide and HI is endothermic.

$$\Delta H^o \text{ (kJ/mol)}$$

+376   +151   -213   -297   +17

Using values for bond dissociation enthalpies, calculate the enthalpy for each of the most likely chain propagation steps. Abstraction of hydrogen by an iodine radical (an iodine atom) is endothermic by 79 kJ/mol. The activation energy for this step is approximately a few kJ/mol greater than 79 kJ/mol. Given this large activation energy and the fact that the overall reaction is endothermic, it does not occur as written.

chain propagation

$$\Delta H^o \text{ (kJ/mol)}$$

+376   +•I   -297   +79

+151   -213   -62

Problem 21.44 Although most alkanes react with chlorine by a radical chain mechanism when reaction is initiated by light or heat, benzene fails to react under the same conditions. Benzene cannot be converted to chlorobenzene by treatment with chlorine in the presence of light or heat.

(a) Explain why benzene fails to react under these conditions. (Consult Appendix 3 for relevant bond dissociation enthalpies).

chain propagation

$$\Delta H^o \text{ (kcal/mol)}$$

+472   +•Cl   -431   +41

+247   -405   +•Cl   -158

-117

As can be seen in the above equations, the rate-determining step is the abstraction of a benzene hydrogen atom by a chlorine radical. This process is endothermic by 41 kJ/mol because the C-H bonds are relatively strong (472 kJ/mol). Therefore, even though the entire process is exothermic by 117 kJ/mol, the reaction does not proceed because of the highly endothermic rate-determining step.

(b) Explain why the bond dissociation enthalpy of a C-H bond in benzene is significantly greater than that in alkanes.

The benzene carbon atoms are *sp²* hybridized, so the C-H bonds are derived from *sp²* hybridized orbitals. Compared with alkanes that use *sp³* hybridized orbitals, the *sp²* orbitals of benzene have greater *s* character and thus the electrons are held closer to the nucleus. This has the effect of increasing C-H bond strength in benzene relative to alkanes.

<u>Problem 21.45</u> Following is an equation for hydroperoxidation of cumene.

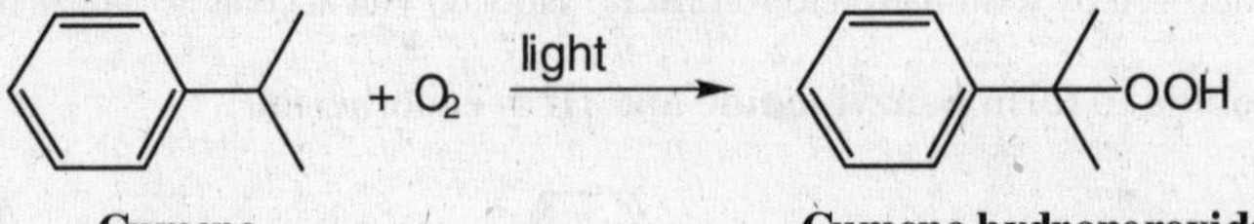

Cumene   **Cumene hydroperoxide**

Propose a radical chain mechanism for this reaction. Assume that initiation is by an unspecified radical, R•.

The stability of the benzyl radical, especially with the added methyl groups, facilitates the reaction with the radical initiator according to the following mechanism.

Step 1: **Inititiation**

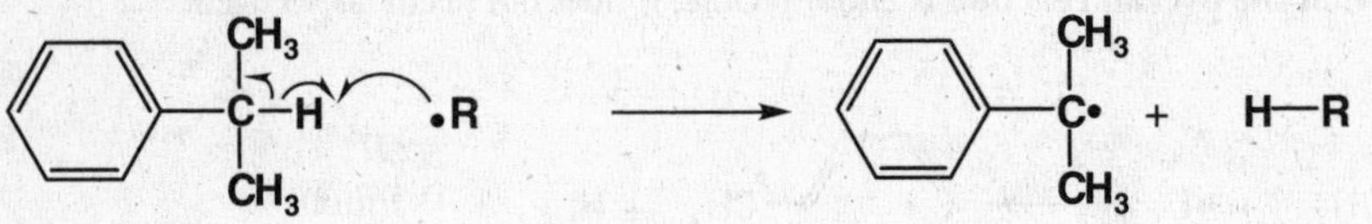

Step 2: **Propagation**

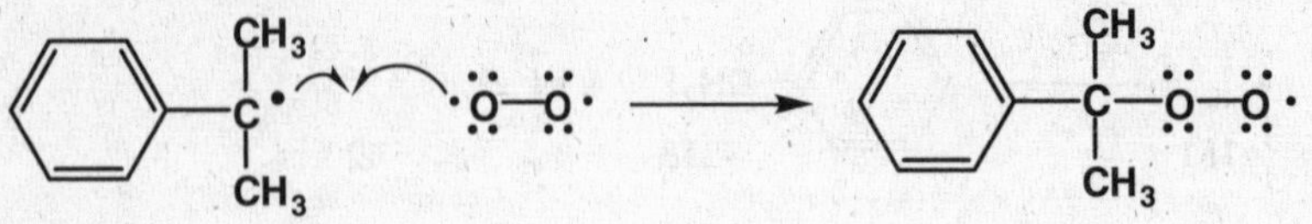

Step 3: **2nd Propagation Step**

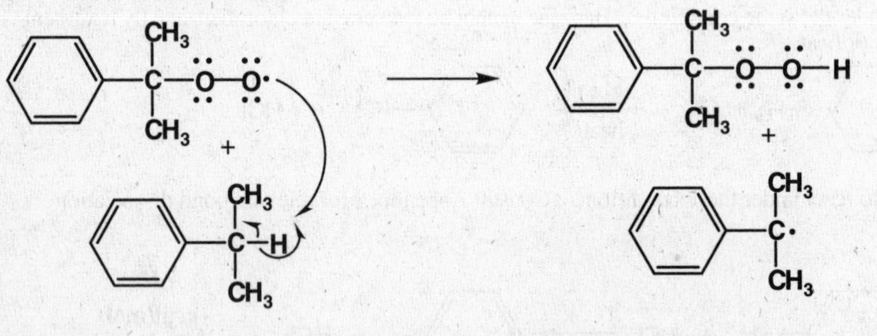

<u>Problem 21.46</u> Para-substituted benzyl halides undergo reaction with methanol by an S_N1 mechanism to give a benzyl ether. Account for the following order of reactivity under these conditions.

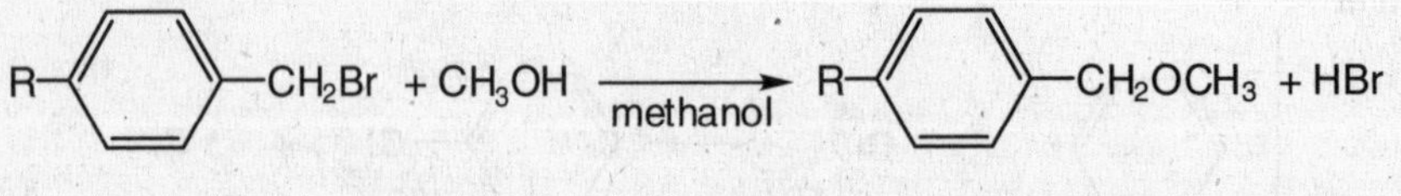

Rate of S_N1 reaction: $R = CH_3O > CH_3 > H > NO_2$

A reasonable S_N1 mechanism is shown below:

Step 1: Break a bond to give stable molecules or ions. **This is the rate determining step.**

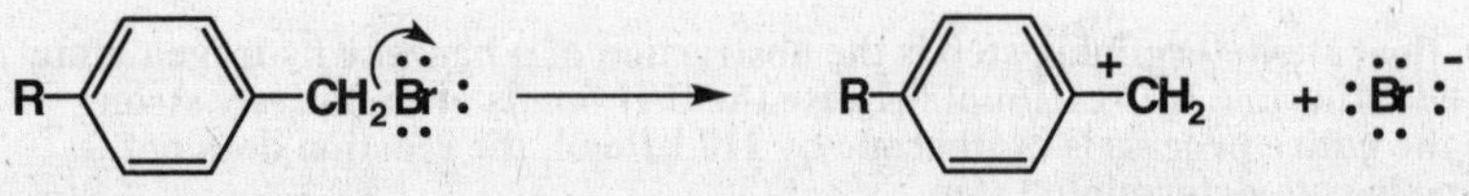

Step 2: Make a bond between a nucleophile and an electrophile.

Step 3: Take a proton away.

The rate-determining step in this process is formation of the benzylic cation, so anything that stabilizes the benzylic cation will speed up the S_N1 reaction. Electron-donating groups such as the methoxy group are stabilizing to a benzylic cation. Electron-withdrawing groups such as the nitro group are destabilizing to a benzylic cation. Thus, the groups are listed in order from most electron-donating to most electron-withdrawing.

<u>Problem 21.47</u> When warmed in dilute sulfuric acid, 1-phenyl-1,2-propanediol undergoes dehydration and rearrangement to give 2-phenylpropanal.

1-Phenyl-1,2-propanediol
(racemic)

2-Phenylpropanal
(racemic)

(a) Propose a mechanism for this example of a pinacol rearrangement (Section 10.7).

Step 1: Add a proton.

Step 2: Break a bond to give stable molecules or ions.

Step 3: 1,2 Shift. The new cation is stabilized by resonance delocalization.

Step 4: Take a proton away.

(b) Account for the fact that 2-phenylpropanal is formed rather than its constitutional isomer, 1-phenyl-1-propanone.

$$O$$
$$C_6H_5-\overset{\overset{\displaystyle O}{\|}}{C}-CH_2-CH_3 \qquad\qquad C_6H_5-\overset{\overset{\displaystyle CH_3}{|}}{CH}-\underset{\underset{\displaystyle O}{\|}}{C}-H$$

1-Phenyl-1-propanone   **2-Phenylpropanal**

In the observed reaction that leads to the aldehyde product, protonation of the benzylic hydroxyl followed by loss of H_2O gives a benzylic carbocation (Step 2). 1-Phenyl-1-propanone would result from protonation and loss of the other -OH group, but that process involves a less stable secondary carbocation so it is not observed.

Problem 21.48 In the chemical synthesis of DNA and RNA, hydroxyl groups are normally converted to triphenylmethyl (trityl) ethers to protect the hydroxyl group from reaction with other reagents.

$$RCH_2OH + Ph_3CCl \xrightarrow[\text{amine}]{\text{tertiary}} RCH_2OCPh_3 + HCl \quad \overset{\text{neutralized by the}}{\text{tertiary amine}}$$

Triphenylmethyl   a triphenylmethyl ether
chloride (Trityl chloride)   (a trityl ether)

Triphenylmethyl ethers are stable to aqueous base, but are rapidly cleaved in aqueous acid.

$$RCH_2OCPh_3 + H_2O \xrightarrow{H^+} RCH_2OH + Ph_3COH$$

(a) Why are triphenylmethyl ethers so readily hydrolyzed by aqueous acid?

The triphenylmethyl ethers are hydrolyzed according to the following mechanism.

Step 1: Add a proton.

Step 2: Break a bond to give stable molecules or ions. **Note that the cation produced in this step is a highly resonance-stabilized benzylic cation due to the three adjacent phenyl rings.**

Step 3: Make a bond between a nucleophile and an electrophile.

Step 4: Take a proton away.

(b) How might the structure of the triphenylmethyl group be modified in order to increase or decrease its acid sensitivity?

Electron-releasing substituents like methoxy groups stabilize the triphenylmethyl cation and thereby increase the sensitivity of these triphenylmethyl ethers to acid. On the other hand, electron-withdrawing groups like nitro groups destabilize the triphenylmethyl cation and thereby decrease the sensitivity of these triphenylmethyl ethers to acid.

Synthesis

Problem 21.49 Using ethylbenzene as the only aromatic starting material, show how to synthesize the following compounds. In addition to ethylbenzene, use any other necessary organic or inorganic chemicals. Any compound already synthesized in one part of this problem may then be used to make any other compound in the problem.

(a)   [structure: benzene ring with —COOH]

Oxidation of ethylbenzene using H_2CrO_4 gives benzoic acid.

[structure: benzene with —CH₂CH₃] $\xrightarrow{H_2CrO_4}$ [structure: benzene with —C(=O)OH]

(b)   [structure: benzene ring with —CH(Br)CH₃, marked with *]

(racemic)

Bromination of the benzylic position using bromine at elevated temperature or NBS in the presence of peroxide. The reaction involves a radical chain mechanism. A racemic mixture is produced because the product is chiral.

[structure: benzene with —CH₂CH₃] + Br_2 $\xrightarrow{heat}$ [structure: benzene with —CHCH₃ bearing Br and *] + HBr

(c)   [structure: benzene ring with —CH=CH₂]

Dehydrohalogenation of the alkyl bromide from (b), brought about by a strong base such as KOH.

[structure: benzene with —CHCH₃ bearing Br and *] + KOH $\xrightarrow[ethanol]{E2}$ [structure: benzene with —CH=CH₂] + KBr + H_2O

(d)   [structure: benzene ring with —CH(OH)CH₃, marked with *]

(racemic)

Acid-catalyzed hydration of a carbon-carbon double bond of (c). The same reaction may also be brought about by oxymercuration followed by reduction with $NaBH_4$. A racemic mixture is produced because the product contains a chiral center.

[structure: benzene with —CH=CH₂] + H_2O $\xrightarrow{H_2SO_4}$ [structure: benzene with —CHCH₃ bearing OH and *]

(e)

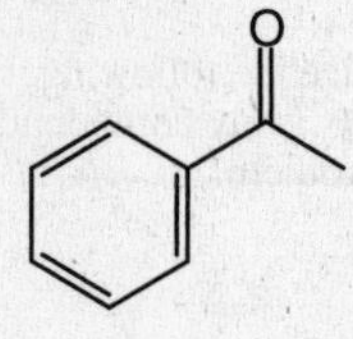

This oxidation is best brought about using the more selective oxidizing agent pyridinium chlorochromate (PCC) to avoid over oxidation to benzoic acid as in part (a).

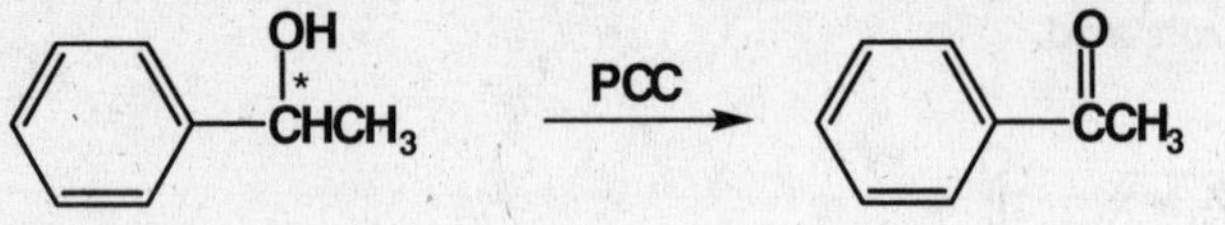

(f)

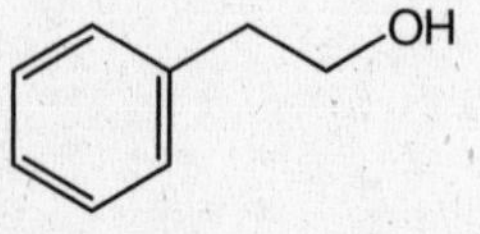

Hydroboration/oxidation of styrene from part (c).

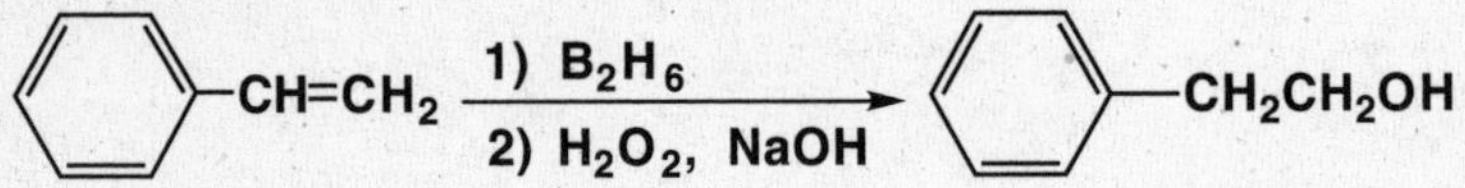

(g)

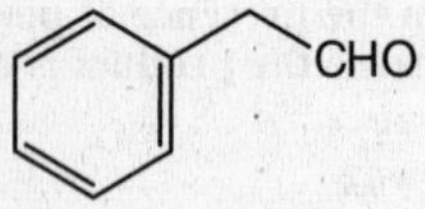

Oxidation of the primary alcohol of (f) using pyridinium chlorochromate.

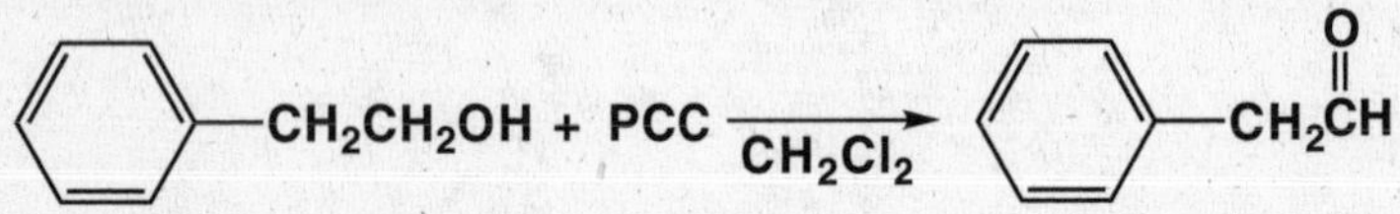

(h)

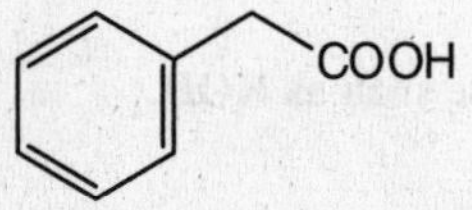

Oxidation of the primary alcohol of (f) using chromic acid. The same product may be formed by similar oxidation of the aldehyde (g). Note how care must be taken to prevent over-oxidation to benzoic acid.

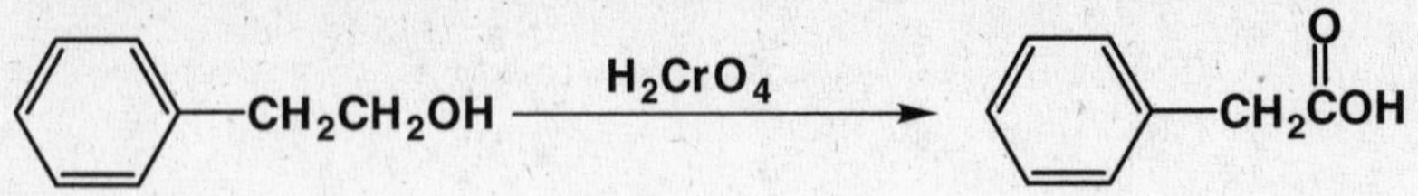

(i)

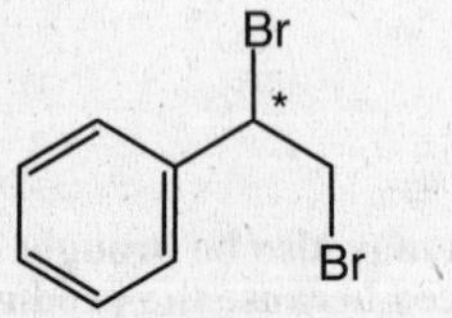

(racemic)

Addition of bromine to the carbon-carbon double bond of styrene from part (c). A racemic mixture is produced because the product contains a chiral center.

(j)

A double dehydrohalogenation of product (i) using sodium amide as the base.

$$\text{Br Br}$$
$$\text{CHCH}_2 \xrightarrow[\text{2.H}_2\text{O}]{\text{1. 3 NaNH}_2} \quad \text{C}\equiv\text{CH}$$

(k)

The terminal alkyne from (j) is deprotonated with sodium amide to produce the anionic species that reacts with allyl chloride to produce the desired product.

$$\text{C}\equiv\text{CH} \xrightarrow{\text{NaNH}_2} \text{C}\equiv\text{C:}^-\text{Na}^+ \xrightarrow[\text{S}_\text{N}2]{\text{ClCH}_2\text{CH}=\text{CH}_2}$$

$$\text{C}\equiv\text{CCH}_2\text{CH}=\text{CH}_2$$

(l)   C≡C(CH$_2$)$_5$CH$_3$

The deprotonated terminal alkyne from (k) reacts with hexyl chloride to produce the desired product.

$$\text{C}\equiv\text{C:}^-\text{Na}^+ \xrightarrow[\text{S}_\text{N}2]{\text{ClCH}_2(\text{CH}_2)_4\text{CH}_3} \quad \text{C}\equiv\text{C(CH}_2)_5\text{CH}_3$$

(m)

$$\begin{array}{c} \text{H} \qquad (\text{CH}_2)_5\text{CH}_3 \\ \text{C}=\text{C} \\ \text{Ph} \qquad\quad \text{H} \end{array}$$

The alkyne produced in part (l) is reduced with sodium metal in liquid ammonia to produce the desired *trans* alkene.

$$\text{C}\equiv\text{C(CH}_2)_5\text{CH}_3 \xrightarrow[\text{NH}_3\,(\textit{l})]{\text{2 Na}} \begin{array}{c} \text{H} \qquad (\text{CH}_2)_5\text{CH}_3 \\ \text{C}=\text{C} \\ \text{Ph} \qquad\quad \text{H} \end{array}$$

(n)

$$\begin{array}{c} \text{Ph} \qquad (\text{CH}_2)_5\text{CH}_3 \\ \text{C}=\text{C} \\ \text{H} \qquad\quad \text{H} \end{array}$$

The alkyne produced in part (l) is reduced with hydrogen and the Lindlar catalyst to produce the desired *cis* alkene.

$$\text{C}\equiv\text{C(CH}_2)_5\text{CH}_3 \xrightarrow[\substack{\text{Lindlar} \\ \text{Catalyst}}]{\text{H}_2} \begin{array}{c} \text{Ph} \qquad (\text{CH}_2)_5\text{CH}_3 \\ \text{C}=\text{C} \\ \text{H} \qquad\quad \text{H} \end{array}$$

Problem 21.50 Show how to convert 1-phenylpropane into the following compounds. In addition to this starting material, use any necessary inorganic reagents. Any compound synthesized in one part of this problem may then be used to make any other compound in the problem.

(a) C_6H_5 —— (with Br, *)
   (racemic)

Radical chain bromination of 1-phenylpropane. Bromination is highly regioselective for the benzylic position. A racemic mixture is produced because the product contains a chiral center.

C_6H_5 —— + Br_2 $\xrightarrow{\text{heat}}$ C_6H_5 —— (with Br, *) + HBr

(b) C_6H_5 —— (with OH, *)
   (racemic)

Dehydrohalogenation of product (a) using KOH or other strong base will give an alkene. Acid-catalyzed hydration of the alkene will give the desired alcohol. The reaction is highly regioselective because of the stability of the benzylic carbocation formed by protonation of the alkene. Alternatively, oxymercuration followed by reduction with $NaBH_4$ forms the same secondary alcohol.

C_6H_5 —— (with Br, *) + KOH $\longrightarrow$ C_6H_5 —— $\xrightarrow[H_2SO_4]{H_2O}$ C_6H_5 —— (with OH, *)

(c) C_6H_5 —— (with O)

Oxidation of the secondary alcohol of (b) using pyridinium chlorochromate (PCC) as the oxidizing agent.

C_6H_5 —— (with OH, *) $\xrightarrow{\text{PCC}}$ C_6H_5 —— (with O)

(d) C_6H_5 —— (with Cl, *, *, Cl)
   (racemic)

Addition of chlorine by electrophilic addition to the alkene formed in the first reaction of answer (b). Starting with the *trans* alkene, the two stereoisomers shown are formed as a racemic mixture.

C_6H_5 —— + Cl_2 $\longrightarrow$ C_6H_5 (with Cl, H, H, Cl) + C_6H_5 (with H, Cl, Cl, H)

(e) 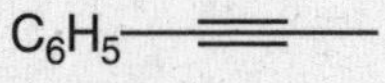 C_6H_5—≡—

Double dehydrohalogenation of product (d) using sodium amide as the base.

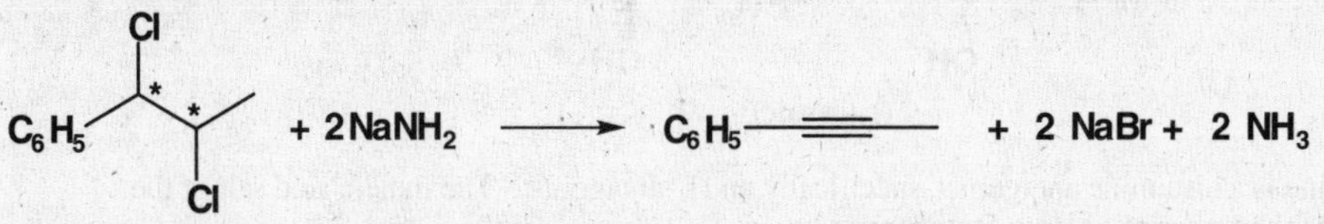

C_6H_5 —CHCl—CHCl—CH$_3$ + 2 NaNH$_2$ ⟶ C_6H_5—≡— + 2 NaBr + 2 NH$_3$

(f) 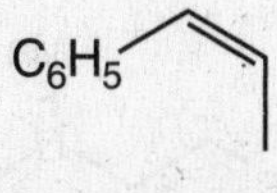 C_6H_5

Catalytic reduction of the alkyne using the Lindlar catalyst to reduce the alkyne of part (e) to the *cis* alkene.

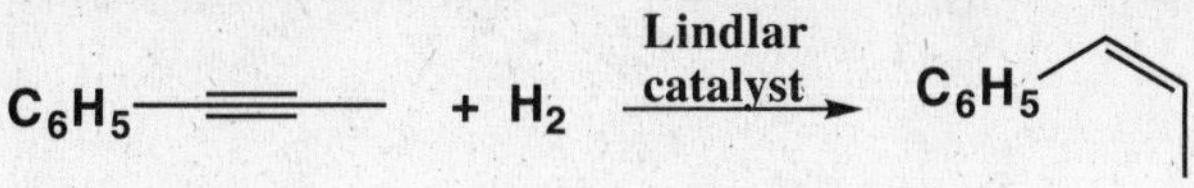

C_6H_5—≡— + H$_2$ $\xrightarrow{\text{Lindlar catalyst}}$ C_6H_5

(g) 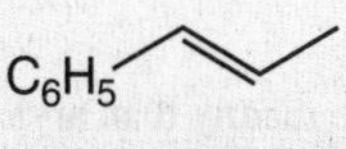 C_6H_5

The *trans* alkene will be the primary product generated upon dehydrohalogenation of the 1-bromo-1-phenylpropane produced in (a).

C_6H_5—CHBr— + KOH ⟶ C_6H_5

Alternatively, the alkyne from part (e) can be treated with Na in ammonia to give the *trans* alkene.

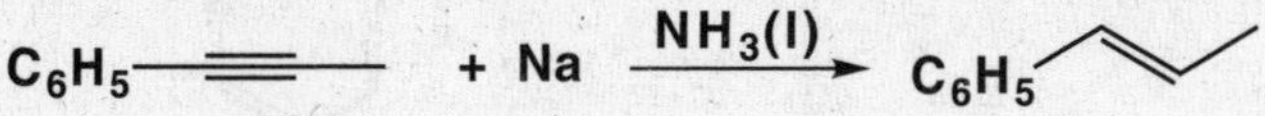

C_6H_5—≡— + Na $\xrightarrow{\text{NH}_3(\text{l})}$ C_6H_5

(h) 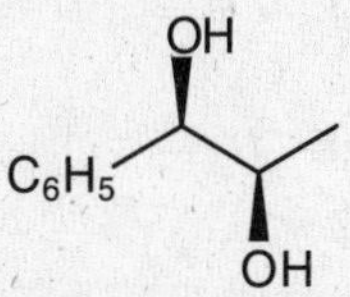 C_6H_5 —(OH)—(OH)—

Treatment of the *E* alkene from part (g) with OsO$_4$/H$_2$O$_2$ will give the desired product because the syn stereoselectivity of the reaction will place both OH groups on the same face of the double bond. Note that the product of this reaction will be a racemic mixture, only one enantiomer of which is shown above.

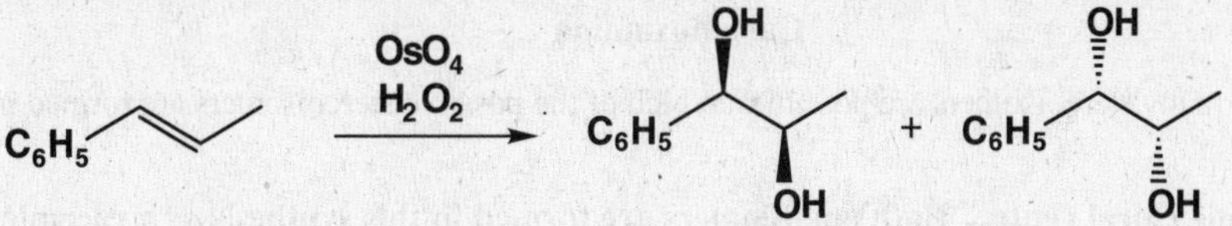

C_6H_5 $\xrightarrow[\text{H}_2\text{O}_2]{\text{OsO}_4}$ C_6H_5—(OH)—(OH)— + C_6H_5—(OH)—(OH)—

(racemic)

(i) 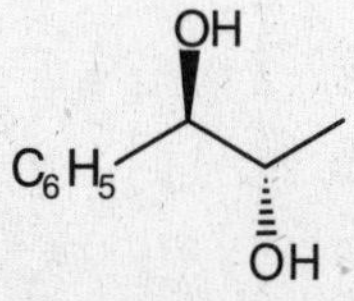 C_6H_5 —(OH)—(OH)—

Oxidation of the alkene from part (g) by first creating an epoxide, followed by opening in acid or base gives the glycol with the *trans* configuration because the anti stereoselectivity of the epoxide ring opening will place the OH groups on opposite faces of the double bond. Note that a racemic mixture is formed, only one enantiomer of which is shown above.

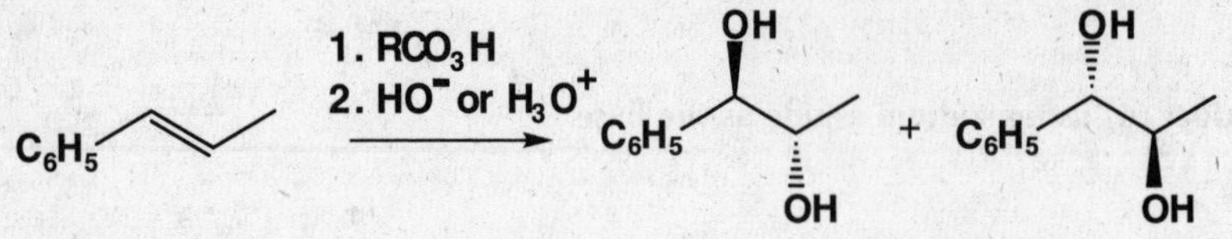

1. RCO₃H
2. HO⁻ or H₃O⁺

C_6H_5 ... C_6H_5 + C_6H_5

(racemic)

<u>Problem 21.51</u> Carbinoxamine is a histamine antagonist, specifically an H_1-antagonist. The maleic acid salt of the levorotatory isomer is sold as the prescription drug Rotoxamine.

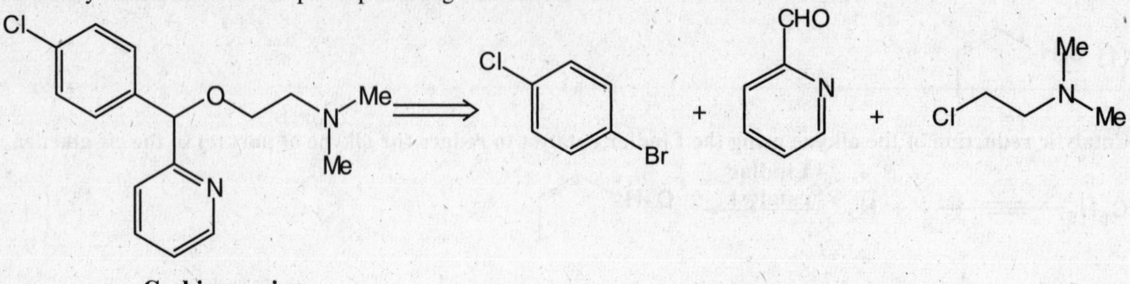

Carbinoxamine

(a) Propose a synthesis of carbinoxamine. (*Note*: Aryl bromides form Grignard reagents much more readily than aryl chlorides.)

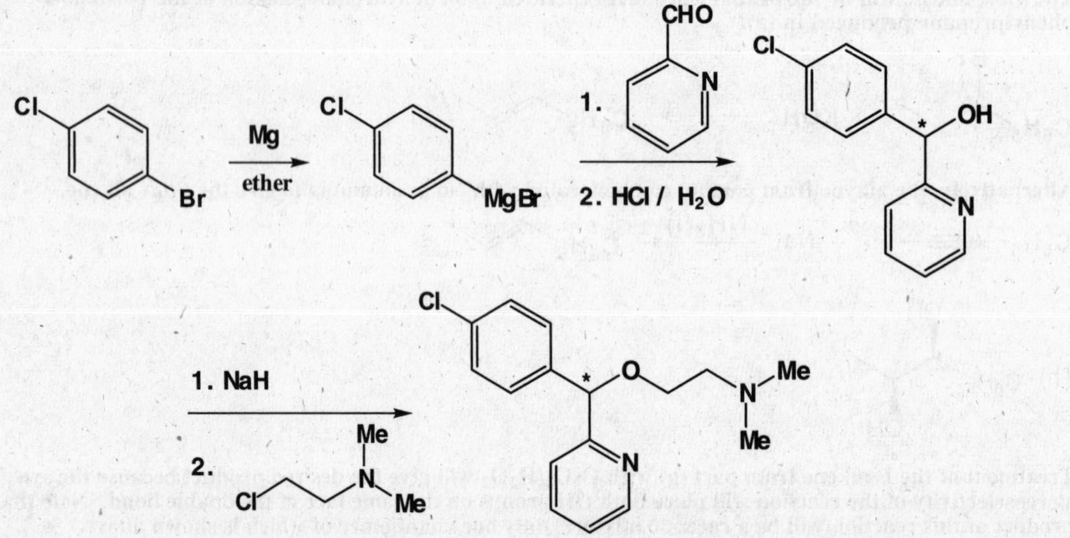

Mg
ether

1.
2. HCl / H₂O

1. NaH
2.

Carbinoxamine

(b) Is carbinoxamine chiral? If so how many stereoisomers are possible? Which of the possible stereoisomers are formed in this synthesis?

Yes, carbinoxamine is chiral with one chiral center. Both enantiomers are formed in this synthesis as a racemic mixture.

<u>Problem 21.52</u> Cromolyn sodium, developed in the 1960s, is used to prevent allergic reactions primarily affecting the lungs, as for example exercise-induced emphysema. It is thought to block the release of histamine, which prevents the sequence of events leading to swelling, itching, and constriction of bronchial tubes. Cromolyn sodium is synthesized in the following series of steps. Treatment of one mole of epichlorohydrin (Section 11.10) with two moles of 2,6-dihydroxyacetophenone in the presence of base gives I. Treatment of I with two moles of diethyl oxalate in the presence of sodium ethoxide gives a diester II. Saponification of the diester with aqueous NaOH gives cromolyn sodium.

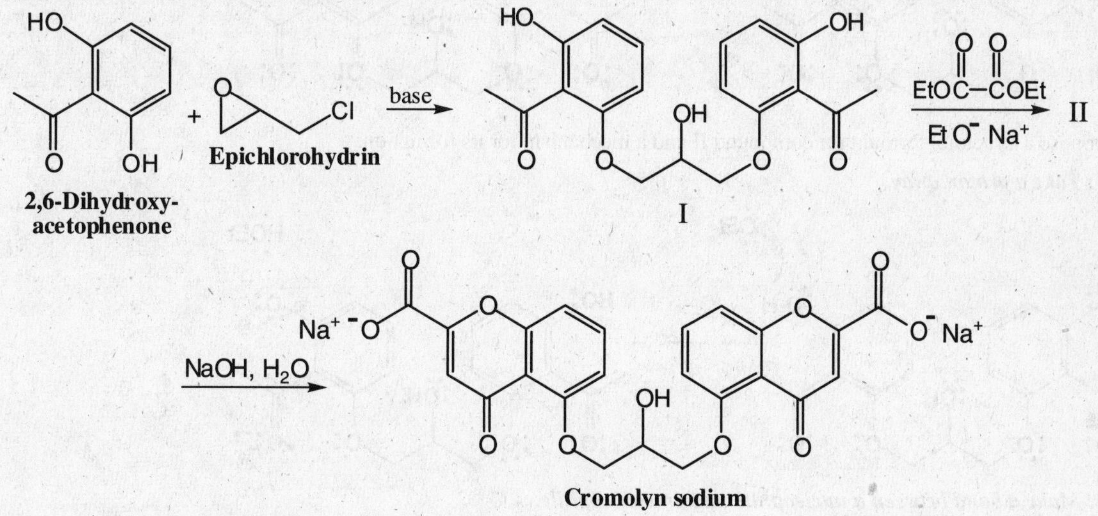

2,6-Dihydroxy-acetophenone

Epichlorohydrin

I

II

Cromolyn sodium

(a) Propose a mechanism for the formation of compound I.

Step 1: Take a proton away.

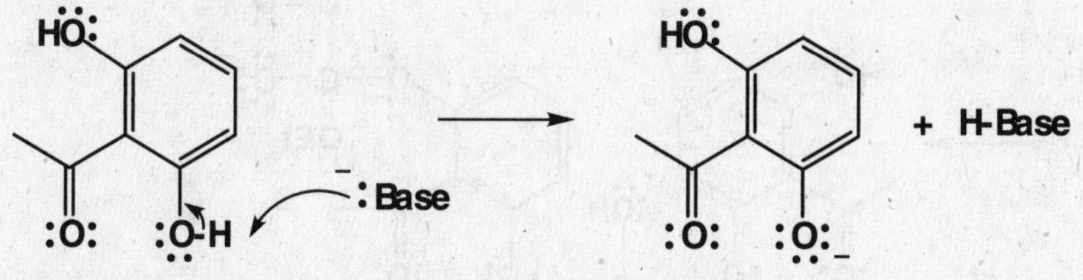

Step 2: Make a bond between a nucleophile and an electrophile and simultaneously break a bond to give stable molecules or ions.

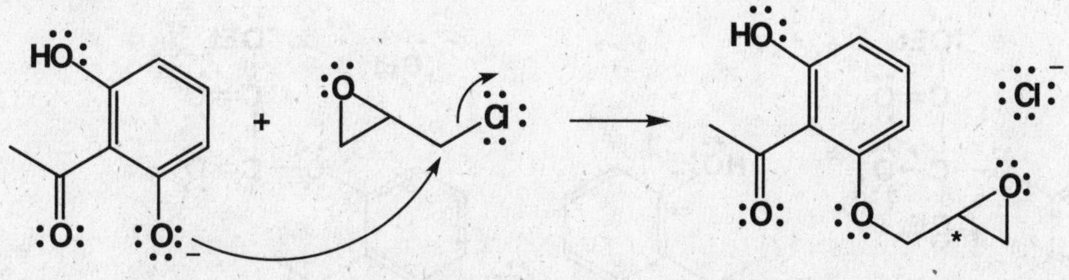

Step 3: Make a bond between a nucleophile and an electrophile and simultaneously break a bond to give stable molecules or ions.

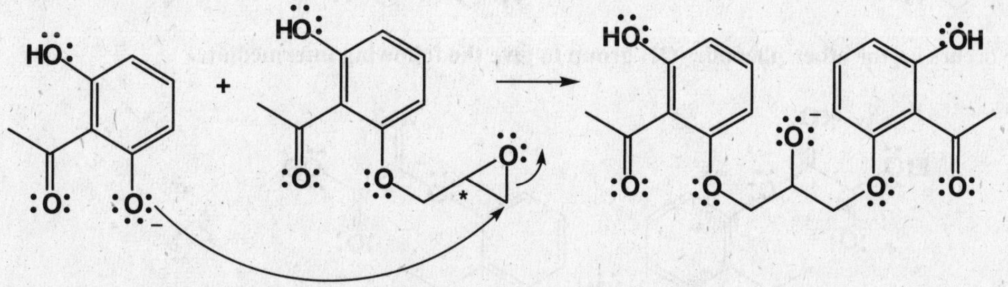

Step 4: Add a proton.

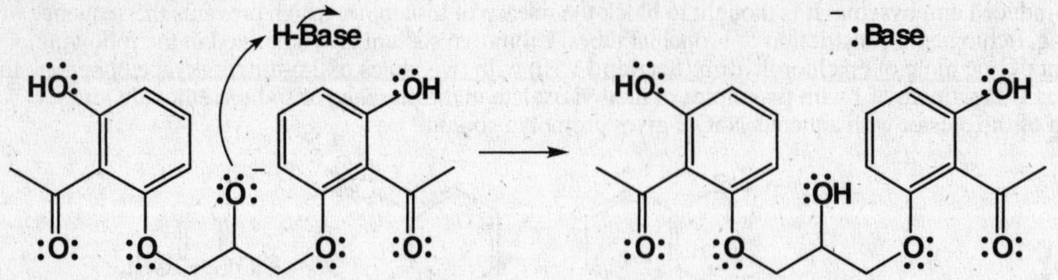

(b) Propose a structural formula for compound II and a mechanism for its formation.

Step 1: Take a proton away.

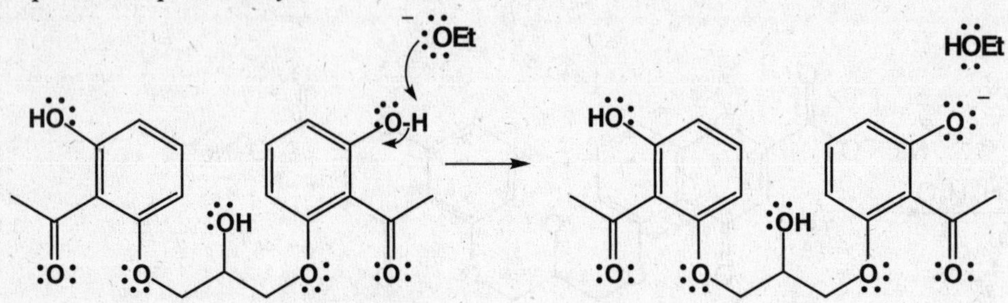

Step 2: Make a bond between a nucleophile and an electrophile.

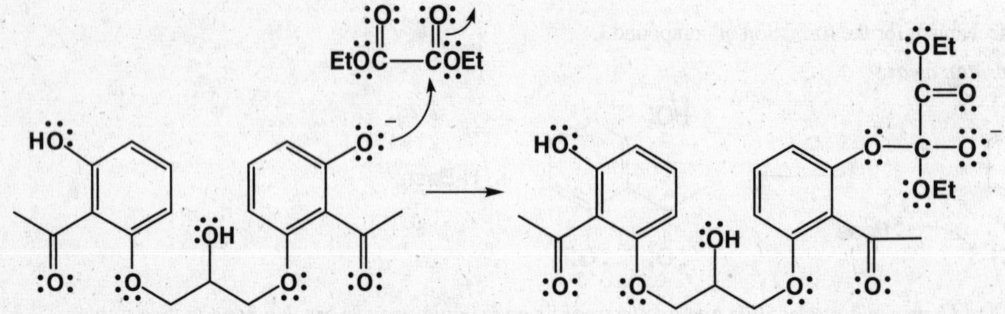

Step 3: Break a bond to give stable molecules or ions.

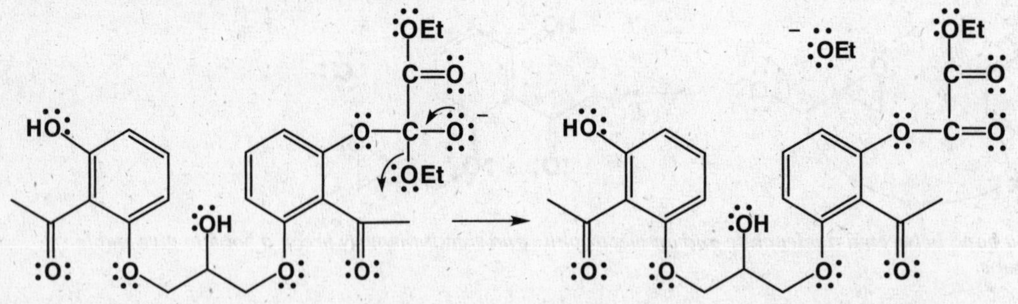

The same sequence occurs on the other phenolic -OH group to give the following intermediate.

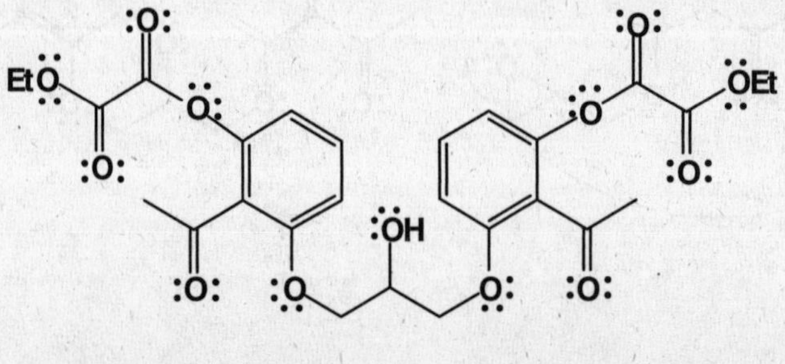

Step 4: Take a proton away. The methyl ketone is deprotonated to give an enolate

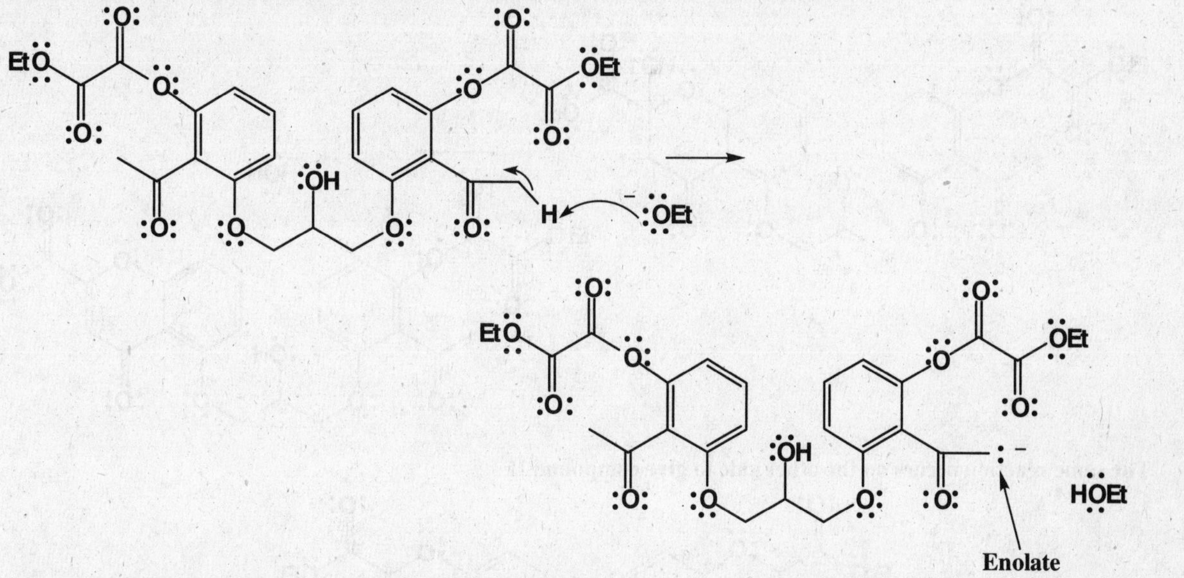

Enolate

Step 5: Make a bond between a nucleophile and an electrophile. The enolate then undergoes an aldol reaction.

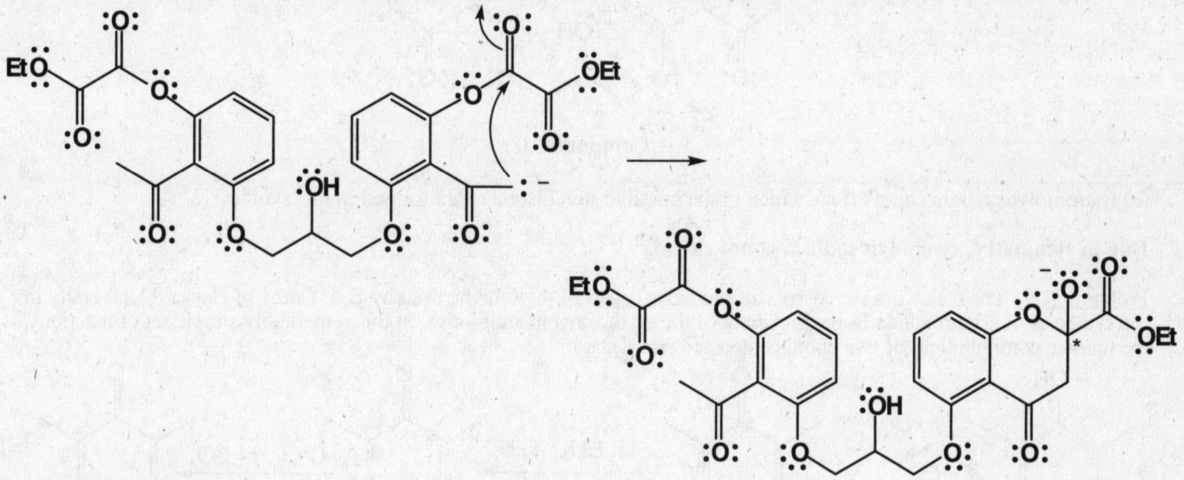

Step 6: Add a proton.

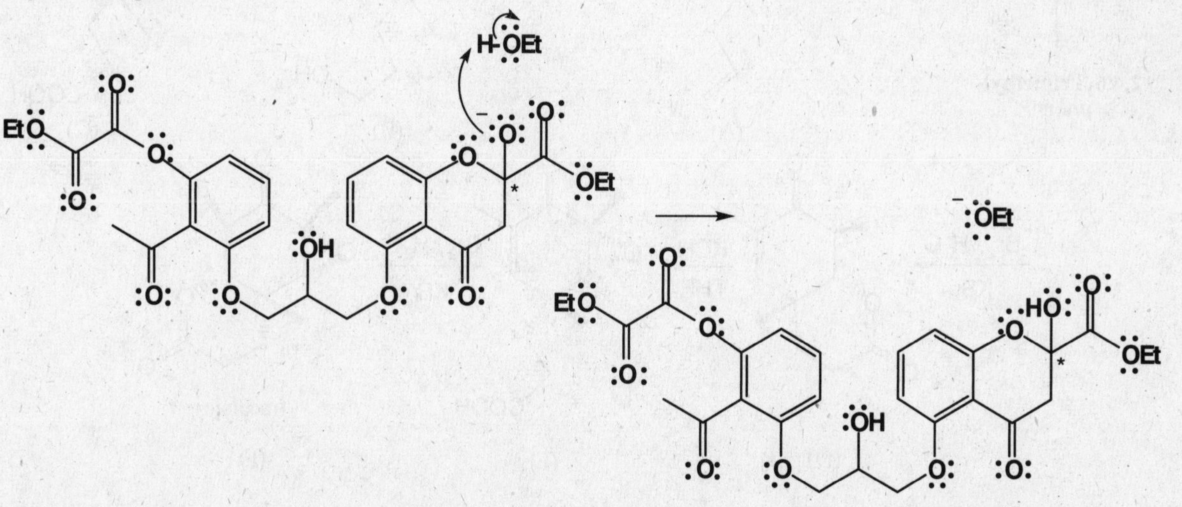

Step 7: Dehydration. Dehydration, which actually involves several steps, completes the process

The same reaction occurs on the other side to give compound II

Compound II

(c) Is cromolyn sodium chiral? If so, which of the possible stereoisomers are formed in this synthesis?

Due to symmetry, cromolyn sodium is not chiral.

<u>Problem 21.53</u> The following stereospecific synthesis is part of the scheme used by E. J. Corey of Harvard University in the synthesis of erythronolide B, the precursor of the erythromycin antibiotics. In this remarkably simple set of reactions, the relative configurations of five chiral centers are established.

(a) Propose a mechanism for the conversion of 2,4,6-trimethylphenol to compound A.

Step 1: Take a proton away.

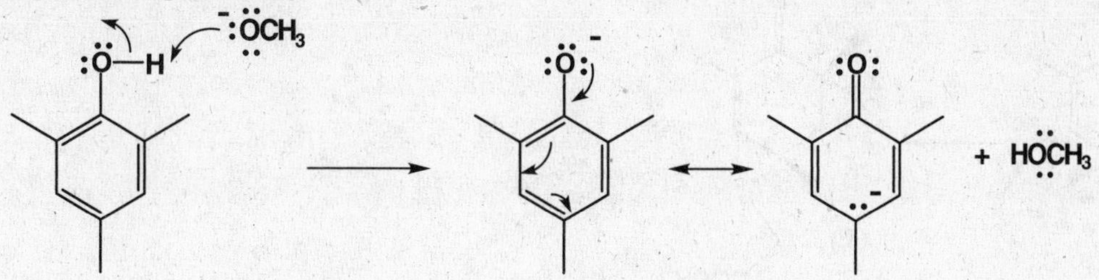

Resonance stabilized

The phenoxide intermediate is resonance stabilized, and a total of four contributing structures can be drawn. For the purposes of this mechanism, the most important is the one shown above in which the negative charge is placed on the ring carbon atom para to the oxygen atom. This partial negative charge is nucleophilic enough to take part in an S_N2 reaction as shown.

Step 2: Make a bond between a nucleophile and an electrophile and simultaneously break a bond to give stable molecules or ions.

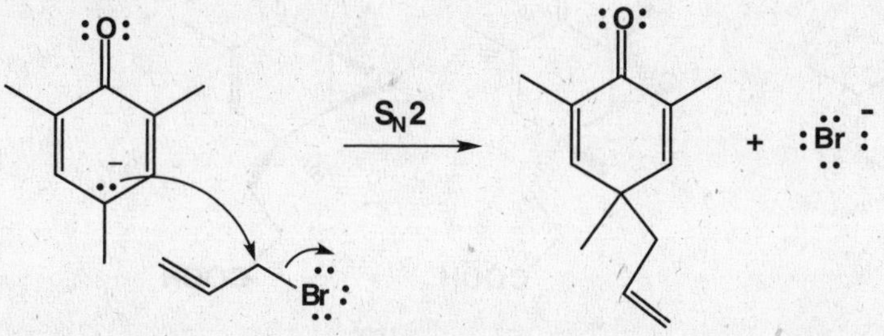

(b) Account for the stereoselectivity and regioselectivity of the three steps in the conversion of compound C to compound F.

In the conversion of C to D (as well as E to F), the mechanism involves electrophilic attack of bromine on a double bond of the ring, leading to a bromonium ion intermediate. The bromonium ion intermediate is attacked by the appended carboxyl group to complete the reaction. The bromonium ion intermediate dictates the anti stereoselectivity observed, and the length of the carboxylic acid chain leads to the formation of the six-membered ring.

In the conversion of (D) to (E), the mechanism is a more complicated. The carbonyl of the lactone ester is attacked by hydroxide to generate a tetrahedral addition intermediate that collapses by expelling an alkoxide that is set up to attack from the backside of the C-Br bond in an intramolecular S_N2 step to give the observed stereochemistry as shown below.

Step 1: Make a bond between a nucleophile and an electrophile.

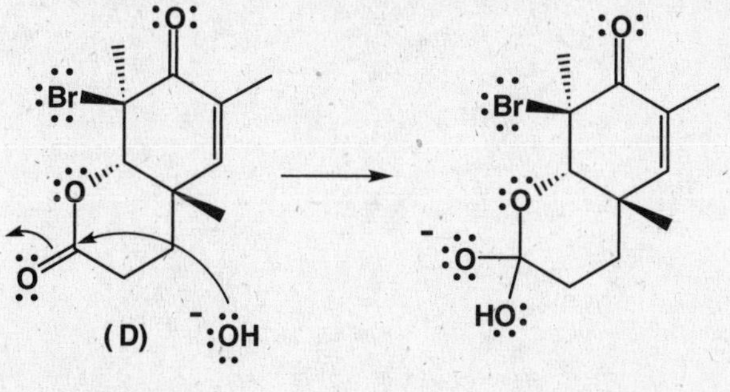

Step 2: Make a bond between a nucleophile and an electrophile and simultaneously break a bond to give stable molecules or ions.

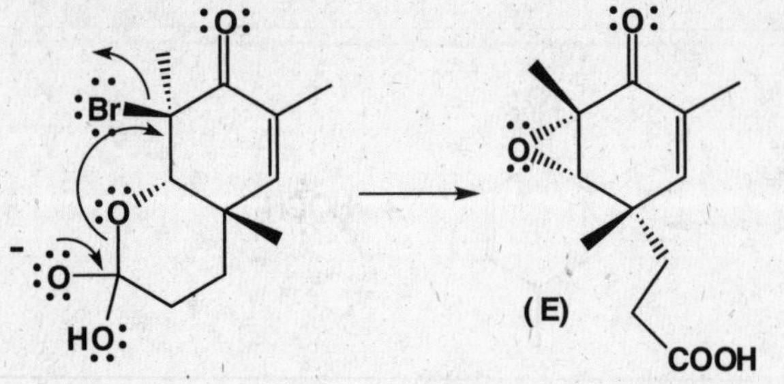

(E)

(c) Is compound F produced in this synthesis as a single enantiomer or as a racemic mixture? Explain.

Even though only one structure is drawn in each step, in reality compounds D, E, and F are produced as racemic mixtures. Compound C is achiral, and there are no chiral reagents being used, so all chiral molecules produced subsequently must be made as racemic mixtures. When molecules have this many chiral centers, it often helps to use molecular models when analyzing stereochemistry. Both enantiomers of D, E, and F are drawn below.

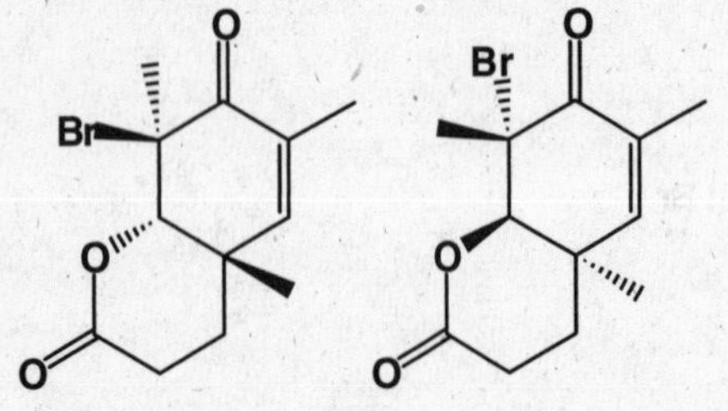

(D)

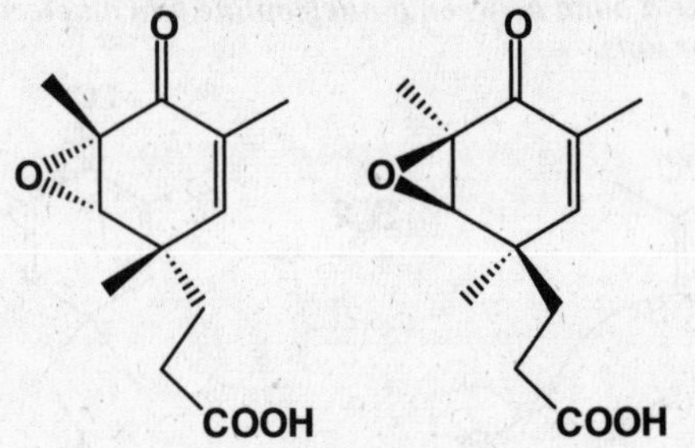

(E)

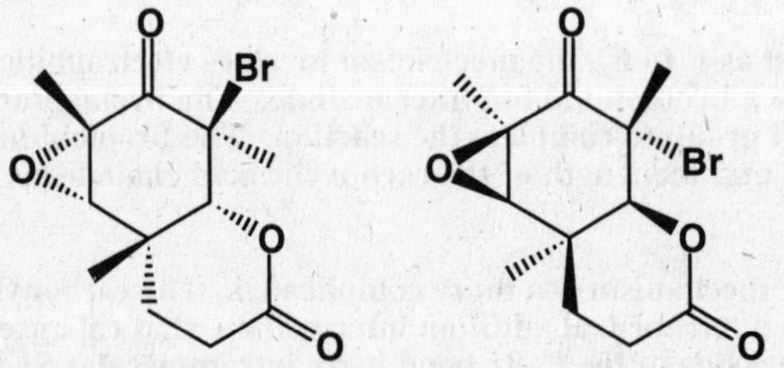

(F)

<u>Problem 21.54</u> Following is an outline of one of the first syntheses of the antidepressant fluoxetine (Prozac).

(A)   (B)   (C)   (D)

(E)   (F)

An *N*-substituted   **Fluoxetine**
carbamic acid

(a) Propose a reagent for the conversion of (A) to (B).

Conversion of (A) to (B) involves a Michael addition (Section 19.8) of dimethylamine to the α,β-unsaturated ketone.

(b) Propose a reagent for the conversion of (B) to (C).

Conversion of (B) to (C) is a two-electron reduction of the carbonyl group of the ketone to a secondary alcohol. Suitable reducing agents are H_2 in the presence of a transition metal catalyst such as Pt or Pd (Section 16.11), or a metal hydride reduction using $NaBH_4$ (Section 16.11).

(c) Propose a reagent for the conversion of (C) to (D).

To convert (C) to (D), treat the secondary alcohol group with thionyl chloride, $SOCl_2$ (Section 10.5).

(d) Propose a mechanism for the conversion of (E) to (F). The reagent used in this synthesis is ethyl chloroformate. The other product of this conversion is chloromethane, CH_3Cl. Your mechanism should show how the CH_3Cl is formed.

The first step in the conversion of (E) to (F) is nucleophilic addition of the tertiary amine nitrogen to the carbonyl group of ethyl chloroformate to give a tetrahedral carbonyl addition intermediate followed by its collapse to give a quaternary amide with positive charge on the nitrogen of the amide. Nucleophilic attack of chloride (an S_N2 reaction, Section 8.3) on one of the two methyl groups and displacement of nitrogen gives the ethyl carbamide and chloromethane.

Step 1: Make a bond between a nucleophile and an electrophile.

Tetrahedral addition intermediate

Step 2: Break a bond to give stable molecules or ions.

Tetrahedral addition intermediate

Step 3: Make a bond between a nucleophile and an electrophile and simultaneously break a bond to give stable molecules or ions.

(e) Propose a reagent or reagents to bring about the conversion of (F) to fluoxetine. Note that the bracketed intermediate formed in this step is an *N*-substituted carbamic acid. Such compounds are unstable and break down to carbon dioxide and an amine.

(F) is converted to fluoxetine by base-promoted hydrolysis of the ester (Section 18.4) using aqueous NaOH to give the *N*-substituted carbamic acid followed by decarboxylation of the carbamic acid.

(f) Is fluoxetine chiral? If so, which of the possible stereoisomers are formed in this synthesis?

Fluoxetine is chiral, because it has one chiral center. Both enantiomers are produced in the above synthesis.

Enantiomers of fluoxetine

Problem 21.55 Following is a synthesis for the antiarrhythmic drug bidisomide. The symbol Bn is an abbreviation for the benzyl group, $C_6H_5CH_2$-.

(a) Propose mechanisms for the conversion of (A) to (B) and of (B) to (C). What is the function of sodium amide in each reaction?

An α-hydrogen of a nitrile is weakly acidic, with a pK_a of approximately 25. Treatment of the nitrile with sodium amide, a very strong base, gives an anion, which then displaces chlorine by an S$_N$2 reaction (Section 8.3) to give (B). The same mechanism applies to the conversion of (B) to (C).

(b) Why is it necessary to incorporate the benzyl group on the chloroamine used to convert (B) to (C)?

The secondary amine of (C) is protected in benzylation to prevent its reaction during the acid-catalyzed hydrolysis of the cyano group to an amide. There is the possibility that the secondary amine will undergo an intramolecular reaction with the amide group to give a cyclic, five-membered lactam.

(c) Propose a reagent or reagents for the removal of the benzyl group in the conversion of (D) to (E).

The benzyl protecting group is removed by hydrogenolysis using H$_2$/Pt (Section 21.5).

(d) Propose a reagent for the conversion of (E) to bidisomide.

Treatment of the secondary amine with either acetic anhydride or acetyl chloride (Section 18.6) gives bidisomide.

(e) Is bidisomide chiral? If so, which of the possible stereoisomers are formed in this synthesis?

Bidosomide has one chiral center so a racemic mixture of enantiomers is produced in this synthesis.

Enantiomers of bidisomide

Problem 21.56   A finding that opened a route to β-blockers was the discovery that β-blocking activity is retained if an oxygen atom is interposed between the aromatic ring and the side chain. To see this difference, compare the structures of labetalol (Problem 22.55) and propranolol. Thus, alkylation of phenoxide ions can be used as a way to introduce this side chain. The first of this new class of drugs was propranolol.

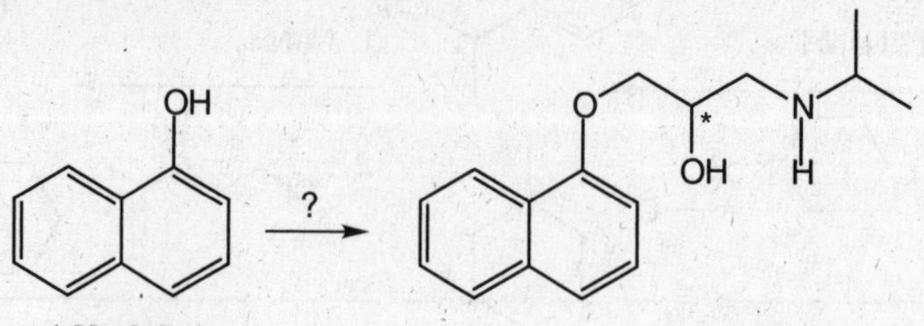

**1-Naphthol
(β-Naphthol)**

Propranolol

(a) Show how propanolol can be synthesized from 1-naphthol, epichlorohydrin (Section 11.10), and isopropylamine.

Epichlorohydrin is useful as a synthetic intermediate because it reacts with two different nucleophiles, in this case a phenolic oxygen atom and an amine nitrogen atom. The synthesis can be carried out by reacting epichlorohydrin with the phenolic oxygen nucleophile first.

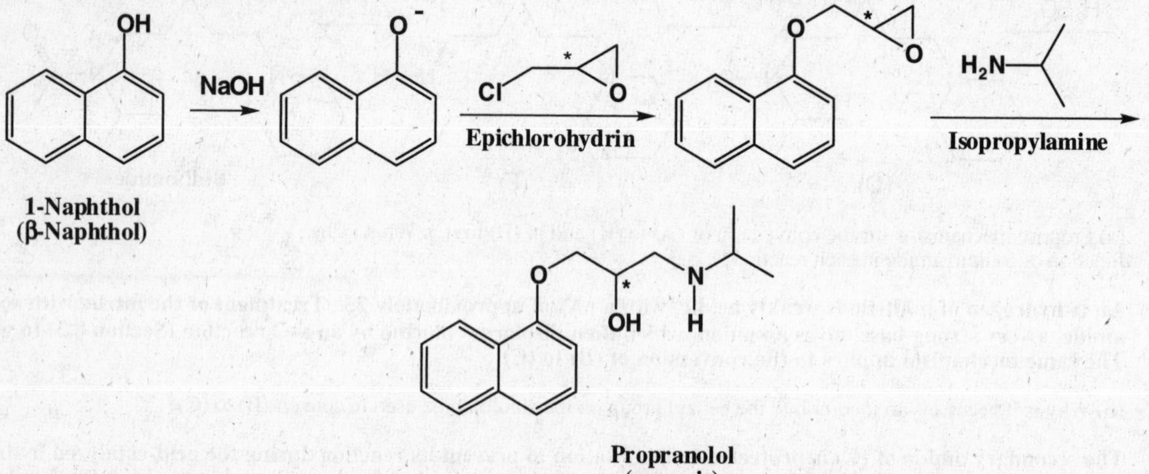

**1-Naphthol
(β-Naphthol)**   **Epichlorohydrin**   **Isopropylamine**

**Propranolol
(racemic)**

Alternatively, the epichlorohydrin can be reacted with the amine first.

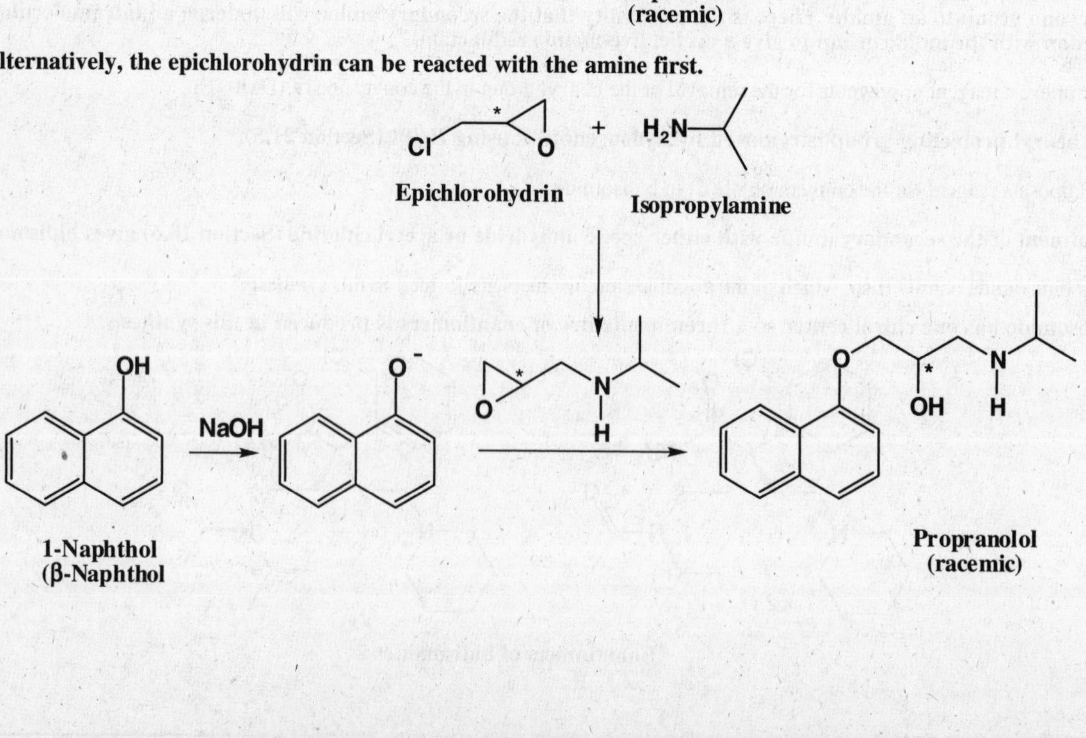

Epichlorohydrin   Isopropylamine

**1-Naphthol
(β-Naphthol**   **Propranolol
(racemic)**

(b) Is propranolol chiral? If so, which of the possible stereoisomers are formed in this synthesis?

Propranolol has one chiral center so a racemic mixture of enantiomers is produced in this synthesis.

Enantiomers of propranolol

<u>Problem 21.57</u> Side effects of propranolol (Problem 21.56) include disturbances of the central nervous system (CNS) such as fatigue, sleep disturbances (including insomnia and nightmares), and depression. Pharmaceutical companies wondered if this drug could be redesigned to eliminate or at least reduce these side effects. Propranolol, it was reasoned, enters the CNS by passive diffusion because of the lipidlike character of its naphthalene ring. The challenge, then, was to design a more hydrophilic drug that does not cross the blood-brain barrier but still retains a β-adrenergic antagonist property. A product of this research is atenolol, a potent β-adrenergic blocker that is hydrophilic enough that it crosses the blood-brain barrier to only a very limited extent. Atenolol is now one of the most widely used β-blockers.

Atenolol

Isopropyl-amine

Epichloro-hydrin

4-Hydroxyphenylacetic acid

(a) Given this retrosynthetic analysis, propose a synthesis for atenolol from the three named starting materials.

Following ester formation and conversion to the amide, the epichlorohydrin can be reacted with the phenolic oxygen followed by the isopropylamine.

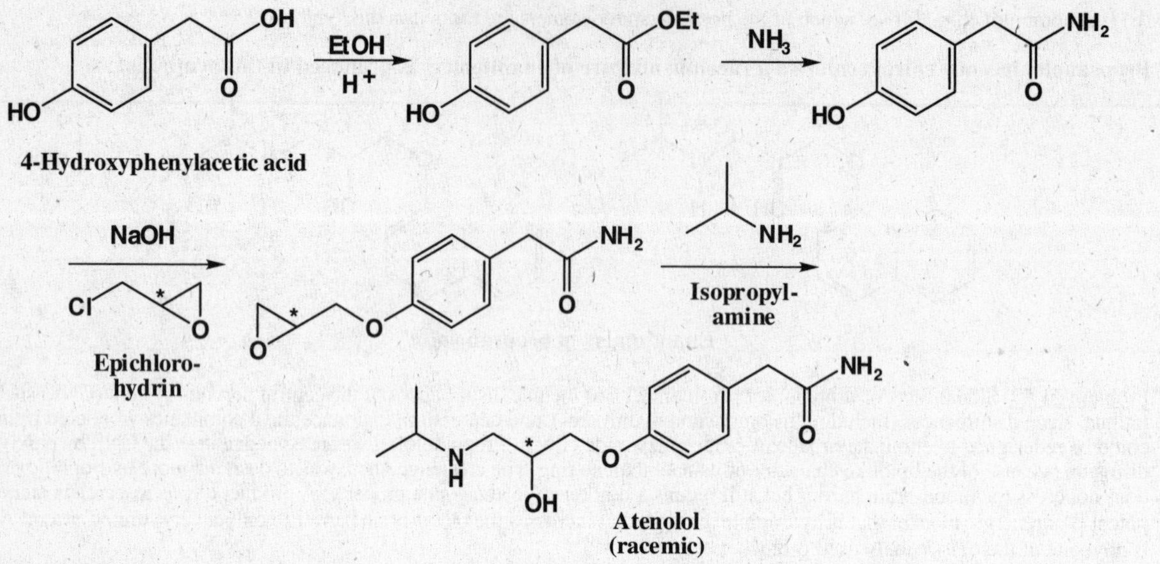

4-Hydroxyphenylacetic acid

**Atenolol
(racemic)**

Alternatively, the epichlorohydrin can react with the isopropylamine first.

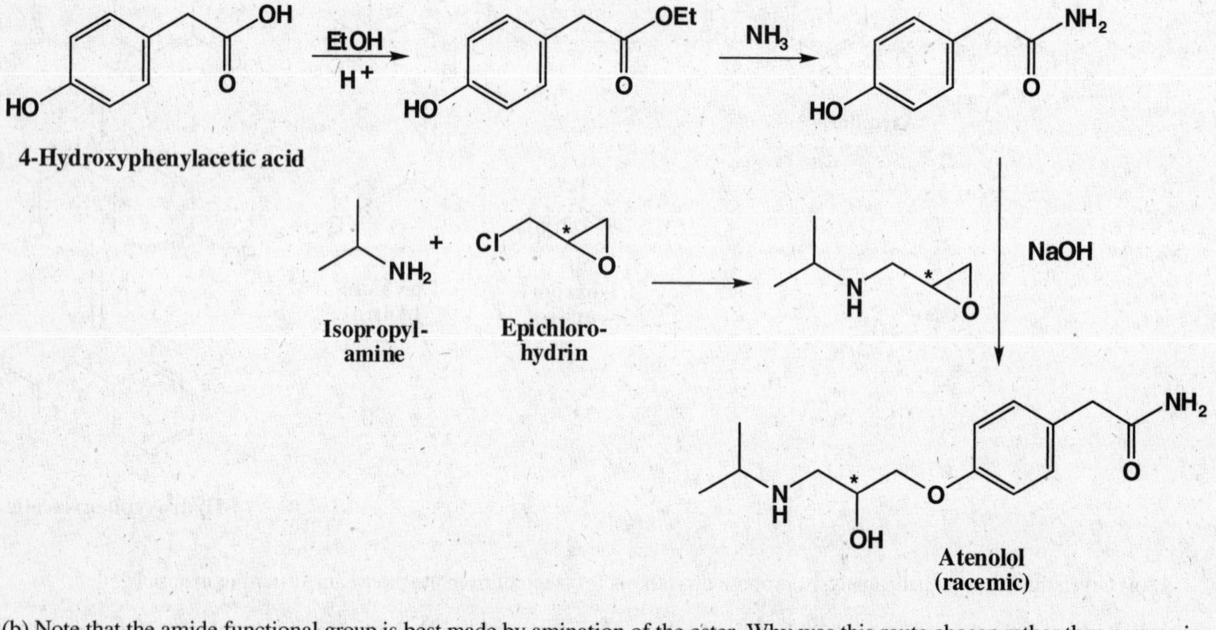

4-Hydroxyphenylacetic acid

**Isopropyl-
amine**

**Epichloro-
hydrin**

**Atenolol
(racemic)**

(b) Note that the amide functional group is best made by amination of the ester. Why was this route chosen rather than conversion of the carboxylic acid to its acid chloride and then treatment of the acid chloride with ammonia?

The phenolic -OH group would have reacted with an acid chloride, but it does not react with the ethyl ester. By using the ethyl ester route, the phenolic -OH does not need to be protected during the synthesis.

(c) Is atenolol chiral? If so, which of the possible stereoisomers are formed in this synthesis?

Atenolol has one chiral center so a racemic mixture of enantiomers is produced in this synthesis.

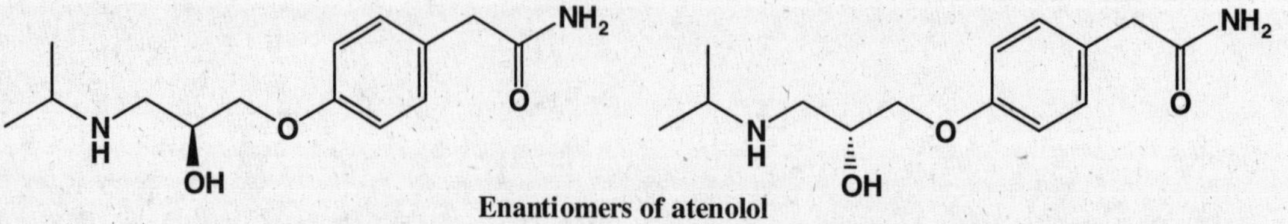

Enantiomers of atenolol

<u>Problem 21.58</u> In certain clinical situations, there is need for an injectable β-blocker with a short biological half-life. The clue to development of such a drug was taken from the structure of atenolol, whose corresponding carboxylic acid (the product of hydrolysis of its amide) has no β-blocking activity. Substitution of an ester for the amide group and lengthening the carbon side chain by one methylene group resulted in esmolol. Its ester group is hydrolyzed quite rapidly to a carboxyl group by serum esterases under physiological conditions. This hydrolysis product has no β-blocking activity. Propose a synthesis for esmolol from 4-hydroxycinnamic acid, epichlorohydrin, and isopropylamine.

(a) Propose a synthesis for esmolol from 4-hydroxycinnamic acid, epichlorohydrin, and isopropylamine.

The double bond of 4-hydroxycinnamic acid is reduced then the carboxyl group is converted to the methyl ester before reaction with epichlorohydrin. Note that NaOH is not used in the reaction of the phenol hydroxyl group with epichlorohydrin in this case, as the methyl ester would be hydrolyzed. Reaction with isopropylamine completes the synthesis.

Alternatively, the epichlorohydrin can react with the isopropylamine first.

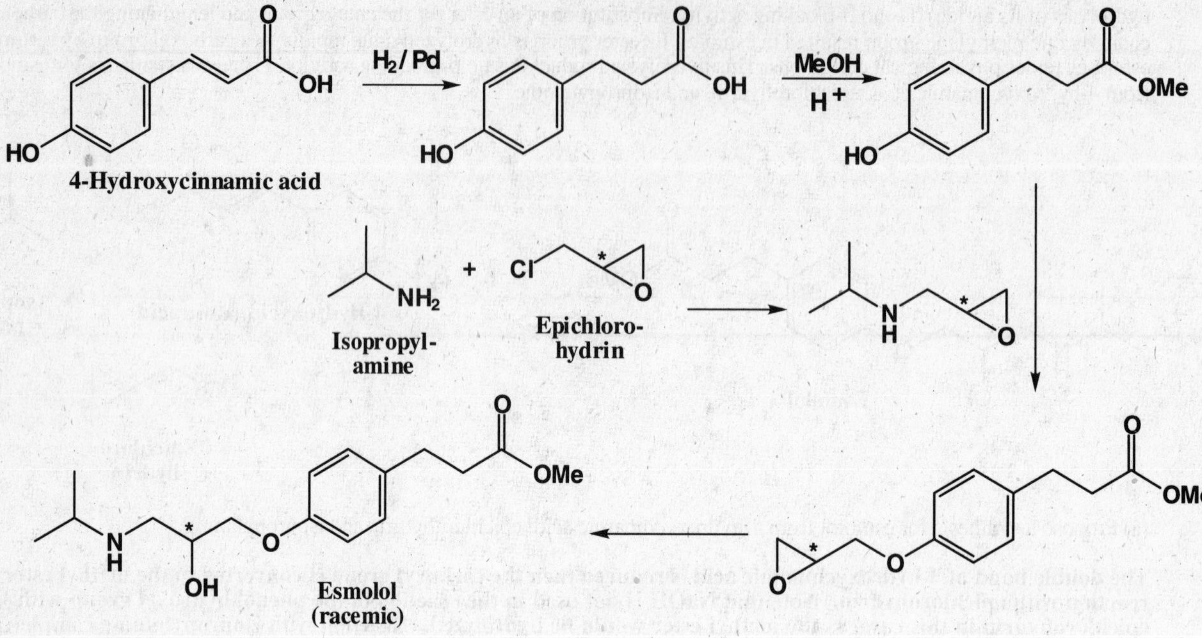

4-Hydroxycinnamic acid

Isopropyl-amine + Epichloro-hydrin

Esmolol
(racemic)

(b) Is esmolol chiral? If so, which of the possible stereoisomers are formed in this synthesis?

Esmolol has one chiral center so a racemic mixture of enantiomers is produced in this synthesis.

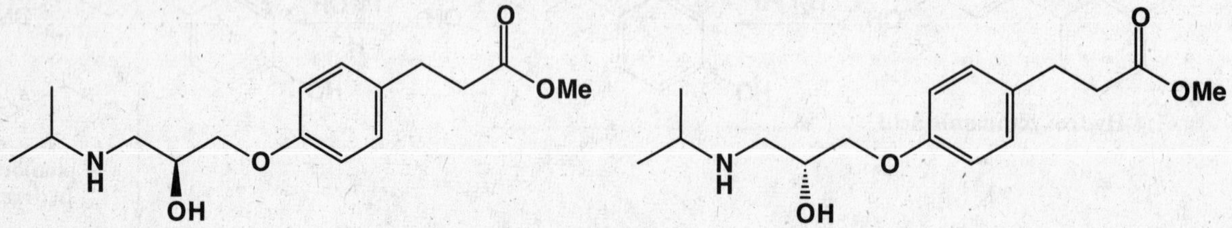

Enantiomers of esmolol

Problem 21.59 Following is an outline of a synthesis of the bronchodilator carbuterol, a beta-2 adrenergic blocker with high selectivity for airway smooth muscle receptors.

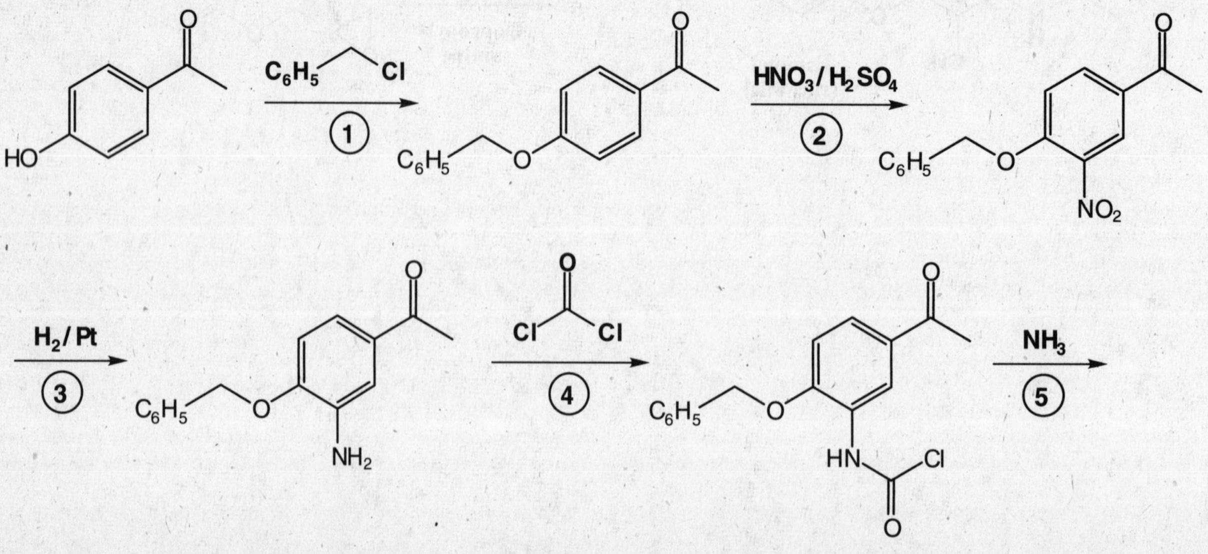

Carbuterol

(a) Propose reagents to bring about each step.

Reagents are listed over the arrows on the scheme above.

Step 1: Treatment of the substituted phenol with NaOH converts it to its sodium salt; in effect the phenolic -OH, a poor nucleophile, is converted to -O $^-$, a good nucleophile (Section 9.4).

Step 2: The position of nitration of the aromatic ring is determined by the RO- group, which is an activating group and ortho, para directing (Section 22.2).

Step 3: Reduction of the nitro group to a primary amino group is accomplished using hydrogen in the presence of a transition metal catalyst (Section 22.1).

Step 4: Treatment of the amine with one equivalent of phosgene (the diacid chloride of carbonic acid) gives an amide (Section 18.6). The product of this reaction is difunctional; it is both an amide and an acid chloride.

Step 5: Treatment of the acid chloride with ammonia gives an amide (Section 18.6), which is actually an *N*-substituted urea. Urea is H_2N-CO-NH_2.

Step 6: The ketone is brominated on its α-carbon by treatment with bromine (Section 16.12).

Step 7: Treatment of the α-bromoketone with *tert*-butylamine results in nucleophilic displacement of bromine (Section 9.4).

Step 8: The benzyl protecting group is removed by treatment with H_2/Pt in a reaction called hydrogenolysis (Section 21.5). Care must be taken to avoid reducing the ketone function.

Step 9: Reduction of the ketone group by sodium borohydride (Section 16.11) completes the synthesis of carbuterol.

Depending on experimental conditions, the product of reaction (4) may lose HCl to give an isocyanate; that is, a compound containing an -N=C=O group. Addition of NH_3 to the isocyanate group followed by intramolecular proton transfer gives the same *N*-substituted urea shown in the problem.

(b) Why is it necessary to add the benzyl group, PhCH$_2$-, as a blocking group in Step 1?

The benzyl blocking group is needed because the nucleophilic phenol hydroxyl group would interfere with step (5) by reacting with the acyl chloride/isocyanate produced in step (4).

(c) Suggest a structural relationship between carbuterol and ephedrine.

The structural features of carbuterol that resemble ephedrine are shown in bold.

(d) Is carbuterol chiral? If so, which of the possible stereoisomers are formed in this synthesis?

Carbuterol has one chiral center so a racemic mixture of enantiomers is produced in this synthesis.

Enantiomers of carbuterol

Problem 21.60 Following is a synthesis for albuterol (Proventil), currently one of the most widely used inhalation bronchodilators.

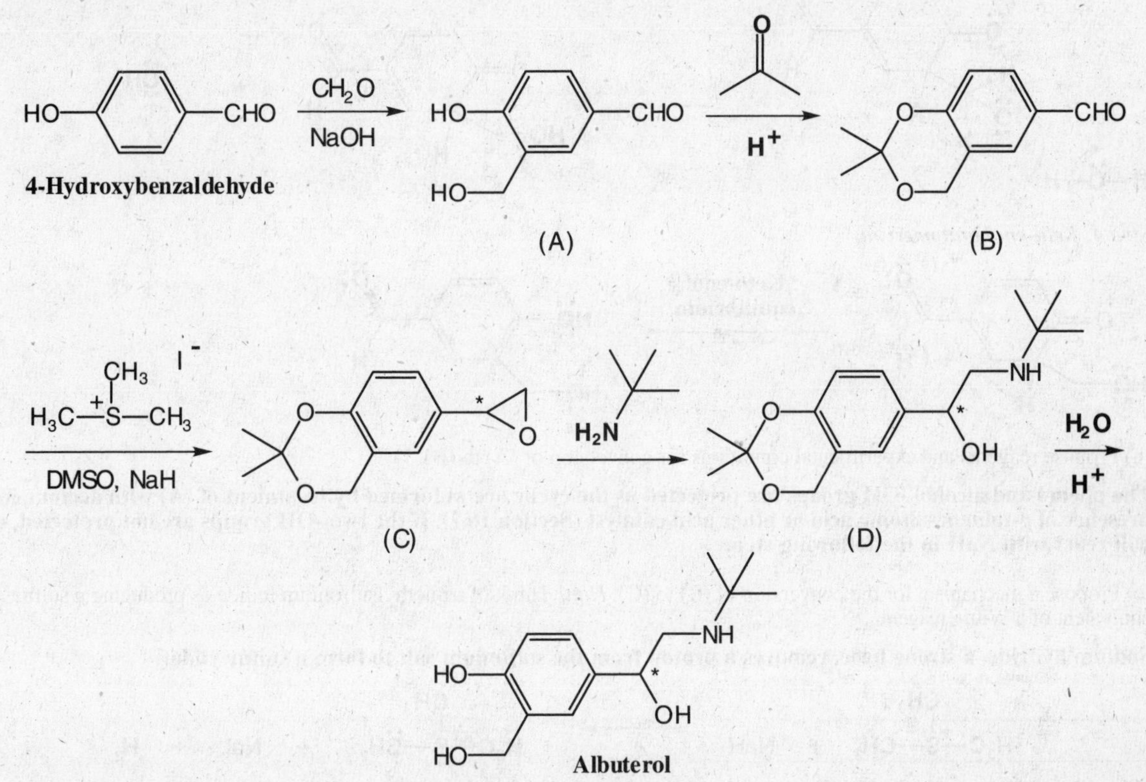

4-Hydroxybenzaldehyde

(A)

(B)

(C)

(D)

Albuterol

(a) Propose a mechanism for conversion of 4-hydroxybenzaldehyde to (A):

This reaction, which results in hydroxymethylation of the aromatic ring, is similar in mechanism to the Kolbe carboxylation (Section 21.4). Both mechanisms begin with an acid-base reaction using NaOH to form a phenoxide ion, followed by addition of the α-carbon of the phenoxide ion to a carbonyl group. In the case of hydroxymethylation, the α-carbon adds to the carbonyl group (step 2) of formaldehyde. Proton transfer in step 3 followed by keto-enol tautomerism (Section 16.9) gives the hydroxymethylated phenol.

Step 1: Take a proton away.

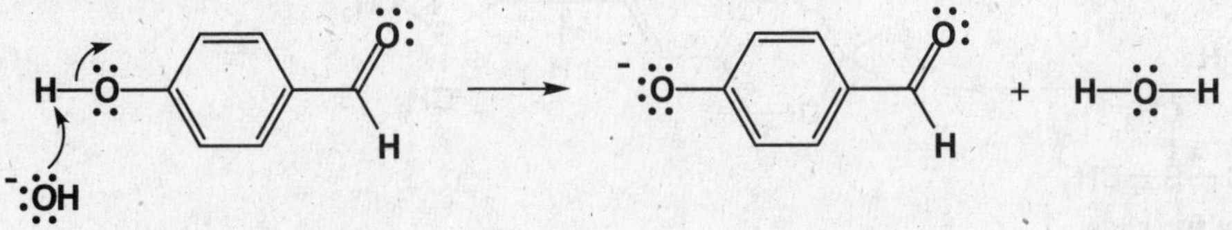

Step 2: Electrophilic addition.

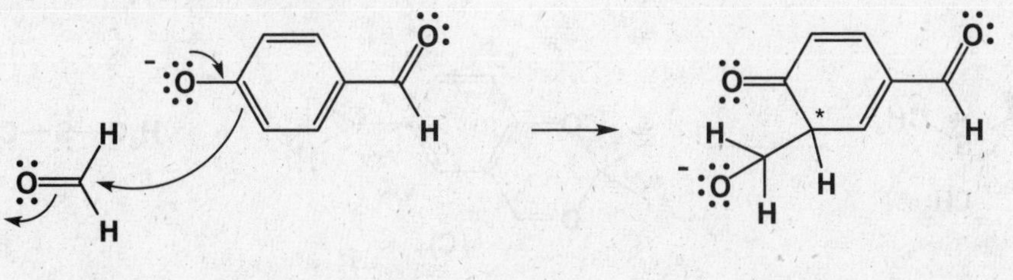

Step 3: Add a proton.

Step 4: Keto-enol tautomerism.

(b) Propose reagents and experimental conditions for conversion of (A) to (B).

The phenol and alcohol -OH groups are protected as the cyclic acetal formed by treatment of (A) with acetone in the presence of *p*-toluenesulfonic acid or other acid catalyst (Section 16.7). If the two -OH groups are not protected, each will react with NaH in the following step.

(c) Propose a mechanism for the conversion of (B) to (C). *Hint:* Think of trimethylsulfonium iodide as producing a sulfur equivalent of a Wittig reagent.

Sodium hydride, a strong base, removes a proton from the sulfonium salt to form a sulfur ylide.

Attack of the sulfur ylide on the carbonyl group of the aldehyde, like the attack of a Wittig reagent on a carbonyl group, gives a betaine. The fate of a sulfur betaine is different from that of a phosphorus betaine because sulfur cannot expand its valence shell to accommodate two more electrons. Instead oxygen of the sulfur betaine attacks the adjacent CH$_2$ group in an internal S$_N$2 reaction and displaces the sulfur group, a good leaving group, to give the epoxide (A).

Step 1: Make a bond between a nucleophile and an electrophile.

Step 2: Make a bond between a nucleophile and an electrophile and simultaneously break a bond to give stable molecules or ions.

(d) Propose reagents and experimental conditions for the conversion of (C) to (D).

Treatment of the epoxide with *tert*-butylamine opens the epoxide ring by nucleophilic displacement of oxygen (Section 11.9).

(e) Propose reagents and experimental conditions for the conversion of (D) to albuterol.

Acid-catalyzed removal of the acetal protecting group (Section 16.7) gives albuterol.

(f) Is albuterol chiral? If so, which of the possible stereoisomers are formed in this synthesis?

Albuterol has one chiral center so a racemic mixture of enantiomers is produced in this synthesis.

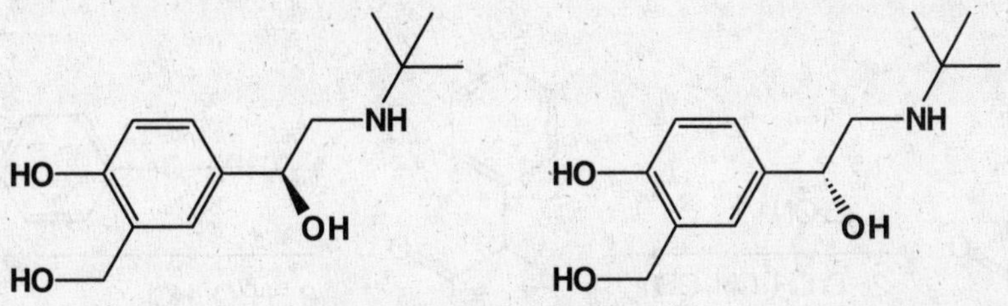

Enantiomers of albuterol

Problem 21.61 Estrogens are female sex hormones, the most potent of which is β-estradiol.

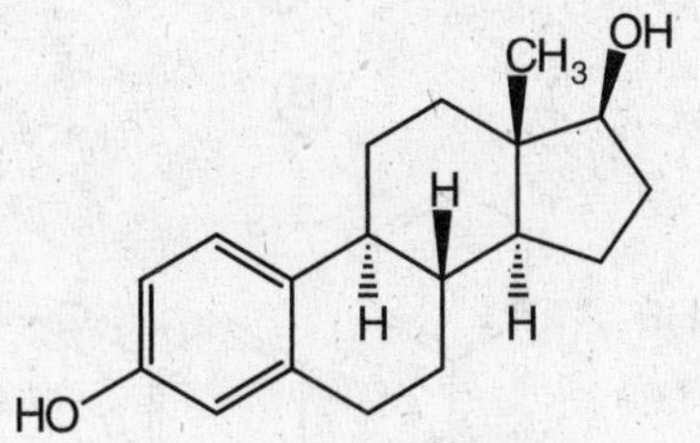

β-Estradiol

In recent years, chemists have focused on designing and synthesizing molecules that bind to estrogen receptors. One target of this research has been nonsteroidal estrogen antagonists, compounds that interact with estrogen receptors and block the effects of both endogenous and exogenous estrogens. A feature common to one type of nonsteroidal estrogen antagonist is the presence of a 1,2-diphenylethylene with one of the benzene rings bearing a dialkylaminoethoxyl substituent. The first nonsteroidal estrogen antagonist of this type to achieve clinical importance was tamoxifen, now an important drug in the treatment of breast cancer. Tamoxifen has the Z configuration as shown here.

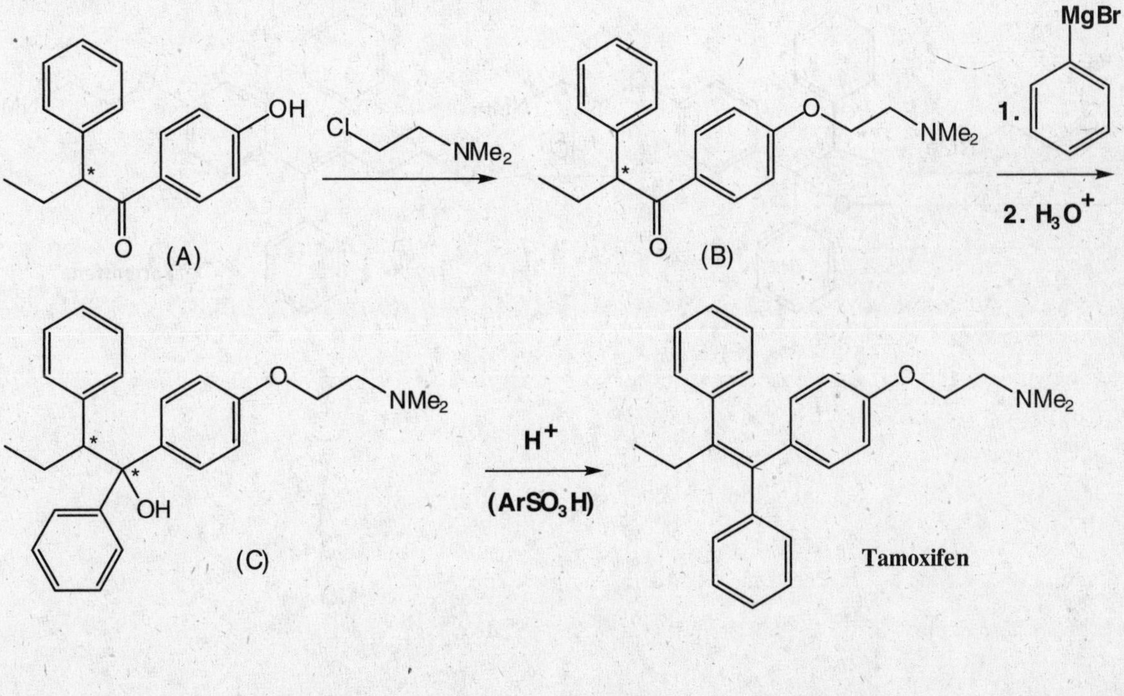

Propose reagents for the conversion of (A) to tamoxifen. *Note:* The final step in this synthesis gives a mixture of *E* and *Z* isomers.

The new carbon-carbon bond is formed by a Grignard reaction (Section 16.5) with phenylmagnesium bromide. Acid-catalyzed dehydration of the tertiary benzylic alcohol gives tamoxifen. One such acid catalyst is *p*-toluenesulfonic acid.

<u>Problem 21.62</u> Following is a synthesis for toremifene, a nonsteroidal estrogen antagonist whose structure is closely related to that of tamoxifen.

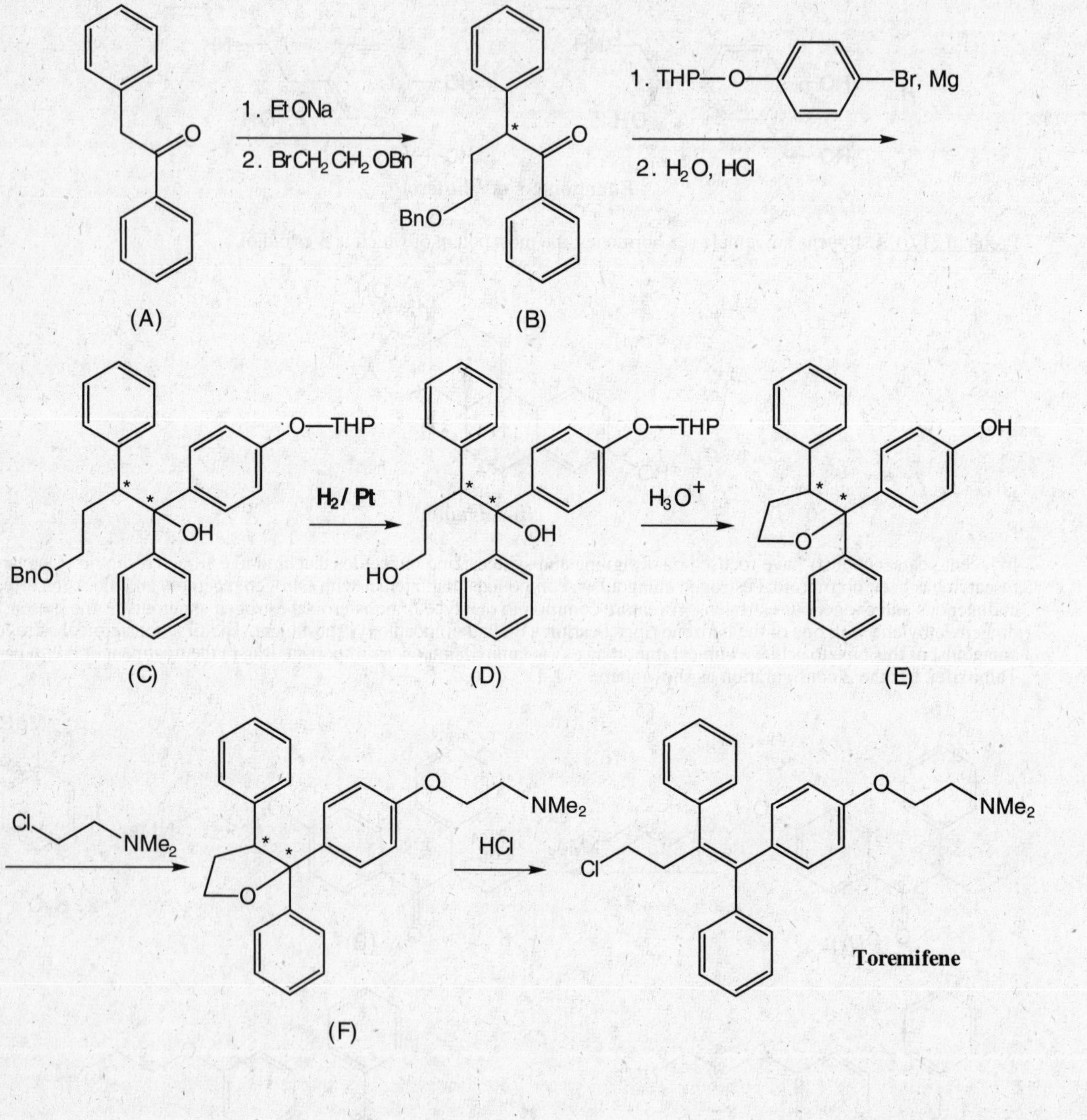

(A)   (B)

(C)   (D)   (E)

(F)

Toremifene

(a) This synthesis makes use of two blocking groups, the benzyl (Bn) group and the tetrahydropyranyl (THP) group. Draw a structural formula of each group, and describe the experimental conditions under which it is attached and removed.

Following are the structural formulas for each blocking group.

The tetrahydropyranyl (THP) group   The benzyl (Bn) group

For how the THP group is added, the conditions under which it is stable, and how it is removed, see Section 16.7. For the same type of information for the benzyl group, see Section 21.5.

(b) Discuss the chemical logic behind the use of each blocking group in this synthesis.

The THP group is necessary to protect the Grignard reagent from being destroyed by reaction between it and the acidic proton of the phenol. Similarly, the benzyl group is necessary to prevent destruction of the Grignard reagent by the acidic proton of the primary alcohol. The benzyl group is selectively removed by hydrogenolysis (Section 21.5) and the THP group is selectively removed by acid-catalyzed hydrolysis (Section 16.7).

(c) Propose a mechanism for the conversion of (D) to (E).

In this mechanism, the two benzene rings are represented by Ph. Step 3 is initiated by proton transfer to the tertiary alcohol to give an oxonium ion followed by loss of water to give a resonance-stabilized tertiary benzylic carbocation. A Lewis acid-base reaction between this cation (a Lewis acid) and the nearby primary alcohol (a Lewis base) gives the cyclic, five-membered ether.

Step 1: Add a proton.

Step 2: Break a bond to give stable molecules or ions.

Step 3: Make a bond between a nucleophile and an electrophile.

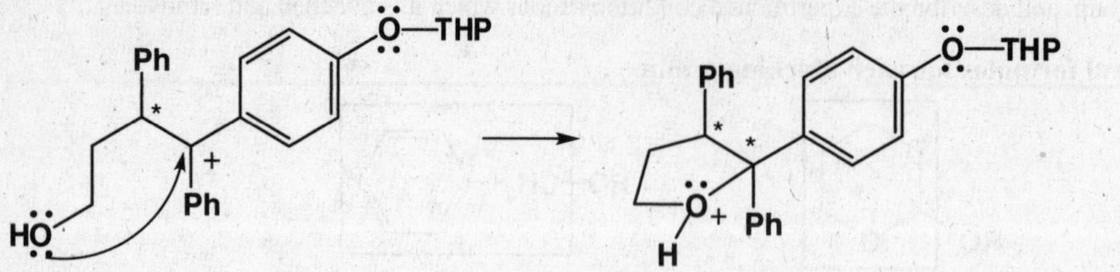

Step 4: Take a proton away.

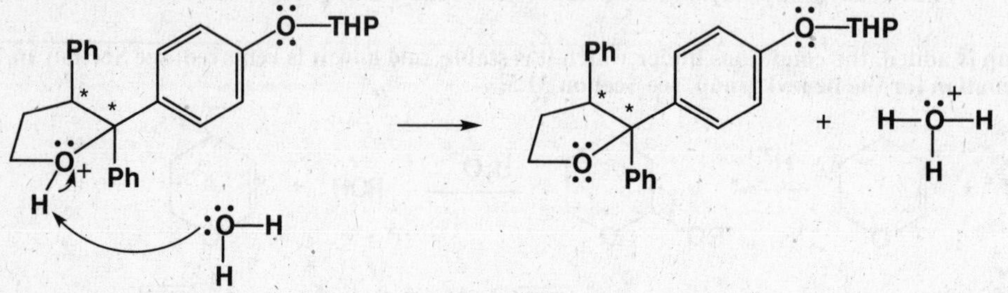

Step 5: Hydrolysis of THP ether. The THP ether is hydrolyzed in acid in a process that involves several steps.

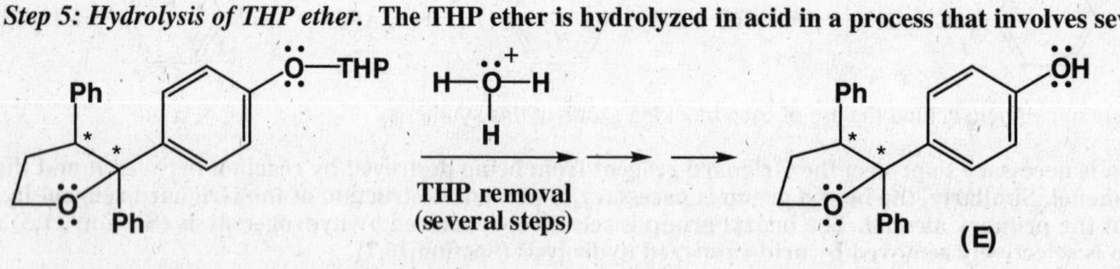

THP removal
(several steps)

(d) Propose a mechanism for the conversion of (F) to toremifene.

Opening of the five-membered cyclic ether begins by proton transfer from HCl to give an oxonium ion. From here, there are several different mechanisms by which toremifene might be formed. One of these is proton transfer from a benzylic position to chloride ion to cleave the ether and open the five-membered ring. This is followed by proton transfer to the primary alcohol followed by nucleophilic displacement of water by chloride ion to give toremifene.

Step 1: Add a proton.

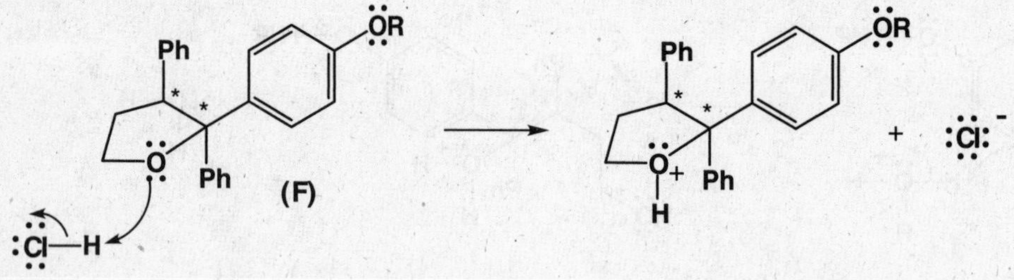

Step 2: Take a proton away and simultaneously break a bond to give stable molecules or ions. This step is actually and E2 reaction.

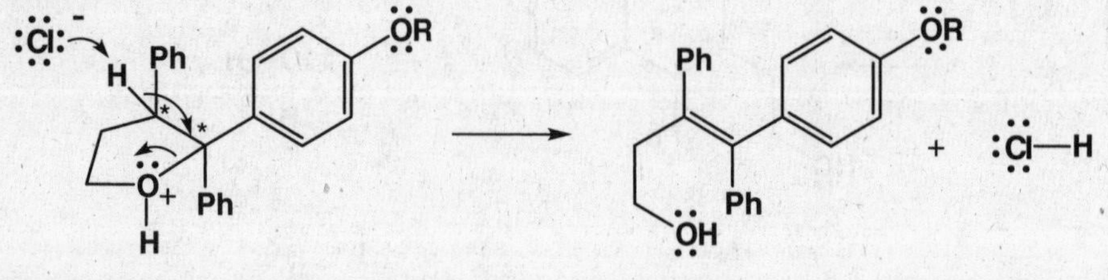

Step 3: Add a proton.

Step 4: Make a bond between a nucleophile and an electrophile and simultaneously break a bond to give stable molecules or ions.

(e) Is toremifene chiral? If so, which of the possible stereoisomers are formed in this synthesis?

No, toremifene is not chiral.

Queen angelfish *Holacanthus ciliaris*
Tobago, West Indies

<u>Problem 22.1</u> Write the stepwise mechanism for sulfonation of benzene by hot, concentrated sulfuric acid. In this reaction, the electrophile is SO_3 formed as shown in the following equation.

$$H_2SO_4 \rightleftharpoons SO_3 + H_2O$$

In sulfonation of benzene, the electrophile is sulfur trioxide. In Step 1, reaction of benzene with the electrophile yields a resonance-stabilized cation. In Step 2, this intermediate loses a proton to complete the reaction: water is shown as the base accepting the proton.

Step 1: Electrophilic addition (aromatic).

(A resonance-stabilized cation intermediate)

Step 2: Take a proton away.

Note that in more strongly acidic conditions, commonly referred to as fuming sulfuric acid, the electrophile can be in the protonated form, namely HSO_3^+, instead of SO_3.

<u>Problem 22.2</u> Write structural formulas for the products you expect from Friedel-Crafts alkylation or acylation of benzene with each compound.

(a)

(b)

(c)

(Rearrangement occurred)

(Not chiral)

<u>Problem 22.3</u> Write a mechanism for the formation of *tert*-butylbenzene from benzene and *tert*-butyl alcohol in the presence of phosphoric acid.

A reasonable mechanism for this reaction involves formation of the *tert*-butyl cation, that then reacts with benzene via electrophilic aromatic substitution.

Step 1: Add a proton.

Step 2: Break a bond to give stable molecules or ions.

Step 3: Electrophilic addition (aromatic).

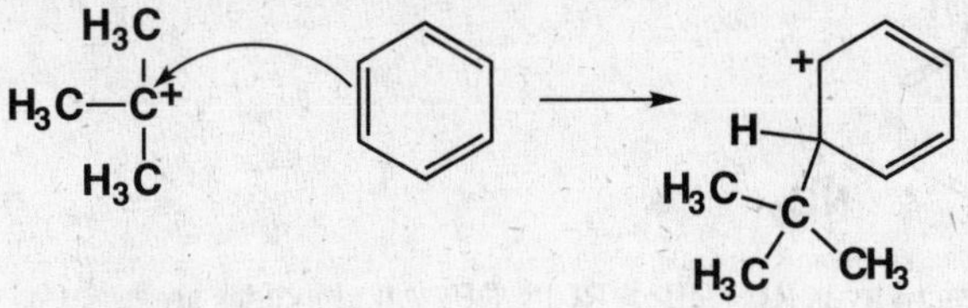

Step 4: Take a proton away.

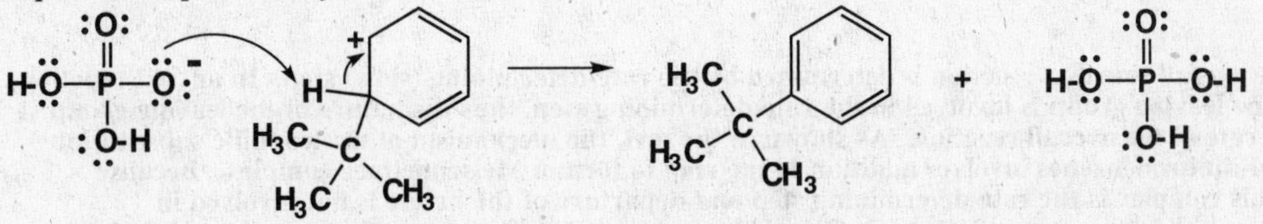

Problem 22.4 Draw structural formulas for the product of nitration of each compound. Where you predict ortho-para substitution, show both products.

(a)

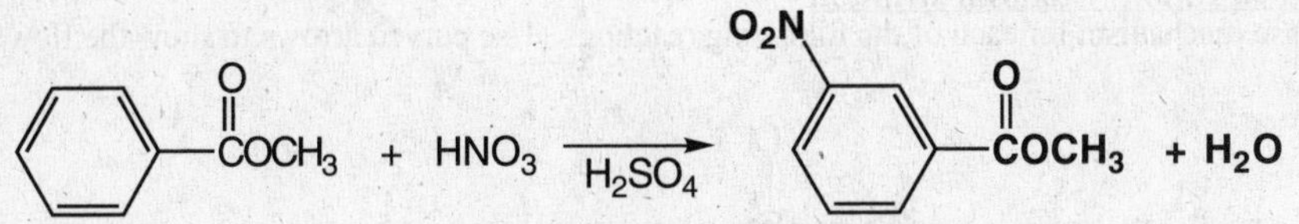

The carboxymethyl group is meta directing and deactivating.

(b)

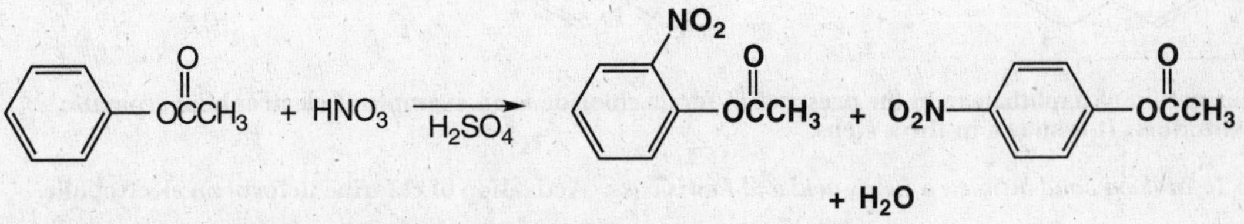

The acetoxy group is ortho/para directing and activating.

Problem 22.5 Predict the major product(s) of each electrophilic aromatic substitution.

When there is more than one substituent on the ring, the predominant product is determined by the more activating (less deactivating) substituent. In addition, steric hindrance precludes reaction between two substituents that are meta with respect to each other.

(a)

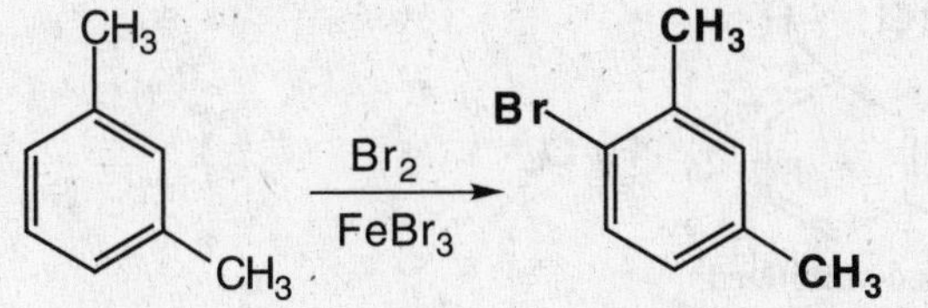

(b)

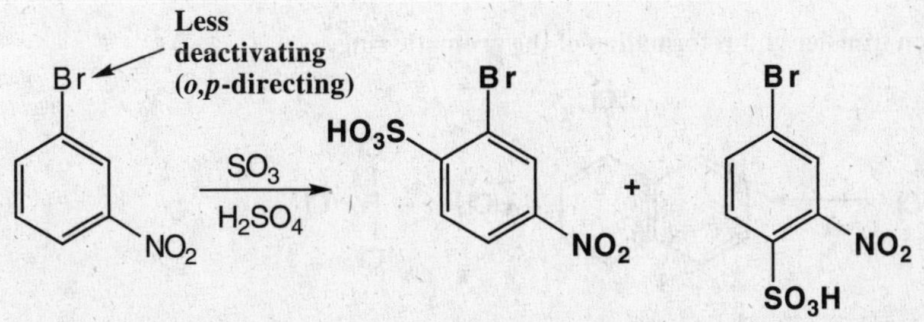

(c)

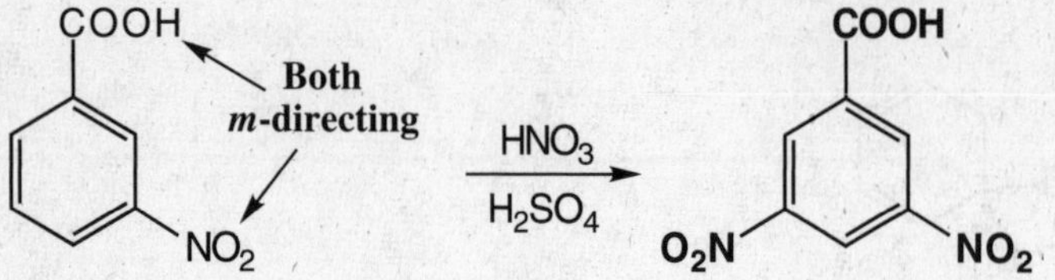

Problem 22.6 In S_N2 reactions of alkyl halides, the order of reactivity is RI > RBr > RCl > RF. Alkyl iodides are considerably more reactive than alkyl fluorides, often by factors as great as 10^6. All 1-halo-2,4-dinitrobenzenes, however, react at approximately the same rate in nucleophilic aromatic substitutions. Account for this difference in relative reactivities.

Recall that the overall rate of a reaction is determined by the rate-determining (slow) step. In an S_N2 reaction, departure of the leaving group is involved in the rate-determining step, thus the nature of the leaving group influences the rate of the overall reaction. As shown in the text, the mechanism of nucleophilic substitution with 1-halo-2,4-dinitrobenzenes involves addition to the ring to form a Meisenheimer complex. Because *formation* of this complex is the rate-determining step and departure of the halide is not involved in Meisenheimer complex formation, the nature of the halogen has little influence on the overall rate of the process.

Electrophilic Aromatic Substitution: Monosubstitution

Problem 22.7 Write a stepwise mechanism for each of the following reactions. Use curved arrows to show the flow of electrons in each step.

(a)

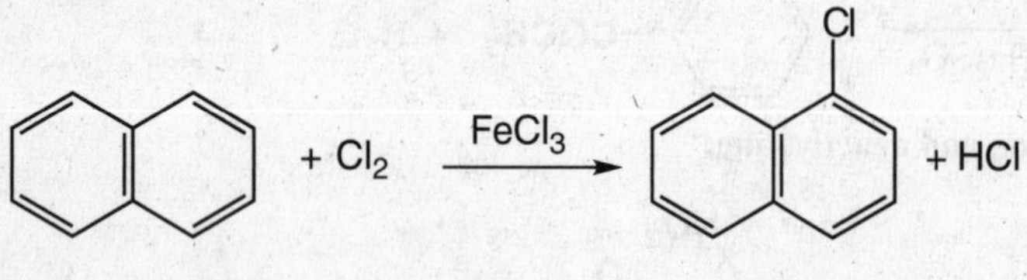

Chlorination of naphthalene in the presence of ferric chloride is an example of electrophilic aromatic substitution. It is shown in three steps.

Step 1: Make a bond between a Lewis acid and Lewis base. **Activation of chlorine to form an electrophile.**

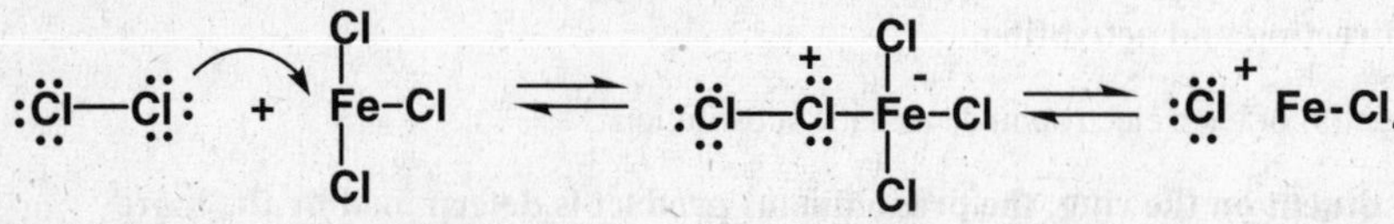

Step 2: Electrophilic addition (aromatic). **Reaction of the electrophile with the aromatic ring, a nucleophile.**

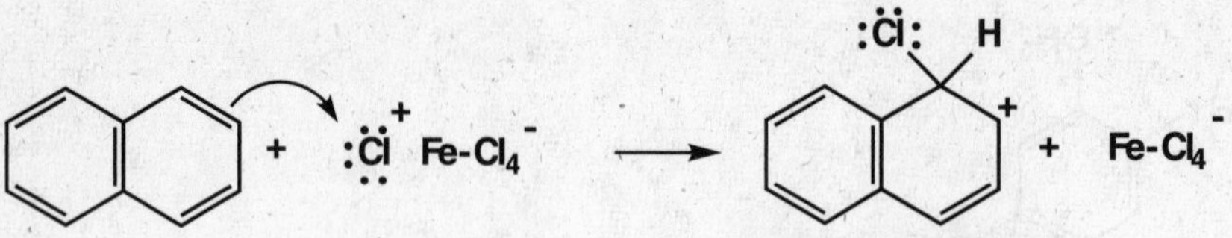

A resonance-stabilized
cation intermediate

Step 3: Take a proton away. **Proton transfer and reformation of the aromatic ring.**

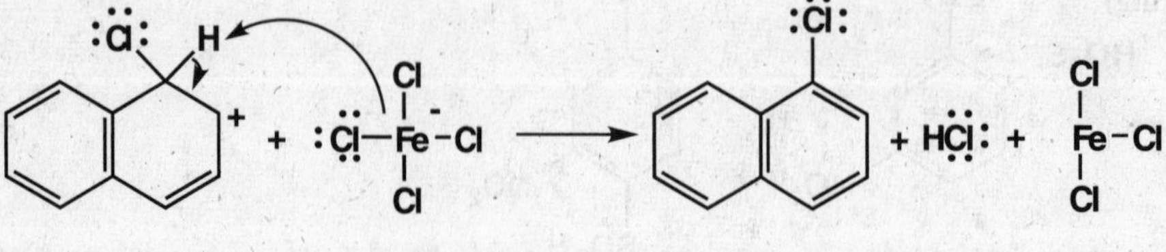

(b)

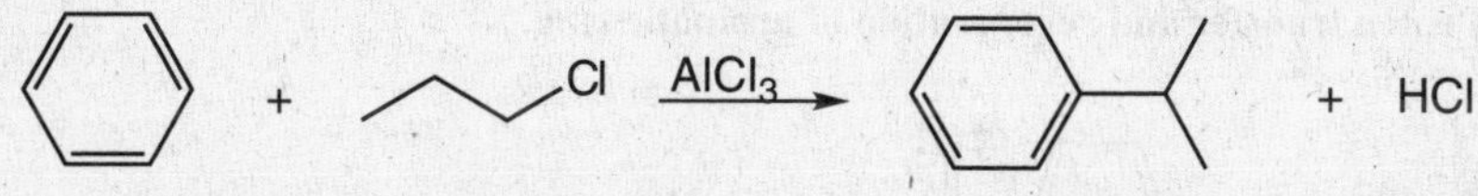

The reaction of benzene with 1-chloropropane in the presence of aluminum chloride involves initial formation of a complex between 1-chloropropane and aluminum chloride, and its rearrangement to an isopropyl cation. This cationic species is the electrophile that undergoes further reaction with benzene.

Step 1: *Make a bond between a Lewis acid and Lewis base.* Formation of a complex between 1-chloropropane (a Lewis base) and aluminum chloride (a Lewis acid).

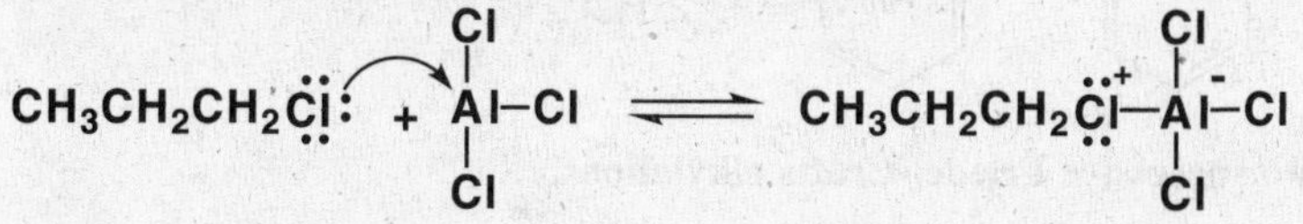

Step 2: *1,2 Shift and break a bond to give stable molecules or ions.* Rearrangement to form an isopropyl cation.

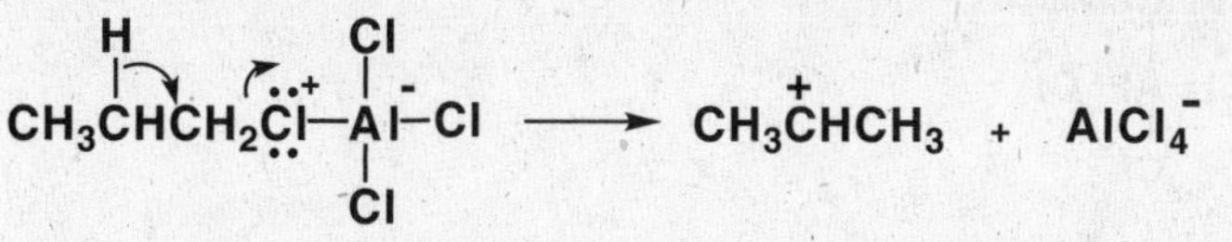

Step 3: *Electrophilic addition (aromatic).* Electrophilic attack on the aromatic ring.

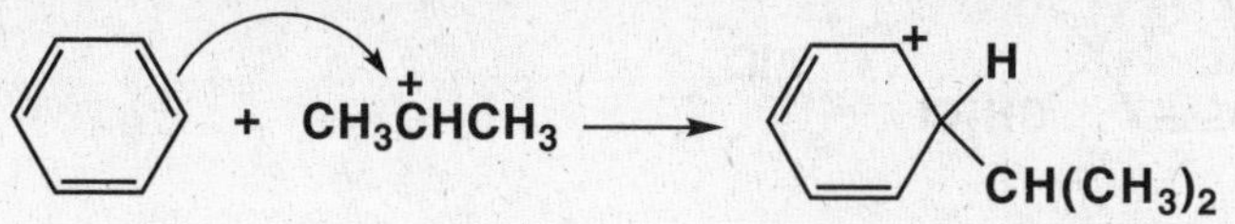

Resonance-stabilized
cation intermediate

Step 4: *Take a proton away.* Proton transfer to regenerate the aromatic ring.

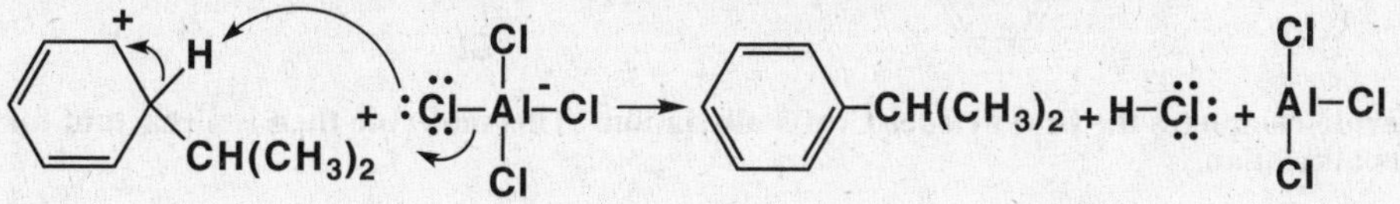

(c)

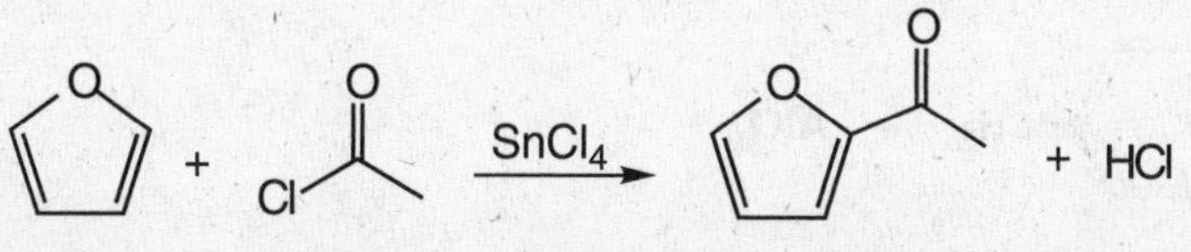

Friedel-Crafts acylation of furan involves electrophilic attack by an acylium ion.
Step 1: *Make a bond between a Lewis acid and Lewis base.* Formation of resonance-stabilized acylium ion.

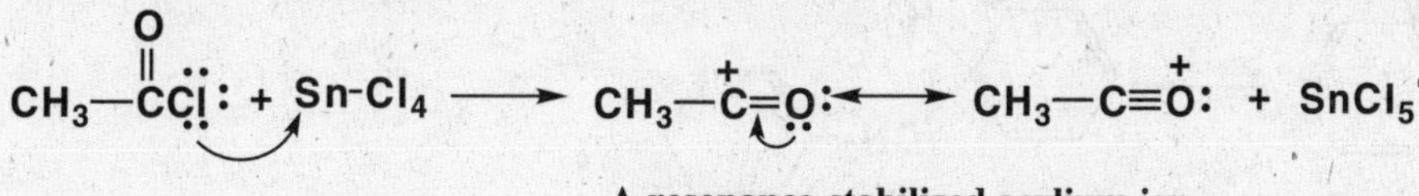

A resonance-stabilized acylium ion

Step 2: *Electrophilic addition (aromatic).* Reaction of the acylium ion (an electrophile) and furan (a nucleophile) to form a resonance-stabilized cation.

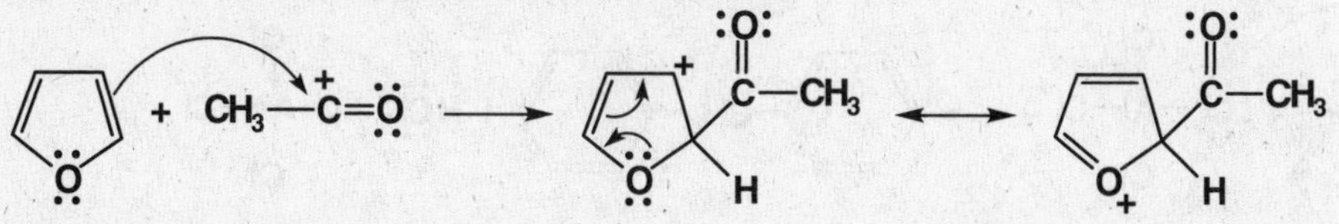

Step 3: *Take a proton away.* Proton transfer and regeneration of aromatic ring.

(d)   2 ⬡ + CH_2Cl_2 $\xrightarrow{AlCl_3}$ ⬡—CH_2—⬡ + 2 HCl

Formation of diphenylmethane involves two successive Friedel-Crafts alkylations.

Step 1: *Make a bond between a Lewis acid and Lewis base.*

Step 2: *Electrophilic addition (aromatic).*

Step 3: *Take a proton away.*

Formation of benzyl chloride completes the first Friedel-Crafts alkylation. This molecule then is a reactant for the second Friedel-Crafts alkylation.

Step 4: *Make a bond between a Lewis acid and Lewis base.*

Step 5: *Electrophilic addition (aromatic).*

Step 6: *Take a proton away.*

Problem 22.8 Pyridine undergoes electrophilic aromatic substitution preferentially at the 3 position as illustrated by the synthesis of 3-nitropyridine.

Pyridine   **3-Nitropyridine**

Under these acidic conditions, the species undergoing nitration is not pyridine but its conjugate acid. Write resonance contributing structures for the intermediate formed by attack of NO_2^+ at the 2, 3, and 4 positions of the conjugate acid of pyridine. From examination of these intermediates, offer an explanation for preferential nitration at the 3 position.

Pyridine is a base, and in the presence of a nitric acid-sulfuric acid mixture, it is protonated. It is the protonated form (i.e. conjugate acid) that must be attacked by the electrophile $^+NO_2$. For nitration at the 3-position, the additional positive charge in the cation intermediate may be delocalized on three carbon atoms of the pyridine ring. None of the contributing structures, however, places both positive charges on the same atom.

For nitration at the 4-position or the 2-position, the additional positive charge in the cation intermediate is also delocalized on three atoms of the pyridine ring, but one of these contributing structures has a charge of +2 on nitrogen. This situation is thus less stable than that which occurs for nitration at the 3-position.

A very poor contributing
structure because it places a
charge of +2 on nitrogen

A very poor contributing
structure because it places a
charge of +2 on nitrogen

<u>Problem 22.9</u> Pyrrole undergoes electrophilic aromatic substitution preferentially at the 2 position as illustrated by the synthesis of 2-nitropyrrole.

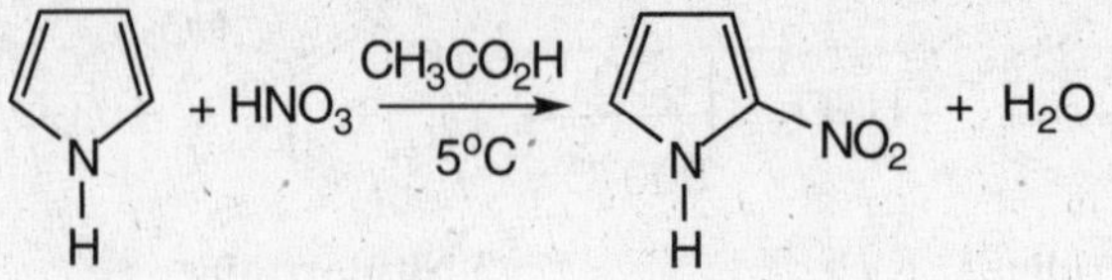

Pyrrole   **2-Nitropyrrole**

Write resonance contributing structures for the intermediate formed by attack of $^{+}NO_2$ at the 2 and 3 positions of pyrrole. From examination of these intermediates, offer an explanation for preferential nitration at the 2 position.

Pyrrole is nitrated under considerably milder conditions than pyridine. For nitration at the 2-position, the positive charge on the cation intermediate is delocalized over three atoms of the pyrrole ring whereas for nitration at the 3-position, it is delocalized over only two atoms. The intermediate with the greater degree of delocalization of charge has a lower activation energy for its formation and hence is formed at a faster rate.

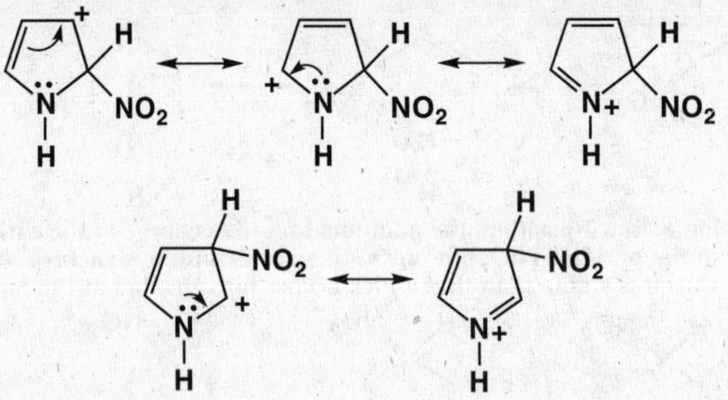

<u>Problem 22.10</u> Addition of *m*-xylene to the strongly acidic solvent HF/SbF_5 at -45° C gives a new species, which shows ^{1}H-NMR resonances at δ 2.88 (3H), 3.00 (3H), 4.67 (2H), 7.93 (1 H), 7.83 (1H), and 8.68 (1H). Assign a structure to the species giving this spectrum.

The strong acid results in protonation of the aromatic ring to create a positively charged species as shown below.

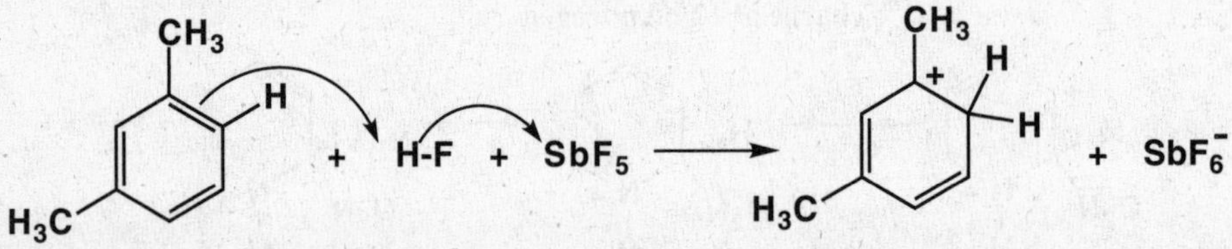

<u>Problem 22.11</u> Addition of *tert*-butylbenzene to the strongly acidic solvent HF/SbF_5 followed by aqueous work-up gives benzene. Propose a mechanism for this dealkylation reaction. What is the other product of the reaction?

A reasonable mechanism for this reaction involves protonation of the aromatic ring, followed by heterolytic bond cleavage and release of the *tert*-butyl cation. After addition of water, *tert*-butyl alcohol will be the other product of the reaction.

Step 1: Electrophilic addition (aromatic)-add a proton

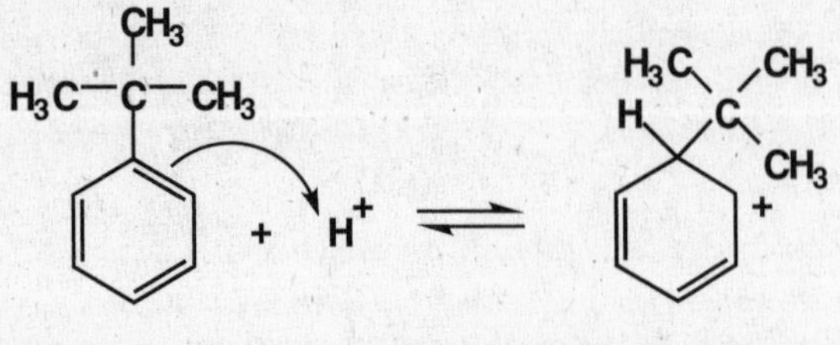

Step 2: Break a bond to give stable molecules or ions.

Step 3: Make a bond between a nucleophile and an electrophile.

Step 4: Take a proton away.

Problem 22.12 What product do you predict from the reaction of SCl_2 with benzene in the presence of $AlCl_3$? What product results if diphenyl ether is treated with SCl_2 and $AlCl_3$?

The Lewis acid, $AlCl_3$, facilitates departure of one of the chlorine atoms from SCl_2, and the resulting electrophile takes part in an electrophilic aromatic substitution reaction to create the -SCl derivative that then reacts again to create diphenyl sulfide. The diphenyl sulfide is activated compared to benzene, so this will react further to generate a polymeric product as shown.

If diphenyl ether were treated in a similar manner, the resulting polymeric species will have alternating ether and thioether functions.

Problem 22.13 Other groups besides H^+ can act as leaving groups in electrophilic aromatic substitution. One of the best leaving groups is the trimethylsilyl group (Me_3Si-). For example, treatment of $Me_3SiC_6H_5$ with CF_3CO_2D rapidly forms DC_6H_5. What are the properties of a silicon-carbon bond that allows you to predict this kind of reactivity?

Based on simple electronegativities, the C-Si bond is polarized such that a partial positive charge is on the Si atom. Furthermore, the heterolytic bond cleavage is facilitated because the $(CH_3)_3Si^+$ cation is so stable.

Disubstitution and Polysubstitution

Problem 22.14 The following groups are ortho-para directors. Draw a contributing structure for the resonance-stabilized aryl cation formed during electrophilic aromatic substitution that shows the role of each group in stabilizing the intermediate by further delocalizing its positive charge.

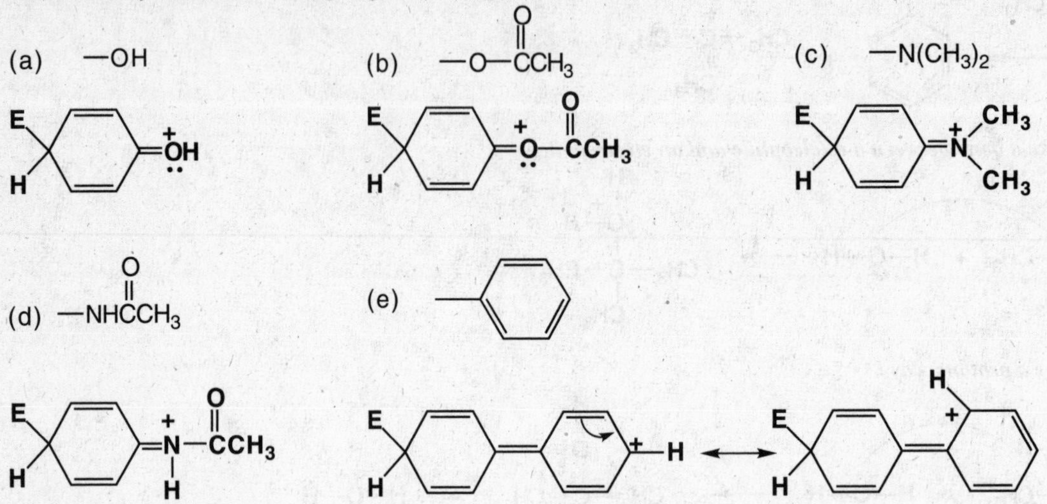

(a) —OH

(b) —O—$\overset{\overset{\displaystyle O}{\|}}{C}$CH₃

(c) —N(CH₃)₂

(d) —NH$\overset{\overset{\displaystyle O}{\|}}{C}$CH₃

(e)

Problem 22.15 Predict the major product or products from treatment of each compound with HNO₃/H₂SO₄.

When there is more than one substituent on a ring, the predominant product is derived from the orientation preference of the more activating substituent.

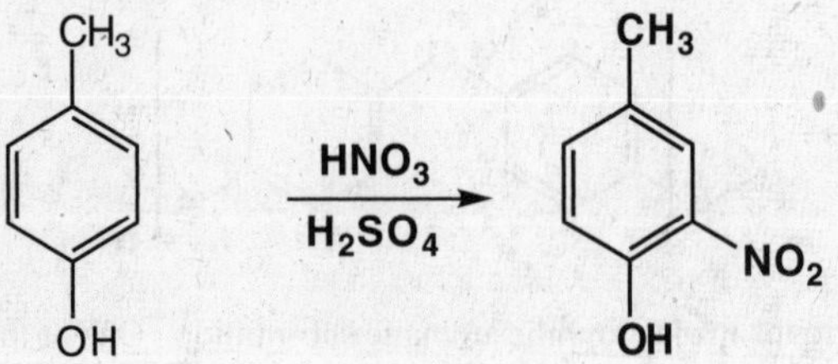

(a) $\xrightarrow[\text{H}_2\text{SO}_4]{\text{HNO}_3}$

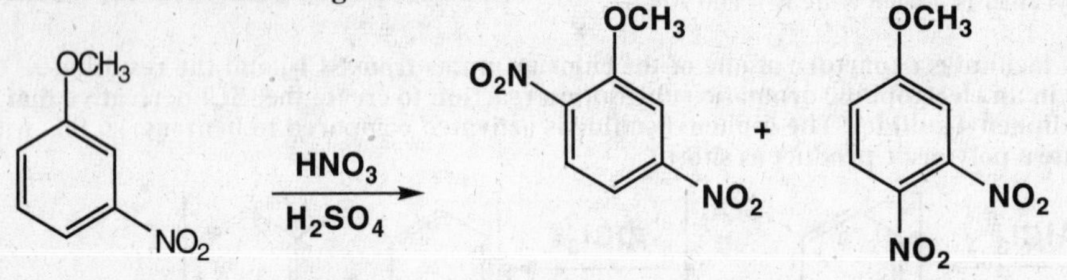

(b) $\xrightarrow[\text{H}_2\text{SO}_4]{\text{HNO}_3}$

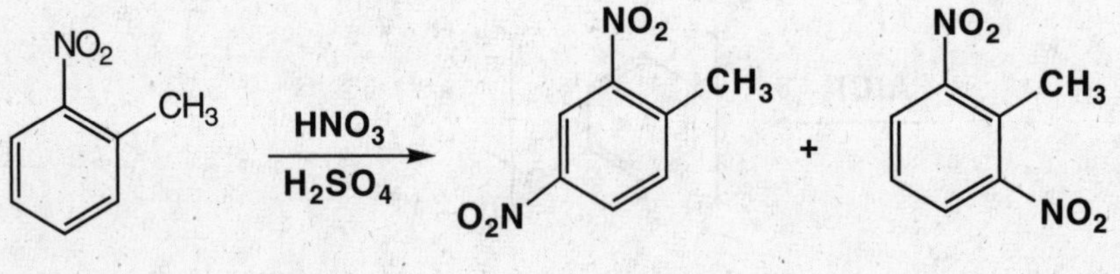

(c) $\xrightarrow[\text{H}_2\text{SO}_4]{\text{HNO}_3}$

(d)

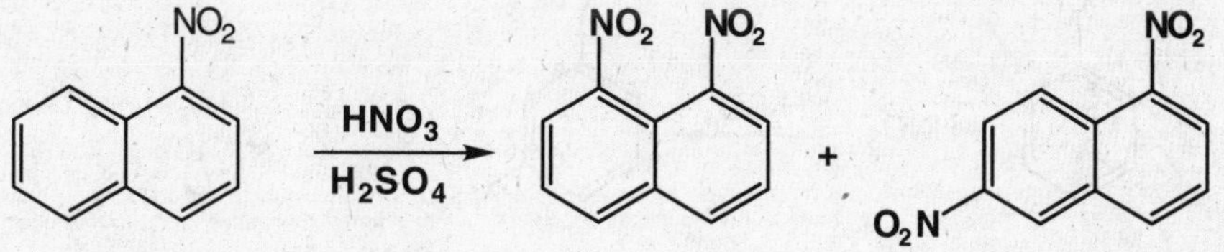

Nitration occurs as shown above, because the ring without the nitro group is less deactivated. For reaction at the positions shown, none of the contributing structures place the positive charge adjacent to the existing nitro group. For example, the following five contributing structures can be drawn for the intermediate leading to the first product.

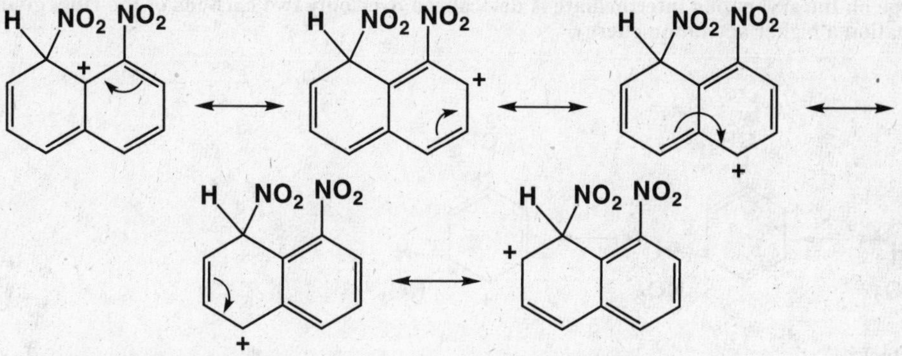

A similar set of contributing structures drawn for reaction at an alternative position includes one that places the positive charge adjacent to the existing nitro group. This structure makes a negligible contribution to the resonance hybrid, so the positive charge is less delocalized and this intermediate is less stable than that for the preferred substitution positions.

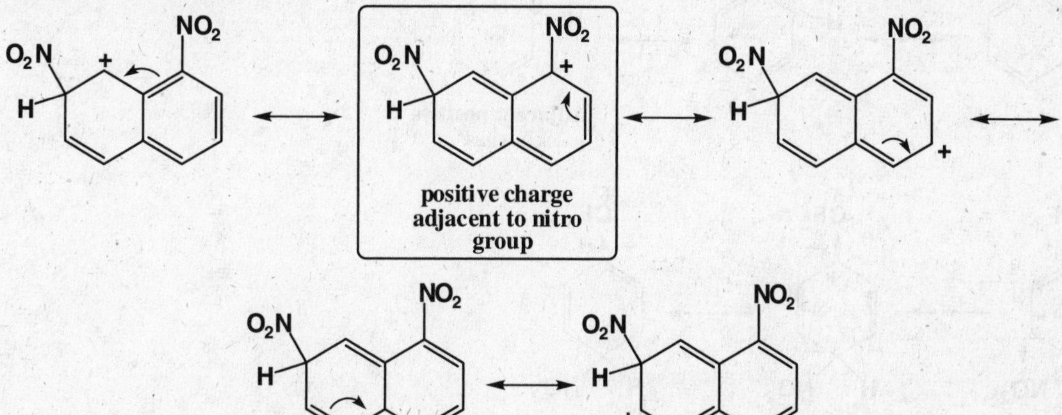

positive charge
adjacent to nitro
group

<u>Problem 22.16</u> How do you account for the fact that *N*-phenylacetamide (acetanilide) is less reactive toward electrophilic aromatic substitution than aniline?

N-Phenylacetamide
(Acetanilide)

Aniline

The unshared pair of electrons on the nitrogen atom of acetanilide is involved in a resonance interaction with the carbonyl group of the amide, and, therefore, less available for stabilization of an aryl cation intermediate compared to aniline.

<u>Problem 22.17</u> Propose an explanation for the fact that the trifluoromethyl group is almost exclusively meta directing.

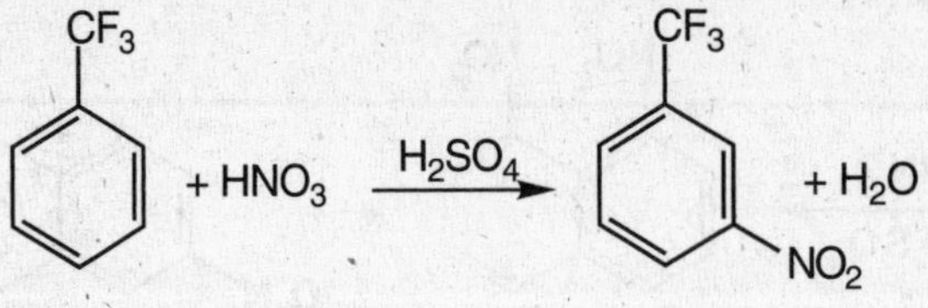

Shown below are contributing structures for meta and para attack of the electrophile. For meta attack, three contributing structures can be drawn and all make approximately equal contributions to the hybrid. Three contributing structures can also be drawn for ortho/para attack, one of which places a positive charge on carbon bearing the trifluoromethyl group; this structure makes only a negligible contribution to the hybrid. Thus, for meta attack, the positive charge on the aryl cation intermediate can be delocalized almost equally over three atoms of the ring giving this cation's formation a lower activation energy. For ortho/para attack, the positive charge on the aryl cation intermediate is delocalized over only two carbons of the ring, giving this cation's formation a higher activation energy.

meta <u>attack:</u>

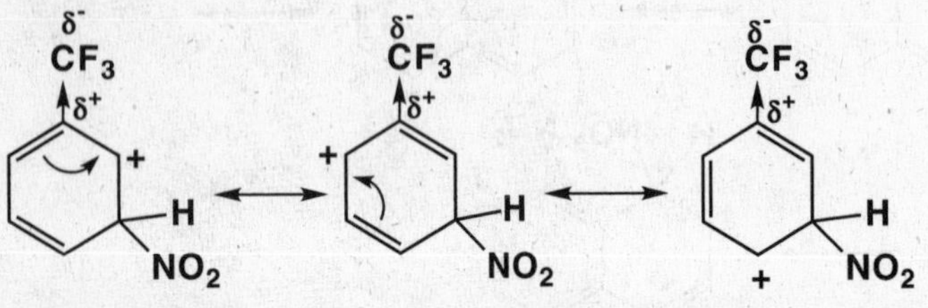

ortho/para <u>attack:</u>

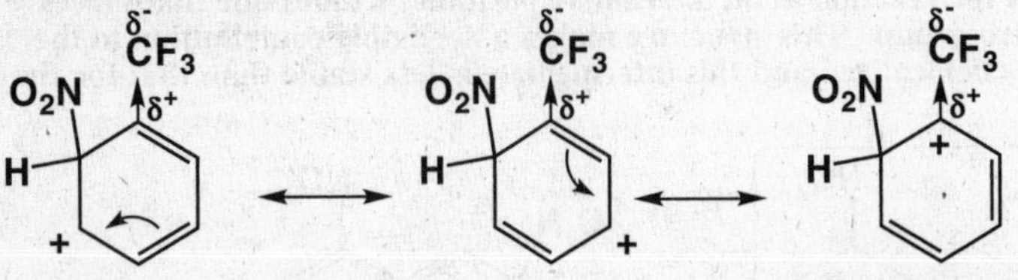

Adjacent positive charges

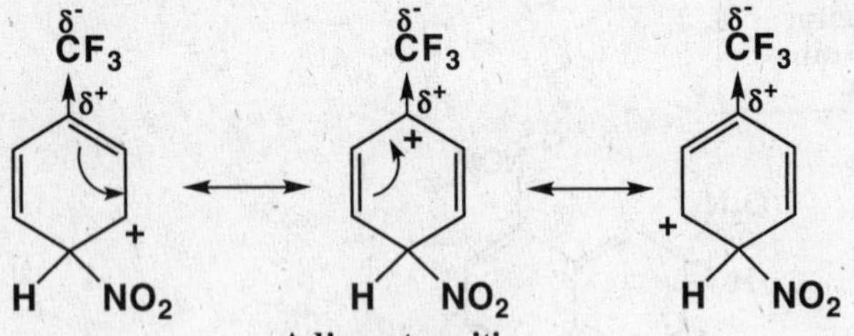

Adjacent positive charges

<u>Problem 22.18</u> Suggest a reason why the nitroso group, -N=O, is ortho-para directing although the nitro group, -NO$_2$, is meta directing.

Like other ortho-para directing groups, the -N=O group has a lone pair of electrons that can stabilize an adjacent positive charge of ortho or para attack via the resonance structures shown below. Without a lone pair of electrons on nitrogen, the nitro group cannot take part in similar stabilization. In fact, an adjacent positive charge is *destabilized* by the electron-withdrawing nature of the nitro group and the partial positive charge on the nitro N atom.

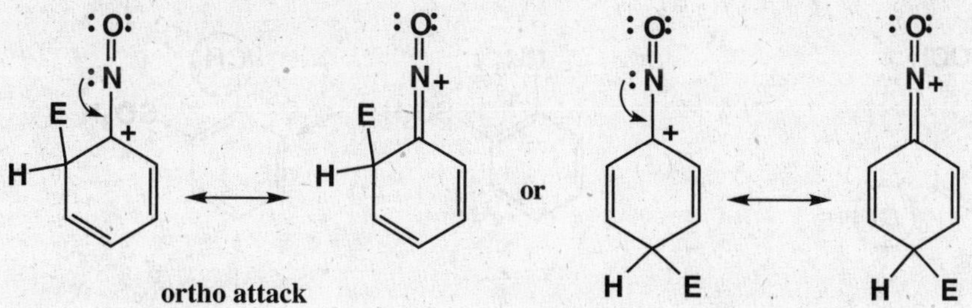

:O:
N
E
H
+
N+
or
H E

ortho attack

para attack

Problem 22.19 Arrange the compounds in each set in order of decreasing reactivity (fastest to slowest) toward electrophilic aromatic substitution.

(a)
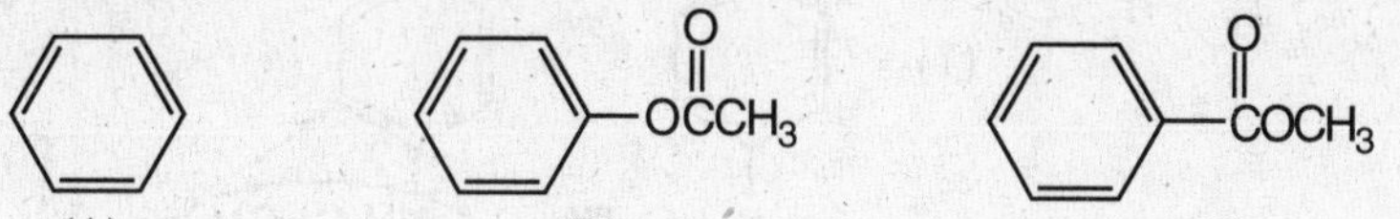

OCCH3
COCH3

(A) (B) (C) **B>A>C**

(b)
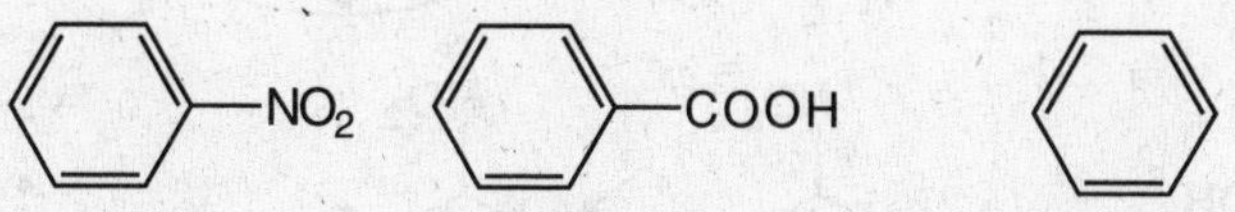

NO2
COOH

(A) (B) (C) **C>B>A**

(c)
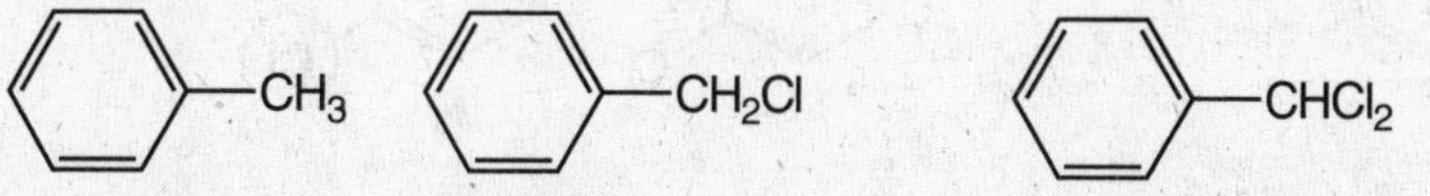

CH3
CH2Cl
CHCl2

(A) (B) (C) **A>B>C**

(d)
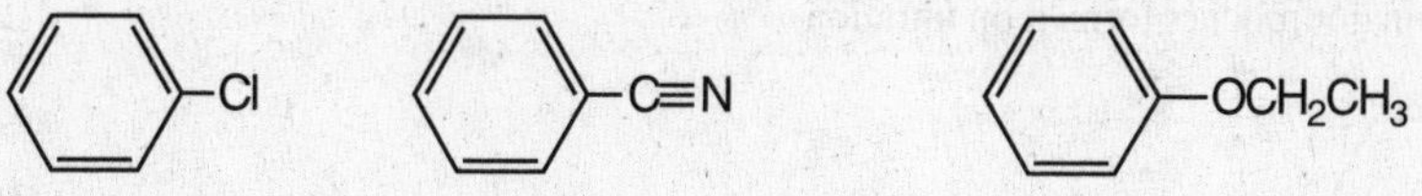

Cl
C≡N
OCH2CH3

(A) (B) (C) **C>A>B**

(e)
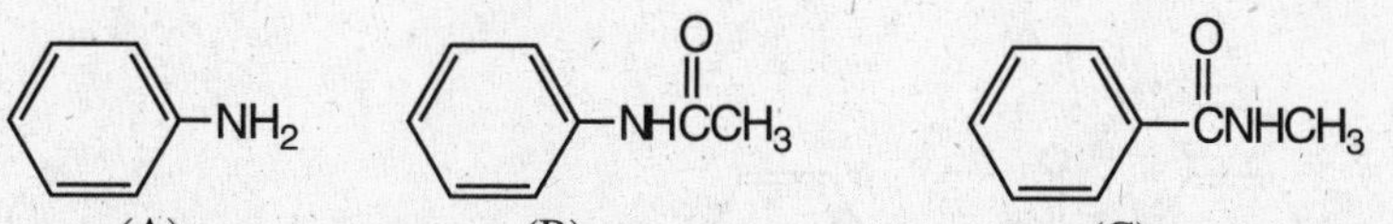

NH2
NHCCH3
CNHCH3

(A) (B) (C) **A>B>C**

Problem 22.20 For each compound, indicate which group on the ring is the more strongly activating and then draw a structural formula of the major product formed by nitration of the compound.

In the following structures, the more strongly activating group is circled and arrows show the position(s) of nitration. Where both ortho and para nitration is possible, two arrows are shown. A broken arrow shows a product formed in only negligible amounts.

(a)
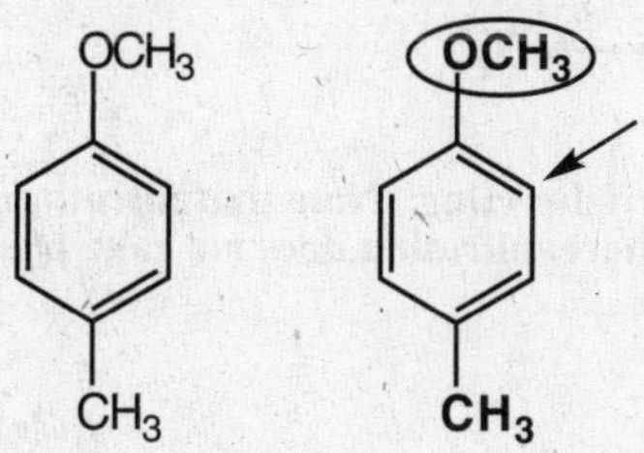

OCH3
CH3
OCH3
CH3

(b)
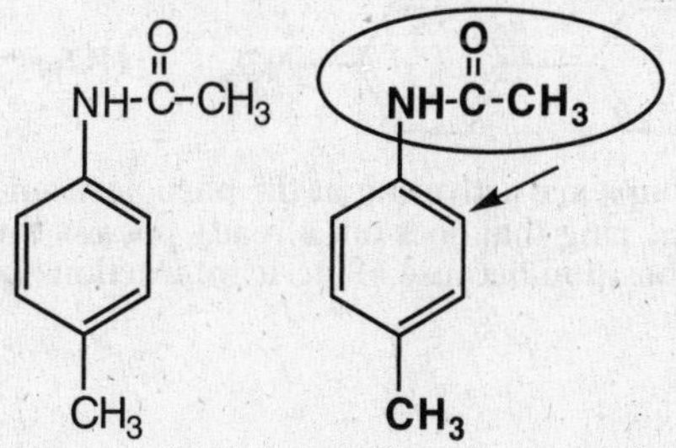

O
NH-C-CH3
CH3
O
NH-C-CH3
CH3

(c)

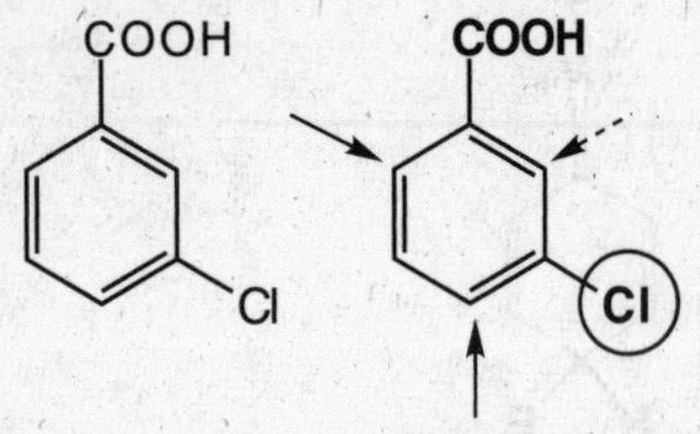

(d)

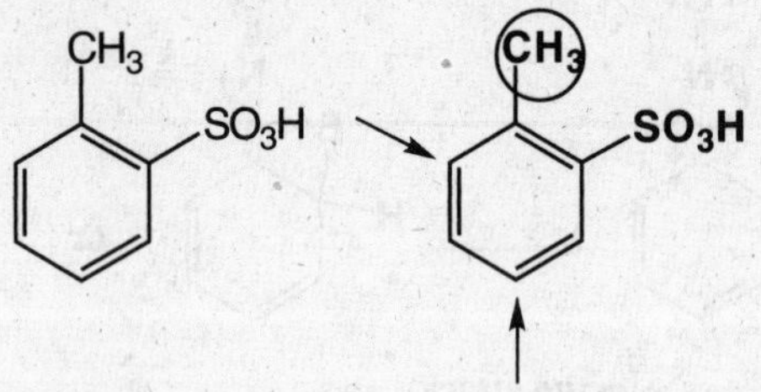

(e)

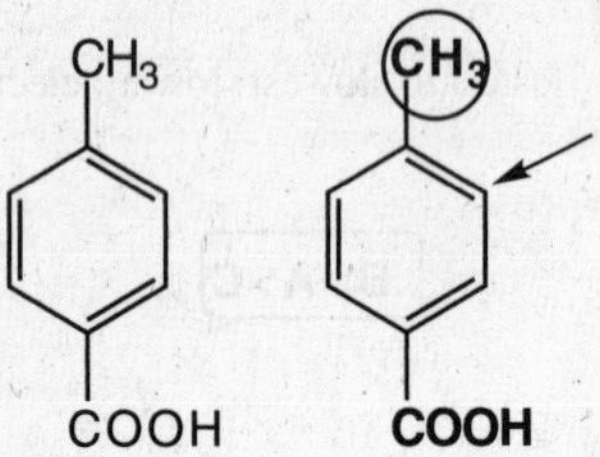

(f)

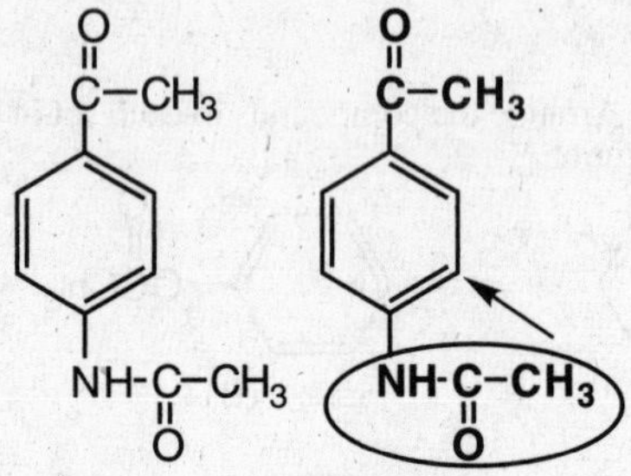

(g)

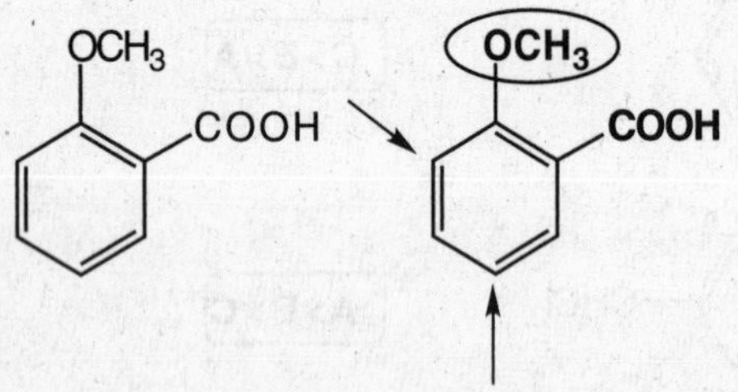

(h)

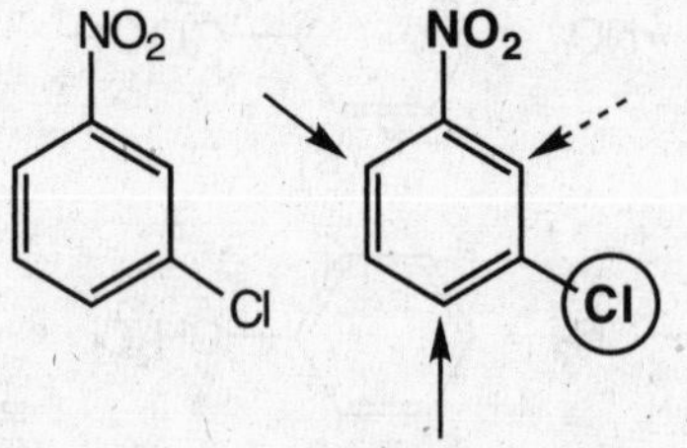

Problem 22.21 The following molecules each contain two rings. Which ring in each undergoes electrophilic aromatic substitution more readily? Draw the major product formed on nitration.

(a)

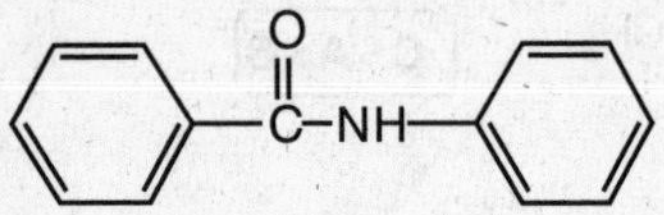

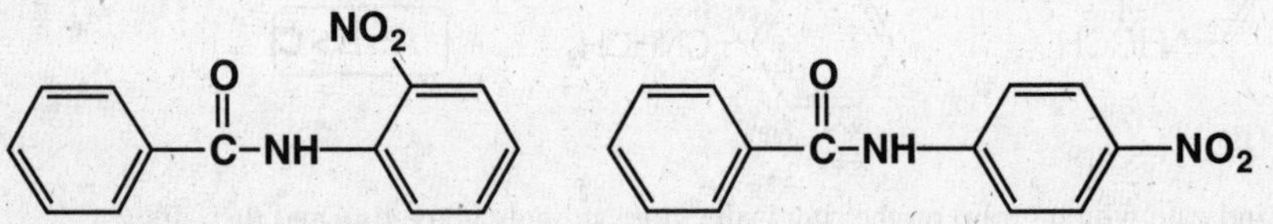

The nitrogen side of the amide group is more activating, so nitration produces the ortho/para products on this ring.

(b)

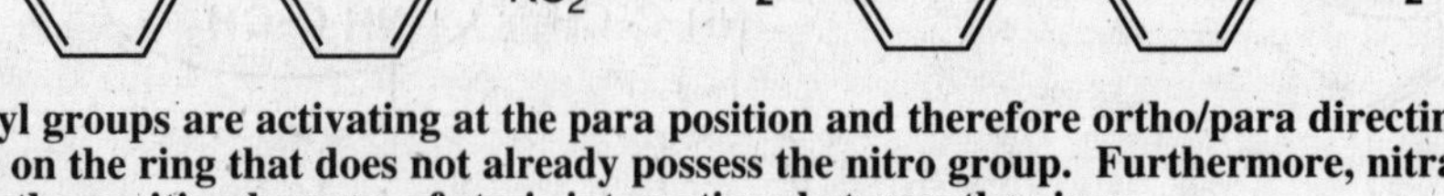

Phenyl groups are activating at the para position and therefore ortho/para directing. Note that nitration takes place on the ring that does not already possess the nitro group. Furthermore, nitration does not take place at the ortho position because of steric interactions between the rings.

(c)

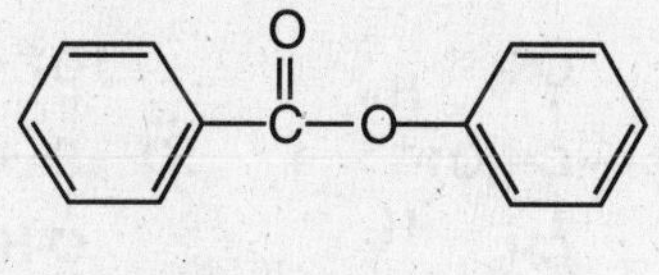

The oxygen side of the ester is more activating, so ortho/para nitration takes place on this ring.

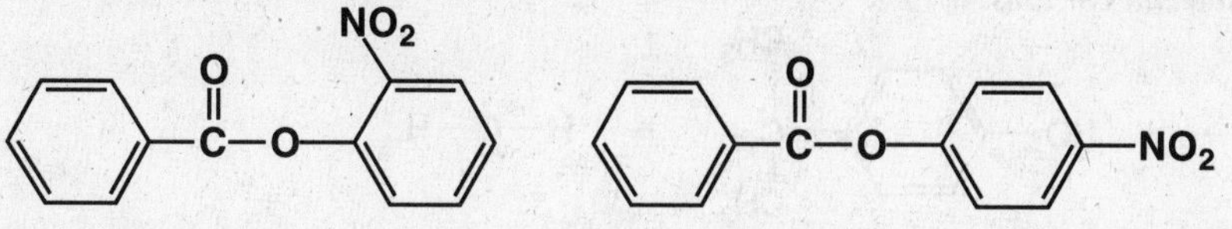

<u>Problem 22.22</u> Reaction of phenol with acetone in the presence of an acid catalyst gives a compound known as bisphenol A, which is used in the production of epoxy and polycarbonate resins. (Section 29.5). Propose a mechanism for the formation of bisphenol A.

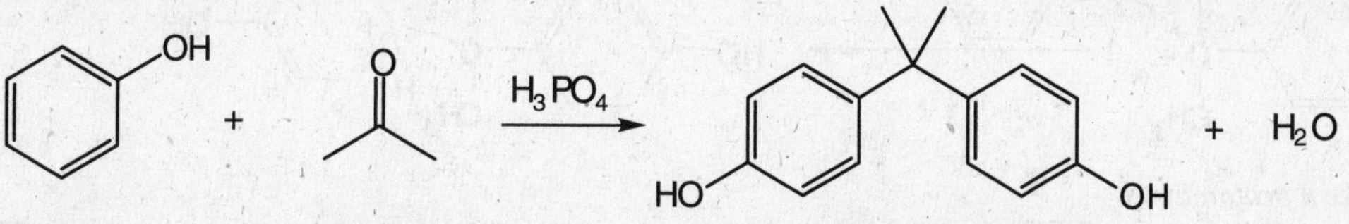

| Phenol | Acetone | Bisphenol A |

The reaction begins by proton transfer from phosphoric acid to acetone to form its conjugate acid which may be written as a hybrid of two contributing structures. The conjugate acid of acetone is an electrophile and reacts with phenol at the para position by electrophilic aromatic substitution to give 2-(4-hydroxyphenyl)-2-propanol. Protonation of the tertiary alcohol in this molecule and departure of water gives a resonance-stabilized cation that reacts with a second molecule of phenol to give bisphenol A.

Step 1: Add a proton.

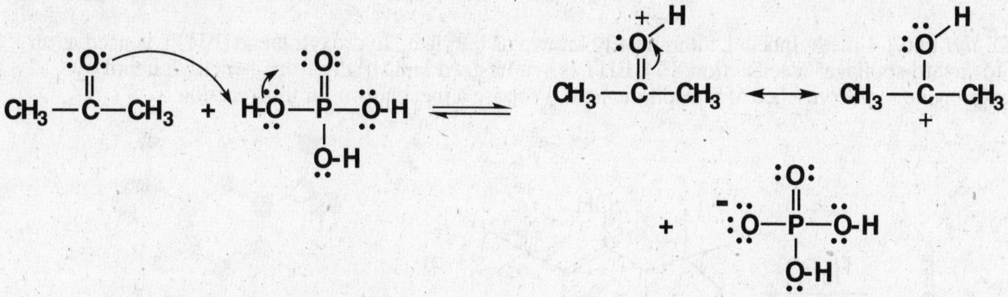

Step 2: Electrophilic addition (aromatic).

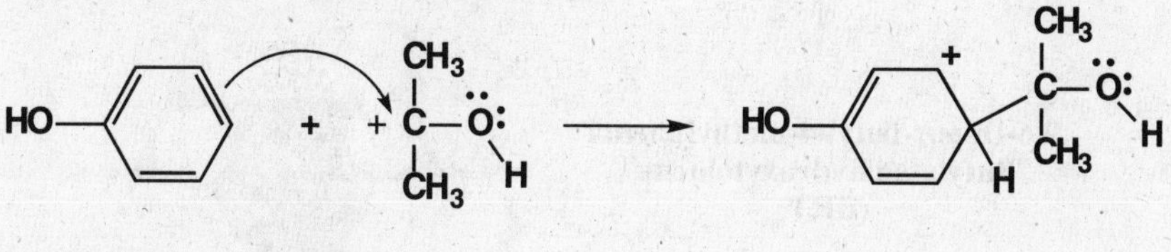

Step 3: Take a proton away.

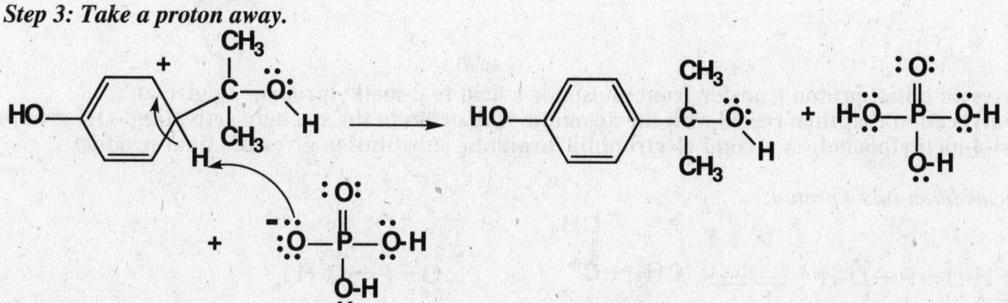

Step 4: Add a proton.

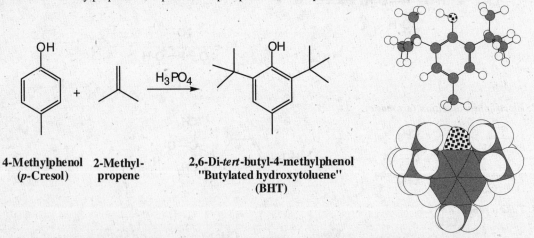

Step 5: Break a bond to give stable molecules or ions.

Step 6: Electrophilic addition (aromatic).

Step 7: Take a proton away.

Bisphenol A

<u>Problem 22.23</u> 2,6-Di-*tert*-butyl-4-methylphenol, alternatively known as butylated hydroxytoluene (BHT), is used as an antioxidant in foods to "retard spoilage" (see Section 8.7). BHT is synthesized industrially from 4-methylphenol by reaction with 2-methylpropene in the presence of phosphoric acid. Propose a mechanism for this reaction.

4-Methylphenol   **2-Methyl-**   **2,6-Di-*tert*-butyl-4-methylphenol**
(*p*-Cresol)   **propene**   **"Butylated hydroxytoluene"**
                                                              (BHT)

The reaction involves an initial proton transfer from phosphoric acid to 2-methylpropene to give an electrophilic *tert*-butyl cation that then reacts with the aromatic ring ortho to the strongly activating -OH group to form 2-*tert*-butyl-4-methylphenol. A second electrophilic aromatic substitution gives the final product.

Step 1: Electrophilic addition-add a proton.

Step 2: Electrophilic addition (aromatic).

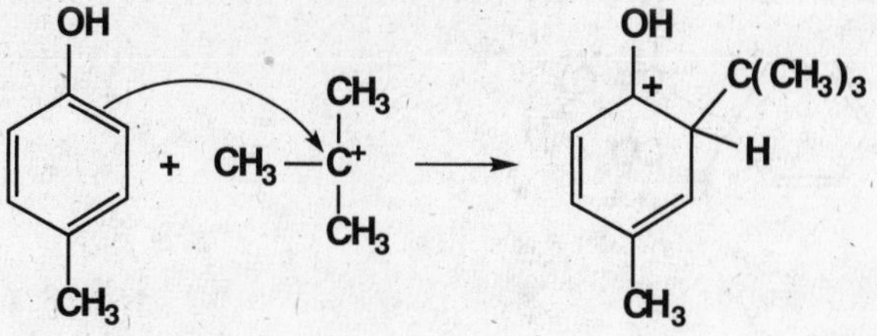

Step 3: Take a proton away.

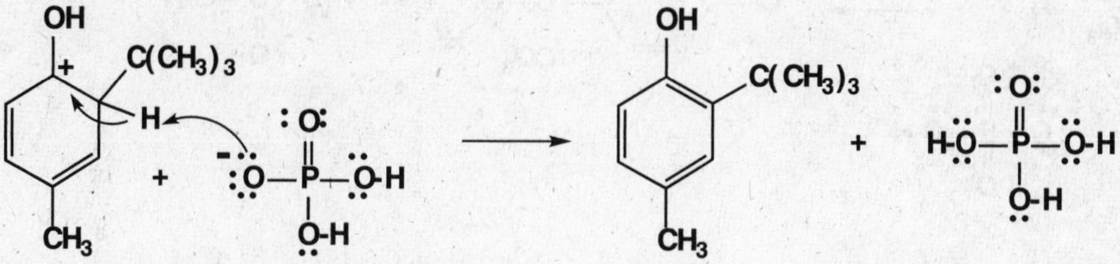

Step 4: Electrophilic addition (aromatic).

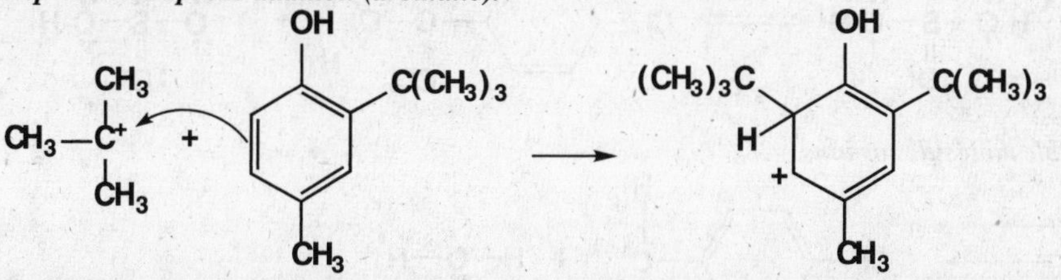

Step 5: Take a proton away.

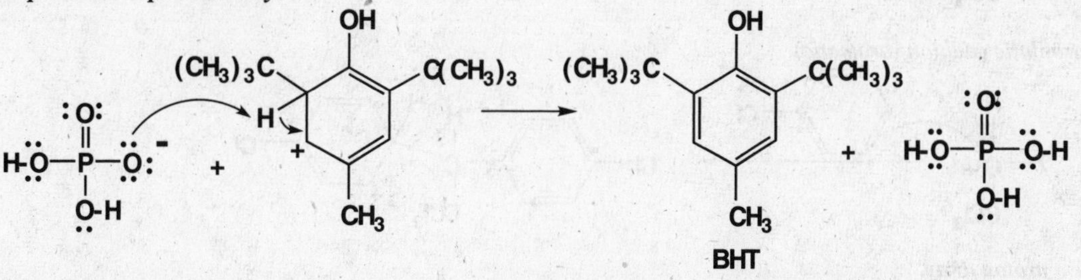

BHT

Problem 22.24 The insecticide DDT is prepared by the following route. Suggest a mechanism for this reaction. The abbreviation DDT is derived from the common name **d**ichloro**d**iphenyl**t**richloroethane.

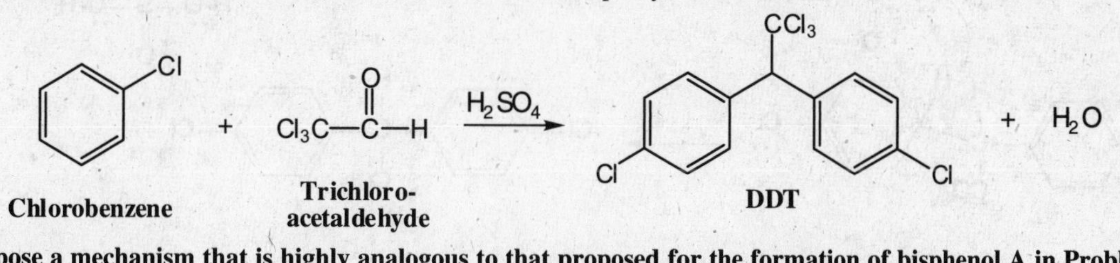

$$Chlorobenzene + Cl_3C-\overset{O}{\overset{\|}{C}}-H \xrightarrow{H_2SO_4} DDT + H_2O$$

Chlorobenzene

**Trichloro-
acetaldehyde**

DDT

Propose a mechanism that is highly analogous to that proposed for the formation of bisphenol A in Problem 22.22 above.

Step 1: Add a proton.

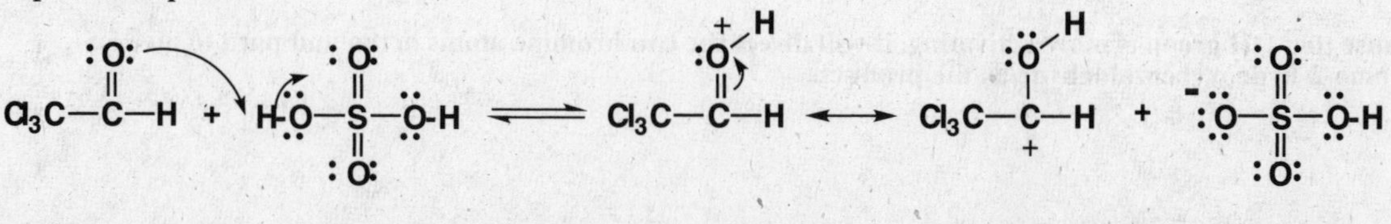

Step 2: Electrophilic addition (aromatic).

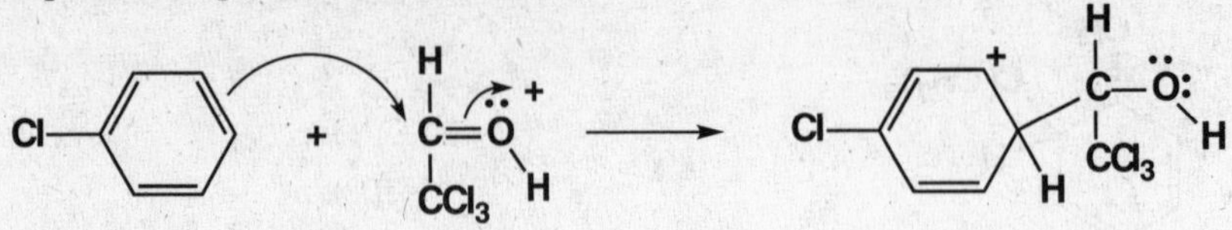

Step 3: Take a proton away.

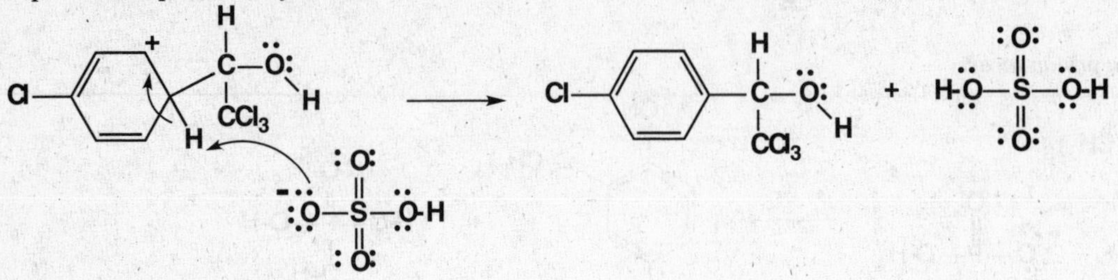

Step 4: Add a proton.

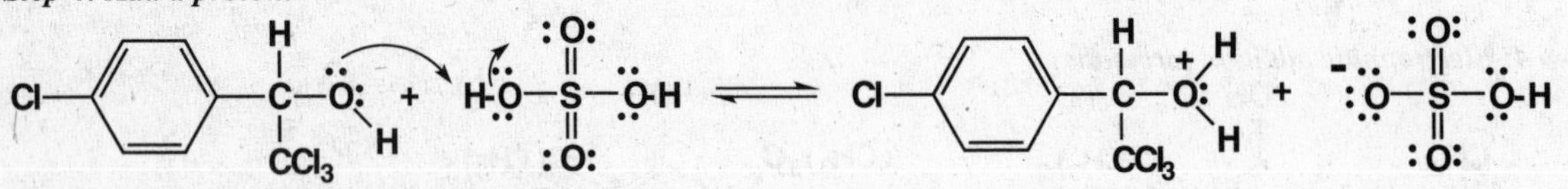

Step 5: Break a bond to give stable molecules or ions.

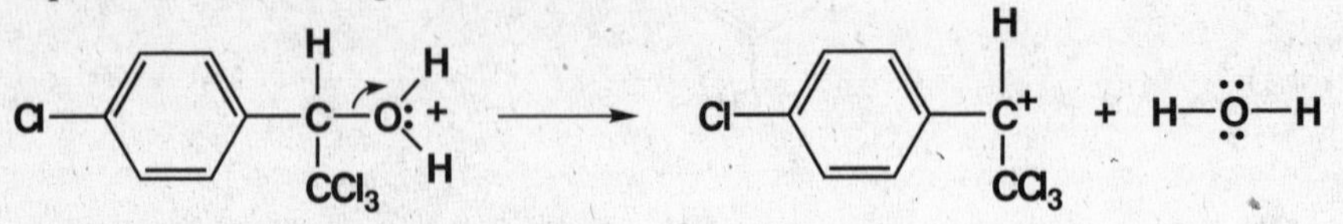

Step 6: Electrophilic addition (aromatic).

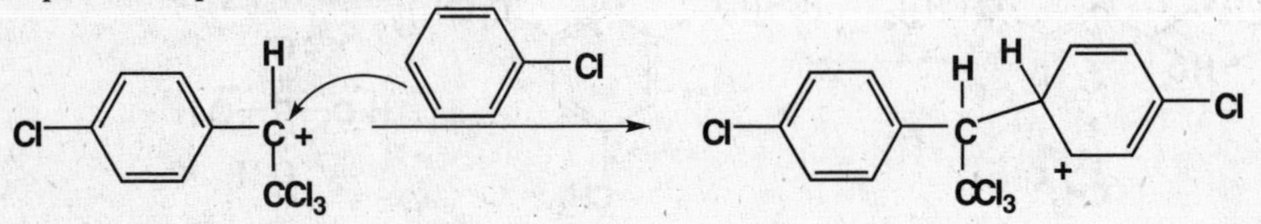

Step 7: Take a proton away.

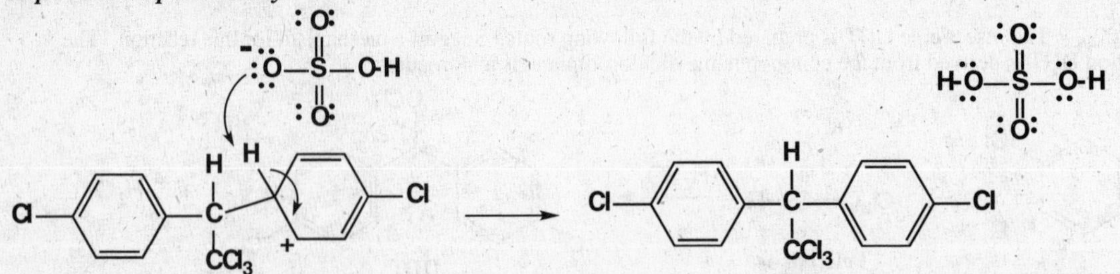

DDT

Problem 22.25 Treatment of salicylaldehyde (2-hydroxybenzaldehyde) with bromine in glacial acetic acid at 0°C gives a compound of molecular formula $C_7H_4Br_2O_2$ which is used as a topical fungicide and antibacterial agent. Propose a structural formula for this compound.

Because the -OH group is more activating, it will direct the two bromine atoms ortho and para to give 3,5-dibromo-2-hydroxybenzaldehyde as the product.

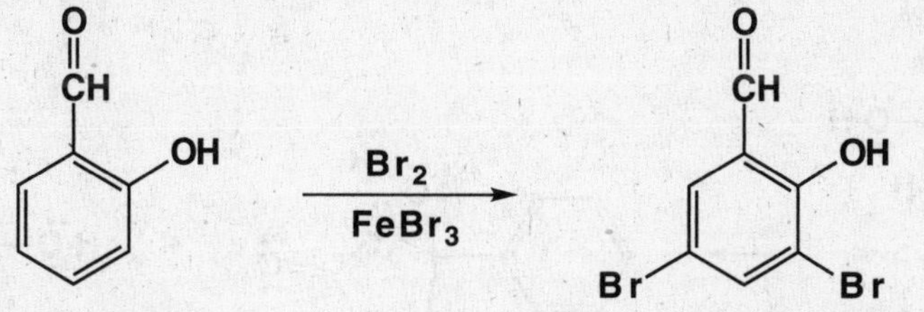

**2-Hydroxybenzaldehyde
(Salicylaldehyde)**

**3,5-Dibromo-2-hydroxybenzaldehyde
(3,5-Dibromosalicylaldehyde)
($C_7H_4Br_2O_2$)**

<u>Problem 22.26</u> Propose a synthesis for 3,5-dibromo-2-hydroxybenzoic acid (3,5-dibromosalicylic acid) from phenol.

A Kolbe reaction followed by bromination leads to 3,5-dibromo-2-hydroxybenzoic acid.

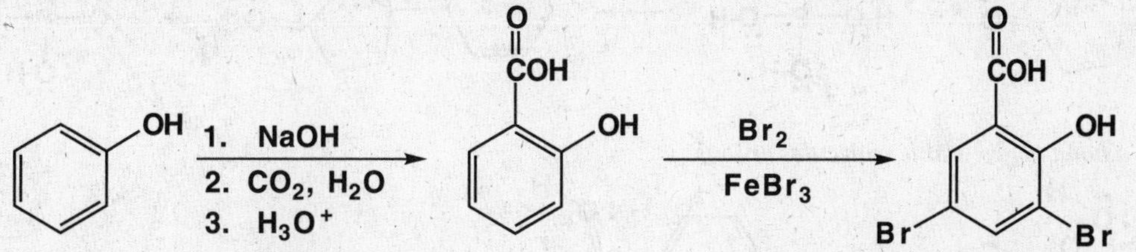

**(Kolbe
synthesis)**

**3,5-Dibromo-2-hydroxybenzoic acid
(3,5-Dibromosalicylic acid)**

<u>Problem 22.27</u> Treatment of benzene with succinic anhydride in the presence of polyphosphoric acid gives the following γ-ketoacid. Propose a mechanism for this reaction.

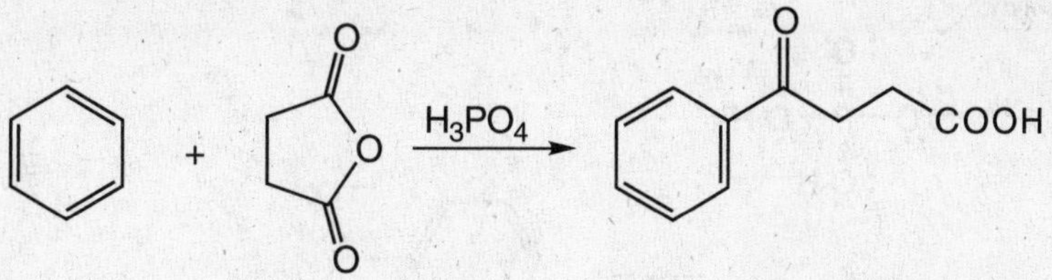

Succinic anhydride   4-Oxo-4-phenylbutanoic acid

Step 1: Add a proton.

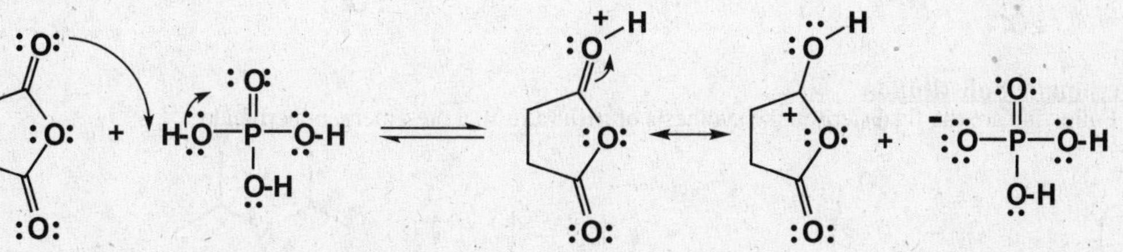

Step 2: Electrophilic addition (aromatic).

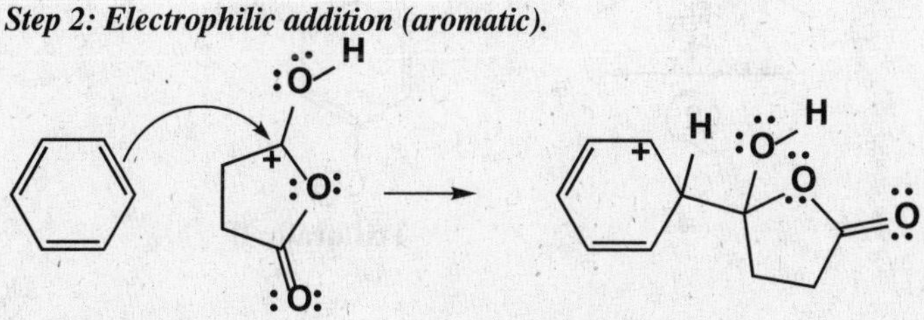

Step 3: Take a proton away.

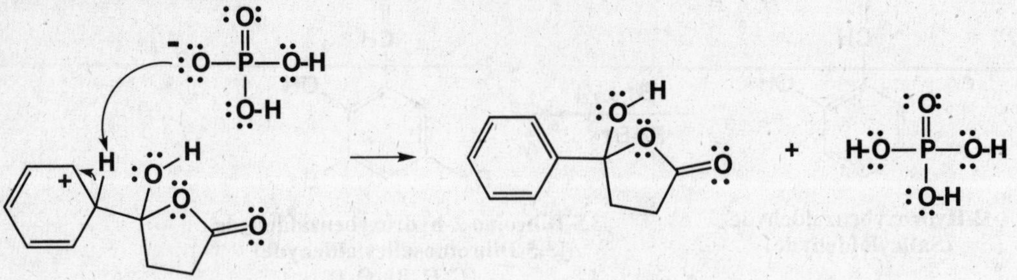

Step 4: Add a proton.

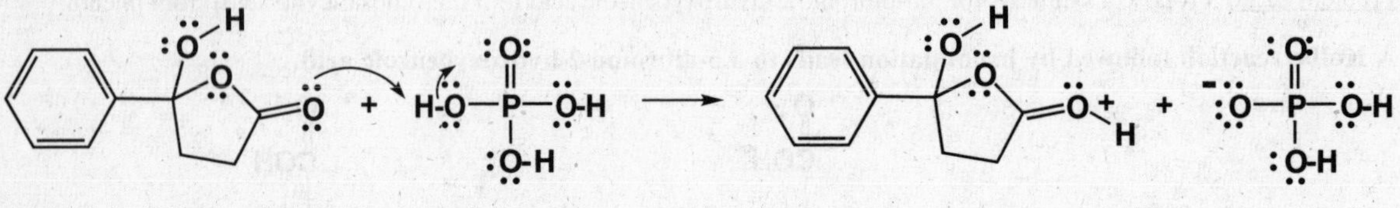

Step 5: Break a bond to give stable molecules or ions.

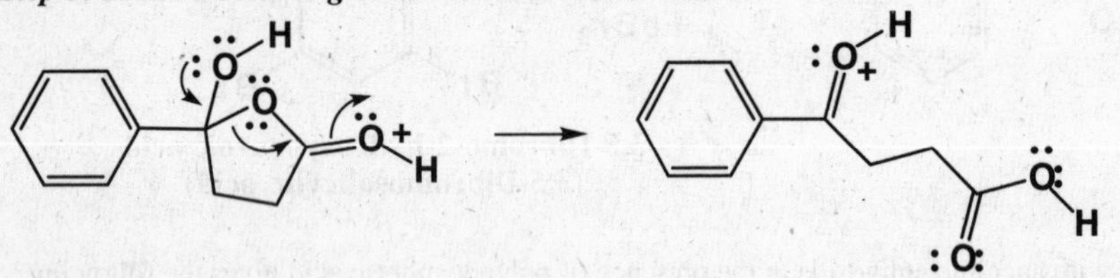

Step 6: Take a proton away.

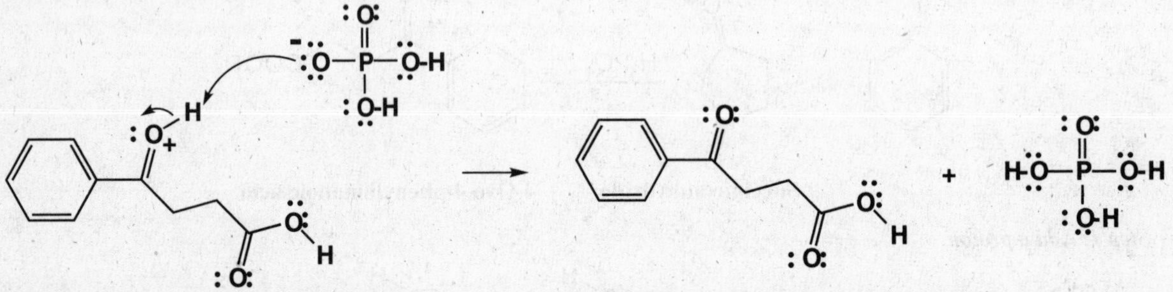

Nucleophilic Aromatic Substitution

Problem 22.28 Following are the final steps in the synthesis of trifluralin B, a pre-emergent herbicide.

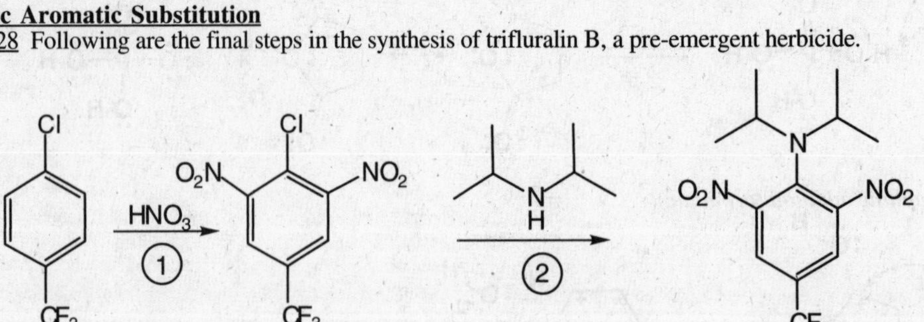

Trifluralin B

(a) Account for the orientation of nitration in Step (1).

The trifluoromethyl group is strongly deactivating and meta directing (Problem 22.17). Chlorine is weakly deactivating and ortho/para directing. The combination directs the incoming nitro groups ortho to the chlorine atom.

(b) Propose a mechanism for the substitution reaction in Step (2).

The second step of this transformation is an example of nucleophilic aromatic substitution. In the following structural formulas, the propyl group of dipropylamine is abbreviated R.

Step 1: Make a bond between a nucleophile and an electrophile (aromatic).

A Meisenheimer complex

Step 2: Take a proton away.

(-H⁺)

Step 3: Break a bond to give stable molecules or ions.

Trifluralin B

Problem 22.29 A problem in dyeing fabrics is the degree of fastness of the dye to the fabric. Many of the early dyes were surface dyes; that is, they did not bond to the fabric, with the result that they tend to wash off after repeated laundering. Indigo, for example, which gives the blue color to blue jeans, is a surface dye. Color fastness can be obtained by bonding a dye to the fabric. The first such dyes were the so-called reactive dyes, developed in the 1930s for covalent bonding dyes containing $-NH_2$ groups to cotton, wool, and silk fabrics. In the first stage of the first developed method for reactive dyeing, the dye is treated with cyanuryl chloride, which links the two through the amino group of the dye. The remaining chlorines are then displaced by -OH groups of cotton (cellulose) or $-NH_2$ groups of wool or silk (both proteins).

Cyanuryl chloride   Dye-NH_2 →   A reactive dye   HO-cotton →   Dye covalently bonded to cotton

Propose a mechanism for the displacement of a chlorine from cyanuryl chloride by (a) the NH_2 group of a dye and (b) by an -OH group of cotton.

In each reaction, the chlorine atoms are replaced by nucleophiles, either the amine groups of the dye or the hydroxyl groups of the cotton fibers. The mechanism will be nucleophilic aromatic substitution, analogous to that presented in problem 22.28 above, involving a Meisenheimer complex intermediate. Note that the chloro groups here can be considered the nitrogen analog of an acid chloride.

Syntheses

Problem 22.30 Show how to convert toluene to these compounds.

(a)

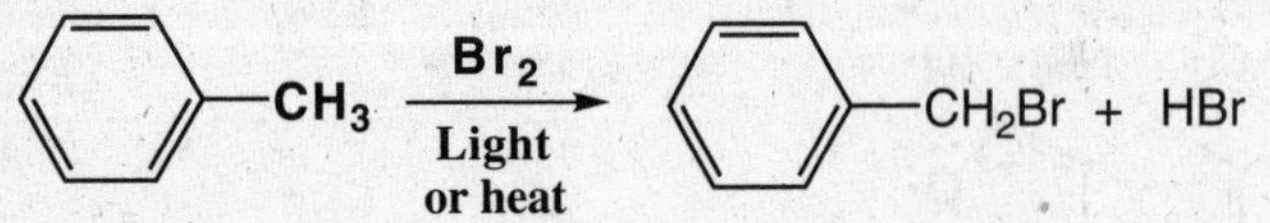

(b)

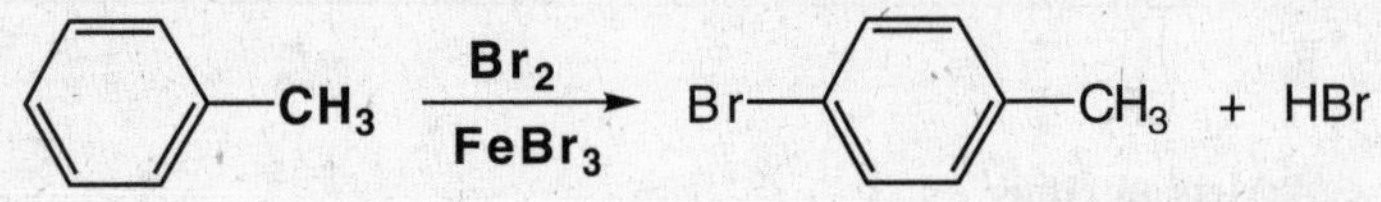

Problem 22.31 Show how to prepare each compound from 1-phenyl-1-propanone.

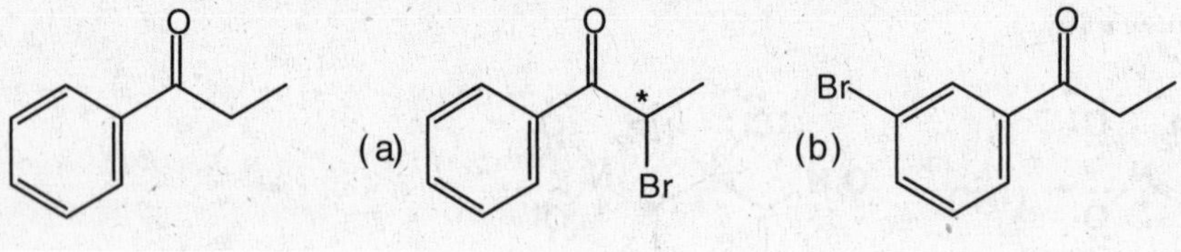

1-Phenyl-1-propanone   (racemic)

Product (a) is the result of acid catalyzed α-halogenation of a ketone (see chapter 16), while product (b) is the result of electrophilic aromatic substitution. In part (a) a chiral center is created, so the a racemic mixture of enantiomers will be produced.

(a)

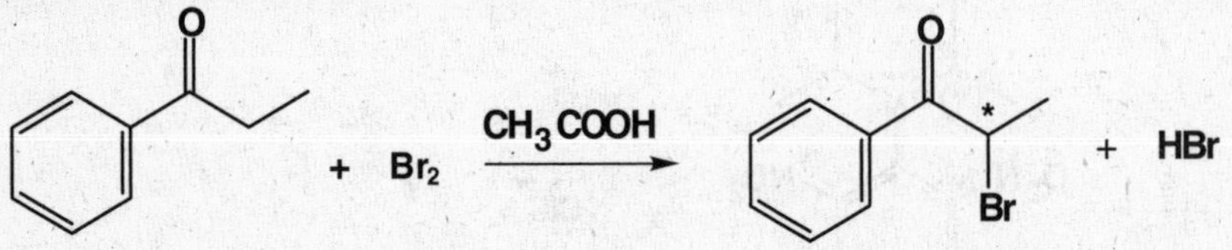

(b)

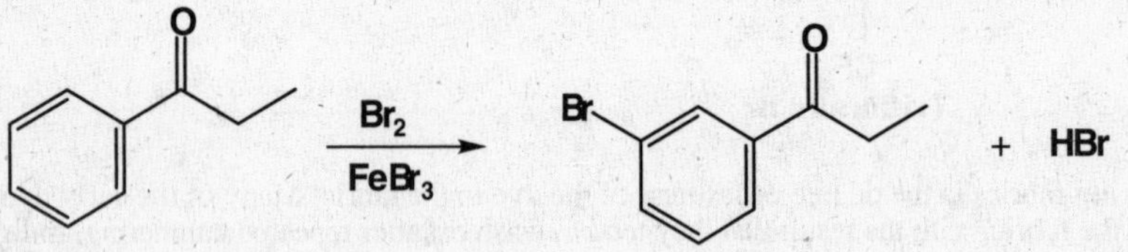

Problem 22.32 Show how to convert toluene to (a) 2,4-dinitrobenzoic acid and (b) 3,5-dinitrobenzoic acid.

Methyl is ortho/para directing. Therefore, toluene can be nitrated twice then oxidized with chromic acid to convert the methyl group into the carboxyl group.

(a)

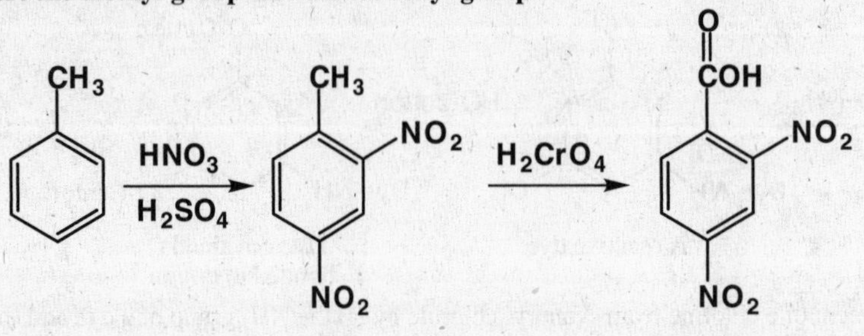

(b)

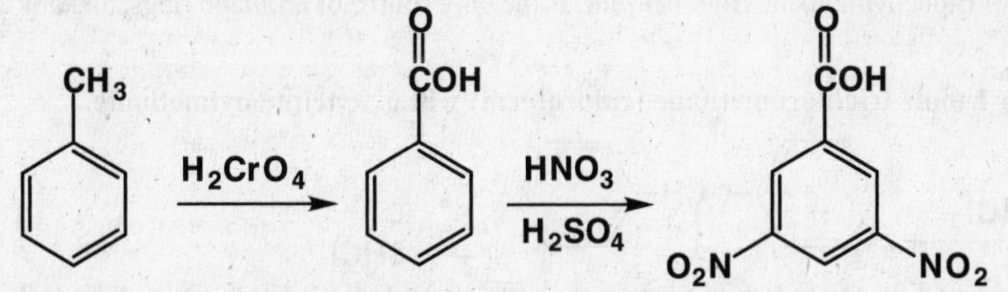

The reaction sequence is very similar to the last one, except now the order of the reactions is reversed because the carboxylic acid group is a meta director.

Problem 22.33 Show reagents and conditions to bring about the following conversions.

(a)

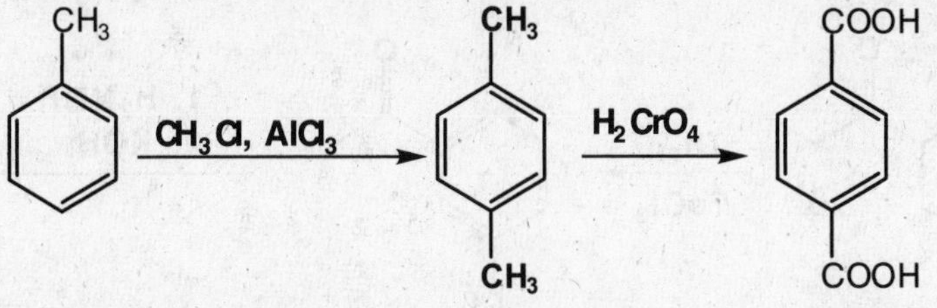

(b)

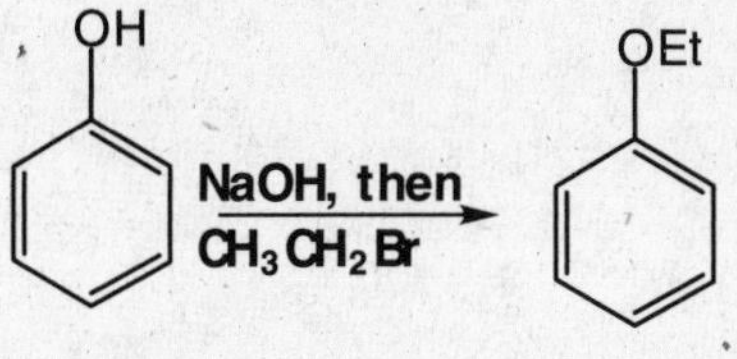

(c)

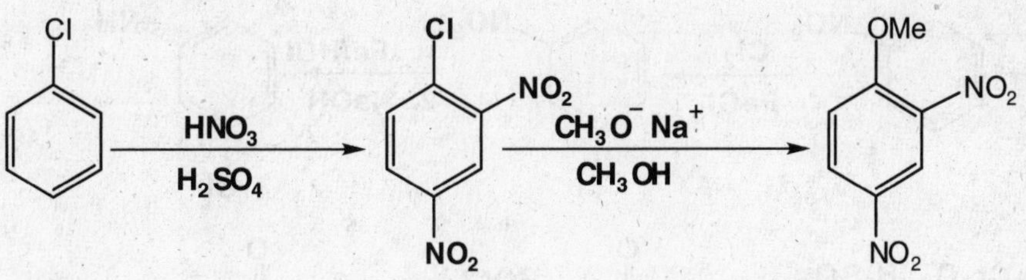

(d)

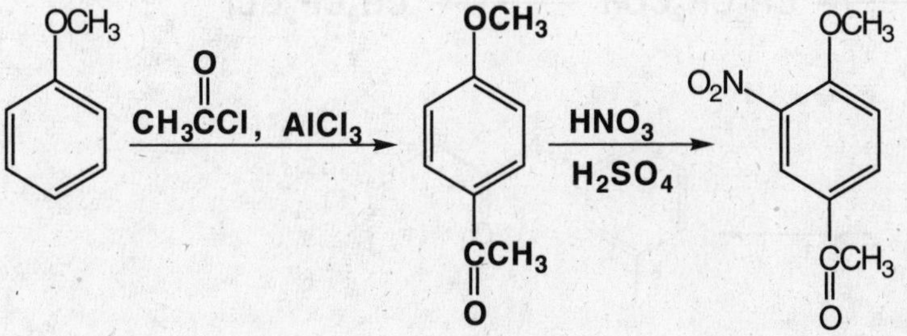

(e)

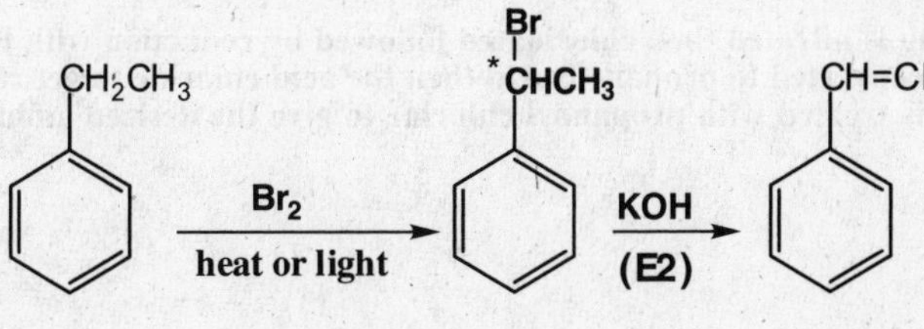

<u>Problem 22.34</u> Propose a synthesis of triphenylmethane from benzene as the only source of aromatic rings, and any other necessary reagents:

Reaction of 3 moles benzene with 1 mole trichloromethane (chloroform) will give triphenylmethane.

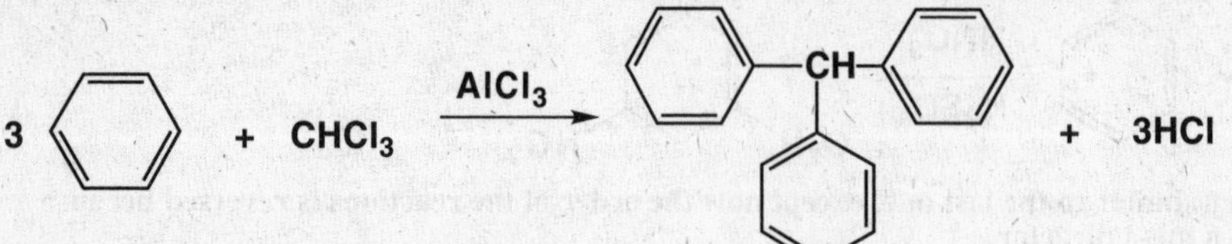

<u>Problem 22.35</u> Propose a synthesis for each compound from benzene.

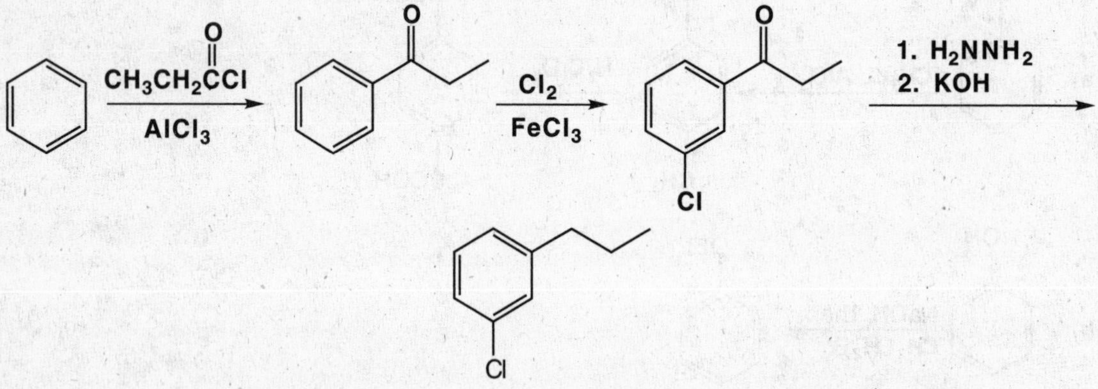

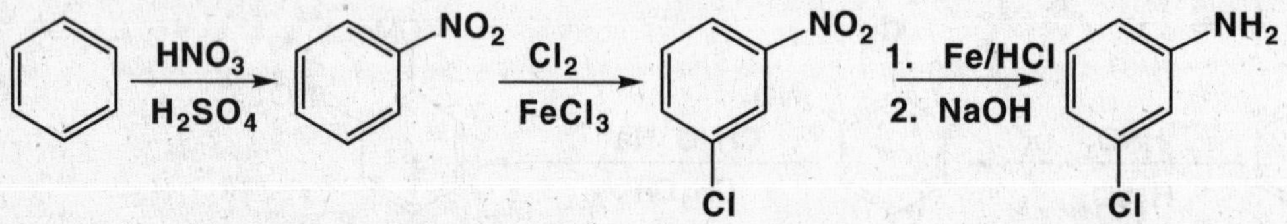

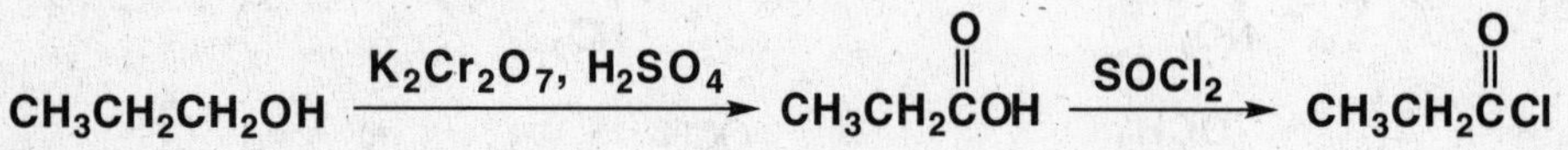

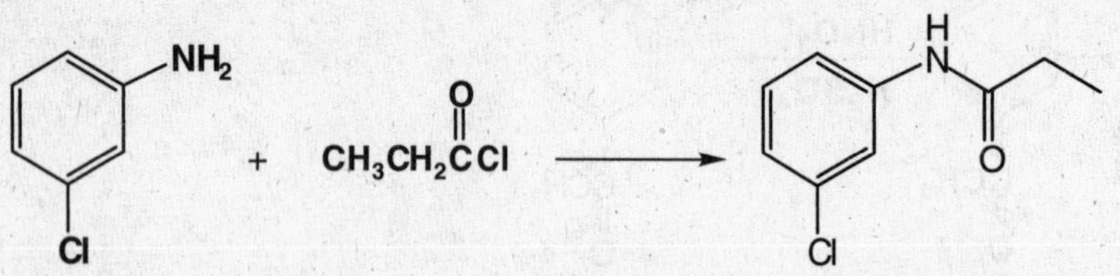

In the above synthesis scheme, the benzene ring is nitrated then chlorinated followed by reduction with Fe/HCl to give the 3-chloroaniline. The 1-propanol is converted to propanoic acid then the acid chloride to get ready for the final step in which the 3-chloroaniline is treated with propanoyl chloride to give the desired amide product.

Problem 22.36 The first widely used herbicide for the control of weeds was 2,4-dichlorophenoxyacetic acid (2,4-D). Show how this compound might be synthesized from phenol and chloroacetic acid by way of the given chlorinated phenol intermediate.

2,4-Dichlorophenoxyacetic acid
(2,4-D)

Chloroacetic
acid

Phenol

1. 2 eq. NaOH, H₂O
2. ClCH₂COOH
3. HCl, H₂O

Problem 22.37 Phenol is the starting material for the synthesis of 2,3,4,5,6-pentachlorophenol, known alternatively as pentachlorophenol or more simply as penta. At one time, penta was widely used as a wood preservative for decks, siding, and outdoor wood furniture. Draw the structural formula for pentachlorophenol, and describe its synthesis from phenol.

"Penta" is synthesized industrially from phenol. Because the -OH group is strongly activating, reaction of phenol with the first few atoms of chlorine occurs readily. However, after three chlorine atoms are on the ring, it is deactivated enough that harsher conditions are usually required to produce the final product. For this reason, the synthesis of 2,3,4,5,6-pentachlorophenol is usually broken up into two steps.

Cl_2
$FeCl_3$

Cl_2
$FeCl_3$
(harsher
conditions)

2,4,6-Trichlorophenol   2,3,4,5,6-Pentachlorophenol

Problem 22.38 Starting with benzene, toluene, or phenol as the only sources of aromatic rings, show how to synthesize the following. Assume in all syntheses that mixtures of ortho-para products can be separated into the desired isomer.
(a) 1-Bromo-3-nitrobenzene

Nitro is meta directing; bromine is ortho/para directing. Therefore, to have the two substituents meta to each other, carry out nitration first followed by bromination.

HNO_3
H_2SO_4

Br_2
$FeBr_3$

Benzene   Nitrobenzene   1-Bromo-3-nitrobenzene

(b) 1-Bromo-4-nitrobenzene

Reverse the order of steps from part (a). Nitro is meta directing; bromine is ortho-para directing. Therefore, to have the two substituents para to each other, carry out bromination first followed by nitration.

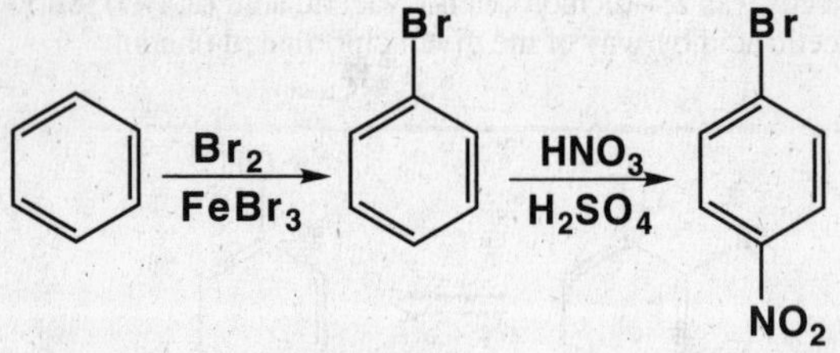

Benzene   Bromobenzene   1-Bromo-4-nitrobenzene

(c) 2,4,6-Trinitrotoluene (TNT)

The methyl group is ortho/para directing. Therefore, nitrate toluene three successive times then be very careful with the product!

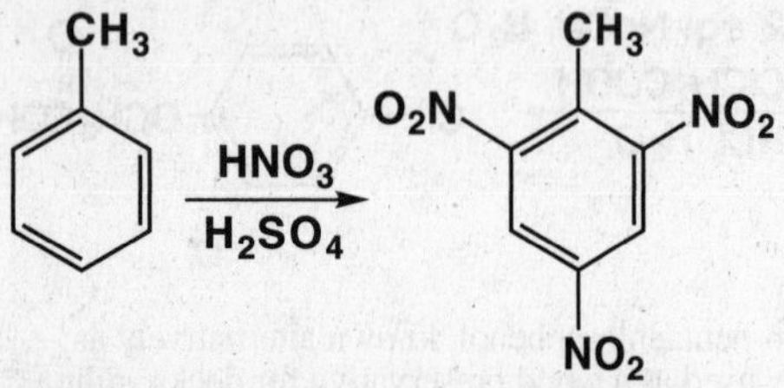

Toluene   2,4,6-Trinitrotoluene

(d) *m*-Chlorobenzoic acid

The carboxyl group and chlorine atom are meta to each other, an orientation best accomplished by chlorination of benzoic acid (the carboxyl group is meta directing). Oxidation of toluene with chromic acid gives benzoic acid. Treatment of benzoic acid with chlorine in the presence of ferric chloride gives the desired product.

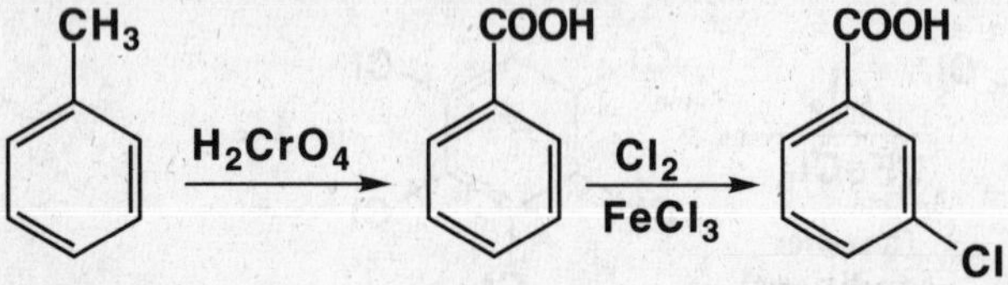

Toluene   Benzoic acid   3-Chlorobenzoic acid

(e) *p*-Chlorobenzoic acid

Start with toluene. The methyl group is weakly activating and directs chlorination to the ortho/para positions. Separate the desired para isomer and then oxidize the methyl group to a carboxyl group using chromic acid.

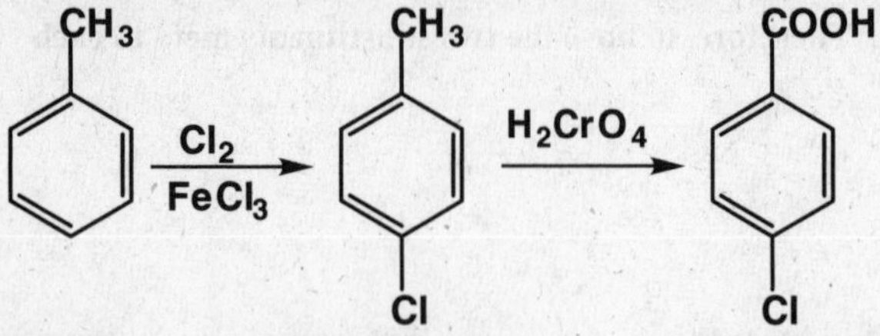

**Toluene   4-Chlorotoluene   4-Chlorobenzoic
                                          acid**

(f) *p*-Dichlorobenzene

Treatment of benzene with chlorine in the presence of aluminum chloride gives chlorobenzene. The chlorine atom is ortho/para directing. Treatment with chlorine in the presence of aluminum chloride a second time gives 1,4-dichlorobenzene.

Benzene   Chlorobenzene   _p_-Dichlorobenzene

(g) _m_-Nitrobenzenesulfonic acid

Both the sulfonic acid group and the nitro group are meta directors. Therefore, the two electrophilic aromatic substitution reactions may be carried out in either order. The sequence shown is nitration followed by sulfonation.

Benzene   Nitrobenzene   _m_-Nitrobenzenesulfonic acid

<u>Problem 22.39</u> 3,5-Dibromo-4-hydroxybenzenesulfonic acid is used as a disinfectant. Propose a synthesis of this compound from phenol.

**3,5-Dibromo-4-hydroxy-
benzenesulfonic acid**

<u>Problem 22.40</u> Propose a synthesis for 3,5-dichloro-2-methoxybenzoic acid starting from phenol.

**3,5-Dichloro-2-methoxy-
benzoic acid**

Note that in the methylation step, the methyl iodide could have reacted with either the phenoxide or carboxylate groups. Recall that phenoxide is a stronger base and also a stronger nucleophile, so the product shown is the predominant one. Dimethyl sulfate can be used in place of methyl iodide to create the methyl ether.

Problem 22.41 The following compound used in perfumery has a violet-like scent. Propose a synthesis of this compound from benzene.

4-Isopropylacetophenone

The isopropyl group is weakly activating and ortho-para directing; the carbonyl of the acetyl group is deactivating and meta directing. Therefore, start with benzene, convert it to isopropylbenzene (cumene) and then carry out a Friedel-Crafts acylation using acetyl chloride in the presence of aluminum chloride. Alkylation of benzene can be accomplished using 2-halopropane, 2-propanol, or propene, each in the presence of an appropriate catalyst.

C_6H_6 + CH_3CHCH_3 with Cl, $AlCl_3$

C_6H_6 + CH_3CHCH_3 with OH, H_2SO_4 →   **Isopropylbenzene (Cumene)** CH_3CCl (C=O), $AlCl_3$ →   **4-Isopropyl-acetophenone**

C_6H_6 + $CH_3CH=CH_2$, H_3PO_4

C_6H_6 + $CH_3CH_2CH_2Cl$, $AlCl_3$

Problem 22.42 Cancer of the prostrate is the second leading cause of cancer deaths among American males, exceeded only by lung cancer. One treatment of prostrate cancer is based on the fact that testosterone and androsterone (both androgens) enhance the proliferation of prostrate tumors. The drug flutamide (an antiandrogen) reduces the level of androgens in target tissues and is currently used to prevent and treat prostate cancer.

Flutamide (Eulexin)   ⟹   **Trifluoromethyl benzene**

Propose a synthesis of flutamide from (trifluoromethyl)benzene.

The $-CF_3$ group is meta directing because it is so electron withdrawing. Nitration followed by reduction gives the amine, which is acylated using 2-methylpropanoyl chloride. The amide group is activating and ortho/para directing, so a final nitration gives a good yield of the desired product flutamide. Note that in this last step, the ortho/para directing amide group take precedence because it is activating, while the $-CF_3$ group is deactivating.

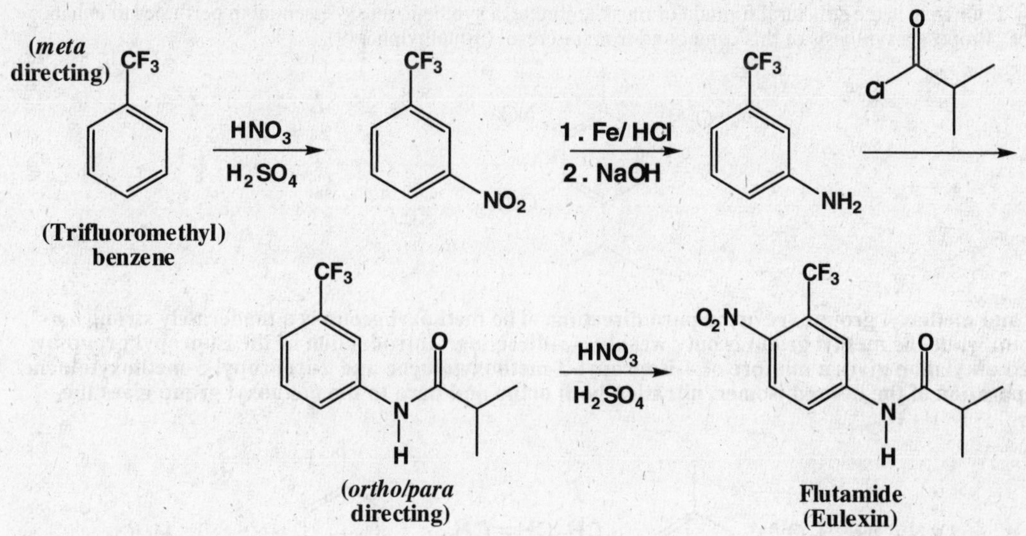

(Trifluoromethyl) benzene

(ortho/para directing)

Flutamide (Eulexin)

<u>Problem 22.43</u> The compound 4-isobutylacetophenone is needed for the synthesis of ibuprofen (See Chemical Connections: "Ibuprofen: The Evolution of an Industrial Synthesis" in Chapter 19). Propose a synthesis of 4-isobutylacetophenone from benzene and any other necessary reagents.

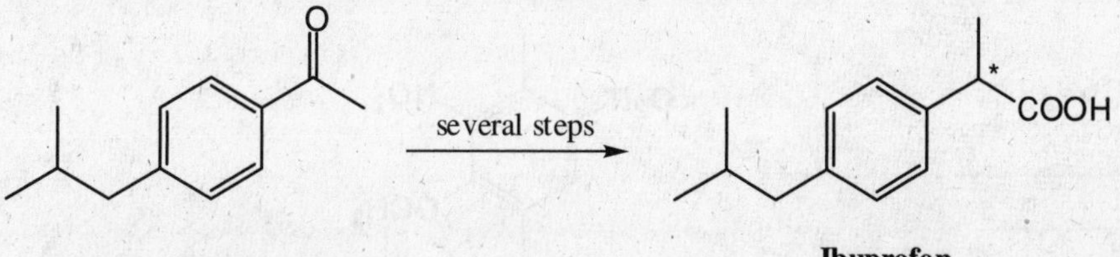

4-Isobutylacetophenone

Ibuprofen (racemic)

Begin with a Friedel-Crafts acylation using 2-methylpropanoyl chloride followed by a Wolff-Kishner reduction. The 2-methylpropyl group is ortho/para directing, so a second Friedel-Crafts acylation using acetyl chloride gives a good yield of 4-isobutylacetophenone.

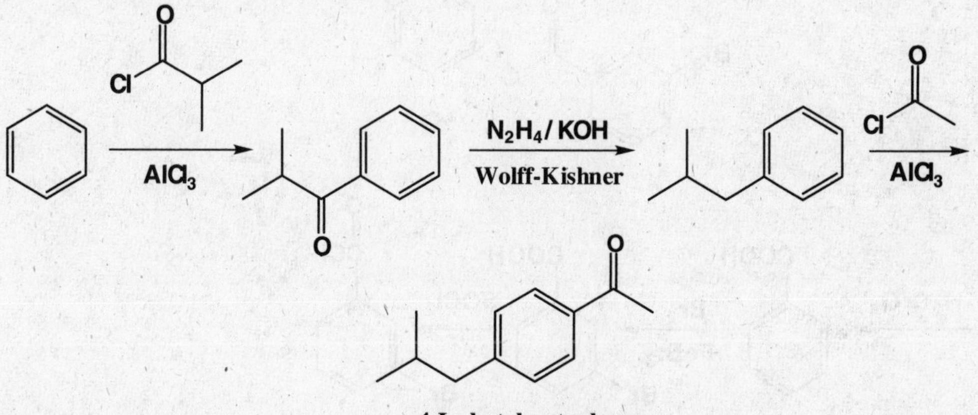

4-Isobutylacetophenone

<u>Problem 22.44</u> Following is the structural formula of musk ambrette, a synthetic musk, essential in perfumes to enhance and retain odor. Propose a synthesis of this compound from *m*-cresol (3-methylphenol).

Both methyl and methoxyl groups are ortho/para directing. The methoxyl group is a moderately strong *o,p*-directing group while the methyl group is only weakly *o,p*-directing. Introduction of the isopropyl group by acid-catalyzed alkylation gives a mixture of 4-isopropyl-3-methoxytoluene and 2-isopropyl-5-methoxytoluene. Following separation of the desired isomer, nitration both ortho and para to the methoxyl group gives the product.

In the first step, dimethyl sulfate can be used in place of methyl iodide to create the methyl ether.

<u>Problem 22.45</u> Propose a synthesis of this compound starting from toluene and phenol.

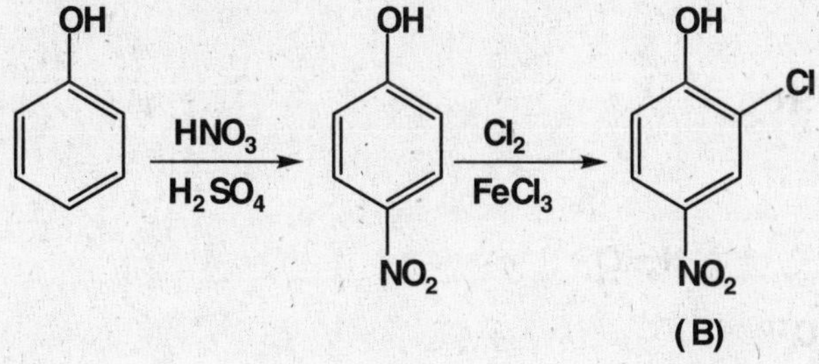

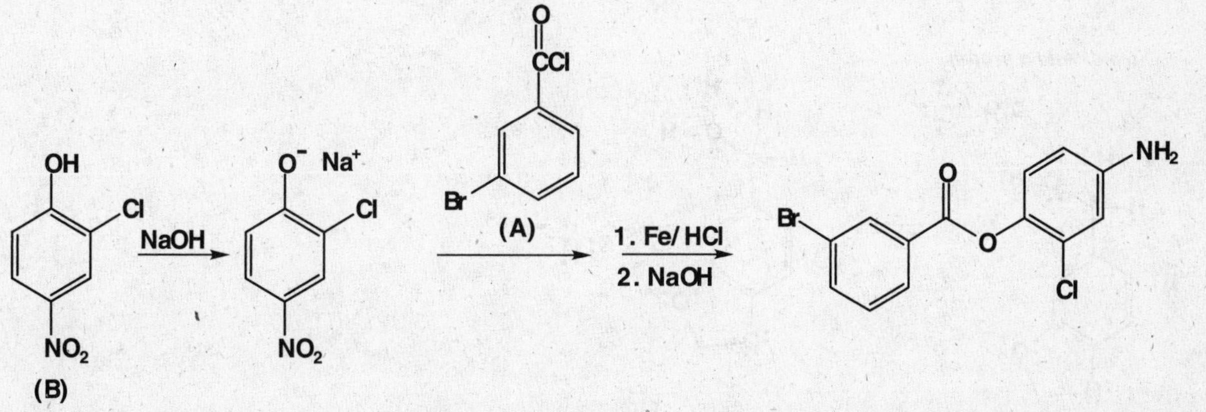

Problem 22.46 When certain aromatic compounds are treated with formaldehyde, CH_2O, and HCl, the CH_2Cl group is introduced onto the ring. This reaction is known as chloromethylation.

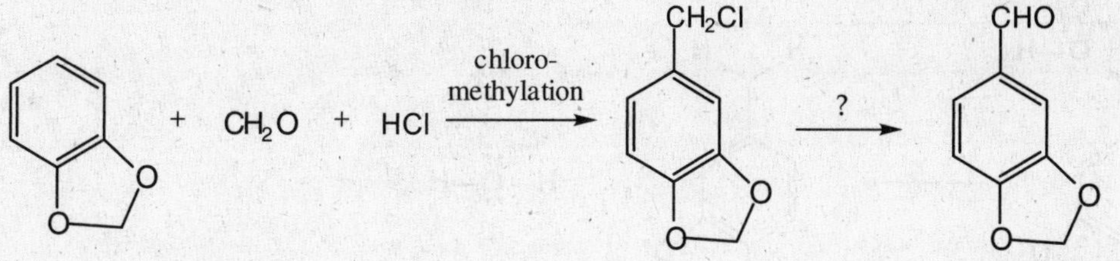

Piperonal

(a) Propose a mechanism for this example of chloromethylation.

The most likely mechanism involves an acid-catalyzed electrophilic aromatic substitution followed by replacement of the resulting -OH group with Cl via a benzyl cation intermediate. The acetal in the bottom of the molecule is likely attacked by the acid, but formaldehyde in the reaction mixture drives the equilibrium to acetal formation.

Step 1: Add a proton.

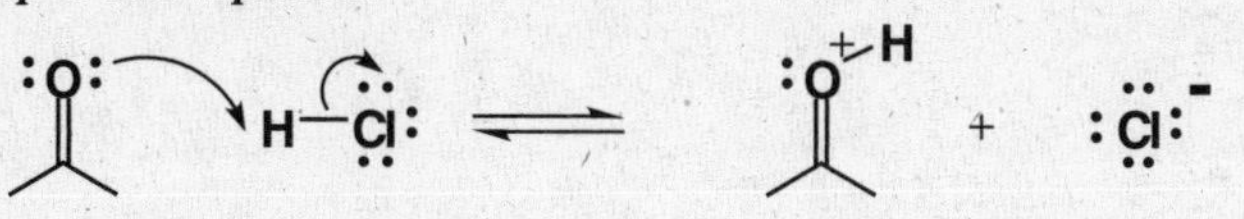

Step 2: Electrophilic addition (aromatic).

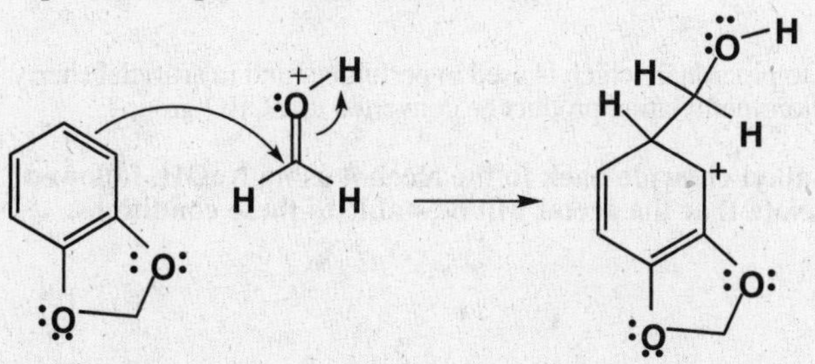

Step 3: Take a proton away.

Step 4: Add a proton.

Step 5: Break a bond to give stable molecules or ions.

Step 6: Make a bond between a nucleophile and an electrophile.

(b) The product of this chloromethylation can be converted to piperonal, which is used in perfumery and in artificial cherry and vanilla flavors. How might the CH₂Cl group of the chloromethylation product be converted to a CHO group?

A simple two-step procedure would be to convert the alkyl chloride back to the alcohol using NaOH, followed by oxidation with PCC to give the desired aldehyde. Note that the acetal will be stable to these conditions.

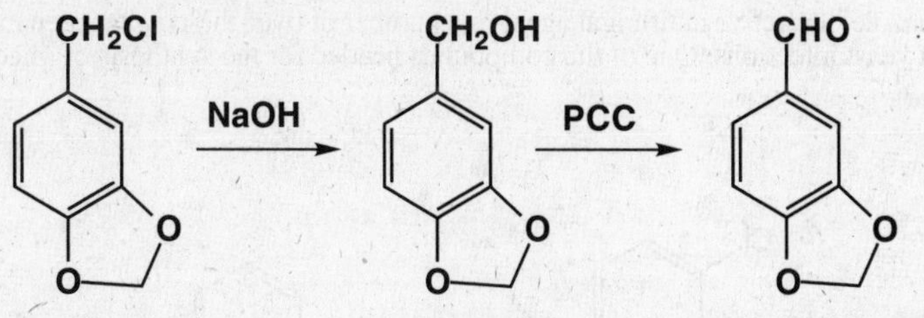

Problem 22.47 Following is a retrosynthetic analysis for the acaracide (killing mites and ticks) and fungicide dinocap.

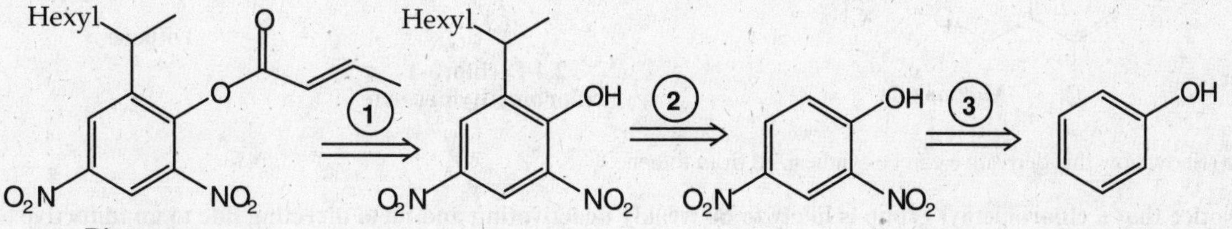

Dinocap

(a) Given this analysis, propose a synthesis for dinocap.

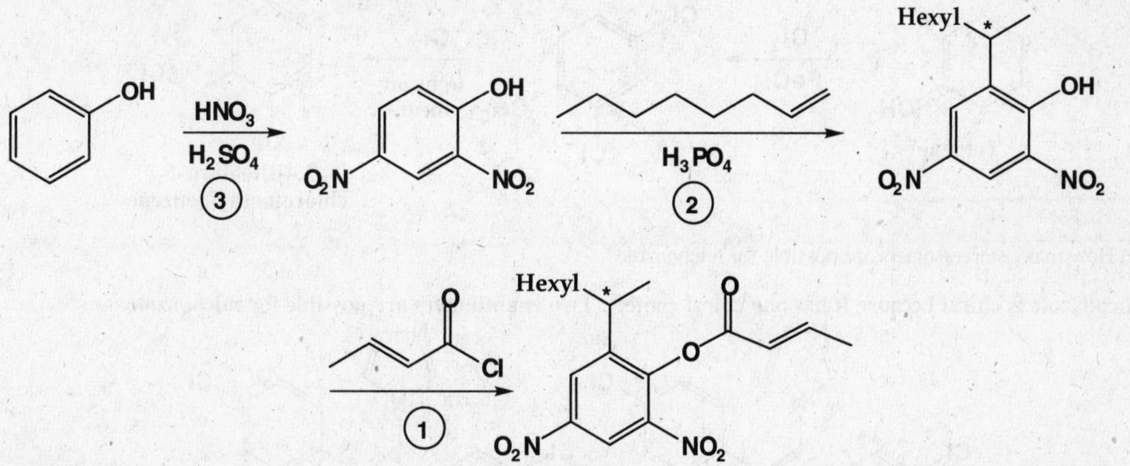

In the above synthesis scheme, the acid-catalyzed alkylation step is carried out on an aromatic ring that contains two nitro groups. Ordinarily, this would render the ring too deactivated to take part in the reaction. However, the presence of the highly activating -OH group adjacent to the site of attack is enough to allow reaction.

(b) Is dinocap chiral? If so, which of the possible stereoisomers are formed in this synthesis?

Dinocap is chiral because it has one chiral center. Both enantiomers are produced in the above synthesis.

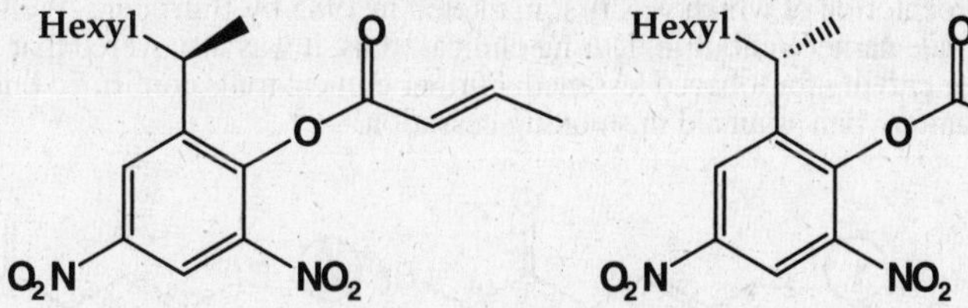

Enantiomers of dinocap

Problem 22.48 Following is the structure of miconazole, the active antifungal agent in a number of over-the-counter preparations, including Monistat, which are used to treat vaginal yeast infections. One of the compounds needed for the synthesis of miconazole is the trichloro derivative of toluene shown on its right.

Miconazole

2,4-Dichloro-1-chloromethylbenzene

Toluene

(a) Show how this derivative can be synthesized from toluene.

Notice that a chloromethyl group is likely to be weakly deactivating and meta directing due to an inductive effect. Therefore, the best synthetic strategy would be to chlorinate the aromatic ring ortho and para to the methyl group first, then halogenate the benzylic position.

Toluene   **2,4-Dichloro-1-chloromethylbenzene**

(b) How many stereoisomers are possible for miconazole?

Miconazole is chiral because it has one chiral center. Two enantiomers are possible for miconazole.

Enantiomers of miconazole

Problem 22.49 Bupropion, the hydrochloride of which was first marketed in 1985 by Burroughs-Wellcome, now GlaxoWellcome, is an antidepressant sold under the trade name Wellbutrin. During clinical trials, it was discovered that smokers, after one to two weeks on the drug, reported that their craving for tobacco lessened. Further clinical trials confirmed this finding, and the drug was marketed in 1997 under the trade name Zyban as an aid in smoking cessation.

Bupropion

(a) Given this retrosynthetic analysis, propose a synthesis for bupropion.

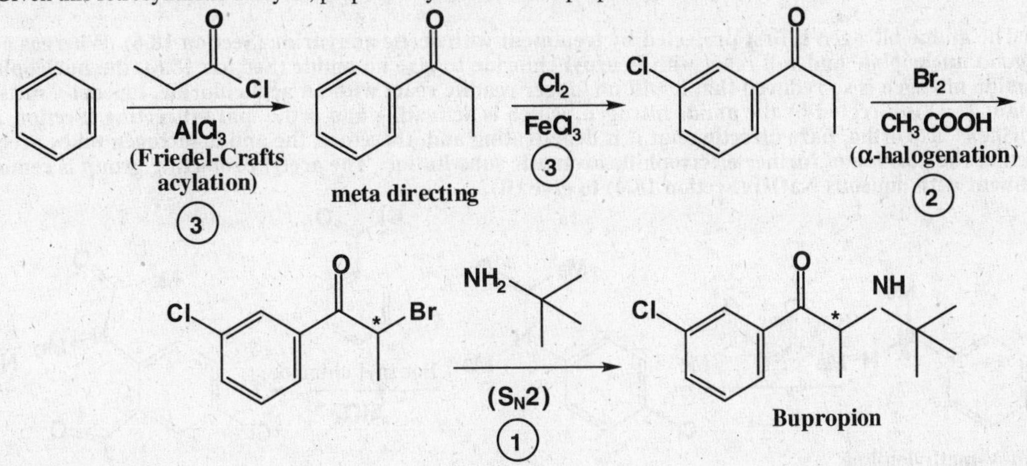

Note that the Friedel-Crafts acylation was carried out first because that produces the desired meta orientation of the chlorine atom and propanoyl group. One chiral center is created in this reaction sequence, so a racemic mixture of enantiomers is produced.

(b) Is bupropion chiral? If so, how many of the possible stereoisomers are formed in this synthesis?

Bupropion is chiral because it has one chiral center. Both enantiomers are produced in the above synthesis.

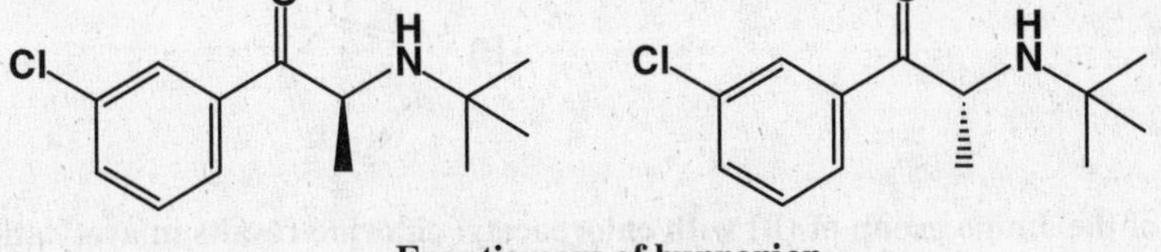

Enantiomers of bupropion

Problem 22.50 Diazepam, better known as Valium, is a central nervous system (CNS) sedative/hypnotic. As a sedative, it diminishes activity and excitement and thereby has a calming effect. In 1976, based on the number of new and refilled prescriptions processed, diazepam was the most prescribed drug in the United States.

Following is a retrosynthetic analysis for a synthesis of diazepam. Note that the formation of compound B involves a Friedel-Crafts acylation. In this reaction it is necessary to protect the 2° amine by prior treatment with acetic anhydride. The acetyl-protecting group is then removed by treatment with aqueous NaOH followed by careful acidification with HCl.

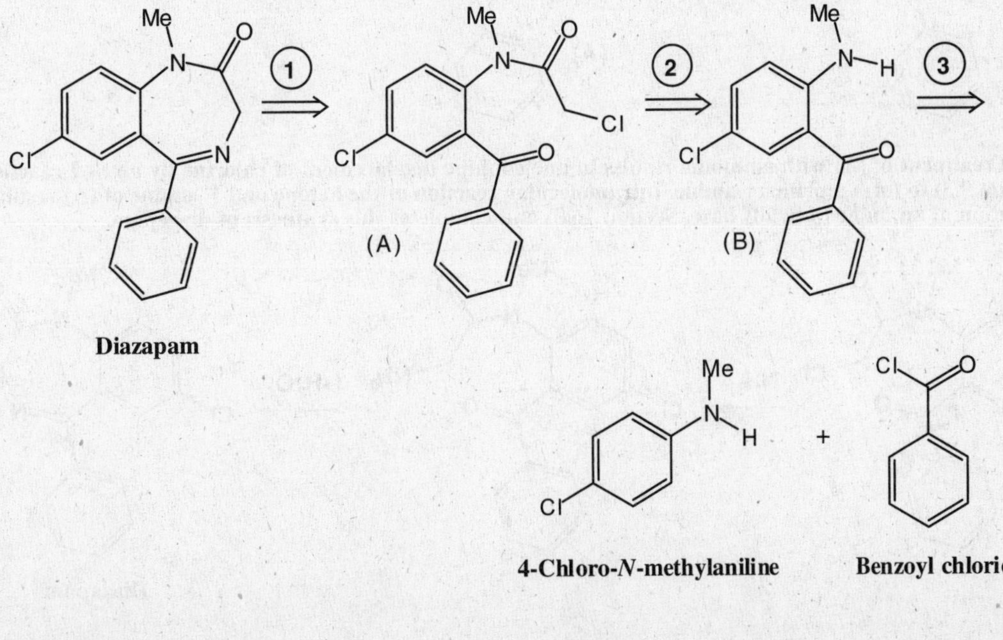

Diazapam

4-Chloro-N-methylaniline   **Benzoyl chloride**

(a) Given this retrosynthetic analysis, propose a synthesis of diazapam.

Step (3) The amine nitrogen is first protected by treatment with acetic anhydride (Section 18.6). Whereas a 2° amine is a good nucleophile and will react with benzoyl chloride to give an amide (Section 18.6), the nucleophilicity of an amide nitrogen is so reduced that it will no longer readily react with an acid chloride. Friedel-Crafts acylation (Section 21.1) is directed by the amide nitrogen, which is activating and ortho, para directing (Section 21.2). Chlorine is also ortho, para directing but it is deactivating and, therefore, the amide nitrogen takes precedence in directing the position of further electrophilic aromatic substitution. The acetyl protecting group is removed by treatment with aqueous NaOH (Section 18.4) to give (B).

Step (2) Treatment of the amino group of (B) with chloroacetyl chloride results in acetylation of the amino group to form the amide group of (A) (Section 18.6).

Step (1) Treatment of (A) with ammonia results in nucleophilic displacement of chlorine by an S$_N$2 reaction (Section 9.3) to form a primary amine. Intramolecular reaction of the ketone and 1° amine of (A) results in formation of an imine (a Schiff base, Section 16.8) and completes this synthesis of diazepam.

(b) Is diazapam chiral? If so, how many of the possible stereoisomers are formed in this synthesis?

Diazapam is not chiral.

Problem 22.51 The antidepressant amitriptyline inhibits the re-uptake of norepinephrine and serotonin from the synaptic cleft. Because the re-uptake of these neurotransmitters is inhibited, their effects are potentiated; they remain available to interact with serotonin and norepinephrine receptor sites longer and continue to cause excitation of serotonin- and norepinephrine-mediated neural pathways. Following is a synthesis for amitriptyline.

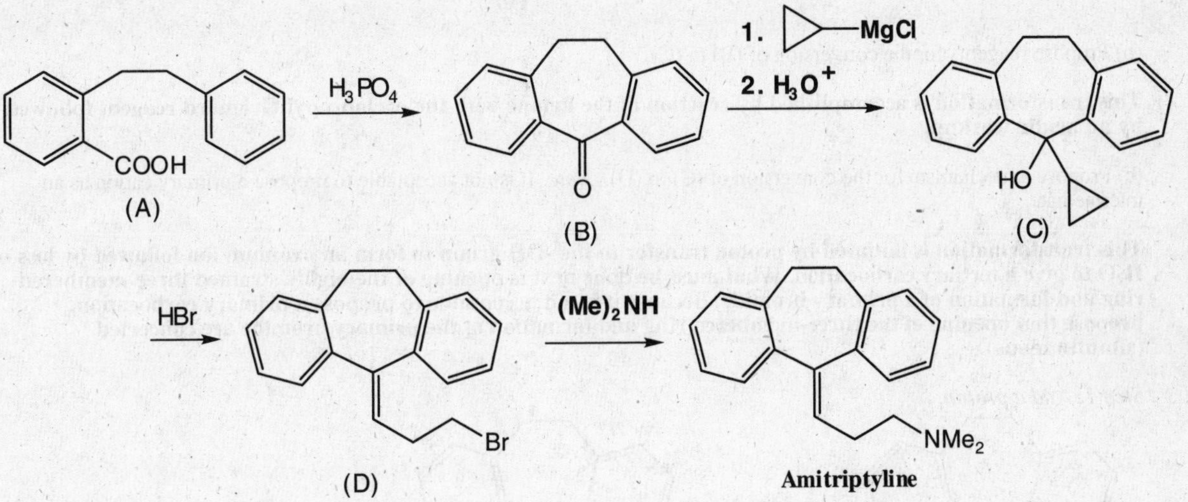

(a) Propose a mechanism for the conversion of (A) to (B).

This step is a type of electrophilic aromatic substitution resulting in acylation of the aromatic ring. The electrophile is an acylium ion.

Step 1: Add a proton.

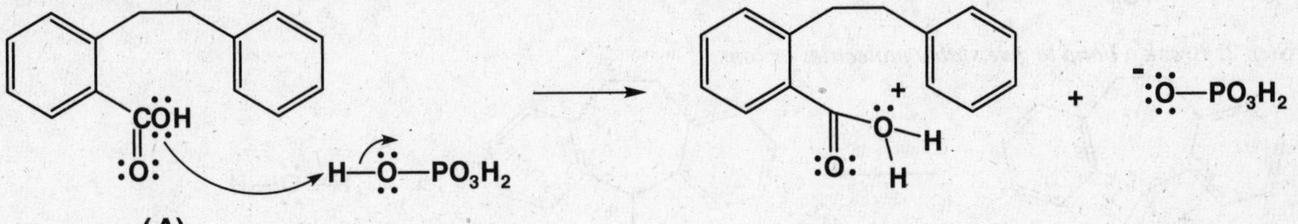

Step 2: Break a bond to give stable molecules or ions.

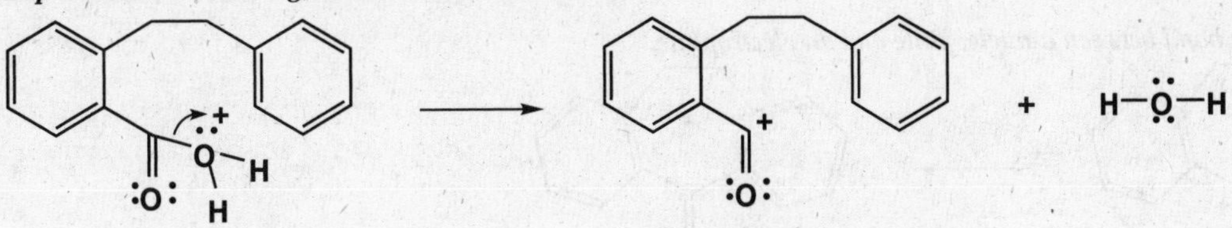

An acylium ion

Step 3: Electrophilic addition (aromatic).

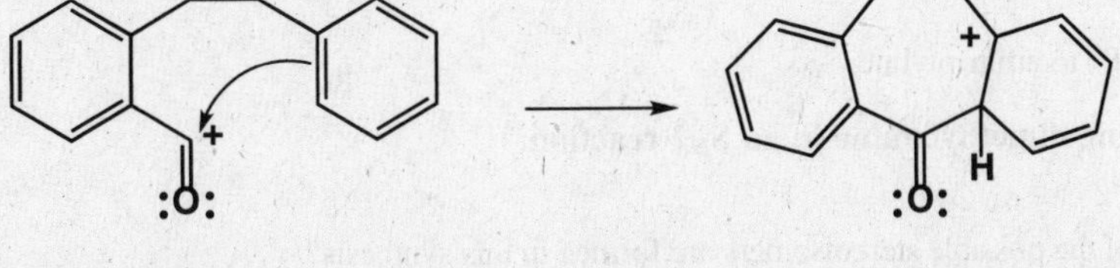

Step 4: Take a proton away.

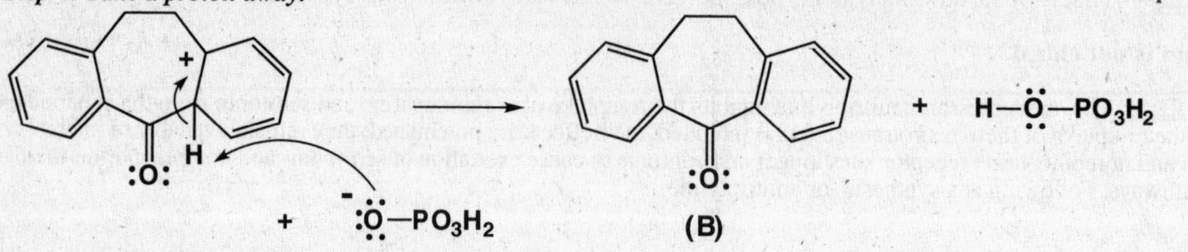

(b) Propose reagents for the conversion of (B) to (C).

This transformation is accomplished by reaction of the ketone with the cyclopropyl Grignard reagent followed by an acidic workup.

(c) Propose a mechanism for the conversion of (C) to (D). *Note:* It is not acceptable to propose a primary cation as an intermediate.

This transformation is initiated by proton transfer to the -OH group to form an oxonium ion followed by loss of H_2O to give a tertiary carbocation. What must be done next is opening of the highly strained three-membered ring and formation of a primary bromide. Because it is not acceptable to propose a primary carbocation, propose that opening of the three-membered ring and formation of the primary bromide are concerted (simultaneous).

Step 1: Add a proton.

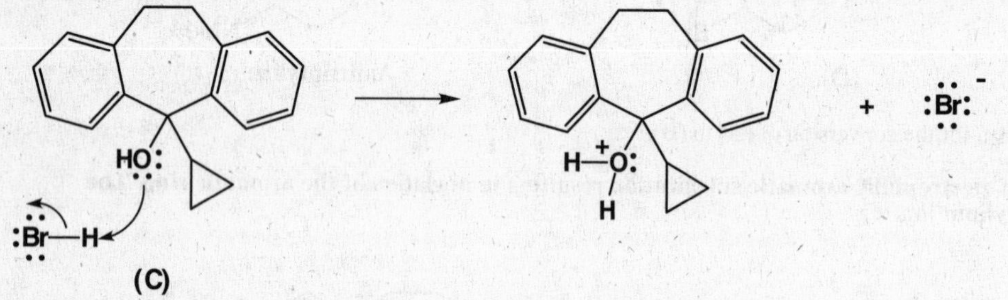

Step 2: Break a bond to give stable molecules or ions.

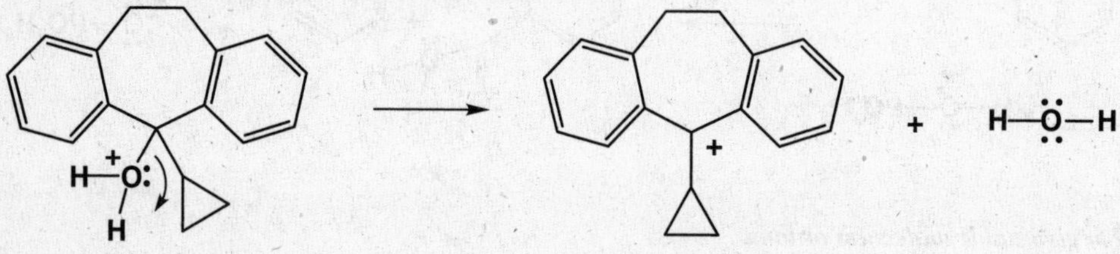

Step 3: Make a bond between a nucleophile and an electrophile.

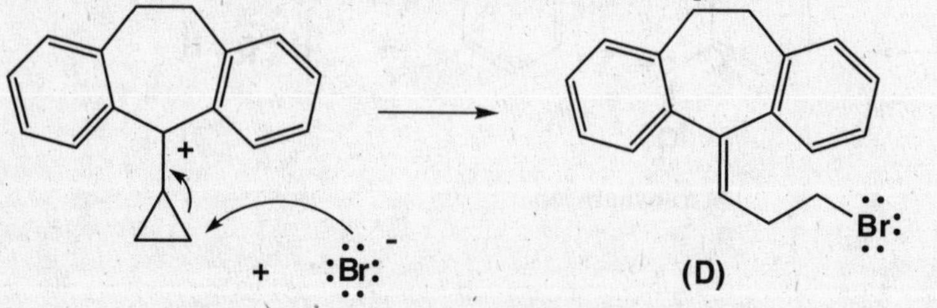

(d) Propose a reagent for the conversion of (D) to amitriptyline.

This transformation is accomplished using dimethylamine in an S_N2 reaction.

(e) Is amitripyline chiral? If so, how many of the possible stereoisomers are formed in this synthesis?

Amitriptyline is not chiral.

<u>Problem 22.52</u> Show how the antidepressant vanlafaxine (Effexor) can be synthesized from these readily available starting materials. Is venlafaxine chiral? If so, how many of the possible stereoisomers are formed in this synthesis?

Step (1) Synthesis begins with a Friedel-Crafts acylation (Section 22.1) of anisole.
Step (2) Nucleophilic displacement of the primary chloride by dimethylamine (S_N2, Section 9.3) gives the tertiary amine.
Step (3) Metal hydride reduction of the ketone (Section 16.11) gives the secondary alcohol. Catalytic reduction using H_2 and a transition metal catalyst such as Pd or Pt (Section 16.11) may also be used to reduce the ketone.
Step (4) Treatment of the secondary alcohol with thionyl chloride (Section 10.5) converts the alcohol to a chloride.
Step (5) The final step in the synthesis is a Grignard reaction with cyclohexanone followed by treatment of the magnesium alkoxide salt with aqueous acid (Section 16.5).
Venlafaxine has two chiral centers, so a racemic mixture of the four stereoisomers is made in this synthesis.

<u>Problem 22.53</u> One potential synthesis of the antiinflammatory and analgesic drug nabumetone is chloromethylation (Problem 22.46) of 2-methoxynaphthalene followed by an acetoacetic ester synthesis (Section 19.6).

(a) Account for the regioselectivity of chloromethylation at carbon 6 rather than at carbons 5 or 7.

Step 1: Add a proton. **Chloromethylation is initiated by proton transfer to formaldehyde to give a resonance-stabilized cation.**

Step 2: Electrophilic addition (aromatic). **This cation is sufficiently electrophilic to react with 2-methoxynaphthalene to give a resonance-stabilized cation. Comparing contributing structures for attack and positions 6 and 7 of the aromatic ring, we see that if attack occurs at carbon 6, oxygen of the methoxyl group can participate in stabilizing the cation. If attack occurs at carbon 7, oxygen of the methoxyl cannot participate in stabilizing the cation. Because cation 6 is more stable, there is a lower activation energy for its formation and it is the preferred cation intermediate.**

Step 3: Take a proton away. **Proton transfer from the cation to chloride ion completes the electrophilic aromatic substitution and gives a primary alcohol.**

Step 4: Add a proton. **Chloromethylation is completed by proton transfer to oxygen of the CH₂OH group to give an oxonium ion, loss of H_2O to give a resonance-stabilized benzylic carbocation, and finally reaction of the benzylic carbocation with chloride ion.**

Step 5: Break a bond to give stable molecules or ions.

Step 6: Make a bond between a nucleophile and an electrophile.

(b) Show steps in the acetoacetic ester synthesis by which the synthesis of nabumetone is completed.

The individual steps are shown on the scheme at the beginning of the problem.

<u>Problem 22.54</u> The analgesic, soporific, and euphoriant properties of the dried juice obtained from unripe seed pods of the opium poppy *Papaver somniferum* have been known for centuries. By the beginning of the 19th century, the active principle, morphine, had been isolated and its structure determined. Even though morphine is one of modern medicine's most effective painkillers, it has two serious disadvantages. First, it is addictive. Second, it depresses the respiratory control center of the central nervous system. Large doses of morphine (or heroin which is *N*-acetylmorphine) can lead to death by respiratory failure.

Morphine   **Morphinan**

(+/-) = **Racemethorphan**
(+) = **Dextromethorphan**
(-) = **Levomethorphan**

For these reasons, chemists have sought to produce painkillers related in structure to morphine, but without these serious disadvantages. One strategy has been to modify the carbon-nitrogen skeleton of morphine in the hope of producing medications equally effective but with reduced side effects. One target of this synthetic effort was morphinan, the bare morphine skeleton. Among the compounds thus synthesized, racemethorphan (the racemic mixture) and levomethorphan (the levorotatory enantiomer) proved to be very potent analgesics. Interestingly, the dextrorotatory enantiomer, dextromethorphan, has no analgesic activity. It does, however, show approximately the same antitussive (cough suppressing) activity as morphine and is, therefore, used extensively in cough remedies.

Following is a synthesis of racemethorphan.

Racemethorphan

(a) Propose a reagent for the conversion of (A) to (B).

Lithium aluminum hydride reduction of a nitrile gives a primary amine (Section 18.10).

(b) Propose a reagent for the conversion of (B) to (C).

The reagent is 4-methoxybenzoyl chloride. Reaction of the primary amine and acid chloride gives an amide (Section 18.6).

(c) Propose a mechanism for the conversion of (C) to (D).

Step 1: Add a proton.

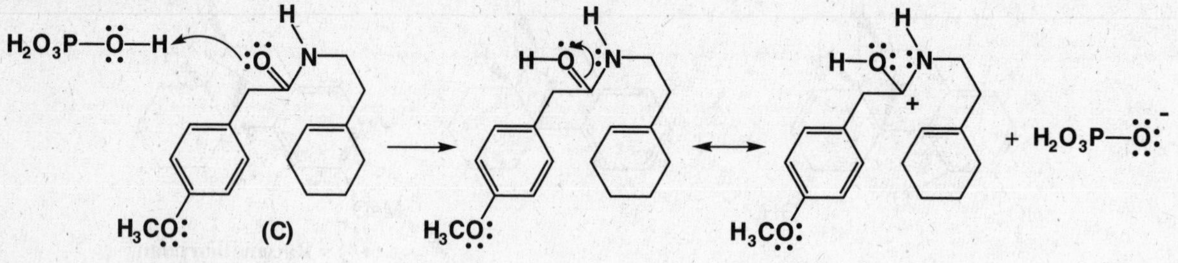

Step 2: Electrophilic addition.

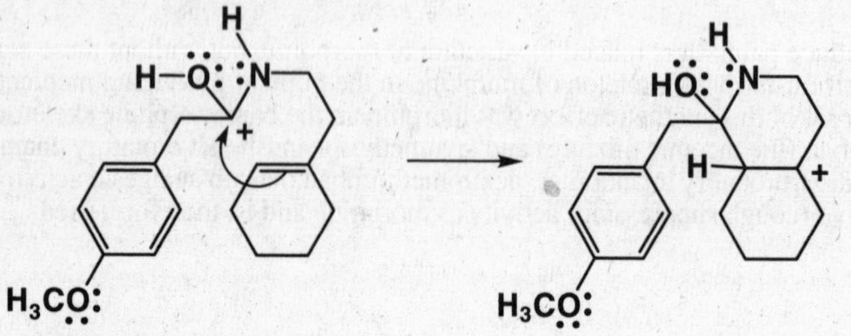

Step 3: Take a proton away.

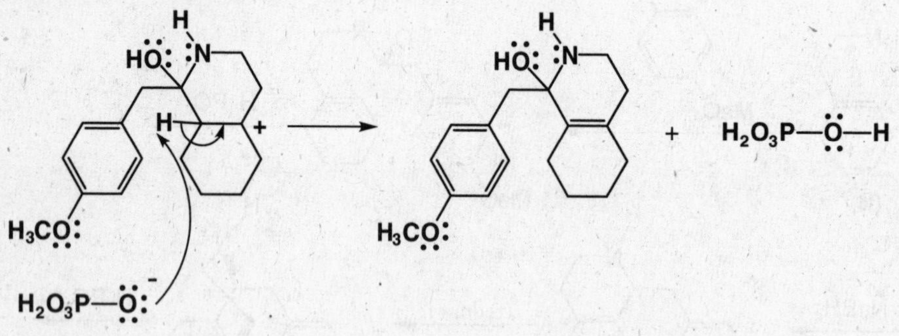

Step 4: Dehydration **This process actually requires several steps.**

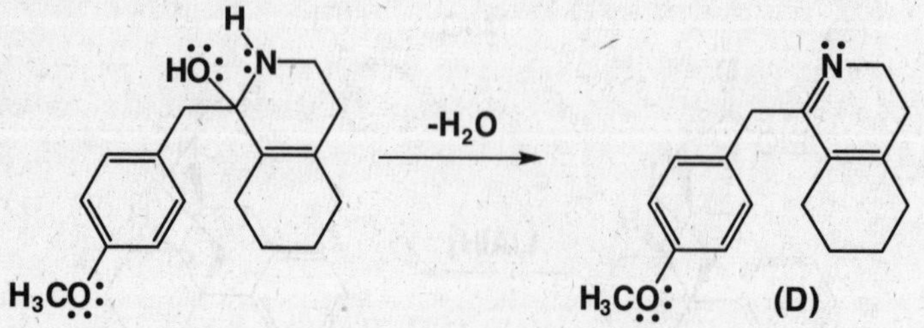

(d) Propose a mechanism for the conversion of (E) to (F).

Step 1: Electrophilic addition-add a proton.

Step 2: Electrophilic addition (aromatic).

Step 3: Take a proton away.

(e) Propose a reagent for the conversion of (F) to (G).

The reagent is ethyl chloroformate.

(f) Propose a reagent for the conversion of (G) to racemethorphan.

The reagent is lithium aluminum hydride.

<u>Problem 22.55</u> Following is the structural formula of the antihypertensive drug labetalol, a nonspecific β-adrenergic blocker with vasodilating activity. Members of this class have received enormous clinical attention because of their effectiveness in treating hypertension (high blood pressure), migraine headaches, glaucoma, ischemic heart disease, and certain cardiac arrhythmias. This retrosynthetic analysis involves disconnects to the α-haloketone (B) and the amine (C). Each is in turn derived from a simpler, readily available precursor.

Salicylic acid

(a) Given this retrosynthetic analysis, propose a synthesis for labetalol from salicylic acid and benzyl chloride. *Note:* The conversion of salicylic acid to (E) involves a Friedel-Crafts acylation in which the phenolic -OH must be protected by treatment with acetic anhydride to prevent the acylation of the -OH group. The protecting group is later removed by treatment with KOH followed by acidification.

Step (7) Treatment of ethyl acetoacetate with sodium ethoxide followed by treatment of the anion of the β-ketoester with benzyl chloride (Section 19.6) gives (F).

Step (6) Base-promoted hydrolysis of the ethyl ester to give a sodium carboxylate salt, acidification to give the β-ketoacid, and heating to bring about decarboxylation (Section 19.6) gives (D).

Step (5) Reductive amination of the ketone (Section 16.8) using NH₃ followed by hydrogenation gives (C).

Step (4) Fischer esterification (Section 17.7) of salicylic acid followed by ammonolysis of the ester (Section 18.6) is carried out first here. Acetylation using acetic anhydride (Section 18.5) is used to protect the phenolic -OH group. Friedel-Crafts acylation using acetyl chloride (Section 21.1) followed by base-promoted hydrolysis of an ester then acidification (Section 18.4) gives (E).

Step (3) Chlorination on the α-carbon of a ketone (Section 16.12) gives (B).

Step (2) Nucleophilic displacement of chlorine on (B) by the primary amino group of (C) gives (A).

Step (1) Reduction of the ketone by sodium borohydride (Section 16.11) gives the product.

(b) Labetalol has two chiral centers and, as produced in this synthesis, is a racemic mixture of the four possible stereoisomers. The active stereoisomer is dilevalol, which has the *R,R* configuration at its chiral centers. Draw a structural formula of dilevalol showing the configuration of each chiral center.

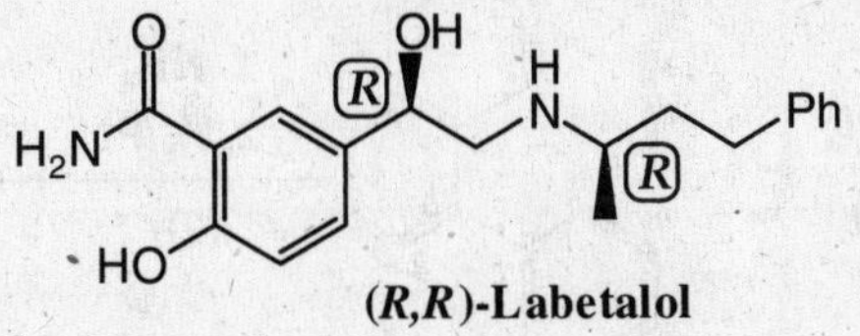

(*R,R*)-Labetalol

<u>Problem 22.56</u> Propose a synthesis for the antihistamine *p*-methyldiphenhydramine, given this retrosynthetic analysis. Is *p*-methyldiphenhydramine chiral? If so, how many of the possible stereoisomers are formed in this synthesis?

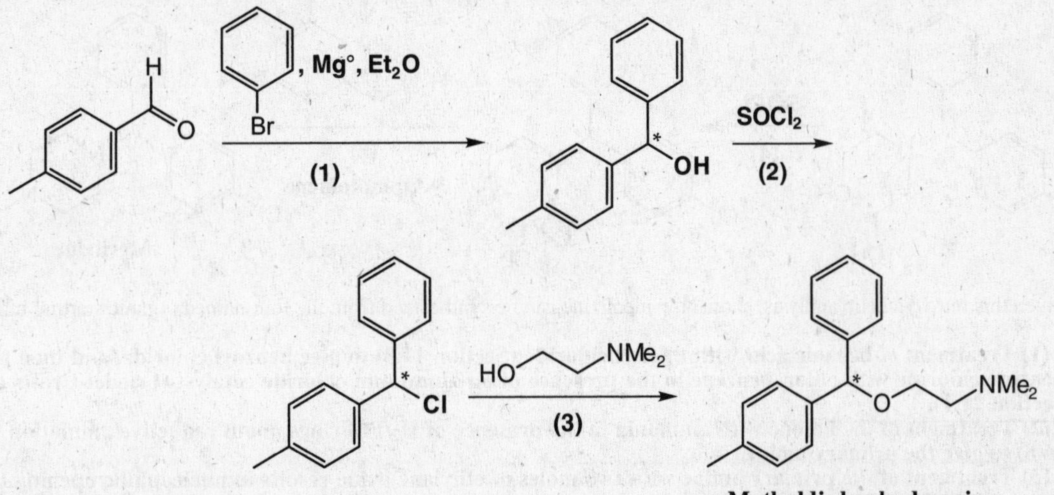

p-Methyldiphenhydramine

Step (1) Conversion of bromobenzene to a Grignard reagent followed by its treatment with 4-methylbenzaldehyde and workup in aqueous acid gives the secondary diaryl alcohol (Section 16.5).

Step (2) Reaction of the secondary alcohol with thionyl chloride (Section 10.5) gives a secondary benzylic chloride.

Step (3) Treatment of the secondary benzylic chloride with the β-aminoalcohol results in solvolysis via a resonance-stabilized benzylic carbocation to give the desired product (Section 9.3).

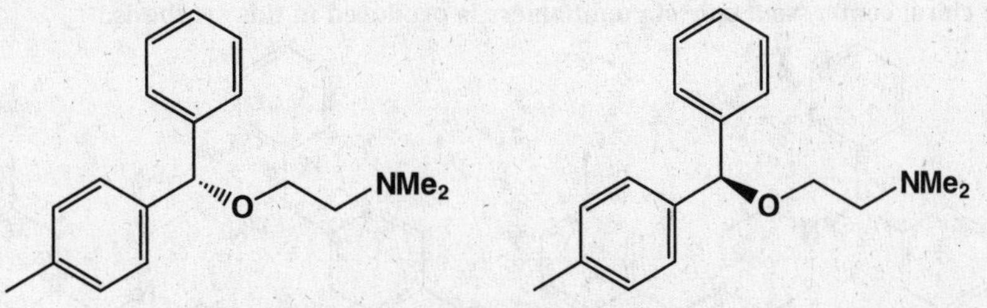

Enantiomers of *p*-methyldiphenhydramine

Problem 22.57 Meclizine is an antiemetic (it helps prevent or at least lessen the throwing up associated with motion sickness, including seasickness). Among the names of its over-the-counter preparations are Bonine, Sea-Legs, Antivert, and Navicalm.

(a) Given this retrosynthetic analysis, show how meclizine can be synthesized from the four named organic starting materials.

Step (1) Treatment of benzoic acid with thionyl chloride (Section 17.8) to give benzoyl chloride, and then treatment of benzoyl chloride with chlorobenzene in the presence of an aluminum chloride catalyst (Friedel-Crafts acylation, Section 21.1).

Step (2) Treatment of the ketone with ammonia in the presence of H₂/Pt brings about reductive amination (Section 16.8) to give the primary amine.

Step (3) Treatment of the primary amine with two moles of ethylene oxide results in nucleophilic opening of the epoxide (Section 11.9).

Step (4) Treatment of the diol with two moles of thionyl chloride converts each primary alcohol to the β-chloroamine (Section 10.5).

Step (5) The β-chloroamine is a nitrogen mustard (Section 9.11) and reacts with the primary amine to give meclizine.

(b) Is meclizine chiral? If so, how many of the possible stereoisomers are formed in this synthesis?

Meclizine has one chiral center, so a pair of enantiomers is produced in this synthesis.

Enantiomers of meclizine

Problem 22.58 Spasmolytol, as its name suggests, is an antispasmodic. Given this retrosynthetic analysis, propose a synthesis for spasmolytol from salicylic acid, ethylene oxide, and diethylamine.

Step (1) Bromination of salicylic acid is controlled by the ortho,para-directing phenolic -OH group (Section 22.1).

Step (2) In order for the phenolic -OH to be converted to an anion and become the nucleophile reacting with dimethyl sulfate, the carboxyl group must be blocked. This is accomplished by Fischer esterification (Section 17.7).

Step (3) Treatment of the phenolic hydroxyl group with sodium hydroxide converts the -OH group (a poor nucleophile) into an -O⁻ group, a good nucleophile. Treatment of this anion with methyl iodide (Section 9.3) gives the methyl ether.

Step (4) Reduction of the ester group with lithium aluminum hydride gives the primary alcohol (Section 18.10).

Step (5) Treatment of ethylene oxide with dimethylamine results in nucleophilic opening of the epoxide ring and gives a β-aminoalcohol (Section 11.9).

Step (6) Treatment of the primary alcohol with thionyl chloride (Section 10.5) gives a primary alkyl chloride.

Step (7) Ionization of the primary chloride assisted by neighboring group participation (Section 9.11) gives a cyclic ammonium ion. Reaction of the cyclic ammonium ion with the primary alcohol gives spasmolytol.

Problem 22.59 Among the first antipsychotic drugs for the treatment of schizophrenia was haloperidol (Haldol), a competitive inhibitor of dopamine receptor sites in the central nervous system.

Haloperidol

(a) Given the this retrosynthetic analysis, propose a synthesis for haloperidol.

Step (1) This reaction is a type of electrophilic aromatic substitution (Section 22.1) that results in acylation of the aromatic ring. For a mechanism of this type of acylation, review the solution to Problem 22.27.

Step (2) To protect the ketone group from reduction in Step 3, it is protected as the cyclic hemiacetal by treatment with ethylene glycol in the presence of p-toluenesulfonic acid (Section 16.7).

Step (3) The carboxyl group is converted to its acid chloride by treatment with thionyl chloride (Section 17.8).

Step (4) Treatment of the acid chloride with ammonia gives the primary amide (Section 18.6).

Step (5) LiAlH₄ reduction of the primary amide gives a primary amine (Section 18.10)

Step (6) Michael addition (Section 19.8) of two moles of methyl propenoate gives the diester.

Step (7) Dieckmann condensation (Section 19.3) using sodium ethoxide gives a β-ketoester.

Step (8) Base-promoted hydrolysis of the ethyl ester followed by acidification with aqueous HCl followed by mild heating results in hydrolysis and decarboxylation of the carboxyl group (Section 17.9).

Step (9) Bromides are more reactive toward Mg° than chlorides (Section 15.1). Therefore, treatment of 4-bromochlorobenzene with one mole of Mg° gives 4-chlorophenylmagnesium bromide. Reaction of this Grignard reagent with the ketone gives a magnesium alkoxide (Section 16.5). Treatment of the Grignard salt with aqueous acid accomplishes two things. The aqueous acid hydrolyzes the magnesium alkoxide to an alcohol (Section 16.5) and removes the ketone protecting group (Section 16.7).

(b) Is haloperidol chiral? If so, how many of the possible stereoisomers are formed in this synthesis?

Haloperidol is not chiral

Problem 22.60 A newer generation of antipsychotics, among them clozapine, are now used to treat the symptoms of schizophrenia. These drugs are more effective than earlier drugs in improving patient response in the areas of social withdrawal, apathy, memory, comprehension, and judgment. They also produce fewer side effects such as seizures and tardive dyskinesia (involuntary body movements). In the following synthesis of clozapine, Step 1 is an Ulmann coupling, a type of nucleophilic aromatic substitution that uses a copper catalyst.

**2,5-Dichloro-
nitrobenzene**

**2-Aminobenzoic acid
(Anthranilic acid)**

(A)

(B)

(C)   **Clozapine**

(a) Show how you might bring about formation of the amide in step 2.

The carboxyl group can be converted to an ester by Fischer esterification (Section 17.7) followed by treatment of the ester with the cyclic secondary amine (Section 18.6).

(b) Propose a reagent for step 3.

Treatment of the nitro group with H_2/Pt or other transition metal catalyst reduces it to a primary aromatic amine (Section 21.1).

(c) Propose a mechanism for step 4.

Review the mechanism in Section 16.8 for the formation of an imine (Schiff base).

(d) Is cloxapine chiral? If so, how many of the possible stereoisomers are formed in this synthesis?

Clozapine is not chiral.

<u>Problem 22.61</u> Proparacaine is one of a class of -caine local anesthetics.

Proparacaine

(a) Given this retrosynthetic analysis, propose a synthesis of proparacaine from 4-hydroxybenzoic acid.

Step (4) Fischer esterification (Section 17.7). Treatment of the substituted phenol with NaOH followed by 1-bromopropane gives the phenyl ether.

Step (3) Nitration of the aromatic ring is directed by the activating and ortho, para direction propoxy group (Section 22.2).

Step (2) Base-promoted hydrolysis of the ethyl ester followed by acidification gives the carboxylic acid (Section 18.5).

Step (1) Fischer esterification (Section 17.7) using the β-aminoalcohol gives the ester. The β-amino alcohol is derived by treating ethylene oxide with diethylamine. Catalytic reduction of the nitro group gives the primary aromatic amine (Section 22.1).

(b) Is proparacaine chiral? If so, how many of the possible stereoisomers are formed in this synthesis?

Proparacaine is not chiral.

Pyramid butterflyfish *Hemitaurichthys polyepis*
Lanai, Hawaii

CHAPTER 23
Solutions to the Problems

<u>Problem 23.1</u> Identify all carbon chiral centers in coniine, nicotine, and cocaine.

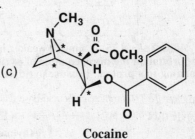

(a)

(b)

(c)

(S)-(+)-Coniine

(S)-(-)-Nicotine

Cocaine

<u>Problem 23.2</u> Write structural formulas for these amines.
(a) 2-Methyl-1-propanamine

(b) Cyclohexanamine

(c) (R)-2-Butanamine

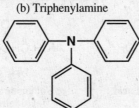

<u>Problem 23.3</u> Write structural formulas for these amines.
(a) Isobutylamine

(b) Triphenylamine

(c) Diisopropylamine

<u>Problem 23.4</u> Write IUPAC and, where possible, common names for each compound.

(a)

(b)

(c)

**(S)-2-Amino-3-phenylpropanoic
acid
(L-Phenylalanine)**

**4-Aminobutanoic acid
(γ-Aminobutyric acid)**

**2,2-Dimethylpropanamine
(Neopentylamine)**

<u>Problem 23.5</u> Predict the position of equilibrium for this acid-base reaction.

$$CH_3NH_3^+ \; + \; H_2O \; \rightleftharpoons \; CH_3NH_2 \; + \; H_3O^+$$

pK_a 10.64   pK_a -1.74

(weaker
acid)   (stronger
acid)

Because equilibrium favors formation of the weaker acid, equilibrium will be to the left as shown.

<u>Problem 23.6</u> Select the stronger acid from each pair of compounds.

(a) 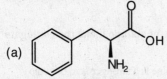   or

(A)   (B)

4-Nitroaniline (pK_b 13.0) is a weaker base than 4-methylaniline (pK_b 8.92). The decreased basicity of 4-nitroaniline is due to the electron-withdrawing effect of the *para* nitro group. Because 4-nitroaniline is the weaker base, its conjugate acid (A) is the stronger acid.

(b)

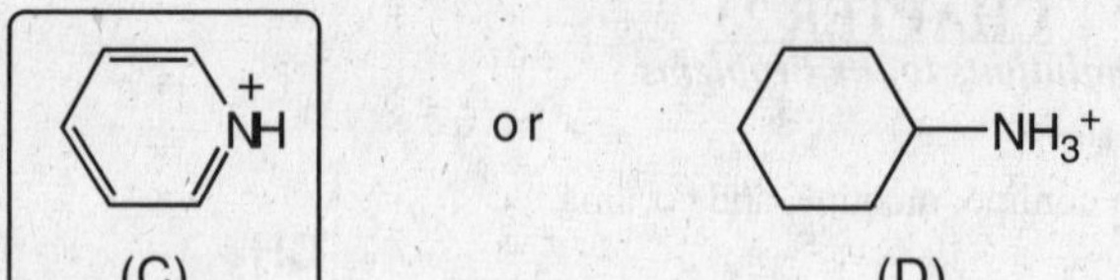

(C)   or   (D)

Pyridine (pK$_b$ 8.75) is a much weaker base than cyclohexanamine (pK$_b$ 3.34). The lone pair of electrons in the sp^2 orbital on the nitrogen atom of pyridine (C) has more s character, so these electrons are less available for bonding to a proton. Because pyridine is a weaker base, its conjugate acid, (C), is the stronger acid

<u>Problem 23.7</u> Complete each acid-base reaction and name the salt formed.

(a) (Et)$_3$N + HCl ⟶ **(Et)$_3$NH$^+$ Cl$^-$**
                          Triethylammonium chloride

(b)

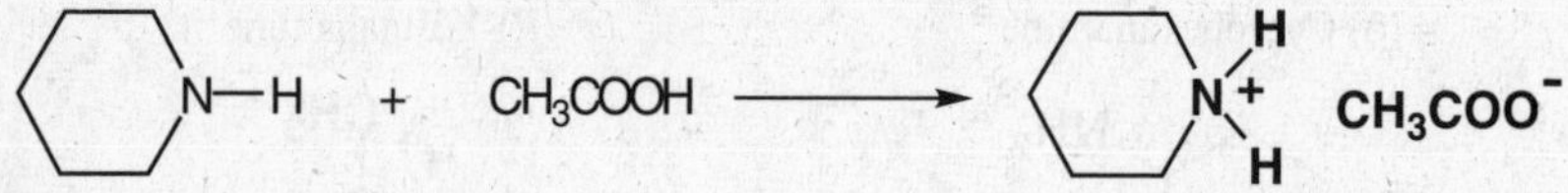

Piperidinium acetate

<u>Problem 23.8</u> Following are structural formulas for propanoic acid and the conjugate acids of isopropylamine and alanine, along with pK$_a$ values for each functional group:

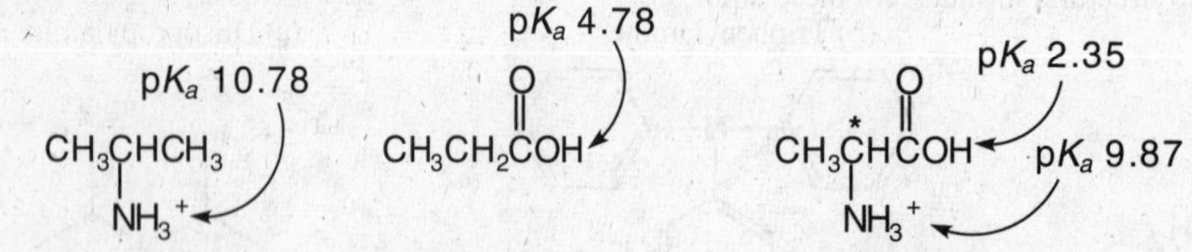

Conjugate acid of   **Propanoic acid**   Conjugate acid
isopropylamine   of alanine

(a) How do you account for the fact that the -NH$_3$$^+$ group of the conjugate acid of alanine is a stronger acid than the -NH$_3$$^+$ group of the conjugate acid of isopropylamine?

The electron-withdrawing properties of the carboxyl group adjacent to the amine of alanine make the conjugate acid of alanine more acidic than the conjugate acid of isopropylamine.

(b) How do you account for the fact that the -COOH group of the conjugate acid of alanine is a stronger acid than the -COOH group of propanoic acid?

The -NH$_3$$^+$ group is electron-withdrawing, so the adjacent -COOH group is made more acidic by an inductive effect. This situation is analogous to the electron-withdrawing effects of halogens adjacent to carboxylic acids in molecules such as chloroacetic acid, which has a pK$_a$ of 2.86. In addition, deprotonation of the carboxylic acid function of alanine results in formation of an overall neutral zwitterion. Thus, the carboxylate form of alanine can be thought of as being neutralized by the adjacent positively-charged ammonium ion.

<u>Problem 23.9</u> In what way(s) might the results of the separation and purification procedure outlined in Example 23.9 be different if the following conditions exist?
(a) Aqueous NaOH is used in place of aqueous NaHCO$_3$?

If NaOH is used in place of aqueous NaHCO$_3$, then the phenol will be deprotonated along with the carboxylic acid, so they will be isolated together in fraction A.

(b) The starting mixture contains an aromatic amine, ArNH$_2$, rather than an aliphatic amine, RNH$_2$?

If the starting mixture contains an aromatic amine, ArNH$_2$, rather than an aliphatic amine, RNH$_2$, then the results will be the same. The aromatic amine will still be protonated by the HCl wash, and deprotonated by the NaOH treatment.

Problem 23.10 Show how to bring about each conversion in good yield. In addition to the given starting material, use any other reagents as necessary.

(a)

(b)

(racemic)　　　　　　　　　　　　　　(racemic)

The stereochemistry of the product will be determined by the stereochemistry of the starting material; if the product is racemic, the starting material must have been racemic as well.

Problem 23.11 How might you bring about this conversion?

The synthesis begins with an aldol reaction using nitromethane, followed by reduction to give a β-aminoalcohol that then undergoes the ring expansion reaction.

Problem 23.12 Show how to convert toluene to 3-hydroxybenzoic acid using the same set of reactions as in Example 23.12, but changing the order in which two or more of the steps are carried out.

The key to this question is that the methyl group is converted to a meta-directing carboxyl group before the nitration reaction. This leads to the desired product with the hydroxy group in the 3 position.
Step (1) Oxidation at a benzylic carbon (Section 20.6A) can be brought about using chromic acid to give benzoic acid.
Step (2) Nitration of the aromatic ring using HNO_3 in H_2SO_4. The meta-directing carboxyl group gives predominantly the desired 3-nitrobenzoic acid product.
Step (3) Reduction of the nitro group to 3-aminobenzoic acid can be brought about using H_2 in the presence of Ni or other transition metal catalyst. Alternatively, it can be brought about using Zn, Sn, or Fe metal in aqueous HCl.
Step (4) Reaction of the aromatic amine with HNO_2 followed by heating in water gives 3-hydroxybenzoic acid.

Problem 23.13 Starting with 3-nitroaniline, show how to prepare the following compounds.
(a) 3-Nitrophenol

(b) 3-Bromoaniline

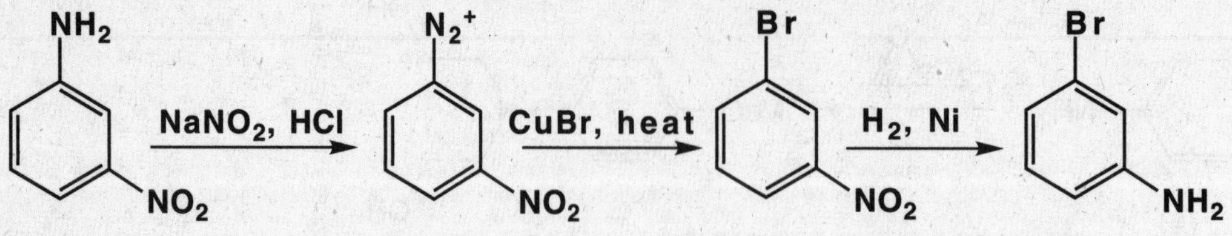

(c) 1,3-Dihydroxybenzene (resorcinol)

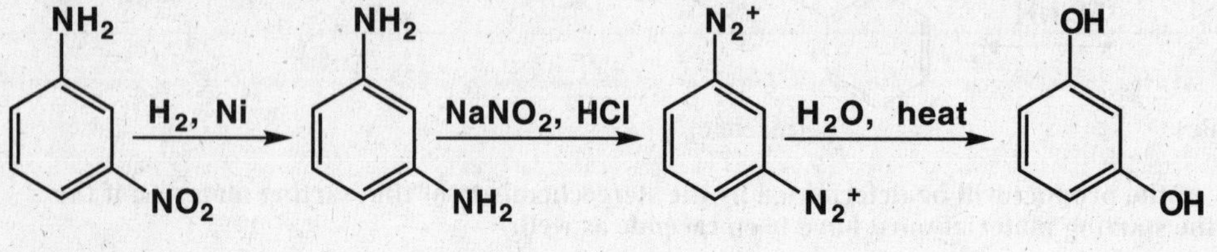

(d) 3-Fluoroaniline

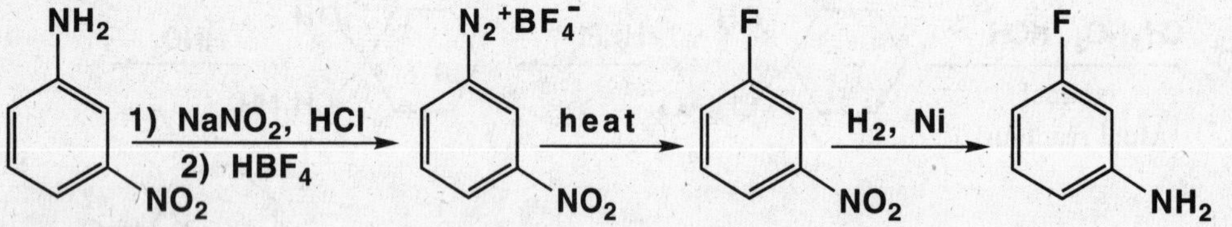

(e) 3-Fluorophenol

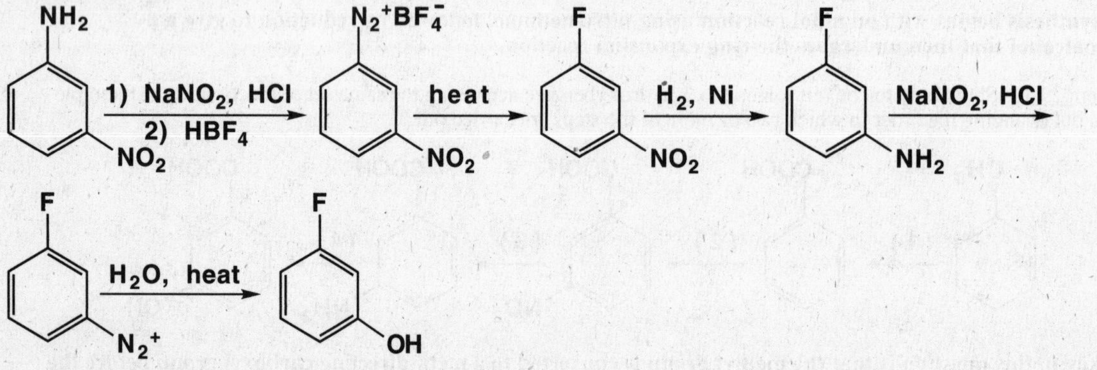

(f) 3-Hydroxybenzonitrile

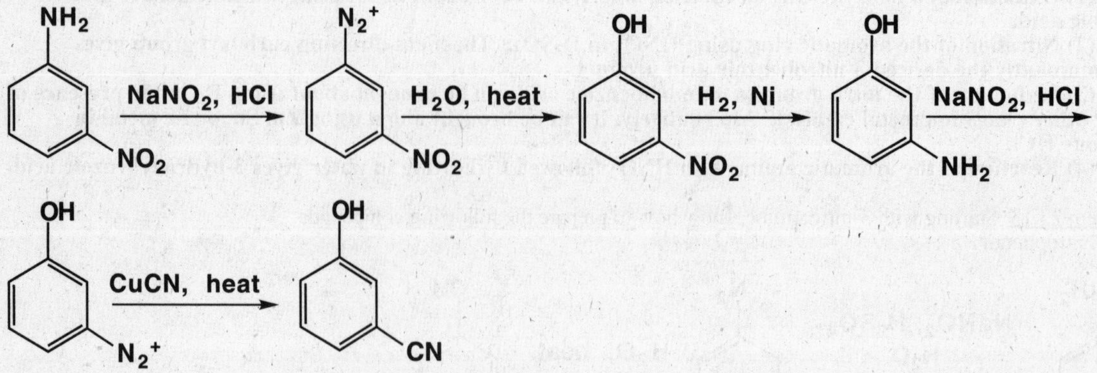

Problem 23.14 The procedure of methylation of amines and thermal decomposition of quaternary ammonium hydroxides was first reported by Hofmann in 1851, but its value as a means of structure determination was not appreciated until 1881 when he published a report of its use in determining the structure of piperidine. Following are the results obtained by Hofmann:

$$C_5H_{11}N \xrightarrow[\substack{\text{1. } CH_3I \text{ (excess), } K_2CO_3 \\ \text{2. } Ag_2O, H_2O \\ \text{3. heat}}]{} C_7H_{15}N \xrightarrow[\substack{\text{4. } CH_3I \text{ (excess), } K_2CO_3 \\ \text{5. } Ag_2O, H_2O \\ \text{6. heat}}]{} CH_2=CHCH_2CH=CH_2$$

Piperidine   **(A)**   **1,4-Pentadiene**

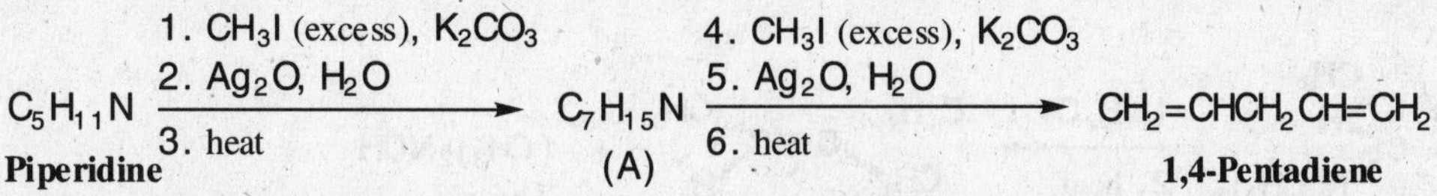

(a) Show that these results are consistent with the structure of piperidine (Section 23.1).

As shown below, the structure of piperidine is consistent with the formulas given, as well as the final product.

1. CH_3I (excess)
2. Ag_2O, H_2O
3. heat

4. CH_3I (excess)
5. Ag_2O, H_2O
6. heat

$$CH_2=CHCH_2CH=CH_2$$

1,4-Pentadiene

$C_5H_{11}N$

Piperidine

$C_7H_{15}N$

(A)

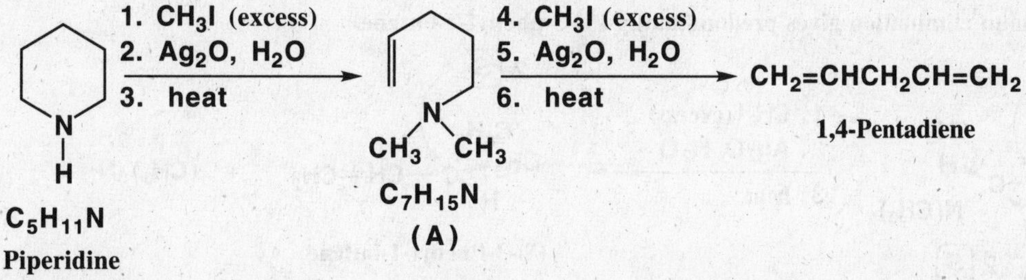

(b) Propose two additional structural formulas (excluding stereoisomers) for $C_5H_{11}N$ that are also consistent with the results obtained by Hofmann.

The following two molecules also have structures that are consistent with the formulas given, as well as the final product. Remember that in Hofmann eliminations, the least substituted alkene is formed predominantly.

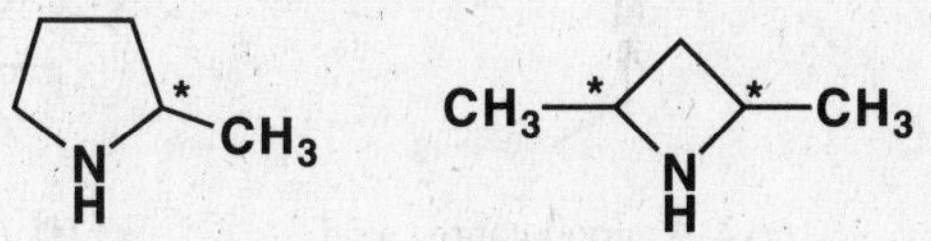

Problem 23.15 In Example 23.15, you considered the product of Cope elimination from the 2R,3S stereoisomer of 2-dimethylamino-3-phenylbutane. What is the product of Cope elimination from the following stereoisomers? What is the product of Hofmann elimination from each stereoisomer?
(a) (2S, 3R) stereoisomer?

The Cope elimination gives predominantly (*E*)-2-phenyl-2-butene.

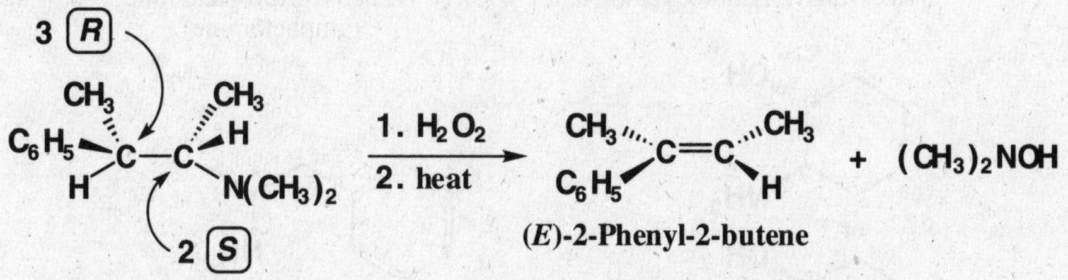

1. H_2O_2
2. heat

$$+ (CH_3)_2NOH$$

(*E*)-2-Phenyl-2-butene

The Hofmann elimination gives predominantly (*R*)-3-phenyl-1-butene.

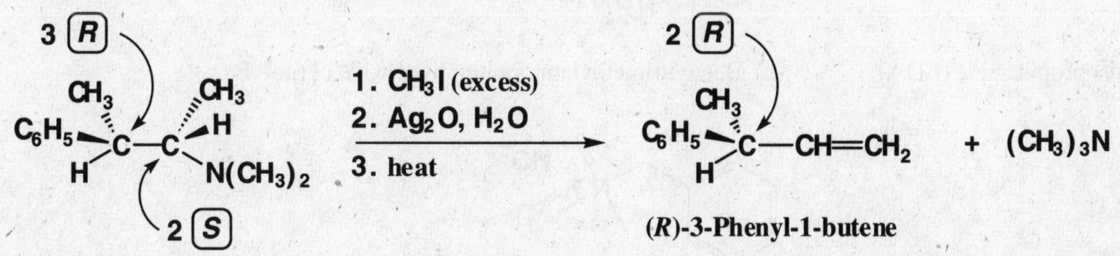

1. CH_3I (excess)
2. Ag_2O, H_2O
3. heat

$$+ (CH_3)_3N$$

(*R*)-3-Phenyl-1-butene

(b) (2S, 3S) stereoisomer?

The Cope elimination gives predominantly (Z)-2-phenyl-2-butene.

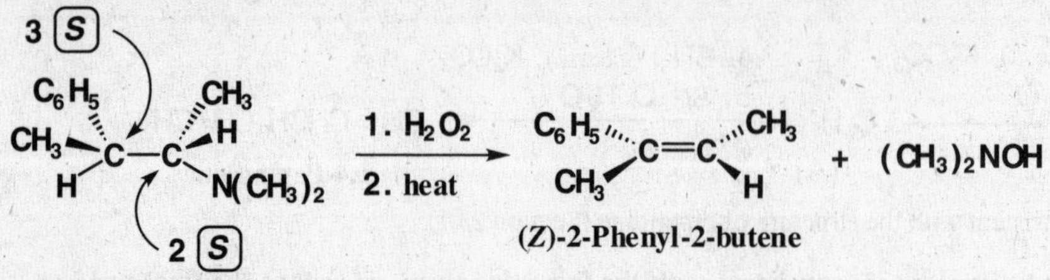

(Z)-2-Phenyl-2-butene

The Hofmann elimination gives predominantly (S)-3-phenyl-1-butene.

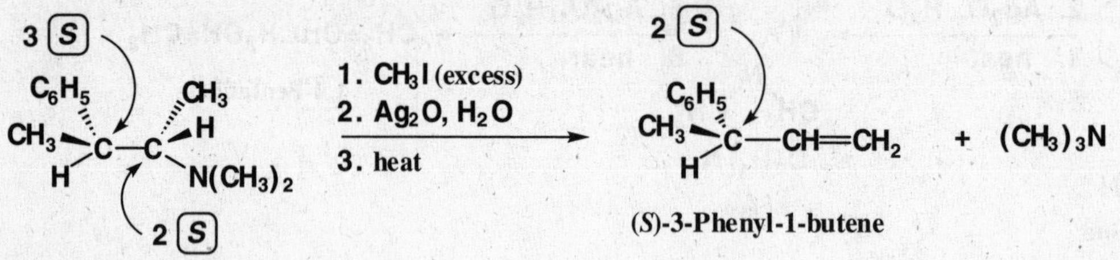

(S)-3-Phenyl-1-butene

Structure and Nomenclature
Problem 23.16 Draw structural formulas for each amine and amine derivative.

(a) *N,N*-Dimethylaniline

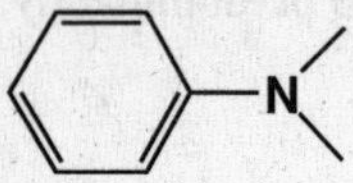

(b) Triethylamine

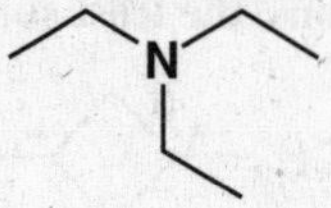

(c) *tert*-Butylamine

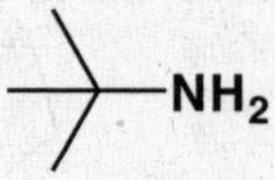

(d) 1,4-Benzenediamine

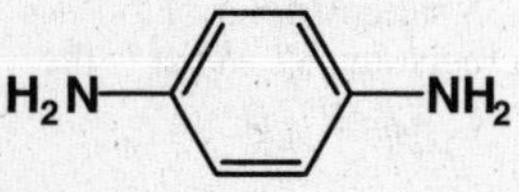

(e) 4-Aminobutanoic acid

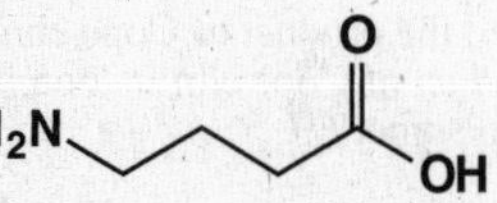

(f) (*R*)-2-Butanamine

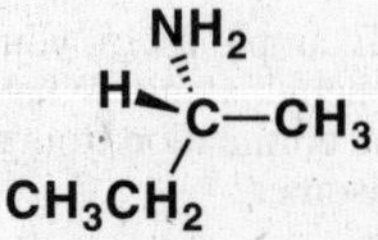

(g) Benzylamine

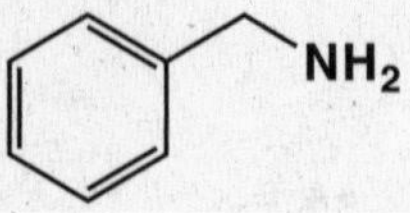

(h) *trans*-2-Aminocyclohexanol

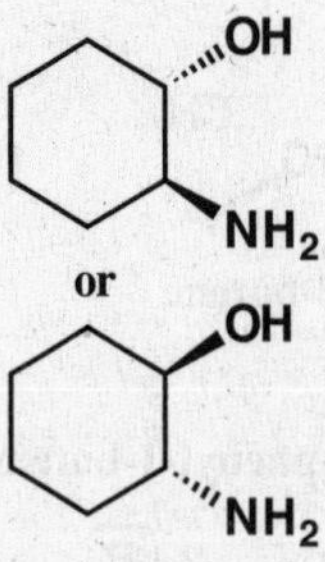

(i) 1-Phenyl-2-propanamine
     (amphetamine)

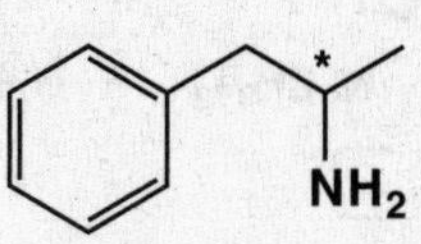

(j) Lithium diisopropylamide (LDA)

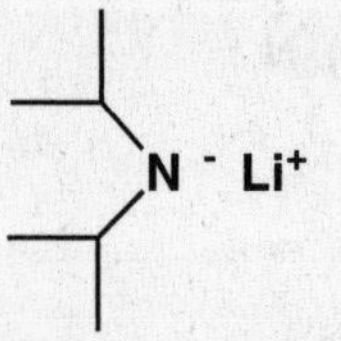

(k) Benzyltrimethylammonium hydroxide (Triton B)

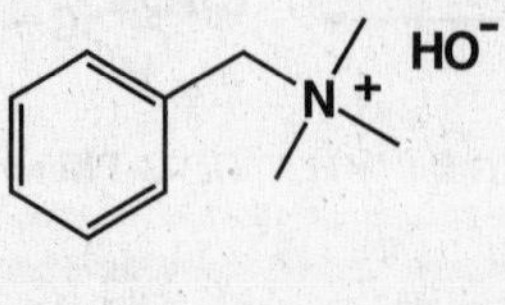

<u>Problem 23.17</u> Give an acceptable name for these compounds.

(a)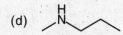

3,4-Dimethoxyaniline

(b)

1-Aminomethylcyclohexanol

(c)

1-Naphthylamine

(d)

Methylpropylamine

(e) —NH$_3^+$ Cl$^-$

Aniline hydrochloride

(f) —N$_2^+$ Cl$^-$

Benzenediazonium chloride

(g)

(R)-2-Hexanamine

(h)

3-Pyridinecarboxylic acid

<u>Problem 23.18</u> Classify each amine as primary, secondary, or tertiary; as aliphatic or aromatic.

(a) **Primary aliphatic amine** ←—

Heterocyclic aromatic amine ←—

Serotonin
(a neurotransmitter)

Primary aromatic amine —

(b)

Benzocaine
(a topical anesthetic)

Tertiary aliphatic amine

(c)

Secondary aromatic amine

Chloroquine
(an antimalarial, racemic)

Heterocyclic aromatic amine

<u>Problem 23.19</u> Epinephrine is a hormone secreted by the adrenal medulla. Among its actions, it is a bronchodilator. Albuterol, sold under several trade names, including Proventil and Salbumol, is one of the most effective and widely prescribed antiasthma drugs. The *R* enantiomer of albuterol is 68 times more effective in the treatment of asthma than the *S* enantiomer.

Secondary aliphatic amine   **Secondary aliphatic amine**

(*R*)-Epinephrine   **(*R*)-Albuterol**
(Adrenaline)

(a) Classify each as a primary, secondary, or tertiary amine.
(b) Compare the similarities and differences between their structural formulas.

The parts of the molecules that are identical are indicated in bold on the above structures. As far as differences are concerned, epinephrine possesses a second hydroxyl group on the aromatic ring and a methyl group on the amine, while (*R*)-albuterol has a hydroxymethyl group on the ring and a *tert*-butyl group on the amine.

<u>Problem 23.20</u> Draw the structural formula for a compound with the given molecular formula.
(a) A 2° arylamine, C_7H_9N (b) A 3° arylamine, $C_8H_{11}N$ (c) A 1° aliphatic amine, C_7H_9N

(d) A chiral 1° amine, $C_4H_{11}N$ (e) A 3° heterocyclic amine, $C_6H_{11}N$ (f) A trisubstituted 1° arylamine, $C_9H_{13}N$

(Other isomers are possible)

(g) A chiral quaternary ammonium salt, $C_6H_{16}NCl$

Problem 23.21 Morphine and its *O*-methylated derivative codeine are among the most effective pain killers known. However, they possess two serious drawbacks: they are addictive, and repeated use induces a tolerance to the drug. Increasingly larger doses become necessary; these doses can lead to respiratory arrest. Many morphine analogs have been prepared in an effort to find drugs that are equally effective as pain killers but that have less risk of physical dependence and potential for abuse. Following are several of these.

R = H; **Morphine**

R = CH₃; **Codeine**

Meperidine
(Demerol)

Methadone

Propoxyphene
(Darvon)

(a) List the structural features common to each of these molecules.

Each of the above molecules contains a tertiary amine and a phenyl ring that is three *sp³* carbons away from the nitrogen atom of the amine.

Quaternary
carbon

(b) The Beckett-Casey rules are a set of empirical rules to predict the structure of molecules that bind to morphine receptors and act as analgesics. According to these rules, to provide an effective morphine-like analgesia, a molecule must have (1) an aromatic ring attached to (2) a quaternary carbon and (3) a nitrogen at a distance equal to two carbon-carbon single bond lengths from the quaternary center. Show that these structural requirements are present in the molecules given in this problem.

By inspection of the structures, it can be seen that all of these three structural requirements are present in the molecules mentioned in this problem.

Problem 23.22 Following is a structural formula of desosamine, a sugar component of several macrolide antibiotics, including erythromycins. The configuration shown is that of the natural or D isomer. Erythromycin is produced by a strain of *streptomyces erytheus* found in a soil sample from the Phillipine Archipelago.

Desosamine

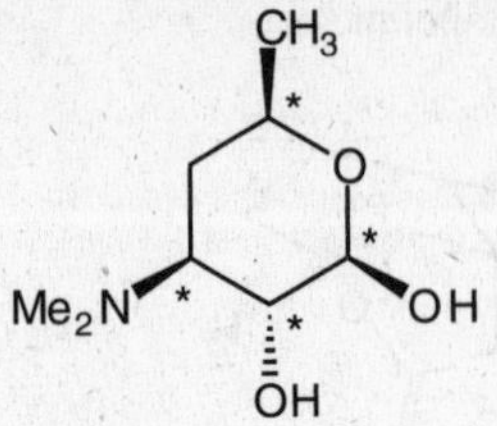

(a) Name all the functional groups in desosamine.

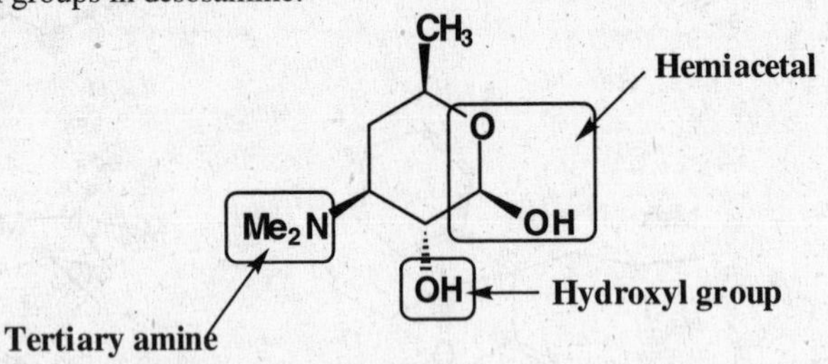

Hemiacetal

Tertiary amine

Hydroxyl group

(b) How many chiral centers are present in desosamine? How many stereoisomers are possible for it? How many pairs of enantiomers are possible for it?

Desosamine has 4 chiral centers, marked with asterisks on the uppermost structure. There are 2 x 2 x 2 x 2 = 16 possible stereoisomers as 8 pairs of enantiomers.

(c) Draw alternative chair conformations for desosamine. In each, label which groups are equatorial and which are axial.

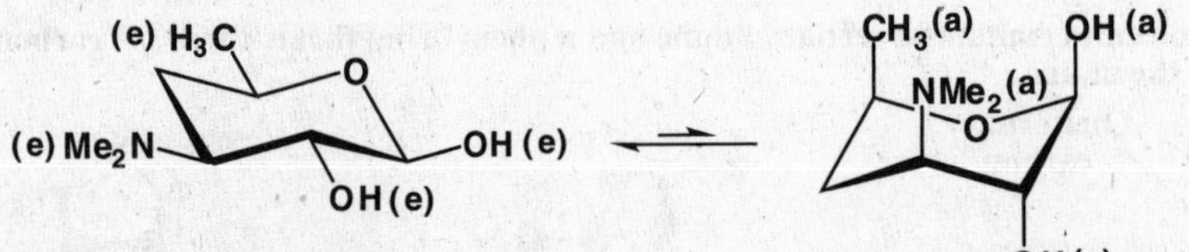

(e) H₃C   **(e) Me₂N**   **OH (e)**   **OH(e)**

CH₃(a) **OH (a)** **NMe₂(a)** **O** **OH(a)**

**More stable
(all equatorial)**

(d) Which of the alternative chair conformations for desosamine is the more stable? Explain.

The chair structure on the left is by far the more stable because all of the groups are equatorial, thereby minimizing non-bonded interaction strain.

Spectroscopy

Problem 23.23 Account for the formation of the base peaks in these mass spectra.
(a) Isobutylmethylamine, *m/z* 30

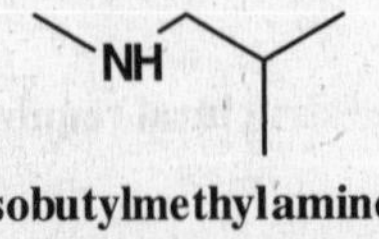

NH

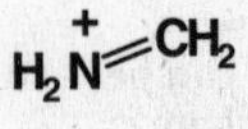

H₂N⁺=CH₂

Isobutylmethylamine   *m/z* = 30

As shown above, the peak at *m/z* = 30 is the result of the characteristic β-cleavage reaction often observed as the base peak in the mass spectra of amines.

(b) Diethylamine, *m/z* 58

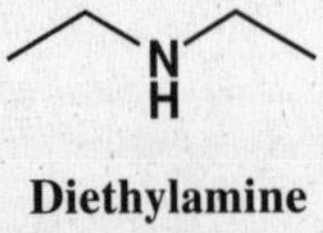

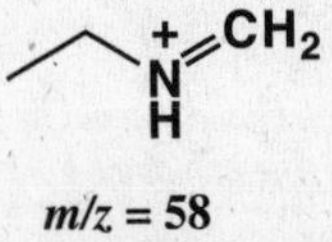

Diethylamine   *m/z* = 58

As shown above, the peak at *m/z* = 58 is the result of the characteristic β-cleavage reaction often observed as the base peak in the mass spectra of amines.

<u>Problem 23.24</u> Propose a structural formula for compound (A), molecular formula $C_5H_{13}N$, given its IR and ^{1}H-NMR spectra.

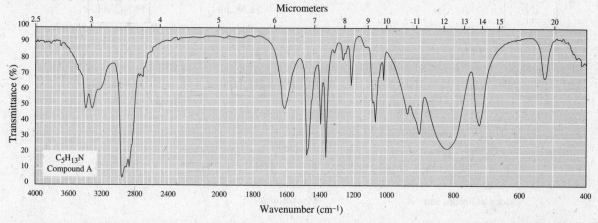

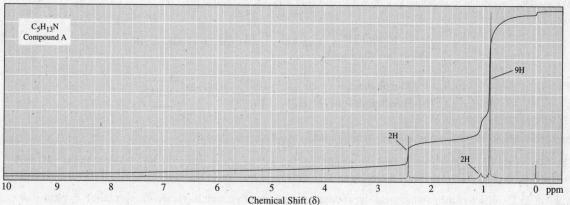

$$\underset{0.86}{\text{H}_3\text{CCCH}_2\text{NH}_2}$$

Compound A
1-Amino-2,2-dimethylpropane
(Neopentylamine)

The presence of two absorptions near 3300 to 3400 cm^{-1} in the IR spectrum indicates that compound A is a primary amine. Furthermore, in the ^{1}H-NMR, there are two sharp singlets at δ 0.86 and δ 2.40 integrating to 9H and 2H, respectively. The above structure is consistent with the index of hydrogen deficiency of zero, because there are no rings or π bonds.

Basicity of Amines
<u>Problem 23.25</u> Select the stronger base from each pair of compounds.

(a) [benzene ring with NH₂ and CH₃ substituents] or [benzene ring with CH₂NH₂]   (b) [benzene ring with NH₂ and O₂N substituents] or [benzene ring with NH₂ and H₃C substituents]

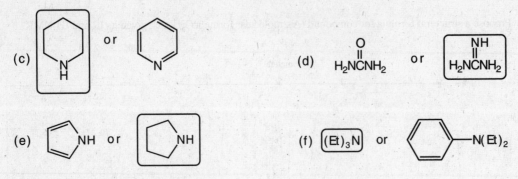

Problem 23.26 The pK_a of morpholine is 8.33.

$$\text{Morpholinium ion} + H_2O \rightleftharpoons \text{Morpholine} + H_3O^+ \qquad pK_a = 8.33$$

(a) Calculate the ratio of morpholine to morpholinium ion in aqueous solution at pH 7.0.

$$K_a = \frac{[\text{Morpholine}][H^+]}{[\text{Morpholinium Ion}]} = 10^{-8.33} \qquad \text{At pH 7.0} \quad [H^+] = 10^{-7}$$

$$\frac{[\text{Morpholine}]}{[\text{Morpholinium Ion}]} = \frac{10^{-8.33}}{[H^+]} = \frac{10^{-8.33}}{10^{-7.0}} = 10^{-1.33} \boxed{= 0.047}$$

(b) At what pH are the concentrations of morpholine and morpholinium ion equal?

The concentrations of morpholine and morpholinium ion will be equal when the pK_a is equal to the pH, that is, at pH 8.33.

Problem 23.27 Which of the two nitrogens in pyridoxamine (a form of vitamin B_6) is the stronger base? Explain your reasoning.

More basic site

Pyridoxamine
(Vitamin B_6)

The nitrogen atom of the primary amine is more basic than the pyridine nitrogen atom. This is because the primary amine nitrogen atom is sp^3 hybridized, while the pyridine nitrogen is sp^2 hybridized. The sp^2 hybridized nitrogen atom has a greater percentage s character, so the electrons are held closer to the nucleus and are less available for interactions with protons.

Problem 23.28 Epibatidine, a colorless oil isolated from the skin of the Ecuadorian poison frog *Epipedobates tricolor* has several times the analgesic potency of morphine. It is the first chlorine-containing, nonopioid (nonmorphine-like in structure) analgesic ever isolated from a natural source.

Epibatidine

(a) Which of the two nitrogen atoms of epibatidine is the more basic?

The nitrogen atom of the secondary amine is more basic than the pyridine nitrogen atom. This is because the secondary amine nitrogen atom is *sp³* hybridized, while the pyridine nitrogen is *sp²* hybridized. The *sp²* hybridized nitrogen atom has a greater percentage *s* character, so the electrons are held tighter and are less available for interactions with protons.

(b) Mark all chiral centers in this molecule.

The three chiral centers are marked with an asterisk (*).

Problem 23.29 Aniline (conjugate acid pK_a 4.63) is a considerably stronger base than diphenylamine (pK_a 0.79). Account for these marked differences.

$$C_6H_5NH_2 \qquad\qquad (C_6H_5)_2NH$$

Aniline	Diphenylamine
Conjugate acid pK_a p 4.63	pK 0.79

Diphenylamine is a weaker base because the nitrogen lone pair is delocalized by interaction with the π system of both aromatic rings, as opposed to aniline in which the lone pair on nitrogen is only delocalized by interaction with a single aromatic ring. This delocalization (resonance stabilization) cannot take place when the amine is protonated. Thus, the diphenylammonium ion is a stronger acid because loss of the proton results in greater resonance stabilization (delocalization into two aromatic rings instead of one).

Problem 23.30 Complete the following acid-base reactions and predict the direction of equilibrium (to the right or to the left) for each. Justify your prediction by citing values of pK_a for the stronger and weaker acid in each equilibrium. For values of acid ionization constants, consult Table 23.2 (Acid Strengths, pK_a, of the Conjugate Acids of Selected Amines), and Appendix 2 (Acid Ionization Constants for the Major Classes of Organic Acids). Where no ionization constants are given, make the best estimate from the information given in the reference tables and sections.

In all cases, the equilibrium favors formation of the weaker acid (higher pK_a) and weaker base (higher pK_b). Recall that $pK_a + pK_b = 14$ for any conjugate acid-base pair.

Equilibrium lies to the right, because the acetate anion and pyridinium ion are the weaker acid and base, respectively.

Equilibrium lies to the right, because the phenoxide anion and the triethylammonium species are the weaker base and acid, respectively.

(c) PhC≡CH   +   NH₃ ⇌ PhC≡C⁻   +   NH₄⁺

    Phenylacetylene   Ammonia

    pK_a ~25 pK_b 4.74 pK_b ~ -11 pK_a 9.26

Equilibrium lies to the left, because the alkyne and ammonia are the weaker acid and base, respectively.

(d) PhC≡CH   +   iPr₂N⁻ Li⁺ ⟶ PhC≡C⁻ Li⁺   +   iPr₂NH

    Phenylacetylene   Lithium
                diisopropylamide
                  (LDA)

    pK_a ~25 pK_b ~ -23 pK_b ~ -11 pK_a ~ 40

Equilibrium lies to the right, because the alkyne anion and the amine are the weaker base and acid, respectively.

(e) PhCO⁻ Na⁺   +   Et₃NH⁺ Cl⁻ ⇌ PhCOH   +   Et₃N   +   NaCl

    Sodium   Triethylammonium
    benzoate   chloride
    pK_b 9.81 pK_a 10.75 pK_a 4.19 pK_b 3.25

Equilibrium lies to the left, because the carboxylate anion and the triethylammonium ion are the weaker base and acid, respectively.

(f)

    1-Phenyl-2-propanamine   2-Hydroxy-
    (Amphetamine)   propanoic acid
                    (Lactic acid)

    pK_b ~3 pK_a 3.08 pK_a ~11 pK_b 9.92

Equilibrium lies to the right, because the 1-phenyl-2-propanammonium ion and the carboxylate ion are the weaker acid and base, respectively.

(g)

    Amphetamine   Sodium
    hydrochloride   bicarbonate
    pK_a ~11 pK_b 7.63 pK_b ~3 pK_a 6.36

Equilibrium lies to the left, because the amphetamine ammonium ion and the bicarbonate ion are the weaker acid and base, respectively.

(h)

                     Tetraethylammonium
    Phenol   hydroxide
    pK_a 9.95 pK_b -1.7 pK_b 4.05 pK_a 15.7

Equilibrium lies to the right, because phenoxide and water are the weaker base and acid, respectively.

<u>Problem 23.31</u> Quinuclidine and triethylamine are both tertiary amines. Quinuclidine, however, is a considerably stronger base than triethylamine. Stated alternatively, the conjugate acid of quinuclidine is a considerably weaker acid than the conjugate acid of triethylamine. Propose an explanation for these differences in acidity/basicity.

Quinuclidine
(pK_a 10.6)

Triethylamine
(pK_a 8.6)

The protonated form of quinuclidine is more compact than protonated triethylamine, because the alkyl groups of quinuclidine are "tied back" allowing it to be solvated better. For this reason, protonated quinuclidine is a weaker acid, so quinuclidine is a stronger base.

<u>Problem 23.32</u> Suppose that you have a mixture of these three compounds. Devise a chemical procedure based on their relative acidity or basicity to separate and isolate each in pure form.

4- Nitrotoluene
(*p*-Nitrotoluene)

4-Methylaniline
(*p*-Toluidine)

4-Methylphenol
(*p*-Cresol)

These molecules can be separated by extraction into different aqueous solutions. First, the mixture is dissolved in an organic solvent such as ether in which all three compounds are soluble. Then, the ether solution is extracted with dilute aqueous HCl. Under these conditions, 4-methylaniline (a weak base) is converted to its protonated form and dissolves in the aqueous solution. The aqueous solution is separated, treated with dilute NaOH, the water-insoluble 4-methylaniline precipitates, and is recovered. The ether solution containing the other two components is then treated with dilute aqueous NaOH. Under these conditions, 4-methylphenol (a weak acid) is converted to its phenoxide ion and dissolves in the aqueous solution. Acidification of this aqueous solution with dilute HCl forms water-insoluble 4-methylphenol that is then isolated. Evaporation of the remaining ether solution gives the 4-nitrotoluene, which is neither acidic or basic.

Preparation of Amines
<u>Problem 23.33</u> Propose a synthesis of 1-hexanamine from the following:
(a) A bromoalkane of six carbon atoms

(b) A bromoalkane of five carbon atoms

Problem 23.34 Show how to convert each starting material into benzylamine in good yield.

(a) Benzaldehyde → (NH₃, -H₂O) → imine → (H₂/Ni) → benzylamine

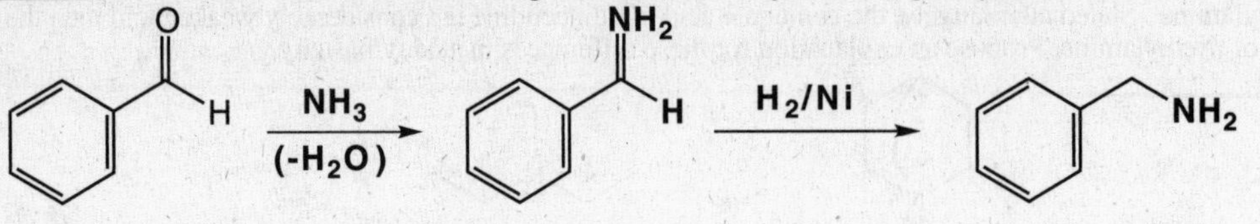

(b) N-benzylacetamide → (NaOH) → benzylamine + $CH_3CO^- Na^+$

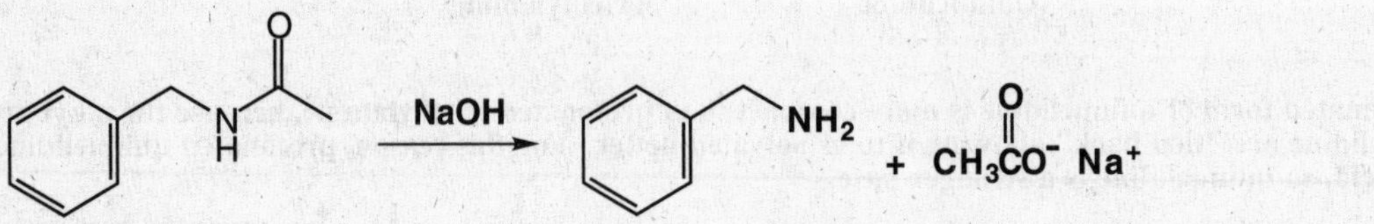

(c) benzyl alcohol → ($SOCl_2$) → benzyl chloride → (excess NH_3) → benzylamine

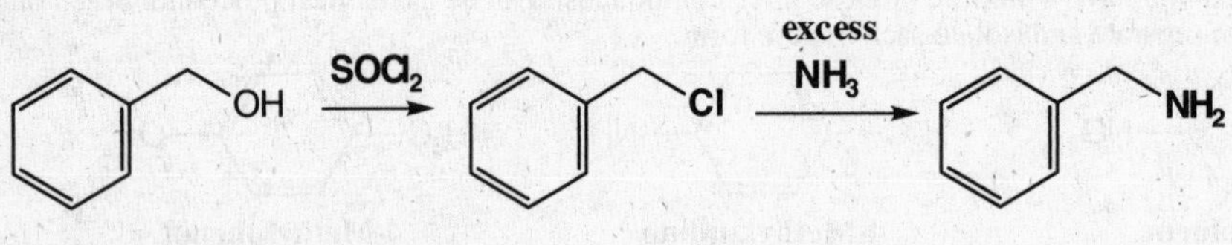

(d) benzyl chloride → (excess NH_3) → benzylamine

or

benzyl chloride → (1. N_3^-, 2. $LiAlH_4$, 3. H_2O) → benzylamine

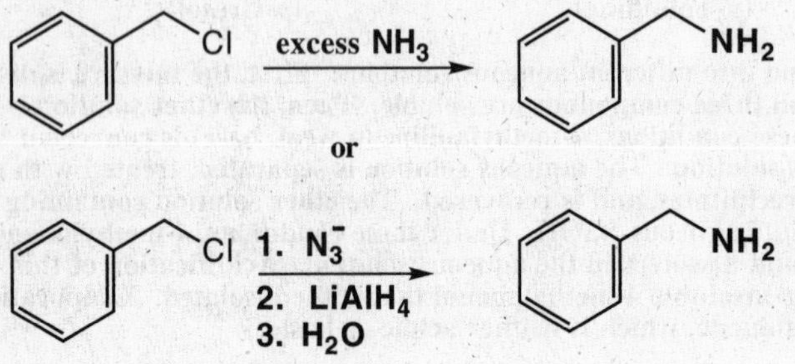

(e) benzoic acid → (1. $SOCl_2$, 2. NH_3) → benzamide → (1. $LiAlH_4$, 2. H_2O) → benzylamine

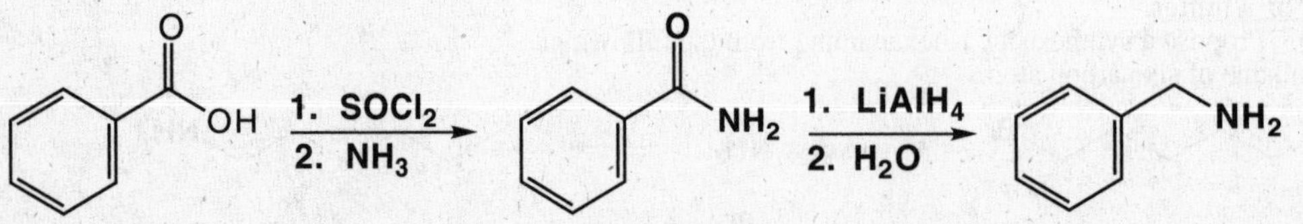

(f) ethyl benzoate → (NH_3) → benzamide → (1. $LiAlH_4$, 2. H_2O) → benzylamine

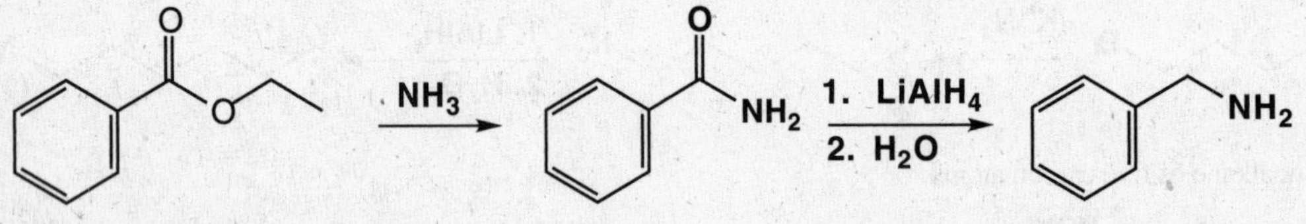

Reactions of Amines

Problem 23.35 Treating trimethylamine with 2-chloroethyl acetate gives acetylcholine as its chloride. Acetylcholine is a neurotransmitter. Propose a structural formula for this quaternary ammonium salt and a mechanism for its formation.

$$(Me)_3N \ + \ CH_3COCH_2CH_2Cl \longrightarrow CH_3COCH_2CH_2N(CH_3)_3 \quad Cl^-$$

$$C_7H_{16}ClNO_2$$

Acetylcholine chloride

Acetylcholine is formed through an S_N2 reaction between trimethylamine and the alkyl chloride, giving the structure shown.

Problem 23.36 *N*-Nitrosamines by themselves are not significant carcinogens. However, they are activated in the liver by a class of iron-containing enzymes (members of the cytochrome P-450 family). Activation involves the oxidation of a C-H bond next to the amine nitrogen to a C-OH group. Show how this hydroxylation product can be transformed into an alkyldiazonium ion, an active carcinogen, in the presence of an acid catalyst.

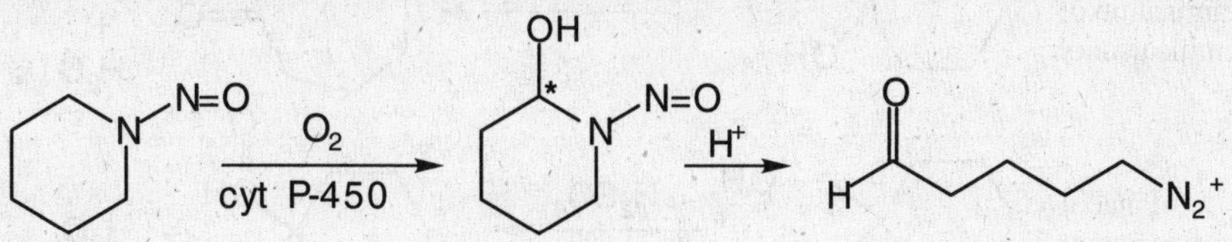

**N-Nitroso-
piperidine**

O_2
cyt P-450

**2-Hydroxy-N-
nitrosopiperidine**

H^+

**An alkyl diazonium ion
(a carcinogen)**

Step 1: Break a bond to give stable molecules or ions-add a proton-take a proton away. These three steps occur in rapic succession to give the open chain molecule shown.

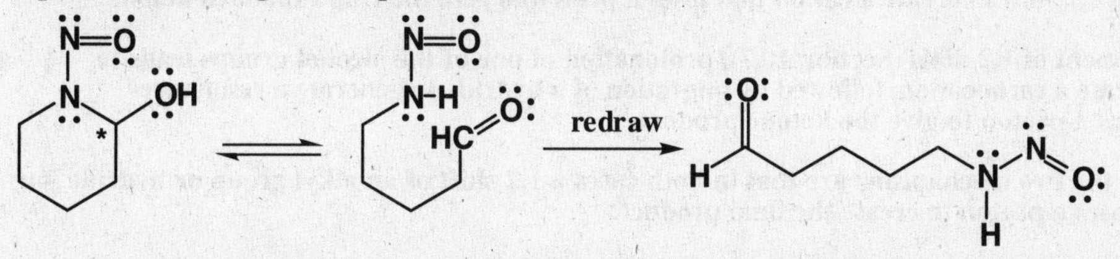

redraw

Step 2: Add a proton. Note that steps 2 and 3 represent an analog of keto-enol tautomerization.

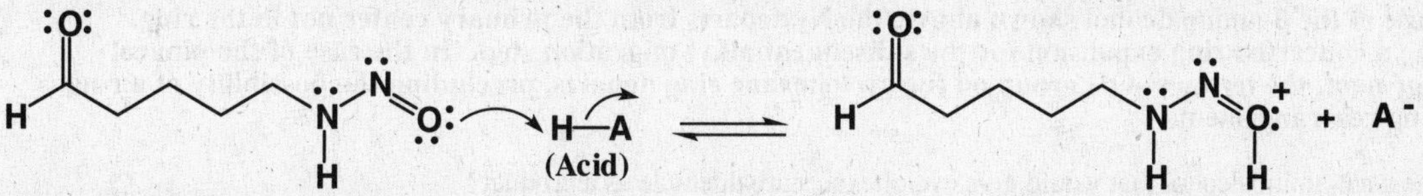

H—A
(Acid)

+ A⁻

Step 3: Take a proton away.

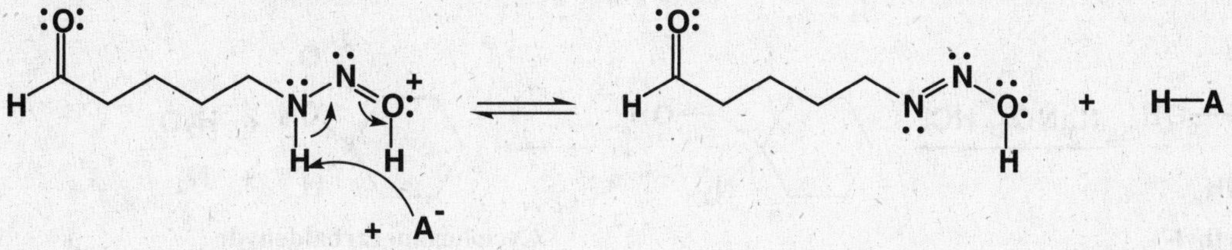

+ H—A

+ A⁻

Step 4: Add a proton.

H—A

+ A⁻

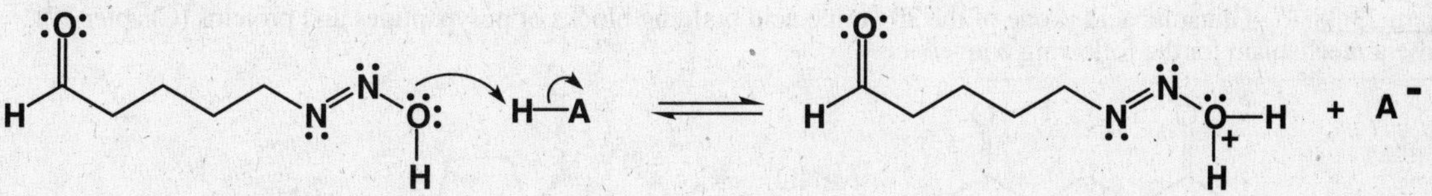

Step 5: Break a bond to give stable molecules or ions.

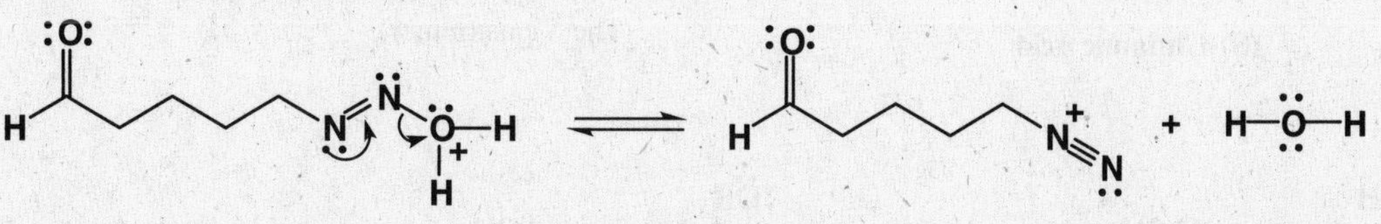

+ H—O—H

Problem 23.37 Marked similarities exist between the mechanism of nitrous acid deamination of β-aminoalcohols and the pinacol rearrangement. Following are examples of each.

Nitrous acid
deamination of
a β-aminoalcohol: [structure] —NH₂ $\xrightarrow{NaNO_2, HCl}$ [cycloheptanone] $N_2 + H_2O$

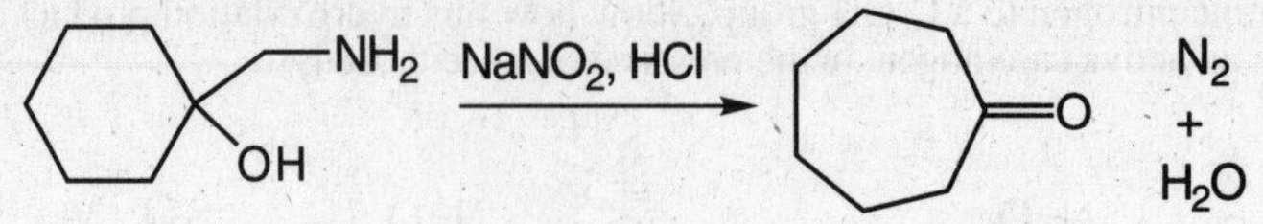

Pinacol
rearrangement: [structure] —OH $\xrightarrow{H_2SO_4}$ [aldehyde] $+ H_2O$

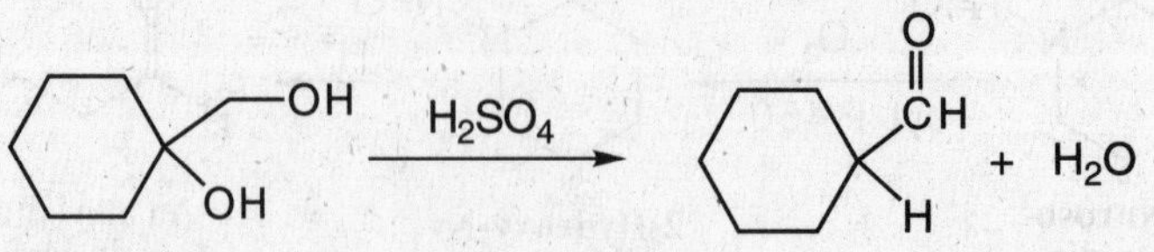

(a) Analyze the mechanism of each rearrangement and list their similarities.

The nitrous acid deamination of a β-aminoalcohol (Section 23.8D) involves formation of a diazonium ion that loses N₂ in concert with a 1,2 shift to create a cation that loses a proton to give the ring expanded ketone.

In the pinacol rearrangement of 1,2 diols (Section 10.7), protonation of one of the alcohol groups leads to departure of water to create a carbocation, followed by migration of a hydride to generate a resonance-stabilized cation that loses a proton to give the ketone product.

The similarities between the two mechanisms are that in both cases a 1,2 shift of an alkyl group or hydride ion produces a cation, that loses a proton to create the final product.

(b) Why does the first reaction, but not the second, give ring expansion?

In the case of the β-aminoalcohol shown above, the N₂ departs from the primary center not in the ring, requiring a concerted ring expansion for the subsequent alkyl migration step. In the case of the pinacol rearrangement, the tertiary -OH group on the cyclohexane ring departs, precluding the possibility of a ring expanding rearrangement.

(c) Suggest a β-aminoalcohol that would give cyclohexanecarbaldehyde as a product?

1-Hydroxymethylcyclohexanamine would undergo reaction to give cyclohexanecarbaldehyde as shown.

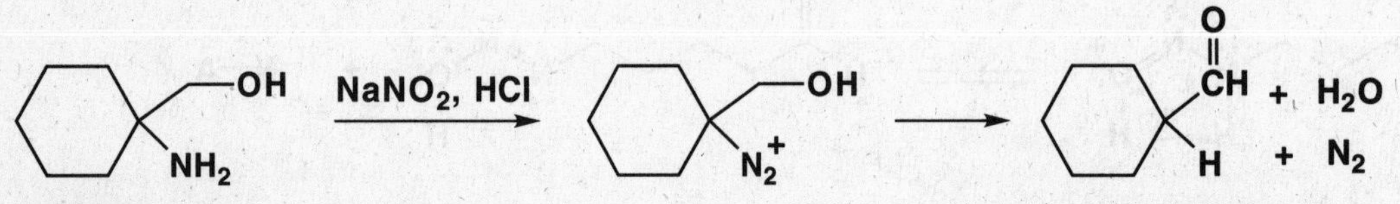

[structure] —OH $\xrightarrow{NaNO_2, HCl}$ [structure] —OH $\longrightarrow$ [structure] $+ H_2O$
—NH₂   —N₂⁺   CH, H $+ N_2$

1-Hydroxymethyl-
cyclohexanamine   **Cyclohexanecarbaldehyde**

Problem 23.38 (*S*)-Glutamic acid is one of the 20 amino acid building blocks of polypeptides and proteins (Chapter 27). Propose a mechanism for the following conversion.

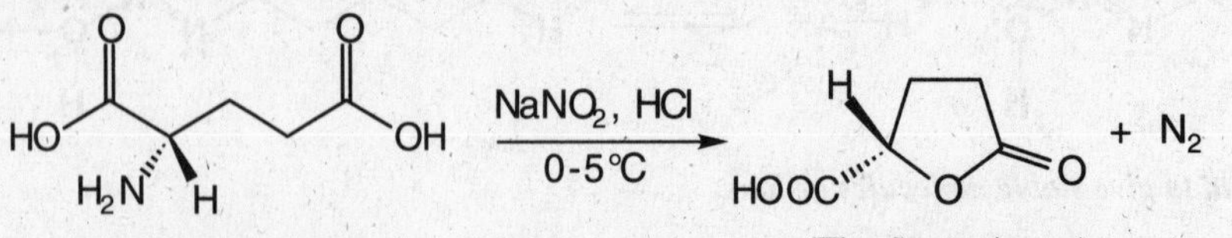

[structure of (S)-Glutamic acid] $\xrightarrow[0-5\,°C]{NaNO_2, HCl}$ [structure] $+ N_2$

(S)-Glutamic acid   (The *S* enantiomer)

Several Steps:

[structure] $\xrightarrow{NaNO_2, HCl}$ [structure]

(S)-Glutamic acid

Step 2: Make a bond between a nucleophile and an electrophile and simultaneously break a bond to give stable molecules or ions. The intriguing part of the mechanism is that there is no racemization. This means that it is unlikely that N_2 departs leaving a free cation behind. Such a cation would be sp^2 hybridized, would be planar and thus would lead to racemization. Therefore, propose the adjacent carbonyl group reacts to form a short-lived three membered ring intermediate that then reacts with the other carboxyl group to make the more stable, five-membered ring intermediate. The chiral center undergoes double inversion, resulting in net retention of stereochemistry.

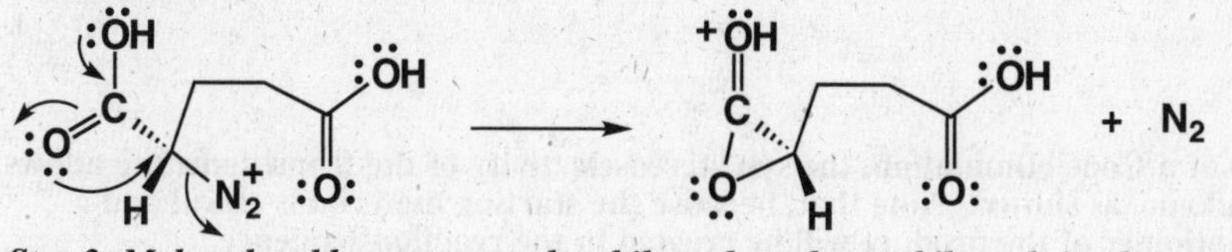

Step 3: *Make a bond between a nucleophile and an electrophile and simultaneously break a bond to give stable molecules or ions.*

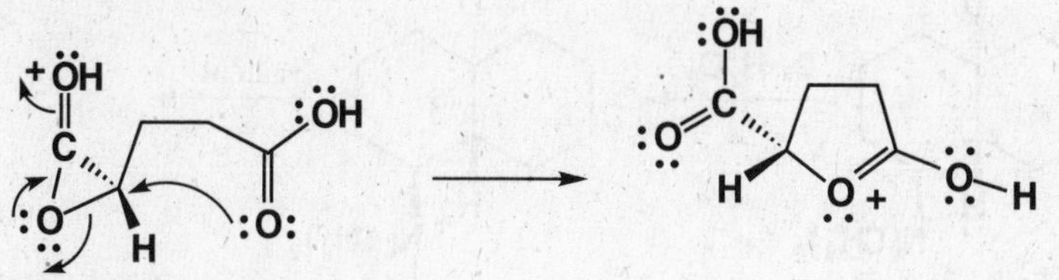

Step 4: *Take a proton away.*

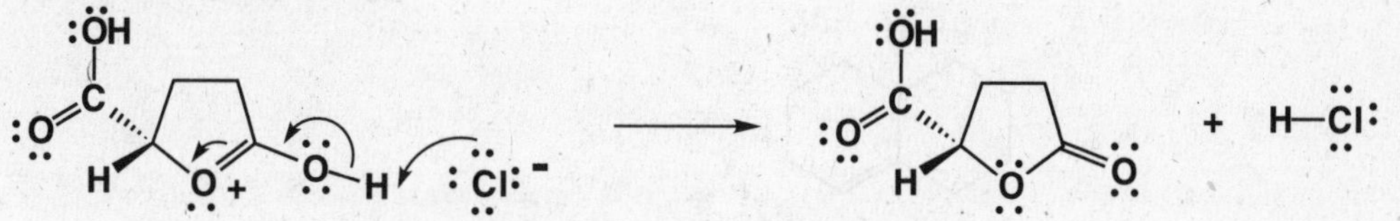

<u>Problem 23.39</u> The following sequence of methylation and Hofmann elimination was used in the determination of the structure of this bicyclic amine. Compound B is a mixture of two isomers.

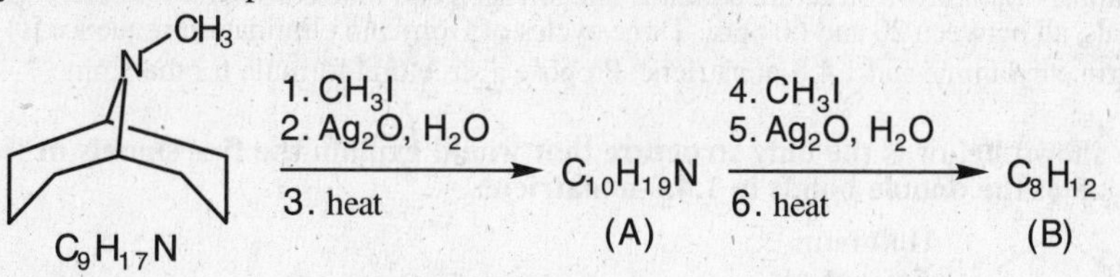

(a) Propose structural formulas for compounds (A) and (B).

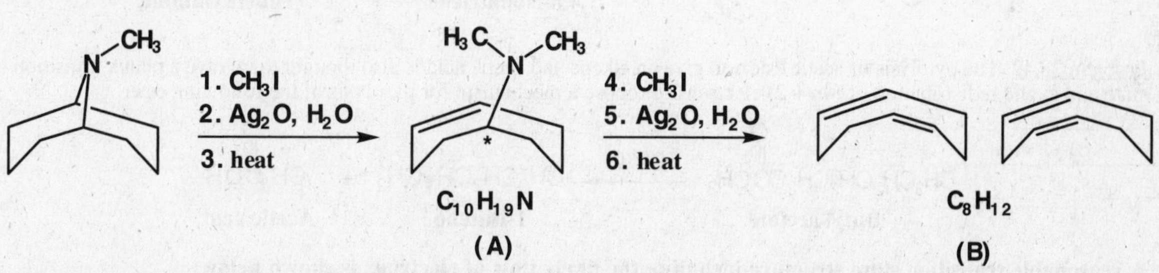

(b) Suppose you were given the structural formula of compound B but only the molecular formulas for compound A and the starting bicyclic amine. Given this information, is it possible, working backward, to arrive at an unambiguous structural formula for compound A? For the bicyclic amine?

Simply knowing the structural formula of compound (B) does not unambiguously establish the structure of compound (A), because the amine could be bonded at either end of what is a double bond in compound (B). Given this, the starting bicyclic amine cannot be unambiguously identified either, because two different structures are possible. One structure contains the bridging nitrogen atom in a [3.3.1] ring system (shown above), and the other structure contains the bridging nitrogen atom in a [4.2.1] ring system.

<u>Problem 23.40</u> Propose a structural formula for the compound, $C_{10}H_{16}$, and account for its formation.

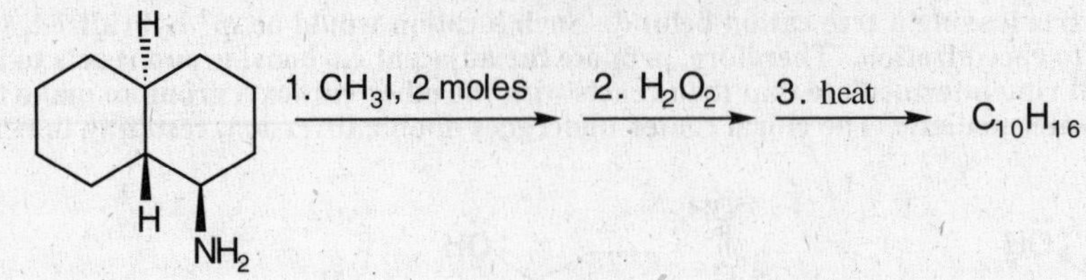

Note how, in the following example of a Cope elimination, the syn stereoselectivity of the transition state allows formation of the more substituted alkene, as shown. Note that, because the starting material is chiral and a single enantiomer, only a single enantiomer of the product will be created in the reaction sequence.

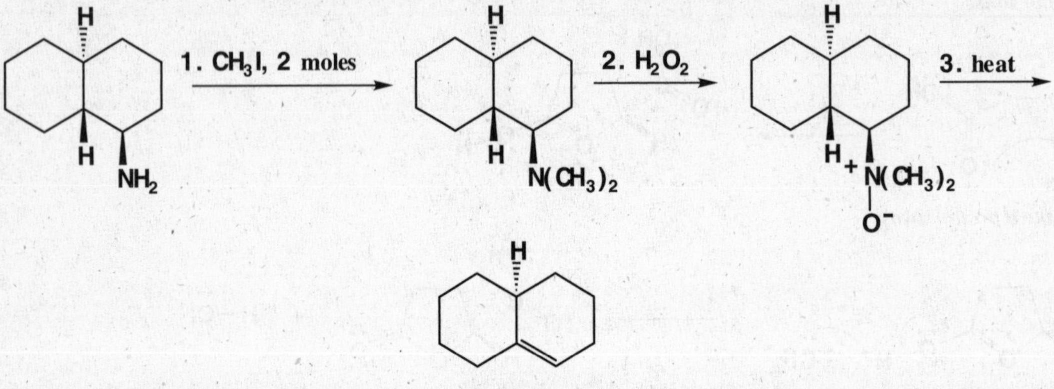

$C_{10}H_{16}$

Problem 23.41 An amine of unknown structure contains one nitrogen and nine carbon atoms. The ^{13}C-NMR spectrum shows only five signals, all between 20 and 60 ppm. Three cycles of Hofmann elimination sequence [(1) CH_3I; (2) Ag_2O, H_2O; (3) heat] give trimethylamine and 1,4,8-nonatriene. Propose a structural formula for the amine.

The bicyclic amine shown below is the only structure that would explain the five signals in the ^{13}C spectrum as well as the location of the double bonds in 1,4,8-nonatriene.

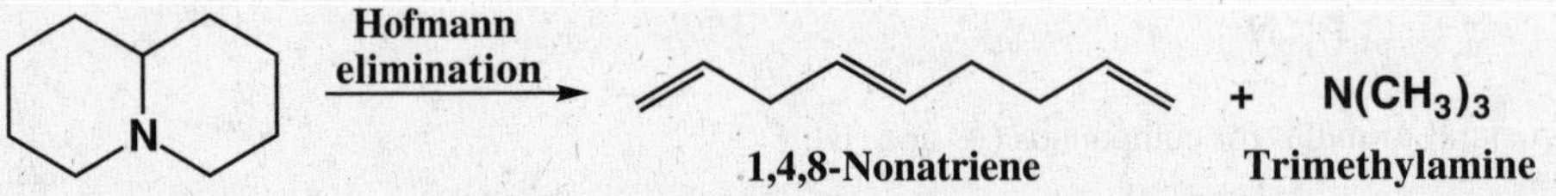

1,4,8-Nonatriene **Trimethylamine**

Problem 23.42 The pyrolysis of acetic esters to give an alkene and acetic acid is also thought to involve a planar transition state and cyclic redistribution of (4*n* + 2) electrons. Propose a mechanism for pyrolysis of the following ester.

$$CH_3CH_2CH_2CH_2O\overset{O}{\overset{\|}{C}}CH_3 \xrightarrow{500°C} CH_3CH_2CH{=}CH_2 + CH_3\overset{O}{\overset{\|}{C}}OH$$

Butyl acetate **1-Butene** **Acetic acid**

A reasonable transition state structure including the likely flow of electrons is shown below:

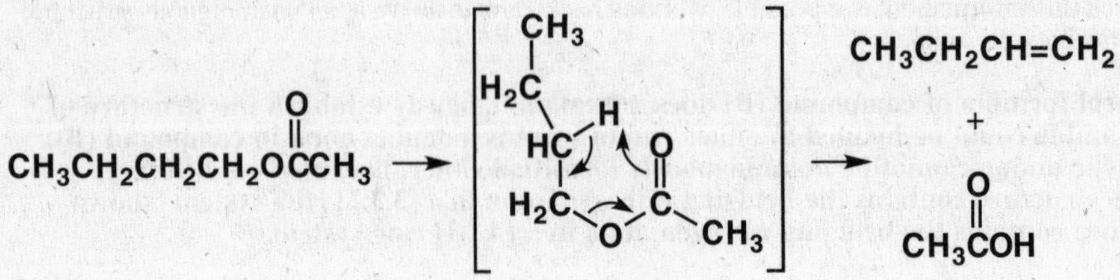

Synthesis

<u>Problem 23.43</u> Propose steps for the following conversions using a reaction of a diazonium salt in at least one step of each conversion.

(a) Toluene to 4-methylphenol (*p*-cresol)

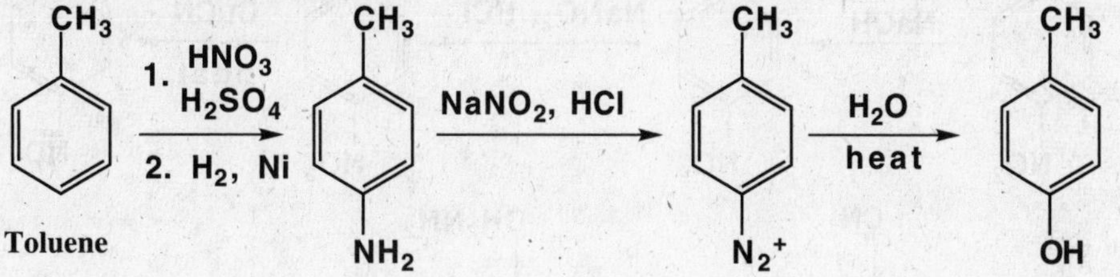

Toluene

(b) Nitrobenzene to 3-bromophenol

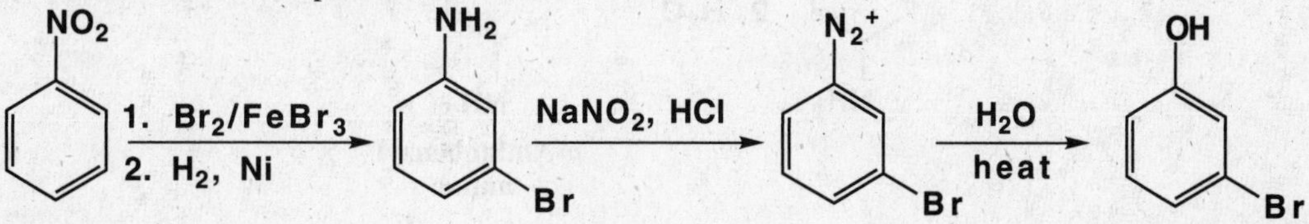

Nitrobenzene   **3-Bromophenol**

(c) Toluene to *p*-cyanobenzoic acid

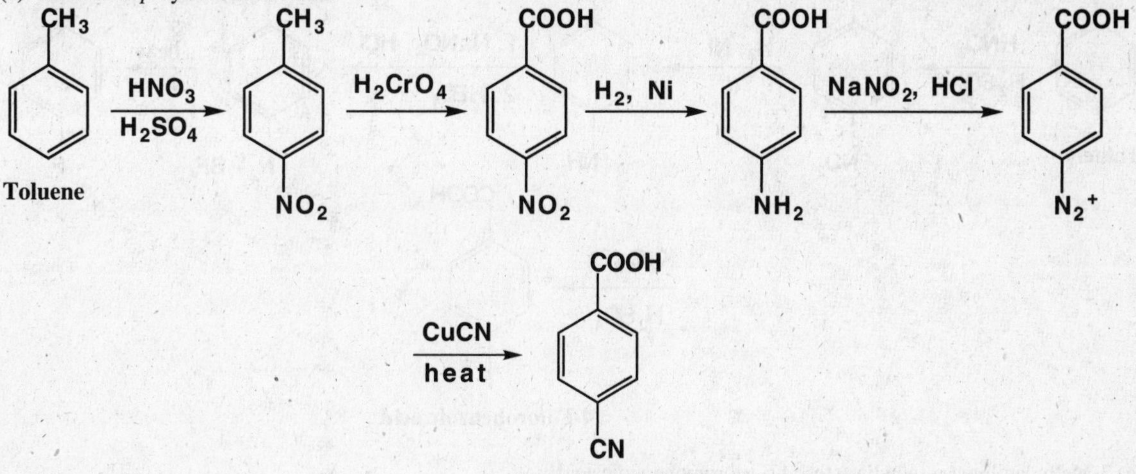

Toluene

p-Cyanobenzoic acid

(d) Phenol to *p*-iodoanisole

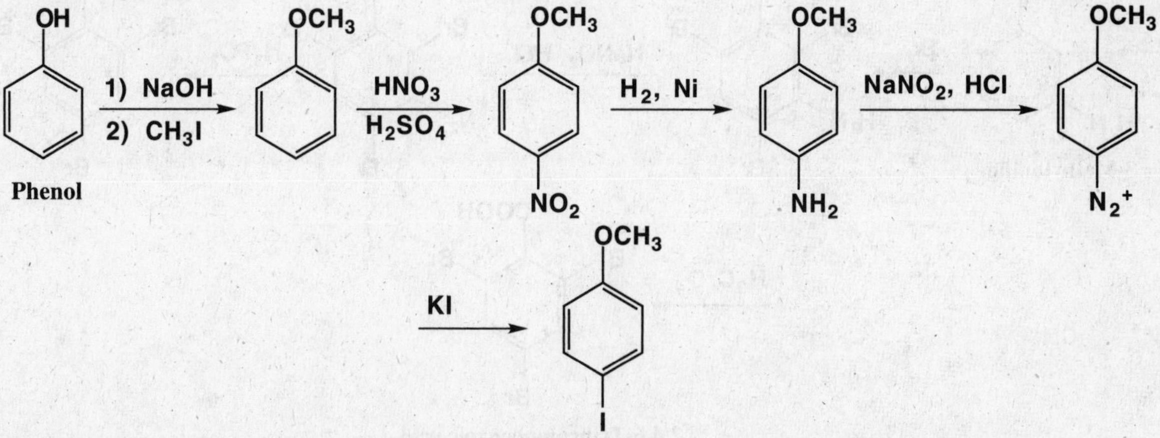

Phenol

p-Iodoanisole

Note that dimethyl sulfate could be used in place of CH₃I to give the methyl ether in the first step of the synthesis.

(e) Acetanilide to *p*-aminobenzylamine

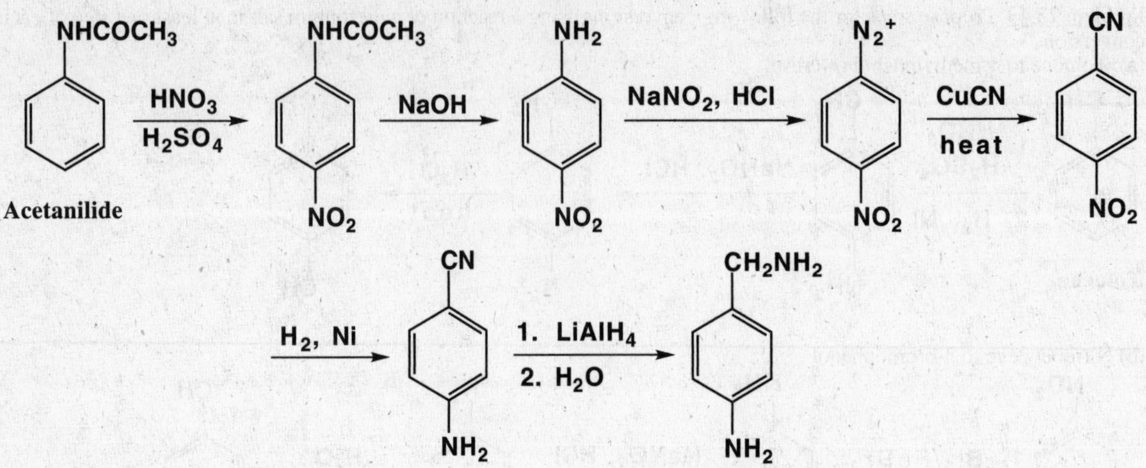

Acetanilide

p-Aminobenzyl-amine

(f) Toluene to 4-fluorobenzoic acid

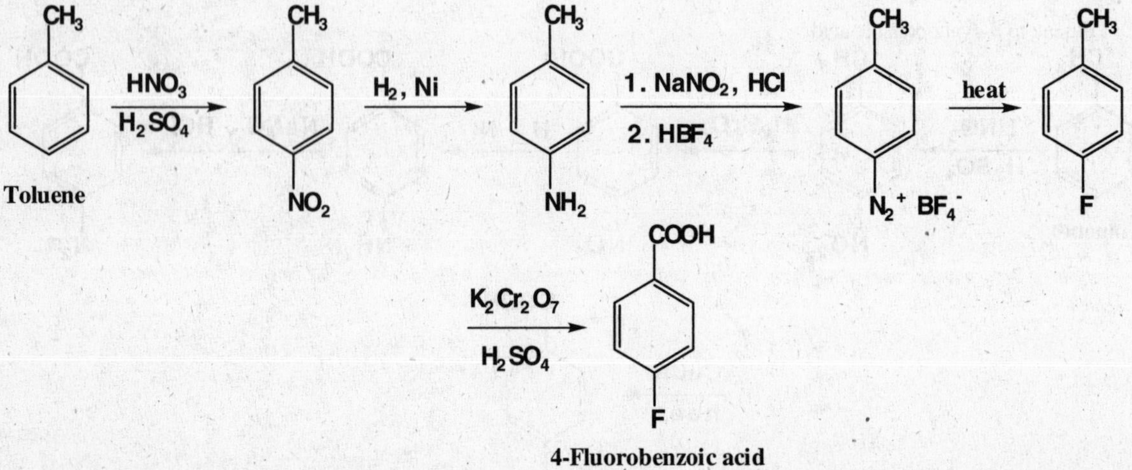

Toluene

4-Fluorobenzoic acid

(g) 3-Methylaniline (*m*-toluidine) to 2,4,6-tribromobenzoic acid

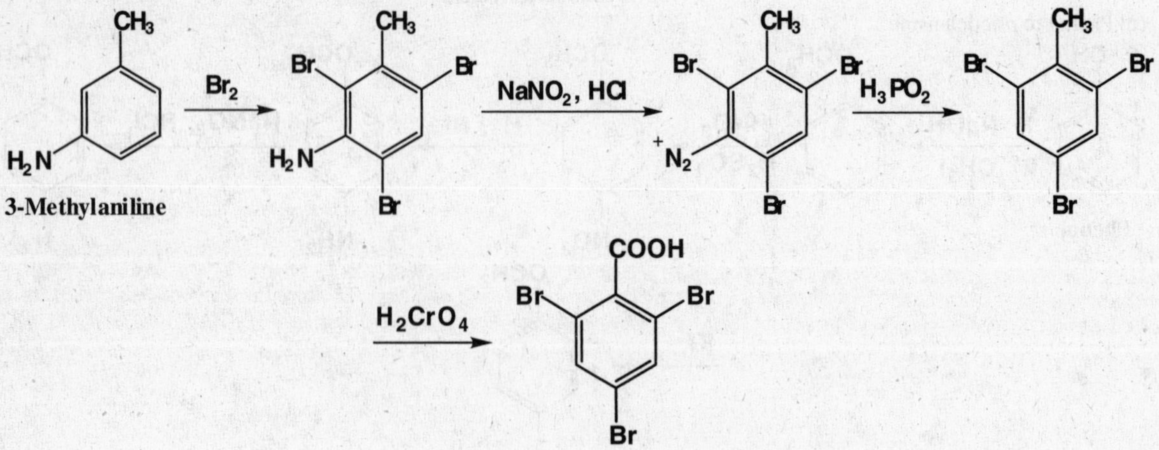

3-Methylaniline

2,4,6-Tribromobenzoic acid

Problem 23.44 Show how to bring about each step in this synthesis of the herbicide propranil.

Propanil

The reagents that can be used in each step are listed below:
Step (1) Aromatic chlorination reaction using Cl_2 and $FeCl_3$.
Step (2) Aromatic nitration reaction using HNO_3 and H_2SO_4.
Step (3) Another aromatic chlorination reaction using Cl_2 and $FeCl_3$. The new Cl atom is directed to the correct position by the groups already present on the ring.
Step (4) This reduction can be carried out using H_2 and a transition metal catalyst.
Step (5) Propanoic acid is first converted to the acid chloride by treatment with $SOCl_2$ and then reacted with the 3,4-dichloroaniline to give the desired propranil.

<u>Problem 23.45</u> Show how to bring about each step in the following synthesis.

The reagents that can be used in each step are listed below:
Step (1) Aromatic nitration reaction using HNO_3 and H_2SO_4.
Step (2) This reduction can be carried out using H_2 and a transition metal catalyst.
Step (3) This transformation can be accomplished by first turning the amino group into a diazonium salt by reaction with $NaNO_2$ and HCl, followed by a Sandmeyer reaction using CuCN and heat.
Step (4) This reduction can be carried out using $LiAlH_4$ followed by H_2O.

<u>Problem 23.46</u> Show how to bring about this synthesis.

<u>Problem 23.47</u> Show how to bring about each step in the following synthesis.

The reagents that can be used in each step are listed below:

Step (1) Reaction with NaOH to produce the phenoxide, followed by treatment with CH₃I.

Step (2) Aromatic nitration reaction using HNO₃ and H₂SO₄.

Step (3) This reduction can be carried out using H₂ and a transition metal catalyst.

Step (4) This transformation can be accomplished by first turning the amino group into a diazonium salt by reaction with NaNO₂ and HCl, followed by a Sandmeyer reaction using CuCN and heat.

Step (5) This reduction can be carried out using (1) H₂ and a transition metal or (2) LiAlH₄ followed by H₂O.

Problem 23.48 Methylparaben is used as a preservative in foods, beverages, and cosmetics. Provide a synthesis of this compound from toluene.

Methyl *p*-hydroxybenzoate
(Methylparaben)

Methyl *p*-hydroxybenzoate
(Methylparaben)

<u>Problem 23.49</u> Given the following retrosynthetic analysis, show how to synthesize the following tertiary amine as a racemic mixture from benzene and any necessary reagents.

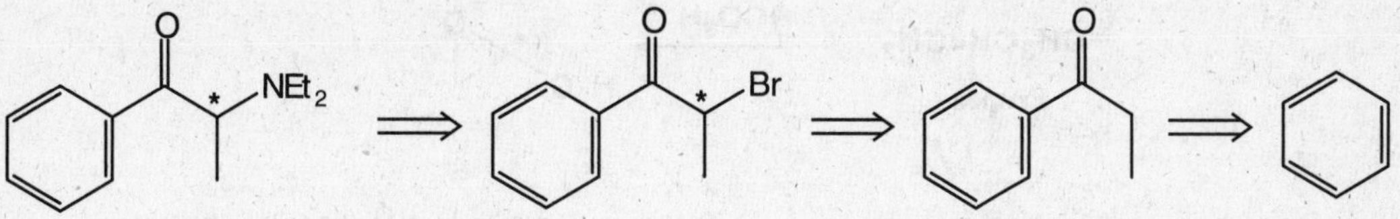

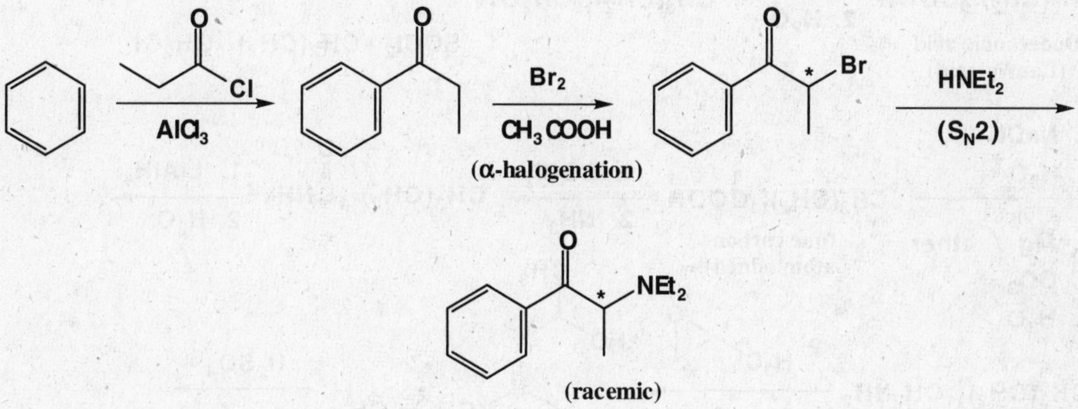

(racemic)

<u>Problem 23.50</u> *N*-Substituted morpholines are a building block in many drugs. Show how to synthesize *N*-methylmorpholine given this retrosynthetic analysis.

Methylamine   Ethylene oxide

N-Methylmorpholine

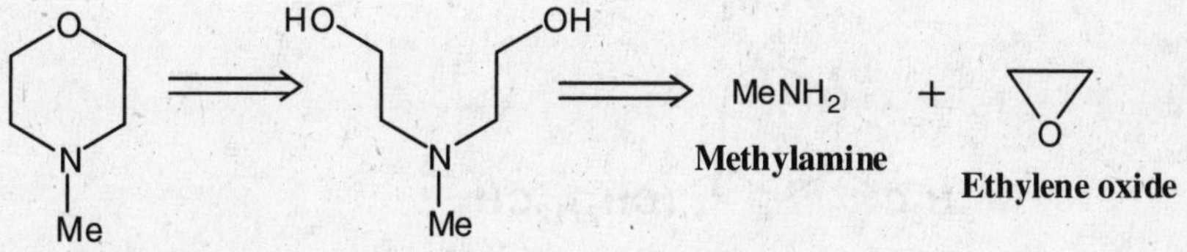

The key step in this synthesis is the last one, in which acid is used to form the morpholine ring. The reaction involves protonation of one of the hydroxyl groups, that then leaves as water. The fact that a six-membered ring is being formed facilitates this reaction.

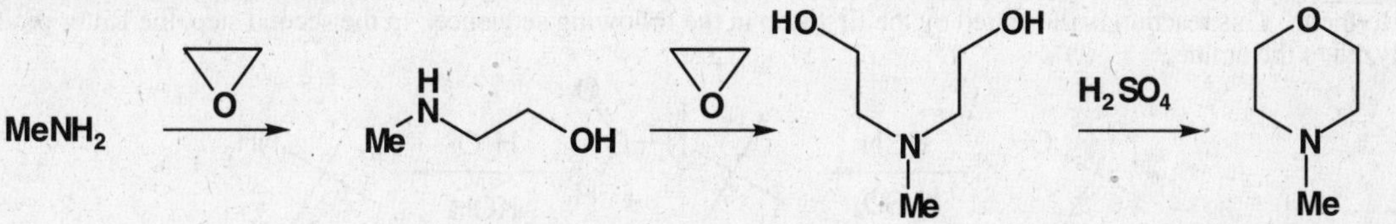

N-Methylmorpholine

<u>Problem 23.51</u> Propose a synthesis for the systemic agricultural fungicide tridemorph from dodecanoic acid (lauric acid) and propene. How many stereoisomers are possible for tridemorph?

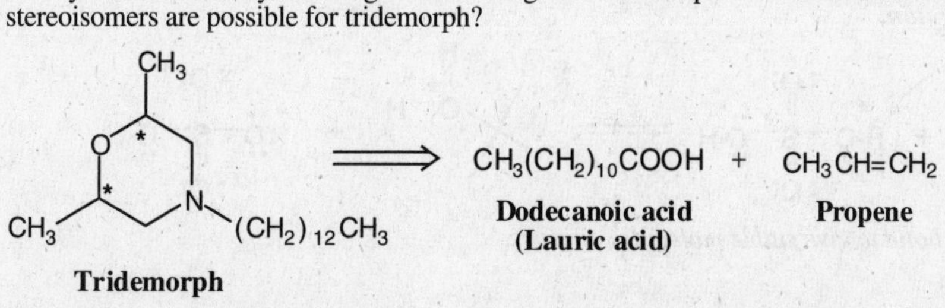

$CH_3(CH_2)_{10}COOH$ + $CH_3CH=CH_2$

Dodecanoic acid   Propene
(Lauric acid)

Tridemorph

There are two keys to this synthesis. The first is to recognize that the morpholine ring is constructed from an acid-catalyzed reaction between two alcohol groups analogous to that used in problem 23.50. Note that the required diol can be prepared from reaction of an aliphatic amine and the epoxide derived from propene. The second key of the

synthesis is to notice that dodecanoic acid must be lengthened by one carbon atom before it is turned into an amine function.

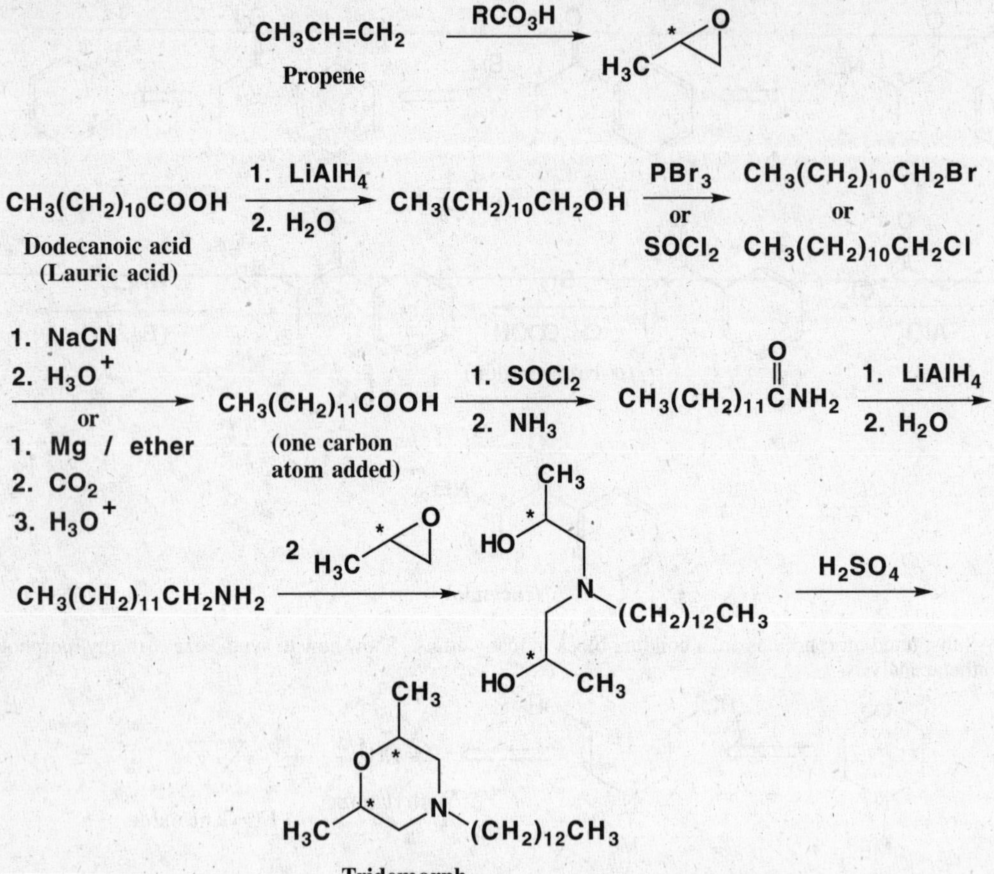

$$CH_3CH=CH_2 \xrightarrow{\quad RCO_3H \quad}$$

Propene

$$CH_3(CH_2)_{10}COOH \xrightarrow[\quad 2.\ H_2O \quad]{\quad 1.\ LiAlH_4 \quad} CH_3(CH_2)_{10}CH_2OH \xrightarrow[SOCl_2 \quad CH_3(CH_2)_{10}CH_2Cl]{PBr_3 \quad CH_3(CH_2)_{10}CH_2Br}$$

Dodecanoic acid
(Lauric acid)

$$\xrightarrow[\substack{1.\ Mg\ /\ ether \\ 2.\ CO_2 \\ 3.\ H_3O^+}]{\substack{1.\ NaCN \\ 2.\ H_3O^+ \\ or}} CH_3(CH_2)_{11}COOH \xrightarrow[2.\ NH_3]{1.\ SOCl_2} CH_3(CH_2)_{11}\overset{O}{\overset{\|}{C}}NH_2 \xrightarrow[2.\ H_2O]{1.\ LiAlH_4}$$

(one carbon atom added)

$$CH_3(CH_2)_{11}CH_2NH_2 \xrightarrow{\quad 2\ H_3C \overset{*}{\triangle} O \quad} \text{(product)} \xrightarrow{\quad H_2SO_4 \quad}$$

Tridemorph

Note that Tridemorph has two chiral centers. Assuming racemic epoxide is used as starting material, a racemic mixture of three stereoisomers (one meso form and two enantiomers) will be formed as the product of the above reaction sequence.

<u>Problem 23.52</u> The Ritter reaction is especially valuable for the synthesis of 3° alkanamines. In fact, there are few alternative routes to them. This reaction is illustrated by the first step in the following sequence. In the second step, the Ritter product is hydrolyzed to the amine.

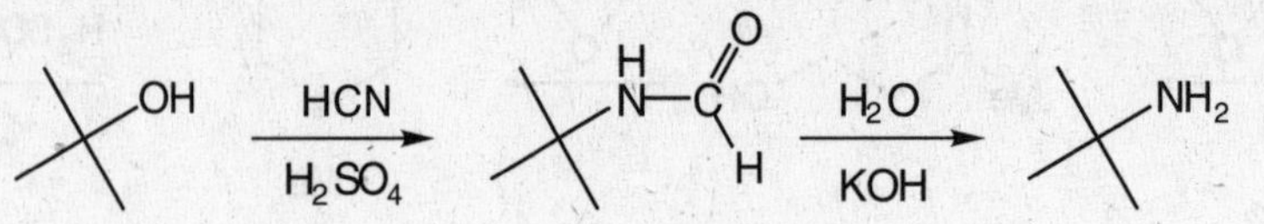

$$\text{(t-BuOH)} \xrightarrow[H_2SO_4]{HCN} \text{(Ritter product)} \xrightarrow[KOH]{H_2O} \text{(t-BuNH}_2)$$

Ritter product

(a) Propose a mechanism for the Ritter reaction.

Step 1: Add a proton.

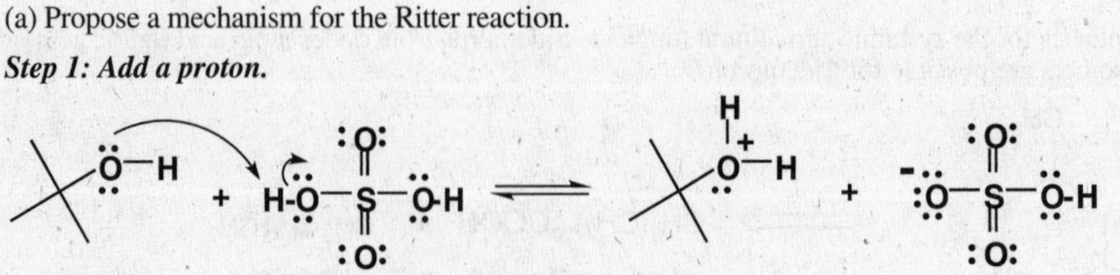

Step 2: Break a bond to give stable molecules or ions.

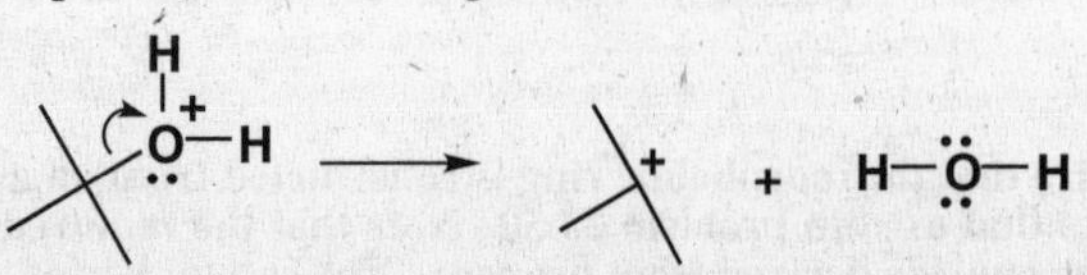

Step 3: Make a bond between a nucleophile and an electrophile.

Step 4: Make a bond between a nucleophile and an electrophile.

Step 5: Take a proton away.

Step 6: Keto-enol tautomerism.

(b) What is the product of a Ritter reaction using acetonitrile, CH_3CN, instead of HCN followed by reduction of the Ritter product with lithium aluminum hydride?

Problem 23.53 Several diamines are building blocks for the synthesis of pharmaceuticals and agrochemicals. Show how both 1,3-propanediamine and 1,4-butanediamine can be prepared from acrylonitrile

1,3-Propanediamine   **1,4-Butanediamine**   **Acrylonitrile**

Acrylonitrile is a Michael acceptor that can react with amines (ammonia) or the cyanide anion.

Problem 23.54 Given the following retrosynthetic analysis, show how the intravenous anesthetic 2,6-diisopropylphenol (Propofol) can be synthesized from phenol.

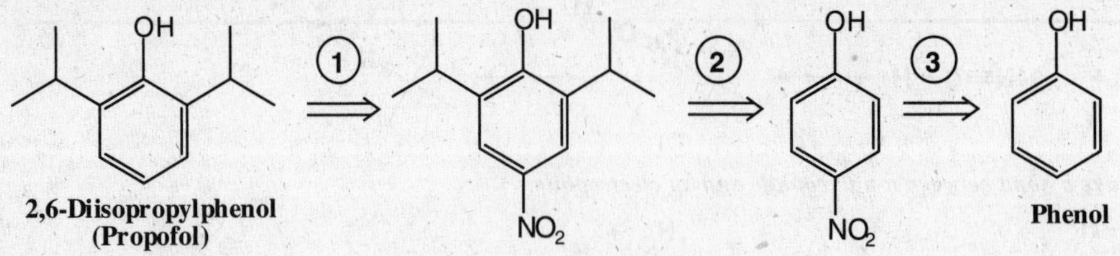

**2,6-Diisopropylphenol
(Propofol)**

Phenol

In the following synthetic scheme, note how the overall yield of product is increased by having the temporary nitro group insure that the alkylation will occur only in the desired positions ortho to the methyl group.

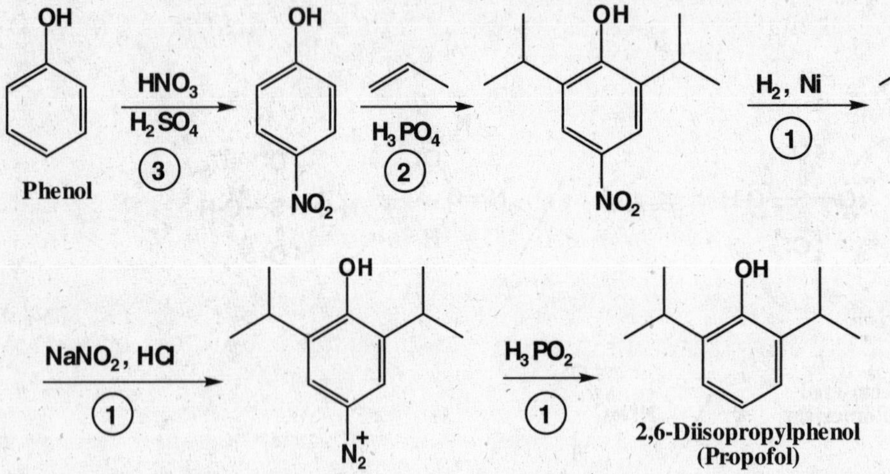

Phenol

**2,6-Diisopropylphenol
(Propofol)**

Problem 23.55 Following is a retrosynthetic analysis for propoxyphene, the hydrochloride salt of which is Darvon. The naphthalenesulfonic acid salt of propoxyphene is Darvon-N. The configuration of the carbon in Darvon bearing the hydroxyl group is *S*, and the configuration of the other chiral center is *R*. Its enantiomer has no analgesic properties, but it is used as a cough suppressant.

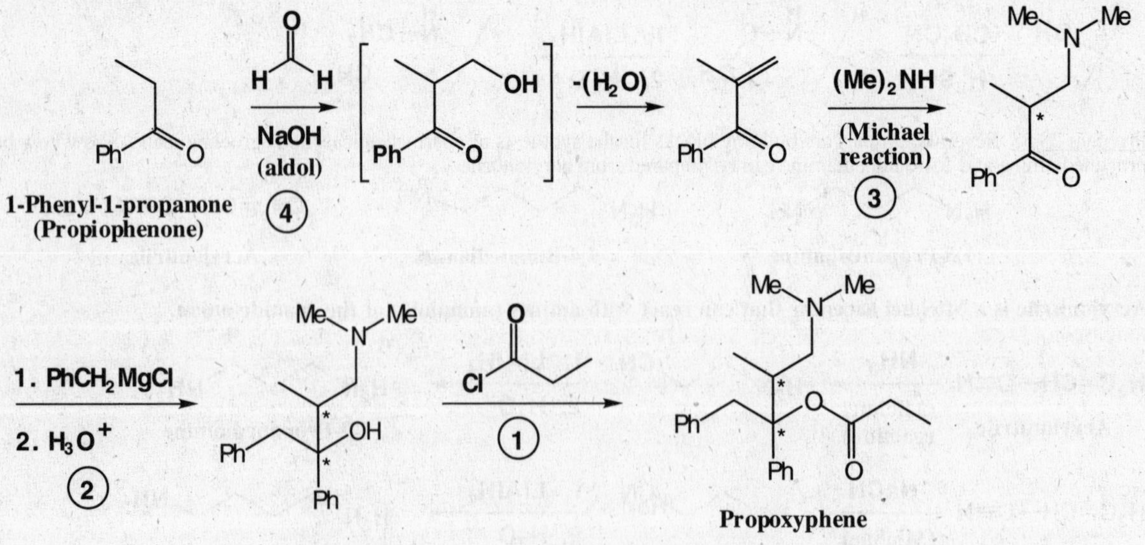

**1-Phenyl-1-propanone
(Propiophenone)**

Propoxyphene

(a) Propose a synthesis for propoxyphene from 1-phenyl-1-propanone and any other necessary reagents.

Step (4) A crossed aldol reaction using formaldehyde followed by dehydration gives the α,β-unsaturated ketone (Section 19.2).
Step (3) A Michael reaction using dimethylamine (Section 19.8) gives the second key intermediate.
Step (2) Reaction of the ketone with a benzyl Grignard reagent (Section 16.5) gives the alcohol product.
Step (1) Reaction with propanoyl chloride (Section 18.5) gives propoxyphene ether.

(b) Is propoxyphene chiral? If so, which of the possible stereoisomers are formed in this synthesis?

Propoxyphene has two chiral centers, so a racemic mixture of four stereoisomers is produced in this synthesis.

<u>Problem 23.56</u> Following is a retrosynthetic analysis for ibutilide, a drug used to treat cardiac arrhythmia. In this scheme, Hept is an abbreviation for the 1-heptyl group.

(a) Propose a synthesis for ibutilide starting with aniline, methanesulfonyl chloride, succinic anhydride, and *N*-ethyl-1-heptanamine.

Step (5) Treatment of a primary or secondary amine with methanesulfonyl chloride (Section 18.1) gives a sulfonamide.
Step (4) This reaction is a type of electrophilic aromatic substitution (Section 22.1) that results in acylation of the aromatic ring. For a mechanism of this type of acylation, review the solution to Problem 22.27.
Step (3) Treatment of the carboxylic acid with thionyl chloride gives an acid chloride (Section 17.8).
Step (2) Treatment of the acid chloride with *N*-ethyl-1-heptanamine gives (a secondary amine) gives an amide (Section 18.6).
Step (1) Reduction with lithium aluminum hydride gives a secondary alcohol (Section 16.11) and amine (Section 18.10).

(b) Is ibutilide chiral? If so, which of the possible stereoisomers are formed in this synthesis?

Ibutilide has one chiral center, so it is produced as a racemic mixture of enantiomers.

Enantiomers of ibutilide

<u>Problem 23.57</u> Propose a synthesis for the antihistamine histapyrrodine.

Step (1) Nucleophilic opening of the epoxide (Section 11.9) gives an aminoalcohol.
Step (2) Treatment of the alcohol with thionyl chloride (Section 10.5) gives an alkyl chloride.
Step (3) Treatment of benzoic acid with thionyl chloride (Section 17.8) gives benzoyl chloride.
Step (4) Reaction of benzoyl chloride and aniline (Section 18.6) gives *N*-phenylbenzamide.
Step (5) LiAlH₄ reduction of the amide (Section 18.10) gives a secondary amine.
Step (6) Because of participation by the neighboring nitrogen atom (Section 9.11), this primary chloride
        undergoes reaction with the secondary amino group of *N*-methylbenzylamine to give histapyrrodine.

Problem 23.58 Following is a retrosynthesis for the coronary vasodilator ganglefene.

Ganglefene

(a) Propose a synthesis for ganglefene from 4-hydroxybenzoic acid and 3-methyl-3-buten-2-one.

Step (3) Michael addition (Section 19.8) of diethylamine to this α,β-unsaturated ketone gives a β-aminoketone. Sodium borohydride reduction of the ketone gives a secondary alcohol (Section 16.11).

Step (2) A carboxylic acid (pK_a 4-5) is a stronger acid than a phenol (pK_a 9-10). Therefore, if 4-hydroxybenzoic acid were treated with one mole of NaOH, the carboxyl proton would be removed preferentially. Treatment of this anion with 1-bromo-2-methylpropane (isobutyl bromide) would then give an isobutyl ester. To ensure that alkylation takes place on the phenolic oxygen, first convert the carboxyl group to an ethyl ester by Fischer esterification (Section 17.7). Then treat this ester with one mole of NaOH followed by isobutyl bromide to form the phenolic ether. Finally remove the carboxyl protecting group by base-promoted hydrolysis (Section 18.4) followed by acidification to give 4-isobutoxybenzoic acid.

Step (1) Fischer esterification (Section 17.7) of the carboxylic acid with the secondary alcohol gives ganglefene. Alternatively, the carboxyl group may be converted to its acid chloride using thionyl chloride (Section 17.8) and the acid chloride reaction with the alcohol to give ganglefene.

(b) Is ganglefene chiral? If so, which of the possible stereoisomers are formed in this synthesis?

Ganglefene has two chiral centers, so it is produced as a racemic mixture of four stereoisomers.

<u>Problem 23.59</u> Moxisylyte, an α-adrenergic blocker, is used as a peripheral vasodilator. Propose a synthesis for this compound from thymol, which occurs in the volatile oils of members of the thyme family. Thymol is made industrially from *m*-cresol.

Step (1) A phenol may be alkylated using an alkene and a phosphoric acid catalyst (Section 21.4), an alcohol and a phosphoric acid catalyst (Section 21.4), or an alkyl halide and aluminum chloride (Friedel-Crafts alkylation (Section 21.1C).

Step (2) Nitration (Section 22.1) is directing by the strongly activating phenolic -OH group.

Step (3) Reaction with this chloramine gives the ether intermediate (Section 9.11).

Step (4) Catalytic reduction of the nitro group (Section 22.1) gives a primary aromatic amine.

Steps (5 and 6) Treatment of the primary aromatic amine with nitrous acid gives an arenediazonium salt (Section 23.8). Warming this salt in water results in evolution of N_2 and formation of a phenol.

Step (7) Treatment of the phenol with acetic anhydride (Section 18.5) gives moxisylyte.

<u>Problem 23.60</u> Propose a synthesis of the local anesthetic ambucaine from 4-nitrosalicylic acid, ethylene oxide, diethylamine, and 1-bromobutane.

Steps (1-3) The synthetic problem here is to alkylate the phenolic -OH group while retaining the -COOH group. This can be accomplished by protection of the -COOH group as its methyl ester (Fischer esterification, Section 17.7), alkylation the phenolic -OH group using NaOH and 1-bromobutane, and finally base-promoted hydrolysis of the ethyl ester followed by acidification with aqueous HCl followed to regenerate the carboxyl group (Section 18.4).

Step (4) Fischer esterification (Section 17.7) using the β-aminoalcohol gives the ester. The β-amino alcohol is derived by treating ethylene oxide with diethylamine.

Step (5) Catalytic reduction of the nitro group gives the primary aromatic amine (Section 21.1) and completes the synthesis of ambucaine.

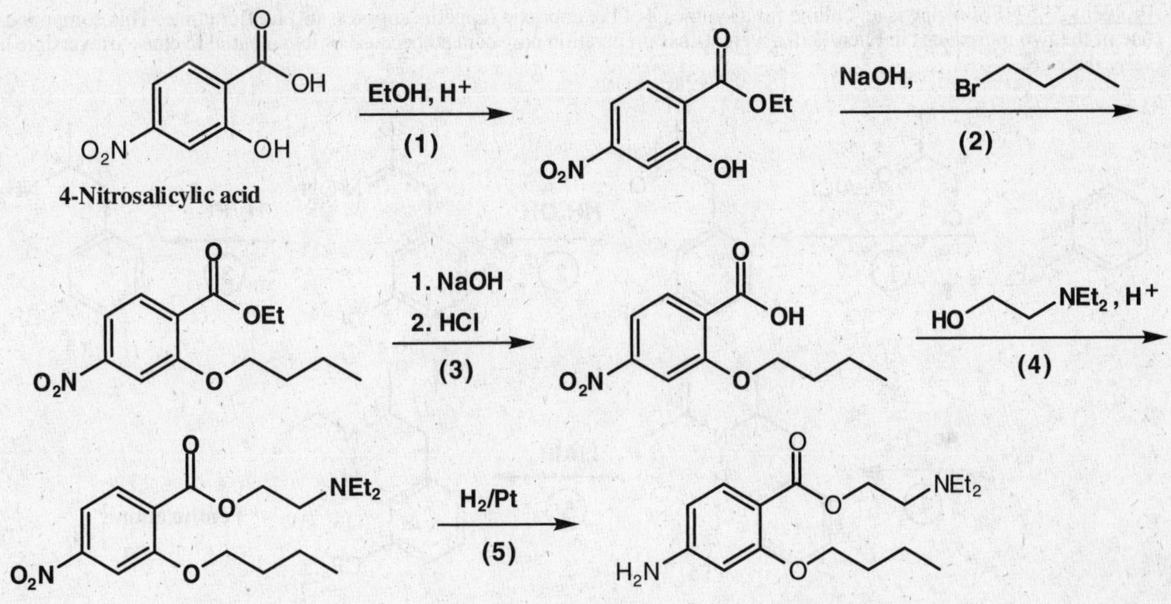

4-Nitrosalicylic acid

Ambucaine

<u>Problem 23.61</u> Given this retrosynthetic analysis, propose a synthesis for the local anesthetic hexylcaine.

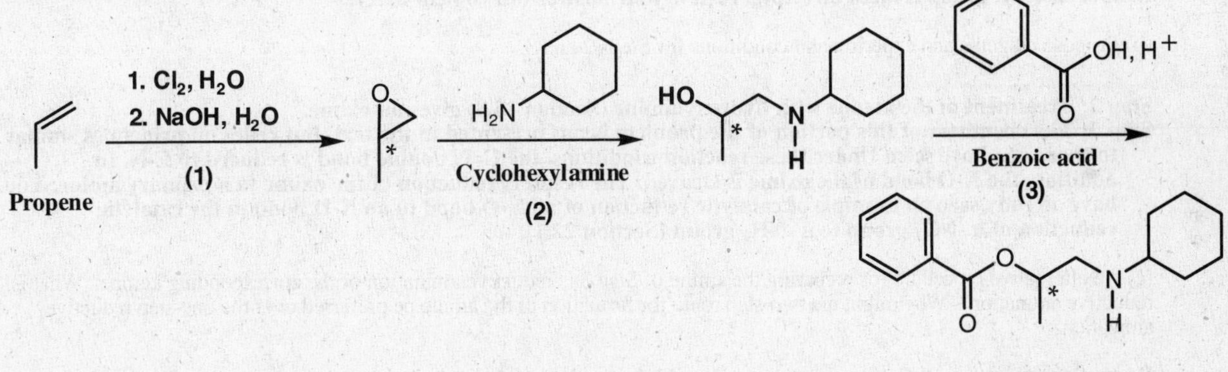

Propene

Cyclohexylamine

Benzoic acid

Hexylcaine

Step 1: This synthesis of methyloxirane (propylene oxide) is described in Section 11.8.
Step 2: Treatment of the epoxide with cyclohexylamine results in regioselective nucleophilic opening of the epoxide ring (Section 11.9).
Step 3: Fischer esterification with benzoic acid gives hexylcaine (Section 17.7). Note that the acid chloride method can not be used here because of secondary amine of the β-aminoalcohol is a stronger nucleophile that the hydroxyl group.

<u>Problem 23.62</u> Following is an outline for a synthesis of the anorexic (appetite suppressant) fenfluramine. This compound was one of the two ingredients in Phen-Fen, a weight-loss preparation now banned because of its potential to cause irreversible heart valve damage.

(a) Propose reagents and conditions for Step 1. Account for the fact that the -CF₃ group is meta directing.

This step is a Friedel-Crafts acylation (Section 22.1) using an α-chloroketone. For the reason the trifluoromethyl group is meta directing, review your answer to Problem 22.17.

(b) Propose reagents and experimental conditions for Steps 2 and 3.

Step 2: Treatment of the ketone with hydroxylamine (Section 16.8) gives an oxime.
Step 3: The chemistry of this portion of the problem is not presented in the text, but relies on principles similar to those you have seen. Under these reaction conditions, the C=N double bond is reduced to C-N. In addition, the N-O bond of the oxime is cleaved. The result is reduction of the oxime to a primary amine. You have already seen an example of catalytic reduction of an N-O bond to an N-H bond in the catalytic reduction of a -NO₂ group to a -NH₂ group (Section 22.1).

(c) An alternative procedure for preparing the amine of Step 3 is reductive amination of the corresponding ketone. What is reductive amination? Why might the two-step route for formation of the amine be preferred over the one-step reductive amination?

Reductive amination refers to the process in which an amine is added along with hydrogen and a transition metal. An imine is formed temporarily, then the C=N bond is reduced by hydrogenation. A potential problem with the reductive amination procedure is that the imides are not that stable, and during the reduction step some imine can revert to the ketone, which is reduced, lowering the yield. The oxime is more stable, so with this procedure, it is easier improve the yield of the overall process.

(d) Propose reagents for Steps 4 and 5.

Step 4: Treatment of the primary amine with acetic anhydride (Section 18.6) gives the amide.
Step 5: Lithium aluminum hydride reduction of the amide (Section 18.10) gives the secondary amine and completes the synthesis of fenfluramine.

(e) Is fenfluramine chiral? If so, which of the possible stereoisomers are formed in this synthesis?

Enantiomers of fenfluramine

<u>Problem 23.63</u> Following is a series of anorexics (appetite suppressants). As you study their structures, you will surely be struck by the sets of characteristic structural features.

(a) Knowing what you do about the synthesis of amines, including the Ritter reaction (Problem 23.52), suggest a synthesis for each compound.

An important observation is that the amine of amphetamine is on a secondary carbon and can be synthesized from an alkyl halide using the azide route. Amphetamine can be used as starting material for several other derivatives (b),(d),(f),(g). Others are on tertiary carbons and can be synthesized by the Ritter reaction (c),(h),(i). In most cases, reaction of a phenyl Grignard reagent with an epoxide will give the required alcohol for Ritter reactions (c),(h),(i), or conversion to an alkyl chloride with SOCl$_2$ for the azide route (a). In other cases, various methods such as reduction of amides (b),(d), Michael reactions (f), or reductive aminations (g) are used to avoid overalkylation of the primary amine of amphetamine.

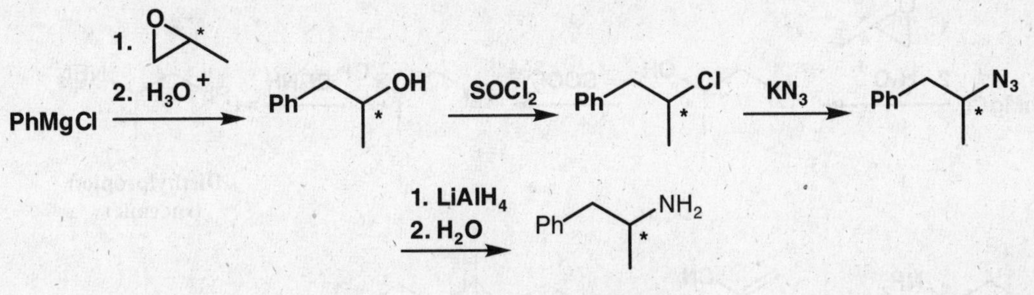

**Amphetamine
(racemic)**

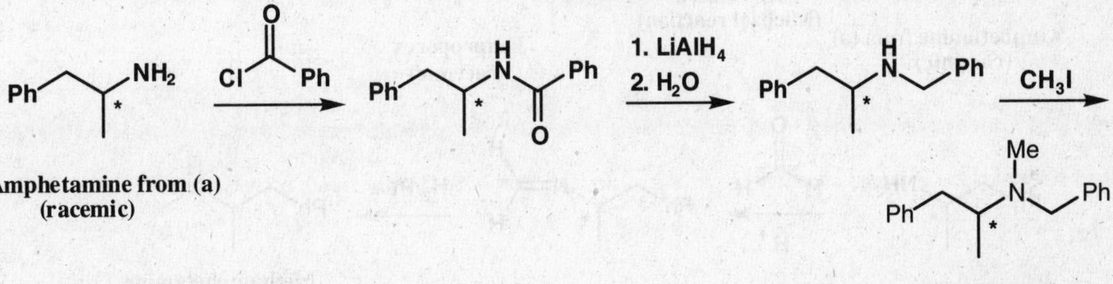

**Amphetamine from (a)
(racemic)**

**Benzphetamine
(racemic)**

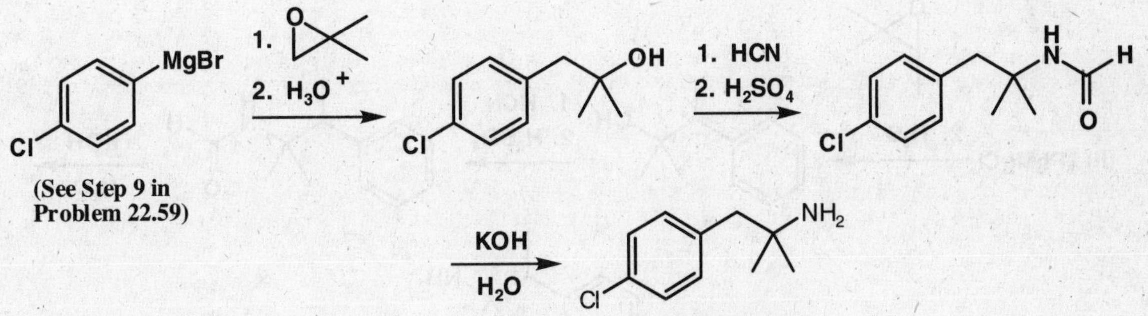

**(See Step 9 in
Problem 22.59)**

Chlorphentermine

(d) **Amphetamine from (a)**
(racemic)

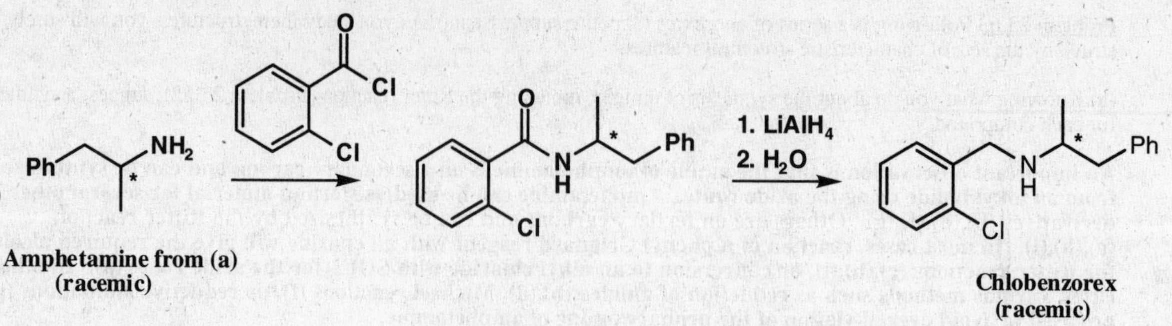

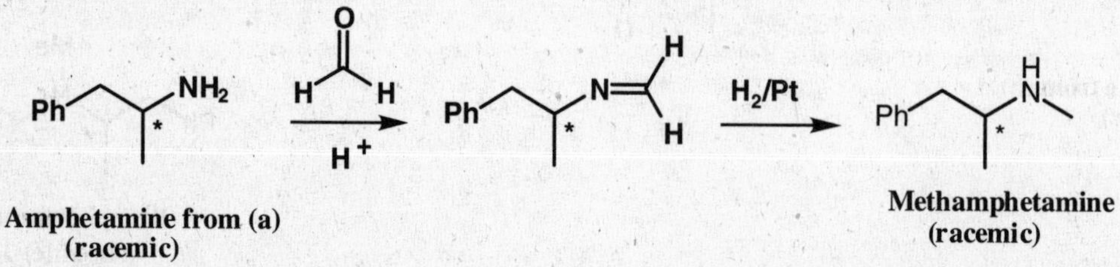

Chlobenzorex
(racemic)

(e) **PhMgCl**

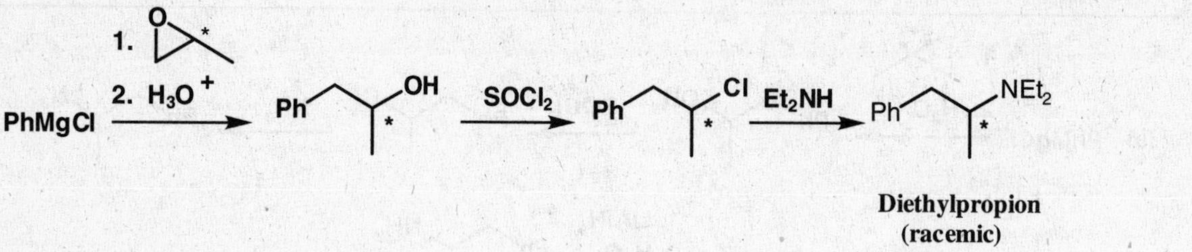

1. (epoxide) *
2. H_3O^+

Ph — OH *

$SOCl_2$

Ph — Cl *

Et_2NH

Ph — NEt_2 *

Diethylpropion
(racemic)

(f) **Amphetamine from (a)**
(racemic)

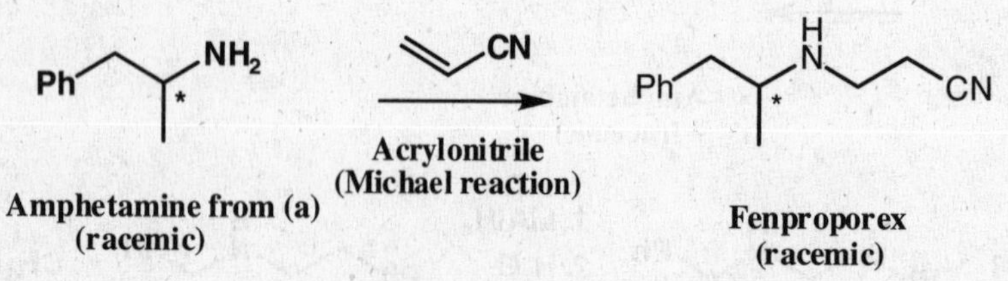

CN

Acrylonitrile
(Michael reaction)

Ph — N — CN

Fenproporex
(racemic)

(g) **Amphetamine from (a)**
(racemic)

H^+

Ph — N=

H_2/Pt

Ph — N — H

Methamphetamine
(racemic)

(h) **PhMgCl**

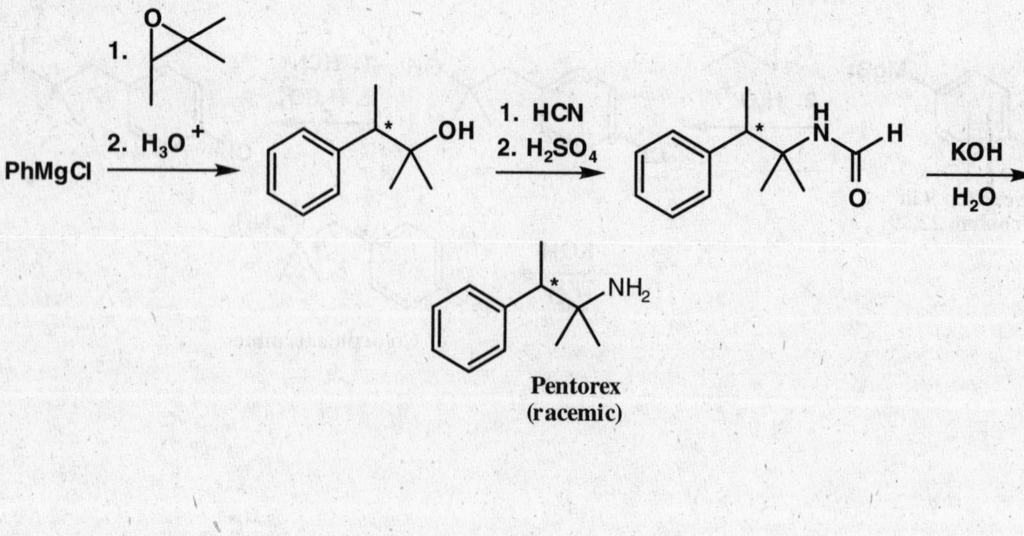

1. (epoxide)
2. H_3O^+

— OH

1. HCN
2. H_2SO_4

— N — H — O

KOH / H_2O

— NH_2

Pentorex
(racemic)

(i) 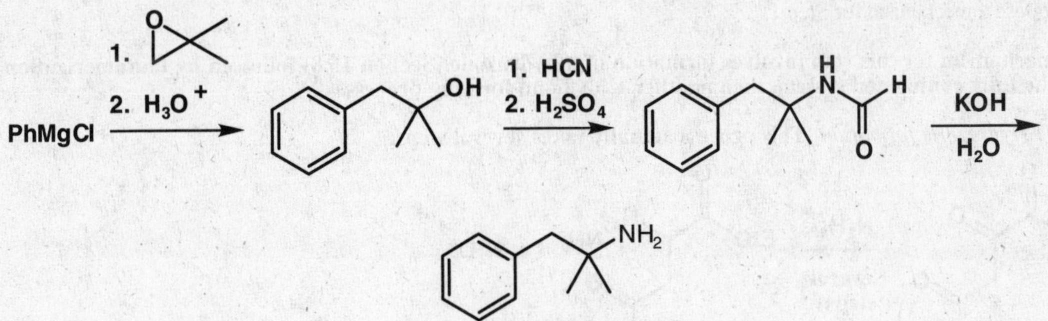 PhMgCl

Phentermine

(b) Which of these compounds are chiral?

As labeled on the structures above, (a), (b), (d), (e), (f), (g), and (h) are chiral because they each have one chiral center. They are produced as racemic mixtures in the proposed syntheses.

Problem 23.64 The drug sildefanil, sold under the trade name Viagra, is a potent inhibitor of phosphodiesterase V (PDE V), an enzyme found in high levels in the corpus carvenosum of the penis. Inhibitors of this enzyme enhance vascular smooth muscle relaxation and are used for treatment of male impotence. Following is an outline for a synthesis of sildefanil.

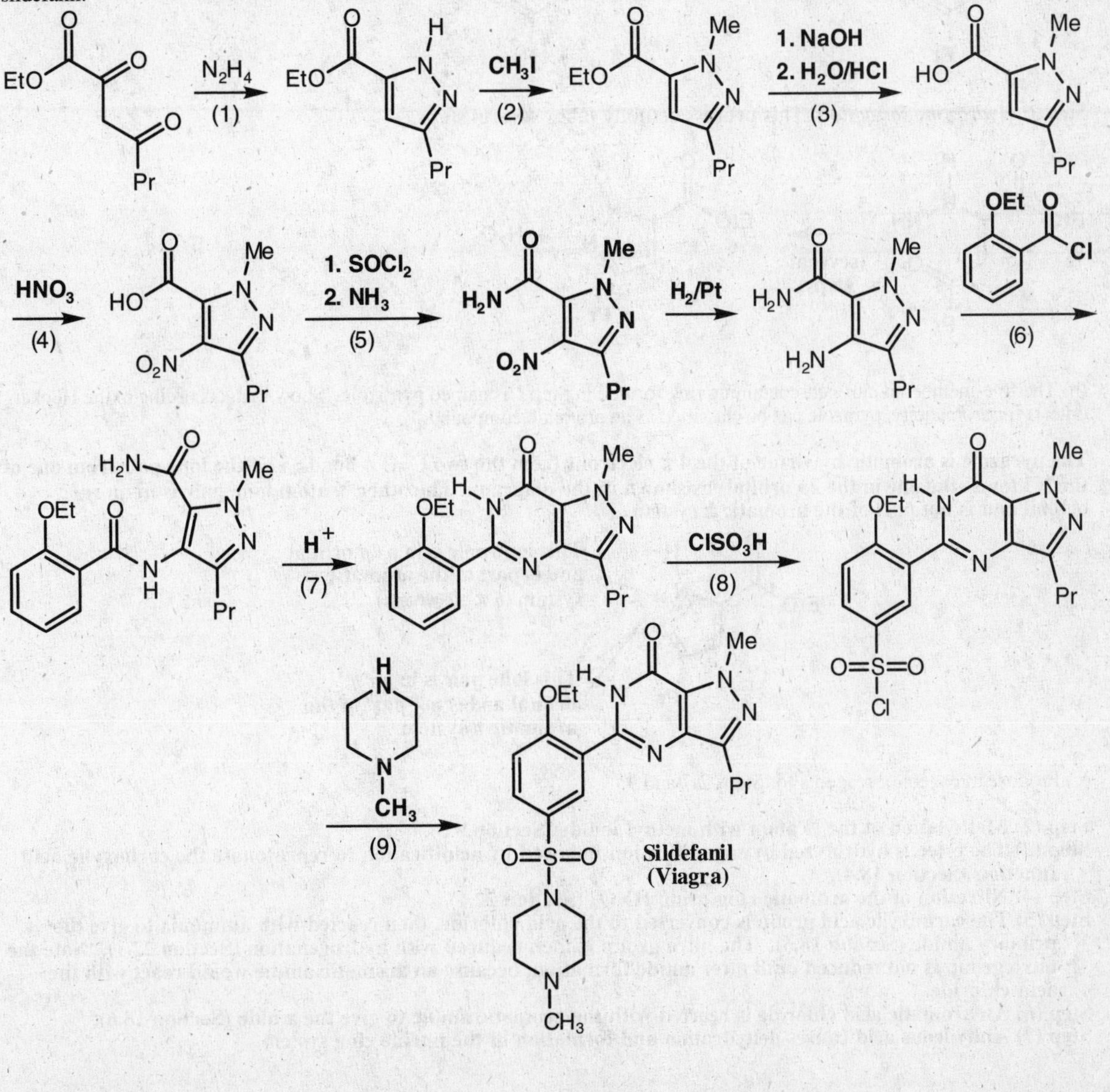

Sildefanil (Viagra)

(a) Propose a mechanism for Step 1.

The mechanism for this step involves formation of a hydrazone (Section 16.8) followed by tautomerization to give the fully conjugated system, then another C=N bond forming process.

Step 1: Hydrazone formation **This process actually takes several steps.**

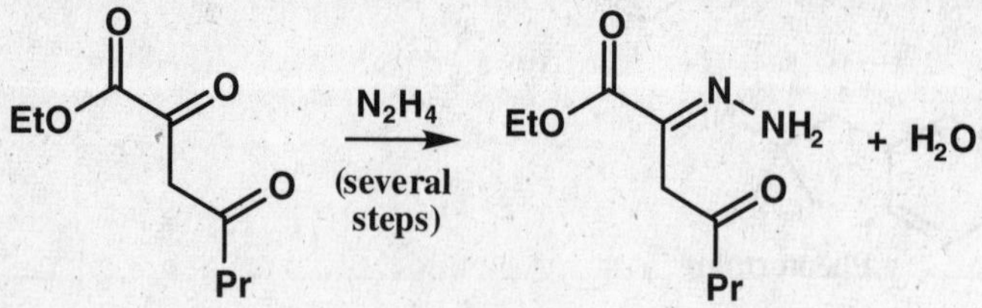

Step 2: Keto-enol tautomerism. **This process actually takes several steps.**

Step 3: Hydrazone formation **This process actually takes several steps.**

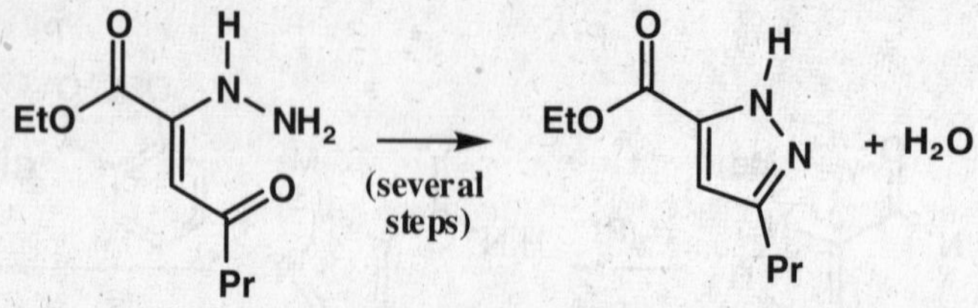

(b) The five-membered nitrogen-containing ring formed in Step 1 is named pyrazole. Show that, according to the Hückel criteria for aromaticity, pyrazole can be classified as an aromatic compound.

The pyrazole is aromatic by virtue of the 4 π electrons from the two C=C π bonds, and the lone pair from one of the N atoms (the one in the 2p orbital) as shown in the diagram. The other N atom lone pair is in an sp^2 orbital and is not part of the aromatic π system.

This lone pair is in a 2p orbtial and is part of the aromatic π system (6 π electrons)

This lone pair is in an sp^2 orbital and is not part of the aromatic π system

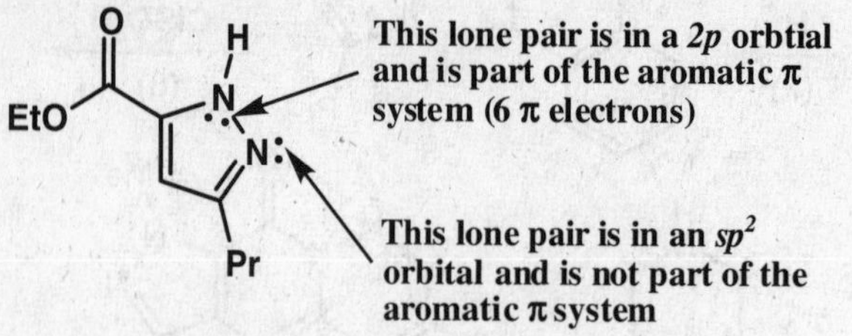

(c) Propose a reagent or reagents for Steps 2-7 and 9.

Step (2) Methylation of the N atom with methyl iodide (Section 9.3).
Step (3) The ester is hydrolyzed by saponification followed by acidification to reprotonate the carboxylic acid function (Section 18.4).
Step (4) Nitration of the aromatic ring using HNO_3 (Section 22.1).
Step (5) The carboxylic acid group is converted to the acid chloride, then reacted with ammonia to give the primary amide (Section 18.6). The nitro group is then reduced with hydrogenation (Section 22.1). Note the nitro group is not reduced until after amide formation, because an aromatic amine would react with the acid chloride.
Step (6) An aromatic acid chloride is reacted with the aromatic amine to give the amide (Section 18.6).
Step (7) Anhydrous acid causes dehydration and formation of the purine ring system.

Step (9) Reaction of the sulfonyl chloride with the appropriate amine completes the synthesis and gives Sildefanil.

(d) Show how the reagent for Step 6 can be prepared from salicylic acid (2-hydroxybenzoic acid). Salicylic acid, the starting material for the synthesis of aspirin and a number of other pharmaceuticals, is readily available by the Kolbe carboxylation of phenol (Section 21.4E).

The synthetic problem here is to alkylate the phenolic -OH group while retaining the -COOH group. There are two solutions to this problem. The first solution involves protection the -COOH group as its methyl ester (Fischer esterification, Section 17.7), alkylation the phenolic -OH group using NaOH and ethyl bromide, and finally base-promoted hydrolysis of the methyl ester followed by acidification with aqueous HCl followed to regenerate the carboxyl group (Section 18.4). The second solution is to alkylate both the phenolic -OH and the -COOH groups using ethyl bromide in the presence of sodium carbonate, and then hydrolyze the ester. Both synthetic strategies have been used to alkylate the -OH group of salicylic acid and substituted salicylic acids. For the present synthesis, the acid is converted to the acid chloride with SOCl$_2$ (Section 17.8).

(e) Chlorosufonic acid, ClSO$_3$H, the reagent used in Step 8 is not described in the text. Given what you have studied about other types of electrophilic aromatic substitutions (Section 22.1), propose a mechanism for the reaction in Step 8.

Step 1: *Hydrolysis* This process actually takes several steps.

Step 2: Electrophilic addition (aromatic).

Step 3: Take a proton away.

(f) Propose a structural formula for the reagent used in Step 9 and show how it can be prepared from methylamine and ethylene oxide.

CH₃NH   +   2   [epoxide]
Methylamine   **Ethylene oxide**

(g) Is sildefanil chiral? If so, which of the possible stereoisomers are formed in this synthesis?

No, sildefanil is not chiral.

Problem 23.65 Radiopaque imaging agents are substances administered either orally or intravenously that absorb x-rays more strongly than body material. One of the best known of these is barium sulfate, the key ingredient in the so-called barium cocktail for imaging of the gastrointestinal tract. Among other x-ray contrast media are the so-called triiodoaromatics. You can get some idea of the imaging for which they are used from the following selection of trade names: Angiografin, Gastrografin, Cardiografin, Cholegrafin, Renografin, and Urografin. Following is a synthesis for diatrizoic acid from benzoic acid.

Diatrizoic acid

(a) Provide reagents and experimental conditions for steps (1), (2), (3), and (5).

Steps (1) and (2) are nitration reactions (Section 22.1). The -COOH group is meta directing. Note the Step (2) will require some heating because the ring is relatively deactivated.
Step (3) Reduction of the nitro groups using hydrogenation (Section 22.1).
Step (5) Amide formation using excess acetyl chloride (Section 18.6).

(b) Iodine monochloride, ICl, a black crystalline solid with a mp of 27.2°C and a bp of 97°C, is prepared by mixing equimolar amounts of I₂ and Cl₂. Propose a mechanism for the iodination of 3,5-diaminobenzoic acid by this reagent.

This reaction is a standard electrophilic aromatic substitution reaction (Section 22.1). The iodine monochloride reacts with the I because the electrophilic atom because Cl is the more electronegative of the two.

Problem 23.66 Show how the synthetic scheme developed in problem 23.65 can be modified to synthesize this triiodobenzoic acid x-ray contrast agent.

Only mononitration would be used, and a diacid chloride would be used in the final step. The other steps are the same. However, in the final coupling step two equivalents of the aromatic amine would be reacted with a single equivalent of the diacid chloride.

Iodipamide

Problem 23.67 A diuretic is a compound that causes increased urination and thereby reduces fluid volume in the body. An important use of diuretics in clinical medicine is in the reduction of the fluid build up, particularly in the lungs, that is associated with congestive heart failure. It is also used as an antihypertensive; that is, to reduce blood pressure. Furosemide, an exceptionally potent diuretic, is prescribed under 30 or more trade names, the best known of which is Lasix. The synthesis of furosemide begins with treatment of 2,4-dichlorobenzoic acid with chlorosulfonic acid in a reaction called chlorosulfonation. The product of this reaction is then treated with ammonia, followed by heating with furfurylamine.

2,4-Dichlorobenzoic acid

Furosemide

(a) Propose a synthesis of 2,4-dichlorobenzoic acid from toluene.

2,4-Dichlorobenzoic acid

(b) Propose a mechanism for the chlorosulfonation reaction in Step (1).

See the answer to Problem 23.64 (e) for this mechanism.

(c) Propose a mechanism for Step (3).

This is an nucleophilic aromatic substitution reaction. The aromatic ring is extremely electron deficient by virtue of the four electron withdrawing groups. Note that the last two steps could occur in either order.

Step 1: Make a bond between a nucleophile and an electrophile (aromatic).

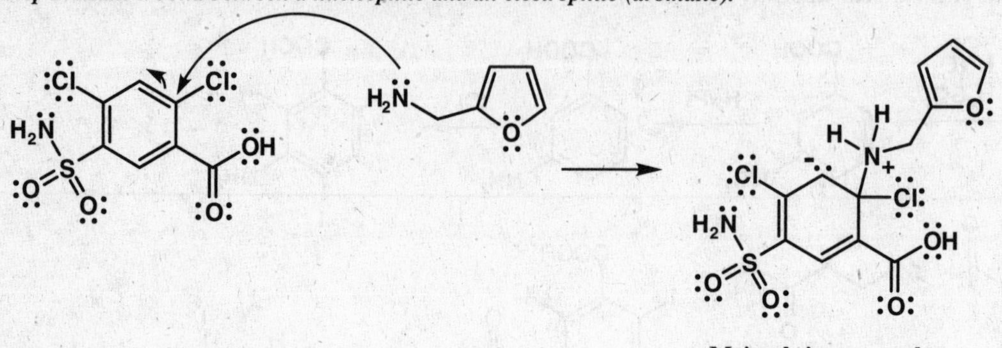

Meisenheimer complex

Step 2: Break a bond to give stable molecules or ions.

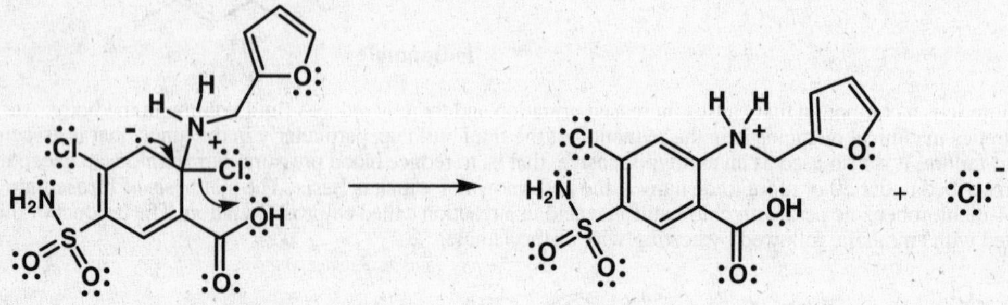

Step 3: Take a proton away.

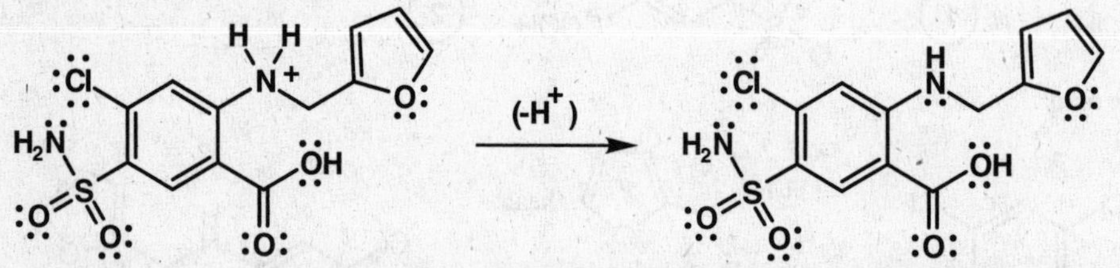

Furosemide

(d) Is furosamide chiral? If so, which of the possible stereoisomers are formed in this synthesis?

No, furosamide is not chiral.

Problem 23.68 Among the newer generation diuretics is bumetanide, prescribed under several trade names, including Bumex and Fordiuran. Following is an outline of a synthesis of this drug.

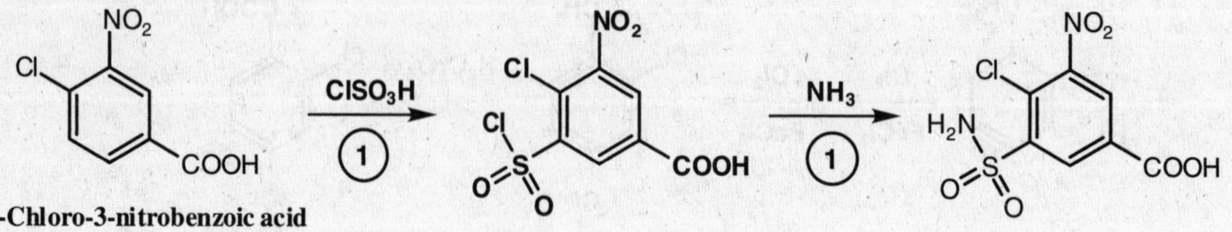

4-Chloro-3-nitrobenzoic acid

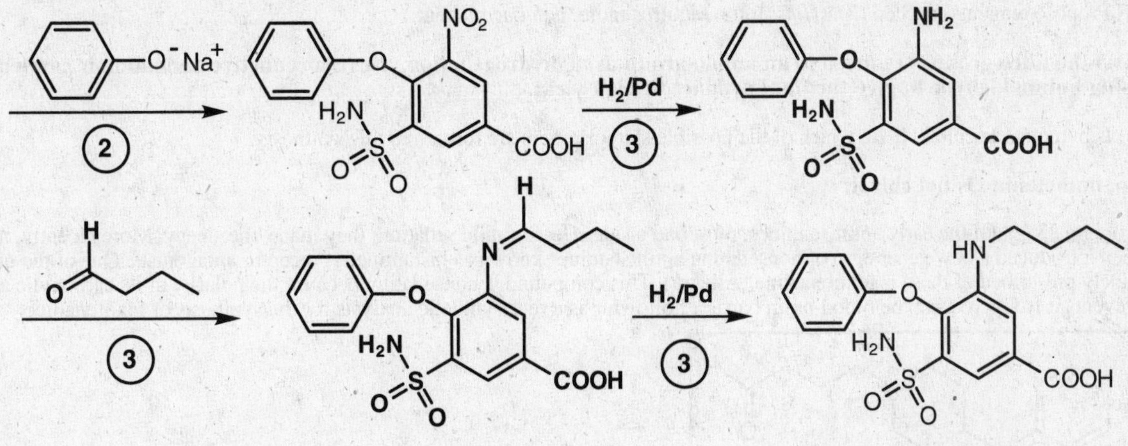

Bumetanide

(a) Propose a synthesis of 4-chloro-3-nitrobenzoic acid from toluene.

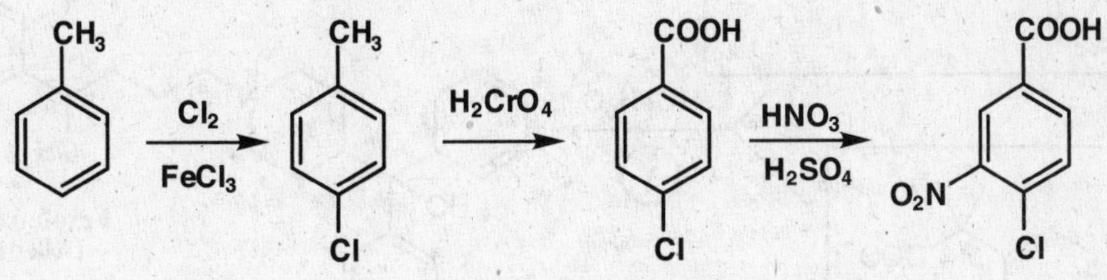

(b) Propose reagents for Step (1). (*Hint:* It requires more than one reagent.)

Chlorosulfonation followed by reaction with ammonia gives the desired sulfonamide.

(c) Propose a mechanism for reaction (2).

This is an nucleophilic aromatic substitution reaction. The aromatic ring is extremely electron deficient by virtue of the four electron withdrawing groups.

Step 1: Make a bond between a nucleophile and an electrophile (aromatic).

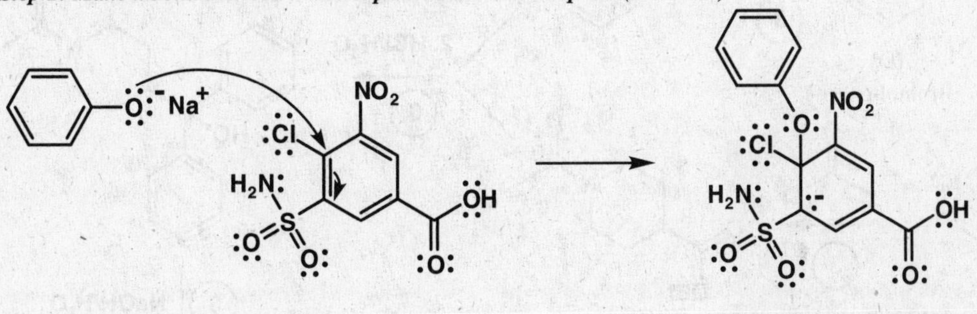

Meisenheimer complex

Step 2: Break a bond to give stable molecules or ions.

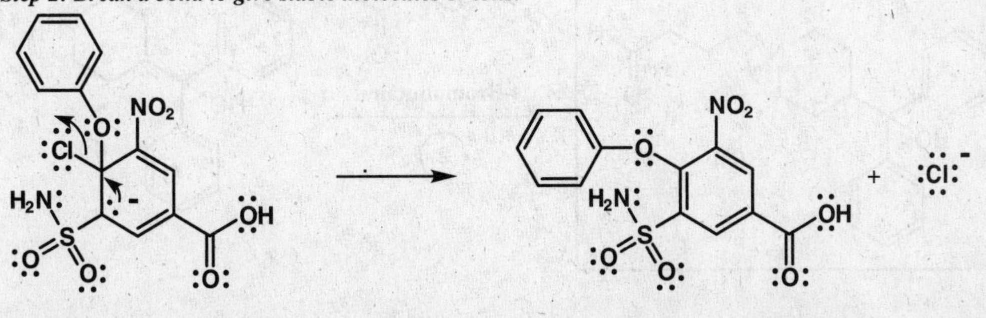

(d) Propose reagents for Step (3). (*Hint:* It too requires more than one reagent.)

First the nitro group is reduced to an amino group using hydrogenation. Next, a reductive amination is carried out using butanaldehyde to give the final product in high yield.

(e) Is bumetanide chiral? If so, which of the possible stereoisomers are formed in this synthesis?

No, bumetanide is not chiral.

<u>Problem 23.69</u> Of the early antihistamines, most had a side effect of mild sedation; they made one sleepy. More recently, there has been introduced a new generation of nonsedating antihistamines known as histamine H_1 receptor antagonists. One of the most widely prescribed of these is fexofenadine (Allegra). This compound is nonsedating because the polarity of its carboxylic anion prevents it from crossing the blood-brain barrier. Following is a retrosynthetic analysis for the synthesis of fexofenadine.

(4-Bromophenyl)ethanenitrile
(A)

(*Note:* The organolithium reagent C cannot be made directly from B because the presence of the carboxyl group in B would lead to intermolecular destruction of the reagent by an acid-base reaction. In practice, B is first converted to its sodium salt by treatment with sodium hydride, NaH, and then the organolithium reagent is prepared.)

(a) Given this retrosynthetic analysis, propose a synthesis for fexofenadine from the four named starting materials.

Step (1) The organolithium reagent reacts with the aldehyde function followed by an aqueous workup to give the racemic alcohol product (fexofenadine) (Section 15.1).

Step (2) The 4-bromobutanal reacts with the secondary amine of the piperidine group in an S_N2 reaction to give (H) (Section 9.3).

Step (3) This conversion requires multiple steps. The piperidine amino group must be protected as the acetamide in preparation for the Grignard reaction (Section 18.9). If left unprotected, it would become deprotonated by the basic Grignard reagent (Section 15.1). The Grignard reaction is made from bromobenzene by reaction with Mg° in ether. The Grignard reagent adds twice to the ester function to give the tertiary alcohol (Section 18.9). The acetamide group is removed in aqueous base (Section 18.4). Note acid is not used here as the tertiary alcohol is sensitive to acid (β-elimination).

Step (4) The organolithium reagent is prepared by first removing the carboxylic acid proton with NaH, followed by reaction with Li° to give (C).

Step (5) 4-(bromophenyl)ethanamide is alkylated with CH_3I two times at the position α to the nitrile through formation of an enolate using NaH or other strong base (Section 19.1). The nitrile is hydrolyzed in aqueous acid to give the carboxylic acid (Section 18.4).

(b) Is fexofenadine chiral? If so, which of the possible stereoisomers are formed in this synthesis?

Fexofenadine has one chiral center, so a racemic mixture of enantiomers is produced in this synthesis.

Enantiomers of fexofenadine (Allegra)

Problem 23.70 Sotalol is a β-adrenergic blocker used to treat certain types of cardiac arrhythmias. Its hydrochloride salt is marketed under several trade names, including Betapace. Following is a retrosynthetic analysis.

(a) Propose a synthesis for sotalol from aniline.

Step (1) Reduction of the ketone (A) using NaBH₄ gives the racemic alcohol, sotalol (Section 16.11).

Step (2) Reaction of (B) with isopropyl amine gives intermediate (A) through an S_N2 reaction (Section 9.3).

Step (3) This transformation can be accomplished in high yield using a combination of a Friedel-Crafts acylation with acetyl chloride (Section 22.1) followed by α-halogenation in acid using Br₂ (Section 16.12).

Step (4) Reaction of aniline with mesyl chloride (MeSO₂Cl) gives (C).

(b) Is sotalol chiral? If so, which if the possible stereoisomers are formed in this synthesis?

Sotalol has one chiral center, so a racemic mixture of enantiomers is produced in this synthesis.

Enantiomers of sotalol

CHAPTER 24
Solutions to the Problems

Problem 24.1 Show how you might prepare each compound by a Heck reaction using methyl 2-propenoate as the starting alkene.

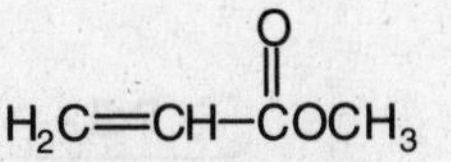

Methyl 2-propenoate
(Methyl acrylate)

(a)

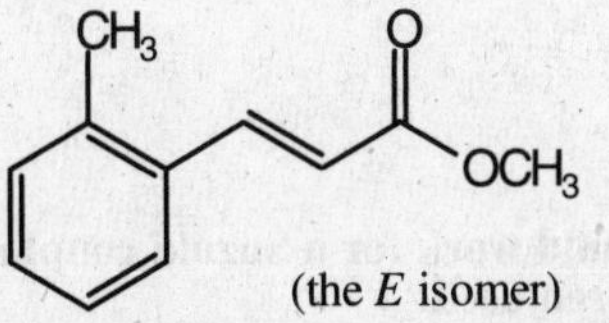

(the *E* isomer)

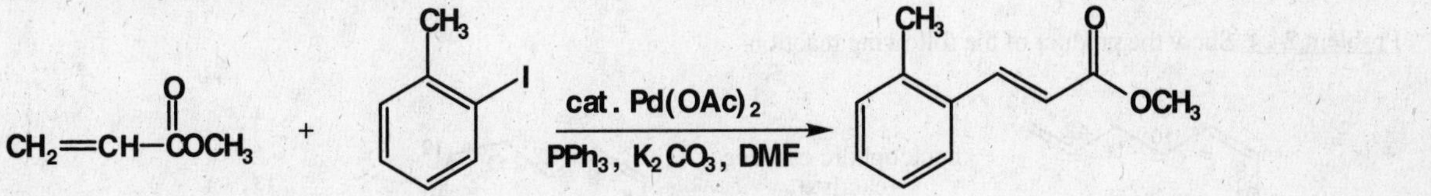

(b)

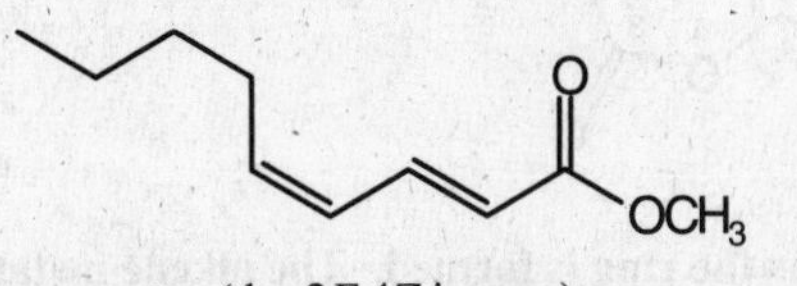

(the 2*E*,4*Z* isomer)

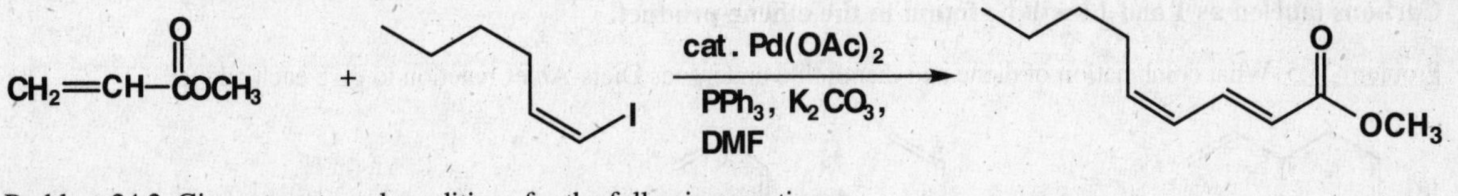

Problem 24.2 Give reagents and conditions for the following reaction.

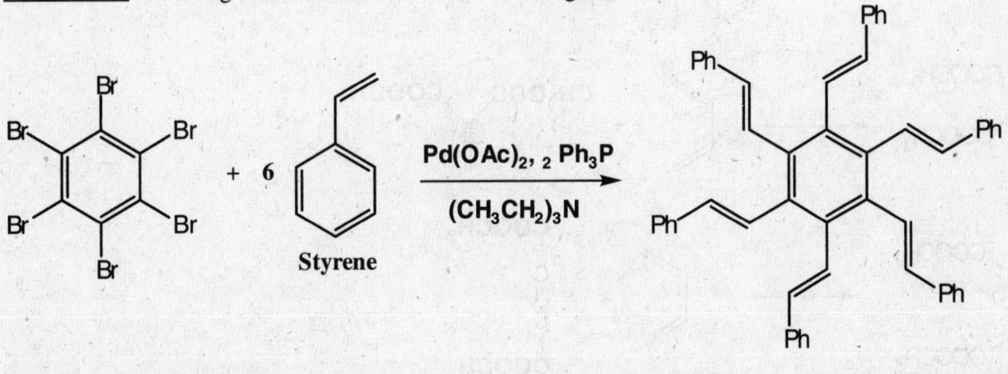

Styrene

Reaction of hexabromobenzene with 6 equivalents of styrene under the normal Heck conditions gives the desired compound.

Problem 24.3 Show how the following compound can be prepared from starting materials containing eight carbons or less.

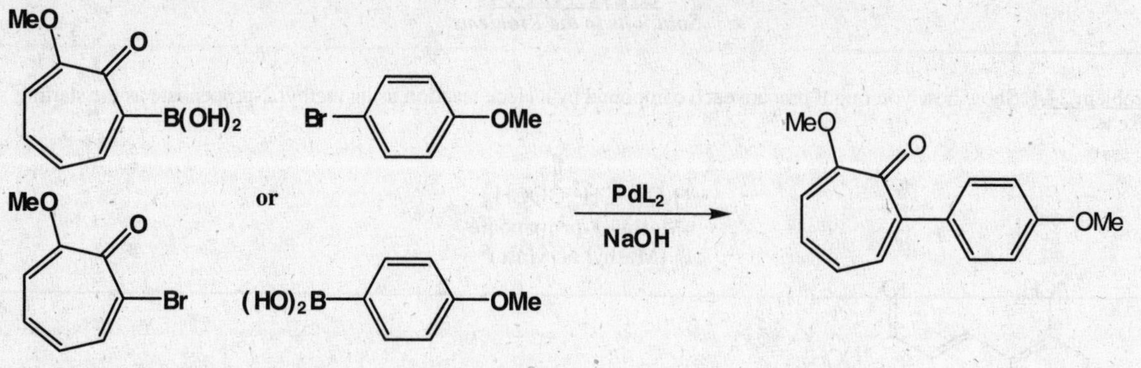

Because both pieces are aryl, there are two combinations of reagents that would work for a Suzuki coupling to make the biaryl product. Note that a borane or boric ester could also have been used.

Problem 24.4 Show the product of the following reaction.

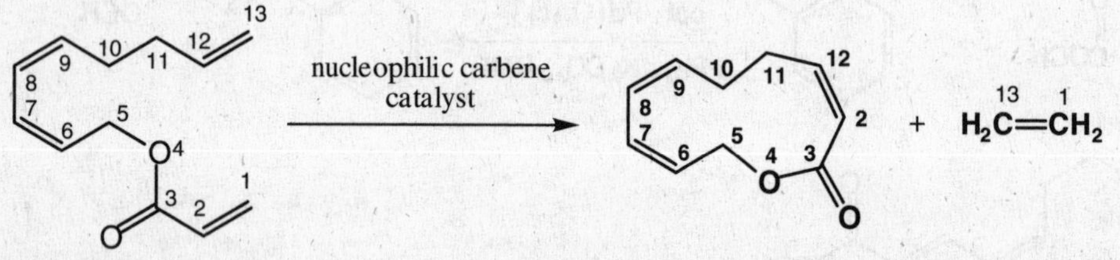

Numbers were added to help you keep track of the different atoms as the ring is formed. The alkene metathesis reaction will generate an eleven-membered ring by forming a bond between carbon atoms labeled as 2 and 12. Carbons labeled as 1 and 13 will be found in the ethene product.

Problem 24.5 What combination of diene and dienophile undergoes Diels-Alder reaction to give each adduct.

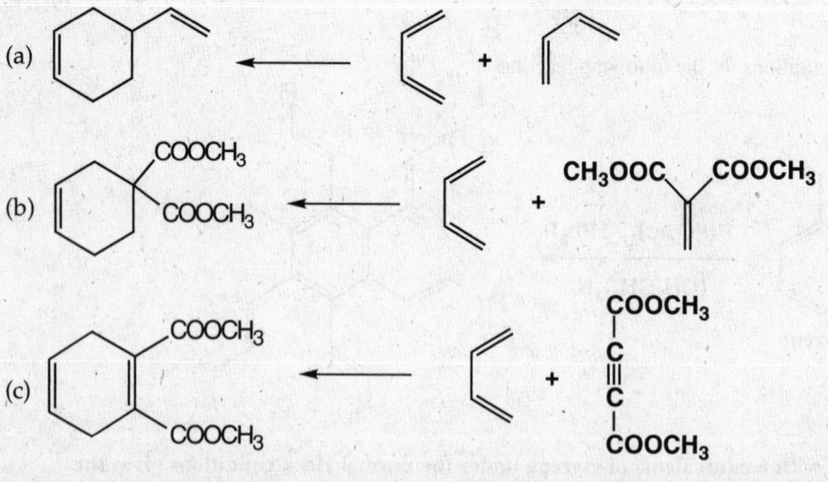

In each part of this problem, the diene is 1,3-butadiene. In (a), the dienophile is one of the double bonds of a second molecule of 1,3-butadiene. In part (b), the dienophile is a 1,1-disubstituted alkene and in part (c), it is a disubstituted alkyne.

<u>Problem 24.6</u> Which molecules can function as dienes in Diels-Alder reactions? Explain your reasoning.

(a) 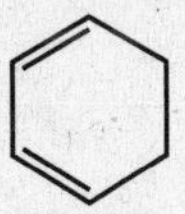   (b) 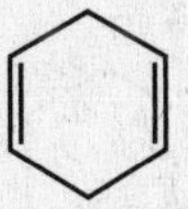   (c)

To function as a diene, the double bonds must be conjugated and able to assume an s-*cis* conformation. Both (a) 1,3-cyclohexadiene and (c) 1,3-cyclopentadiene can function as dienes. The double bonds in (b) 1,4-cyclohexadiene are not conjugated and, therefore, this molecule cannot function as a diene.

<u>Problem 24.7</u> What diene and dienophile might you use to prepare the following racemic Diels-Alder adduct?

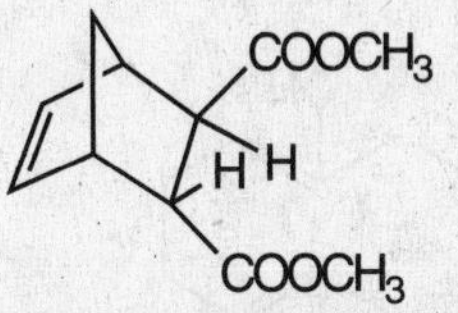

Use cyclopentadiene and the *E*-alkene. Note that two new chiral centers are formed in the reaction so the product is actually a racemic mixture of the two enantiomers having a *trans* arrangement of the ester groups.

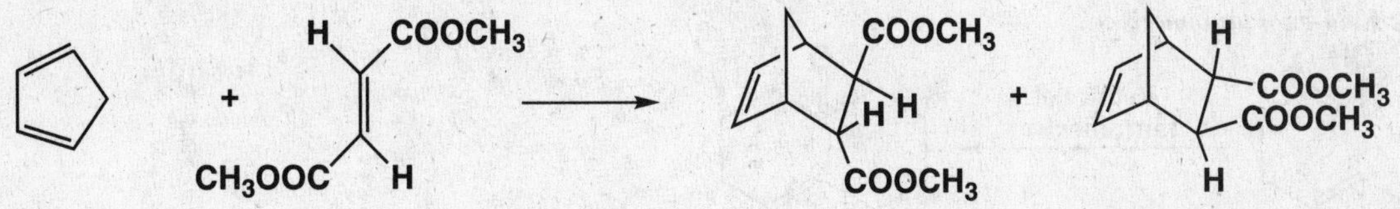

<u>Problem 24.8</u> Show how to synthesize allyl phenyl ether and 2-butenyl phenyl ether from phenol and appropriate alkenyl halides.

Prepare each by a Williamson ether synthesis. Treat phenol with sodium hydroxide to form sodium phenoxide followed by treatment with the appropriate alkenyl chloride.

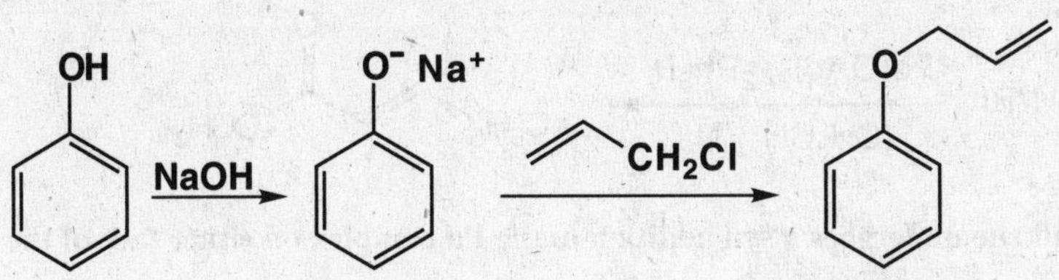

Allyl phenyl ether

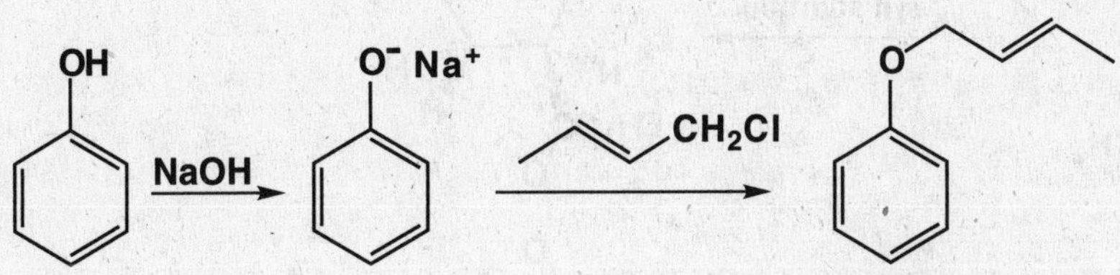

2-Butenyl phenyl ether

Problem 24.9 Propose a mechanism for the following Cope rearrangement.

Cope rearrangement gives an enol that then undergoes keto-enol tautomerism (Section 15.11B) to give the observed product.

Step 1: Pericyclic reaction.

Step 2: Keto-enol tautomerism.

The Heck Reaction
Problem 24.10 As has been demonstrated in the text, when the starting alkene has CH_2 as its terminal group, the Heck reaction is highly stereoselective for formation of the *E* isomer as illustrated in this example. Here, the benzene ring is abbreviated C_6H_5-. Show how the mechanism proposed in the text allows you to account for this stereoselectivity.

In the second step of the mechanism, the alkene undergoes a syn addition to the Pd complex on either face of the alkene.

In the third step, the central bond of each enantiomer rotates to minimize torsional strain between the large -$COOCH_3$ and -C_6H_5 groups.

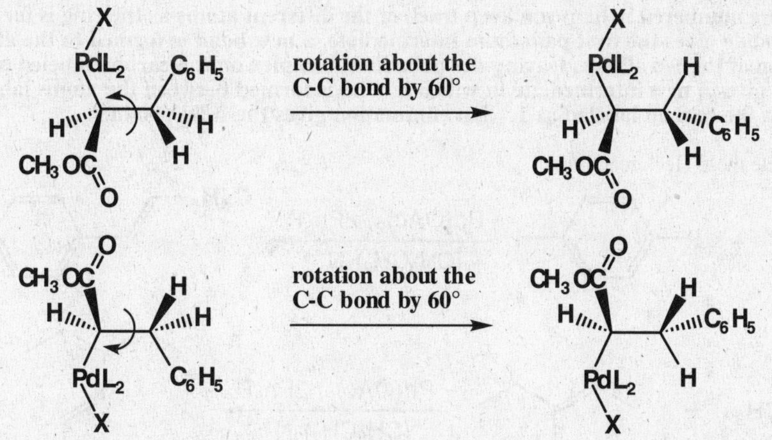

With the bond rotated in this conformation, the *E* product will be produced in the syn elimination step from both enantiomers. The key here is that the syn elimination occurs preferentially from the conformation that places the bulky -COOCH$_3$ and -C$_6$H$_5$ groups far apart from each other, because this is the lower energy conformation.

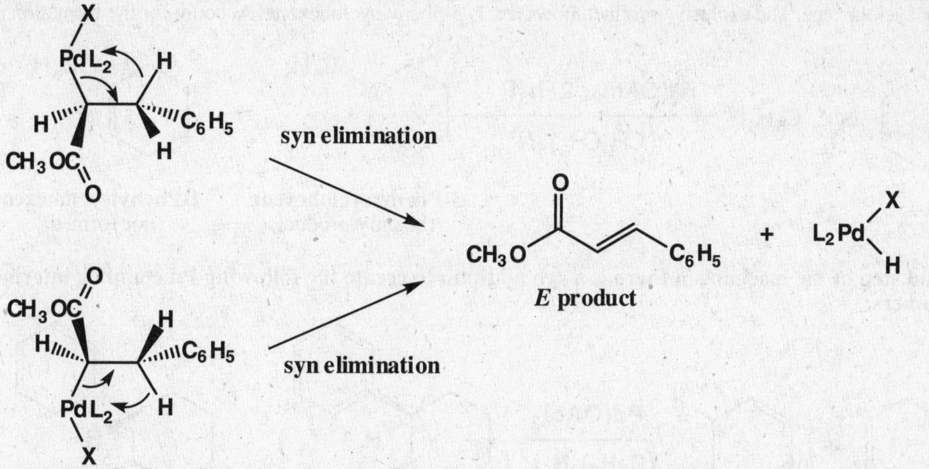

Problem 24.11 The following reaction involves two sequential Heck reactions. Draw structural formulas for each organopalladium intermediate formed in the sequence and show how the final product is formed. Note from the molecular formula given under each structural formula that this conversion corresponds to a loss of H and I from the starting material. Acetonitrile, CH$_3$CN, is the solvent.

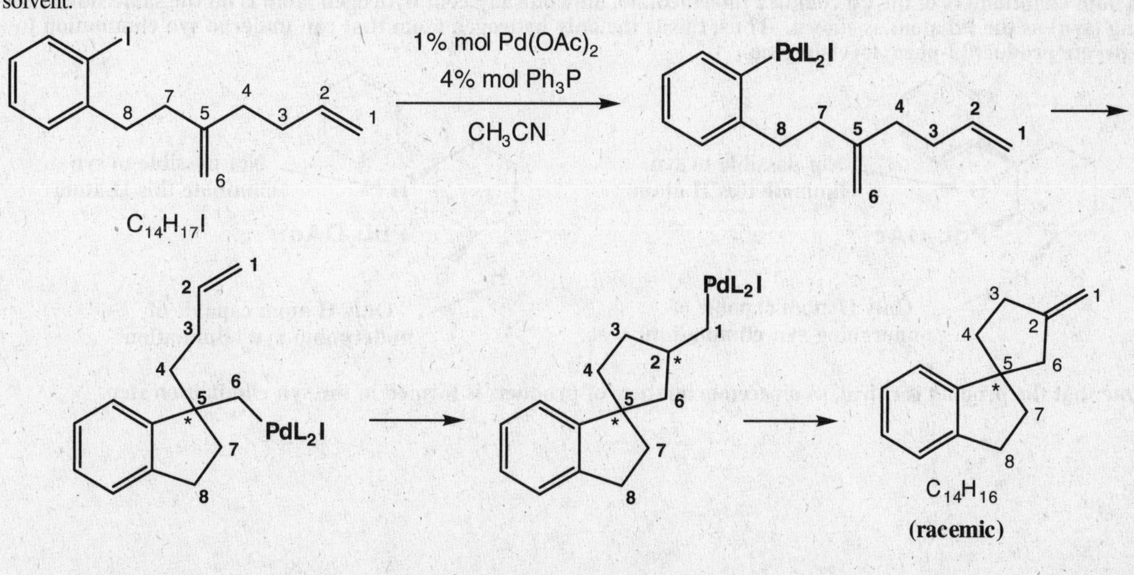

Key carbon atoms were numbered to help you keep track of the different atoms as the ring is formed. Oxidative addition of the aryl iodide gives the first palladium intermediate, a new bond is formed to the atom labeled as 5 through a syn addition of the 5-6 alkene, leaving the palladium complex on the carbon labeled as 6. Reaction with the other alkene gives a new intermediate in which a bond is formed between the atoms labeled as 6 and 2, with the palladium on the carbon labeled as 1. Syn elimination gives the final product.

Problem 24.12 Complete these Heck reactions.

(a) $2 C_6H_5CH=CH_2$ + I—⟨ ⟩—I →[$Pd(OAc)_2, 2Ph_3P$][$(CH_3CH_2)_3N$] [product structure]

(b) $CH_2=CH-\overset{O}{\overset{\|}{C}}OCH_3$ + [alkene structure] →[$Pd(OAc)_2, 2Ph_3P$][$(CH_3CH_2)_3N$] [product structure]

Problem 24.13 Treatment of cyclohexene with iodobenzene under the conditions of the Heck reaction might be expected to give 1-phenylcyclohexene. The exclusive product, however, is 3-phenylcyclohexene. Account for the formation of this product.

[reaction scheme]

+ C_6H_5I →[$Pd(OAc)_2, 2Ph_3P$][$(CH_3CH_2)_3N$] [3-phenylcyclohexene structure] + [1-phenylcyclohexene structure]

3-Phenylcyclohexene   **1-Phenylcyclohexene**
(the only product)   (not formed)

In the second step of the mechanism there is a syn addition to create the following Pd complex intermediate as two enantiomers.

[reaction scheme]

+ [iodobenzene] →[$Pd(OAc)_2$][$(C_2H_5)_3N$] [Pd complex structure with PdL_2OAc] + [Pd complex enantiomer structure with PdL_2OAc]

In both enantiomers of the Pd complex intermediate, only one adjacent hydrogen atom is on the same side of the ring (syn) as the Pd atom as shown. Thus, this is the only hydrogen atom that can undergo syn elimination to generate product (3-phenylcyclohexene).

[two structures with annotations]

Not possible to syn eliminate this H atom   Not possible to syn eliminate this H atom

PdL_2OAc PdL_2OAc

Only H atom capable of undergoing syn elimination   Only H atom capable of undergoing syn elimination

Note that the product is chiral, so a racemic mixture of products is formed in the syn elimination step.

syn elimination

syn elimination

**Formed as a
racemic mixture**

<u>Product 24.14</u> Account for the formation of the product and for the *cis* stereochemistry of its ring junction. (The function of silver carbonate is to enhance the rate of reaction.)

COOMe

$\xrightarrow{\text{Pd(OAc)}_2,\ \text{PPh}_3}{\text{Ag}_2\text{CO}_3,\ \text{CH}_3\text{CN}}$

COOMe

H

86%

The Pd complex could add to either side of the double bond in the syn addition step of the mechanism. However, in this case, the aryl iodide is primarily oriented so that it is adjacent to the bottom face of the reactive double bond. Here "bottom" refers to the side opposite the -COOMe group.

Reactive double bond

COOMe

Top face

Bottom face

**Oriented toward bottom
face of double bond**

Because of the "bottom face" orientation, the syn addition step occurs primarily from the bottom face, leading to the observed *cis* ring junction.

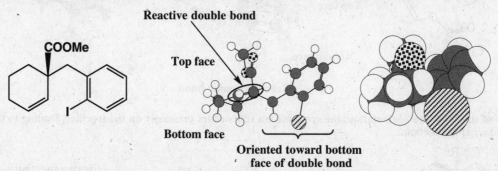

COOMe

$\xrightarrow{\text{Pd(OAc)}_2,\ \text{PPh}_3}{\text{Ag}_2\text{CO}_3,\ \text{CH}_3\text{CN}}$

COOMe

H

H

H

PdL₂OAc

Cis ring junction
derived from syn
addition to bottom
face of double bond

Syn elimination of this complex gives the predominant product.

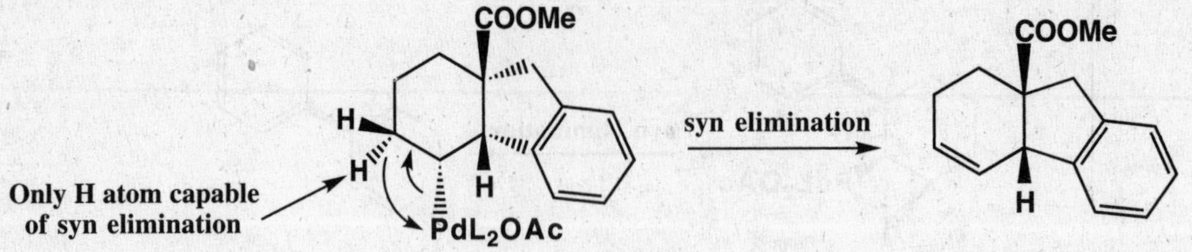

Only H atom capable of syn elimination

syn elimination

<u>Problem 24.15</u> Account for the formation of the following product, including the *cis* stereochemistry at the ring junction.

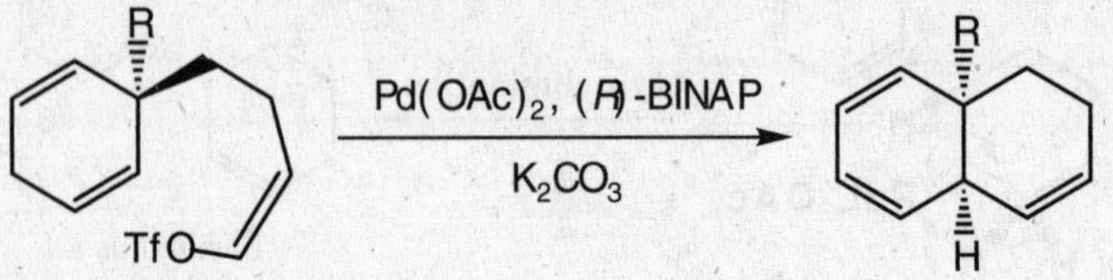

$$Pd(OAc)_2, (R)\text{-}BINAP$$
$$K_2CO_3$$

Tf is an abbreviation for the trifluoromethanesulfonate group, $CF_3SO_2^-$, which can be used in the place of a halogen atom in a Heck reaction. The Pd complex could add to either side of the double bond in the syn addition step of the mechanism. However, in this case, the reactive TfO group is primarily oriented so that is adjacent to the top face of the reactive double bond. Here "top" refers to the side opposite the R group. Note that in the following molecular models a methyl group was used to illustrate the location of the "R" group in the problem.

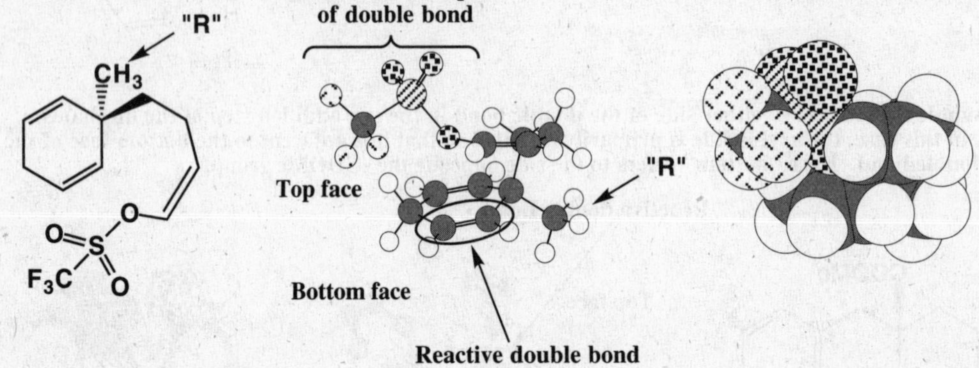

Reactive TfO group oriented toward top face of double bond

"R"

CH₃

Top face

Bottom face

Reactive double bond

"R"

Because of the "top face" orientation, the syn addition step occurs primarily on the top face, leading to the observed *cis* ring junction.

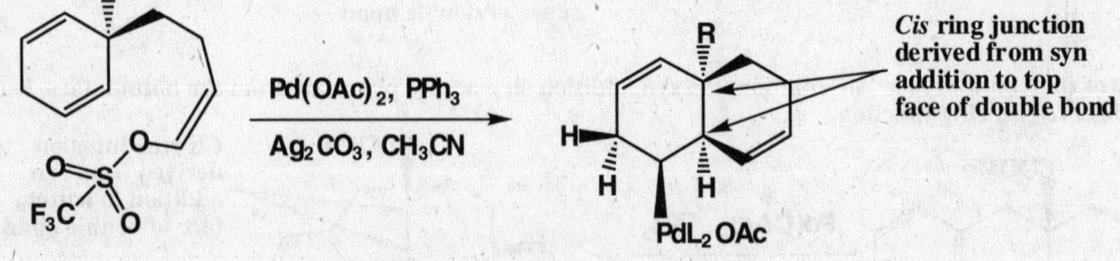

$$Pd(OAc)_2, PPh_3$$
$$Ag_2CO_3, CH_3CN$$

Cis ring junction derived from syn addition to top face of double bond

Syn elimination of this complex gives the predominant product.

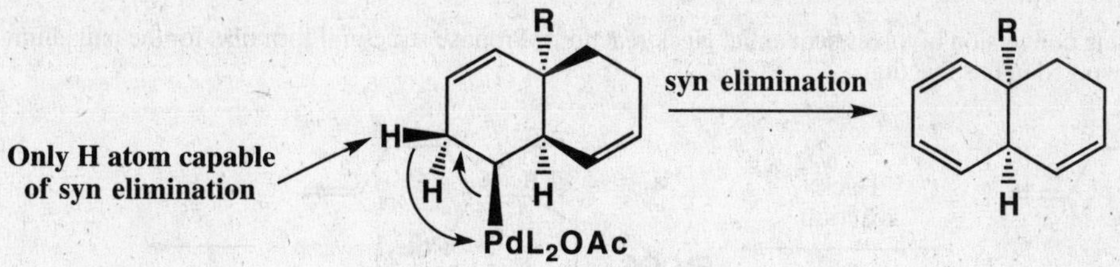

Only H atom capable of syn elimination → **syn elimination** →

<u>Problem 24.16</u> The aryl diene undergoes sequential Heck reactions to give a product with the molecular formula $C_{15}H_{18}$. Propose a structural formula for this product.

Key carbon atoms have been numbered to help you keep track of the atoms taking part in ring forming reactions.

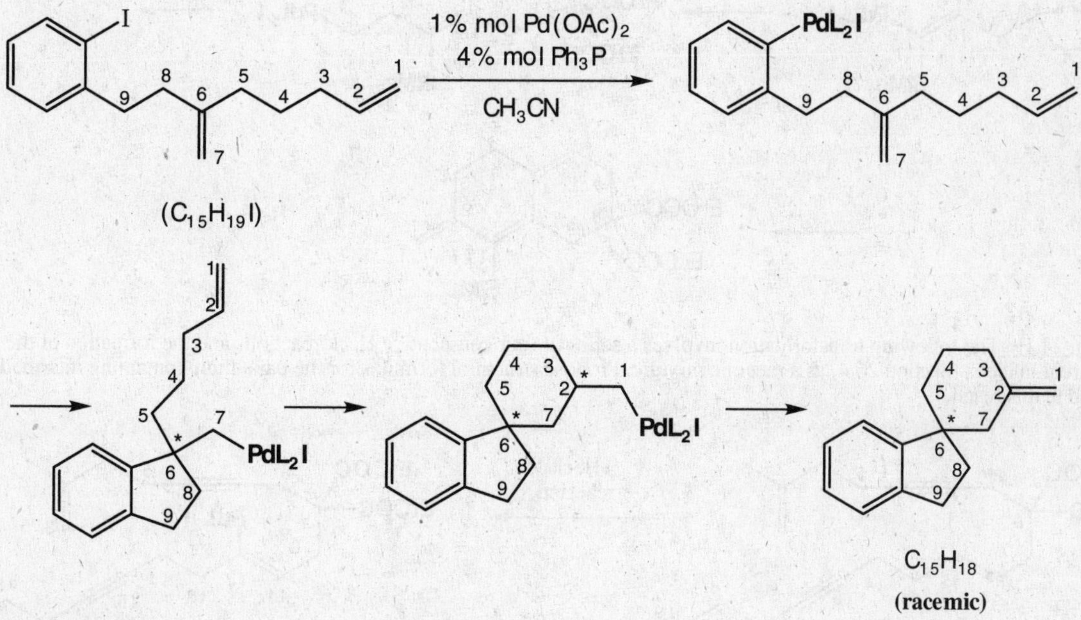

1% mol $Pd(OAc)_2$
4% mol Ph_3P
CH_3CN

$(C_{15}H_{19}I)$

PdL_2I

$C_{15}H_{18}$
(racemic)

<u>Problem 24.17</u> Heck reactions take place with alkynes as well as alkenes. The following conversion involves an intramolecular Heck reaction followed by an intermolecular Heck. Propose structural formulas for the palladium-containing intermediates involved in this reaction.

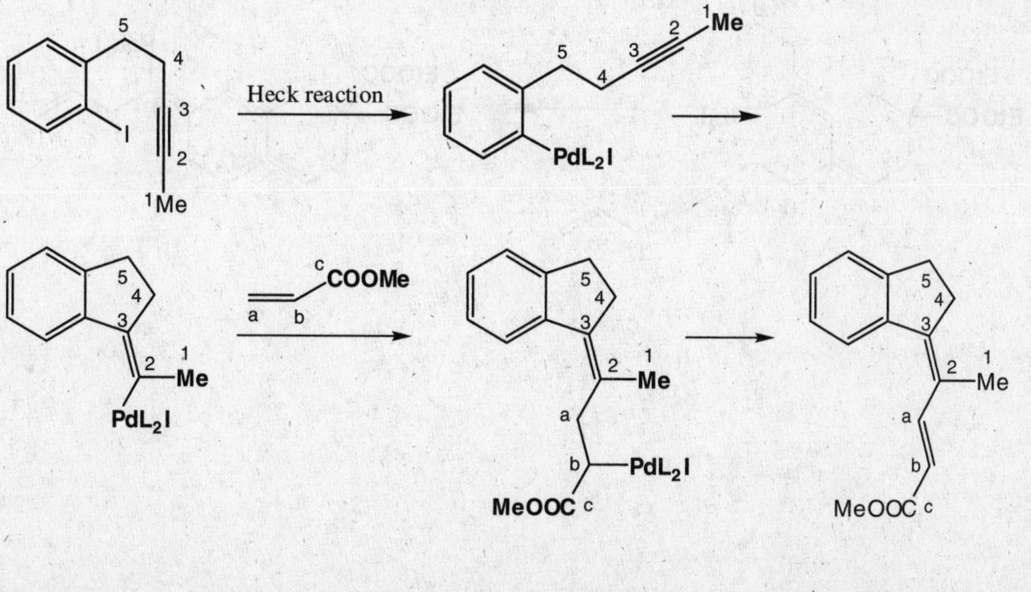

Heck reaction

PdL_2I

PdL_2I

$COOMe$

PdL_2I

$MeOOC$ $MeOOC$

Problem 24.18 The following conversion involves sequential Heck reactions. Propose structural formulas for the palladium-containing intermediates involved in this reaction.

Problem 24.19 The following transformation involves a series of four consecutive Heck reactions and the formation of the four-ring steroid nucleus (Section 26.4) as a racemic mixture. Propose structural formulas for the palladium-containing intermediates involved in this reaction.

Problem 24.20 Suggest reagents and the other fragment that could be used to carry out the indicated conversion.

In this Suzuki coupling reaction, the following aryl iodide is used to construct the desired biaryl. Note that the aryl chloride group does not interfere, because chlorides are generally unreactive in Suzuki coupling reactions. The base could be a variety of things including carbonate or trialkylamine. Note that an aryl bromide could be used in place of the aryl iodide.

Problem 24.21 Show how the following compound could be prepared by a Suzuki reaction.

Either combination of aryl boric acid and aryl halide could be used to prepare the desired compound using a Suzuki reaction.

Problem 24.22 Show the sequence of Heck reactions by which the following conversion takes place. Note from the molecular formula given under each structural formula that this conversion corresponds to a loss of H and I from the starting material.

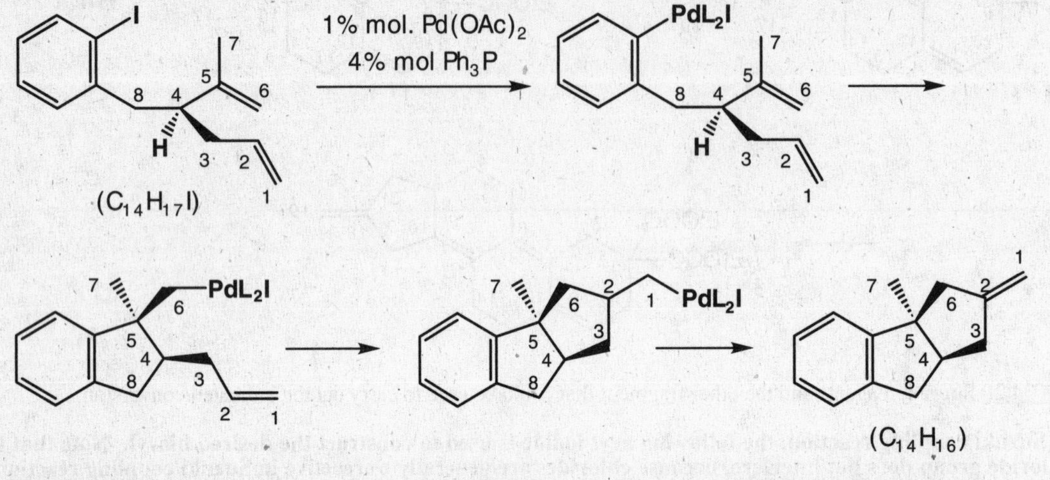

In the following sequence, note how the stereochemistry at the chiral center labeled as 4 controls the stereochemistry at the carbon labeled as 5. A common theme in modern organic synthesis is that the presence of one chiral center is used to control the stereochemistry of chiral centers produced in later synthetic steps. Although a complete explanation for the stereospecificity of these reactions can be quite complex, one contributing factor is likely to be the conformational preference of the starting material. Rotation about the carbon-4 carbon-5 bond yields two conformational isomers that would give rise to the two stereoisomer products as shown. The conformer on the right is the more stable, and this leads to the observed predominant product stereoisomer.

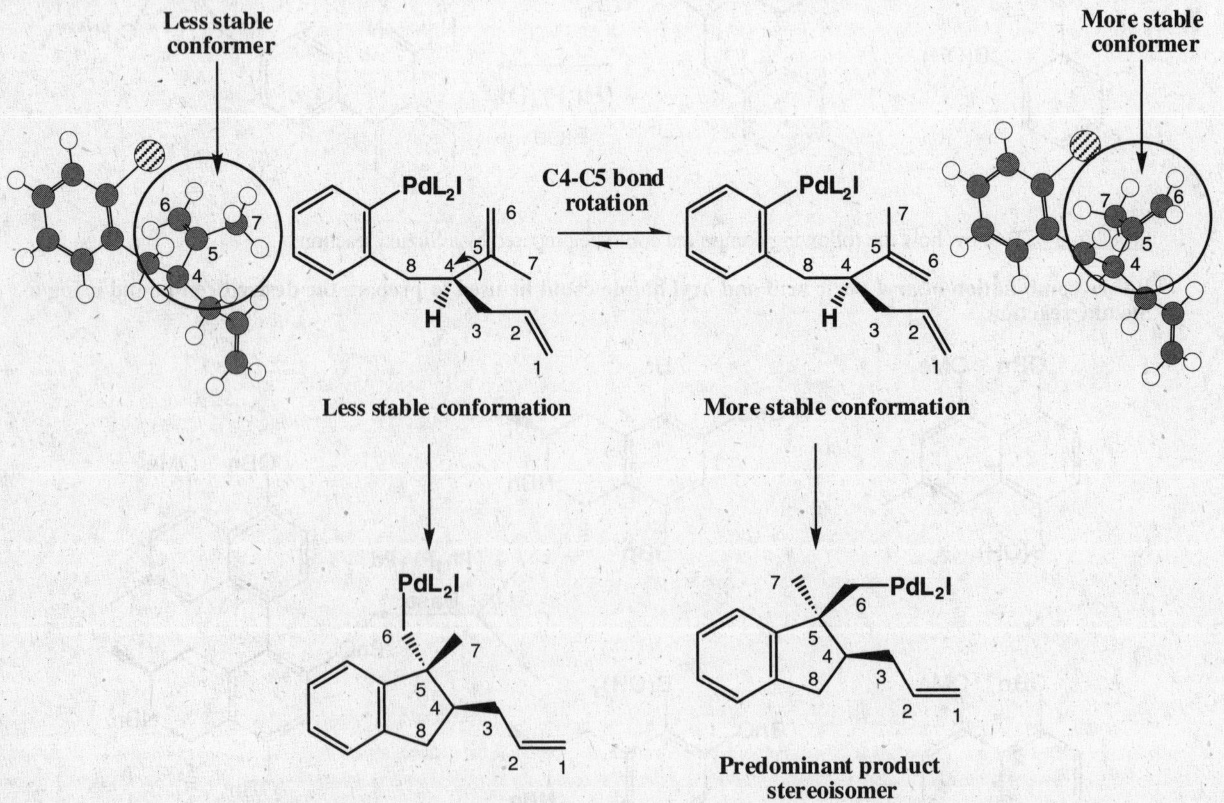

Less stable conformer

More stable conformer

C4-C5 bond rotation

Less stable conformation

More stable conformation

Predominant product stereoisomer

Problem 24. 23 The cyclic ester (lactone) Exaltolide has a musk-like fragrance and is used as a fixative in perfumery. Show how this compound could be synthesized from the indicated starting material. Give the structure of R.

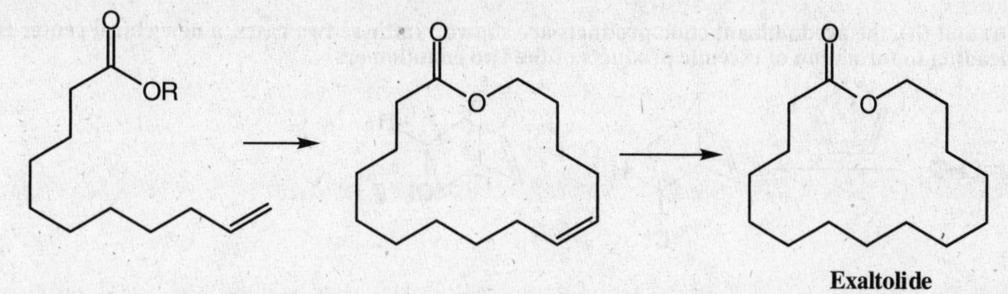

Exaltolide

Like many syntheses, the key step is the carbon-carbon bond-forming step. Here propose an alkene metathesis reaction, dictating the ester has the structure shown. The numbers and letters were added to the structures to help you keep track of the atoms during the ring forming process.

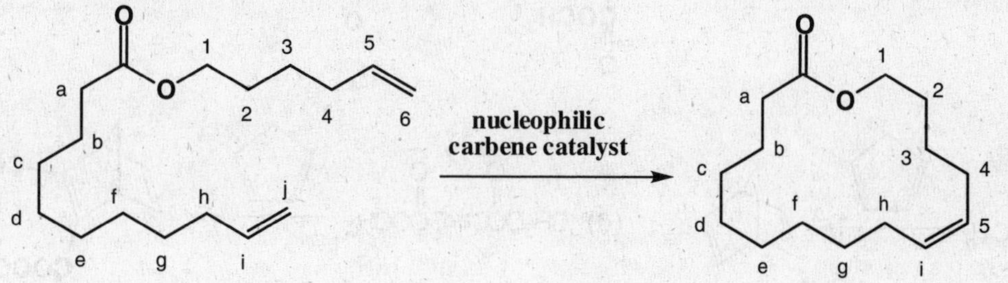

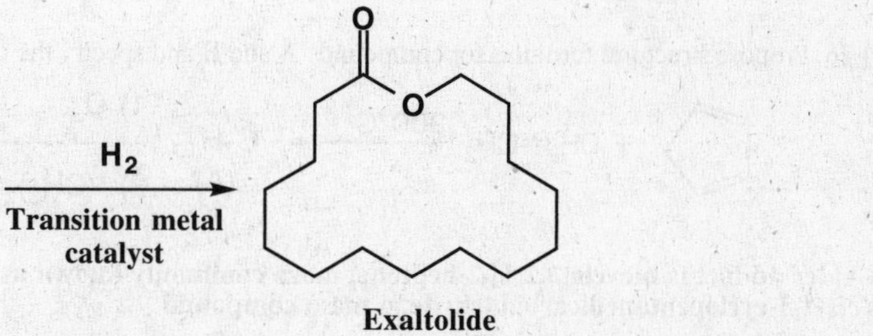

nucleophilic
carbene catalyst

H_2

**Transition metal
catalyst**

Exaltolide

Problem 24.24 Predict the product of each alkene metathesis reaction using a Ru-nucleophilic carbene catalyst.

(a)   5 mole % Ru catalyst
CH_2Cl_2, 40°C, 30 min   + $CH_2=CH_2$

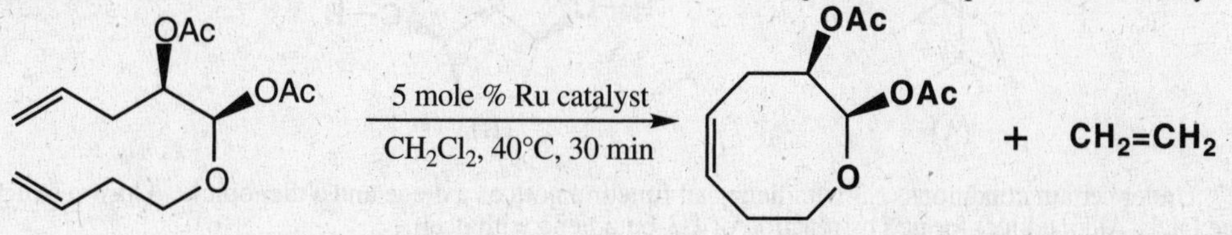

(b)   5 mole % Ru catalyst
CH_2Cl_2, 40°C, 30 min   + $CH_2=CH_2$

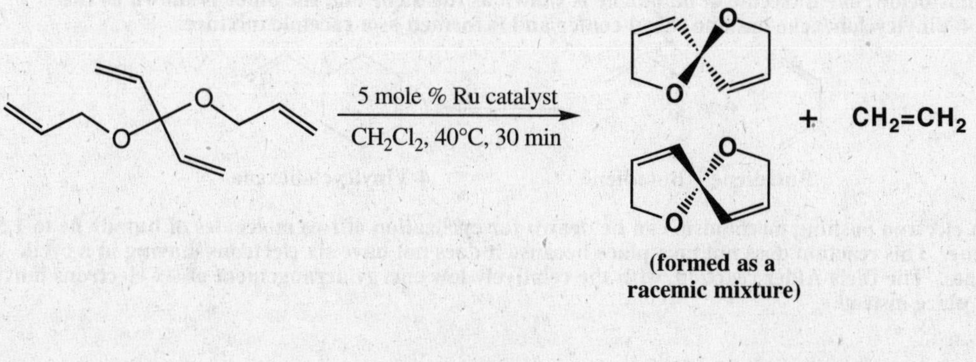

**(formed as a
racemic mixture)**

Diels-Alder Reaction

<u>Problem 24.25</u> Draw structural formulas for the products of reaction of cyclopentadiene with each dienophile.

In both (a) and (b), the predominant endo products are shown. In these two cases, a new chiral center is created, leading to formation of racemic products of the two enantiomers.

(a) $CH_2=CHCl$

(b) $CH_2=CHCOCH_3$

(c) $HC\equiv CH$

(d) $CH_3OCC\equiv CCOCH_3$

<u>Problem 24.26</u> Propose structural formulas for compounds A and B and specify the configuration of compound B.

$+ CH_2=CH_2 \xrightarrow{200°C} C_7H_{10} \xrightarrow[\text{2) } (CH_3)_2S]{\text{1) } O_3} C_7H_{10}O_2$

(A) (B)

The Diels-Alder adduct is bicyclo[2.2.1]-2-heptene, more commonly known as norbornene. The oxidation product is *cis*-1,3-cyclopentanedicarbaldehyde, a meso compound.

(A)

(B)

<u>Problem 24.27</u> Under certain conditions, 1,3-butadiene can function both as a diene and a dienophile. Draw a structural formula for the Diels-Alder adduct formed by reaction of 1,3-butadiene with itself.

In the formulas below, one molecule of butadiene is shown as the diene and the other is shown as the dienophile. 4-Vinylcyclohexene has one chiral center and is formed as a racemic mixture.

Butadiene   Butadiene   4-Vinylcyclohexene

Note that an electron pushing mechanism can be drawn for cyclization of two molecules of butadiene to 1,5-cyclooctadiene. This reaction does not take place because it does not have six electrons moving in a cyclic *trans*ition state. The Diels-Alder reaction, with the relatively low energy arrangement of six electrons moving in a ring takes place instead.

1,5-Cyclooctadiene

<u>Problem 24.28</u> 1,3-Butadiene is a gas at room temperature and a requires gas-handling apparatus to use in a Diels-Alder reaction. Butadiene sulfone is a convenient substitute for gaseous 1,3-butadiene. This sulfone is a solid at room temperature (mp 66°C) and, when heated above its boiling point of 110°C, decomposes by a reverse Diels-Alder reaction to give s-*cis*-1,3-butadiene and sulfur dioxide. Draw a Lewis structure for butadiene sulfone and SO_2; then, and show by curved arrows the path of this reaction, which resembles a reverse Diels-Alder reaction.

Butadiene sulfone   **Butadiene**   **Sulfur dioxide**

The electron flow in this decomposition is the reverse of that observed in a Diels-Alder reaction.

Butadiene sulfone   **Butadiene**   **Sulfur dioxide**

<u>Problem 24.29</u> The following triene undergoes an intramolecular Diels-Alder reaction to give the product shown. Show how the carbon skeleton of the triene must be coiled to give this product, and show by curved arrows the redistribution of electron pairs that takes place to give the product.

Locate the position of the carbon-carbon double bond in the product and realize that the two carbons of this double bond were carbons 2 and 3 of the conjugated diene. The other two carbon atoms making up this six-membered ring were from the dienophile. Note that two new chiral centers are formed in the product, and because the starting trienone is not chiral, the product is actually a racemic mixture of the two *trans* isomers. The *trans* geometry of the ring fusion (atoms "4" and "9") is a consequence of the concerted nature of the reaction.

this part was
the diene

the dienophile
was here

the triene was coiled in this
manner so as to align the
diene and dienophile properly

Problem 24.30 The following trienone undergoes an intramolecular Diels-Alder reaction to give a bicyclic product. Propose a structural formula for the product. Account for the observation that the Diels-Alder reaction given in this problem takes place under milder conditions (at lower temperature) than the analogous Diels-Alder reaction shown in problem 24.29?

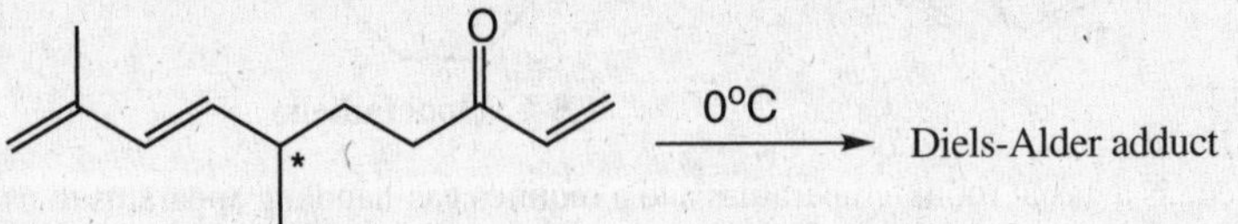

0°C → Diels-Alder adduct

Follow the arrangement of diene and dienophile illustrated in the previous problem.

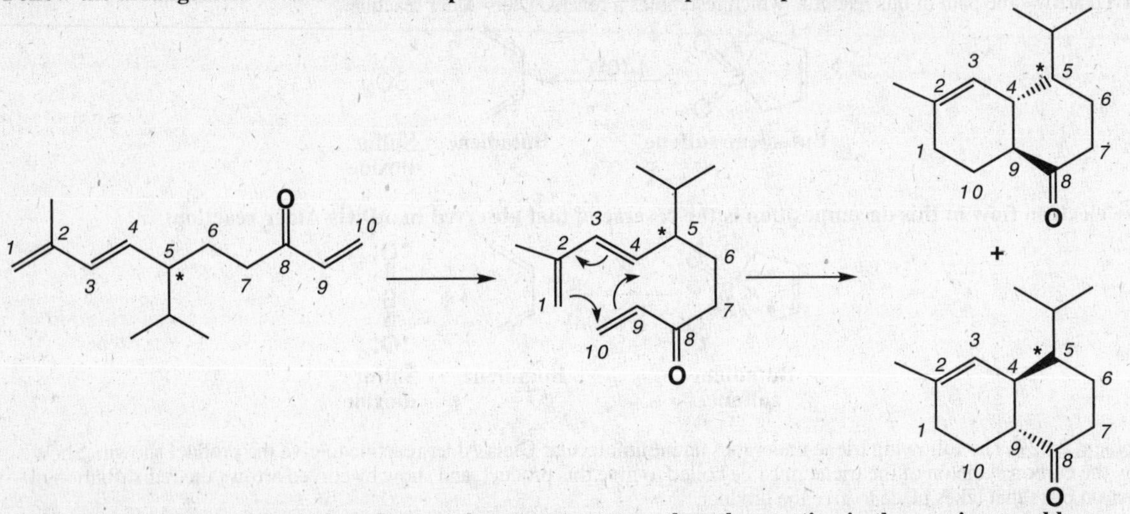

The reaction in this problem takes place at a lower temperature than the reaction in the previous problem, because the dienophile portion of this molecule is more electron deficient by virtue of the carbonyl group. An electron deficient dienophile is a better dienophile, so the reaction is accelerated. Note how the starting material is chiral and presumably racemic, while two new chiral centers are created in the product. The concerted nature of the reaction will lead to a *trans* geometry at the ring fusion (atoms "4" and "9"), and the third chiral center (atom "5") will lead to a mixture of stereoisomers. If the starting material is racemic, then the product mixture will be racemic as well.

Problem 24.31 The following compound undergoes an intramolecular Diels-Alder reaction to give a bicyclic product. Propose a structural formula for the product.

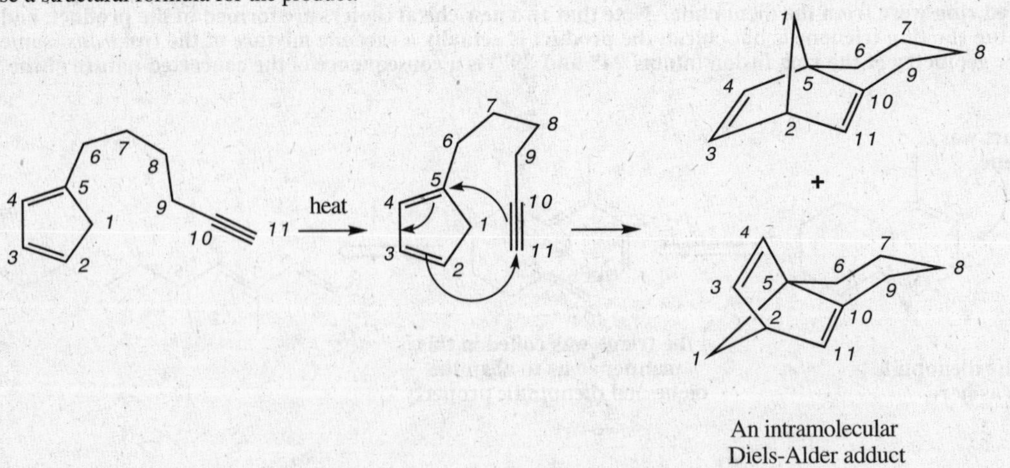

heat →

An intramolecular
Diels-Alder adduct

The product has two new chiral centers, but because of the bicyclic structure there are only two possible enantiomers. The different enantiomers arise from reaction on either face of the diene. Only one enantiomer is drawn below as a three-dimensional model.

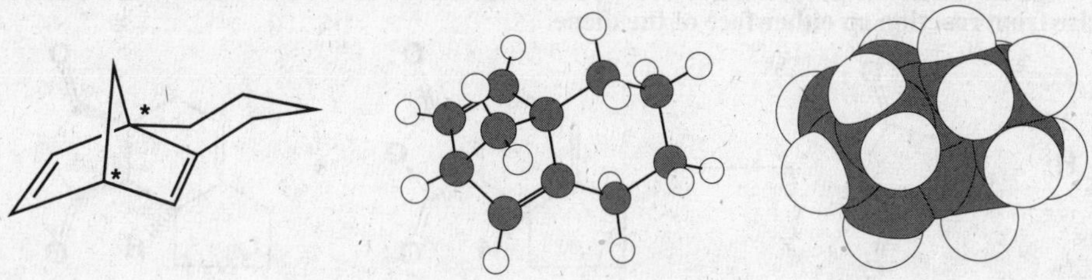

Problem 24.32 Draw a structural formula for the product of this Diels-Alder reaction, including the stereochemistry of the product.

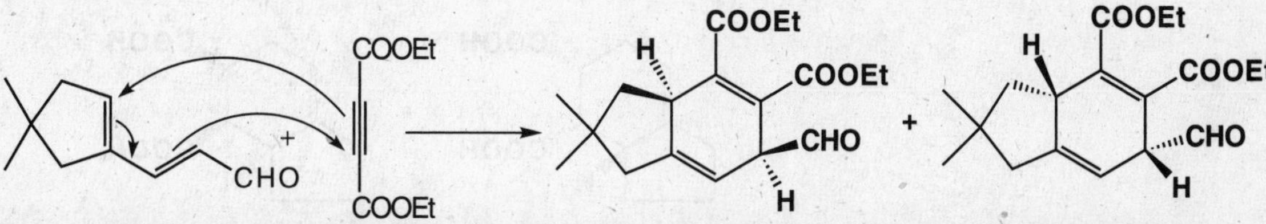

Two new chiral centers are created in the reaction. The concerted nature of the reaction dictates that the addition leads to the two possible *cis* enantiomers. Note that the products are actually a racemic mixture of the two *cis* enantiomers, because neither the diene or dienophile has a chiral center that would favor formation of one enantiomer over the other. The different enantiomers arise from reaction on either face of the diene.

Problem 24.33 Following is a retrosynthetic analysis for the dicarboxylic acid shown on the left.

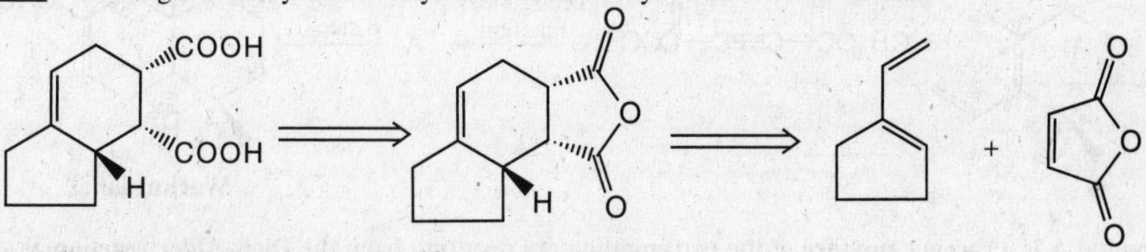

(a) Propose a synthesis of the diene from cyclopentanone and acetylene.

The acetylene is deprotonated, then the resulting alkyne anion is reacted with cyclopentanone to create the alcohol following an acid work-up. The alcohol is dehydrated to give the cyclopentene derivative and hydrogenation over the Lindlar catalyst completes the synthesis.

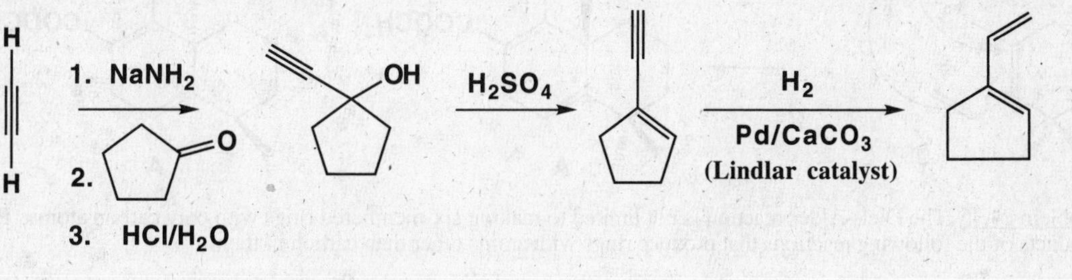

(b) Rationalize the stereochemistry of the target dicarboxylic acid.

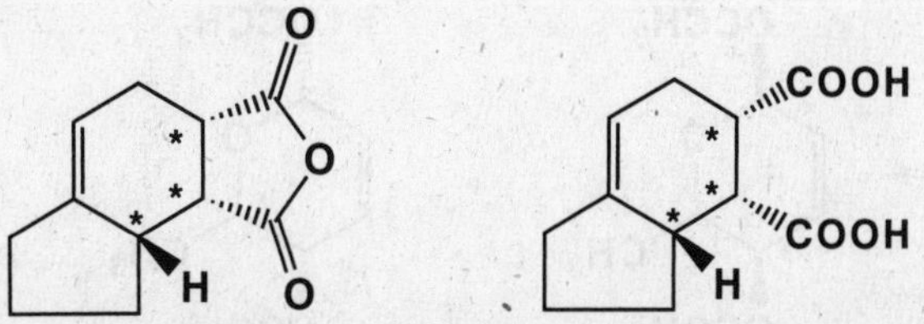

The Diels-Alder adduct and subsequent diacid each have three chiral centers as indicated by asterisks in the above structures. Like all Diels-Alder reactions, the dienophile and diene add in a concerted fashion, fixing the two carbonyls *cis* to each other as well as fixing the relative configuration of the third chiral center. Note that the Diels-Alder product and thus the diacid are actually a racemic mixture of enantiomers, because neither the diene or dienophile has a chiral center that would favor formation of one enantiomer over the other. Only one of

the enantiomers was shown in the problem, however, both enantiomers are shown below. The different enantiomers arise from reaction on either face of the diene.

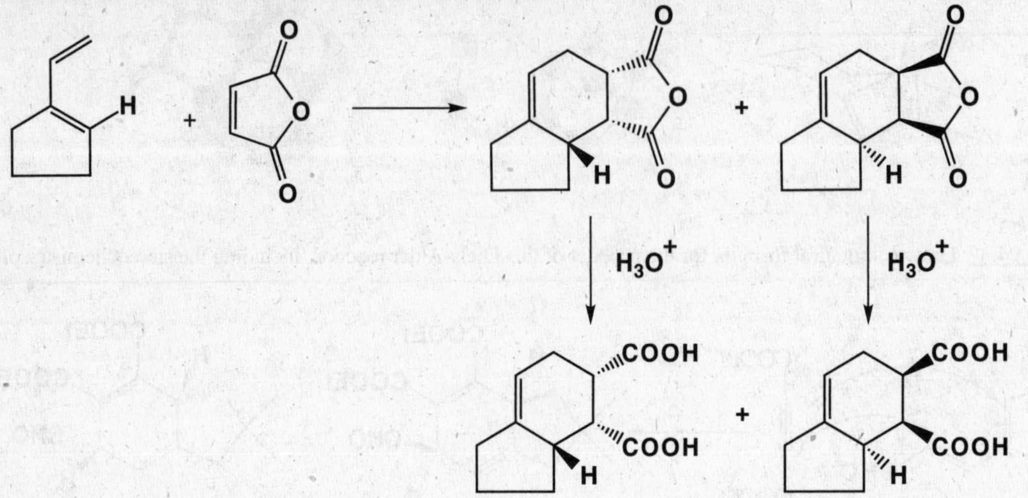

Problem 24.34 One of the published syntheses of warburganal begins with the following Diels-Alder reaction. Propose a structure for Compound A.

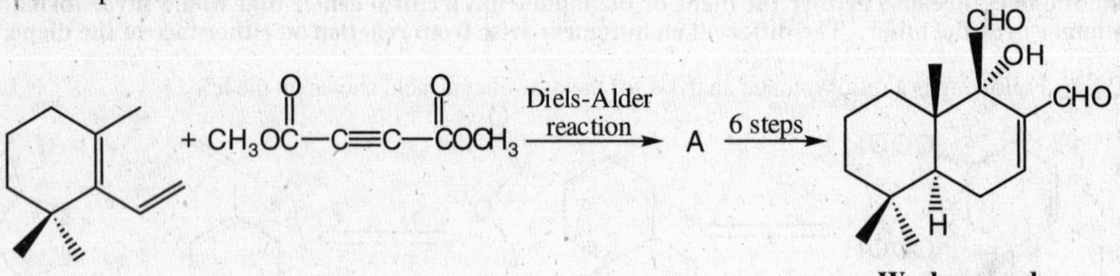

Warburganal

Compound A is a racemic mixture of the two enantiomers resulting from the Diels-Alder reaction as shown. The different enantiomers arise from reaction on either face of the diene.

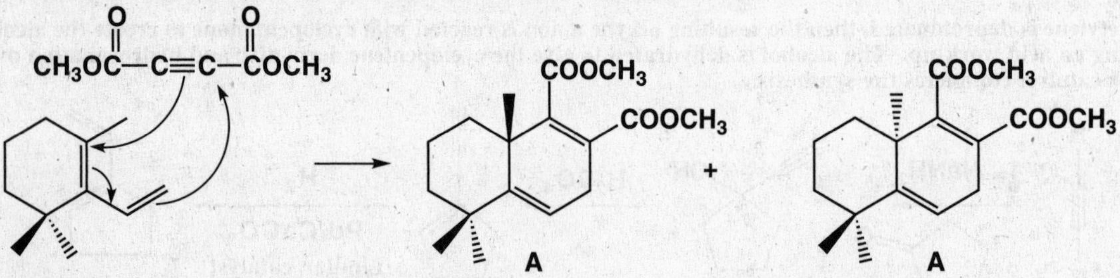

Problem 24.35 The Diels-Alder reaction is not limited to making six-membered rings with only carbon atoms. Predict the products of the following reactions that produce rings with atoms other than carbon in them.

(a)

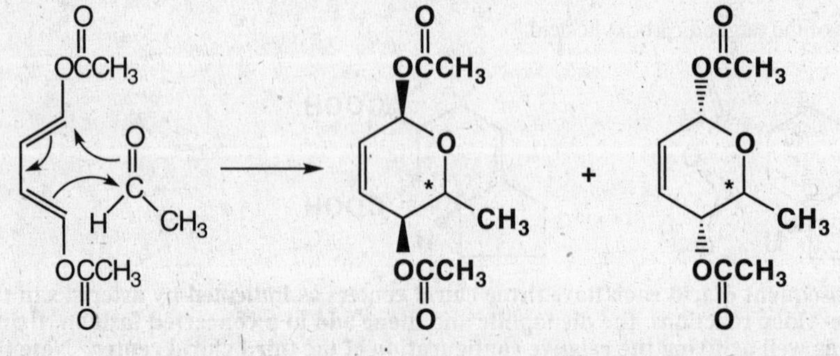

Note that in the above products, three new chiral centers are created. The two ester groups will be *cis* to each other in all products, but the third chiral center could be either *cis* or *trans* relative to these leading to a complex racemic mixture of stereoisomers.

(b)

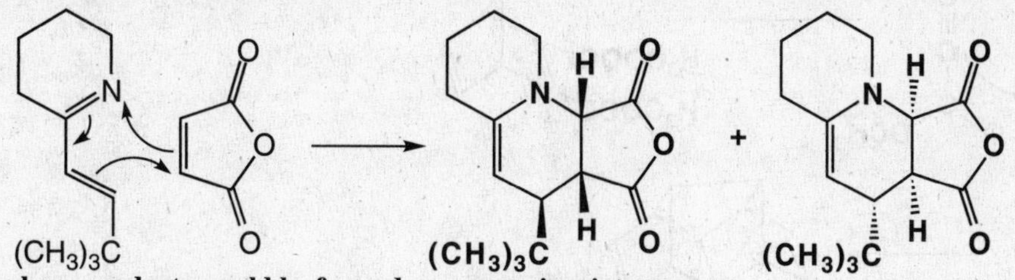

The above products would be formed as a racemic mixture.

(c)

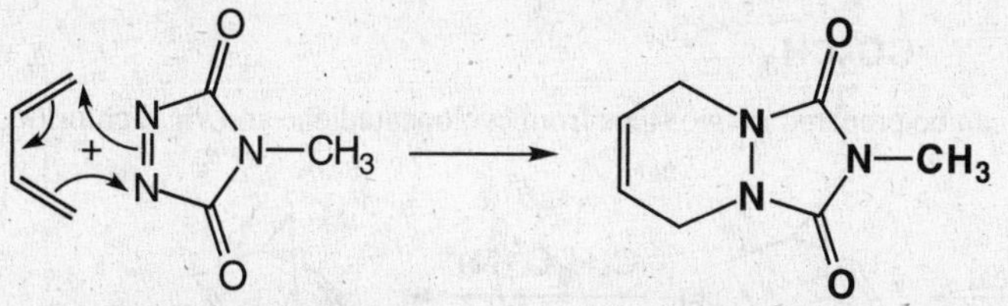

Because of symmetry in the molecule, no new chiral centers are formed.

(d)

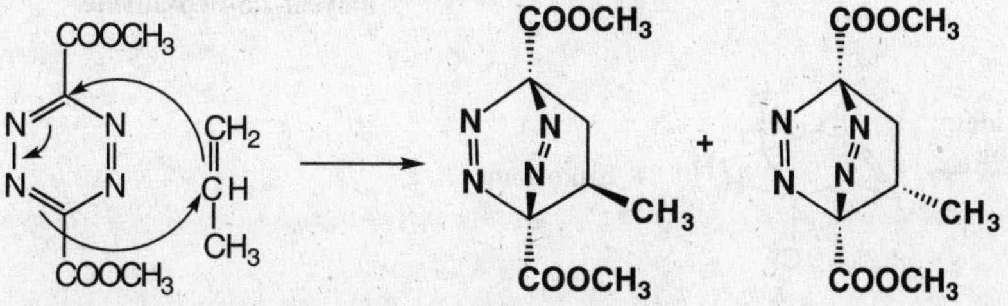

Because of symmetry in the product, only one chiral center is formed. The product enantiomers will be formed as a racemic mixture.

(e)

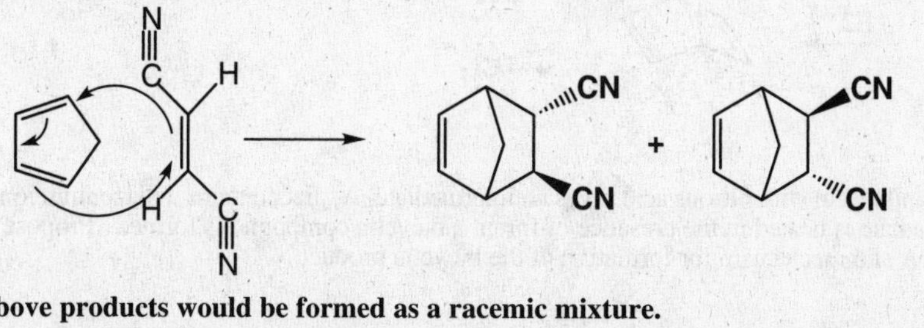

The above products would be formed as a racemic mixture.

Problem 24.36 The first step in a synthesis of dodecahedrane involves a Diels-Alder reaction between the cyclopentadiene derivative (1) and dimethyl acetylenedicarboxylate (2). Show how these two molecules react to form the dodecahedrane synthetic intermediate (3).

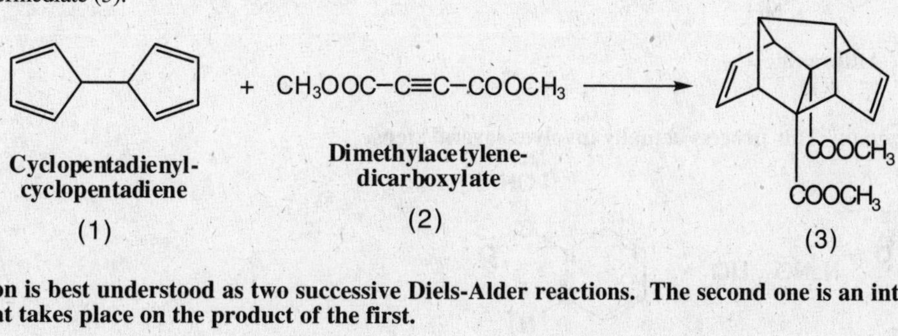

Cyclopentadienyl-
cyclopentadiene

(1)

Dimethylacetylene-
dicarboxylate

(2)

(3)

This reaction is best understood as two successive Diels-Alder reactions. The second one is an intramolecular reaction that takes place on the product of the first.

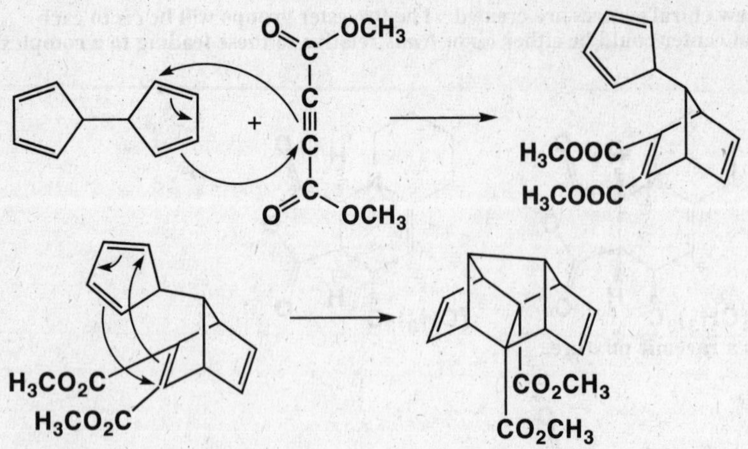

<u>Problem 24.37</u> Bicyclo[2.2.1]-2,5-heptadiene can be prepared in two steps from cyclopentadiene and vinyl chloride. Provide a mechanism for each step.

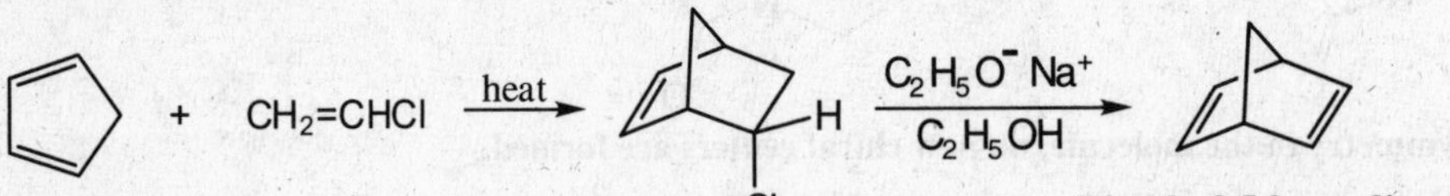

Bicyclo-2,5-heptadiene

Step 1: Pericyclic reaction.

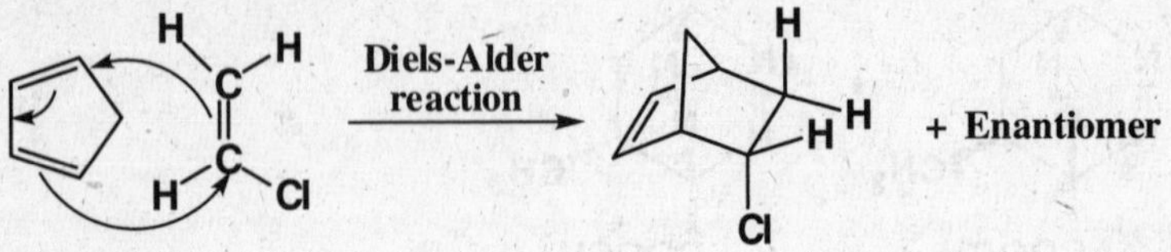

Step : Take a proton away and simultaneously break a bond to give stable molecules or ions (E2).

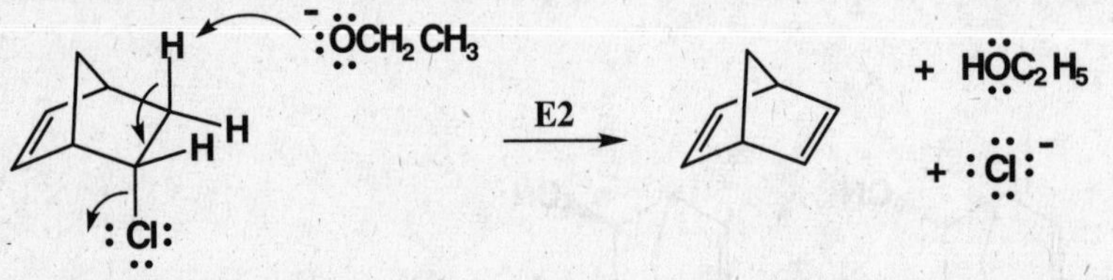

<u>Problem 24.38</u> Treatment of anthranilic acid with nitrous acid gives an intermediate, A, that contains a diazonium ion and a carboxyl group. When this intermediate is heated in the presence of furan, a bicyclic compound is formed. Propose a structural formula for intermediate A and mechanism for formation of the bicyclic product.

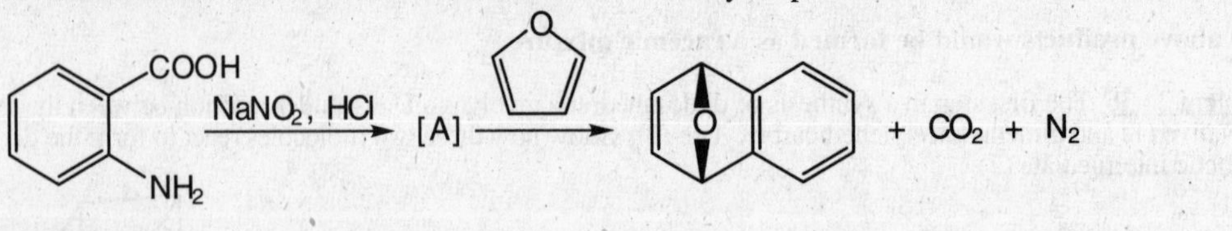

**Anthranilic
acid**

Step 1: Diazotization. This process actually involves several steps.

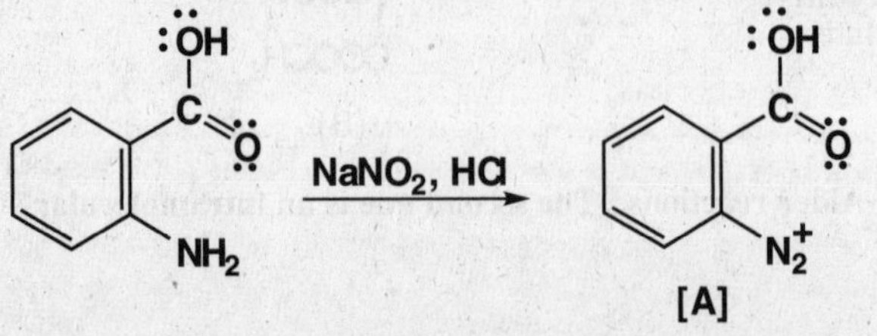

Step 2: Make a bond between a nucleophile and an electrophile and simultaneously break a bond to give stable molecules or ions.

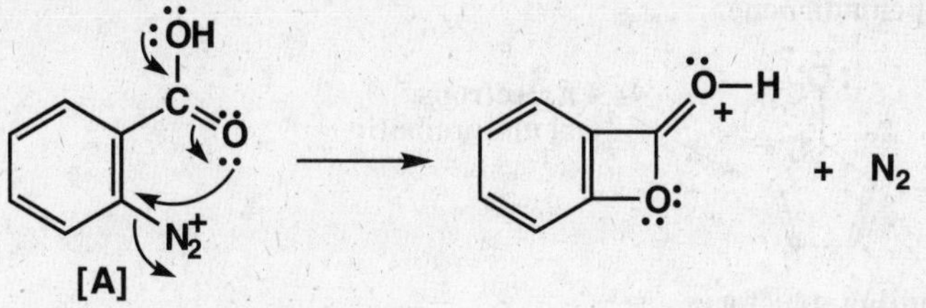

Step 3: Take a proton away.

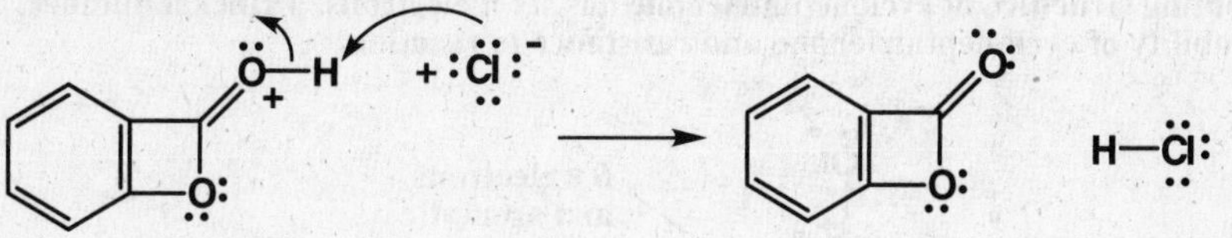

Step 4: Break a bond to give stable molecules or ions.

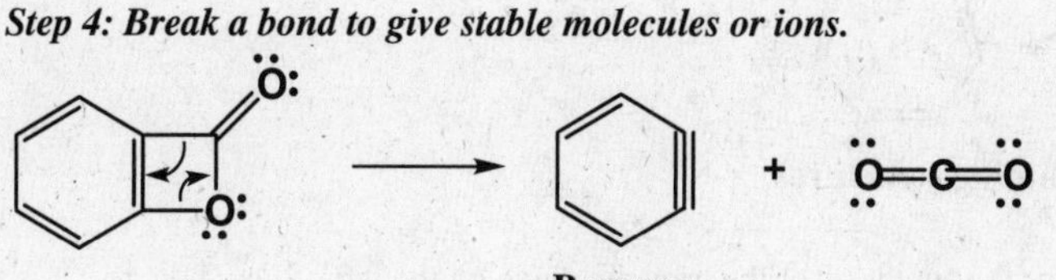

Benzyne

Step 5: Pericyclic reaction.

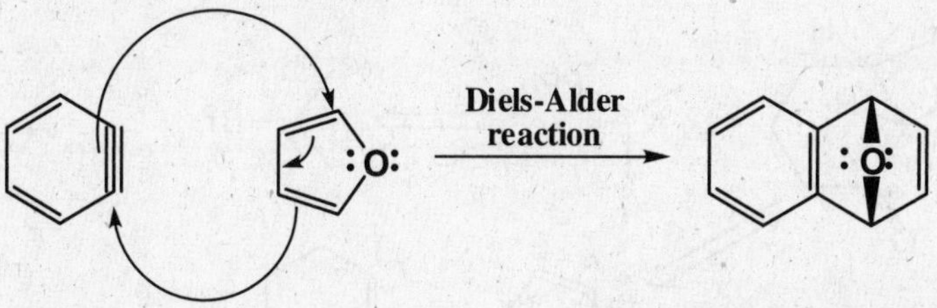

Diels-Alder reaction

The product is a meso compound.

<u>Problem 24.39</u> All attempts to synthesize cyclopentadienone yield only a Diels-Alder adduct. Cycloheptatrienone, however, has been prepared by several methods and is stable. *Hint:* consider important resonance contributing structures.

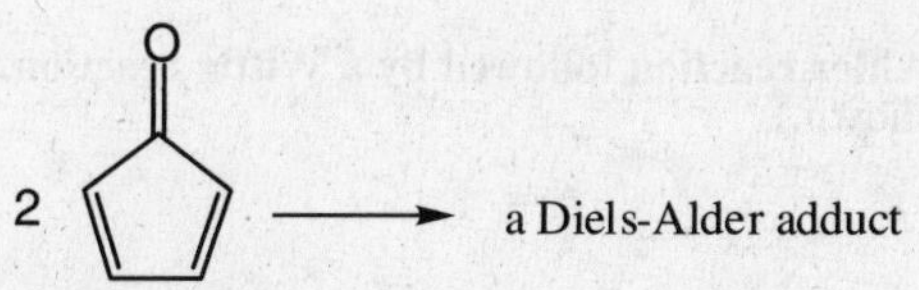

2 **Cyclopentadienone** ⟶ a Diels-Alder adduct

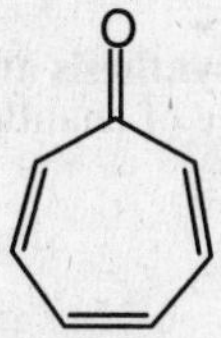

Cycloheptatrienone

(a) Draw a structural formula for the Diels-Alder adduct formed by cyclopentadienone.

The Diels-Alder adduct is a racemic mixture of two enantiomers as shown.

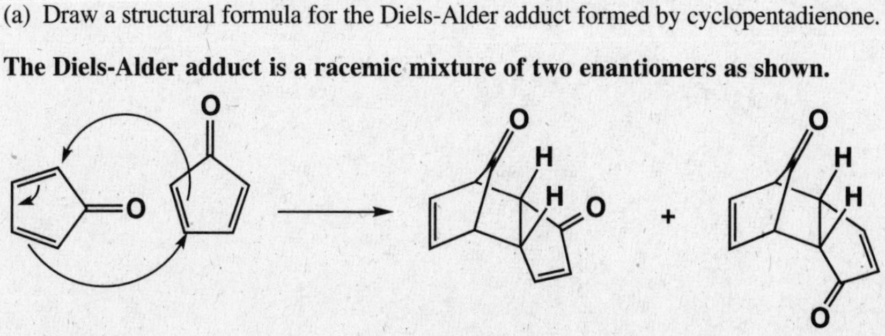

(b) How do you account for the marked difference in stability of these two ketones?

A major contributing structure for cyclopentadienone has only four π electrons, an anti-aromatic number, thereby explaining the extreme reactivity of cyclopentadienone.

4 π electrons and antiaromatic

Cyclopentadienone contributing structures

On the other hand, a major contributing structure of cycloheptatrienone has six π electrons, a Hückel number, thereby explaining the enhanced stability of cycloheptatrienone and resistance to reaction.

6 π electrons and aromatic

Cycloheptatrienone contributing structures

Problem 24.40 Following is a retrosynthetic scheme for the synthesis of the tricyclic diene on the left. Show how to accomplish this synthesis from 2-bromopropane, cyclopentadiene, and 2-cyclohexanone.

The key elements of this synthesis are a Diels-Alder reaction followed by a Wittig reaction. Note the product is actually a racemic mixture of enantiomers as shown.

<u>Problem 24.41</u> Claisen rearrangement of an allyl phenyl ether with substituent groups in both ortho positions leads to formation of a para-substituted product. Propose a mechanism for the following rearrangement.

Propose rearrangement of the allyl group to the ortho position to form a substituted cyclohexadienone, as in the normal Claisen rearrangement. This intermediate, however, cannot undergo keto-enol tautomerism to become an aromatic compound. A second rearrangement, this time of the allyl group to the para position, gives a substituted cyclohexadienone that can undergo keto-enol tautomerism to give an aromatic compound.

Step 1: Pericyclic reaction.

Step 2: Pericyclic reaction.

Step 3: Keto-enol tautomerism.

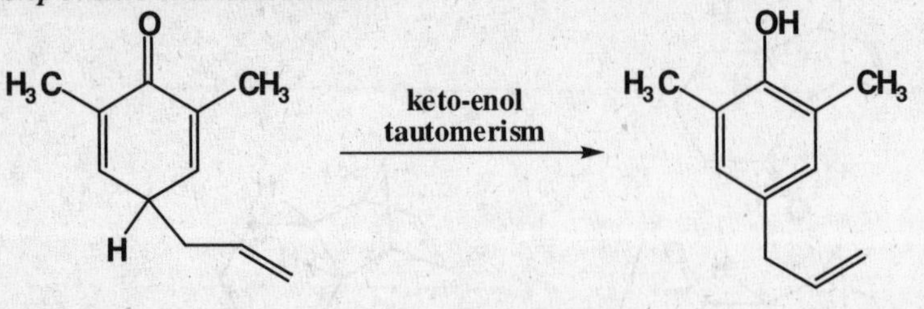

Problem 24.42 Following are three examples of Cope rearrangements of 1,5-dienes. Show that each product can be formed in a single step by a mechanism involving redistribution of six electrons in a cyclic transition state.

Keep in mind as you do these problems that the Cope rearrangement involves flow of three pairs of electrons in a cyclic, six-membered transition state. Bonding in the six carbon atoms participating in the reaction is C=C-C-C-C=C (double-single-single-single-double). Find that combination of carbon atoms and you are well on your way. Numbers have been assigned to the atoms to keep track of the bonds made and broken. It will help to use molecular models when answering these questions.

(a)

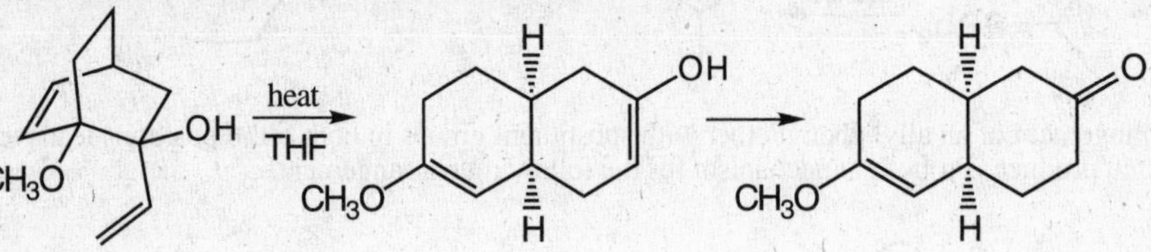

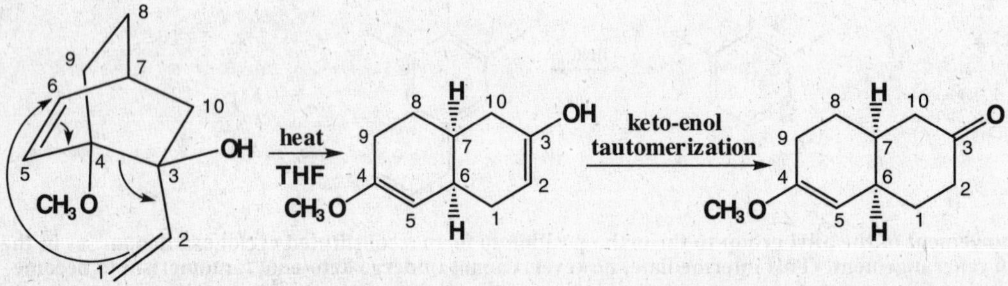

In the above product, two new chiral centers are formed, but only one enantiomer is produced due to the stereochemistry of the starting material. Recall that when a reactant is chiral, it is possible to create a single enantiomer of a product.

(b)

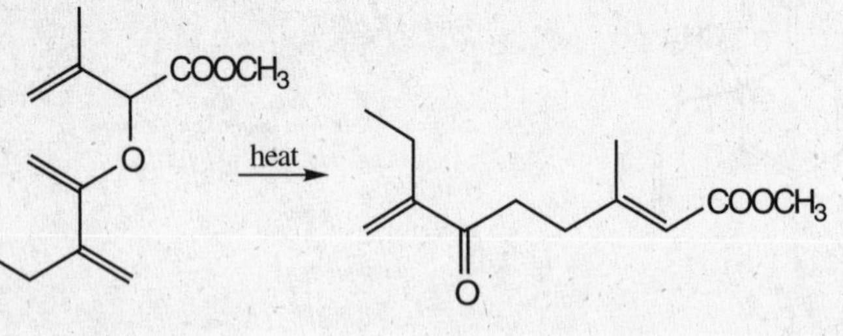

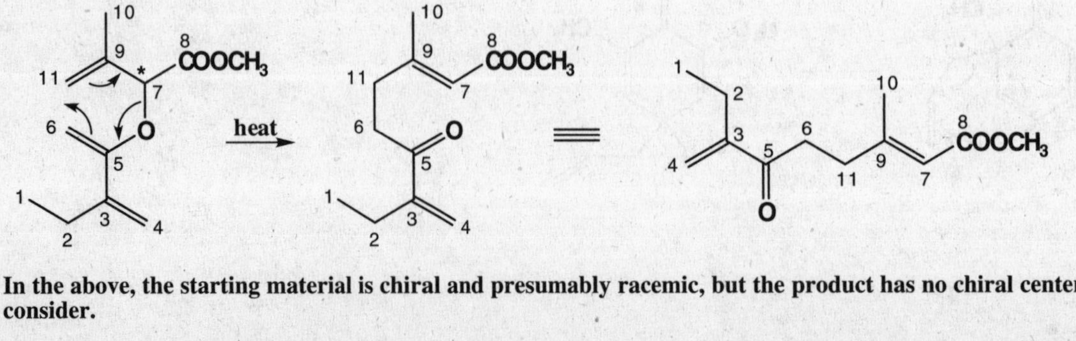

In the above, the starting material is chiral and presumably racemic, but the product has no chiral centers to consider.

(c)

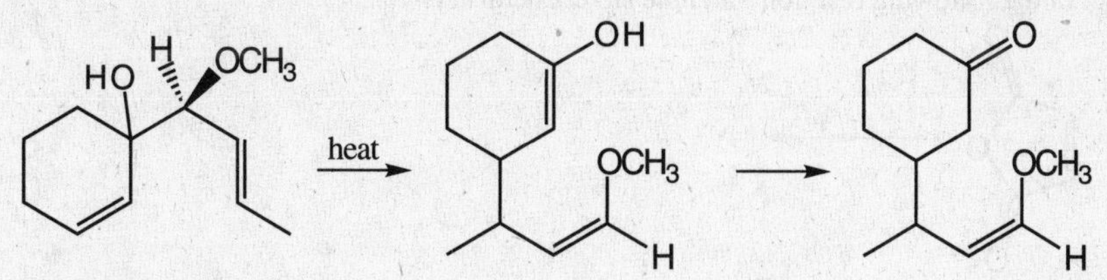

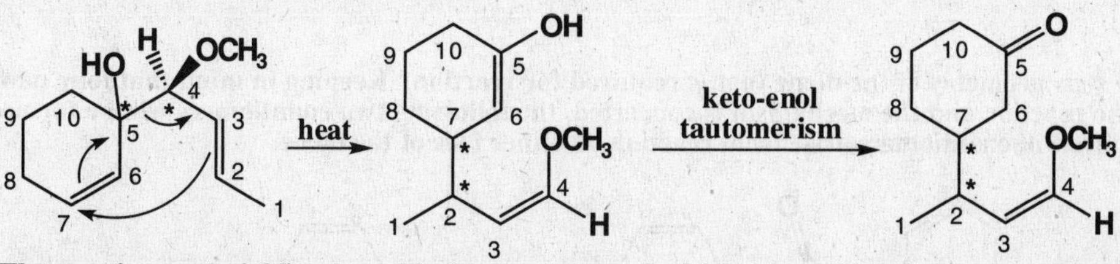

The starting material has two chiral centers, one of which (atom "4") is shown explicitly, while the other (atom "5") is not and thus presumably racemic. As a result, the product of the reaction, which has two new chiral centers created, will be a mixture of stereoisomers.

<u>Problem 24.43</u> The following transformation is an example of the Carroll reaction, named after the English chemist, M. F. Carroll, who first reported it. Propose a mechanism for this reaction.

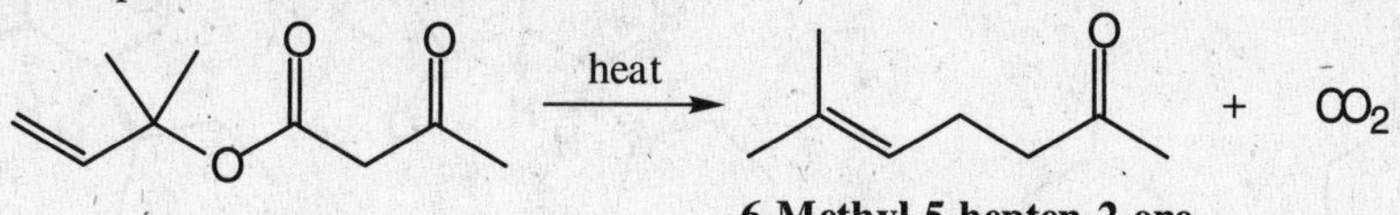

6-Methyl-5-hepten-2-one

Step 1: Keto-enol tautomerism. The reaction begins with formation of an enol. Note that this enol is stabilized by conjugation to the adjacent carbonyl.

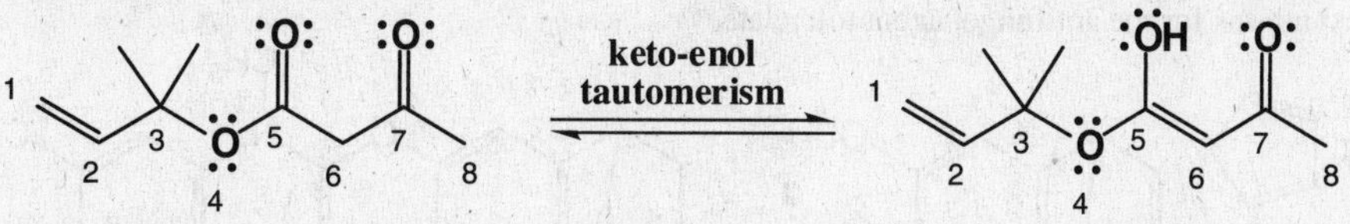

Step 2: Pericyclic reaction. The rearrangement step involves the familiar six-membered ring transition state with three pairs of electrons moving around the ring.

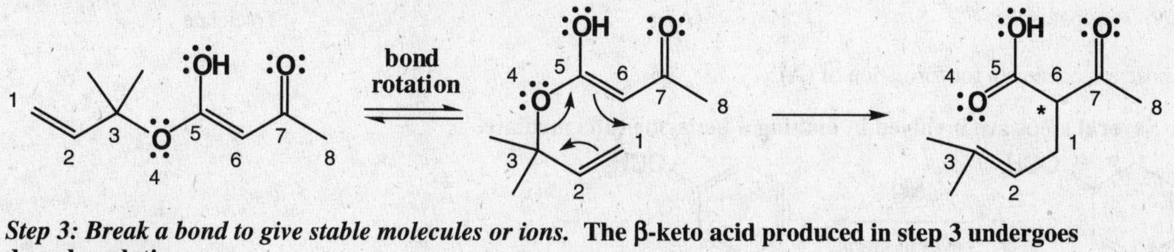

Step 3: Break a bond to give stable molecules or ions. The β-keto acid produced in step 3 undergoes decarboxylation.

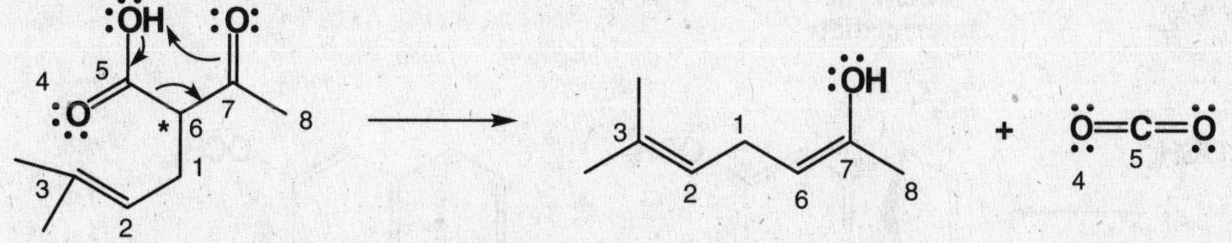

Step 4: Keto-enol tautomerism. Tautomerismcompletes the reaction.

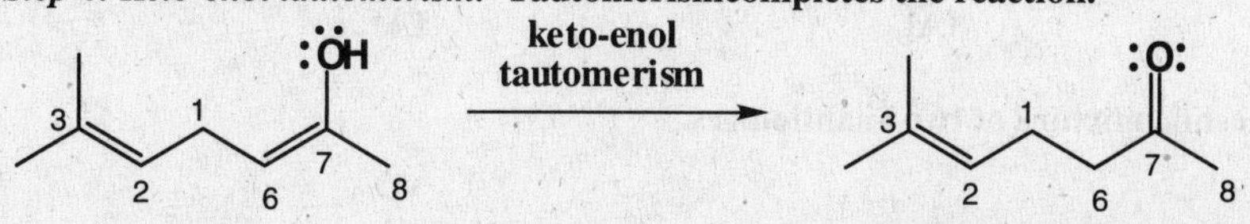

6-Methyl-5-hepten-2-one

<u>Problem 24.44</u> Show the product of the following reaction. Include stereochemistry.

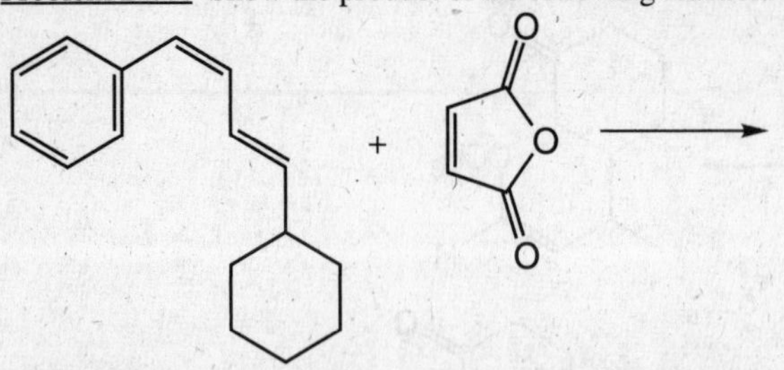

It is helpful to draw the s-*cis* geometry of the diene that is required for reaction. Keeping in mind that four new chiral centers are created in the reaction and the mechanism is concerted, the following two enantiomers will be formed as a racemic mixture. The different enantiomers arise from reaction on either face of the diene.

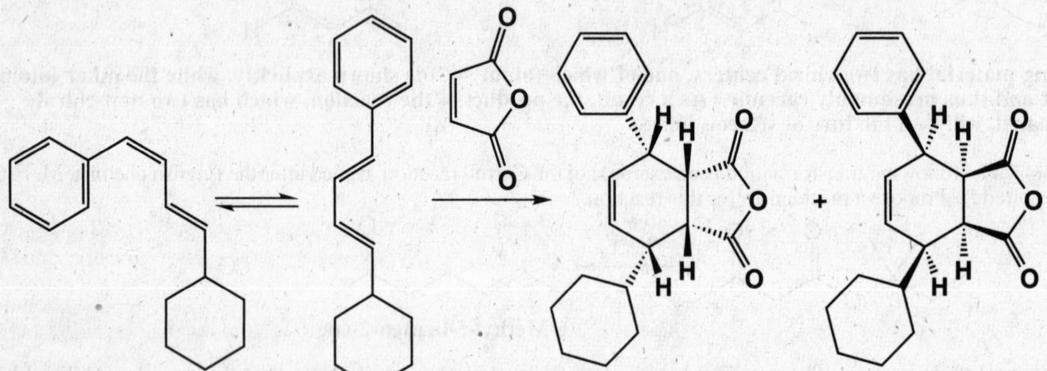

<u>Problem 24.45</u> Following is a synthesis for the antifungal agent tolciclate.

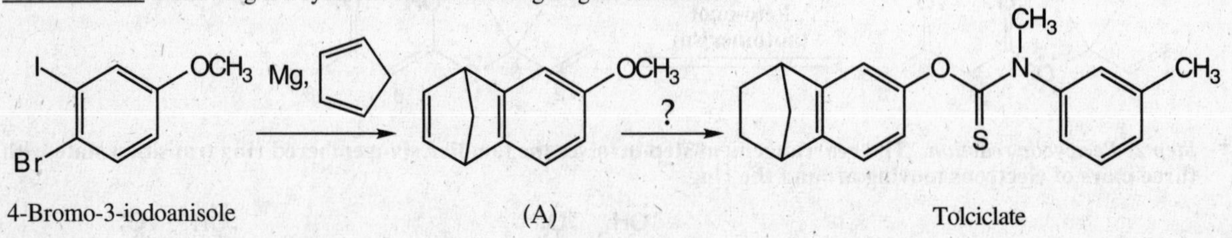

4-Bromo-3-iodoanisole   (A)   Tolciclate

(a) Propose a mechanism for formation of (A).

Step 1: **Several steps are involved in making a benzyme intermediate.**

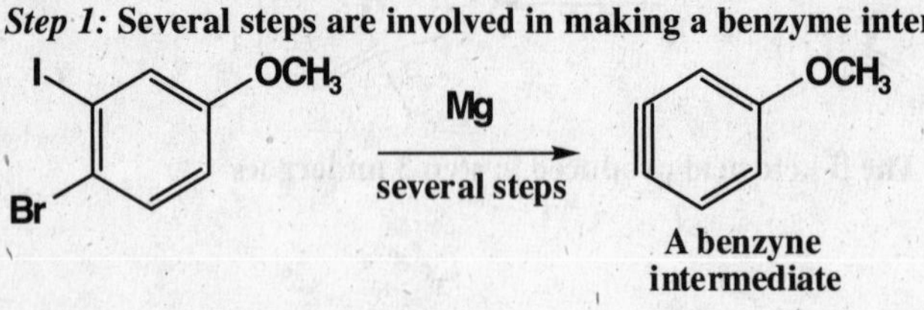

A benzyne
intermediate

Step 2: Pericyclic reaction.

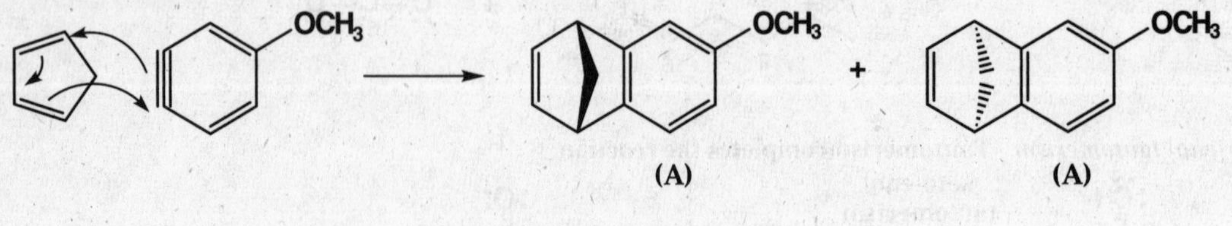

(A)   (A)

Note that (A) will be formed as a racemic mixture of two enantiomers.

(b) Show how (A) can be converted to tolciclate. Use 3-methyl-*N*-methylaniline as the source of the amine nitrogen and thiophosgene, Cl₂C=S as the source of the C=S group.

(A)

3-methyl-*N*-methylaniline

Tolciclate

Note that each compound involved with this reaction sequence is actually a racemic mixture of enantiomers.

Problem 24.46 Ascaridole is a natural product from Oil of Chenopodium (from *chenopodium ambrosioides,* also called American wormseed, Mexican tea, epazote [from Nahuatl words for skunk and sweat!] or ambrosia; the herb is used in seasoning in Yucatán cuisine) that has been used to treat intestinal worms. After World War II, the German population was near starvation, and intestinal worms were a major problem. G. O. Schenck (then in Heidelberg, later Göttingen, and then Mühlheim) devised a remarkable industrial-scale synthesis of this compound in a bombed-out lot in Heidelberg. Using chlorophyll extracted from spinach, he used large carboys with a methanol solution of α-terpinene (isolated from natural oils such as cardamom) with air bubbling and sunlight to produce large amounts of this compound.

α-Terpinene   **Ascaridole**

Suggest a mechanism for this reaction.

Step 1: **The chlorophyll and light convert the triplet ground state (³O₂) of the oxygen molecule into the excited state singlet state (¹O₂).**

O₂ in the ground state (triplet, ³O₂))   **O₂ in the excited state (singlet, ¹O₂)**

Step 2: Pericyclic reaction. **The singlet oxygen (³O₂) undergoes a Diels-Alder reaction with the α-terpinene diene as shown to create the product ascaridole.**

α-Terpinene   **Ascaridole**

<u>Problem 24.47</u> The following transformation can be accomplished by two reactions we have studied in this chapter. Name the type of reaction used in each step.

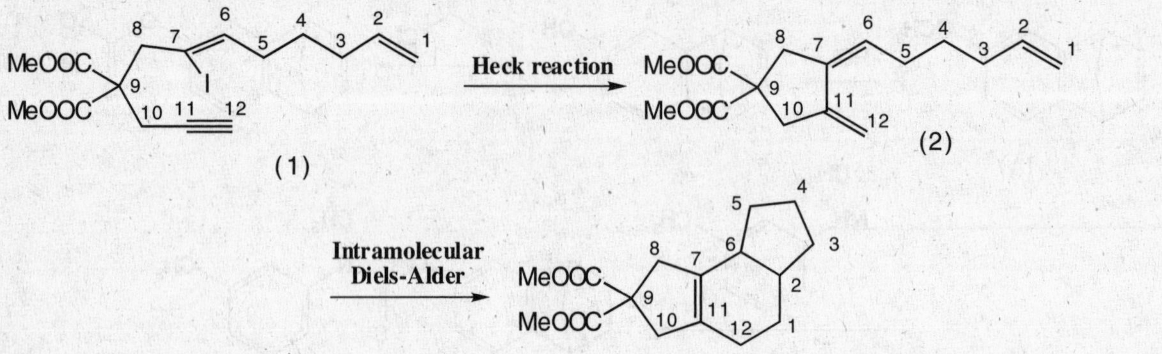

(1)

Heck reaction

(2)

Intramolecular Diels-Alder

This molecule is made from an intramolecular Heck reaction followed by an intramolecular Diels-Alder reaction. The Heck reaction creates intermediate (2) and makes a bond between the carbons labeled as 7 and 11.

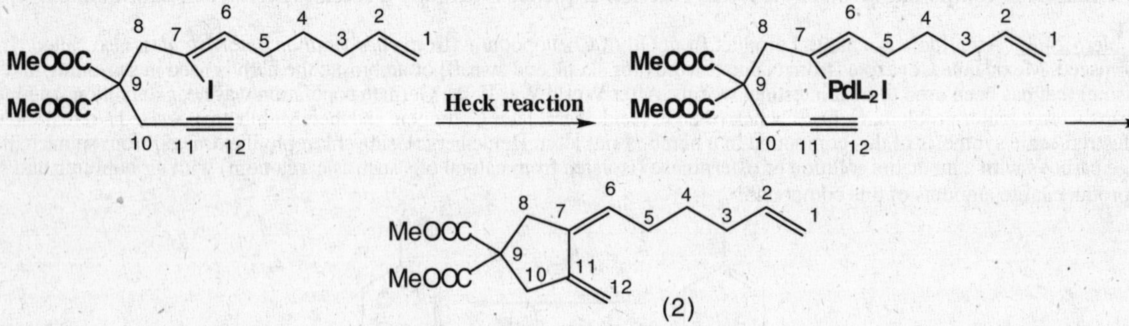

Heck reaction

(2)

The Diels-Alder reaction takes place between the diene and terminal alkene so that new bonds are made between the carbons labeled as 6 and 2 as well as between 12 and 1.

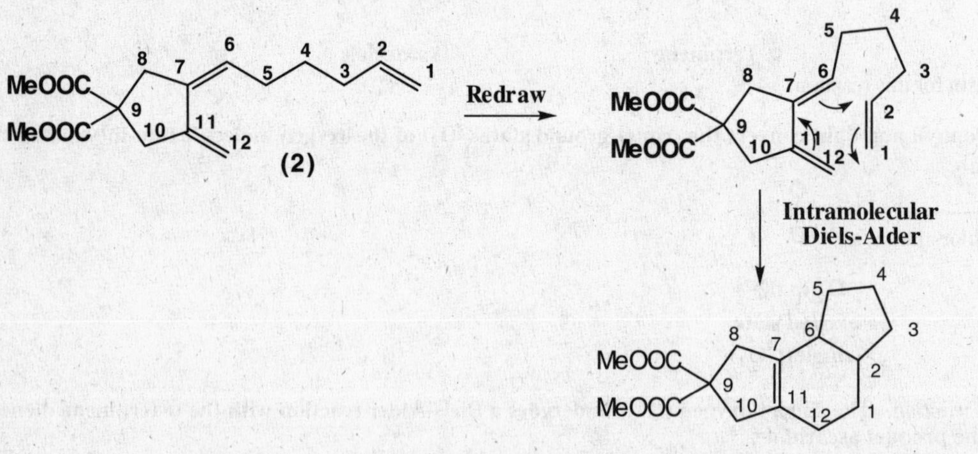

(2)

Redraw

Intramolecular Diels-Alder

Synthesis

The following problems are based on relatively recent total syntheses of important natural products. Many such syntheses are outlined in compendia of synthetic reactions. Particularly valuable in preparing these problems were *Classics in Total Synthesis,* K. C. Nicolaou and E. J. Sorensen, Wiley-VCH Weinheim, New York, Basel, Cambridge, Tokyo, 1996; *Classics in Total Synthesis II,* K. C. Nicolaou and S. A. Snyder, Wiley-VCH Verlag GMBH, Weinheim (2003).

Problem 24.48 Following is an outline of the stereospecific synthesis of the "Corey lactone." Professor E. J. Corey (Harvard University) describes it in this way. "The first general synthetic route to all the known prostaglandins was developed by way of bicycloheptene intermediates. The design was guided by the requirements that the route be versatile enough to allow the synthesis of many analogs and also allow early resolution. This synthesis has been used on a large scale and in laboratories throughout the world; it has been applied to the production of countless prostaglandin analogs." Corey was awarded the 1990 Nobel Prize for chemistry for the development of retrosynthetic analysis for synthetic production of complex molecules. See E. J. Corey and Xue-Min Cheng, *The Logic of Chemical Synthesis,* John Wiley & Sons, New York, 1989, p. 255. For the structure of the prostaglandins, see Section 26.3. *Note:* The wavy lines in compound C indicate that the stereochemistry of -Cl and -CN groups was not determined. (The conversion of (D) to (E) involves an oxidation of the ketone group to a lactone by the Baeyer-Villiger reaction, which we have not studied in this text.)

(a) What is the function of sodium hydride, NaH, in the first step? What is the pKa of cyclopentadiene? How do you account for its remarkable acidity?

The function of the NaH is as a strong, irreversible base to deprotonate the cyclopentadiene (A) and convert it into its anion. This transforms it into a better nucleophile to set up the S$_N$2 reaction with MeOCH$_2$Cl. The pKa of cyclopentadiene is 16.0 (Section 21.2), which makes it about as acidic as alcohols. Its acidity is due to the aromaticity of the cyclopentadienyl anion.

(b) By what type of reaction is (B) converted to (C)?

The conversion of (B) to (C) is a Diels-Alder reaction (Section 24.6).

(c) What is the function of the carbon dioxide added to the reaction mixture in Step 2 of the conversion of (E) to (F)? *Hint:* What happens when carbon dioxide is dissolved in water? Why not just use HCl?

Step 1 in the conversion of (E) to (F) is base-promoted hydrolysis of the lactone; the product is a carboxylate anion. The function of carbon dioxide is to acidify the alkaline solution and thereby convert the -COO- group to a COOH group.

(d) The tributyltin hydride, $(Bu)_3SnH$, used in the conversion of (H) to (I) reacts via a radical chain reaction; the first step involves a reaction with an radical initiator to form $(Bu)_3Sn$. Suggest a mechanism for the rest of the reaction.

The reaction is a radical chain process in which the I atom is replaced with H.

Step 1

$$(Bu)_3SnH \xrightarrow{\text{radical initiator}} (Bu)_3Sn \cdot \; + \; H \cdot$$

Step 2

Step 3

(e) The Corey lactone contains four chiral centers with the relative configurations shown. In what step or steps in this synthesis is the configuration of each chiral center determined? Propose a mechanism to account for the observed stereospecificity of the relevant steps.

Carbons 1, 2, 4, and 5 of the Corey lactone are chiral centers.

(K)
(Corey lactone)

Following is an outline of this synthesis showing only the structural formulas for steps in which chiral centers are created or undergo reaction. The carbons of the five-membered ring of the final lactone are numbered 1 through 5; carbons 1, 2, 4, and 5 of the ring are chiral centers. Retaining this numbering pattern and working back toward starting materials, you will see that the four chiral centers arise from carbons 1, 2, 4, and 5 of cyclopentadiene!

(E)

(F)

(G)

$(H) \longrightarrow (I) \longrightarrow (J) \longrightarrow$

(K)

(Corey lactone)

The configurations of chiral centers 1, 4, and 5 are determined in the Diels-Alder reaction that forms (C). The configuration of chiral center 2 is established in the conversion of (E) to (F). As background for understanding this stereo- and regiochemistry, review Section 6.3 on the addition of Cl_2/H_2O and Br_2/H_2O to a cycloalkene to form a halohydrin. In this step of the Corey lactone synthesis, I_2 reacts with the carbon-carbon double from the top, less hindered side of the ring to give a cyclic iodinium ion. Given the *trans* and coplanar regiospecificity of opening of this cyclic intermediate and the fact that the side-chain carboxyl group is on the bottom side of the ring, OH of the carboxyl group attacks the iodinium ion intermediate from the bottom side of the ring, thus establishing the configuration of carbon 2.

(f) Compound (F) was resolved using (+)-ephedrine. Following is the structure of (-)-ephedrine, the naturally occurring stereoisomer. What is meant by "resolution" and what is the rationale for using a chiral, enantiomerically pure amine for the resolution of (F)?

Ephedrine $[\alpha]_D^{21}$ -41

Resolution (Section 3.8) is the separation of a racemic mixture into its enantiomers, each in pure form. Enantiomerically pure ephedrine reacts with racemic (F) to form a pair of diastereomeric salts. Because diastereomeric salts are different compounds (Section 3.8), they have different chemical and physical properties and can be separated by physical means, most commonly fractional crystallization. The separated diastereomeric salts are then treated with HCl or other mineral acid to give the individual enantiomeric carboxylic acids.

(g) You have not studied the Baeyer-Villiger reaction (D to E). The mechanism involves nucleophilic reaction of the peroxyacid with the carbonyl, followed by a rearrangement much like that involved in the hydroboration reaction (Section 6.4). Write a mechanism for this reaction.

As stated in the problem, the mechanism begins with a nucleophilic attack of the peracid on the carbonyl.

Step 1: Make a bond between a nucleophile and an electrophile.

Step 2: Take a proton away. A proton is transferred to base, in this case the HCO_3^- ion.

Step 3: 1,2 Shift-break a bond to give stable molecules or ions. The key step in a Baeyer-Villager oxidation is the migration of an alkyl group onto the terminal peracid oxygen atom to create an ester and carboxylate. Many previous experiments have established that tertiary alkyl groups will move in preference to a $-CH_2-$ group, explaining why the major product has the bridgehead carbon attached to the -O- oxygen.

(E)

Note: By resolving at this stage, one half of the material is discarded. A more efficient route would be to have an earlier resolution; in fact, Corey solved this problem in a very elegant way by using an enantioselective Diels-Alder with the alkene in the form of an acrylate ester of enantiomerically pure 8-phenylmenthol. Asymmetric induction gave a product with a diastereoselectivity of 97 : 3. So rather than resolving, he was able to get the correct stereoisomer directly!

<u>Problem 24.49</u> Chapman's (O. L. Chapman, then at Iowa State, later UCLA) classic total synthesis of (±)-Carpanone is so remarkably simple that it is used as an undergraduate laboratory preparation. It is modeled on a possible biosynthetic route for this lignan-derived natural product. Phenol oxidations figure prominently in many such biosyntheses of natural products. In one step, this reaction creates no less than five contiguous chiral centers, all in the correct relative configuration.

(±)-Carpanone

(a) Give a mechanism for the first step of the reaction and explain why it goes in the direction it does.

The base is equilibrating the double bond. Although the phenolic -OH is the most acidic site on the molecule, in strong base, other sites are deprotonated as well. In this case, the benzylic position can be deprotonated to give an anion that is resonance stabilized and is reprotonated on the terminal carbon. Because the product *trans* alkene is conjugated to the aromatic ring, it is more stable than the starting material and predominates under the reaction conditions.

Step 1: Take a proton away.

Step 2: Add a proton.

(b) The oxidation step uses a palladium salt. Suggest a mechanism for this coupling, which you have not encountered. (*Hint:* Do not concern yourself with the role of the metal except as an acceptor of electrons.)

The Pd(II) metal ion is thought to coordinate two molecules through the phenolic oxygen atoms. A rearrangement of electrons, including addition of two electrons to the Pd ion, gives the coupled product and a Pd(0) metal center.

(c) The third step is spontaneous. Give a mechanism for this reaction and show how it accounts for the stereochemistry of the final product.

This step is an intramolecular Diels-Alder reaction, which is best seen after rotation about the bond of carbons labeled as 11 and 13. The Diels-Alder reaction fixes the stereochemistry at the carbons labeled as 8, 15, and 16. In particular, the chiral centers of the carbons labeled as 9 and 11 fix the geometry of the molecule such that the dienophile (carbons labeled as 15 and 16) approaches from above the diene (atoms labeled as 1, 2, 7, and 8). Given this geometry, the H atom on 3, the H atom on 15 and the O atom on 15 are all fixed pointing downward.

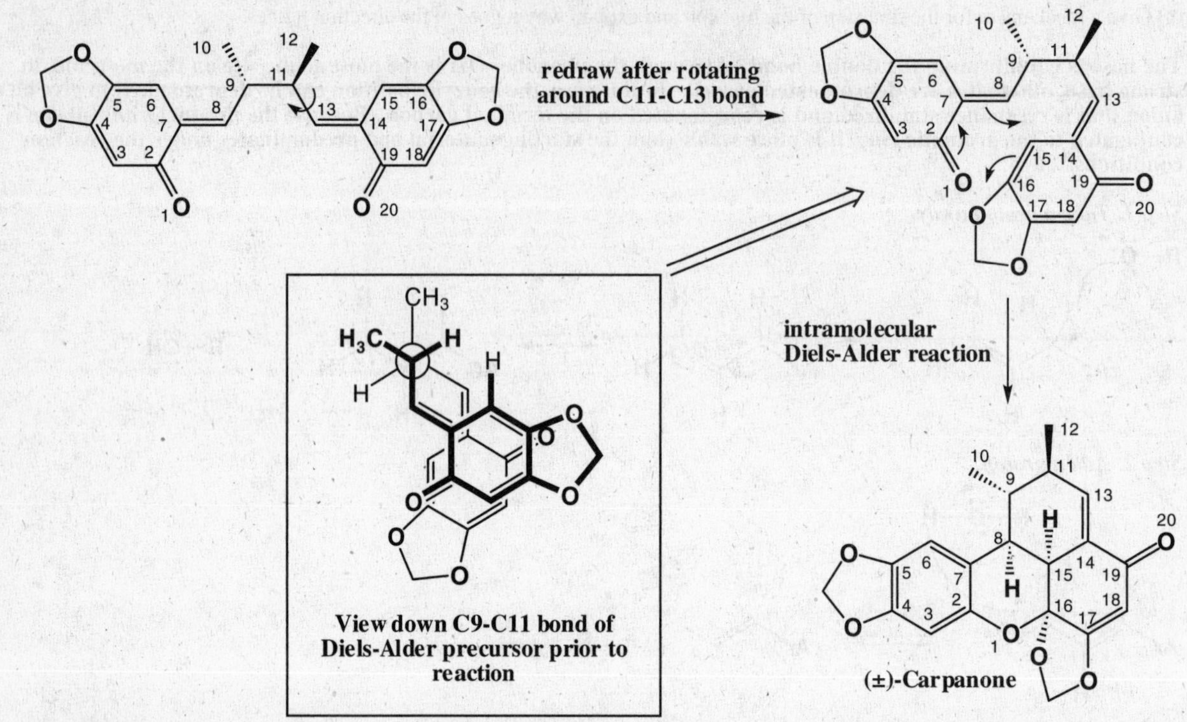

redraw after rotating around C11-C13 bond

intramolecular Diels-Alder reaction

View down C9-C11 bond of Diels-Alder precursor prior to reaction

(±)-Carpanone

(d) Would you expect the product to be racemic or a single enantiomer?

The phenolic coupling will yield a racemic mixture of enantiomers, which in turn, will lead to a racemic mixture of enantiomers following the Diels-Alder reaction.

Enantiomers of Diels-Alder precursor

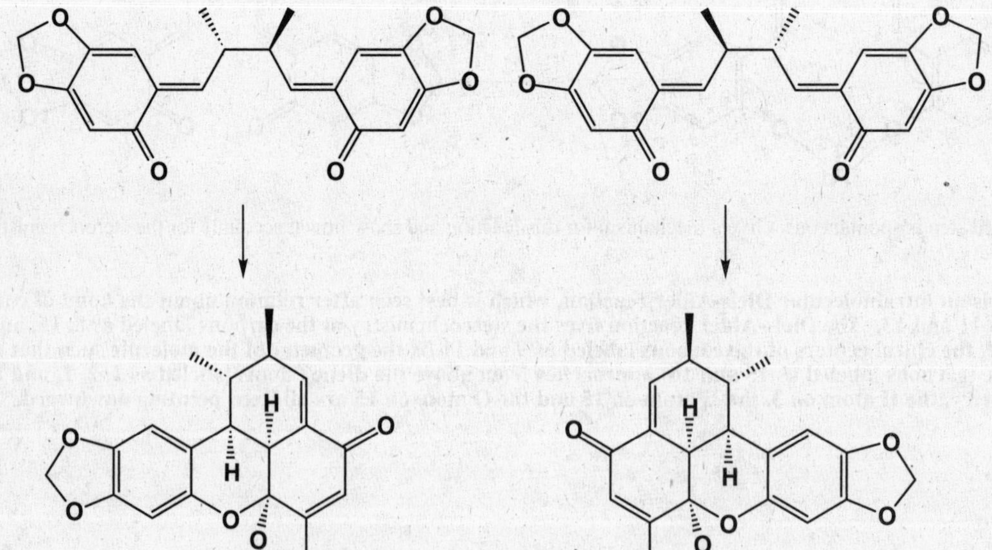

Enantiomers of Carpanone

<u>Problem 24.50</u> Gilvocarcin M is isolated from *Streptomyces* strains and has strong antitumor activity.

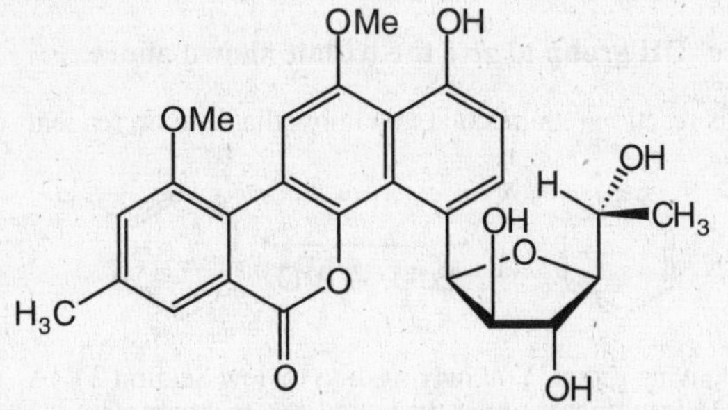

Suzuki and coworkers were able to carry out the total synthesis of naturally occurring (-)-gilvocarcin M. Their synthesis included the following steps. (The wavy line means stereochemistry is unspecified or a mixture.) The stereochemistry of the product appears to be counterintuitive (apparent attack from the more hindered side). The reason is that the reaction involves initial O-alkylation followed by a rearrangement that need not concern us.

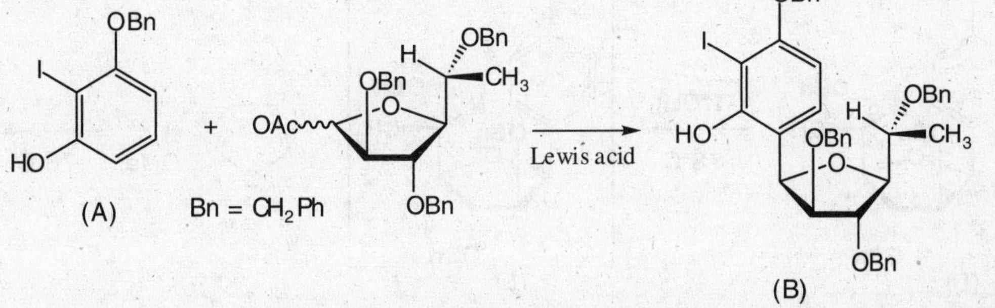

(a) This reaction gives both high regioselectivity and stereoselectivity. What other products might have been expected?

One might expect the other anomer, that is, the molecule with the aryl ring in the down position. By controlling conditions of the reactions carefully, especially solvent, temperature, and the nature of the Lewis acid, a good yield of the desired anomer was obtained and the product shown below was only a minor component of the product mixture.

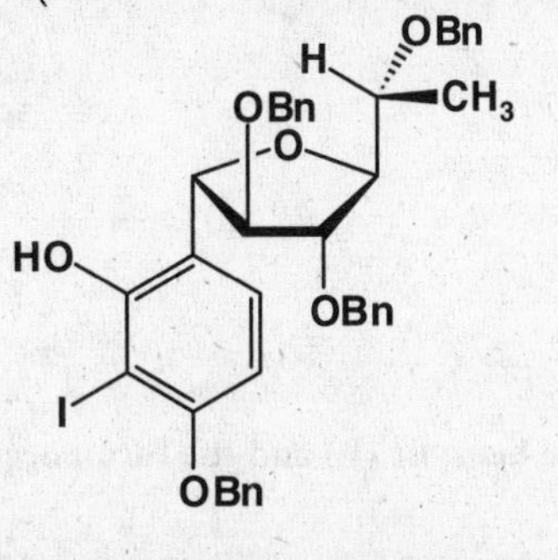

Alternative stereoisomer that might have been expected.

The next step involves triflation and treatment with butyl lithium.

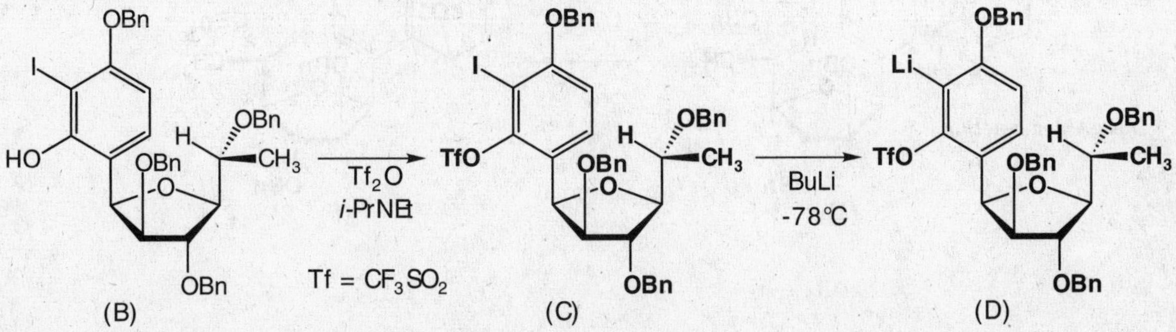

(b) Give a structure for (C).

The Tf₂O reacts with the free -OH group to give the triflate shown above.

(c) Give a structure for (D). This reaction requires that you know that lithium reagents can interchange with aryl halides:

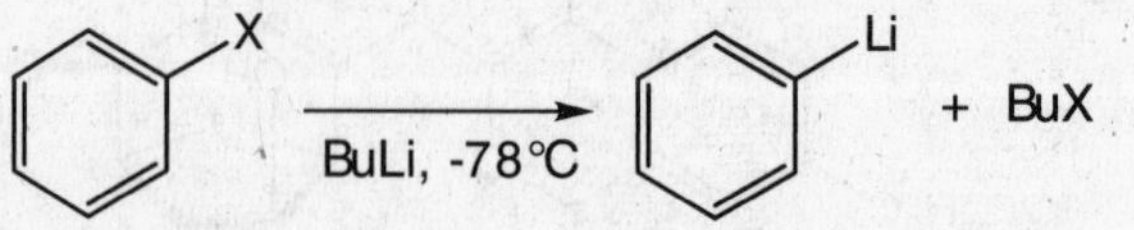

Recall that OTf is an excellent leaving group. You may wish to review Section 24.3A. The reaction yielding (D) is carried out in the presence of 2-methoxyfuran. (D) decomposes under the conditions to a compound (E) that instantly reacts with the furan to give (F).
(d) Give a structure for E and the mechanism of (D) to (E).
(e) Give a mechanism for (E) to (F).

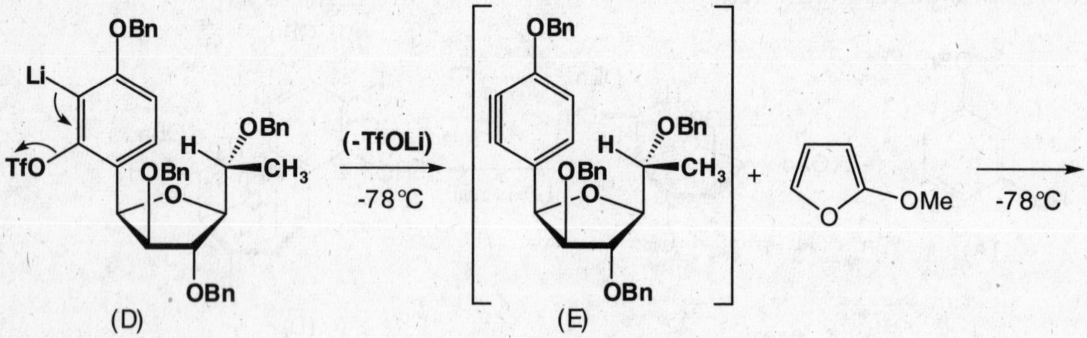

Benzyne intermediate

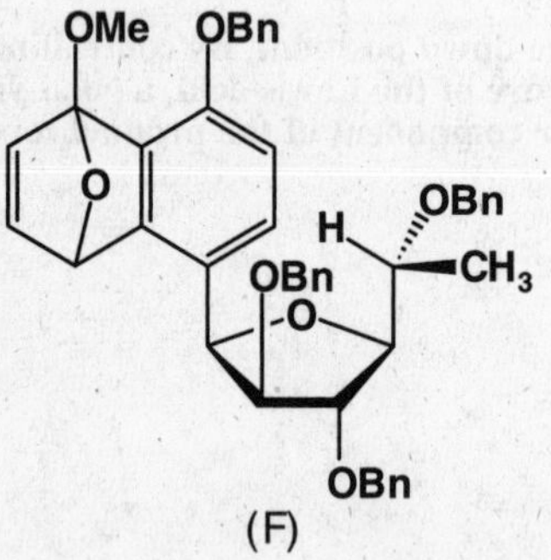

The reaction of the triple bond of the benzyne (E) and the furan occurs through a Diels-Alder reaction.

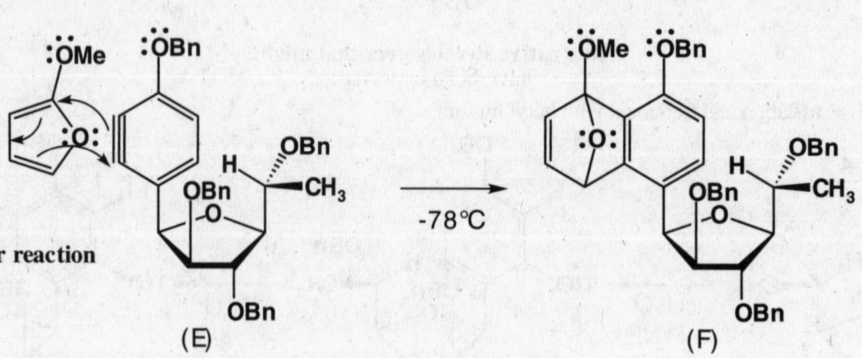

Diels-Alder reaction

(F) is unstable and undergoes ring opening on workup to give (G).

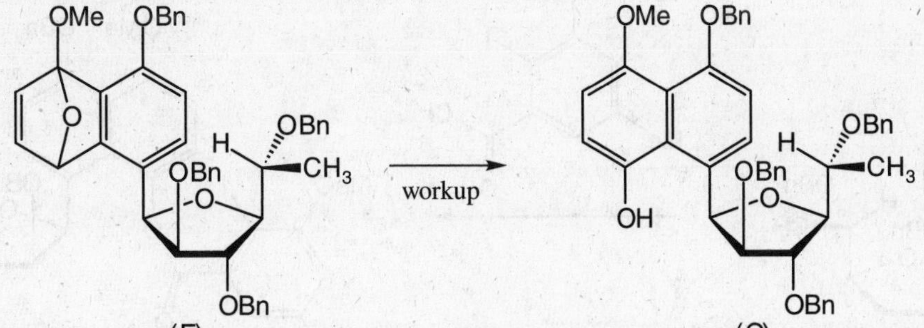

(F)   → workup   (G)

(f) Give a mechanism for (F) to (G).

In the workup, HCl is added to effect the ring opening according to the following mechanism.

Step 1: Add a proton.

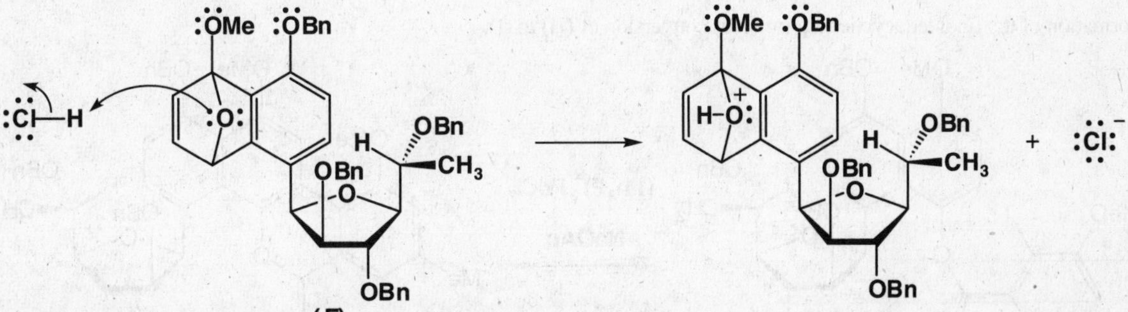

(F)

Step 2: Break a bond to give stable molecules or ions.

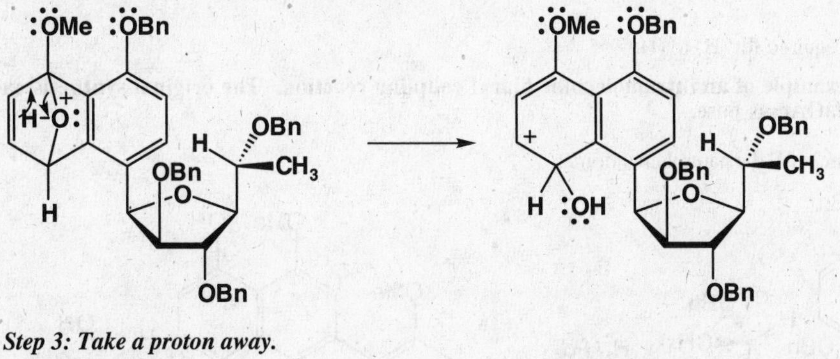

Step 3: Take a proton away.

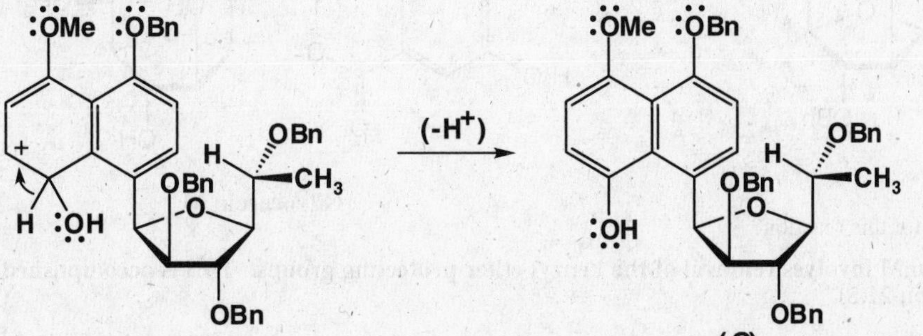

$(-H^+)$ (G)

The next step involves conversion of (G) to (H).

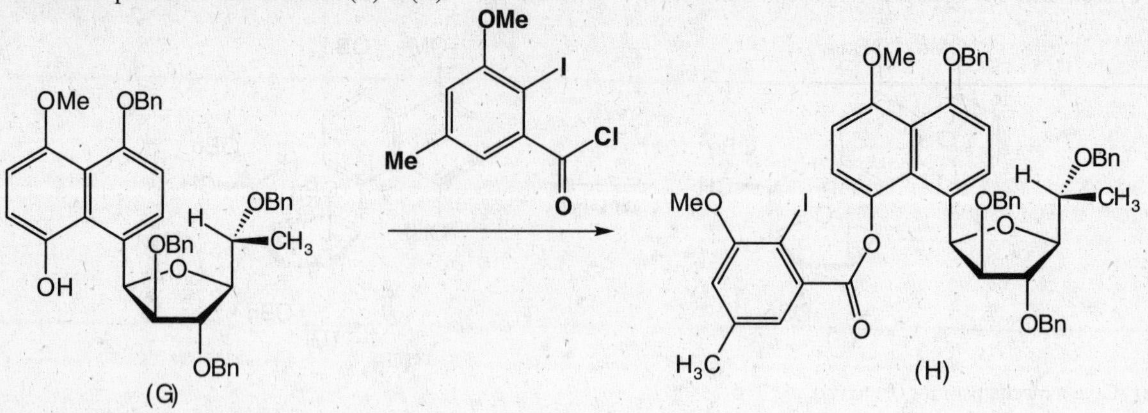

(g) Give reagents and conditions required for (G) to (H).

As shown above, this conversion involves an acid chloride reacting with the free -OH group.

Formation of the final tetracyclic ring involves conversion of (H) to (I).

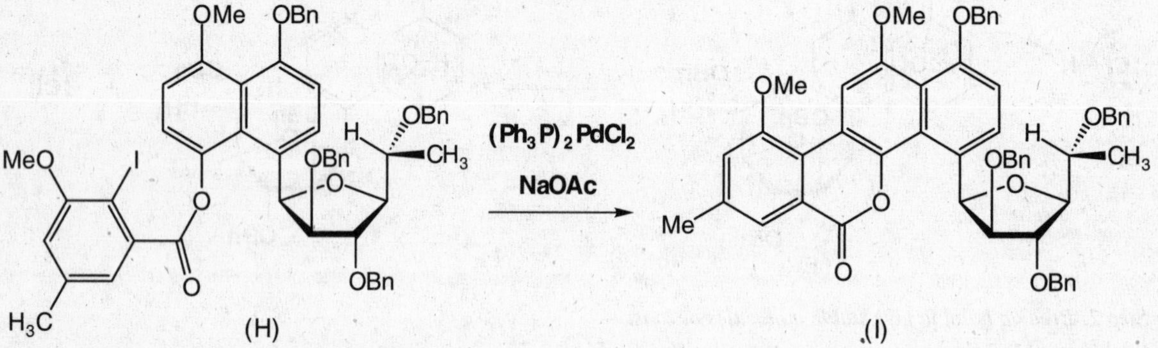

(h) Give reagents and conditions required for (H) to (I).

Conversion of (H) to (I) is an example of an intramolecular biaryl coupling reaction. The original synthesis calls for the use of (Ph₃P)₂PdCl₂ and NaOAc as base.

I is then is converted to (-)-gilvocarcin M, the natural enantiomer.

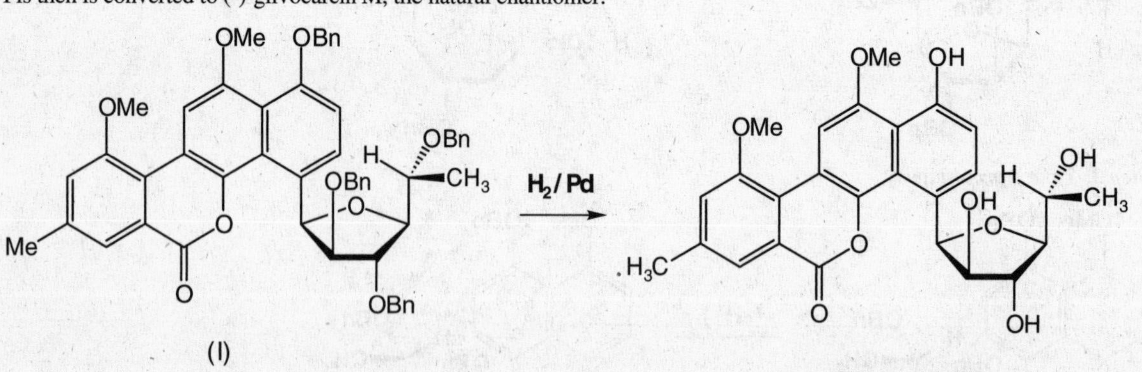

Gilvocarcin M

(i) What reagents could be used for this reaction?

Conversion of I to Gilvocarcin M involves removal of the benzyl ether protecting groups. This is accomplished using hydrogenolysis (H₂/Pd, Section 21.5).

(j) Comment on the probable source of the chiral centers in this synthesis. Note that the chirality was not created in any of the reaction steps. You can find a possible readily available and inexpensive source (see Chapter 25, "Carbohydrates").

A carbohydrate is a great chiral starting material. They are available from nature as single enantiomers, and often contain highly functionalized backbones with several chiral centers. The synthesis of (-)-Gilvocarcin M utilized D-fucofuranosyl acetate as the chiral carbohydrate starting material.

(k) Given reactions that are later in the sequence, why is it necessary to protect some of the OH groups as the benzyl ether? What side reactions would occur without this protection? Starting with OH groups, how would you add these protecting groups?

Conversion of (G) to (H) involves reaction of the one free -OH group with an acid chloride. If the other -OH groups were not protected as benzyl ether groups, an undesirable mixture of isomers would be formed. Benzyl ether protecting groups are added using PhCH₂Br in the presence of base such as potassium carbonate (K₂CO₃).

Problem 24.51 Vancomycin is an important antibiotic. It is isolated from the bacterium *Streptomyces orientalis* and functions by inhibiting bacterial mucopeptide synthesis. It is a last line of defense against the resistant Staph organisms that are now common in hospitals.

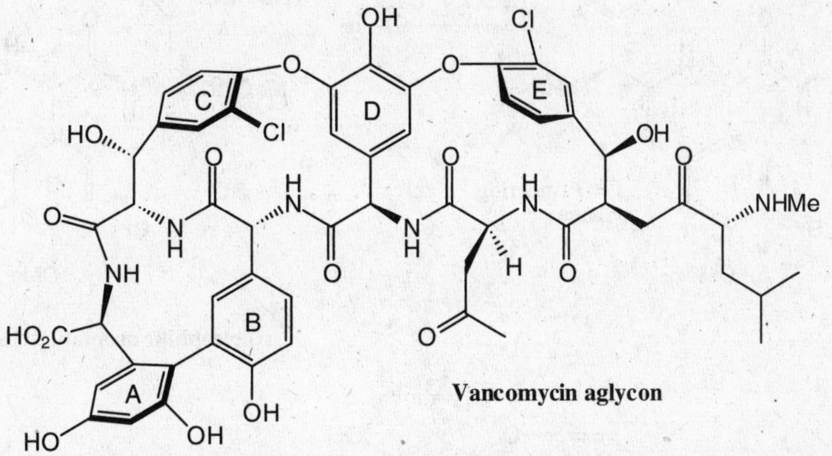

Vancomycin aglycon

In 1999, Professor Dale Boger (The Scripps Research Institute) reported a synthesis of vancomycin **aglycon** (aglycon = lacking a sugar) involving the following steps, among others. Compound I was prepared from simple starting materials by a series of steps involving forming amide bonds. (a) Suggest reasonable precursors and show how the bonds could be formed (the actual reagents used have not been introduced, but they work in a similar way to those you know).

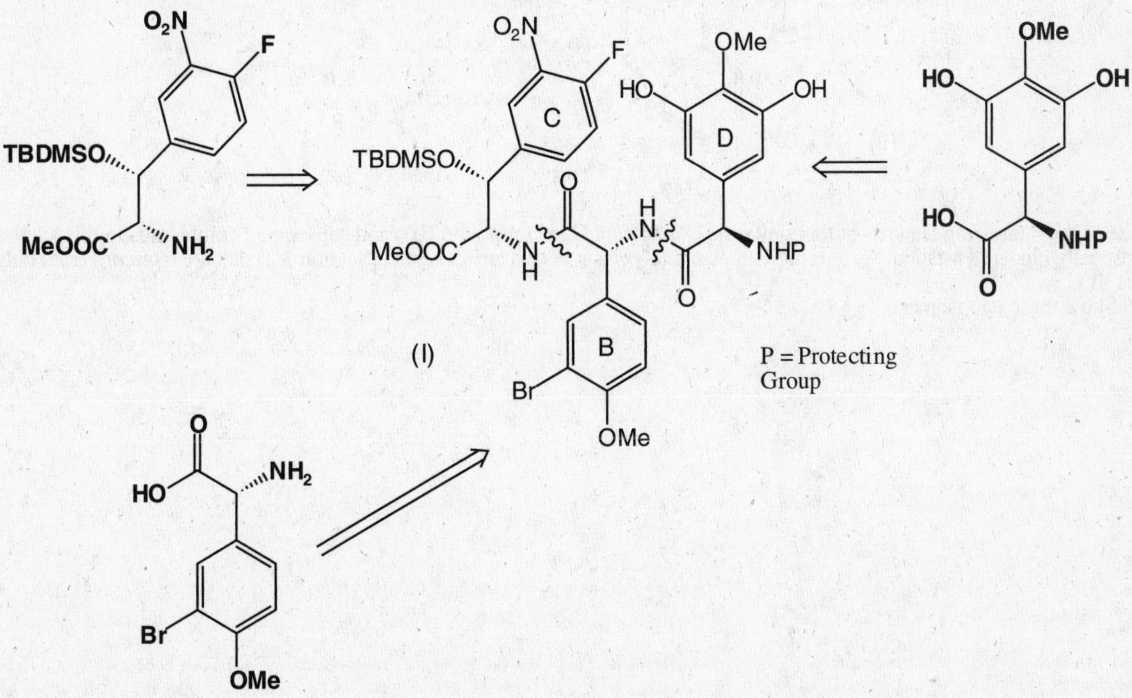

(I)

P = Protecting Group

(I) was then converted into (II).
(b) Give reagents for this reaction and suggest the mechanism.

This transformation is accomplished by base, in the actual synthesis, a combination of K$_2$CO$_3$ and CaCO$_3$ was used. The phenolic -OH group is deprotonated in the base, allowing a nucleophilic aromatic substitution reaction that creates an ether linkage between the C and D rings.

One of the interesting features of this synthesis is that Ring C in compound (II) (and subsequent compounds in this synthesis) has extremely hindered rotation. As a result, compound (II) exists as two atropisomers (Section 3.2) that are interconverted only at 140°C.
(c) Show these two isomers.

(II) was then converted to (III).
(d) Suggest reagents to accomplish this transformation.

This is a two-step process that first involves reduction of the nitro group to an amine with H₂/Pd, followed by diazotization (RONO) then a Sandmeyer reaction with CuCl.

(III) was then converted to (IV).
(e) Suggest reagents and the ring A fragment that could be used for this reaction.

This is a Suzuki coupling as shown.

Closure of an amide link between the amine on ring A (after removal of the protecting group) and the carbomethoxy group above it led to a precursor of vancomycin.
(f) Show the ring closure reaction of the deprotected free amino group and its mechanism.

The methyl ester and protecting group were both removed, then a carbodiimide reagent was used to form the amide bond. Carbodiimide reagents are among the most important methods for amide bond formation, and are discussed in Section 27.5. The reaction utilizes a free carboxylic acid and amine.

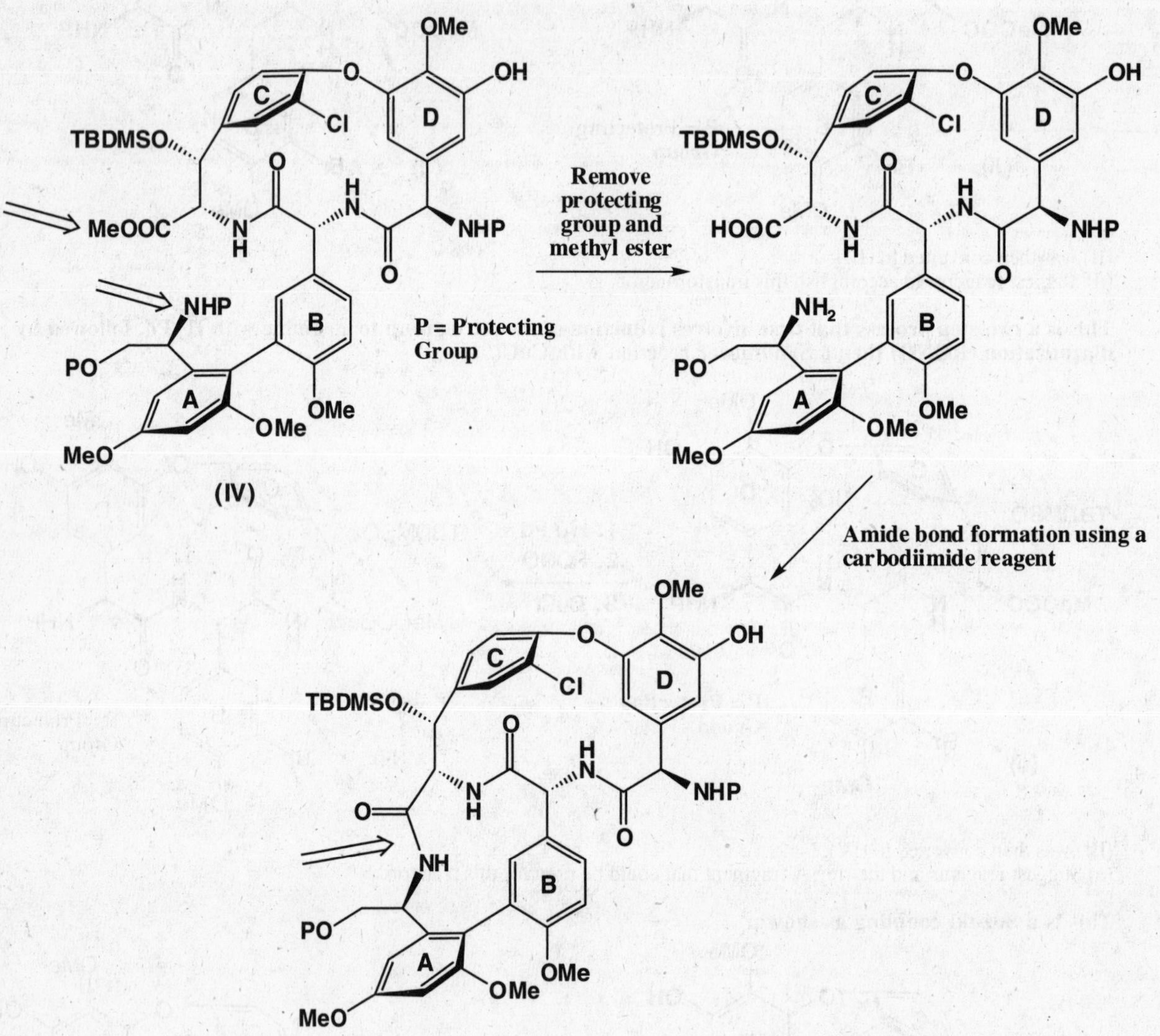

(IV)

P = Protecting Group

Remove protecting group and methyl ester

Amide bond formation using a carbodiimide reagent

Another interesting feature of this synthesis is that ring A and B also form atropisomers. These can be converted into a 3:1 mixture of the desired and undesired atropisomers on heating at 120°C.

(g) Draw these atropisomers and show that only one can be converted to vancomycin. The synthesis of the aglycon was completed by functional manipulation and addition of ring E by chemistry similar to that detailed earlier. Yet another set of atropisomers (this time of ring E) was formed! However, this one was more easily equilibrated than the others; model studies had shown that the activation barrier for this set of atropisomers should be lower that of the others.

This atropisomer does *not* lead to Vancomycin

This atropisomer leads to Vancomycin

Problem 24.52 E. J. Corey's 1964 total synthesis of α-caryophyllene (essence of cloves) solves a number of problems of construction of unusual-sized rings.

α-Caryophyllene

The first step uses an efficient photochemical [2 + 2] reaction. The desired stereochemistry and regiochemistry had been predicted based on model reactions.

(a) [2 + 2] Reactions are quite common in photochemical reactions. Would this reaction be predicted to occur in the ground state?

A [2 + 2] reaction would not be predicted to occur in the ground state. Reactions such as the Diels-Alder that occur in the ground state have a Hückel number of electrons involved, in the case of the Diels-Alder reaction that would be [4 + 2] = 6 electrons.

The next steps follow. Basic alumina is a chromatography support that will often act as a base catalyst.

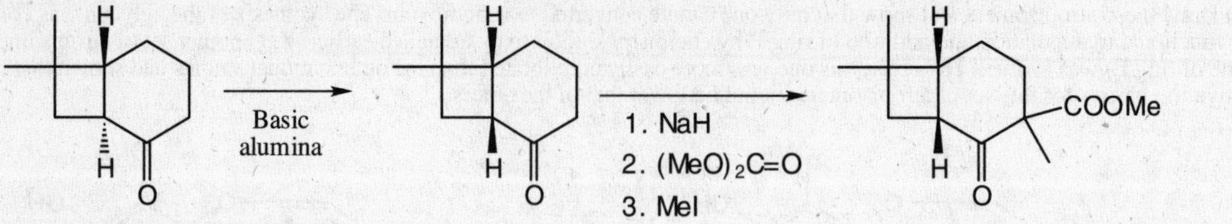

(b) What is the mechanism of the first step?

This first step involves an enolate formation, followed by reprotonation on the opposite face (top face as drawn) of the molecule.

Step 1: Take a proton away.

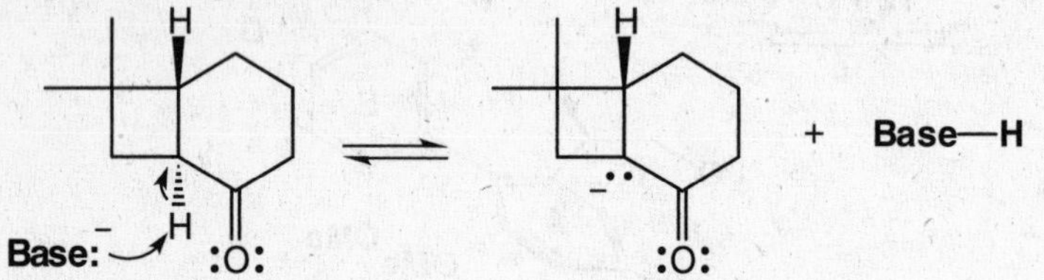

Step 2: Add a proton.

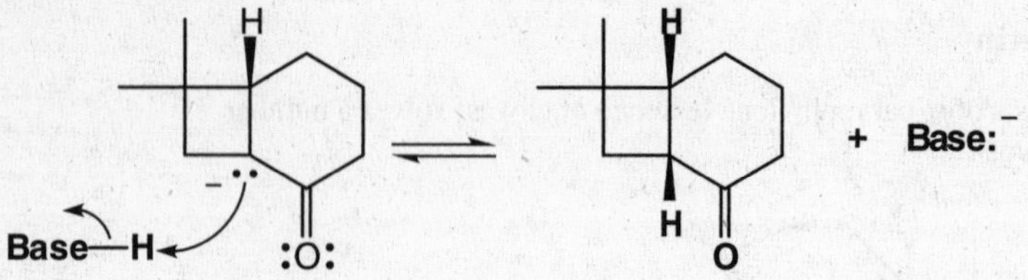

(c) What is the mechanism of the second step?

There are several individual steps involved, and these are presented in an abbreviated form.

Step 1: Take a proton away.

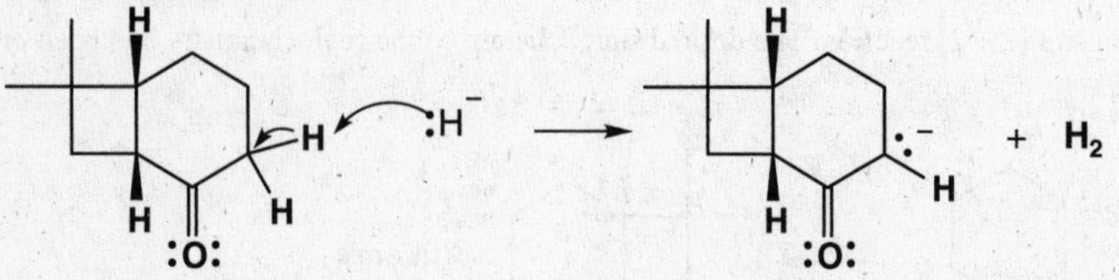

An interesting feature of this system is that the β-ketoester is in equilibrium with the enol as shown.

Step 2: Make a bond between a nucleophile and an electrophile and keto-enol tautomerism.

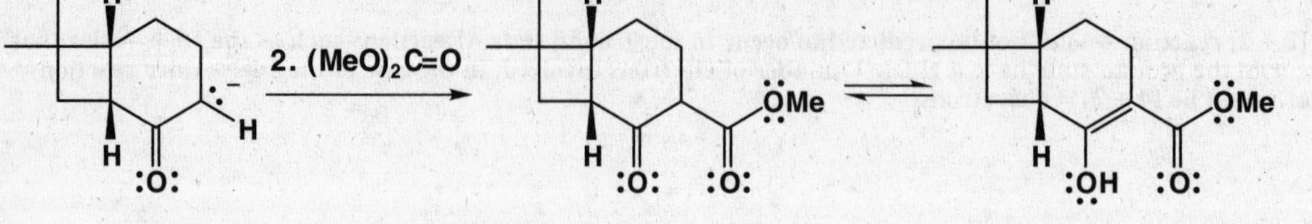

Step 3: Make a bond between a nucleophile and an electrophile and simultaneously break a bond to give stable molecules or ions. No base is required in the methylation step because MeI reacts directly with the enol to give the desired product.

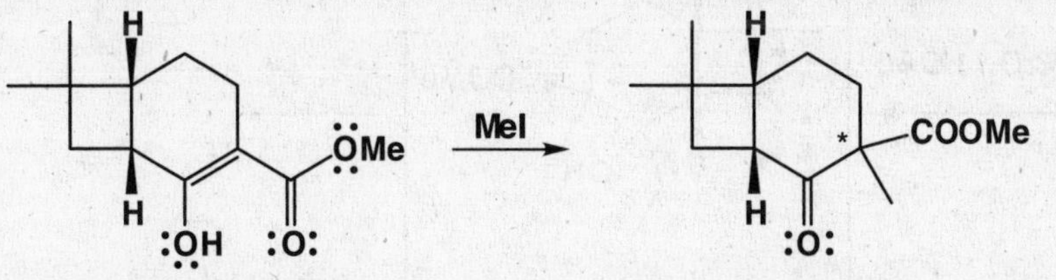

(d) Look at later steps in the synthesis. Does the stereochemistry of the added carbomethoxy group matter?

Because this chiral center is lost in a later ring expansion step, the stereochemistry of the added carbomethoxy group does not matter.

The next steps are shown here.

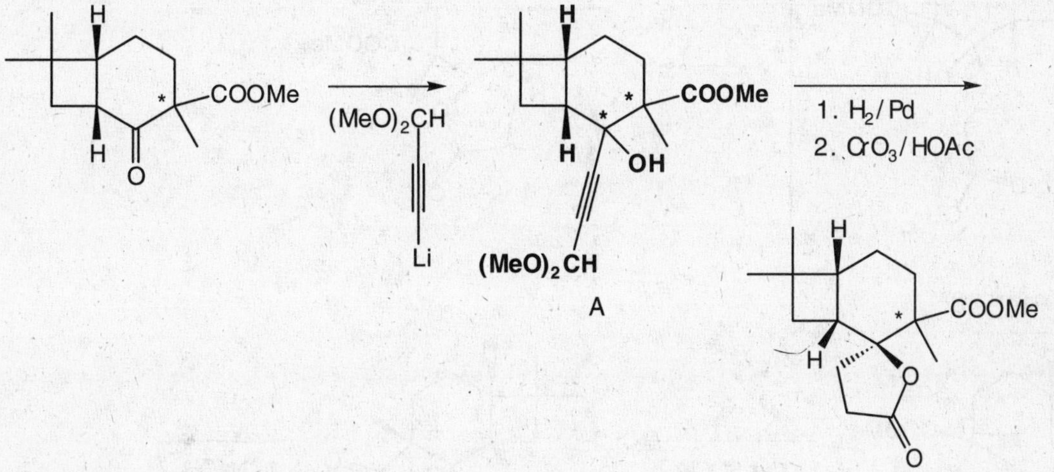

(e) What is the structure of compound (A)?

Compound A is the result of addition of the acetylide anion to the ketone, to give the corresponding alcohol.

(f) Give a mechanism for the formation of the cyclized product.

First, the alkyne is hydrogenated.

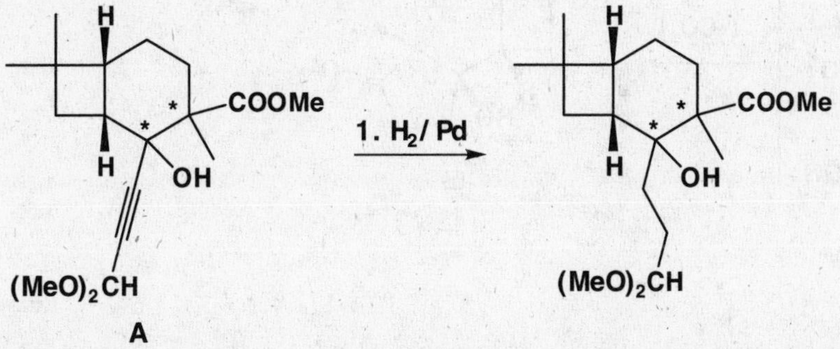

Next, in acidic solution, the dimethyl acetal is removed to reveal the aldehyde, which is immediately oxidized to the carboxylic acid. Under these acidic conditions, the lactone is formed from the carboxylic acid and alcohol functions.

Here are the next steps.

(g) Give a mechanism for the first step. (*Hint:* Attack on the lactone carbonyl may be the first step.)

A reasonable mechanism has an initial deprotonation of an H atom α to the lactone carbonyl to generate an enolate intermediate.

Step 1: Take a proton away.

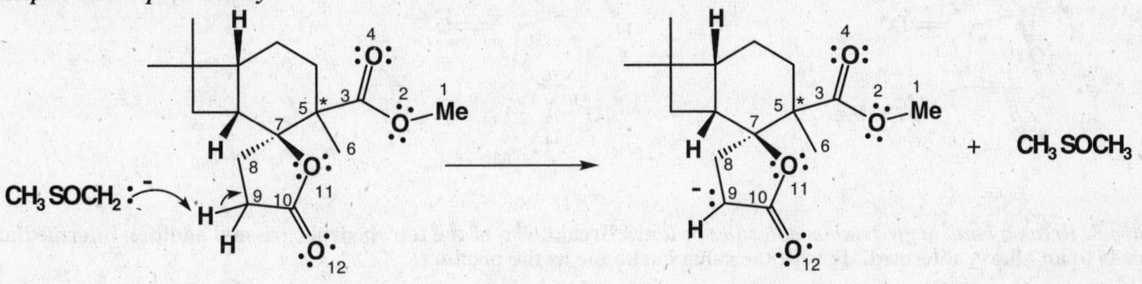

Enolate intermediate

The enolate intermediate can be redrawn from a different angle to emphasize proximity with the methyl ester carbonyl.

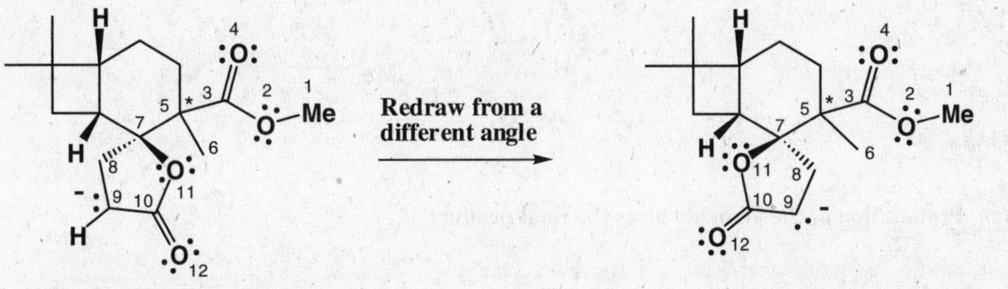

Redraw from a different angle

Enolate intermediate   **Enolate intermediate**

Step 2: Make a bond between a nucleophile and an electrophile. **Attack of the enolate onto the carbon labeled as 3 in a Dieckmann type reaction leads to a bicyclic intermediate.**

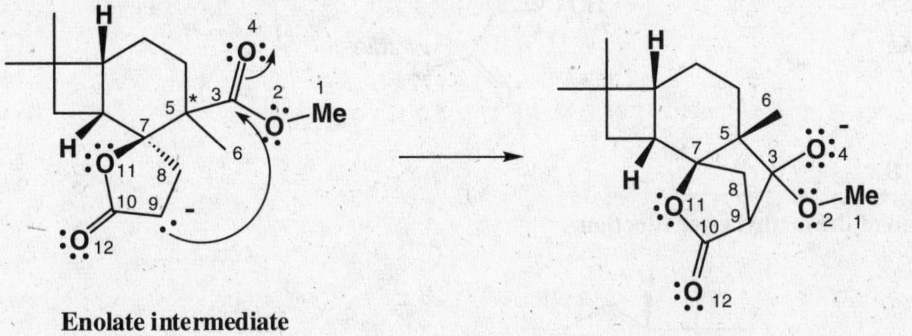

Enolate intermediate

Step 3: Break a bond to give stable molecules or ions. **Loss of methoxide generates a bicyclic keto lactone.**

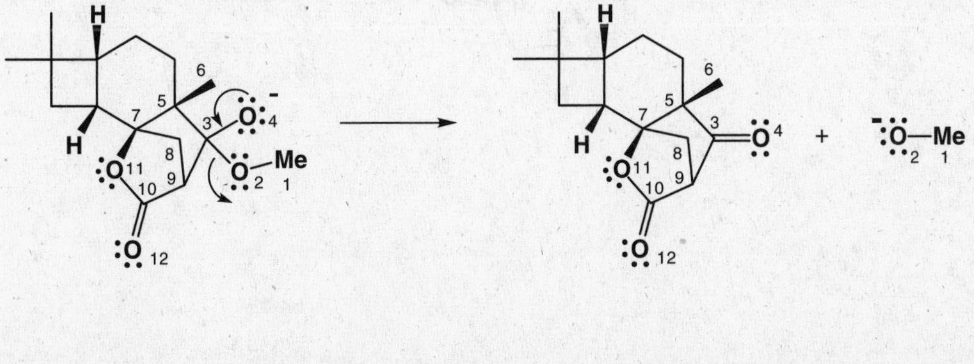

Step 4: Make a bond between a nucleophile and an electrophile. **Methoxide can then attack the lactone carbonyl.**

Step 5: Break a bond to give stable molecules or ions. **Breakdown of the tetrahedral carbonyl addition intermediate leads to an alkoxy intermediate with the same backbone as the product.**

Step 6: Add a proton. **Protonation of the alkoxide gives the final product.**

(h) Give a structure for product (B).

B results from decarboxylation of the methyl ester function.

B

The following two steps are next.

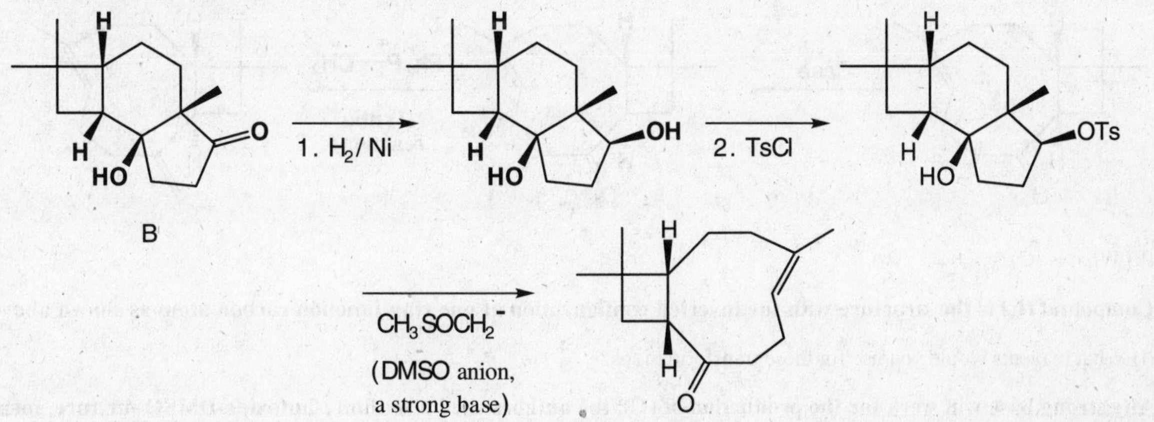
H
HO
O
B
1. H_2/ Ni
OH
2. TsCl
OTs
$CH_3SOCH_2^-$
(DMSO anion,
a strong base)

(i) Show the reactions of (B).

As shown in the above scheme, the hydrogenation reduces the ketone to an alcohol. In the next step, only the secondary alcohol reacts with the TsCl because the other -OH group is tertiary and is less reactive due to steric hindrance.

(j) Write a mechanism for the ring-opening reaction. *Hint:* Note the presence of an acidic proton and a good leaving group in the molecule.

Step 1: Take a proton away. **For the following, the atoms were renumbered compared to earlier parts of this question for clarity. A reasonable mechanism involves a first step in which the -OH group on the carbon labeled as 9 is deprotonated by the basic DMSO anion.**

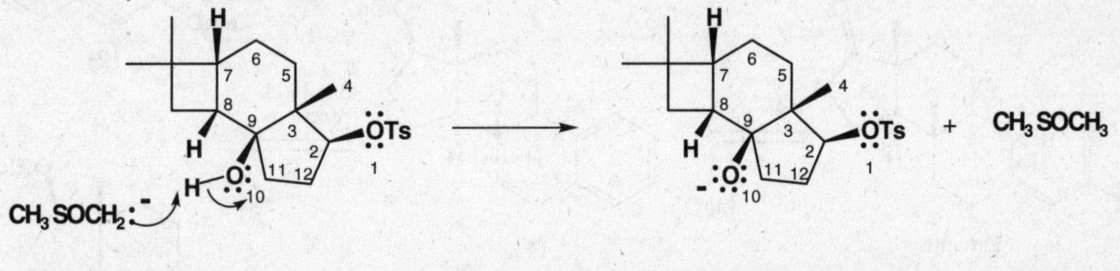
OTs
CH_3SOCH_2:
+
CH_3SOCH_3

Step 2: 1,2-Shift-break a bond to give stable molecules or ions. **The key ring expansion step involves formation of a ketone along with loss of the TsO⁻ leaving group. The bond between the carbons labeled as 3 and 9 is broken, and a new double bond is formed between the carbons labeled as 2 and 3 in the process.**

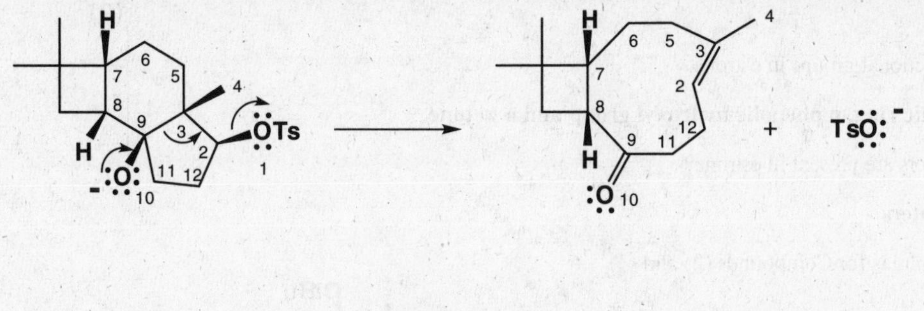
OTs
+
TsO:

The synthesis was completed by the following steps.

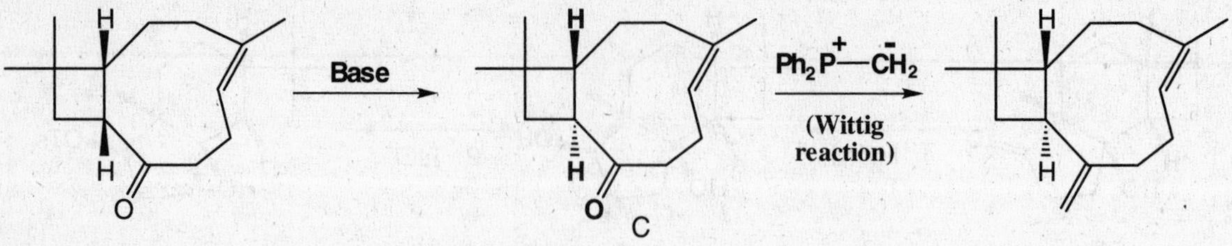

(k) What is (C)?

Compound (C) is the structure with the inverted configuration at one ring junction carbon atom as shown above.

(l) What reagents would you use for these transformations?

Any strong base will work for the production of (C), the authors used a sodium *t*-butoxide-DMSO mixture, meaning the DMSO anion was probably the important base in the reaction. The second reaction is a Wittig reaction involving the methylene Wittig reagent.

Problem 24.53   Over the past several decades, chemists have developed a number of synthetic methodologies for the synthesis of steroid hormones. One of these, developed by Lutz Tietze at the Institut für Organische Chemie der Georg-August-Universität, Göttingen, Germany, used a double Heck reaction to create ring B of the steroid nucleus. As shown in the following retrosynthetic analysis, a key intermediate in his synthesis is Compound (1). Two Heck reaction disconnects of this intermediate give compounds (2) and (3). Compound (2) contains the aromatic ring that becomes ring A of estrone. Compound (3) contains the fused five- and six-membered rings that become rings C and D of estrone.

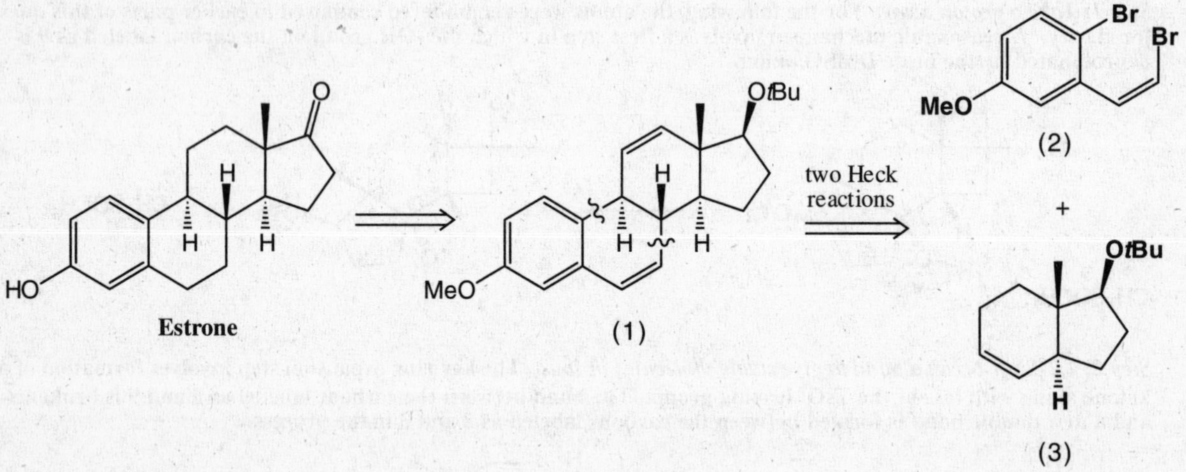

(a) Name the types of functional groups in estrone.

Estrone has one aromatic ring, a phenolic hydroxyl group and a ketone.

(b) How many chiral centers are present in estrone?

Estrone has 5 chiral centers.

(c) Propose structural formulas for Compounds (2) and (3).

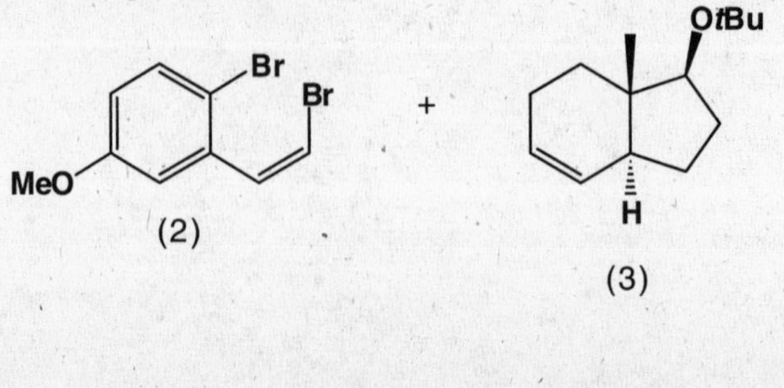

(d) Show how your proposals for Compounds (2) and (3) can be converted to Compound (1). *Note:* In the course of developing this synthesis, Tietze discovered that vinylic bromides and iodides are more reactive in Heck reactions than aryl bromides and iodides.

The key to this question is realizing that both an aryl halide and vinyl halide need to be on the same molecule (2), and that the vinyl halide would have to react first in a Heck reaction, followed by reaction of the aryl halide. As Tietze discovered that vinylic bromides and iodides are more reactive in Heck reactions than aryl bromides and iodides, therefore the best way to control the chemistry was to use a molecule with a relatively reactive vinyl halide, and a much less reactive aryl halide. Work with model systems indicated that aa dibromide had the right combination differential reactivity, yet high enough overall reactivity to give high yields in both steps.

The first Heck reaction takes place with the vinyl bromide as shown.

First Heck reaction

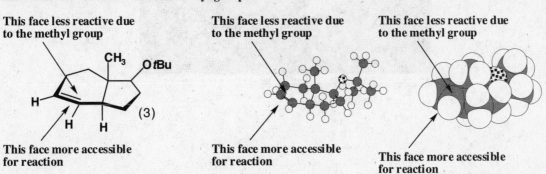

The second Heck reaction takes place with the aryl bromide. Here, a special Pd catalyst was used, presumably to compensate for the lower reactivity of the aryl bromide, and also to help assure the proper stereochemistry of the product.

Second Heck reaction

(e) In the course of the double Heck reactions, two new chiral centers are created. Assume that in Compound (3), the precursor to rings C and D of estrone, the fusion of rings C and D is *trans* and that the angular methyl group is above the plane of the ring. Given this stereochemistry, predict the stereochemistry of Compound (1) formed by the double Heck reaction.

The methyl group of (3) extends above the plane of the ring and represents a barrier to reaction on this face. Thus, both Heck reactions take place from underneath, forcing both H atoms at the ring junction upward, that is, the same side of the steroid skeleton as the methyl group.

This face less reactive due to the methyl group

This face less reactive due to the methyl group

This face less reactive due to the methyl group

(3)

This face more accessible for reaction

This face more accessible for reaction

This face more accessible for reaction

Facing upward

OtBu

A · B · C · D

MeO

(1)

(f) To convert (1) to estrone, the *tert*-butyl ether on ring D must be converted to a ketone. How might this transformation be accomplished?

This transformation is accomplished by cleaving the *tert*-butyl ether in acid, followed by oxidation of the resulting alcohol function to the ketone.

OtBu

MeO

(1)

$\xrightarrow{\text{Acid}}$

OH

MeO

$\xrightarrow{\text{Oxidation}}$

O

MeO

Garibaldi *Hypsypops rubicundus*
Catalina Island, California

CHAPTER 25
Solutions to the Problems

<u>Problem 25.1</u> Draw Fischer projections for all 2-ketopentoses. Which are D-2-ketopentoses, which are L-2-ketopentoses, and which are enantiomers?

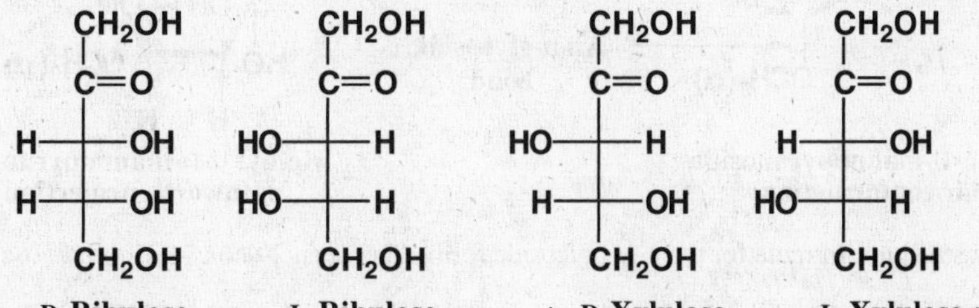

D-Ribulose   L-Ribulose   D-Xylulose   L-Xylulose

a pair of enantiomers   a pair of enantiomers

<u>Problem 25.2</u> Mannose exists in aqueous solution as a mixture of α-D-mannopyranose and β-D-mannopyranose. Draw Haworth projections for these molecules.

D-Mannose differs in configuration from D-glucose at carbon-2. Therefore, the alpha and beta forms of D-mannopyranose differ from those of alpha and beta D-glucopyranoses only in the orientation of the -OH on carbon-2. Following are Haworth projections for these compounds.

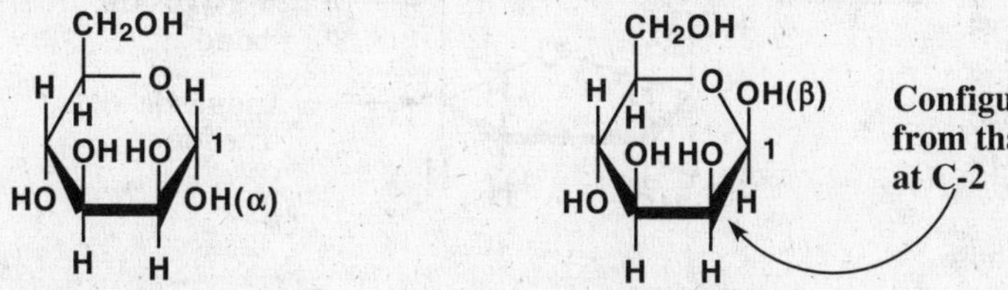

Configuration differs from that of D-glucose at C-2

α-D-Mannopyranose
(α-D-Mannose)

β-D-Mannopyranose
(β-D-Mannose)

<u>Problem 25.3</u> Draw chair conformations for α-D-mannopyranose and β-D-mannopyranose. Label the anomeric carbon in each.

D-Mannose differs in configuration from D-glucose at carbon-2. Draw chair conformations for the alpha and beta forms of D-glucopyranose and then invert the configuration of the -OH on carbon-2. For reference, the open chain form of D-mannose is also drawn.

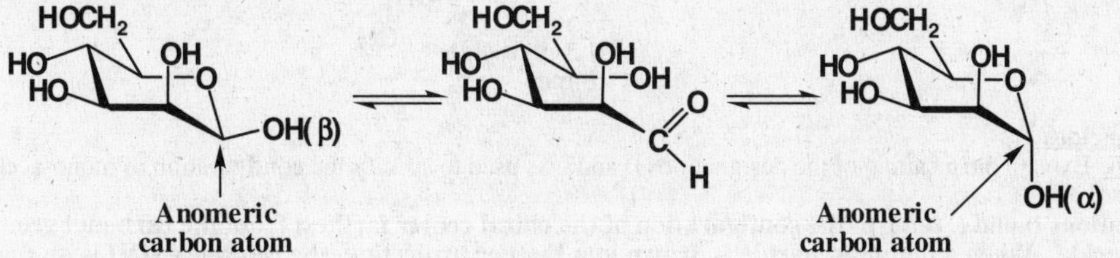

Anomeric carbon atom

Anomeric carbon atom

β-D-Mannopyranose
(β-D-Mannose)

D-Mannose

α-D-Mannopyranose
(α-D-Mannose)

Problem 25.4 Draw a Haworth projection and a chair conformation for methyl α-D-mannopyranoside (methyl α-D-mannoside). Label the anomeric carbon and the glycosidic bond.

Methyl α-D-mannopyranoside
(Chair conformation)

Methyl α-D-mannopyranoside
(Haworth projection)

Anomeric
carbon atom

An α-glycosidic
bond

Problem 25.5 Draw a structural formula for the β-N-glycoside formed between 2-deoxy-D-ribofuranose and adenine.

Following are structural formulas for adenine, the monosaccharide hemiacetal, and the N-glycoside.

Adenine

A β-N-glycosidic
bond

Anomeric
carbon

+ H₂O

β-2-Deoxy-D-Ribofuranose

Problem 25.6 Draw a chair conformation for the α form of a disaccharide in which two units of D-glucopyranose are joined by a β-1,3-glycosidic bond.

β-1,3-glycosidic
bond

Monosaccharides

Problem 25.7 Explain the meaning of the designations D and L as used to specify the configuration of monosaccharides.

The designations D and L refer to the configuration of the chiral center farthest from the carbonyl group of the monosaccharide. When a monosaccharide is drawn in a Fischer projection, the reference -OH is on the right in a D-monosaccharide and on the left in an L-monosaccharide. Note that the conventions D and L specify the configuration at one and only one of however many chiral centers there are in a particular monosaccharide.

Problem 25.8 How many chiral centers are present in D-glucose? In D-ribose?

The number of chiral centers present depends on whether you consider the open chain or cyclic forms of the molecules. In each case, an additional chiral center is present in the cyclic forms due to the anomeric carbon atom. For D-glucose there are 4 (open chain) and 5 chiral centers (cyclic form), while for D-ribose there are 3 (open chain) and 4 (cyclic form) chiral centers.

D-Glucose

D-Ribose

Problem 25.9 Which carbon of an aldopentose determines whether the pentose has a D or L configuration?

The stereochemistry at this
carbon determines
configuration for
aldopentoses

The reference point for determining D or L configuration is the chiral center farthest from the carbonyl group. This chiral center is always the next to the last carbon on the chain.

Problem 25.10 How many aldooctoses are possible? How many D-aldooctoses are possible?

The number of different aldooctoses possible is determined by the number of chiral centers in the open chain form of the molecule. There are 6 chiral centers in the open chain form of aldooctoses so there are $2^6 = 64$ possible. Half of those will have the D configuration, so there are 32 possible D-aldooctoses.

Problem 25.11 Which compounds are D-monosaccharides. Which are L-monosaccharides?

Compounds (a) and (c) are D-monosaccharides, and compound (b) is an L-monosaccharide.

Problem 25.12 Write Fischer projections for L-ribose and L-arabinose.

L-Ribose and L-arabinose are the mirror images of D-ribose and D-arabinose, respectively. The most common error in answering this question is to start with the Fischer projection for the D sugar and then invert the configuration of carbon-4 only. While the monosaccharide thus drawn is an L-sugar, it is not the correct one. All of the chiral centers must be changed to draw the true enantiomers.

D-Ribose   **L-Ribose**   **D-Arabinose**   **L-Arabinose**

Problem 25.13 What is the meaning of the prefix *deoxy-* as it is used in carbohydrate chemistry?

The prefix *deoxy-* means "without oxygen".

Problem 25.14 Give L-fucose (Chemical Connections: "A, B, AB, and O Blood group Substances") a name incorporating the prefix *deoxy-* that shows its relationship to galactose.

A systemic name for L-fucose is 6-*deoxy*-L-galactose.

Problem 25.15 2,6-Dideoxy-D-altrose, known alternatively as D-digitoxose, is a monosaccharide obtained on hydrolysis of digitoxin, a natural product extracted from foxglove (*Digitalis purpurea*). Digitoxin is used in cardiology to reduce pulse rate, regularize heart rhythm, and strengthen heart beat. Draw the structural formula of 2,6-dideoxy-D-altrose.

2,6-Dideoxy-D-altrose
(D-Digitoxose)

The Cyclic Structure of Monosaccharides

Problem 25.16 Define the term *anomeric carbon*. In glucose, which carbon is the anomeric carbon?

The new chiral center created from the carbonyl in forming the cyclic form of a carbohydrate is called the anomeric carbon.

β-D-Glucose

Problem 25.17 Define the terms (a) *pyranose* and (b) *furanose*.

The six-membered hemiacetal ring formed by a carbohydrate is called a pyranose. The five-membered hemiacetal ring formed by a carbohydrate is called a furanose.

Problem 25.18 Which is the anomeric carbon in a 2-ketohexose?

A 2-ketohexose forms a five-membered ring, and the anomeric carbon is the new chiral center created from the carbonyl.

D-Sorbose
(a 2-ketohexose)

Problem 25.19 Are α-D-glucose and β-D-glucose enantiomers? Explain.

They are not enantiomers because only the anomeric carbon atom is different. The other chiral centers are the same, thus α-D-glucose and β-D-glucose are diastereomers.

Problem 25.20 Convert each Haworth projection to an open-chain form and then to a Fischer projection. Name the monosaccharide you have drawn.

(a)

D-Allose

(b)

D-Altrose

Problem 25.21   Convert each chair conformation to an open-chain form and then to a Fischer projection. Name the monosaccharide you have drawn.

(a)

D-Galactose

(b)

D-Allose

Problem 25.22 Explain the phenomenon of mutarotation with reference to carbohydrates. By what means is it detected?

Any monosaccharide of four or more carbons can exist in an open-chain form and two or more cyclic hemiacetal (i.e. furanose or pyranose) forms, each having a different specific rotation. The specific rotation of an aqueous solution, measured with a polarimeter, of any one form changes until an equilibrium value is reached, representing an equilibrium concentration of the different forms. Mutarotation is the change in specific rotation toward an equilibrium value.

Problem 25.23 The specific rotation of α-D-glucose is +112.2.
(a) What is the specific rotation of α-L-glucose?

α-L-Glucose is the enantiomer of α-D-glucose so its specific rotation will be -112.2.

(b) When α-D-glucose is dissolved in water, the specific rotation of the solution changes from +112.2 to +52.7. Does the specific rotation of α-L-glucose also change when it is dissolved in water? If so, to what value does it change?

The specific rotation of α-D-glucose changes because it mutarotates until equilibrium is reached between α- and β-D-glucose. The same thing will happen with α-L-glucose, but because it is the enantiomer, the equilibrium specific rotation will be -52.7.

Reactions of Monosaccharides

Problem 25.24 Draw Fischer projections for the product(s) formed by reaction of D-galactose with the following. In addition, state whether each product is optically active or inactive.

(a) NaBH₄ in H₂O   (b) H₂/Pt   (c) HNO₃, warm   (d) Br₂, H₂O/CaCO₃

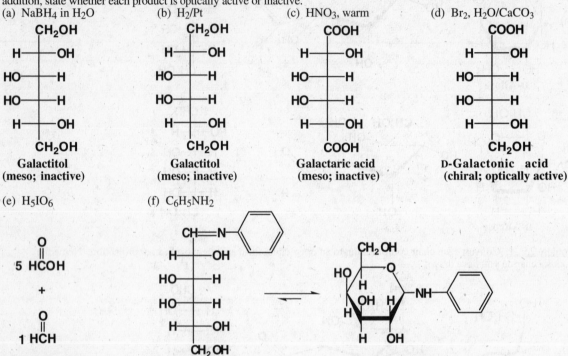

Galactitol
(meso; inactive)

Galactitol
(meso; inactive)

Galactaric acid
(meso; inactive)

D-Galactonic acid
(chiral; optically active)

(e) H₅IO₆   (f) C₆H₅NH₂

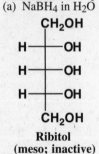

Formic acid and formaldehyde
(achiral; inactive)

A Schiff base of D-galactose

The cyclic form predominates
(chiral; optically active)

Problem 25.25 Repeat Problem 25.24 using D-ribose.

(a) NaBH₄ in H₂O   (b) H₂/Pt   (c) HNO₃, warm   (d) Br₂, H₂O/CaCO₃

Ribitol
(meso; inactive)

Ribitol
(meso; inactive)

D-Ribaric acid
(meso; inactive)

D-Ribonic acid
(chiral; optically active)

(e) H_5IO_6

$$4 \ \underset{\substack{| \\ \text{OH}}}{\overset{\substack{\text{O} \\ \|}}{\text{HC}}}$$

+

$$1 \ \overset{\substack{\text{O} \\ \|}}{\text{HCH}}$$

**Formic acid and
formaldehyde
(achiral; inactive)**

(f) $C_6H_5NH_2$

**A Schiff base of
D-ribose**

**The cyclic form
predominates
(chiral; optically active)**

Problem 25.26 An important technique for establishing relative configurations among isomeric aldoses and ketoses is to convert both terminal carbon atoms to the same functional group. This can be done either by selective oxidation or reduction. As a specific example, nitric acid oxidation of D-erythrose gives meso-tartaric acid (Section 3.4B). Similar oxidation of D-threose gives (2S,3S)-tartaric acid. Given this information and the fact that D-erythrose and D-threose are diastereomers, draw Fischer projections for D-erythrose and D-threose. Check your answers against Table 25.1.

D-Erythrose   **meso-Tartaric acid**   **D-Threose**   **D-Tartaric acid**

Problem 25.27 There are four D-aldopentoses (Table 25.1). If each is reduced with $NaBH_4$, which yield optically active alditols? Which yield optically inactive alditols?

D-Ribose and D-xylose yield different achiral (meso) alditols. D-Arabinose and D-lyxose yield the same chiral alditol.

D-Ribose   **Ribitol (meso)**   **D-Xylose**   **Xylitol (meso)**

D-Arabinose   **D-Arabinitol**   **D-Lyxose**

$$[\alpha]^{25}_D = -32$$

Problem 25.28 Name the two alditols formed by $NaBH_4$ reduction of D-fructose?

D-Glucitol and D-mannitol. Each differs in configuration only at carbon-2.

Problem 25.29 One pathway for the metabolism of D-glucose 6-phosphate is its enzyme-catalyzed conversion to D-fructose 6-phosphate. Show that this transformation can be accomplished as two enzyme-catalyzed keto-enol tautomerisms.

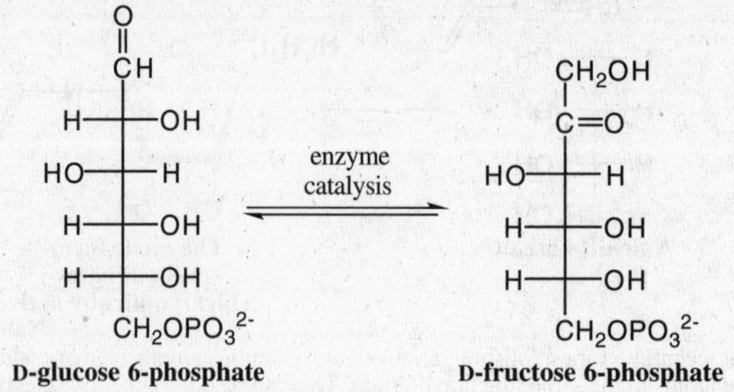

D-glucose 6-phosphate   **D-fructose 6-phosphate**

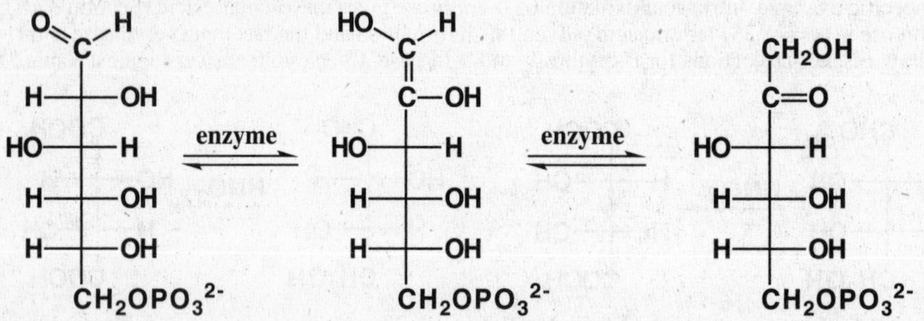

Problem 25.30 L-Fucose, one of several monosaccharides commonly found in the surface polysaccharides of animal cells, is synthesized biochemically from D-mannose in the following eight steps:

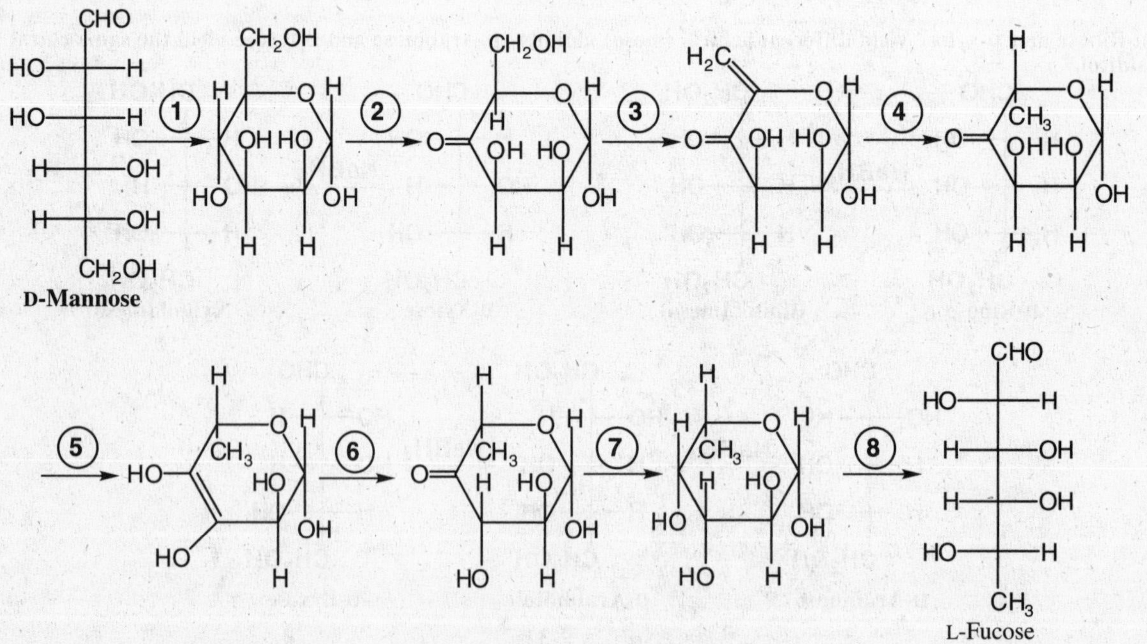

(a) Describe the type of reaction (that is, oxidation, reduction, hydration, dehydration, and so on) involved in each step.

Following is the type of reaction in each step.
(1) Formation of a cyclic hemiacetal from a carbonyl group and a secondary alcohol.
(2) A two-electron oxidation of a secondary alcohol to a ketone.
(3) Dehydration of a β-hydroxyketone to an α,β-unsaturated ketone.
(4) A two-electron reduction of a carbon-carbon double bond to a carbon-carbon single bond.
(5) Keto-enol tautomerism of an α-hydroxyketone to form an enediol.

(6) **Keto-enol tautomerism of an enediol to form an α-hydroxyketone.**
(7) **A two-electron reduction of a ketone to a secondary alcohol.**
(8) **Opening of a cyclic hemiacetal to form an aldehyde and an alcohol.**

(b) Explain why this monosaccharide derived from D-mannose now belongs to the L series.

It is the configuration at carbon-5 of this aldohexose that determines whether it is of the D-series or of the L-series. The result of steps 3 and 4 is inversion of configuration at carbon-5 and, therefore, conversion of a D-aldohexose to an L-aldohexose.

Problem 25.31 What is the difference in meaning between the terms glycosidic bond and glucosidic bond?

A glycosidic bond is the bond from the anomeric carbon of a glycoside to an -OR group. A glucosidic bond is a glycosidic bond that yields glucose upon hydrolysis.

Problem 25.32 Treatment of methyl β-D-glucopyranoside with benzaldehyde forms a six-membered cyclic acetal. Draw the most stable conformation of this acetal. Identify each new chiral center in the acetal.

Problem 25.33 Vanillin (4-hydroxy-3-methoxybenzaldehyde), the principal component of vanilla, occurs in vanilla beans and other natural sources as a β-D-glucopyranoside. Draw a structural formula for this glycoside, showing the D-glucose unit as a chair conformation.

Vanillin

Problem 25.34 Hot water extracts of ground willow and poplar bark are an effective pain reliever. Unfortunately, the liquid is so bitter that most persons refuse it. The pain reliever in these infusions is salicin, a β-glycoside of D-glucopyranose and the phenolic -OH group of 2-(hydroxymethyl)phenol. Draw a structural formula for salicin, showing the glucose ring as a chair conformation.

Problem 25.35 Draw structural formulas for the products formed by hydrolysis at pH 7.4 (the pH of blood plasma) of all ester, thioester, amide, anhydride, and glycoside groups in acetyl coenzyme A. Name as many of the products as you can.

Acetyl coenzyme A
(Acetyl-CoA)

Following are the smaller molecules formed by hydrolysis of each amide, ester, glycosidic, and anhydride bond. They are arranged to correspond roughly to their location from left to right in acetyl CoA.

CH_3COH
Acetic acid

$HOCCH_2CH_2NH_2$
3-Aminopropanoic acid

$2 HPO_4^{2-}$
Phosphate

Adenine

$HSCH_2CH_2NH_2$
2-Aminoethanethiol

$HOCCHCCH_2OH$
2,4-Dihydroxy-3,3-dimethylbutanoic acid

β-D-Ribofuranose

HPO_4^{2-}
Phosphate

The molecule formed by amide formation between 3-aminopropanoic acid and 2,4-dihydroxy-3,3-dimethylbutanoic acid is given the special name pantothenic acid. Pantothenoic acid is a vitamin, most commonly contained in vitamin pills as calcium pantothenate. Its minimum daily requirement (MDR) has not yet been determined.

$HOCCH_2CH_2NHCCHCCH_2OH$
Pantothenic acid

Disaccharides and Oligosaccharides

Problem 25.36 In making candy or sugar syrups, sucrose is boiled in water with a little acid, such as lemon juice. Why does the product mixture taste sweeter than the starting sucrose solution?

Sucrose is a disaccharide composed of the monosaccharides glucose and fructose linked through a glycosidic bond. The acid catalyzes hydrolysis of the glycosidic bond, and the monomeric glucose and fructose are more soluble than sucrose itself. Because of this, the syrups and candy have higher concentrations of these sugars than is possible with sucrose. As a result, the candy and syrup taste sweeter. Furthermore, fructose actually tastes sweeter than sucrose, having a relative sweetness of 174, compared with 100 for sucrose. Thus, converting the sucrose into fructose increases the sweetness of the mixture.

Problem 25.37 Trehalose is found in young mushrooms and is the chief carbohydrate in the blood of certain insects. Trehalose is a disaccharide consisting of two D-monosaccharide units each joined to the other by an α-1,1-glycosidic bond.

Trehalose

(a) Is trehalose a reducing sugar?

Trehalose is not a reducing sugar because each anomeric carbon is involved in formation of the glycosidic bond.

(b) Does trehalose undergo mutarotation?

It will not undergo mutarotation for the reason given in (a).

(c) Name the two monosaccharide units of which trehalose is composed.

Both monosaccharide units are D-Glucose.

Problem 25.38 The trisaccharide raffinose occurs principally in cottonseed meal.

Raffinose

(a) Name the three monosaccharide units in raffinose.

The three monosaccharide units in raffinose, from top to bottom, are D-galactose, D-glucose, and D-fructose.

(b) Describe each glycosidic bond in this trisaccharide.

Reading from left to right, they are D-galactopyranose joined by an α-1,6-glycosidic bond to D-glucopyranose and then D-glucopyranose, in turn, joined by an α-1,2-glycosidic bond to β-D-fructofuranose.

(c) Is raffinose a reducing sugar?

No, it is not a reducing sugar.

(d) With how many moles of periodic acid will raffinose react?

Raffinose will react with 5 mol of HIO$_4$; 2 mol for the D-galactopyranose ring, 2 mol for the D-glucopyranose ring, and 1 mol for the D-fructofuranose ring.

Problem 25.39 Amygdalin is a toxic component in the pits of bitter almonds, peaches and apricots.

(a) Name the two monosaccharide units in amygdalin and describe the glycoside bond by which they are joined.

Both of the monosaccharide units in amygdalin are D-glucose. In both cases they have a β-glycosidic bond, the bond between the β-D-glucopyranose units being a β-1,6-glycosidic bond.

(b) Account for the fact that on hydrolysis of amygdalin in warm aqueous acid liberates benzaldehyde and HCN.

Hydrolysis of the glycosidic bond in aqueous acid gives D-glucose and the cyanohydrin of benzaldehyde. This cyanohydrin is in equilibrium with benzaldehyde and HCN.

Problem 25.40 Following is a structural formula for stachyose, a water-soluble tetrasaccharide component of many plants including lentils and soybeans. Humans cannot digest stachyose, and its accumulation leads to distension of the gut and flatulence.

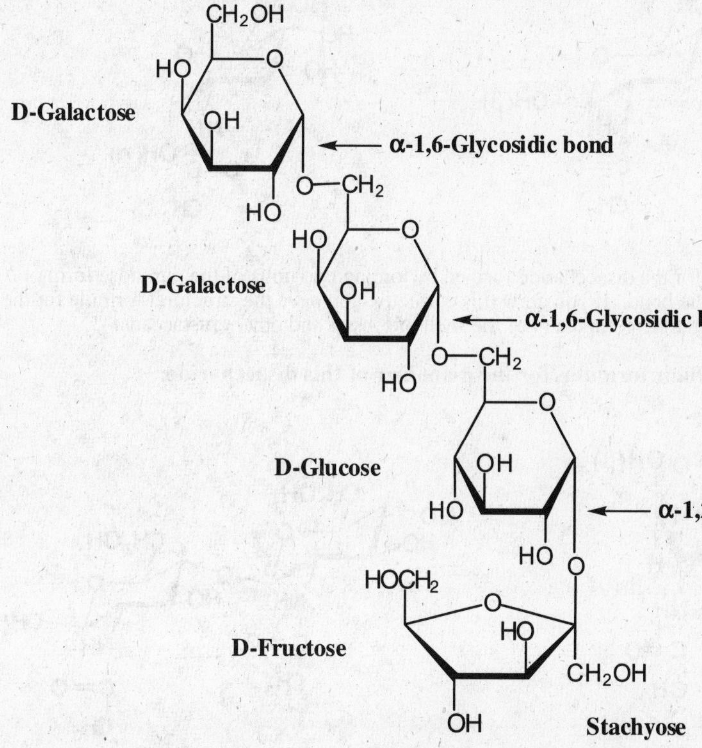

Stachyose

(a) Name each monosaccharide unit in stachyose D-monosaccharide or an L-monosaccharide.

Stachyose is composed of two D-galactose, one D-glucose, and one D-fructose monosaccharides.

(b) Describe each glycosidic bond in stachyose.

There are two α-1,6 and one α-1,2 glycosidic bonds in stachyose.

Polysaccharides
Problem 25.41 What is the difference in structure between oligo- and polysaccharides?

The general term oligosaccharide is often used for carbohydrates that contain from four to ten monosaccharide units. Carbohydrates containing larger numbers of monosaccharide units are called polysaccharides.

Problem 25.42 Why is cellulose insoluble in water?

The polysaccharide chains of cellulose are held together through extensive networks of hydrogen bonds. The chains are held together so tightly that water molecules cannot effectively solvate them.

Problem 25.43 Consider *N*-acetyl-D-glucosamine (Section 25.1D).

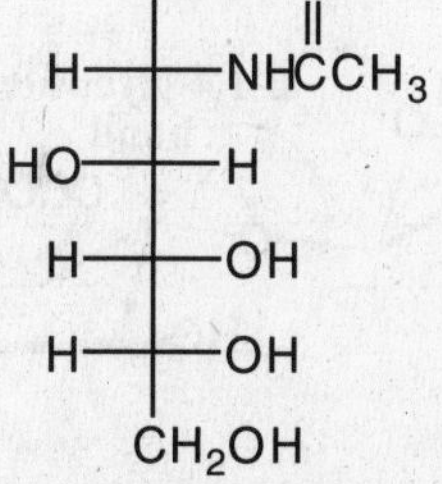

N-**Acetyl-D-glucosamine**

(a) Draw a chair conformation for the α- and β-pyranose forms of this monosaccharide.

To draw the α- and β-pyranose forms, invert configuration at carbon-1.

(b) Draw a chair conformation for the disaccharide formed by joining two units of the pyranose form of *N*-acetyl-D-glucosamine by a β-1,4-glycosidic bond. If you drew this correctly, you have the structural formula for the repeating dimer of chitin, the structural polysaccharide component of the shell of lobster and other crustaceans.

Following are Haworth and chair formulas for the β-anomer of this disaccharide.

Problem 25.44 Propose structural formulas for the following polysaccharides:
(a) Alginic acid, isolated from seaweed, is used as a thickening agent in ice cream and other foods. Alginic acid is a polymer of D-mannuronic acid in the pyranose form joined by β-1,4-glycosidic bonds.

D-Mannuronic acid   D-Galacturonic acid

Following is the chair conformation for repeating disaccharide units of alginic acid.

Alginic acid

(b) Pectic acid is the main component of pectin, which is responsible for the formation of jellies from fruits and berries. Pectic acid is a polymer of D-galacturonic acid in the pyranose form joined by α-1,4-glycosidic bonds.

Following is the chair conformation for repeating disaccharide units of pectic acid.

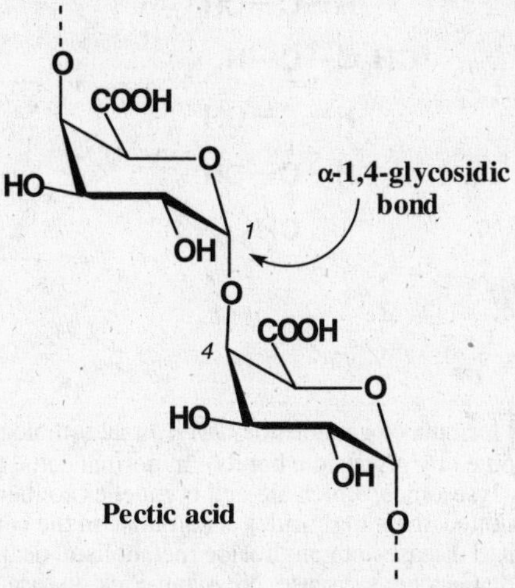

Pectic acid

Problem 25.45 Digitalis is a preparation made from the dried seeds and leaves of the purple foxglove, *Digitalis purpurea* , a plant native to southern and central Europe and cultivated in the United States. The preparation is a mixture of several active components, including digitalin. Digitalis is used in medicine to increase the force of myocardial contraction and as a conduction depressant to decrease heart rate (the heart pumps more forcefully but less often).

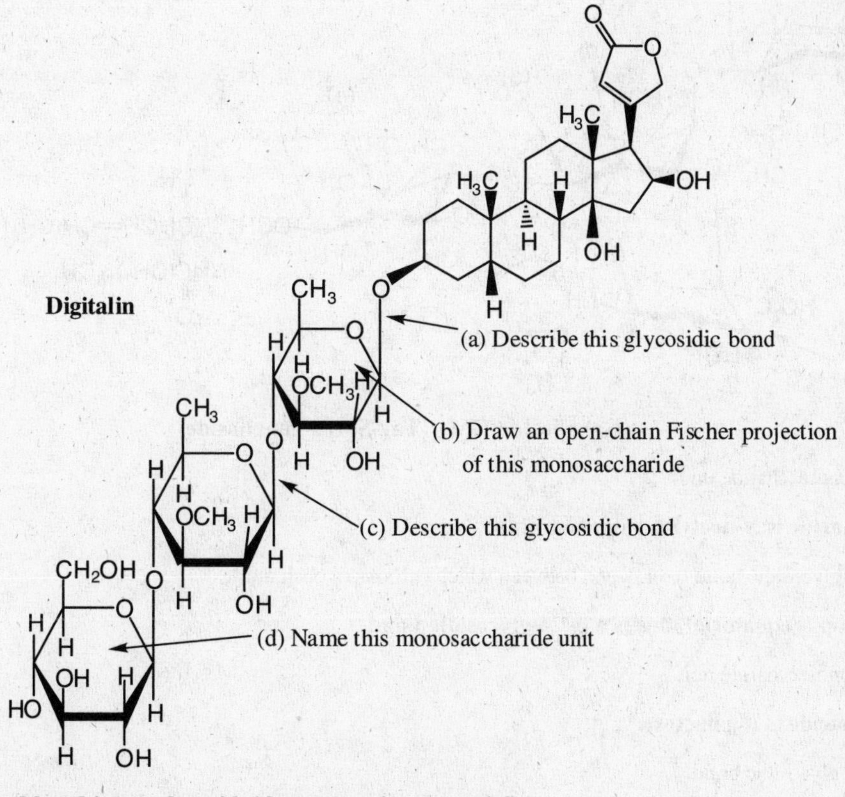

Digitalin

(a) Describe this glycosidic bond

(b) Draw an open-chain Fischer projection of this monosaccharide

(c) Describe this glycosidic bond

(d) Name this monosaccharide unit

a) The indicated bond is a β-glycoside(the oxygen is equatorial).

b) The first monosaccharide corresponds to the following Fischer projection.

$$
\begin{array}{c}
\mathrm{O}{=}\mathrm{C}{-}\mathrm{H} \\
\mathrm{H}{-}\mathrm{C}{-}\mathrm{OH} \\
\mathrm{CH_3O}{-}\mathrm{C}{-}\mathrm{H} \\
\mathrm{H}{-}\mathrm{C}{-}\mathrm{OH} \\
\mathrm{H}{-}\mathrm{C}{-}\mathrm{OH} \\
\mathrm{CH_3}
\end{array}
$$

c) This bond is a β-1,4-glycosidic bond.

d) This monosaccharide is glucose.

Problem 25.46 Following is the structural formula of ganglioside GM_2, a macromolecular glycolipid (meaning that it contains lipid and monosaccharide units joined by glycosidic bonds). In normal cells, this and other gangliosides are synthesized continuously and degraded by lysosomes, which are cell organelles containing digestive enzymes. If pathways for the degradation of gangliosides are inhibited, the gangliosides accumulate in the central nervous system causing all sorts of life-threatening consequences. In inherited diseases of ganglioside metabolism, death usually occurs at an early age. Diseases of ganglioside metabolism include Gaucher's disease, Niemann-Pick disease, and Tay-Sachs disease. Tay-Sachs disease is a hereditary defect that is transmitted as an autosomal recessive gene. The concentration of ganglioside GM_2 is abnormally high in this disease because the enzyme responsible for catalyzing the hydrolysis of glycosidic bond (b) is absent.

Ganglioside GM_2 (Tay-Sachs ganglioside)

(a) Name this monosaccharide unit.

This monosaccharide is *N*-acetyl-D-galactosamine

(b) Describe this glycosidic bond (α or β, and between which carbons of each unit).

Because the group is equatorial, this is a β-1,4-glycosidic bond.

(c) Name this monosaccharide unit.

This monosaccharide is D-galactose.

(d) Describe this glycosidic bond.

This is also a β-1,4-glycosidic bond.

(e) Name this monosaccharide unit.

This monosaccharide is D-glucose

(f) Describe this glycosidic bond.

This is a β-glycosidic bond.

(g) This unit is *N*-acetylneuraminic acid, the most abundant member of a family of amino sugars containing nine or more carbons and distributed widely throughout the animal kingdom. Draw the open-chain form of this amino sugar. Do not be concerned with the configuration of the five chiral centers in the open-chain form.

Problem 25.47 Hyaluronic acid acts as a lubricant in the synovial fluid of joints. In rheumatoid arthritis, inflammation breaks hyaluronic acid down to smaller molecules. Under these conditions, what happens to the lubricating power of the synovial fluid?

Smaller molecules are less effective lubricants because they are less able to prevent contact between joint components compared to longer molecules.

Problem 25.48 The anticlotting property of heparin is partly due to the negative charges it carries.
(a) Identify the functional groups that provide the negative charges.

Carboxylates and sulfonates are the functional groups that provide the negative charges of heparin.

(b) Which type of heparin is a better anticoagulant, one with a high or a low degree of polymerization?

Highly polymerized heparin is a better anticoagulant because it is more effective at binding to antithrombin III, its enzyme target.

Problem 25.49 Keratin sulfate is an important component of the cornea of the eye. Following is the repeating unit of this acidic polysaccharide.

(a) From what monosaccharides or derivatives of monosaccharides is keratin sulfate made?

Keratin sulfate is made from D-Galactose and *N*-acetyl-D-glucosamine.

(b) Describe the glycosidic bond in this repeating disaccharide unit.

This is a β-1,4-glycosidic bond

(c) What is the net charge on this repeating disaccharide unit at pH 7.0?

The repeating disaccharide unit has a single negative charge at pH 7.0 due to the sulfonate group.

Problem 25.50 Following is a chair conformation for the repeating disaccharide unit in chondroitin 6-sulfate. This biopolymer acts as the flexible connecting matrix between the tough protein filaments in cartilage. It is available as a dietary supplement, often combined with D-glucosamine sulfate. Some believe this combination can strengthen and improve joint flexibility.

(a) From what two monosaccharide units is the repeating disaccharide unit of chondroitin 6-sulfate derived?

The two monosaccharide units are D-glucuronic acid and *N*-acetyl-D-galactosamine.

(b) Describe the glycosidic bond between the two units.

The bond is a β-1,3-glycosidic bond.

Problem 25.51 Following is a structural formula for the repeating disaccharide unit of dermatan sulfate. Dermatan sulfate is a component of the extracellular matrix of the skin.

Repeating disaccharide unit
of **Dermatan Sulfate**

(a) Name the monosaccharide from which each unit of this disaccharide is derived.

The repeating disaccharide is composed of glucouronic acid and *N*-acetyl galactosamine sulfate units.

(b) Describe the glycoside bonds in dermatan sulfate.

There are β-1,3 and β-1,4-glycosidic bonds linking the units of Dermatan sulfate.

CHAPTER 26
Solutions to the Problems

<u>Problem 26.1</u> (a) How many constitutional isomers are possible for a triglyceride containing one molecule each of palmitic acid, oleic acid, and stearic acid?

There are three constitutional isomers possible, the difference being which fatty acid is in the middle of the molecule:

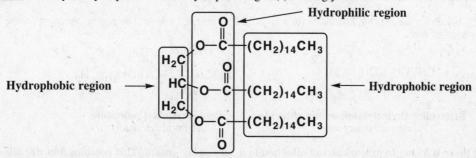

Each of these molecules has one chiral center as indicated by the asterisk, so each constitutional isomer shown above can exist as a pair of enantiomers. Thus there are 2 x 3 = 6 total isomers possible. Note that for oleic acid, the carbon-carbon double bond is assumed to have the Z (_cis_) configuration only.

(b) Which of these constitutional isomers are chiral?

They are all chiral because they each have a chiral center at the middle carbon of the glycerol moiety.

<u>Problem 26.2</u> Define the term *hydrophobic*.

A good working definition of the term hydrophobic is "dislikes water". Hydrophobic materials are nonpolar, and do not dissolve in water. Because they are not solvated, hydrophobic surfaces of molecules aggregate in polar solvents like water. This aggregation behavior of hydrophobic surfaces is an extremely important phenomenon that underlies a number of important aspects of biological chemistry such as membrane formation, surfactant action, protein folding, and protein-protein recognition.

<u>Problem 26.3</u> Identify the hydrophobic and the hydrophilic region(s) of a triglyceride.

Note that the vast majority of the molecule is hydrophobic, thus explaining why triglycerides are so hydrophobic overall.

<u>Problem 26.4</u> Explain why the melting points of unsaturated fatty acids are lower than those of saturated fatty acids.

When fatty acids pack together better, the attractive dispersion forces between molecules are stronger, thereby increasing the melting point. Saturated fatty acids can adopt a much more regular, compact structure compared to unsaturated fatty acids that have a kink induced by the _cis_ double bond. The more regular compact saturated fatty acids can pack together better, so their melting points are higher.

<u>Problem 26.5</u> Which would you expect to have the higher melting point, glyceryl trioleate or glyceryl trilinoleate?

***Cis* double bonds disrupt chain packing and thus lower melting points. The triglyceride with fewer *cis* double bonds will have the higher melting point. Each oleic acid unit has only one *cis* double bond, while each linoleic acid has two (Please see Table 26.1). Glycerol trioleate will have the higher melting point.**

Problem 26.6 Draw a structural formula for methyl linoleate. Be certain to show the correct configuration of groups about each carbon-carbon double bond.

Methyl linoleate

Problem 26.7 Explain why coconut oil is a liquid triglyceride, even though most of its fatty acid components are saturated.

Triglycerides having fatty acids with shorter chains have lower melting points. As can be seen in Table 26.2, coconut oil is 45% lauric acid. Lauric acid is only a C12 fatty acid, so coconut oil has a melting point that is low enough to make it a liquid near room temperature.

Problem 26.8 It is common now to see "contains no tropical oils" on cooking oil labels, meaning that the oil contains no palm or coconut oil. What is the difference between the composition of tropical oils and that of vegetable oils, such as corn oil, soybean oil, and peanut oil?

Tropical oils are considerably richer in low-molecular-weight saturated fatty acids.

Problem 26.9 What is meant by the term *hardening* as applied to fats and oils?

The term "hardening" refers to the process of catalytic hydrogenation using H_2 and a transition metal with polyunsaturated plant oils. By removing the Z double bonds, the reduction reaction allows the fatty acids to pack together better and thus the triacylglycerols become more solid.

Problem 26.10 How many moles of H_2 are used in the catalytic hydrogenation of one mole of a triglyceride derived from glycerol and equal portions of stearic acid, linoleic acid, and arachidonic acid?

One molecule of H_2 is used per double bond in the triglyceride. Stearic acid does not have any double bonds, linoleic acid has 2 and arachidonic acid has 4 double bonds, respectively. Thus, 2 + 4 = 6 mole of H_2 will be used per mole of the triglyceride.

Problem 26.11 Characterize the structural features necessary to make a good synthetic detergent.

A good synthetic detergent should have a long hydrocarbon tail and a very polar group at one end. This combination will allow for the production of micelle structures in aqueous solution that will dissolve hydrophobic dirt such as grease and oil. The very polar group should not form insoluble salts with the ions normally found in hard water such as Ca(II), Mg(II), and Fe(III).

Problem 26.12 Following are structural formulas for a cationic detergent and a neutral detergent. Account for the detergent properties of each.

Benzyldimethyloctylammonium chloride
(a cationic detergent)

Pentaerythrityl palmitate
(a neutral detergent)

In each case there is a long hydrocarbon tail attached to a very polar group. This combination will allow for the production of micelle structures in aqueous solution that will dissolve nonpolar, hydrophobic dirt such as grease and oil. In the case of benzyldimethyloctylammonium chloride, the polar group is the positively-charged ammonium group, while for the pentaerythrityl palmitate the polar group is composed of the triol functions.

Problem 26.13 Identify some of the detergents used in shampoos and dish washing solutions. Are they primarily anionic, neutral, or cationic detergents.

Most detergents in shampoos and dish washing solutions are anionic detergents such as sodium lauryl sulfate.

Sodium Lauryl Sulfate

Problem 26.14 Show how to convert palmitic acid (hexadecanoic acid) into the following:
(a) Ethyl palmitate

$$CH_3(CH_2)_{14}\overset{\overset{\displaystyle O}{\|}}{C}OH \;+\; CH_3CH_2OH \;\xrightarrow{H^+}\; CH_3(CH_2)_{14}\overset{\overset{\displaystyle O}{\|}}{C}OCH_2CH_3$$

Ethyl palmitate

(b) Palmitoyl chloride

$$CH_3(CH_2)_{14}\overset{\overset{\displaystyle O}{\|}}{C}OH \;+\; SOCl_2 \;\longrightarrow\; CH_3(CH_2)_{14}\overset{\overset{\displaystyle O}{\|}}{C}Cl$$

Palmitoyl chloride

(c) 1-Hexadecanol (cetyl alcohol)

$$CH_3(CH_2)_{14}\overset{\overset{\displaystyle O}{\|}}{C}OH \;\xrightarrow[\;\textbf{2) }H_2O\;]{\textbf{1) LiAlH}_4\textbf{, ether or THF}}\; CH_3(CH_2)_{14}CH_2OH$$

1-Hexadecanol
(Cetyl alcohol)

(d) 1-Hexadecamine

$$CH_3(CH_2)_{14}\overset{\overset{\displaystyle O}{\|}}{C}OH \;+\; SOCl_2 \;\longrightarrow\; CH_3(CH_2)_{14}\overset{\overset{\displaystyle O}{\|}}{C}Cl \;\xrightarrow{NH_3}\; CH_3(CH_2)_{14}\overset{\overset{\displaystyle O}{\|}}{C}NH_2$$

$$\xrightarrow[\;\textbf{2. }H_2O\;]{\textbf{1. LiAlH}_4\textbf{, ether or THF}}\; CH_3(CH_2)_{14}CH_2NH_2$$

1-Hexadecanamine

(e) *N,N*-Dimethylhexadecanamide

$$CH_3(CH_2)_{14}\overset{\overset{\displaystyle O}{\|}}{C}OH \;+\; SOCl_2 \;\longrightarrow\; CH_3(CH_2)_{14}\overset{\overset{\displaystyle O}{\|}}{C}Cl \;\xrightarrow{HN(CH_3)_2}\; CH_3(CH_2)_{14}\overset{\overset{\displaystyle O}{\|}}{C}N(CH_3)_2$$

***N,N*-Dimethylhexadecanamide**

Problem 26.15 Palmitic acid (hexadecanoic acid) is the source of the hexadecyl (cetyl) group in the following compounds. Each is a mild surface-acting germicide and fungicide and is used as a topical antiseptic and disinfectant. They are examples of quaternary ammonium detergents, commonly called "quats."

Cetylpyridinium chloride   **Benzylcetyldimethylammonium chloride**

(a) Cetylpyridinium chloride is prepared by treating pyridine with 1-chlorohexadecane (cetyl chloride). Show how to convert palmitic acid to cetyl chloride.

$$CH_3(CH_2)_{14}\overset{\overset{\displaystyle O}{\|}}{C}OH \;\xrightarrow[\;\textbf{2) }H_2O\;]{\textbf{1) LiAlH}_4\textbf{, ether or THF}}\; CH_3(CH_2)_{14}CH_2OH \;\xrightarrow{SOCl_2}\; CH_3(CH_2)_{14}CH_2Cl$$

1-Chlorohexadecane
(Cetyl chloride)

(b) Benzylcetyldimethylammonium chloride is prepared by treating benzyl chloride with *N,N*-dimethyl-1-hexadecanamine. Show how this tertiary amine can be prepared from palmitic acid.

$$CH_3(CH_2)_{14}\overset{O}{\overset{\|}{C}}OH + SOCl_2 \longrightarrow CH_3(CH_2)_{14}\overset{O}{\overset{\|}{C}}Cl \xrightarrow{HN(CH_3)_2} CH_3(CH_2)_{14}\overset{O}{\overset{\|}{C}}N(CH_3)_2$$

$$\xrightarrow[\text{2) } H_2O]{\text{1) } LiAlH_4, \text{ ether or THF}} CH_3(CH_2)_{14}CH_2N(CH_3)_2$$

N,N-Dimethyl-1-hexadecanamine

Problem 26.16 Lipases are enzymes that catalyze the hydrolysis of esters, especially esters of glycerol. Because enzymes are chiral catalysts, they catalyze the hydrolysis of only one enantiomer of a racemic mixture. For example, porcine pancreatic lipase catalyzes the hydrolysis of only one enantiomer of the following racemic epoxyester. Calculate the number of grams of epoxyalcohol that can be obtained from 100 g of racemic epoxyester by this method.

A racemic mixture $\xrightarrow[\text{lipase}]{H_2O, \text{ OH}^-}$ This enantiomer is recovered unhydrolyzed   This epoxyalcohol is obtained in pure form

The molecular weight for the epoxyester starting material ($C_8H_{14}O_3$) is 158 g/mol, while the molecular weight of the epoxyalcohol product ($C_3H_6O_2$) is 74 g/mol. Because the starting material is racemic, there is actually only 100 g/2 or 50 g of the starting material epoxyester enantiomer that will be converted to product. As a result, the number of grams of the epoxyalcohol product is calculated as:

$$\frac{(50 \text{ g})(74 \text{ g} / \text{mol})}{(158 \text{ g} / \text{mol})} = 23.4 \text{ g}$$

Thus, 23.4 g of epoxyalcohol will be produced in this enzyme-catalyzed reaction.

Prostaglandins

Problem 26.17 Examine the structure of PGF$_{2\alpha}$. Identify all chiral centers, all double bonds about which *cis,trans* isomerism is possible, and state the number of stereoisomers possible for a molecule of this structure.

These double bonds are indicated by the arrows.

cis-trans isomerization possible

PGF$_{2\alpha}$

There are 2^5 stereoisomers possible and 2^2 *cis-trans* isomers possible for a grand total of 32 x 4 = 128 possible stereoisomers.

Problem 26.18 Following is the structure of unoprostone, a compound patterned after the natural prostaglandins (Section 26.3). Rescula, the isopropyl ester of unoprostone, is an antiglaucoma drug used to treat ocular hypertension. Compare the structural formula of this synthetic prostaglandin with that of PGF$_{2\alpha}$.

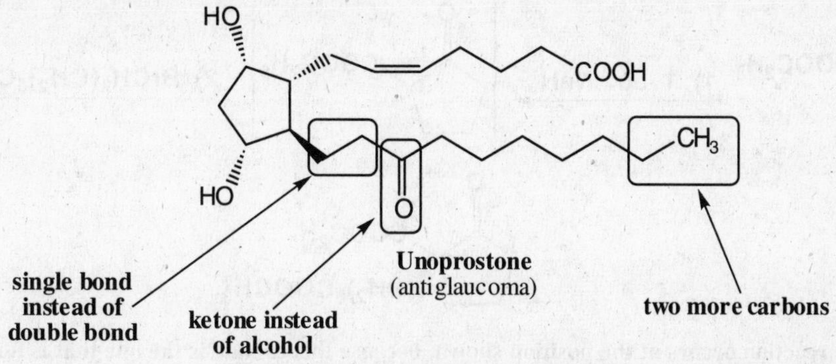

single bond instead of double bond

ketone instead of alcohol

Unoprostone (antiglaucoma)

two more carbons

Unoprostone has two more carbons, a ketone instead of an alcohol group, and one less double bond compared to PGF$_{2\alpha}$.

Problem 26.19 Doxaprost, an orally active bronchodilator patterned after the natural prostaglandins (Section 26.3), is synthesized in the following series of reactions starting with ethyl 2-oxocyclopentanecarboxylate. Except for the Nef reaction in Step 8, we have seen examples of all other types of reactions involved in this synthesis.

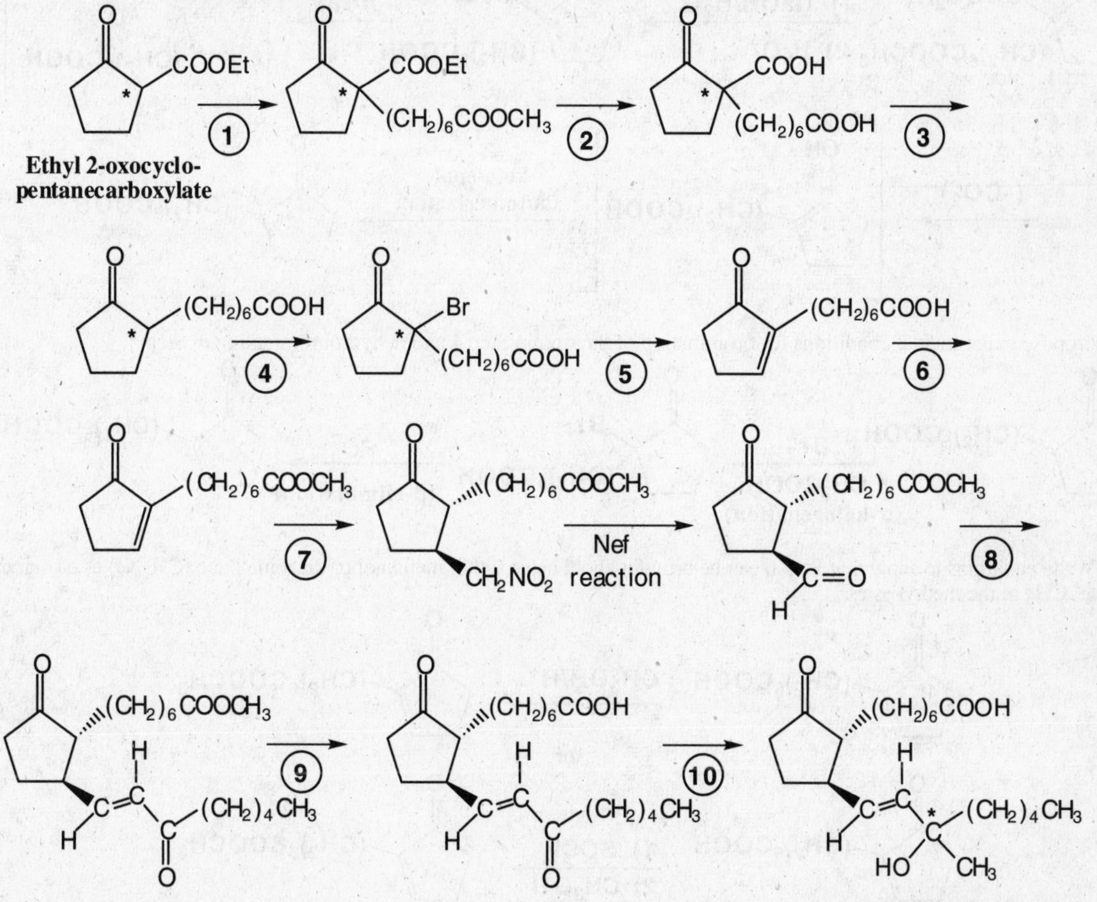

Ethyl 2-oxocyclo-pentanecarboxylate

Nef reaction

Doxaprost (an orally active bronchodilator)

Note that there are actually two *trans* isomers produced as racemic mixtures following reactions (7) - (11), although only one is drawn.

(a) Propose a set of experimental conditions to bring about the alkylation in Step 1. Account for the regioselectivity of the alkylation, that is, that it takes place on the carbon between the two carbonyl groups rather than on the other side of the ketone carbonyl.

The alkylation reaction occurs at the position shown, because this enolate is the one that is formed predominantly due to the stabilization provided by both of the adjacent carbonyl functions.

(b) Propose experimental conditions to bring about Steps 2 and 3.

(c) Propose experimental conditions for bromination of the ring in Step 4 and dehydrobromination in Step 5.

(d) Write equations to show that Step 6 can be brought about using either methanol or diazomethane (CH_2N_2) as a source of the -CH_3 in the methyl ester.

(e) Describe experimental conditions to bring about Step 7 and account for the fact that the *trans* isomer is formed in this step.

This Michael reaction leads to a *trans* product because this minimizes nonbonded interaction strain between the nitromethyl and methyl hexanoate groups. Note that two new chiral centers are produced in the reaction, and the product is actually a racemic mixture of the two *trans* enantiomers.

(f) Step 9 is done by a Wittig reaction. Suggest a structural formula for a Wittig reagent that gives the product shown.

(g) Name the type of reaction involved in Step 9.

Step 9 is an ester hydrolysis reaction.

(h) Step 10 can best be described as a Grignard reaction with methylmagnesium bromide under very carefully controlled conditions. In addition to the observed reaction, what other Grignard reactions might take place in Step 10?

The carboxyl group will be deprotonated by the first equivalent of methylmagnesium bromide. As far as other Grignard reactions go, the cyclopentyl ketone might also take part in a Grignard reaction as shown:

(i) Assuming that the two side chains on the cyclopentanone ring are *trans*, how many stereoisomers are possible from this synthetic sequence?

There are actually two enantiomers of the *trans* product produced during the synthesis, although only one is drawn in the problem. The Grignard reaction also gives a new chiral center, so there are a total of four stereoisomers produced in this synthesis.

Steroids
Problem 26.20 Draw the structural formula for the product formed by treatment of cholesterol with H_2/Pd; with Br_2.

**Approach from
bottom face preferred**

Cholesterol structural formula and conformational formula

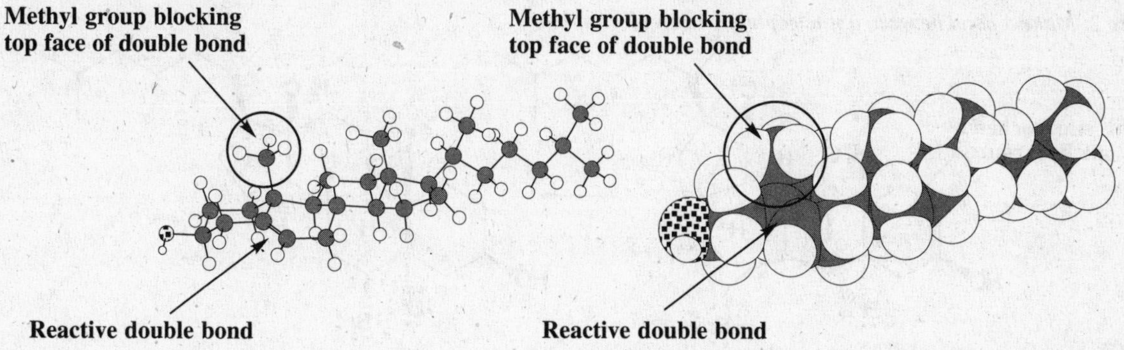

Methyl group blocking top face of double bond

Methyl group blocking top face of double bond

Reactive double bond

Reactive double bond

As can be seen in the models above, the methyl group blocks the top face of the reactive double bond. Thus, when predicting the stereochemistry of reaction at this position, predict that the reagents will approach from the bottom face.

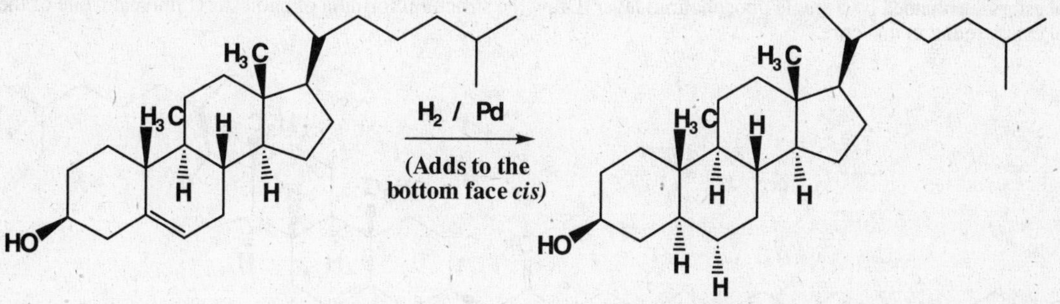

H_2 / Pd

(Adds to the bottom face *cis*)

Predominant stereoisomer

For the reaction with bromine, the bromonium ion is formed on the bottom face. According to the mechanism of the reaction, the Br- nucleophile must approach from the top face to complete the reaction. This will occur at the position farthest away from the methyl group, leading to the single stereoisomer shown. This stereochemistry is also preferred because the bromide anion must add *anti* and coplanar to give a diaxial product.

Step 1: Electrophilic addition.

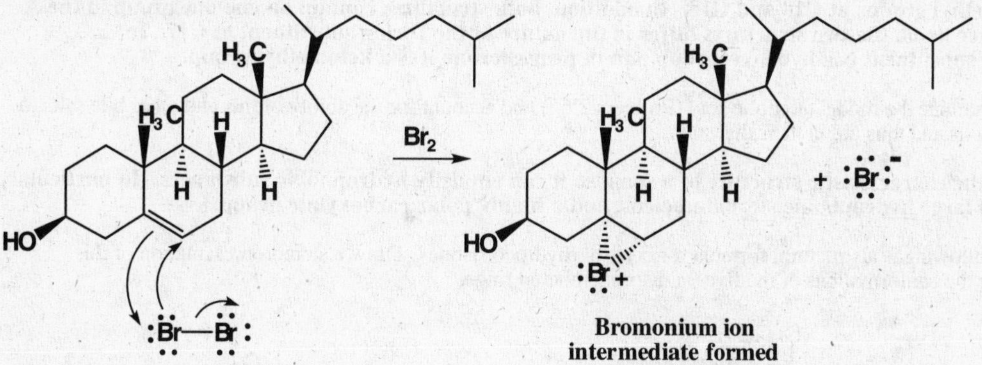

Br$_2$

Bromonium ion intermediate formed on bottom face

Step 2: Make a bond between a nucleophile and an electrophile.

Predominant stereoisomer (diaxial bromine atoms)

Problem 26.21 Both low-density lipoproteins (LDL) and high-density lipoproteins (HDL) consist of a core of triacylglycerols and cholesterol esters surrounded by a single phospholipid layer. Draw the structural formula of cholesteryl linoleate, one of the cholesterol esters found in this core.

Problem 26.22 Examine the structural formulas of testosterone (a male sex hormone) and progesterone (a female sex hormone). What are the similarities in structure between the two? What are the differences?

Overall, these structures are remarkably similar. Both steroids contain the standard four ring steroid structure with the axial methyl groups at C10 and C13. In addition, both structures contain an ene-one group in the A ring. On the other hand, the two structures differ in the nature of the D ring substituent at C17. In testosterone, the substituent is a hydroxyl group and in progesterone it is a ketomethyl group.

Problem 26.23 Examine the model of cholic acid (Problem 2.65) and account for the ability of this and other bile salts to emulsify fats and oils and thus aid in their digestion.

Cholic acid has the characteristic structure of a soap, so it can emulsify hydrophobic substances. In particular, cholic acid has a large hydrophobic steroid nucleus, and a highly polar carboxylate group.

Problem 26.24 Following is a structural formula for cortisol (hydrocortisone). Draw a stereorepresentation of this molecule showing the conformations of the five- and six-membered rings.

Cortisol
(Hydrocortisone)
structural formula

Cortisol
(Hydrocortisone)
conformational formula

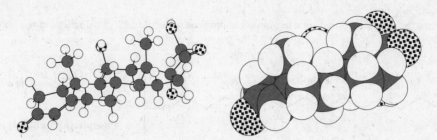

Problem 26.25 Much of our understanding of conformational analysis has arisen from studies on the reactions of rigid steroid nuclei. For example, the concept of *trans*-diaxial ring opening of epoxycyclohexanes was proposed to explain the stereoselective reactions seen with steroidal epoxides. Predict the product when each of the following steroidal epoxides is treated with LiAlH₄;

(a)

(b)

(c)

(d)

The idea here is that the hydride reagent attacks such that the epoxide opens to give a diaxial product. This dictates that only one product is observed for each reaction, the one that is shown below:

(a)

(b)

(c)

(d)

Phospholipids

Problem 26.26 Draw the structural formula of a lecithin containing one molecule each of palmitic acid and linoleic acid.

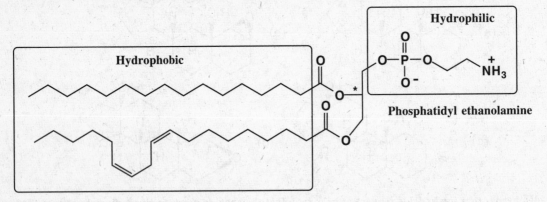

Problem 26.27 Identify the hydrophobic and hydrophilic region(s) of a phospholipid.

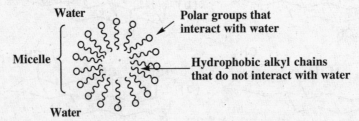

Problem 26.28 The hydrophobic effect is one of the most important noncovalent forces directing the self-assembly of biomolecules in aqueous solution. The hydrophobic effect arises from tendencies of biomolecules (1) to arrange polar groups so that they interact with the aqueous environment by hydrogen bonding and (2) to arrange nonpolar groups so that they are shielded from the aqueous environment. Show how the hydrophobic effect is involved in directing the following.
(a) Formation of micelles by soaps and detergents.

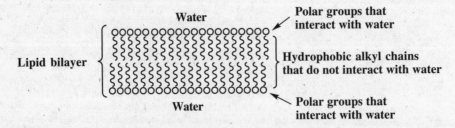

In micelles, the hydrophobic hydrocarbon tails are associated with each other to form the hydrophobic interior, while the polar groups are associated with each other on the outside surface where they interact with water.

(b) Formation of lipid bilayers by phospholipids.

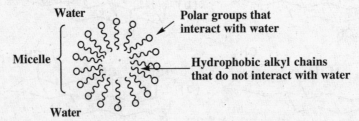

In lipid bilayers, the hydrophobic hydrocarbon tails are associated with each other to form the hydrophobic inner layer, while the polar head groups are associated with each other on both outside surfaces where they interact with water.

Problem 26.29 How does the presence of unsaturated fatty acids contribute to the fluidity of biological membranes?

Biological membranes must have the proper fluidity to function correctly. Fatty acid components of membranes add the *cis* double bonds of the unsaturated fatty acids that interfere with chain packing, increasing fluidity of the membrane. Organisms modulate membrane fluidity by controlling the kind of fatty acids in their membranes. For example, organisms living in colder environments tend to have more unsaturated fatty acids to insure membrane fluidity at low temperature.

Problem 26.30 Lecithins can act as emulsifying agents. The lecithin of egg yolk, for example, is used to make mayonnaise. Identify the hydrophobic part(s) and the hydrophilic part(s) of a lecithin. Which parts interact with the oils used in making mayonnaise? Which parts interact with the water?

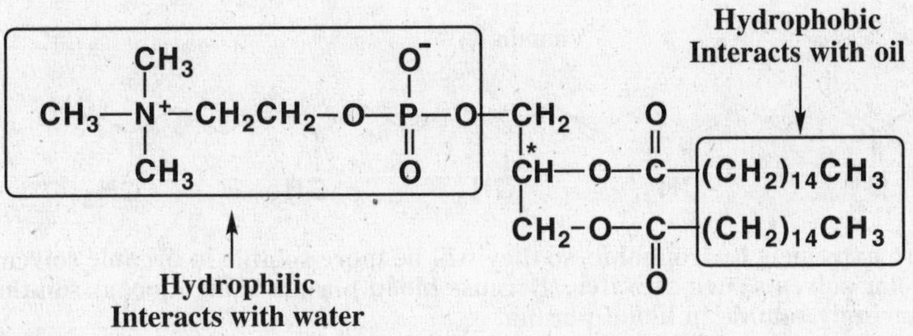

Fat-soluble Vitamins

Problem 26.31 Examine the structural formula of vitamin A and state the number of *cis-trans* isomers possible for this molecule.

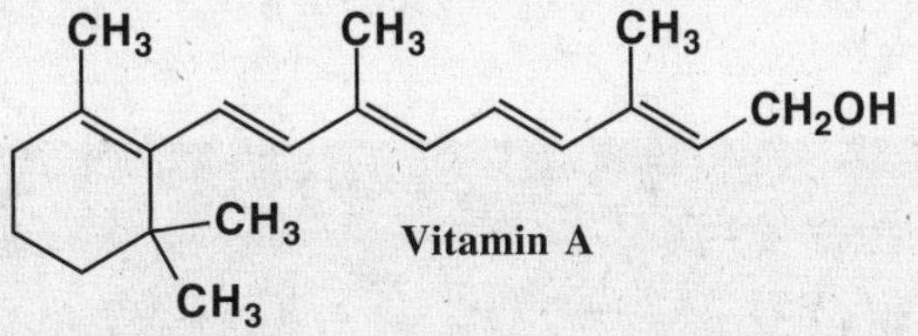

As shown in the structure above, vitamin A has four double bonds that can be either *cis* or *trans*, thus there are 2^4 or 16 possible *cis-trans* isomers. Note that the double bond in the ring cannot have *cis-trans* isomers.

Problem 26.32 The form of vitamin A present in many food supplements is vitamin A palmitate. Draw the structural formula of this molecule.

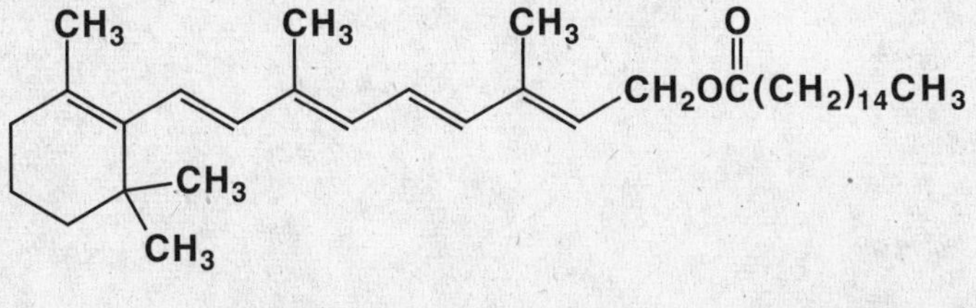

Problem 26.33 Examine the structural formulas of vitamin A, 1,25-dihydroxy-D_3, vitamin E, and vitamin K_1 (Section 26.6). Do you expect them to be more soluble in water or in dichloromethane? Do you expect them to be soluble in blood plasma?

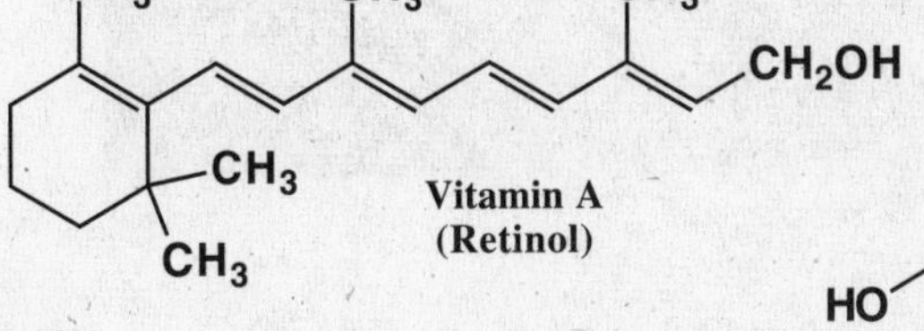

Vitamin A (Retinol)

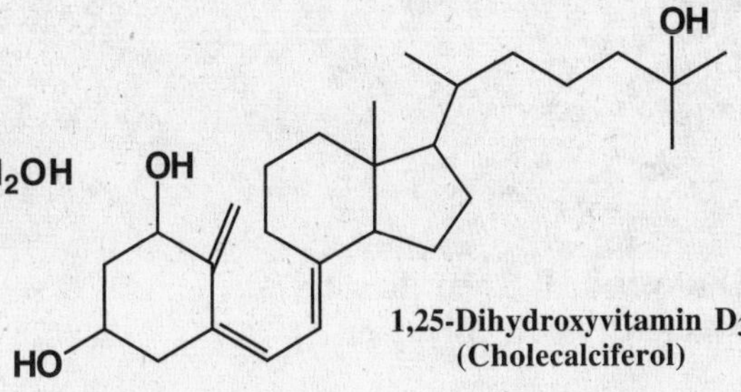

1,25-Dihydroxyvitamin D_3 (Cholecalciferol)

Vitamin E
(α-Tocopherol)

Vitamin K₁

All of these structures are extremely hydrophobic, so they will be more soluble in organic solvents such as dichloromethane than polar solvents such as water. Because blood plasma is an aqueous solution, these vitamins will only be sparingly soluble in blood plasma.

Dusky batfish *Platax pinnatus*
Moorea, French Polynesia

CHAPTER 27
Solutions to the Problems

<u>Problem 27.1</u> Of the 20 protein-derived amino acids shown in Table 27.1, which contain (a) no chiral center, (b) two chiral centers.

The only amino acid with no chiral centers is glycine (Gly, G). Both isoleucine (Ile, I) and threonine (Thr, T) have two chiral centers as shown with asterisks in the structures below.

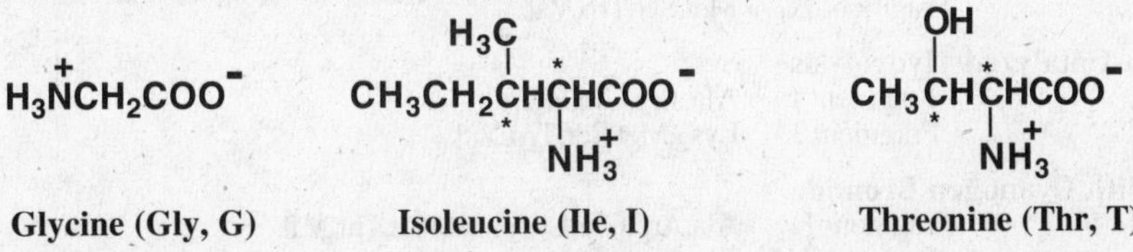

$$H_3\overset{+}{N}CH_2COO^- \qquad CH_3CH_2\overset{*}{C}H\overset{*}{C}HCOO^- \qquad CH_3\overset{*}{C}H\overset{*}{C}HCOO^-$$

Glycine (Gly, G)   Isoleucine (Ile, I)   Threonine (Thr, T)

<u>Problem 27.2</u> The isoelectric point of histidine is 7.64. Toward which electrode does histidine migrate on paper electrophoresis at pH 7.0?

An amino acid will have at least a partial positive charge at any pH that is below its isoelectric point. A pH of 7.0 is below the isoelectric point of histidine (7.64), so it will have a partial positive charge. Therefore, at this pH histidine migrates toward the negative electrode.

<u>Problem 27.3</u> Describe the behavior of a mixture of glutamic acid, arginine, and valine on paper electrophoresis at pH 6.0.

The pI's for glutamic acid, arginine, and valine are 3.08, 10.76, and 6.00, respectively. Therefore, at pH 6.0 glutamic acid is negatively charged, arginine is positively charged, and valine is neutral. Thus, on paper electrophoresis, glutamic acid will migrate toward the positive electrode, arginine will migrate toward the negative electrode, and valine will not move.

<u>Problem 27.4</u> Draw a structural formula for Lys-Phe-Ala. Label the *N*-terminal amino acid and the *C*-terminal amino acid. What is the net charge on this tripeptide at pH 6.0?

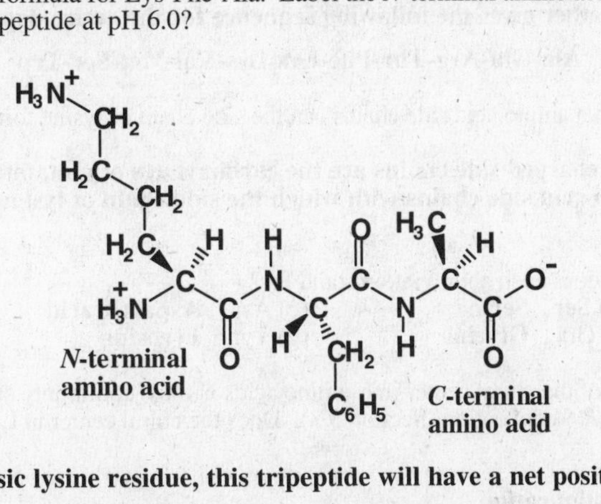

Due to the presence of the basic lysine residue, this tripeptide will have a net positive charge at pH 6.0

<u>Problem 27.5</u> Which of these tripeptides are hydrolyzed by trypsin? By chymotrypsin?
(a) Tyr-Gln-Val   (b) Thr-Phe-Ser   (c) Thr-Ser-Phe

Based on the substrate specificities listed in Table 27.3, trypsin will not cleave any of these tripeptides because there are no arginine or lysine residues. On the other hand, chymotrypsin will cleave peptides (a) and (b) between the Tyr-Gln and Phe-Ser residues, respectively, but not (c).

<u>Problem 27.6</u> Deduce the amino acid sequence of an undecapeptide (11 amino acids) from the experimental results shown in the accompanying table.

Experimental Procedure	Amino Acid Composition
Undecapeptide	Ala,Arg,Glu,Lys$_2$,Met,Phe,Ser,Thr,Trp,Val
Edman degradation	Ala
Trypsin-Catalyzed Hydrolysis	
Fragment E	Ala,Glu,Arg
Fragment F	Thr,Phe,Lys
Fragment G	Lys
Fragment H	Met,Ser,Trp,Val
Chymotrypsin-Catalyzed Hydrolysis	
Fragment I	Ala,Arg,Glu,Phe,Thr
Fragment J	Lys$_2$,Met,Ser,Trp,Val
Reaction with Cyanogen Bromide	
Fragment K	Ala,Arg,Glu,Lys$_2$,Met,Phe,Thr,Val
Fragment L	Trp,Ser

Based on the Edman degradation result, alanine (Ala) is the *N*-terminal residue of the peptide.
Fragment E must have Arg on the *C*-terminal end because it is a peptide produced by trypsin cleavage. Because we know Ala is the *N*-terminal residue, this means fragment E must be of the sequence Ala-Glu-Arg.
There must be two lysine residues or an arginine and a lysine residue adjacent to each other based on the appearance of a single lysine residue as Fragment G. Because Fragment J has two lysines and no arginine residues, the two lysine residues must be adjacent to each other.
Methionine must be the third to the last residue, because CNBr treatment created fragment L that is only Ser and Trp. In addition, Trp and Ser must be the last two residues. Combining this information with the knowledge that there are two lysine residues adjacent to each other indicates the Fragment J is of the sequence Lys-Lys-Val-Met-Ser-Trp. Note that Val must come after the two Lys residues because of Fragment H. In addition, Trp has to be on the *C*-terminus or a residue would have been cleaved off by chymotrypsin.
Phenylalanine must be on the *C* terminus of Fragment I because it results from chymotrypsin cleavage. We already know that Fragment I must start with Ala-Glu-Arg, so the entire sequence of Fragment I must be Ala-Glu-Arg-Thr-Phe.
Putting Fragments I and J together gives the following sequence for the entire peptide:

Ala-Glu-Arg-Thr-Phe-Lys-Lys-Val-Met-Ser-Trp

<u>Problem 27.7</u> At pH 7.4, with what amino acid side chains can the side chain of lysine form salt linkages.

At pH 7.4, the only negatively charged side chains are the carboxylates of glutamic acid and aspartic acid. Therefore, these are the amino acid side chains with which the side chain of lysine can form a salt linkage.

<u>**Amino Acids**</u>
<u>Problem 27.8</u> What amino acids does each abbreviation stand for?
(a) Phe   **Phenylalanine**   (b) Ser   **Serine**   (c) Asp   **Aspartic acid**   (d) Gln   **Glutamine**
(e) His   **Histidine**   (f) Gly   **Glycine**   (g) Tyr   **Tyrosine**

<u>Problem 27.9</u> The configuration of the chiral center in α-amino acids is most commonly specified using the D,L convention. It can also be identified using the *R,S* convention (Section 3.3). Does the chiral center in L-serine have the *R* or the *S* configuration?

The chiral center has the *S* configuration.

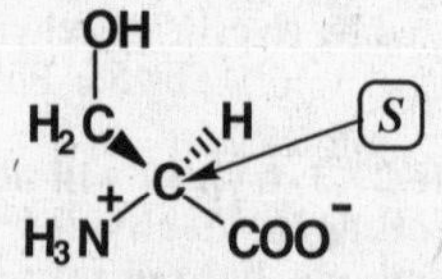

L-Serine

<u>Problem 27.10</u> Assign an *R* or *S* configuration to the chiral center in each amino acid.

(a) L-Phenylalanine   (b) L-Glutamic acid   (c) L-Methionine

L-Phenylalanine   L-Glutamic acid   L-Methionine

All of these L amino acids have the *S* configuration.

<u>Problem 27.11</u> The amino acid threonine has two chiral centers. The stereoisomer found in proteins has the configuration 2S,3R about the two chiral centers. Draw (a) a Fischer projection of this stereoisomer and (b) a three-dimensional representation.

L-Threonine

<u>Problem 27.12</u> Define the term *zwitterion*.

A zwitterion refers to the situation in which an acidic and a basic group in the same molecule react with each other to form a dipolar ion or internal salt. A zwitterion has no net charge; it contains one positive charge and one negative charge.

<u>Problem 27.13</u> Draw zwitterion forms of these amino acids.

(a) Valine   (b) Phenylalanine   (c) Glutamine

L-Valine   L-Phenylalanine   L-Glutamine

<u>Problem 27.14</u> Why are Glu and Asp often referred to as acidic amino acids.

The side chains of glutamic acid (Glu) and aspartic acid (Asp) have carboxyl groups, so these amino acids are referred to as acidic amino acids.

<u>Problem 27.15</u> Why is Arg often referred to as a basic amino acid? Which two other amino acids are also basic amino acids?

The guanidino group of arginine (Arg) is strongly basic, so this amino acid is referred to as a basic amino acid. Note that this means arginine is positively charged at neutral pH. Lysine (Lys) and histidine (His) are also referred to as basic amino acids because their side chains contain a primary amine and imidazole functions, respectively.

<u>Problem 27.16</u> What is the meaning of the alpha as it is used in α-amino acid?

The alpha in α-amino acids indicates that the amino group is on the carbon atom that is α to the carboxyl group.

<u>Problem 27.17</u> Several β-amino acids exist. There is a unit of β-alanine, for example, contained within the structure of coenzyme A (Problem 25.35). Write the structural formula of β-alanine.

$$H_3\overset{+}{N}CH_2CH_2COO^-$$

β-Alanine

<u>Problem 27.18</u> Although only L-amino acids occur in proteins, D-amino acids are often a part of the metabolism of lower organisms. The antibiotic actinomycin D, for example, contains a unit of D-valine, and the antibiotic bacitracin A contains units of D-asparagine and D-glutamic acid. Draw Fischer projections and three-dimensional representations for these three D-amino acids.

D-Valine

D-Asparagine

D-Glutamic acid

<u>Problem 27.19</u> Histamine is synthesized from one of the 20 protein-derived amino acids. Suggest which amino acid is its biochemical precursor and the type of organic reaction(s) involved in its biosynthesis (for example, oxidation, reduction, decarboxylation, nucleophilic substitution).

Histamine

Histidine

Histamine is derived from the amino acid histidine and is the result of a biosynthetic decarboxylation reaction. Note how both histamine and histidine are drawn in the form present at basic pH.

<u>Problem 27.20</u> As discussed in the Chemical Connections: "Vitamin K, Blood Clotting, and Basicity," in Section 26.6D, vitamin K participates in carboxylation of glutamic acid residues of the blood-clotting protein prothrombin.
(a) Write a structural formula for γ-carboxyglutamate.

γ-Carboxyglutamate

(b) Account for the fact that the presence of γ-carboxyglutamate escaped detection for many years; on routine amino acid analyses, only glutamic acid was detected.

This amino acid was not detected because it is a β-dicarboxylic acid and therefore easily decarboxylated under conditions of routine amino acid analysis.

Problem 27.21 Both norepinephrine and epinephrine are synthesized from the same protein-derived amino acid. From which amino acid are they synthesized and what types of reactions are involved in their biosynthesis?

(a)

Norepinephrine

(b)

Epinephrine (Adrenaline)

L-Tyrosine

Norepinephrine and epinephrine are derived from the amino acid tyrosine. In both cases, biosynthesis of these molecules involves decarboxylation, aromatic hydroxylation (a two-electron oxidation) ortho to the original aromatic -OH group, and hydroxylation (a second two-electron oxidation) of the benzylic methylene group. Epinephrine is also methylated on the α-amino group. Note how all of the molecules in the problem are drawn in the form present at basic pH.

Problem 27.22 From which amino acid are serotonin and melatonin synthesized and what types of reactions are involved in their biosynthesis?

(a)

Serotonin

(b)

Melatonin

L-Tryptophan

Serotonin and melatonin are derived from the amino acid tryptophan. In both cases, biosynthesis of these molecules involves decarboxylation. In the case of serotonin there is also an aromatic hydroxylation (a two-electron oxidation). For melatonin the phenolic -OH group of seratonin is methylated and the primary amino group is acetylated. Note how all of the molecules in the problem are drawn in the form present at basic pH.

Acid-Base Behavior of Amino Acids

Problem 27.23 Draw a structural formula for the form of each amino acid most prevalent at pH 1.0.

(a) Threonine

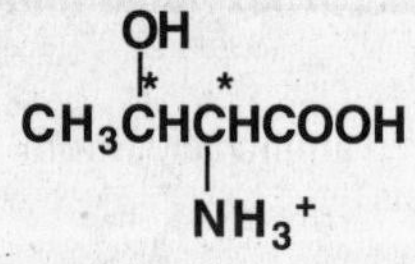

$$\underset{\underset{NH_3^+}{|}}{CH_3\overset{OH}{\underset{*}{\underset{|}{C}}HC}HCOOH}$$

(b) Arginine

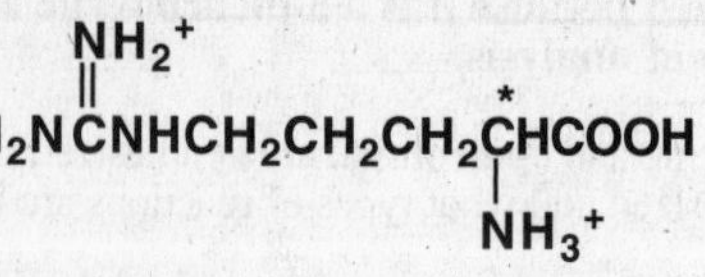

$$H_2N\overset{\overset{+}{NH_2}}{\underset{\parallel}{C}}NHCH_2CH_2CH_2\overset{*}{\underset{\underset{NH_3^+}{|}}{C}}HCOOH$$

(c) Methionine

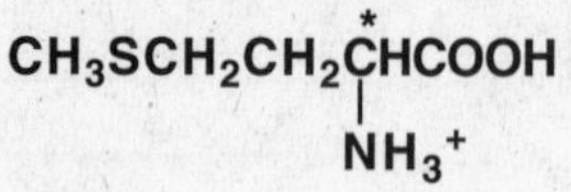

$$CH_3SCH_2CH_2\overset{*}{\underset{\underset{NH_3^+}{|}}{C}}HCOOH$$

(d) Tyrosine

$$HO-\!\!\!\!\bigcirc\!\!\!\!-CH_2\overset{*}{\underset{\underset{NH_3^+}{|}}{C}}HCOOH$$

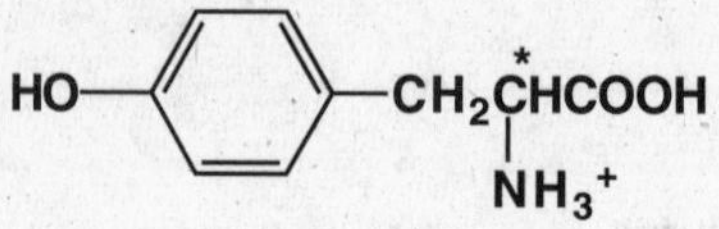

Problem 27.24 Draw a structural formula for the form of each amino acid most prevalent at pH 10.0.

(a) Leucine

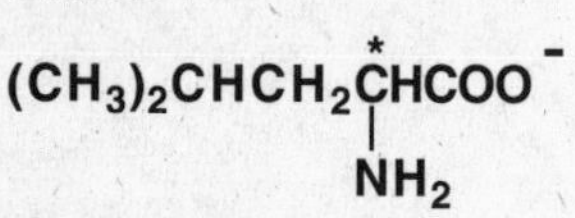

$$(CH_3)_2CHCH_2\overset{*}{\underset{\underset{NH_2}{|}}{C}}HCOO^-$$

(b) Valine

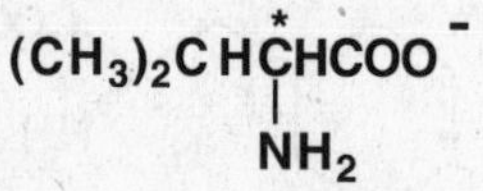

$$(CH_3)_2CH\overset{*}{\underset{\underset{NH_2}{|}}{C}}HCOO^-$$

(c) Proline

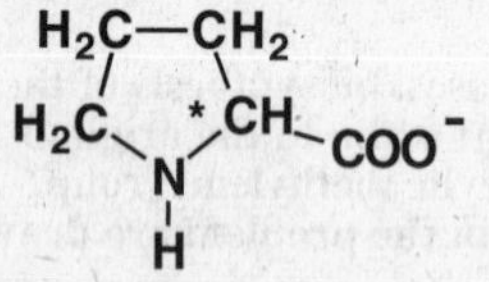

$$\begin{array}{c} H_2C\!-\!CH_2 \\ H_2C\diagdown\quad\diagup\overset{*}{C}H\!-\!COO^- \\ \underset{H}{N} \end{array}$$

(d) Aspartic acid

$$^-OOCCH_2\overset{*}{\underset{\underset{NH_2}{|}}{C}}HCOO^-$$

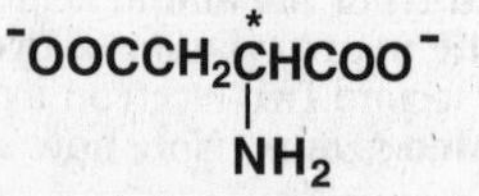

Problem 27.25 Write the zwitterion form of alanine and show its reaction with the following:

(a) 1 mol NaOH

$$CH_3\overset{*}{\underset{\underset{NH_3^+}{|}}{C}}HCOO^- \;+\; 1\text{ mole }NaOH \longrightarrow CH_3\overset{*}{\underset{\underset{NH_2}{|}}{C}}HCOO^-$$

(b) 1 mol HCl

$$CH_3\overset{*}{\underset{\underset{NH_3^+}{|}}{C}}HCOO^- \;+\; 1\text{ mole }HCl \longrightarrow CH_3\overset{*}{\underset{\underset{NH_3^+}{|}}{C}}HCOOH$$

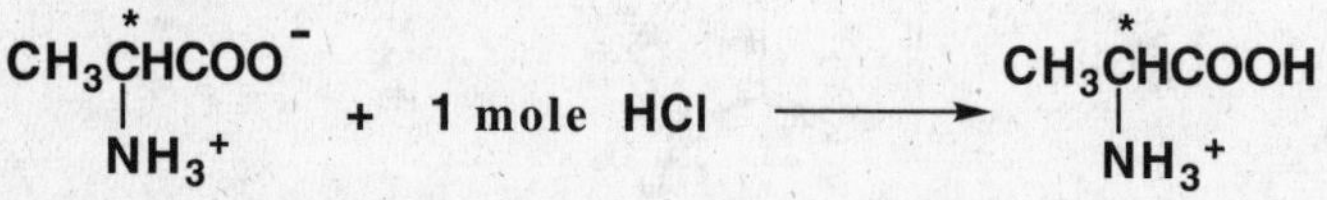

Problem 27.26 Write the form of lysine most prevalent at pH 1.0 and then show its reaction with the following. Consult Table 27.2 for pK_a values of the ionizable groups in lysine.

At pH 1.0, the most prevalent form of lysine has both amino groups as well as the carboxylic acid group protonated and a total charge of +2 as shown in the following structure.

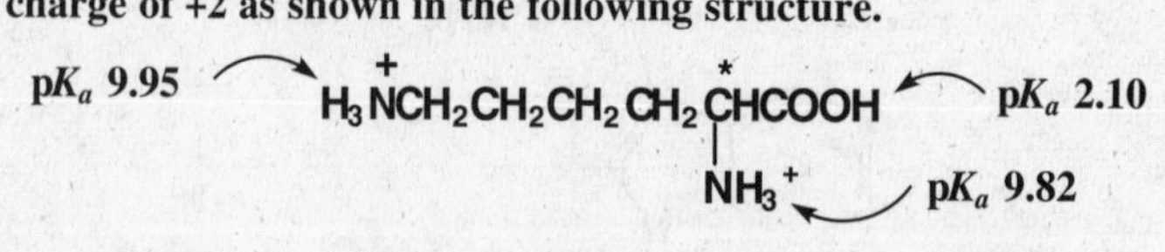

$$pK_a\;9.95 \quad \overset{+}{H_3}NCH_2CH_2CH_2CH_2\overset{*}{\underset{\underset{NH_3^+}{|}}{C}}HCOOH \quad pK_a\;2.10 \qquad pK_a\;9.82$$

(a) 1 mol NaOH

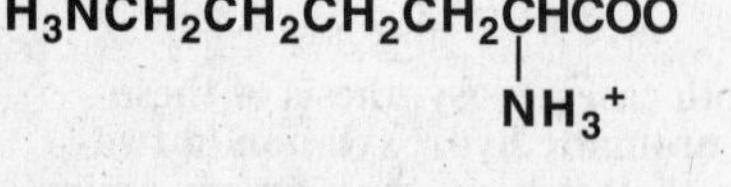

$$\overset{+}{H_3}NCH_2CH_2CH_2CH_2\overset{*}{\underset{\underset{NH_3^+}{|}}{C}}HCOO^-$$

(b) 2 mol NaOH

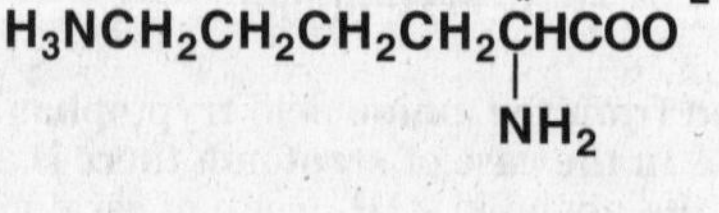

$$\overset{+}{H_3}NCH_2CH_2CH_2CH_2\overset{*}{\underset{\underset{NH_2}{|}}{C}}HCOO^-$$

(c) 3 mol NaOH

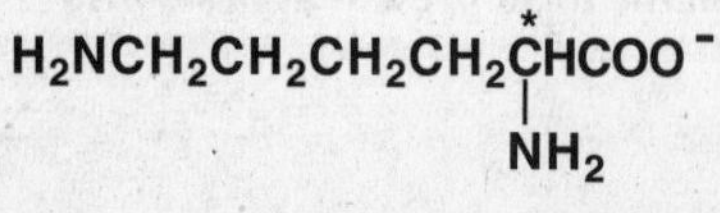

$$H_2NCH_2CH_2CH_2CH_2\overset{*}{\underset{\underset{NH_2}{|}}{C}}HCOO^-$$

<u>Problem 27.27</u> Write the form of aspartic acid most prevalent at pH 1.0 and then show its reaction with the following. Consult Table 27.2 for pK_a values of the ionizable groups in aspartic acid.

At pH 1.0, the most prevalent form of aspartic acid has both carboxylic acid groups as well as the amino group protonated and a total charge of +1 as shown in the following structure.

$$pK_a\ 3.86 \quad \longrightarrow \quad HOOCCH_2\overset{*}{C}HCOOH \quad \longleftarrow \quad pK_a\ 2.10$$
$$NH_3{}^+ \quad \longleftarrow \quad pK_a\ 9.82$$

(a) 1 mol NaOH

$$HOOCCH_2\overset{*}{C}HCOO^-$$
$$NH_3{}^+$$

(b) 2 mol NaOH

$$^-OOCCH_2\overset{*}{C}HCOO^-$$
$$NH_3{}^+$$

(c) 3 mol NaOH

$$^-OOCCH_2\overset{*}{C}HCOO^-$$
$$NH_2$$

<u>Problem 27.28</u> Given pK_a values for ionizable groups from Table 27.2, sketch curves for the titration of (a) glutamic acid with NaOH, and (b) histidine with NaOH.

Glutamic acid has pK_a values of 2.10, 4.07, and 9.47 so the titration curve would look something like the following:

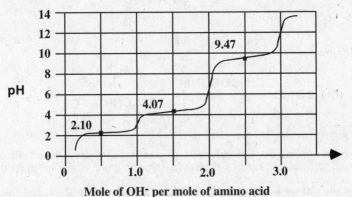

Histidine has pK_a values of 1.77, 6.10, and 9.18 so the titration curve would look something like the following:

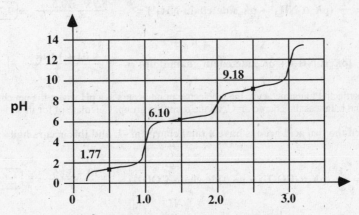

Problem 27.29 Draw a structural formula for the product formed when alanine is treated with the following reagents.
(a) Aqueous NaOH

(b) Aqueous HCl

(c) CH₃CH₂OH, H₂SO₄

(d) (CH₃CO)₂O, CH₃CO₂Na

Problem 27.30 For lysine and arginine, the isoelectric point, pI, occurs at a pH where the net charge on the nitrogen-containing groups is +1 and balances the charge of -1 on the α–carboxyl group. Calculate pI for these amino acids.

The pI will occur when the nitrogen-containing groups have a total charge of +1, and this occurs halfway between their respective pK_a values.
For lysine:

$$pI = \frac{1}{2} \ (pK_a\alpha\text{-}NH_3{}^+ + pK_a \text{side chain-}NH_3{}^+) = \quad \frac{8.95 + 10.53}{2} \quad = \boxed{9.74}$$

For arginine:

$$pI = \frac{1}{2} \ (pK_a\alpha\text{-}NH_3{}^+ + pK_a \text{side chain guanidium}) = \quad \frac{9.04 + 12.48}{2} \quad = \boxed{10.76}$$

Problem 27.31 For aspartic and glutamic acids, the isoelectric point occurs at a pH where the net charge on the two carboxyl groups is -1 and balances the charge of +1 on the α–amino group. Calculate pI for these amino acids.

The pI will occur when the two acid groups have a total charge of -1, and this occurs halfway between their respective pK_a values.
For aspartic acid:

$$pI = \frac{1}{2} \ (pK_a\alpha\text{-}COOH + pK_a \text{side chain-}COOH) = \frac{2.10 + 3.86}{2} \quad = \boxed{2.98}$$

For glutamic acid:

$$pI = \frac{1}{2} \ (pK_a\alpha\text{-}COOH + pK_a \text{side chain-}COOH) = \frac{2.10 + 4.07}{2} \quad = \boxed{3.08}$$

Problem 27.32 Account for the fact that the isoelectric point of glutamine (pI 5.65) is higher than the isoelectric point of glutamic acid (pI 3.08).

Amino acids have no net charge at their pI. For this to happen with glutamic acid, the net charge on the α-carboxyl and side chain carboxyl groups must be -1 to balance the +1 charge of the α-amino group. This will occur at a pI = (1/2)(2.10 + 4.07) = 3.08. The amide side chain of glutamine is already neutral near neutral pH, so the pI of the amino acid is determined by the values for the only ionizable groups, namely the α-carboxyl

group and α-amino groups, according to the equation pI = (1/2)(2.17 + 9.03) = 5.6. This value is near that of the other amino acids with non-ionizable functional groups on their side chains.

Problem 27.33 Enzyme-catalyzed decarboxylation of glutamic acid gives 4-aminobutanoic acid (Section 27.1D). Estimate the pI of 4-aminobutanoic acid.

There is little if any inductive effect operating between the amino and carboxyl groups of 4-aminobutanoic acid because there are three methylene groups between them. Thus, the pK_a of the amino group of 4-aminobutanoic acid is like that of a simple amino group, near 10.0. Similarly, the pK_a of the carboxyl group is like that of a simple carboxyl group, near 4.5. Given these estimates for the pK_a values, the pI would be:

$$pI = \frac{1}{2}\,(pK_a\text{-COOH} + pK_a\ \alpha\text{-NH}_2) = \frac{4.5 + 10.0}{2} = \boxed{7.25}$$

Problem 27.34 Guanidine and the guanidino group present in arginine are two of the strongest organic bases known. Account for their basicity.

The guanidino group is strongly basic because of resonance stabilization of the protonated guanidinium ion as shown below:

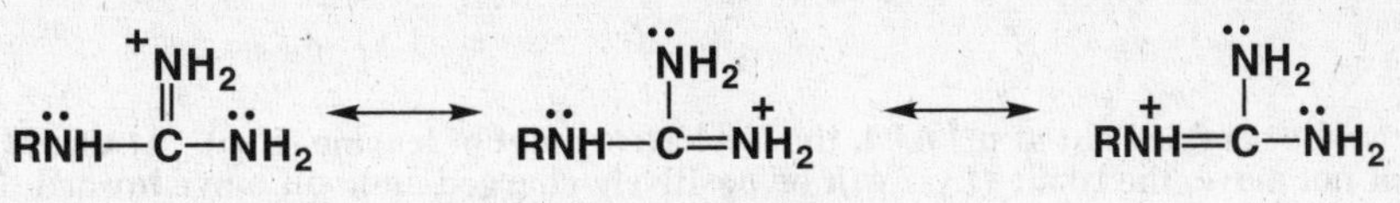

R = H or alkyl group

Problem 27.35 At pH 7.4, the pH of blood plasma, do the majority of protein-derived amino acids bear a net negative charge or a net positive charge? Explain.

The majority of amino acids have a pI near 5 or 6, so they will bear a net negative charge at pH 7.4.

Problem 27.36 Do the following molecules migrate to the cathode or to the anode on electrophoresis at the specified pH?

The key to determining which way the molecules migrate is to estimate the net charge on the molecules at the given pH. Molecules with a net positive charge will migrate toward the negative electrode and molecules with a net negative charge will migrate toward the positive electrode. Molecules at a pH below their isoelectric point (Table 27.2) have a net positive charge, molecules at a pH above their isoelectric point have a net negative charge, and molecules at a pH that equals their isoelectric point have no net charge.

(a) Histidine at pH 6.8

pI = 7.64, so at pH 6.8 histidine has a net positive charge and migrates toward the negative electrode (cathode).

(b) Lysine at pH 6.8

pI = 9.74, so at pH 6.8 lysine has a net positive charge and migrates toward the negative electrode (cathode).

(c) Glutamic acid at pH 4.0

pI = 3.08, so at pH 4.0 glutamic acid has a net negative charge and migrates toward the positive electrode (anode).

(d) Glutamine at pH 4.0

pI = 5.65, so at pH 4.0 glutamine has a net positive charge and migrates toward the negative electrode (cathode).

(e) Glu-Ile-Val at pH 6.0

The glutamic acid residue has a carboxyl group that will be largely deprotonated at pH 6.0, so the overall molecule will have a net negative charge and will migrate toward the positive electrode (anode).

(f) Lys-Gln-Tyr at pH 6.0

The lysine residue has an amino group on its side chain that will be protonated at pH 6.0, so the molecule will have a net positive charge and will migrate toward the negative electrode (cathode).

Problem 27.37 At what pH would you carry out an electrophoresis to separate the amino acids in each mixture?

Recall that an amino acid below its isoelectric point will have some degree of positive charge, an amino acid above its isoelectric point will have some degree of negative charge, and an amino acid at its isoelectric point will have no net charge.

(a) Ala, His, Lys

Electrophoresis could be carried out at pH 7.64, the isoelectric point of histidine (His). At this pH, the histidine is neutral and would not move, the lysine (Lys) will be positively charged and will move toward the negative electrode, and the alanine (Ala) will be slightly negatively charged and will move toward the positive electrode.

(b) Glu, Gln, Asp

Electrophoresis could be carried out at pH 3.08, the isoelectric point of glutamic acid (Glu). At this pH, the glutamic acid is neutral and would not move, the glutamine (Gln) will be positively charged and will move toward the negative electrode, and the aspartic acid (Asp) will be slightly negatively charged and will move toward the positive electrode.

(c) Lys, Leu, Tyr

Electrophoresis could be carried out at pH 6.04, the isoelectric point of leucine (Leu). At this pH, the leucine is neutral and would not move, the lysine (Lys) will be positively charged and will move toward the negative electrode, and the tyrosine (Tyr) will be slightly negatively charged and will move toward the positive electrode.

Problem 27.38 Examine the amino acid sequence of human insulin (Figure 27.16), and list each Asp, Glu, His, Lys, and Arg in this molecule. Do you expect human insulin to have an isoelectric point nearer that of the acidic amino acids (pI 2.0 - 3.0), the neutral amino acids (pI 5.5 - 6.5), or the basic amino acids (pI 9.5 - 11.0)?

A listing of the amino acids present are shown below:

aspartic acid (Asp)	**0**
glutamic acid (Glu)	**4**
histidine (His)	**2**
lysine (Lys)	**1**
arginine (Arg)	**1**

The charge will only be neutral when there are four positively charged residues to neutralize the four negative charges of the carboxylates from the four Glu residues. For this to happen, the Lys, Arg, and both His residues must be positively charged. Because the imidazole of His is not protonated until the pH is below 6, the entire molecule will only be neutral around this pH. Thus, insulin is expected to have an isoelectric point nearer to that of the neutral amino acids. Its measured isoelectric point is 5.30 to 5.35.

Problem 27.39 A chemically modified guanidino group is present in cimetidine (Tagamet), a widely prescribed drug for the control of gastric acidity and peptic ulcers. Cimetidine reduces gastric acid secretion by inhibiting the interaction of histamine with gastric H2 receptors. In the development of this drug, a cyano group was added to the substituted guanidino group to alter its basicity. Do you expect this modified guanidino group to be more basic or less basic than the guanidino group of arginine? Explain.

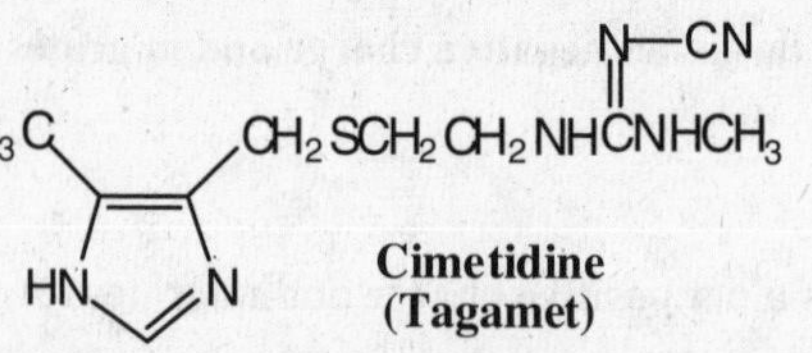

**Cimetidine
(Tagamet)**

A cyano group is electron-withdrawing. As a result, the guanidino function will have less electron density available to interact with a proton, and will be less basic than a similar guanidino group without the cyano attached.

Problem 27.40 Draw a structural formula for the product formed when alanine is treated with the following reagents.

(a)

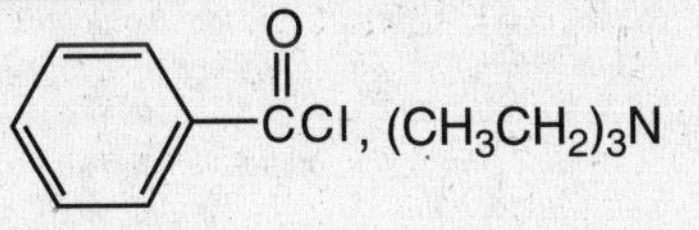

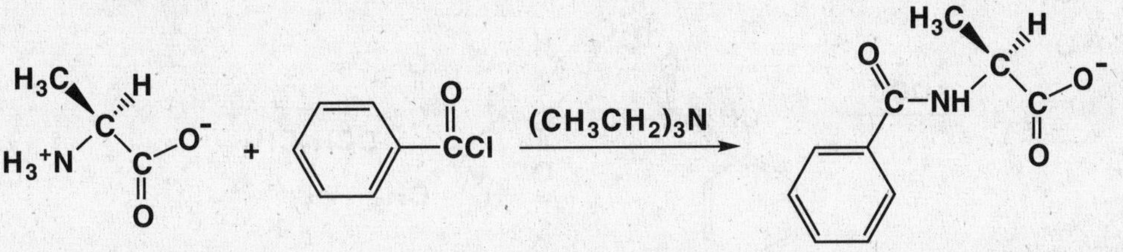

(b)

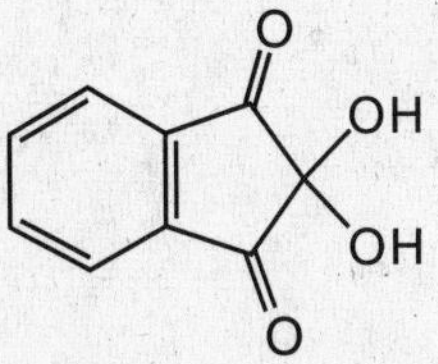

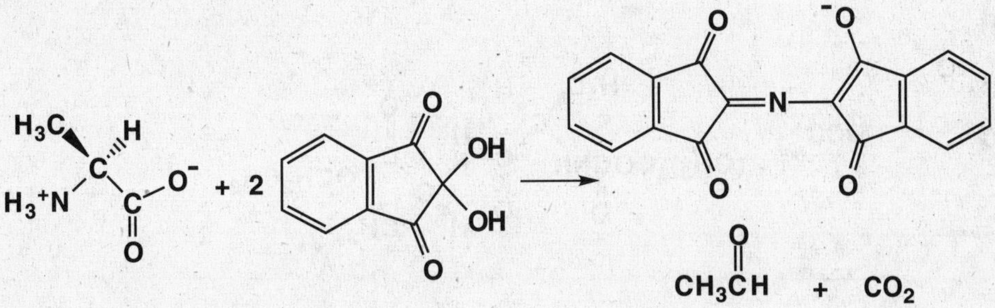

$$CH_3CH \quad + \quad CO_2$$

(c)

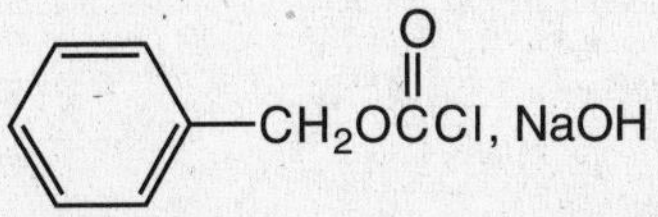

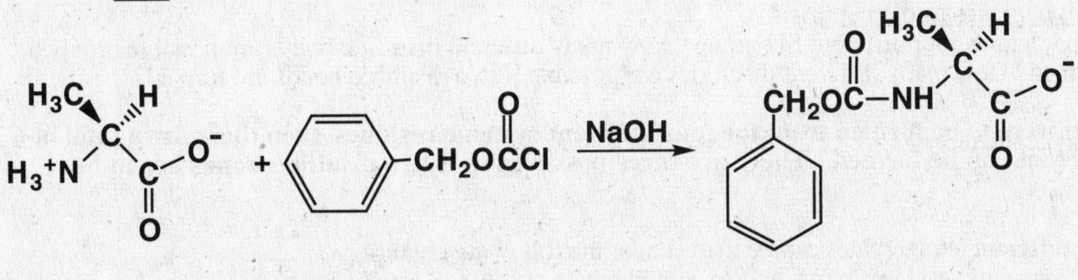

(d) $(CH_3)_3COCOCOC(CH_3)_3$, NaOH

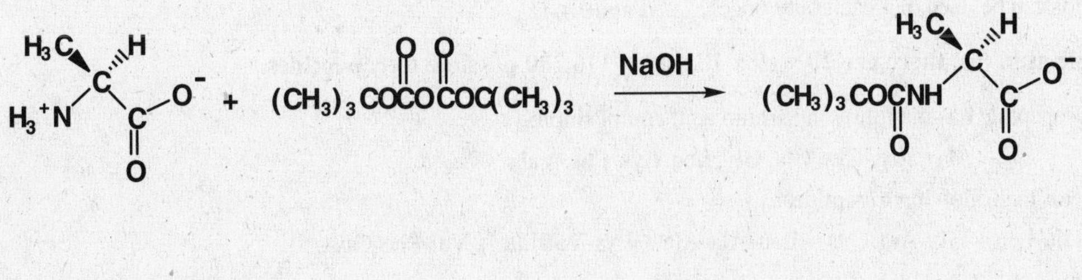

(e) Product from (c) + L-alanine ethyl ester + DCC

(f) Product from (d) + L-alanine ethyl ester + DCC

Primary Structure of Polypeptides and Proteins

Problem 27.41 If a protein contains four different SH groups, how many different disulfide bonds are possible if only a single disulfide bond is formed? How many different disulfides are possible if two disulfide bonds are formed?

If only one disulfide bond were to be formed from the four different cysteine residues, then there are a total of 6 different disulfide bonds that can be formed. There are three possibilities if two disulfide bonds are to be formed.

Problem 27.42 How many different tetrapeptides can be made under the following conditions?
(a) The tetrapeptide contains one unit each of Asp, Glu, Pro, and Phe.

There could be any of the four residues in the first position, any of the remaining three amino acids in the second position and so on. Thus, there are 4 x 3 x 2 x 1 = 24 possible tetrapeptides.

(b) All 20 amino acids can be used, but each only once.

Using the same logic as in (a), there are 20 x 19 x 18 x 17 = 116,280 possible tetrapeptides.

Problem 27.43 A decapeptide has the following amino acid composition:

$$Ala_2, Arg, Cys, Glu, Gly, Leu, Lys, Phe, Val$$

Partial hydrolysis yields the following tripeptides:

$$Cys\text{-}Glu\text{-}Leu + Gly\text{-}Arg\text{-}Cys + Leu\text{-}Ala\text{-}Ala + Lys\text{-}Val\text{-}Phe + Val\text{-}Phe\text{-}Gly$$

One round of Edman degradation yields a lysine phenylthiohydantoin. From this information, deduce the primary structure of this decapeptide.

Due to the Edman degradation result, the Lys residue must be at the *N*-terminus. Given this information, the rest of the peptide sequence is deduced because of overlap among the tripeptide sequences as shown below.

The complete peptide is:
Lys-Val-Phe-Gly-Arg-Cys-Glu-Leu-Ala-Ala

The peptides fit as follows:
Lys-Val-Phe
Val-Phe-Gly
Gly-Arg-Cys
Cys-Glu-Leu
Leu-Ala-Ala

<u>Problem 27.44</u> Following is the primary structure of glucagon a polypeptide hormone of 29 amino acids. Glucagon is produced in the α-cells of the pancreas and helps maintain blood glucose levels in a normal concentration range.

```
       1               5              10             15
   His-Ser-Glu-Gly-Thr-Phe-Thr-Ser-Asp-Tyr-Ser-Lys-Tyr-Leu-Asp-Ser-Arg-Arg-

                  20             25          29
        Ala-Gln-Asp-Phe-Val-Gln-Trp-Leu-Met-Asn-Thr
```

Which peptide bonds are hydrolyzed when this polypeptide is treated with each reagent?
(a) Phenyl isothiocyanate

This reagent only hydrolyzes the *N*-terminal amino acid, so the His-Ser bond would be hydrolyzed. The site of cleavage is indicated by the j.

```
     1               5              10             15
   His]j[Ser-Glu-Gly-Thr-Phe-Thr-Ser-Asp-Tyr-Ser-Lys-Tyr-Leu-Asp-Ser-Arg-

                  20             25          29
        Arg-Ala-Gln-Asp-Phe-Val-Gln-Trp-Leu-Met-Asn-Thr
```

(b) Chymotrypsin

Chymotrypsin catalyzes the hydrolysis of the peptide bonds that are located on the carboxyl side of phenylalanine, tyrosine, and tryptophan residues. The sites of cleavage are indicated by the j.

```
     1               5              10             15
   His-Ser-Glu-Gly-Thr-Phe]j[Thr-Ser-Asp-Tyr]j[Ser-Lys-Tyr]j[Leu-Asp-Ser-

                  20             25          29
        Arg-Arg-Ala-Gln-Asp-Phe]j[Val-Gln-Trp]j[Leu-Met-Asn-Thr
```

(c) Trypsin

Trypsin catalyzes the hydrolysis of the peptide bonds that are located on the carboxyl side of arginine and lysine residues. The sites of cleavage are indicated by the j.

```
     1               5              10             15
   His-Ser-Glu-Gly-Thr-Phe-Thr-Ser-Asp-Tyr-Ser-Lys]j[Tyr-Leu-Asp-Ser-

                  20             25          29
        Arg]j[Arg]j[Ala-Gln-Asp-Phe-Val-Gln-Trp-Leu-Met-Asn-Thr
```

(d) Br-CN

Cyanogen bromide cleaves on the *C*-terminal side of methionine residues. The site of cleavage is indicated by the j.

```
     1               5              10             15
   His-Ser-Glu-Gly-Thr-Phe-Thr-Ser-Asp-Tyr-Ser-Lys-Tyr-Leu-Asp-Ser-

                  20             25          29
        Arg-Arg-Ala-Gln-Asp-Phe-Val-Gln-Trp-Leu-Met]j[Asn-Thr
```

Problem 27.45 A tetradecapeptide (14 amino acid residues) gives the following peptide fragments on partial hydrolysis. From this information, deduce the primary structure of this polypeptide. Fragments are grouped according to size.

Pentapeptide Fragments	Tetrapeptide Fragments
Phe-Val-Asn-Gln-His	Gln-His-Leu-Cys
His-Leu-Cys-Gly-Ser	His-Leu-Val-Glu
Gly-Ser-His-Leu-Val	Leu-Val-Glu-Ala

The complete peptide is:
Phe-Val-Asn-Gln-His-Leu-Cys-Gly-Ser-His-Leu-Val-Glu-Ala

The peptides fit as follows:
        Phe-Val-Asn-Gln-His
                Gln-His-Leu-Cys
                His-Leu-Cys-Gly-Ser
                        Gly-Ser-His-Leu-Val
                              His-Leu-Val-Glu
                                Leu-Val-Glu-Ala

Problem 27.46 Draw a structural formula of these tripeptides. Mark each peptide bond, the *N*-terminal amino acid, and the *C*-terminal amino acid.
(a) Phe-Val-Asn   (b) Leu-Val-Gln

Problem 27.47 Estimate the pI of each tripeptide on Problem 27.46.

These pI values can be estimated by using the pK_a of the amino group for the *N*-terminal amino acid, and the pK_a of the carboxylic acid group for the *C*-terminal amino acid. Using the values for the appropriate amino groups and carboxylic acid groups listed in table 27.2 leads to the values of pI = 1/2(9.24 + 2.02) = 5.63 and pI = 1/2(9.76 + 2.17) = 5.96 for (a) Phe-Val-Asn and (b) Leu-Val-Gln, respectively.

Problem 27.48 Glutathione (G-SH), one of the most common tripeptides in animals, plants, and bacteria, is a scavenger of oxidizing agents. In reacting with oxidizing agents, glutathione is converted to G-S-S-G.

Glutathione

(a) Name the amino acids in this tripeptide.

The amino acids in glutathione are glutamic acid (Glu), cysteine (Cys), and glycine (Gly).

(b) What is unusual about the peptide bond formed by the *N*-terminal amino acid?

The *N*-terminal glutamic acid is linked to the next residue by an amide bond between the carboxyl group of the side chain, not the α-carboxyl group.

(c) Write a balanced half-reaction for the reaction of two molecules of glutathione to form a disulfide bond. Is glutathione a biological oxidizing agent or a biological reducing agent?

$$2 \text{ G-SH} \longrightarrow \text{G-S-S-G} + 2\text{H}^+ + 2\text{e}^-$$

The glutathione is oxidized in this process, so it is a biological reducing agent.

(d) Write a balanced equation for reaction of glutathione with molecular oxygen, O_2, to form G-S-S-G and H_2O. Is molecular oxygen oxidized or reduced in this process?

$$4 \text{ G-SH} + O_2 \longrightarrow 2 \text{ G-S-S-G} + 2 H_2O$$

Molecular oxygen is reduced in this process.

<u>Problem 27.49</u> Following iare a structural formula and a ball-and-stick model for the artificial sweetener aspartame. Each amino acid has the L configuration.

Aspartame

(a) Name the two amino acids in this molecule.

Aspartame is composed of aspartic acid (Asp) attached via a peptide bond to the methyl ester of phenylalanine (Phe).

(b) Estimate the isoelectric point of aspartame?

These pI value can be estimated by using the pK_a of the amino group for the N-terminal amino acid, and the pK_a of the carboxylic acid group for aspartic acid residue. Using the values listed in table 27.2 leads to the values of pI = 1/2(9.82 + 3.86) = 6.84.

(c) Draw structural formulas for the products of hydrolysis of aspartame in 1 M HCl.

Aspartame

Problem 27.50 2,4-Dinitrofluorobenzene, very often known as Sanger's reagent after the English chemist Frederick Sanger who popularized its use, reacts selectively with the N-terminal amino group of a polypeptide chain. Sanger was awarded the 1958 Nobel Prize for chemistry for his work in determining the primary structure of bovine insulin. One of the few persons to be awarded two Nobel Prizes, he also shared the 1980 award in chemistry with American chemists, Paul Berg and Walter Gilbert, for the development of chemical and biological analyses of DNAs.

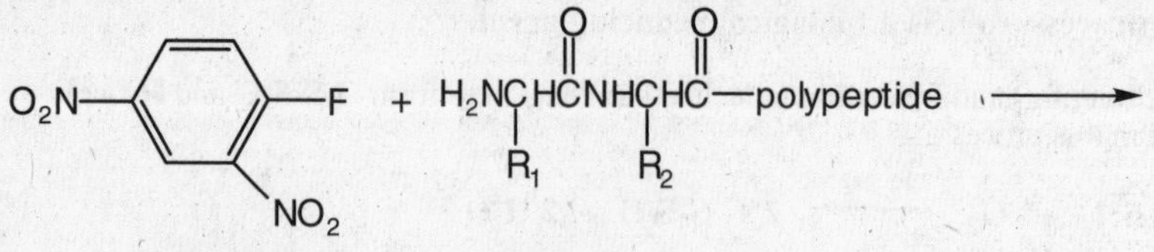

2,4-Dinitro-fluorobenzene   (*N*-terminal end of a polypeptide chain)

Following reaction with 2,4-dinitrofluorobenzene, all amide bonds of the polypeptide chain are hydrolyzed, and the amino acid labeled with a 2,4-dinitrophenyl group is separated by either paper or column chromatography and identified.

(a) Write a structural formula for the product formed by treatment of the *N*-terminal amino group with Sanger's reagent and propose a mechanism for its formation.

Step 1: Make a bond between a nucleophile and an electrophile (aromatic).

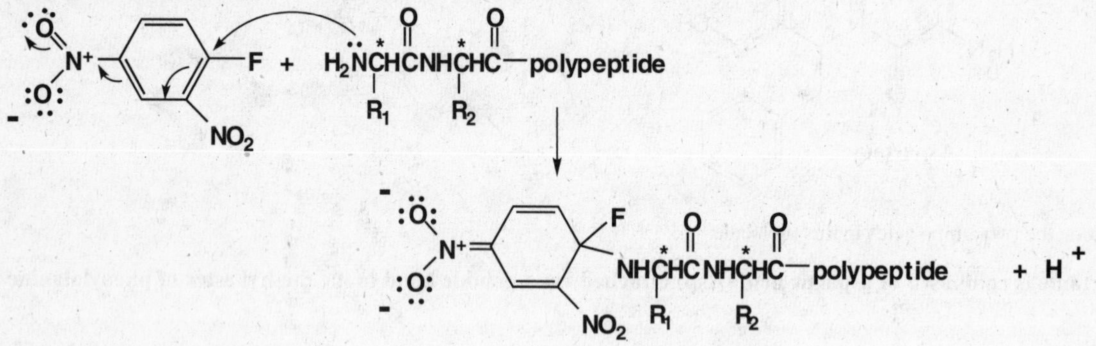

Meisenheimer complex

Step 2: Break a bond to give stable molecules or ions.

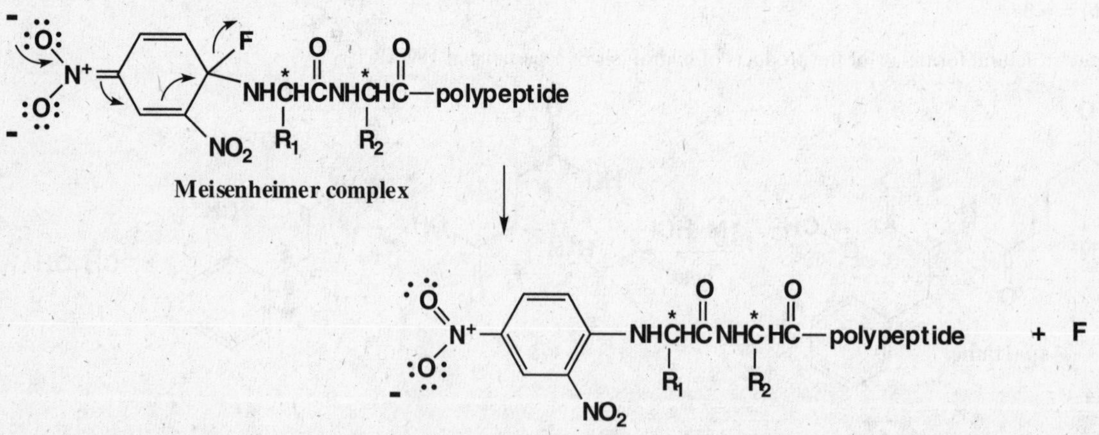

Meisenheimer complex

(b) When bovine insulin is treated with Sanger's reagent followed by hydrolysis of all peptide bonds, two labeled amino acids are detected: glycine and phenylalanine. What conclusions can be drawn from this information about the primary structure of bovine insulin?

This result indicates that insulin is actually composed of two polypeptide chains, so there are two *N*-terminal residues.

(c) Compare and contrast the structural information that can be obtained from use of Sanger's reagent with that from use of the Edman degradation.

Sanger's reagent only allows identification of the *N*-terminal amino acid, while the Edman degradation can sequentially determine the sequence of amino acids at the *N*-terminus.

Synthesis of Polypeptides

Problem 27.51 In a variation of the Merrifield solid-phase peptide synthesis, the amino group is protected by a fluorenylmethoxycarbonyl (FMOC) group. This protecting group is removed by treatment with a weak base such as the secondary amine, piperidine. Write a balanced equation and propose a mechanism for this deprotection.

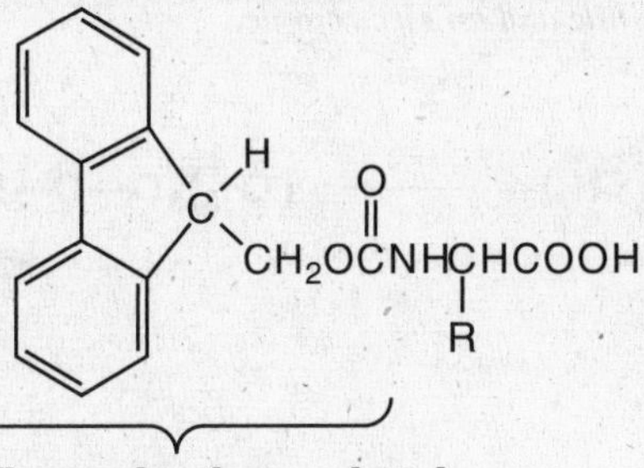

Fluorenylmethoxycarbonyl
(FMOC) group

The key to the mechanism is that the FMOC group is unusually acidic for a hydrocarbon, because deprotonation generates a relatively stable dibenzocyclopentadienyl anion. The stability of this anion is analogous to cyclopentadiene itself. The following is a reasonable mechanism for the removal of FMOC protecting groups in the presence of a weak base such as piperidine.

Step : Take a proton away.

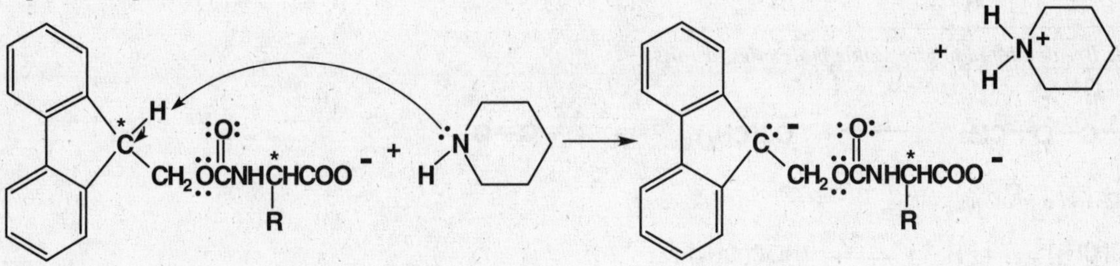

Step 2: Break a bond to give stable molecules or ions.

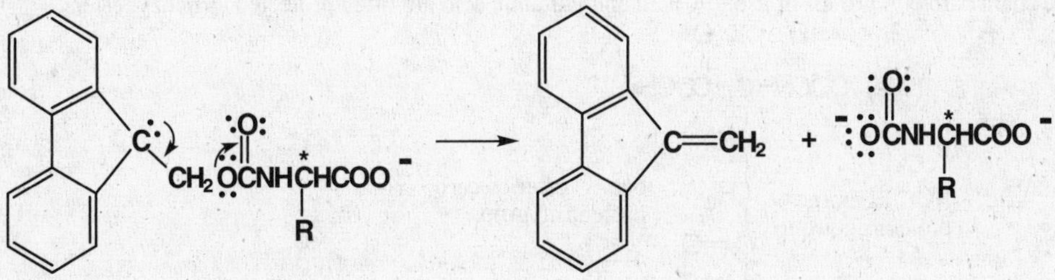

Step 3: Add a proton.

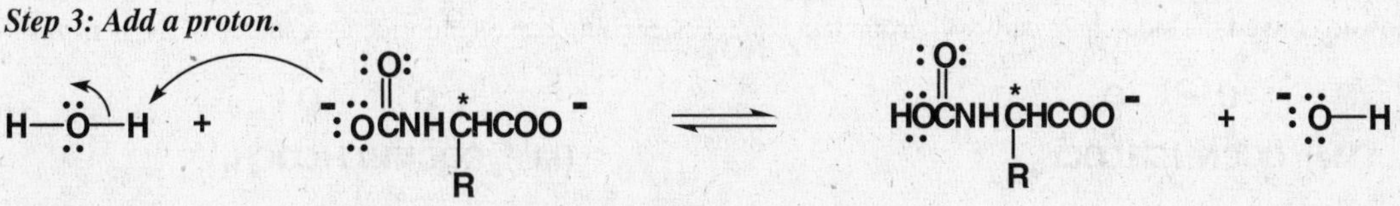

Step 4: Decarboxylation. **This process actually involves several steps.**

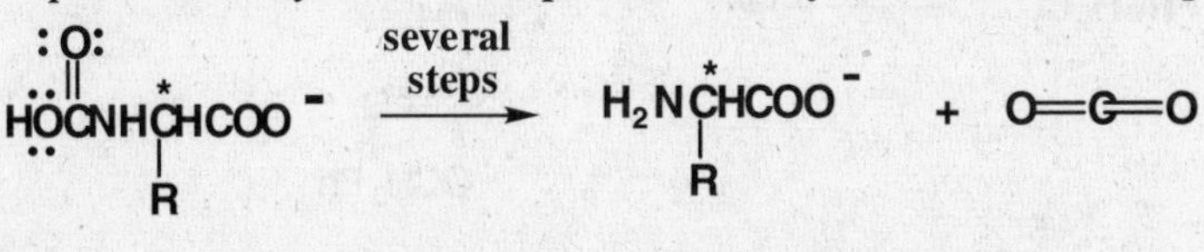

Problem 27.52 The BOC-protecting group may be added by treatment of an amino acid with di-*tert*-butyl dicarbonate as shown in the following reaction sequence.   Propose a mechanism to account for formation of these products.

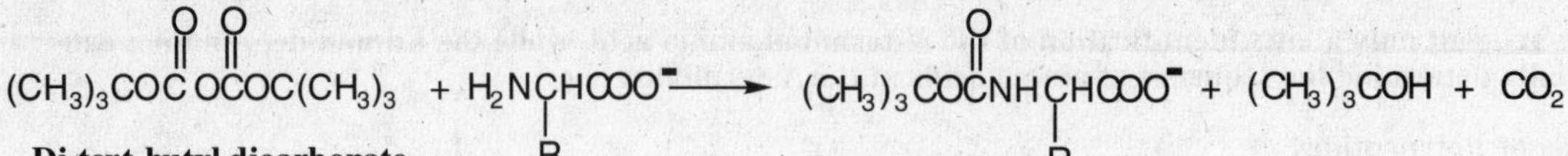

$$(CH_3)_3COC\ OCOC(CH_3)_3 + H_2NCHCOO^- \longrightarrow (CH_3)_3COCNHCHCOO^- + (CH_3)_3COH + CO_2$$

Di-tert-butyl dicarbonate
R

R
BOC-amino acid

Step 1: Make a bond between a nucleophile and an electrophile.

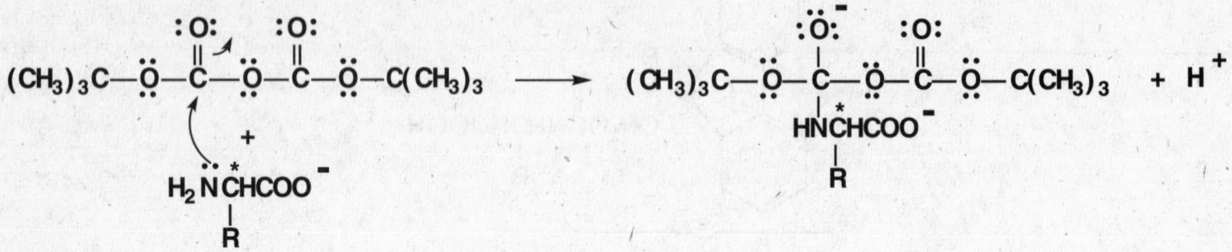

Step 2: Break a bond to give stable molecules or ions.

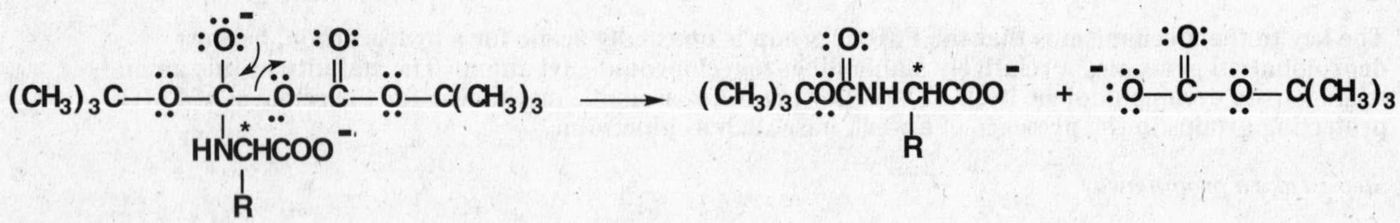

Step 3: Break a bond to give stable molecules or ions.

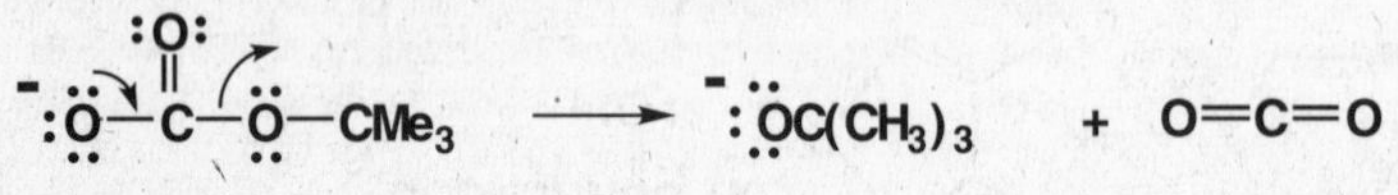

Step 4: Add a proton.

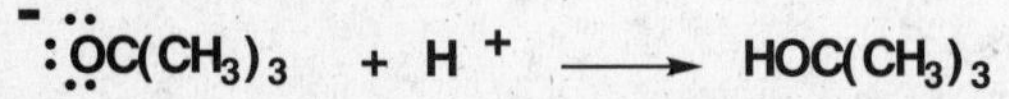

$$:\overset{..}{\underset{..}{O}}C(CH_3)_3 + H^+ \longrightarrow HOC(CH_3)_3$$

Problem 27.53 The side chain carboxyl groups of aspartic acid and glutamic acid are often protected as benzyl esters.

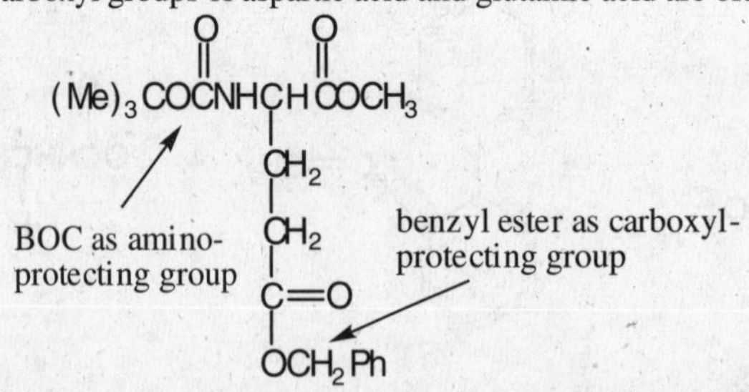

BOC as amino-protecting group

benzyl ester as carboxyl-protecting group

(a) Show how to convert the side-chain carboxyl group to a benzyl ester using benzyl chloride as a source of the benzyl group.

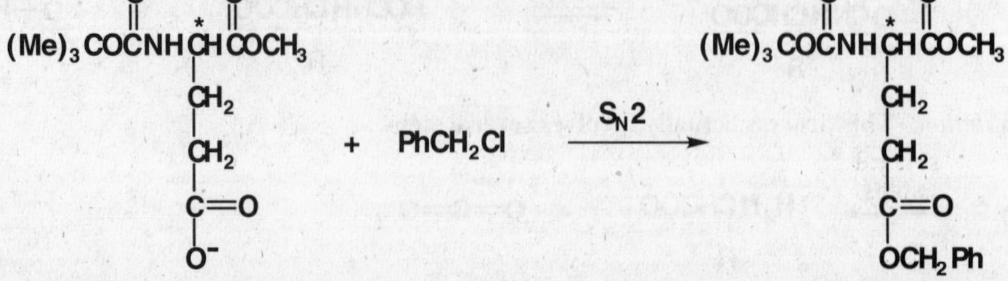

(b) How do you deprotect the side-chain carboxyl under mild conditions without removing the BOC protecting group at the same time?

The benzyl esters can be removed under very mild conditions by hydrogenolysis using hydrogen in the presence of a transition metal catalyst.

Three-Dimensional Shapes of Polypeptides and Proteins

Problem 27.54 Examine the α-helix conformation. Are amino acid side chains arranged all inside the helix, all outside the helix, or is their arrangement random?

All of the amino acid side chains extend outside the helix.

Problem 27.55 Distinguish between intermolecular and intramolecular hydrogen bonding between the backbone groups on polypeptide chains. In what type of secondary structure do you find intermolecular hydrogen bonds? In what type do you find intramolecular hydrogen bonding?

Substantial intermolecular hydrogen bonding is possible with β-sheet secondary structures, while only minimal intermolecular hydrogen bonding is possible with α-helix secondary structures. This is because the entire edges of β-sheets have backbone hydrogen bonding groups exposed, while α-helices have only the hydrogen bonding groups exposed on the ends. Both β-sheets and α-helices are stabilized by extensive intramolecular hydrogen bonding.

Problem 27.56 Many plasma proteins found in an aqueous environment are globular in shape. Which amino acid side chains would you expect to find on the surface of a globular protein and in contact with the aqueous environment? Which would you expect to find inside, shielded from the aqueous environment? Explain.
(a) Leu   (b) Arg   (c) Ser   (d) Lys   (e) Phe

In general, charged or hydrophilic amino acids are exposed to the aqueous solution on the surface of a globular protein. Thus, (b) Arg, (c) Ser, and (d) Lys will be on the surface. The hydrophobic amino acids (a) Leu and (e) Phe will generally be inside the protein, shielded from the aqueous environment.

Problem 27.57 Denaturation of a protein is a physical change, the most readily observable result of which is loss of biological activity. Denaturation stems from changes in secondary, tertiary, and quaternary structure through disruption of noncovalent interactions including hydrogen bonding and hydrophobic interactions. Two common denaturing agents are sodium dodecyl sulfate (SDS) and urea. What kinds of noncovalent interactions might each reagent disrupt?

The SDS disrupts the hydrophobic forces that keep the non-polar residues in the interior of the structure away from the aqueous solvent and the urea disrupts the hydrogen bonds of the protein that stabilize higher order structure.

French angelfish *Pomacanthus paru*
Tobago, West Indies

CHAPTER 28
Solutions to the Problems

Problem 28.1 Draw a structural formula for each nucleotide.
(a) 2'-Deoxythymidine 5'-monophosphate

(b) 2'-Deoxythymidine 3'-monophosphate

Problem 28.2 Draw a structural formula for the section of DNA that contains the base sequence CTG and is phosphorylated on the 3' end only.

Problem 28.3 Write the complementary DNA base sequence for 5'-CCGTACGA-3'.

The complementary sequence would be 3'-GGCATGCT-5'

<u>Problem 28.4</u> Here is a portion of the nucleotide sequence in phenylalanine tRNA.

3'-ACCACCUGCUCAGGCCUU-5'

Write the nucleotide sequence of its DNA complement.

Remember that the base uracil (U) in RNA is complementary to adenine (A) in DNA. The complement DNA sequence of the above RNA sequence would be:

5'-TGGTGGACGAGTCCGGAA-3'.

<u>Problem 28.5</u> The following section of DNA codes for oxytocin, a polypeptide hormone.

3'-ACG-ATA-TAA-GTT-TTA-ACG-GGA-GAA-CCA-ACT-5'

(a) Write the base sequence of the mRNA synthesized from this section of DNA.

The base sequence of the mRNA synthesized from this section of DNA would be:

5'-UGC-UAU-AUU-CAA-AAU-UGC-CCU-CUU-GGU-UGA-3'

(b) Given the sequence of bases in part (a), write the primary structure of oxytocin.

The primary sequence of oxytocin would be:

Amino terminus- Cys-Tyr-Ile-Gln-Asn-Cys-Pro-Leu-Gly -Carboxyl terminus

Note how the last codon, UGA, does not code for an amino acid, but rather is the stop signal.

<u>Problem 28.6</u> The following is another section of the bovine rhodopsin gene. Which of the endonucleases given in Example 28.6 will catalyze cleavage of this section.

FnuDII   HpaII

5'-ACGTCGGGTCGTCGTCCTCT⎡CGCG⎤GTGGTGAGTCTT⎡CCGG⎤CTCTTCT-3'

The F<i>nu</i>DII and H<i>pa</i>II cleavage sites are shown on the sequence above.

PROBLEMS
Nucleosides and Nucleotides

<u>Problem 28.7</u> A pioneer of designing and synthesizing antimetabolites that could destroy cancer cells was George Hitchings at Burroughs Wellcome Co. In 1942 he initiated a program to discover DNA antimetabolites, and in 1948 he and Gertrude Elion synthesized 6-mercaptopurine, a successful drug for treating acute leukemia. Another DNA antimetabolite synthesized by Hutchings and Elion was 6-thioguanine. Hitchings and Elion along with Sir James W. Black won the 1988 Nobel Prize for Physiology or Medicine for their discoveries of "important principles of drug treatment". In each drug, the oxygen at carbon 6 of the parent molecule is replaced by divalent sulfur. Draw structural formulas for the enethiol (the sulfur equivalent of an enol) forms of 6-mercaptopurine and 6-thioguanine.

6-Mercaptopurine   6-Thioguanine

The enethiol forms are shown below:

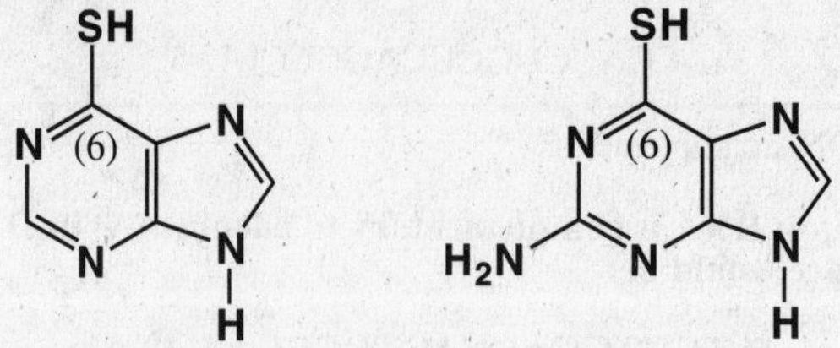

Problem 28.8 Following are structural formulas for cytosine and thymine. Draw two additional tautomeric forms for cytosine and three additional tautomeric forms for thymine.

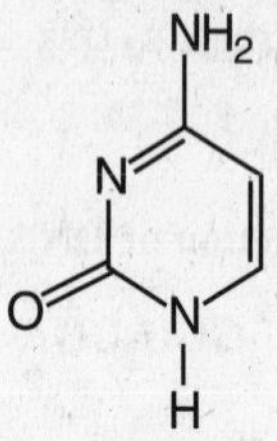

Cytosine (C) 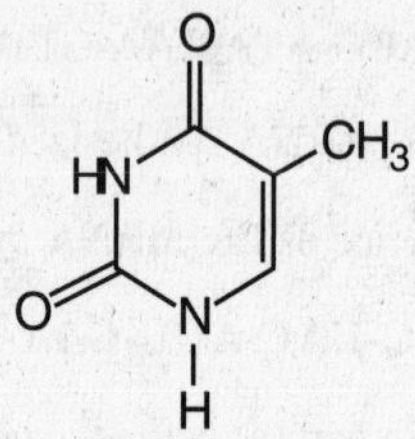   **Thymine (T)**

Three additional tautomeric forms for cytosine are:

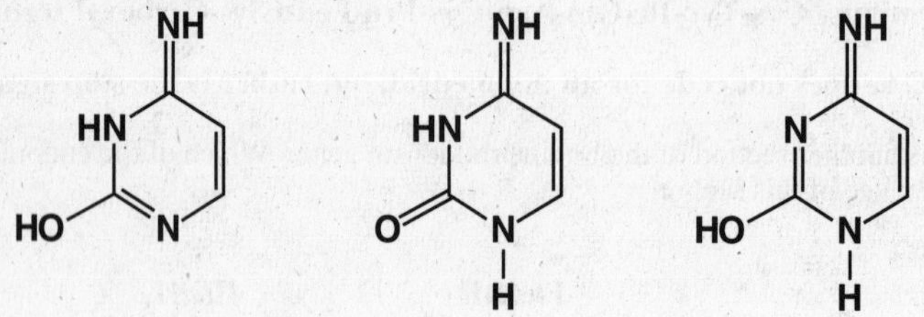

Four additional tautomeric forms for thymine are:

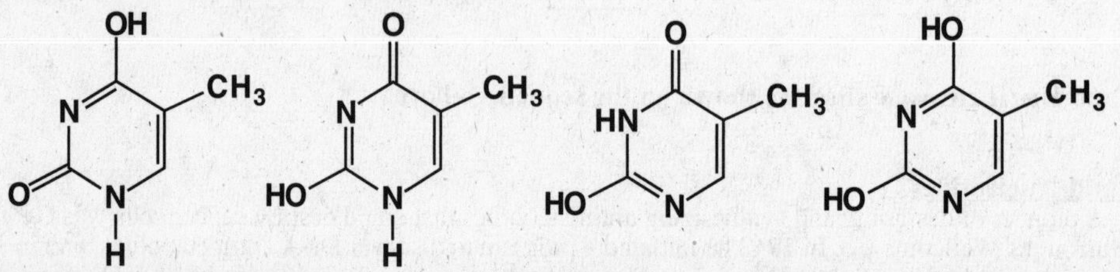

Problem 28.9 Draw a structural formula for a nucleoside composed of the following.
(a) α-D-Ribose and adenine   (b) β-2-Deoxy-D-ribose and cytosine

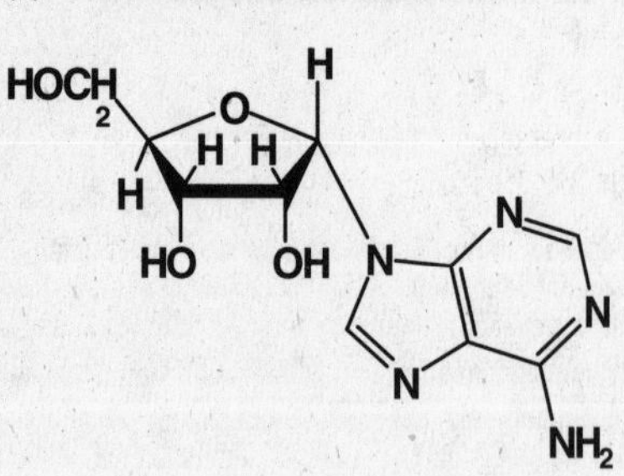

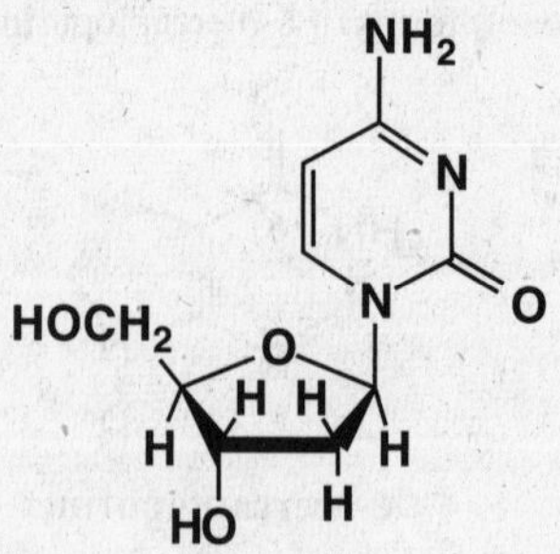

<u>Problem 28.10</u> Nucleosides are stable in water and in dilute base. In dilute acid, however, the glycosidic bond of a nucleoside undergoes hydrolysis to give a pentose and a heterocyclic aromatic amine base. Propose a mechanism for this acid-catalyzed hydrolysis.

Acid-catalyzed glycosidic bond hydrolysis is most pronounced for purine nucleosides. A reasonable mechanism involves protonation of the heterocyclic base to create a good leaving group that is displaced by water to produce the product pentose and free base. The reaction of guanosine in acid (H-A) is shown below.

Step 1: Add a proton.

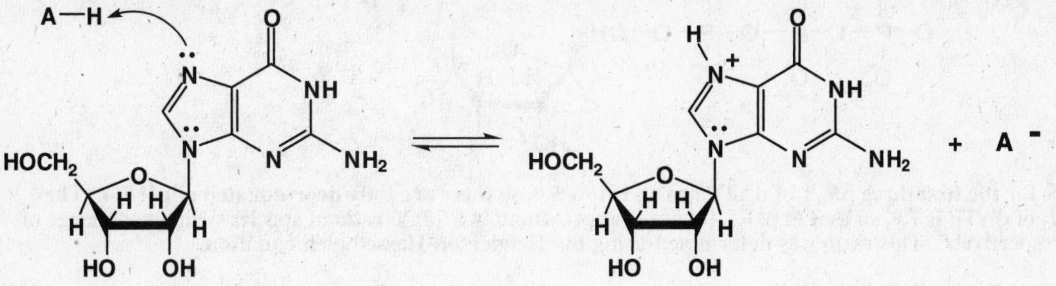

Step 2: Break a bond to give stable molecules or ions.

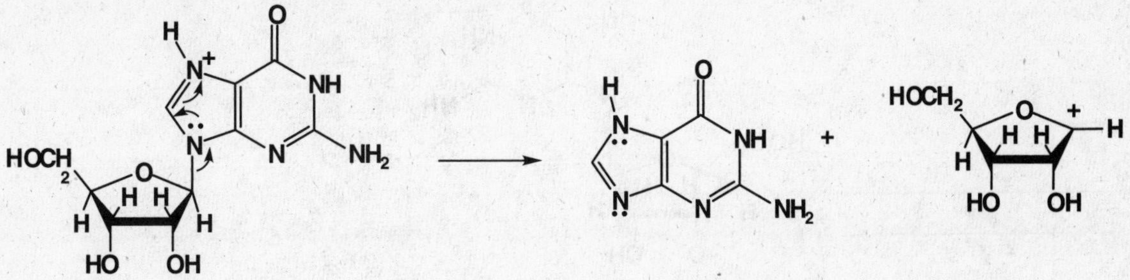

Step 3: Make a bond between a nucleophile and an electrophile.

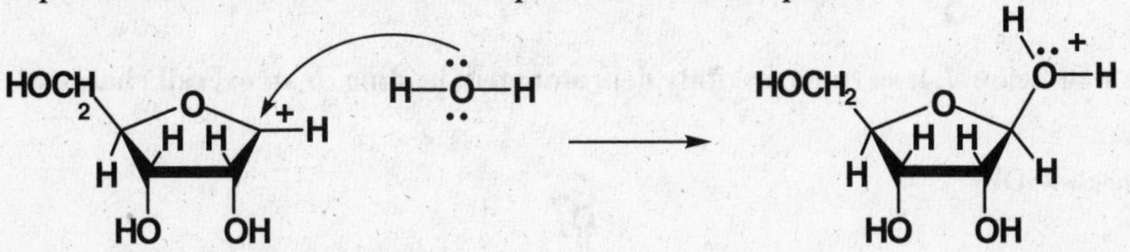

Step 4: Take a proton away.

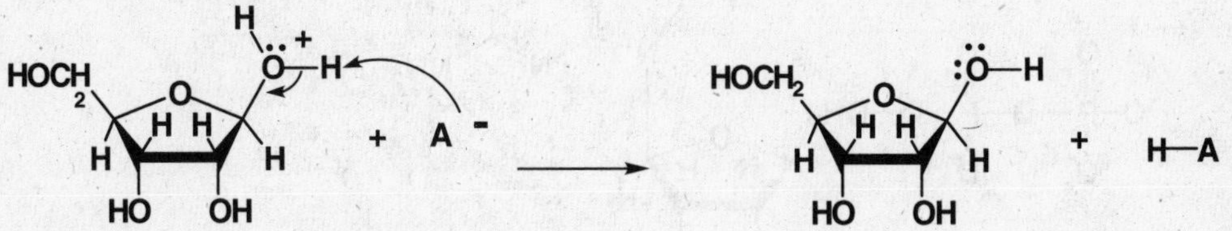

<u>Problem 28.11</u> Explain the difference in structure between a nucleoside and a nucleotide.

A nucleoside consists of a D-ribose or 2'-deoxy-D-ribose bonded to an heterocylic aromatic base by a β-*N*-glycosidic bond. A nucleotide is a nucleoside that has one or more molecules of phosphoric acid esterified at an -OH group of the monosaccharide, usually at the 3' and/or 5' -OH group.

Problem 28.12 Draw a structural formula for each nucleotide, and estimate its net charge at pH 7.4, the pH of blood plasma.

(a) 2'-Deoxyadenosine 5'-triphosphate (dATP)

The values for the first three pK_a's of dATP are all below 5.0, so these are fully deprotonated at pH 7.4. The fourth pK_a of dATP is 7.0, so that at pH 7.4 there is approximately a 70:30 ratio of species with a net charge of -4 or -3, respectively. This ratio was determined using the Henderson-Hasselbalch equation.

(b) Guanosine 3'-monophosphate (GMP)

The two pK_a values for GMP are well below 7.4, so these are fully deprotonated, leading to an overall charge of -2.

(c) 2'-Deoxyguanosine 5'-diphosphate (dGDP)

The values for the first two pK_a's of dGDP are all below 5.0, so these are fully deprotonated at pH 7.4. The third pK_a of dATP is 6.7, so that at pH 7.4 there is approximately a 83:17 ratio of species with a net charge of -3 or -2, respectively. This ratio was determined using the Henderson-Hasselbalch equation.

Problem 28.13 Cyclic-AMP, first isolated in 1959, is involved in many diverse biological processes as a regulator of metabolic and physiological activity. In it, a single phosphate group is esterified with both the 3' and 5' hydroxyls of adenosine. Draw a structural formula of cyclic-AMP.

Cyclic-AMP

The Structure of DNA

Problem 28.14 Why are deoxyribonucleic acids called acids? What are the acidic groups in their structure?

Deoxyribonucleic acids are called acids because the phosphodiester groups of the backbone are acidic. At neutral pH, they are fully deprotonated, leading to the anionic nature of DNA.

Problem 28.15 Human DNA contains approximately 30.4% A. Estimate the percentages of G, C, and T and compare them with the values presented in Table 28.1.

The A residues must be paired with T residues, so estimate that there is also 30.4% T. A and T must therefore account for 30.4% + 30.4% = 60.8% of the bases. That leaves (100% - 60.8%) / 2 = 39.2% / 2 = 19.6% each for G and C. In Table 28.1, there is actually slightly less T than expected, so there is also slightly more G and C than expected.

Problem 28.16 Draw a structural formula of the DNA tetranucleotide 5'-A-G-C-T-3'. Estimate the net charge on this tetranucleotide at pH = 7.0. What is the complementary tetranucleotide to this sequence?

As shown in the preceding structure, there is a net charge of -5 on this tetranucleotide at pH 7.0. This oligonucleotide is self-complementary, that is the complementary oligonucleotide also has the sequence 5'-A-G-C-T-3'.

Problem 28.17 List the postulates of the Watson-Crick model of DNA secondary structure.

Major postulates of the Watson-Crick model are that:
1) A molecule of DNA consists of two antiparallel strands coiled in a right handed manner about the same axis, thereby creating a double helix.
2) The bases project inward toward the helix axis.

3) The bases are paired through hydrogen bonding, with a purine paired to a pyrimidine so that each base pair is of the same size and shape.
4) In particular A pairs with T and G pairs with C.
5) The paired bases are stacked one on top of another in the interior of the double helix.
6) There is a distance of 0.34 nm between adjacent stacked paired bases.
7) There are ten paired bases per turn of the helix, and these are slightly offset from each other. The slight offset provides two grooves of different dimensions along the helix, the so-called major and minor grooves.

Problem 28.18 The Watson-Crick model is based on certain experimental observations of base composition and molecular dimensions. Describe these observations and show how the Watson-Crick model accounts for each.

Chargaff found that in different organisms, the amount of A always equals the amount of T and the amount of G always equals the amount of C, even though different organisms have different ratios of A to G. The base-pairing postulates of the Watson-Crick model fully explain the observed ratios of bases. The geometry of the Watson-Crick model also accounts perfectly for the periodicity observed in the X-ray diffraction data.

Problem 28.19 If you read J. D. Watson's account of the discovery of the structure of DNA, *The Double Helix,* you will find that for a time in their model-building studies, he and Crick were using alternative (and incorrect, at least in terms of their final model of the double helix) tautomeric structures for some of the heterocyclic bases.
(a) Write at least one alternative tautomeric structure for adenine.

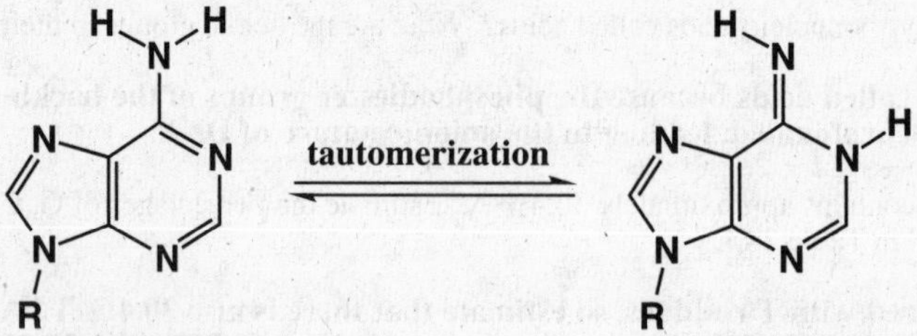

Adenine   **tautomerization**   **Tautomeric adenine**

(b) Would this structure still base-pair with thymine, or would it now base-pair more efficiently with a different base and if so, with what base?

This tautomeric form of adenine would not be able to base pair with thymine, but it would be able to form a reasonable base pair with cytosine, as shown below:

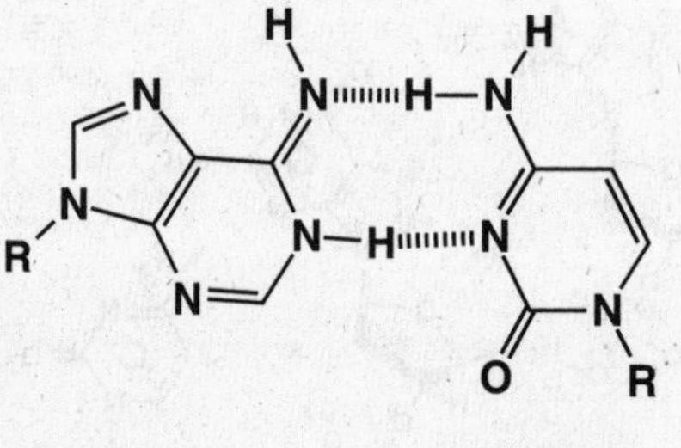

Cytosine

Problem 28.20 Compare the α-helix of proteins and the double helix of DNA in these ways.
(a) The units that repeat in the backbone of the polymer chain.

The α-helix of a protein is composed of amino acids, so the repeating unit of the backbone is a carboxyl group bonded to a tetrahedral carbon atom and a nitrogen atom. Of course, the carboxyl group and nitrogen atoms are linked in amide bonds. The repeating unit of the double helix in DNA is a 2'-deoxy-D-ribose unit linked in 3'-5' phosphodiester bonds.

(b) The projection in space of substituents along the backbone (the R groups in the case of amino acids; purine and pyrimidine bases in the case of double-stranded DNA) relative to the axis of the helix.

The α-helix of a protein has the R groups pointed out away from the helix axis. The DNA bases of the double helix are pointed inward, toward the helix axis.

Problem 28.21 Discuss the role of the hydrophobic interactions in stabilizing the following.
(a) Double-stranded DNA

In the DNA double helix, the relatively hydrophobic bases are stacked on the inside, surrounded by the relatively hydrophilic sugar-phosphate backbone that is on the outside of the structure. The stacking of the hydrophobic bases minimizes contact with water.

(b) Lipid bilayers

In lipid bilayers, the hydrophobic hydrocarbon tails are associated with each other to form the hydrophobic inner layer, while the polar head groups are associated with each other on both outside surfaces.

(c) Soap micelles

In micelles, the hydrophobic hydrocarbon tails are associated with each other to form the hydrophobic interior, while the polar groups are associated with each other on the outside surface.

Problem 28.22 Name the type of covalent bond(s) joining monomers in these biopolymers.
(a) Polysaccharides   (b) Polypeptides   (c) Nucleic acids

Polysaccharides have glycosidic linkages, polypeptides have amide linkages and nucleic acids have phosphodiester linkages between the monomers, respectively.

Problem 28.23 In terms of hydrogen bonding, which is more stable, an A-T base pair or a G-C base pair?

A G-C base pair is held together by 3 hydrogen bonds, while an A-T base pair is held together by only two hydrogen bonds. Thus, in terms of hydrogen bonds alone, a G-C base pair is more stable than an A-T base pair.

Problem 28.24 At elevated temperatures, nucleic acids become denatured, that is, they unwind into single-stranded DNA. Account for the observation that the higher the G-C content of a nucleic acid, the higher the temperature required for its thermal denaturation.

G-C base pairs have three hydrogen bonds between them, while A-T base pairs have only two. Thus, the G-C base pairs are held together with stronger overall attractive forces and require higher temperatures to denature.

Problem 28.25 Write the DNA complement for 5'-ACCGTTAAT-3'. Be certain to label which is the 5' end and which is the 3' end of the complement strand.

The complementary sequence is 3'-TGGCAATTA-5'

Problem 28.26 Write the DNA complement for 5'-TCAACGAT-3'.

The complementary sequence is 3'-AGTTGCTA-5'

Ribonucleic Acids (RNA)
Problem 28.27 Compare the degree of hydrogen bonding in the base pair A-T found in DNA with that in the base pair A-U found in RNA.

The only difference between uracil (U) and thymine (T) is the presence of a methyl group at the 5 position of thymine, that is absent in uracil. As can be seen in the structures, the presence or absence of this methyl group has very little influence on hydrogen bonding.

Problem 28.28 Compare DNA and RNA in these ways:
(a) Monosaccharide units

DNA contains 2'-deoxy-D-ribose units, while RNA contains D-ribose units.

(b) Principal purine and pyrimidine bases

DNA		RNA	
Purines	Pyrimidines	Purines	Pyrimidines
Adenine	Thymine	Adenine	Uracil
Guanine	Cytosine	Guanine	Cytosine

(c) Primary structure

The monosaccharide unit in DNA is 2'-deoxy-D-ribose, the monosaccharide in RNA is D-ribose. The bases are the same between the two types of nucleic acids, except thymine is found in DNA while uracil is found in RNA. DNA is usually double stranded and RNA is primarily single stranded. In both DNA and RNA, the primary sequence consists of linear chains of the nucleic acids linked by phosphodiester bonds involving the 3' and 5' hydroxyl groups of the monosaccharide units.

(d) Location in the cell

DNA is found in cell nuclei, while the bulk of RNA occurs as ribosome particles in the cytoplasm.

(e) Function in the cell

DNA serves to store and transmit genetic information, and RNA is primarily involved with the transcription and translation of that genetic information during the production of proteins.

Problem 28.29 What type of RNA has the shortest lifetime in cells?

Messenger RNA has the shortest lifetime in cells, usually on the order of a few minutes or less. This short lifetime is thought to allow for very tight control over how much protein is produced in a cell at any one time.

Problem 28.30 Write the mRNA complement for 5'-ACCGTTAAT-3'. Be certain to label which is the 5' end and which is the 3' end of the mRNA strand.

The mRNA complement would be 3'-UGGCAAUUA-5'

Problem 28.31 Write the mRNA complement for 5'-TCAACGAT-3'.

The mRNA complement would be 3'-AGUUGCUA-5'

The Genetic Code

Problem 28.32 What does it mean to say that the genetic code is degenerate?

The genetic code is referred to as degenerate because more than one codon can code for the same amino acid. This is because there are 64 different codons, but only twenty amino acids and a stop signal for which coding is needed.

Problem 28.33 Write the mRNA codons for the following.
(a) Valine   **GUU, GUC, GUA, GUG**   (b) Histidine   **CAU, CAC**
(c) Glycine   **GGU, GGC, GGA, GGG**

Problem 28.34 Aspartic acid and glutamic acid have carboxyl groups on their side chains and are called acidic amino acids. Compare the codons for these two amino acids.

All of the codons for these two acidic amino acids begin with GA. The codons for aspartic acid are GAU and GAC, while the codons for glutamic acid are GAA and GAG.

Problem 28.35 Compare the structural formulas of the aromatic amino acids phenylalanine and tyrosine. Compare also the codons for these two amino acids.

L-Phenylalanine   L-Tyrosine

Phenylalanine has a phenyl group while tyrosine has a phenol group. The mRNA codons for phenylalanine are UUU and UUC, while the mRNA codons for tyrosine are UAU and UAC.

<u>Problem 28.36</u> Glycine, alanine, and valine are classified as nonpolar amino acids. Compare their codons What similarities do you find? What differences do you find?

Glycine	Alanine	Valine
GGU	GCU	GUU
GGC	GCC	GUC
GGA	GCA	GUA
GGG	GCG	GUG

All of these amino acids have four mRNA codons, all codons start with G, and in each case, the first two bases of the codon are identical for a given amino acid. This makes the last base irrelevant.

<u>Problem 28.37</u> Codons in the set CUU, CUC, CUA, and CUG all code for the amino acid leucine. In this set, the first and second bases are identical, and the identity of the third base is irrelevant. For what other sets of codons is the third base also irrelevant, and for what amino acid(s) does each set code?

The third base is also irrelevant for GUX (valine), GCX (alanine), GGX (glycine), ACX (threonine), CCX (proline), CGX (arginine), and UCX (serine). In the preceding codons, X stands for any of the bases.

<u>Problem 28.38</u> Compare the codons with a pyrimidine, either U or C, as the second base. Do the majority of the amino acids specified by these codons have hydrophobic or hydrophilic side chains?

The majority of amino acids with a pyrimidine in the second position of their codons are hydrophobic. This set contains phenylalanine, leucine, isoleucine, methionine, valine, proline, and alanine. Only serine and threonine have a pyrimidine in the second position and also have a somewhat hydrophilic side chain.

<u>Problem 28.39</u> Compare the codons with a purine, either A or G, as the second base. Do the majority of the amino acids specified by these codons have hydrophilic or hydrophobic side chains?

The majority of amino acids with a purine in the second position of their codons are hydrophilic. This set contains histidine, glutamine, asparagine, lysine, aspartic acid, glutamic acid, arginine, cysteine, and serine. Only glycine and tryptophan are not hydrophilic, while tyrosine is a special case that is aromatic with a polar group.

<u>Problem 28.40</u> What polypeptide is coded for by this mRNA sequence?

5'-GCU-GAA-GUC-GAG-GUG-UGG-3'

This mRNA codes for the following polypeptide:

Amino terminus- Ala-Glu-Val-Glu-Val-Trp -Carboxyl terminus.

<u>Problem 28.41</u> The alpha chain of human hemoglobin has 141 amino acids in a single polypeptide chain. Calculate the minimum number of bases on DNA necessary to code for the alpha chain. Include in your calculation the bases necessary for specifying termination of polypeptide synthesis.

The minimum number of bases needed for the alpha chain of human hemoglobin must code for the 141 amino acids as well as three extra bases for the stop codon. Therefore, the minimum number of bases that will be required is (3 x 141) + (1 x 3) = 426 bases.

<u>Problem 28.42</u> In HbS, the human hemoglobin found in individuals with sickle-cell anemia, glutamic acid at position 6 in the beta chain is replaced by valine.
(a) List the two codons for glutamic acid and the four codons for valine.

The two mRNA codons for glutamic acid are GAA and GAG, while the four mRNA codons for valine are GUU, GUC, GUA, and GUG.

(b) Show that one of the glutamic acid codons can be converted to a valine codon by a single substitution mutation, that is, by changing one letter in one codon.

Both of the glutamic acid codons can be converted to valine by replacing the central A with a U residue.

CHAPTER 29
Solutions to the Problems

Problem 29.1 Given the following structure, determine the polymer's repeat unit, redraw the structure using the simplified parenthetical notation, and name the polymer.

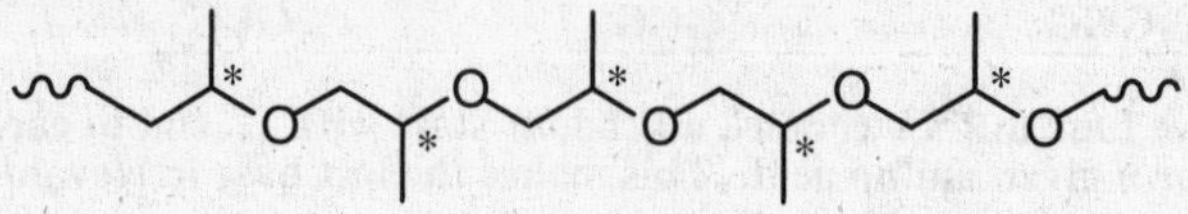

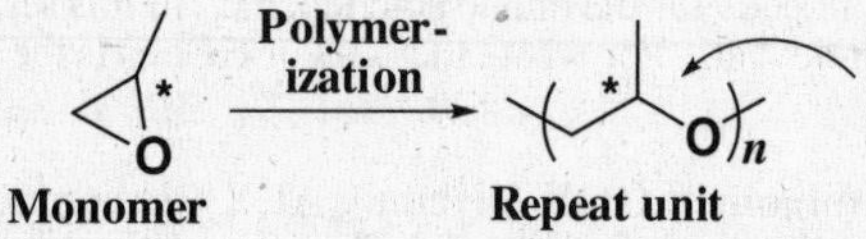

Monomer **Polymer-ization** → **Repeat unit**

This polymer is derived from propylene oxide and, therefore, named poly(propylene oxide)

Problem 29.2 Write the repeating unit of the polymer formed from the following reaction and propose a mechanism for its formation.

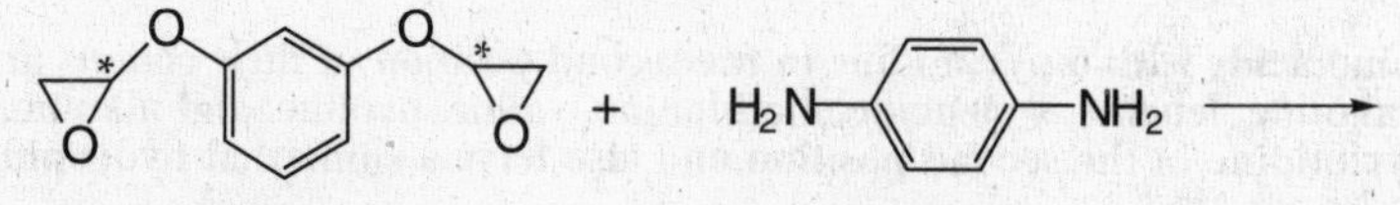

A diepoxide   A diamine

Following is the structural formula of the repeat unit of this polymer.

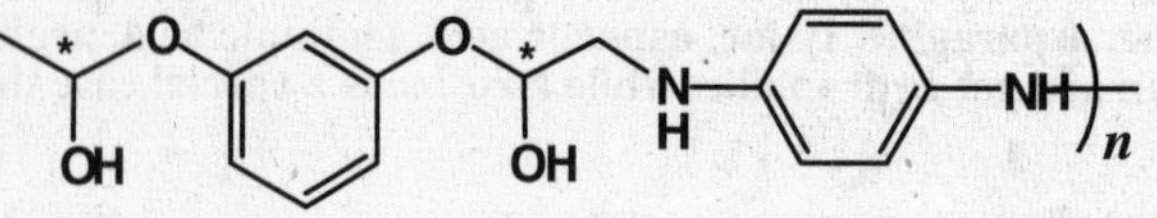

As a mechanism, propose nucleophilic attack of the amine on the less hindered carbon of the epoxide followed by proton transfer from nitrogen to oxygen.

Step 1: Make a bond between a nucleophile and an electrophile and simultaneously break a bond to give stable molecules or ions. Nucleophilic ring opening of the epoxide.

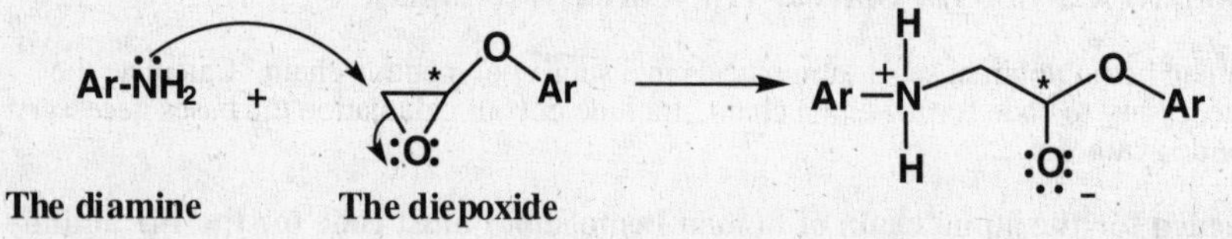

The diamine   The diepoxide

Step 2: Add a proton-take a proton away. Proton transfer from nitrogen to oxygen.

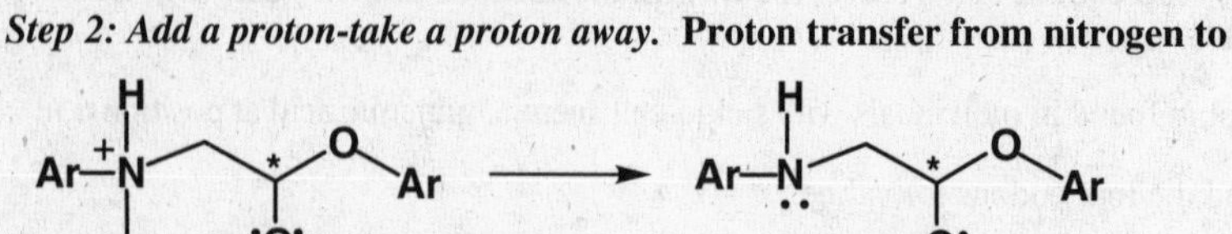

Problem 29.3 Show how to prepare polybutadiene that is terminated at both ends with primary alcohol groups.

Treat 1,3-butadiene with two moles of lithium metal to form a dianion followed by addition of monomer units as in Example 29.3 to form a living polymer. Cap the active end groups by treatment of the living polymer with ethylene oxide followed by aqueous acid.

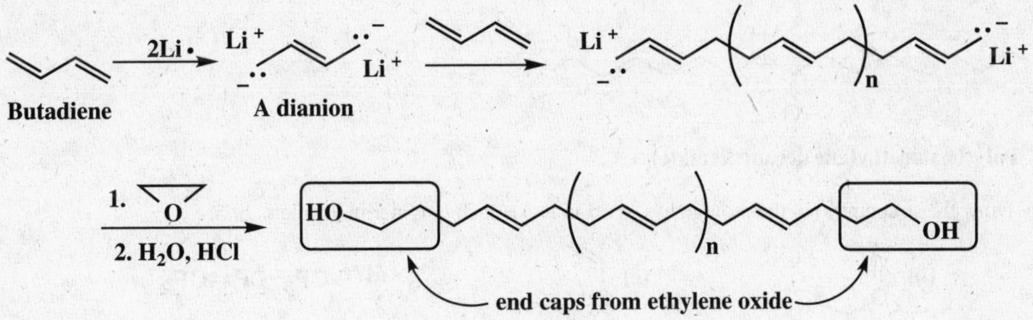

Butadiene → 2Li· → **A dianion** → (butadiene chain)

1. (ethylene oxide)
2. H_2O, HCl

→ HO—(chain)—OH

end caps from ethylene oxide

Problem 29.4 Write a mechanism for the polymerization of methyl vinyl ether initiated by 2-chloro-2-phenylpropane and $SnCl_4$. Label the initiation, propagation, and termination steps.

The mechanism for this cationic polymerization is similar to that shown in Example 29.4 for the cationic polymerization of 2-methylpropene. Treatment of 2-chloro-2-phenylpropane with $SnCl_4$ forms the 2-phenyl-2-propyl cation, the initiating cation. The termination step shown here is loss of H^+ from the end of the polymer chain to form a carbon-carbon double bond.

Initiation:

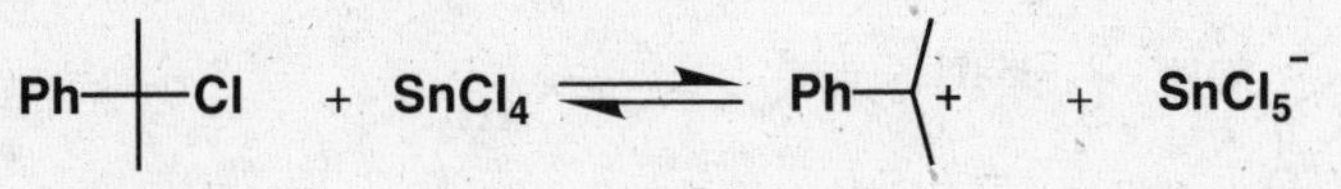

Ph—Cl + $SnCl_4$ ⇌ Ph—(+) + $SnCl_5^-$

2-Chloro-2-phenylpropane

Propagation:

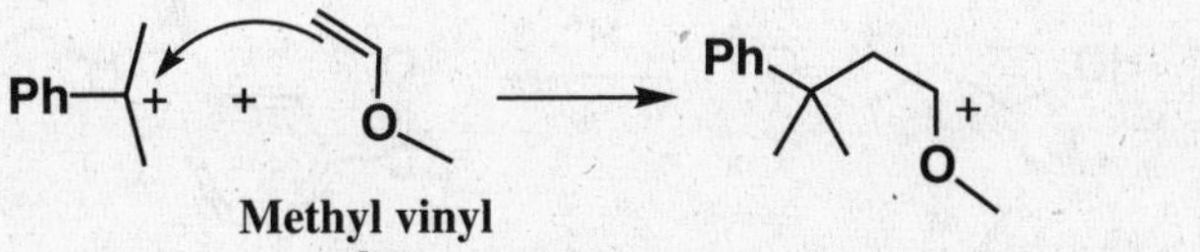

Ph—(+) + (Methyl vinyl ether) → Ph—(—O)(+)

Methyl vinyl ether

Ph—(—O)(+) + n (O) → Ph—(—O—)$_n$(+)

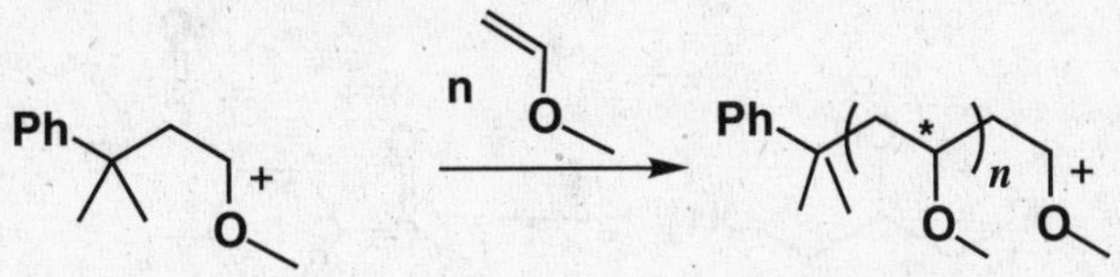

Termination:

Ph—(—O—)$_n$—O—(+) H → $SnCl_5^-$ → Ph—(—O—)$_n$—O= + HCl + $SnCl_4$

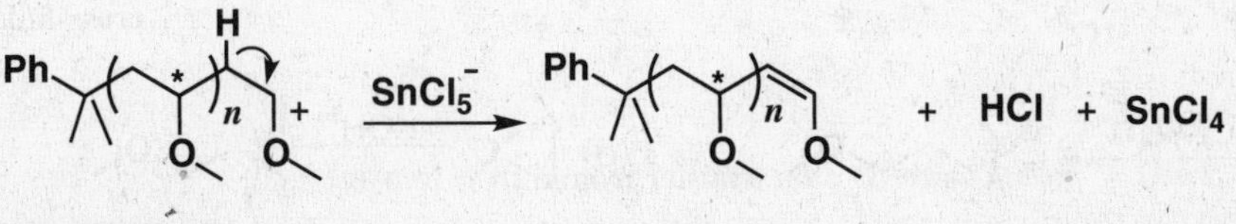

Structure and Nomenclature
Problem 29.5 Name the following polymers.

(a) 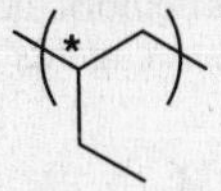 **Poly(1-butene)**

(b) 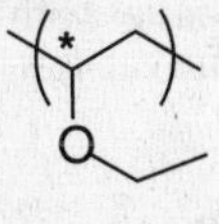 **Poly(ethyl vinyl ether)**

(c) 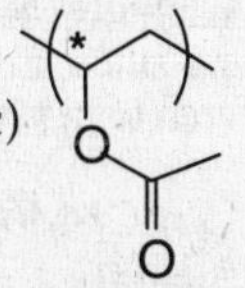 **Poly(vinyl acetate)**

(d) 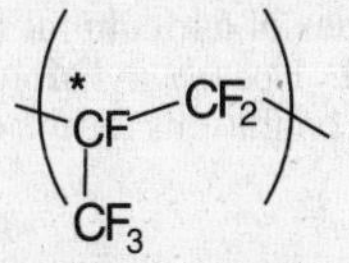 **Poly(perfluoroethylene)**

(e) 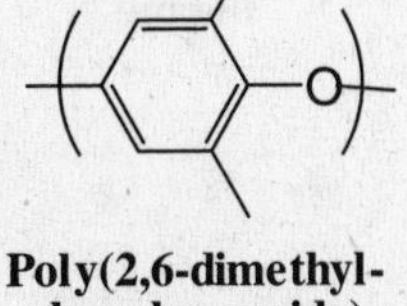 **Poly(2,6-dimethyl-phenylene oxide)**

(f) 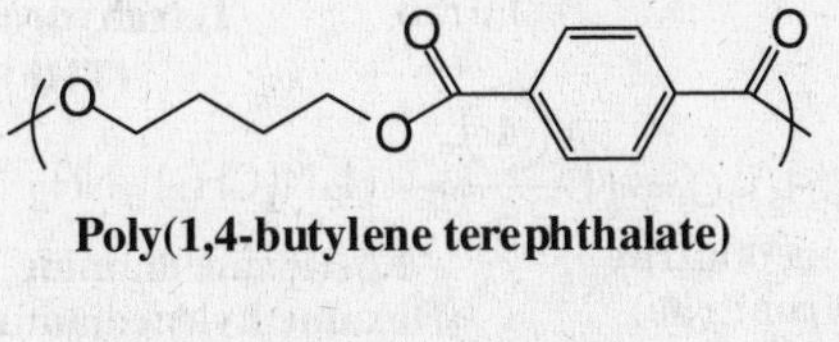 **Poly(1,4-butylene terephthalate)**

(g) 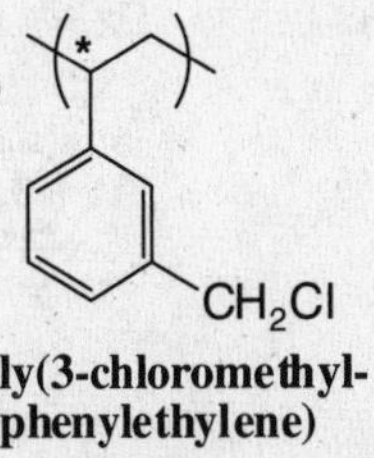 **Poly(3-chloromethyl-phenylethylene)**

(h)

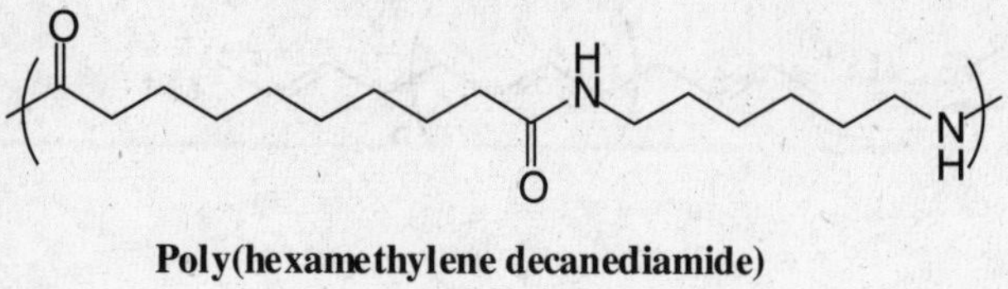

Poly(hexamethylene decanediamide)

<u>Problem 29.6</u> Draw the structure(s) of the monomer(s) used to make each polymer in Problem 29.5.

(a)

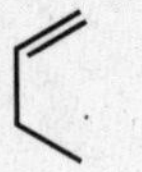

(b)

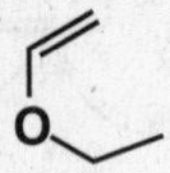

(c)

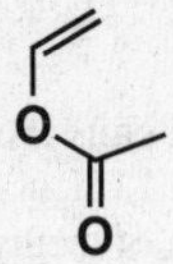

(d) CF$_3$–CF=CF$_2$

(e)

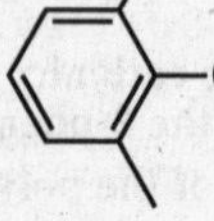

(f)

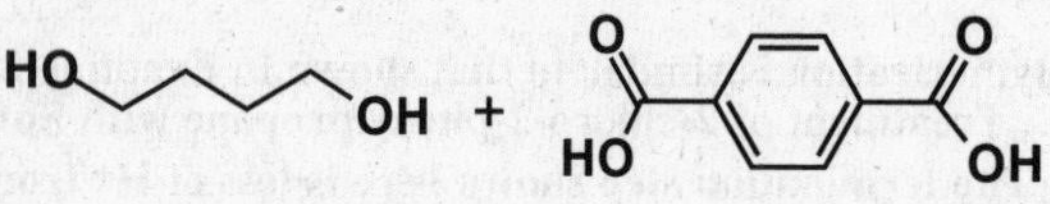

(g)

(h)

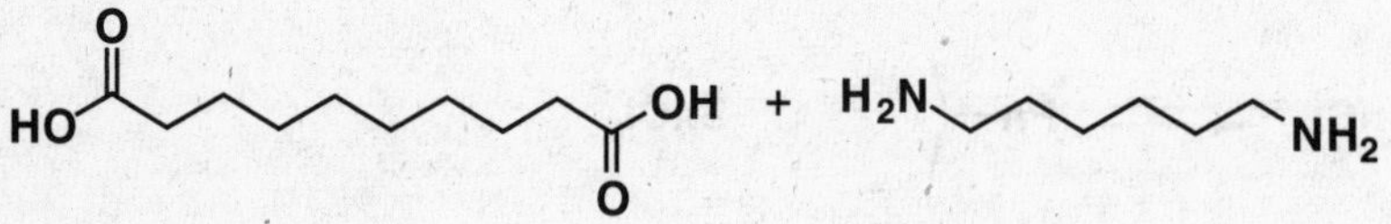

<u>Problem 29.7</u> Draw the structure of the polymer formed in the following reactions.

(a)

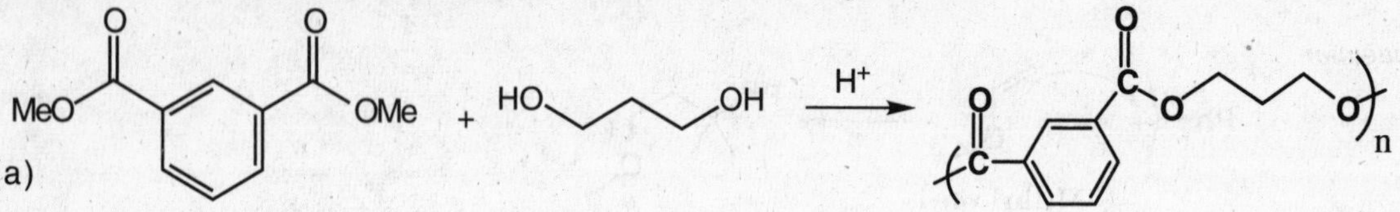

(b) 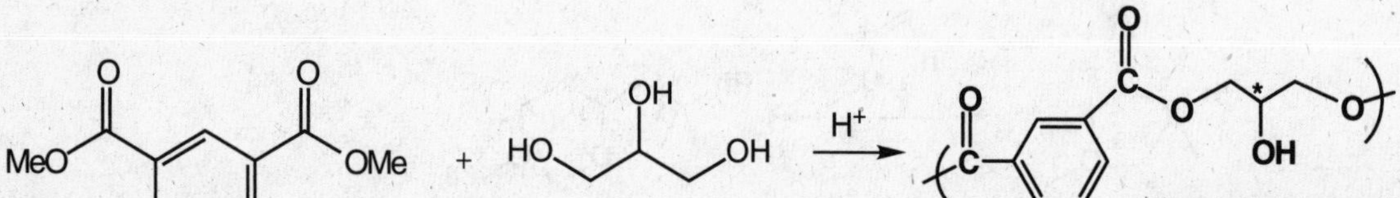   **Note: there may also be cross-linking**

(c)

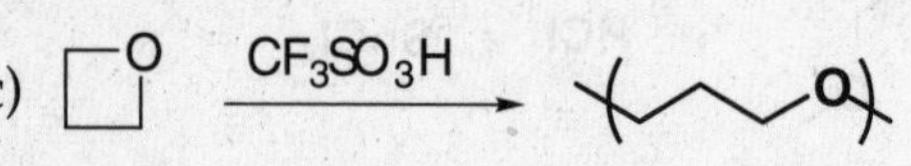

(d)

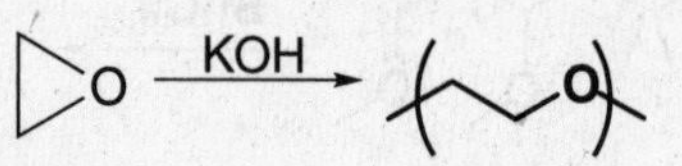

<u>Problem 29.8</u> At one time, a raw material for the production of hexamethylenediamine was the pentose-based polysaccharides of agricultural wastes, such as oat hulls. Treatment of these wastes with sulfuric acid or hydrochloric acid gives furfural. Decarbonylation of furfural over a zinc-chromium-molybdenum catalyst gives furan. Propose reagents and experimental conditions for the conversion of furan to hexamethylenediamine.

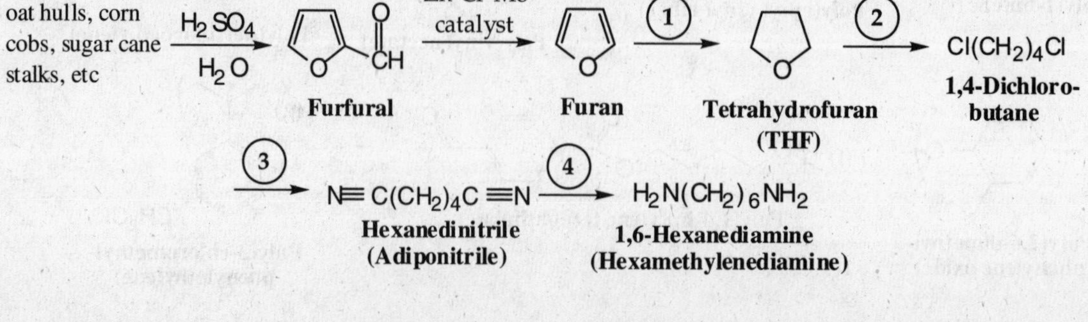

Step 1: Catalytic hydrogenation using H_2 over a transition metal catalyst.
Step 2: Cleavage of the ether using concentrated HCl at elevated temperature.
Step 3: Treatment of the dihalide with NaCN, by an S_N2 pathway.
Step 4: Catalytic hydrogenation of the cyano groups using H_2 over a transition metal catalyst.

Problem 29.9   Another raw material for the production of hexamethylenediamine is butadiene derived from thermal and catalytic cracking of petroleum. Propose reagents and experimental conditions for the conversion of butadiene to hexamethylenediamine.

$$CH_2=CHCH=CH_2 \xrightarrow{\text{(1)}} CICH_2CH=CHCH_2CI \xrightarrow{\text{(2)}} N\equiv CCH_2CH=CHCH_2C\equiv N \xrightarrow{\text{(3)}}$$

Butadiene   **1,4-Dichloro-2-butene**   **3-Hexenedinitrile**

$$H_2N(CH_2)_6NH_2$$

1,6-Hexanediamine
(Hexamethylenediamine).

Step 1: 1,4- Addition of Cl_2 to the conjugated diene.
Step 2: Treatment of the dihalide with NaCN, by an S_N2 pathway.
Step 3: Catalytic hydrogenation of the cyano groups and the carbon-carbon double bond using H_2 over a transition metal catalyst.

Problem 29.10 Propose reagents and experimental conditions for the conversion of butadiene to adipic acid.

1,3 Butadiene   **Hexanedioic acid**
                                      (Adipic acid)

See Problem 29.9 for the conversion of butadiene to 3-hexenedinitrile. (3) Catalytic hydrogenation of the carbon-carbon double bond in 3-hexenedinitrile followed by (4) hydrolysis of the cyano groups in aqueous acid gives adipic acid.

$$CH_2=CHCH=CH_2 \xrightarrow{\text{(1)}} CICH_2CH=CHCH_2CI \xrightarrow{\text{(2)}} N\equiv CCH_2CH=CHCH_2C\equiv N \xrightarrow{\text{(3)}}$$

Butadiene   **1,4-Dichloro-2-butene**   **3-Hexenedinitrile**

$$N\equiv C(CH_2)_4C\equiv N \xrightarrow{\text{(4)}} HOOC(CH_2)_4COOH$$
Hexanedinitrile   **Hexanedioic acid**
(Adiponitrile)   **(Adipic acid)**

Problem 29.11 Polymerization of 2-chloro-1,3-butadiene under Ziegler-Natta conditions gives a synthetic elastomer called neoprene. All carbon-carbon double bonds in the polymer chain have the *trans* configuration. Draw the repeat unit in neoprene.

Problem 29.12 Poly(ethylene terephthalate) (PET) can be prepared by this reaction. Propose a mechanism for the step-growth reaction in this polymerization.

Dimethyl terephthalate   **Ethylene glycol**   **Poly(ethylene terephthalate)**   **Methanol**

Propose addition of a hydroxyl group to a carbonyl carbon of dimethyl terephthalate to form a tetrahedral carbonyl addition intermediate, followed by its collapse to give an ester bond of the polymer plus methanol etc. This is an example of transesterification.

Step 1: Make a bond between a nucleophile and an electrophile.

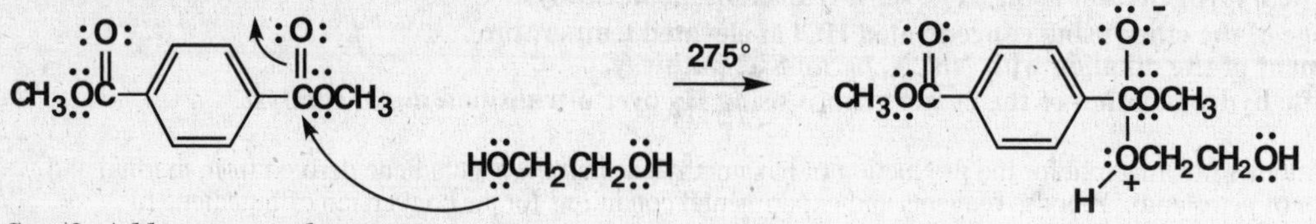

$275°$

Step 2: Add a proton-take a proton away.

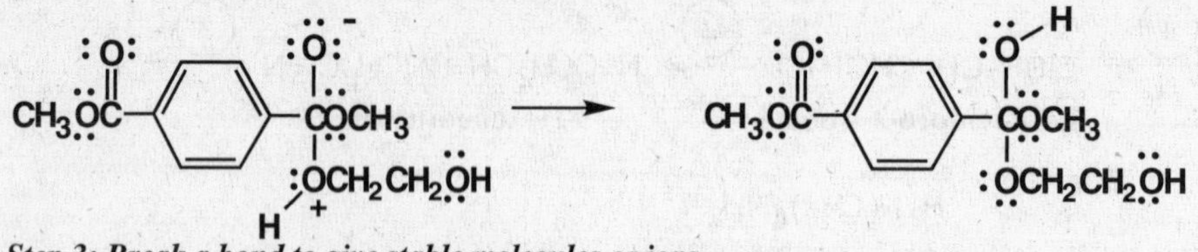

Step 3: Break a bond to give stable molecules or ions.

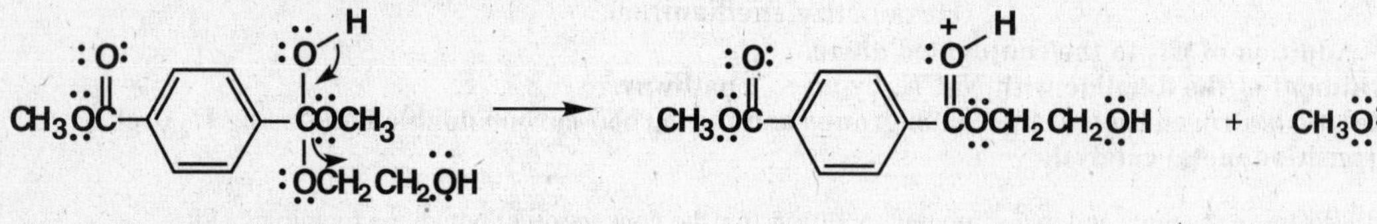

Step 4: Take a proton away.

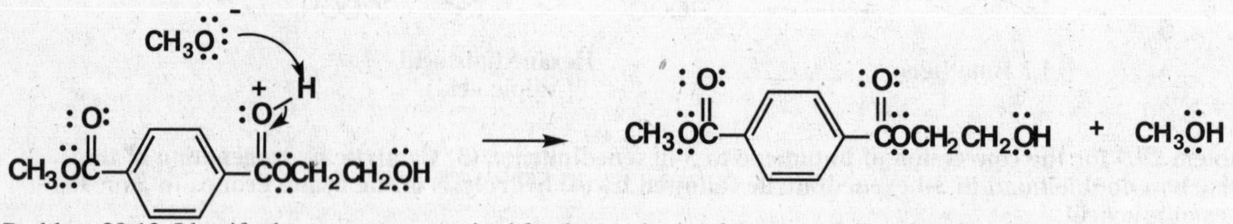

<u>Problem 29.13</u> Identify the monomers required for the synthesis of these step-growth polymers.

(a)

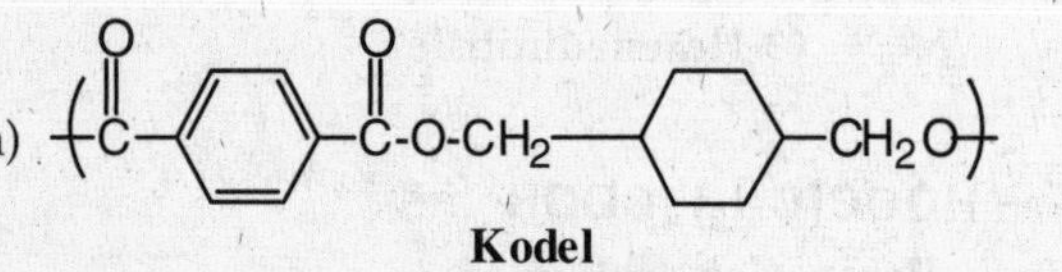

Kodel
(a polyester)

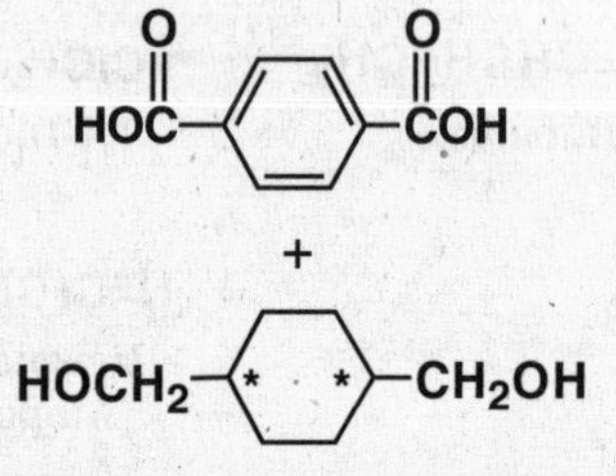

HO_2C — — CO_2H

+

$HOCH_2$ — * — * — CH_2OH

(b)

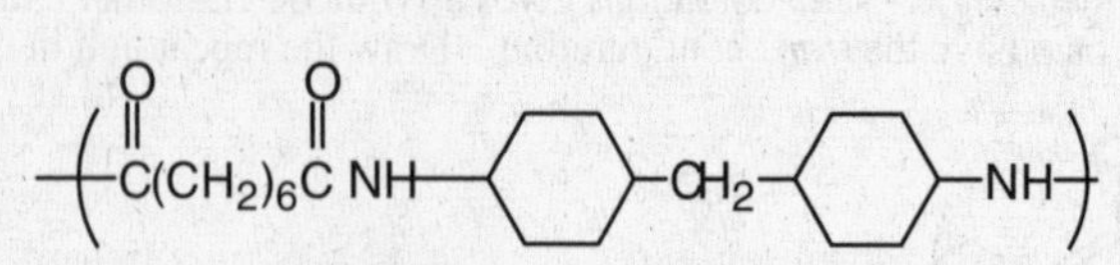

Quiana
(a polyamide)

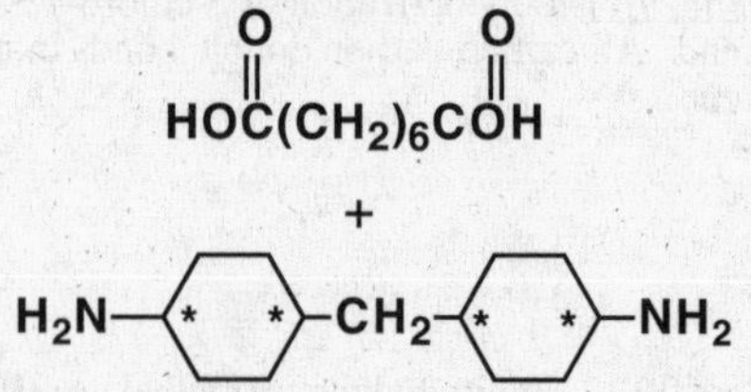

$HOC(CH_2)_6COH$

+

H_2N — * — * — CH_2 — * — * — NH_2

Problem 29.14 Nomex, another aromatic polyamide (compare aramid) is prepared by polymerization of 1,3-benzenediamine and the diacid chloride of 1,3-benzenedicarboxylic acid. The physical properties of the polymer make it suitable for high strength, high temperature applications such as parachute cords and jet aircraft tires. Draw a structural formula for the repeating unit of Nomex.

1,3-Benzenediamine   **1,3-Benzene-dicarbonyl chloride**   polymerization → Nomex

Following is the repeat unit in Nomex.

Problem 29.15 Caprolactam, the monomer from which nylon 6 is synthesized, is prepared from cyclohexanone in two steps. In step 1, cyclohexanone is treated with hydroxylamine to form cyclohexanone oxime. Treatment of the oxime with concentrated sulfuric acid in Step 2 gives caprolactam by a reaction called a Beckmann rearrangement. Propose a mechanism for the conversion of cyclohexanone oxime to caprolactam.

Cyclohexanone   **Cyclohexanone oxime**   **Caprolactam**

The mechanism is shown divided into five steps.

    Step 1: Add a proton. **Proton transfer from H_3O^+ to the oxygen atom of the oxime generates an oxonium ion, which converts OH, a poor leaving group, into OH_2, a better leaving group.**

    Step 2: 1,2 Shift-break a bond to give stable molecules or ions. **Migration of the electron pair of an adjacent carbon-carbon bond to nitrogen accompanies departure of H_2O.**

Step 3: Make a bond between a nucleophile and an electrophile. **Reaction of the carbocation from step 2 with H_2O to give an oxonium ion.**

Step 4: Take a proton away. **Proton transfer to solvent gives the enol of an amide.**

Step 5: Keto-enol tautomerism. **Keto-enol tautomerism of the enol form of the amide gives caprolactam.**

<u>Problem 29.16</u> Nylon 6,10 is prepared by polymerization of a diamine and a diacid chloride. Draw the structural formula for each reactant and for the repeat unit in this polymer.

The six-carbon diamine is 1,6-hexanediamine and the ten-carbon diacid chloride is decanedioyl chloride.

<u>Problem 29.17</u> Polycarbonates (Section 29.5C) are also formed by using a nucleophilic aromatic substitution route (Section 22.3B) involving aromatic difluoro monomers and carbonate ion.   Propose a mechanism for this reaction.

Review Section 21.3B for the addition-elimination mechanism of nucleophilic aromatic substitution.

Step 1: Make a bond between a nucleophile and an electrophile (aromatic). **Nucleophilic addition of carbonate ion to the aromatic ring at the carbon bearing the fluorine atom forms a Meisenheimer complex.**

Step 2: Break a bond to give stable molecules or ions. **Collapse of the Meisenheimer complex with ejection for fluoride ion.**

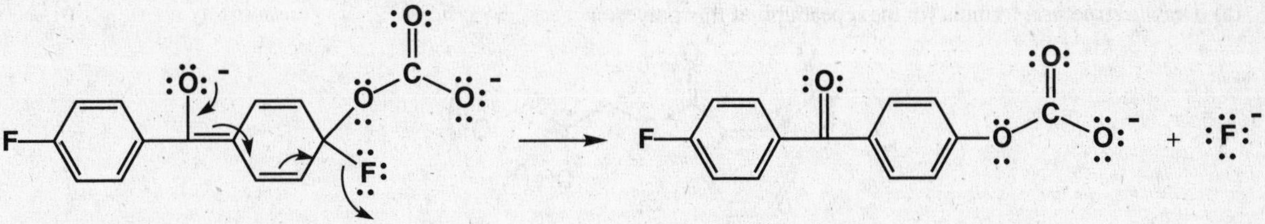

Problem 29.18 Propose a mechanism for the formation of this polyphenylurea. To simplify your presentation of the mechanism, consider the reaction of one -NCO group with one -NH₂ group.

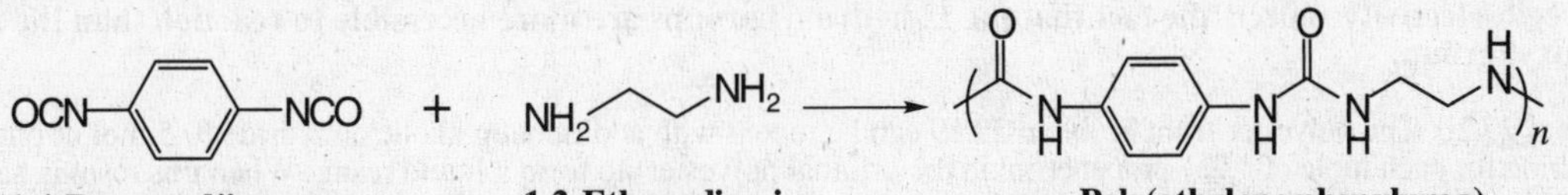

1,4-Benzenediisocyanate	1,2-Ethanediamine	Poly(ethylene phenylurea)

Propose addition of an amino group to the C=O of the isocyanate group to give an enol, followed by keto-enol tautomerism of the enol to give a disubstituted urea.

Step 1: Make a bond between a nucleophile and an electrophile.

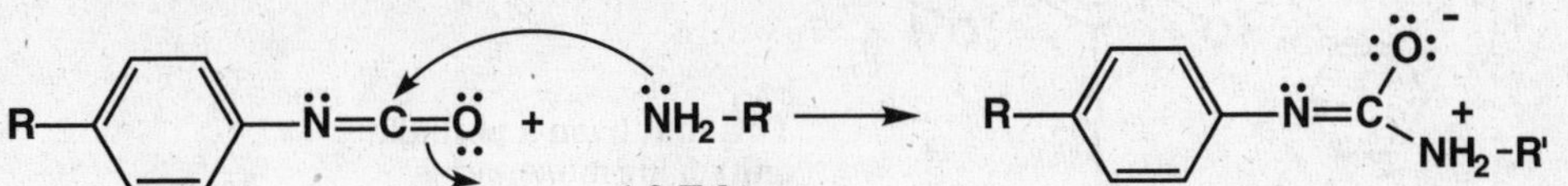

1,4-Benzene diisocyanate   1,2-Ethane-diamine

Step 2: Add a proton-take a proton away.

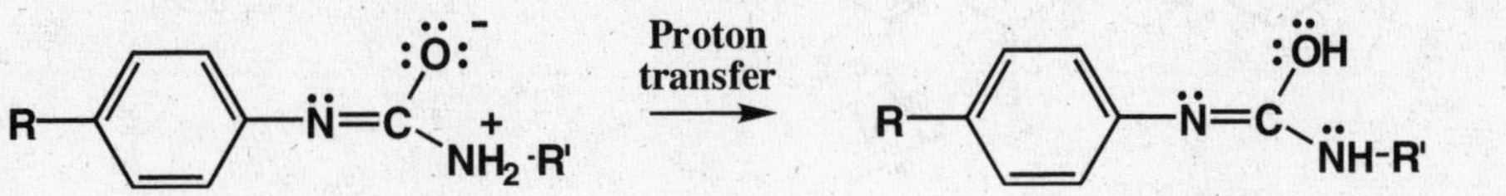

Proton transfer

An enol

Step 3: Keto-enol tautomerism.

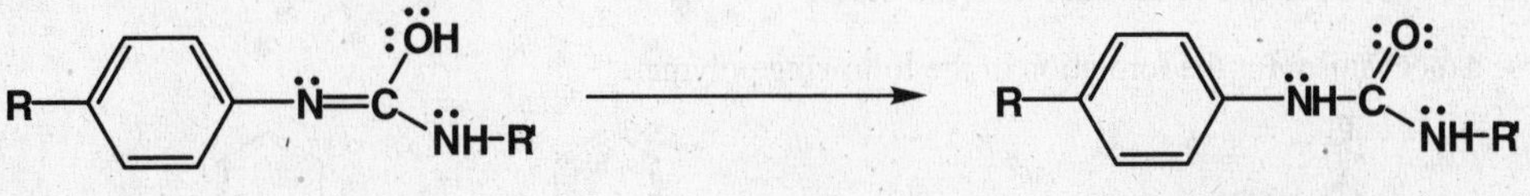

A disubstituted urea

Problem 29.19 When equal molar amounts of phthalic anhydride and 1,2,3-propanetriol are heated, they form an amorphous polyester. Under these conditions, polymerization is regioselective for the primary hydroxyl groups of the triol.

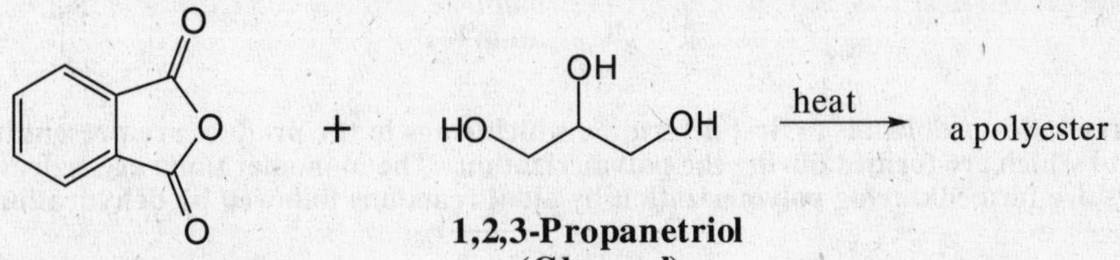

heat

a polyester

Phthalic anhydride   1,2,3-Propanetriol (Glycerol)

(a) Draw a structural formula for the repeat unit of this polyester

(b) Account for the regioselective reaction with the primary hydroxyl groups only.

The regioselectivity reflects the fact that the 1° hydroxyl groups are more accessible to reaction than the 2° hydroxyl group.

Problem 29.20 The polyester from Problem 29.19 can be mixed with additional phthalic anhydride (0.5 mol of phthalic anhydride for each mole of 1,2,3-propanetriol in the original polyester) to form a liquid resin. When this resin is heated, it forms a hard, insoluble, thermosetting polyester called glyptal.
(a) Propose a structure for the repeat unit in glyptal.

The polymer described in Problem 29.19 becomes cross linked as shown.

This unit from a phthalic anhydride monomer is the cross-linking unit.

(b) Account for the fact that glyptal is a thermosetting plastic.

Because of the extensive cross linking, the individual polymer chains can no longer be made to flow and, therefore, the polymer cannot be made to assume a liquid state.

Problem 29.21 Propose a mechanism for the formation of the following polymer.

One way to attack this problem is to first determine which rings in the product are present in the original monomers, and which are formed during the polymerization. The monomer units are redrawn here to show that new rings are formed during polymerization by aldol reactions followed by dehydration and by imine formation.

**aldol reaction
here followed
by dehydration**

imine formation here

base | -2nH$_2$O

**These rings formed by the
combination of aldol/dehydration and
imine formation**

<u>Problem 29.22</u> Draw the structural formula of the polymer resulting from base-catalyzed polymerization of each compound. Would you expect the polymers to be optically active? (S)-(+)-lactide is the dilactone formed from two molecules of (S)-(+)-lactic acid.

(a)

(S)-(+)-lactide

**each chiral center has the S
configuration; the polymer
is optically active**

(b)

(R)-Propylene oxide

**each chiral center has the R
configuration; the polymer
is optically active**

<u>Problem 29.23</u> Poly(3-hydroxybutanoic acid), a biodegradable polyester, is an insoluble, opaque material that is difficult to process into shapes. In contrast, the copolymer of 3-hydroxybutanoic acid and 3-hydroxyoctanoic acid is a transparent polymer that shows good solubility in a number of organic solvents. Explain the difference in properties between these two polymers in terms of their structure.

Poly(3-hydroxybutanoic acid)

**Poly(3-hydroxybutanoic acid -
3-hydroxyoctanoic acid) copolymer**

The polymer chains of poly(3-hydroxybutanoic acid) can assume a highly ordered arrangement with a high degree of crystallinity, hence its insolubility and its opaque character. In contrast, the polymer chains of the copolymer of 3-hydroxybutanoic acid and 3-hydroxyoctanoic acid have bulky five-carbon chains that effectively prevent polymer chains from assuming any regular ordered structure. As a result, the polymer has little crystalline character, that is, it is an amorphous material with little crystalline character to reflect light.

<u>Problem 29.24</u> How might you determine experimentally if a particular polymerization is propagating by a step-growth or a chain-growth mechanism?

Analyze the distribution of polymer molecular weights as a function of degree of polymerization. As discussed in the introduction to Section 29.5, high molecular weight polymers are not produced until very late in step-growth polymerization, typically past 99% conversion of monomers to polymers. Given the mechanism of chain-growth polymerization, high molecular weight polymer molecules are produced very early and continuously in the polymerization process.

<u>Problem 29.25</u> Draw a structural formula for the polymer formed in the following reactions.

(a) (b)

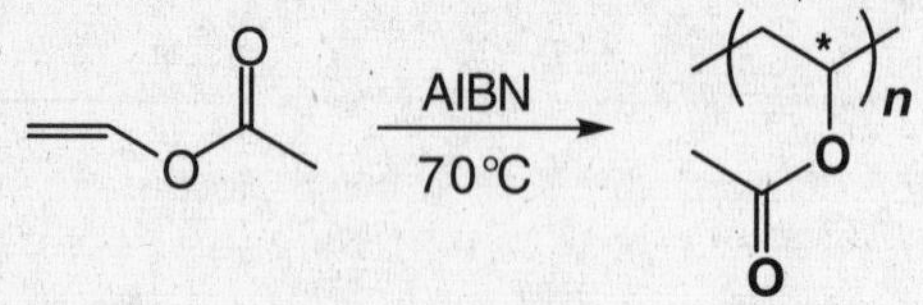

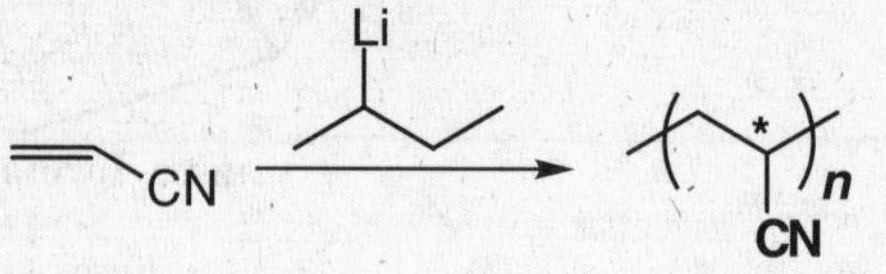

<u>Problem 29.26</u> Select the monomer in each pair that is more reactive toward cationic polymerization.

The more reactive monomer in each pair is the one forming the more stable carbocation. The first structure in each pair forms the more stable carbocation.

(a)

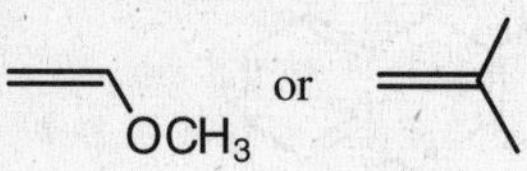

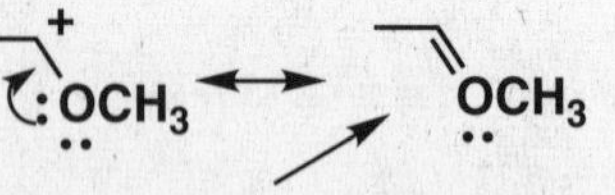

More important contributing structure; carbon and oxygen have complete valence shells

(b)

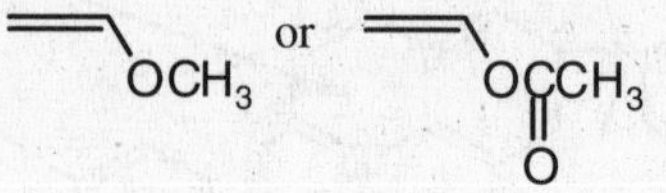

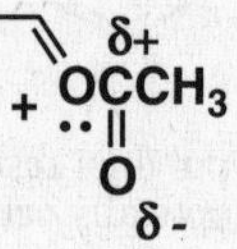

This structure makes little contribution to the hybrid because of adjacent positive and partial positive charges.

(c)

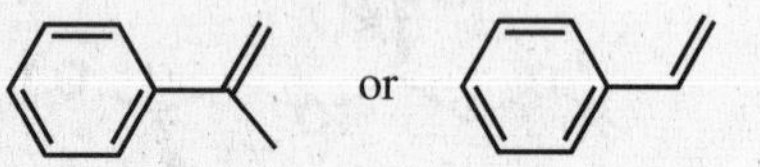

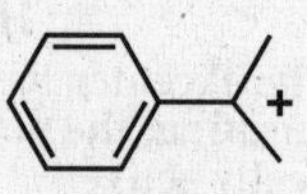

A 3° benzylic carbocation

(d)

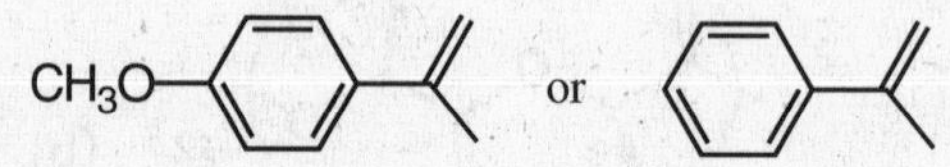

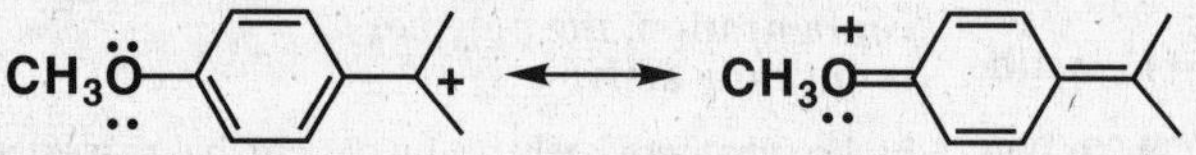

<u>Problem 29.27</u> Polymerization of vinyl acetate gives poly(vinyl acetate). Hydrolysis of this polymer in aqueous sodium hydroxide gives the useful water-soluble polymer poly(vinyl alcohol). Draw the repeat units of both poly(vinyl acetate) and poly(vinyl alcohol).

Following are structural formulas for the monomer and repeat unit of each polymer.

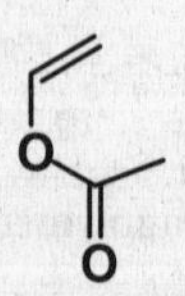

Vinyl acetate 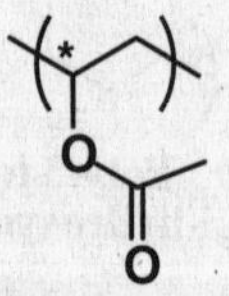   Poly(vinyl acetate) 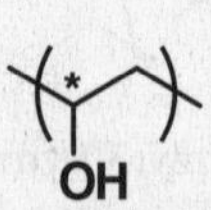   Poly(vinyl alcohol)

<u>Problem 29.28</u> Benzoquinone can be used to inhibit radical polymerizations. This compound reacts with a radical intermediate, R•, to form a less reactive radical that does not participate in chain propagation steps and, thus, breaks the chain.

Draw a series of contributing structures for this less reactive radical and account for its stability.

This radical can be represented as a hybrid of five contributing structures; three place the single electron on carbon atoms of the ring, and two place it on the oxygen atoms bonded to the ring. This radical is stabilized by the significant degree of delocalization of the single electron.

<u>Problem 29.29</u> Following is the structural formula of a section of polypropylene derived from three units of propylene monomer.

Polypropylene

Draw structural formulas for comparable sections of the following.
(a) Poly(vinyl chloride)   (b) Polytetrafluoroethylene   (c) Poly(methyl methacrylate) (Plexiglas)

$-CF_2CF_2-CF_2CF_2-CF_2CF_2-$

(d) Poly(1,1-dichloroethylene)

$-CH_2CCl_2-CH_2CCl_2-CH_2CCl_2-$

<u>Problem 29.30</u> Low-density polyethylene (LDPE) has a higher degree of chain branching than high-density polyethylene (HDPE). Explain the relationship between chain branching and density.

Unbranched polyethylene packs more efficiently into compact structures which have more mass per unit volume than structures formed by packing of branched polyethylene chains. Therefore, unbranched polyethylene has a higher density than branched-chain polyethylene.

Problem 29.31 We saw how intramolecular chain transfer in radical polymerization of ethylene creates a four-carbon branch on a polyethylene chain. What branch is created by a comparable intramolecular chain transfer during radical polymerization of styrene.

A six-membered transition state leading to 1,5-hydrogen abstraction

This four-carbon branch is created

Problem 29.32 Compare the densities of low-density polyethylene (LDPE) and high-density polyethylene (HDPE) with the densities of the liquid alkanes listed in Table 2.5. How might you account for the differences between them?

Given in the table are densities of several liquid alkanes reported in Table 2.5, plus densities for pentadecane, eicosane, and tricosane. As you can see, densities for these unbranched alkanes reach a maximum in the range 0.77-0.79 g/mL, which is significantly less than the density of both LDPE and HDPE. From this data, we conclude that both LDPE and HDPE pack more efficiently (have greater mass per unit volume) than their lower molecular weight counterparts.

Alkane	Formula	Density (g/mL)
Pentane	C_5H_{12}	0.626
Heptane	C_7H_{16}	0.684
Decane	$C_{10}H_{22}$	0.730
Pentadecane	$C_{15}H_{32}$	0.769
Eicosane	$C_{20}H_{42}$	0.789
Tricosane	$C_{30}H_{62}$	0.779
LDPE	(CH_2)	0.91-0.94
HDPE	(CH_2)	0.96

Problem 29.33 Natural rubber is the all *cis* polymer of 2-methyl-1,3-butadiene (isoprene).

Poly(2-methyl-1,3-butadiene)
(Polyisoprene)

(a) Draw a structural formula for the repeat unit of natural rubber.

(b) Draw a structural formula of the product of oxidation of natural rubber by ozone followed by a workup in the presence of $(CH_3)_2S$. Name each functional group present in this product.

Aldehyde   **Ketone**

4-Oxopentanal

(c) The smog prevalent in many major metropolitan areas contains oxidizing agents, including ozone. Account for the fact that this type of smog attacks natural rubber (automobile tires and the like) but does not attack polyethylene or polyvinyl chloride.

Polyethylene and poly(vinyl chloride) do not contain carbon-carbon double bonds, which are susceptible to attack by oxidizing agents such as ozone.

(d) Account for the fact that natural rubber is an elastomer but the synthetic all-*trans* isomer is not.

The natural cis isomer is kinked by virtue of the *cis* bond geometry while the all *trans* synthetic rubber has a more uniform staggered polymer chain. The *trans* synthetic rubber chains can thus pack together better making it more rigid compared to natural rubber.

Problem 29.34 Radical polymerization of styrene gives a linear polymer. Radical polymerization of a mixture of styrene and 1,4-divinylbenzene gives a cross linked network polymer of the type shown in Figure 29.1. Show by drawing structural formulas how incorporation of a few percent 1,4-divinylbenzene in the polymerization mixture gives a cross linked polymer.

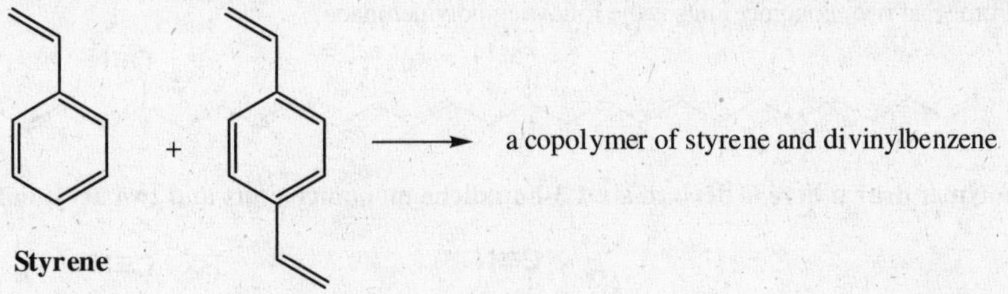

Styrene

1,4-Divinylbenzene

a copolymer of styrene and divinylbenzene

Drawn here is a section of the copolymer showing cross linking by one molecule of 1,4-divinylbenzene. Benzene rings derived from, PhCH=CH₂, are shown as Ph.

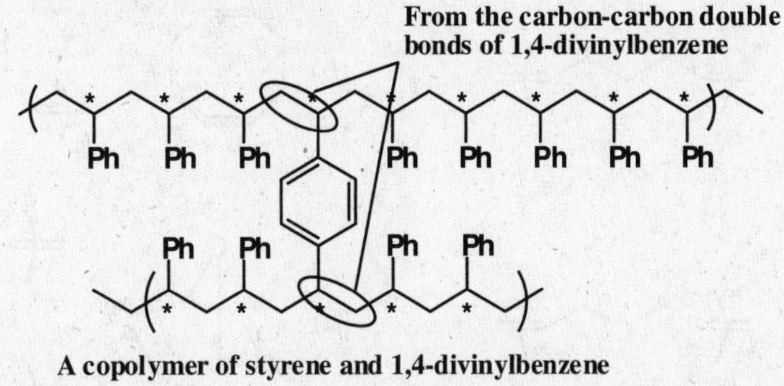

From the carbon-carbon double bonds of 1,4-divinylbenzene

A copolymer of styrene and 1,4-divinylbenzene

Problem 29.35 One common type of cation exchange resin is prepared by polymerization of a mixture containing styrene and 1,4-divinylbenzene (Problem 29.34). The polymer is then treated with concentrated sulfuric acid to sulfonate a majority of the aromatic rings in the polymer.
(a) Show the product of sulfonation of each benzene ring.

The following is a structural formula for a section of the polymer. Structural formulas for only the sulfonated rings are written in full; unsulfonated benzene rings are shown as Ph.

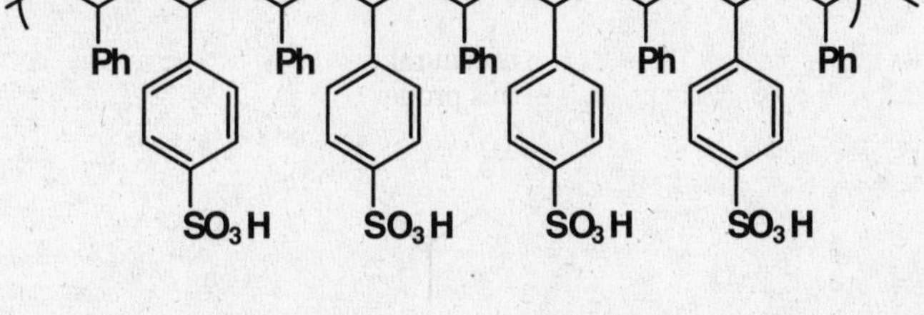

(b) Explain how this sulfonated polymer can act as a cation exchange resin.

The resin is shown in the acid or protonated form. When functioning as a cation exchange resin, cations displace H+ and become bound to the negatively charged -SO3- groups.

<u>Problem 29.36</u> The most widely used synthetic rubber is a copolymer of styrene and butadiene called SB rubber. Ratios of butadiene to styrene used in polymerization vary depending on the end use of the polymer. The ratio used most commonly in the preparation of SB rubber for use in automobile tires is 1 mole styrene to 3 moles butadiene. Draw a structural formula of a section of the polymer formed from this ratio of reactants. Assume that all carbon-carbon double bonds in the polymer chain are in the *cis* configuration.

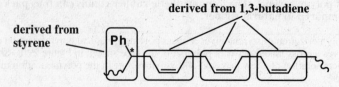

<u>Problem 29.37</u> From what two monomer units is the following polymer made?

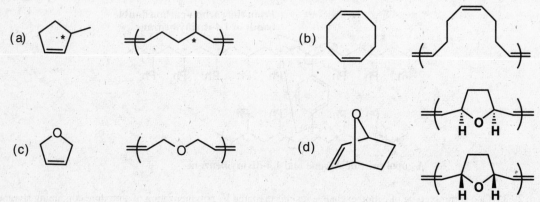

The section of polymer drawn here is derived six 1,3-butadiene monomer units and two acrylonitrile monomer units.

<u>Problem 29.38</u> Draw the structure of the polymer formed from ring-opening metathesis polymerization (ROMP) of each monomer.

(a)

(b)

(c)

(d)